United Nations

Vatican City

Russia

Kazakhstan

Mongolia

North Korea

Japan

Georgia

Uzbekistan

Kirghizia

China

South Korea

Azerbaijan

Turkmenistan

Tajikistan

Armenia

Iraq

Bhutan

Taiwan, Republic of China

Kuwait

Afghanistan

Pakistan

Nepal

Syria

Qatar

Iran

Bangladesh

Hong Kong

Lebanon

Saudi Arabia

Bahrain

Burma

Laos

Philippines

Jordan

India

Yemen

United Arab Emirates

Thailand

Vietnam

Ethiopia

Oman

Cambodia

Malaysia

Brunei

Kiribati

Djibouti

Sri Lanka

Uganda

Somalia

Maldive Islands

Singapore

Nauru

Rwanda

Indonesia

Solomon Is.

Tuvalu

Burundi

Kenya

Seychelles

Papua New Guinea

Western Samoa

Tanzania

Comoros

Vanuatu

Fiji

Malawi

Mauritius

Zimbabwe

Mozambique

Madagascar

Australia

Tonga

Swaziland

New Zealand

COPYRIGHT GEORGE PHILIP & SON. LTD.

OXFORD

CONCISE
ATLAS
OF THE
WORLD

CONTENTS

© 1993 Reed International Books Limited
Cartography by Philip's

Reed International Books Limited
81 Fulham Road, London SW3 6RB, England and
Auckland, Melbourne, Singapore and Toronto.

Published in the United States of America by
Oxford University Press, Inc.,
200 Madison Avenue,
New York, N.Y. 10016, U.S.A.

Oxford is a registered trademark of Oxford University Press

Library of Congress Cataloging-in-Publication Data
Concise atlas of the world
p. cm,
Includes indexes,
ISBN 0-19-521024-7
1.Atlases. I.Oxford University Press
G1021.C74 1993 <G&M>
912-dc20   93-3811
                                        CIP
                                        MAP

Printing (last digit): 9 8 7 6 5 4 3 2

Printed in China on acid free paper.

WORLD STATISTICS: COUNTRIES

This alphabetical list includes the principal countries and territories of the world. If a territory is not completely independent, then the country it is associated with is named. The area figures give the total area of land, inland water and ice. Units for areas and populations are thousands. The annual income is the Gross National Product per capita in US dollars. The figures are the latest available, usually 1991.

Country/Territory	Area km^2 Thousands	Area $miles^2$ Thousands	Population Thousands	Capital	Annual Income US $
Afghanistan	652	252	16,433	Kabul	450
Albania	28.8	11.1	3,250	Tirana	1,000
Algeria	2,382	920	24,960	Algiers	2,020
American Samoa (US)	0.20	0.08	39	Pago Pago	6,000
Andorra	0.45	0.17	52	Andorre-la-Vella	-
Angola	1,247	481	10,020	Luanda	620
Anguilla (UK)	0.09	0.04	8	The Valley	-
Antigua & Barbuda	0.44	0.17	77	St John's	4,770
Argentina	2,767	1,068	32,322	Buenos Aires	2,780
Armenia	29.8	11.5	3,416	Yerevan	2,150
Aruba (Neths.)	0.19	0.07	60	Oranjestad	6,000
Australia	7,687	2,968	17,086	Canberra	16,590
Austria	83.9	32.4	7,712	Vienna	20,380
Azerbaijan	86.6	33.4	7,451	Baku	1,670
Azores (Port.)	2.2	0.87	260	Ponta Delgada	-
Bahamas	13.9	5.4	253	Nassau	11,720
Bahrain	0.68	0.26	503	Manama	6,910
Bangladesh	144	56	115,594	Dacca	220
Barbados	0.43	0.17	255	Bridgetown	6,630
Belau (US)	0.46	0.18	15	Koror	-
Belgium	30.5	11.8	9,845	Brussels	19,300
Belize	23	8.9	188	Belmopan	2,050
Belorussia	207.8	80.1	10,374	Minsk	3,110
Benin	113	43	4,736	Porto-Novo	380
Bermuda (UK)	0.05	0.02	61	Hamilton	25,000
Bhutan	47	18.1	1,517	Thimphu	180
Bolivia	1,099	424	7,400	La Paz/Sucre	650
Bosnia-Herzegovina	51.2	19.8	4,364	Sarajevo	-
Botswana	582	225	1,291	Gaborone	2,590
Brazil	8,512	3,286	153,322	Brasilia	2,920
Brit. Antarctic Terr. (UK)	1,709	660	0.3	Stanley	-
Brit. Ind. Ocean Terr. (UK)	0.08	0.03	3	-	-
Brunei	5.8	2.2	266	Bandar Seri Begawan	6,000
Bulgaria	111	43	9,011	Sofia	1,840
Burkina Faso	274	106	9,001	Ouagadougou	350
Burma (Myanmar)	677	261	41,675	Rangoon	500
Burundi	27.8	10.7	5,438	Bujumbura	210
Cambodia	181	70	8,246	Phnom Penh	200
Cameroon	475	184	11,834	Yaoundé	940
Canada	9,976	3,852	26,522	Ottawa	21,260
Canary Is. (Spain)	7.3	2.8	1,700	Las Palmas/Santa Cruz	-
Cape Verde Is.	4	1.6	370	Praia	750
Cayman Is. (UK)	0.26	0.10	27	Georgetown	-
Central African Republic	623	241	3,039	Bangui	390
Chad	1,284	496	5,679	Ndjamena	220
Chile	757	292	13,386	Santiago	2,160
China	9,597	3,705	1,139,060	Beijing (Peking)	370
Colombia	1,139	440	32,987	Bogotá	1,280
Comoros	2.2	0.86	551	Moroni	500
Congo	342	132	2,271	Brazzaville	1,120
Costa Rica	51.1	19.7	2,994	San José	1,930
Croatia	56.5	21.8	4,784	Zagreb	-
Cuba	111	43	10,609	Havana	3,000
Cyprus	9.3	3.6	702	Nicosia	8,640
Czech Republic	78.9	30.4	10,299	Prague	2,370
Denmark	43.1	16.6	5,140	Copenhagen	23,660
Djibouti	23.2	9	409	Djibouti	1,000
Dominica	0.75	0.29	83	Roseau	2,440
Dominican Republic	48.7	18.8	7,170	Santo Domingo	950
Ecuador	284	109	10,782	Quito	1,020
Egypt	1,001	387	53,153	Cairo	620
El Salvador	21	8.1	5,252	San Salvador	1,070
Equatorial Guinea	28.1	10.8	348	Malabo	330
Estonia	44.7	17.3	1,600	Tallinn	3,830
* Ethiopia	1,222	472	50,974	Addis Ababa	120
Falkland Is. (UK)	12.2	4.7	2	Stanley	-
Faroe Is. (Den.)	1.4	0.54	47	Tórshavn	23,660
Fiji	18.3	7.1	765	Suva	1,830
Finland	338	131	4,986	Helsinki	24,400
France	552	213	56,440	Paris	20,600
French Guiana (Fr.)	90	34.7	99	Cayenne	2,500
French Polynesia (Fr.)	4	1.5	206	Papeete	6,000
Gabon	268	103	1,172	Libreville	3,780
Gambia, The	11.3	4.4	861	Banjul	360
Georgia	69.7	26.9	5,571	Tbilisi	1,640
Germany	357	138	79,479	Berlin	17,000
Ghana	239	92	15,028	Accra	400
Gibraltar (UK)	0.007	0.003	31	-	4,000
Greece	132	51	10,269	Athens	6,230
Greenland (Den.)	2,176	840	57	Godthåb	6,000
Grenada	0.34	0.13	85	St George's	2,180
Guadeloupe (Fr.)	1.7	0.66	344	Basse-Terre	7,000
Guam (US)	0.55	0.21	119	Agana	6,000
Guatemala	109	42	9,197	Guatemala City	930
Guinea	246	95	5,756	Conakry	450
Guinea-Bissau	36.1	13.9	965	Bissau	190
Guyana	215	83	796	Georgetown	290
Haiti	27.8	10.7	6,486	Port-au-Prince	370
Honduras	112	43	5,105	Tegucigalpa	570
Hong Kong (UK)	1.1	0.40	5,801	-	13,200
Hungary	93	35.9	10,344	Budapest	2,690
Iceland	103	40	255	Reykjavik	22,580
India	3,288	1,269	843,931	Delhi	330
Indonesia	1,905	735	179,300	Jakarta	610
Iran	1,648	636	58,031	Tehran	2,320
Iraq	438	169	18,920	Baghdad	2,000
Ireland	70.3	27.1	3,523	Dublin	10,780
Israel	27	10.3	4,659	Jerusalem	11,330
Italy	301	116	57,663	Rome	18,580
Ivory Coast	322	125	11,998	Abidjan	690
Jamaica	11	4.2	2,420	Kingston	1,380
Japan	378	146	123,537	Tokyo	26,920
Jordan	89.2	34.4	4,009	Amman	1,120
Kazakhstan	2,717	1,049	17,104	Alma Ata	2,470
Kenya	580	224	24,032	Nairobi	340
Kirghizia	198.5	76.6	4,568	Bishkek	1,550
Kiribati	0.72	0.28	66	Tarawa	750
Korea, North	121	47	21,773	Pyongyang	900
Korea, South	99	38.2	43,302	Seoul	6,340
Kuwait	17.8	6.9	2,143	Kuwait City	16,380
Laos	237	91	4,139	Vientiane	230
Latvia	63.1	24.4	2,700	Riga	3,410
Lebanon	10.4	4	2,701	Beirut	2,000
Lesotho	30.4	11.7	1,774	Maseru	580
Liberia	111	43	2,607	Monrovia	500
Libya	1,760	679	4,545	Tripoli	5,800
Liechtenstein	0.16	0.06	29	Vaduz	33,000
Lithuania	65.2	25.2	3,751	Vilnius	2,710
Luxembourg	2.6	1	384	Luxembourg	31,080
Macau (Port.)	0.02	0.006	479	-	2,000
Macedonia	25.3	9.8	2,174	Skopje	-
Madagascar	587	227	11,197	Antananarivo	210
Malawi	118	46	8,556	Lilongwe	230
Malaysia	330	127	17,861	Kuala Lumpur	2,490
Maldives	0.30	0.12	215	Malé	460
Mali	1,240	479	8,156	Bamako	280
Malta	0.32	0.12	354	Valletta	6,850
Martinique (Fr.)	1.1	0.42	341	Fort-de-France	4,000
Mauritania	1,025	396	2,050	Nouakchott	510
Mauritius	1.9	0.72	1,075	Port Louis	2,420
Mexico	1,958	756	86,154	Mexico City	2,870
Micronesia, Fed. States (US)	0.70	0.27	103	Kolonia	-
Moldavia	33.7	13	4,458	Kishinev	2,170
Monaco	0.002	0.0001	29	-	20,000
Mongolia	1,567	605	2,190	Ulan Bator	400
Montserrat (UK)	0.10	0.04	13	Plymouth	-
Morocco	447	172	25,061	Rabat	1,030
Mozambique	802	309	15,656	Maputo	70
Namibia	824	318	1,781	Windhoek	1,120
Nauru	0.02	0.008	10	Domaneab	-
Nepal	141	54	18,916	Katmandu	180
Netherlands	41.9	16.2	15,019	Amsterdam	18,560
Neths. Antilles (Neths.)	0.99	0.38	189	Willemstad	6,000
New Caledonia (Fr.)	19	7.3	168	Nouméa	4,000
New Zealand	269	104	3,429	Wellington	12,140
Nicaragua	130	50	3,871	Managua	340
Niger	1,267	489	7,732	Niamey	300
Nigeria	924	357	108,542	Lagos/Abuja	290
Norway	324	125	4,242	Oslo	24,160
Oman	212	82	1,502	Muscat	5,220
Pakistan	796	307	112,050	Islamabad	400
Panama	77.1	29.8	2,418	Panama City	2,180
Papua New Guinea	463	179	3,699	Port Moresby	820
Paraguay	407	157	4,277	Asunción	1,210
Peru	1,285	496	22,332	Lima	1,020
Philippines	300	116	61,480	Manila	740
Poland	313	121	38,180	Warsaw	1,830
Portugal	92.4	35.7	10,525	Lisbon	5,620
Puerto Rico (US)	8.9	3.4	3,599	San Juan	6,330
Qatar	11	4.2	368	Doha	15,860
Réunion (Fr.)	2.5	0.97	599	St-Denis	4,000
Romania	238	92	23,200	Bucharest	1,340
Russia	17,075	6,592	149,527	Moscow	3,220
Rwanda	26.3	10.2	7,181	Kigali	260
St Christopher/Nevis	0.36	0.14	44	Basseterre	3,960
St Lucia	0.62	0.24	151	Castries	2,500
St Pierre & Miquelon (Fr.)	0.24	0.09	7	St-Pierre	-
St Vincent/Grenadines	0.39	0.15	116	Kingstown	1,730
San Marino	0.06	0.02	24	San Marino	-
São Tomé & Príncipe	0.96	0.37	121	São Tomé	350
Saudi Arabia	2,150	830	14,870	Riyadh	7,070
Senegal	197	76	7,327	Dakar	720
Seychelles	0.46	0.18	67	Victoria	5,110
Sierra Leone	71.7	27.7	4,151	Freetown	210
Singapore	0.62	0.24	3,003	Singapore	12,890
Slovak Republic	49	18.9	5,269	Bratislava	1,650
Slovenia	20.3	7.8	1,963	Ljubljana	-
Solomon Is.	28.9	11.2	321	Honiara	560
Somalia	638	246	7,497	Mogadishu	150
South Africa	1,221	471	35,282	Pretoria	2,530
Spain	505	195	38,959	Madrid	12,460
Sri Lanka	65.6	25.3	16,993	Colombo	500
Sudan	2,506	967	25,204	Khartoum	400
Surinam	163	63	422	Paramaribo	3,610
Swaziland	17.4	6.7	768	Mbabane	1,060
Sweden	450	174	8,618	Stockholm	25,490
Switzerland	41.3	15.9	6,712	Bern	33,510
Syria	185	71	12,116	Damascus	1,110
Taiwan	36	13.9	20,300	Taipei	6,600
Tajikistan	143.1	55.2	5,680	Dushanbe	1,050
Tanzania	945	365	25,635	Dar es Salaam	100
Thailand	513	198	57,196	Bangkok	1,580
Togo	56.8	21.9	3,531	Lomé	410
Tokelau (NZ)	0.01	0.005	2	Nukunonu	-
Tonga	0.75	0.29	95	Nuku'alofa	1,100
Trinidad & Tobago	5.1	2	1,227	Port of Spain	3,620
Tunisia	164	63	8,180	Tunis	1,510
Turkey	779	301	57,326	Ankara	1,820
Turkmenistan	488.1	188.5	3,838	Ashkhabad	1,700
Turks & Caicos Is. (UK)	0.43	0.17	10	Grand Turk	-
Tuvalu	0.03	0.01	10	Funafuti	600
Uganda	236	91	18,795	Kampala	160
Ukraine	603.7	233.1	51,940	Kiev	2,340
United Arab Emirates	83.6	32.3	1,589	Abu Dhabi	19,860
United Kingdom	243.3	94	54,889	London	16,750
United States	9,373	3,619	249,975	Washington	22,560
Uruguay	177	68	3,094	Montevideo	2,860
Uzbekistan	447.4	172.7	21,627	Tashkent	1,350
Vanuatu	12.2	4.7	147	Port Vila	1,120
Vatican City	0.0004	0.0002	1	-	-
Venezuela	912	352	19,735	Caracas	2,610
Vietnam	332	127	66,200	Hanoi	300
Virgin Is. (UK)	0.15	0.06	13	Road Town	-
Virgin Is. (US)	0.34	0.13	117	Charlotte Amalie	12,000
Western Sahara (Mor.)	266	103	179	El Aaiún	-
Western Samoa	2.8	1.1	164	Apia	930
Yemen	528	204	11,282	Sana	540
Yugoslavia	102.3	39.5	10,642	Belgrade	2,940
Zaïre	2,345	906	35,562	Kinshasa	230
Zambia	753	291	8,073	Lusaka	420
Zimbabwe	391	151	9,369	Harare	620

** Eritrea formally declared full independence from Ethiopia on 24th May 1993*

World Statistics: Physical Dimensions

Each topic list is divided into continents and within a continent the items are listed in size order. The order of the continents is as in the atlas. The bottom part of many of the lists are selective. The world top ten are shown in square brackets; in the case of mountains this has not been done because the world top 30 are all in Asia. The figures are rounded as appropriate.

World, Continents, Oceans

	km²	miles²	%
The World	509,450,000	196,672,000	-
Land	149,450,000	57,688,000	29.3
Water	360,000,000	138,984,000	70.7
Asia	44,500,000	17,177,000	29.8
Africa	30,302,000	11,697,000	20.3
North America	24,241,000	9,357,000	16.2
South America	17,793,000	6,868,000	11.9
Antarctica	14,100,000	5,443,000	9.4
Europe	9,957,000	3,843,000	6.7
Australia & Oceania	8,557,000	3,303,000	5.7
Pacific Ocean	179,679,000	69,356,000	49.9
Atlantic Ocean	92,373,000	35,657,000	25.7
Indian Ocean	73,917,000	28,532,000	20.5
Arctic Ocean	14,090,000	5,439,000	3.9

Mountains

		m	ft
Europe			
Mont Blanc	*France/Italy*	4,807	15,771
Monte Rosa	*Italy/Switzerland*	4,634	15,203
Dom	*Switzerland*	4,545	14,911
Weisshorn	*Switzerland*	4,505	14,780
Matterhorn/Cervino	*Italy/Switzerland*	4,478	14,691
Mt Maudit	*France/Italy*	4,465	14,649
Finsteraarhorn	*Switzerland*	4,275	14,025
Aletschhorn	*Switzerland*	4,182	13,720
Jungfrau	*Switzerland*	4,158	13,642
Barre des Ecrins	*France*	4,103	13,461
Gran Paradiso	*Italy*	4,061	13,323
Piz Bernina	*Italy/Switzerland*	4,052	13,294
Ortles	*Italy*	3,899	12,792
Monte Viso	*Italy*	3,841	12,602
Grossglockner	*Austria*	3,797	12,457
Mulhacén	*Spain*	3,478	11,411
Pico de Aneto	*Spain*	3,404	11,168
Etna	*Italy*	3,340	10,958
Galdhöpiggen	*Norway*	2,469	8,100
Hvannadalshnúkur	*Iceland*	2,119	6,952
Ben Nevis	*UK*	1,343	4,406
Asia			
Everest	*China/Nepal*	8,848	29,029
Godwin Austen (K2)	*China/Kashmir*	8,611	28,251
Kanchenjunga	*India/Nepal*	8,598	28,208
Lhotse	*China/Nepal*	8,516	27,939
Makalu	*China/Nepal*	8,481	27,824
Cho Oyu	*China/Nepal*	8,201	26,906
Dhaulagiri	*Nepal*	8,172	26,811
Manaslu	*Nepal*	8,156	26,758
Nanga Parbat	*Kashmir*	8,126	26,660
Annapurna	*Nepal*	8,078	26,502
Gasherbrum	*China/Kashmir*	8,068	26,469
Broad Peak	*India*	8,051	26,414
Gosainthan	*China*	8,012	26,286
Disteghil Sar	*Kashmir*	7,885	25,869
Nuptse	*Nepal*	7,879	25,849
Elbrus	*Russia*	5,633	18,481
Fuji-san	*Japan*	3,776	12,388
Pidurutalagala	*Sri Lanka*	2,524	8,281
Africa			
Kilimanjaro	*Tanzania*	5,895	19,340
Mt Kenya	*Kenya*	5,199	17,057
Ruwenzori	*Uganda/Zaïre*	5,109	16,762
Ras Dashan	*Ethiopia*	4,620	15,157
Meru	*Tanzania*	4,565	14,977
Karisimbi	*Rwanda/Zaïre*	4,507	14,787
Mt Elgon	*Kenya/Uganda*	4,321	14,176
Batu	*Ethiopia*	4,307	14,130
Gughe	*Ethiopia*	4,200	13,779
Toubkal	*Morocco*	4,165	13,665
Oceania			
Puncak Jaya	*Indonesia*	5,029	16,499
Puncak Mandala	*Indonesia*	4,760	15,617
Puncak Trikora	*Indonesia*	4,750	15,584
Mt Wilhelm	*Papua New Guinea*	4,508	14,790
Mauna Kea	*USA (Hawaii)*	4,208	13,806
Mauna Loa	*USA (Hawaii)*	4,169	13,678
Mt Cook	*New Zealand*	3,753	12,313
Mt Kosciusko	*Australia*	2,230	7,316
North America			
Mt McKinley	*USA (Alaska)*	6,194	20,321
Mt Logan	*Canada*	6,050	19,849
Citlaltepetl	*Mexico*	5,700	18,701
Mt St Elias	*USA/Canada*	5,489	18,008
Popocatepetl	*Mexico*	5,452	17,887
Mt Foraker	*USA (Alaska)*	5,304	17,401
Ixtaccihuatl	*Mexico*	5,286	17,342
Lucania	*USA (Alaska)*	5,226	17,145
Mt Steele	*Canada*	5,011	16,440
Mt Bona	*USA (Alaska)*	5,005	16,420
South America			
Aconcagua	*Argentina*	6,960	22,834
Illimani	*Bolivia*	6,882	22,578
Bonete	*Argentina*	6,872	22,546
Ojos del Salado	*Argentina/Chile*	6,863	22,516
Tupungato	*Argentina/Chile*	6,800	22,309
Pissis	*Argentina*	6,779	22,241
Mercedario	*Argentina/Chile*	6,770	22,211
Huascaran	*Peru*	6,768	22,204
Llullaillaco	*Argentina/Chile*	6,723	22,057
Nudo de Cachi	*Argentina*	6,720	22,047
Antarctica			
Vinson Massif		4,897	16,066

Ocean Depths

	m	ft
Atlantic Ocean		
Puerto Rico (Milwaukee) Deep [7]	9,200	30,183
Cayman Trench [10]	7,680	25,197
Gulf of Mexico	5,203	17,070
Mediterranean Sea	5,121	16,801
Black Sea	2,211	7,254
North Sea	310	1,017
Baltic Sea	294	965
Hudson Bay	111	364
Indian Ocean		
Java Trench	7,450	24,442
Red Sea	2,266	7,434
Persian Gulf	73	239
Pacific Ocean		
Mariana Trench [1]	11,022	36,161
Tonga Trench [2]	10,822	35,505
Japan Trench [3]	10,554	34,626
Kuril Trench [4]	10,542	34,586
Mindanao Trench [5]	10,497	34,439
Kermadec Trench [6]	10,047	32,962
Peru-Chile Trench [8]	8,050	26,410
Aleutian Trench [9]	7,822	25,662
Middle American Trench	6,662	21,857
Arctic Ocean		
Molloy Deep	5,608	18,399

Land Lows

		m	ft
Caspian Sea	*Europe*	-28	-92
Dead Sea	*Asia*	-400	-1,312
Lake Assal	*Africa*	-156	-512
Lake Eyre North	*Oceania*	-16	-52
Death Valley	*N. America*	-86	-282
Valdés Peninsula	*S. America*	-40	-131

Rivers

		km	miles
Europe			
Volga	*Caspian Sea*	3,700	2,300
Danube	*Black Sea*	2,850	1,770
Ural	*Caspian Sea*	2,535	1,574
Dnieper	*Volga*	2,285	1,420
Kama	*Volga*	2,030	1,260
Don	*Volga*	1,990	1,240
Petchora	*Arctic Ocean*	1,790	1,110
Dniester	*Black Sea*	1,400	870
Rhine	*North Sea*	1,320	820
Elbe	*North Sea*	1,145	710
Vistula	*Baltic Sea*	1,090	675
Loire	*Atlantic Ocean*	1,020	635
W. Dvina	*Baltic Sea*	1,019	633
Asia			
Yangtze [3]	*Pacific Ocean*	6,380	3,960
Yenisey-Angara [5]	*Arctic Ocean*	5,550	3,445
Ob-Irtysh [6]	*Arctic Ocean*	5,410	3,360
Hwang Ho [7]	*Pacific Ocean*	4,840	3,005
Amur [9]	*Pacific Ocean*	4,510	2,800
Mekong [10]	*Pacific Ocean*	4,500	2,795
Lena	*Arctic Ocean*	4,400	2,730
Irtysh	*Ob*	4,250	2,640
Yenisey	*Arctic Ocean*	4,090	2,540
Ob	*Arctic Ocean*	3,680	2,285
Indus	*Indian Ocean*	3,100	1,925
Brahmaputra	*Indian Ocean*	2,900	1,800
Syr Darya	*Aral Sea*	2,860	1,775
Salween	*Indian Ocean*	2,800	1,740
Euphrates	*Indian Ocean*	2,700	1,675
Vilyuy	*Lena*	2,650	1,645
Kolyma	*Arctic Ocean*	2,600	1,615
Amu Darya	*Aral Sea*	2,540	1,575
Ural	*Caspian Sea*	2,535	1,575
Ganges	*Indian Ocean*	2,510	1,560
Si Kiang	*Pacific Ocean*	2,100	1,305
Irrawaddy	*Indian Ocean*	2,010	1,250
Tigris	*Indian Ocean*	1,900	1,180
Africa			
Nile [1]	*Mediterranean Sea*	6,670	4,140
Zaïre/Congo [8]	*Atlantic Ocean*	4,670	2,900
Niger	*Atlantic Ocean*	4,180	2,595
Zambezi	*Indian Ocean*	2,740	1,700
Oubangi/Uele	*Zaïre*	2,250	1,400
Kasai	*Zaïre*	1,950	1,210
Shaballe	*Indian Ocean*	1,930	1,200
Orange	*Atlantic Ocean*	1,860	1,155
Australia			
Murray-Darling	*Indian Ocean*	3,720	2,310
Darling	*Murray*	3,070	1,905
Murray	*Indian Ocean*	2,575	1,600
Murrumbidgee	*Murray*	1,690	1,050
North America			
Mississippi-Missouri [4]	*Gulf of Mexico*	6,020	3,740
Mackenzie	*Arctic Ocean*	4,240	2,630
Mississippi	*Gulf of Mexico*	3,780	2,350
Missouri	*Mississippi*	3,725	2,310
Yukon	*Pacific Ocean*	3,185	1,980
Rio Grande	*Gulf of Mexico*	3,030	1,880
Arkansas	*Mississippi*	2,340	1,450
Colorado	*Pacific Ocean*	2,330	1,445
Red	*Mississippi*	2,040	1,270
Columbia	*Pacific Ocean*	1,950	1,210
Saskatchewan	*Lake Winnipeg*	1,940	1,205
Snake	*Columbia*	1,670	1,040
South America			
Amazon [2]	*Atlantic Ocean*	6,430	3,990
Paraná-Plate	*Atlantic Ocean*	4,000	2,480
Purus	*Amazon*	3,350	2,080
Madeira	*Amazon*	3,200	1,990
São Francisco	*Atlantic Ocean*	2,900	1,800
Paraná	*Plate*	2,800	1,740
Tocantins	*Atlantic Ocean*	2,640	1,640
Paraguay	*Paraná*	2,550	1,580
Orinoco	*Atlantic Ocean*	2,500	1,550
Pilcomayo	*Paraná*	2,500	1,550

Lakes

		km²	miles²
Europe			
Lake Ladoga	*Russia*	18,400	7,100
Lake Onega	*Russia*	9,700	3,700
Saimaa system	*Finland*	8,000	3,100
Asia			
Caspian Sea [1]	*Asia*	371,000	143,000
Aral Sea [6]	*Kazakh./Uzbek.*	36,000	13,900
Lake Baykal [9]	*Russia*	31,500	12,200
Tonlé Sap	*Cambodia*	20,000	7,700
Lake Balkhash	*Kazakhstan*	18,500	7,100
Africa			
Lake Victoria [3]	*E. Africa*	68,000	26,000
Lake Tanganyika [7]	*C. Africa*	33,000	13,000
Lake Malawi [10]	*E. Africa*	29,000	11,000
Lake Chad	*C. Africa*	25,000	9,700
Australia			
Lake Eyre	*Australia*	9,000	3,500
North America			
Lake Superior [2]	*Canada/USA*	82,200	31,700
Lake Huron [4]	*Canada/USA*	59,600	23,000
Lake Michigan [5]	*USA*	58,000	22,400
Great Bear Lake [8]	*Canada*	31,500	12,200
Great Slave Lake	*Canada*	28,700	11,100
Lake Erie	*Canada/USA*	25,700	9,900
Lake Winnipeg	*Canada*	24,400	9,400
Lake Ontario	*Canada/USA*	19,500	7,500
South America			
Lake Titicaca	*Bolivia/Peru*	8,200	3,200

Islands

		km²	miles²
Europe			
Great Britain [8]	*UK*	229,880	88,700
Iceland	*Atlantic Ocean*	103,000	39,800
Ireland	*Ireland/UK*	84,400	32,600
Asia			
Borneo [3]	*S. E. Asia*	737,000	284,000
Sumatra [6]	*Indonesia*	425,000	164,000
Honshu [7]	*Japan*	230,000	88,800
Celebes	*Indonesia*	189,000	73,000
Java	*Indonesia*	126,700	48,900
Luzon	*Philippines*	104,700	40,400
Mindanao	*Philippines*	95,000	36,700
Hokkaido	*Japan*	78,400	30,300
Sakhalin	*Russia*	76,400	29,500
Sri Lanka	*Indian Ocean*	65,600	25,300
Africa			
Madagascar [4]	*Indian Ocean*	587,000	226,600
Oceania			
New Guinea [2]	*Indon./Pap. NG*	780,000	301,080
New Zealand (S.)	*New Zealand*	150,500	58,100
New Zealand (N.)	*New Zealand*	114,400	44,200
Tasmania	*Australia*	67,800	26,200
North America			
Greenland [1]	*Greenland*	2,175,600	839,800
Baffin Is. [5]	*Canada*	508,000	196,100
Victoria Is. [9]	*Canada*	212,200	81,900
Ellesmere Is. [10]	*Canada*	212,000	81,800
Cuba	*Cuba*	114,500	44,200
Newfoundland	*Canada*	96,000	37,100
Hispaniola	*Atlantic Ocean*	76,200	29,400
South America			
Tierra del Fuego	*Argentina/Chile*	47,000	18,100

MAP PROJECTIONS

MAP PROJECTIONS

A map projection is the systematic depiction on a plane surface of the imaginary lines of latitude or longitude from a globe of the earth. This network of lines is called the graticule and forms the framework upon which an accurate depiction of the earth is made. The map graticule, which is the basis of any map, is constructed sometimes by graphical means, but often by using mathematical formulae to give the intersections of the graticule plotted as x and y co-ordinates. The choice between projections is based upon which properties the cartographer wishes the map to possess, the map scale and also the extent of the area to be mapped. Since the globe is three dimensional, it is not possible to depict its surface on a two dimensional plane without distortion. Preservation of one of the basic properties listed below can only be secured at the expense of the others and the choice of projection is often a compromise solution.

Correct Area

In these projections the areas from the globe are to scale on the map. For example, if you look at the diagram at the top right, areas of 10° x 10° are shown from the equator to the poles. The proportion of this area at the extremities are approximately 11:1. An equal area projection will retain that proportion in its portrayal of those areas. This is particularly useful in the mapping of densities and distributions. Projections with this property are termed **Equal Area, Equivalent or Homolographic.**

Correct Distance

In these projections the scale is correct along the meridians, or in the case of the Azimuthal Equidistant scale is true along any line drawn from the centre of the projection. They are called **Equidistant**.

Correct Shape

This property can only be true within small areas as it is achieved only by having a uniform scale distortion along both x and y axes of the projection. The projections are called **Conformal** or **Orthomorphic.**

In order to minimise the distortions at the edges of some projections, central portions of them are often selected for atlas maps. Below are listed some of the major types of projection.

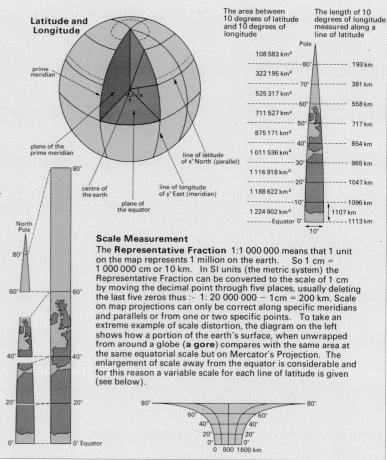

Latitude and Longitude

prime meridian

plane of the prime meridian

centre of the earth

plane of the equator

line of latitude of x° North (parallel)

line of longitude of y° East (meridian)

The area between 10 degrees of latitude and 10 degrees of longitude

Pole	
108 583 km²	80°
322 195 km²	70°
525 317 km²	60°
711 527 km²	50°
875 171 km²	40°
1 011 536 km²	30°
1 116 918 km²	20°
1 188 622 km²	10°
1 224 902 km²	
Equator 0°	

The length of 10 degrees of longitude measured along a line of latitude

80°	193 km
70°	381 km
60°	558 km
50°	717 km
40°	854 km
30°	965 km
20°	1047 km
10°	1096 km
	1107 km
	1113 km
10°	

North Pole

80°

80°

60° 60°

40° 40°

20° 20°

0° 0° Equator

Scale Measurement

The **Representative Fraction** 1:1 000 000 means that 1 unit on the map represents 1 million on the earth. So 1 cm = 1 000 000 cm or 10 km. In SI units (the metric system) the Representative Fraction can be converted to the scale of 1 cm by moving the decimal point through five places, usually deleting the last five zeros thus :- 1: 20 000 000 – 1cm = 200 km. Scale on map projections can only be correct along specific meridians and parallels or from one or two specific points. To take an extreme example of scale distortion, the diagram on the left shows how a portion of the earth's surface, when unwrapped from around a globe (**a gore**) compares with the same area at the same equatorial scale but on Mercator's Projection. The enlargement of scale away from the equator is considerable and for this reason a variable scale for each line of latitude is given (see below).

80° 80°

60° 60°

40° 40°

20° 20°

0° 0°

0 800 1600 km

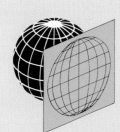

AZIMUTHAL OR ZENITHAL PROJECTIONS

These are constructed by the projection of part of the graticule from the globe onto a plane tangential to any single point on it. This plane may be tangential to the equator (**equatorial case**), the poles (**polar case**) or any other point (**oblique case**). Any straight line drawn from the point at which the plane touches the globe is the shortest distance from that point and is known as a **great circle**. In its **Gnomonic** construction *any* straight line on the map is a great circle, but there is great exaggeration towards the edges and this reduces its general uses. There are five different ways of transferring the graticule onto the plane and these are shown on the right. The central diagram below shows how the graticules vary, using the polar case as the example.

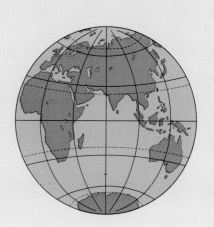

Equidistant Equal-Area Orthographic Gnomonic Stereographic (conformal)

Oblique Case

The plane touches the globe at any point between the equator and poles. The oblique orthographic uses the distortion in azimuthal projections away from the centre to give a graphic depiction of the earth as seen from any desired point in space. It can also be used in both Polar and Equatorial cases. It is used not only for the earth but also for the moon and planets.

Polar Case

The polar case is the simplest to construct and the diagram below shows the differing effects of all five methods of construction comparing their coverage, distortion etc., using North America as the example.

Equatorial Case

The example shown here is Lambert's Equivalent Azimuthal. It is the only projection which is both equal area and where bearing is true from the centre.

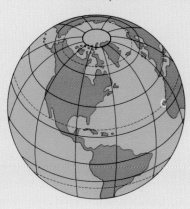

Stereographic

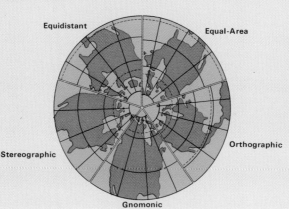

Equidistant Equal-Area

Gnomonic

Orthographic

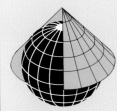

CONICAL PROJECTIONS

These use the projection of the graticule from the globe onto a cone which is tangential to a line of latitude (termed the **standard parallel**). This line is always an arc and scale is always true along it. Because of its method of construction it is used mainly for depicting the temperate latitudes around the standard parallel i.e. where there is least distortion. To reduce the distortion and include a larger range of latitudes, the projection may be constructed with the cone bisecting the surface of the globe so that there are two standard parallels each of which is true to scale. The distortion is thus spread more evenly between the two chosen parallels.

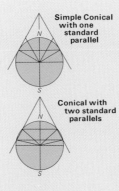

Simple Conical with one standard parallel

Conical with two standard parallels

Bonne
This is a modification of the simple conic whereby the true scale along the meridians is sacrificed to enable the accurate representation of areas. However scale is true along each parallel but shapes are distorted at the edges.

Simple Conic
Scale is correct not only along the standard parallel but also along all meridians. The selection of the standard parallel used is crucial because of the distortion away from it. The projection is usually used to portray regions or continents at small scales.

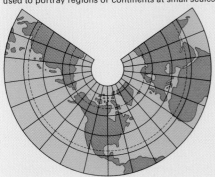

Lambert's Conformal Conic
This projection uses two standard parallels but instead of being equal area as Albers, it is Conformal. Because it has comparatively small distortion, direction and distances can be readily measured and it is therefore used for some navigational charts.

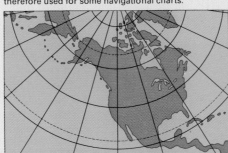

Albers Conical Equal Area
This projection uses two standard parallels and once again the selection of the two specific ones relative to the land area to be mapped is very important. It is equal area and is especially useful for large land masses oriented East-West, for example the U.S.A.

CYLINDRICAL AND OTHER WORLD PROJECTIONS

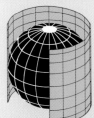

This group of projections are those which permit the whole of the Earth's surface to be depicted on one map. They are a very large group of projections and the following are only a few of them. Cylindrical projections are constructed by the projection of the graticule from the globe onto a cylinder tangential to the globe. In the examples shown here the cylinder touches the equator, but it can be moved through 90° so it touches the poles - this is called the **Transverse Aspect**. If the cylinder is twisted so that it touches anywhere between the equator and poles it is called the **Oblique Aspect**. Although cylindrical projections can depict all the main land masses, there is considerable distortion of shape and area towards the poles. One cylindrical projection, **Mercator** overcomes this shortcoming by possessing the unique navigational property that any straight drawn on it is a line of constant bearing (**loxodrome**), i.e. a straight line route on the globe crosses the parallels and meridians on the map at the same angles as on the globe. It is used for maps and charts between 15° either side of the equator. Beyond this enlargement of area is a serious drawback, although it is used for navigational charts at all latitudes.

Cylindrical with two standard parallels

Simple Cylindrical

Mercator

Mollweide
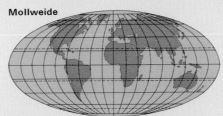

Sanson-Flamsteed

Hammer
This is not a cylindrical projection, but is developed from the Lambert Azimuthal Equal Area by doubling all the East-West distances along the parallels from the central meridian. Like both Sanson-Flamsteed and Mollweide it is distorted towards its edges but has curved parallels to lessen the distortion.

Mollweide and Sanson-Flamsteed
Both of these projections are termed **pseudo-cylindrical**. They are basically cylindrical projections where parallels have been progressively shortened and drawn to scale towards the poles. This allows them to overcome the gross distortions exhibited by the ordinary cylindrical projections and they are in fact Equal Area, Mollweide's giving a slightly better shape. To improve the shape of the continents still further they, like some other projections can be **Interrupted** as can be seen below, but this is at the expense of contiguous sea areas. These projections can have any central meridian and so can be 'centred' on the Atlantic, Pacific, Asia, America etc. In this form both projections are suitable for any form of mapping statistical distributions.

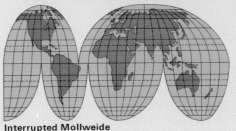

Interrupted Mollweide

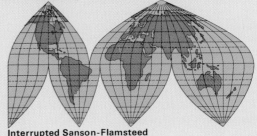

Interrupted Sanson-Flamsteed

User Guide

Organization of the atlas

Prepared in accordance with the highest standards of international cartography to provide accurate and detailed representation of the earth, the atlas is made up of four separate sections and is organized with ease of use in mind.

The first section of the atlas consists of up-to-date world geographical and demographical statistics, graphics on map projections intended to help the reader understand how cartographers create and use map projections, and this user guide.

The second section of the atlas, the 32-page United States Maps section, has blue page borders and offers comprehensive coverage of the United States and its outlying areas, with climate and agricultural maps, politically colored maps with some topographical detail, maps of major urban areas, and a 16-page index with longitude and latitude coordinates.

The third section of the atlas, the 32-page Introduction to World Geography section, consists of thematic maps, graphs, and charts on a range of geographical and demographical topics, and a subject index.

The fourth and final section of the atlas, the 96-page World Maps section, has gray page borders and covers the earth continent by continent in the classic sequence adopted by cartographers since the 16th century. This section begins with Europe, then Asia, Africa, Australia and Oceania, North America, and South America. For each continent, there are maps at a variety of scales: first, physical relief maps and political maps of the whole continent, then large scale maps of the most important or densely populated areas.

The governing principle is that by turning the pages of the World Maps section, the reader moves steadily from north to south through each continent, with each map overlapping its neighbors. Immediately following the maps in the World Maps section is the comprehensive index to the maps, which contains 44,000 entries of both place names and geographical features. The index provides the latitude and longitude coordinates as well as letters and numbers, so that locating any site can be accomplished with speed and accuracy.

Map presentation

All of the maps in the atlas are drawn with north at the top (except for the map of the Arctic Ocean and the map of Antarctica). The maps in the United States Maps section and the World Maps section contain the following information in their borders: the map title; the scale; the projection used; the degrees of latitude and longitude; and on the physical relief maps, a height and depth reference panel identifying the colors used for each layer of contouring. In addition to this information, the maps in the World Maps section also contain locator diagrams which show the area covered, the page numbers for adjacent maps, and the letters and numbers used in the index for locating place names and geographical features.

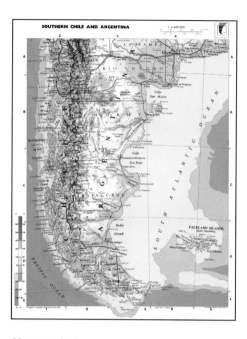

Map symbols

Each map contains a vast amount of detail which is conveyed clearly and accurately by the use of symbols. Points and circles of varying sizes locate and identify the relative importance of towns and cities; different styles of type are employed for administrative, geographical and regional place names. A variety of pictorial symbols denote landscape features such as glaciers, marshes and reefs, and man-made structures including roads, railroads, airports, canals and dams. International borders are shown by red lines. Where neighboring countries are in dispute, the maps show the *de facto* boundary between nations, regardless of the legal or historical situation. The symbols are explained on the first page of each of the map sections.

Map scales

The scale of each map is given in the numerical form known as the representative fraction. The first figure is always one, signifying one unit of distance on the map; the second figure, usually in millions, is the number by which the map unit must be multiplied to give the equivalent distance on

LARGE SCALE		
1: 1 000 000	1 cm = 10 km	1 inch = 16 miles
1: 2 500 000	1 cm = 25 km	1 inch = 39.5 miles
1: 5 000 000	1 cm = 50 km	1 inch = 79 miles
1: 6 000 000	1 cm = 60 km	1 inch = 95 miles
1: 8 000 000	1 cm = 80 km	1 inch = 126 miles
1: 10 000 000	1 cm = 100 km	1 inch = 158 miles
1: 15 000 000	1 cm = 150 km	1 inch = 237 miles
1: 20 000 000	1 cm = 200 km	1 inch = 316 miles
1: 50 000 000	1 cm = 500 km	1 inch = 790 miles
SMALL SCALE		

the earth's surface. Calculations can easily be made in centimeters and kilometers, by dividing the earth units figure by 100 000 (i.e. deleting the last five 0s). Thus 1:1 000 000 means l cm = 10 km. The calculation for inches and miles is more laborious, but 1 000 000

divided by 63 360 (the number of inches in a mile) shows that 1:1 000 000 means approximately 1 inch = 16 miles. The table shown provides distance equivalents for scales down to 1:50 000 000.

Measuring distances

Although each map is accompanied by a scale bar, distances cannot always be measured with confidence because of the distortions involved in portraying the curved surface of the earth on a flat page. As a general rule, the larger the map scale (i.e. the lower the number of earth units in the representative fraction), the more accurate and reliable will be the distance measured. On small scale maps such as those of the world and of entire continents, measurement may only be accurate along the standard parallels, or central axes, and should not be attempted without considering the map projection.

Latitude and longitude

Accurate positioning of individual points on the earth's surface is made possible by reference to the geometrical system of latitude and longitude. Latitude parallels are drawn west–east around the earth and numbered by degrees north and south of the Equator, which is designated 0° of latitude. Longitude meridians are drawn north–south and numbered by degrees east and west of the prime meridian, 0° of longitude, which passes through Greenwich in England. By referring to these coordinates and their subdivisions of minutes (1/60th of a degree) and seconds (1/60th of a minute), any place on earth can be located to within a few hundred yards. Latitude and longitude are indicated by blue lines on the maps; they are straight or curved according to the projection employed. Reference to these lines is the easiest way of determining the relative positions of places on different large scale maps, and for plotting compass directions.

Name forms

For ease of reference, both English and local name forms appear in the atlas. Oceans, seas and countries are shown in English throughout the atlas; country names may be abbreviated to their commonly accepted form (e.g. Germany, not The Federal Republic of Germany). Conventional English forms are also used for place names on the smaller scale maps of the continents. However, local name forms are used on all large scale and regional maps, with the English form given in brackets only for important cities – the large scale map of Eastern Europe and Turkey thus shows Moskva (Moscow). For countries which do not use a Roman script, place names have been transcribed according to the systems adopted by the British and US Geographic Names Authorities. For China, the Pin Yin system has been used, with some more widely known forms appearing in brackets, as with Beijing (Peking). Both English and local names appear in the index to the world maps.

UNITED STATES MAPS

— SETTLEMENTS —

◌ WASHINGTON D.C.　■ Tampa　◉ Fresno　◉ Waterloo　◉ *Ventura*　⊙ *Barstow*　○ *Blythe*　○ *Hope*

Settlement symbols and type styles vary according to the scale of each map and indicate the importance
of towns on the map rather than specific population figures

— ADMINISTRATION —

——— International Boundaries

‒‒‒···· Internal Boundaries

National Parks, Recreation
Areas and Monuments

Country Names
CANADA

Administrative
Area Names

MICHIGAN

— COMMUNICATIONS —

══ Major Highways

⌒ Other Principal Roads

≍ Passes

✈ ⊹ ⊙ Airports and Airfields

⌒ Principal Railroads

‒·‒‒ Railroads
Under Construction

⌒ Other Railroads

⊣‒‒‒⊦ Railroad Tunnels

⊔⊔⊔⊔⊔ Principal Canals

— PHYSICAL FEATURES —

⌒ Perennial Streams

‒‒‒‒‒ Intermittent Streams

⬭ Perennial Lakes
and Reservoirs

⬭ Intermittent Lakes and
Salt Flats

Swamps and Marshes

Permanent Ice
and Glaciers

▲ 8848 Elevations in meters

▼ 8050 Sea Depths in meters

1134 Height of Lake Surface
Above Sea Level
in meters

I meter is approx. 3.3 feet

— CITY MAPS —

In addition to, or instead of, the symbols explained above, the following symbols are used on the city maps between pages 20-29

Urban Areas

⌒ Limited Access Roads

‒·‒·‒ Aqueducts

Woodland and Parks

⌒ Secondary Roads

‒‒‒‒ Ferry Routes

⌒ State Boundaries

✕ Airports

⌒ Canals

⌒ County Boundaries

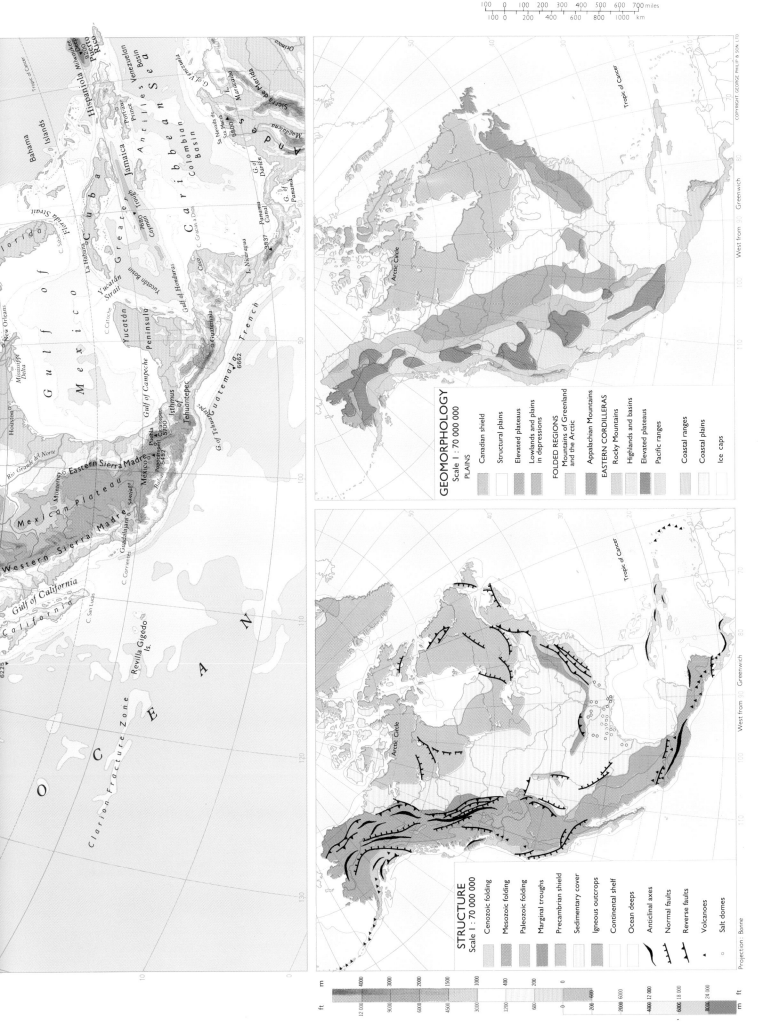

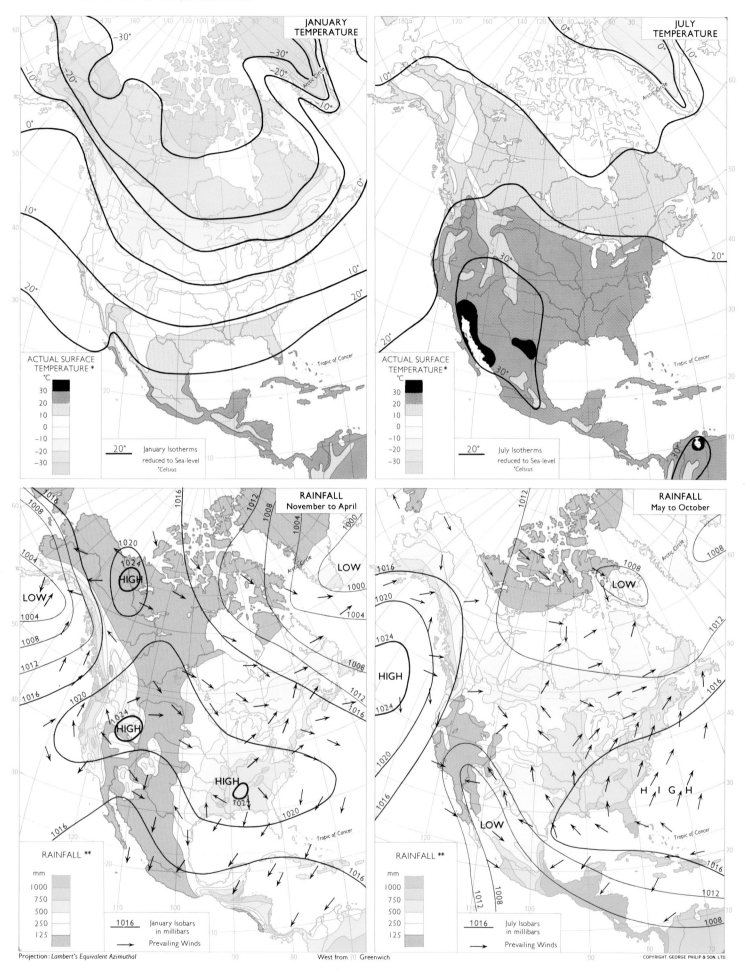

*To convert °C to °F, multiply by 1.8, then add 32   **1 in equals 25.4mm

1:32 000 000

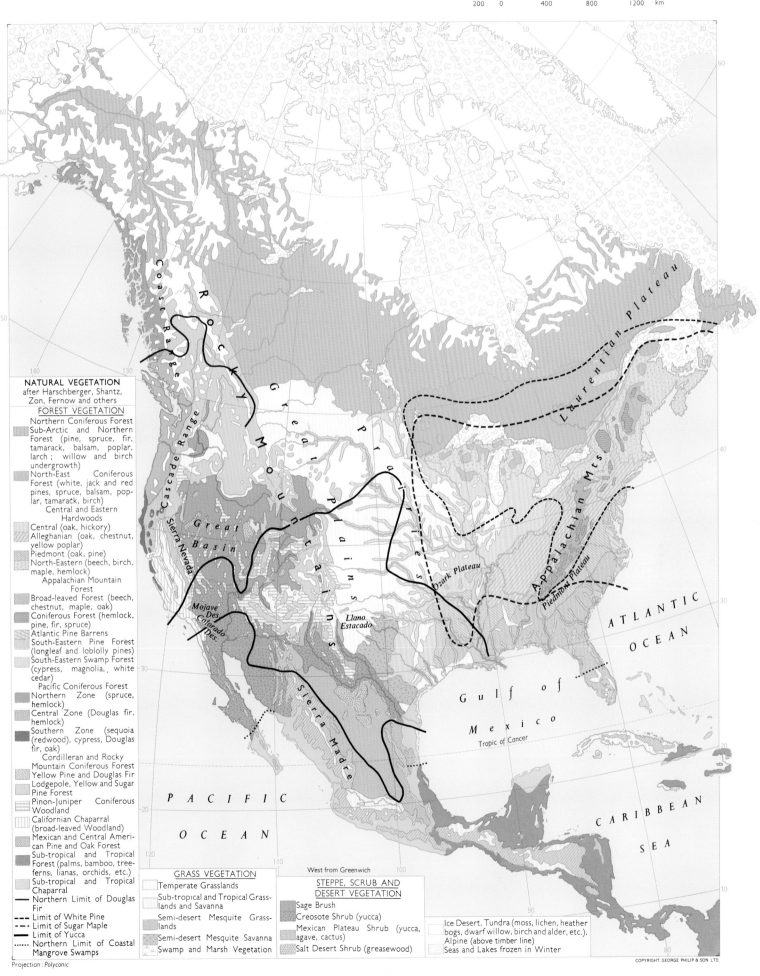

NATURAL VEGETATION
after Harschberger, Shantz,
Zon, Fernow and others

FOREST VEGETATION

Northern Coniferous Forest

Sub-Arctic and Northern Forest (pine, spruce, fir, tamarack, balsam, poplar, larch; willow and birch undergrowth)

North-East Coniferous Forest (white, jack and red pines, spruce, balsam, poplar, tamarack, birch)

Central and Eastern Hardwoods

Central (oak, hickory)

Alleghanian (oak, chestnut, yellow poplar)

Piedmont (oak, pine)

North-Eastern (beech, birch, maple, hemlock)

Appalachian Mountain Forest

Broad-leaved Forest (beech, chestnut, maple, oak)

Coniferous Forest (hemlock, pine, fir, spruce)

Atlantic Pine Barrens

South-Eastern Pine Forest (longleaf and loblolly pines)

South-Eastern Swamp Forest (cypress, magnolia, white cedar)

Pacific Coniferous Forest

Northern Zone (spruce, hemlock)

Central Zone (Douglas fir, hemlock)

Southern Zone (sequoia (redwood), cypress, Douglas fir, oak)

Cordilleran and Rocky Mountain Coniferous Forest

Yellow Pine and Douglas Fir

Lodgepole, Yellow and Sugar Pine Forest

Pinon-Juniper Coniferous Woodland

Californian Chaparral (broad-leaved Woodland)

Mexican and Central American Pine and Oak Forest

Sub-tropical and Tropical Forest (palms, bamboo, tree-ferns; lianas, orchids, etc.)

Sub-tropical and Tropical Chaparral

——— Northern Limit of Douglas Fir

– – – Limit of White Pine

–·–·– Limit of Sugar Maple

━━━ Limit of Yucca

········· Northern Limit of Coastal Mangrove Swamps

GRASS VEGETATION

Temperate Grasslands

Sub-tropical and Tropical Grasslands and Savanna

Semi-desert Mesquite Grasslands

Semi-desert Mesquite Savanna

Swamp and Marsh Vegetation

STEPPE, SCRUB AND DESERT VEGETATION

Sage Brush

Creosote Shrub (yucca)

Mexican Plateau Shrub (yucca, agave, cactus)

Salt Desert Shrub (greasewood)

Ice Desert, Tundra (moss, lichen, heather bogs, dwarf willow, birch and alder, etc.).

Alpine (above timber line)

Seas and Lakes frozen in Winter

West from Greenwich

Projection: Polyconic

HAWAII
1:10 000 000

Projection: Albers' Equal Area with two standard parallels

West from Greenwich

National Capital ★
State Capital ■ ● ● ● ●

1:12 000 000

COPYRIGHT GEORGE PHILIP & SON, LTD.

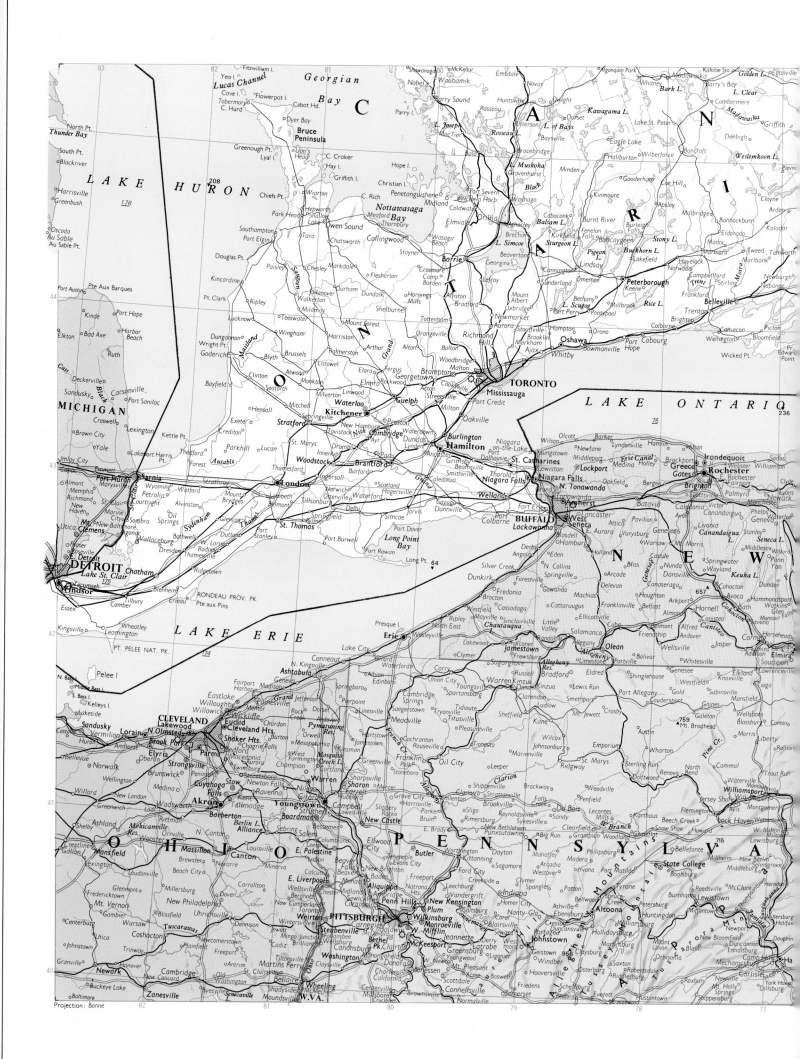

Projection: Bonne

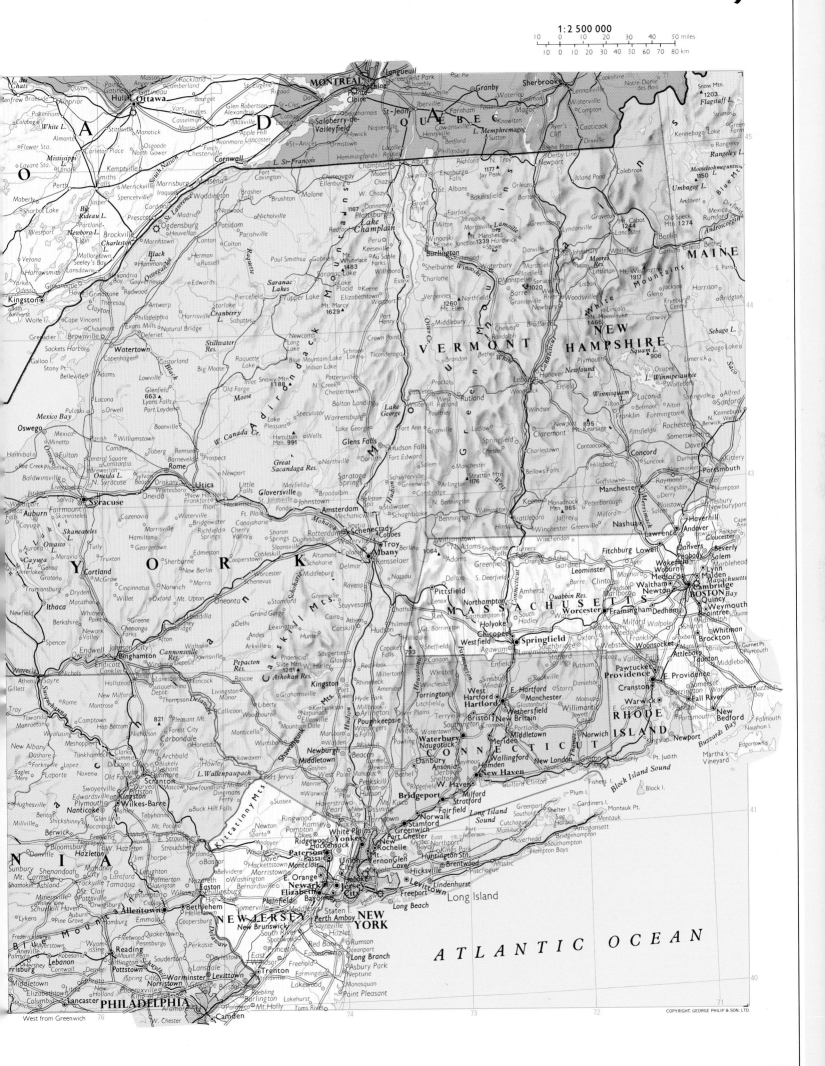

1 : 2 500 000

10   0   10   20   30   40   50 miles
10 0 10 20 30 40 50 60 70 80 km

ATLANTIC OCEAN

COPYRIGHT. GEORGE PHILIP & SON. LTD.

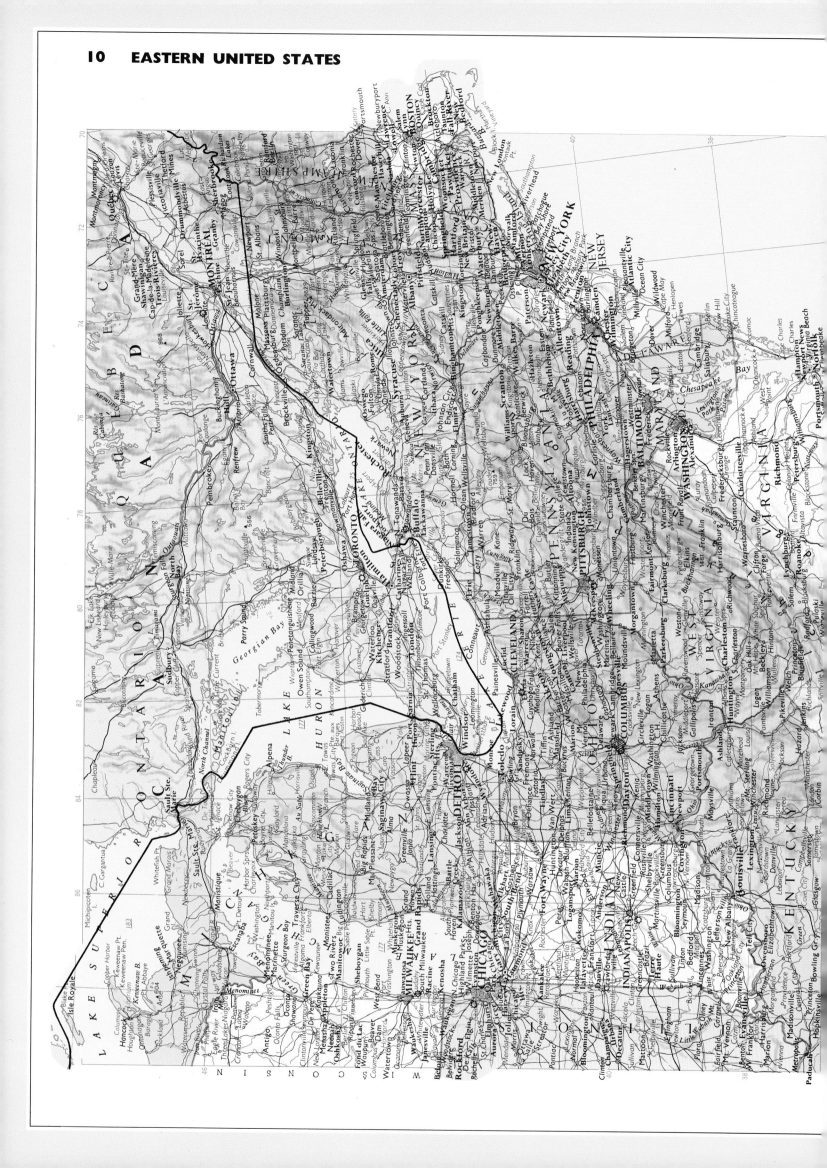

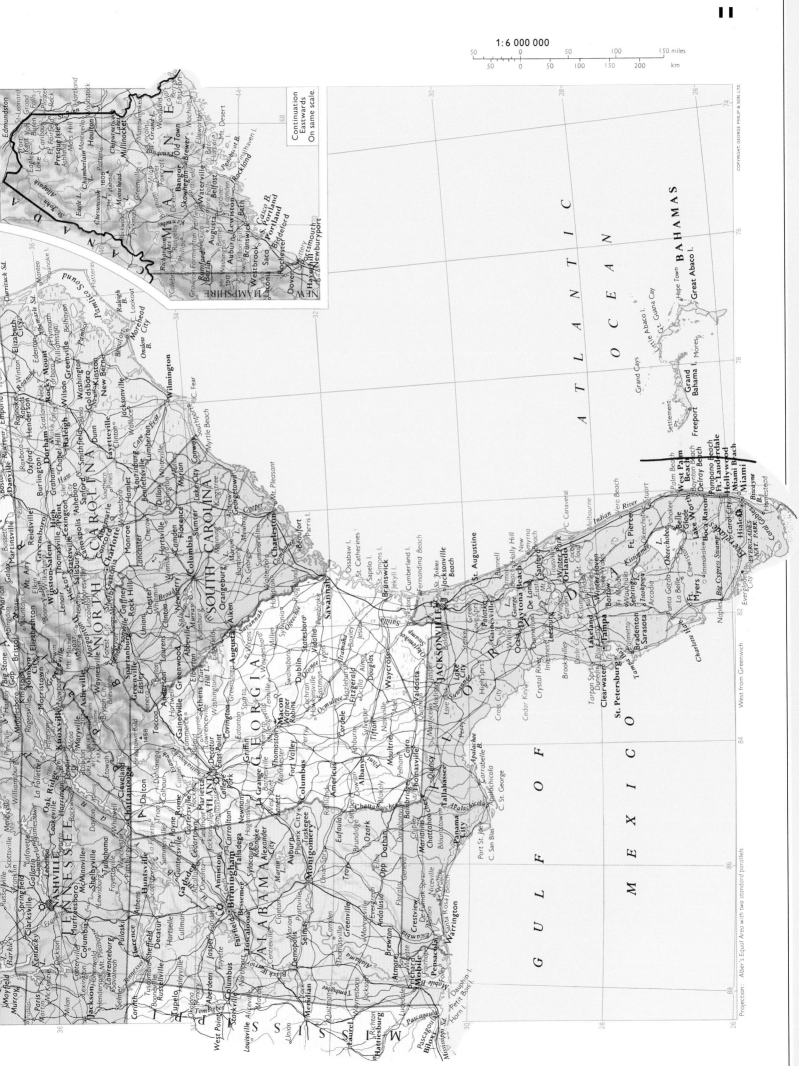

1 : 6 000 000

Continuation
Eastwards
On same scale.

50   50   100   150 miles

50   0   50   100   150   200   km

MAINE

NEW HAMPSHIRE

VERMONT

NORTH CAROLINA

SOUTH CAROLINA

GEORGIA

FLORIDA

ALABAMA

TENNESSEE

MISSISSIPPI

A T L A N T I C

O C E A N

BAHAMAS

Great Abaco I.

Grand
Bahama I.

G U L F   O F

M E X I C O

Projection: Alber's Equal Area with two standard parallels

West from Greenwich

1:6 000 000

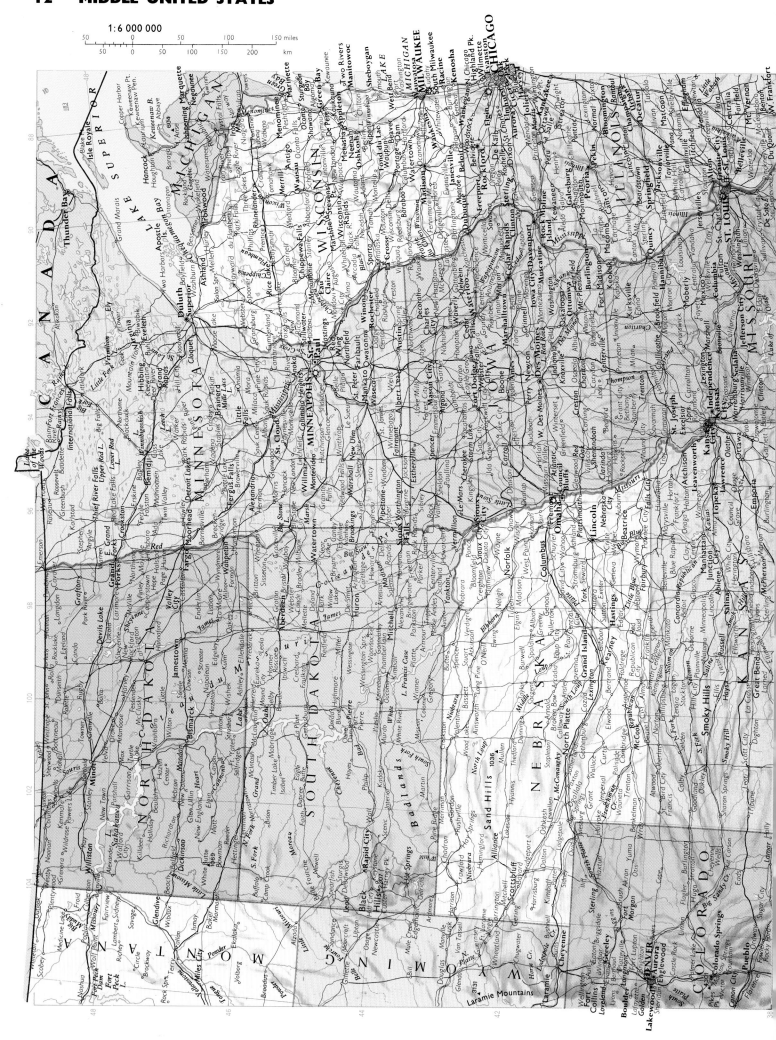

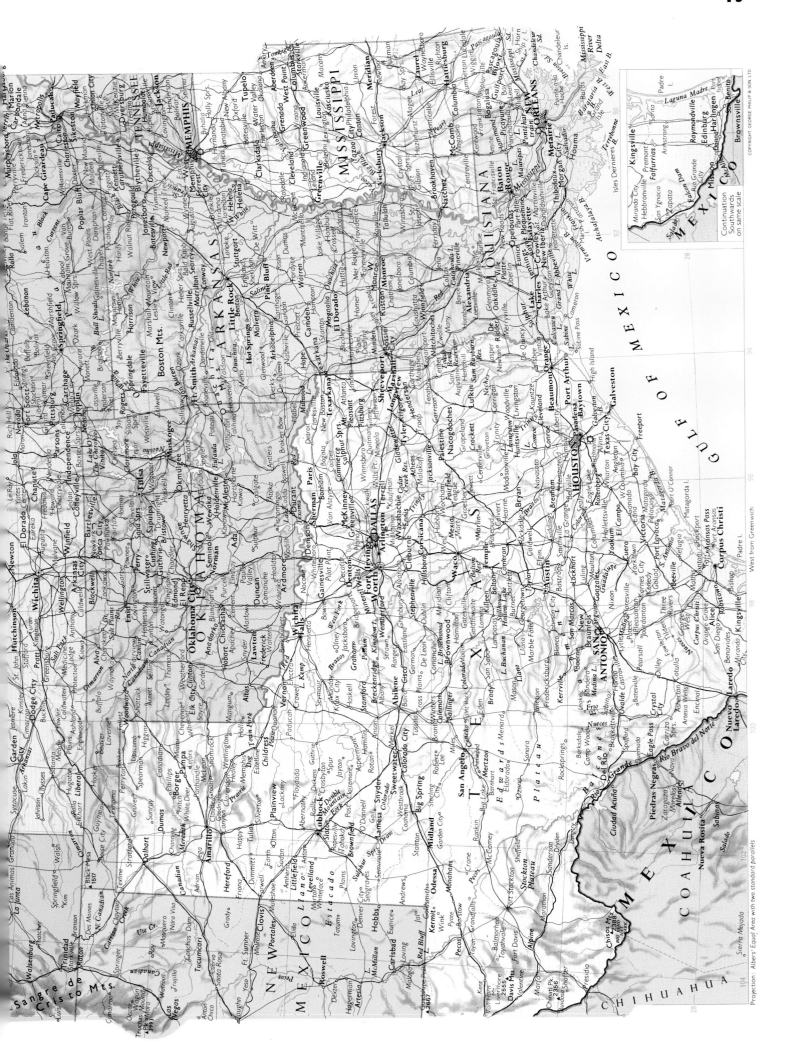

Projection: Albers' Equal Area with two standard parallels

West from Greenwich

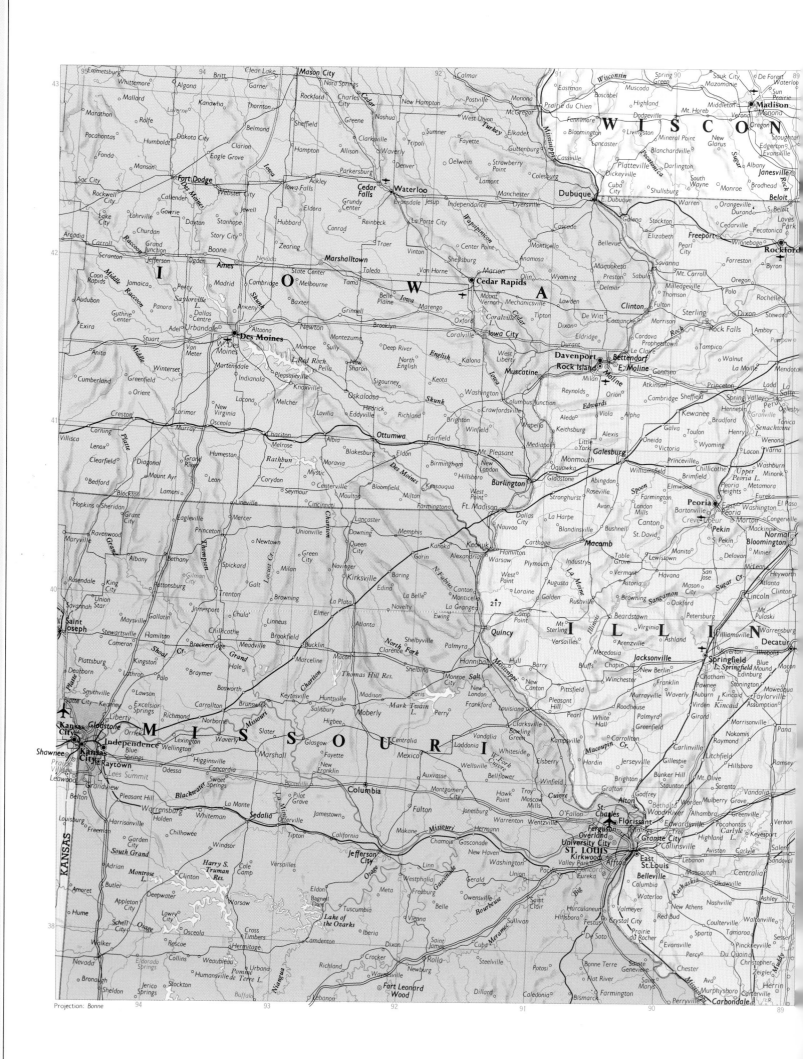

Projection: Bonne

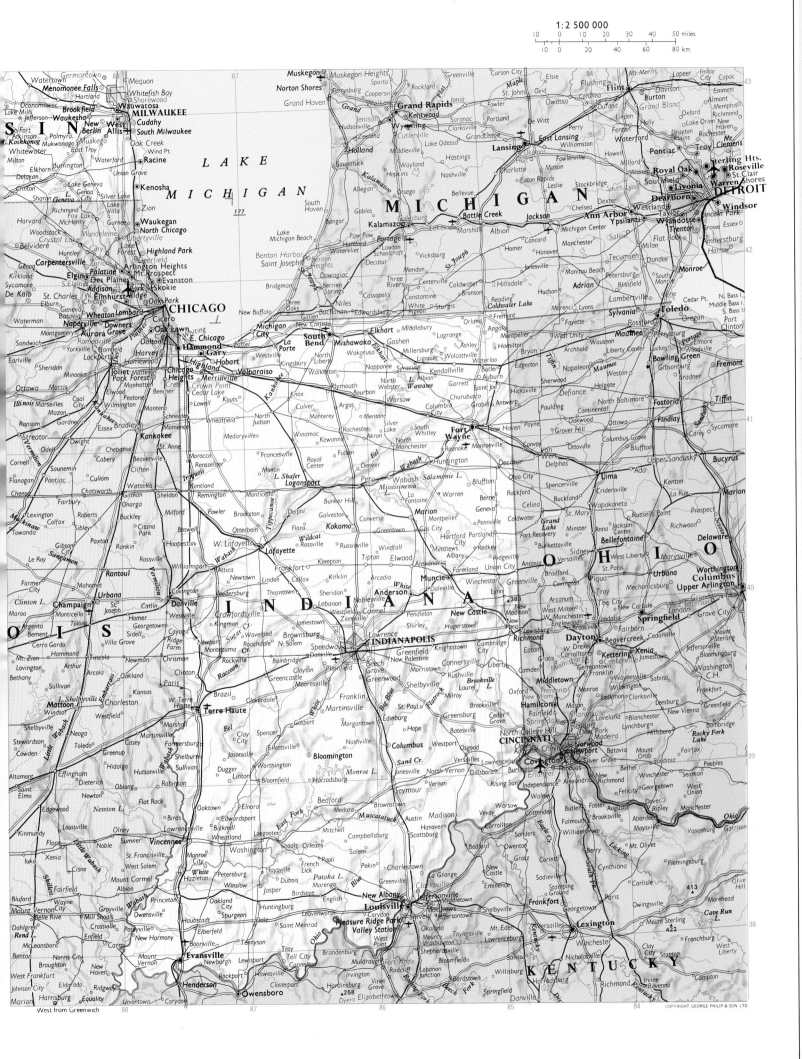

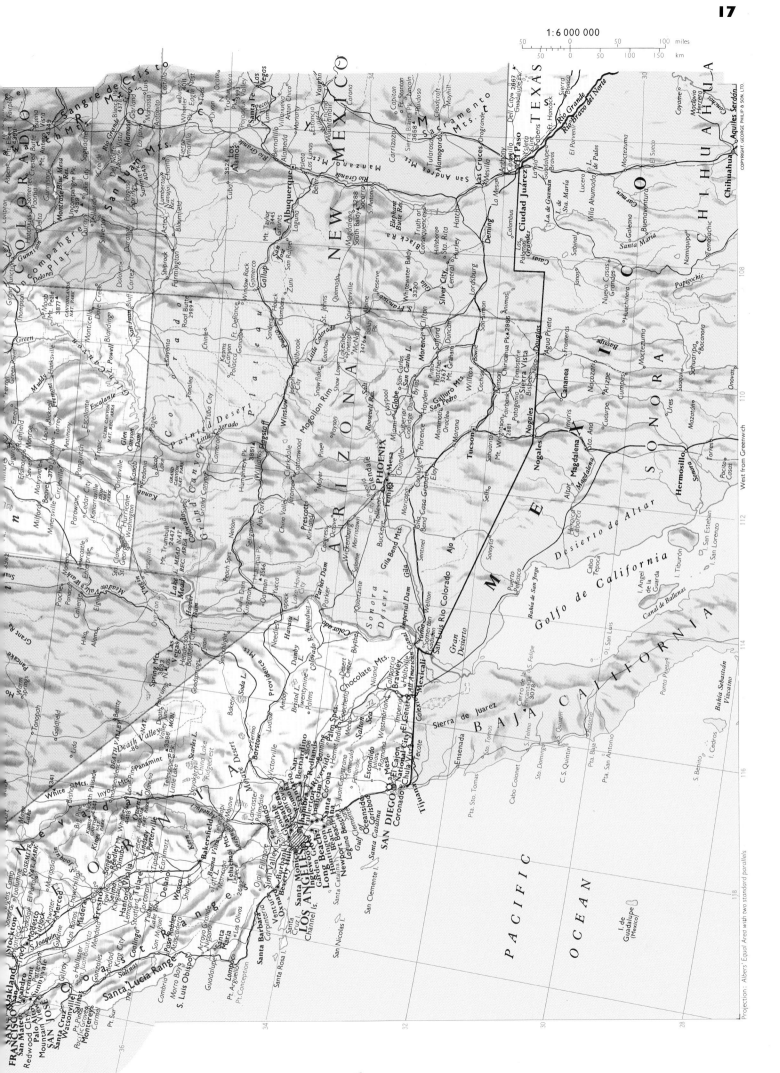

1 : 6 000 000

50   0   50   100 miles
50   0   50   100   150 km

COPYRIGHT GEORGE PHILIP & SON LTD.

West from Greenwich

Projection: Albers' Equal Area with two standard parallels

COLORADO

NEW MEXICO

TEXAS

CHIHUAHUA

ARIZONA

SONORA

MÉXICO

BAJA CALIFORNIA

CALIFORNIA

NEVADA

UTAH

Sangre de Cristo Mts.
San Juan Mts.

Colorado Plateau
Painted Desert

Grand Canyon

Mogollon Rim

Sonora Desert

Gran Desierto

Desierto de Altar

Golfo de California

Santa Lucia Range

Death Valley

PACIFIC

OCEAN

SAN FRANCISCO

LOS ANGELES

PHOENIX

SAN DIEGO

Tijuana
Mexicali

Ciudad Juárez

El Paso

Albuquerque

Santa Fe

Tucson

Nogales

Hermosillo

Las Vegas

Lake Mead

Rio Grande
Rio Bravo del Norte

I. de Guadalupe
(Mexico)

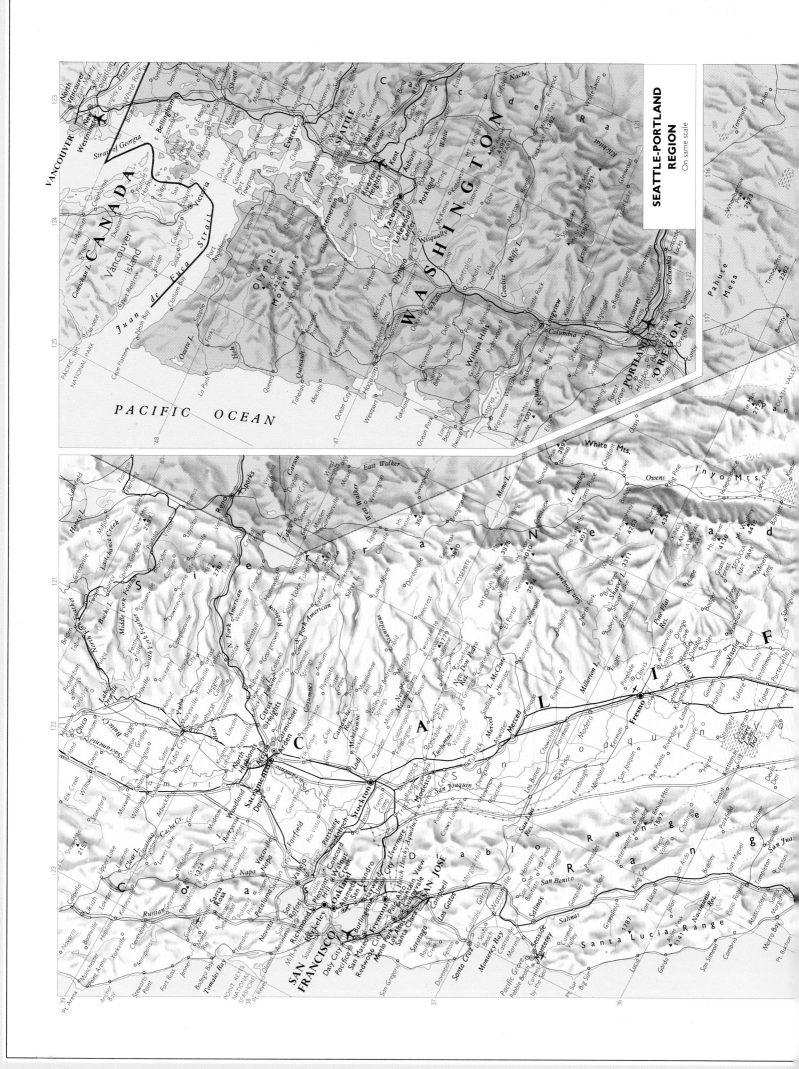

SEATTLE-PORTLAND
REGION
On same scale

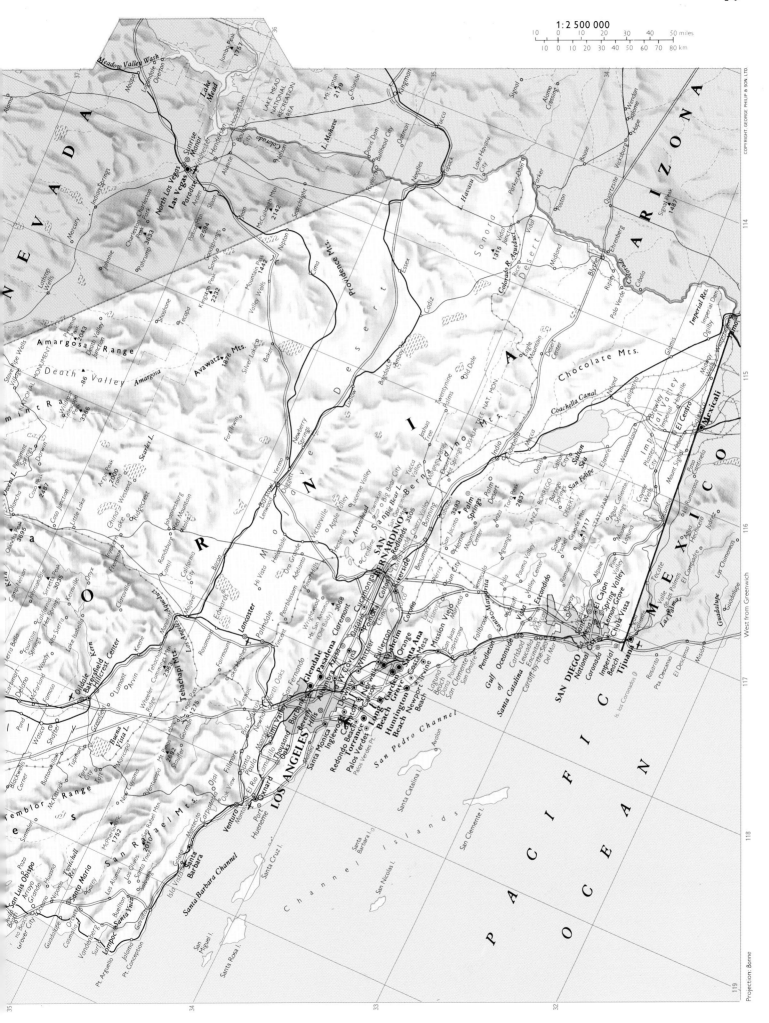

1 : 2 500 000

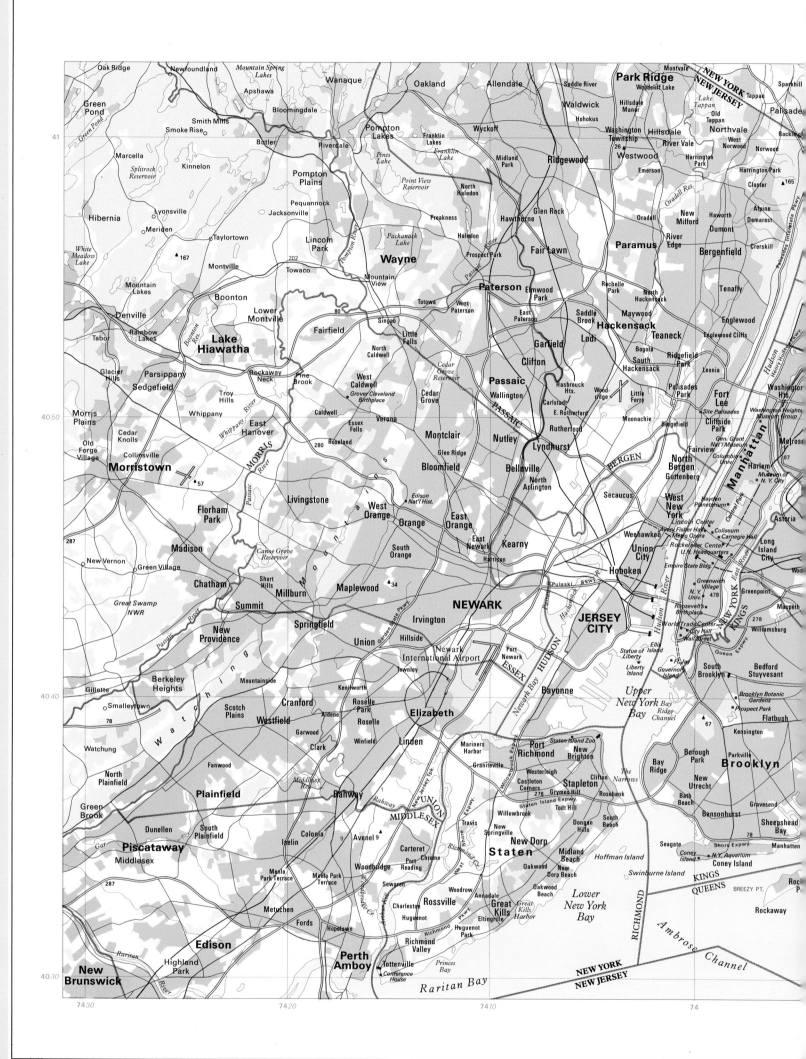

1 : 250 000

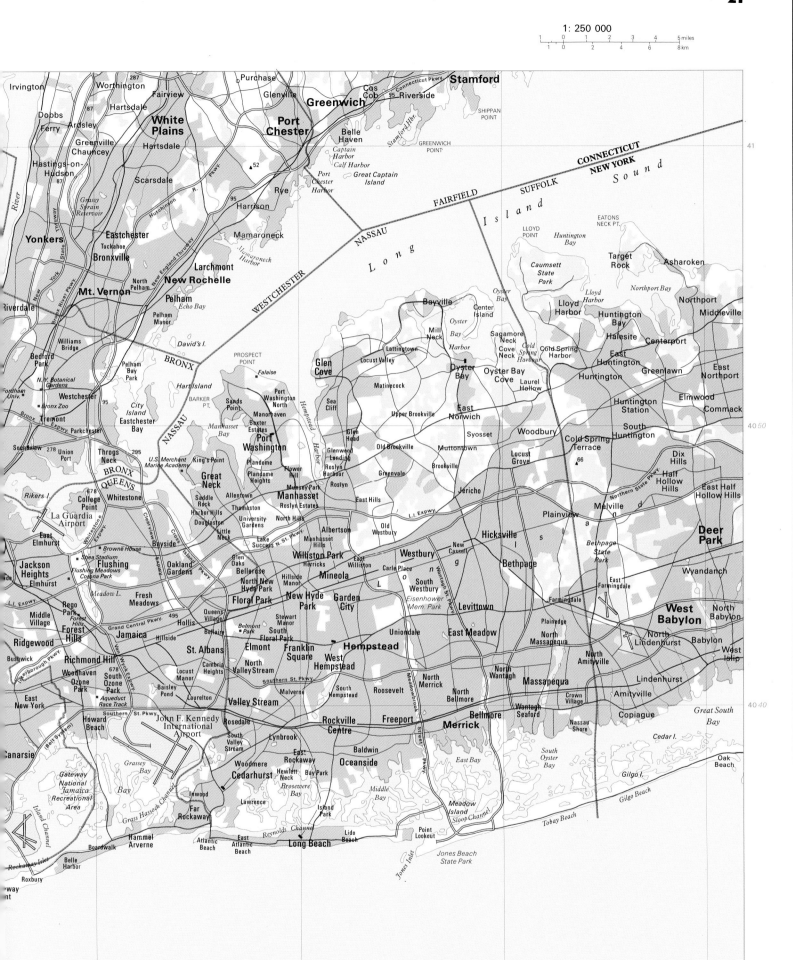

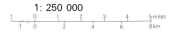

1: 250 000

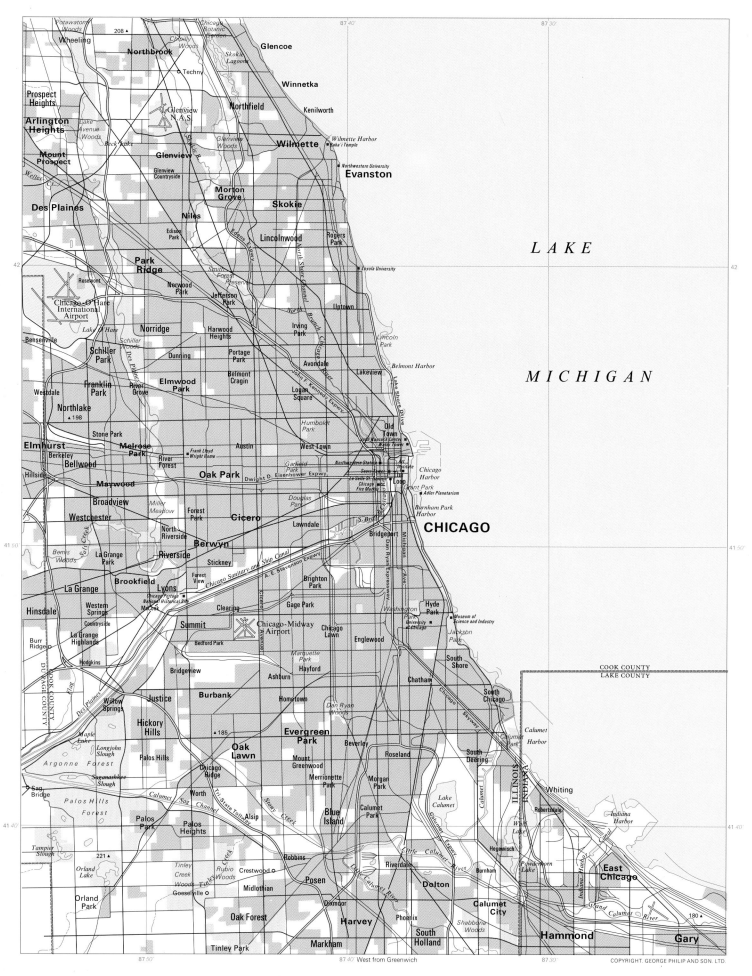

LAKE

MICHIGAN

CHICAGO

87 50'   87 40' West from Greenwich   87 30'

1: 250 000

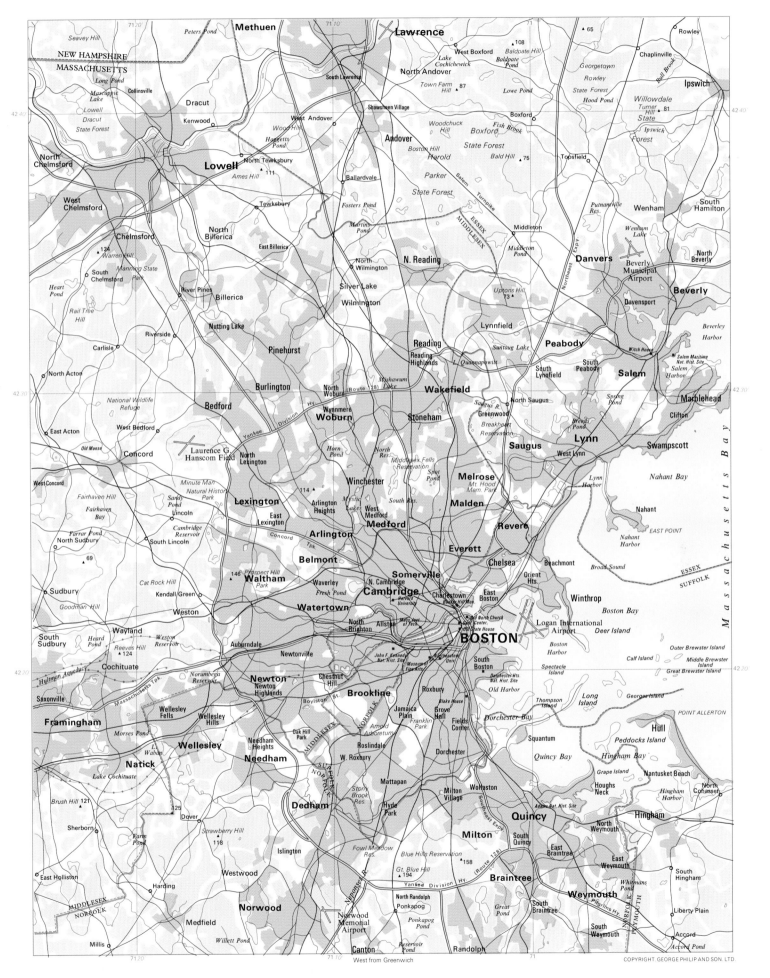

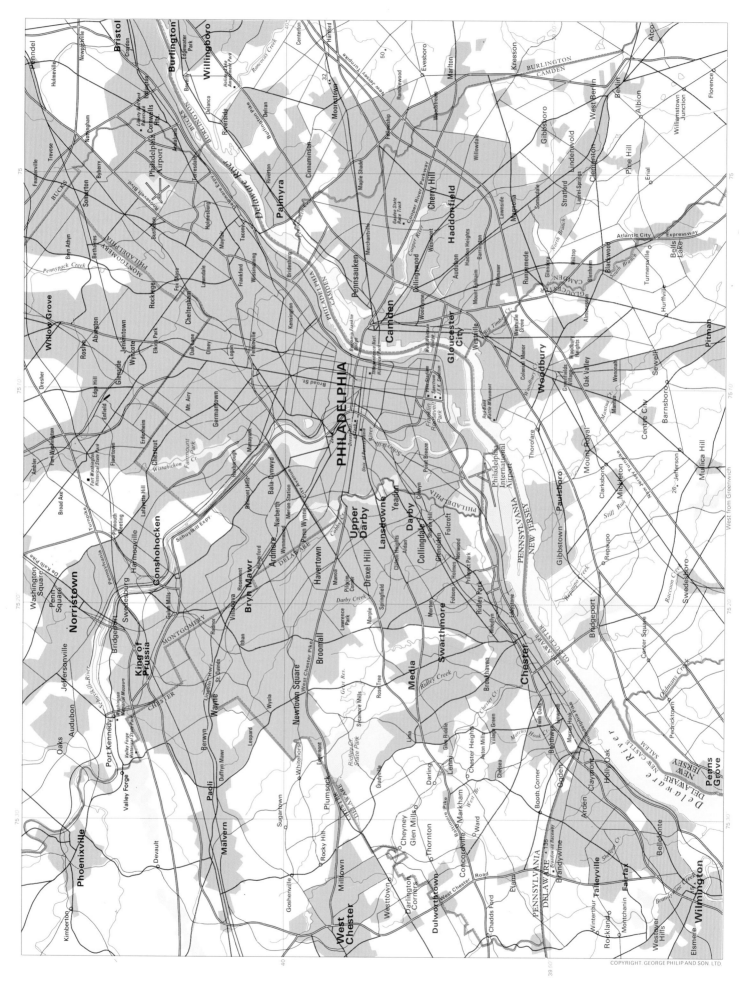

1: 250 000

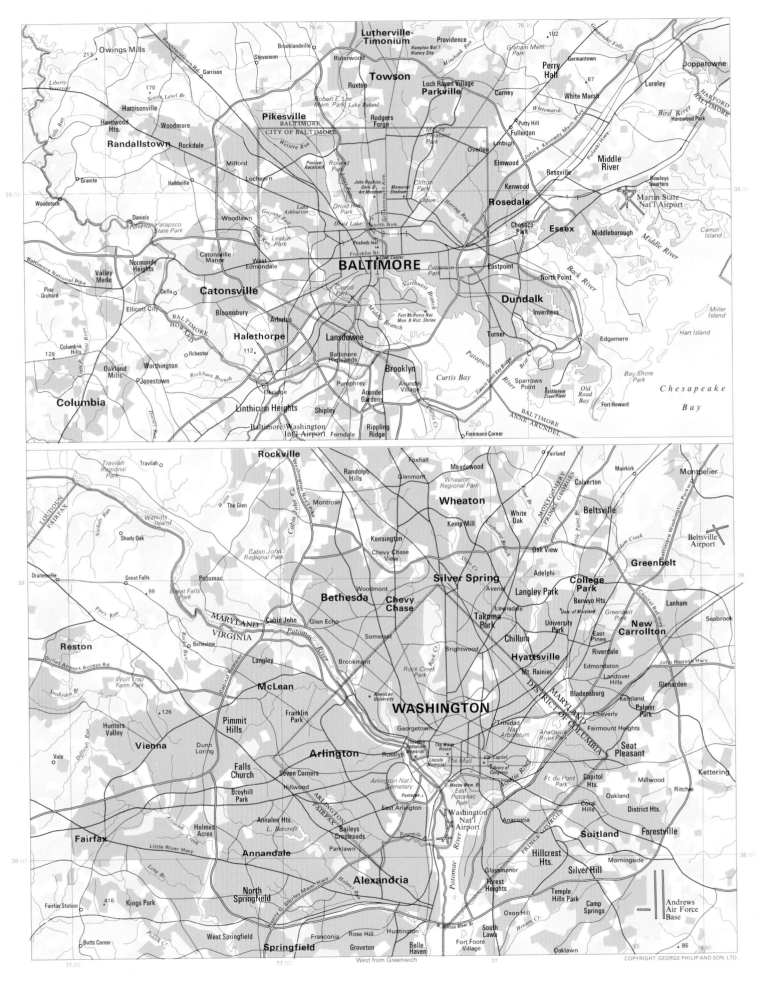

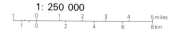

1: 250 000

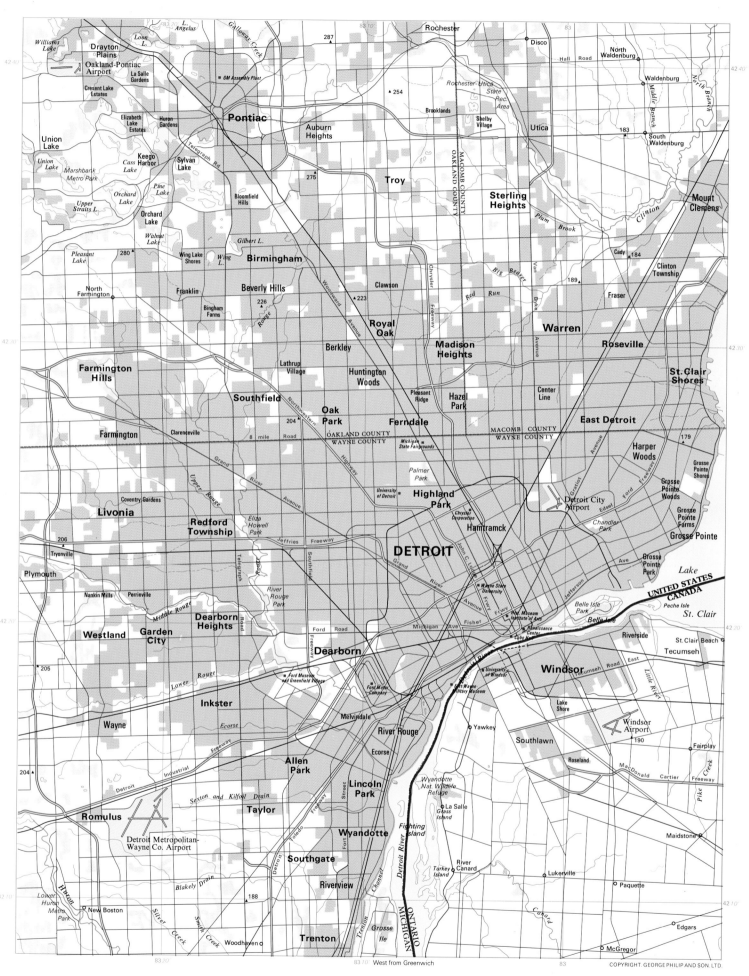

West from Greenwich

1: 250 000

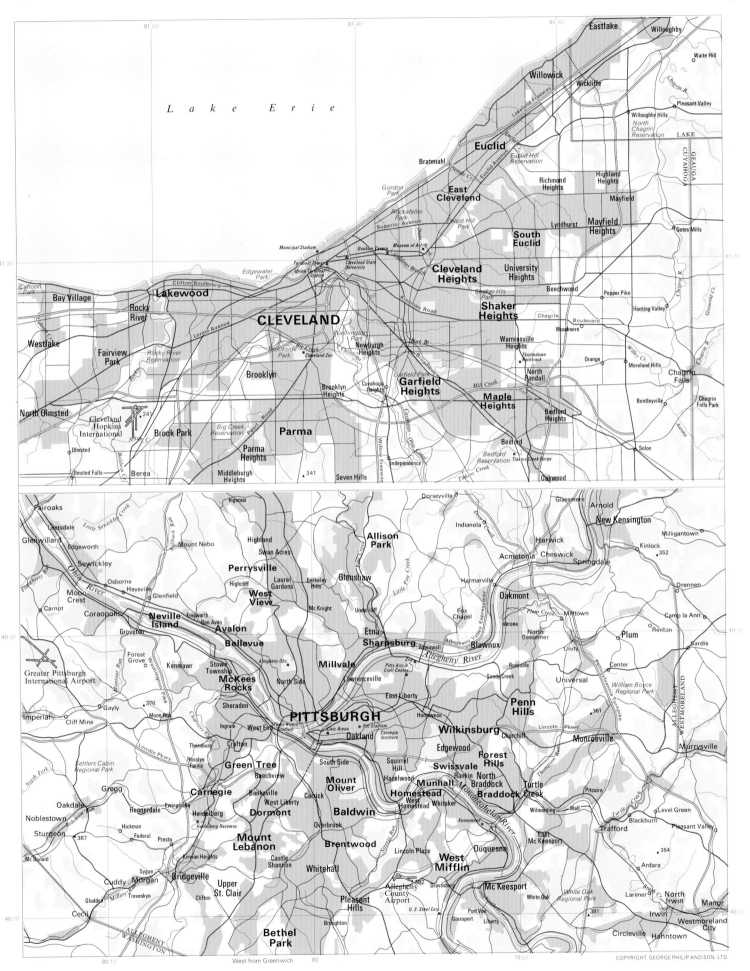

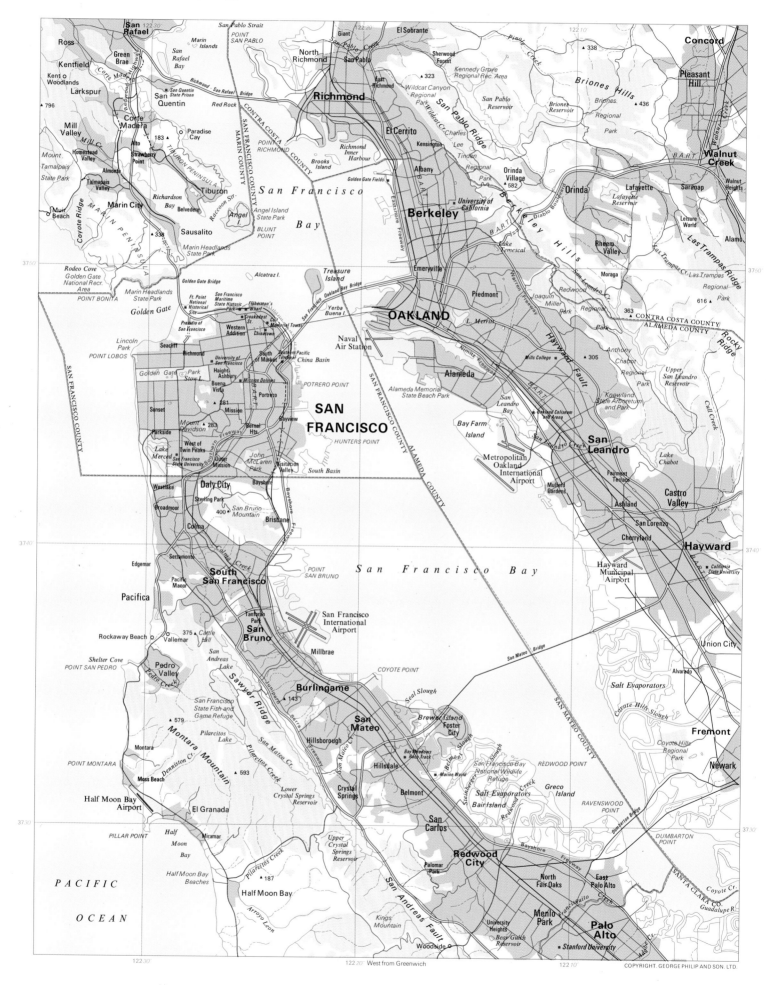

1 : 250 000

PACIFIC

OCEAN

West from Greenwich

1:250 000

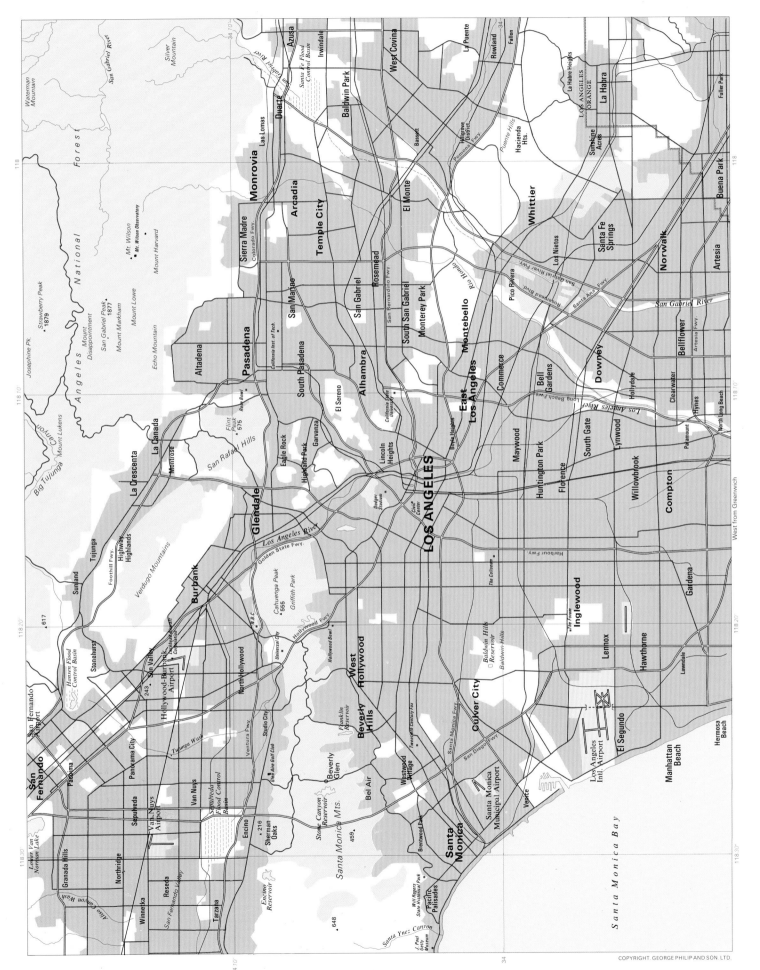

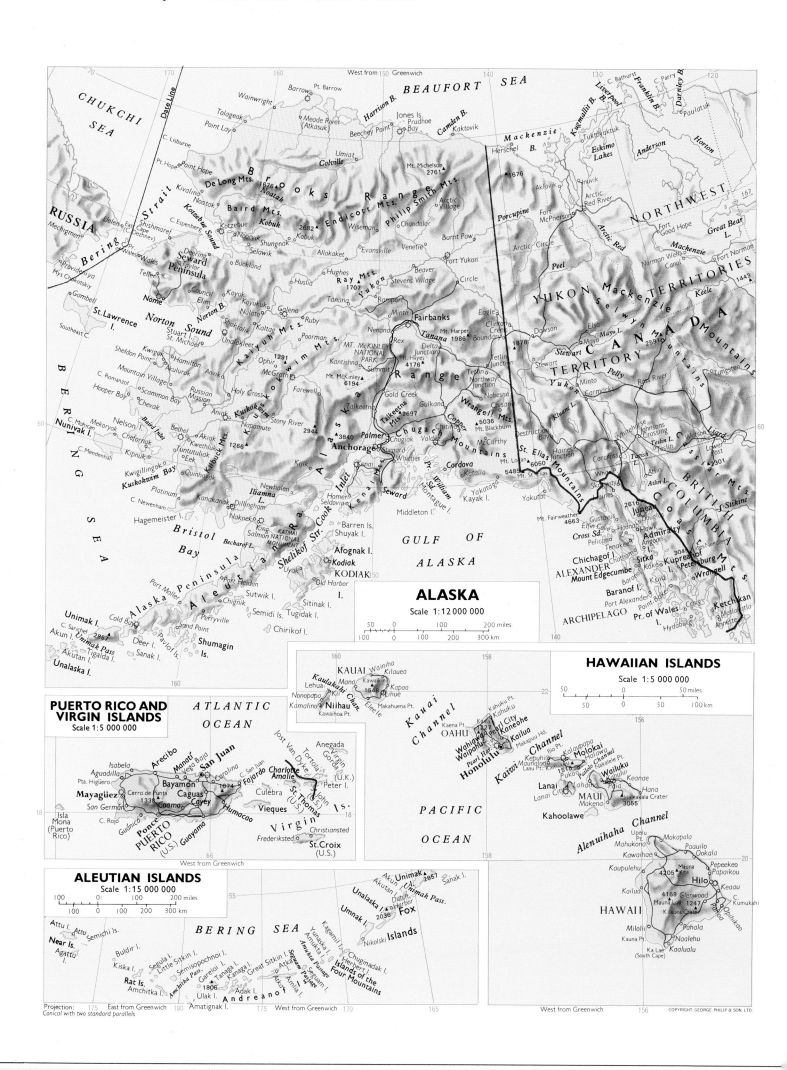

ALASKA
Scale 1:12 000 000

HAWAIIAN ISLANDS
Scale 1:5 000 000

PUERTO RICO AND VIRGIN ISLANDS
Scale 1:5 000 000

ALEUTIAN ISLANDS
Scale 1:15 000 000

Projection:
Conical with two standard parallels

COPYRIGHT. GEORGE PHILIP & SON. LTD.

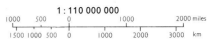

1 : 110 000 000

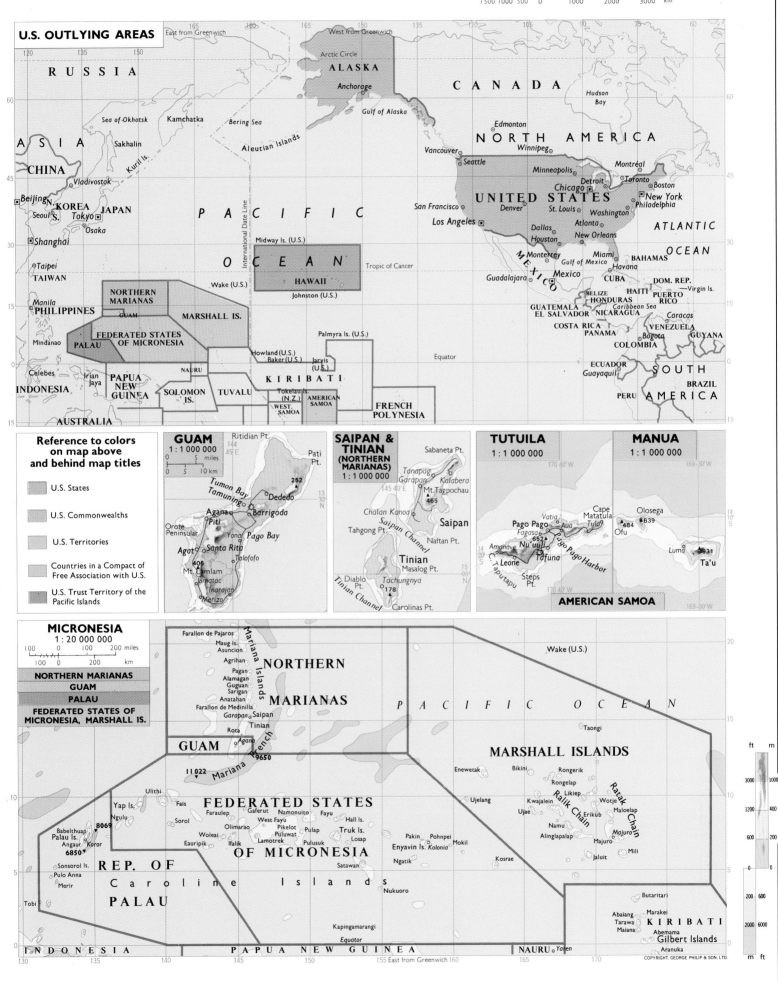

U.S. OUTLYING AREAS

RUSSIA
ASIA
CHINA
Sea of Okhotsk
Kamchatka
Sakhalin
Bering Sea
Aleutian Islands
Kuril
Vladivostok
Beijing
N. KOREA
Seoul S.
Tokyo
JAPAN
Osaka
Shanghai
Taipei
TAIWAN
Manila
PHILIPPINES
Mindanao
Celebes
Irian Jaya
INDONESIA
AUSTRALIA

ALASKA
Anchorage
Arctic Circle
Gulf of Alaska
CANADA
Hudson Bay
NORTH AMERICA
Edmonton
Vancouver
Winnipeg
Seattle
Minneapolis
Montréal
Detroit
Toronto
Boston
Chicago
UNITED STATES
New York
Philadelphia
San Francisco
Denver
St. Louis
Washington
Los Angeles
Dallas
Atlanta
ATLANTIC
Houston
New Orleans
OCEAN
Monterrey
Miami
BAHAMAS
Gulf of Mexico
Havana
Guadalajara
Mexico
CUBA
DOM. REP.
Virgin Is.
BELIZE
HAITI
PUERTO
HONDURAS
RICO
GUATEMALA
Caribbean Sea
EL SALVADOR NICARAGUA
Caracas
COSTA RICA
VENEZUELA
PANAMA
Bogota
GUYANA
COLOMBIA
ECUADOR
SOUTH
Guayaquil
BRAZIL
PERU
AMERICA

PACIFIC
OCEAN
International Date Line
Midway Is. (U.S.)
Tropic of Cancer
HAWAII
Wake (U.S.)
Johnston (U.S.)
NORTHERN
MARIANAS
GUAM
MARSHALL IS.
FEDERATED STATES
OF MICRONESIA
PALAU
Palmyra Is. (U.S.)
Howland (U.S.)
Baker (U.S.)
Jarvis (U.S.)
Equator
NAURU
KIRIBATI
PAPUA
NEW
GUINEA
SOLOMON
IS.
TUVALU
Tokelau Is.
(N.Z.)
WEST.
SAMOA
AMERICAN
SAMOA
FRENCH
POLYNESIA

Reference to colors on map above and behind map titles

- U.S. States
- U.S. Commonwealths
- U.S. Territories
- Countries in a Compact of Free Association with U.S.
- U.S. Trust Territory of the Pacific Islands

GUAM
1 : 1 000 000
0   5   miles
0   5   10 km
Ritidian Pt.
Pati Pt.
252
Tumon Bay
Tamuning
Dededo
Aganа
Barrigada
Piti
Orote Peninsular
Yona
Pago Bay
Agat
Santa Rita
Talofofo
406
Mt. Lamlam
Umatac
Inarajan
Merizo

SAIPAN & TINIAN (NORTHERN MARIANAS)
1 : 1 000 000
Sabaneta Pt.
Tanapag
Garapan
Kalabera
Mt. Tagpochau
465
Chalan Kanoa
Saipan Channel
Saipan
Tahgong Pt.
Naftan Pt.
Tinian
Masalog Pt.
Diablo Pt.
Tachungnya
178
Tinian Channel
Carolinas Pt.

TUTUILA
1 : 1 000 000
Vatia
Cape Matatula
Pago Pago
Aua
Tula
Fagasa
652
Amanave
Nu'uuli
Tafuna
C. Taputapu
Leone
Steps Pt.
484
Ofu
Olosega
839
Luma
931
Ta'u

MANUA
1 : 1 000 000

AMERICAN SAMOA

MICRONESIA
1 : 20 000 000
100   0   100   200 miles
100   0   200   km

NORTHERN MARIANAS
GUAM
PALAU
FEDERATED STATES OF MICRONESIA, MARSHALL IS.

Farallon de Pajaros
Maug Is.
Asuncion
Agrihan
NORTHERN
Pagan
Alamagan
Guguan
Sarigan
MARIANAS
Anatahan
Farallon de Medinilla
Garapan
Saipan
Tinian
Rota
GUAM
Agana
9650
11 022
Mariana Trench
Wake (U.S.)
PACIFIC   OCEAN
Taongi
MARSHALL ISLANDS
Enewetak
Bikini
Rongerik
Rongelap
Ujelang
Likiep
Wotje
Ralik Chain
Ratak Chain
Kwajalein
Erikub
Maloelap
Ujae
Namu
Majuro
Alinglapalap
Majuro
Mili
Jaluit
Ulithi
Yap Is.
Fais
FEDERATED STATES
Ngulu
Sorol
Faraulep
Gaferut
Namonuito
Fayu
West Fayu
Hall Is.
8069
Olimarao
Pikelot
Pulap
Pakin
Pohnpei
Babelthuap
Palau Is.
Woleai
Lamotrek
Puluwat
Truk Is.
Losap
Kolonia
Mokil
Angaur
Koror
Eauripik
Ifalik
Pulusuk
Enyavin Is.
6850
OF MICRONESIA
Ngatik
Kosrae
Sonsorol Is.
Satawan
Pulo Anna
REP. OF
Merir
Caroline   Islands
Nukuoro
Butaritari
Tobi
PALAU
Abaiang
Marakei
Tarawa
KIRIBATI
Kapingamarangi
Maiana
Gilbert Islands
Equator
Abemama
INDONESIA
PAPUA   NEW   GUINEA
NAURU
Yaren
Aranuka

ft   m
3000   1000
1200   400
600   200
0   0
200   600
2000   6000
m   ft

INDEX

UNITED STATES & OUTLYING AREAS

This index lists all the place names which appear on the large scale maps of the United States and outlying areas (pages which precede this index). Place names for the rest of the world can be found in the World Maps Index at the end of the atlas.

The number in dark type which follows each name in the index refers to the page number on which the place or feature is located. The geographical coordinates which follow the page number give the latitude and longitude of each place. The first coordinate indicates the latitude – the distance north or south of the Equator. The second coordinate indicates the longitude – the distance east or west of the Greenwich Meridian. Both latitude and longitude are measured in degrees and minutes (there are 60 minutes in a degree). Rivers are indexed to their mouths or confluences. A solid square ■ follows the name of a country, while an open square □ signifies that the name is a state. An arrow ⟶ follows the name of a river.

The alphabetic order of names composed of two or more words is governed by the first word and then by the second. Names composed of a proper name (Alaska) and a description (Gulf of) are positioned alphabetically by the proper name. All names beginning St. are alphabetized under Saint and those beginning Mc under Mac.

Abbreviations used in the index

Ala. — Alabama
Amer. — America, American
Ariz. — Arizona
Ark. — Arkansas
B. — Bay
C. — Cape
Calif. — California
Chan. — Channel
Colo. — Colorado
Conn. — Connecticut
Cr. — Creek
D.C. — District of Columbia
Del. — Delaware
Dist. — District
E. — East, Eastern
Fla. — Florida
G. — Gulf
Ga. — Georgia
Gt. — Great
I.(s) — Island(s)
Ill. — Illinois
Ind. — Indiana
Kans. — Kansas
Ky. — Kentucky
L. — Lake
La. — Louisiana
Ld. — Land
Mass. — Massachusetts
Md. — Maryland
Mich. — Michigan
Minn. — Minnesota
Miss. — Mississippi
Mo. — Missouri
Mont. — Montana
Mt.(s) — Mountain(s)
N. — North, Northern
N.B. — New Brunswick
N.C. — North Carolina
N. Dak. — North Dakota
N.H. — New Hampshire
N.J. — New Jersey
N. Mex. — New Mexico
N.Y. — New York
Nat. Mon. — National Monument
Nat. Park — National Park
Nat. Rec. Area. — National Recreation Area
Nebr. — Nebraska
Nev. — Nevada
Okla. — Oklahoma
Oreg. — Oregon
Pa. — Pennsylvania
Pac. Oc. — Pacific Ocean
Pass. — Passage
Pen. — Peninsula
Pk. — Peak
Pt. — Point
R. — Rio, River
R.I. — Rhode Island
Ra.(s) — Range(s)
Res. — Reserve, Reservoir, Reservation
S.C. — South Carolina
S. Dak. — South Dakota
Sa. — Serra, Sierra
Sd. — Sound
St. — Saint
Ste. — Sainte
Str. — Strait
Tenn. — Tennessee
Tex. — Texas
U.S.A. — United States of America
Va. — Virginia
Vt. — Vermont
Wash. — Washington
W. — West, Western
W. Va. — West Virginia
Wis. — Wisconsin
Wyo. — Wyoming

A

Abbaye, Pt., *Mich.* **10** 46 58N 88 8W
Abbeville, *La.* **13** 29 58N 92 8W
Abbeville, *S.C.* **11** 34 11N 82 23W
Abbotsford, *Wis.* **12** 44 57N 90 19W
Aberdeen, *Ala.* **11** 33 49N 88 33W
Aberdeen, *Idaho* **16** 42 57N 112 50W
Aberdeen, *S. Dak.* **12** 45 28N 98 29W
Aberdeen, *Wash.* **18** 46 59N 123 50W
Abernathy, *Tex.* **13** 33 50N 101 51W
Abert, L., *Oreg.* **16** 42 38N 120 14W
Abilene, *Kans.* **12** 38 55N 97 13W
Abilene, *Tex.* **13** 32 28N 99 43W
Abingdon, *Ill.* **14** 40 48N 90 24W
Abingdon, *Va.* **11** 36 43N 81 59W
Absaroka Range, *Wyo.* **16** 44 45N 109 50W
Accomac, *Va.* **10** 37 43N 75 40W
Ackerman, *Miss.* **13** 33 19N 89 11W
Ada, *Minn.* **12** 47 18N 96 31W
Ada, *Okla.* **13** 34 46N 96 41W
Adams, *Mass.* **9** 42 38N 73 7W
Adams, *N.Y.* **9** 43 49N 76 1W
Adams, *Wis.* **12** 43 57N 89 49W
Adams Mt., *Wash.* **18** 46 12N 121 30W
Adel, *Ga.* **11** 31 8N 83 25W
Adelanto, *Calif.* **19** 34 35N 117 22W
Adin, *Calif.* **16** 41 12N 120 57W
Adirondack Mts., *N.Y.* **9** 44 0N 74 0W
Admiralty I., *Alaska* **30** 57 30N 134 30W
Admiralty Inlet, *Wash.* **16** 48 8N 122 58W
Adrian, *Mich.* **15** 41 54N 84 2W
Adrian, *Tex.* **13** 35 16N 102 40W
Affton, *Miss.* **14** 38 33N 90 20W
Afognak I., *Alaska* **30** 58 15N 152 30W
Afton, *N.Y.* **9** 42 14N 75 32W
Agana, *Guam* **31** 13 28N 144 45 E
Agattu I., *Alaska* **30** 52 25N 172 30 E
Agua Caliente Springs, *Calif.* **19** 32 56N 116 19W
Aguadilla, *Puerto Rico* **30** 18 26N 67 10W
Aguanga, *Calif.* **19** 33 27N 116 51W
Aiken, *S.C.* **11** 33 34N 81 43W
Ainsworth, *Nebr.* **12** 42 33N 99 52W
Aitkin, *Minn.* **12** 46 32N 93 42W
Ajo, *Ariz.* **17** 32 22N 112 52W
Akiak, *Alaska* **30** 60 55N 161 13W
Akron, *Colo.* **12** 40 10N 103 13W
Akron, *Ohio* **8** 41 5N 81 31W
Akulurak, *Alaska* **30** 62 40N 164 35W
Akun I., *Alaska* **30** 54 11N 165 32W
Akutan I., *Alaska* **30** 54 7N 165 55W
Alabama □ **11** 33 0N 87 0W
Alabama ⟶, *Ala.* **11** 31 8N 87 57W
Alameda, *Calif.* **28** 37 46N 122 15W
Alameda, *N. Mex.* **17** 35 11N 106 37W
Alameda County, *Calif.* **28** 37 40N 122 10W
Alamo, *Nev.* **19** 36 21N 115 10W
Alamogordo, *N. Mex.* **17** 32 54N 105 57W
Alamosa, *Colo.* **17** 37 28N 105 52W
Alaska □ **30** 64 0N 154 0W
Alaska, G. of, *Pac. Oc.* **30** 58 0N 145 0W
Alaska Peninsula, *Alaska* **30** 56 0N 159 0W
Alaska Range, *Alaska* **30** 62 50N 151 0W
Alava, C., *Wash.* **16** 48 10N 124 44W
Albany, *Ga.* **11** 31 35N 84 10W
Albany, *Minn.* **12** 45 38N 94 34W
Albany, *N.Y.* **9** 42 39N 73 45W
Albany, *Oreg.* **16** 44 38N 123 6W
Albany, *Tex.* **13** 32 44N 99 18W
Albemarle, *N.C.* **11** 35 21N 80 11W
Albemarle Sd., *N.C.* **11** 36 5N 76 0W
Albert Lea, *Minn.* **12** 43 39N 93 22W
Albia, *Iowa* **14** 41 2N 92 48W
Albion, *Idaho* **16** 42 25N 113 35W
Albion, *Mich.* **15** 42 15N 84 45W
Albion, *Nebr.* **12** 41 42N 98 0W
Albuquerque, *N. Mex.* **17** 35 5N 106 39W
Alcatraz I., *Calif.* **28** 37 49N 122 25W
Alcoa, *Tenn.* **11** 35 48N 83 59W
Alcova, *Wyo.* **16** 42 34N 106 43W
Alder, *Mont.* **16** 45 19N 112 6W
Alder Pk., *Calif.* **18** 35 53N 121 22W
Aledo, *Ill.* **14** 41 12N 90 45W
Alenuihaha Channel, *Hawaii* **30** 20 30N 156 0W
Aleutian Is., *Pac. Oc.* **30** 52 0N 175 0W
Aleutian Ra., *Alaska* **30** 55 0N 155 0W
Alexander, *N. Dak.* **12** 47 51N 103 39W
Alexander Arch., *Alaska* **30** 56 0N 136 0W
Alexander City, *Ala.* **11** 32 56N 85 58W
Alexandria, *Ind.* **15** 40 16N 85 41W
Alexandria, *La.* **13** 31 18N 92 27W
Alexandria, *Minn.* **12** 45 53N 95 22W
Alexandria, *S. Dak.* **12** 43 39N 97 47W
Alexandria, *Va.* **25** 38 49N 77 6W
Alexandria Bay, *N.Y.* **9** 44 20N 75 55W
Alfred, *Maine* **9** 43 29N 70 43W
Algoma, *Wis.* **10** 44 36N 87 26W
Algona, *Iowa* **14** 43 4N 94 14W
Alhambra, *Calif.* **29** 34 5N 118 9W
Alice, *Tex.* **13** 27 45N 98 5W
Aliceville, *Ala.* **11** 33 8N 88 9W
Aliquippa, *Pa.* **8** 40 37N 80 15W
All American Canal, *Calif.* **17** 32 45N 115 15W
Allakaket, *Alaska* **30** 66 34N 152 39W
Allegan, *Mich.* **15** 42 32N 85 51W
Allegheny ⟶, *Pa.* **8** 40 27N 80 1W
Allegheny Plateau, *Va.* **10** 38 0N 80 0W
Allen Park, *Mich.* **26** 42 14N 83 12W
Allentown, *Pa.* **9** 40 37N 75 29W
Alliance, *Nebr.* **12** 42 6N 102 52W
Alliance, *Ohio* **8** 40 55N 81 6W
Allison Park, *Pa.* **27** 40 33N 79 56W
Alma, *Ga.* **11** 31 33N 82 28W
Alma, *Kans.* **12** 39 1N 96 17W
Alma, *Mich.* **10** 43 23N 84 39W
Alma, *Nebr.* **12** 40 6N 99 22W
Alma, *Wis.* **12** 44 20N 91 55W
Almanor, L., *Calif.* **16** 40 14N 121 9W
Alpaugh, *Calif.* **18** 35 53N 119 29W
Alpena, *Mich.* **10** 45 4N 83 27W
Alpine, *Ariz.* **17** 33 51N 109 9W
Alpine, *Calif.* **19** 32 50N 116 46W
Alpine, *Tex.* **13** 30 22N 103 40W
Alta Sierra, *Calif.* **19** 35 42N 118 33W
Altadena, *Calif.* **29** 34 11N 118 8W
Altamaha ⟶, *Ga.* **11** 31 20N 81 20W
Altamont, *N.Y.* **9** 42 43N 74 3W
Altavista, *Va.* **10** 37 6N 79 17W
Alton, *Ill.* **14** 38 53N 90 11W
Altoona, *Pa.* **8** 40 31N 78 24W
Alturas, *Calif.* **16** 41 29N 120 32W
Altus, *Okla.* **13** 34 38N 99 20W
Alva, *Okla.* **13** 36 48N 98 40W
Alvarado, *Tex.* **13** 32 24N 97 13W
Alvin, *Tex.* **13** 29 26N 95 15W
Alzada, *Mont.* **12** 45 2N 104 25W
Amargosa ⟶, *Calif.* **19** 36 14N 116 51W
Amargosa Range, *Calif.* **19** 36 20N 116 45W
Amarillo, *Tex.* **13** 35 13N 101 50W
Amatignak I., *Alaska* **30** 51 16N 179 6W
Amboy, *Calif.* **19** 34 33N 115 45W
Amchitka I., *Alaska* **30** 51 32N 179 0 E
Amchitka Pass., *Alaska* **30** 51 30N 179 0W
American Falls, *Idaho* **16** 42 47N 112 51W
American Falls Reservoir, *Idaho* **16** 42 47N 112 52W
American Samoa ■, *Pac. Oc.* **31** 14 20S 170 40W
Americus, *Ga.* **11** 32 4N 84 14W
Ames, *Iowa* **14** 42 2N 93 37W
Amesbury, *Mass.* **9** 42 51N 70 56W
Amherst, *Mass.* **9** 42 23N 72 31W
Amherst, *Tex.* **13** 34 1N 102 25W
Amite, *La.* **13** 30 44N 90 30W
Amlia I., *Alaska* **30** 52 4N 173 30W
Amory, *Miss.* **11** 33 59N 88 29W
Amsterdam, *N.Y.* **9** 42 56N 74 11W
Amukta I., *Alaska* **30** 52 30N 171 16W
Anaconda, *Mont.* **16** 46 8N 112 57W
Anacortes, *Wash.* **18** 48 30N 122 37W
Anadarko, *Okla.* **13** 35 4N 98 15W
Anaheim, *Calif.* **19** 33 50N 117 55W
Anamoose, *N. Dak.* **12** 47 53N 100 15W
Anamosa, *Iowa* **14** 42 7N 91 17W
Anatone, *Wash.* **16** 46 8N 117 8W
Anchorage, *Alaska* **30** 61 13N 149 54W
Andalusia, *Ala.* **11** 31 18N 86 29W
Anderson, *Calif.* **16** 40 27N 122 18W
Anderson, *Ind.* **15** 40 10N 85 41W
Anderson, *Mo.* **13** 36 39N 94 27W
Anderson, *S.C.* **11** 34 31N 82 39W
Andover, *Mass.* **23** 42 39N 71 7W
Andreanof Is., *Alaska* **30** 52 0N 178 0W
Andrews, *S.C.* **11** 33 27N 79 34W
Andrews, *Tex.* **13** 32 19N 102 33W
Andrews Air Force Base, *Md.* **25** 38 48N 76 52W
Anegada I., *Virgin Is.* **30** 18 45N 64 20W
Angeles National Forest, *Calif.* **29** 34 15N 118 5W
Angels Camp, *Calif.* **18** 38 4N 120 32W
Angleton, *Tex.* **13** 29 10N 95 26W
Angola, *Ind.* **15** 41 38N 85 0W
Angoon, *Alaska* **30** 57 30N 134 35W
Aniak, *Alaska* **30** 61 35N 159 32W
Animas, *N. Mex.* **17** 31 57N 108 48W
Ann, C., *Mass.* **9** 42 38N 70 35W
Ann Arbor, *Mich.* **15** 42 17N 83 45W
Anna, *Ill.* **13** 37 28N 89 15W
Annandale, *Va.* **25** 38 50N 77 12W
Annapolis, *Md.* **10** 38 59N 76 30W
Annette, *Alaska* **30** 55 2N 131 35W
Anniston, *Ala.* **11** 33 39N 85 50W
Annville, *Pa.* **9** 40 20N 76 31W
Anoka, *Minn.* **12** 45 12N 93 23W
Ansley, *Nebr.* **12** 41 18N 99 23W
Anson, *Tex.* **13** 32 45N 99 54W
Ansonia, *Conn.* **9** 41 21N 73 5W
Antero, Mt., *Colo.* **17** 38 41N 106 15W
Anthony, *Kans.* **13** 37 9N 98 2W
Anthony, *N. Mex.* **17** 32 0N 106 36W
Antigo, *Wis.* **12** 45 9N 89 9W
Antimony, *Utah* **17** 38 7N 112 0W
Antioch, *Calif.* **18** 38 1N 121 48W

Blackfoot

Antler, N. Dak. 12 48 59N 101 17W
Antlers, Okla. 13 34 14N 95 37W
Anton, Tex. 13 33 49N 102 10W
Anton Chico, N. Mex. 17 35 12N 105 9W
Antonito, Colo. 17 37 5N 106 0W
Anvik, Alaska 30 62 39N 160 13W
Anza, Calif. 19 33 35N 116 39W
Apache, Okla. 13 34 54N 98 22W
Apalachee B., Fla. 11 30 0N 84 0W
Apalachicola, Fla. 11 29 43N 84 59W
Apalachicola →, Fla. 11 29 43N 84 58W
Apostle Is., Wis. 12 47 0N 90 40W
Appalachian Mts., Va. 10 38 0N 80 0W
Apple Valley, Calif. 19 34 32N 117 14W
Appleton, Wis. 10 44 16N 88 25W
Aransas Pass, Tex. 13 27 55N 97 9W
Arapahoe, Nebr. 12 40 18N 99 54W
Arbuckle, Calif. 18 39 1N 122 3W
Arcadia, Calif. 29 34 7N 118 1W
Arcadia, Fla. 11 27 13N 81 52W
Arcadia, La. 13 32 33N 92 55W
Arcadia, Nebr. 12 41 25N 99 8W
Arcadia, Wis. 12 44 15N 91 30W
Arcata, Calif. 16 40 52N 124 5W
Archbald, Pa. 9 41 30N 75 32W
Arco, Idaho 16 43 38N 113 18W
Arctic Village, Alaska 30 68 8N 145 32W
Ardmore, Okla. 13 34 10N 97 8W
Ardmore, Pa. 24 40 0N 75 19W
Ardmore, S. Dak. 12 43 1N 103 40W
Arecibo, Puerto Rico 30 18 29N 66 43W
Argonne Forest, Ill. 22 41 42N 87 53W
Arguello, Pt., Calif. 19 34 35N 120 39W
Argus Pk., Calif. 19 35 52N 117 26W
Argyle, Minn. 12 48 20N 96 49W
Arivaca, Ariz. 17 31 37N 111 25W
Arizona □ 17 34 0N 112 0W
Arkadelphia, Ark. 13 34 7N 93 4W
Arkansas □ 13 35 0N 92 30W
Arkansas →, Ark. 13 33 47N 91 4W
Arkansas City, Kans. 13 37 4N 97 2W
Arlee, Mont. 16 47 10N 114 5W
Arlington, Mass. 23 42 24N 71 9W
Arlington, Oreg. 16 45 43N 120 12W
Arlington, S. Dak. 12 44 22N 97 8W
Arlington, Va. 25 38 53N 77 7W
Arlington, Wash. 18 48 12N 122 8W
Arlington Heights, Ill. 22 42 5N 87 54W
Arlington National Cemetery, D.C. 25 38 52N 77 4W
Armour, S. Dak. 12 43 19N 98 21W
Armstrong, Tex. 13 26 56N 97 47W
Arnett, Okla. 13 36 8N 99 46W
Arnold, Calif. 18 38 15N 120 20W
Arnold, Nebr. 12 41 26N 100 12W
Arrow Rock Res., Idaho 16 43 45N 115 50W
Arrowhead, L., Calif. 19 34 16N 117 10W
Arroyo Grande, Calif. 19 35 7N 120 35W
Artesia, Calif. 29 33 51N 118 5W
Artesia, N. Mex. 13 32 51N 104 24W
Artesia Wells, Tex. 13 28 17N 99 17W
Artesian, S. Dak. 12 44 1N 97 55W
Arundel Gardens, Md. 25 39 12N 76 37W
Arvada, Wyo. 16 44 39N 106 8W
Arvin, Calif. 19 35 12N 118 50W
Asbury Park, N.J. 9 40 13N 74 1W
Ash Fork, Ariz. 17 35 13N 112 29W
Ash Grove, Mo. 13 37 19N 93 35W
Ashburn, Ga. 11 31 43N 83 39W
Asheboro, N.C. 11 35 43N 79 49W
Asherton, Tex. 13 28 27N 99 46W
Asheville, N.C. 11 35 36N 82 33W
Ashford, Wash. 16 46 46N 122 2W
Ashland, Kans. 13 37 11N 99 46W
Ashland, Ky. 10 38 28N 82 38W
Ashland, Maine 11 46 38N 68 24W
Ashland, Mont. 16 45 36N 106 16W
Ashland, Nebr. 12 41 3N 96 23W
Ashland, Ohio 8 40 52N 82 19W
Ashland, Oreg. 16 42 12N 122 43W
Ashland, Pa. 9 40 45N 76 22W
Ashland, Va. 10 37 46N 77 29W
Ashland, Wis. 12 46 35N 90 53W
Ashley, N. Dak. 12 46 2N 99 22W
Ashtabula, Ohio 8 41 52N 80 47W
Ashton, Idaho 16 44 4N 111 27W
Asotin, Wash. 16 46 20N 117 3W
Aspen, Colo. 17 39 11N 106 49W
Aspermont, Tex. 13 33 8N 100 14W
Astoria, Oreg. 18 46 11N 123 50W
Atascadero, Calif. 17 35 32N 120 44W
Atchafalaya B., La. 13 29 25N 91 25W
Atchison, Kans. 12 39 34N 95 7W
Athens, Ala. 11 34 48N 86 58W
Athens, Ga. 11 33 57N 83 23W
Athens, N.Y. 9 42 16N 73 49W
Athens, Ohio 10 39 20N 82 6W
Athens, Pa. 9 41 57N 76 31W
Athens, Tenn. 11 35 27N 84 36W
Athens, Tex. 13 32 12N 95 51W
Atka, Alaska 30 52 5N 174 40W
Atkasuk, Alaska 30 70 30N 157 20W
Atkinson, Nebr. 12 42 32N 98 59W
Atlanta, Ga. 11 33 45N 84 23W
Atlanta, Tex. 13 33 7N 94 10W
Atlantic, Iowa 12 41 24N 95 1W
Atlantic City, N.J. 10 39 21N 74 27W
Atmore, Ala. 11 31 2N 87 29W
Atoka, Okla. 13 34 23N 96 8W
Atolia, Calif. 19 35 19N 117 37W
Attalla, Ala. 11 34 1N 86 6W
Attica, Ind. 15 40 18N 87 15W
Attleboro, Mass. 9 41 57N 71 17W
Attu, Alaska 30 52 56N 173 15 E
Atwater, Calif. 18 37 21N 120 37W
Atwood, Kans. 12 39 48N 101 3W
Au Sable →, Mich. 10 44 25N 83 20W
Auberry, Calif. 18 37 7N 119 29W
Auburn, Ala. 11 32 36N 85 29W
Auburn, Calif. 18 38 54N 121 4W
Auburn, Ind. 15 41 22N 85 4W
Auburn, N.Y. 9 42 56N 76 34W
Auburn, Nebr. 12 40 23N 95 51W
Auburndale, Fla. 11 28 4N 81 48W
Audubon, Iowa 14 41 43N 94 56W
Augusta, Ark. 13 35 17N 91 22W
Augusta, Ga. 11 33 28N 81 58W
Augusta, Kans. 13 37 41N 96 59W
Augusta, Maine 11 44 19N 69 47W
Augusta, Mont. 16 47 30N 112 24W
Augusta, Wis. 12 44 41N 91 7W
Aukum, Calif. 18 38 34N 120 43W
Ault, Colo. 12 40 35N 104 44W
Aurora, Colo. 12 39 44N 104 52W
Aurora, Ill. 15 41 45N 88 19W
Aurora, Mo. 13 36 58N 93 43W
Aurora, Nebr. 12 40 52N 98 0W
Austin, Minn. 12 43 40N 92 58W
Austin, Nev. 16 39 30N 117 4W
Austin, Tex. 13 30 17N 97 45W
Avalon, Calif. 19 33 21N 118 20W
Avalon, Pa. 27 40 30N 80 4W
Avawatz Mts., Calif. 19 35 40N 116 30W
Avenal, Calif. 18 36 0N 120 8W
Avery, Idaho 16 47 15N 115 49W
Avila Beach, Calif. 19 35 11N 120 44W
Avon, S. Dak. 12 43 0N 98 4W
Aztec, N. Mex. 17 36 49N 107 59W
Azusa, Calif. 29 34 7N 117 54W

B

Babb, Mont. 16 48 51N 113 27W
Bad →, S. Dak. 12 44 21N 100 22W
Bad Axe, Mich. 8 43 48N 83 0W
Bad Lands, S. Dak. 12 43 40N 102 10W
Badger, Calif. 18 36 38N 119 1W
Bagdad, Calif. 19 34 35N 115 53W
Baggs, Wyo. 16 41 2N 107 39W
Bagley, Minn. 12 47 32N 95 24W
Bainbridge, Ga. 11 30 55N 84 35W
Bainbridge, N.Y. 9 42 18N 75 29W
Bainville, Mont. 12 48 8N 104 13W
Baird, Tex. 13 32 24N 99 24W
Baird Inlet, Alaska 30 60 50N 164 18W
Baird Mts., Alaska 30 67 0N 160 0W
Baker, Calif. 19 35 16N 116 4W
Baker, Mont. 12 46 22N 104 17W
Baker, Oreg. 16 44 47N 117 50W
Baker Mt., Wash. 16 48 50N 121 49W
Bakersfield, Calif. 19 35 23N 119 1W
Bald Knob, Ark. 13 35 19N 91 34W
Baldwin, Fla. 11 30 18N 81 59W
Baldwin, Mich. 10 43 54N 85 51W
Baldwin, Pa. 27 40 23N 79 58W
Baldwin Park, Calif. 29 34 5N 117 57W
Baldwinsville, N.Y. 9 43 10N 76 20W
Baldy Peak, Ariz. 17 33 54N 109 34W
Ballinger, Tex. 13 31 45N 99 57W
Balmorhea, Tex. 13 30 59N 103 45W
Balta, N. Dak. 12 48 10N 100 2W
Baltimore, Md. 25 39 17N 76 37W
Baltimore Washington International Airport, Md. 25 39 11N 76 41W
Bamberg, S.C. 11 33 18N 81 2W
Bandera, Tex. 13 29 44N 99 5W
Bangor, Maine 11 44 48N 68 46W
Bangor, Pa. 9 40 52N 75 13W
Banning, Calif. 19 33 56N 116 53W
Bar Harbor, Maine 11 44 23N 68 13W
Baraboo, Wis. 12 43 28N 89 45W
Baraga, Mich. 12 46 47N 88 30W
Baranof I., Alaska 30 57 0N 135 0W
Barataria B., La. 13 29 20N 89 55W
Barberton, Ohio 8 41 0N 81 39W
Barbourville, Ky. 11 36 52N 83 53W
Bardstown, Ky. 15 37 49N 85 28W
Barksdale, Tex. 13 29 44N 100 2W
Barnesville, Ga. 11 33 3N 84 9W
Barneveld, N.Y. 9 43 16N 75 14W
Barnhart, Tex. 13 31 8N 101 10W
Barnsville, Minn. 12 46 43N 96 28W
Barques, Pt. Aux, Mich. 10 44 4N 82 58W
Barre, Mass. 9 42 25N 72 6W
Barre, Vt. 9 44 12N 72 30W
Barren Is., Alaska 30 58 45N 152 0W
Barrington, R.I. 9 41 44N 71 18W
Barrow, Alaska 30 71 18N 156 47W
Barrow Pt., Alaska 30 71 24N 156 29W
Barstow, Calif. 19 34 54N 117 1W
Barstow, Tex. 13 31 28N 103 24W
Bartlesville, Okla. 13 36 45N 95 59W
Bartlett, Calif. 18 36 29N 118 2W
Bartlett, Tex. 13 30 48N 97 26W
Bartow, Fla. 11 27 54N 81 50W
Basin, Wyo. 16 44 23N 108 2W
Bassett, Nebr. 12 42 35N 99 32W
Bassett, Va. 11 36 46N 79 59W
Bastrop, Tex. 13 30 7N 97 19W
Batavia, N.Y. 8 43 0N 78 11W
Batesburg, S.C. 11 33 54N 81 33W
Batesville, Ark. 13 35 46N 91 39W
Batesville, Miss. 13 34 19N 89 57W
Batesville, Tex. 13 28 58N 99 37W
Bath, Maine 11 43 55N 69 49W
Bath, N.Y. 8 42 20N 77 19W
Baton Rouge, La. 13 30 27N 91 11W
Battle Creek, Mich. 15 42 19N 85 11W
Battle Lake, Minn. 12 46 17N 95 43W
Battle Mountain, Nev. 16 40 38N 116 56W
Baudette, Minn. 12 48 43N 94 36W
Baxley, Ga. 11 31 47N 82 21W
Baxter Springs, Kans. 13 37 2N 94 44W
Bay City, Mich. 10 43 36N 83 54W
Bay City, Oreg. 16 45 31N 123 53W
Bay City, Tex. 13 28 59N 95 58W
Bay Minette, Ala. 11 30 53N 87 46W
Bay St. Louis, Miss. 13 30 19N 89 20W
Bay Springs, Miss. 13 31 59N 89 17W
Bay Village, Ohio 27 41 29N 81 53W
Bayamón, Puerto Rico 30 18 24N 66 10W
Bayard, Nebr. 12 41 45N 103 20W
Bayfield, Wis. 12 46 49N 90 49W
Bayonne, N.J. 20 40 40N 74 6W
Baytown, Tex. 13 29 43N 94 59W
Beach, N. Dak. 12 46 58N 104 0W
Beacon, N.Y. 9 41 30N 73 58W
Bear L., Utah 16 41 59N 111 21W
Bearcreek, Mont. 16 45 11N 109 6W
Beardstown, Ill. 14 40 1N 90 26W
Bearpaw Mts., Mont. 16 48 12N 109 30W
Beatrice, Nebr. 12 40 16N 96 45W
Beatty, Nev. 18 36 54N 116 46W
Beaufort, N.C. 11 34 43N 76 40W
Beaufort, S.C. 11 32 26N 80 40W
Beaumont, Calif. 19 33 56N 116 58W
Beaumont, Tex. 13 30 5N 94 6W
Beaver, Alaska 30 66 22N 147 24W
Beaver, Okla. 13 36 49N 100 31W
Beaver, Utah 17 38 17N 112 38W
Beaver City, Nebr. 12 40 8N 99 50W
Beaver Dam, Wis. 12 43 28N 88 50W
Beaver Falls, Pa. 8 40 46N 80 20W
Beaver I., Mich. 10 45 40N 85 33W
Becharof L., Alaska 30 57 56N 156 23W
Beckley, W. Va. 10 37 47N 81 11W
Bedford, Ind. 15 38 52N 86 29W
Bedford, Iowa 14 40 40N 94 44W
Bedford, Mass. 23 42 29N 71 15W
Bedford, Ohio 8 41 23N 81 32W
Bedford, Va. 10 37 20N 79 31W
Beech Grove, Ind. 15 39 44N 86 3W
Beechey Point, Alaska 30 70 27N 149 18W
Beeville, Tex. 13 28 24N 97 45W
Belen, N. Mex. 17 34 40N 106 46W
Belfast, Maine 11 44 26N 69 1W
Belfield, N. Dak. 12 46 53N 103 12W
Belfry, Mont. 16 45 9N 109 1W
Belgrade, Mont. 16 45 47N 111 11W
Belhaven, N.C. 11 35 33N 76 37W
Bell Gardens, Calif. 29 33 58N 118 9W
Bellaire, Ohio 8 40 1N 80 45W
Belle Fourche, S. Dak. 12 44 40N 103 51W
Belle Fourche →, S. Dak. 12 44 26N 102 18W
Belle Glade, Fla. 11 26 41N 80 40W
Belle Plaine, Iowa 14 41 54N 92 17W
Belle Plaine, Minn. 12 44 37N 93 46W
Bellefontaine, Ohio 15 40 22N 83 46W
Bellefonte, Pa. 8 40 55N 77 47W
Belleville, Ill. 14 38 31N 89 59W
Belleville, Kans. 12 39 50N 97 38W
Belleville, N.J. 20 40 48N 74 9W
Belleville, N.Y. 9 43 46N 76 10W
Bellevue, Idaho 16 43 28N 114 16W
Bellevue, Pa. 27 40 29N 80 3W
Bellflower, Calif. 29 33 53N 118 8W
Bellingham, Wash. 18 48 46N 122 29W
Bellmore, N.Y. 21 40 39N 73 31W
Bellows Falls, Vt. 9 43 8N 72 27W
Bellville, Tex. 13 29 57N 96 15W
Bellwood, Ill. 22 41 53N 87 53W
Belmont, Mass. 23 42 24N 71 10W
Beloit, Kans. 12 39 28N 98 6W
Beloit, Wis. 14 42 31N 89 2W
Belton, S.C. 11 34 31N 82 30W
Belton, Tex. 13 31 3N 97 28W
Belton Res., Tex. 13 31 8N 97 32W
Beltsville, Md. 25 39 2N 76 55W
Belvidere, Ill. 15 42 15N 88 50W
Belvidere, N.J. 9 40 50N 75 5W
Belzoni, Miss. 13 33 11N 90 29W
Bemidji, Minn. 12 47 28N 94 53W
Benavides, Tex. 13 27 36N 98 25W
Bend, Oreg. 16 44 4N 121 19W
Benicia, Calif. 18 38 3N 122 9W
Benkelman, Nebr. 12 40 3N 101 32W
Bennettsville, S.C. 11 34 37N 79 41W
Bennington, N.H. 9 43 0N 71 55W
Benson, Ariz. 17 31 58N 110 18W
Benton, Ark. 13 34 34N 92 35W
Benton, Calif. 18 37 48N 118 32W
Benton, Ill. 14 38 0N 88 55W
Benton Harbor, Mich. 15 42 6N 86 27W
Beowawe, Nev. 16 40 35N 116 29W
Berea, Ky. 10 37 34N 84 17W
Bergenfield, N.J. 20 40 55N 73 58W
Bering Sea, Pac. Oc. 30 58 0N 171 0 E
Bering Strait, Alaska 30 65 30N 169 0W
Berkeley, Calif. 28 37 52N 122 17W
Berkeley Springs, W. Va. 10 39 38N 78 14W
Berkley, Mich. 26 42 29N 83 11W
Berlin, Md. 10 38 20N 75 13W
Berlin, N.H. 9 44 28N 71 11W
Berlin, Wis. 10 43 58N 88 57W
Bernado, N. Mex. 17 34 30N 106 53W
Bernalillo, N. Mex. 17 35 18N 106 33W
Berryville, Ark. 13 36 22N 93 34W
Berthold, N. Dak. 12 48 19N 101 44W
Berthoud, Colo. 12 40 19N 105 5W
Bertrand, Nebr. 12 40 32N 99 38W
Berwick, Pa. 9 41 3N 76 14W
Berwyn, Ill. 22 41 50N 87 47W
Bessemer, Ala. 11 33 24N 86 58W
Bessemer, Mich. 12 46 29N 90 3W
Bethany, Mo. 14 40 16N 94 2W
Bethel, Alaska 30 60 48N 161 45W
Bethel Park, Pa. 27 40 19N 80 1W
Bethesda, Md. 25 38 59N 77 6W
Bethlehem, Pa. 9 40 37N 75 23W
Bethpage, N.Y. 21 40 45N 73 29W
Beulah, N. Dak. 12 47 16N 101 47W
Beverly, Mass. 23 42 33N 70 52W
Beverly, Wash. 16 46 50N 119 56W
Beverly Hills, Calif. 29 34 5N 118 24W
Beverly Hills, Mich. 26 42 31N 83 15W
Bicknell, Ind. 15 38 47N 87 19W
Bicknell, Utah 17 38 20N 111 33W
Biddeford, Maine 11 43 30N 70 28W
Bieber, Calif. 16 41 7N 121 8W
Big Bear City, Calif. 19 34 16N 116 51W
Big Bear Lake, Calif. 19 34 15N 116 56W
Big Belt Mts., Mont. 16 46 30N 111 25W
Big Bend National Park, Tex. 13 29 20N 103 5W
Big Black →, Miss. 13 32 3N 91 4W
Big Blue →, Kans. 12 39 35N 96 34W
Big Creek, Calif. 18 37 11N 119 14W
Big Cypress Swamp, Fla. 11 26 12N 81 10W
Big Falls, Minn. 12 48 12N 93 48W
Big Fork →, Minn. 12 48 31N 93 43W
Big Horn Mts. = Bighorn Mts., Wyo. 16 44 30N 107 30W
Big Lake, Tex. 13 31 12N 101 28W
Big Moose, N.Y. 9 43 49N 74 58W
Big Muddy Cr. →, Mont. 12 48 8N 104 36W
Big Pine, Calif. 18 37 10N 118 17W
Big Piney, Wyo. 16 42 32N 110 7W
Big Rapids, Mich. 10 43 42N 85 29W
Big Sable Pt., Mich. 10 44 3N 86 1W
Big Sandy, Mont. 16 48 11N 110 7W
Big Sandy Cr. →, Colo. 12 38 7N 102 29W
Big Sioux →, S. Dak. 12 42 29N 96 27W
Big Spring, Tex. 13 32 15N 101 28W
Big Springs, Nebr. 12 41 4N 102 5W
Big Stone City, S. Dak. 12 45 18N 96 28W
Big Stone Gap, Va. 11 36 52N 82 47W
Big Stone L., Minn. 12 45 30N 96 35W
Big Sur, Calif. 18 36 15N 121 48W
Big Timber, Mont. 16 45 50N 109 57W
Bigfork, Mont. 16 48 4N 114 4W
Bighorn, Mont. 16 46 10N 107 27W
Bighorn →, Mont. 16 46 10N 107 28W
Bighorn Mts., Wyo. 16 44 30N 107 30W
Bikini Atoll, Pac. Oc. 31 12 0N 167 30 E
Bill, Wyo. 12 43 14N 105 16W
Billerica, Mass. 23 42 33N 71 15W
Billings, Mont. 16 45 47N 108 30W
Biloxi, Miss. 13 30 24N 88 53W
Bingham, Maine 11 45 3N 69 53W
Bingham Canyon, Utah 16 40 32N 112 9W
Binghamton, N.Y. 9 42 6N 75 55W
Bird City, Kans. 12 39 45N 101 32W
Birmingham, Ala. 11 33 31N 86 48W
Birmingham, Mich. 26 42 33N 83 13W
Bisbee, Ariz. 17 31 27N 109 55W
Biscayne B., Fla. 11 25 40N 80 12W
Bishop, Calif. 18 37 22N 118 24W
Bishop, Tex. 13 27 35N 97 48W
Bismarck, N. Dak. 12 46 48N 100 47W
Bison, S. Dak. 12 45 31N 102 28W
Bitter Creek, Wyo. 16 41 33N 108 33W
Bitterroot →, Mont. 16 46 52N 114 7W
Bitterroot Range, Idaho 16 46 0N 114 20W
Bitterwater, Calif. 18 36 23N 121 0W
Biwabik, Minn. 12 47 32N 92 21W
Black →, Ark. 13 35 38N 91 20W
Black →, Wis. 12 43 57N 91 22W
Black Hills, S. Dak. 12 44 0N 103 45W
Black L., Mich. 10 45 28N 84 16W
Black Mesa, Okla. 13 36 58N 102 58W
Black Range, N. Mex. 17 33 15N 107 50W
Black River Falls, Wis. 12 44 18N 90 51W
Black Warrior →, Ala. 11 32 32N 87 51W
Blackburn, Mt., Alaska 30 61 44N 143 26W
Blackduck, Minn. 12 47 44N 94 33W
Blackfoot, Idaho 16 43 11N 112 21W
Blackfoot →, Mont. 16 46 52N 113 53W

Blackfoot River Reservoir

Blackfoot River Reservoir, *Idaho* **16** 43 0N 111 43W
Blacksburg, *Va.* **10** 37 14N 80 25W
Blackstone, *Va.* **10** 37 4N 78 0W
Blackwell, *Okla.* **13** 36 48N 97 17W
Blackwells Corner, *Calif.* **19** 35 37N 119 47W
Blackwood, *N.J.* **24** 39 48N 75 4W
Blaine, *Wash.* **18** 48 59N 122 45W
Blair, *Nebr.* **12** 41 33N 96 8W
Blake Pt., *Mich.* **12** 48 11N 88 25W
Blakely, *Ga.* **11** 31 23N 84 56W
Blanca Peak, *Colo.* **17** 37 35N 105 29W
Blanchard, *Okla.* **13** 35 8N 97 39W
Blanco, *Tex.* **13** 30 6N 98 25W
Blanco, C., *Oreg.* **16** 42 51N 124 34W
Blanding, *Utah* **17** 37 37N 109 29W
Block I., *R.I.* **9** 41 11N 71 35W
Bloomer, *Wis.* **12** 45 6N 91 29W
Bloomfield, *Iowa* **14** 40 45N 92 25W
Bloomfield, *N.J.* **20** 40 48N 74 12W
Bloomfield, *N. Mex.* **17** 36 43N 107 59W
Bloomfield, *Nebr.* **12** 42 36N 97 39W
Bloomingdale, *N.J.* **20** 41 0N 74 15W
Bloomington, *Ill.* **14** 40 28N 89 0W
Bloomington, *Ind.* **15** 39 10N 86 32W
Bloomsburg, *Pa.* **9** 41 0N 76 27W
Blossburg, *Pa.* **8** 41 41N 77 4W
Blountstown, *Fla.* **11** 30 27N 85 3W
Blue Island, *Ill.* **10** 41 40N 87 40W
Blue Island, *Ill.* **22** 41 40N 87 40W
Blue Lake, *Calif.* **16** 40 53N 123 59W
Blue Mesa Reservoir, *Colo.* **17** 38 28N 107 20W
Blue Mts., *Oreg.* **16** 45 15N 119 0W
Blue Mts., *Pa.* **9** 40 30N 76 30W
Blue Rapids, *Kans.* **12** 39 41N 96 39W
Blue Ridge Mts., *N.C.* **11** 36 30N 80 15W
Bluefield, *Va.* **10** 37 15N 81 17W
Bluff, *Utah* **17** 37 17N 109 33W
Bluffton, *Ind.* **15** 40 44N 85 11W
Blunt, *S. Dak.* **12** 44 31N 99 59W
Bly, *Oreg.* **16** 42 24N 121 3W
Blythe, *Calif.* **19** 33 37N 114 36W
Boca Raton, *Fla.* **11** 26 21N 80 5W
Boerne, *Tex.* **13** 29 47N 98 44W
Bogalusa, *La.* **13** 30 47N 89 52W
Bogata, *Tex.* **13** 33 28N 95 13W
Boise, *Idaho* **16** 43 37N 116 13W
Boise City, *Okla.* **13** 36 44N 102 31W
Bolivar, *Mo.* **13** 37 37N 93 25W
Bolivar, *Tenn.* **13** 35 12N 89 0W
Bonham, *Tex.* **13** 33 35N 96 11W
Bonne Terre, *Mo.* **14** 37 55N 90 33W
Bonners Ferry, *Idaho* **16** 48 42N 116 19W
Bonsall, *Calif.* **19** 33 16N 117 14W
Booker, *Tex.* **13** 36 27N 100 32W
Boone, *Iowa* **14** 42 4N 93 53W
Boone, *N.C.* **11** 36 13N 81 41W
Booneville, *Ark.* **13** 35 8N 93 55W
Booneville, *Miss.* **11** 34 39N 88 34W
Boonville, *Ind.* **15** 38 3N 87 16W
Boonville, *Mo.* **14** 38 58N 92 44W
Boonville, *N.Y.* **9** 43 29N 75 20W
Borah Peak, *Idaho* **16** 44 8N 113 47W
Borger, *Tex.* **13** 35 39N 101 24W
Boron, *Calif.* **19** 35 0N 117 39W
Borrego Springs, *Calif.* **19** 33 15N 116 23W
Bossier City, *La.* **13** 32 31N 93 44W
Boston, *Mass.* **23** 42 21N 71 3W
Boswell, *Okla.* **13** 34 2N 95 52W
Bottineau, *N. Dak.* **12** 48 50N 100 27W
Boulder, *Colo.* **12** 40 1N 105 17W
Boulder, *Mont.* **16** 46 14N 112 7W
Boulder City, *Nev.* **19** 35 59N 114 50W
Boulder Creek, *Calif.* **18** 37 7N 122 7W
Boulder Dam = Hoover Dam, *Ariz.* **19** 36 1N 114 44W
Boundary, *Alaska* **30** 64 4N 141 6W
Boundary Peak, *Nev.* **18** 37 51N 118 21W
Bountiful, *Utah* **16** 40 53N 111 53W
Bovill, *Idaho* **16** 46 51N 116 24W
Bowbells, *N. Dak.* **12** 48 48N 102 15W
Bowdle, *S. Dak.* **12** 45 27N 99 39W
Bowie, *Ariz.* **17** 32 19N 109 29W
Bowie, *Tex.* **13** 33 34N 97 51W
Bowling Green, *Ky.* **10** 36 59N 86 27W
Bowling Green, *Ohio* **15** 41 23N 83 39W
Bowman, *N. Dak.* **12** 46 11N 103 24W
Boyce, *La.* **13** 31 23N 92 40W
Boyne City, *Mich.* **10** 45 13N 85 1W
Boynton Beach, *Fla.* **11** 26 32N 80 4W
Bozeman, *Mont.* **16** 45 41N 111 2W
Brackettville, *Tex.* **13** 29 19N 100 25W
Braddock, *Pa.* **27** 40 24N 79 51W
Bradenton, *Fla.* **11** 27 30N 82 34W
Bradford, *Pa.* **8** 41 58N 78 38W
Bradley, *Ark.* **13** 33 6N 93 39W
Bradley, *Calif.* **18** 35 52N 120 48W
Bradley, *S. Dak.* **12** 45 5N 97 39W
Brady, *Tex.* **13** 31 9N 99 20W
Brainerd, *Minn.* **12** 46 22N 94 12W
Braintree, *Mass.* **23** 42 12N 71 0W
Brandywine, *Del.* **24** 39 49N 75 32W
Branford, *Conn.* **9** 41 17N 72 49W
Branson, *Colo.* **13** 37 1N 103 53W
Branson, *Mo.* **13** 36 39N 93 13W
Brasstown Bald, *Ga.* **11** 34 53N 83 49W
Brattleboro, *Vt.* **9** 42 51N 72 34W
Brawley, *Calif.* **19** 32 59N 115 31W
Brazil, *Ind.* **15** 39 32N 87 8W
Brazos →, *Tex.* **13** 28 53N 95 23W
Breckenridge, *Colo.* **16** 39 29N 106 3W
Breckenridge, *Minn.* **12** 46 16N 96 35W
Breckenridge, *Tex.* **13** 32 45N 98 54W
Bremerton, *Wash.* **18** 47 34N 122 38W
Brenham, *Tex.* **13** 30 10N 96 24W
Brentwood, *Pa.* **27** 40 22N 79 59W
Breton Sd., *La.* **13** 29 35N 89 15W
Brevard, *N.C.* **11** 35 14N 82 44W
Brewer, *Maine* **11** 44 48N 68 46W
Brewer, Mt., *Calif.* **18** 36 44N 118 28W
Brewster, *N.Y.* **9** 41 23N 73 37W
Brewster, *Wash.* **16** 48 6N 119 47W
Brewton, *Ala.* **11** 31 7N 87 4W
Bridgehampton, *N.Y.* **9** 40 56N 72 19W
Bridgeport, *Calif.* **18** 38 15N 119 14W
Bridgeport, *Conn.* **9** 41 11N 73 12W
Bridgeport, *Nebr.* **12** 41 40N 103 6W
Bridgeport, *Pa.* **24** 39 48N 75 21W
Bridgeport, *Tex.* **13** 33 13N 97 45W
Bridger, *Mont.* **16** 45 18N 108 55W
Bridgeton, *N.J.* **10** 39 26N 75 14W
Bridgeville, *Pa.* **27** 21 0N 80 6W
Bridgewater, *Mass.* **9** 41 59N 70 58W
Bridgewater, *S. Dak.* **12** 43 33N 97 30W
Briggsdale, *Colo.* **12** 40 38N 104 20W
Brigham City, *Utah* **16** 41 31N 112 1W
Brighton, *Colo.* **12** 39 59N 104 49W
Brinkley, *Ark.* **13** 34 53N 91 12W
Bristol, *Conn.* **9** 41 40N 72 57W
Bristol, *Pa.* **24** 40 6N 74 53W
Bristol, *R.I.* **9** 41 40N 71 16W
Bristol, *S. Dak.* **12** 45 21N 97 45W
Bristol, *Tenn.* **11** 36 36N 82 11W
Bristol B., *Alaska* **30** 58 0N 160 0W
Bristol L., *Calif.* **17** 34 23N 116 50W
Bristow, *Okla.* **13** 35 50N 96 23W
Britton, *S. Dak.* **12** 45 48N 97 45W
Broad →, *S.C.* **11** 34 1N 81 4W
Broadus, *Mont.* **12** 45 27N 105 25W
Broadview, *Ill.* **22** 41 51N 87 52W
Brockport, *N.Y.* **8** 43 13N 77 56W
Brockton, *Mass.* **9** 42 5N 71 1W
Brockway, *Mont.* **12** 47 18N 105 45W
Brogan, *Oreg.* **16** 44 15N 117 31W
Broken Bow, *Nebr.* **12** 41 24N 99 38W
Broken Bow, *Okla.* **13** 34 2N 94 44W
Bronte, *Tex.* **13** 31 53N 100 18W
Bronx, *N.Y.* **21** 40 50N 73 52W
Bronxville, *N.Y.* **21** 40 56N 73 49W
Brook Park, *Ohio* **27** 41 24N 81 48W
Brookfield, *Ill.* **22** 41 48N 87 50W
Brookfield, *Mo.* **14** 39 47N 93 4W
Brookhaven, *Miss.* **13** 31 35N 90 26W
Brookings, *Oreg.* **16** 42 3N 124 17W
Brookings, *S. Dak.* **12** 44 19N 96 48W
Brookline, *Mass.* **23** 42 19N 71 7W
Brooklyn, *Md.* **25** 39 13N 76 35W
Brooklyn, *N.Y.* **20** 40 37N 73 57W
Brooklyn, *Ohio* **27** 41 26N 81 44W
Brooks Ra., *Alaska* **30** 68 40N 147 0W
Brooksville, *Fla.* **11** 28 33N 82 23W
Brookville, *Ind.* **15** 39 25N 85 1W
Broomall, *Pa.* **24** 39 58N 75 22W
Brothers, *Oreg.* **16** 43 49N 120 36W
Browerville, *Minn.* **12** 46 5N 94 52W
Brownfield, *Tex.* **13** 33 11N 102 17W
Browning, *Mont.* **16** 48 34N 113 1W
Brownsville, *Oreg.* **16** 44 24N 122 59W
Brownsville, *Tenn.* **13** 35 36N 89 16W
Brownsville, *Tex.* **13** 25 54N 97 30W
Brownwood, *Tex.* **13** 31 43N 98 59W
Brownwood, L., *Tex.* **13** 31 51N 98 35W
Brundidge, *Ala.* **11** 31 43N 85 49W
Bruneau, *Idaho* **16** 42 53N 115 48W
Bruneau →, *Idaho* **16** 42 56N 115 57W
Brunswick, *Ga.* **11** 31 10N 81 30W
Brunswick, *Maine* **11** 43 55N 69 58W
Brunswick, *Md.* **10** 39 19N 77 38W
Brunswick, *Mo.* **14** 39 26N 93 8W
Brush, *Colo.* **12** 40 15N 103 37W
Bryan, *Ohio* **15** 41 28N 84 33W
Bryan, *Tex.* **13** 30 40N 96 22W
Bryant, *S. Dak.* **12** 44 35N 97 28W
Bryn Mawr, *Pa.* **24** 40 1N 75 19W
Bryson City, *N.C.* **11** 35 26N 83 27W
Buchanan, L., *Tex.* **13** 30 45N 98 25W
Buchon, Pt., *Calif.* **18** 35 15N 120 54W
Buckeye, *Ariz.* **17** 33 22N 112 35W
Buckhannon, *W. Va.* **10** 39 0N 80 8W
Buckland, *Alaska* **30** 65 59N 161 8W
Buckley, *Wash.* **16** 47 10N 122 2W
Bucklin, *Kans.* **13** 37 33N 99 38W
Bucyrus, *Ohio* **15** 40 48N 82 59W
Buellton, *Calif.* **19** 34 37N 120 12W
Buena Park, *Calif.* **29** 33 51N 118 1W
Buena Vista, *Colo.* **17** 38 51N 106 8W
Buena Vista, *Va.* **10** 37 44N 79 21W
Buena Vista L., *Calif.* **19** 35 12N 119 18W
Buffalo, *Mo.* **14** 37 39N 93 6W
Buffalo, *N.Y.* **8** 42 53N 78 53W
Buffalo, *Okla.* **13** 36 50N 99 38W
Buffalo, *S. Dak.* **12** 45 35N 103 33W
Buffalo, *Wyo.* **16** 44 21N 106 42W
Buford, *Ga.* **11** 34 10N 84 0W
Buhl, *Idaho* **16** 42 36N 114 46W
Buhl, *Minn.* **12** 47 30N 92 46W
Buldir I., *Alaska* **30** 52 21N 175 56 E
Bull Shoals L., *Ark.* **13** 36 22N 92 35W
Bunker Hill Monument, *Mass.* **23** 42 21N 71 3W
Bunkie, *La.* **13** 30 57N 92 11W
Bunnell, *Fla.* **11** 29 28N 81 16W
Buras, *La.* **13** 29 22N 89 32W
Burbank, *Calif.* **29** 34 11N 118 18W
Burbank, *Ill.* **22** 41 44N 87 47W
Burkburnett, *Tex.* **13** 34 6N 98 34W
Burke, *Idaho* **16** 47 31N 115 49W
Burley, *Idaho* **16** 42 32N 113 48W
Burlingame, *Calif.* **28** 37 35N 122 22W
Burlington, *Colo.* **12** 39 18N 102 16W
Burlington, *Iowa* **14** 40 49N 91 14W
Burlington, *Kans.* **12** 38 12N 95 45W
Burlington, *Mass.* **23** 42 30N 71 13W
Burlington, *N.C.* **11** 36 6N 79 26W
Burlington, *N.J.* **24** 40 4N 74 54W
Burlington, *Vt.* **9** 44 29N 73 12W
Burlington, *Wash.* **18** 48 28N 122 20W
Burlington, *Wis.* **10** 42 41N 88 17W
Burnet, *Tex.* **13** 30 45N 98 14W
Burney, *Calif.* **16** 40 53N 121 40W
Burns, *Oreg.* **16** 43 35N 119 3W
Burns, *Wyo.* **12** 41 12N 104 21W
Burnt Paw, *Alaska* **30** 67 2N 142 43W
Burwell, *Nebr.* **12** 41 47N 99 8W
Bushnell, *Ill.* **12** 40 33N 90 31W
Bushnell, *Nebr.* **12** 41 14N 103 54W
Butler, *Mo.* **14** 38 16N 94 20W
Butler, *Pa.* **8** 40 52N 79 54W
Butte, *Mont.* **16** 46 0N 112 32W
Butte, *Nebr.* **12** 42 58N 98 51W
Buttonwillow, *Calif.* **19** 35 24N 119 28W
Buzzards Bay, *Mass.* **9** 41 45N 70 37W
Byers, *Colo.* **12** 39 43N 104 14W
Byhalia, *Miss.* **13** 34 52N 89 41W
Bylas, *Ariz.* **17** 33 8N 110 7W

C

Cabazon, *Calif.* **19** 33 55N 116 47W
Cabinet Mts., *Mont.* **16** 48 0N 115 30W
Cabool, *Mo.* **13** 37 7N 92 6W
Caddo, *Okla.* **13** 34 7N 96 16W
Cadillac, *Mich.* **10** 44 15N 85 24W
Caguas, *Puerto Rico* **30** 18 14N 66 2W
Cairo, *Ga.* **11** 30 52N 84 13W
Cairo, *Ill.* **13** 37 0N 89 11W
Calais, *Maine* **11** 45 11N 67 17W
Calcasieu L., *La.* **13** 29 55N 93 18W
Caldwell, *Idaho* **16** 43 40N 116 41W
Caldwell, *Kans.* **13** 37 2N 97 37W
Caldwell, *Tex.* **13** 30 32N 96 42W
Calexico, *Calif.* **19** 32 40N 115 30W
Calhoun, *Ga.* **11** 34 30N 84 57W
Caliente, *Nev.* **17** 37 37N 114 31W
California, *Mo.* **14** 38 38N 92 34W
California □ **17** 37 30N 119 30W
California, University of, *Calif.* **28** 37 52N 122 15W
California City, *Calif.* **19** 35 10N 117 55W
California Hot Springs, *Calif.* **19** 35 51N 118 41W
Calipatria, *Calif.* **19** 33 8N 115 31W
Calistoga, *Calif.* **18** 38 35N 122 35W
Callaway, *Nebr.* **12** 41 18N 99 56W
Calumet, *Mich.* **10** 47 14N 88 27W
Calumet City, *Ill.* **22** 41 37N 87 32W
Calvert, *Tex.* **13** 30 59N 96 40W
Calwa, *Calif.* **18** 36 42N 119 46W
Camanche Reservoir, *Calif.* **18** 38 14N 121 1W
Camarillo, *Calif.* **19** 34 13N 119 2W
Camas, *Wash.* **18** 45 35N 122 24W
Camas Valley, *Oreg.* **16** 43 2N 123 40W
Cambria, *Calif.* **18** 35 34N 121 5W
Cambridge, *Idaho* **16** 44 34N 116 41W
Cambridge, *Mass.* **23** 42 22N 71 6W
Cambridge, *Md.* **10** 38 34N 76 5W
Cambridge, *Minn.* **12** 45 34N 93 13W
Cambridge, *N.Y.* **9** 43 2N 73 22W
Cambridge, *Nebr.* **12** 40 17N 100 10W
Cambridge, *Ohio* **8** 40 2N 81 35W
Camden, *Ala.* **11** 31 59N 87 17W
Camden, *Ark.* **13** 33 35N 92 50W
Camden, *Maine* **11** 44 13N 69 4W
Camden, *N.J.* **24** 39 56N 75 7W
Camden, *S.C.* **11** 34 16N 80 36W
Camden, B., *Alaska* **30** 70 30N 145 0W
Camdenton, *Mo.* **14** 38 1N 92 45W
Cameron, *Ariz.* **17** 35 53N 111 25W
Cameron, *La.* **13** 29 48N 93 20W
Cameron, *Mo.* **14** 39 44N 94 14W
Cameron, *Tex.* **13** 30 51N 96 59W
Camino, *Calif.* **18** 38 44N 120 41W
Camp Crook, *S. Dak.* **12** 45 33N 103 59W
Camp Nelson, *Calif.* **19** 36 8N 118 39W
Camp Wood, *Tex.* **13** 29 40N 100 1W
Campbell, *Calif.* **18** 37 17N 121 57W
Campbellsville, *Ky.* **10** 37 21N 85 20W
Canadian, *Tex.* **13** 35 55N 100 23W
Canadian →, *Okla.* **13** 35 28N 95 3W
Canandaigua, *N.Y.* **8** 42 54N 77 17W
Canarsie, *N.Y.* **21** 40 38N 73 53W
Canaveral, C., *Fla.* **11** 28 27N 80 32W
Canby, *Calif.* **16** 41 27N 120 52W
Canby, *Minn.* **12** 44 43N 96 16W
Canby, *Oreg.* **18** 45 16N 122 42W
Cando, *N. Dak.* **12** 48 32N 99 12W
Cannon Ball →, *N. Dak.* **12** 46 20N 100 38W
Canon City, *Colo.* **12** 38 27N 105 14W
Cantil, *Calif.* **19** 35 18N 117 58W
Canton, *Ga.* **11** 34 14N 84 29W
Canton, *Ill.* **14** 40 33N 90 2W
Canton, *Miss.* **13** 32 37N 90 2W
Canton, *Mo.* **14** 40 8N 91 32W
Canton, *N.Y.* **9** 44 36N 75 10W
Canton, *Ohio* **8** 40 48N 81 23W
Canton, *Okla.* **13** 36 3N 98 35W
Canton, *S. Dak.* **12** 43 18N 96 35W
Canton L., *Okla.* **13** 36 6N 98 35W
Canutillo, *Tex.* **17** 31 55N 106 36W
Canyon, *Tex.* **13** 34 59N 101 55W
Canyon, *Wyo.* **16** 44 43N 110 36W
Canyonlands National Park, *Utah* **17** 38 15N 110 0W
Canyonville, *Oreg.* **16** 42 56N 123 17W
Cape Charles, *Va.* **10** 37 16N 76 1W
Cape Fear →, *N.C.* **11** 33 53N 78 1W
Cape Girardeau, *Mo.* **13** 37 19N 89 32W
Cape May, *N.J.* **10** 38 56N 74 56W
Capitan, *N. Mex.* **17** 33 35N 105 35W
Capitola, *Calif.* **18** 36 59N 121 57W
Carbondale, *Colo.* **16** 39 24N 107 13W
Carbondale, *Ill.* **14** 37 44N 89 13W
Carbondale, *Pa.* **9** 41 35N 75 30W
Cardiff-by-the-Sea, *Calif.* **19** 33 1N 117 17W
Carey, *Idaho* **16** 43 19N 113 57W
Carey, *Ohio* **15** 40 57N 83 23W
Caribou, *Maine* **11** 46 52N 68 1W
Carlin, *Nev.* **16** 40 43N 116 7W
Carlinville, *Ill.* **14** 39 17N 89 53W
Carlisle, *Pa.* **8** 40 12N 77 12W
Carlsbad, *Calif.* **19** 33 10N 117 21W
Carlsbad, *N. Mex.* **13** 32 25N 104 14W
Carlyle, *Ill.* **12** 38 37N 89 22W
Carmel, *N.Y.* **9** 41 26N 73 41W
Carmel-by-the-Sea, *Calif.* **18** 36 33N 121 55W
Carmel Valley, *Calif.* **18** 36 29N 121 43W
Carmi, *Ill.* **15** 38 5N 88 10W
Carmichael, *Calif.* **18** 38 38N 121 19W
Carnegie, *Pa.* **27** 40 24N 80 5W
Caro, *Mich.* **10** 43 29N 83 24W
Carol City, *Fla.* **11** 25 56N 80 16W
Caroline Is., *Pac. Oc.* **31** 8 0N 150 0 E
Carpinteria, *Calif.* **19** 34 24N 119 31W
Carrabelle, *Fla.* **11** 29 51N 84 40W
Carrington, *N. Dak.* **12** 47 27N 99 8W
Carrizo Cr. →, *N. Mex.* **13** 36 55N 103 55W
Carrizo Springs, *Tex.* **13** 28 31N 99 52W
Carrizozo, *N. Mex.* **17** 33 38N 105 53W
Carroll, *Iowa* **14** 42 4N 94 52W
Carrollton, *Ga.* **11** 33 35N 85 5W
Carrollton, *Ill.* **12** 39 18N 90 24W
Carrollton, *Ky.* **15** 38 41N 85 11W
Carrollton, *Mo.* **14** 39 22N 93 30W
Carson, *N. Dak.* **12** 46 25N 101 34W
Carson City, *Nev.* **18** 39 10N 119 46W
Carson Sink, *Nev.* **16** 39 50N 118 25W
Cartersville, *Ga.* **11** 34 10N 84 48W
Carthage, *Ark.* **13** 34 4N 92 33W
Carthage, *Ill.* **14** 40 25N 91 8W
Carthage, *Mo.* **13** 37 11N 94 19W
Carthage, *S. Dak.* **12** 44 10N 97 43W
Carthage, *Tex.* **13** 32 9N 94 20W
Caruthersville, *Mo.* **13** 36 11N 89 39W
Casa Grande, *Ariz.* **17** 32 53N 111 45W
Cascade, *Idaho* **16** 44 31N 116 2W
Cascade, *Mont.* **16** 47 16N 111 42W
Cascade Locks, *Oreg.* **18** 45 40N 121 54W
Cascade Ra., *Wash.* **18** 47 0N 121 30W
Cashmere, *Wash.* **16** 47 31N 120 28W
Casmalia, *Calif.* **19** 34 50N 120 32W
Casper, *Wyo.* **16** 42 51N 106 19W
Cass City, *Mich.* **10** 43 36N 83 11W
Cass Lake, *Minn.* **12** 47 23N 94 37W
Casselton, *N. Dak.* **12** 46 54N 97 13W
Cassville, *Mo.* **13** 36 41N 93 52W
Castaic, *Calif.* **19** 34 30N 118 38W
Castle Dale, *Utah* **16** 39 13N 111 1W
Castle Rock, *Colo.* **12** 39 22N 104 51W
Castle Rock, *Wash.* **18** 46 17N 122 54W
Castro Valley, *Calif.* **28** 37 42N 122 4W
Castroville, *Calif.* **18** 36 46N 121 45W
Castroville, *Tex.* **13** 29 21N 98 53W
Cat I., *Miss.* **13** 30 14N 89 6W
Catahoula L., *La.* **13** 31 31N 92 7W
Cathlamet, *Wash.* **18** 46 12N 123 23W
Catlettsburg, *Ky.* **10** 38 25N 82 36W
Catonsville, *Md.* **25** 39 16N 76 43W
Catskill, *N.Y.* **9** 42 14N 73 52W
Catskill Mts., *N.Y.* **9** 42 10N 74 25W
Cavalier, *N. Dak.* **12** 48 48N 97 37W
Cave City, *Ky.* **10** 37 8N 85 58W
Cayey, *Puerto Rico* **30** 18 7N 66 10W
Cayuga L., *N.Y.* **9** 42 41N 76 41W
Cedar →, *Iowa* **14** 41 17N 91 21W
Cedar City, *Utah* **17** 37 41N 113 4W
Cedar Creek Reservoir, *Tex.* **13** 32 11N 96 4W

Crestline

Cedar Falls, *Iowa* **14** 42 32N 92 27W
Cedar Key, *Fla.* **11** 29 8N 83 2W
Cedar Rapids, *Iowa* **14** 41 59N 91 40W
Cedarhurst, *N.Y.* **21** 40 37N 73 42W
Cedartown, *Ga.* **11** 34 1N 85 15W
Cedarville, *Calif.* **16** 41 32N 120 10W
Celina, *Ohio* **15** 40 33N 84 35W
Cement, *Okla.* **13** 34 56N 98 8W
Center, *N. Dak.* **12** 47 7N 101 18W
Center, *Tex.* **13** 31 48N 94 11W
Centerfield, *Utah* **17** 39 8N 111 49W
Centerville, *Calif.* **18** 36 44N 119 30W
Centerville, *Iowa* **14** 40 44N 92 52W
Centerville, *S. Dak.* **12** 43 7N 96 58W
Centerville, *Tenn.* **11** 35 47N 87 28W
Centerville, *Tex.* **13** 31 16N 95 59W
Central, *N. Mex.* **17** 32 47N 108 9W
Central City, *Ky.* **10** 37 18N 87 7W
Central City, *Nebr.* **12** 41 7N 98 0W
Centralia, *Ill.* **14** 38 32N 89 8W
Centralia, *Mo.* **14** 39 13N 92 8W
Centralia, *Wash.* **18** 46 43N 122 58W
Centreville, *Ala.* **11** 32 57N 87 8W
Centreville, *Miss.* **13** 31 5N 91 4W
Ceres, *Calif.* **18** 37 35N 120 57W
Cerro de Punta, Mt., *Puerto Rico* **30** 18 10N 67 0W
Chadron, *Nebr.* **12** 42 50N 103 0W
Chalfant, *Calif.* **18** 37 32N 118 21W
Challis, *Idaho* **16** 44 30N 114 14W
Chama, *N. Mex.* **17** 36 54N 106 35W
Chamberlain, *S. Dak.* **12** 43 49N 99 20W
Chambers, *Ariz.* **17** 35 11N 109 26W
Chambersburg, *Pa.* **10** 39 56N 77 40W
Champaign, *Ill.* **15** 40 7N 88 15W
Champlain, *N.Y.* **9** 44 59N 73 27W
Champlain, L., *N.Y.* **9** 44 40N 73 20W
Chandalar, *Alaska* **30** 67 30N 148 35W
Chandeleur Is., *La.* **13** 29 55N 88 57W
Chandeleur Sd., *La.* **13** 29 55N 89 0W
Chandler, *Ariz.* **17** 33 18N 111 50W
Chandler, *Okla.* **13** 35 42N 96 53W
Channel Is., *Calif.* **19** 33 40N 119 15W
Channing, *Mich.* **10** 46 9N 88 5W
Channing, *Tex.* **13** 35 41N 102 20W
Chanute, *Kans.* **13** 37 41N 95 27W
Chapel Hill, *N.C.* **11** 35 55N 79 4W
Chariton →, *Mo.* **14** 39 19N 92 58W
Charles, C., *Va.* **10** 37 7N 75 58W
Charles City, *Iowa* **14** 43 4N 92 41W
Charles Town, *W. Va.* **10** 39 17N 77 52W
Charleston, *Ill.* **10** 39 30N 88 10W
Charleston, *Miss.* **13** 34 1N 90 4W
Charleston, *Mo.* **13** 36 55N 89 21W
Charleston, *S.C.* **11** 32 46N 79 56W
Charleston, *W. Va.* **10** 38 21N 81 38W
Charlestown, *Ind.* **15** 38 27N 85 40W
Charlevoix, *Mich.* **10** 45 19N 85 16W
Charlotte, *Mich.* **15** 42 34N 84 50W
Charlotte, *N.C.* **11** 35 13N 80 51W
Charlotte Amalie, *Virgin Is.* **30** 18 21N 64 56W
Charlotte Harbor, *Fla.* **11** 26 50N 82 10W
Charlottesville, *Va.* **10** 38 2N 78 30W
Charlton, *Iowa* **12** 40 59N 93 20W
Chatfield, *Minn.* **12** 43 51N 92 11W
Chatham, *Alaska* **30** 57 30N 135 0W
Chatham, *La.* **13** 32 18N 92 27W
Chatham, *N.J.* **20** 40 44N 74 23W
Chatham, *N.Y.* **9** 42 21N 73 36W
Chattahoochee →, *Ga.* **11** 30 54N 84 57W
Chattanooga, *Tenn.* **11** 35 3N 85 19W
Cheboygan, *Mich.* **10** 45 39N 84 29W
Checotah, *Okla.* **13** 35 28N 95 31W
Chefornak, *Alaska* **30** 60 13N 164 12W
Chehalis, *Wash.* **18** 46 40N 122 58W
Chelan, *Wash.* **16** 47 51N 120 1W
Chelan, L., *Wash.* **16** 48 11N 120 30W
Chelmsford, *Mass.* **23** 42 35N 71 20W
Chelsea, *Mass.* **23** 42 23N 71 1W
Chelsea, *Okla.* **13** 36 32N 95 26W
Cheltenham, *Pa.* **24** 40 3N 75 6W
Chemult, *Oreg.* **16** 43 14N 121 47W
Chenango Forks, *N.Y.* **9** 42 15N 75 51W
Cheney, *Wash.* **16** 47 30N 117 35W
Chequamegon B., *Mich.* **12** 46 40N 90 30W
Cheraw, *S.C.* **11** 34 42N 79 53W
Cherokee, *Iowa* **12** 42 45N 95 33W
Cherokee, *Okla.* **13** 36 45N 98 21W
Cherokees, Lake O' The, *Okla.* **13** 36 28N 95 2W
Cherry Creek, *Nev.* **16** 39 54N 114 53W
Cherry Hill, *N.J.* **24** 39 54N 75 1W
Cherry Valley, *Calif.* **19** 33 59N 116 57W
Cherryvale, *Kans.* **13** 37 16N 95 33W
Chesapeake, *Va.* **10** 36 50N 76 17W
Chesapeake B., *Va.* **10** 38 0N 76 10W
Chester, *Calif.* **16** 40 19N 121 14W
Chester, *Ill.* **14** 37 55N 89 49W
Chester, *Mont.* **16** 48 31N 110 58W
Chester, *Pa.* **24** 39 51N 75 22W
Chester, *S.C.* **11** 34 43N 81 12W
Chesuncook L., *Maine* **11** 46 0N 69 21W
Chevy Chase, *Md.* **25** 38 59N 77 4W
Chewelah, *Wash.* **16** 48 17N 117 43W
Cheyenne, *Okla.* **13** 35 37N 99 40W
Cheyenne, *Wyo.* **12** 41 8N 104 49W
Cheyenne →, *S. Dak.* **12** 44 41N 101 18W
Cheyenne Wells, *Colo.* **12** 38 49N 102 21W
Cheyney, *Pa.* **24** 39 55N 75 32W
Chicago, *Ill.* **22** 41 52N 87 38W
Chicago, University of, *Ill.* **22** 41 47N 87 35W
Chicago Heights, *Ill.* **15** 41 30N 87 38W
Chicago Midway Airport, *Ill.* **22** 41 46N 87 44W
Chicago O'Hare International Airport, *Ill.* **22** 41 58N 87 54W
Chickasha, *Okla.* **13** 35 3N 97 58W
Chico, *Calif.* **18** 39 44N 121 50W
Chicopee, *Mass.* **9** 42 9N 72 37W
Chignik, *Alaska* **30** 56 18N 158 24W
Childress, *Tex.* **13** 34 25N 100 13W
Chillicothe, *Ill.* **14** 40 55N 89 29W
Chillicothe, *Mo.* **14** 39 48N 93 33W
Chillicothe, *Ohio* **10** 39 20N 82 59W
Chillum, *Md.* **25** 38 57N 76 58W
Chilton, *Wis.* **10** 44 2N 88 10W
China Lake, *Calif.* **19** 35 44N 117 37W
Chinati Peak, *Tex.* **13** 29 57N 104 29W
Chincoteague, *Va.* **10** 37 56N 75 23W
Chinle, *Ariz.* **17** 36 9N 109 33W
Chino, *Calif.* **19** 34 1N 117 41W
Chino Valley, *Ariz.* **17** 34 45N 112 27W
Chinook, *Mont.* **16** 48 35N 109 14W
Chipley, *Fla.* **11** 30 47N 85 32W
Chippewa →, *Wis.* **12** 44 25N 92 5W
Chippewa Falls, *Wis.* **12** 44 56N 91 24W
Chiricahua Peak, *Ariz.* **17** 31 51N 109 18W
Chirikof I., *Alaska* **30** 55 50N 155 40W
Chisos Mts., *Tex.* **13** 29 5N 103 15W
Chitina, *Alaska* **30** 61 31N 144 26W
Cholame, *Calif.* **18** 35 44N 120 18W
Choteau, *Mont.* **16** 47 49N 112 11W
Chowchilla, *Calif.* **18** 37 7N 120 16W
Chugach Mts., *Alaska* **30** 60 45N 147 0W
Chugiak, *Alaska* **30** 61 24N 149 29W
Chuginadak I., *Alaska* **30** 52 50N 169 45W
Chugwater, *Wyo.* **12** 41 46N 104 50W
Chula Vista, *Calif.* **19** 32 39N 117 5W
Cicero, *Ill.* **22** 41 51N 87 45W
Cimarron, *Kans.* **13** 37 48N 100 21W
Cimarron, *N. Mex.* **13** 36 31N 104 55W
Cimarron →, *Okla.* **13** 36 10N 96 17W
Cincinnati, *Ohio* **15** 39 6N 84 31W
Circle, *Alaska* **30** 65 50N 144 4W
Circle, *Mont.* **12** 47 25N 105 35W
Circleville, *Ohio* **10** 39 36N 82 57W
Circleville, *Utah* **17** 38 10N 112 16W
Cisco, *Tex.* **13** 32 23N 98 59W
Clairemont, *Tex.* **13** 33 9N 100 44W
Clanton, *Ala.* **11** 32 51N 86 38W
Claraville, *Calif.* **19** 35 24N 118 20W
Clare, *Mich.* **10** 43 49N 84 46W
Claremont, *Calif.* **19** 34 6N 117 43W
Claremont, *N.H.* **9** 43 23N 72 20W
Claremore, *Okla.* **13** 36 19N 95 36W
Clarendon, *Ark.* **13** 34 42N 91 19W
Clarendon, *Tex.* **13** 34 56N 100 53W
Clarinda, *Iowa* **12** 40 44N 95 2W
Clarion, *Iowa* **14** 42 44N 93 44W
Clark, *S. Dak.* **12** 44 53N 97 44W
Clark Fork, *Idaho* **16** 48 9N 116 11W
Clark Fork →, *Idaho* **16** 48 9N 116 15W
Clark Hill Res., *Ga.* **11** 33 45N 82 20W
Clarkdale, *Ariz.* **17** 34 46N 112 3W
Clark's Fork →, *Wyo.* **16** 45 39N 108 43W
Clarks Summit, *Pa.* **9** 41 30N 75 42W
Clarksburg, *W. Va.* **10** 39 17N 80 30W
Clarksdale, *Miss.* **13** 34 12N 90 35W
Clarkston, *Wash.* **16** 46 25N 117 3W
Clarksville, *Ark.* **13** 35 28N 93 28W
Clarksville, *Tenn.* **11** 36 32N 87 21W
Clarksville, *Tex.* **13** 33 37N 95 3W
Clatskanie, *Oreg.* **18** 46 6N 123 12W
Claude, *Tex.* **13** 35 7N 101 22W
Clay, *Calif.* **18** 38 17N 121 10W
Clay Center, *Kans.* **12** 39 23N 97 8W
Claymont, *Del.* **24** 39 48N 75 28W
Claypool, *Ariz.* **17** 33 25N 110 51W
Clayton, *Idaho* **16** 44 16N 114 24W
Clayton, *N. Mex.* **13** 36 27N 103 11W
Cle Elum, *Wash.* **16** 47 12N 120 56W
Clear L., *Calif.* **18** 39 2N 122 47W
Clear Lake, *S. Dak.* **12** 44 45N 96 41W
Clear Lake, *Wash.* **16** 48 27N 122 15W
Clear Lake Reservoir, *Calif.* **16** 41 56N 121 5W
Clearfield, *Pa.* **10** 41 2N 78 27W
Clearfield, *Utah* **16** 41 7N 112 2W
Clearmont, *Wyo.* **16** 44 38N 106 23W
Clearwater, *Fla.* **11** 27 58N 82 48W
Clearwater Mts., *Idaho* **16** 46 5N 115 20W
Cleburne, *Tex.* **13** 32 21N 97 23W
Cleveland, *Miss.* **13** 33 45N 90 43W
Cleveland, *Ohio* **27** 41 29N 81 42W
Cleveland, *Okla.* **13** 36 19N 96 28W
Cleveland, *Tenn.* **11** 35 10N 84 53W
Cleveland, *Tex.* **13** 30 21N 95 5W
Cleveland Heights, *Ohio* **27** 41 29N 81 35W
Cleveland Hopkins International Airport, *Ohio* **27** 41 24N 81 51W
Clewiston, *Fla.* **11** 26 45N 80 56W
Clifton, *Ariz.* **17** 33 3N 109 18W
Clifton, *N.J.* **20** 40 52N 74 8W
Clifton, *Tex.* **13** 31 47N 97 35W
Clifton Forge, *Va.* **10** 37 49N 79 50W
Clinch →, *Tenn.* **11** 35 53N 84 29W
Clingmans Dome, *Tenn.* **11** 35 34N 83 30W
Clint, *Tex.* **17** 31 35N 106 14W
Clinton, *Ark.* **13** 35 36N 92 28W
Clinton, *Ill.* **12** 40 9N 88 57W
Clinton, *Ind.* **15** 39 40N 87 24W
Clinton, *Iowa* **14** 41 51N 90 12W
Clinton, *Mass.* **9** 42 25N 71 41W
Clinton, *Mo.* **14** 38 22N 93 46W
Clinton, *N.C.* **11** 35 0N 78 22W
Clinton, *Okla.* **13** 35 31N 98 58W
Clinton, *S.C.* **11** 34 29N 81 53W
Clinton, *Tenn.* **11** 36 6N 84 8W
Clintonville, *Wis.* **12** 44 37N 88 46W
Cloquet, *Minn.* **12** 46 43N 92 28W
Cloud Peak, *Wyo.* **16** 44 23N 107 11W
Cloudcroft, *N. Mex.* **17** 32 58N 105 45W
Cloverdale, *Calif.* **18** 38 48N 123 1W
Clovis, *Calif.* **18** 36 49N 119 42W
Clovis, *N. Mex.* **13** 34 24N 103 12W
Coachella, *Calif.* **19** 33 41N 116 10W
Coahoma, *Tex.* **13** 32 18N 101 18W
Coalgate, *Okla.* **13** 34 32N 96 13W
Coalinga, *Calif.* **18** 36 9N 120 21W
Coalville, *Utah* **16** 40 55N 111 24W
Coamo, *Puerto Rico* **30** 18 5N 66 22W
Coast Ranges, *Calif.* **18** 39 0N 123 0W
Coatesville, *Pa.* **10** 39 59N 75 50W
Cobleskill, *N.Y.* **9** 42 41N 74 29W
Cobre, *Nev.* **16** 41 7N 114 24W
Cochise, *Ariz.* **17** 32 7N 109 55W
Cochran, *Ga.* **11** 32 23N 83 21W
Cocoa, *Fla.* **11** 28 21N 80 44W
Cody, *Wyo.* **16** 44 32N 109 3W
Cœur d'Alene, *Idaho* **16** 47 45N 116 51W
Cœur d'Alene L., *Idaho* **16** 47 32N 116 48W
Coffeyville, *Kans.* **13** 37 2N 95 37W
Cohagen, *Mont.* **16** 47 3N 106 37W
Cohoes, *N.Y.* **9** 42 46N 73 42W
Cokeville, *Wyo.* **16** 42 5N 110 57W
Colby, *Kans.* **12** 39 24N 101 3W
Coldwater, *Kans.* **13** 37 16N 99 20W
Colebrook, *N.H.* **9** 44 54N 71 30W
Coleman, *Tex.* **13** 31 50N 99 26W
Coleville, *Calif.* **18** 38 34N 119 30W
Colfax, *La.* **13** 31 31N 92 42W
Colfax, *Wash.* **16** 46 53N 117 22W
Collbran, *Colo.* **17** 39 14N 107 58W
College Park, *Ga.* **11** 33 40N 84 27W
College Park, *Md.* **25** 38 59N 76 55W
Collingdale, *Pa.* **24** 39 54N 75 16W
Colome, *S. Dak.* **12** 43 16N 99 43W
Colonial Heights, *Va.* **10** 37 15N 77 25W
Colorado □ **17** 39 30N 105 30W
Colorado →, *N. Amer.* **17** 31 45N 114 40W
Colorado →, *Tex.* **13** 28 36N 95 59W
Colorado City, *Tex.* **13** 32 24N 100 52W
Colorado Plateau, *Ariz.* **17** 37 0N 111 0W
Colorado River Aqueduct, *Calif.* **19** 34 17N 114 10W
Colorado Springs, *Colo.* **12** 38 50N 104 49W
Colton, *Calif.* **19** 34 4N 117 20W
Colton, *Wash.* **16** 46 34N 117 8W
Columbia, *La.* **13** 32 6N 92 5W
Columbia, *Md.* **25** 39 12N 76 50W
Columbia, *Miss.* **13** 31 15N 89 50W
Columbia, *Mo.* **14** 38 57N 92 20W
Columbia, *Pa.* **9** 40 2N 76 30W
Columbia, *S.C.* **11** 34 0N 81 2W
Columbia, *Tenn.* **11** 35 37N 87 2W
Columbia →, *Oreg.* **16** 46 15N 124 5W
Columbia, District of □ **10** 38 55N 77 0W
Columbia Basin, *Wash.* **16** 46 45N 119 5W
Columbia Falls, *Mont.* **16** 48 23N 114 11W
Columbia Heights, *Minn.* **12** 45 3N 93 15W
Columbus, *Ga.* **11** 32 28N 84 59W
Columbus, *Ind.* **15** 39 13N 85 55W
Columbus, *Kans.* **13** 37 10N 94 50W
Columbus, *Miss.* **11** 33 30N 88 25W
Columbus, *Mont.* **16** 45 38N 109 15W
Columbus, *N. Dak.* **12** 48 54N 102 47W
Columbus, *N. Mex.* **17** 31 50N 107 38W
Columbus, *Nebr.* **12** 41 26N 97 22W
Columbus, *Ohio* **15** 39 58N 83 0W
Columbus, *Tex.* **13** 29 42N 96 33W
Columbus, *Wis.* **12** 43 21N 89 1W
Colusa, *Calif.* **18** 39 13N 122 1W
Colville, *Wash.* **16** 48 33N 117 54W
Colville →, *Alaska* **30** 70 25N 150 30W
Comanche, *Okla.* **13** 34 22N 97 58W
Comanche, *Tex.* **13** 31 54N 98 36W
Combahee →, *S.C.* **11** 32 30N 80 31W
Commerce, *Calif.* **29** 34 0N 118 9W
Commerce, *Ga.* **11** 34 12N 83 28W
Commerce, *Tex.* **13** 33 15N 95 54W
Compton, *Calif.* **29** 33 53N 118 14W
Conception, Pt., *Calif.* **19** 34 27N 120 28W
Conchas Dam, *N. Mex.* **13** 35 22N 104 11W
Concho, *Ariz.* **17** 34 28N 109 36W
Concho →, *Tex.* **13** 31 34N 99 43W
Concord, *Calif.* **28** 37 57N 122 3W
Concord, *Mass.* **23** 42 27N 71 20W
Concord, *N.C.* **11** 35 25N 80 35W
Concord, *N.H.* **9** 43 12N 71 32W
Concordia, *Kans.* **12** 39 34N 97 40W
Concordville, *Pa.* **24** 39 53N 75 31W
Concrete, *Wash.* **16** 48 32N 121 45W
Conde, *S. Dak.* **12** 45 9N 98 6W
Condon, *Oreg.* **16** 45 14N 120 11W
Congress, *Ariz.* **17** 34 9N 112 51W
Conneaut, *Ohio* **8** 41 57N 80 34W
Connecticut □ **9** 41 30N 72 45W
Connecticut →, *Conn.* **9** 41 16N 72 20W
Connell, *Wash.* **16** 46 40N 118 52W
Connellsville, *Pa.* **8** 40 1N 79 35W
Connersville, *Ind.* **15** 39 39N 85 8W
Conrad, *Mont.* **16** 48 10N 111 57W
Conroe, *Tex.* **13** 30 19N 95 27W
Conshohocken, *Pa.* **24** 40 4N 75 18W
Contact, *Nev.* **16** 41 46N 114 45W
Contoocook, *N.H.* **9** 43 13N 71 45W
Conway, *Ark.* **13** 35 5N 92 26W
Conway, *N.H.* **9** 43 59N 71 7W
Conway, *S.C.* **11** 33 51N 79 3W
Cook, *Minn.* **12** 47 49N 92 39W
Cook Inlet, *Alaska* **30** 60 0N 152 0W
Cookeville, *Tenn.* **11** 36 10N 85 30W
Coolidge, *Ariz.* **17** 32 59N 111 31W
Coolidge Dam, *Ariz.* **17** 33 0N 110 20W
Cooper, *Tex.* **13** 33 23N 95 42W
Cooper →, *S.C.* **11** 32 50N 79 56W
Cooperstown, *N. Dak.* **12** 47 27N 98 8W
Cooperstown, *N.Y.* **9** 42 42N 74 56W
Coos Bay, *Oreg.* **16** 43 22N 124 13W
Cope, *Colo.* **12** 39 40N 102 51W
Copper →, *Alaska* **30** 60 18N 145 3W
Copper Center, *Alaska* **30** 61 58N 145 18W
Copper Harbor, *Mich.* **10** 47 28N 87 53W
Copperopolis, *Calif.* **18** 37 58N 120 38W
Coquille, *Oreg.* **16** 43 11N 124 11W
Coral Gables, *Fla.* **11** 25 45N 80 16W
Corbin, *Ky.* **10** 36 57N 84 6W
Corcoran, *Calif.* **18** 36 6N 119 33W
Cordele, *Ga.* **11** 31 58N 83 47W
Cordell, *Okla.* **13** 35 17N 98 59W
Cordova, *Ala.* **11** 33 46N 87 11W
Cordova, *Alaska* **30** 60 33N 145 45W
Corinth, *Miss.* **11** 34 56N 88 31W
Corinth, *N.Y.* **9** 43 15N 73 49W
Cornell, *Wis.* **12** 45 10N 91 9W
Corning, *Ark.* **13** 36 25N 90 35W
Corning, *Calif.* **16** 39 56N 122 11W
Corning, *Iowa* **14** 40 59N 94 44W
Corning, *N.Y.* **8** 42 9N 77 3W
Corona, *Calif.* **19** 33 53N 117 34W
Corona, *N. Mex.* **17** 34 15N 105 36W
Coronado, *Calif.* **19** 32 41N 117 11W
Corpus Christi, *Tex.* **13** 27 47N 97 24W
Corpus Christi, L., *Tex.* **13** 28 2N 97 52W
Corrigan, *Tex.* **13** 31 0N 94 52W
Corry, *Pa.* **8** 41 55N 79 39W
Corsicana, *Tex.* **13** 32 6N 96 28W
Corte Madera, *Calif.* **28** 37 55N 122 30W
Cortez, *Colo.* **17** 37 21N 108 35W
Cortland, *N.Y.* **9** 42 36N 76 11W
Corvallis, *Oreg.* **16** 44 34N 123 16W
Corydon, *Iowa* **14** 40 46N 93 19W
Coshocton, *Ohio* **8** 40 16N 81 51W
Coso Junction, *Calif.* **19** 36 3N 117 57W
Coso Pk., *Calif.* **19** 36 13N 117 44W
Costa Mesa, *Calif.* **19** 33 38N 117 55W
Costilla, *N. Mex.* **17** 36 59N 105 32W
Cosumnes →, *Calif.* **18** 38 16N 121 26W
Coteau des Prairies, *S. Dak.* **12** 45 20N 97 50W
Coteau du Missouri, *N. Dak.* **12** 47 0N 100 0W
Cottage Grove, *Oreg.* **16** 43 48N 123 3W
Cottonwood, *Ariz.* **17** 34 45N 112 1W
Cotulla, *Tex.* **13** 28 26N 99 14W
Coudersport, *Pa.* **8** 41 46N 78 1W
Coulee City, *Wash.* **16** 47 37N 119 17W
Coulterville, *Calif.* **18** 37 43N 120 12W
Council, *Alaska* **30** 64 55N 163 45W
Council, *Idaho* **16** 44 44N 116 26W
Council Bluffs, *Iowa* **12** 41 16N 95 52W
Council Grove, *Kans.* **12** 38 40N 96 29W
Courtland, *Calif.* **18** 38 20N 121 34W
Coushatta, *La.* **13** 32 1N 93 21W
Covington, *Ga.* **11** 33 36N 83 51W
Covington, *Ky.* **15** 39 5N 84 31W
Covington, *Okla.* **13** 36 18N 97 35W
Covington, *Tenn.* **13** 35 34N 89 39W
Cozad, *Nebr.* **12** 40 52N 99 59W
Craig, *Alaska* **30** 55 29N 133 9W
Craig, *Colo.* **16** 40 31N 107 33W
Crandon, *Wis.* **12** 45 34N 88 54W
Crane, *Oreg.* **16** 43 25N 118 35W
Crane, *Tex.* **13** 31 24N 102 21W
Cranford, *N.J.* **20** 40 39N 74 19W
Cranston, *R.I.* **9** 41 47N 71 26W
Crater L., *Oreg.* **16** 42 56N 122 6W
Crawford, *Nebr.* **12** 42 41N 103 25W
Crawfordsville, *Ind.* **15** 40 2N 86 54W
Crazy Mts., *Mont.* **16** 46 12N 110 20W
Creede, *Colo.* **17** 37 51N 106 56W
Creighton, *Nebr.* **12** 42 28N 97 54W
Cresbard, *S. Dak.* **12** 45 10N 98 57W
Crescent, *Okla.* **13** 35 57N 97 36W
Crescent, *Oreg.* **16** 43 28N 121 42W
Crescent City, *Calif.* **16** 41 45N 124 12W
Crested Butte, *Colo.* **17** 38 52N 106 59W
Crestline, *Calif.* **19** 34 14N 117 18W

Creston

Creston, Calif. 18 35 32N 120 33W
Creston, Iowa 14 41 4N 94 22W
Creston, Wash. 16 47 46N 118 31W
Crestview, Calif. 18 37 46N 118 58W
Crestview, Fla. 11 30 46N 86 34W
Crete, Nebr. 12 40 38N 96 58W
Crockett, Tex. 13 31 19N 95 27W
Crooked →, Oreg. 16 44 32N 121 16W
Crookston, Minn. 12 47 47N 96 37W
Crookston, Nebr. 12 42 56N 100 45W
Crooksville, Ohio 10 39 46N 82 6W
Crosby, Minn. 12 46 29N 93 58W
Crosbyton, Tex. 13 33 40N 101 14W
Cross City, Fla. 11 29 38N 83 7W
Cross Plains, Tex. 13 32 8N 99 11W
Cross Sound, Alaska 30 58 0N 135 0W
Crossett, Ark. 13 33 8N 91 58W
Croton-on-Hudson, N.Y. 9 41 12N 73 55W
Crow Agency, Mont. 16 45 36N 107 28W
Crowell, Tex. 13 33 59N 99 43W
Crowley, La. 13 30 13N 92 22W
Crowley, L., Calif. 18 37 35N 118 42W
Crown Point, Ind. 15 41 25N 87 22W
Crows Landing, Calif. 18 37 23N 121 6W
Crystal City, Mo. 14 38 13N 90 23W
Crystal City, Tex. 13 28 41N 99 50W
Crystal Falls, Mich. 10 46 5N 88 20W
Crystal River, Fla. 11 28 54N 82 35W
Crystal Springs, Miss. 13 31 59N 90 21W
Cuba, N. Mex. 17 36 1N 107 4W
Cudahy, Wis. 15 42 58N 87 52W
Cuero, Tex. 13 29 6N 97 17W
Cuervo, N. Mex. 13 35 2N 104 25W
Culbertson, Mont. 12 48 9N 104 31W
Culebra, Isla de,
Puerto Rico 30 18 19N 65 18W
Cullman, Ala. 11 34 11N 86 51W
Culpeper, Va. 10 38 30N 78 0W
Culver City, Calif. 29 34 1N 118 23W
Cumberland, Md. 10 39 39N 78 46W
Cumberland, Wis. 12 45 32N 92 1W
Cumberland →, Tenn. 11 36 15N 87 0W
Cumberland I., Ga. 11 30 50N 81 25W
Cumberland Plateau,
Tenn. 11 36 0N 85 0W
Cummings Mt., Calif. 19 35 2N 118 34W
Currant, Nev. 16 38 51N 115 32W
Current →, Ark. 13 36 15N 90 55W
Currie, Nev. 16 40 16N 114 45W
Currituck Sd., N.C. 11 36 20N 75 52W
Curtis, Nebr. 12 40 38N 100 31W
Cushing, Okla. 13 35 59N 96 46W
Custer, S. Dak. 12 43 46N 103 36W
Cut Bank, Mont. 16 48 38N 112 20W
Cuthbert, Ga. 11 31 46N 84 48W
Cutler, Calif. 18 36 31N 119 17W
Cuyahoga Falls, Ohio 8 41 8N 81 29W
Cynthiana, Ky. 15 38 23N 84 18W

D

Dade City, Fla. 11 28 22N 82 11W
Daggett, Calif. 19 34 52N 116 52W
Dahlonega, Ga. 11 34 32N 83 59W
Dakota City, Nebr. 12 42 25N 96 25W
Dalhart, Tex. 13 36 4N 102 31W
Dallas, Oreg. 16 44 55N 123 19W
Dallas, Tex. 13 32 47N 96 49W
Dalton, Ga. 11 34 46N 84 58W
Dalton, Mass. 9 42 28N 73 11W
Dalton, Nebr. 12 41 25N 102 58W
Daly City, Calif. 28 37 42N 122 26W
Dana, Mt., Calif. 18 37 54N 119 12W
Danbury, Conn. 9 41 24N 73 28W
Danby L., Calif. 17 34 13N 115 5W
Danforth, Maine 11 45 40N 67 52W
Daniel, Wyo. 16 42 52N 110 4W
Danielson, Conn. 9 41 48N 71 53W
Dannemora, N.Y. 9 44 43N 73 44W
Dansville, N.Y. 8 42 34N 77 42W
Danvers, Mass. 23 42 34N 70 56W
Danville, Ill. 15 40 8N 87 37W
Danville, Ky. 15 37 39N 84 46W
Danville, Va. 11 36 36N 79 23W
Darby, Mont. 16 46 1N 114 11W
Darby, Pa. 24 39 55N 75 16W
Dardanelle, Ark. 13 35 13N 93 9W
Dardanelle, Calif. 18 38 20N 119 50W
Darlington, S.C. 11 34 18N 79 52W
Darlington, Wis. 14 42 41N 90 7W
Darrington, Wash. 16 48 15N 121 36W
Darwin, Calif. 19 36 15N 117 35W
Dauphin I., Ala. 11 30 15N 88 11W
Davenport, Calif. 18 37 1N 122 12W
Davenport, Iowa 14 41 32N 90 35W
Davenport, Wash. 16 47 39N 118 9W
David City, Nebr. 12 41 15N 97 8W
Davis, Calif. 18 38 33N 121 44W
Davis Dam, Ariz. 19 35 11N 114 34W
Davis Mts., Tex. 13 30 50N 103 55W
Dawson, Ga. 11 31 46N 84 27W
Dawson, N. Dak. 12 46 52N 99 45W
Dayton, Ohio 10 39 45N 84 12W
Dayton, Tenn. 11 35 30N 85 1W
Dayton, Wash. 16 46 19N 117 59W
Daytona Beach, Fla. 11 29 13N 81 1W
Dayville, Oreg. 16 44 28N 119 32W
De Funiak Springs, Fla. 11 30 43N 86 7W
De Kalb, Ill. 15 41 56N 88 46W
De Land, Fla. 11 29 2N 81 18W
De Leon, Tex. 13 32 7N 98 32W
De Long Mts., Alaska 30 68 30N 163 0W
De Pere, Wis. 10 44 27N 88 4W
De Queen, Ark. 13 34 2N 94 21W
De Quincy, La. 13 30 27N 93 26W
De Ridder, La. 13 30 51N 93 17W
De Smet, S. Dak. 12 44 23N 97 33W
De Soto, Mo. 14 38 8N 90 34W
De Tour Village, Mich. 10 46 0N 83 56W
De Witt, Ark. 13 34 18N 91 20W
Deadwood, S. Dak. 12 44 23N 103 44W
Dearborn, Mich. 26 42 19N 83 10W
Dearborn Heights, Mich. 26 42 20N 83 17W
Death Valley, Calif. 19 36 15N 116 50W
Death Valley Junction,
Calif. 19 36 20N 116 25W
Death Valley National
Monument, Calif. 19 36 45N 117 15W
Decatur, Ala. 11 34 36N 86 59W
Decatur, Ga. 11 33 47N 84 18W
Decatur, Ill. 14 39 51N 88 57W
Decatur, Ind. 15 40 50N 84 56W
Decatur, Tex. 13 33 14N 97 35W
Decorah, Iowa 12 43 18N 91 48W
Dedham, Mass. 23 42 15N 71 10W
Deer I., Alaska 30 54 55N 162 18W
Deer Lodge, Mont. 16 46 24N 112 44W
Deer Park, N.Y. 21 40 46N 73 19W
Deer Park, Wash. 16 47 57N 117 28W
Deer River, Minn. 12 47 20N 93 48W
Deering, Alaska 30 66 4N 162 42W
Defiance, Ohio 15 41 17N 84 22W
Del Mar, Calif. 19 32 58N 117 16W
Del Norte, Colo. 17 37 41N 106 21W
Del Rio, Tex. 13 29 22N 100 54W
Delano, Calif. 19 35 46N 119 15W
Delavan, Wis. 15 42 38N 88 39W
Delaware, Ohio 15 40 18N 83 4W
Delaware □ 10 39 0N 75 20W
Delaware →, Del. 10 39 15N 75 20W
Delhi, N.Y. 9 42 17N 74 55W
Dell City, Tex. 17 31 56N 105 12W
Dell Rapids, S. Dak. 12 43 50N 96 43W
Delphi, Ind. 15 40 36N 86 41W
Delphos, Ohio 15 40 51N 84 21W
Delray Beach, Fla. 11 26 28N 80 4W
Delta, Colo. 17 38 44N 108 4W
Delta, Utah 16 39 21N 112 35W
Deming, N. Mex. 17 32 16N 107 46W
Demopolis, Ala. 11 32 31N 87 50W
Denair, Calif. 18 37 32N 120 48W
Denison, Iowa 12 42 1N 95 21W
Denison, Tex. 13 33 45N 96 33W
Denton, Mont. 16 47 19N 109 57W
Denton, Tex. 13 33 13N 97 8W
Denver, Colo. 12 39 44N 104 59W
Denver City, Tex. 13 32 58N 102 50W
Deposit, N.Y. 9 42 4N 75 25W
Derby, Conn. 9 41 19N 73 5W
Dernieres, Isles, La. 13 29 2N 90 50W
Des Moines, Iowa 14 41 35N 93 37W
Des Moines, N. Mex. 13 36 46N 103 50W
Des Moines →, Iowa 12 40 23N 91 25W
Des Plaines, Ill. 22 42 2N 87 54W
Deschutes →, Oreg. 16 45 38N 120 55W
Desert Center, Calif. 19 33 43N 115 24W
Desert Hot Springs, Calif. 19 33 58N 116 30W
Detour, Pt., Mich. 10 45 40N 86 40W
Detroit, Mich. 26 42 20N 83 2W
Detroit, Tex. 13 33 40N 95 16W
Detroit City Airport,
Mich. 26 42 24N 83 0W
Detroit Lakes, Minn. 12 46 49N 95 51W
Detroit-Wayne Airport,
Mich. 26 42 13N 83 20W
Devils Den, Calif. 18 35 46N 119 58W
Devils Lake, N. Dak. 12 48 7N 98 52W
Dexter, Mo. 13 36 48N 89 57W
Dexter, N. Mex. 13 33 12N 104 22W
Diablo, Mt., Calif. 18 37 53N 121 56W
Diablo Range, Calif. 18 37 20N 121 25W
Diamond Mts., Nev. 16 39 50N 115 30W
Diamond Springs, Calif. 18 38 42N 120 49W
Diamondville, Wyo. 16 41 47N 110 32W
Dickinson, N. Dak. 12 46 53N 102 47W
Dickson, Tenn. 11 36 5N 87 23W
Dickson City, Pa. 9 41 29N 75 40W
Dierks, Ark. 13 34 7N 94 1W
Dighton, Kans. 12 38 29N 100 28W
Dilley, Tex. 13 28 40N 99 10W
Dillingham, Alaska 30 59 3N 158 28W
Dillon, Mont. 16 45 13N 112 38W
Dillon, S.C. 11 34 25N 79 22W
Dimmitt, Tex. 13 34 33N 102 19W
Dingmans Ferry, Pa. 9 41 13N 74 55W
Dinosaur National
Monument, Colo. 16 40 30N 108 45W
Dinuba, Calif. 18 36 32N 119 23W
Disappointment, C.,
Wash. 16 46 18N 124 5W
Divide, Mont. 16 45 45N 112 45W
Dixon, Calif. 18 38 27N 121 49W
Dixon, Ill. 14 41 50N 89 29W
Dixon, Mont. 16 47 19N 114 19W
Dixon, N. Mex. 17 36 12N 105 53W
Dodge Center, Minn. 12 44 2N 92 52W
Dodge City, Kans. 13 37 45N 100 1W
Dodgeville, Wis. 14 42 58N 90 8W
Dodson, Mont. 16 48 24N 108 15W
Doland, S. Dak. 12 44 54N 98 6W
Dolores, Colo. 17 37 28N 108 30W
Dolores →, Utah 17 38 49N 109 17W
Dolton, Ill. 22 41 37N 87 35W
Donaldsonville, La. 13 30 6N 90 59W
Donalsonville, Ga. 11 31 3N 84 53W
Doniphan, Mo. 13 36 37N 90 50W
Donna, Tex. 13 26 9N 98 4W
Dormont, Pa. 27 40 23N 80 2W
Dorris, Calif. 16 41 58N 121 55W
Dos Palos, Calif. 18 36 59N 120 37W
Dothan, Ala. 11 31 13N 85 24W
Douglas, Alaska 30 58 17N 134 24W
Douglas, Ariz. 17 31 21N 109 33W
Douglas, Ga. 11 31 31N 82 51W
Douglas, Wyo. 12 42 45N 105 24W
Douglasville, Ga. 11 33 45N 84 45W
Dove Creek, Colo. 17 37 46N 108 54W
Dover, Del. 10 39 10N 75 32W
Dover, N.H. 9 43 12N 70 56W
Dover, N.J. 9 40 53N 74 34W
Dover, Ohio 8 40 32N 81 29W
Dover-Foxcroft, Maine 11 45 11N 69 13W
Dover Plains, N.Y. 9 41 43N 73 35W
Dowagiac, Mich. 15 41 59N 86 6W
Downey, Calif. 29 33 56N 118 8W
Downey, Idaho 16 42 26N 112 7W
Downieville, Calif. 18 39 34N 120 50W
Doylestown, Pa. 9 40 21N 75 10W
Drain, Oreg. 16 43 40N 123 19W
Drake, N. Dak. 12 47 55N 100 23W
Drexel Hill, Pa. 24 39 56N 75 18W
Driggs, Idaho 16 43 44N 111 6W
Drummond, Mont. 16 46 40N 113 9W
Drumright, Okla. 13 35 59N 96 36W
Dryden, Tex. 13 30 3N 102 7W
Du Bois, Pa. 8 41 8N 78 46W
Du Quoin, Ill. 14 38 1N 89 14W
Duanesburg, N.Y. 9 42 45N 74 11W
Duarte, Calif. 29 34 8N 117 57W
Dublin, Ga. 11 32 32N 82 54W
Dublin, Tex. 13 32 5N 98 21W
Dubois, Idaho 16 44 10N 112 14W
Dubuque, Iowa 14 42 30N 90 41W
Duchesne, Utah 16 40 10N 110 24W
Duckwall, Mt., Calif. 18 37 58N 120 7W
Duluth, Minn. 12 46 47N 92 6W
Dulworthtown, Pa. 24 39 54N 75 33W
Dumas, Ark. 13 33 53N 91 29W
Dumas, Tex. 13 35 52N 101 58W
Duncan, Ariz. 17 32 43N 109 6W
Duncan, Okla. 13 34 30N 97 57W
Dundalk, Md. 25 39 16N 76 30W
Dunedin, Fla. 11 28 1N 82 47W
Dunkirk, N.Y. 8 42 29N 79 20W
Dunlap, Iowa 12 41 51N 95 36W
Dunmore, Pa. 9 41 25N 75 38W
Dunn, N.C. 11 35 19N 78 37W
Dunnellon, Fla. 11 29 3N 82 28W
Dunning, Nebr. 12 41 50N 100 6W
Dunseith, N. Dak. 12 48 50N 100 3W
Dunsmuir, Calif. 16 41 13N 122 16W
Dupree, S. Dak. 12 45 4N 101 35W
Dupuyer, Mont. 16 48 13N 112 30W
Duquesne, Pa. 27 40 22N 79 52W
Durand, Mich. 15 42 55N 83 59W
Durango, Colo. 17 37 16N 107 53W
Durant, Okla. 13 33 59N 96 25W
Durham, N.C. 11 35 59N 78 54W
Duryea, Pa. 9 41 20N 75 45W
Dutch Harbor, Alaska 30 53 53N 166 32W
Dwight, Ill. 15 41 5N 88 26W
Dyersburg, Tenn. 13 36 3N 89 23W

E

Eads, Colo. 12 38 29N 102 47W
Eagle, Alaska 30 64 47N 141 12W
Eagle, Colo. 16 39 39N 106 50W
Eagle Butte, S. Dak. 12 45 0N 101 10W
Eagle Grove, Iowa 14 42 40N 93 54W
Eagle L., Calif. 16 40 39N 120 45W
Eagle L., Maine 11 46 20N 69 22W
Eagle Lake, Tex. 13 29 35N 96 20W
Eagle Nest, N. Mex. 17 36 33N 105 16W
Eagle Pass, Tex. 13 28 43N 100 30W
Eagle Pk., Calif. 18 38 10N 119 25W
Eagle River, Wis. 12 45 55N 89 15W
Earle, Ark. 13 35 16N 90 28W
Earlimart, Calif. 19 35 53N 119 16W
Earth, Tex. 13 34 14N 102 24W
Easley, S.C. 11 34 50N 82 36W
East B., La. 13 29 0N 89 15W
East Chicago, Ind. 22 41 38N 87 26W
East Cleveland, Ohio 27 41 32N 81 35W
East Detroit, Mich. 26 42 27N 82 58W
East Grand Forks, Minn. 12 47 56N 97 1W
East Greenwich, R.I. 9 41 40N 71 27W
East Hartford, Conn. 9 41 46N 72 39W
East Helena, Mont. 16 46 35N 111 56W
East Jordan, Mich. 10 45 10N 85 7W
East Lansing, Mich. 15 42 44N 84 29W
East Liverpool, Ohio 8 40 37N 80 35W
East Los Angeles, Calif. 29 34 1N 118 10W
East Meadow, N.Y. 21 40 42N 73 31W
East Orange, N.J. 20 40 46N 74 11W
East Point, Ga. 11 33 41N 84 27W
East Providence, R.I. 9 41 49N 71 23W
East St. Louis, Ill. 14 38 37N 90 9W
East Stroudsburg, Pa. 9 41 1N 75 11W
East Tawas, Mich. 10 44 17N 83 29W
East Walker →, Nev. 18 38 52N 119 10W
Eastchester, N.Y. 21 40 57N 73 49W
Eastlake, Ohio 27 41 38N 81 28W
Eastland, Tex. 13 32 24N 98 49W
Eastman, Ga. 11 32 12N 83 11W
Easton, Md. 10 38 47N 76 5W
Easton, Pa. 9 40 41N 75 13W
Easton, Wash. 18 47 14N 121 11W
Eastport, Maine 11 44 56N 67 0W
Eaton, Colo. 12 40 32N 104 42W
Eatonton, Ga. 11 33 20N 83 23W
Eatontown, N.J. 9 40 19N 74 4W
Eau Claire, Wis. 12 44 49N 91 30W
Eden, N.C. 11 36 29N 79 53W
Eden, Tex. 13 31 13N 99 51W
Eden, Wyo. 16 42 3N 109 26W
Edenton, N.C. 11 36 4N 76 39W
Edgar, Nebr. 12 40 22N 97 58W
Edgartown, Mass. 9 41 23N 70 31W
Edgefield, S.C. 11 33 47N 81 56W
Edgeley, N. Dak. 12 46 22N 98 43W
Edgemont, S. Dak. 12 43 18N 103 50W
Edina, Mo. 14 40 10N 92 11W
Edinburg, Tex. 13 26 18N 98 10W
Edison, N.J. 20 40 31N 74 22W
Edmeston, N.Y. 9 42 42N 75 15W
Edmond, Okla. 13 35 39N 97 29W
Edmonds, Wash. 18 47 49N 122 23W
Edna, Tex. 13 28 59N 96 39W
Edwards, Calif. 19 34 55N 117 51W
Edwards Plateau, Tex. 13 30 45N 101 20W
Edwardsville, Pa. 9 41 15N 75 56W
Eek, Alaska 30 60 14N 162 2W
Effingham, Ill. 15 39 7N 88 33W
Egeland, N. Dak. 12 48 38N 99 6W
Ekalaka, Mont. 12 45 53N 104 33W
El Cajon, Calif. 19 32 48N 116 58W
El Campo, Tex. 13 29 12N 96 16W
El Centro, Calif. 19 32 48N 115 34W
El Cerrito, Calif. 28 37 54N 122 18W
El Dorado, Ark. 13 33 12N 92 40W
El Dorado, Kans. 13 37 49N 96 52W
El Granada, Calif. 28 37 30N 122 28W
El Monte, Calif. 29 34 3N 118 1W
El Paso, Tex. 17 31 45N 106 29W
El Paso Robles, Calif. 18 35 38N 120 41W
El Portal, Calif. 18 37 41N 119 47W
El Reno, Okla. 13 35 32N 97 57W
El Rio, Calif. 19 34 14N 119 10W
El Segundo, Calif. 29 33 55N 118 24W
Elba, Ala. 11 31 25N 86 4W
Elbert, Mt., Colo. 17 39 7N 106 27W
Elberta, Mich. 10 44 37N 86 14W
Elberton, Ga. 11 34 7N 82 52W
Eldon, Mo. 14 38 21N 92 35W
Eldora, Iowa 14 42 22N 93 5W
Eldorado, Ill. 15 37 49N 88 26W
Eldorado, Tex. 13 30 52N 100 36W
Eldorado Springs, Mo. 14 37 52N 94 1W
Electra, Tex. 13 34 2N 98 55W
Eleele, Hawaii 30 21 54N 159 35W
Elephant Butte Reservoir,
N. Mex. 17 33 9N 107 11W
Elfin Cove, Alaska 30 58 12N 136 22W
Elgin, Ill. 15 42 2N 88 17W
Elgin, N. Dak. 12 46 24N 101 51W
Elgin, Nebr. 12 41 59N 98 5W
Elgin, Nev. 17 37 21N 114 32W
Elgin, Oreg. 16 45 34N 117 55W
Elgin, Tex. 13 30 21N 97 22W
Elida, N. Mex. 13 33 57N 103 39W
Elim, Alaska 30 64 37N 162 15W
Elizabeth, N.J. 20 40 39N 74 12W
Elizabeth City, N.C. 11 36 18N 76 14W
Elizabethton, Tenn. 11 36 21N 82 13W
Elizabethtown, Ky. 10 37 42N 85 52W
Elizabethtown, Pa. 9 40 9N 76 36W
Elk City, Okla. 13 35 25N 99 25W
Elk Grove, Calif. 18 38 25N 121 22W
Elk River, Idaho 16 46 47N 116 11W
Elk River, Minn. 12 45 18N 93 35W
Elkhart, Ind. 15 41 41N 85 58W
Elkhart, Kans. 13 37 0N 101 54W
Elkhorn →, Nebr. 12 41 8N 96 19W
Elkin, N.C. 11 36 15N 80 51W
Elkins, W. Va. 10 38 55N 79 51W
Elko, Nev. 16 40 50N 115 46W
Ellendale, N. Dak. 12 46 0N 98 32W
Ellensburg, Wash. 16 46 59N 120 34W
Ellenville, N.Y. 9 41 43N 74 24W
Ellinwood, Kans. 12 38 21N 98 35W
Ellis, Kans. 12 38 56N 99 34W
Ellisville, Miss. 13 31 36N 89 12W
Ellsworth, Kans. 12 38 44N 98 14W
Ellwood City, Pa. 8 40 52N 80 17W
Elma, Wash. 18 47 0N 123 25W

Georgetown

Georgetown, *S.C.* 11 33 23N 79 17W
Georgetown, *Tex.* 13 30 38N 97 41W
Georgia □ 11 32 50N 83 15W
Geraldine, *Mont.* 16 47 36N 110 16W
Gering, *Nebr.* 12 41 50N 103 40W
Gerlach, *Nev.* 16 40 39N 119 21W
Gettysburg, *Pa.* 10 39 50N 77 14W
Gettysburg, *S. Dak.* 12 45 1N 99 57W
Geyser, *Mont.* 16 47 16N 110 30W
Giant Forest, *Calif.* 18 36 36N 118 43W
Gibbon, *Nebr.* 12 40 45N 98 51W
Giddings, *Tex.* 13 30 11N 96 56W
Gila →, *Ariz.* 17 32 43N 114 33W
Gila Bend, *Ariz.* 17 32 57N 112 43W
Gila Bend Mts., *Ariz.* 17 33 10N 113 0W
Gillette, *Wyo.* 12 44 18N 105 30W
Gilmer, *Tex.* 13 32 44N 94 57W
Gilroy, *Calif.* 18 37 1N 121 34W
Girard, *Kans.* 13 37 31N 94 51W
Glacier Park, *Mont.* 16 48 30N 113 18W
Glacier Peak, *Wash.* 16 48 7N 121 7W
Gladewater, *Tex.* 13 32 33N 94 56W
Gladstone, *Mich.* 10 45 51N 87 1W
Gladwin, *Mich.* 10 43 59N 84 29W
Glasco, *Kans.* 12 39 22N 97 50W
Glasco, *N.Y.* 9 42 3N 73 57W
Glasgow, *Ky.* 10 37 0N 85 55W
Glasgow, *Mont.* 16 48 12N 106 38W
Glastonbury, *Conn.* 9 41 43N 72 37W
Glen Canyon Dam, *Ariz.* 17 36 57N 111 29W
Glen Canyon National Recreation Area, *Utah* 17 37 15N 111 0W
Glen Cove, *N.Y.* 21 40 52N 73 38W
Glen Lyon, *Pa.* 9 41 10N 76 5W
Glen Ullin, *N. Dak.* 12 46 49N 101 50W
Glencoe, *Ill.* 22 42 7N 87 44W
Glencoe, *Minn.* 12 44 46N 94 9W
Glendale, *Ariz.* 17 33 32N 112 11W
Glendale, *Calif.* 29 34 9N 118 14W
Glendale, *Oreg.* 16 42 44N 123 26W
Glendive, *Mont.* 12 47 7N 104 43W
Glendo, *Wyo.* 12 42 30N 105 2W
Glenmora, *La.* 13 30 59N 92 35W
Glenns Ferry, *Idaho* 16 42 57N 115 18W
Glenrock, *Wyo.* 16 42 52N 105 52W
Glens Falls, *N.Y.* 9 43 19N 73 39W
Glenshaw, *Pa.* 27 40 32N 79 58W
Glenview, *Ill.* 22 42 4N 87 48W
Glenville, *W. Va.* 10 38 56N 80 50W
Glenwood, *Ark.* 13 34 20N 93 33W
Glenwood, *Hawaii* 30 19 29N 155 9W
Glenwood, *Iowa* 12 41 3N 95 45W
Glenwood, *Minn.* 12 45 39N 95 23W
Glenwood Springs, *Colo.* 16 39 33N 107 19W
Globe, *Ariz.* 17 33 24N 110 47W
Gloucester, *Mass.* 9 42 37N 70 40W
Gloucester City, *N.J.* 24 39 53N 75 7W
Gloversville, *N.Y.* 9 43 3N 74 21W
Gogebic, L., *Mich.* 12 46 30N 89 35W
Golconda, *Nev.* 16 40 58N 117 30W
Gold Beach, *Oreg.* 16 42 25N 124 25W
Gold Creek, *Alaska* 30 62 46N 149 41W
Gold Hill, *Oreg.* 16 42 26N 123 3W
Golden, *Colo.* 12 39 42N 105 15W
Golden Gate, *Calif.* 16 37 54N 122 30W
Golden Gate, *Calif.* 28 37 48N 122 30W
Golden Gate Bridge, *Calif.* 28 37 49N 122 28W
Goldendale, *Wash.* 16 45 49N 120 50W
Goldfield, *Nev.* 17 37 42N 117 14W
Goldsboro, *N.C.* 11 35 23N 77 59W
Goldsmith, *Tex.* 13 31 59N 102 37W
Goldthwaite, *Tex.* 13 31 27N 98 34W
Goleta, *Calif.* 19 34 27N 119 50W
Goliad, *Tex.* 13 28 40N 97 23W
Gonzales, *Calif.* 18 36 30N 121 26W
Gonzales, *Tex.* 13 29 30N 97 27W
Gooding, *Idaho* 16 42 56N 114 43W
Goodland, *Kans.* 12 39 21N 101 43W
Goodnight, *Tex.* 13 35 2N 101 11W
Goodsprings, *Nev.* 17 35 50N 115 26W
Goose L., *Calif.* 16 41 56N 120 26W
Gorda, *Calif.* 18 35 53N 121 26W
Gordon, *Nebr.* 12 42 48N 102 12W
Gorman, *Calif.* 19 34 47N 118 51W
Gorman, *Tex.* 13 32 12N 98 41W
Goshen, *Calif.* 18 36 21N 119 25W
Goshen, *Ind.* 15 41 35N 85 50W
Goshen, *N.Y.* 9 41 24N 74 20W
Gothenburg, *Nebr.* 12 40 56N 100 10W
Gowanda, *N.Y.* 8 42 28N 78 56W
Grace, *Idaho* 16 42 35N 111 44W
Graceville, *Minn.* 12 45 34N 96 26W
Grady, *N. Mex.* 13 34 49N 103 19W
Grafton, *N. Dak.* 12 48 25N 97 25W
Graham, *N.C.* 11 36 5N 79 25W
Graham, *Tex.* 13 33 6N 98 35W
Graham, Mt., *Ariz.* 17 32 42N 109 52W
Granada, *Colo.* 13 38 4N 102 19W
Granbury, *Tex.* 13 32 27N 97 47W
Grand →, *Mo.* 14 39 23N 93 7W
Grand →, *S. Dak.* 12 45 40N 100 45W
Grand Canyon, *Ariz.* 17 36 3N 112 9W
Grand Canyon National Park, *Ariz.* 17 36 15N 112 30W
Grand Coulee, *Wash.* 16 47 57N 119 0W
Grand Coulee Dam, *Wash.* 16 47 57N 118 59W
Grand Forks, *N. Dak.* 12 47 55N 97 3W
Grand Haven, *Mich.* 15 43 4N 86 13W
Grand I., *Mich.* 10 46 31N 86 40W
Grand Island, *Nebr.* 12 40 55N 98 21W
Grand Isle, *La.* 13 29 14N 90 0W
Grand Junction, *Colo.* 17 39 4N 108 33W
Grand L., *La.* 13 29 55N 92 47W
Grand Lake, *Colo.* 16 40 15N 105 49W
Grand Marais, *Mich.* 10 46 40N 85 59W
Grand Rapids, *Mich.* 15 42 58N 85 40W
Grand Rapids, *Minn.* 12 47 14N 93 31W
Grand Teton, *Idaho* 16 43 54N 111 50W
Grand Valley, *Colo.* 16 39 27N 108 3W
Grande, Rio →, *Tex.* 13 25 58N 97 9W
Grandfalls, *Tex.* 13 31 20N 102 51W
Grandview, *Wash.* 16 46 15N 119 54W
Granger, *Wash.* 16 46 21N 120 11W
Granger, *Wyo.* 16 41 35N 109 58W
Grangeville, *Idaho* 16 45 56N 116 7W
Granite City, *Ill.* 14 38 42N 90 9W
Granite Falls, *Minn.* 12 44 49N 95 33W
Granite Mt., *Calif.* 19 33 5N 116 28W
Granite Peak, *Mont.* 16 45 10N 109 48W
Grant, *Nebr.* 12 40 53N 101 42W
Grant, Mt., *Nev.* 16 38 34N 118 48W
Grant City, *Mo.* 14 40 29N 94 25W
Grant Range, *Nev.* 17 38 30N 115 25W
Grants, *N. Mex.* 17 35 9N 107 52W
Grants Pass, *Oreg.* 16 42 26N 123 19W
Grantsburg, *Wis.* 12 45 47N 92 41W
Grantsville, *Utah* 16 40 36N 112 28W
Granville, *N. Dak.* 12 48 16N 100 47W
Granville, *N.Y.* 10 43 24N 73 16W
Grapeland, *Tex.* 13 31 30N 95 29W
Grass Range, *Mont.* 16 47 0N 109 0W
Grass Valley, *Calif.* 18 39 13N 121 4W
Grass Valley, *Oreg.* 16 45 22N 120 47W
Grayling, *Mich.* 10 44 40N 84 43W
Grays Harbor, *Wash.* 16 46 59N 124 1W
Grays L., *Idaho* 16 43 4N 111 26W
Great Barrington, *Mass.* 9 42 12N 73 22W
Great Basin, *Nev.* 16 40 0N 117 0W
Great Bend, *Kans.* 12 38 22N 98 46W
Great Bend, *Pa.* 9 41 58N 75 45W
Great Falls, *Mont.* 16 47 30N 111 17W
Great Kills, *N.Y.* 20 40 32N 74 9W
Great Neck, *N.Y.* 21 40 48N 73 44W
Great Plains, *N. Amer.* 2 47 0N 105 0W
Great Salt L., *Utah* 16 41 15N 112 40W
Great Salt Lake Desert, *Utah* 16 40 50N 113 30W
Great Salt Plains L., *Okla.* 13 36 45N 98 8W
Great Sitkin I., *Alaska* 30 52 3N 176 6W
Great Smoky Mts. Nat. Pk., *Tenn.* 11 35 40N 83 40W
Greater Pittsburgh International Airport, *Pa.* 27 40 29N 80 13W
Greeley, *Colo.* 12 40 25N 104 42W
Greeley, *Nebr.* 12 41 33N 98 32W
Green →, *Ky.* 10 37 54N 87 30W
Green →, *Utah* 17 38 11N 109 53W
Green B., *Wis.* 10 45 0N 87 30W
Green Bay, *Wis.* 10 44 31N 88 0W
Green Cove Springs, *Fla.* 11 29 59N 81 42W
Green River, *Utah* 17 38 59N 110 10W
Green Tree, *Pa.* 27 40 25N 80 4W
Greenbelt, *Md.* 25 39 0N 76 52W
Greenbush, *Minn.* 12 48 42N 96 11W
Greencastle, *Ind.* 15 39 38N 86 52W
Greene, *N.Y.* 9 42 20N 75 46W
Greenfield, *Calif.* 18 36 19N 121 15W
Greenfield, *Calif.* 19 35 15N 119 0W
Greenfield, *Ind.* 15 39 47N 85 46W
Greenfield, *Iowa* 14 41 18N 94 28W
Greenfield, *Mass.* 9 42 35N 72 36W
Greenfield, *Mo.* 13 37 25N 93 51W
Greenport, *N.Y.* 9 41 6N 72 22W
Greensboro, *Ga.* 11 33 35N 83 11W
Greensboro, *N.C.* 11 36 4N 79 48W
Greensburg, *Ind.* 15 39 20N 85 29W
Greensburg, *Kans.* 13 37 36N 99 18W
Greensburg, *Pa.* 8 40 18N 79 33W
Greenville, *Ala.* 11 31 50N 86 38W
Greenville, *Calif.* 18 40 8N 120 57W
Greenville, *Ill.* 14 38 53N 89 25W
Greenville, *Maine* 11 45 28N 69 35W
Greenville, *Mich.* 15 43 11N 85 15W
Greenville, *Miss.* 13 33 24N 91 4W
Greenville, *N.C.* 11 35 37N 77 23W
Greenville, *Ohio* 15 40 6N 84 38W
Greenville, *Pa.* 8 41 24N 80 23W
Greenville, *S.C.* 11 34 51N 82 24W
Greenville, *Tenn.* 11 36 13N 82 51W
Greenville, *Tex.* 13 33 8N 96 7W
Greenwich, *Conn.* 21 41 1N 73 38W
Greenwich, *N.Y.* 9 43 5N 73 30W
Greenwood, *Miss.* 13 33 31N 90 11W
Greenwood, *S.C.* 11 34 12N 82 10W
Gregory, *S. Dak.* 12 43 14N 99 20W
Grenada, *Miss.* 13 33 47N 89 49W
Grenora, *N. Dak.* 12 48 37N 103 56W
Gresham, *Oreg.* 18 45 30N 122 26W
Greybull, *Wyo.* 16 44 30N 108 3W
Gridley, *Calif.* 18 39 22N 121 42W
Griffin, *Ga.* 11 33 15N 84 16W
Grinnell, *Iowa* 14 41 45N 92 43W
Groesbeck, *Tex.* 13 30 48N 96 31W
Groom, *Tex.* 13 35 12N 101 6W
Grosse Pointe, *Mich.* 26 42 23N 82 54W
Groton, *Conn.* 9 41 21N 72 5W
Groton, *S. Dak.* 12 45 27N 98 6W
Grouse Creek, *Utah* 16 41 42N 113 53W
Groveland, *Calif.* 18 37 50N 120 14W
Grover City, *Calif.* 19 35 7N 120 37W
Groveton, *N.H.* 9 44 36N 71 31W
Groveton, *Tex.* 13 31 4N 95 8W
Grundy Center, *Iowa* 14 42 22N 92 47W
Gruver, *Tex.* 13 36 16N 101 24W
Guadalupe, *Calif.* 19 34 59N 120 33W
Guadalupe →, *Tex.* 13 28 27N 96 47W
Guadalupe Peak, *Tex.* 17 31 50N 104 52W
Guam ■, *Pac. Oc.* 31 13 27N 144 45 E
Guánica, *Puerto Rico* 30 17 58N 66 55W
Guayama, *Puerto Rico* 30 17 59N 66 7W
Guernsey, *Wyo.* 12 42 19N 104 45W
Gueydan, *La.* 13 30 2N 92 31W
Guilford, *Maine* 11 45 10N 69 23W
Gulfport, *Miss.* 13 30 22N 89 6W
Gulkana, *Alaska* 30 62 16N 145 23W
Gunnison, *Colo.* 17 38 33N 106 56W
Gunnison, *Utah* 16 39 9N 111 49W
Gunnison →, *Colo.* 17 39 4N 108 35W
Guntersville, *Ala.* 11 34 21N 86 18W
Gurdon, *Ark.* 13 33 55N 93 9W
Gustavus, *Alaska* 30 58 25N 135 44W
Gustine, *Calif.* 18 37 16N 121 0W
Guthrie, *Okla.* 13 35 53N 97 25W
Guttenberg, *Iowa* 14 42 47N 91 6W
Guymon, *Okla.* 13 36 41N 101 29W
Gwinn, *Mich.* 10 46 19N 87 27W

H

Hackensack, *N.J.* 20 40 53N 74 3W
Haddonfield, *N.J.* 24 39 53N 75 2W
Hagemeister I., *Alaska* 30 58 39N 160 54W
Hagerman, *N. Mex.* 13 33 7N 104 20W
Hagerstown, *Md.* 10 39 39N 77 43W
Hailey, *Idaho* 16 43 31N 114 19W
Haines, *Alaska* 30 59 14N 135 26W
Haines, *Oreg.* 16 44 55N 117 56W
Haines City, *Fla.* 11 28 7N 81 38W
Halawa, *Hawaii* 30 21 9N 156 47W
Haleakala Crater, *Hawaii* 30 20 43N 156 16W
Halethorpe, *Md.* 25 39 14N 76 41W
Haleyville, *Ala.* 11 34 14N 87 37W
Half Moon B., *Calif.* 28 37 29N 122 28W
Half Moon Bay, *Calif.* 28 37 27N 122 25W
Hallettsville, *Tex.* 13 29 27N 96 57W
Halliday, *N. Dak.* 12 47 21N 102 20W
Hallstead, *Pa.* 9 41 58N 75 45W
Halstad, *Minn.* 12 47 21N 96 50W
Hamburg, *Ark.* 13 33 14N 91 48W
Hamburg, *Iowa* 12 40 36N 95 39W
Hamburg, *Pa.* 9 40 33N 75 59W
Hamden, *Conn.* 9 41 23N 72 54W
Hamilton, *Alaska* 30 62 54N 163 53W
Hamilton, *Mo.* 12 39 45N 93 59W
Hamilton, *Mont.* 16 46 15N 114 10W
Hamilton, *N.Y.* 9 42 50N 75 33W
Hamilton, *Ohio* 15 39 24N 84 34W
Hamilton, *Tex.* 13 31 42N 98 7W
Hamlet, *N.C.* 11 34 53N 79 42W
Hamlin, *Tex.* 13 32 53N 100 8W
Hammond, *Ind.* 22 41 35N 87 29W
Hammond, *La.* 13 30 30N 90 28W
Hammonton, *N.J.* 10 39 39N 74 48W
Hampton, *Ark.* 13 33 32N 92 28W
Hampton, *Iowa* 14 42 45N 93 13W
Hampton, *N.H.* 9 42 57N 70 50W
Hampton, *S.C.* 11 32 52N 81 7W
Hampton, *Va.* 10 37 2N 76 21W
Hamtramck, *Mich.* 26 42 23N 83 4W
Hana, *Hawaii* 30 20 45N 155 59W
Hancock, *Mich.* 12 47 8N 88 35W
Hancock, *Minn.* 12 45 30N 95 48W
Hancock, *N.Y.* 9 41 57N 75 17W
Hanford, *Calif.* 18 36 20N 119 39W
Hankinson, *N. Dak.* 12 46 4N 96 54W
Hanksville, *Utah* 17 38 22N 110 43W
Hannaford, *N. Dak.* 12 47 19N 98 11W
Hannah, *N. Dak.* 12 48 58N 98 42W
Hannibal, *Mo.* 14 39 42N 91 22W
Hanover, *N.H.* 9 43 42N 72 17W
Hanover, *Pa.* 10 39 48N 76 59W
Happy, *Tex.* 13 34 45N 101 52W
Happy Camp, *Calif.* 16 41 48N 123 23W
Harbor Beach, *Mich.* 10 43 51N 82 39W
Harbor Springs, *Mich.* 10 45 26N 85 0W
Hardin, *Mont.* 16 45 44N 107 37W
Hardman, *Oreg.* 16 45 10N 119 41W
Hardy, *Ark.* 13 36 19N 91 29W
Harlan, *Iowa* 12 41 39N 95 19W
Harlan, *Ky.* 11 36 51N 83 19W
Harlem, *Mont.* 16 48 32N 108 47W
Harlem, *N.Y.* 20 40 48N 73 56W
Harlingen, *Tex.* 13 26 12N 97 42W
Harlowton, *Mont.* 16 46 26N 109 50W
Harney Basin, *Oreg.* 16 43 30N 119 0W
Harney L., *Oreg.* 16 43 14N 119 8W
Harney Peak, *S. Dak.* 12 43 52N 103 32W
Harper, Mt., *Alaska* 30 64 14N 143 51W
Harper Woods, *Mich.* 26 42 26N 82 56W
Harriman, *Tenn.* 11 35 56N 84 33W
Harrisburg, *Ill.* 15 37 44N 88 32W
Harrisburg, *Nebr.* 12 41 33N 103 44W
Harrisburg, *Oreg.* 16 44 16N 123 10W
Harrisburg, *Pa.* 8 40 16N 76 53W
Harrison, *Ark.* 13 36 14N 93 7W
Harrison, *Idaho* 16 47 27N 116 47W
Harrison, *Nebr.* 12 42 41N 103 53W
Harrison Bay, *Alaska* 30 70 40N 151 0W
Harrisonburg, *Va.* 10 38 27N 78 52W
Harrisonville, *Mo.* 14 38 39N 94 21W
Harrisville, *Mich.* 8 44 39N 83 17W
Hart, *Mich.* 10 43 42N 86 22W
Hartford, *Conn.* 9 41 46N 72 41W
Hartford, *Ky.* 10 37 27N 86 55W
Hartford, *S. Dak.* 12 43 38N 96 57W
Hartford, *Wis.* 12 43 19N 88 22W
Hartford City, *Ind.* 15 40 27N 85 22W
Hartselle, *Ala.* 11 34 27N 86 56W
Hartshorne, *Okla.* 13 34 51N 95 34W
Hartsville, *S.C.* 11 34 23N 80 4W
Hartwell, *Ga.* 11 34 21N 82 56W
Harvard University, *Mass.* 23 42 22N 71 7W
Harvey, *Ill.* 22 41 36N 87 39W
Harvey, *N. Dak.* 12 47 47N 99 56W
Harwood Heights, *Ill.* 22 41 57N 87 47W
Haskell, *Okla.* 13 35 50N 95 40W
Haskell, *Tex.* 13 33 10N 99 44W
Hastings, *Mich.* 15 42 39N 85 17W
Hastings, *Minn.* 12 44 44N 92 51W
Hastings, *Nebr.* 12 40 35N 98 23W
Hatch, *N. Mex.* 17 32 40N 107 9W
Hatteras, C., *N.C.* 11 35 14N 75 32W
Hattiesburg, *Miss.* 13 31 20N 89 17W
Havana, *Ill.* 14 40 18N 90 4W
Havasu, L., *Ariz.* 19 34 18N 114 28W
Haverhill, *Mass.* 9 42 47N 71 5W
Haverstraw, *N.Y.* 9 41 12N 73 58W
Havertown, *Pa.* 24 39 58N 75 18W
Havre, *Mont.* 16 48 33N 109 41W
Haw →, *N.C.* 11 35 36N 79 3W
Hawaii □ 30 19 30N 156 30W
Hawaii I., *Pac. Oc.* 30 20 0N 155 0W
Hawaiian Is., *Pac. Oc.* 30 20 30N 156 0W
Hawarden, *Iowa* 12 43 0N 96 29W
Hawkinsville, *Ga.* 11 32 17N 83 28W
Hawley, *Minn.* 12 46 53N 96 19W
Hawthorne, *Calif.* 29 33 54N 118 21W
Hawthorne, *Nev.* 16 38 32N 118 38W
Haxtun, *Colo.* 12 40 39N 102 38W
Hay Springs, *Nebr.* 12 42 41N 102 41W
Hayden, *Ariz.* 17 33 0N 110 47W
Hayden, *Colo.* 16 40 30N 107 16W
Hayes, *S. Dak.* 12 44 23N 101 1W
Haynesville, *La.* 13 32 58N 93 8W
Hays, *Kans.* 12 38 53N 99 20W
Hayward, *Calif.* 28 37 40N 122 4W
Hayward, *Wis.* 12 46 1N 91 29W
Hazard, *Ky.* 10 37 15N 83 12W
Hazel Park, *Mich.* 26 42 28N 83 5W
Hazelton, *N. Dak.* 12 46 29N 100 17W
Hazen, *N. Dak.* 12 47 18N 101 38W
Hazen, *Nev.* 16 39 34N 119 3W
Hazlehurst, *Ga.* 11 31 52N 82 36W
Hazlehurst, *Miss.* 13 31 52N 90 24W
Hazleton, *Pa.* 9 40 57N 75 59W
Healdsburg, *Calif.* 18 38 37N 122 52W
Healdton, *Okla.* 13 34 14N 97 29W
Hearne, *Tex.* 13 30 53N 96 36W
Heart →, *N. Dak.* 12 46 46N 100 50W
Heavener, *Okla.* 13 34 53N 94 36W
Hebbronville, *Tex.* 13 27 18N 98 41W
Heber Springs, *Ark.* 13 35 30N 92 2W
Hebgen L., *Mont.* 16 44 52N 111 20W
Hebron, *N. Dak.* 12 46 54N 102 3W
Hebron, *Nebr.* 12 40 10N 97 35W
Hecla, *S. Dak.* 12 45 53N 98 9W
Hedley, *Tex.* 13 34 52N 100 39W
Helena, *Ark.* 13 34 32N 90 36W
Helena, *Mont.* 16 46 36N 112 2W
Helendale, *Calif.* 19 34 44N 117 19W
Helper, *Utah* 16 39 41N 110 51W
Hemet, *Calif.* 19 33 45N 116 58W
Hemingford, *Nebr.* 12 42 19N 103 4W
Hemphill, *Tex.* 13 31 20N 93 51W
Hempstead, *N.Y.* 21 40 42N 73 37W
Hempstead, *Tex.* 13 30 6N 96 5W
Henderson, *Ky.* 15 37 50N 87 35W
Henderson, *N.C.* 11 36 20N 78 25W
Henderson, *Nev.* 19 36 2N 114 59W
Henderson, *Tenn.* 11 35 26N 88 38W
Henderson, *Tex.* 13 32 9N 94 48W
Hendersonville, *N.C.* 11 35 19N 82 28W
Henlopen, C., *Del.* 10 38 48N 75 6W
Hennessey, *Okla.* 13 36 6N 97 54W
Henrietta, *Tex.* 13 33 49N 98 12W
Henry, *Ill.* 14 41 7N 89 22W
Henryetta, *Okla.* 13 35 27N 95 59W
Heppner, *Oreg.* 16 45 21N 119 33W
Herbert I., *Alaska* 30 52 45N 170 7W
Hereford, *Tex.* 13 34 49N 102 24W
Herington, *Kans.* 12 38 40N 96 57W
Herkimer, *N.Y.* 9 43 0N 74 59W
Herman, *Minn.* 12 45 49N 96 9W
Hermann, *Mo.* 12 38 42N 91 27W
Hermiston, *Oreg.* 16 45 51N 119 17W

Keewatin

Hernandez, *Calif.* **18** 36 24N 120 46W
Hernando, *Miss.* **13** 34 50N 90 0W
Herreid, *S. Dak.* **12** 45 50N 100 4W
Herrin, *Ill.* **14** 37 48N 89 2W
Hesperia, *Calif.* **19** 34 25N 117 18W
Hetch Hetchy Aqueduct, *Calif.* **18** 37 29N 122 19W
Hettinger, *N. Dak.* **12** 46 0N 102 42W
Hi Vista, *Calif.* **19** 34 45N 117 46W
Hialeah, *Fla.* **11** 25 50N 80 17W
Hiawatha, *Kans.* **12** 39 51N 95 32W
Hiawatha, *Utah* **16** 39 29N 111 1W
Hibbing, *Minn.* **12** 47 25N 92 56W
Hickory, *N.C.* **11** 35 44N 81 21W
Hickory Hills, *Ill.* **22** 41 43N 87 50W
Hicksville, *N.Y.* **21** 40 46N 73 30W
Higgins, *Tex.* **13** 36 7N 100 2W
High Island, *Tex.* **13** 29 34N 94 24W
High Point, *N.C.* **11** 35 57N 80 0W
High Springs, *Fla.* **11** 29 50N 82 36W
Highland Park, *Ill.* **15** 42 11N 87 48W
Highland Park, *Mich.* **26** 42 24N 83 6W
Highmore, *S. Dak.* **12** 44 31N 99 27W
Hiko, *Nev.* **18** 37 32N 115 14W
Hill City, *Idaho* **16** 43 18N 115 3W
Hill City, *Kans.* **12** 39 22N 99 51W
Hill City, *Minn.* **12** 46 59N 93 36W
Hill City, *S. Dak.* **12** 43 56N 103 35W
Hillcrest Heights, *Md.* **25** 38 50N 76 57W
Hillman, *Mich.* **10** 45 4N 83 54W
Hillsboro, *Kans.* **12** 38 21N 97 12W
Hillsboro, *N. Dak.* **12** 47 26N 97 3W
Hillsboro, *N.H.* **9** 43 7N 71 54W
Hillsboro, *N. Mex.* **17** 32 55N 107 34W
Hillsboro, *Oreg.* **18** 45 31N 122 59W
Hillsboro, *Tex.* **13** 32 1N 97 8W
Hillsdale, *Mich.* **15** 41 56N 84 38W
Hillsdale, *N.J.* **20** 41 0N 74 2W
Hillsdale, *N.Y.* **9** 42 11N 73 30W
Hilo, *Hawaii* **30** 19 44N 155 5W
Hinckley, *Utah* **16** 39 20N 112 40W
Hingham, *Mass.* **23** 42 14N 70 54W
Hingham, *Mont.* **16** 48 33N 110 25W
Hinsdale, *Ill.* **22** 41 47N 87 56W
Hinsdale, *Mont.* **16** 48 24N 107 5W
Hinton, *W. Va.* **10** 37 40N 80 54W
Hobart, *Okla.* **13** 35 1N 99 6W
Hobbs, *N. Mex.* **13** 32 42N 103 8W
Hoboken, *N.J.* **20** 40 44N 74 3W
Hogansville, *Ga.* **11** 33 10N 84 55W
Hogeland, *Mont.* **16** 48 51N 108 40W
Hohenwald, *Tenn.* **11** 35 33N 87 33W
Hoisington, *Kans.* **12** 38 31N 98 47W
Holbrook, *Ariz.* **17** 34 54N 110 10W
Holden, *Utah* **16** 39 6N 112 16W
Holdenville, *Okla.* **13** 35 5N 96 24W
Holdrege, *Nebr.* **12** 40 26N 99 23W
Holland, *Mich.* **15** 42 47N 86 7W
Hollidaysburg, *Pa.* **8** 40 26N 78 24W
Hollis, *Okla.* **13** 34 41N 99 55W
Hollister, *Calif.* **18** 36 51N 121 24W
Hollister, *Idaho* **16** 42 21N 114 35W
Holly, *Colo.* **12** 38 3N 102 7W
Holly Hill, *Fla.* **11** 29 16N 81 3W
Holly Springs, *Miss.* **13** 34 46N 89 27W
Hollywood, *Calif.* **17** 34 7N 118 25W
Hollywood, *Fla.* **11** 26 1N 80 9W
Holton, *Kans.* **12** 39 28N 95 44W
Holtville, *Calif.* **19** 32 49N 115 23W
Holy Cross, *Alaska* **30** 62 12N 159 46W
Holyoke, *Colo.* **12** 40 35N 102 18W
Holyoke, *Mass.* **9** 42 12N 72 37W
Homedale, *Idaho* **16** 43 37N 116 56W
Homer, *Alaska* **30** 59 39N 151 33W
Homer, *La.* **13** 32 48N 93 4W
Homestead, *Fla.* **11** 25 28N 80 29W
Homestead, *Oreg.* **16** 45 2N 116 51W
Homestead, *Pa.* **27** 40 24N 79 55W
Hominy, *Okla.* **13** 36 25N 96 24W
Hondo, *Tex.* **13** 29 21N 99 9W
Honey L., *Calif.* **18** 40 15N 120 19W
Honolulu, *Hawaii* **30** 21 19N 157 52W
Hood, Mt., *Oreg.* **16** 45 23N 121 42W
Hood River, *Oreg.* **16** 45 43N 121 31W
Hoodsport, *Wash.* **18** 47 24N 123 9W
Hooker, *Okla.* **13** 36 52N 101 13W
Hoonah, *Alaska* **30** 58 7N 135 27W
Hooper Bay, *Alaska* **30** 61 32N 166 6W
Hoopeston, *Ill.* **15** 40 28N 87 40W
Hoover Dam, *Ariz.* **19** 36 1N 114 44W
Hop Bottom, *Pa.* **9** 41 42N 75 46W
Hope, *Ark.* **13** 33 40N 93 36W
Hope, *N. Dak.* **12** 47 19N 97 43W
Hope, Pt., *Alaska* **30** 68 20N 166 50W
Hopkins, *Mo.* **14** 40 33N 94 49W
Hopkinsville, *Ky.* **11** 36 52N 87 29W
Hopland, *Calif.* **18** 38 58N 123 7W
Hoquiam, *Wash.* **18** 46 59N 123 53W
Horn I., *Miss.* **11** 30 14N 88 39W
Hornbeck, *La.* **13** 31 20N 93 24W
Hornbrook, *Calif.* **16** 41 55N 122 33W
Hornell, *N.Y.* **8** 42 20N 77 40W
Hornitos, *Calif.* **18** 37 30N 120 14W
Horse Creek, *Wyo.* **12** 41 57N 105 10W
Horton, *Kans.* **12** 39 40N 95 32W
Hosmer, *S. Dak.* **12** 45 34N 99 28W
Hot Creek Range, *Nev.* **16** 38 40N 116 20W
Hot Springs, *Ark.* **13** 34 31N 93 3W
Hot Springs, *S. Dak.* **12** 43 26N 103 29W
Hotchkiss, *Colo.* **17** 38 48N 107 43W
Houck, *Ariz.* **17** 35 20N 109 10W
Houghton, *Mich.* **12** 47 7N 88 34W
Houghton L., *Mich.* **10** 44 21N 84 44W
Houlton, *Maine* **11** 46 8N 67 51W
Houma, *La.* **13** 29 36N 90 43W
Houston, *Mo.* **13** 37 22N 91 58W
Houston, *Tex.* **13** 29 46N 95 22W
Howard, *Kans.* **13** 37 28N 96 16W
Howard, *S. Dak.* **12** 44 1N 97 32W
Howe, *Idaho* **16** 43 48N 113 0W
Howell, *Mich.* **15** 42 36N 83 56W
Hualapai Peak, *Ariz.* **17** 35 5N 113 54W
Huasna, *Calif.* **19** 35 6N 120 24W
Hubbard, *Tex.* **13** 31 51N 96 48W
Hudson, *Mich.* **15** 41 51N 84 21W
Hudson, *N.Y.* **9** 42 15N 73 46W
Hudson, *Wis.* **12** 44 58N 92 45W
Hudson, *Wyo.* **16** 42 54N 108 35W
Hudson →, *N.Y.* **9** 40 42N 74 2W
Hudson Falls, *N.Y.* **9** 43 18N 73 35W
Hughes, *Alaska* **30** 66 3N 154 15W
Hugo, *Colo.* **12** 39 8N 103 28W
Hugoton, *Kans.* **13** 37 11N 101 21W
Hull, *Mass.* **23** 42 18N 70 54W
Humacao, *Puerto Rico* **30** 18 9N 65 50W
Humble, *Tex.* **13** 29 59N 93 18W
Humboldt, *Iowa* **14** 42 44N 94 13W
Humboldt, *Tenn.* **13** 35 50N 88 55W
Humboldt →, *Nev.* **16** 39 59N 118 36W
Hume, *Calif.* **18** 36 48N 118 54W
Humphreys, Mt., *Calif.* **18** 37 17N 118 40W
Humphreys Peak, *Ariz.* **17** 35 21N 111 41W
Hunter, *N. Dak.* **12** 47 12N 97 13W
Hunter, *N.Y.* **9** 42 13N 74 13W
Huntingburg, *Ind.* **15** 38 18N 86 57W
Huntingdon, *Pa.* **8** 40 30N 78 1W
Huntington, *Ind.* **15** 40 53N 85 30W
Huntington, *N.Y.* **21** 40 52N 73 25W
Huntington, *Oreg.* **16** 44 21N 117 16W
Huntington, *Utah* **16** 39 20N 110 58W
Huntington, *W. Va.* **10** 38 25N 82 27W
Huntington Beach, *Calif.* **19** 33 40N 118 5W
Huntington Park, *Calif.* **29** 33 58N 118 13W
Huntington Woods, *Mich.* **26** 42 28N 83 10W
Huntsville, *Ala.* **11** 34 44N 86 35W
Huntsville, *Tex.* **13** 30 43N 95 33W
Hurley, *N. Mex.* **17** 32 42N 108 8W
Hurley, *Wis.* **12** 46 27N 90 11W
Huron, *Calif.* **18** 36 12N 120 6W
Huron, *S. Dak.* **12** 44 22N 98 13W
Huron, L., *Mich.* **8** 44 30N 82 40W
Hurricane, *Utah* **17** 37 11N 113 17W
Huslia, *Alaska* **30** 65 41N 156 24W
Hutchinson, *Kans.* **13** 38 5N 97 56W
Hutchinson, *Minn.* **12** 44 54N 94 22W
Huttig, *Ark.* **13** 33 2N 92 11W
Hyannis, *Nebr.* **12** 42 0N 101 46W
Hyattsville, *Md.* **25** 38 57N 76 58W
Hydaburg, *Alaska* **30** 55 12N 132 50W
Hyndman Peak, *Idaho* **16** 43 45N 114 8W
Hyrum, *Utah* **16** 41 38N 111 51W
Hysham, *Mont.* **16** 46 18N 107 14W

I

Ida Grove, *Iowa* **12** 42 21N 95 28W
Idabel, *Okla.* **13** 33 54N 94 50W
Idaho □ **16** 45 0N 115 0W
Idaho City, *Idaho* **16** 43 50N 115 50W
Idaho Falls, *Idaho* **16** 43 30N 112 2W
Idaho Springs, *Colo.* **16** 39 45N 105 31W
Idria, *Calif.* **18** 36 25N 120 41W
Iliamna L., *Alaska* **30** 59 30N 155 0W
Iliff, *Colo.* **12** 40 45N 103 4W
Ilio Pt., *Hawaii* **30** 21 13N 157 16W
Ilion, *N.Y.* **9** 43 1N 75 2W
Illinois □ **14** 40 15N 89 30W
Illinois →, *Ill.* **14** 38 58N 90 28W
Imbler, *Oreg.* **16** 45 28N 117 58W
Imlay, *Nev.* **16** 40 40N 118 9W
Immokalee, *Fla.* **11** 26 25N 81 25W
Imperial, *Calif.* **19** 32 51N 115 34W
Imperial, *Nebr.* **12** 40 31N 101 39W
Imperial Beach, *Calif.* **19** 32 35N 117 8W
Imperial Dam, *Ariz.* **19** 32 55N 114 25W
Independence, *Calif.* **18** 36 48N 118 12W
Independence, *Iowa* **14** 42 28N 91 54W
Independence, *Kans.* **13** 37 14N 95 42W
Independence, *Mo.* **14** 39 6N 94 25W
Independence, *Oreg.* **16** 44 51N 123 11W
Independence Mts., *Nev.* **16** 41 20N 116 0W
Indian →, *Fla.* **11** 27 59N 80 34W
Indiana, *Pa.* **8** 40 37N 79 9W
Indiana □ **15** 40 0N 86 0W
Indianapolis, *Ind.* **15** 39 46N 86 9W
Indianola, *Iowa* **14** 41 22N 93 34W
Indianola, *Miss.* **13** 33 27N 90 39W
Indio, *Calif.* **19** 33 43N 116 13W
Inglewood, *Calif.* **29** 33 57N 118 19W
Ingomar, *Mont.* **16** 46 35N 107 23W
Inkom, *Idaho* **16** 42 48N 112 15W
Inkster, *Mich.* **26** 42 17N 83 16W
Interior, *S. Dak.* **12** 43 44N 101 59W
International Falls, *Minn.* **12** 48 36N 93 25W
Inverness, *Fla.* **11** 28 50N 82 20W
Inyo Mts., *Calif.* **17** 36 40N 118 0W
Inyokern, *Calif.* **19** 35 39N 117 49W
Iola, *Kans.* **13** 37 55N 95 24W
Ione, *Calif.* **18** 38 21N 120 56W
Ione, *Wash.* **16** 48 45N 117 25W
Ionia, *Mich.* **15** 42 59N 85 4W
Iowa □ **12** 42 18N 93 30W
Iowa City, *Iowa* **14** 41 40N 91 32W
Iowa Falls, *Iowa* **14** 42 31N 93 16W
Ipswich, *Mass.* **23** 42 41N 70 50W
Ipswich, *S. Dak.* **12** 45 27N 99 2W
Iron Mountain, *Mich.* **10** 45 49N 88 4W
Iron River, *Mich.* **12** 46 6N 88 39W
Ironton, *Mo.* **13** 37 36N 90 38W
Ironton, *Ohio* **10** 38 32N 82 41W
Ironwood, *Mich.* **12** 46 27N 90 9W
Irvine, *Ky.* **15** 37 42N 83 58W
Irvington, *N.Y.* **20** 40 42N 74 13W
Isabel, *S. Dak.* **12** 45 24N 101 26W
Isabela, *Puerto Rico* **30** 18 30N 67 2W
Ishpeming, *Mich.* **10** 46 29N 87 40W
Isla Vista, *Calif.* **19** 34 25N 119 53W
Island Falls, *Maine* **11** 46 1N 68 16W
Island Pond, *Vt.* **9** 44 49N 71 53W
Isle Royale, *Mich.* **12** 48 0N 88 54W
Isleta, *N. Mex.* **17** 34 55N 106 42W
Isleton, *Calif.* **18** 38 10N 121 37W
Ismay, *Mont.* **12** 46 30N 104 48W
Istokpoga, L., *Fla.* **11** 27 23N 81 17W
Ithaca, *N.Y.* **9** 42 27N 76 30W
Ivanhoe, *Calif.* **18** 36 23N 119 13W

J

Jackman, *Maine* **11** 45 35N 70 17W
Jacksboro, *Tex.* **13** 33 14N 98 15W
Jackson, *Ala.* **11** 31 31N 87 53W
Jackson, *Calif.* **18** 38 21N 120 46W
Jackson, *Ky.* **10** 37 33N 83 23W
Jackson, *Mich.* **15** 42 15N 84 24W
Jackson, *Minn.* **12** 43 37N 95 1W
Jackson, *Miss.* **13** 32 18N 90 12W
Jackson, *Mo.* **13** 37 23N 89 40W
Jackson, *Ohio* **10** 39 3N 82 39W
Jackson, *Tenn.* **11** 35 37N 88 49W
Jackson, *Wyo.* **16** 43 29N 110 46W
Jackson Heights, *N.Y.* **21** 40 44N 73 53W
Jackson L., *Wyo.* **16** 43 52N 110 36W
Jacksonville, *Ala.* **11** 33 49N 85 46W
Jacksonville, *Calif.* **18** 37 52N 120 24W
Jacksonville, *Fla.* **11** 30 20N 81 39W
Jacksonville, *Ill.* **14** 39 44N 90 14W
Jacksonville, *N.C.* **11** 34 45N 77 26W
Jacksonville, *Oreg.* **16** 42 19N 122 57W
Jacksonville, *Tex.* **13** 31 58N 95 17W
Jacksonville Beach, *Fla.* **11** 30 17N 81 24W
Jacob Lake, *Ariz.* **17** 36 43N 112 13W
Jal, *N. Mex.* **13** 32 7N 103 12W
Jalama, *Calif.* **19** 34 29N 120 29W
Jaluit I., *Pac. Oc.* **31** 6 0N 169 30 E
Jamaica, *N.Y.* **21** 40 42N 73 48W
James →, *S. Dak.* **12** 42 52N 97 18W
Jamestown, *Ky.* **10** 36 59N 85 4W
Jamestown, *N. Dak.* **12** 46 54N 98 42W
Jamestown, *N.Y.* **8** 42 6N 79 14W
Jamestown, *Tenn.* **11** 36 26N 84 56W
Janesville, *Wis.* **14** 42 41N 89 1W
Jasper, *Ala.* **11** 33 50N 87 17W
Jasper, *Fla.* **11** 30 31N 82 57W
Jasper, *Minn.* **12** 43 51N 96 24W
Jasper, *Tex.* **13** 30 56N 94 1W
Jay, *Okla.* **13** 36 25N 94 48W
Jayton, *Tex.* **13** 33 15N 100 34W
Jean, *Nev.* **19** 35 47N 115 20W
Jeanerette, *La.* **13** 29 55N 91 40W
Jefferson, *Iowa* **14** 42 1N 94 23W
Jefferson, *Tex.* **13** 32 46N 94 21W
Jefferson, *Wis.* **15** 43 0N 88 48W
Jefferson, Mt., *Nev.* **16** 38 51N 117 0W
Jefferson, Mt., *Oreg.* **16** 44 41N 121 48W
Jefferson City, *Mo.* **14** 38 34N 92 10W
Jefferson City, *Tenn.* **11** 36 7N 83 30W
Jeffersonville, *Ind.* **15** 38 17N 85 44W
Jena, *La.* **13** 31 41N 92 8W
Jenkins, *Ky.* **10** 37 10N 82 38W
Jennings, *La.* **13** 30 13N 92 40W
Jermyn, *Pa.* **9** 41 31N 75 31W
Jerome, *Ariz.* **17** 34 45N 112 7W
Jersey City, *N.J.* **20** 40 42N 74 4W
Jersey Shore, *Pa.* **8** 41 12N 77 15W
Jerseyville, *Ill.* **14** 39 7N 90 20W
Jesup, *Ga.* **11** 31 36N 81 53W
Jetmore, *Kans.* **13** 38 4N 99 54W
Jewett, *Tex.* **13** 31 22N 96 9W
Jewett City, *Conn.* **9** 41 36N 72 0W
Johannesburg, *Calif.* **19** 35 22N 117 38W
John Day, *Oreg.* **16** 44 25N 118 57W
John Day →, *Oreg.* **16** 45 44N 120 39W
John F.Kennedy International Airport, *N.Y.* **21** 40 38N 73 46W
John H. Kerr Reservoir, *N.C.* **11** 36 36N 78 18W
Johnson, *Kans.* **13** 37 34N 101 45W
Johnson City, *N.Y.* **9** 42 7N 75 58W
Johnson City, *Tenn.* **11** 36 19N 82 21W
Johnson City, *Tex.* **13** 30 17N 98 25W
Johnsondale, *Calif.* **19** 35 58N 118 32W
Johnstown, *N.Y.* **9** 43 0N 74 22W
Johnstown, *Pa.* **8** 40 20N 78 55W
Joliet, *Ill.* **15** 41 32N 88 5W
Jolon, *Calif.* **18** 35 58N 121 9W
Jonesboro, *Ark.* **13** 35 50N 90 42W
Jonesboro, *Ill.* **13** 37 27N 89 16W
Jonesboro, *La.* **13** 32 15N 92 43W
Jonesport, *Maine* **11** 44 32N 67 37W
Joplin, *Mo.* **13** 37 6N 94 31W
Joppatowne, *Md.* **25** 39 24N 76 21W
Jordan, *Mont.* **16** 47 19N 106 55W
Jordan Valley, *Oreg.* **16** 42 59N 117 3W
Joseph, *Oreg.* **16** 45 21N 117 14W
Joseph City, *Ariz.* **17** 34 57N 110 20W
Joshua Tree, *Calif.* **19** 34 8N 116 19W
Joshua Tree National Monument, *Calif.* **19** 33 55N 116 0W
Jourdanton, *Tex.* **13** 28 55N 98 33W
Judith →, *Mont.* **16** 47 44N 109 39W
Judith, Pt., *R.I.* **9** 41 22N 71 29W
Judith Gap, *Mont.* **16** 46 41N 109 45W
Julesburg, *Colo.* **12** 40 59N 102 16W
Julian, *Calif.* **19** 33 4N 116 38W
Junction, *Tex.* **13** 30 29N 99 46W
Junction, *Utah* **17** 38 14N 112 13W
Junction City, *Kans.* **12** 39 2N 96 50W
Junction City, *Oreg.* **16** 44 13N 123 12W
Juneau, *Alaska* **30** 58 18N 134 25W
Juniata →, *Pa.* **8** 40 30N 77 40W
Juntura, *Oreg.* **16** 43 45N 118 5W
Justice, *Ill.* **22** 41 44N 87 49W

K

Ka Lae, *Hawaii* **30** 18 55N 155 41W
Kaala, *Hawaii* **30** 21 31N 158 9W
Kadoka, *S. Dak.* **12** 43 50N 101 31W
Kaena Pt., *Hawaii* **30** 21 35N 158 17W
Kagamil I., *Alaska* **30** 53 0N 169 43W
Kahoka, *Mo.* **14** 40 25N 91 44W
Kahoolawe, *Hawaii* **30** 20 33N 156 37W
Kahuku Pt., *Hawaii* **30** 21 43N 157 59W
Kahului, *Hawaii* **30** 20 54N 156 28W
Kailua Kona, *Hawaii* **30** 19 39N 155 59W
Kaiwi Channel, *Hawaii* **30** 21 15N 157 30W
Kaiyuh Mts., *Alaska* **30** 64 30N 158 0W
Kake, *Alaska* **30** 56 59N 133 57W
Kaktovik, *Alaska* **30** 70 8N 143 38W
Kalama, *Wash.* **18** 46 1N 122 51W
Kalamazoo, *Mich.* **15** 42 17N 85 35W
Kalamazoo →, *Mich.* **15** 42 40N 86 10W
Kalaupapa, *Hawaii* **30** 21 12N 156 59W
Kalispell, *Mont.* **16** 48 12N 114 19W
Kalkaska, *Mich.* **10** 44 44N 85 11W
Kamalino, *Hawaii* **30** 21 50N 160 14W
Kamiah, *Idaho* **16** 46 14N 116 2W
Kanab, *Utah* **17** 37 3N 112 32W
Kanab →, *Ariz.* **17** 36 24N 112 38W
Kanaga I., *Alaska* **30** 51 45N 177 22W
Kanakanak, *Alaska* **30** 59 0N 158 58W
Kanarraville, *Utah* **17** 37 32N 113 11W
Kanawha →, *W. Va.* **10** 38 50N 82 9W
Kane, *Pa.* **8** 41 40N 78 49W
Kaneohe, *Hawaii* **30** 21 25N 157 48W
Kankakee, *Ill.* **15** 41 7N 87 52W
Kankakee →, *Ill.* **15** 41 23N 88 15W
Kannapolis, *N.C.* **11** 35 30N 80 37W
Kansas □ **12** 38 30N 99 0W
Kansas →, *Kans.* **12** 39 7N 94 37W
Kansas City, *Kans.* **14** 39 7N 94 38W
Kansas City, *Mo.* **14** 39 6N 94 35W
Kantishna, *Alaska* **30** 63 31N 151 5W
Kapaa, *Hawaii* **30** 22 5N 159 19W
Karlstad, *Minn.* **12** 48 35N 96 31W
Karnes City, *Tex.* **13** 28 53N 97 54W
Kaskaskia →, *Ill.* **14** 37 58N 89 57W
Katalla, *Alaska* **30** 60 12N 144 31W
Katmai National Park, *Alaska* **30** 58 20N 155 0W
Kauai, *Hawaii* **30** 22 3N 159 30W
Kauai Channel, *Hawaii* **30** 21 45N 158 50W
Kaufman, *Tex.* **13** 32 35N 96 19W
Kaukauna, *Wis.* **10** 44 17N 88 17W
Kaupulehu, *Hawaii* **30** 19 43N 155 53W
Kawaihae, *Hawaii* **30** 20 3N 155 50W
Kawaihoa Pt., *Hawaii* **30** 21 47N 160 12W
Kawaikimi, *Hawaii* **30** 22 5N 159 29W
Kayak I., *Alaska* **30** 59 56N 144 23W
Kaycee, *Wyo.* **16** 43 43N 106 38W
Kayenta, *Ariz.* **17** 36 44N 110 15W
Kaysville, *Utah* **16** 41 2N 111 56W
Keaau, *Hawaii* **30** 19 37N 155 2W
Keams Canyon, *Ariz.* **17** 35 49N 110 12W
Keanae, *Hawaii* **30** 20 52N 156 9W
Kearney, *Nebr.* **12** 40 42N 99 5W
Kearny, *N.J.* **20** 40 45N 74 9W
Keeler, *Calif.* **18** 36 29N 117 52W
Keene, *Calif.* **19** 35 13N 118 33W
Keene, *N.H.* **9** 42 56N 72 17W
Keewatin, *Minn.* **12** 47 24N 93 5W

Keller

Keller, Wash. 16 48 5N 118 41W
Kellogg, Idaho 16 47 32N 116 7W
Kelso, Wash. 18 46 9N 122 54W
Kemmerer, Wyo. 16 41 48N 110 32W
Kemp, L., Tex. 13 33 46N 99 9W
Kenai, Alaska 30 60 33N 151 16W
Kenai Mts., Alaska 30 60 0N 150 0W
Kendallville, Ind. 15 41 27N 85 16W
Kendrick, Idaho 16 46 37N 116 39W
Kenedy, Tex. 13 28 49N 97 51W
Kenmare, N. Dak. 12 48 41N 102 5W
Kennebec, S. Dak. 12 43 54N 99 52W
Kennett, Mo. 13 36 14N 90 3W
Kennewick, Wash. 16 46 12N 119 7W
Kenosha, Wis. 15 42 35N 87 49W
Kensington, Kans. 12 39 46N 99 2W
Kent, Ohio 8 41 9N 81 22W
Kent, Oreg. 16 45 12N 120 42W
Kent, Tex. 13 31 4N 104 13W
Kentfield, Calif. 28 37 57N 122 33W
Kentland, Ind. 15 40 46N 87 27W
Kenton, Ohio 15 40 39N 83 37W
Kentucky □ 10 37 0N 84 0W
Kentucky →, Ky. 15 38 41N 85 11W
Kentucky L., Ky. 11 37 1N 88 16W
Kentwood, La. 13 31 0N 90 30W
Kentwood, La. 13 30 56N 90 31W
Keokuk, Iowa 14 40 24N 91 24W
Kepuhi, Hawaii 30 21 10N 157 10W
Kerman, Calif. 18 36 43N 120 4W
Kermit, Tex. 13 31 52N 103 6W
Kern →, Calif. 19 35 16N 119 18W
Kernville, Calif. 19 35 45N 118 26W
Kerrville, Tex. 13 30 3N 99 8W
Ketchikan, Alaska 30 55 21N 131 39W
Ketchum, Idaho 16 43 41N 114 22W
Kettle Falls, Wash. 16 48 37N 118 3W
Kettleman City, Calif. 18 36 1N 119 58W
Kevin, Mont. 16 48 45N 111 58W
Kewanee, Ill. 14 41 14N 89 56W
Kewaunee, Wis. 10 44 27N 87 31W
Keweenaw B., Mich. 10 47 0N 88 15W
Keweenaw Pen., Mich. 10 47 30N 88 0W
Keweenaw Pt., Mich. 10 47 25N 87 43W
Keyser, W. Va. 10 39 26N 78 59W
Keystone, S. Dak. 12 43 54N 103 25W
Kijik, Alaska 30 60 20N 154 20W
Kilauea, Hawaii 30 22 13N 159 25W
Kilauea Crater, Hawaii 30 19 25N 155 17W
Kilbuck Mts., Alaska 30 60 30N 160 0W
Kilgore, Tex. 13 32 23N 94 53W
Killdeer, N. Dak. 12 47 26N 102 48W
Killeen, Tex. 13 31 7N 97 44W
Kim, Colo. 13 37 15N 103 21W
Kimball, Nebr. 12 41 14N 103 40W
Kimball, S. Dak. 12 43 45N 98 57W
Kimberly, Idaho 16 42 32N 114 22W
King City, Calif. 18 36 13N 121 8W
King of Prussia, Pa. 24 40 5N 75 22W
Kingfisher, Okla. 13 35 52N 97 56W
Kingman, Ariz. 19 35 12N 114 4W
Kingman, Kans. 13 37 39N 98 7W
Kings →, Calif. 18 36 3N 119 50W
Kings Canyon National Park, Calif. 18 36 50N 118 40W
Kings Mountain, N.C. 11 35 15N 81 20W
King's Peak, Utah 16 40 46N 110 27W
Kingsburg, Calif. 18 36 31N 119 33W
Kingsley, Iowa 12 42 35N 95 58W
Kingsport, Tenn. 11 36 33N 82 33W
Kingston, N.Y. 9 41 56N 73 59W
Kingston, Pa. 9 41 16N 75 54W
Kingston, R.I. 9 41 29N 71 30W
Kingstree, S.C. 11 33 40N 79 50W
Kingsville, Tex. 13 27 31N 97 52W
Kinsley, Kans. 13 37 55N 99 25W
Kinston, N.C. 11 35 16N 77 35W
Kiowa, Kans. 13 37 1N 98 29W
Kiowa, Okla. 13 34 43N 95 54W
Kipnuk, Alaska 30 59 56N 164 3W
Kirkland, Ariz. 17 34 25N 112 43W
Kirksville, Mo. 14 40 12N 92 35W
Kiska I., Alaska 30 51 59N 177 30 E
Kissimmee, Fla. 11 28 18N 81 24W
Kissimmee →, Fla. 11 27 9N 80 52W
Kit Carson, Colo. 12 38 46N 102 48W
Kittanning, Pa. 8 40 49N 79 31W
Kittatinny Mts., N.J. 9 41 0N 75 0W
Kittery, Maine 11 43 5N 70 45W
Kivalina, Alaska 30 67 44N 164 33W
Klamath →, Calif. 16 41 33N 124 5W
Klamath Falls, Oreg. 16 42 13N 121 46W
Klamath Mts., Calif. 16 41 20N 123 0W
Klein, Mont. 16 46 24N 108 33W
Klickitat, Wash. 16 45 49N 121 9W
Knights Ferry, Calif. 18 37 50N 120 40W
Knights Landing, Calif. 18 38 48N 121 43W
Knox, Ind. 15 41 18N 86 37W
Knox City, Tex. 13 33 25N 99 49W
Knoxville, Iowa 14 41 19N 93 6W
Knoxville, Tenn. 11 35 58N 83 55W
Kobuk, Alaska 30 66 55N 156 52W
Kobuk →, Alaska 30 66 55N 157 0W
Kodiak, Alaska 30 57 47N 152 24W
Kodiak I., Alaska 30 57 30N 152 45W
Kokomo, Ind. 15 40 29N 86 8W
Konawa, Okla. 13 34 58N 96 45W
Kooskia, Idaho 16 46 9N 115 59W
Koror, Pac. Oc. 31 7 20N 134 28 E
Kosciusko, Miss. 13 33 4N 89 35W
Kotzebue, Alaska 30 66 53N 162 39W
Kotzebue Sound, Alaska 30 66 20N 163 0W
Kountze, Tex. 13 30 22N 94 19W
Koyuk, Alaska 30 64 56N 161 9W
Koyukuk →, Alaska 30 64 55N 157 32W
Kremmling, Colo. 16 40 4N 106 24W
Kualakahi Chan, Hawaii 30 22 2N 159 53W
Kuiu I., Alaska 30 57 45N 134 10W
Kulm, N. Dak. 12 46 18N 98 57W
Kumukahi, C., Hawaii 30 19 31N 154 49W
Kupreanof I., Alaska 30 56 50N 133 30W
Kuskokwim →, Alaska 30 60 5N 162 25W
Kuskokwim B., Alaska 30 59 45N 162 25W
Kuskokwim Mts., Alaska 30 62 30N 156 0W
Kwethluk, Alaska 30 60 49N 161 26W
Kwigillingok, Alaska 30 59 51N 163 8W
Kwiguk, Alaska 30 62 46N 164 30W
Kyburz, Calif. 18 38 47N 120 18W

L

La Barge, Wyo. 16 42 16N 110 12W
La Belle, Fla. 11 26 46N 81 26W
La Canada, Calif. 29 34 12N 118 12W
La Conner, Wash. 16 48 23N 122 30W
La Crescenta, Calif. 29 34 13N 118 14W
La Crosse, Kans. 12 38 32N 99 18W
La Crosse, Wis. 12 43 48N 91 15W
La Fayette, Ga. 11 34 42N 85 17W
La Follette, Tenn. 11 36 23N 84 7W
La Grande, Oreg. 16 45 20N 118 5W
La Grange, Calif. 18 37 42N 120 27W
La Grange, Ga. 11 33 2N 85 2W
La Grange, Ill. 22 41 48N 87 53W
La Grange, Ky. 10 38 25N 85 23W
La Grange, Tex. 13 29 54N 96 52W
La Guardia Airport, N.Y. 21 40 46N 73 52W
La Habra, Calif. 29 33 56N 117 57W
La Harpe, Ill. 14 40 35N 90 58W
La Jara, Colo. 17 37 16N 105 58W
La Junta, Colo. 13 37 59N 103 33W
La Mesa, Calif. 19 32 46N 117 3W
La Mesa, N. Mex. 17 32 7N 106 42W
La Moure, N. Dak. 12 46 21N 98 18W
La Pine, Oreg. 16 43 40N 121 30W
La Plant, S. Dak. 12 45 9N 100 39W
La Porte, Ind. 15 41 36N 86 43W
La Push, Wash. 18 47 55N 124 38W
La Salle, Ill. 14 41 20N 89 6W
La Selva Beach, Calif. 18 36 56N 121 51W
Laau Pt., Hawaii 30 21 6N 157 19W
Lac du Flambeau, Wis. 12 45 58N 89 53W
Lackawanna, N.Y. 8 42 50N 78 50W
Lacona, N.Y. 9 43 39N 76 10W
Laconia, N.H. 9 43 32N 71 28W
Lacrosse, Wash. 16 46 51N 117 58W
Ladysmith, Wis. 12 45 28N 91 12W
Lafayette, Colo. 12 39 58N 105 12W
Lafayette, Ind. 15 40 25N 86 54W
Lafayette, La. 13 30 14N 92 1W
Lafayette, Tenn. 11 36 31N 86 2W
Laguna, N. Mex. 17 35 2N 107 25W
Laguna Beach, Calif. 19 33 33N 117 47W
Lahaina, Hawaii 30 20 53N 156 41W
Lahontan Reservoir, Nev. 16 39 28N 119 4W
Lake Alpine, Calif. 18 38 29N 120 0W
Lake Andes, S. Dak. 12 43 9N 98 32W
Lake Anse, Mich. 10 46 42N 88 25W
Lake Arthur, La. 13 30 5N 92 41W
Lake Charles, La. 13 30 14N 93 13W
Lake City, Colo. 17 38 2N 107 19W
Lake City, Fla. 11 30 11N 82 38W
Lake City, Iowa 14 42 16N 94 44W
Lake City, Mich. 10 44 20N 85 13W
Lake City, Minn. 12 44 27N 92 16W
Lake City, S.C. 11 33 52N 79 45W
Lake George, N.Y. 9 43 26N 73 43W
Lake Havasu City, Ariz. 19 34 27N 114 22W
Lake Hiawatha, N.J. 20 40 52N 74 22W
Lake Hughes, Calif. 19 34 41N 118 26W
Lake Isabella, Calif. 19 35 38N 118 28W
Lake Mead National Recreation Area, Ariz. 19 36 15N 114 30W
Lake Mills, Iowa 12 43 25N 93 32W
Lake Providence, La. 13 32 48N 91 10W
Lake Village, Ark. 13 33 20N 91 17W
Lake Wales, Fla. 11 27 54N 81 35W
Lake Worth, Fla. 11 26 37N 80 3W
Lakeland, Fla. 11 28 3N 81 57W
Lakeside, Ariz. 17 34 9N 109 58W
Lakeside, Calif. 19 32 52N 116 55W
Lakeside, Nebr. 12 42 3N 102 26W
Lakeview, Oreg. 16 42 11N 120 21W
Lakewood, Colo. 12 39 44N 105 5W
Lakewood, N.J. 9 40 6N 74 13W
Lakewood, Ohio 27 41 29N 81 49W
Lakin, Kans. 13 37 57N 101 15W
Lakota, N. Dak. 12 48 2N 98 21W
Lamar, Colo. 12 38 5N 102 37W
Lamar, Mo. 13 37 30N 94 16W
Lambert, Mont. 12 47 41N 104 37W
Lame Deer, Mont. 16 45 37N 106 40W
Lamesa, Tex. 13 32 44N 101 58W
Lamont, Calif. 19 35 15N 118 55W
Lampasas, Tex. 13 31 4N 98 11W
Lamy, N. Mex. 17 35 29N 105 53W
Lanai City, Hawaii 30 20 50N 156 55W
Lanai I., Hawaii 30 20 50N 156 55W
Lancaster, Calif. 19 34 42N 118 8W
Lancaster, Ky. 10 37 37N 84 35W
Lancaster, N.H. 9 44 29N 71 34W
Lancaster, Pa. 9 40 2N 76 19W
Lancaster, S.C. 11 34 43N 80 46W
Lancaster, Wis. 14 42 51N 90 43W
Lander, Wyo. 16 42 50N 108 44W
Lanesboro, Pa. 9 41 57N 75 34W
Lanett, Ala. 11 32 52N 85 12W
Langdon, N. Dak. 12 48 45N 98 22W
Langley Park, Md. 25 38 59N 76 58W
Langlois, Oreg. 16 42 56N 124 27W
Langtry, Tex. 13 29 49N 101 34W
Lansdale, Pa. 9 40 14N 75 17W
Lansdowne, Md. 25 39 14N 76 39W
Lansdowne, Pa. 24 39 56N 75 15W
Lansford, Pa. 9 40 50N 75 53W
Lansing, Mich. 15 42 44N 84 33W
Laona, Wis. 10 45 34N 88 40W
Lapeer, Mich. 15 43 3N 83 19W
Laporte, Pa. 9 41 25N 76 30W
Laramie, Wyo. 12 41 19N 105 35W
Laramie Mts., Wyo. 12 42 0N 105 30W
Larchmont, N.Y. 21 40 55N 73 44W
Laredo, Tex. 13 27 30N 99 30W
Larimore, N. Dak. 12 47 54N 97 38W
Larkspur, Calif. 28 37 56N 122 32W
Larned, Kans. 12 38 11N 99 6W
Las Animas, Colo. 12 38 4N 103 13W
Las Cruces, N. Mex. 17 32 19N 106 47W
Las Vegas, N. Mex. 17 35 36N 105 13W
Las Vegas, Nev. 19 36 10N 115 9W
Lassen Pk., Wash. 16 40 29N 121 31W
Lathrop Wells, Nev. 19 36 39N 116 24W
Laton, Calif. 18 36 26N 119 41W
Laurel, Miss. 13 31 41N 89 8W
Laurel, Mont. 16 45 40N 108 46W
Laurens, S.C. 11 34 30N 82 1W
Laurinburg, N.C. 11 34 47N 79 28W
Laurium, Mich. 10 47 14N 88 27W
Lava Hot Springs, Idaho 16 42 37N 112 1W
Laverne, Okla. 13 36 43N 99 54W
Lawrence, Kans. 12 38 58N 95 14W
Lawrence, Mass. 23 42 43N 71 7W
Lawrenceburg, Ind. 15 39 6N 84 52W
Lawrenceburg, Tenn. 11 35 14N 87 20W
Lawrenceville, Ga. 11 33 57N 83 59W
Laws, Calif. 18 37 24N 118 20W
Lawton, Okla. 13 34 37N 98 25W
Laytonville, Calif. 16 39 41N 123 29W
Le Mars, Iowa 12 42 47N 96 10W
Le Roy, Kans. 13 38 5N 95 38W
Le Sueur, Minn. 12 44 28N 93 55W
Lead, S. Dak. 12 44 21N 103 46W
Leadville, Colo. 17 39 15N 106 18W
Leaf →, Miss. 13 30 59N 88 44W
Leakey, Tex. 13 29 44N 99 46W
Leamington, Utah 16 39 32N 112 17W
Leavenworth, Kans. 12 39 19N 94 55W
Leavenworth, Wash. 16 47 36N 120 40W
Lebanon, Ind. 15 40 3N 86 28W
Lebanon, Kans. 12 39 49N 98 33W
Lebanon, Ky. 10 37 34N 85 15W
Lebanon, Mo. 14 37 41N 92 40W
Lebanon, Oreg. 16 44 32N 122 55W
Lebanon, Pa. 9 40 20N 76 26W
Lebanon, Tenn. 11 36 12N 86 18W
Lebec, Calif. 19 34 50N 118 52W
Lee Vining, Calif. 18 37 58N 119 7W
Leech L., Minn. 12 47 10N 94 24W
Leedey, Okla. 13 35 52N 99 21W
Leeds, Ala. 11 33 33N 86 33W
Leesburg, Fla. 11 28 49N 81 53W
Leesville, La. 13 31 9N 93 16W
Lefors, Tex. 13 35 26N 100 48W
Lehi, Utah 16 40 24N 111 51W
Lehighton, Pa. 9 40 50N 75 43W
Lehua I., Hawaii 30 22 1N 160 6W
Leland, Miss. 13 33 24N 90 54W
Lemhi Ra., Idaho 16 44 30N 113 30W
Lemmon, S. Dak. 12 45 57N 102 10W
Lemon Grove, Calif. 19 32 45N 117 2W
Lemoore, Calif. 18 36 18N 119 46W
Lennox, Calif. 29 33 56N 118 21W
Lenoir, N.C. 11 35 55N 81 32W
Lenoir City, Tenn. 11 35 48N 84 16W
Lenora, Kans. 12 39 37N 100 0W
Lenox, Mass. 9 42 22N 73 17W
Lenwood, Calif. 19 34 53N 117 7W
Leola, S. Dak. 12 45 43N 98 56W
Leominster, Mass. 9 42 32N 71 46W
Leon, Iowa 14 40 44N 93 45W
Leonardtown, Md. 10 38 17N 76 38W
Leoti, Kans. 12 38 29N 101 21W
Leslie, Ark. 13 35 50N 92 34W
Leucadia, Calif. 19 33 4N 117 18W
Levan, Utah 16 39 33N 111 52W
Levelland, Tex. 13 33 35N 102 23W
Levittown, N.Y. 21 40 43N 73 31W
Levittown, Pa. 9 40 9N 74 51W
Lewellen, Nebr. 12 41 20N 102 9W
Lewes, Del. 10 38 46N 75 9W
Lewis Range, Mont. 16 48 5N 113 5W
Lewisburg, Pa. 8 40 58N 76 54W
Lewisburg, Tenn. 11 35 27N 86 48W
Lewiston, Idaho 16 46 25N 117 1W
Lewiston, Maine 11 44 6N 70 13W
Lewistown, Mont. 16 47 4N 109 26W
Lewistown, Pa. 8 40 36N 77 34W
Lexington, Ill. 15 40 39N 88 47W
Lexington, Ky. 15 38 3N 84 30W
Lexington, Mass. 23 42 26N 71 13W
Lexington, Miss. 13 33 7N 90 3W
Lexington, Mo. 14 39 11N 93 52W
Lexington, N.C. 11 35 49N 80 15W
Lexington, Nebr. 12 40 47N 99 45W
Lexington, Oreg. 16 45 27N 119 42W
Lexington, Tenn. 11 35 39N 88 24W
Lexington Park, Md. 10 38 16N 76 27W
Libby, Mont. 16 48 23N 115 33W
Liberal, Kans. 13 37 3N 100 55W
Liberal, Mo. 13 37 34N 94 31W
Liberty, Mo. 14 39 15N 94 25W
Liberty, Tex. 13 30 3N 94 48W
Lida, Nev. 17 37 28N 117 30W
Lihue, Hawaii 30 21 59N 159 23W
Lima, Mont. 16 44 38N 112 36W
Lima, Ohio 15 40 44N 84 6W
Limon, Colo. 12 39 16N 103 41W
Lincoln, Ill. 14 40 9N 89 22W
Lincoln, Kans. 12 39 3N 98 9W
Lincoln, Maine 11 45 22N 68 30W
Lincoln, N. Mex. 17 33 30N 105 23W
Lincoln, Nebr. 12 40 49N 96 41W
Lincoln Park, Mich. 26 42 14N 83 9W
Lincolnton, N.C. 11 35 29N 81 16W
Lincolnwood, Ill. 22 42 1N 87 45W
Lind, Wash. 16 46 58N 118 37W
Linden, Calif. 18 38 1N 121 5W
Linden, N.J. 20 40 38N 74 14W
Linden, Tex. 13 33 1N 94 22W
Lindsay, Calif. 18 36 12N 119 5W
Lindsay, Okla. 13 34 50N 97 38W
Lindsborg, Kans. 12 38 35N 97 40W
Lingle, Wyo. 12 42 8N 104 21W
Linthicum Heights, Md. 25 39 12N 76 41W
Linton, Ind. 15 39 2N 87 10W
Linton, N. Dak. 12 46 16N 100 14W
Lipscomb, Tex. 13 36 14N 100 16W
Lisbon, N. Dak. 12 46 27N 97 41W
Lisburne, C., Alaska 30 68 53N 166 13W
Litchfield, Conn. 9 41 45N 73 11W
Litchfield, Ill. 14 39 11N 89 39W
Litchfield, Minn. 12 45 8N 94 32W
Little Belt Mts., Mont. 16 46 40N 110 45W
Little Blue →, Kans. 12 39 42N 96 41W
Little Colorado →, Ariz. 17 36 12N 111 48W
Little Falls, Minn. 12 45 59N 94 22W
Little Falls, N.Y. 9 43 3N 74 51W
Little Fork →, Minn. 12 48 31N 93 35W
Little Humboldt →, Nev. 16 41 1N 117 43W
Little Lake, Calif. 19 35 56N 117 55W
Little Missouri →, N. Dak. 12 47 36N 102 25W
Little Red →, Ark. 13 35 11N 91 27W
Little Rock, Ark. 13 34 45N 92 17W
Little Sable Pt., Mich. 10 43 38N 86 33W
Little Sioux →, Iowa 12 41 48N 96 4W
Little Snake →, Colo. 16 40 27N 108 26W
Little Wabash →, Ill. 15 37 55N 88 5W
Littlefield, Tex. 13 33 55N 102 20W
Littlefork, Minn. 12 48 24N 93 34W
Littleton, N.H. 9 44 18N 71 46W
Live Oak, Fla. 11 30 18N 82 59W
Livermore, Calif. 18 37 41N 121 47W
Livermore, Mt., Tex. 13 30 38N 104 11W
Livingston, Calif. 18 37 23N 120 43W
Livingston, Mont. 16 45 40N 110 34W
Livingston, N.J. 20 40 47N 74 18W
Livingston, Tex. 13 30 43N 94 56W
Livonia, Mich. 26 42 24N 83 22W
Llano, Tex. 13 30 45N 98 41W
Llano →, Tex. 13 30 39N 98 26W
Llano Estacado, Tex. 13 33 30N 103 0W
Loa, Utah 17 38 24N 111 39W
Lock Haven, Pa. 8 41 8N 77 28W
Lockeford, Calif. 18 38 10N 121 9W
Lockhart, Tex. 13 29 53N 97 40W
Lockney, Tex. 13 34 7N 101 27W
Lockport, N.Y. 8 43 10N 78 42W
Lodge Grass, Mont. 16 45 19N 107 22W
Lodgepole, Nebr. 12 41 9N 102 38W
Lodgepole Cr. →, Wyo. 12 41 20N 104 30W
Lodi, Calif. 18 38 8N 121 16W
Lodi, N.J. 20 40 52N 74 5W
Logan, Kans. 12 39 40N 99 34W
Logan, Ohio 10 39 32N 82 25W
Logan, Utah 16 41 44N 111 50W
Logan, W. Va. 10 37 51N 81 59W
Logan International Airport, Mass. 23 42 21N 71 0W
Logansport, Ind. 15 40 45N 86 22W
Logansport, La. 13 31 58N 94 0W
Lolo, Mont. 16 46 45N 114 5W
Loma, Mont. 16 47 56N 110 30W
Loma Linda, Calif. 19 34 3N 117 16W
Lometa, Tex. 13 31 13N 98 24W
Lompoc, Calif. 19 34 38N 120 28W
London, Ky. 10 37 8N 84 5W
London, Ohio 15 39 53N 83 27W
Lone Pine, Calif. 18 36 36N 118 4W
Long Beach, Calif. 19 33 47N 118 11W
Long Beach, N.Y. 21 40 35N 73 40W
Long Beach, Wash. 18 46 21N 124 3W

Millerton

Long Branch, *N.J.* **9** 40 18N 74 0W
Long Creek, *Oreg.* **16** 44 43N 119 6W
Long I., *N.Y.* **9** 40 45N 73 30W
Long Island Sd., *N.Y.* **9** 41 10N 73 0W
Long Pine, *Nebr.* **12** 42 32N 99 42W
Longmont, *Colo.* **12** 40 10N 105 6W
Longview, *Tex.* **13** 32 30N 94 44W
Longview, *Wash.* **18** 46 8N 122 57W
Lonoke, *Ark.* **13** 34 47N 91 54W
Lookout, C., *N.C.* **11** 34 35N 76 32W
Lorain, *Ohio* **8** 41 28N 82 11W
Lordsburg, *N. Mex.* **17** 32 21N 108 43W
Los Alamos, *Calif.* **19** 34 44N 120 17W
Los Alamos, *N. Mex.* **17** 35 53N 106 19W
Los Altos, *Calif.* **18** 37 23N 122 7W
Los Angeles, *Calif.* **29** 34 3N 118 13W
Los Angeles Aqueduct, *Calif.* **19** 35 22N 118 5W
Los Angeles International Airport, *Calif.* **29** 33 56N 118 23W
Los Banos, *Calif.* **18** 37 4N 120 51W
Los Lunas, *N. Mex.* **17** 34 48N 106 44W
Los Olivos, *Calif.* **19** 34 40N 120 7W
Loudon, *Tenn.* **11** 35 45N 84 20W
Louisa, *Ky.* **10** 38 7N 82 36W
Louisiana, *Mo.* **14** 39 27N 91 3W
Louisiana □ **13** 30 50N 92 0W
Louisville, *Ky.* **15** 38 15N 85 46W
Louisville, *Miss.* **13** 33 7N 89 3W
Loup City, *Nebr.* **12** 41 17N 98 58W
Loveland, *Colo.* **12** 40 24N 105 5W
Lovell, *Wyo.* **16** 44 50N 108 24W
Lovelock, *Nev.* **16** 40 11N 118 28W
Loving, *N. Mex.* **13** 32 17N 104 6W
Lovington, *N. Mex.* **13** 32 57N 103 21W
Lowell, *Mass.* **23** 42 38N 71 16W
Lower L., *Calif.* **16** 41 16N 120 2W
Lower Lake, *Calif.* **18** 38 55N 122 37W
Lower Red L., *Minn.* **12** 47 58N 95 0W
Lowville, *N.Y.* **9** 43 47N 75 29W
Lubbock, *Tex.* **13** 33 35N 101 51W
Lucedale, *Miss.* **11** 30 56N 88 35W
Lucerne Valley, *Calif.* **19** 34 27N 116 57W
Ludington, *Mich.* **10** 43 57N 86 27W
Ludlow, *Calif.* **19** 34 43N 116 10W
Ludlow, *Vt.* **9** 43 24N 72 42W
Lufkin, *Tex.* **13** 31 21N 94 44W
Luling, *Tex.* **13** 29 41N 97 39W
Luma, *Amer. Samoa* **31** 14 15S 169 32W
Lumberton, *Miss.* **13** 31 0N 89 27W
Lumberton, *N.C.* **11** 34 37N 79 0W
Lumberton, *N. Mex.* **17** 36 56N 106 56W
Lund, *Nev.* **16** 38 52N 115 0W
Luning, *Nev.* **16** 38 30N 118 11W
Luray, *Va.* **10** 38 40N 78 28W
Lusk, *Wyo.* **12** 42 46N 104 27W
Lutherville-Timonium, *Md.* **25** 39 25N 76 36W
Luverne, *Minn.* **12** 43 39N 96 13W
Lyman, *Wyo.* **16** 41 20N 110 18W
Lynchburg, *Va.* **10** 37 25N 79 9W
Lynden, *Wash.* **18** 48 57N 122 27W
Lyndhurst, *N.J.* **20** 40 49N 74 8W
Lynn, *Mass.* **23** 42 28N 70 57W
Lynwood, *Calif.* **29** 33 55N 118 12W
Lyons, *Colo.* **12** 40 14N 105 16W
Lyons, *Ga.* **11** 32 12N 82 19W
Lyons, *Ill.* **22** 41 48N 87 49W
Lyons, *Kans.* **12** 38 21N 98 12W
Lyons, *N.Y.* **8** 43 5N 77 0W
Lytle, *Tex.* **13** 29 14N 98 48W

M

Mabton, *Wash.* **16** 46 13N 120 0W
McAlester, *Okla.* **13** 34 56N 95 46W
McAllen, *Tex.* **13** 26 12N 98 14W
McCall, *Idaho* **16** 44 55N 116 6W
McCamey, *Tex.* **13** 31 8N 102 14W
McCammon, *Idaho* **16** 42 39N 112 12W
McCarthy, *Alaska* **30** 61 26N 142 56W
McCloud, *Calif.* **16** 41 15N 122 8W
McClure, L., *Calif.* **18** 37 35N 120 16W
McClusky, *N. Dak.* **12** 47 29N 100 27W
McComb, *Miss.* **13** 31 15N 90 27W
McConaughy, L., *Nebr.* **12** 41 14N 101 40W
McCook, *Nebr.* **12** 40 12N 100 38W
McDermitt, *Nev.* **16** 41 59N 117 43W
McFarland, *Calif.* **19** 35 41N 119 14W
McGehee, *Ark.* **13** 33 38N 91 24W
McGill, *Nev.* **16** 39 23N 114 47W
McGregor, *Iowa* **14** 43 1N 91 11W
Machias, *Maine* **11** 44 43N 67 28W
McIntosh, *S. Dak.* **12** 45 55N 101 21W
Mackay, *Idaho* **16** 43 55N 113 37W
McKees Rocks, *Pa.* **27** 40 28N 80 3W
McKeesport, *Pa.* **27** 40 21N 79 51W
McKenzie, *Tenn.* **11** 36 8N 88 31W
McKenzie →, *Oreg.* **16** 44 7N 123 6W
Mackinaw City, *Mich.* **10** 45 47N 84 44W
McKinley, Mt., *Alaska* **30** 63 4N 151 0W
McKinney, *Tex.* **13** 33 12N 96 37W
McLaughlin, *S. Dak.* **12** 45 49N 100 49W
McLean, *Tex.* **13** 35 14N 100 36W
McLean, *Va.* **25** 38 56N 77 10W
McLeansboro, *Ill.* **15** 38 6N 88 32W
McLoughlin, Mt., *Oreg.* **16** 42 27N 122 19W
McMillan, L., *N. Mex.* **13** 32 36N 104 21W
McMinnville, *Oreg.* **16** 45 13N 123 12W
McMinnville, *Tenn.* **11** 35 41N 85 46W
McNary, *Ariz.* **17** 34 4N 109 51W
Macomb, *Ill.* **14** 40 27N 90 40W
Macon, *Ga.* **11** 32 51N 83 38W
Macon, *Miss.* **11** 33 7N 88 34W
Macon, *Mo.* **14** 39 44N 92 28W
McPherson, *Kans.* **12** 38 22N 97 40W
McPherson Pk., *Calif.* **19** 34 53N 119 53W
McVille, *N. Dak.* **12** 47 46N 98 11W
Madera, *Calif.* **18** 36 57N 120 3W
Madill, *Okla.* **13** 34 6N 96 46W
Madison, *Fla.* **11** 30 28N 83 25W
Madison, *Ind.* **15** 38 44N 85 23W
Madison, *N.J.* **20** 40 45N 74 25W
Madison, *Nebr.* **12** 41 50N 97 27W
Madison, *S. Dak.* **12** 44 0N 97 7W
Madison, *Wis.* **14** 43 4N 89 24W
Madison →, *Mont.* **16** 45 56N 111 31W
Madison Heights, *Mich.* **26** 42 29N 83 6W
Madisonville, *Ky.* **10** 37 20N 87 30W
Madisonville, *Tex.* **13** 30 57N 95 55W
Madras, *Oreg.* **16** 44 38N 121 8W
Madre, Laguna, *Tex.* **13** 27 0N 97 30W
Magdalena, *N. Mex.* **17** 34 7N 107 15W
Magee, *Miss.* **13** 31 52N 89 44W
Magnolia, *Ark.* **13** 33 16N 93 14W
Magnolia, *Miss.* **13** 31 9N 90 28W
Mahanoy City, *Pa.* **9** 40 49N 76 9W
Mahnomen, *Minn.* **12** 47 19N 95 58W
Mahukona, *Hawaii* **30** 20 11N 155 52W
Maine □ **11** 45 20N 69 0W
Makapuu Hd., *Hawaii* **30** 21 19N 157 39W
Makena, *Hawaii* **30** 20 39N 156 27W
Malad City, *Idaho* **16** 42 12N 112 15W
Malaga, *N. Mex.* **13** 32 14N 104 4W
Malakoff, *Tex.* **13** 32 10N 96 1W
Malden, *Mass.* **23** 42 26N 71 3W
Malden, *Mo.* **13** 36 34N 89 57W
Malheur →, *Oreg.* **16** 44 4N 116 59W
Malheur L., *Oreg.* **16** 43 20N 118 48W
Malibu, *Calif.* **19** 34 2N 118 41W
Malone, *N.Y.* **9** 44 51N 74 18W
Malta, *Idaho* **16** 42 18N 113 22W
Malta, *Mont.* **16** 48 21N 107 52W
Malvern, *Ark.* **13** 34 22N 92 49W
Malvern, *Pa.* **24** 40 2N 75 31W
Mammoth, *Ariz.* **17** 32 43N 110 39W
Mana, *Hawaii* **30** 22 2N 159 47W
Manasquan, *N.J.* **9** 40 8N 74 3W
Manassa, *Colo.* **17** 37 11N 105 56W
Manatí, *Puerto Rico* **30** 18 26N 66 29W
Mancelona, *Mich.* **10** 44 54N 85 4W
Manchester, *Conn.* **9** 41 47N 72 31W
Manchester, *Ga.* **11** 32 51N 84 37W
Manchester, *Iowa* **14** 42 29N 91 27W
Manchester, *Ky.* **10** 37 9N 83 46W
Manchester, *N.H.* **9** 42 59N 71 28W
Mandan, *N. Dak.* **12** 46 50N 100 54W
Mangum, *Okla.* **13** 34 53N 99 30W
Manhasset, *N.Y.* **21** 40 47N 73 39W
Manhattan, *Kans.* **12** 39 11N 96 35W
Manhattan, *N.Y.* **20** 40 48N 73 57W
Manhattan Beach, *Calif.* **29** 33 53N 118 25W
Manila, *Utah* **16** 40 59N 109 43W
Manistee, *Mich.* **10** 44 15N 86 19W
Manistee →, *Mich.* **10** 44 15N 86 21W
Manistique, *Mich.* **10** 45 57N 86 15W
Manitou Is., *Mich.* **10** 45 8N 86 0W
Manitou Springs, *Colo.* **12** 38 52N 104 55W
Manitowoc, *Wis.* **10** 44 5N 87 40W
Mankato, *Kans.* **12** 39 47N 98 13W
Mankato, *Minn.* **12** 44 10N 94 0W
Manning, *S.C.* **11** 33 42N 80 13W
Mannington, *W. Va.* **10** 39 32N 80 21W
Mansfield, *La.* **13** 32 2N 93 43W
Mansfield, *Mass.* **9** 42 2N 71 13W
Mansfield, *Ohio* **8** 40 45N 82 31W
Mansfield, *Pa.* **8** 41 48N 77 5W
Mansfield, *Wash.* **16** 47 49N 119 38W
Manteca, *Calif.* **18** 37 48N 121 13W
Manteo, *N.C.* **11** 35 55N 75 40W
Manti, *Utah* **16** 39 16N 111 38W
Manton, *Mich.* **10** 44 25N 85 24W
Manua Is., *Amer. Samoa* **31** 14 13S 169 35W
Manville, *Wyo.* **12** 42 47N 104 37W
Many, *La.* **13** 31 34N 93 29W
Manzano Mts., *N. Mex.* **17** 34 40N 106 20W
Maple Heights, *Ohio* **27** 41 25N 81 33W
Mapleton, *Oreg.* **16** 44 2N 123 52W
Maplewood, *N.J.* **20** 40 43N 74 16W
Maquoketa, *Iowa* **14** 42 4N 90 40W
Marana, *Ariz.* **17** 32 27N 111 13W
Marathon, *N.Y.* **9** 42 27N 76 2W
Marathon, *Tex.* **13** 30 12N 103 15W
Marble Falls, *Tex.* **13** 30 35N 98 16W
Marblehead, *Mass.* **23** 42 29N 70 51W
Marengo, *Iowa* **14** 41 48N 92 4W
Marfa, *Tex.* **13** 30 19N 104 1W
Mariana Trench, *Pac. Oc.* **31** 13 0N 145 0 E
Marianna, *Ark.* **13** 34 46N 90 46W
Marianna, *Fla.* **11** 30 46N 85 14W
Marias →, *Mont.* **16** 47 56N 110 30W
Maricopa, *Ariz.* **17** 33 4N 112 3W
Maricopa, *Calif.* **19** 35 4N 119 24W
Marietta, *Ga.* **11** 33 57N 84 33W
Marietta, *Ohio* **10** 39 25N 81 27W
Marin City, *Calif.* **28** 37 52N 122 30W
Marina, *Calif.* **18** 36 41N 121 48W
Marine City, *Mich.* **10** 42 43N 82 30W
Marinette, *Wis.* **10** 45 6N 87 38W
Marion, *Ala.* **11** 32 38N 87 19W
Marion, *Ill.* **14** 37 44N 88 56W
Marion, *Ind.* **15** 40 32N 85 40W
Marion, *Iowa* **14** 42 2N 91 36W
Marion, *Kans.* **12** 38 21N 97 1W
Marion, *Mich.* **10** 44 6N 85 9W
Marion, *N.C.* **11** 35 41N 82 1W
Marion, *Ohio* **15** 40 35N 83 8W
Marion, *S.C.* **11** 34 11N 79 24W
Marion, *Va.* **11** 36 50N 81 31W
Marion, L., *S.C.* **11** 33 28N 80 10W
Mariposa, *Calif.* **18** 37 29N 119 58W
Marked Tree, *Ark.* **13** 35 32N 90 25W
Markham, *Ill.* **22** 41 35N 87 41W
Markleeville, *Calif.* **18** 38 42N 119 47W
Marksville, *La.* **13** 31 8N 92 4W
Marlboro, *Mass.* **9** 42 19N 71 33W
Marlin, *Tex.* **13** 31 18N 96 54W
Marlow, *Okla.* **13** 34 39N 97 58W
Marmarth, *N. Dak.* **12** 46 18N 103 54W
Marple, *Pa.* **24** 39 56N 75 21W
Marquette, *Mich.* **10** 46 33N 87 24W
Marsh I., *La.* **13** 29 34N 91 53W
Marsh L., *Minn.* **12** 45 5N 96 0W
Marshall, *Ark.* **13** 35 55N 92 38W
Marshall, *Mich.* **15** 42 16N 84 58W
Marshall, *Minn.* **12** 44 25N 95 45W
Marshall, *Mo.* **14** 39 7N 93 12W
Marshall, *Tex.* **13** 32 33N 94 23W
Marshall Is. ■, *Pac. Oc.* **31** 9 0N 171 0 E
Marshalltown, *Iowa* **14** 42 3N 92 55W
Marshfield, *Mo.* **13** 37 15N 92 54W
Marshfield, *Wis.* **12** 44 40N 90 10W
Mart, *Tex.* **13** 31 33N 96 50W
Martha's Vineyard, *Mass.* **9** 41 25N 70 38W
Martin, *S. Dak.* **12** 43 11N 101 44W
Martin, *Tenn.* **13** 36 21N 88 51W
Martin L., *Ala.* **11** 32 41N 85 55W
Martin State National Airport, *Md.* **25** 39 19N 76 25W
Martinez, *Calif.* **18** 38 1N 122 8W
Martinsburg, *W. Va.* **10** 39 27N 77 58W
Martinsville, *Ind.* **15** 39 26N 86 25W
Martinsville, *Va.* **11** 36 41N 79 52W
Maryland □ **10** 39 0N 76 30W
Marysvale, *Utah* **17** 38 27N 112 14W
Marysville, *Calif.* **18** 39 9N 121 35W
Marysville, *Kans.* **12** 39 51N 96 39W
Marysville, *Ohio* **15** 40 14N 83 22W
Maryville, *Tenn.* **11** 35 46N 83 58W
Mason, *Nev.* **18** 38 56N 119 8W
Mason, *Tex.* **13** 30 45N 99 14W
Mason City, *Iowa* **14** 43 9N 93 12W
Massachusetts □ **9** 42 30N 72 0W
Massachusetts B., *Mass.* **9** 42 20N 70 50W
Massapequa, *N.Y.* **21** 40 41N 73 28W
Massena, *N.Y.* **9** 44 56N 74 54W
Massillon, *Ohio* **8** 40 48N 81 32W
Matagorda, *Tex.* **13** 28 42N 95 58W
Matagorda B., *Tex.* **13** 28 40N 96 0W
Matagorda I., *Tex.* **13** 28 15N 96 30W
Mathis, *Tex.* **13** 28 6N 97 50W
Mattawamkeag, *Maine* **11** 45 32N 68 21W
Mattituck, *N.Y.* **9** 40 59N 72 32W
Maui, *Hawaii* **30** 20 48N 156 20W
Maumee, *Ohio* **15** 41 34N 83 39W
Maumee →, *Ohio* **15** 41 42N 83 28W
Mauna Kea, *Hawaii* **30** 19 50N 155 28W
Mauna Loa, *Hawaii* **30** 19 30N 155 35W
Maupin, *Oreg.* **16** 45 11N 121 5W
Maurepas, L., *La.* **13** 30 15N 90 30W
Mauston, *Wis.* **12** 43 48N 90 5W
Max, *N. Dak.* **12** 47 49N 101 18W
Mayagüez, *Puerto Rico* **30** 18 12N 67 9W
Maybell, *Colo.* **16** 40 31N 108 5W
Mayer, *Ariz.* **17** 34 24N 112 14W
Mayfield, *Ky.* **11** 36 44N 88 38W
Mayfield Heights, *Ohio* **27** 41 31N 81 28W
Mayhill, *N. Mex.* **17** 32 53N 105 29W
Maysville, *Ky.* **15** 38 39N 83 46W
Mayville, *N. Dak.* **12** 47 30N 97 20W
Maywood, *Calif.* **29** 33 59N 118 12W
Maywood, *Ill.* **22** 41 52N 87 52W
McGrath, *Alaska* **30** 62 58N 155 40W
Mead, L., *Ariz.* **19** 36 1N 114 44W
Meade, *Kans.* **13** 37 17N 100 20W
Meade River = Atkasuk, *Alaska* **30** 70 30N 157 20W
Meadow Valley Wash →, *Nev.* **19** 36 40N 114 34W
Meadville, *Pa.* **8** 41 39N 80 9W
Meares, C., *Oreg.* **16** 45 37N 124 0W
Mecca, *Calif.* **19** 33 34N 116 5W
Mechanicsburg, *Pa.* **8** 40 13N 77 1W
Mechanicville, *N.Y.* **9** 42 54N 73 41W
Medford, *Mass.* **23** 42 25N 71 7W
Medford, *Oreg.* **16** 42 19N 122 52W
Medford, *Wis.* **12** 45 9N 90 20W
Media, *Pa.* **24** 39 55N 75 23W
Medical Lake, *Wash.* **16** 47 34N 117 41W
Medicine Bow, *Wyo.* **16** 41 54N 106 12W
Medicine Bow Pk., *Wyo.* **16** 41 21N 106 19W
Medicine Bow Ra., *Wyo.* **16** 41 10N 106 25W
Medicine Lake, *Mont.* **12** 48 30N 104 30W
Medicine Lodge, *Kans.* **13** 37 17N 98 35W
Medina, *N. Dak.* **12** 46 54N 99 18W
Medina, *N.Y.* **8** 43 13N 78 23W
Medina, *Ohio* **8** 41 8N 81 52W
Medina →, *Tex.* **13** 29 16N 98 29W
Medina L., *Tex.* **13** 29 32N 98 56W
Meeker, *Colo.* **16** 40 2N 107 55W
Meeteetse, *Wyo.* **16** 44 9N 108 52W
Mekoryok, *Alaska* **30** 60 20N 166 20W
Melbourne, *Fla.* **11** 28 5N 80 37W
Mellen, *Wis.* **12** 46 20N 90 40W
Mellette, *S. Dak.* **12** 45 9N 98 30W
Melrose, *Mass.* **23** 42 27N 71 2W
Melrose, *N. Mex.* **13** 34 26N 103 38W
Melrose Park, *Ill.* **22** 41 53N 87 53W
Melstone, *Mont.* **16** 46 36N 107 52W
Memphis, *Tenn.* **13** 35 8N 90 3W
Memphis, *Tex.* **13** 34 44N 100 33W
Mena, *Ark.* **13** 34 35N 94 15W
Menard, *Tex.* **13** 30 55N 99 47W
Menasha, *Wis.* **10** 44 13N 88 26W
Mendenhall, C., *Alaska* **30** 59 45N 166 10W
Mendocino, *Calif.* **16** 39 19N 123 48W
Mendocino, C., *Calif.* **16** 40 26N 124 25W
Mendota, *Calif.* **18** 36 45N 120 23W
Mendota, *Ill.* **14** 41 33N 89 7W
Menlo Park, *Calif.* **28** 37 26N 122 11W
Menominee, *Mich.* **10** 45 6N 87 37W
Menominee →, *Wis.* **10** 45 6N 87 36W
Menomonie, *Wis.* **12** 44 53N 91 55W
Mer Rouge, *La.* **13** 32 47N 91 48W
Merced, *Calif.* **18** 37 18N 120 29W
Merced Pk., *Calif.* **18** 37 36N 119 24W
Meredith, L., *Tex.* **13** 35 43N 101 33W
Meriden, *Conn.* **9** 41 32N 72 48W
Meridian, *Idaho* **16** 43 37N 116 24W
Meridian, *Miss.* **11** 32 22N 88 42W
Meridian, *Tex.* **13** 31 56N 97 39W
Merkel, *Tex.* **13** 32 28N 100 1W
Merrick, *N.Y.* **21** 40 39N 73 32W
Merrill, *Oreg.* **16** 42 1N 121 36W
Merrill, *Wis.* **12** 45 11N 89 41W
Merriman, *Nebr.* **12** 42 55N 101 42W
Merryville, *La.* **13** 30 45N 93 33W
Mertzon, *Tex.* **13** 31 16N 100 49W
Mesa, *Ariz.* **17** 33 25N 111 50W
Meshoppen, *Pa.* **9** 41 36N 76 3W
Mesick, *Mich.* **10** 44 24N 85 43W
Mesilla, *N. Mex.* **17** 32 16N 106 48W
Mesquite, *Nev.* **17** 36 47N 114 6W
Metairie, *La.* **13** 29 58N 90 10W
Metaline Falls, *Wash.* **16** 48 52N 117 22W
Methuen, *Mass.* **23** 42 43N 71 12W
Metlakatla, *Alaska* **30** 55 8N 131 35W
Metropolis, *Ill.* **13** 37 9N 88 44W
Metropolitan Oakland International Airport, *Calif.* **28** 37 43N 122 13W
Mexia, *Tex.* **13** 31 41N 96 29W
Mexico, *Mo.* **14** 39 10N 91 53W
Miami, *Ariz.* **17** 33 24N 110 52W
Miami, *Fla.* **11** 25 47N 80 11W
Miami, *Tex.* **13** 35 42N 100 38W
Miami →, *Ohio* **10** 39 20N 84 40W
Miami Beach, *Fla.* **11** 25 47N 80 8W
Miamisburg, *Ohio* **15** 39 38N 84 17W
Michelson, Mt., *Alaska* **30** 69 20N 144 20W
Michigan □ **10** 44 0N 85 0W
Michigan, L., *Mich.* **10** 44 0N 87 0W
Michigan City, *Ind.* **15** 41 43N 86 54W
Micronesia, Federated States of ■, *Pac. Oc.* **31** 11 0N 160 0 E
Middle Alkali L., *Calif.* **16** 41 27N 120 5W
Middle Loup →, *Nebr.* **12** 41 17N 98 24W
Middle River, *Md.* **25** 39 21N 76 26W
Middleburg, *N.Y.* **9** 42 36N 74 20W
Middleport, *Ohio* **10** 39 0N 82 3W
Middlesboro, *Ky.* **11** 36 36N 83 43W
Middlesex, *N.J.* **9** 40 36N 74 30W
Middleton I., *Alaska* **30** 59 26N 146 20W
Middletown, *Conn.* **9** 41 34N 72 39W
Middletown, *N.Y.* **9** 41 27N 74 25W
Middletown, *Ohio* **15** 39 31N 84 24W
Middletown, *Pa.* **9** 40 12N 76 44W
Midland, *Mich.* **10** 43 37N 84 14W
Midland, *Tex.* **13** 32 0N 102 3W
Midlothian, *Tex.* **13** 32 30N 97 0W
Midwest, *Wyo.* **16** 43 25N 106 16W
Milaca, *Minn.* **12** 45 45N 93 39W
Milan, *Mo.* **14** 40 12N 93 7W
Milan, *Tenn.* **11** 35 55N 88 46W
Milbank, *S. Dak.* **12** 45 13N 96 38W
Miles, *Tex.* **13** 31 36N 100 11W
Miles City, *Mont.* **12** 46 25N 105 51W
Milford, *Conn.* **9** 41 14N 73 3W
Milford, *Del.* **10** 38 55N 75 26W
Milford, *Mass.* **9** 42 8N 71 31W
Milford, *Pa.* **9** 41 19N 74 48W
Milford, *Utah* **17** 38 24N 113 1W
Milk →, *Mont.* **16** 48 4N 106 19W
Mill City, *Oreg.* **16** 44 45N 122 29W
Mill Valley, *Calif.* **28** 37 54N 122 32W
Millburn, *N.J.* **20** 40 43N 74 19W
Mille Lacs L., *Minn.* **12** 46 15N 93 39W
Milledgeville, *Ga.* **11** 33 5N 83 14W
Millen, *Ga.* **11** 32 48N 81 57W
Miller, *S. Dak.* **12** 44 31N 98 59W
Millersburg, *Pa.* **8** 40 32N 76 58W
Millerton, *N.Y.* **9** 41 57N 73 31W

Millerton L.

Millerton L., *Calif.* **18** 37 1N 119 41W
Millinocket, *Maine* **11** 45 39N 68 43W
Milltown, *Pa.* **24** 39 57N 75 32W
Millvale, *Pa.* **27** 40 28N 79 59W
Millville, *N.J.* **10** 39 24N 75 2W
Millwood L., *Ark.* **13** 33 42N 93 58W
Milnor, *N. Dak.* **12** 46 16N 97 27W
Milolii, *Hawaii* **30** 19 11N 155 55W
Milton, *Calif.* **18** 38 3N 120 51W
Milton, *Fla.* **11** 30 38N 87 3W
Milton, *Mass.* **23** 42 14N 71 2W
Milton, *Pa.* **8** 41 1N 76 51W
Milton-Freewater, *Oreg.* **16** 45 56N 118 23W
Milwaukee, *Wis.* **15** 43 2N 87 55W
Milwaukie, *Oreg.* **18** 45 27N 122 38W
Mina, *Nev.* **17** 38 24N 118 7W
Minden, *La.* **13** 32 37N 93 17W
Mineola, *N.Y.* **21** 40 44N 73 38W
Mineola, *Tex.* **13** 32 40N 95 29W
Mineral King, *Calif.* **18** 36 27N 118 36W
Mineral Wells, *Tex.* **13** 32 48N 98 7W
Minersville, *Pa.* **9** 40 41N 76 16W
Minersville, *Utah* **17** 38 13N 112 56W
Minetto, *N.Y.* **9** 43 24N 76 28W
Minidoka, *Idaho* **16** 42 45N 113 29W
Minneapolis, *Kans.* **12** 39 8N 97 42W
Minneapolis, *Minn.* **12** 44 59N 93 16W
Minnesota □ **12** 46 0N 94 15W
Minot, *N. Dak.* **12** 48 14N 101 18W
Minto, *Alaska* **30** 64 53N 149 11W
Minturn, *Colo.* **16** 39 35N 106 26W
Mirando City, *Tex.* **13** 27 26N 99 0W
Mishawaka, *Ind.* **15** 41 40N 86 11W
Mission, *S. Dak.* **12** 43 18N 100 39W
Mission, *Tex.* **13** 26 13N 98 20W
Mississippi □ **13** 33 0N 90 0W
Mississippi →, *La.* **13** 29 9N 89 15W
Mississippi River Delta,
La. **13** 29 10N 89 15W
Mississippi Sd., *Miss.* **13** 30 20N 89 0W
Missoula, *Mont.* **16** 46 52N 114 1W
Missouri □ **12** 38 25N 92 30W
Missouri →, *Mo.* **12** 38 49N 90 7W
Missouri Valley, *Iowa* **12** 41 34N 95 53W
Mitchell, *Ind.* **15** 38 44N 86 28W
Mitchell, *Nebr.* **12** 41 57N 103 49W
Mitchell, *Oreg.* **16** 44 34N 120 9W
Mitchell, *S. Dak.* **12** 43 43N 98 2W
Mitchell, Mt., *N.C.* **11** 35 46N 82 16W
Moab, *Utah* **17** 38 35N 109 33W
Moberly, *Mo.* **14** 39 25N 92 26W
Mobile, *Ala.* **11** 30 41N 88 3W
Mobile B., *Ala.* **11** 30 30N 88 0W
Mobridge, *S. Dak.* **12** 45 32N 100 26W
Moclips, *Wash.* **18** 47 14N 124 13W
Modena, *Utah* **17** 37 48N 113 56W
Modesto, *Calif.* **18** 37 39N 121 0W
Mohall, *N. Dak.* **12** 48 46N 101 31W
Mohawk →, *N.Y.* **9** 42 47N 73 41W
Mohican, C., *Alaska* **30** 60 12N 167 25W
Mojave, *Calif.* **19** 35 3N 118 10W
Mojave Desert, *Calif.* **19** 35 0N 116 30W
Mokelumne →, *Calif.* **18** 38 13N 121 28W
Mokelumne Hill, *Calif.* **18** 38 18N 120 43W
Moline, *Ill.* **14** 41 30N 90 31W
Molokai, *Hawaii* **30** 21 8N 157 0W
Monahans, *Tex.* **13** 31 36N 102 54W
Mondovi, *Wis.* **12** 44 34N 91 40W
Monessen, *Pa.* **8** 40 9N 79 54W
Monett, *Mo.* **13** 36 55N 93 55W
Monmouth, *Ill.* **14** 40 55N 90 39W
Mono L., *Calif.* **18** 38 1N 119 1W
Monolith, *Calif.* **19** 35 7N 118 22W
Monroe, *Ga.* **11** 33 47N 83 43W
Monroe, *La.* **13** 32 30N 92 7W
Monroe, *Mich.* **15** 41 55N 83 24W
Monroe, *N.C.* **11** 34 59N 80 33W
Monroe, *Utah* **17** 38 38N 112 7W
Monroe, *Wis.* **14** 42 36N 89 38W
Monroe City, *Mo.* **14** 39 39N 91 44W
Monroeville, *Ala.* **11** 31 31N 87 20W
Monroeville, *Pa.* **27** 40 26N 79 46W
Monrovia, *Calif.* **29** 34 9N 118 1W
Montague, *Calif.* **16** 41 44N 122 32W
Montague I., *Alaska* **30** 60 0N 147 30W
Montalvo, *Calif.* **19** 34 15N 119 12W
Montana □ **16** 47 0N 110 0W
Montauk, *N.Y.* **9** 41 3N 71 57W
Montauk Pt., *N.Y.* **9** 41 4N 71 52W
Montclair, *N.J.* **20** 40 49N 74 12W
Monte Vista, *Colo.* **17** 37 35N 106 9W
Montebello, *Calif.* **29** 34 1N 118 8W
Montecito, *Calif.* **19** 34 26N 119 40W
Montello, *Wis.* **12** 43 48N 89 20W
Monterey, *Calif.* **18** 36 37N 121 55W
Monterey B., *Calif.* **18** 36 45N 122 0W
Monterey Park, *Calif.* **29** 34 3N 118 7W
Montesano, *Wash.* **18** 46 59N 123 36W
Montevideo, *Minn.* **12** 44 57N 95 43W
Montezuma, *Iowa* **14** 41 35N 92 32W
Montgomery, *Ala.* **11** 32 23N 86 19W
Montgomery, *W. Va.* **10** 38 11N 81 19W
Monticello, *Ark.* **13** 33 38N 91 47W
Monticello, *Fla.* **11** 30 33N 83 52W
Monticello, *Ind.* **15** 40 45N 86 46W
Monticello, *Iowa* **14** 42 15N 91 12W
Monticello, *Ky.* **11** 36 50N 84 51W
Monticello, *Minn.* **12** 45 18N 93 48W
Monticello, *Miss.* **13** 31 33N 90 7W
Monticello, *N.Y.* **9** 41 39N 74 42W
Monticello, *Utah* **17** 37 52N 109 21W
Montour Falls, *N.Y.* **8** 42 21N 76 51W
Montpelier, *Idaho* **16** 42 19N 111 18W
Montpelier, *Md.* **25** 39 3N 76 50W
Montpelier, *Ohio* **15** 41 35N 84 37W
Montpelier, *Vt.* **9** 44 16N 72 35W
Montrose, *Colo.* **17** 38 29N 107 53W
Montrose, *Pa.* **9** 41 50N 75 53W
Moorcroft, *Wyo.* **12** 44 16N 104 57W
Moorefield, *W. Va.* **10** 39 5N 78 59W
Mooresville, *N.C.* **11** 35 35N 80 48W
Moorhead, *Minn.* **12** 46 53N 96 45W
Moorpark, *Calif.* **19** 34 17N 118 53W
Moose Lake, *Minn.* **12** 46 27N 92 46W
Moosehead L., *Maine* **11** 45 38N 69 40W
Moosup, *Conn.* **9** 41 43N 71 53W
Mora, *Minn.* **12** 45 53N 93 18W
Mora, *N. Mex.* **17** 35 58N 105 20W
Moran, *Kans.* **13** 37 55N 95 10W
Moran, *Wyo.* **16** 43 53N 110 37W
Moravia, *Iowa* **14** 40 53N 92 49W
Moreau →, *S. Dak.* **12** 45 18N 100 43W
Morehead, *Ky.* **15** 38 11N 83 26W
Morehead City, *N.C.* **11** 34 43N 76 43W
Morenci, *Ariz.* **17** 33 5N 109 22W
Morgan, *Utah* **16** 41 2N 111 41W
Morgan City, *La.* **13** 29 42N 91 12W
Morgan Hill, *Calif.* **18** 37 8N 121 39W
Morganfield, *Ky.* **10** 37 41N 87 55W
Morganton, *N.C.* **11** 35 45N 81 41W
Morgantown, *W. Va.* **10** 39 38N 79 57W
Morongo Valley, *Calif.* **19** 34 3N 116 37W
Morrilton, *Ark.* **13** 35 9N 92 44W
Morris, *Ill.* **15** 41 22N 88 26W
Morris, *Minn.* **12** 45 35N 95 55W
Morrison, *Ill.* **14** 41 49N 89 58W
Morristown, *Ariz.* **17** 33 51N 112 37W
Morristown, *N.J.* **20** 40 48N 74 26W
Morristown, *S. Dak.* **12** 45 56N 101 43W
Morristown, *Tenn.* **11** 36 13N 83 18W
Morro Bay, *Calif.* **18** 35 22N 120 51W
Morton, *Tex.* **13** 33 44N 102 46W
Morton, *Wash.* **18** 46 34N 122 17W
Morton Grove, *Ill.* **22** 42 2N 87 45W
Moscow, *Idaho* **16** 46 44N 117 0W
Moses Lake, *Wash.* **16** 47 8N 119 17W
Mosquero, *N. Mex.* **13** 35 47N 103 58W
Mott, *N. Dak.* **12** 46 23N 102 20W
Moulton, *Tex.* **13** 29 35N 97 9W
Moultrie, *Ga.* **11** 31 11N 83 47W
Moultrie, L., *S.C.* **11** 33 20N 80 5W
Mound City, *Mo.* **12** 40 7N 95 14W
Mound City, *S. Dak.* **12** 45 44N 100 4W
Moundsville, *W. Va.* **8** 39 55N 80 44W
Mount Airy, *N.C.* **11** 36 31N 80 37W
Mount Angel, *Oreg.* **16** 45 4N 122 48W
Mount Carmel, *Ill.* **15** 38 25N 87 46W
Mount Clemens, *Mich.* **8** 42 35N 82 53W
Mount Clemens, *Mich.* **26** 42 35N 82 53W
Mount Desert I., *Maine* **11** 44 21N 68 20W
Mount Dora, *Fla.* **11** 28 48N 81 38W
Mount Edgecumbe,
Alaska **30** 57 3N 135 21W
Mount Hope, *W. Va.* **10** 37 54N 81 10W
Mount Horeb, *Wis.* **14** 43 1N 89 44W
Mount Laguna, *Calif.* **19** 32 52N 116 25W
Mount Lebanon, *Pa.* **27** 40 22N 80 2W
Mount McKinley
National Park, *Alaska* **30** 63 30N 150 0W
Mount Morris, *N.Y.* **8** 42 44N 77 52W
Mount Oliver, *Pa.* **27** 40 24N 79 59W
Mount Pleasant, *Iowa* **14** 40 58N 91 33W
Mount Pleasant, *Mich.* **10** 43 36N 84 46W
Mount Pleasant, *S.C.* **11** 32 47N 79 52W
Mount Pleasant, *Tenn.* **11** 35 32N 87 12W
Mount Pleasant, *Tex.* **13** 33 9N 94 58W
Mount Pleasant, *Utah* **16** 39 33N 111 27W
Mount Pocono, *Pa.* **9** 41 7N 75 22W
Mount Prospect, *Ill.* **22** 42 3N 87 55W
Mount Rainier National
Park, *Wash.* **18** 46 55N 121 50W
Mount Royal, *N.J.* **24** 39 48N 75 13W
Mount Shasta, *Calif.* **16** 41 19N 122 19W
Mount Sterling, *Ill.* **14** 39 59N 90 45W
Mount Sterling, *Ky.* **15** 38 4N 83 56W
Mount Vernon, *Ind.* **15** 38 17N 88 57W
Mount Vernon, *N.Y.* **21** 40 54N 73 49W
Mount Vernon, *Ohio* **8** 40 23N 82 29W
Mount Vernon, *Wash.* **18** 48 25N 122 20W
Mount Wilson
Observatory, *Calif.* **29** 34 13N 118 4W
Mountain Center, *Calif.* **19** 33 42N 116 44W
Mountain City, *Nev.* **16** 41 50N 115 58W
Mountain City, *Tenn.* **11** 36 29N 81 48W
Mountain Grove, *Mo.* **13** 37 8N 92 16W
Mountain Home, *Ark.* **13** 36 20N 92 23W
Mountain Home, *Idaho* **16** 43 8N 115 41W
Mountain Iron, *Minn.* **12** 47 32N 92 37W
Mountain View, *Ark.* **13** 35 52N 92 7W
Mountain View, *Calif.* **18** 37 23N 122 5W
Mountain Village, *Alaska* **30** 62 5N 163 43W
Mountainair, *N. Mex.* **17** 34 31N 106 15W
Muddy Cr. →, *Utah* **17** 38 24N 110 42W
Mule Creek, *Wyo.* **12** 43 19N 104 8W
Muleshoe, *Tex.* **13** 34 13N 102 43W
Mullen, *Nebr.* **12** 42 3N 101 1W
Mullens, *W. Va.* **10** 37 35N 81 23W
Mullin, *Tex.* **13** 31 33N 98 40W
Mullins, *S.C.* **11** 34 12N 79 15W
Mulvane, *Kans.* **13** 37 29N 97 15W
Muncie, *Ind.* **15** 40 12N 85 23W
Munday, *Tex.* **13** 33 27N 99 38W
Munhall, *Pa.* **27** 40 24N 79 54W
Munising, *Mich.* **10** 46 25N 86 40W
Murdo, *S. Dak.* **12** 43 53N 100 43W
Murfreesboro, *Tenn.* **11** 35 51N 86 24W
Murphy, *Idaho* **16** 43 13N 116 33W
Murphys, *Calif.* **18** 38 8N 120 28W
Murphysboro, *Ill.* **14** 37 46N 89 20W
Murray, *Ky.* **11** 36 37N 88 19W
Murray, *Utah* **16** 40 40N 111 53W
Murray, L., *S.C.* **11** 34 3N 81 13W
Murrieta, *Calif.* **19** 33 33N 117 13W
Murrysville, *Pa.* **27** 40 25N 79 41W
Muscatine, *Iowa* **14** 41 25N 91 3W
Muskegon, *Mich.* **15** 43 14N 86 16W
Muskegon →, *Mich.* **10** 43 14N 86 21W
Muskegon Heights, *Mich.* **15** 43 12N 86 16W
Muskogee, *Okla.* **13** 35 45N 95 22W
Musselshell →, *Mont.* **16** 47 21N 107 57W
Myerstown, *Pa.* **9** 40 22N 76 19W
Myrtle Beach, *S.C.* **11** 33 42N 78 53W
Myrtle Creek, *Oreg.* **16** 43 1N 123 17W
Myrtle Point, *Oreg.* **16** 43 4N 124 8W
Mystic, *Conn.* **9** 41 21N 71 58W
Myton, *Utah* **16** 40 12N 110 4W

N

Naalehu, *Hawaii* **30** 19 4N 155 35W
Nabesna, *Alaska* **30** 62 22N 143 0W
Naches, *Wash.* **16** 46 44N 120 42W
Nacimiento Reservoir,
Calif. **18** 35 46N 120 53W
Naco, *Ariz.* **17** 31 20N 109 57W
Nacogdoches, *Tex.* **13** 31 36N 94 39W
Nakalele Pt., *Hawaii* **30** 21 2N 156 35W
Naknek, *Alaska* **30** 58 44N 157 1W
Nampa, *Idaho* **16** 43 34N 116 34W
Nanticoke, *Pa.* **9** 41 12N 76 0W
Napa, *Calif.* **18** 38 18N 122 17W
Napa →, *Calif.* **18** 38 10N 122 19W
Napamute, *Alaska* **30** 61 30N 158 45W
Napanoch, *N.Y.* **9** 41 44N 74 22W
Naples, *Fla.* **11** 26 8N 81 48W
Napoleon, *N. Dak.* **12** 46 30N 99 46W
Napoleon, *Ohio* **15** 41 23N 84 8W
Nara Visa, *N. Mex.* **13** 35 37N 103 6W
Narrows, The, *N.Y.* **20** 40 37N 74 3W
Nashua, *Iowa* **14** 42 57N 92 32W
Nashua, *Mont.* **16** 48 8N 106 22W
Nashua, *N.H.* **9** 42 45N 71 28W
Nashville, *Ark.* **13** 33 57N 93 51W
Nashville, *Ga.* **11** 31 12N 83 15W
Nashville, *Tenn.* **11** 36 10N 86 47W
Nassau, *N.Y.* **9** 42 31N 73 37W
Natchez, *Miss.* **13** 31 34N 91 24W
Natchitoches, *La.* **13** 31 46N 93 5W
Natick, *Mass.* **23** 42 16N 71 21W
National City, *Calif.* **19** 32 41N 117 6W
Natoma, *Kans.* **12** 39 11N 99 2W
Navajo Reservoir,
N. Mex. **17** 36 48N 107 36W
Navasota, *Tex.* **13** 30 23N 96 5W
Neah Bay, *Wash.* **18** 48 22N 124 37W
Near Is., *Alaska* **30** 53 0N 172 0 E
Nebraska □ **12** 41 30N 99 30W
Nebraska City, *Nebr.* **12** 40 41N 95 52W
Necedah, *Wis.* **12** 44 2N 90 4W
Neches →, *Tex.* **13** 29 58N 93 51W
Needham, *Mass.* **23** 42 16N 71 13W
Needles, *Calif.* **19** 34 51N 114 37W
Neenah, *Wis.* **10** 44 11N 88 28W
Negaunee, *Mich.* **10** 46 30N 87 36W
Neihart, *Mont.* **16** 47 0N 110 44W
Neilton, *Wash.* **16** 47 25N 123 53W
Neligh, *Nebr.* **12** 42 8N 98 2W
Nelson, *Ariz.* **17** 35 31N 113 19W
Nelson I., *Alaska* **30** 60 40N 164 40W
Nenana, *Alaska* **30** 64 34N 149 5W
Neodesha, *Kans.* **13** 37 25N 95 41W
Neosho, *Mo.* **13** 36 52N 94 22W
Neosho →, *Okla.* **13** 36 48N 95 18W
Nephi, *Utah* **16** 39 43N 111 50W
Neptune, *N.J.* **9** 40 13N 74 2W
Neuse →, *N.C.* **11** 35 6N 76 29W
Nevada, *Mo.* **14** 37 51N 94 22W
Nevada □ **16** 39 0N 117 0W
Nevada, Sierra, *Calif.* **16** 39 0N 120 30W
Nevada City, *Calif.* **18** 39 16N 121 1W
Neville Island, *Pa.* **27** 40 30N 80 6W
New Albany, *Ind.* **15** 38 18N 85 49W
New Albany, *Miss.* **13** 34 29N 89 0W
New Albany, *Pa.* **9** 41 36N 76 27W
New Bedford, *Mass.* **9** 41 38N 70 56W
New Bern, *N.C.* **11** 35 7N 77 3W
New Boston, *Tex.* **13** 33 28N 94 25W
New Braunfels, *Tex.* **13** 29 42N 98 8W
New Britain, *Conn.* **9** 41 40N 72 47W
New Brunswick, *N.J.* **9** 40 30N 74 27W
New Carrollton, *Md.* **25** 38 58N 76 53W
New Castle, *Ind.* **15** 39 55N 85 22W
New Castle, *Pa.* **8** 41 0N 80 21W
New City, *N.Y.* **9** 41 9N 73 59W
New Cuyama, *Calif.* **19** 34 57N 119 38W
New Don Pedro
Reservoir, *Calif.* **18** 37 43N 120 24W
New Dorp, *N.Y.* **20** 40 34N 74 8W
New England, *N. Dak.* **12** 46 32N 102 52W
New Hampshire □ **9** 44 0N 71 30W
New Hampton, *Iowa* **14** 43 3N 92 19W
New Haven, *Conn.* **9** 41 18N 72 55W
New Hyde Park, *N.Y.* **21** 40 43N 73 39W
New Iberia, *La.* **13** 30 1N 91 49W
New Jersey □ **9** 40 0N 74 30W
New Kensington, *Pa.* **8** 40 34N 79 46W
New Kensington, *Pa.* **27** 40 34N 79 46W
New Lexington, *Ohio* **10** 39 43N 82 13W
New London, *Conn.* **9** 41 22N 72 6W
New London, *Minn.* **12** 45 18N 94 56W
New London, *Wis.* **12** 44 23N 88 45W
New Madrid, *Mo.* **13** 36 36N 89 32W
New Meadows, *Idaho* **16** 44 58N 116 18W
New Melones L., *Calif.* **18** 37 57N 120 31W
New Mexico □ **17** 34 30N 106 0W
New Milford, *Conn.* **9** 41 35N 73 25W
New Milford, *Pa.* **9** 41 52N 75 44W
New Orleans, *La.* **13** 29 58N 90 4W
New Philadelphia, *Ohio* **8** 40 30N 81 27W
New Plymouth, *Idaho* **16** 43 58N 116 49W
New Providence, *N.J.* **20** 40 42N 74 23W
New Richmond, *Wis.* **12** 45 7N 92 32W
New Roads, *La.* **13** 30 42N 91 26W
New Rochelle, *N.Y.* **21** 40 55N 73 45W
New Rockford, *N. Dak.* **12** 47 41N 99 8W
New Salem, *N. Dak.* **12** 46 51N 101 25W
New Smyrna Beach, *Fla.* **11** 29 1N 80 56W
New Town, *N. Dak.* **12** 47 59N 102 30W
New Ulm, *Minn.* **12** 44 19N 94 28W
New York, *N.Y.* **20** 40 42N 74 0W
New York □ **9** 43 0N 75 0W
Newark, *Del.* **10** 39 41N 75 46W
Newark, *N.J.* **20** 40 43N 74 10W
Newark, *N.Y.* **8** 43 3N 77 6W
Newark, *Ohio* **8** 40 3N 82 24W
Newark International
Airport, *N.J.* **20** 40 41N 74 10W
Newaygo, *Mich.* **10** 43 25N 85 48W
Newberg, *Oreg.* **16** 45 18N 122 58W
Newberry, *Mich.* **10** 46 21N 85 30W
Newberry, *S.C.* **11** 34 17N 81 37W
Newberry Springs, *Calif.* **19** 34 50N 116 41W
Newburgh, *N.Y.* **9** 41 30N 74 1W
Newburyport, *Mass.* **9** 42 49N 70 53W
Newcastle, *Wyo.* **12** 43 50N 104 11W
Newell, *S. Dak.* **12** 44 43N 103 25W
Newenham, C., *Alaska* **30** 58 39N 162 11W
Newhalen, *Alaska* **30** 59 43N 154 54W
Newhall, *Calif.* **19** 34 23N 118 32W
Newkirk, *Okla.* **13** 36 53N 97 3W
Newman, *Calif.* **18** 37 19N 121 1W
Newmarket, *N.H.* **9** 43 5N 70 56W
Newnan, *Ga.* **11** 33 23N 84 48W
Newport, *Ark.* **13** 35 37N 91 16W
Newport, *Ky.* **15** 39 5N 84 30W
Newport, *N.H.* **9** 43 22N 72 10W
Newport, *Oreg.* **16** 44 39N 124 3W
Newport, *R.I.* **9** 41 29N 71 19W
Newport, *Tenn.* **11** 35 58N 83 11W
Newport, *Vt.* **9** 44 56N 72 13W
Newport, *Wash.* **16** 48 11N 117 3W
Newport Beach, *Calif.* **19** 33 37N 117 56W
Newport News, *Va.* **10** 36 59N 76 25W
Newton, *Iowa* **14** 41 42N 93 3W
Newton, *Mass.* **23** 42 19N 71 13W
Newton, *Miss.* **13** 32 19N 89 10W
Newton, *N.C.* **11** 35 40N 81 13W
Newton, *N.J.* **9** 41 3N 74 45W
Newton, *Tex.* **13** 30 51N 93 46W
Newtown Square, *Pa.* **24** 39 59N 75 24W
Nezperce, *Idaho* **16** 46 14N 116 14W
Niagara, *Mich.* **10** 45 45N 88 0W
Niagara Falls, *N.Y.* **8** 43 5N 79 4W
Niceville, *Fla.* **11** 30 31N 86 30W
Nicholasville, *Ky.* **15** 37 53N 84 34W
Nichols, *N.Y.* **9** 42 1N 76 22W
Nicholson, *Pa.* **9** 41 37N 75 47W
Niihau, *Hawaii* **30** 21 54N 160 9W
Nikolski, *Alaska* **30** 52 56N 168 52W
Niland, *Calif.* **19** 33 14N 115 31W
Niles, *Ill.* **22** 42 1N 87 48W
Niles, *Ohio* **8** 41 11N 80 46W
Niobrara, *Nebr.* **12** 42 45N 98 2W
Niobrara →, *Nebr.* **12** 42 46N 98 3W
Nipomo, *Calif.* **19** 35 3N 120 29W
Nixon, *Tex.* **13** 29 16N 97 46W
Noatak, *Alaska* **30** 67 34N 162 58W
Noatak →, *Alaska* **30** 68 0N 161 0W
Noblesville, *Ind.* **15** 40 3N 86 1W
Nocona, *Tex.* **13** 33 47N 97 44W
Noel, *Mo.* **13** 36 33N 94 29W
Nogales, *Ariz.* **17** 31 20N 110 56W
Nome, *Alaska* **30** 64 30N 165 25W
Nonopapa, *Hawaii* **30** 21 50N 160 15W
Noonan, *N. Dak.* **12** 48 54N 103 1W
Noorvik, *Alaska* **30** 66 50N 161 3W
Norco, *Calif.* **19** 33 56N 117 33W
Norfolk, *Nebr.* **12** 42 2N 97 25W
Norfolk, *Va.* **10** 36 51N 76 17W
Norfork Res., *Ark.* **13** 36 13N 92 15W
Normal, *Ill.* **14** 40 31N 88 59W
Norman, *Okla.* **13** 35 13N 97 26W
Norridge, *Ill.* **22** 41 57N 87 49W

Perry

Norris, *Mont.* **16** 45 34N 111 41W
Norristown, *Pa.* **24** 40 7N 75 20W
North Adams, *Mass.* **9** 42 42N 73 7W
North Bend, *Oreg.* **16** 43 24N 124 14W
North Bergen, *N.J.* **20** 40 48N 74 0W
North Berwick, *Maine* .. **9** 43 18N 70 44W
North Billerica, *Mass.* .. **23** 42 35N 71 16W
North Braddock, *Pa.* **27** 40 25N 79 51W
North Canadian →, *Okla.* **13** 35 16N 95 31W
North Carolina □ **11** 35 30N 80 0W
North Chelmsford, *Mass.* **23** 42 38N 71 23W
North Chicago, *Ill.* **15** 42 19N 87 51W
North Dakota □ **12** 47 30N 100 15W
North Fork, *Calif.* **18** 37 14N 119 21W
North Las Vegas, *Nev.* .. **19** 36 12N 115 7W
North Loup →, *Nebr.* .. **12** 41 17N 98 24W
North Olmsted, *Ohio* .. **27** 41 24N 81 55W
North Palisade, *Calif.* .. **18** 37 6N 118 31W
North Platte, *Nebr.* **12** 41 8N 100 46W
North Platte →, *Nebr.* .. **12** 41 7N 100 42W
North Powder, *Oreg.* **16** 45 2N 117 55W
North Reading, *Mass.* .. **23** 42 34N 71 5W
North Richmond, *Calif.* . **28** 37 57N 122 22W
North Springfield, *Va.* .. **25** 38 48N 77 12W
North Tonawanda, *N.Y.* **8** 43 2N 78 53W
North Truchas Pk., *N. Mex.* **17** 36 0N 105 30W
North Vernon, *Ind.* **15** 39 0N 85 38W
Northampton, *Mass.* **9** 42 19N 72 38W
Northampton, *Pa.* **9** 40 41N 75 30W
Northbridge, *Mass.* **9** 42 9N 71 39W
Northbrook, *Ill.* **22** 42 7N 87 53W
Northern Marianas ■, *Pac. Oc.* .... **31** 17 0N 145 0 E
Northfield, *Ill.* **22** 42 5N 87 44W
Northfield, *Minn.* **12** 44 27N 93 9W
Northlake, *Ill.* **22** 41 54N 87 53W
Northome, *Minn.* **12** 47 52N 94 17W
Northport, *Ala.* **11** 33 14N 87 35W
Northport, *Mich.* **10** 45 8N 85 37W
Northport, *Wash.* **16** 48 55N 117 48W
Northway, *Alaska* **30** 62 58N 141 56W
Northwood, *Iowa* **12** 43 27N 93 13W
Northwood, *N. Dak.* **12** 47 44N 97 34W
Norton, *Kans.* **12** 39 50N 99 53W
Norton B., *Alaska* **30** 64 45N 161 15W
Norton Sd., *Alaska* **30** 63 50N 164 0W
Norwalk, *Calif.* **29** 33 54N 118 4W
Norwalk, *Conn.* **9** 41 7N 73 22W
Norwalk, *Ohio* **8** 41 15N 82 37W
Norway, *Mich.* **10** 45 47N 87 55W
Norwich, *Conn.* **9** 41 31N 72 5W
Norwich, *N.Y.* **9** 42 32N 75 32W
Norwood, *Mass.* **23** 42 11N 71 13W
Nottoway →, *Va.* **10** 36 33N 76 55W
Novato, *Calif.* **18** 38 6N 122 35W
Noxen, *Pa.* **9** 41 25N 76 4W
Noxon, *Mont.* **16** 48 0N 115 43W
Nueces →, *Tex.* **13** 27 51N 97 30W
Nulato, *Alaska* **30** 64 43N 158 6W
Nunivak I., *Alaska* **30** 60 10N 166 30W
Nutley, *N.J.* **20** 40 49N 74 9W
Nyack, *N.Y.* **9** 41 5N 73 55W
Nyssa, *Oreg.* **16** 43 53N 117 0W

O

Oacoma, *S. Dak.* **12** 43 48N 99 24W
Oahe, L., *S. Dak.* **12** 44 27N 100 24W
Oahe Dam, *S. Dak.* **12** 44 27N 100 24W
Oahu, *Hawaii* **30** 21 28N 157 58W
Oak Creek, *Colo.* **16** 40 16N 106 57W
Oak Forest, *Ill.* **22** 41 36N 87 44W
Oak Harbor, *Wash.* **18** 48 18N 122 39W
Oak Hill, *W. Va.* **10** 37 59N 81 9W
Oak Lawn, *Ill.* **22** 41 42N 87 44W
Oak Park, *Ill.* **22** 41 52N 87 46W
Oak Park, *Mich.* **26** 42 27N 83 11W
Oak Ridge, *Tenn.* **11** 36 1N 84 16W
Oak View, *Calif.* **19** 34 24N 119 18W
Oakdale, *Calif.* **18** 37 46N 120 51W
Oakdale, *La.* **13** 30 49N 92 40W
Oakes, *N. Dak.* **12** 46 8N 98 6W
Oakesdale, *Wash.* **16** 47 8N 117 15W
Oakhurst, *Calif.* **18** 37 19N 119 40W
Oakland, *Calif.* **28** 37 48N 122 17W
Oakland, *N.J.* **20** 41 2N 74 13W
Oakland, *Oreg.* **16** 43 25N 123 18W
Oakland City, *Ind.* **15** 38 20N 87 21W
Oakland Pontiac Airport, *Mich.* **26** 42 40N 83 24W
Oakley, *Idaho* **16** 42 15N 113 53W
Oakley, *Kans.* **12** 39 8N 100 51W
Oakmont, *Pa.* **27** 40 31N 79 50W
Oakridge, *Oreg.* **16** 43 45N 122 28W
Oasis, *Calif.* **19** 33 28N 116 6W
Oasis, *Nev.* **18** 37 29N 117 55W
Oatman, *Ariz.* **19** 35 1N 114 19W
Oberlin, *Kans.* **12** 39 49N 100 32W
Oberlin, *La.* **13** 30 37N 92 46W
Ocala, *Fla.* **11** 29 11N 82 8W
Ocanomowoc, *Wis.* **12** 43 7N 88 30W
Ocate, *N. Mex.* **13** 36 11N 105 3W
Ocean City, *N.J.* **10** 39 17N 74 35W
Ocean Park, *Wash.* **18** 46 30N 124 3W
Oceano, *Calif.* **19** 35 6N 120 37W
Oceanside, *Calif.* **19** 33 12N 117 23W
Oceanside, *N.Y.* **21** 40 38N 73 37W
Ocilla, *Ga.* **11** 31 36N 83 15W
Ocmulgee →, *Ga.* **11** 31 58N 82 33W
Oconee →, *Ga.* **11** 31 58N 82 33W
Oconto, *Wis.* **10** 44 53N 87 52W
Oconto Falls, *Wis.* **10** 44 52N 88 9W
Octave, *Ariz.* **17** 34 10N 112 43W
Odessa, *Tex.* **13** 31 52N 102 23W
Odessa, *Wash.* **16** 47 20N 118 41W
O'Donnell, *Tex.* **13** 32 58N 101 50W
Oelrichs, *S. Dak.* **12** 43 11N 103 14W
Oelwein, *Iowa* **12** 42 41N 91 55W
Ofu, *Amer. Samoa* **31** 14 11S 169 41W
Ogallala, *Nebr.* **12** 41 8N 101 43W
Ogden, *Iowa* **14** 42 2N 94 2W
Ogden, *Utah* **16** 41 13N 111 58W
Ogdensburg, *N.Y.* **9** 44 42N 75 30W
Ogeechee →, *Ga.* **11** 31 50N 81 3W
Ohio □ **10** 40 15N 82 45W
Ohio →, *Ohio* **10** 36 59N 89 8W
Oil City, *Pa.* **8** 41 26N 79 42W
Oildale, *Calif.* **19** 35 25N 119 1W
Ojai, *Calif.* **19** 34 27N 119 15W
Okanogan, *Wash.* **16** 48 22N 119 35W
Okanogan →, *Wash.* .. **16** 48 6N 119 44W
Okeechobee, *Fla.* **11** 27 15N 80 50W
Okeechobee, L., *Fla.* **11** 27 0N 80 50W
Okefenokee Swamp, *Ga.* **11** 30 40N 82 20W
Oklahoma □ **13** 35 20N 97 30W
Oklahoma City, *Okla.* .. **13** 35 30N 97 30W
Okmulgee, *Okla.* **13** 35 37N 95 58W
Okolona, *Miss.* **13** 34 0N 88 45W
Ola, *Ark.* **13** 35 2N 93 13W
Olancha, *Calif.* **19** 36 17N 118 1W
Olancha Pk., *Calif.* **19** 36 15N 118 7W
Olathe, *Kans.* **12** 38 53N 94 49W
Old Baldy Pk. = San Antonio, Mt., *Calif.* .. **19** 34 17N 117 38W
Old Dale, *Calif.* **19** 34 8N 115 47W
Old Forge, *N.Y.* **9** 43 43N 74 58W
Old Forge, *Pa.* **9** 41 22N 75 45W
Old Harbor, *Alaska* **30** 57 12N 153 18W
Old Town, *Maine* **11** 44 56N 68 39W
Olean, *N.Y.* **8** 42 5N 78 26W
Olema, *Calif.* **18** 38 3N 122 47W
Olney, *Ill.* **15** 38 44N 88 5W
Olney, *Tex.* **13** 33 22N 98 45W
Olosega, *Amer. Samoa* **31** 14 11S 169 38W
Olton, *Tex.* **13** 34 11N 102 8W
Olympia, *Wash.* **18** 47 3N 122 53W
Olympic Mts., *Wash.* **18** 47 55N 123 45W
Olympic Nat. Park, *Wash.* **18** 47 48N 123 30W
Olympus, Mt., *Wash.* .. **18** 47 48N 123 43W
Omaha, *Nebr.* **12** 41 17N 95 58W
Omak, *Wash.* **16** 48 25N 119 31W
Onaga, *Kans.* **12** 39 29N 96 10W
Onalaska, *Wis.* **12** 43 53N 91 14W
Onamia, *Minn.* **12** 46 4N 93 40W
Onancock, *Va.* **10** 37 43N 75 45W
Onawa, *Iowa* **12** 42 2N 96 6W
Onaway, *Mich.* **10** 45 21N 84 14W
Oneida, *N.Y.* **9** 43 6N 75 39W
Oneida L., *N.Y.* **9** 43 12N 75 54W
O'Neill, *Nebr.* **12** 42 27N 98 39W
Oneonta, *Ala.* **11** 33 57N 86 28W
Oneonta, *N.Y.* **9** 42 27N 75 4W
Onida, *S. Dak.* **12** 44 42N 100 4W
Onslow B., *N.C.* **11** 34 20N 77 15W
Ontario, *Calif.* **19** 34 4N 117 39W
Ontario, *Oreg.* **16** 44 2N 116 58W
Ontario, L., *N. Amer.* **8** 43 20N 78 0W
Ontonagon, *Mich.* **12** 46 52N 89 19W
Onyx, *Calif.* **19** 35 41N 118 14W
Ookala, *Hawaii* **30** 20 1N 155 17W
Opelousas, *La.* **13** 30 32N 92 5W
Opheim, *Mont.* **16** 48 51N 106 24W
Ophir, *Alaska* **30** 63 10N 156 31W
Opp, *Ala.* **11** 31 17N 86 16W
Oracle, *Ariz.* **17** 32 37N 110 46W
Orange, *Calif.* **19** 33 47N 117 51W
Orange, *Mass.* **9** 42 35N 72 19W
Orange, *N.J.* **20** 40 46N 74 13W
Orange, *Tex.* **13** 30 6N 93 44W
Orange, *Va.* **10** 38 15N 78 7W
Orange Cove, *Calif.* **18** 36 38N 119 19W
Orange Grove, *Tex.* **13** 27 58N 97 56W
Orangeburg, *S.C.* **11** 33 30N 80 52W
Orcutt, *Calif.* **19** 34 52N 120 27W
Orderville, *Utah* **17** 37 17N 112 38W
Ordway, *Colo.* **12** 38 13N 103 46W
Oregon, *Ill.* **14** 42 1N 89 20W
Oregon □ **16** 44 0N 121 0W
Oregon City, *Oreg.* **18** 45 21N 122 36W
Orem, *Utah* **16** 40 19N 111 42W
Orinda, *Calif.* **28** 37 52N 122 10W
Orland, *Calif.* **18** 39 45N 122 12W
Orland Park, *Ill.* **22** 41 37N 87 52W
Orlando, *Fla.* **11** 28 33N 81 23W
Ormond Beach, *Fla.* **11** 29 17N 81 3W
Oro Grande, *Calif.* **19** 34 36N 117 20W
Orogrande, *N. Mex.* **17** 32 24N 106 5W
Oroville, *Calif.* **18** 39 31N 121 33W
Oroville, *Wash.* **16** 48 56N 119 26W
Osage, *Iowa* **12** 43 17N 92 49W
Osage, *Wyo.* **12** 43 59N 104 25W
Osage →, *Mo.* **14** 38 35N 91 57W
Osage City, *Kans.* **12** 38 38N 95 50W
Osawatomie, *Kans.* **12** 38 31N 94 57W
Osborne, *Kans.* **12** 39 26N 98 42W
Osceola, *Ark.* **13** 35 42N 89 58W
Osceola, *Iowa* **14** 41 2N 93 46W
Oscoda, *Mich.* **8** 44 26N 83 20W
Oshkosh, *Nebr.* **12** 41 24N 102 21W
Oshkosh, *Wis.* **12** 44 1N 88 33W
Oskaloosa, *Iowa* **14** 41 18N 92 39W
Ossabaw I., *Ga.* **11** 31 50N 81 5W
Ossining, *N.Y.* **9** 41 10N 73 55W
Oswego, *N.Y.* **9** 43 27N 76 31W
Othello, *Wash.* **16** 46 50N 119 10W
Otis, *Colo.* **12** 40 9N 102 58W
Ottawa, *Ill.* **15** 41 21N 88 51W
Ottawa, *Kans.* **12** 38 37N 95 16W
Ottumwa, *Iowa* **14** 41 1N 92 25W
Ouachita →, *La.* **13** 31 38N 91 49W
Ouachita, L., *Ark.* **13** 34 34N 93 12W
Ouachita Mts., *Ark.* **13** 34 40N 94 25W
Ouray, *Colo.* **17** 38 1N 107 40W
Outlook, *Mont.* **12** 48 53N 104 47W
Overlea, *Md.* **25** 39 21N 76 33W
Overton, *Nev.* **19** 36 33N 114 27W
Ovid, *Colo.* **12** 40 58N 102 23W
Owatonna, *Minn.* **12** 44 5N 93 14W
Owego, *N.Y.* **9** 42 6N 76 16W
Owens →, *Calif.* **18** 36 32N 117 59W
Owens L., *Calif.* **19** 36 26N 117 57W
Owensboro, *Ky.* **15** 37 46N 87 7W
Owensville, *Mo.* **14** 38 21N 91 30W
Owings Mills, *Md.* **25** 39 25N 76 48W
Owosso, *Mich.* **15** 43 0N 84 10W
Owyhee, *Nev.* **16** 41 57N 116 6W
Owyhee →, *Oreg.* **16** 43 49N 117 2W
Owyhee, L., *Oreg.* **16** 43 38N 117 14W
Oxford, *Miss.* **13** 34 22N 89 31W
Oxford, *N.C.* **11** 36 19N 78 35W
Oxford, *Ohio* **15** 39 31N 84 45W
Oxnard, *Calif.* **19** 34 12N 119 11W
Oyster Bay, *N.Y.* **21** 40 52N 73 31W
Ozark, *Ala.* **11** 31 28N 85 39W
Ozark, *Ark.* **13** 35 29N 93 50W
Ozark, *Mo.* **13** 37 1N 93 12W
Ozark Plateau, *Mo.* **13** 37 20N 91 40W
Ozarks, L. of the, *Mo.* .. **14** 38 12N 92 38W
Ozona, *Tex.* **13** 30 43N 101 12W

P

Paauilo, *Hawaii* **30** 20 2N 155 22W
Pacific Grove, *Calif.* **18** 36 38N 121 56W
Pacifica, *Calif.* **28** 37 38N 122 29W
Padre I., *Tex.* **13** 27 10N 97 25W
Paducah, *Ky.* **10** 37 5N 88 37W
Paducah, *Tex.* **13** 34 1N 100 18W
Page, *Ariz.* **17** 36 57N 111 27W
Page, *N. Dak.* **12** 47 10N 97 34W
Pago Pago, *Amer. Samoa* **31** 14 16S 170 43W
Pagosa Springs, *Colo.* . **17** 37 16N 107 1W
Pahala, *Hawaii* **30** 19 12N 155 29W
Pahoa, *Hawaii* **30** 19 30N 154 57W
Pahokee, *Fla.* **11** 26 50N 80 40W
Pahrump, *Nev.* **19** 36 12N 115 59W
Pahute Mesa, *Nev.* **18** 37 20N 116 45W
Paia, *Hawaii* **30** 20 54N 156 22W
Paicines, *Calif.* **18** 36 44N 121 17W
Pailolo Channel, *Hawaii* **30** 21 0N 156 40W
Painesville, *Ohio* **8** 41 43N 81 15W
Paint Rock, *Tex.* **13** 31 31N 99 55W
Painted Desert, *Ariz.* **17** 36 0N 111 0W
Paintsville, *Ky.* **10** 37 49N 82 48W
Paisley, *Oreg.* **16** 42 42N 120 32W
Pala, *Calif.* **19** 33 22N 117 5W
Palacios, *Tex.* **13** 28 42N 96 13W
Palatka, *Fla.* **11** 29 39N 81 38W
Palau ■, *Pac. Oc.* .... **31** 7 30N 134 30 E
Palermo, *Calif.* **16** 39 26N 121 33W
Palestine, *Tex.* **13** 31 46N 95 38W
Palisade, *Nebr.* **12** 40 21N 101 7W
Palisades, *N.Y.* **20** 41 1N 73 55W
Palm Beach, *Fla.* **11** 26 43N 80 2W
Palm Desert, *Calif.* **19** 33 43N 116 22W
Palm Springs, *Calif.* **19** 33 50N 116 33W
Palmdale, *Calif.* **19** 34 35N 118 7W
Palmer, *Alaska* **30** 61 36N 149 7W
Palmer Lake, *Colo.* **12** 39 7N 104 55W
Palmerton, *Pa.* **9** 40 48N 75 37W
Palmetto, *Fla.* **11** 27 31N 82 34W
Palmyra, *Mo.* **14** 39 48N 91 32W
Palmyra, *N.J.* **24** 40 0N 75 1W
Palo Alto, *Calif.* **28** 37 26N 122 8W
Palos Heights, *Ill.* **22** 41 40N 87 47W
Palos Hills Forest, *Ill.* .. **22** 41 40N 87 52W
Palos Verdes, *Calif.* **19** 33 48N 118 23W
Palos Verdes, Pt., *Calif.* . **19** 33 43N 118 26W
Palouse, *Wash.* **16** 46 55N 117 4W
Pamlico →, *N.C.* **11** 35 20N 76 28W
Pamlico Sd., *N.C.* **11** 35 20N 76 0W
Pampa, *Tex.* **13** 35 32N 100 58W
Pana, *Ill.* **14** 39 23N 89 5W
Panaca, *Nev.* **17** 37 47N 114 23W
Panama City, *Fla.* **11** 30 10N 85 40W
Panamint Range, *Calif.* . **19** 36 20N 117 20W
Panamint Springs, *Calif.* **19** 36 20N 117 28W
Pancake Range, *Nev.* .. **17** 38 30N 115 50W
Panguitch, *Utah* **17** 37 50N 112 26W
Panhandle, *Tex.* **13** 35 21N 101 23W
Paola, *Kans.* **12** 38 35N 94 53W
Paoli, *Pa.* **24** 40 2N 75 28W
Paonia, *Colo.* **17** 38 52N 107 36W
Papaikou, *Hawaii* **30** 19 47N 155 6W
Paradise, *Mont.* **16** 47 23N 114 48W
Paradise Valley, *Nev.* .. **16** 41 30N 117 32W
Paragould, *Ark.* **13** 36 3N 90 29W
Paramus, *N.J.* **20** 40 56N 74 2W
Paris, *Idaho* **16** 42 14N 111 24W
Paris, *Ky.* **15** 38 13N 84 15W
Paris, *Tenn.* **11** 36 18N 88 19W
Paris, *Tex.* **13** 33 40N 95 33W
Parish, *N.Y.* **9** 43 25N 76 8W
Park City, *Utah* **16** 40 39N 111 30W
Park Falls, *Wis.* **12** 45 56N 90 27W
Park Range, *Colo.* **16** 40 0N 106 30W
Park Rapids, *Minn.* **12** 46 55N 95 4W
Park Ridge, *Ill.* **22** 42 0N 87 50W
Park Ridge, *N.J.* **20** 41 2N 74 2W
Park River, *N. Dak.* **12** 48 24N 97 45W
Parker, *Ariz.* **19** 34 9N 114 17W
Parker, *S. Dak.* **12** 43 24N 97 8W
Parker Dam, *Ariz.* **19** 34 18N 114 8W
Parkersburg, *W. Va.* **10** 39 16N 81 34W
Parkfield, *Calif.* **18** 35 54N 120 26W
Parkston, *S. Dak.* **12** 43 24N 97 59W
Parkville, *Md.* **25** 39 23N 76 34W
Parma, *Idaho* **16** 43 47N 116 57W
Parma, *Ohio* **27** 41 24N 81 43W
Parma Heights, *Ohio* ... **27** 41 23N 81 45W
Parowan, *Utah* **17** 37 51N 112 50W
Parris I., *S.C.* **11** 32 20N 80 41W
Parshall, *N. Dak.* **12** 47 57N 102 8W
Parsons, *Kans.* **13** 37 20N 95 16W
Pasadena, *Calif.* **29** 34 9N 118 8W
Pasadena, *Tex.* **13** 29 43N 95 13W
Pascagoula, *Miss.* **13** 30 21N 88 33W
Pascagoula →, *Miss.* .. **13** 30 23N 88 37W
Pasco, *Wash.* **16** 46 14N 119 6W
Paso Robles, *Calif.* **17** 35 38N 120 41W
Passaic, *N.J.* **20** 40 51N 74 9W
Patagonia, *Ariz.* **17** 31 33N 110 45W
Patchogue, *N.Y.* **9** 40 46N 73 1W
Pateros, *Wash.* **16** 48 3N 119 54W
Paterson, *N.J.* **20** 40 54N 74 10W
Pathfinder Reservoir, *Wyo.* **16** 42 28N 106 51W
Patten, *Maine* **11** 46 0N 68 38W
Patterson, *Calif.* **18** 37 28N 121 8W
Patterson, *La.* **13** 29 42N 91 18W
Patterson, Mt., *Calif.* ... **18** 38 29N 119 20W
Paullina, *Iowa* **12** 42 59N 95 41W
Pauls Valley, *Okla.* **13** 34 44N 97 13W
Paulsboro, *N.J.* **24** 39 49N 75 14W
Pauma Valley, *Calif.* **19** 33 16N 116 58W
Pavlof Is., *Alaska* **30** 55 30N 161 30W
Pawhuska, *Okla.* **13** 36 40N 96 20W
Pawling, *N.Y.* **9** 41 34N 73 36W
Pawnee, *Okla.* **13** 36 20N 96 48W
Pawnee City, *Nebr.* **12** 40 7N 96 9W
Pawtucket, *R.I.* **9** 41 53N 71 23W
Paxton, *Ill.* **15** 40 27N 88 6W
Paxton, *Nebr.* **12** 41 7N 101 21W
Payette, *Idaho* **16** 44 5N 116 56W
Paynesville, *Minn.* **12** 45 23N 94 43W
Payson, *Ariz.* **17** 34 14N 111 20W
Payson, *Utah* **16** 40 3N 111 44W
Pe Ell, *Wash.* **18** 46 34N 123 18W
Peabody, *Mass.* **23** 42 32N 70 57W
Peach Springs, *Ariz.* **17** 35 32N 113 25W
Peale, Mt., *Utah* **17** 38 26N 109 14W
Pearblossom, *Calif.* **19** 34 30N 117 55W
Pearl →, *Miss.* **13** 30 11N 89 32W
Pearl City, *Hawaii* **30** 21 24N 157 59W
Pearl Harbor, *Hawaii* ... **30** 21 21N 157 57W
Pearsall, *Tex.* **13** 28 54N 99 6W
Pease →, *Tex.* **13** 34 12N 99 2W
Pebble Beach, *Calif.* **18** 36 34N 121 57W
Pecos, *Tex.* **13** 31 26N 103 30W
Pecos →, *Tex.* **13** 29 42N 101 22W
Pedro Valley, *Calif.* **28** 37 35N 122 28W
Peekskill, *N.Y.* **9** 41 17N 73 55W
Pekin, *Ill.* **14** 40 35N 89 40W
Pelham, *Ga.* **11** 31 8N 84 9W
Pelham, *N.Y.* **21** 40 54N 73 46W
Pelican, *Alaska* **30** 57 58N 136 14W
Pella, *Iowa* **14** 41 25N 92 55W
Pembina, *N. Dak.* **12** 48 58N 97 15W
Pembine, *Wis.* **10** 45 38N 87 59W
Pembroke, *Ga.* **11** 32 8N 81 37W
Pend Oreille →, *Wash.* **16** 49 4N 117 37W
Pend Oreille L., *Idaho* .. **16** 48 10N 116 21W
Pendleton, *Calif.* **19** 33 16N 117 23W
Pendleton, *Oreg.* **16** 45 40N 118 47W
Penn Hills, *Pa.* **27** 40 27N 79 50W
Penn Yan, *N.Y.* **8** 42 40N 77 3W
Pennsauken, *N.J.* **24** 39 57N 75 5W
Pennsylvania □ **10** 40 45N 77 30W
Pensacola, *Fla.* **11** 30 25N 87 13W
Peoria, *Ariz.* **17** 33 35N 112 14W
Peoria, *Ill.* **14** 40 42N 89 36W
Perham, *Minn.* **12** 46 36N 95 34W
Perris, *Calif.* **19** 33 47N 117 14W
Perry, *Fla.* **11** 30 7N 83 35W
Perry, *Ga.* **11** 32 28N 83 44W
Perry, *Iowa* **14** 41 51N 94 6W

Perry

Perry, *Maine* **11** 44 58N 67 5W
Perry, *Okla.* **13** 36 17N 97 14W
Perry Hall, *Md.* **25** 39 24N 76 28W
Perrysville, *Pa.* **27** 40 32N 80 1W
Perryton, *Tex.* **13** 36 24N 100 48W
Perryville, *Alaska* **30** 55 55N 159 9W
Perryville, *Mo.* **14** 37 43N 89 52W
Perth Amboy, *N.J.* **20** 40 30N 74 16W
Peru, *Ill.* **14** 41 20N 89 8W
Peru, *Ind.* **15** 40 45N 86 4W
Peshtigo, *Mich.* **10** 45 4N 87 46W
Petaluma, *Calif.* **18** 38 14N 122 39W
Peterborough, *N.H.* **9** 42 53N 71 57W
Petersburg, *Alaska* **30** 56 48N 132 58W
Petersburg, *Ind.* **15** 38 30N 87 17W
Petersburg, *Va.* **10** 37 14N 77 24W
Petersburg, *W. Va.* **10** 39 1N 79 5W
Petit Bois I., *Miss.* **11** 30 12N 88 26W
Petoskey, *Mich.* **10** 45 22N 84 57W
Phelps, *N.Y.* **8** 42 58N 77 3W
Phelps, *Wis.* **12** 46 4N 89 5W
Phenix City, *Ala.* **11** 32 28N 85 0W
Philadelphia, *Miss.* **13** 32 46N 89 7W
Philadelphia, *Pa.* **24** 39 58N 75 10W
Philadelphia Airport, *Pa.* **24** 40 4N 75 1W
Philadelphia International Airport, *Pa.* **24** 39 52N 75 14W
Philip, *S. Dak.* **12** 44 2N 101 40W
Philip Smith Mts., *Alaska* **30** 68 0N 146 0W
Philipsburg, *Mont.* **16** 46 20N 113 18W
Phillips, *Tex.* **13** 35 42N 101 22W
Phillips, *Wis.* **12** 45 42N 90 24W
Phillipsburg, *Kans.* **12** 39 45N 99 19W
Phillipsburg, *N.J.* **9** 40 42N 75 12W
Philmont, *N.Y.* **9** 42 15N 73 39W
Philomath, *Oreg.* **16** 44 32N 123 22W
Phoenix, *Ariz.* **17** 33 27N 112 4W
Phoenix, *N.Y.* **9** 43 14N 76 18W
Phoenixville, *Pa.* **24** 40 7N 75 31W
Picayune, *Miss.* **13** 30 32N 89 41W
Pico Rivera, *Calif.* **29** 33 59N 118 5W
Piedmont, *Ala.* **11** 33 55N 85 37W
Piedmont Plateau, *S.C.* **11** 34 0N 81 30W
Pierce, *Idaho* **16** 46 30N 115 48W
Pierre, *S. Dak.* **12** 44 22N 100 21W
Pigeon, *Mich.* **10** 43 50N 83 16W
Piggott, *Ark.* **13** 36 23N 90 11W
Pikes Peak, *Colo.* **12** 38 50N 105 3W
Pikesville, *Md.* **25** 39 22N 76 41W
Pikeville, *Ky.* **10** 37 29N 82 31W
Pilot Point, *Tex.* **13** 33 24N 96 58W
Pilot Rock, *Oreg.* **16** 45 29N 118 50W
Pima, *Ariz.* **17** 32 54N 109 50W
Pimmit Hills, *Va.* **25** 38 54N 77 12W
Pinckneyville, *Ill.* **14** 38 5N 89 23W
Pine, *Ariz.* **17** 34 23N 111 27W
Pine Bluff, *Ark.* **13** 34 13N 92 1W
Pine City, *Minn.* **12** 45 50N 92 59W
Pine Flat L., *Calif.* **18** 36 50N 119 20W
Pine Ridge, *S. Dak.* **12** 43 2N 102 33W
Pine River, *Minn.* **12** 46 43N 94 24W
Pine Valley, *Calif.* **19** 32 50N 116 32W
Pinecrest, *Calif.* **18** 38 12N 120 1W
Pinedale, *Calif.* **18** 36 50N 119 48W
Pinehurst, *Mass.* **23** 42 31N 71 12W
Pinetop, *Ariz.* **17** 34 8N 109 56W
Pinetree, *Wyo.* **16** 43 42N 105 52W
Pineville, *Ky.* **11** 36 46N 83 42W
Pineville, *La.* **13** 31 19N 92 26W
Pinnacles, *Calif.* **18** 36 33N 121 19W
Pinon Hills, *Calif.* **19** 34 26N 117 39W
Pinos, Mt., *Calif.* **19** 34 49N 119 8W
Pinos Pt., *Calif.* **17** 36 38N 121 57W
Pioche, *Nev.* **17** 37 56N 114 27W
Pipestone, *Minn.* **12** 44 0N 96 19W
Piqua, *Ohio* **15** 40 9N 84 15W
Piru, *Calif.* **19** 34 25N 118 48W
Piscataway, *N.J.* **20** 40 34N 74 27W
Pismo Beach, *Calif.* **19** 35 9N 120 38W
Pittsburg, *Kans.* **13** 37 25N 94 42W
Pittsburg, *Tex.* **13** 33 0N 94 59W
Pittsburgh, *Pa.* **27** 40 26N 79 59W
Pittsfield, *Ill.* **14** 39 36N 90 49W
Pittsfield, *Mass.* **9** 42 27N 73 15W
Pittsfield, *N.H.* **9** 43 18N 71 20W
Pittston, *Pa.* **9** 41 19N 75 47W
Pixley, *Calif.* **18** 35 58N 119 18W
Placerville, *Calif.* **18** 38 44N 120 48W
Plain Dealing, *La.* **13** 32 54N 93 42W
Plainfield, *N.J.* **20** 40 36N 74 24W
Plainfield, *N.J.* **9** 40 37N 74 25W
Plains, *Kans.* **13** 37 16N 100 35W
Plains, *Mont.* **16** 47 28N 114 53W
Plains, *Tex.* **13** 33 11N 102 50W
Plainview, *Nebr.* **12** 42 21N 97 47W
Plainview, *Tex.* **13** 34 11N 101 43W
Plainville, *Kans.* **12** 39 14N 99 18W
Plainwell, *Mich.* **10** 42 27N 85 38W
Planada, *Calif.* **18** 37 16N 120 19W
Plankinton, *S. Dak.* **12** 43 43N 98 29W
Plano, *Tex.* **13** 33 1N 96 42W
Plant City, *Fla.* **11** 28 1N 82 7W
Plaquemine, *La.* **13** 30 17N 91 14W
Plateau du Coteau du Missouri, *N. Dak.* **12** 47 9N 101 5W
Platinum, *Alaska* **30** 59 1N 161 49W
Platte, *S. Dak.* **12** 43 23N 98 51W
Platte →, *Mo.* **14** 39 16N 94 50W
Platteville, *Colo.* **12** 40 13N 104 49W
Plattsburgh, *N.Y.* **9** 44 42N 73 28W
Plattsmouth, *Nebr.* **12** 41 1N 95 53W
Pleasant Hill, *Calif.* **28** 37 56N 122 4W
Pleasant Hill, *Mo.* **14** 38 47N 94 16W
Pleasant Hills, *Pa.* **27** 40 20N 79 58W
Pleasanton, *Tex.* **13** 28 58N 98 29W
Pleasantville, *N.J.* **10** 39 24N 74 32W
Plentywood, *Mont.* **12** 48 47N 104 34W
Plum I., *N.Y.* **9** 41 11N 72 12W
Plummer, *Idaho* **16** 47 20N 116 53W
Plymouth, *Calif.* **18** 38 29N 120 51W
Plymouth, *Ind.* **15** 41 21N 86 19W
Plymouth, *Mass.* **9** 41 57N 70 40W
Plymouth, *N.C.* **11** 35 52N 76 43W
Plymouth, *Pa.* **9** 41 14N 75 57W
Plymouth, *Wis.* **10** 43 45N 87 59W
Plymouth Meeting, *Pa.* **24** 40 6N 75 17W
Pocahontas, *Ark.* **13** 36 16N 90 58W
Pocahontas, *Iowa* **14** 42 44N 94 40W
Pocatello, *Idaho* **16** 42 52N 112 27W
Pocomoke City, *Md.* **10** 38 5N 75 34W
Pohnpei, *Pac. Oc.* **31** 6 55N 158 10 E
Point Baker, *Alaska* **30** 56 21N 133 37W
Point Hope, *Alaska* **30** 68 21N 166 47W
Point Lay, *Alaska* **30** 69 46N 163 3W
Point Pleasant, *W. Va.* **10** 38 51N 82 8W
Pointe-à-la Hache, *La.* **13** 29 35N 89 55W
Pojoaque Valley, *N. Mex.* **17** 35 54N 106 1W
Polacca, *Ariz.* **17** 35 50N 110 23W
Pollock, *S. Dak.* **12** 45 55N 100 17W
Polo, *Ill.* **14** 41 59N 89 35W
Polson, *Mont.* **16** 47 41N 114 9W
Pomeroy, *Ohio* **10** 39 2N 82 2W
Pomeroy, *Wash.* **16** 46 28N 117 36W
Pomona, *Calif.* **19** 34 4N 117 45W
Pompano Beach, *Fla.* **11** 26 14N 80 8W
Pompeys Pillar, *Mont.* **16** 45 59N 107 57W
Pompton Plains, *N.J.* **20** 40 58N 74 18W
Ponca, *Nebr.* **12** 42 34N 96 43W
Ponca City, *Okla.* **13** 36 42N 97 5W
Ponce, *Puerto Rico* **30** 18 1N 66 37W
Ponchatoula, *La.* **13** 30 26N 90 26W
Pond, *Calif.* **19** 35 43N 119 20W
Pontchartrain L., *La.* **13** 30 5N 90 5W
Pontiac, *Ill.* **15** 40 53N 88 38W
Pontiac, *Mich.* **26** 42 38N 83 17W
Poorman, *Alaska* **30** 64 5N 155 48W
Poplar, *Mont.* **12** 48 7N 105 12W
Poplar Bluff, *Mo.* **13** 36 46N 90 24W
Poplarville, *Miss.* **13** 30 51N 89 32W
Porcupine →, *Alaska* **30** 66 34N 145 19W
Port Alexander, *Alaska* **30** 56 15N 134 38W
Port Allegany, *Pa.* **8** 41 48N 78 17W
Port Allen, *La.* **13** 30 27N 91 12W
Port Angeles, *Wash.* **18** 48 7N 123 27W
Port Aransas, *Tex.* **13** 27 50N 97 4W
Port Arthur, *Tex.* **13** 29 54N 93 56W
Port Austin, *Mich.* **8** 44 3N 83 1W
Port Chester, *N.Y.* **21** 41 0N 73 40W
Port Clinton, *Ohio* **15** 41 31N 82 56W
Port Gibson, *Miss.* **13** 31 58N 90 59W
Port Heiden, *Alaska* **30** 56 55N 158 41W
Port Henry, *N.Y.* **9** 44 3N 73 28W
Port Hueneme, *Calif.* **19** 34 7N 119 12W
Port Huron, *Mich.* **10** 42 58N 82 26W
Port Isabel, *Tex.* **13** 26 5N 97 12W
Port Jefferson, *N.Y.* **9** 40 57N 73 3W
Port Jervis, *N.Y.* **9** 41 22N 74 41W
Port Lavaca, *Tex.* **13** 28 37N 96 38W
Port O'Connor, *Tex.* **13** 28 26N 96 24W
Port Orchard, *Wash.* **18** 47 32N 122 38W
Port Orford, *Oreg.* **16** 42 45N 124 30W
Port Richmond, *N.Y.* **20** 40 38N 74 7W
Port St. Joe, *Fla.* **11** 29 49N 85 18W
Port Sanilac, *Mich.* **8** 43 26N 82 33W
Port Townsend, *Wash.* **18** 48 7N 122 45W
Port Washington, *N.Y.* **21** 40 49N 73 41W
Port Washington, *Wis.* **10** 43 23N 87 53W
Portage, *Wis.* **12** 43 33N 89 28W
Portageville, *Mo.* **13** 36 26N 89 42W
Portales, *N. Mex.* **13** 34 11N 103 20W
Porterville, *Calif.* **18** 36 4N 119 1W
Porthill, *Idaho* **16** 48 59N 116 30W
Portland, *Conn.* **9** 41 34N 72 38W
Portland, *Maine* **11** 43 39N 70 16W
Portland, *Mich.* **15** 42 52N 84 54W
Portland, *Oreg.* **18** 45 32N 122 37W
Portola, *Calif.* **18** 39 49N 120 28W
Portsmouth, *N.H.* **9** 43 5N 70 45W
Portsmouth, *Ohio* **10** 38 44N 82 57W
Portsmouth, *R.I.* **9** 41 36N 71 15W
Portsmouth, *Va.* **10** 36 50N 76 18W
Post, *Tex.* **13** 33 12N 101 23W
Post Falls, *Idaho* **16** 47 43N 116 57W
Poteau, *Okla.* **13** 35 3N 94 37W
Poteet, *Tex.* **13** 29 2N 98 35W
Potomac →, *Md.* **10** 38 0N 76 23W
Potsdam, *N.Y.* **9** 44 40N 74 59W
Potter, *Nebr.* **12** 41 13N 103 19W
Pottstown, *Pa.* **9** 40 15N 75 39W
Pottsville, *Pa.* **9** 40 41N 76 12W
Poughkeepsie, *N.Y.* **9** 41 42N 73 56W
Poulsbo, *Wash.* **18** 47 44N 122 39W
Poway, *Calif.* **19** 32 58N 117 2W
Powder →, *Mont.* **12** 46 45N 105 26W
Powder River, *Wyo.* **16** 43 2N 106 59W
Powell, *Wyo.* **16** 44 45N 108 46W
Powell L., *Utah* **17** 36 57N 111 29W
Powers, *Mich.* **10** 45 41N 87 32W
Powers, *Oreg.* **16** 42 53N 124 4W
Powers Lake, *N. Dak.* **12** 48 34N 102 39W
Pozo, *Calif.* **19** 35 20N 120 24W
Prairie →, *Tex.* **13** 34 30N 99 23W
Prairie City, *Oreg.* **16** 44 28N 118 43W
Prairie du Chien, *Wis.* **14** 43 3N 91 9W
Pratt, *Kans.* **13** 37 39N 98 44W
Prattville, *Ala.* **11** 32 28N 86 29W
Premont, *Tex.* **13** 27 22N 98 7W
Prentice, *Wis.* **12** 45 33N 90 17W
Prescott, *Ariz.* **17** 34 33N 112 28W
Prescott, *Ark.* **13** 33 48N 93 23W
Presho, *S. Dak.* **12** 43 54N 100 3W
Presidio, *Tex.* **13** 29 34N 104 22W
Presque Isle, *Maine* **11** 46 41N 68 1W
Preston, *Idaho* **16** 42 6N 111 53W
Preston, *Minn.* **12** 43 40N 92 5W
Preston, *Nev.* **16** 38 55N 115 4W
Price, *Utah* **16** 39 36N 110 49W
Prichard, *Ala.* **11** 30 44N 88 5W
Priest →, *Idaho* **16** 48 12N 116 54W
Priest L., *Idaho* **16** 48 35N 116 52W
Priest Valley, *Calif.* **18** 36 10N 120 39W
Prince of Wales, C., *Alaska* **30** 65 36N 168 5W
Prince of Wales I., *Alaska* **30** 55 47N 132 50W
Prince William Sd., *Alaska* **30** 60 40N 147 0W
Princeton, *Ill.* **14** 41 23N 89 28W
Princeton, *Ind.* **15** 38 21N 87 34W
Princeton, *Ky.* **10** 37 7N 87 53W
Princeton, *Mo.* **14** 40 24N 93 35W
Princeton, *N.J.* **9** 40 21N 74 39W
Princeton, *W. Va.* **10** 37 22N 81 6W
Prineville, *Oreg.* **16** 44 18N 120 51W
Prospect Heights, *Ill.* **22** 42 6N 87 54W
Prosser, *Wash.* **16** 46 12N 119 46W
Protection, *Kans.* **13** 37 12N 99 29W
Providence, *Ky.* **10** 37 24N 87 46W
Providence, *R.I.* **9** 41 49N 71 24W
Providence Mts., *Calif.* **17** 35 10N 115 15W
Provo, *Utah* **16** 40 14N 111 39W
Prudhoe Bay, *Alaska* **30** 70 18N 148 22W
Pryor, *Okla.* **13** 36 19N 95 19W
Pueblo, *Colo.* **12** 38 16N 104 37W
Puerco →, *N. Mex.* **17** 34 22N 107 50W
Puerto Rico ■, *W. Indies* **30** 18 15N 66 45W
Puget Sound, *Wash.* **16** 47 50N 122 30W
Pukoo, *Hawaii* **30** 21 4N 156 48W
Pulaski, *N.Y.* **9** 43 34N 76 8W
Pulaski, *Tenn.* **11** 35 12N 87 2W
Pulaski, *Va.* **10** 37 3N 80 47W
Pullman, *Wash.* **16** 46 44N 117 10W
Punta Gorda, *Fla.* **11** 26 56N 82 3W
Punxsatawney, *Pa.* **8** 40 57N 78 59W
Purcell, *Okla.* **13** 35 1N 97 22W
Putnam, *Conn.* **9** 41 55N 71 55W
Puyallup, *Wash.* **18** 47 12N 122 18W
Pyote, *Tex.* **13** 31 32N 103 8W
Pyramid L., *Nev.* **16** 40 1N 119 35W
Pyramid Pk., *Calif.* **19** 36 25N 116 37W

Q

Quakertown, *Pa.* **9** 40 26N 75 21W
Quanah, *Tex.* **13** 34 18N 99 44W
Quartzsite, *Ariz.* **19** 33 40N 114 13W
Queens, *N.Y.* **21** 40 42N 73 50W
Quemado, *N. Mex.* **17** 34 20N 108 30W
Quemado, *Tex.* **13** 28 58N 100 35W
Questa, *N. Mex.* **17** 36 42N 105 36W
Quincy, *Calif.* **18** 39 56N 120 57W
Quincy, *Fla.* **11** 30 35N 84 34W
Quincy, *Ill.* **12** 39 56N 91 23W
Quincy, *Mass.* **23** 42 14N 71 0W
Quincy, *Wash.* **16** 47 22N 119 56W
Quinhagak, *Alaska* **30** 59 45N 161 54W
Quitman, *Ga.* **11** 30 47N 83 34W
Quitman, *Miss.* **11** 32 2N 88 44W
Quitman, *Tex.* **13** 32 48N 95 27W

R

Racine, *Wis.* **15** 42 41N 87 51W
Radford, *Va.* **10** 37 8N 80 34W
Rahway, *N.J.* **20** 40 36N 74 17W
Rainier, *Wash.* **18** 46 53N 122 41W
Rainier, Mt., *Wash.* **18** 46 52N 121 46W
Raleigh, *N.C.* **11** 35 47N 78 39W
Raleigh B., *N.C.* **11** 34 50N 76 15W
Ralls, *Tex.* **13** 33 41N 101 24W
Ramona, *Calif.* **19** 33 2N 116 52W
Rampart, *Alaska* **30** 65 30N 150 10W
Ranchester, *Wyo.* **16** 44 54N 107 10W
Randallstown, *Md.* **25** 39 21N 76 46W
Randolph, *Mass.* **9** 42 10N 71 2W
Randolph, *Utah* **16** 41 40N 111 11W
Rangeley, *Maine* **9** 44 58N 70 39W
Rangely, *Colo.* **16** 40 5N 108 48W
Ranger, *Tex.* **13** 32 28N 98 41W
Rankin, *Tex.* **13** 31 13N 101 56W
Rantoul, *Ill.* **15** 40 19N 88 9W
Rapid City, *S. Dak.* **12** 44 5N 103 14W
Rapid River, *Mich.* **10** 45 55N 86 58W
Rat Islands, *Alaska* **30** 52 0N 178 0 E
Raton, *N. Mex.* **13** 36 54N 104 24W
Ravena, *N.Y.* **9** 42 28N 73 49W
Ravenna, *Nebr.* **12** 41 1N 98 55W
Ravenswood, *W. Va.* **10** 38 57N 81 46W
Rawlins, *Wyo.* **16** 41 47N 107 14W
Ray, *N. Dak.* **12** 48 21N 103 10W
Ray Mts., *Alaska* **30** 66 0N 152 0W
Raymond, *Calif.* **18** 37 13N 119 54W
Raymond, *Wash.* **18** 46 41N 123 44W
Raymondville, *Tex.* **13** 26 29N 97 47W
Rayne, *La.* **13** 30 14N 92 16W
Rayville, *La.* **13** 32 29N 91 46W
Reading, *Mass.* **23** 42 31N 71 5W
Reading, *Pa.* **9** 40 20N 75 56W
Red →, *La.* **13** 31 1N 91 45W
Red →, *N. Dak.* **12** 49 0N 97 15W
Red Bank, *N.J.* **9** 40 21N 74 5W
Red Bluff, *Calif.* **16** 40 11N 122 15W
Red Bluff L., *N. Mex.* **13** 31 54N 103 55W
Red Cloud, *Nebr.* **12** 40 5N 98 32W
Red Lake Falls, *Minn.* **12** 47 53N 96 16W
Red Lodge, *Mont.* **16** 45 11N 109 15W
Red Mountain, *Calif.* **19** 35 37N 117 38W
Red Oak, *Iowa* **12** 41 1N 95 14W
Red Rock, L., *Iowa* **14** 41 22N 92 59W
Red Slate Mt., *Calif.* **18** 37 31N 118 52W
Red Wing, *Minn.* **12** 44 34N 92 31W
Redding, *Calif.* **16** 40 35N 122 24W
Redfield, *S. Dak.* **12** 44 53N 98 31W
Redford Township, *Mich.* **26** 42 23N 83 17W
Redlands, *Calif.* **19** 34 4N 117 11W
Redmond, *Oreg.* **16** 44 17N 121 11W
Redwood City, *Calif.* **28** 37 29N 122 13W
Redwood Falls, *Minn.* **12** 44 32N 95 7W
Reed City, *Mich.* **10** 43 53N 85 31W
Reeder, *N. Dak.* **12** 46 7N 102 57W
Reedley, *Calif.* **18** 36 36N 119 27W
Reedsburg, *Wis.* **12** 43 32N 90 0W
Reedsport, *Oreg.* **16** 43 42N 124 6W
Refugio, *Tex.* **13** 28 18N 97 17W
Reidsville, *N.C.* **11** 36 21N 79 40W
Reinbeck, *Iowa* **14** 42 19N 92 36W
Reno, *Nev.* **18** 39 31N 119 48W
Renovo, *Pa.* **8** 41 20N 77 45W
Rensselaer, *Ind.* **15** 40 57N 87 9W
Rensselaer, *N.Y.* **9** 42 38N 73 45W
Renton, *Wash.* **18** 47 29N 122 12W
Republic, *Mich.* **10** 46 25N 87 59W
Republic, *Wash.* **16** 48 39N 118 44W
Republican →, *Kans.* **12** 39 4N 96 48W
Republican City, *Nebr.* **12** 40 6N 99 13W
Reserve, *N. Mex.* **17** 33 43N 108 45W
Reston, *Va.* **25** 38 57N 77 20W
Revere, *Mass.* **23** 42 25N 71 1W
Rex, *Alaska* **30** 64 10N 149 20W
Rexburg, *Idaho* **16** 43 49N 111 47W
Reyes, Pt., *Calif.* **18** 38 0N 123 0W
Rhinelander, *Wis.* **12** 45 38N 89 25W
Rhode Island □ **9** 41 40N 71 30W
Rice Lake, *Wis.* **12** 45 30N 91 44W
Rich Hill, *Mo.* **13** 38 6N 94 22W
Richardton, *N. Dak.* **12** 46 53N 102 19W
Richey, *Mont.* **12** 47 39N 105 4W
Richfield, *Idaho* **16** 43 3N 114 9W
Richfield, *Utah* **17** 38 46N 112 5W
Richland, *Ga.* **11** 32 5N 84 40W
Richland, *Oreg.* **16** 44 46N 117 10W
Richland, *Wash.* **16** 46 17N 119 18W
Richland Center, *Wis.* **12** 43 21N 90 23W
Richlands, *Va.* **10** 37 6N 81 48W
Richmond, *Calif.* **28** 37 56N 122 22W
Richmond, *Ind.* **15** 39 50N 84 53W
Richmond, *Ky.* **15** 37 45N 84 18W
Richmond, *Mo.* **12** 39 17N 93 58W
Richmond, *Tex.* **13** 29 35N 95 46W
Richmond, *Utah* **16** 41 56N 111 48W
Richmond, *Va.* **10** 37 33N 77 27W
Richmond Hill, *N.Y.* **21** 40 41N 73 49W
Richton, *Miss.* **11** 31 16N 88 56W
Richwood, *W. Va.* **10** 38 14N 80 32W
Ridgecrest, *Calif.* **19** 35 38N 117 40W
Ridgeland, *S.C.* **11** 32 29N 80 59W
Ridgewood, *N.J.* **20** 40 59N 74 6W
Ridgewood, *N.Y.* **21** 40 42N 73 52W
Ridgway, *Pa.* **8** 41 25N 78 44W
Rifle, *Colo.* **16** 39 32N 107 47W
Rigby, *Idaho* **16** 43 40N 111 55W
Riggins, *Idaho* **16** 45 25N 116 19W
Riley, *Oreg.* **16** 43 32N 119 28W
Rimrock, *Wash.* **18** 46 38N 121 10W
Ringling, *Mont.* **16** 46 16N 110 49W
Rio Grande →, *Tex.* **13** 25 57N 97 9W
Rio Grande City, *Tex.* **13** 26 23N 98 49W
Rio Vista, *Calif.* **18** 38 10N 121 42W
Ripley, *Tenn.* **13** 35 45N 89 32W
Ripon, *Calif.* **18** 37 44N 121 7W
Ripon, *Wis.* **10** 43 51N 88 50W
Rison, *Ark.* **13** 33 58N 92 11W
Ritzville, *Wash.* **16** 47 8N 118 23W
River Rouge, *Mich.* **26** 42 16N 83 8W
Riverdale, *Calif.* **18** 36 26N 119 52W
Riverdale, *N.Y.* **21** 40 54N 73 54W
Riverhead, *N.Y.* **9** 40 55N 72 40W
Riverside, *Calif.* **19** 33 59N 117 22W

Selma

Riverside, *Ill.* **22** 41 49N 87 48W
Riverside, *N.J.* **24** 40 2N 74 58W
Riverside, *Wyo.* **16** 41 13N 106 47W
Riverton, *Wyo.* **16** 43 2N 108 23W
Riverview, *Mich.* **26** 42 10N 83 11W
Roanoke, *Ala.* **11** 33 9N 85 22W
Roanoke, *Va.* **10** 37 16N 79 56W
Roanoke →, *N.C.* **11** 35 57N 76 42W
Roanoke I., *Ala.* **11** 35 55N 75 40W
Roanoke Rapids, *N.C.* **11** 36 28N 77 40W
Robert Lee, *Tex.* **13** 31 54N 100 29W
Roberts, *Idaho* **16** 43 43N 112 8W
Robstown, *Tex.* **13** 27 47N 97 40W
Rochelle, *Ill.* **14** 41 56N 89 4W
Rochester, *Ind.* **15** 41 4N 86 13W
Rochester, *Minn.* **12** 44 1N 92 28W
Rochester, *N.H.* **9** 43 18N 70 59W
Rochester, *N.Y.* **8** 43 10N 77 37W
Rock Hill, *S.C.* **11** 34 56N 81 1W
Rock Island, *Ill.* **14** 41 30N 90 34W
Rock Rapids, *Iowa* **12** 43 26N 96 10W
Rock River, *Wyo.* **16** 41 44N 105 58W
Rock Springs, *Mont.* **16** 46 49N 106 15W
Rock Springs, *Wyo.* **16** 41 35N 109 14W
Rock Valley, *Iowa* **12** 43 12N 96 18W
Rockdale, *Tex.* **13** 30 39N 97 0W
Rockford, *Ill.* **14** 42 16N 89 6W
Rocklake, *N. Dak.* **12** 48 47N 99 15W
Rockland, *Idaho* **16** 42 34N 112 53W
Rockland, *Maine* **11** 44 6N 69 7W
Rockland, *Mich.* **12** 46 44N 89 11W
Rockmart, *Ga.* **11** 34 0N 85 3W
Rockport, *Mo.* **12** 40 25N 95 31W
Rockport, *Tex.* **13** 28 2N 97 3W
Rocksprings, *Tex.* **13** 30 1N 100 13W
Rockville, *Conn.* **9** 41 52N 72 28W
Rockville, *Md.* **25** 39 4N 77 9W
Rockville Center, *N.Y.* **21** 40 39N 73 38W
Rockwall, *Tex.* **13** 32 56N 96 28W
Rockwell City, *Iowa* **14** 42 24N 94 38W
Rockwood, *Tenn.* **11** 35 52N 84 41W
Rocky Ford, *Colo.* **12** 38 3N 103 43W
Rocky Mount, *N.C.* **11** 35 57N 77 48W
Rocky Mts., *N. Amer.* **2** 39 0N 106 0W
Rocky River, *Ohio* **27** 41 28N 81 50W
Roebling, *N.J.* **9** 40 7N 74 47W
Rogers, *Ark.* **13** 36 20N 94 7W
Rogers City, *Mich.* **10** 45 25N 83 49W
Rogerson, *Idaho* **16** 42 13N 114 36W
Rogersville, *Tenn.* **11** 36 24N 83 1W
Rogue →, *Oreg.* **16** 42 26N 124 26W
Rohnert Park, *Calif.* **18** 38 16N 122 40W
Rojo, Cabo, *Puerto Rico* **30** 17 56N 67 12W
Rolette, *N. Dak.* **12** 48 40N 99 51W
Rolla, *Kans.* **13** 37 7N 101 38W
Rolla, *Mo.* **14** 37 57N 91 46W
Rolla, *N. Dak.* **12** 48 52N 99 37W
Romanzof C., *Alaska* **30** 61 49N 166 6W
Rome, *Ga.* **11** 34 15N 85 10W
Rome, *N.Y.* **9** 43 13N 75 27W
Romney, *W. Va.* **10** 39 21N 78 45W
Romulus, *Mich.* **26** 42 13N 83 23W
Ronan, *Mont.* **16** 47 32N 114 6W
Ronceverte, *W. Va.* **10** 37 45N 80 28W
Roof Butte, *Ariz.* **17** 36 28N 109 5W
Roosevelt, *Minn.* **12** 48 48N 95 6W
Roosevelt, *Utah* **16** 40 18N 109 59W
Roosevelt Res., *Ariz.* **17** 33 46N 111 0W
Ropesville, *Tex.* **13** 33 26N 102 9W
Rosalia, *Wash.* **16** 47 14N 117 22W
Rosamond, *Calif.* **19** 34 52N 118 10W
Roscoe, *S. Dak.* **12** 45 27N 99 20W
Roscommon, *Mich.* **10** 44 30N 84 35W
Roseau, *Minn.* **12** 48 51N 95 46W
Rosebud, *Tex.* **13** 31 4N 96 59W
Roseburg, *Oreg.* **16** 43 13N 123 20W
Rosedale, *Md.* **25** 39 19N 76 32W
Rosedale, *Miss.* **13** 33 51N 91 2W
Rosemead, *Calif.* **29** 34 4N 118 4W
Rosenberg, *Tex.* **13** 29 34N 95 49W
Roseville, *Calif.* **18** 38 45N 121 17W
Roseville, *Mich.* **26** 42 30N 82 57W
Ross, *Calif.* **28** 37 58N 122 33W
Ross L., *Wash.* **16** 48 44N 121 4W
Rossville, *N.Y.* **20** 40 32N 74 12W
Roswell, *N. Mex.* **13** 33 24N 104 32W
Rotan, *Tex.* **13** 32 51N 100 28W
Round Mountain, *Nev.* **16** 38 43N 117 4W
Roundup, *Mont.* **16** 46 27N 108 33W
Rouses Point, *N.Y.* **9** 44 59N 73 22W
Roxboro, *N.C.* **11** 36 24N 78 59W
Roy, *Mont.* **16** 47 20N 108 58W
Roy, *N. Mex.* **13** 35 57N 104 12W
Royal Oak, *Mich.* **26** 42 30N 83 9W
Ruby, *Alaska* **30** 64 45N 155 30W
Ruby L., *Nev.* **16** 40 10N 115 28W
Ruby Mts., *Nev.* **16** 40 30N 115 20W
Rudyard, *Mich.* **10** 46 14N 84 36W
Rugby, *N. Dak.* **12** 48 22N 100 0W
Ruidosa, *Tex.* **13** 29 59N 104 41W
Ruidoso, *N. Mex.* **17** 33 20N 105 41W
Rumford, *Maine* **9** 44 33N 70 33W
Rushford, *Minn.* **12** 43 49N 91 46W
Rushville, *Ill.* **14** 40 7N 90 34W
Rushville, *Ind.* **15** 39 37N 85 27W
Rushville, *Nebr.* **12** 42 43N 102 28W
Russell, *Kans.* **12** 38 54N 98 52W
Russellville, *Ala.* **11** 34 30N 87 44W
Russellville, *Ark.* **13** 35 17N 93 8W
Russellville, *Ky.* **11** 36 51N 86 53W
Russian Mission, *Alaska* **30** 61 47N 161 19W
Ruston, *La.* **13** 32 32N 92 38W
Ruth, *Nev.* **16** 39 17N 114 59W
Rye Patch Reservoir, *Nev.* **16** 40 28N 118 19W
Ryegate, *Mont.* **16** 46 18N 109 15W

S

Sabinal, *Tex.* **13** 29 19N 99 28W
Sabine →, *La.* **13** 29 59N 93 47W
Sabine L., *La.* **13** 29 53N 93 51W
Sabine Pass, *Tex.* **13** 29 44N 93 54W
Sac City, *Iowa* **14** 42 25N 95 0W
Saco, *Maine* **11** 43 30N 70 27W
Saco, *Mont.* **16** 48 28N 107 21W
Sacramento, *Calif.* **18** 38 35N 121 29W
Sacramento →, *Calif.* **18** 38 3N 121 56W
Sacramento Mts., *N. Mex.* **17** 32 30N 105 30W
Safford, *Ariz.* **17** 32 50N 109 43W
Sag Harbor, *N.Y.* **9** 41 0N 72 18W
Saginaw, *Mich.* **10** 43 26N 83 56W
Saginaw B., *Mich.* **10** 43 50N 83 40W
Saguache, *Colo.* **17** 38 5N 106 8W
Sahuarita, *Ariz.* **17** 31 57N 110 58W
St. Albans, *N.Y.* **21** 40 42N 73 44W
St. Albans, *Vt.* **9** 44 49N 73 5W
St. Albans, *W. Va.* **10** 38 23N 81 50W
St. Anthony, *Idaho* **16** 43 58N 111 41W
St. Augustine, *Fla.* **11** 29 54N 81 19W
St. Catherines I., *Ga.* **11** 31 40N 81 10W
St. Charles, *Ill.* **15** 41 54N 88 19W
St. Charles, *Mo.* **14** 38 47N 90 29W
St. Clair, *Pa.* **9** 40 43N 76 12W
St. Clair Shores, *Mich.* **26** 42 29N 82 54W
St. Cloud, *Fla.* **11** 28 15N 81 17W
St. Cloud, *Minn.* **12** 45 34N 94 10W
St. Croix, *Virgin Is.* **30** 17 45N 64 45W
St. Croix →, *Wis.* **12** 44 45N 92 48W
St. Croix Falls, *Wis.* **12** 45 24N 92 38W
St. Elias, Mt., *Alaska* **30** 60 18N 140 56W
St. Francis, *Kans.* **12** 39 47N 101 48W
St. Francis →, *Ark.* **13** 34 38N 90 36W
St. Francisville, *La.* **13** 30 47N 91 23W
St. George, *S.C.* **11** 33 11N 80 35W
St. George, *Utah* **17** 37 6N 113 35W
St. George, C., *Fla.* **11** 29 40N 85 5W
St. Helena, *Calif.* **16** 38 30N 122 28W
St. Helens, *Oreg.* **18** 45 52N 122 48W
St. Ignace, *Mich.* **10** 45 52N 84 44W
St. Ignatius, *Mont.* **16** 47 19N 114 6W
St. James, *Minn.* **12** 43 59N 94 38W
St. John, *Kans.* **13** 38 0N 98 46W
St. John, *N. Dak.* **12** 48 57N 99 43W
St. John →, *Maine* **11** 45 12N 66 5W
St. John I., *Virgin Is.* **30** 18 20N 64 42W
St. Johns, *Ariz.* **17** 34 30N 109 22W
St. Johns, *Mich.* **15** 43 0N 84 33W
St. Johns →, *Fla.* **11** 30 24N 81 24W
St. Johnsbury, *Vt.* **9** 44 25N 72 1W
St. Johnsville, *N.Y.* **9** 43 0N 74 43W
St. Joseph, *La.* **13** 31 55N 91 14W
St. Joseph, *Mich.* **15** 42 6N 86 29W
St. Joseph, *Mo.* **14** 39 46N 94 50W
St. Joseph →, *Mich.* **15** 42 7N 86 29W
St. Lawrence I., *Alaska* **30** 63 30N 170 30W
St. Louis, *Mich.* **10** 43 25N 84 36W
St. Louis, *Mo.* **14** 38 37N 90 12W
St. Louis →, *Minn.* **12** 47 15N 92 45W
St. Maries, *Idaho* **16** 47 19N 116 35W
St. Martinville, *La.* **13** 30 7N 91 50W
St. Marys, *Pa.* **8** 41 26N 78 34W
St. Michael, *Alaska* **30** 63 29N 162 2W
St. Paul, *Minn.* **12** 44 57N 93 6W
St. Paul, *Nebr.* **12** 41 13N 98 27W
St. Peter, *Minn.* **12** 44 20N 93 57W
St. Petersburg, *Fla.* **11** 27 46N 82 39W
St. Regis, *Mont.* **16** 47 18N 115 6W
St. Thomas I., *Virgin Is.* **30** 18 20N 64 55W
Ste. Genevieve, *Mo.* **14** 37 59N 90 2W
Saipan, *Pac. Oc.* **31** 15 12N 145 45 E
Sakakawea, L., *N. Dak.* **12** 47 30N 101 25W
Salamanca, *N.Y.* **8** 42 10N 78 43W
Salem, *Ind.* **15** 38 36N 86 6W
Salem, *Mass.* **23** 42 30N 70 55W
Salem, *Mo.* **13** 37 39N 91 32W
Salem, *N.J.* **10** 39 34N 75 28W
Salem, *Ohio* **8** 40 54N 80 52W
Salem, *Oreg.* **16** 44 56N 123 2W
Salem, *S. Dak.* **12** 43 44N 97 23W
Salem, *Va.* **10** 37 18N 80 3W
Salina, *Kans.* **12** 38 50N 97 37W
Salinas, *Calif.* **18** 36 40N 121 39W
Salinas →, *Calif.* **18** 36 45N 121 48W
Saline →, *Ark.* **13** 33 10N 92 8W
Saline →, *Kans.* **12** 38 52N 97 30W
Salisbury, *Md.* **10** 38 22N 75 36W
Salisbury, *N.C.* **11** 35 40N 80 29W
Sallisaw, *Okla.* **13** 35 28N 94 47W
Salmon, *Idaho* **16** 45 11N 113 54W
Salmon →, *Idaho* **16** 45 51N 116 47W
Salmon Falls, *Idaho* **16** 42 48N 114 59W
Salmon River Mts., *Idaho* **16** 45 0N 114 30W
Salome, *Ariz.* **19** 33 47N 113 37W
Salt →, *Ariz.* **17** 33 23N 112 19W
Salt Fork Arkansas →, *Okla.* **13** 36 36N 97 3W
Salt Lake City, *Utah* **16** 40 45N 111 53W
Salton City, *Calif.* **19** 33 29N 115 51W
Salton Sea, *Calif.* **19** 33 15N 115 45W
Saltville, *Va.* **10** 36 53N 81 46W
Saluda →, *S.C.* **11** 34 1N 81 4W
Salvador, L., *La.* **13** 29 43N 90 15W
Salyersville, *Ky.* **10** 37 45N 83 4W
Sam Rayburn Reservoir, *Tex.* **13** 31 4N 94 5W
San Andreas, *Calif.* **18** 38 12N 120 41W
San Andres Mts., *N. Mex.* **17** 33 0N 106 30W
San Angelo, *Tex.* **13** 31 28N 100 26W
San Anselmo, *Calif.* **18** 37 59N 122 34W
San Antonio, *N. Mex.* **17** 33 55N 106 52W
San Antonio, *Tex.* **13** 29 25N 98 30W
San Antonio →, *Tex.* **13** 28 30N 96 54W
San Antonio, Mt., *Calif.* **19** 34 17N 117 38W
San Ardo, *Calif.* **18** 36 1N 120 54W
San Augustine, *Tex.* **13** 31 30N 94 7W
San Benito, *Tex.* **13** 26 8N 97 38W
San Benito →, *Calif.* **18** 36 53N 121 34W
San Benito Mt., *Calif.* **18** 36 22N 120 37W
San Bernardino, *Calif.* **19** 34 7N 117 19W
San Blas, C., *Fla.* **11** 29 40N 85 21W
San Bruno, *Calif.* **28** 37 37N 122 24W
San Carlos, *Ariz.* **17** 33 21N 110 27W
San Carlos, *Calif.* **28** 37 30N 122 16W
San Carlos L., *Ariz.* **17** 33 11N 110 32W
San Clemente, *Calif.* **19** 33 26N 117 37W
San Clemente I., *Calif.* **19** 32 53N 118 29W
San Diego, *Calif.* **19** 32 43N 117 9W
San Diego, *Tex.* **13** 27 46N 98 14W
San Felipe →, *Calif.* **19** 33 12N 115 49W
San Fernando, *Calif.* **29** 34 17N 118 26W
San Francisco, *Calif.* **28** 37 46N 122 25W
San Francisco →, *Ariz.* **17** 32 59N 109 22W
San Francisco Bay, *Calif.* **28** 37 40N 122 15W
San Francisco International Airport, *Calif.* **28** 37 37N 122 22W
San Gabriel, *Calif.* **29** 34 5N 118 5W
San Germán, *Puerto Rico* **30** 18 5N 67 3W
San Gorgonio Mt., *Calif.* **19** 34 7N 116 51W
San Gregorio, *Calif.* **18** 37 20N 122 23W
San Jacinto, *Calif.* **19** 33 47N 116 57W
San Joaquin, *Calif.* **18** 36 36N 120 11W
San Joaquin →, *Calif.* **18** 38 4N 121 51W
San Joaquin Valley, *Calif.* **18** 37 20N 121 0W
San Jose, *Calif.* **18** 37 20N 121 53W
San Jose →, *N. Mex.* **17** 34 25N 106 45W
San Juan, *Dom. Rep.* **30** 18 49N 71 12W
San Juan, *Puerto Rico* **30** 18 28N 66 7W
San Juan →, *Utah* **17** 37 16N 110 26W
San Juan, C., *Puerto Rico* **30** 18 23N 65 37W
San Juan Bautista, *Calif.* **18** 36 51N 121 32W
San Juan Capistrano, *Calif.* **19** 33 30N 117 40W
San Juan Cr. →, *Calif.* **18** 35 40N 120 22W
San Juan Mts., *Colo.* **17** 37 30N 107 0W
San Leandro, *Calif.* **28** 37 44N 122 9W
San Lucas, *Calif.* **18** 36 8N 121 1W
San Luis, *Colo.* **17** 37 12N 105 25W
San Luis Obispo, *Calif.* **19** 35 17N 120 40W
San Luis Reservoir, *Calif.* **18** 37 4N 121 5W
San Marcos, *Tex.* **13** 29 53N 97 56W
San Marino, *Calif.* **29** 34 7N 118 5W
San Mateo, *Calif.* **28** 37 34N 122 20W
San Mateo Bridge, *Calif.* **28** 37 36N 122 11W
San Mateo County, *Calif.* **28** 37 30N 122 20W
San Miguel, *Calif.* **18** 35 45N 120 42W
San Miguel I., *Calif.* **19** 34 2N 120 23W
San Nicolas I., *Calif.* **19** 33 15N 119 30W
San Onofre, *Calif.* **19** 33 22N 117 34W
San Pedro →, *Ariz.* **17** 32 59N 110 47W
San Pedro Channel, *Calif.* **19** 33 30N 118 25W
San Quentin, *Calif.* **28** 37 56N 122 29W
San Rafael, *Calif.* **28** 37 58N 122 31W
San Rafael, *N. Mex.* **17** 35 7N 107 53W
San Rafael Mt., *Calif.* **19** 34 41N 119 52W
San Saba, *Tex.* **13** 31 12N 98 43W
San Simeon, *Calif.* **18** 35 39N 121 11W
San Simon, *Ariz.* **17** 32 16N 109 14W
San Ygnacio, *Tex.* **13** 27 3N 99 26W
Sanak I., *Alaska* **30** 54 25N 162 40W
Sand Point, *Alaska* **30** 55 20N 160 30W
Sand Springs, *Okla.* **13** 36 9N 96 7W
Sanders, *Ariz.* **17** 35 13N 109 20W
Sanderson, *Tex.* **13** 30 9N 102 24W
Sandpoint, *Idaho* **16** 48 17N 116 33W
Sandusky, *Mich.* **8** 43 25N 82 50W
Sandusky, *Ohio* **8** 41 27N 82 42W
Sandy Cr. →, *Wyo.* **16** 41 51N 109 47W
Sanford, *Fla.* **11** 28 48N 81 16W
Sanford, *Maine* **9** 43 27N 70 47W
Sanford, *N.C.* **11** 35 29N 79 10W
Sanger, *Calif.* **18** 36 42N 119 33W
Sangre de Cristo Mts., *N. Mex.* **13** 37 0N 105 0W
Santa Ana, *Calif.* **19** 33 46N 117 52W
Santa Barbara, *Calif.* **19** 34 25N 119 42W
Santa Barbara Channel, *Calif.* **19** 34 15N 120 0W
Santa Barbara I., *Calif.* **19** 33 29N 119 2W
Santa Catalina, Gulf of, *Calif.* **19** 33 10N 117 50W
Santa Catalina I., *Calif.* **19** 33 23N 118 25W
Santa Clara, *Calif.* **18** 37 21N 121 57W
Santa Clara, *Utah* **17** 37 8N 113 39W
Santa Cruz, *Calif.* **18** 36 58N 122 1W
Santa Cruz I., *Calif.* **19** 34 1N 119 43W
Santa Fe, *N. Mex.* **17** 35 41N 105 57W
Santa Fe Springs, *Calif.* **29** 33 56N 118 3W
Santa Lucia Range, *Calif.* **18** 36 0N 121 20W
Santa Margarita, *Calif.* **18** 35 23N 120 37W
Santa Margarita →, *Calif.* **19** 33 13N 117 23W
Santa Maria, *Calif.* **19** 34 57N 120 26W
Santa Monica, *Calif.* **29** 34 1N 118 29W
Santa Rita, *N. Mex.* **17** 32 48N 108 4W
Santa Rosa, *Calif.* **18** 38 26N 122 43W
Santa Rosa, *N. Mex.* **13** 34 57N 104 41W
Santa Rosa I., *Calif.* **19** 33 58N 120 6W
Santa Rosa I., *Fla.* **11** 30 20N 86 50W
Santa Rosa Range, *Nev.* **16** 41 45N 117 40W
Santa Ynez →, *Calif.* **19** 35 41N 120 36W
Santa Ynez Mts., *Calif.* **19** 34 30N 120 0W
Santa Ysabel, *Calif.* **19** 33 7N 116 40W
Santaquin, *Utah* **16** 39 59N 111 47W
Sapelo I., *Ga.* **11** 31 25N 81 12W
Sapulpa, *Okla.* **13** 35 59N 96 5W
Saranac Lake, *N.Y.* **9** 44 20N 74 8W
Sarasota, *Fla.* **11** 27 20N 82 32W
Saratoga, *Calif.* **18** 37 16N 122 2W
Saratoga, *Wyo.* **16** 41 27N 106 49W
Saratoga Springs, *N.Y.* **9** 43 5N 73 47W
Sargent, *Nebr.* **12** 41 39N 99 22W
Sarichef C., *Alaska* **30** 54 38N 164 59W
Sarita, *Tex.* **13** 27 13N 97 47W
Sarles, *N. Dak.* **12** 48 58N 99 0W
Satanta, *Kans.* **13** 37 26N 100 59W
Satilla →, *Ga.* **11** 30 59N 81 29W
Saugerties, *N.Y.* **9** 42 5N 73 57W
Saugus, *Mass.* **23** 42 28N 71 0W
Sauk Centre, *Minn.* **12** 45 44N 94 57W
Sauk Rapids, *Minn.* **12** 45 35N 94 10W
Sault Ste. Marie, *Mich.* **10** 46 30N 84 21W
Sausalito, *Calif.* **28** 37 51N 122 28W
Savage, *Mont.* **12** 47 27N 104 21W
Savanna, *Ill.* **14** 42 5N 90 8W
Savannah, *Ga.* **11** 32 5N 81 6W
Savannah, *Mo.* **14** 39 56N 94 50W
Savannah, *Tenn.* **11** 35 14N 88 15W
Savannah →, *Ga.* **11** 32 2N 80 53W
Sawatch Mts., *Colo.* **17** 38 30N 106 30W
Sayre, *Okla.* **13** 35 18N 99 38W
Sayre, *Pa.* **9** 41 59N 76 32W
Scammon Bay, *Alaska* **30** 61 51N 165 35W
Scenic, *S. Dak.* **12** 43 47N 102 33W
Schell Creek Ra., *Nev.* **16** 39 15N 114 30W
Schenectady, *N.Y.* **9** 42 49N 73 57W
Schiller Park, *Ill.* **22** 41 56N 87 52W
Schofield, *Wis.* **12** 44 54N 89 36W
Schurz, *Nev.* **16** 38 57N 118 49W
Schuyler, *Nebr.* **12** 41 27N 97 4W
Schuylkill Haven, *Pa.* **9** 40 37N 76 11W
Scioto →, *Ohio* **10** 38 44N 83 1W
Scobey, *Mont.* **12** 48 47N 105 25W
Scotia, *Calif.* **16** 40 29N 124 6W
Scotia, *N.Y.* **9** 42 50N 73 58W
Scotland, *S. Dak.* **12** 43 9N 97 43W
Scotland Neck, *N.C.* **11** 36 8N 77 25W
Scott City, *Kans.* **12** 38 29N 100 54W
Scottsbluff, *Nebr.* **12** 41 52N 103 40W
Scottsboro, *Ala.* **11** 34 40N 86 2W
Scottsburg, *Ind.* **15** 38 41N 85 47W
Scottsville, *Ky.* **11** 36 45N 86 11W
Scottville, *Mich.* **10** 43 58N 86 17W
Scranton, *Pa.* **9** 41 25N 75 40W
Seaford, *Del.* **10** 38 39N 75 37W
Seagraves, *Tex.* **13** 32 57N 102 34W
Sealy, *Tex.* **13** 29 47N 96 9W
Searchlight, *Nev.* **19** 35 28N 114 55W
Searcy, *Ark.* **13** 35 15N 91 44W
Searles L., *Calif.* **19** 35 44N 117 21W
Sears Tower, *Ill.* **22** 41 52N 87 37W
Seaside, *Calif.* **18** 36 37N 121 50W
Seaside, *Oreg.* **18** 46 0N 123 56W
Seat Pleasant, *Md.* **25** 38 53N 76 53W
Seattle, *Wash.* **18** 47 36N 122 20W
Sebastopol, *Calif.* **18** 38 24N 122 49W
Sebewaing, *Mich.* **10** 43 44N 83 27W
Sebring, *Fla.* **11** 27 30N 81 27W
Sedalia, *Mo.* **14** 38 42N 93 14W
Sedan, *Kans.* **13** 37 8N 96 11W
Sedro-Woolley, *Wash.* **18** 48 30N 122 14W
Seguam I., *Alaska* **30** 52 19N 172 30W
Seguam Pass, *Alaska* **30** 52 0N 172 30W
Seguin, *Tex.* **13** 29 34N 97 58W
Segula I., *Alaska* **30** 52 0N 177 50 E
Seiling, *Okla.* **13** 36 9N 98 56W
Selah, *Wash.* **16** 46 39N 120 32W
Selawik, *Alaska* **30** 66 36N 160 0W
Selby, *S. Dak.* **12** 45 31N 100 2W
Selden, *Kans.* **12** 39 33N 100 34W
Seldovia, *Alaska* **30** 59 26N 151 43W
Selfridge, *N. Dak.* **12** 46 2N 100 56W
Seligman, *Ariz.* **17** 35 20N 112 53W
Sells, *Ariz.* **17** 31 55N 111 53W
Selma, *Ala.* **11** 32 25N 87 1W
Selma, *Calif.* **18** 36 34N 119 37W
Selma, *N.C.* **11** 35 32N 78 17W

Selmer

Selmer, Tenn. 11 35 10N 88 36W
Seminoe Reservoir, Wyo. 16 42 9N 106 55W
Seminole, Okla. 13 35 14N 96 41W
Seminole, Tex. 13 32 43N 102 39W
Semisopochnoi I., Alaska 30 51 55N 179 36 E
Senatobia, Miss. 13 34 37N 89 58W
Seneca, Oreg. 16 44 8N 118 58W
Seneca, S.C. 11 34 41N 82 57W
Seneca Falls, N.Y. 9 42 55N 76 48W
Seneca L., N.Y. 8 42 40N 76 54W
Sentinel, Ariz. 17 32 52N 113 13W
Sequim, Wash. 18 48 5N 123 6W
Sequoia National Park, Calif. 18 36 30N 118 30W
Settlement Pt., Bahamas 11 26 40N 79 0W
Sevier, Utah 17 38 39N 112 11W
Sevier →, Utah 17 39 4N 113 6W
Sevier L., Utah 16 38 54N 113 9W
Seward, Alaska 30 60 7N 149 27W
Seward, Nebr. 12 40 55N 97 6W
Seward Pen., Alaska 30 65 0N 164 0W
Seymour, Conn. 9 41 24N 73 4W
Seymour, Ind. 15 38 58N 85 53W
Seymour, Tex. 13 33 35N 99 16W
Seymour, Wis. 10 44 31N 88 20W
Shafter, Calif. 19 35 30N 119 16W
Shafter, Tex. 13 29 49N 104 18W
Shaker Heights, Ohio 27 41 28N 81 33W
Shakopee, Minn. 12 44 48N 93 32W
Shaktolik, Alaska 30 64 30N 161 15W
Shamokin, Pa. 9 40 47N 76 34W
Shamrock, Tex. 13 35 13N 100 15W
Shandon, Calif. 18 35 39N 120 23W
Shaniko, Oreg. 16 45 0N 120 45W
Sharon, Mass. 9 42 7N 71 11W
Sharon, Pa. 8 41 14N 80 31W
Sharon Springs, Kans. 12 38 54N 101 45W
Sharpsburg, Pa. 27 40 29N 79 56W
Shasta, Mt., Calif. 16 41 25N 122 12W
Shasta L., Calif. 16 40 43N 122 25W
Shattuck, Okla. 13 36 16N 99 53W
Shaver L., Calif. 18 37 9N 119 18W
Shawano, Wis. 10 44 47N 88 36W
Shawnee, Okla. 13 35 20N 96 55W
Sheboygan, Wis. 10 43 46N 87 45W
Sheffield, Ala. 11 34 46N 87 41W
Sheffield, Mass. 9 42 5N 73 21W
Sheffield, Tex. 13 30 41N 101 49W
Shelburne Falls, Mass. 9 42 36N 72 45W
Shelby, Mich. 10 43 37N 86 22W
Shelby, Mont. 16 48 30N 111 51W
Shelby, N.C. 11 35 17N 81 32W
Shelbyville, Ill. 15 39 24N 88 48W
Shelbyville, Ind. 15 39 31N 85 47W
Shelbyville, Tenn. 11 35 29N 86 28W
Sheldon, Iowa 12 43 11N 95 51W
Sheldon Point, Alaska 30 62 32N 164 52W
Shelikof Strait, Alaska 30 57 30N 155 0W
Shelton, Conn. 9 41 19N 73 5W
Shelton, Wash. 18 47 13N 123 6W
Shenandoah, Iowa 12 40 46N 95 22W
Shenandoah, Pa. 9 40 49N 76 12W
Shenandoah, Va. 10 38 29N 78 37W
Shenandoah →, Va. 10 39 19N 77 44W
Sheridan, Ark. 13 34 19N 92 24W
Sheridan, Wyo. 16 44 48N 106 58W
Sherman, Tex. 13 33 40N 96 35W
Sherwood, N. Dak. 12 48 57N 101 38W
Sherwood, Tex. 13 31 18N 100 45W
Sheyenne, N. Dak. 12 47 50N 99 7W
Sheyenne →, N. Dak. 12 47 2N 96 50W
Ship I., Miss. 13 30 13N 88 55W
Shippensburg, Pa. 8 40 3N 77 31W
Shiprock, N. Mex. 17 36 47N 108 41W
Shishmaref, Alaska 30 66 15N 166 4W
Shoshone, Calif. 19 35 58N 116 16W
Shoshone, Idaho 16 42 56N 114 25W
Shoshone L., Wyo. 16 44 22N 110 43W
Shoshone Mts., Nev. 16 39 20N 117 25W
Shoshoni, Wyo. 16 43 14N 108 7W
Show Low, Ariz. 17 34 15N 110 2W
Shreveport, La. 13 32 31N 93 45W
Shumagin Is., Alaska 30 55 7N 159 45W
Shungnak, Alaska 30 66 52N 157 9W
Shuyak I., Alaska 30 58 31N 152 30W
Sibley, Iowa 12 43 24N 95 45W
Sibley, La. 13 32 33N 93 18W
Sidney, Mont. 12 47 43N 104 9W
Sidney, N.Y. 9 42 19N 75 24W
Sidney, Nebr. 12 41 8N 102 59W
Sidney, Ohio 15 40 17N 84 9W
Sierra Blanca, Tex. 17 31 11N 105 22W
Sierra Blanca Peak, N. Mex. 17 33 23N 105 49W
Sierra City, Calif. 18 39 34N 120 38W
Sierra Madre, Calif. 29 34 9N 118 3W
Sigurd, Utah 17 38 50N 111 58W
Sikeston, Mo. 13 36 53N 89 35W
Siler City, N.C. 11 35 44N 79 28W
Siloam Springs, Ark. 13 36 11N 94 32W
Silsbee, Tex. 13 30 21N 94 11W
Silver City, N. Mex. 17 32 46N 108 17W
Silver City, Nev. 16 39 15N 119 48W
Silver Cr. →, Oreg. 16 43 16N 119 13W
Silver Creek, N.Y. 8 42 33N 79 10W
Silver Hill, Md. 25 38 49N 76 55W
Silver L., Calif. 18 38 39N 120 6W
Silver L., Calif. 19 35 21N 116 7W
Silver Lake, Oreg. 16 43 8N 121 3W
Silver Spring, Md. 25 39 0N 77 1W
Silverton, Colo. 17 37 49N 107 40W
Silverton, Tex. 13 34 28N 101 19W
Silvies →, Oreg. 16 43 34N 119 2W
Simi Valley, Calif. 19 34 16N 118 47W
Simmler, Calif. 19 35 21N 119 59W
Sinclair, Wyo. 16 41 47N 107 7W
Sinton, Tex. 13 28 2N 97 31W
Sioux City, Iowa 12 42 30N 96 24W
Sioux Falls, S. Dak. 12 43 33N 96 44W
Sirretta Pk., Calif. 19 35 56N 118 19W
Sisseton, S. Dak. 12 45 40N 97 3W
Sisters, Oreg. 16 44 18N 121 33W
Sitka, Alaska 30 57 3N 135 20W
Skagway, Alaska 30 59 28N 135 19W
Skokie, Ill. 22 42 2N 87 42W
Skowhegan, Maine 11 44 46N 69 43W
Skunk →, Iowa 14 40 42N 91 7W
Skykomish, Wash. 16 47 42N 121 22W
Slaton, Tex. 13 33 26N 101 39W
Sleepy Eye, Minn. 12 44 18N 94 43W
Slidell, La. 13 30 17N 89 47W
Sloansville, N.Y. 9 42 45N 74 22W
Sloughhouse, Calif. 18 38 26N 121 12W
Smith Center, Kans. 12 39 47N 98 47W
Smithfield, N.C. 11 35 31N 78 21W
Smithfield, Utah 16 41 50N 111 50W
Smithville, Tex. 13 30 1N 97 10W
Smoky Hill →, Kans. 12 39 4N 96 48W
Snake →, Wash. 16 46 12N 119 2W
Snake Range, Nev. 16 39 0N 114 20W
Snake River Plain, Idaho 16 42 50N 114 0W
Snelling, Calif. 18 37 31N 120 26W
Snohomish, Wash. 18 47 55N 122 6W
Snow Hill, Md. 10 38 11N 75 24W
Snowflake, Ariz. 17 34 30N 110 5W
Snowshoe Pk., Mont. 16 48 13N 115 41W
Snowville, Utah 16 41 58N 112 43W
Snyder, Okla. 13 34 40N 98 57W
Snyder, Tex. 13 32 44N 100 55W
Soap Lake, Wash. 16 47 23N 119 29W
Socorro, N. Mex. 17 34 4N 106 54W
Soda L., Calif. 17 35 10N 116 4W
Soda Springs, Idaho 16 42 39N 111 36W
Sodus, N.Y. 8 43 14N 77 4W
Soledad, Calif. 18 36 26N 121 20W
Solomon, N. Fork →, Kans. 12 39 29N 98 26W
Solomon, S. Fork →, Kans. 12 39 25N 99 12W
Solon Springs, Wis. 12 46 22N 91 49W
Solvang, Calif. 19 34 36N 120 8W
Solvay, N.Y. 9 43 3N 76 13W
Somers, Mont. 16 48 5N 114 13W
Somerset, Colo. 17 38 56N 107 28W
Somerset, Ky. 10 37 5N 84 36W
Somerset, Mass. 9 41 47N 71 8W
Somerton, Ariz. 17 32 36N 114 43W
Somerville, Mass. 23 42 23N 71 5W
Somerville, N.J. 9 40 35N 74 38W
Sonora, Calif. 18 37 59N 120 23W
Sonora, Tex. 13 30 34N 100 39W
South Baldy, N. Mex. 17 33 59N 107 11W
South Bend, Ind. 15 41 41N 86 15W
South Bend, Wash. 18 46 40N 123 48W
South Boston, Va. 11 36 42N 78 54W
South C. = Ka Lae, Hawaii 30 18 55N 155 41W
South Cape, Hawaii 30 18 58N 155 24 E
South Carolina □ 11 34 0N 81 0W
South Charleston, W. Va. 10 38 22N 81 44W
South Dakota □ 12 44 15N 100 0W
South Euclid, Ohio 27 41 31N 81 32W
South Fork →, Mont. 16 47 54N 113 15W
South Fork, American →, Calif. 18 38 45N 121 5W
South Gate, Calif. 29 33 56N 118 12W
South Haven, Mich. 15 42 24N 86 16W
South Holland, Ill. 22 41 36N 87 36W
South Loup →, Nebr. 12 41 4N 98 39W
South Milwaukee, Wis. 15 42 55N 87 52W
South Pasadena, Calif. 29 34 7N 118 8W
South Pass, Wyo. 16 42 20N 108 58W
South Pittsburg, Tenn. 11 35 1N 85 42W
South Platte →, Nebr. 12 41 7N 100 42W
South River, N.J. 9 40 27N 74 23W
South San Francisco, Calif. 28 37 39N 122 24W
South Sioux City, Nebr. 12 42 28N 96 24W
Southampton, N.Y. 9 40 53N 72 23W
Southbridge, Mass. 9 42 5N 72 2W
Southeast C., Alaska 30 62 56N 169 39W
Southern Pines, N.C. 11 35 11N 79 24W
Southfield, Mich. 26 42 28N 83 15W
Southgate, Mich. 26 42 11N 83 12W
Southington, Conn. 9 41 36N 72 53W
Southold, N.Y. 9 41 4N 72 26W
Southport, N.C. 11 33 55N 78 1W
Spalding, Nebr. 12 41 42N 98 22W
Spanish Fork, Utah 16 40 7N 111 39W
Sparks, Nev. 18 39 32N 119 45W
Sparta, Ga. 11 33 17N 82 58W
Sparta, Wis. 12 43 56N 90 49W
Spartanburg, S.C. 11 34 56N 81 57W
Spearfish, S. Dak. 12 44 30N 103 52W
Spearman, Tex. 13 36 12N 101 12W
Spenard, Alaska 30 61 11N 149 55W
Spencer, Idaho 16 44 22N 112 11W
Spencer, Iowa 12 43 9N 95 9W
Spencer, N.Y. 9 42 13N 76 30W
Spencer, Nebr. 12 42 53N 98 42W
Spencer, W. Va. 10 38 48N 81 21W
Spirit Lake, Idaho 16 47 58N 116 52W
Spofford, Tex. 13 29 10N 100 25W
Spokane, Wash. 16 47 40N 117 24W
Spooner, Wis. 12 45 50N 91 53W
Sprague, Wash. 16 47 18N 117 59W
Sprague River, Oreg. 16 42 27N 121 30W
Spray, Oreg. 16 44 50N 119 48W
Spring City, Utah 16 39 29N 111 30W
Spring Mts., Nev. 17 36 0N 115 45W
Spring Valley, Minn. 12 43 41N 92 23W
Springdale, Ark. 13 36 11N 94 8W
Springdale, Wash. 16 48 4N 117 45W
Springer, N. Mex. 13 36 22N 104 36W
Springerville, Ariz. 17 34 8N 109 17W
Springfield, Colo. 13 37 24N 102 37W
Springfield, Ill. 14 39 48N 89 39W
Springfield, Mass. 9 42 6N 72 35W
Springfield, Mo. 13 37 13N 93 17W
Springfield, N.J. 20 40 42N 74 18W
Springfield, Ohio 15 39 55N 83 49W
Springfield, Oreg. 16 44 3N 123 1W
Springfield, Tenn. 11 36 31N 86 53W
Springfield, Va. 25 38 46N 77 10W
Springfield, Vt. 9 43 18N 72 29W
Springvale, Maine 9 43 28N 70 48W
Springville, Calif. 18 36 8N 118 49W
Springville, N.Y. 8 42 31N 78 40W
Springville, Utah 16 40 10N 111 37W
Spur, Tex. 13 33 28N 100 52W
Stafford, Kans. 13 37 58N 98 36W
Stafford Springs, Conn. 9 41 57N 72 18W
Stamford, Conn. 9 41 3N 73 32W
Stamford, Tex. 13 32 57N 99 48W
Stamps, Ark. 13 33 22N 93 30W
Stanberry, Mo. 12 40 13N 94 35W
Standish, Mich. 10 43 59N 83 57W
Stanford, Mont. 16 47 9N 110 13W
Stanislaus →, Calif. 18 37 40N 121 14W
Stanley, Idaho 16 44 13N 114 56W
Stanley, N. Dak. 12 48 19N 102 23W
Stanley, Wis. 12 44 58N 90 56W
Stanton, Tex. 13 32 8N 101 48W
Staples, Minn. 12 46 21N 94 48W
Stapleton, N.Y. 20 40 36N 74 5W
Stapleton, Nebr. 12 41 29N 100 31W
Starke, Fla. 11 29 57N 82 7W
Starkville, Colo. 13 37 8N 104 30W
Starkville, Miss. 11 33 28N 88 49W
State College, Pa. 8 40 48N 77 52W
Staten Island, N.Y. 20 40 34N 74 9W
Statesboro, Ga. 11 32 27N 81 47W
Statesville, N.C. 11 35 47N 80 53W
Statue of Liberty, N.J. 20 40 41N 74 2W
Stauffer, Calif. 19 34 45N 119 3W
Staunton, Ill. 14 39 1N 89 47W
Staunton, Va. 10 38 9N 79 4W
Steamboat Springs, Colo. 16 40 29N 106 50W
Steele, N. Dak. 12 46 51N 99 55W
Steelton, Pa. 8 40 14N 76 50W
Steelville, Mo. 14 37 58N 91 22W
Stephen, Minn. 12 48 27N 96 53W
Stephenville, Tex. 13 32 13N 98 12W
Sterling, Colo. 12 40 37N 103 13W
Sterling, Ill. 14 41 48N 89 42W
Sterling, Kans. 12 38 13N 98 12W
Sterling City, Tex. 13 31 51N 101 0W
Sterling Heights, Mich. 26 42 35N 83 3W
Steubenville, Ohio 8 40 22N 80 37W
Stevens Point, Wis. 12 44 31N 89 34W
Stevens Village, Alaska 30 66 1N 149 6W
Stigler, Okla. 13 35 15N 95 8W
Stillwater, Minn. 12 45 3N 92 49W
Stillwater, N.Y. 9 42 55N 73 41W
Stillwater, Okla. 13 36 7N 97 4W
Stillwater Range, Nev. 16 39 50N 118 5W
Stilwell, Okla. 13 35 49N 94 38W
Stockett, Mont. 16 47 21N 111 10W
Stockton, Calif. 18 37 58N 121 17W
Stockton, Kans. 12 39 26N 99 16W
Stockton, Mo. 14 37 42N 93 48W
Stoneham, Mass. 23 42 29N 71 5W
Stony River, Alaska 30 61 47N 156 35W
Storm Lake, Iowa 12 42 39N 95 13W
Stove Pipe Wells Village, Calif. 19 36 35N 117 11W
Strasburg, N. Dak. 12 46 8N 100 10W
Stratford, Calif. 18 36 11N 119 49W
Stratford, Conn. 9 41 12N 73 8W
Stratford, Tex. 13 36 20N 102 4W
Strathmore, Calif. 18 36 9N 119 4W
Stratton, Colo. 12 39 19N 102 36W
Strawberry Reservoir, Utah 16 40 8N 111 9W
Strawn, Tex. 13 32 33N 98 30W
Streator, Ill. 15 41 8N 88 50W
Streeter, N. Dak. 12 46 39N 99 21W
Stromsburg, Iowa 12 41 7N 97 36W
Stroudsburg, Pa. 9 40 59N 75 12W
Struthers, Ohio 8 41 4N 80 39W
Stryker, Mont. 16 48 41N 114 46W
Stuart, Fla. 11 27 12N 80 15W
Stuart, Nebr. 12 42 36N 99 8W
Stuart I., Alaska 30 63 55N 164 50W
Sturgeon Bay, Wis. 10 44 50N 87 23W
Sturgis, Mich. 15 41 48N 85 25W
Sturgis, S. Dak. 12 44 25N 103 31W
Stuttgart, Ark. 13 34 30N 91 33W
Stuyvesant, N.Y. 9 42 23N 73 45W
Sudan, Tex. 13 34 4N 102 32W
Suffolk, Va. 10 36 44N 76 35W
Sugar City, Colo. 12 38 14N 103 40W
Suitland, Md. 25 38 50N 76 55W
Sullivan, Ill. 15 39 36N 88 37W
Sullivan, Ind. 15 39 6N 87 24W
Sullivan, Mo. 14 38 13N 91 10W
Sulphur, La. 13 30 14N 93 23W
Sulphur, Okla. 13 34 31N 96 58W
Sulphur Springs, Tex. 13 33 8N 95 36W
Sulphur Springs Draw →, Tex. 13 32 12N 101 36W
Sumatra, Mont. 16 46 37N 107 33W
Summer L., Oreg. 16 42 50N 120 45W
Summerville, Ga. 11 34 29N 85 21W
Summerville, S.C. 11 33 1N 80 11W
Summit, Alaska 30 63 20N 149 7W
Summit, Ill. 22 41 47N 87 47W
Summit, N.J. 20 40 43N 74 21W
Summit Peak, Colo. 17 37 21N 106 42W
Sumner, Iowa 14 42 51N 92 6W
Sumter, S.C. 11 33 55N 80 21W
Sun City, Ariz. 17 33 36N 112 17W
Sun City, Calif. 19 33 42N 117 11W
Sunburst, Mont. 16 48 53N 111 55W
Sunbury, Pa. 9 40 52N 76 48W
Suncook, N.H. 9 43 8N 71 27W
Sundance, Wyo. 12 44 24N 104 23W
Sunnyside, Utah 16 39 34N 110 23W
Sunnyside, Wash. 16 46 20N 120 0W
Sunnyvale, Calif. 18 37 23N 122 2W
Sunray, Tex. 13 36 1N 101 49W
Sunshine Acres, Calif. 29 33 56N 117 59W
Supai, Ariz. 17 36 15N 112 41W
Superior, Ariz. 17 33 18N 111 6W
Superior, Mont. 16 47 12N 114 53W
Superior, Nebr. 12 40 1N 98 4W
Superior, Wis. 12 46 44N 92 6W
Superior, L., N. Amer. 10 47 0N 87 0W
Sur, Pt., Calif. 18 36 18N 121 54W
Surf, Calif. 19 34 41N 120 36W
Susanville, Calif. 16 40 25N 120 39W
Susquehanna →, Pa. 9 39 33N 76 5W
Susquehanna Depot, Pa. 9 41 57N 75 36W
Sussex, N.J. 9 41 13N 74 37W
Sutherland, Nebr. 12 41 10N 101 8W
Sutherlin, Oreg. 16 43 23N 123 19W
Sutter Creek, Calif. 18 38 24N 120 48W
Sutton, Nebr. 12 40 36N 97 52W
Sutwik I., Alaska 30 56 34N 157 12W
Suwannee →, Fla. 11 29 17N 83 10W
Swainsboro, Ga. 11 32 36N 82 20W
Swampscott, Mass. 23 42 28N 70 53W
Swarthmore, Pa. 24 39 54N 75 20W
Sweet Home, Oreg. 16 44 24N 122 44W
Sweetwater, Nev. 18 38 27N 119 9W
Sweetwater, Tex. 13 32 28N 100 25W
Sweetwater →, Wyo. 16 42 31N 107 2W
Swissvale, Pa. 27 40 25N 79 52W
Sylacauga, Ala. 11 33 10N 86 15W
Sylvania, Ga. 11 32 45N 81 38W
Sylvester, Ga. 11 31 32N 83 50W
Syracuse, Kans. 13 37 59N 101 45W
Syracuse, N.Y. 9 43 3N 76 9W

T

Tacoma, Wash. 18 47 14N 122 26W
Taft, Calif. 19 35 8N 119 28W
Taft, Tex. 13 27 59N 97 24W
Tahoe, L., Nev. 18 39 6N 120 2W
Tahoe City, Calif. 18 39 10N 120 9W
Takoma Park, Md. 25 38 58N 77 0W
Talihina, Okla. 13 34 45N 95 3W
Talkeetna, Alaska 30 62 20N 150 6W
Talkeetna Mts., Alaska 30 62 20N 149 0W
Talladega, Ala. 11 33 26N 86 6W
Tallahassee, Fla. 11 30 27N 84 17W
Talleyville, Del. 24 39 48N 75 32W
Tallulah, La. 13 32 25N 91 11W
Tama, Iowa 14 41 58N 92 35W
Tamaqua, Pa. 9 40 48N 75 58W
Tampa, Fla. 11 27 57N 82 27W
Tampa B., Fla. 11 27 50N 82 30W
Tanana, Alaska 30 65 10N 152 4W
Tanana →, Alaska 30 65 10N 151 58W
Taos, N. Mex. 17 36 24N 105 35W
Tappahannock, Va. 10 37 56N 76 52W
Tarboro, N.C. 11 35 54N 77 32W
Tarpon Springs, Fla. 11 28 9N 82 45W
Tarrytown, N.Y. 9 41 4N 73 52W
Tatum, N. Mex. 13 33 16N 103 19W
Tau, W. Samoa 31 14 15S 169 30W
Taunton, Mass. 9 41 54N 71 6W
Tawas City, Mich. 10 44 16N 83 31W
Taylor, Alaska 30 65 40N 164 50W
Taylor, Mich. 26 42 13N 83 15W
Taylor, Nebr. 12 41 46N 99 23W
Taylor, Pa. 9 41 23N 75 43W
Taylor, Tex. 13 30 34N 97 25W
Taylor, Mt., N. Mex. 17 35 14N 107 37W
Taylortown, N.J. 20 40 56N 74 23W
Taylorville, Ill. 14 39 33N 89 18W
Teague, Tex. 13 31 38N 96 17W

Washington

Teaneck, *N.J.* **20** 40 52N 74 1W
Tecopa, *Calif.* **19** 35 51N 116 13W
Tecumseh, *Mich.* **15** 42 0N 83 57W
Tehachapi, *Calif.* **19** 35 8N 118 27W
Tehachapi Mts., *Calif.* **19** 35 0N 118 30W
Tejon Pass, *Calif.* **19** 34 49N 118 53W
Tekamah, *Nebr.* **12** 41 47N 96 13W
Tekoa, *Wash.* **16** 47 14N 117 4W
Telescope Pk., *Calif.* **19** 36 10N 117 5W
Tell City, *Ind.* **15** 37 57N 86 46W
Teller, *Alaska* **30** 65 16N 166 22W
Telluride, *Colo.* **17** 37 56N 107 49W
Temblor Range, *Calif.* **19** 35 20N 119 50W
Temecula, *Calif.* **19** 33 30N 117 9W
Tempe, *Ariz.* **17** 33 25N 111 56W
Temple, *Tex.* **13** 31 6N 97 21W
Temple City, *Calif.* **29** 34 6N 118 2W
Templeton, *Calif.* **18** 35 33N 120 42W
Tenaha, *Tex.* **13** 31 57N 94 15W
Tennessee □ **11** 36 0N 86 30W
Tennessee —►, *Tenn.* **10** 37 4N 88 34W
Tennille, *Ga.* **11** 32 56N 82 48W
Terra Bella, *Calif.* **19** 35 58N 119 3W
Terre Haute, *Ind.* **15** 39 28N 87 25W
Terrebonne B., *La.* **13** 29 5N 90 35W
Terrell, *Tex.* **13** 32 44N 96 17W
Terry, *Mont.* **12** 46 47N 105 19W
Tetlin, *Alaska* **30** 63 8N 142 31W
Tetlin Junction, *Alaska* **30** 63 29N 142 55W
Teton —►, *Mont.* **16** 47 56N 110 31W
Texarkana, *Ark.* **13** 33 26N 94 2W
Texarkana, *Tex.* **13** 33 26N 94 3W
Texas □ **13** 31 40N 98 30W
Texas City, *Tex.* **13** 29 24N 94 54W
Texhoma, *Okla.* **13** 36 30N 101 47W
Texline, *Tex.* **13** 36 23N 103 2W
Texoma, L., *Tex.* **13** 33 50N 96 34W
Thames —►, *Conn.* **9** 41 18N 72 5W
Thatcher, *Ariz.* **17** 32 51N 109 46W
Thatcher, *Colo.* **13** 37 33N 104 7W
Thayer, *Mo.* **13** 36 31N 91 33W
The Dalles, *Oreg.* **16** 45 36N 121 10W
Thedford, *Nebr.* **12** 41 59N 100 35W
Thermopolis, *Wyo.* **16** 43 39N 108 13W
Thibodaux, *La.* **13** 29 48N 90 49W
Thief River Falls, *Minn.* **12** 48 7N 96 10W
Thomas, *Okla.* **13** 35 45N 98 45W
Thomas, *W. Va.* **10** 39 9N 79 30W
Thomaston, *Ga.* **11** 32 53N 84 20W
Thomasville, *Ala.* **11** 31 55N 87 44W
Thomasville, *Ga.* **11** 30 50N 83 59W
Thomasville, *N.C.* **11** 35 53N 80 5W
Thompson, *Utah* **17** 38 58N 109 43W
Thompson —►, *Mo.* **12** 39 46N 93 37W
Thompson Falls, *Mont.* **16** 47 36N 115 21W
Thompson Pk., *Calif.* **16** 41 0N 123 0W
Thousand Oaks, *Calif.* **19** 34 10N 118 50W
Three Forks, *Mont.* **16** 45 54N 111 33W
Three Lakes, *Wis.* **12** 45 48N 89 10W
Three Rivers, *Calif.* **18** 36 26N 118 54W
Three Rivers, *Tex.* **13** 28 28N 98 11W
Three Sisters, *Oreg.* **16** 44 4N 121 51W
Thunder B., *Mich.* **8** 45 0N 83 20W
Tiber Reservoir, *Mont.* **16** 48 19N 111 6W
Tiburon, *Calif.* **28** 37 52N 122 27W
Ticonderoga, *N.Y.* **9** 43 51N 73 26W
Tierra Amarilla, *N. Mex.* **17** 36 42N 106 33W
Tiffin, *Ohio* **15** 41 7N 83 11W
Tifton, *Ga.* **11** 31 27N 83 31W
Tigalda I., *Alaska* **30** 54 6N 165 5W
Tilden, *Nebr.* **12** 42 3N 97 50W
Tilden, *Tex.* **13** 28 28N 98 33W
Tillamook, *Oreg.* **16** 45 27N 123 51W
Tilton, *N.H.* **9** 43 27N 71 36W
Timber Lake, *S. Dak.* **12** 45 26N 101 5W
Timber Mt., *Nev.* **18** 37 6N 116 28W
Tin Mt., *Calif.* **18** 36 50N 117 10W
Tinian, *Pac. Oc.* **31** 15 0N 145 38 E
Tioga, *Pa.* **8** 41 55N 77 8W
Tipton, *Calif.* **18** 36 4N 119 19W
Tipton, *Ind.* **15** 40 17N 86 2W
Tipton, *Iowa* **14** 41 46N 91 8W
Tiptonville, *Tenn.* **13** 36 23N 89 29W
Tishomingo, *Okla.* **13** 34 14N 96 41W
Titusville, *Fla.* **11** 28 37N 80 49W
Titusville, *Pa.* **8** 41 38N 79 41W
Tobyhanna, *Pa.* **9** 41 11N 75 25W
Toccoa, *Ga.* **11** 34 35N 83 19W
Tolageak, *Alaska* **30** 70 2N 162 50W
Toledo, *Ohio* **15** 41 39N 83 33W
Toledo, *Oreg.* **16** 44 37N 123 56W
Toledo, *Wash.* **16** 46 26N 122 51W
Tolleson, *Ariz.* **17** 33 27N 112 16W
Tollhouse, *Calif.* **18** 37 1N 119 24W
Tomah, *Wis.* **12** 43 59N 90 30W
Tomahawk, *Wis.* **12** 45 28N 89 44W
Tomales, *Calif.* **18** 38 15N 122 53W
Tomales B., *Calif.* **18** 38 15N 123 58W
Tombigbee —►, *Ala.* **11** 31 8N 87 57W
Tombstone, *Ariz.* **17** 31 43N 110 4W
Toms Place, *Calif.* **18** 37 34N 118 41W
Toms River, *N.J.* **9** 39 58N 74 12W
Tonalea, *Ariz.* **17** 36 19N 110 56W
Tonasket, *Wash.* **16** 48 42N 119 26W
Tonawanda, *N.Y.* **8** 43 1N 78 53W
Tongue —►, *Mont.* **12** 46 25N 105 52W
Tonkawa, *Okla.* **13** 36 41N 97 18W
Tonopah, *Nev.* **17** 38 4N 117 14W
Tooele, *Utah* **16** 40 32N 112 18W
Topaz, *Calif.* **18** 38 41N 119 30W
Topeka, *Kans.* **12** 39 3N 95 40W
Topock, *Calif.* **19** 34 46N 114 29W
Toppenish, *Wash.* **16** 46 23N 120 19W
Toro Pk., *Calif.* **19** 33 34N 116 24W
Toronto, *Ohio* **8** 40 28N 80 36W
Torrance, *Calif.* **19** 33 50N 118 19W
Torrey, *Utah* **17** 38 18N 111 25W
Torrington, *Conn.* **9** 41 48N 73 7W
Torrington, *Wyo.* **12** 42 4N 104 11W
Tortola, *Virgin Is.* **30** 18 19N 64 45W
Towanda, *Pa.* **9** 41 46N 76 27W
Tower, *Minn.* **12** 47 48N 92 17W
Towner, *N. Dak.* **12** 48 21N 100 25W
Townsend, *Mont.* **16** 46 19N 111 31W
Towson, *Md.* **25** 39 24N 76 36W
Toyah, *Tex.* **13** 31 19N 103 48W
Toyahvale, *Tex.* **13** 30 57N 103 47W
Tracy, *Calif.* **18** 37 44N 121 26W
Tracy, *Minn.* **12** 44 14N 95 37W
Trapper Pk., *Mont.* **16** 45 54N 114 18W
Traverse City, *Mich.* **10** 44 46N 85 38W
Tremonton, *Utah* **16** 41 43N 112 10W
Trenton, *Mo.* **14** 40 5N 93 37W
Trenton, *N.J.* **9** 40 14N 74 46W
Trenton, *Nebr.* **12** 40 11N 101 1W
Trenton, *Tenn.* **13** 35 59N 88 56W
Tres Pinos, *Calif.* **18** 36 48N 121 19W
Tribune, *Kans.* **12** 38 28N 101 45W
Trinidad, *Colo.* **13** 37 10N 104 31W
Trinity, *Tex.* **13** 30 57N 95 22W
Trinity —►, *Calif.* **16** 41 11N 123 42W
Trinity —►, *Tex.* **13** 29 45N 94 43W
Trinity Range, *Nev.* **16** 40 15N 118 45W
Trion, *Ga.* **11** 34 33N 85 19W
Tripp, *S. Dak.* **12** 43 13N 97 58W
Trona, *Calif.* **19** 35 46N 117 23W
Tropic, *Utah* **17** 37 37N 112 5W
Troup, *Tex.* **13** 32 9N 95 7W
Troy, *Ala.* **11** 31 48N 85 58W
Troy, *Idaho* **16** 46 44N 116 46W
Troy, *Kans.* **13** 39 47N 95 5W
Troy, *Mich.* **26** 42 35N 83 9W
Troy, *Mo.* **14** 38 59N 90 59W
Troy, *Mont.* **16** 48 28N 115 53W
Troy, *N.Y.* **9** 42 44N 73 41W
Troy, *Ohio* **15** 40 2N 84 12W
Truckee, *Calif.* **18** 39 20N 120 11W
Trujillo, *N. Mex.* **13** 35 32N 104 42W
Truk,
U.S. Pac. Is. Trust Terr. **31** 7 25N 151 46 E
Trumann, *Ark.* **13** 35 41N 90 31W
Trumbull, Mt., *Ariz.* **17** 36 25N 113 8W
Truth or Consequences,
N. Mex. **17** 33 8N 107 15W
Tryon, *N.C.* **11** 35 13N 82 14W
Tuba City, *Ariz.* **17** 36 8N 111 14W
Tucson, *Ariz.* **17** 32 13N 110 58W
Tucumcari, *N. Mex.* **13** 35 10N 103 44W
Tugidak I., *Alaska* **30** 56 30N 154 40W
Tulare, *Calif.* **18** 36 13N 119 21W
Tulare Lake Bed, *Calif.* **18** 36 0N 119 48W
Tularosa, *N. Mex.* **17** 33 5N 106 1W
Tulia, *Tex.* **13** 34 32N 101 46W
Tullahoma, *Tenn.* **11** 35 22N 86 13W
Tulsa, *Okla.* **13** 36 10N 95 55W
Tumwater, *Wash.* **16** 47 1N 122 54W
Tunica, *Miss.* **13** 34 41N 90 23W
Tunkhannock, *Pa.* **9** 41 32N 75 57W
Tuntutuliak, *Alaska* **30** 60 22N 162 38W
Tuolumne, *Calif.* **18** 37 58N 120 15W
Tuolumne —►, *Calif.* **18** 37 36N 121 13W
Tupelo, *Miss.* **11** 34 16N 88 43W
Tupman, *Calif.* **19** 35 18N 119 21W
Tupper Lake, *N.Y.* **9** 44 14N 74 28W
Turlock, *Calif.* **18** 37 30N 120 51W
Turner, *Mont.* **16** 48 51N 108 24W
Turners Falls, *Mass.* **9** 42 36N 72 33W
Turon, *Kans.* **13** 37 48N 98 26W
Turtle Creek, *Pa.* **27** 40 24N 79 49W
Turtle Lake, *N. Dak.* **12** 47 31N 100 53W
Turtle Lake, *Wis.* **12** 45 24N 92 8W
Tuscaloosa, *Ala.* **11** 33 12N 87 34W
Tuscola, *Ill.* **15** 39 48N 88 17W
Tuscola, *Tex.* **13** 32 12N 99 48W
Tuscumbia, *Ala.* **11** 34 44N 87 42W
Tuskegee, *Ala.* **11** 32 25N 85 42W
Tuttle, *N. Dak.* **12** 47 9N 100 0W
Tutuila, *Amer. Samoa* **31** 14 19S 170 50W
Twain Harte, *Calif.* **18** 38 2N 120 14W
Twentynine Palms, *Calif.* **19** 34 8N 116 3W
Twin Bridges, *Mont.* **16** 45 33N 112 20W
Twin Falls, *Idaho* **16** 42 34N 114 28W
Twin Valley, *Minn.* **12** 47 16N 96 16W
Twisp, *Wash.* **16** 48 22N 120 7W
Two Harbors, *Minn.* **12** 47 2N 91 40W
Two Rivers, *Wis.* **10** 44 9N 87 34W
Tyler, *Minn.* **12** 44 18N 96 8W
Tyler, *Tex.* **13** 32 21N 95 18W

U

U.S.A. = United States
of America ■,
N. Amer. **6** 37 0N 96 0W
Uhrichsville, *Ohio* **8** 40 24N 81 21W
Uinta Mts., *Utah* **16** 40 45N 110 30W
Ukiah, *Calif.* **18** 39 9N 123 13W
Ulak I., *Alaska* **30** 51 22N 178 57W
Ulysses, *Kans.* **13** 37 35N 101 22W
Umatilla, *Oreg.* **16** 45 55N 119 21W
Umiat, *Alaska* **30** 69 22N 152 8W
Uminak I., *Alaska* **30** 53 20N 168 20W
Umnak I., *Alaska* **30** 53 15N 168 20W
Umpqua —►, *Oreg.* **16** 43 40N 124 12W
Unadilla, *N.Y.* **9** 42 20N 75 19W
Unalaska, *Alaska* **30** 53 53N 166 32W
Uncompahgre Peak,
Colo. **17** 38 4N 107 28W
Unimak I., *Alaska* **30** 54 45N 164 0W
Unimak Pass, *Alaska* **30** 54 15N 164 30W
Union, *Miss.* **13** 32 34N 89 7W
Union, *Mo.* **14** 38 27N 91 0W
Union, *N.J.* **20** 40 42N 74 15W
Union, *S.C.* **11** 34 43N 81 37W
Union, Mt., *Ariz.* **17** 34 34N 112 21W
Union City, *Calif.* **28** 37 36N 122 3W
Union City, *N.J.* **20** 40 45N 74 2W
Union City, *Pa.* **8** 41 54N 79 51W
Union City, *Tenn.* **13** 36 26N 89 3W
Union Gap, *Wash.* **16** 46 33N 120 28W
Union Springs, *Ala.* **11** 32 9N 85 43W
Uniontown, *Pa.* **10** 39 54N 79 44W
Unionville, *Mo.* **14** 40 29N 93 1W
United States of
America ■, *N. Amer.* **6** 37 0N 96 0W
University Heights, *Ohio* **27** 41 29N 81 31W
Upolu Pt., *Hawaii* **30** 20 16N 155 52W
Upper Alkali Lake, *Calif.* **16** 41 47N 120 8W
Upper Darby, *Pa.* **24** 39 57N 75 16W
Upper Klamath L., *Oreg.* **16** 42 25N 121 55W
Upper Lake, *Calif.* **18** 39 10N 122 54W
Upper Red L., *Minn.* **12** 48 8N 94 45W
Upper Sandusky, *Ohio* **15** 40 50N 83 17W
Upper St. Clair, *Pa.* **27** 40 21N 80 3W
Upton, *Wyo.* **12** 44 6N 104 38W
Urbana, *Ill.* **15** 40 7N 88 12W
Urbana, *Ohio* **15** 40 7N 83 45W
Utah □ **16** 39 20N 111 30W
Utah, L., *Utah* **16** 40 10N 111 58W
Ute Creek —►, *N. Mex.* **13** 35 21N 103 50W
Utica, *N.Y.* **9** 43 6N 75 14W
Uvalde, *Tex.* **13** 29 13N 99 47W

V

Vacaville, *Calif.* **18** 38 21N 121 59W
Valdez, *Alaska* **30** 61 7N 146 16W
Valdosta, *Ga.* **11** 30 50N 83 17W
Vale, *Oreg.* **16** 43 59N 117 15W
Valentine, *Nebr.* **12** 42 52N 100 33W
Valentine, *Tex.* **13** 30 35N 104 30W
Valier, *Mont.* **16** 48 18N 112 16W
Vallejo, *Calif.* **18** 38 7N 122 14W
Valley Center, *Calif.* **19** 33 13N 117 2W
Valley City, *N. Dak.* **12** 46 55N 98 0W
Valley Falls, *Oreg.* **16** 42 29N 120 17W
Valley Springs, *Calif.* **18** 38 12N 120 50W
Valley Stream, *N.Y.* **21** 40 40N 73 42W
Valparaiso, *Ind.* **15** 41 28N 87 4W
Van Alstyne, *Tex.* **13** 33 25N 96 35W
Van Buren, *Ark.* **13** 35 26N 94 21W
Van Buren, *Maine* **11** 47 10N 67 58W
Van Buren, *Mo.* **13** 37 0N 91 1W
Van Horn, *Tex.* **13** 31 3N 104 50W
Van Tassell, *Wyo.* **12** 42 40N 104 5W
Van Wert, *Ohio* **15** 40 52N 84 35W
Vancouver, *Wash.* **18** 45 38N 122 40W
Vandalia, *Ill.* **14** 38 58N 89 6W
Vandalia, *Mo.* **14** 39 19N 91 29W
Vandenburg, *Calif.* **19** 34 35N 120 33W
Vandergrift, *Pa.* **8** 40 36N 79 34W
Variadero, *N. Mex.* **13** 35 43N 104 17W
Vassar, *Mich.* **10** 43 22N 83 35W
Vaughn, *Mont.* **16** 47 33N 111 33W
Vaughn, *N. Mex.* **17** 34 36N 105 13W
Vega, *Tex.* **13** 35 15N 102 26W
Vega Baja, *Puerto Rico* **30** 18 27N 66 23W
Velva, *N. Dak.* **12** 48 4N 100 56W
Venetie, *Alaska* **30** 67 1N 146 25W
Ventucopa, *Calif.* **19** 34 50N 119 29W
Ventura, *Calif.* **19** 34 17N 119 18W
Verdigre, *Nebr.* **12** 42 36N 98 2W
Vermilion, B., *La.* **13** 29 45N 91 55W
Vermilion L., *Minn.* **12** 47 53N 92 26W
Vermillion, *S. Dak.* **12** 42 47N 96 56W
Vermont □ **9** 44 0N 73 0W
Vernal, *Utah* **16** 40 27N 109 32W
Vernalis, *Calif.* **18** 37 36N 121 17W
Vernon, *Tex.* **13** 34 9N 99 17W
Vero Beach, *Fla.* **11** 27 38N 80 24W
Vicksburg, *Mich.* **15** 42 7N 85 32W
Vicksburg, *Miss.* **13** 32 21N 90 53W
Victor, *Colo.* **12** 38 43N 105 9W
Victoria, *Kans.* **12** 38 52N 99 9W
Victoria, *Tex.* **13** 28 48N 97 0W
Victorville, *Calif.* **19** 34 32N 117 18W
Vidalia, *Ga.* **11** 32 13N 82 25W
Vienna, *Ill.* **13** 37 25N 88 54W
Vienna, *Va.* **25** 38 54N 77 17W
Vieques, Isla de,
Puerto Rico **30** 18 8N 65 25W
Villanueva, *N. Mex.* **17** 35 16N 105 22W
Ville Platte, *La.* **13** 30 41N 92 17W
Villisca, *Iowa* **14** 40 56N 94 59W
Vincennes, *Ind.* **15** 38 41N 87 32W
Vincent, *Calif.* **19** 34 33N 118 11W
Vineland, *N.J.* **10** 39 29N 75 2W
Vinita, *Okla.* **13** 36 39N 95 9W
Vinton, *Iowa* **14** 42 10N 92 1W
Vinton, *La.* **13** 30 11N 93 35W
Virgin —►, *Nev.* **17** 36 28N 114 21W
Virgin Gorda, *Virgin Is.* **30** 18 30N 64 26W
Virgin Is. (British) ■,
W. Indies **30** 18 30N 64 30W
Virginia, *Minn.* **12** 47 31N 92 32W
Virginia □ **10** 37 30N 78 45W
Virginia Beach, *Va.* **10** 36 51N 75 59W
Virginia City, *Mont.* **16** 45 18N 111 56W
Virginia City, *Nev.* **18** 39 19N 119 39W
Viroqua, *Wis.* **12** 43 34N 90 53W
Visalia, *Calif.* **18** 36 20N 119 18W
Vista, *Calif.* **19** 33 12N 117 14W
Volborg, *Mont.* **12** 45 51N 105 41W
Vulcan, *Mich.* **10** 45 47N 87 53W

W

Wabash, *Ind.* **15** 40 48N 85 49W
Wabash —►, *Ill.* **10** 37 48N 88 2W
Wabeno, *Wis.* **10** 45 26N 88 39W
Wabuska, *Nev.* **16** 39 9N 119 11W
Waco, *Tex.* **13** 31 33N 97 9W
Wadena, *Minn.* **12** 46 26N 95 8W
Wadesboro, *N.C.* **11** 34 58N 80 5W
Wadsworth, *Nev.* **16** 39 38N 119 17W
Wagon Mound, *N. Mex.* **13** 36 1N 104 42W
Wagoner, *Okla.* **13** 35 58N 95 22W
Wahiawa, *Hawaii* **30** 21 30N 158 2W
Wahoo, *Nebr.* **12** 41 13N 96 37W
Wahpeton, *N. Dak.* **12** 46 16N 96 36W
Wailuku, *Hawaii* **30** 20 53N 156 30W
Wainiha, *Hawaii* **30** 22 9N 159 34W
Wainwright, *Alaska* **30** 70 38N 160 2W
Waipahu, *Hawaii* **30** 21 23N 158 1W
Waitsburg, *Wash.* **16** 46 16N 118 9W
Wake Forest, *N.C.* **11** 35 59N 78 30W
Wake I., *Pac. Oc.* **31** 19 18N 166 36 E
Wakefield, *Mass.* **23** 42 30N 71 4W
Wakefield, *Mich.* **12** 46 29N 89 56W
Walcott, *Wyo.* **16** 41 46N 106 51W
Walden, *Colo.* **16** 40 44N 106 17W
Walden, *N.Y.* **9** 41 34N 74 11W
Waldport, *Oreg.* **16** 44 26N 124 4W
Waldron, *Ark.* **13** 34 54N 94 5W
Wales, *Alaska* **30** 65 37N 168 5W
Walker, *Minn.* **12** 47 6N 94 35W
Walker L., *Nev.* **16** 38 42N 118 43W
Wall, *S. Dak.* **12** 44 0N 102 8W
Wall Street, *N.Y.* **20** 40 42N 74 0W
Walla Walla, *Wash.* **16** 46 4N 118 20W
Wallace, *Idaho* **16** 47 28N 115 56W
Wallace, *N.C.* **11** 34 44N 77 59W
Wallace, *Nebr.* **12** 40 50N 101 10W
Wallowa, *Oreg.* **16** 45 34N 117 32W
Wallowa Mts., *Oreg.* **16** 45 20N 117 30W
Wallula, *Wash.* **16** 46 5N 118 54W
Walnut Creek, *Calif.* **28** 37 53N 122 3W
Walnut Ridge, *Ark.* **13** 36 4N 90 57W
Walsenburg, *Colo.* **13** 37 38N 104 47W
Walsh, *Colo.* **13** 37 23N 102 17W
Walterboro, *S.C.* **11** 32 55N 80 40W
Walters, *Okla.* **13** 34 22N 98 19W
Waltham, *Mass.* **23** 42 23N 71 13W
Waltman, *Wyo.* **16** 43 4N 107 12W
Walton, *N.Y.* **9** 42 10N 75 8W
Wamego, *Kans.* **12** 39 12N 96 18W
Wapakoneta, *Ohio* **15** 40 34N 84 12W
Wapato, *Wash.* **16** 46 27N 120 25W
Wappingers Falls, *N.Y.* **9** 41 36N 73 55W
Wapsipinicon —►, *Iowa* **14** 41 44N 90 19W
Ward Mt., *Calif.* **18** 37 12N 118 54W
Ware, *Mass.* **9** 42 16N 72 14W
Wareham, *Mass.* **9** 41 46N 70 43W
Warm Springs, *Nev.* **17** 38 10N 116 20W
Warner Mts., *Calif.* **16** 41 40N 120 15W
Warner Robins, *Ga.* **11** 32 37N 83 36W
Warren, *Ark.* **13** 33 37N 92 4W
Warren, *Mich.* **26** 42 31N 83 0W
Warren, *Minn.* **12** 48 12N 96 46W
Warren, *Ohio* **8** 41 14N 80 49W
Warren, *Pa.* **8** 41 51N 79 9W
Warrensburg, *Mo.* **12** 38 46N 93 44W
Warrenton, *Oreg.* **18** 46 10N 123 56W
Warrington, *Fla.* **11** 30 23N 87 17W
Warroad, *Minn.* **12** 48 54N 95 19W
Warsaw, *Ind.* **15** 41 14N 85 51W
Warwick, *R.I.* **9** 41 42N 71 28W
Wasatch Ra., *Utah* **16** 40 30N 111 15W
Wasco, *Calif.* **19** 35 36N 119 20W
Wasco, *Oreg.* **16** 45 36N 120 42W
Waseca, *Minn.* **12** 44 5N 93 30W
Washburn, *N. Dak.* **12** 47 17N 101 2W
Washburn, *Wis.* **12** 46 40N 90 54W
Washington, *D.C.* **25** 38 53N 77 2W
Washington, *Ga.* **11** 33 44N 82 44W
Washington, *Ind.* **15** 38 40N 87 10W
Washington, *Iowa* **14** 41 18N 91 42W
Washington, *Mo.* **14** 38 33N 91 1W
Washington, *N.C.* **11** 35 33N 77 3W

Washington

INTRODUCTION TO WORLD GEOGRAPHY

PLANET EARTH

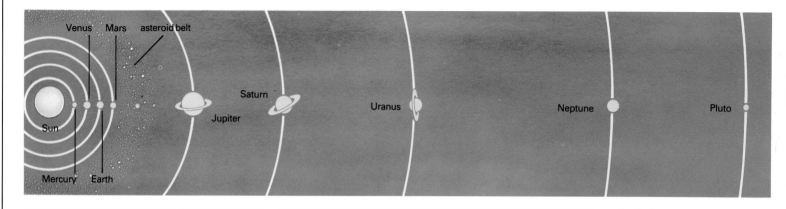

Venus Mars asteroid belt
Sun
Saturn
Jupiter
Uranus
Neptune
Pluto
Mercury Earth

THE SOLAR SYSTEM

A minute part of one of the billions of galaxies (collections of stars) that comprise the Universe, the Solar System lies some 27,000 light-years from the center of our own galaxy, the 'Milky Way'. Thought to be over 4,700 million years old, it consists of a central sun with nine planets and their moons revolving around it, attracted by its gravitational pull. The planets orbit the Sun in the same direction – counter-clockwise when viewed from the Northern Heavens – and almost in the same plane. Their orbital paths, however, vary enormously.

The Sun's diameter is 109 times that of Earth, and the temperature at its core – caused by continuous thermonuclear fusion of hydrogen into helium – is estimated to be 27 million degrees Fahrenheit. It is the Solar System's only source of light and heat.

PROFILE OF THE PLANETS

	Mean distance from Sun (million mi)	Mass (Earth = 1)	Period of orbit	Period of rotation (in days)	Diameter (mi)	Number of known satellites
Mercury	36.4	0.06	88 days	58.67	3,049	0
Venus	67.3	0.8	224.7 days	243.0	7,565	0
Earth	93.5	1.0	365.24 days	0.99	7,973	1
Mars	142.1	0.1	1.88 years	1.02	4,246	2
Jupiter	486.2	317.8	11.86 years	0.41	89,250	16
Saturn	891.9	95.2	29.63 years	0.42	75,000	17
Uranus	1795.2	14.5	83.97 years	0.45	32,500	15
Neptune	2814.2	17.2	164.8 years	0.67	30,250	8
Pluto	3683.9	0.002	248.63 years	6.38	1,500	1

All planetary orbits are elliptical in form, but only Pluto and Mercury follow paths that deviate noticeably from a circular one. Near Perihelion - its closest approach to the Sun - Pluto actually passes inside the orbit of Neptune, an event that last occurred in 1983. Pluto will not regain its station as outermost planet until February 1999.

Northern Spring Equinox
Southern Autumn Equinox

Equinox is one of the two times in the year when day and night are of equal length due to the Sun being overhead at the Equator.

21 March

Northern Summer Solstice

Northern Winter Solstice

21 June

SUN

21 December

Southern Winter Solstice

Southern Summer Solstice

21 September

Solstice is one of the two times in the year when the Sun is overhead at one of the Tropics, 23½° north or south of the equator.

Southern Spring Equinox
Northern Autumn Equinox

21 June
66½°
23½° SHORT NIGHT
0° LONG DAY
Equator
23½° LONG NIGHT
Sun's rays
13½ hours daylight
13½ hours daylight
12 hours daylight
SHORT DAY
Antarctic Circle: 24 hours darkness
S. Pole: 6 months darkness
10½ hours daylight

N. Pole: 6 months daylight
Arctic Circle: 24 hours daylight

21 December
66½°
N. Pole: 6 months darkness
Arctic Circle: 24 hours darkness
10½ hours daylight
12 hours daylight
SHORT DAY
LONG NIGHT
Equator
Tropic 23½° of Cancer
LONG DAY
SHORT NIGHT
0°
23½° Tropic of Capricorn
13½ hours daylight
12 hours daylight
Antarctic Circle: 24 hours daylight
S. Pole: 6 months daylight

THE SEASONS

The Earth revolves around the Sun once a year in a counter-clockwise direction, tilted at a constant angle 66½°. In June, the northern hemisphere is tilted towards the Sun: as a result it receives more hours of sunshine in a day and therefore has its warmest season, summer. By December, the Earth has rotated halfway round the Sun so that the southern hemisphere is tilted towards the Sun and has its summer; the hemisphere that is tilted away from the Sun has winter. On 21 June the Sun is directly overhead at the Tropic of Cancer (23½° N), and this is midsummer in the northern hemisphere. Midsummer in the southern hemisphere occurs on 21 December, when the Sun is overhead at the Tropic of Capricorn (23½° S).

DAY & NIGHT

The Sun appears to rise in the east, reach its highest point at noon, and then set in the west, to be followed by night. In reality it is not the Sun that is moving but the Earth revolving from west to east. Due to the tilting of the Earth the length of day and night varies from place to place and month to month.

At the summer solstice in the northern hemisphere (21 June), the Arctic has total daylight and the Antarctic total darkness. The opposite occurs at the winter solstice (21 December). At the equator, the length of day and night are almost equal all year, at latitude 30° the length of day varies from about 14 hours to 10 hours, and at latitude 50° from about 16 hours to about 8 hours.

TIME

Year: the time taken by the Earth to revolve around the Sun, or 365.24 days.

Month: the approximate time taken by the Moon to revolve around the Earth. The 12 months of the year in fact vary from 28 (29 in a Leap Year) to 31 days.

Week: an artificial period of 7 days, not based on astronomical time.

Day: the time taken by the Earth to complete one rotation on its axis.

Hour: 24 hours make one day. Usually the day is divided into hours AM (ante meridiem or before noon) and PM (post meridiem or after noon), although most timetables now use the 24-hour system, from midnight to midnight.

SUNRISE

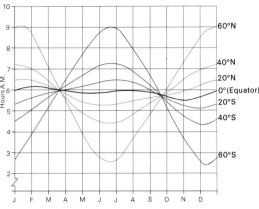

SUNSET

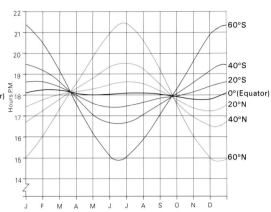

THE MOON

Distance from Earth: 221,463 mi – 252,710 mi; Mean diameter: 2,160 mi; Mass: approx. 1/81 that of Earth; Surface gravity: one sixth of Earth's; Daily range of temperature at lunar equator: 360°F; Average orbital speed: 2,300 mph

PHASES OF THE MOON

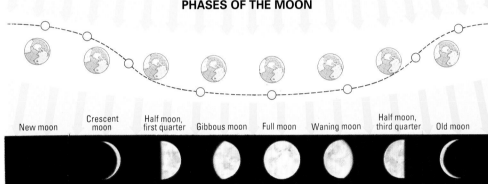

| New moon | Crescent moon | Half moon, first quarter | Gibbous moon | Full moon | Waning moon | Half moon, third quarter | Old moon |

The Moon rotates more slowly than the Earth, making one complete turn on its axis in just over 27 days. Since this corresponds to its period of revolution around the Earth, the Moon always presents the same hemisphere or face to us, and we never see 'the dark side'. The interval between one full Moon and the next (and between new Moons) is about 29¹/₂ days - a lunar month. The apparent changes in the shape of the Moon are caused by its changing position in relation to the Earth; like the planets, it produces no light of its own and shines only by reflecting the rays of the Sun.

ECLIPSES

When the Moon passes between the Sun and the Earth it causes a partial eclipse of the Sun (1) if the Earth passes through the Moon's outer shadow (P), or a total eclipse (2) if the inner cone shadow crosses the Earth's surface. In a lunar eclipse, the Earth's shadow crosses the Moon and, again, provides either a partial or total eclipse. Eclipses of the Sun and the Moon do not occur every month because of the 5° difference between the plane of the Moon's orbit and the plane in which the Earth moves. In the 1990s only 14 lunar eclipses are possible, for example, seven partial and seven total; each is visible only from certain, and variable, parts of the world. The same period witnesses 13 solar eclipses - six partial (or annular) and seven total.

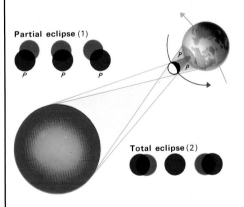

Partial eclipse (1)

Total eclipse (2)

Lunar eclipse

TIDES

The daily rise and fall of the ocean's tides are the result of the gravitational pull of the Moon and that of the Sun, though the effect of the latter is only 46.6% as strong as that of the Moon. This effect is greatest on the hemisphere facing the Moon and causes a tidal 'bulge'. When lunar and solar forces pull together, with Sun, Earth and Moon in line (near new and full Moons), higher 'spring tides' (and lower low tides) occur; when lunar and solar forces are least coincidental with the Sun and Moon at an angle (near the Moon's first and third quarters), 'neap tides' occur, which have a small tidal range.

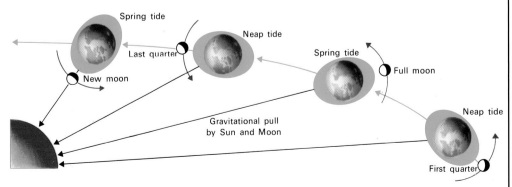

Spring tide

Neap tide

Last quarter

Spring tide

New moon

Full moon

Gravitational pull by Sun and Moon

Neap tide

First quarter

RESTLESS EARTH

THE EARTH'S STRUCTURE

Upper mantle (c. 230 mi)

Crust (average 3-30 mi)

Transition zone (370 mi)

Outer core (1,300 mi)

Lower mantle (1,050 mi)

Inner core (1,650 mi)

CONTINENTAL DRIFT

About 200 million years ago the original Pangaea landmass began to split into two continental groups, which further separated over time to produce the present day configuration.

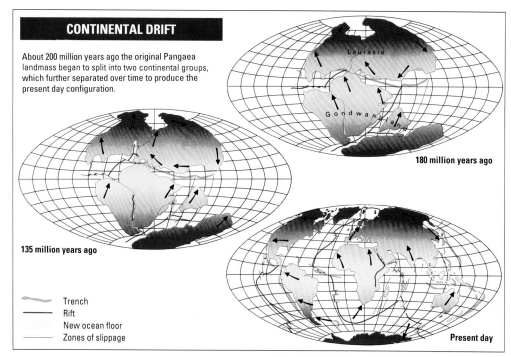

Laurasia

Gondwanaland

180 million years ago

135 million years ago

Present day

Trench
Rift
New ocean floor
Zones of slippage

EARTHQUAKES

Earthquake magnitude is usually rated according to either the Richter or the Modified Mercalli scale, both devised by seismologists in the 1930s. The Richter scale measures absolute earthquake power with mathematical precision: each step upwards represents a tenfold increase in shockwave amplitude. Theoretically, there is no upper limit, but the largest earthquakes measured have been rated at between 8.8 and 8.9. The 12-point Mercalli scale, based on observed effects, is often more meaningful, ranging from I (earthquakes noticed only by seismographs) to XII (total destruction); intermediate points include V (people awakened at night; unstable objects overturned), VII (collapse of ordinary buildings; chimneys and monuments fall) and IX (conspicuous cracks in ground; serious damage to reservoirs).

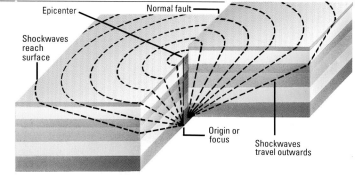

Epicenter
Normal fault
Shockwaves reach surface
Origin or focus
Shockwaves travel outwards

NOTABLE EARTHQUAKES SINCE 1900

Year	Location	Mag.	Deaths
1906	San Francisco, USA	8.3	503
1906	Valparaiso, Chile	8.6	22,000
1908	Messina, Italy	7.5	83,000
1915	Avezzano, Italy	7.5	30,000
1920	Gansu (Kansu), China	8.6	180,000
1923	Yokohama, Japan	8.3	143,000
1927	Nan Shan, China	8.3	200,000
1932	Gansu (Kansu), China	7.6	70,000
1934	Bihar, India/Nepal	8.4	10,700
1935	Quetta, India*	7.5	60,000
1939	Chillan, Chile	8.3	28,000
1939	Erzincan, Turkey	7.9	30,000
1960	Agadir, Morocco	5.8	12,000
1962	Khorasan, Iran	7.1	12,230
1963	Skopje, Yugoslavia	6.0	1,000
1964	Anchorage, Alaska	8.4	131
1968	N. E. Iran	7.4	12,000
1970	N. Peru	7.7	66,794
1972	Managua, Nicaragua	6.2	5,000
1974	N. Pakistan	6.3	5,200
1976	Guatemala	7.5	22,778
1976	Tangshan, China	8.2	650,000
1978	Tabas, Iran	7.7	25,000
1980	El Asnam, Algeria	7.3	20,000
1980	S. Italy	7.2	4,800
1985	Mexico City, Mexico	8.1	4,200
1988	N.W. Armenia	6.8	55,000
1990	N. Iran	7.7	36,000

The highest magnitude recorded on the Richter scale was 8.9, in Japan on 2 March 1933 (2,990 deaths). The most devastating quake ever was in Shaanxi (Shensi) province, central China, on 24 January 1566, when an estimated 830,000 people were killed.

* now Pakistan

DISTRIBUTION OF EARTHQUAKES

Major earthquake zones
Areas experiencing frequent earthquakes

Earthquakes are a series of rapid vibrations originating from the slipping or faulting of parts of the Earth's crust when stresses within build up to breaking point. They usually happen at depths varying from 5 mi to 20 mi. Severe earthquakes cause extensive damage when they take place in populated areas, destroying structures and severing communications. Most initial loss of life occurs due to secondary causes such as falling masonry, fires and flooding.

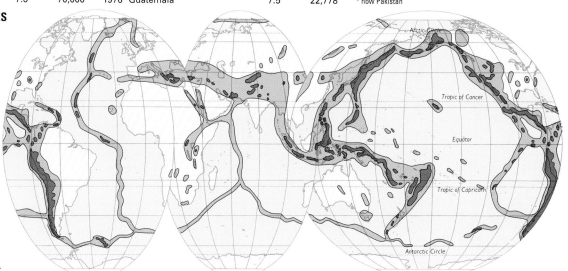

Arctic Circle
Tropic of Cancer
Equator
Tropic of Capricorn
Antarctic Circle

PLATE TECTONICS

The drifting of the continents is a feature unique to Planet Earth. The complementary, almost jigsaw-puzzle fit of the coastlines on each side of the Atlantic Ocean inspired Alfred Wegener's theory of continental drift in 1915. The theory suggested that an ancient super-continent, which Wegener named Pangaea, incorporated all of the Earth's land masses and gradually split up to form today's continents.

The original debate about continental drift was a prelude to a more radical idea: plate tectonics. The basic theory is that the Earth's crust is made up of a series of rigid plates which float on a soft layer of the mantle and are moved about by continental convection currents in the Earth's interior. These plates diverge and converge along margins marked by earthquakes, volcanoes and other seismic activity. Plates diverge from mid-ocean ridges where molten lava pushes upwards and forces the plates apart at a rate of up to 1.5 inches a year; converging plates form either a trench (where the oceanic plate sinks below the lighter continental rock) or mountain ranges (where two continents collide).

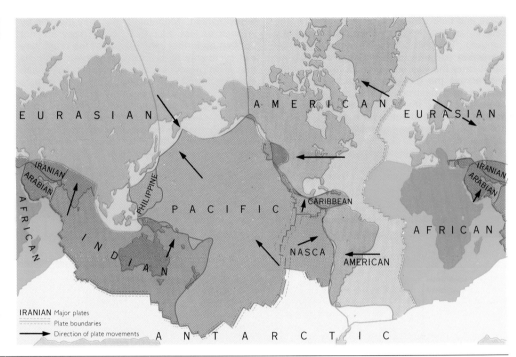

IRANIAN Major plates
- - - - - Plate boundaries
→ Direction of plate movements

VOLCANOES

The word 'volcano' derives from the island of Vulcano off Sicily, in the Mediterranean Sea. In classical times the people of this area thought that Vulcano was the chimney of the forge of Vulcan, blacksmith of the Roman gods. Today volcanoes might be the subject of scientific study but they remain both dramatic and unpredictable, if not exactly supernatural: in 1991 Mount Pinatubo, about 60 miles north of the Philippines capital Manila, suddenly burst into life after more than six centuries of lying dormant.

Most of the world's active volcanoes occur in a belt around the Pacific Ocean, on the edge of the Pacific plate, called the 'ring of fire'. Indonesia has the greatest concentration with 90 volcanoes, 12 of which are active. The most famous, Krakatau, erupted in 1883 with such force that the resulting tidal wave killed 36,000 people and tremors were felt as far away as the English Channel.

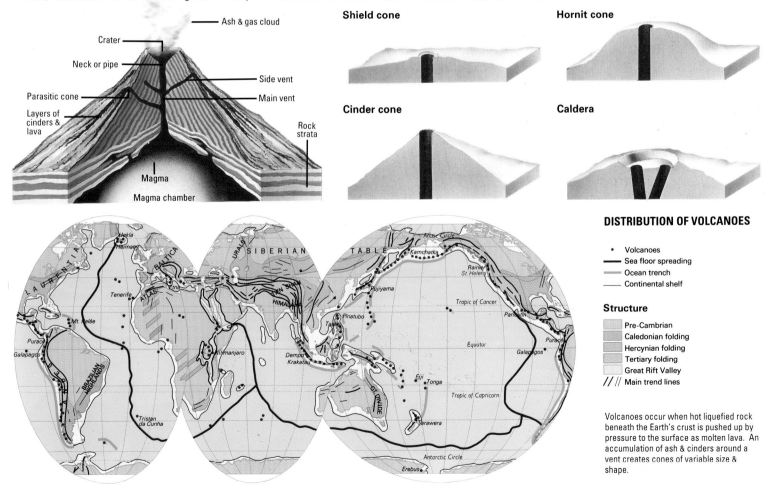

DISTRIBUTION OF VOLCANOES

- • Volcanoes
- ▬ Sea floor spreading
- ▬ Ocean trench
- ▬ Continental shelf

Structure

- Pre-Cambrian
- Caledonian folding
- Hercynian folding
- Tertiary folding
- Great Rift Valley
- /// Main trend lines

Volcanoes occur when hot liquefied rock beneath the Earth's crust is pushed up by pressure to the surface as molten lava. An accumulation of ash & cinders around a vent creates cones of variable size & shape.

OCEANS

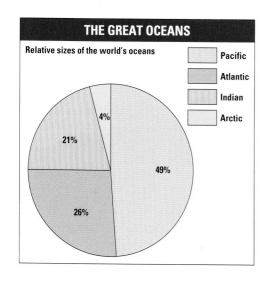
The Earth is a watery planet: more than 70.9% of its surface – almost 140,000,000 square miles – is covered by the oceans and seas. The mighty Pacific alone accounts for nearly 36% of the total, and 49% of the sea area. Gravity holds in around 320 million cubic miles of water, of which over 97% is saline.

The vast underwater world starts in the shallows of the seaside and plunges to depths of more than 36,000 feet. The continental shelf, part of the land mass, drops gently to around 600 feet; here the seabed falls away suddenly at and angle of 3 - 6° - the continental slope. The third stage, called the continental rise, is more gradual with gradients varying from 1 in 100 to 1 in 700. At an average depth of about 16,000 feet there begins the aptly-named abyssal plain, immense submarine expanses far beyond the reach of sunlight, where few creatures are able to survive.

From these plains rise volcanoes which, taken from base to top, rival and even surpass the biggest continental mountains in height. Mount Kea, on Hawaii, reaches a total of 33,400 feet, almost 4,500 feet more than Mount Everest, although scarcely 40% is visible above sea level.

In addition there are underwater mountain chains up to 600 miles across, whose peaks sometimes appear above sea level as islands such as Iceland and Tristan da Cunha.

THE OCEAN DEPTHS

Average and maximum depths of the world's great oceans, in feet

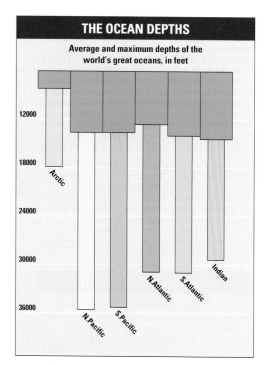

OCEAN CURRENTS

[Cold currents are shown in blue, warm currents in red]

WINTER IN NORTHERN HEMISPHERE

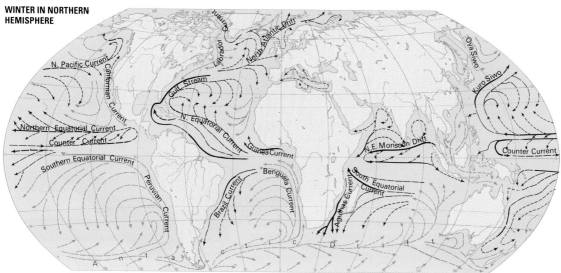

SUMMER IN NORTHERN HEMISPHERE

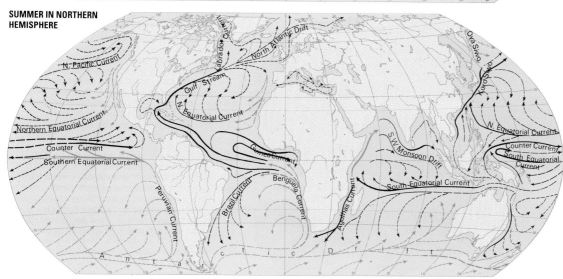

Moving immense quantities of energy as well as billions of tonnes of water every hour, the ocean currents are a vital part of the great heat engine that drives the Earth's climate. They themselves are produced by a twofold mechanism. At the surface, winds push huge masses of water before them; in the deep ocean, below an abrupt temperature gradient that separates the churning surface waters from the still depths, density variations cause slow vertical movements. The pattern of circulation of the great surface currents is determined by the displacement known as the Coriolis effect. As the Earth turns beneath a moving object - whether it is a tennis ball or a vast mass of water - it appears to be deflected to one side. The deflection is most obvious near the equator, where the Earth's surface is spinning eastward at 1,050 mph; currents moving poleward are curved clockwise in the northern hemisphere and counter-clockwise in the southern.

The result is a system of spinning circles known as gyres. The Coriolis effect piles up water on the left of each gyre, creating a narrow, fast-moving stream that is matched by a slower, broader returning current on the right. North and south of the equator, the fastest currents are located in the west and in the east respectively. In each case, warm water moves from the equator and cold water returns to it. Cold currents often bring an upwelling of nutrients with them, supporting the world's most economically important fisheries.

Depending on the prevailing winds, some currents on or near the equator may reverse their direction in the course of the year - a seasonal variation on which Asian monsoon rains depend, and whose occasional failure can bring disaster to millions.

PLATE TECTONICS

The drifting of the continents is a feature unique to Planet Earth. The complementary, almost jigsaw-puzzle fit of the coastlines on each side of the Atlantic Ocean inspired Alfred Wegener's theory of continental drift in 1915. The theory suggested that an ancient super-continent, which Wegener named Pangaea, incorporated all of the Earth's land masses and gradually split up to form today's continents.

The original debate about continental drift was a prelude to a more radical idea: plate tectonics. The basic theory is that the Earth's crust is made up of a series of rigid plates which float on a soft layer of the mantle and are moved about by continental convection currents in the Earth's interior. These plates diverge and converge along margins marked by earthquakes, volcanoes and other seismic activity. Plates diverge from mid-ocean ridges where molten lava pushes upwards and forces the plates apart at a rate of up to 1.5 inches a year; converging plates form either a trench (where the oceanic plate sinks below the lighter continental rock) or mountain ranges (where two continents collide).

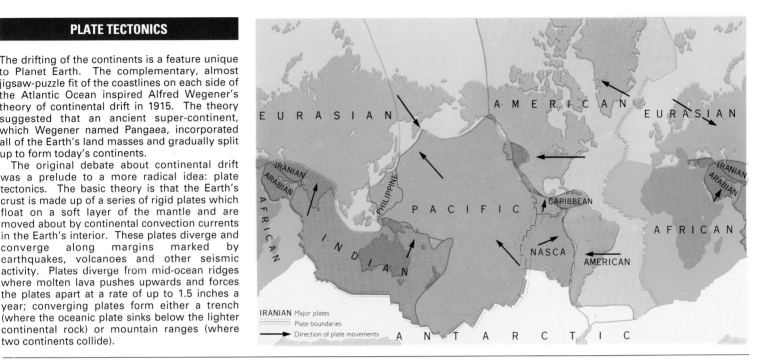

IRANIAN Major plates
Plate boundaries
Direction of plate movements

VOLCANOES

The word 'volcano' derives from the island of Vulcano off Sicily, in the Mediterranean Sea. In classical times the people of this area thought that Vulcano was the chimney of the forge of Vulcan, blacksmith of the Roman gods. Today volcanoes might be the subject of scientific study but they remain both dramatic and unpredictable, if not exactly supernatural: in 1991 Mount Pinatubo, about 60 miles north of the Philippines capital Manila, suddenly burst into life after more than six centuries of lying dormant.

Most of the world's active volcanoes occur in a belt around the Pacific Ocean, on the edge of the Pacific plate, called the 'ring of fire'. Indonesia has the greatest concentration with 90 volcanoes, 12 of which are active. The most famous, Krakatau, erupted in 1883 with such force that the resulting tidal wave killed 36,000 people and tremors were felt as far away as the English Channel.

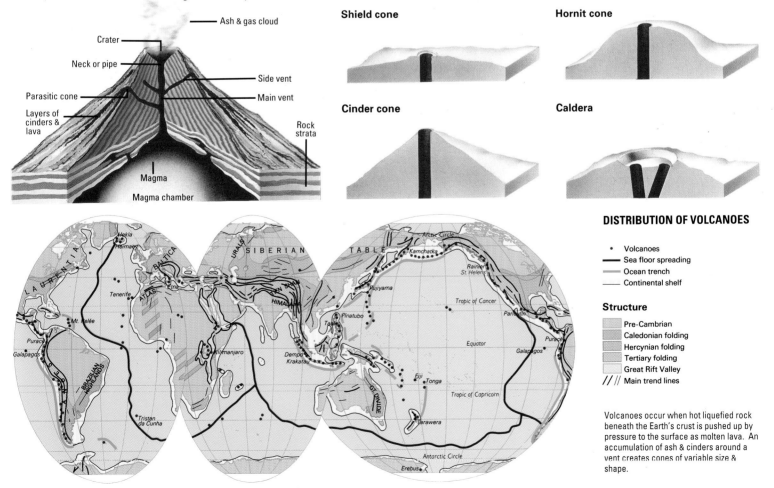

DISTRIBUTION OF VOLCANOES

- Volcanoes
- Sea floor spreading
- Ocean trench
- Continental shelf

Structure

- Pre-Cambrian
- Caledonian folding
- Hercynian folding
- Tertiary folding
- Great Rift Valley
- // /// Main trend lines

Volcanoes occur when hot liquefied rock beneath the Earth's crust is pushed up by pressure to the surface as molten lava. An accumulation of ash & cinders around a vent creates cones of variable size & shape.

LANDSCAPE

Above and below the surface of the oceans, the features of the Earth's crust are constantly changing. The phenomenal forces generated by convection currents in the molten core of our planet carry the vast segments or 'plates' of the crust across the globe in an endless cycle of creation and destruction. A continent may travel little more than an inch each year, yet in the vast span of geological time, this movement throws up giant mountain ranges and creates new land.

Destruction of the landscape, however, begins as soon as it is formed. Wind, water, ice and sea, the main agents of erosion, mount a constant assault that even the hardest rocks can

not withstand. Mountain peaks may dwindle by an inch or less each year, but if they are not uplifted by further movements of the crust they will eventually be reduced to rubble. Water is the most powerful destroyer – it has been estimated that 100 billion tonnes of rock is washed into the oceans every year.

Rivers and glaciers, like the sea itself, generate much of their effect through abrasion – pounding the landscape with the debris they carry with them. But as well as destroying they also create new landscapes, many of them spectacular: vast deltas like the Mississippi and the Nile, or the fiords cut by glaciers in British Columbia, Norway and New Zealand.

THE SPREADING EARTH

THE SPREADING EARTH

The vast ridges that divide the Earth's crust beneath each of the world's oceans mark the boundaries between tectonic plates which are moving gradually in opposite directions. As the plates shift apart, molten magma rises from the Earth's core to seal the rift and the sea floor slowly spreads towards the continental landmasses. The rate of spreading has been calculated by magnetic analysis of the rock at 1.5 inches a year in the North Atlantic. Underwater volcanoes mark the line where the continental rise begins. As the plates meet, much of the denser ocean crust dips beneath the continental plate and melts back to the magma.

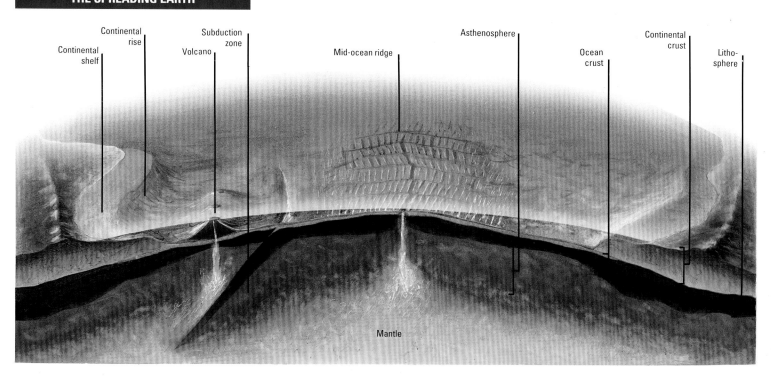

MOUNTAIN BUILDING

Mountains are formed when pressures on the Earth's crust caused by continental drift become so intense that the surface buckles or cracks. This happens most dramatically where two tectonic plates collide : the Rockies, Andes, Alps, Urals and Himalayas resulted from such impacts. These are all known as fold mountains, because they were formed by the compression of the rocks, forcing the surface to bend and fold like a crumpled rug.

The other main building process is when the crust fractures to create faults, allowing rock to be forced upwards in large blocks; or when the pressure of magma within the crust forces the surface to bulge into a dome, or erupts to form a volcano. Large mountain ranges may reveal a combination of those features; the Alps, for example, have been compressed so violently that the folds are fragmented by numerous faults and intrusions of molten rock.

Over millions of years, even the greatest mountain ranges can be reduced by erosion to a rugged landscape known as a peneplain.

Types of fold: Geographers give different names to the degrees of fold thta result from continuing pressure on the rock strata. A simple fold may be symmetric, with even slopes on either side, but as the pressure builds up, one slope becomes steeper and the fold becomes asymmetric. Later, the ridge or 'anticline' at the top of the fold may slide over the lower ground or 'syncline' to form a recumbent fold. Eventually, the rock strata may break under the pressure to form an overthrust and finally a nappe fold.

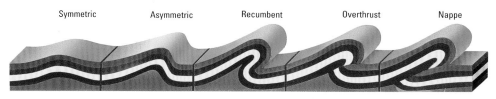

Types of faults: Faults are classified by the direction in which the blocks of rock have moved. A normal fault results when a vertical movement causes the surface to break apart; compression causes a reverse fault. Sideways movement causes shearing, known as a strike-slip fault. When the rock breaks in two places, the central block may be pushed up in a horst fault, or sink in a graben fault.

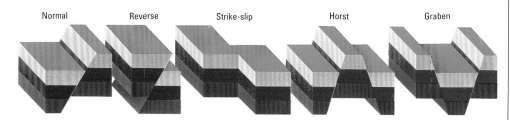

SHAPING FORCES: GLACIERS

Many of the world's most dramatic landscapes have been carved by glaciers. During the Ice Ages of the Pleistocene Era (before 12,000 years ago) up to a third of the land surface was glaciated; even today a tenth is covered in ice. Glaciers are formed from compressed snow called névé accumulating in a valley head or cirque. Slowly the glacier moves downhill scraping away debris from the mountains and valleys through which it passes. The debris, or moraine, adds to the abrasive power of the ice.

The rate of movement can vary from an inch or so to several yards a day; but the end of the glacier may not reach the bottom of the valley: the position of the snout depends on the rate at which the ice melts. Glaciers create numerous distinctive landscape features from arête ridges and pyramidal peaks to ice-dammed lakes and truncated spurs, with the U-shape distinguishing a glacial valley from one cut by a river.

SHAPING FORCES: RIVERS

From their origins as upland rills and streams channeling rainfall, or as springs releasing water that has seeped into the ground, all rivers are incessantly at work cutting and shaping the landscape on their way to the sea. In highland regions their flow may be rapid, pounding rocks and boulders with enough violence to cut deep gorges and V-shaped valleys through softer rocks, or tumble as waterfalls over harder ones.

As they reach more gentle slopes, rivers release some of the pebbles they have carried downstream and flow more slowly, broadening out and raising levees or ridges along their banks by depositing mud and sand. In lowland plains, where the gradient is minimal, the river drifts into meanders, depositing deep layers of sediment especially on the inside of each bend, where the flow is weakest. Here farmers may dig drainage ditches and artificial levees to keep the floodplain dry.

As the river finally reaches the sea, it deposits all its remaining sediments, and estuaries are formed where the tidal currents are strong enough to remove them; if not, the debris creates a delta, through which the river cuts outlet streams known as distributaries.

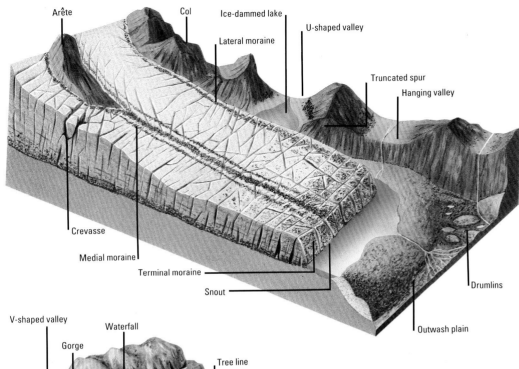

Arête · Col · Ice-dammed lake · U-shaped valley · Lateral moraine · Truncated spur · Hanging valley · Crevasse · Medial moraine · Terminal moraine · Snout · Drumlins · Outwash plain

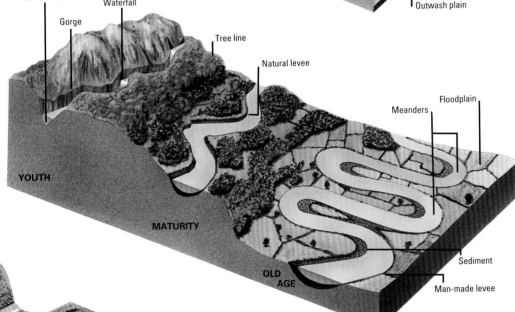

V-shaped valley · Gorge · Waterfall · Tree line · Natural levee · Meanders · Floodplain · YOUTH · MATURITY · OLD AGE · Sediment · Man-made levee

SHAPING FORCES: THE SEA

Under the constant assault from tides and currents, wind and waves, coastlines change faster than most landscape features, both by erosion and by the build-up of sand and pebbles carried by the sea. In severe storms, giant waves pound the shoreline with rocks and boulders, and frequently destroy concrete coastal defenses; but even in quieter conditions, the sea steadily erodes cliffs and headlands, creating new land in the form of sand-dunes, spits and salt marshes.

Where the coastline is formed from soft rocks such as sandstones, debris may fall evenly and be carried away by currents from shelving beaches. In areas with harder rock, the waves may cut steep cliffs and form underwater platforms; eroded debris is deposited as a terrace. Bays are formed when sections of soft rock are carved away between headlands of harder rock. These are then battered by waves from both sides, until the headlands are eventually reduced to rock arches and stacks.

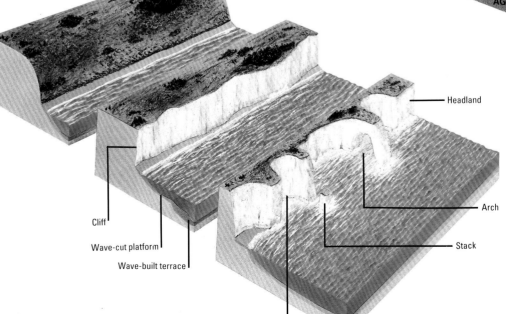

Headland · Cliff · Wave-cut platform · Wave-built terrace · Cove · Arch · Stack

OCEANS

THE GREAT OCEANS

Relative sizes of the world's oceans

Legend:
- Pacific
- Atlantic
- Indian
- Arctic

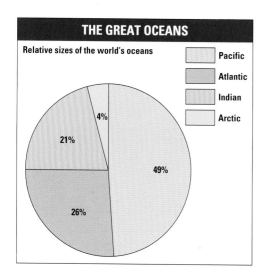

- 49%
- 26%
- 21%
- 4%

In a strict geographical sense there are only three true oceans - the Atlantic, Indian and Pacific. The legendary 'Seven Seas' would require these to be divided at the equator and the addition of the Arctic Ocean - which accounts for less than 4% of the total sea area. The International Hydrographic Bureau does not recognize the Antarctic Ocean (even less the 'Southern Ocean') as a separate entity.

The Earth is a watery planet: more than 70.9% of its surface – almost 140,000,000 square miles – is covered by the oceans and seas. The mighty Pacific alone accounts for nearly 36% of the total, and 49% of the sea area. Gravity holds in around 320 million cubic miles of water, of which over 97% is saline.

The vast underwater world starts in the shallows of the seaside and plunges to depths of more than 36,000 feet. The continental shelf, part of the land mass, drops gently to around 600 feet; here the seabed falls away suddenly at and angle of 3 - 6° - the continental slope. The third stage, called the continental rise, is more gradual with gradients varying from 1 in 100 to 1 in 700. At an average depth of about 16,000 feet there begins the aptly-named abyssal plain, immense submarine expanses far beyond the reach of sunlight, where few creatures are able to survive.

From these plains rise volcanoes which, taken from base to top, rival and even surpass the biggest continental mountains in height. Mount Kea, on Hawaii, reaches a total of 33,400 feet, almost 4,500 feet more than Mount Everest, although scarcely 40% is visible above sea level.

In addition there are underwater mountain chains up to 600 miles across, whose peaks sometimes appear above sea level as islands such as Iceland and Tristan da Cunha.

THE OCEAN DEPTHS

Average and maximum depths of the world's great oceans, in feet

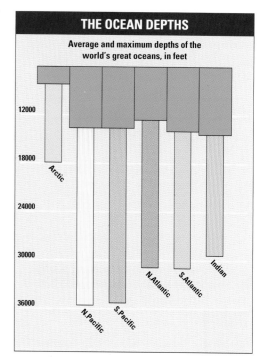

- 12000
- 18000
- 24000
- 30000
- 36000

Arctic, N.Pacific, S.Pacific, N.Atlantic, S.Atlantic, Indian

OCEAN CURRENTS

[Cold currents are shown in blue, warm currents in red]

WINTER IN NORTHERN HEMISPHERE

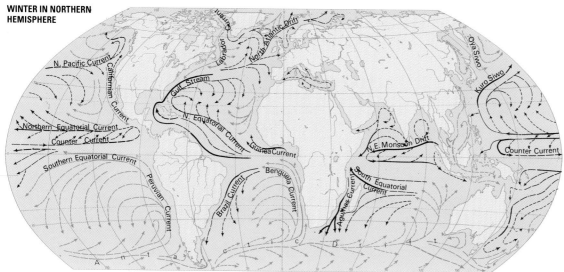

SUMMER IN NORTHERN HEMISPHERE

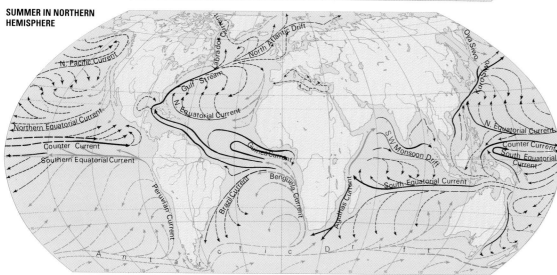

Moving immense quantities of energy as well as billions of tonnes of water every hour, the ocean currents are a vital part of the great heat engine that drives the Earth's climate. They themselves are produced by a twofold mechanism. At the surface, winds push huge masses of water before them; in the deep ocean, below an abrupt temperature gradient that separates the churning surface waters from the still depths, density variations cause slow vertical movements. The pattern of circulation of the great surface currents is determined by the displacement known as the Coriolis effect. As the Earth turns beneath a moving object - whether it is a tennis ball or a vast mass of water - it appears to be deflected to one side. The deflection is most obvious near the equator, where the Earth's surface is spinning eastward at 1,050 mph; currents moving poleward are curved clockwise in the northern hemisphere and counter-clockwise in the southern.

The result is a system of spinning circles known as gyres. The Coriolis effect piles up water on the left of each gyre, creating a narrow, fast-moving stream that is matched by a slower, broader returning current on the right. North and south of the equator, the fastest currents are located in the west and in the east respectively. In each case, warm water moves from the equator and cold water returns to it. Cold currents often bring an upwelling of nutrients with them, supporting the world's most economically important fisheries.

Depending on the prevailing winds, some currents on or near the equator may reverse their direction in the course of the year - a seasonal variation on which Asian monsoon rains depend, and whose occasional failure can bring disaster to millions.

FISHING

Main commercial fishing areas

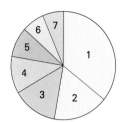

Percentage of world catch

1. North Pacific 36%
2. North Atlantic 17%
3. South Pacific 14%
4. Central Pacific 11%
5. Central Atlantic 9%
6. South Atlantic 7%
7. Indian 6%

Leading fishing nations

Japan 26.5% USSR 22.8% South Korea 8.3% North Korea 4.5% USA 3.3% China 3.1%

World total (1989): 14,143,923 tonnes

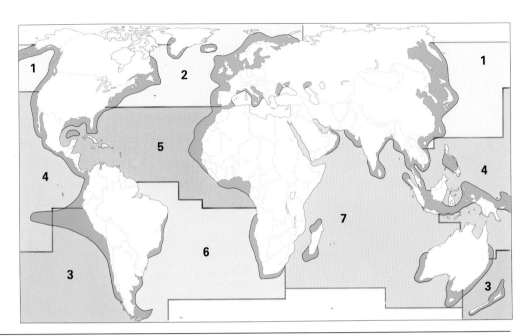

MARINE POLLUTION

Sources of marine oil pollution (1980s)

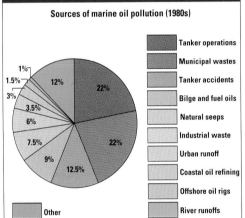

- Tanker operations
- Municipal wastes
- Tanker accidents
- Bilge and fuel oils
- Natural seeps
- Industrial waste
- Urban runoff
- Coastal oil refining
- Offshore oil rigs
- River runoffs
- Other

OIL SPILLS

Major oil spills from tankers & combined carriers

Year	Vessel	Location	Spill (barrels)*	Cause
1979	Atlantic Empress	West Indies	1,890,000	collision
1983	Castillo De Bellver	South Africa	1,760,000	fire
1978	Amoco Cadiz	France	1,628,000	grounding
1988	Odyssey	Canada	1,000,000	fire
1967	Torrey Canyon	UK	909,000	grounding
1972	Sea Star	Gulf of Oman	902,250	collision
1977	Hawaiian Patriot	Hawaiian Is.	742,500	fire
1979	Independenta	Turkey	696,350	collision
1976	Urquiola	Spain	670,000	grounding
1980	Irenes Serenade	Greece	600,000	fire
1989	Khark V	Morocco	560,000	fire

Other sources of major oil spills

Year	Source	Location	Spill (barrels)	Cause
1983	Nowruz oilfield	Persian Gulf	4,250,000 †	war
1979	Ixtoc 1 oilwell	Gulf of Mexico	4,200,000	blow-out
1991	Kuwait	Persian Gulf	2,500,000	war

* 1 barrel = 0.136 tonnes/159 lit./35 Imperial gal./42 US gal.   † estimated

RIVER POLLUTION

Sources of river pollution, USA (1987)

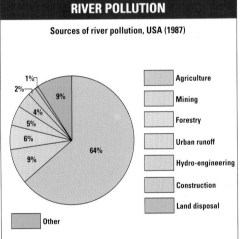

- Agriculture
- Mining
- Forestry
- Urban runoff
- Hydro-engineering
- Construction
- Land disposal
- Other

WATER POLLUTION

Severely polluted sea areas & lakes

Less polluted sea areas & lakes

Areas of frequent oil pollution by shipping

Major oil tanker spills

Major oil rig blow outs

Offshore dumpsites for industrial & municipal waste

Severely polluted rivers & estuaries

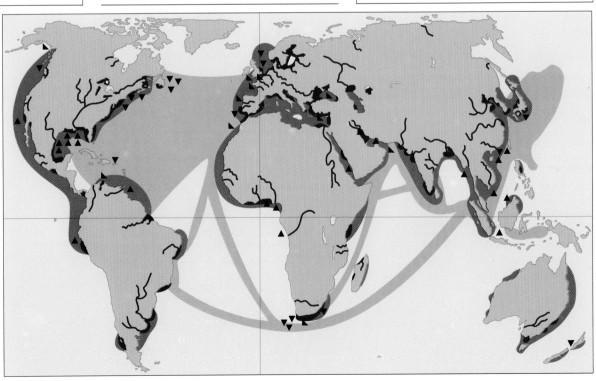

The most notorious tanker spillage of the 1980s occurred when the Exxon Valdez ran aground in Prince William Sound, Alaska in 1989, spilling 267,000 barrels of crude oil close to shore in a sensitive ecological area. This rates as the world's 28th worst spill in terms of volume.

CLIMATE

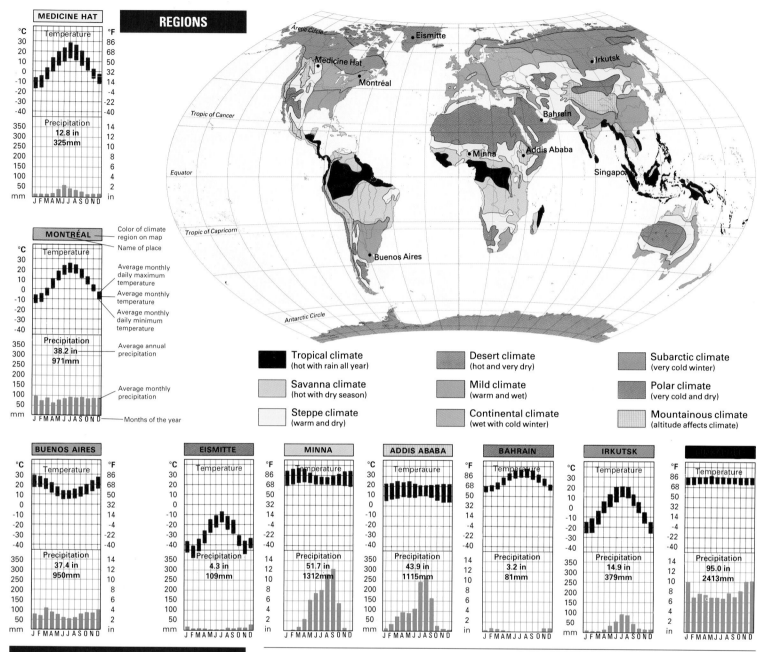

MEDICINE HAT

Temperature / Precipitation 12.8 in 325mm

MONTRÉAL

- Color of climate region on map
- Name of place
- Average monthly daily maximum temperature
- Average monthly temperature
- Average monthly daily minimum temperature
- Average annual precipitation
- Average monthly precipitation
- Months of the year

Temperature / Precipitation 38.2 in 971mm

REGIONS

Tropical climate (hot with rain all year)

Savanna climate (hot with dry season)

Steppe climate (warm and dry)

Desert climate (hot and very dry)

Mild climate (warm and wet)

Continental climate (wet with cold winter)

Subarctic climate (very cold winter)

Polar climate (very cold and dry)

Mountainous climate (altitude affects climate)

BUENOS AIRES — Temperature / Precipitation 37.4 in 950mm

EISMITTE — Temperature / Precipitation 4.3 in 109mm

MINNA — Temperature / Precipitation 51.7 in 1312mm

ADDIS ABABA — Temperature / Precipitation 43.9 in 1115mm

BAHRAIN — Temperature / Precipitation 3.2 in 81mm

IRKUTSK — Temperature / Precipitation 14.9 in 379mm

Temperature / Precipitation 95.0 in 2413mm

CLIMATE RECORDS

Temperature

Highest recorded temperature: Al Aziziyah, Libya, 58°C [136.4°F], 13 Sep. 1922.

Highest mean annual temperature: Dallol, Ethiopia, 34.4°C [94°F], 1960-66.

Longest heatwave: Marble Bar, W. Australia, 162 days over 38°C [100°F], 23 Oct. 1923 - 7 Apr. 1924.

Lowest recorded temperature (outside poles): Verkhoyansk, Siberia, -68°C [-90°F], 6 Feb. 1933. Verkhoyansk also registered the greatest annual range of temperature: - 70°C to 37°C [-94°F to 98°F].

Lowest mean annual temperature: Polus Nedostupnosti, Pole of Cold, Antarctica, -57.8°C [-72°F].

Precipitation

Driest place: Arica, N. Chile, 0.8mm [0.03 in] per year (60-year average).

Longest drought: Calama, N. Chile: no recorded rainfall in 400 years to 1971.

Wettest place (average): Tututendo, Colombia: mean annual rainfall 11,770 mm [463.4 in].

Wettest place (12 months): Cherrapunji, Meghalaya, N.E. India, 26,470 mm [1,040 in], Aug. 1860 to Aug. 1861. Cherrapunji also holds the record for rainfall in one month: 930 mm [37 in] July 1861.

Wettest place (24 hours): Cilaos, Réunion, Indian Ocean, 1,870 mm [73.6 in], 15-16 March 1952.

Heaviest hailstones: Gopalganj, Bangladesh, up to 1.02 kg [2.25 lb], 14 Apr. 1986 (killed 92 people).

Heaviest snowfall (continuous): Bessans, Savoie, France, 1730 mm [68 in] in 19 hours, 5-6 Apr. 1969.

Heaviest snowfall (season/year): Paradise Ranger Station, Mt Rainier, Washington, USA, 31,102 mm [1,224.5 in], 19 Feb. 1971 to 18 Feb. 1972.

Pressure & Winds

Highest barometric pressure: Agata, Siberia, 1,083.8 mb [32 in] at altitude 262 m [862 ft] 31 December 1968.

Lowest barometric pressure: Typhoon Tip, 480 km [300 mls] west of Guam, Pacific Ocean, 870 mb [25.69 in], 12 Oct. 1979.

Highest recorded windspeed: Mt Washington, New Hampshire, USA, 371 kph [231 mph], 12 Apr. 1934. This is three times as strong as hurricane force on the Beaufort Scale.

Windiest place: Commonwealth Bay, George V Coast, Antarctica, where gales reach over 320 kph [200 mph].

WINDCHILL FACTOR

In sub-zero weather, even moderate winds significantly reduce effective temperatures. The chart below shows the windchill effect across a range of speeds. Figures in the pink zone are not dangerous to well-clad people; in the blue zone, the risk of serious frostbite is acute.

	Wind speed (mph)				
	5	15	25	35	45
30°F	27	9	1	-4	-6
25°F	21	2	-7	-12	-14
20°F	16	-5	-15	-20	-22
15°F	12	-11	-22	-27	-30
10°F	7	-18	-29	-35	-38
5°F	0	-25	-36	-43	-46
0°F	-5	-31	-44	-52	-54
-5°F	-10	-38	-51	-58	-62
-10°F	-15	-45	-59	-67	-70
-15°F	-21	-51	-66	-74	-78
-20°F	-26	-58	-74	-82	-85

BEAUFORT WIND SCALE

Named after the 19th-century British naval officer who devised it, the Beaufort Scale assesses wind speed according to its effects. It was originally designed as an aid for sailors, but has since been adapted for use on land.

Scale	Wind speed kph	mph	Effect
0	0-1	0-1	**Calm** Smoke rises vertically
1	1-5	1-3	**Light air** Wind direction shown only by smoke drift
2	6-11	4-7	**Light breeze** Wind felt on face; leaves rustle; vanes moved by wind
3	12-19	8-12	**Gentle breeze** Leaves and small twigs in constant motion; wind extends small flag
4	20-28	13-18	**Moderate** Raises dust and loose paper; small branches move
5	29-38	19-24	**Fresh** Small trees in leaf sway; crested wavelets on inland waters
6	39-49	25-31	**Strong** Large branches move; difficult to use umbrellas; overhead wires whistle
7	50-61	32-38	**Near gale** Whole trees in motion; difficult to walk against wind
8	62-74	39-46	**Gale** Twigs break from trees; walking very difficult
9	75-88	47-54	**Strong gale** Slight structural damage
10	89-102	55-63	**Storm** Trees uprooted; serious structural damage
11	103-117	64-72	**Violent Storm** Widespread damage
12	118+	73+	**Hurricane**

Conversions
°C – (°F - 32) x 5/9;   °F = (°C x 9/5) ı 32;   0°C = 32°F
1 in = 25.4 mm;   1 mm = 0.0394 in;   100 mm = 3.94 in

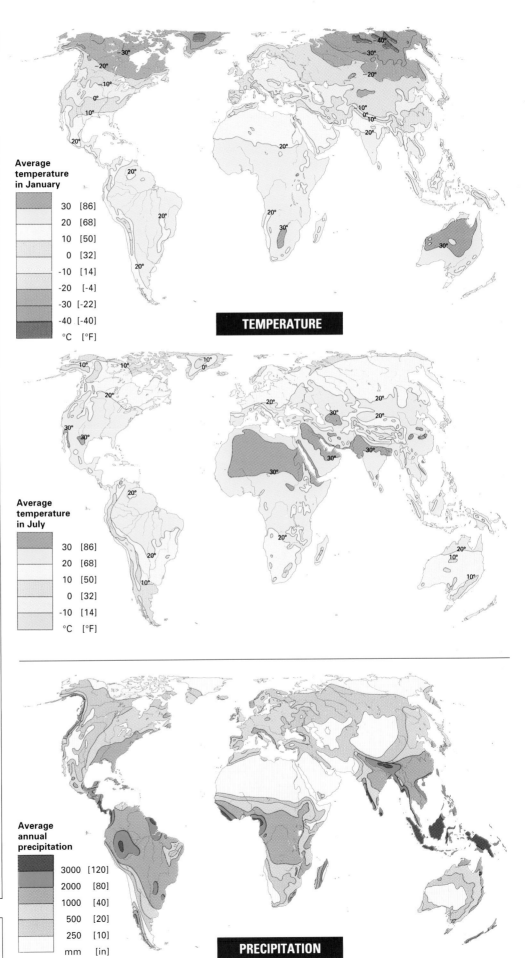

Average temperature in January

30	[86]
20	[68]
10	[50]
0	[32]
-10	[14]
-20	[-4]
-30	[-22]
-40	[-40]
°C	[°F]

TEMPERATURE

Average temperature in July

30	[86]
20	[68]
10	[50]
0	[32]
-10	[14]
°C	[°F]

Average annual precipitation

3000	[120]
2000	[80]
1000	[40]
500	[20]
250	[10]
mm	[in]

PRECIPITATION

WATER

THE HYDROLOGICAL CYCLE

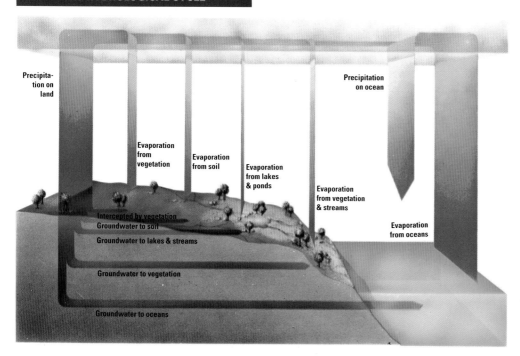

Precipitation on land

Precipitation on ocean

Evaporation from vegetation

Evaporation from soil

Evaporation from lakes & ponds

Evaporation from vegetation & streams

Evaporation from oceans

Intercepted by vegetation
Groundwater to soil
Groundwater to lakes & streams
Groundwater to vegetation
Groundwater to oceans

WATER DISTRIBUTION

The distribution of planetary water, by percentage. Oceans and icecaps together account for more than 99% of the total; the breakdown of the remainder is estimated.

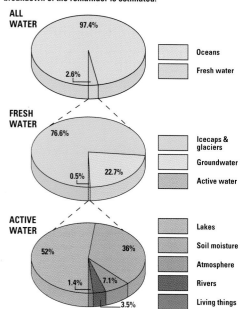

ALL WATER
97.4%
2.6%

- Oceans
- Fresh water

FRESH WATER
76.6%
0.5%
22.7%

- Icecaps & glaciers
- Groundwater
- Active water

ACTIVE WATER
52%
36%
1.4%
7.1%
3.5%

- Lakes
- Soil moisture
- Atmosphere
- Rivers
- Living things

WATER RUNOFF

Annual freshwater runoff by continent in cubic miles

- Asia
- North America
- South America
- Australasia
- Europe
- Africa

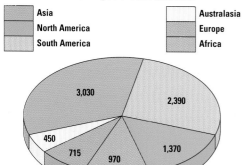

3,030
2,390
450
715
970
1,370

WATER UTILIZATION

- Domestic
- Industrial
- Agriculture

The percentage breakdown of water usage by sector, selected countries (1980s)

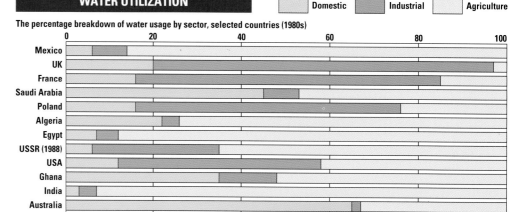

Mexico
UK
France
Saudi Arabia
Poland
Algeria
Egypt
USSR (1988)
USA
Ghana
India
Australia

WATER SUPPLY

Percentage of total population with access to safe drinking water (latest available year, 1980s)

- Over 90%
- 75 - 90%
- 60 - 75%
- 45 - 60%
- 30 - 45%
- Under 30%

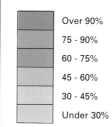

Least well provided countries (rural areas only):

Paraguay	8%	Guinea	15%
Mozambique	12%	Mauritania	17%
Uganda	12%	Malawi	17%
Angola	15%	Morocco	17%

Regional variation in vegetation

- Tundra & mountain vegetation
- Needleleaf evergreen forest
- Mixed needleleaf evergreen & broadleaf deciduous trees
- Broadleaf deciduous woodland
- Mid-latitude grassland
- Evergreen broadleaf & deciduous trees & shrubs
- Semi-desert scrub
- Desert
- Tropical grassland (savanna)
- Tropical broadleaf rainforest & monsoon forest
- Sub-tropical broadleaf & needleleaf forest

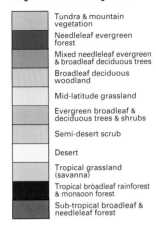

The map shows the natural 'climax vegetation' of regions, as dictated by climate and topography. In most cases, however, agricultural activity has drastically altered the vegetation pattern. Western Europe, for example, lost most of its broadleaf forest many centuries ago, while irrigation has turned some natural semi-desert into productive land.

LAND USE BY CONTINENT

- Forest
- Permanent pasture & rough grazing
- Permanent crops & plantations
- Arable
- Non-productive

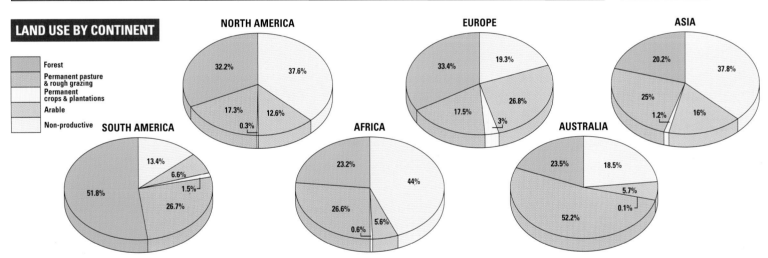

NORTH AMERICA
37.6%, 32.2%, 17.3%, 12.6%, 0.3%

EUROPE
19.3%, 33.4%, 26.8%, 17.5%, 3%

ASIA
20.2%, 37.8%, 25%, 16%, 1.2%

SOUTH AMERICA
13.4%, 6.6%, 1.5%, 51.8%, 26.7%

AFRICA
23.2%, 44%, 26.6%, 5.6%, 0.6%

AUSTRALIA
23.5%, 18.5%, 5.7%, 52.2%, 0.1%

FORESTRY: PRODUCTION

	Forest & woodland (million acres)	Annual production (1980s average, million cubic yards)	
		Fuelwood & charcoal	Industrial roundwood
World	*10,180*	*2,150*	*2,010*
USSR	2,295	112	373
S. America	2,140	285	120
N. America	1,990	202	705
Africa	1,740	504	70
Asia	1,230	968	320
Europe	393	74	383
Australasia	392	12	37

PAPER & BOARD

Top producers (1988)*

USA	69,477
Japan	24,624
Canada	16,638
China	12,645
USSR	10,750

Top exporters (1988)*

Canada	11,420
Finland	7,185
Sweden	6,377
USA	4,294
Germany	3,780

* in thousand tonnes

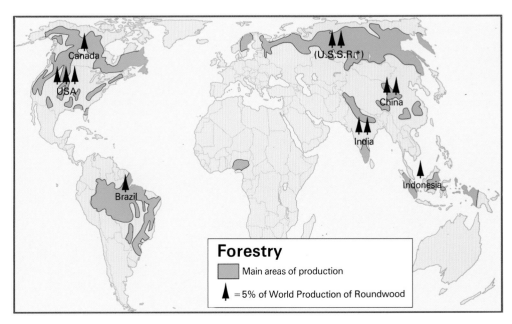

Forestry

- Main areas of production
- ▲ = 5% of World Production of Roundwood

ENVIRONMENT

Humans have always had a dramatic effect on their environment, at least since the invention of agriculture almost 10,000 years ago. Generally, the Earth has accepted human interference without obvious ill effects: the complex systems that regulate the global environment have been able to absorb substantial damage while maintaining a stable and comfortable home for the planet's trillions of lifeforms. But advancing human technology and the rapidly-expanding populations it supports are now threatening to overwhelm the Earth's ability to compensate.

Industrial wastes, acid rainfall, desertification and large-scale deforestation: all combine to create environmental change at a rate far faster than the slow cycles of planetary evolution can accommodate. Equipped with chainsaws and flamethrowers, humans can now destroy more forest in a day than their ancestors could in a century, upsetting the balance between plant and animal, carbon dioxide and oxygen, on which all life ultimately depends.

The fossil fuels that power industrial civilisation have pumped enough carbon dioxide and other so-called greenhouse gases into the atmosphere to make climatic change a near-certainty. Chlorofluorocarbons - CFCs - and other man-made chemicals are rapidly eroding the ozone layer, the atmosphere's screen against ultra-violet radiation.

As a result, the Earth's average temperature has risen by almost 1 F° since the beginning of the 20th century, and is still rising.

CLIMATIC CHANGE

Carbon dioxide emissions in tonnes per person per year (1980s)

High atmospheric concentrations of heat-absorbing gases, especially carbon dioxide, appear to be causing a steady rise in average temperatures worldwide - by as much as 3 F° by the year 2020, according to some estimates. Global warming is likely to bring with it a rise in sea levels that may flood some of the Earth's most densely populated coastlines.

- Over 10 tonnes of CO_2
- 5 - 10 tonnes of CO_2
- 1 - 5 tonnes of CO_2
- Under 1 tonne of CO_2

— Coastal areas in danger of flooding from rising sea levels caused by global warming

Statistics for each of the new republics of the former U.S.S.R. and Yugoslavia are not yet available.

GREENHOUSE POWER

Relative contributions to the Greenhouse Effect by the major heat-absorbing gases in the atmosphere

The chart combines greenhouse potency and volume. Carbon dioxide has a greenhouse potential of only 1 but its concentration of 350 parts per million, makes it predominate. CFC 12 , with 25,000 times the absorption capacity of CO_2, is present only as 0.00044 ppm.

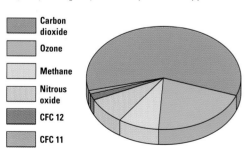

- Carbon dioxide
- Ozone
- Methane
- Nitrous oxide
- CFC 12
- CFC 11

CARBON DIOXIDE

Carbon dioxide released in millions of tonnes (1980s)

Although most of the net increase in atmospheric carbon dioxide comes from fossil fuel combustion, deforestation and changing land use also contribute.

- Fuel burning
- Deforestation

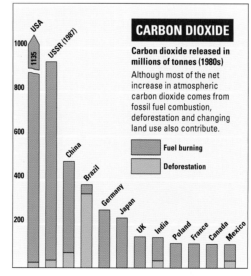

USA 1135
USSR (1987)
China
Brazil
Germany
Japan
UK
India
Poland
France
Canada
Mexico

GLOBAL WARMING

The rise in average temperatures caused by carbon dioxide and other greenhouse gases (1960-2020)

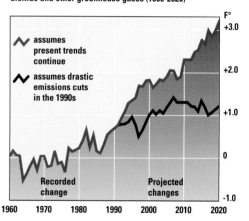

- ‿ assumes present trends continue
- ‿ assumes drastic emissions cuts in the 1990s

Recorded change

Projected changes

THE GREENHOUSE EFFECT

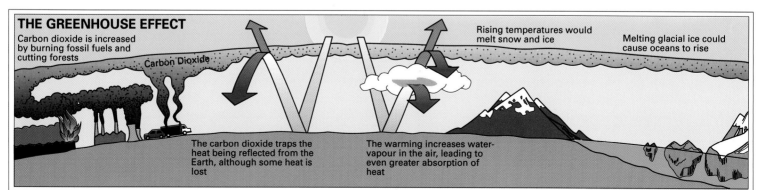

Carbon dioxide is increased by burning fossil fuels and cutting forests

Carbon Dioxide

Rising temperatures would melt snow and ice

Melting glacial ice could cause oceans to rise

The carbon dioxide traps the heat being reflected from the Earth, although some heat is lost

The warming increases water-vapour in the air, leading to even greater absorption of heat

DESERTIFICATION

- Existing deserts
- Areas with a high risk of desertification
- Areas with a moderate risk of desertification
- Former areas of rainforest
- Existing rainforest

DEFORESTATION

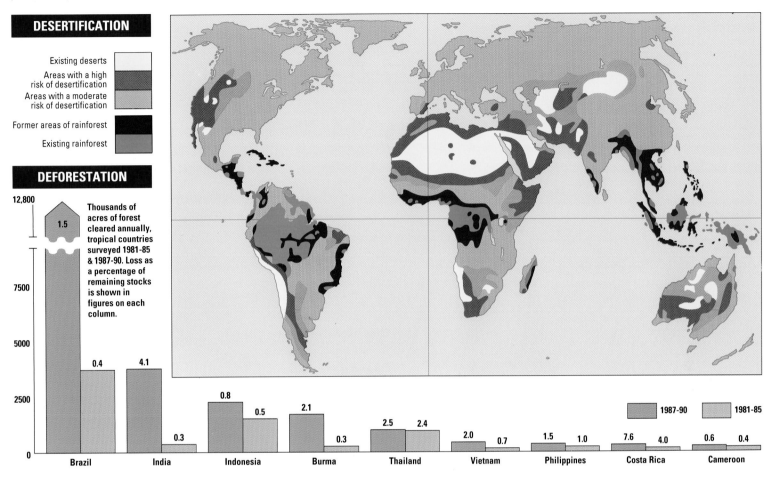

Thousands of acres of forest cleared annually, tropical countries surveyed 1981-85 & 1987-90. Loss as a percentage of remaining stocks is shown in figures on each column.

12,800

1.5

7500

5000

2500

0

| | 1987-90 | 1981-85 |

Brazil	India	Indonesia	Burma	Thailand	Vietnam	Philippines	Costa Rica	Cameroon
1.5 / 0.4	4.1 / 0.3	0.8 / 0.5	2.1 / 0.3	2.5 / 2.4	2.0 / 0.7	1.5 / 1.0	7.6 / 4.0	0.6 / 0.4

DESERTIFICATION

The result of overcultivation, overgrazing, and overcutting of ground cover for firewood, desertification is also caused by faulty irrigation techniques that leave land too saline or alkaline to support viable crops. Changing rainfall patterns or prolonged droughts exacerbate the process. As much as 60% of the world's croplands and rangelands are in some danger, with 15 million acres lost altogether every year and a further 53 million rendered agriculturally worthless. Africa is especially badly hit: in Mali the Sahara advanced more than 200 miles southwards in only 20 years.

DEFORESTATION

The Earth's remaining forests are under attack from three directions: expanding agriculture, logging, and growing consumption of fuelwood, often in combination. Sometimes deforestation is the direct result of government policy, as in the efforts made to resettle the urban poor in some parts of Brazil; just as often, it comes about despite state attempts at conservation. Loggers, licensed or unlicensed, blaze a trail into virgin forest, often destroying twice as many trees as they harvest. Landless farmers follow, burning away most of what remains to plant their crops, completing the destruction.

ACID RAIN

Killing trees, poisoning lakes and rivers and eating away buildings, acid rain is mostly produced by sulfur dioxide emissions from industry, although the burning of savanna lands by African farmers has also caused acid downpours on tropical rainforests. By the late 1980s, acid rain had sterilized 4,000 or more of Sweden's lakes and left 45% of Switzerland's alpine conifers dead or dying, while the monuments of Greece were dissolving in Athens smog. Prevailing wind patterns mean that the acids often fall hundreds of miles from where the original pollutants were discharged.

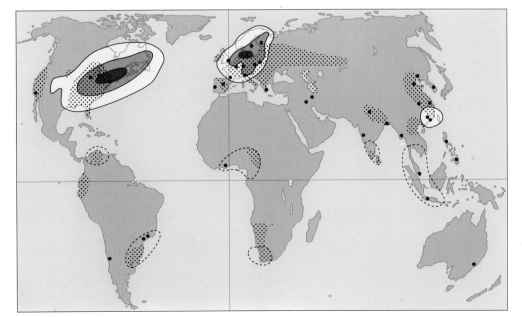

ACID RAIN

Acid rainfall & sources of acidic emissions (1980s)

Acid rain is caused when sulfur & nitrogen oxides in the air combine with water vapor to form sulfuric, nitric & other acids.

Regions where sulfur and nitrogen oxides are released in high concentrations, mainly from fossil fuel combustion.

 Major cities with high levels of air pollution (including nitrogen & sulfur emissions)

Areas of heavy acid deposition

pH numbers indicate acidity, decreasing from a neutral 7. Normal rain, slightly acid from dissolved carbon dioxide, never exceeds a pH of 5.6.

pH less than 4.0 (most acidic)

pH 4.0 to 4.5

pH 4.5 to 5.0

Areas where acid rain is a potential danger

POPULATION

Developed nations such as the US have populations evenly spread across age groups and, usually, a growing proportion of elderly people. Developing nations fall into a pattern somewhere between that of Kenya and the world model: the great majority of their people are in the younger age groups, about to enter their most fertile years. In time, even Kenya's population profile should resemble the world profile, but the transition will come about only after a few more generations of rapid population growth.

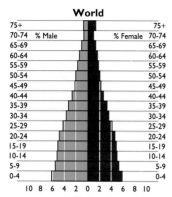

World

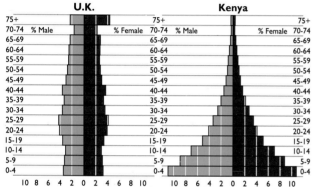

U.K. **Kenya**

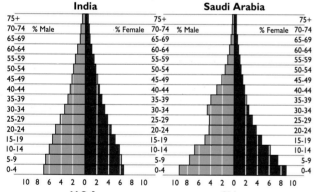

India **Saudi Arabia**

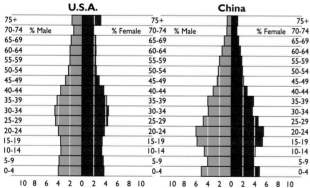

U.S.A. **China**

MOST POPULOUS NATIONS [in millions (1989)]

1.	China	1120	9.	Pakistan	109	17.	Turkey	57
2.	India	812	10.	Bangladesh	107	18.	France	56
3.	USA	250	11.	Mexico	84	19.	Thailand	55
4.	Indonesia	179	12.	Germany	79	20.	Iran	55
5.	Brazil	148	13.	Vietnam	66	21.	Egypt	53
6.	Russia	147	14.	Philippines	60	22.	Ukraine	52
7.	Japan	123	15.	Italy	58	23.	Ethiopia	51
8.	Nigeria	109	16.	UK	57	24.	S. Korea	43

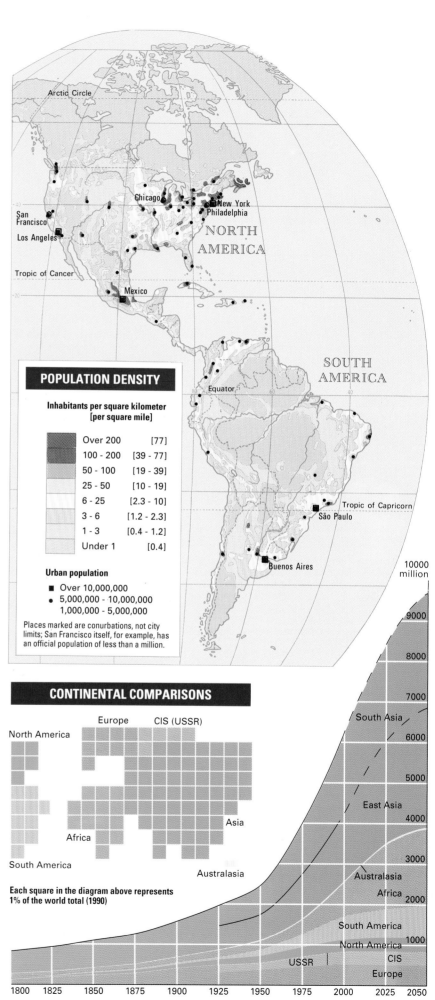

POPULATION DENSITY

Inhabitants per square kilometer [per square mile]

	Over 200	[77]
	100 - 200	[39 - 77]
	50 - 100	[19 - 39]
	25 - 50	[10 - 19]
	6 - 25	[2.3 - 10]
	3 - 6	[1.2 - 2.3]
	1 - 3	[0.4 - 1.2]
	Under 1	[0.4]

Urban population

■ Over 10,000,000
● 5,000,000 - 10,000,000
  1,000,000 - 5,000,000

Places marked are conurbations, not city limits; San Francisco itself, for example, has an official population of less than a million.

CONTINENTAL COMPARISONS

Each square in the diagram above represents 1% of the world total (1990)

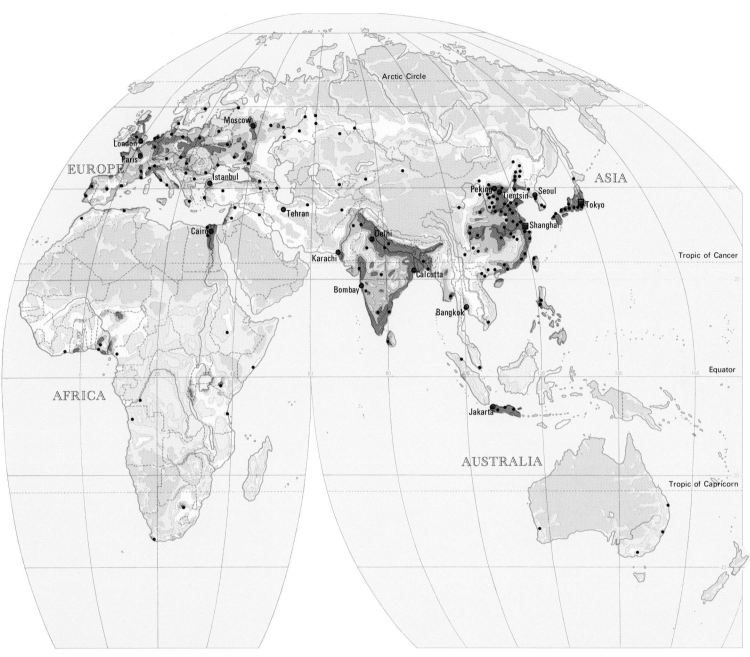

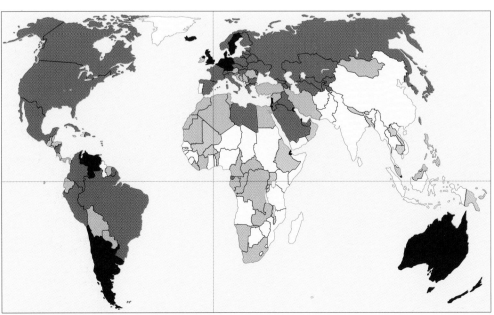

URBAN POPULATION

Percentage of total population living in towns & cities (1990)

Over 80%	
60 - 80%	
40 - 60%	
20 - 40%	
Under 20%	

Most urbanized

Singapore	100%
Belgium	97%
Kuwait	96%
Hong Kong	93%
UK	93%

Least urbanized

Nepal	10%
Burkina Faso	9%
Rwanda	8%
Burundi	7%
Bhutan	5%

*Statistics for each of the new republics of the former
U.S.S.R. and Yugoslavia are not yet available.*

THE HUMAN FAMILY

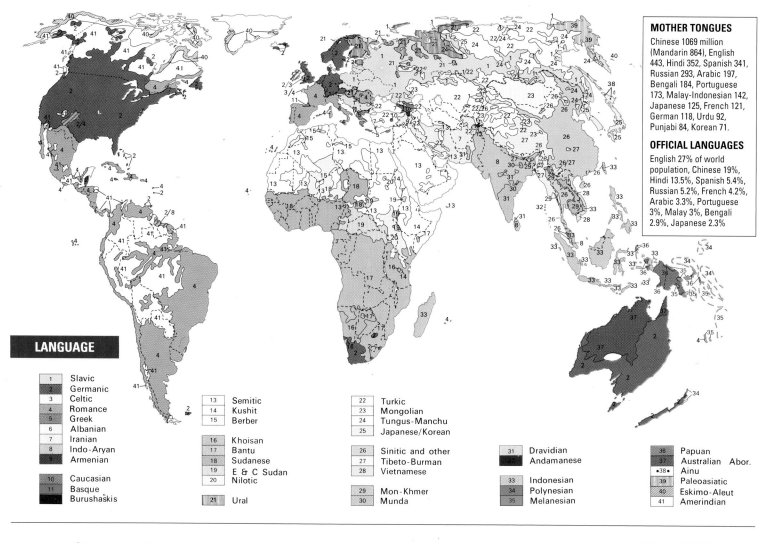

MOTHER TONGUES

Chinese 1069 million (Mandarin 864), English 443, Hindi 352, Spanish 341, Russian 293, Arabic 197, Bengali 184, Portuguese 173, Malay-Indonesian 142, Japanese 125, French 121, German 118, Urdu 92, Punjabi 84, Korean 71.

OFFICIAL LANGUAGES

English 27% of world population, Chinese 19%, Hindi 13.5%, Spanish 5.4%, Russian 5.2%, French 4.2%, Arabic 3.3%, Portuguese 3%, Malay 3%, Bengali 2.9%, Japanese 2.3%

LANGUAGE

1	Slavic
2	Germanic
3	Celtic
4	Romance
5	Greek
6	Albanian
7	Iranian
8	Indo-Aryan
9	Armenian
10	Caucasian
11	Basque
	Burushaskis

13	Semitic
14	Kushit
15	Berber
16	Khoisan
17	Bantu
18	Sudanese
19	E & C Sudan
20	Nilotic
21	Ural

22	Turkic
23	Mongolian
24	Tungus-Manchu
25	Japanese/Korean
26	Sinitic and other
27	Tibeto-Burman
28	Vietnamese
29	Mon-Khmer
30	Munda

31	Dravidian
32	Andamanese
33	Indonesian
34	Polynesian
35	Melanesian

36	Papuan
37	Australian Abor.
•38•	Ainu
39	Paleoasiatic
40	Eskimo-Aleut
41	Amerindian

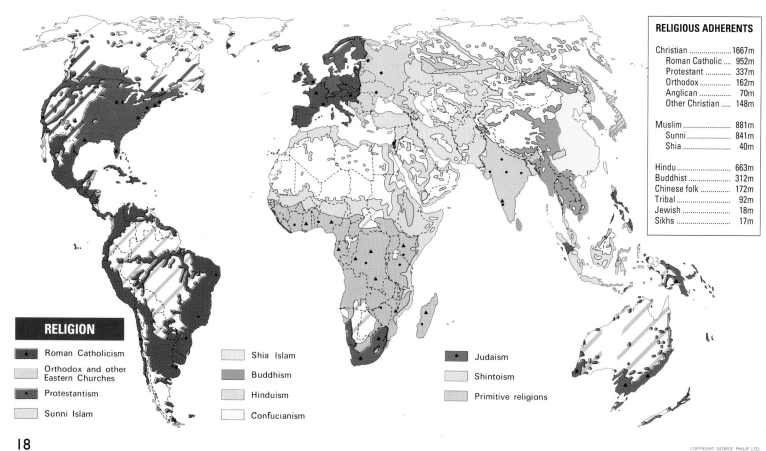

RELIGIOUS ADHERENTS

Christian	1667m
Roman Catholic	952m
Protestant	337m
Orthodox	162m
Anglican	70m
Other Christian	148m
Muslim	881m
Sunni	841m
Shia	40m
Hindu	663m
Buddhist	312m
Chinese folk	172m
Tribal	92m
Jewish	18m
Sikhs	17m

RELIGION

▲	Roman Catholicism
	Orthodox and other Eastern Churches
•	Protestantism
	Sunni Islam

	Shia Islam
	Buddhism
	Hinduism
	Confucianism

•	Judaism
	Shintoism
	Primitive religions

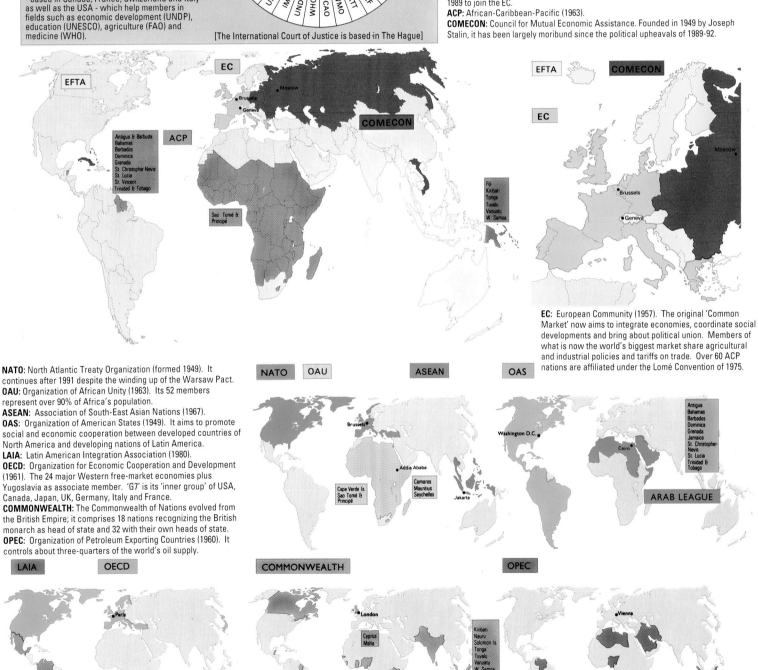

UNITED NATIONS

Created in 1945 to promote peace and cooperation and based in New York, the United Nations is the world's largest international organization, with over 160 members and an annual budget exceeding two billion US dollars. Each member of the General Assembly has one vote, while the permanent members of the 15-nation Security Council - USA, Russia, China, UK and France - hold a veto. The 54 members of the Economic and Social Council are responsible for economic, social, cultural, educational, health and related matters. The Secretariat is the UN's chief administrative arm; the only territory now administered by the Trusteeship Council is Belau (by the USA). The UN has 24 specialized agencies - based in Canada, France, Switzerland and Italy as well as the USA - which help members in fields such as economic development (UNDP), education (UNESCO), agriculture (FAO) and medicine (WHO).

The Secretariat (civil servants who run the UN)

Security Council (tries to keep the peace between countries)

Trusteeship Council (looks after Trust Terr.)

Economic & Social Council (looks after UN agencies)

International Court of Justice

U.N AGENCIES

IAEA ILO FAO UNESCO IMF UNDP OHM ICAO WMO UNICEF GATT UNIDO UNFPA ITU

[The International Court of Justice is based in The Hague]

Membership: There are 13 independent states who are not members of the UN - Andorra, Kiribati, Liechtenstein, N. Korea, S. Korea, Monaco, Nauru, San Marino, Switzerland, Taiwan, Tonga, Tuvalu and Vatican City. By 1992, the successor states of the former USSR had either joined or planned to. There were 51 members in 1945. Official languages are Chinese, English, French, Russian, Spanish and Arabic.

Funding: The UN budget for 1988-1989 was US $ 1,788,746,000. Contributions are assessed by members' ability to pay, with the maximum 25% of the total, the minimum 0.01%. Contributions for 1988-1989 were: USA 25%, Japan 11.38%, USSR 9.99%, W. Germany 8.08%, France 6.25%, UK 4.86%, Italy 3.99%, Canada 3.09%, Spain 1.95%, Netherlands 1.65% (others 23.75%).

Peacekeeping: The UN has been involved in 18 peacekeeping operations worldwide since 1945, five of which (Afghanistan/Pakistan, Iran/Iraq, Angola, Namibia and Honduras) were initiated in 1988-1989. In June 1991 UN personnel totaling over 11,000 were working in eight separate areas.

EFTA: European Free Trade Association (formed in 1960). Portugal left the 'Seven' in 1989 to join the EC.

ACP: African-Caribbean-Pacific (1963).

COMECON: Council for Mutual Economic Assistance. Founded in 1949 by Joseph Stalin, it has been largely moribund since the political upheavals of 1989-92.

EC: European Community (1957). The original 'Common Market' now aims to integrate economies, coordinate social developments and bring about political union. Members of what is now the world's biggest market share agricultural and industrial policies and tariffs on trade. Over 60 ACP nations are affiliated under the Lomé Convention of 1975.

NATO: North Atlantic Treaty Organization (formed 1949). It continues after 1991 despite the winding up of the Warsaw Pact.

OAU: Organization of African Unity (1963). Its 52 members represent over 90% of Africa's population.

ASEAN: Association of South-East Asian Nations (1967).

OAS: Organization of American States (1949). It aims to promote social and economic cooperation between developed countries of North America and developing nations of Latin America.

LAIA: Latin American Integration Association (1980).

OECD: Organization for Economic Cooperation and Development (1961). The 24 major Western free-market economies plus Yugoslavia as associate member. 'G7' is its 'inner group' of USA, Canada, Japan, UK, Germany, Italy and France.

COMMONWEALTH: The Commonwealth of Nations evolved from the British Empire; it comprises 18 nations recognizing the British monarch as head of state and 32 with their own heads of state.

OPEC: Organization of Petroleum Exporting Countries (1960). It controls about three-quarters of the world's oil supply.

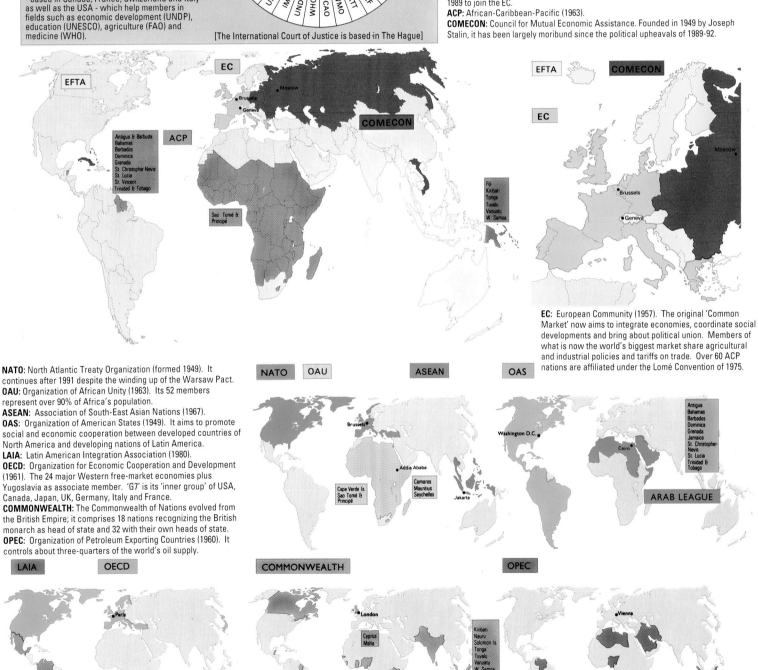

Wealth

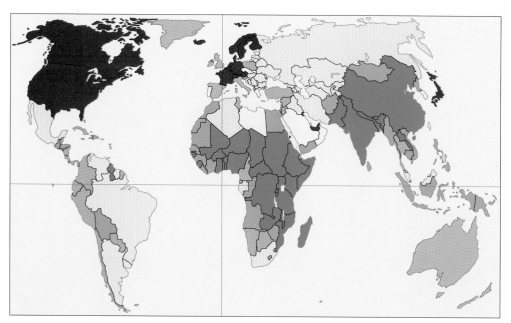

LEVELS OF INCOME

Gross National Product per capita: the value of total production divided by population (1989)

- Over 400% of world average
- 200 - 400% of world average
- 100 - 200% of world average

[World average wealth per person US $3,980]

- 50 - 100% of world average
- 25 - 50% of world average
- 10 - 25% of world average
- Under 10% of world average

Richest countries		Poorest 5 countries	
Switzerland	$30,270	Somalia	$170
Luxembourg	$24,860	Ethiopia	$120
Japan	$23,730	Tanzania	$120
Finland	$22,060	Mozambique	$80

WEALTH CREATION

The Gross National Product (GNP) of the world's largest economies, US $ billion (1989)

1. USA	5,237,707	21. Denmark	105,263
2. Japan	2,920,310	22. Norway	92,097
3. Germany	1,272,959	23. Saudi Arabia	89,986
4. France	1,000,866	24. Indonesia	87,936
5. Italy	871,955	25. South Africa	86,029
6. UK	834,166	26. Turkey	74,731
7. Canada	500,337	27. Argentina	68,780
8. China	393,006	28. Poland	66,974
9. Brazil	375,146	29. Thailand	64,437
10. Spain	358,352	30. Hong Kong	59,202
11. India	287,383	31. Yugoslavia	59,080
12. Australia	242,131	32. Greece	53,626
13. Netherlands	237,451	33. Algeria	53,116
14. Switzerland	197,984	34. Venezuela	47,164
15. South Korea	186,467	35. Israel	44,131
16. Sweden	184,230	36. Portugal	44,058
17. Mexico	170,053	37. Philippines	42,754
18. Belgium	162,026	38. Pakistan	40,134
19. Austria	131,899	39. New Zealand	39,437
20. Finland	109,705	40. Colombia	38,607

There are no accurate figures available for either the USSR or its successor nations.

THE WEALTH GAP

The world's richest & poorest countries, by Gross National Product per capita in US $ (1989)

1. Liechtenstein	33,000	1. Mozambique	80
2. Switzerland	30,270	2. Ethiopia	120
3. Bermuda	25,000	3. Tanzania	120
4. Luxembourg	24,860	4. Laos	170
5. Japan	23,730	5. Nepal	170
6. Finland	22,060	6. Somalia	170
7. Norway	21,850	7. Bangladesh	180
8. Sweden	21,710	8. Malawi	180
9. Iceland	21,240	9. Bhutan	190
10. USA	21,100	10. Chad	190
11. Denmark	20,510	11. Sierra Leone	200
12. Canada	19,020	12. Burundi	220
13. UAE	18,430	13. Gambia	230
14. France	17,830	14. Madagascar	230
15. Austria	17,360	15. Nigeria	250
16. Germany	16,500	16. Uganda	250
17. Belgium	16,390	17. Mali	260
18. Kuwait	16,380	18. Zaire	260
19. Netherlands	16,010	19. Niger	290
20. Italy	15,150	20. Burkina Faso	310

GNP per capita is calculated by dividing a country's Gross National Product by its population. The UK ranks 21st, with US $14,570.

CONTINENTAL SHARES

Shares of population & of wealth (GNP) by continent

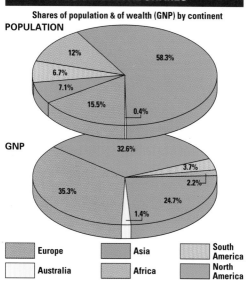

POPULATION: 58.3%, 12%, 6.7%, 7.1%, 15.5%, 0.4%

GNP: 32.6%, 35.3%, 1.4%, 24.7%, 2.2%, 3.7%

- Europe
- Asia
- South America
- Australia
- Africa
- North America

INFLATION

Average annual rate of inflation (1980-1988)

- Over 50%
- 20 - 50%
- 7.5 - 20%
- 0 - 7.5%
- Negative inflation
- No data available

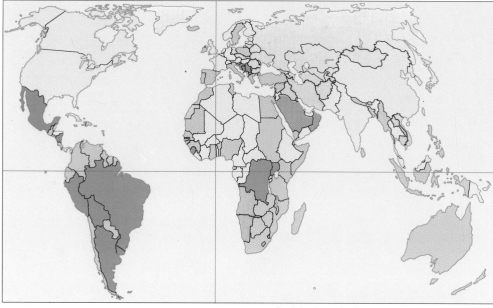

Highest average inflation		Lowest average inflation	
Bolivia	483%	Oman	-6.5%
Argentina	291%	Saudi Arabia	-4.2%
Brazil	189%	Kuwait	-3.9%
Israel	137%	Libya	0.1%
Peru	119%	U.A.E.	0.1%
Uganda	101%	Congo	0.8%
Nicaragua	87%	Gabon	0.9%

Statistics for each of the new republics of the former U.S.S.R. and Yugoslavia are not yet available.

INTERNATIONAL AID

COPYRIGHT GEORGE PHILIP LTD.

Aid provided or received, divided by total population in US $ (latest available year)

PROVIDERS

Over $100 per person	
$20 - 100 per person	
$0 - 20 per person	
No aid given or received	
$0 - 20 per person	
$20 - 100 per person	
Over $100 per person	

RECEIVERS

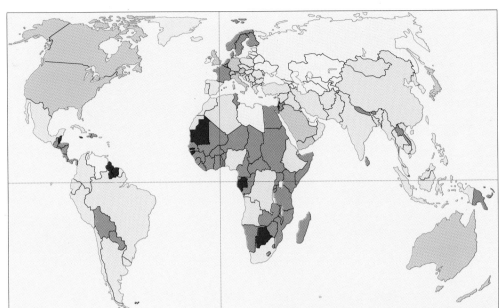

Top 5 providers		Top 5 receivers	
Norway	$218	Seychelles	$515
Sweden	$212	Vanuatu	$333
Denmark	$184	São Tomé & Príncipe	$311
Finland	$141	Samoa	$308
Netherlands	$139	Djibouti	$293

DEBT & AID

International debtors & the aid they receive (1989)

Although aid grants make a vital contribution to many of the world's poorer countries, they are usually dwarfed by the burden of debt that developing economies are expected to repay. In the case of Mozambique, aid amounted to more than 70% of GNP. In 1990, the World Bank rated Mozambique as the world's poorest country; yet debt interest payments came to almost 75 times its entire export earnings.

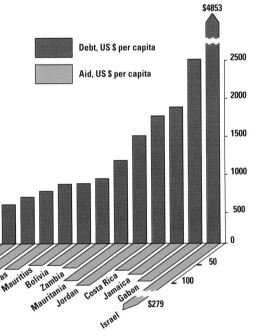

■ Debt, US $ per capita
■ Aid, US $ per capita

$4853

DISTRIBUTION OF SPENDING

Percentage share of household spending

- Food
- Medicine & Education
- Clothing
- Transport
- Energy & Housing
- Other

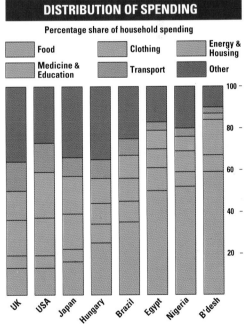

UK USA Japan Hungary Brazil Egypt Nigeria B'desh

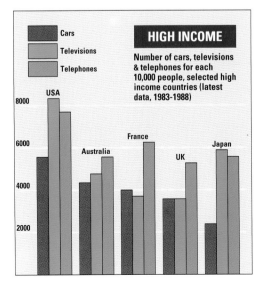

HIGH INCOME

- Cars
- Televisions
- Telephones

Number of cars, televisions & telephones for each 10,000 people, selected high income countries (latest data, 1983-1988)

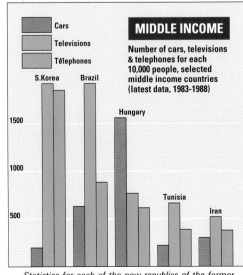

MIDDLE INCOME

- Cars
- Televisions
- Telephones

Number of cars, televisions & telephones for each 10,000 people, selected middle income countries (latest data, 1983-1988)

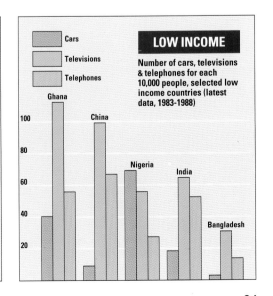

LOW INCOME

- Cars
- Televisions
- Telephones

Number of cars, televisions & telephones for each 10,000 people, selected low income countries (latest data, 1983-1988)

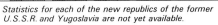

Statistics for each of the new republics of the former U.S.S.R. and Yugoslavia are not yet available.

21

QUALITY OF LIFE

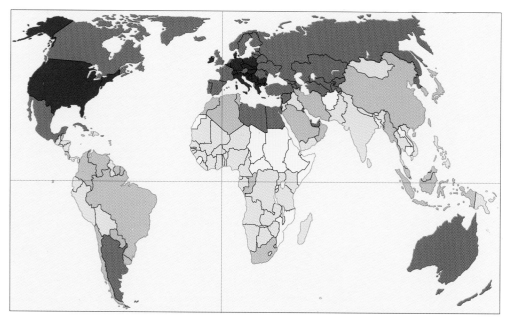

FOOD CONSUMPTION

Average daily food intake per person in calories (1986-88)

- Over 3,500 cal.
- 3,000 - 3,500 cal.
- 2,500 - 3,000 cal.
- 2,000 - 2,500 cal.
- Under 2,000 cal.

Top 5 countries		Bottom 5 countries	
Belgium	3,901 cal.	Bangladesh	1,925 cal.
Greece	3,702 cal.	Rwanda	1,817 cal.
Ireland	3,688 cal.	Sierra Leone	1,813 cal.
Bulgaria	3,650 cal.	Somalia	1,781 cal.
Germany	3,650 cal.	Mozambique	1,604 cal.

[USA 3,645]   [UK 3,256]

HOSPITAL CAPACITY

**Hospital beds available for each 1000 people,
(latest available year 1983-1988)**

Highest capacity		Lowest capacity	
Finland	14.9	Bangladesh	0.2
Sweden	13.2	Nepal	0.2
France	12.9	Ethiopia	0.3
USSR (1986)	12.8	Mauritania	0.4
Netherlands	12.0	Mali	0.5
North Korea	11.7	Burkina Faso	0.6
Switzerland	11.3	Pakistan	0.6
Austria	10.4	Niger	0.7
Czechoslovakia	10.1	Haiti	0.8
Hungary	9.1	Chad	0.8

[UK 8] [USA 5.9]

Although the ratio of people to hospital beds gives a good approximation of a country's health provision, it is not an absolute indicator. Raw numbers may mask inefficiency and other weaknesses: the high availability of beds in North Korea, for example, has not prevented infant mortality rates almost three times as high as in the United Kingdom.

LIFE EXPECTANCY

Years of life expectancy at birth, selected countries (1988-89)

The chart shows combined data for both sexes. On average, women live longer than men worldwide, even in developing countries with high maternal mortality rates. Overall, life expectancy is steadily rising, though the difference between rich and poor nations remains dramatic.

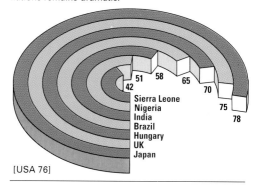

42 51 58 65 70 75 78

Sierra Leone
Nigeria
India
Brazil
Hungary
UK
Japan

[USA 76]

INFECTIOUS DISEASE

**Deaths from infectious disease,
per 100,000 people,
selected countries (latest figures, 1983-88)**

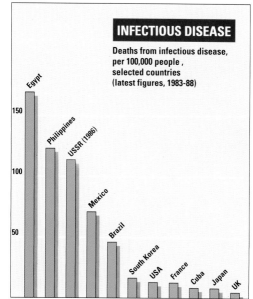

Egypt, Philippines, USSR (1986), Mexico, Brazil, South Korea, USA, France, Cuba, Japan, UK

CHILD MORTALITY

Number of babies who will die before the age of one year, per thousand live births (average 1990-95)

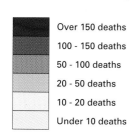

- Over 150 deaths
- 100 - 150 deaths
- 50 - 100 deaths
- 20 - 50 deaths
- 10 - 20 deaths
- Under 10 deaths

Highest child mortality		Lowest child mortality	
Afghanistan	162	Hong Kong	6
Mali	159	Denmark	6
Sierra Leone	143	Japan	5
Guinea-Bissau	140	Iceland	5
Malawi	138	Finland	5

[USA 9]   [UK 8]

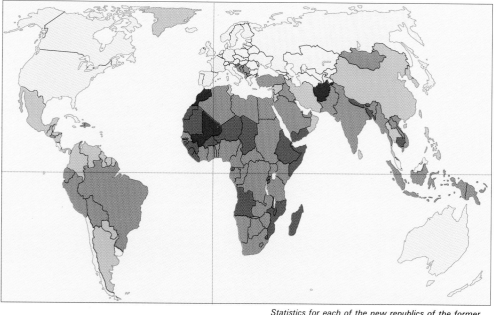

*Statistics for each of the new republics of the former
U.S.S.R. and Yugoslavia are not yet available.*

LITERACY

Percentage of adult population unable to read or write (1988)

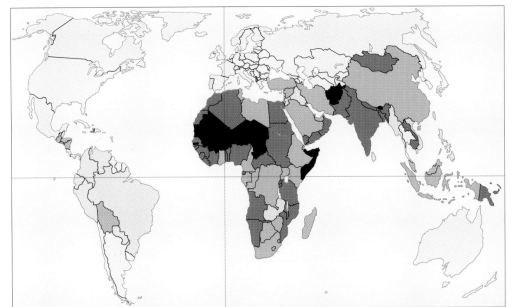

■	Over 75%
▓	50 - 75%
▒	25 - 50%
░	10 - 15%
□	Under 10%

Highest illiteracy

Somalia	88%
Burkina Faso	87%
Niger	86%
Mali	83%
Mauritania	83%
Yemen	82%
Chad	78%
Benin	76%
Nepal	76%

Highest female illiteracy

Yemen	97%
Burkina Faso	94%
Somalia	94%
Niger	91%
Chad	89%
Mali	89%
Nepal	88%
Benin	84%
Guinea	83%

EDUCATION

Percentage of age group in secondary school, selected countries (1987)

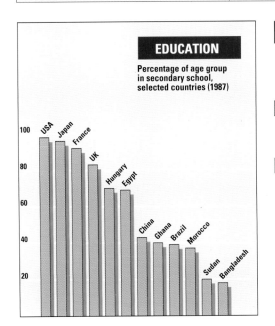

FERTILITY & EDUCATION

Fertility rates compared with female education, selected countries (1988)

░	Fertility rate: number of children borne by average woman
▒	Percentage of female age group in secondary education

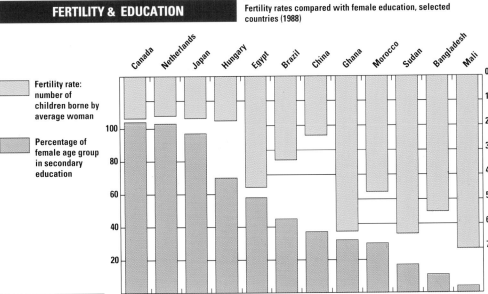

WOMEN IN THE WORKFORCE

Women in paid employment as a percentage of the total workforce (1989)

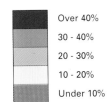

■	Over 40%
▓	30 - 40%
▒	20 - 30%
□	10 - 20%
▓	Under 10%

Highest proportion

Botswana	53%
Burundi	53%
Ghana	51%
Burkina Faso	49%
Sweden	48%
Bulgaria	48%
Barbados	47%
Thailand	47%
Finland	47%

Lowest proportion

UAE	6%
Saudi Arabia	7%
Bangladesh	7%
Qatar	7%
Algeria	8%
Oman	8%
Libya	8%
Jordan	9%
Egypt	9%

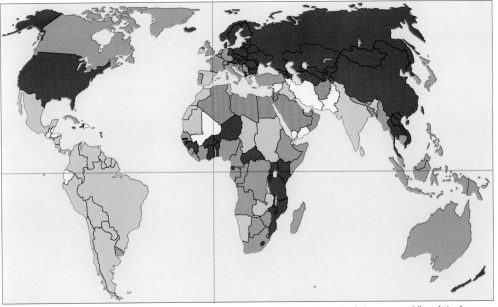

Statistics for each of the new republics of the former U.S.S.R. and Yugoslavia are not yet available.

ENERGY

PRODUCTION

[Each square represents 1% of world energy production]

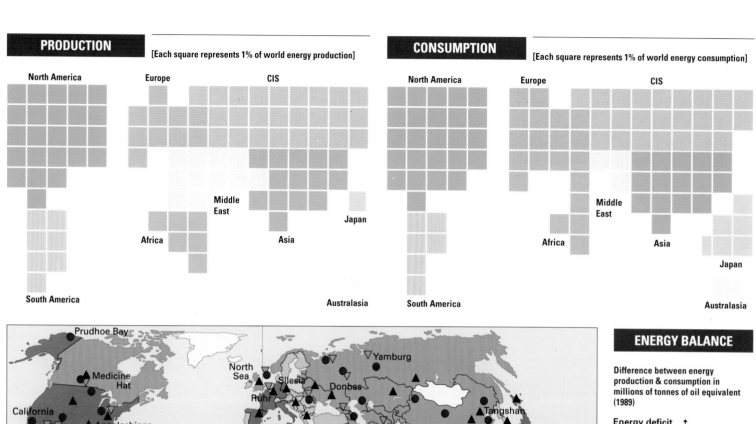

North America   Europe   CIS

Middle East

Africa   Asia   Japan

South America   Australasia

CONSUMPTION

[Each square represents 1% of world energy consumption]

North America   Europe   CIS

Middle East

Africa   Asia

Japan

South America   Australasia

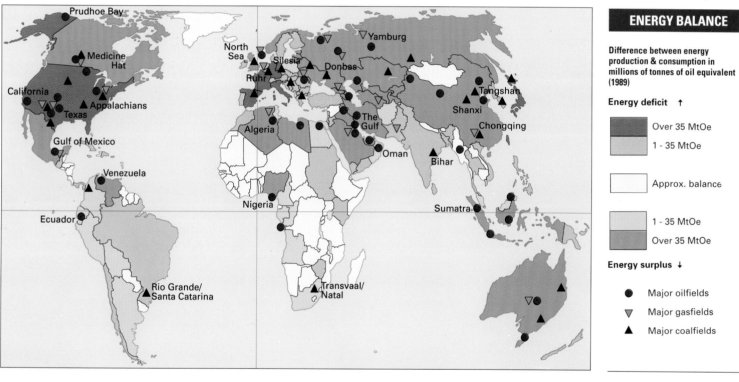

Map labels: Prudhoe Bay, Medicine Hat, California, Appalachians, Texas, Gulf of Mexico, Venezuela, Ecuador, Rio Grande/Santa Catarina, North Sea, Ruhr, Silesia, Donbas, Algeria, Nigeria, The Gulf, Oman, Yamburg, Shanxi, Tangshan, Chongqing, Bihar, Sumatra, Transvaal/Natal

ENERGY BALANCE

Difference between energy production & consumption in millions of tonnes of oil equivalent (1989)

Energy deficit ↑

- Over 35 MtOe
- 1 – 35 MtOe

- Approx. balance

- 1 – 35 MtOe
- Over 35 MtOe

Energy surplus ↓

- ● Major oilfields
- ▽ Major gasfields
- ▲ Major coalfields

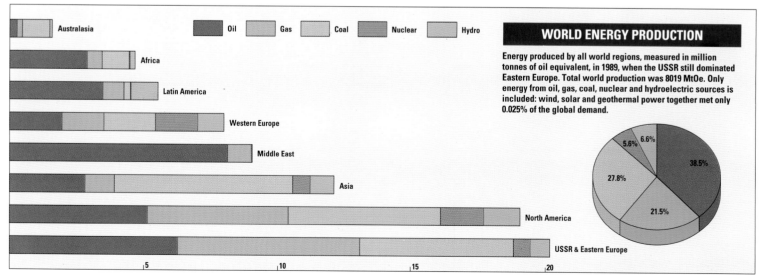

Legend: Oil | Gas | Coal | Nuclear | Hydro

Bar chart regions: Australasia, Africa, Latin America, Western Europe, Middle East, Asia, North America, USSR & Eastern Europe

Scale: 5, 10, 15, 20

WORLD ENERGY PRODUCTION

Energy produced by all world regions, measured in million tonnes of oil equivalent, in 1989, when the USSR still dominated Eastern Europe. Total world production was 8019 MtOe. Only energy from oil, gas, coal, nuclear and hydroelectric sources is included: wind, solar and geothermal power together met only 0.025% of the global demand.

Pie chart: 38.5%, 21.5%, 27.8%, 5.6%, 6.6%

NUCLEAR POWER

Percentage of electricity generated by nuclear power stations, leading nations (1988)

1.	France 70%	11.	Germany (W) 34%
2.	Belgium 66%	12.	Japan 28%
3.	Hungary 49%	13.	Czechoslovakia .. 27%
4.	South Korea 47%	14.	UK 18%
5.	Sweden 46%	15.	USA 17%
6.	Taiwan 41%	16.	Canada 16%
7.	Switzerland 37%	17.	Argentina 12%
8.	Finland 36%	18.	USSR (1989) 11%
9.	Spain 36%	19.	Yugoslavia 6%
10.	Bulgaria 36%	20.	Netherlands 5%

The decade 1980-1990 was a bad time for the nuclear power industry. Major projects regularly ran vastly over-budget, and fears of long-term environmental damage were heavily reinforced by the 1986 Soviet disaster at Chernobyl. Although the number of reactors in service continued to increase throughout the period, orders for new plant shrank dramatically, and most countries cut back on their nuclear programs.

HYDROELECTRICITY

Percentage of electricity generated by hydroelectrical power stations, leading nations (1988)

1.	Paraguay 99.9%	11.	Laos 95.5%
2.	Zambia 99.6%	12.	Nepal 95.2%
3.	Norway............ 99.5%	13.	Iceland............. 94.0%
4.	Congo 99.1%	14.	Uruguay 93.0%
5.	Costa Rica 98.3%	15.	Brazil............... 91.7%
6.	Uganda 98.3%	16.	Albania 87.2%
7.	Rwanda 97.7%	17.	Fiji 81.4%
8.	Malawi 97.6%	18.	Ecuador 80.7%
9.	Zaïre 97.2%	19.	C. African Rep. 80.4%
10.	Cameroon 97.2%	20.	Sri Lanka 80.4%

Countries heavily reliant on hydroelectricity are usually small and non-industrial: a high proportion of hydroelectric power more often reflects a modest energy budget than vast hydroelectric resources. The USA, for instance, produces only 8% of power requirements from hydroelectricity; yet that 8% amounts to more than three times the HEP generated by all of Africa.

ALTERNATIVE ENERGY SOURCES

Solar: Each year the sun bestows upon the Earth almost a million times as much energy as is locked up in all the planet's oil reserves, but only an insignificant fraction is trapped and used commercially. In some experimental installations, mirrors focus the sun's rays on to boilers, whose steam generates electricity by spinning conventional turbines. Solar cells turn sunlight into electricity directly. Efficiencies are still low, but advancing technology could make the sun a major electricity source by 2100.

Wind: Caused by the uneven heating of the spinning Earth, winds are themselves the product of solar energy. Traditional windmills turn wind power into mechanical work; recent models usually generate electricity. But efficient windmills are expensive to build, and suitable locations are few.

Tidal: The energy from tides is potentially enormous, although only a few installations have been built to exploit it. In theory at least, waves and currents could also provide almost unimaginable power, and the thermal differences in the ocean depths are another huge well of potential energy.

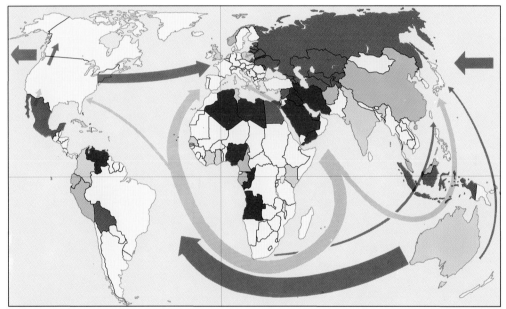

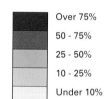

FUEL EXPORTS

Fuels as a percentage of total value of all exports (1986)

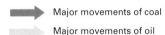

- Over 75%
- 50 - 75%
- 25 - 50%
- 10 - 25%
- Under 10%

Direction of trade

→ Major movements of coal

→ Major movements of oil

CONVERSIONS

For historical reasons, oil is still traded in 'barrels'. The weight and volume equivalents shown below are all based on average density 'Arabian light' crude oil.

The energy equivalents given for a tonne of oil are also somewhat imprecise: oil and coal of different qualities will have varying energy contents, a fact usually reflected in their price on world markets.

1 barrel: 0.136 tonnes/159 liters/35 Imperial gallons/ 42 US gallons. **1 tonne:** 7.33 barrels/1185 liters/ 256 Imperial gallons/261 US gallons. **1 tonne oil:** 1.5 tonnes hard coal/3.0 tonnes lignite/12,000 kWh.

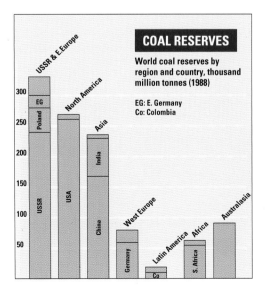

COAL RESERVES

World coal reserves by region and country, thousand million tonnes (1988)

EG: E. Germany
Co: Colombia

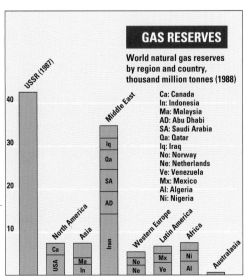

GAS RESERVES

World natural gas reserves by region and country, thousand million tonnes (1988)

Ca: Canada
In: Indonesia
Ma: Malaysia
AD: Abu Dhabi
SA: Saudi Arabia
Qa: Qatar
Iq: Iraq
No: Norway
Ne: Netherlands
Ve: Venezuela
Mx: Mexico
Al: Algeria
Ni: Nigeria

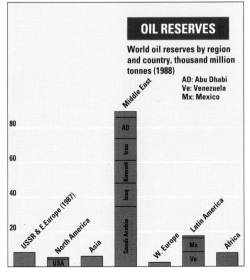

OIL RESERVES

World oil reserves by region and country, thousand million tonnes (1988)

AD: Abu Dhabi
Ve: Venezuela
Mx: Mexico

PRODUCTION

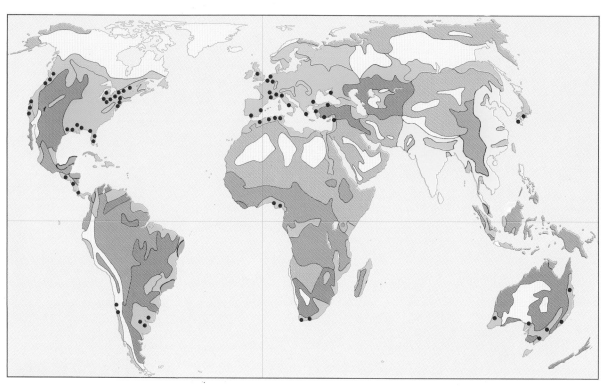

AGRICULTURE

- Nomadic herding
- Primitive subsistence farming
- Intensive subsistence farming
- Tropical plantation agriculture
- Commercial ranching
- Commercial livestock & grain farming
- Commercial fruit growing & market gardening
- Forestry
- Land of little or no agricultural value

STAPLE CROPS

Separate figures for Russia, Ukraine and the other successors of the defunct USSR are not yet available

Wheat
China 16.9% | USSR 16.8% | USA 10.3% | India 10.0% | France 5.9% | Canada 4.5% | Turkey 2.3%

World total (1989): 538,056,000 tonnes

Rice
China 35.4% | India 21.2% | Indonesia 8.6% | Bangladesh 5.3% | Thailand 4.2% | Vietnam 3.6%

World total (1989): 506,291,000 tonnes

Maize
USA 40.7% | China 16.1% | Brazil 5.6% | USSR 3.6% | France 2.7%

World total (1989): 470,318,000 tonnes

Potatoes
USSR 26.0% | Poland 12.4% | China 10.9% | Germany 6.1% | USA 6.0% | India 5.2%

World total (1989): 276,740,000 tonnes

Millet
India 32.8% | China 18.7% | USSR 13.1% | Nigeria 11.5% | Niger 4.2%

World total (1989): 30,512,000 tonnes

Rye
USSR 53.9% | Poland 17.8% | Germany 11.2% | China 2.9% | Canada 2.4%

World total (1989): 34,893,000 tonnes

Soya
USA 48.9% | Brazil 22.4% | China 10.1% | Argentina 5.8%

World total (1989): 107,350,000 tonnes

Cassava
Thailand 15.9% | Brazil 15.8% | Indonesia 11.24% | Nigeria 11.19% | Zaïre 11.1% | Tanzania 4.3% | India 3.6%

World total (1989): 147,500,000 tonnes

SUGARS

Sugar cane
Brazil 22.4% | India 19.7% | Cuba 7.3% | China 5.5% | Mexico 4.0% | Pakistan 3.7% | Thailand 3.6%

World total (1989): 1,007,184,000 tonnes

Sugar beet
USSR 31.9% | Germany 8.8% | France 7.7% | USA 7.6% | Italy 4.9% | Poland 4.7% | Turkey 4.0%

World total (1989): 305,882,000 tonnes

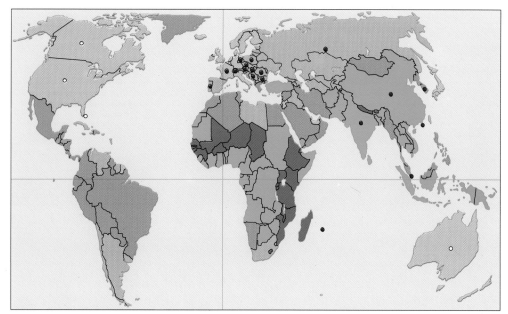

BALANCE OF EMPLOYMENT

Percentage of total workforce employed in agriculture, forestry & fishing (latest available year)

- Over 75%
- 50 - 75%
- 25 - 50%
- 10 - 25%
- Under 10%

- ● Over 25% of total workforce employed in manufacturing

- ○ Over 75% of total workforce employed in service industries (e.g. work in offices, shops, tourism, transport, entertainment & administration)

MINERAL PRODUCTION

Separate figures for Russia, Ukraine and the other successors of the defunct USSR are not yet available

Copper
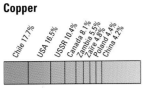
Chile 17.7% | USA 16.5% | USSR 10.4% | Canada 8.1% | Zambia 5.5% | Zaire 4.8% | Poland 4.4% | China 4.2%

World total (1989): 9,100,000 tonnes

Iron
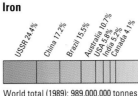
USSR 24.4% | China 17.2% | Brazil 15.5% | Australia 10.7% | USA 5.8% | India 5.2% | Canada 4.1%

World total (1989): 989,000,000 tonnes

Chromium
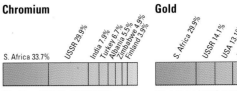
S. Africa 33.7% | USSR 29.9% | India 7.9% | Turkey 6.7% | Albania 5.5% | Zimbabwe 4.9% | Finland 3.9%

World total (1989): 12,700,000 tonnes

Gold
S. Africa 29.9% | USSR 14.1% | USA 13.1% | Australia 10.0% | Canada 4.2% | Brazil 2.4%

World total (1989): 2,026,000 kilograms

Uranium
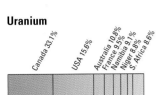
Canada 33.1% | USA 15.6% | Australia 10.8% | France 9.5% | Namibia 9.1% | Niger 8.8% | S. Africa 8.6%

World total (1989): 34,000 tonnes

Lead
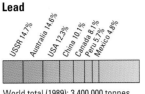
USSR 14.7% | Australia 14.6% | USA 12.3% | China 10.1% | Canada 8.1% | Peru 5.1% | Mexico 4.8%

World total (1989): 3,400,000 tonnes

Tin
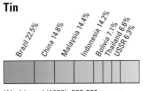
Brazil 22.5% | China 14.8% | Malaysia 14.4% | Indonesia 14.2% | Bolivia 7.1% | Thailand 6.6% | USSR 6.3%

World total (1989): 223,000 tonnes

Manganese
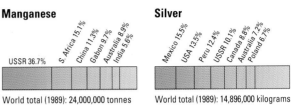
USSR 36.7% | S. Africa 15.1% | China 11.3% | Gabon 9.7% | Australia 8.9% | India 5.6%

World total (1989): 24,000,000 tonnes

Silver
Mexico 15.5% | USA 13.5% | Peru 12.4% | USSR 10.1% | Canada 8.8% | Australia 7.2% | Poland 6.7%

World total (1989): 14,896,000 kilograms

Aluminium
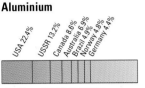
USA 22.4% | USSR 13.2% | Canada 8.6% | Australia 6.9% | Brazil 4.9% | Norway 4.8% | Germany 4.4%

World total (1989): 18,000,000 tonnes

Mercury

USSR 27.3% | China 18.2% | Spain 17.6% | Algeria 12.7% | USA 7.8% | Mexico 6.3% | Turkey 3.7%

World total (1989): 5,500,000 kilograms

Zinc
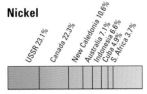
Canada 16.6% | USSR 12.9% | Australia 11.0% | China 8.5% | Peru 8.2% | USA 4.0% | Mexico 3.9%

World total (1989): 7,300,000 tonnes

Nickel
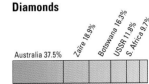
USSR 23.1% | Canada 22.3% | New Caledonia 10.6% | Indonesia 7.1% | Cuba 6.6% | Australia 4.9% | S. Africa 3.7%

World total (1989): 910,000 tonnes

Diamonds
Australia 37.5% | Zaire 18.9% | Botswana 16.3% | USSR 11.8% | S. Africa 9.7%

World total (1989): 96,600,000 carats

MINERAL DISTRIBUTION

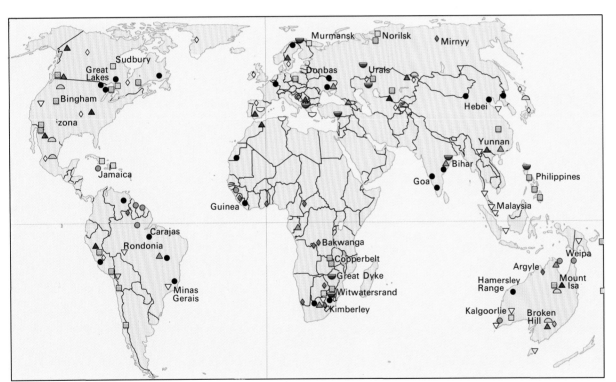

Light metals
- ● Bauxite

Base metals
- ◻ Copper
- ▲ Lead
- ▽ Mercury
- ▽ Tin
- ◇ Zinc

Iron and ferro-alloys
- ● Iron
- ⬗ Chrome
- ▲ Manganese
- ◻ Nickel

Precious metals
- ▽ Gold
- △ Silver

Precious stones
- ◇ Diamonds

Map labels: Murmansk, Norilsk, Mirnyy, Sudbury, Great Lakes, Donbas, Urals, Hebei, Bingham, izona, Yunnan, Bihar, Jamaica, Goa, Philippines, Malaysia, Guinea, Carajas, Rondonia, Bakwanga, Copperbelt, Weipa, Argyle, Mount Isa, Great Dyke, Hamersley Range, Minas Gerais, Witwatersrand, Kimberley, Kalgoorlie, Broken Hill

---

STEEL PRODUCTION
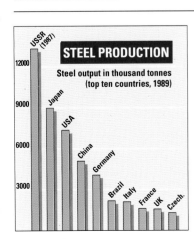
Steel output in thousand tonnes (top ten countries, 1989)

USSR (1987), Japan, USA, China, Germany, Brazil, Italy, France, UK, Czech.

SHIPBUILDING
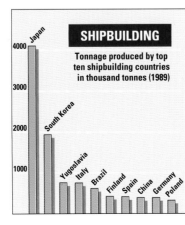
Tonnage produced by top ten shipbuilding countries in thousand tonnes (1989)

Japan, South Korea, Yugoslavia, Italy, Brazil, Finland, Spain, China, Germany, Poland

AUTOMOBILES
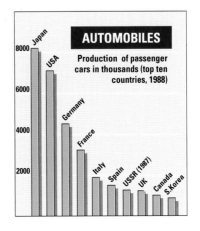
Production of passenger cars in thousands (top ten countries, 1988)

Japan, USA, Germany, France, Italy, Spain, USSR (1987), UK, Canada, S.Korea

COMMERCIAL VEHICLES
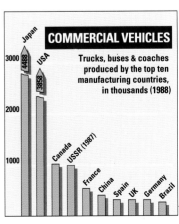
Trucks, buses & coaches produced by the top ten manufacturing countries, in thousands (1988)

Japan 4488, USA 3858, Canada, USSR (1987), France, China, Spain, UK, Germany, Brazil

TRADE

SHARE OF WORLD TRADE

Proportion of total world exports by value (1989)

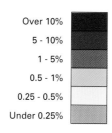

- Over 10%
- 5 - 10%
- 1 - 5%
- 0.5 - 1%
- 0.25 - 0.5%
- Under 0.25%

International trade is dominated by a handful of powerful maritime nations. The members of 'G7', the inner circle of OECD and the top seven countries listed in the diagram below account for more than half the total. The majority of nations – including all but four in Africa – contribute less than one quarter of one per cent to the worldwide total of exports. The EC countries account for 40%, the Pacific Rim nations over 35%.

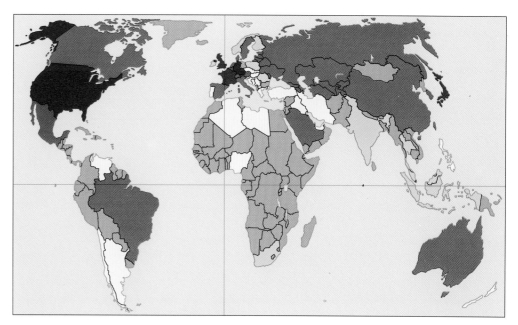

THE GREAT TRADING NATIONS

The imports and exports of the top ten trading nations as a percentage of world trade (1989). Each country's trade in manufactured goods is shown in orange.

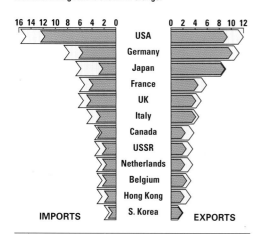

IMPORTS		EXPORTS
USA		
Germany		
Japan		
France		
UK		
Italy		
Canada		
USSR		
Netherlands		
Belgium		
Hong Kong		
S. Korea		

PATTERNS OF TRADE

Thriving international trade is the outward sign of a healthy world economy, the obvious indicator that some countries have goods to sell and others the wherewithal to buy them. Despite local fluctuations, trade throughout the 1980s grew consistently faster than output, increasing in value by almost 50% in the decade 1979-89. It remains dominated by the rich, industrialized countries of the Organization for Economic Development: between them, OECD members account for almost 75% of world imports and exports in most years. OECD dominance is just as marked in the trade in 'invisibles' - a column in the balance sheet that includes among other headings the export of services, interest payments on overseas investments, tourism and even remittances from migrant workers abroad. In the US, invisibles account for almost 40% of all trading income.

However, the size of these great trading economies means that imports and exports usually make up only a fraction of their total wealth: in the case of the famously export-conscious Japanese, trade in goods and services amounts to less than 18% of GDP. In poorer countries, trade - often in a single commodity - may amount to 50% GDP or more.

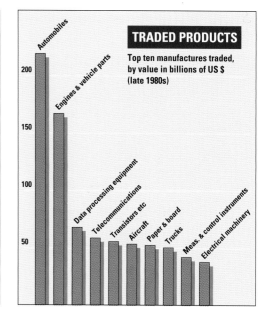

TRADED PRODUCTS

Top ten manufactures traded, by value in billions of US $ (late 1980s)

BALANCE OF TRADE

Value of exports in proportion to the value of imports (1988)

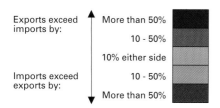

Exports exceed imports by:
- More than 50%
- 10 - 50%
- 10% either side

Imports exceed exports by:
- 10 - 50%
- More than 50%

The total world trade balance should amount to zero, since exports must equal imports on a global scale. In practice, at least $100 billions in exports go unrecorded, leaving the world with an apparent deficit and many countries in a better position than public accounting reveals. However, a favorable trade balance is not necessarily a sign of prosperity: many poorer countries must maintain a high surplus in order to service debts, and do so by restricting imports below the levels needed to sustain successful economies.

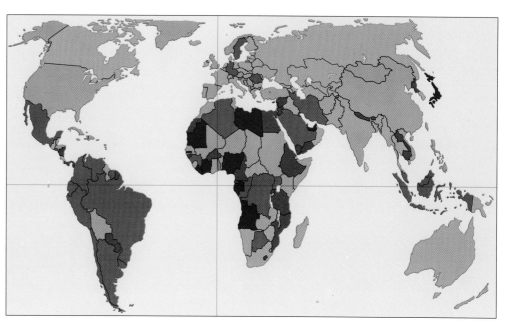

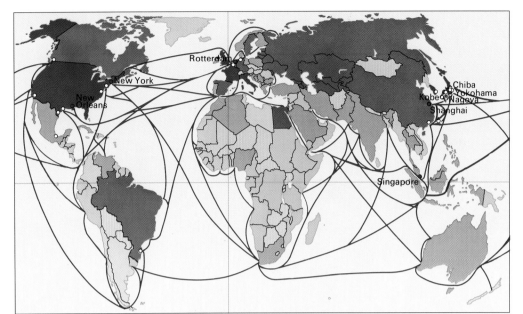

Freight unloaded in millions of tonnes (1988)

- Over 100
- 50 - 100
- 10 - 50
- 5 - 10
- Under 5
- Land locked countries

Major seaports

- ● Over 100 million tonnes per year
- ○ 50-100 million tonnes per year

CARGOES

Type of seaborne freight

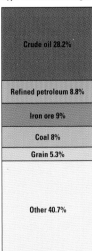

- Crude oil 28.2%
- Refined petroleum 8.8%
- Iron ore 9%
- Coal 8%
- Grain 5.3%
- Other 40.7%

MERCHANT FLEETS

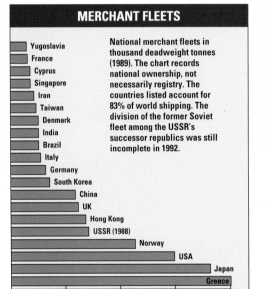

National merchant fleets in thousand deadweight tonnes (1989). The chart records national ownership, not necessarily registry. The countries listed account for 83% of world shipping. The division of the former Soviet fleet among the USSR's successor republics was still incomplete in 1992.

Yugoslavia, France, Cyprus, Singapore, Iran, Taiwan, Denmark, India, Brazil, Italy, Germany, South Korea, China, UK, Hong Kong, USSR (1988), Norway, USA, Japan, Greece

20,000 40,000 60,000 80,000

WORLD SHIPPING

World merchant fleet by type of vessel and deadweight tonnage (1989)

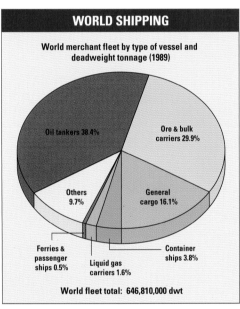

- Oil tankers 38.4%
- Ore & bulk carriers 29.9%
- General cargo 16.1%
- Others 9.7%
- Container ships 3.8%
- Liquid gas carriers 1.6%
- Ferries & passenger ships 0.5%

World fleet total: 646,810,000 dwt

THE GREAT PORTS

The world's ten busiest ports by million tonnes of shipping arrivals (late 1980s)

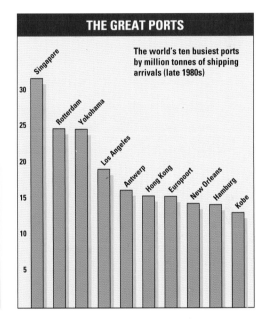

Singapore, Rotterdam, Yokohama, Los Angeles, Antwerp, Hong Kong, Europoort, New Orleans, Hamburg, Kobe

30, 25, 20, 15, 10, 5

DEPENDENCE ON TRADE

Value of exports as a percentage of Gross Domestic Product (1988)

- Over 50%
- 40 - 50%
- 30 - 40%
- 20 - 30%
- 10 - 20%
- Under 10%

- ● Most dependent on industrial exports (over 75% of total exports)
- ● Most dependent on fuel exports (over 75% of total exports)
- ○ Most dependent on mineral & metal exports (over 75% of total exports)

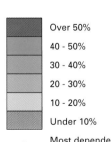

TRAVEL & TOURISM

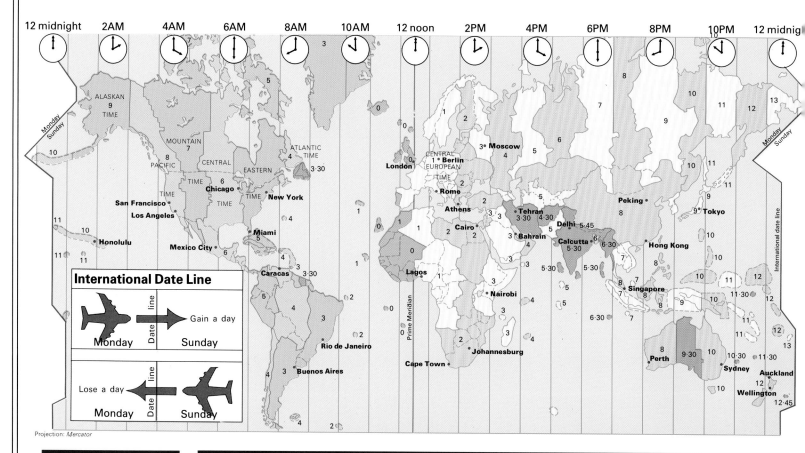

12 midnight | 2AM | 4AM | 6AM | 8AM | 10AM | 12 noon | 2PM | 4PM | 6PM | 8PM | 10PM | 12 midnig

Projection: *Mercator*

International Date Line

Gain a day
Monday — Date line — Sunday

Lose a day
Monday — Date line — Sunday

TIME ZONES

Zones using GMT
(Greenwich Mean Time)

Half-hour zones

Zones slow of GMT

Zones fast of GMT

The time when it is
12 noon at Greenwich

RAIL & ROAD: THE LEADING NATIONS

Total rail network ('000 miles)		Passenger miles per head per year		Total road network ('000 miles)		Vehicle miles per head per year		Vehicle mi per head per year per mi road network	
1. USSR (1986)	154.5	Japan	1,716	1. USA	389.6	USA	7,815	Hong Kong	2,941.1
2. USA	140.9	Switzerland	952	2. Brazil	104.6	Luxembourg	4,993	Kuwait	2,145.9
3. Canada	41.1	Germany (E:'87.)	846	3. USSR (1986)	99.1	Kuwait	4,531	Jordan	825.9
4. India	38.6	Czech'vakia	804	4. India	97.1	France	4,463	UK	581.0
5. China	32.9	Poland	801	5. Japan	69.0	Sweden	4,369	Italy	577.1
6. Germany	25.9	USSR	798	6. China	61.4	Germany	4,253	Germany	527.4
7. Australia	24.6	France	671	7. Australia	53.3	Denmark	4,227	Netherlands	475.9
8. France	21.6	Austria	607	8. Canada	52.8	Austria	4,073	Tunisia	454.0
9. Argentina	21.3	Denmark	586	9. France	50.3	Netherlands	3,740	Iraq	410.8
10. Poland	15.1	Hungary	566	10. Germany	30.9	UK	3,586	Luxembourg	363.3
11. South Africa	14.9	Bulgaria	561	11. Poland	22.6	Canada	3,433	Japan	310.6
12. Brazil	13.8	South Africa	546	12. UK	22.0	Italy	3,032	USA	308.9
13. Mexico	12.5	Italy	451	13. Turkey	20.1	Belgium	3,013	Denmark	306.9
14. Japan	12.4	Sweden	448	14. Spain	19.9	Japan	2,797	Austria	290.6
15. UK	10.4	Germany (W:'88)	400	15. Italy	18.9	South Africa	1,735	France	287.3

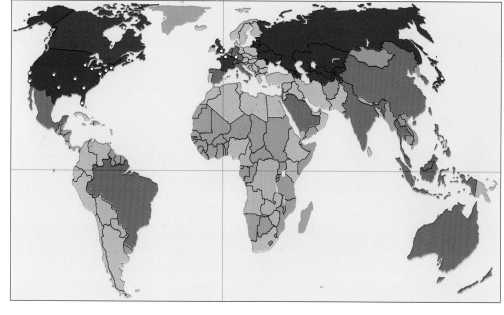

AIR TRAVEL

Passenger miles [the number of passengers, international
& domestic, multiplied by the distance flown by each passenger
from the airport of origin] (1988)

Over 60,000 million

30,000 - 60,000 million

6,000 - 30,000 million

600 - 6,000 million

300 - 600 million

Under 300 million

○ Major airports (handling over
20 million passengers per year)

World's busiest airports (total passengers)		World's busiest airports (international passengers)	
1. Chicago	(O'Hare)	1. London	(Heathrow)
2. Atlanta	(Hatsfield)	2. London	(Gatwick)
3. Los Angeles	(Inter'nl)	3. Frankfurt	(International)
4. Dallas	(Dallas/Ft Worth)	4. New York	(Kennedy)
5. London	(Heathrow)	5. Paris	(De Gaulle)

DESTINATIONS

- ■ Cultural & historical centers
- □ Coastal resorts
- □ Ski resorts
- ▨ Centers of entertainment
- ▩ Places of pilgrimage
- □ Places of great natural beauty

VISITORS TO THE USA

International visitors spending in US $ million (1989)

1.	Japan	7,480
2.	Canada	6,020
3.	Mexico	4,170
4.	UK	4,130
5.	Germany	2,450
6.	France	1,290
7.	Australia	1,120
8.	All others	16,380

A record 38.3 million foreigners visited the US in 1989, about 70% of them on vacation. Between them they spent $ 43 billion.

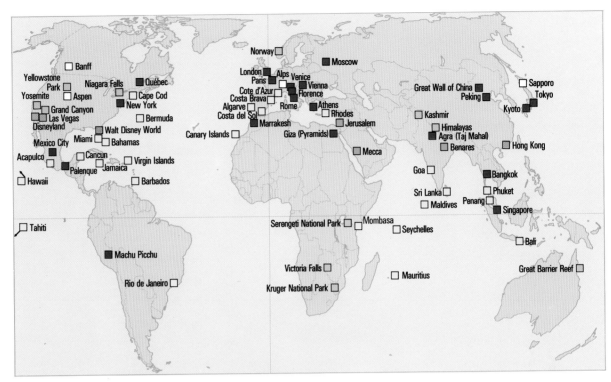

World map with destination markers: Banff, Yellowstone Park, Niagara Falls, Québec, Yosemite, Aspen, Cape Cod, New York, Grand Canyon, Las Vegas, Disneyland, Bermuda, Mexico City, Walt Disney World, Miami, Bahamas, Acapulco, Cancun, Palenque, Jamaica, Virgin Islands, Hawaii, Barbados, Tahiti, Machu Picchu, Rio de Janeiro, Norway, London, Paris, Alps, Venice, Vienna, Cote d'Azur, Florence, Costa Brava, Rome, Athens, Algarve, Rhodes, Costa del Sol, Marrakesh, Jerusalem, Canary Islands, Giza (Pyramids), Mecca, Moscow, Great Wall of China, Peking, Sapporo, Tokyo, Kyoto, Kashmir, Himalayas, Agra (Taj Mahal), Benares, Hong Kong, Goa, Bangkok, Sri Lanka, Phuket, Maldives, Penang, Singapore, Bali, Serengeti National Park, Mombasa, Seychelles, Great Barrier Reef, Victoria Falls, Mauritius, Kruger National Park

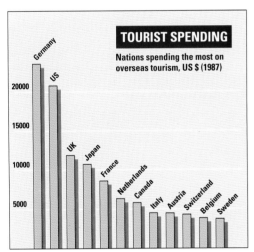

TOURIST SPENDING

Nations spending the most on overseas tourism, US $ (1987)

Bar chart: Germany, US, UK, Japan, France, Netherlands, Canada, Italy, Austria, Switzerland, Belgium, Sweden

IMPORTANCE OF TOURISM

		Arrivals from abroad (1987)	Receipts as % of GDP (1987)
1.	France	36,820,000	1.4%
2.	Spain	32,900,000	5.1%
3.	USA	28,790,000	0.4%
4.	Italy	25,750,000	1.6%
5.	Austria	15,760,000	6.5%
6.	UK	15,445,000	1.5%
7.	Canada	15,040,000	0.9%
8.	Germany	12,780,000	0.7%
9.	Hungary	11,830,000	3.2%
10.	Switzerland	11,600,000	3.1%
11.	China	10,760,000	0.7%
12.	Greece	7,564,000	4.7%

Small economies in attractive areas are often completely dominated by tourism: in some West Indian islands, tourist spending provides over 90% of total income. In cash terms the USA is the world leader: its 1987 earnings exceeded 15 billion dollars, though that sum amounted to only 0.4% of GDP.

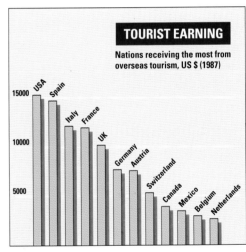

TOURIST EARNING

Nations receiving the most from overseas tourism, US $ (1987)

Bar chart: USA, Spain, Italy, France, UK, Germany, Austria, Switzerland, Canada, Mexico, Belgium, Netherlands

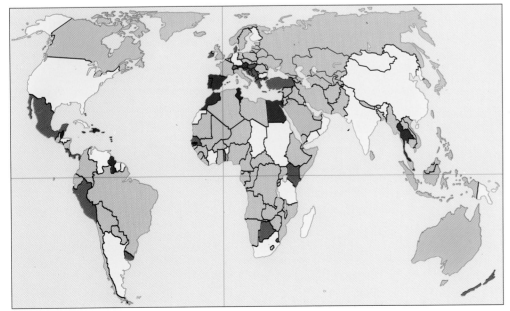

TOURISM

Receipts from tourism as a percentage of Gross National Product (1988)

- ■ Over 10%
- ▨ 5 - 10%
- ▩ 2.5 - 5%
- ▧ 1 - 2.5%
- □ 0.5 - 1%
- ▢ Under 0.5%

Largest share of world spending on tourism

Germany	16%
USA	14%
UK	8%
Japan	7%
France	6%

Largest share of world receipts from tourism

USA	10%
Spain	9%
Italy	8%
France	8%
UK	6%

SUBJECT INDEX

WORLD MAPS

SETTLEMENTS

◌ PARIS   ■ Berne   ◉ Livorno   ◉ Brugge   ◎ Algeciras   ◌ Fréjus   ○ Oberammergau   ○ Thira

Settlement symbols and type styles vary according to the scale of each map and indicate the importance
of towns on the map rather than specific population figures

∴ Ruins or Archæological Sites   ˅ Wells in Desert

ADMINISTRATION

International Boundaries

International Boundaries
(Undefined or Disputed)

Internal Boundaries

National Parks

Country Names

NICARAGUA

Administrative
Area Names

KENT

CALABRIA

International boundaries show the *de facto* situation where there are rival claims to territory

COMMUNICATIONS

Principal Roads

Other Roads

Trails and Seasonal Roads

⋈ Passes

✿ Airfields

Principal Railroads

Railroads
Under Construction

Other Railroads

╡---╞ Railroad Tunnels

Principal Canals

PHYSICAL FEATURES

Perennial Streams

Intermittent Streams

Perennial Lakes

Intermittent Lakes

Swamps and Marshes

Permanent Ice
and Glaciers

▲ 8848 Elevations (m)

▼ 8050 Sea Depths (m)

1134 Height of Lake Surface
Above Sea Level (m)

Projection: *Hammer Equal Area*

ARCTIC OCEAN

10 11 12 13 14 15 16 17 18

Laptev Sea
New Siberian Is.
East Siberian Sea

A

Svalbard
(Norway)
Zemlya Frantsa Iosifa
Novaya Zemlya
Severnaya Zemlya
Ust Port
Tiksi
Verkhoyansk
Nizhne-Kolymsk
Arctic Circle
Anadyr'

Barents Sea
Kara Sea
Nord Kapp
Murmansk

Narvik

gian

Bering Sea
Kamchatka

B

Arkhangelsk
Salekhard
Yenisey
Ob
Vilyuysk
Lena
Yakutsk
Okhotsk
Sea of Okhotsk
Petropavlovsk-Kamchatskiy

FINLAND
Helsinki
St. Peterburg
RUSSIA
Perm
Yekaterinburg
Tomsk
Krasnoyarsk
L.Baykal
Ulan Ude

Sakhalin
C.Lopatka

orth
Oslo
SWEDEN
Stockholm
Yaroslavl
Kazan
Novosibirsk

Komsomolsk
Khabarovsk
Kuril Is.

DENMARK
København
Moskva
Ufa
Chelyabinsk
Omsk
Novokuznetsk
Barnaul
Irtysh
Irkutsk

Amur
Vladivostok
Sapporo
Hakodate

Hamburg
Amsterdam
LATVI
LITH.
BELO.
Samara
Saratov
Orenburg
Karaganda

MONGOLIA
Ulaanbaatar
Harbin
Changchun
Shenyang
N.KOREA
Sea of Japan
JAPAN
Tōkyō

Berlin
POLAND Warszawa
Minsk
Kiyev
KAZAKHSTAN

Beijing
Tianjin
Dalian
S. KOREA
Sŏul
Pusan
Kyoto
Yokohama
Nagoya
Ōsaka

Paris
Praha
Lvov
Kharkov
UKRAINE
Volga
Volgograd
Aral Sea
L.Balkhash
Alma Ata

Taiyuan
Jinan
Qingdao
Kōbe
Kitakyūshū

C

yon
Torino
AUSTRIA
Budapest
Odessa
Astrakhan
Caspian Sea
UZBEKISTAN
Tashkent
KIRGIZIA

CHINA
Lanzhou
Xi'an
Huang
Nanjing
Shanghai

Marseille
Roma
Beograd
ROMANIA
Bucuresti
Black Sea
Grozny
Tbilisi
Samarkand
Dushanbe
TAJ.

Chengdu
Chongqing
Wuhan
Changsha
East China Sea

Barcelona
Sardinia
BULGARIA
Istanbul
Yerevan
Baku
TURKMENISTAN
Ashkhabad
AFGHANISTAN
Kabul

XIZANG (TIBET)
Lhasa
PACIFIC

Valencia
Sicily Athinai
Crete
Izmir TURKEY
Tabriz
Mashhad
Srinagar

Fuzhou

Alger Tunis
TUNISIA
CYPRUS
SYRIA
Halab
Tehrān
Rawalpindi
Lahore
NEPAL Kathmandu
BHU.

Guangzhou
TAIWAN
Tropic of Cancer

GERIA
Ain Salah
Mediterranean Sea
Bayrut
Dimashq
Baghdad
IRAN
Ispahan
Delhi
Agra
Kanpur
LUCKNOW
BANGLA
Dhaka
BURMA
(MYANMAR)

Hong Kong

D

LIBYA
EGYPT
Tel Aviv-Yafo
Al Iskandariya
Amman
Jerusalem
IRAQ
Abadan
Shiraz
PAKISTAN
Karachi
Ahmadabad
INDIA
Calcutta
Mandalay

Huang
South China Sea

NORTHERN MARIANAS
Wake I.
(U.S.)

Banghazi
El Qahira
Aswan
JORDAN
KUWAIT
BAHRAIN QATAR
U.A.E.
Nagpur
Bombay
Pune
Bay of Bengal
Rangoon
Hanoi
Hainan
VIET-

Guam
(U.S.)

NIGER
CHAD
El Khartum
Omdurman
SAUDI ARABIA
Makkah
YEMEN
Arabian Sea
Hyderabad
THAILAND
Bangkok
NAM
Manila
PHILIPPINES
Yap
FEDERATED STATES

MARSHALL IS.

Niamey
Kano
L.Chad
Ndjamena
Aden
Gulf of Aden
Socotra
(Yemen)
Bangalore
Madras
Andaman Is.
(India)
CAMBODIA
Phnom Penh
Phanh Bho Ho Chi Minh
Cebu
BELAU
Caroline Is.
Truk
Ponape

OF MICRONESIA

bougou
NIGERIA
Ibadan
Lagos
SOMALI REP.
DJIBOUTI
Addis Abeba
ETHIOPIA
Lakshadweep Is.
Nicobar Is.
(India)
Colombo
SRI LANKA
(CEYLON)
MALAYSIA
Kuala Lumpur
PEN. MALAYSIA
Kuching
SABAH
BRUNEI

Gilbert Is.

Douala
Yaounde
CENTRAL AFRICAN REPUBLIC
Bangui
L.Turkana
KENYA
MALDIVES
Dondra Hd.
Medan
SINGAPORE
Borneo
NAURU

Guinea
Librevlle
GABON
ZAÏRE (CONGO)
Kisangani
UGANDA
Kampala
Equator
Banjarmasin
Sulawesi
Maluku
Irian Jaya
KIRIBATI

SÃO TOMÉ AND PRINCIPE
Congo
Zaire
Victoria
Nairobi
INDIAN
Palembang
Ujung Pandang
INDONESIA
Rabaul
New Ireland

Brazzaville
Kinshasa
Kananga
BUR.
Mombasa
Chagos Arch.
(Br.)
Jakarta
Jawa
Surabaya
PAPUA NEW GUINEA
New Britain
SOLOMON IS.

Luanda
CABINDA
Kasai
L.Tanganyika
TANZANIA
Zanzibar
Dar es Salaam
Amirante Is.
SEYCHELLES
Bandung
Port Moresby
Louisade Arch.
TUVALU

ANGOLA
Lubumbashi
Aldabra
OCEAN
Cocos
(Keeling Is.)
(Australia)
Timor Sea
Arafura Sea
C.York
Santa Cruz Is.

E

Benguela
COMORO Is.
Christmas I.
(Australia)
Timor
Darwin
VANUATU
Vanua Levu

ZAMBIA
Lusaka
Malawi
MADAGASCAR
MAURITIUS
NORTHERN TERRITORY
Cairns
Viti Levu
Suva
FIJI

NAMIBIA
ZIMBABWE
Bulawayo
Harare
Antananarivo
Rodriguez
Réunion
(Fr.)
Townsville
New Caledonia
(Fr.)

BOTSWANA
Windhoek
(SOUTH)
Zomba
MOZAMBIQUE
Tropic of Capricorn
North West C.
WESTERN
QUEENSLAND
Alice Springs
Rockhampton

WEST
Gaborone
Johannesburg
Pretoria
SW
Maputo
Mozambique Chan.
AUSTRALIA
AUSTRALIA
SOUTH
Brisbane

AFRICA
SOUTH AFRICA
LES.
Durban
Perth
Kalgoorlie-Boulder
AUSTRALIA
NEW SOUTH WALES
Newcastle
Norfolk I.
(Australia)

F

Cape Town
C.of Good Hope
Port Elizabeth
Fremantle
C.Leeuwin
Great Australian Bight
Adelaide
VICTORIA
Melbourne
Darling
Sydney
Canberra
Lord Howe
(Australia)
North C.
Auckland

Amsterdam
(Fr.)
St.Paul
(Fr.)
TASMANIA
Tasman Sea
NEW ZEALAND
Wellington
C.Farewell
Christchurch

Pr.Edward Is.
(South Africa)
Crozet Is.
(Fr.)
Hobart
South I.
Dunedin

Kerguelen
(Fr.)
Stewart I.
Antipodes Is.
(N.Z.)

Bouvet I.
(Norway)
McDonald I.
(Australia)
Heard I.
(Australia)
Macquarie I.
(Australia)
Campbell I.
(N.Z.)
Auckland Is.
(N.Z.)

G

OUTHERN OCEAN

Antarctic Circle
Enderby Land
Wilkes Land
S. Magnetic Pole
Balleny Is.

Maud Land

CTICA
Ross Sea
H

East from Greenwich
20 40 60 80 100 120 140 160 180

10 11 12 13 14 15 16 17 18

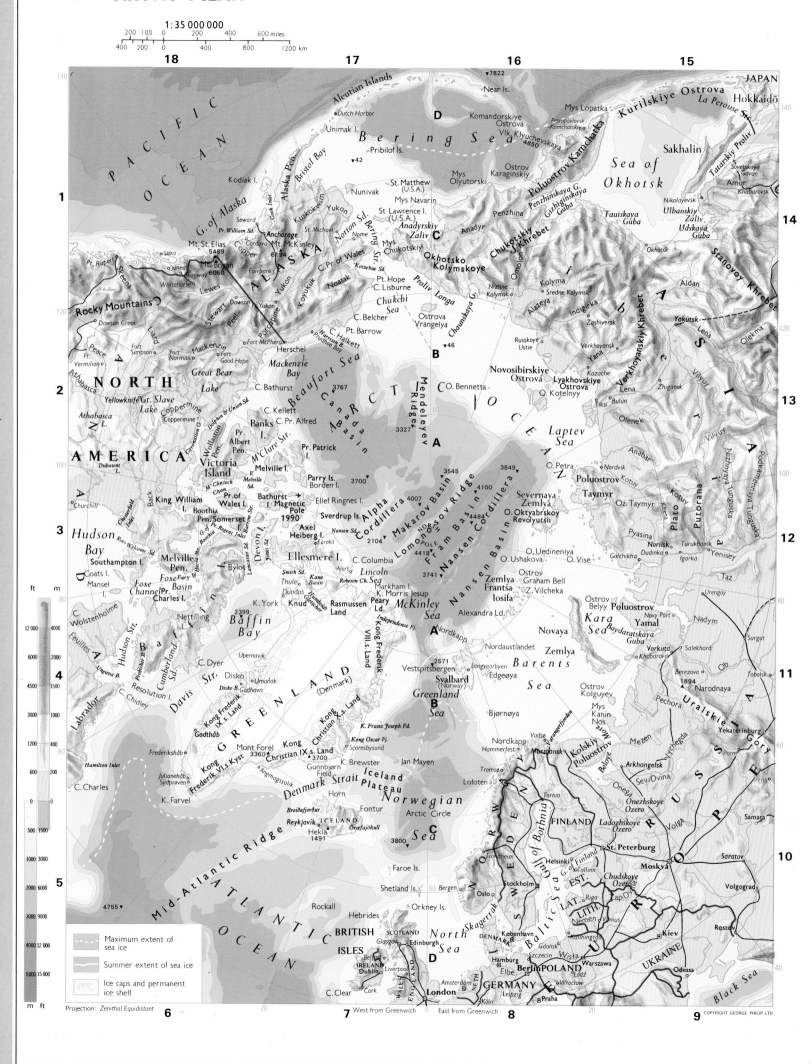

1:35 000 000

Maximum extent of sea ice

Summer extent of sea ice

Ice caps and permanent ice shelf

Projection: Zenithal Equidistant

West from Greenwich

East from Greenwich

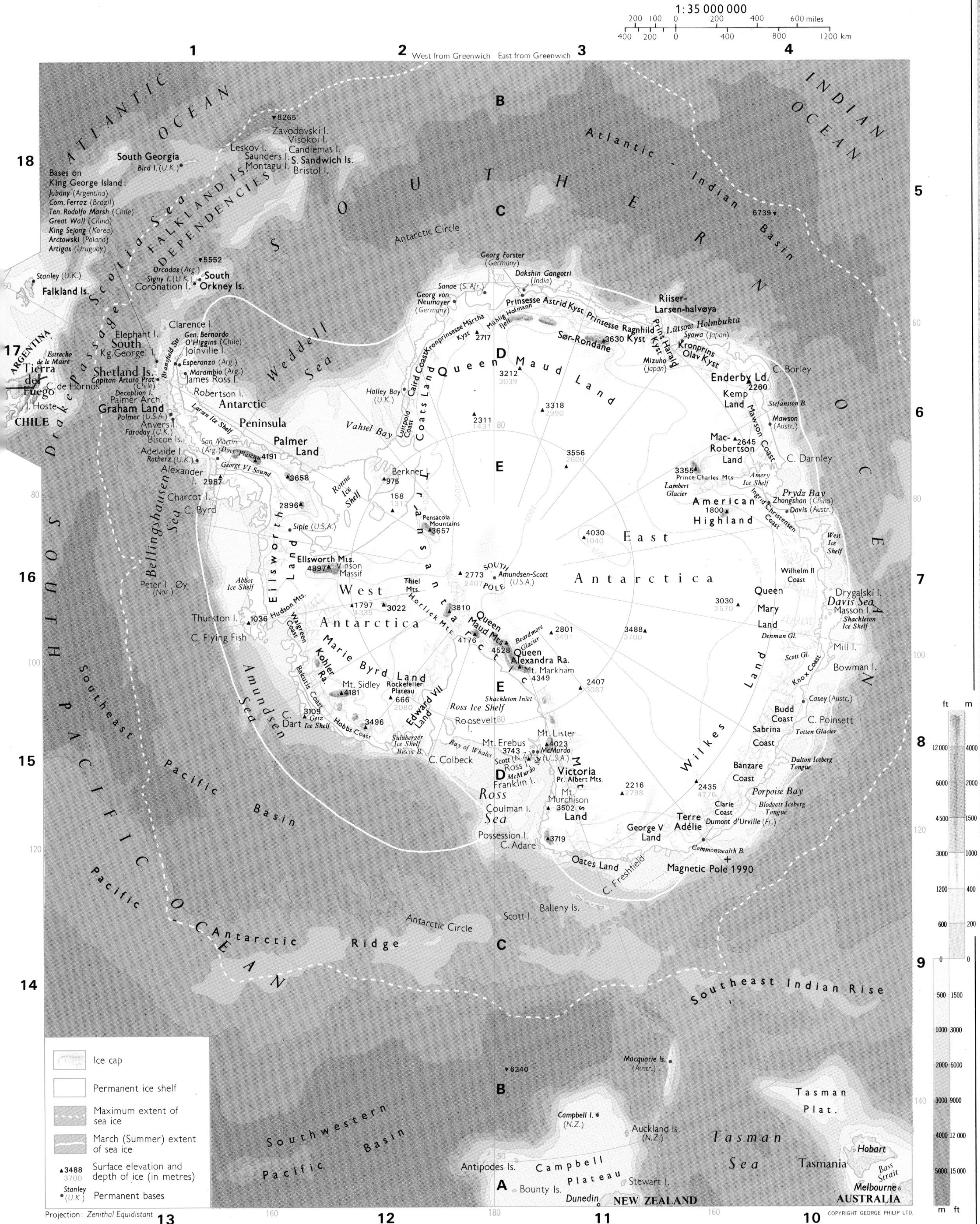

1:35 000 000

200 100 0 200 400 600 miles
400 200 0 400 800 1200 km

1   2 West from Greenwich East from Greenwich 3   4

B

ATLANTIC OCEAN

INDIAN OCEAN

Atlantic – Indian Basin

S O U T H E R N

18

South Georgia
Bird I. (U.K.)

Bases on
King George Island:
Jubany (Argentina)
Com. Ferraz (Brazil)
Ten. Rodolfo Marsh (Chile)
Great Wall (China)
King Sejong (Korea)
Arctowski (Poland)
Artigas (Uruguay)

▼8265
Zavodovski I.
Visokoi I.
Leskov I.   Candlemas I.
Saunders I.
Montagu I.
Bristol I.
S. Sandwich Is.

C

Antarctic Circle

6739▼

5

Stanley (U.K.)
Falkland Is.

FALKLAND IS. DEPENDENCIES

Scotia Sea

▼5552
Orcadas (Arg.)
Signy I. (U.K.)   South
Coronation I.   Orkney Is.

Georg Forster (Germany)
Dakshin Gangotri (India)
Sanae (S. Afr.)
Georg von Neumayer (Germany)

Prinsesse Astrid Kyst   Prinsesse Ragnhild
Kronprinsesse Martha Kyst   Mühlig Holmann fjell   Kyst

Riiser-Larsen-halvøya
Prins Harald Kyst   Lützow Holmbukta

Syowa (Japan)

INDIAN

17

ARGENTINA
Tierra del Fuego
Estrecho de le Maire
C. de Hornos
I. Hoste
CHILE

Elephant I.
South
Kg. George I.
Clarence I.
Gen. Bernardo O'Higgins (Chile)
Joinville I.
Shetland Is.
Capitan Arturo Prat (Chile)
James Ross I.
Deception I.
Palmer Arch.
Graham Land
Palmer (U.S.A.)
Anvers I.
Faraday (U.K.)
Biscoe Is.
Adelaide I.
Rothera (U.K.)
Alexander I.
Charcot I.
C. Byrd

Weddell Sea

Esperanza (Arg.)
Marambio (Arg.)
Robertson I.
Antarctic Peninsula
Larsen Ice Shelf
San Martin (Arg.)
Dyer Plateau 4191
2987
George VI Sound
3658

Halley Bay (U.K.)
Vahsel Bay

Coats Land
Luitpold Coast

Berkner I.
975
158

Queen   Maud   Land
2717
3212
3039

3318
2990

2311
1431   80

3556
2660

Sør-Rondane   3630

Kronprins Olav Kyst
Mizuho (Japan)

Enderby Ld.
2260
Kemp Land
Stefansson B.
Mawson (Austr.)
C. Borley

Mac-Robertson Land
2645

Prince Charles Mts.
3355
Amery Ice Shelf
Lambert Glacier
American Highland
1800
1040

Prydz Bay
Zhongshan (China)
Davis (Austr.)
C. Darnley
Ingrid Christensen Coast

West Ice Shelf

OCEAN

6

D

E

Ronne Ice Shelf

Transantarctic

Pensacola Mountains
3657

4030
1040

East Antarctica

2570

3030

Wilhelm II Coast

Drygalski I.
Davis Sea
Masson I.
Shackleton Ice Shelf

7

16

Bellingshausen Sea

Peter I. Øy (Nor.)

Abbot Ice Shelf

Ellsworth Mts.
4897
Vinson Massif

Hudson Mts.
1797   3022
4335

Siple (U.S.A.)

West Antarctica

2896
131

2773   South Pole
2407
Amundsen-Scott (U.S.A.)

Horlick Mts.
3810
4176

Queen Maud Mts.
4528

3488
3700

2801
3491

Queen Mary Land

Scott Gl.
Knox Coast
Denman Gl.
Mill I.
Bowman I.

100

15

Thurston I.
1036
C. Flying Fish

Walgreen Coast

Marie Byrd Land
Kohler Ra.
4181

Mt. Sidley
Rockefeller Plateau
666
2080

Edward VII Land

Beardmore Glacier
Queen Alexandra Ra.
Mt. Markham
4349

2407
3087

Budd Coast
Sabrina Coast
Totten Glacier
Casey (Austr.)
C. Poinsett

8

Amundsen Sea

C. Dart
Getz Ice Shelf
3109
3496
Hobbs Coast
Sulzberger Ice Shelf
Biscoe B.

Shackleton Inlet
Ross Ice Shelf
Roosevelt I.
80
Bay of Whales

Mt. Erebus
3743
McMurdo
Ross (N.Z.)
McMurdo (U.S.A.)
Scott (N.Z.)
Franklin I.

Mt. Lister
4023

Victoria
Pr. Albert Mts.

2216
2798

2435
4776

Terre Adélie

Banzare Coast
Dalton Iceberg Tongue
Porpoise Bay
Clarie Coast
Blodgett Iceberg Tongue
Dumont d'Urville (Fr.)

120

Pacific Basin

D

Ross Sea
C. Colbeck
Coulman I.
Possession I.
C. Adare

Mt. Murchison
3502
Land
3719

George V Land

Oates Land
C. Freshfield
Commonwealth B.
+ Magnetic Pole 1990

14

Southeast Pacific

Antarctic Ridge

SOUTH   PACIFIC   OCEAN

C

Antarctic Circle

Scott I.
Balleny Is.

Southeast Indian Rise

9

Southwestern

Pacific Basin

B

▼6240

Campbell I. (N.Z.)

Macquarie I. (Austr.)

Tasman Plat.

140

A

Antipodes Is.

Southwestern Pacific

Campbell Plateau
Bounty Is.
Stewart I.
Dunedin   NEW ZEALAND

Auckland Is. (N.Z.)

Tasman Sea

Tasmania
Hobart
Bass Strait
Melbourne
AUSTRALIA

ft   m
12 000   4000
6000   2000
4500   1500
3000   1000
1200   400
600   200
0   0
500   1500
1000   3000
2000   6000
3000   9000
4000   12 000
5000   15 000
m   ft

Legend:
Ice cap
Permanent ice shelf
Maximum extent of sea ice
March (Summer) extent of sea ice
▲3488 / 3700   Surface elevation and depth of ice (in metres)
● Stanley (U.K.)   Permanent bases

Projection: Zenithal Equidistant

13   12   180   11   160   10   COPYRIGHT GEORGE PHILIP LTD.

The Antarctic Treaty was signed in Washington in 1959 so that scientific and technical research could continue unhampered by international politics.

All territorial claims covering land areas south of latitude 60°S have been suspended. Those claims were:

Norwegian claim	45°E – 20°W
Australian claims	45°E – 136°E / 142°E – 160°E
French claim	136°E – 142°E
New Zealand claim	160°E – 150°W
Chilean claim	90°W – 53°W
British claim	80°W – 20°W
Argentine claim	74°W – 53°W

1 : 20 000 000

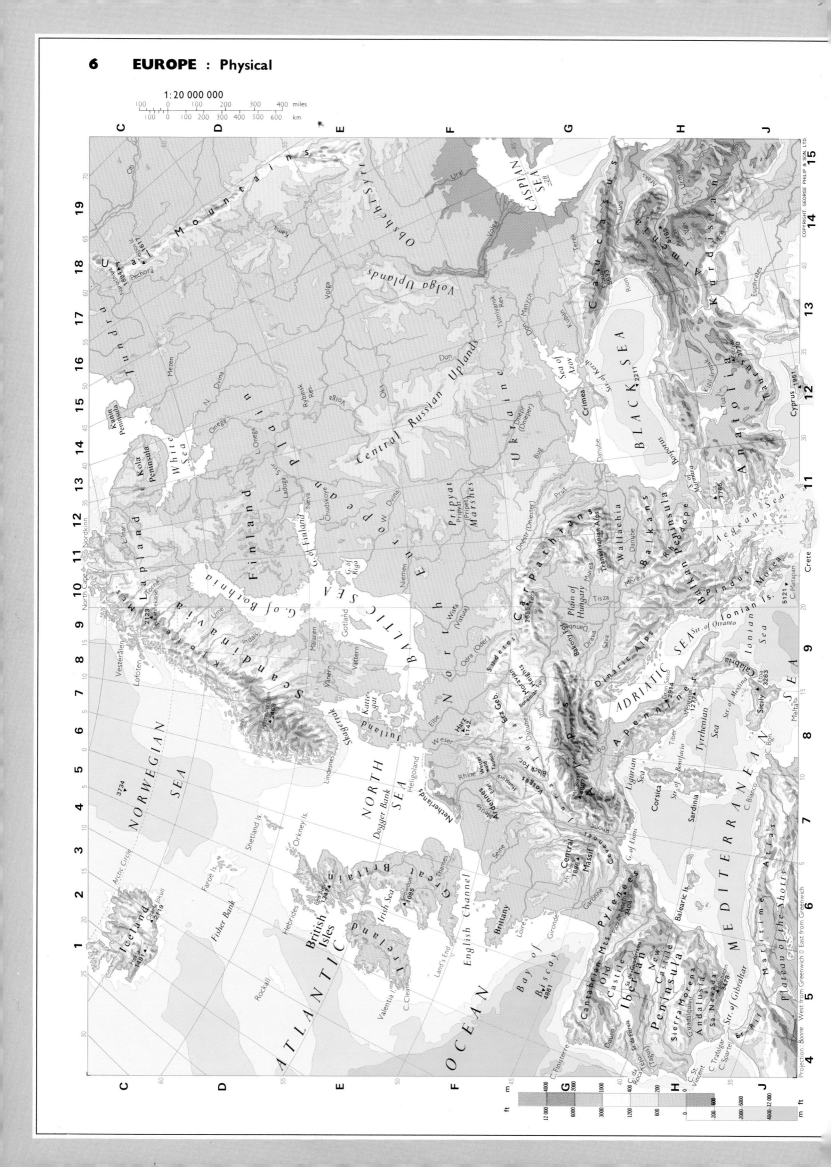

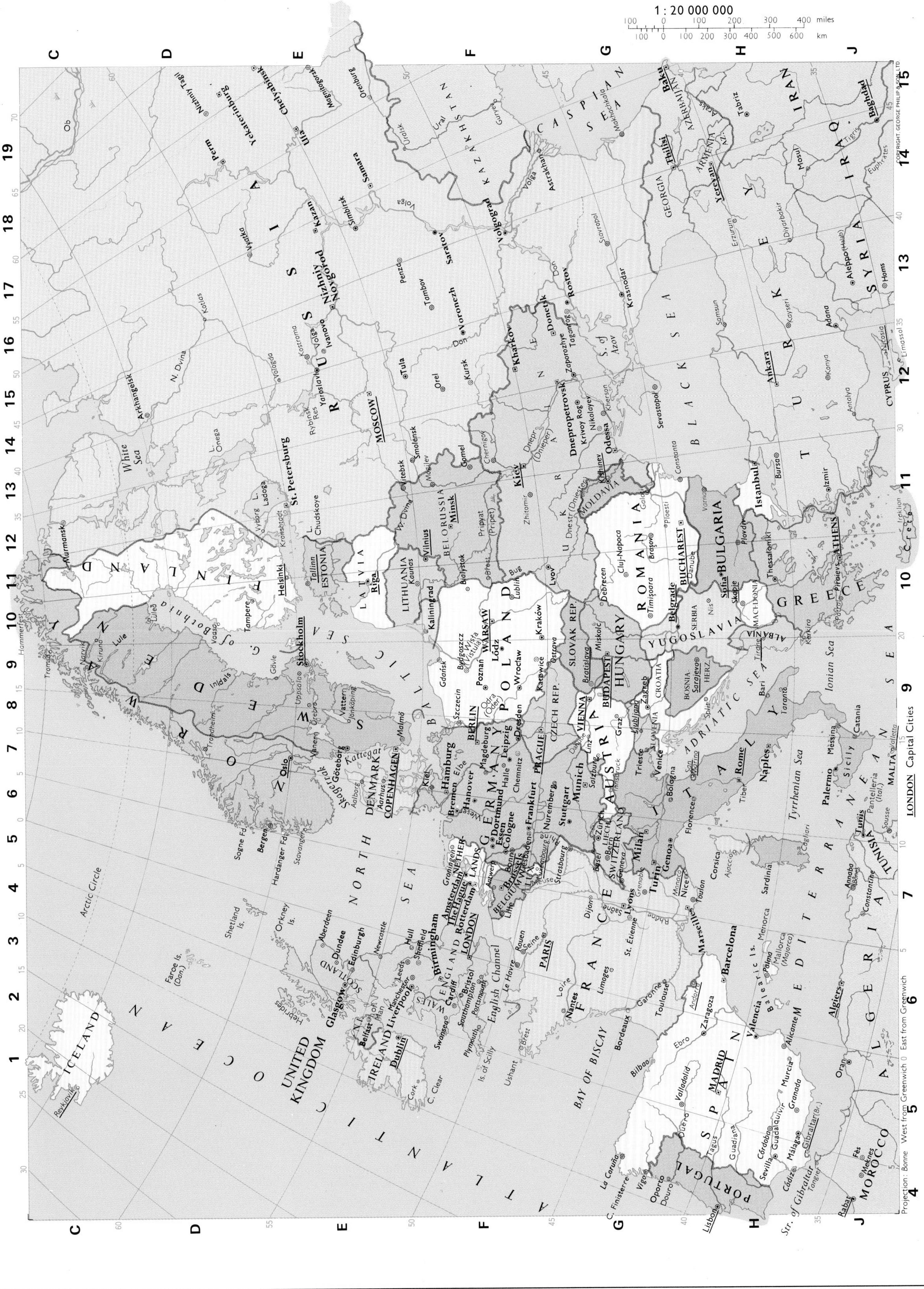

1 : 20 000 000

100   0   100   200   300   400 miles
100   0   100   200   300   400   500   600 km

LONDON Capital Cities

Projection: Bonne   West from Greenwich 0 East from Greenwich   COPYRIGHT GEORGE PHILIP & SON, LTD.

ICELAND
Reykjavik

ATLANTIC OCEAN

Arctic Circle

NORWAY
Tromsö
Narvik
Kiruna
Trondheim
Bergen
Oslo
Sogne Fd.
Hardanger Fd.
Stavanger
Skagerrak

SWEDEN
Luleå
Umeå
Sundsvall
Uppsala
STOCKHOLM
Göteborg
Vänern
Vättern
Jönköping
Kattegat

FINLAND
Hammerfest
Kemi
Oulu
Vaasa
Tampere
HELSINKI

DENMARK
COPENHAGEN
Aarhus
Aalborg
Odense
Kiel

NORTH SEA
BALTIC SEA
G. of Bothnia
White Sea
Murmansk
Arkhangelsk

UNITED KINGDOM
IRELAND
DUBLIN
Belfast
N.I.
SCOTLAND
Glasgow
Edinburgh
Aberdeen
Dundee
Newcastle
ENGLAND
Leeds
Manchester
Liverpool
Sheffield
Birmingham
WALES
Cardiff
Swansea
Bristol
LONDON
Southampton
Portsmouth
Plymouth
Shetland Is.
Orkney Is.
Hebrides
I. of Man
English Channel

NETHERLANDS
AMSTERDAM
The Hague
Rotterdam
Groningen
BELGIUM
BRUSSELS
LUX.

FRANCE
PARIS
Le Havre
Rouen
Nantes
Loire
Bordeaux
Garonne
Toulouse
Lyons
St. Étienne
Marseille
Nice
Strasbourg
Dijon
Limoges
Seine
Rhône
BAY OF BISCAY

GERMANY
BERLIN
Hamburg
Bremen
Hanover
Dortmund
Essen
Cologne
Düsseldorf
Bonn
Frankfurt
Leipzig
Dresden
Halle
Chemnitz
Magdeburg
Stuttgart
Munich
Nuremberg
Elbe

SWITZERLAND
BERN
Zürich
Basel
Geneva

AUSTRIA
VIENNA
Linz
Salzburg
Graz
Innsbruck

ITALY
ROME
Milan
Turin
Genoa
Venice
Bologna
Florence
Naples
Palermo
Sicily
Catania
Messina
Sardinia
Corsica
Bari
Taranto
TYRRHENIAN SEA
ADRIATIC SEA
IONIAN SEA
MEDITERRANEAN SEA

SPAIN
MADRID
Barcelona
Valencia
Zaragoza
Bilbao
Valladolid
Córdoba
Sevilla
Málaga
Granada
Murcia
Alicante
Balearic Is.
Mallorca
Menorca
Ebro
Tagus
Guadiana

PORTUGAL
LISBON
Oporto
Douro
C. Finisterre
Vigo

GIBRALTAR (Br.)
Str. of Gibraltar

MOROCCO
Rabat
Fez
ALGERIA
Algiers
Oran
Constantine
TUNISIA
TUNIS
MALTA
Valletta

POLAND
WARSAW
Kraków
Łódź
Poznań
Wrocław
Gdańsk
Szczecin
Bydgoszcz
Katowice
Lublin
Vistula
Oder

CZECH REP.
PRAGUE
SLOVAK REP.
BRATISLAVA

HUNGARY
BUDAPEST
Miskolc
Debrecen

ROMANIA
BUCHAREST
Cluj-Napoca
Timişoara
Braşov
Ploieşti
Constanţa
Danube

BULGARIA
SOFIA
Plovdiv
Varna

YUGOSLAVIA
BELGRADE
SERBIA
MONTENEGRO
KOSOVO
SKOPJE
MACEDONIA

CROATIA
ZAGREB
SLOVENIA
LJUBLJANA
BOSNIA
HERZ.
SARAJEVO
Split

ALBANIA
Tiranë

GREECE
ATHENS
Thessaloniki
Crete
Ionian Sea

ESTONIA
TALLINN
LATVIA
RIGA
LITHUANIA
VILNIUS
Kaunas
Kaliningrad

BELORUSSIA
MINSK
L. Chudskoye

MOLDAVIA
KISHINEV

RUSSIA
MOSCOW
St. Petersburg
Nizhniy Novgorod
Kazan
Samara
Saratov
Volgograd
Rostov
Voronezh
Tambov
Tula
Orel
Kursk
Smolensk
Vitebsk
Yaroslavl
Kostroma
Ivanovo
Penza
Perm
Vologda
Ufa
Orenburg
Astrakhan
Krasnodar
Stavropol
Volga
Don
Ural
N. Dvina
Onega
Ob
Pechora
Nizhniy Tagil
Chelyabinsk
Yekaterinburg
Murmansk
Kotlas

UKRAINE
KIEV
Kharkov
Donetsk
Dnepropetrovsk
Odessa
Zaporozhye
Krivoy Rog
Nikolayev
Zhitomir
Lvov
Dnepr (Dnieper)
Dniester

CASPIAN SEA
BLACK SEA
SEA OF AZOV
Sevastopol
Simferopol
Novorossiysk

GEORGIA
TBILISI
ARMENIA
YEREVAN
AZERBAIJAN
BAKU

TURKEY
ANKARA
Istanbul
Izmir
Bursa
Adana
Konya
Kayseri
Samsun
Erzurum
Diyarbakir

CYPRUS
NICOSIA
Limassol

SYRIA
Aleppo
Homs

IRAQ
BAGHDAD
Mosul
Tigris
Euphrates

IRAN
Tabriz

KAZAKHSTAN

FINLAND
SWEDEN
NORWAY

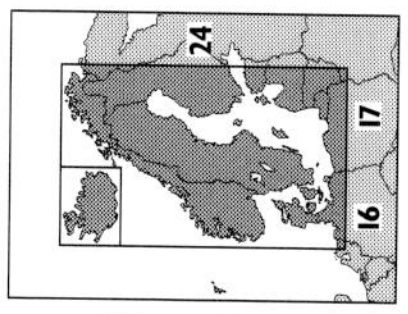

ICELAND
on the same scale
as general map

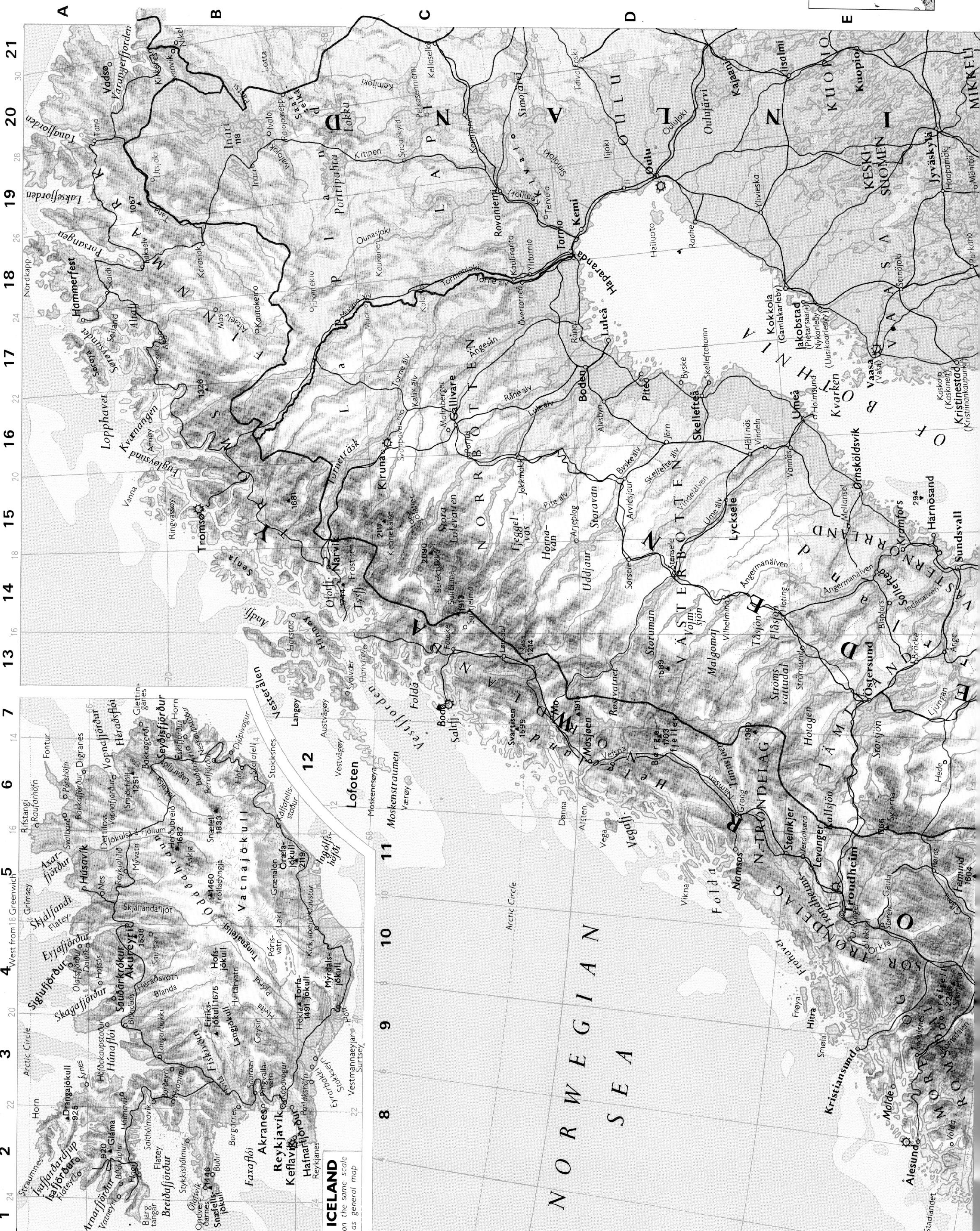

NORWEGIAN SEA

LAPPLAND

FINLAND

NORRBOTTEN

VÄSTERBOTTEN

NORDLAND

N-TRÖNDELAG

SØR-TRÖNDELAG

Lofoten

Vesterålen

Arctic Circle

Reykjavik
Keflavik
Akranes
Hafnarfjörður

Akureyri
Siglufjörður
Sauðárkrókur

Vatnajökull
Hofsjökull
Langjökull
Hekla Torfa 1491 Jökull
Eiríksjökull 1675
Mýrdalsjökull
Öræfajökull 2119

Drangajökull 925

Hammerfest
Vadsø
Varangerfjorden
Tromsø
Narvik
Bodø
Mosjøen
Namsos
Steinkjer
Levanger
Trondheim
Kristiansund
Molde
Ålesund

Kiruna
Gällivare
Luleå
Boden
Piteå
Skellefteå
Umeå
Härnösand
Sundsvall
Östersund

Oulu
Kemi
Tornio
Haparanda
Rovaniemi
Kokkola
Jakobstad
Vaasa
Kuopio
Jyväskylä

Torneträsk
Stora Lulevatten
Storuman
Storsjön
Kallsjön

VAASA
KESKI-SUOMEN
MIKKELI
KUOPIO

Arctic Circle

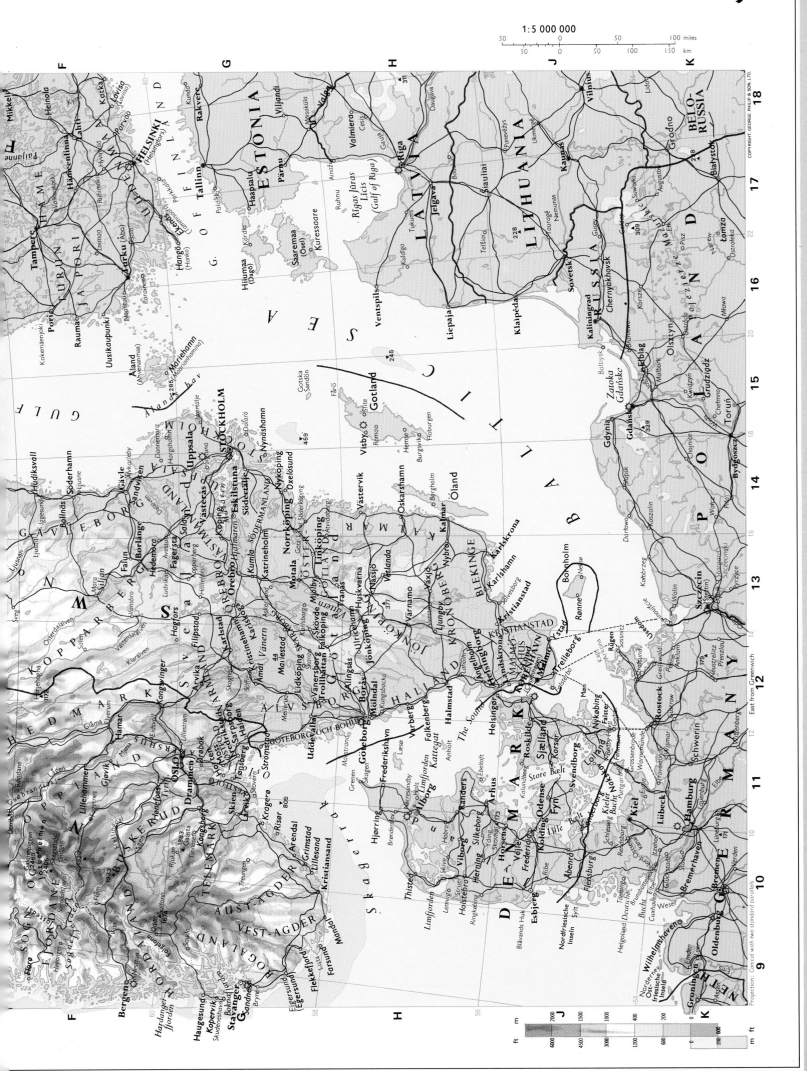

1:5 000 000

50   0   50   100 miles

50   0   50   100   150   km

Projection: Conical with two standard parallels

East from Greenwich

F   **G**   **H**   **J**   **K**

Mikkeli

Heinola

Lahti

Lovisa (Loviisa)

Kotka

Kouvola

HELSINKI (Helsingfors)

Hämeenlinna

TAMPERE

TURUN

PORI

Rauma

Uusikaupunki

Turku (Åbo)

Hangö (Hanko)

Tallinn

ESTONIA

Haapsalu

Pärnu

Hiiumaa (Dagö)

Saaremaa (Ösel)

Kuressaare

Valga

Valmiera

Rīga

Rīgas Jūras Līcis (Gulf of Riga)

LATVIA

Jelgava

Šiauliai

LITHUANIA

Kaunas

Vilnius

Grodno

Białystok

BELO-RUSSIA

RUSSIA

Sovetsk

Klaipēda

Kaliningrad

Chernyakhovsk

Olsztyn

Grudziądz

POLAND

Toruń

Bydgoszcz

Gdańsk

Gdynia

Szczecin (Stettin)

Kołobrzeg

Koszalin

Słupsk

Bornholm

Rønne

BALTIC SEA

Gotland

Visby

Öland

Kalmar

KALMAR

KRONOBERG

Växjö

BLEKINGE

Karlskrona

Karlshamn

Kristianstad

Ystad

Trelleborg

MALMÖ

Lund

Helsingborg

HALLAND

Halmstad

Varberg

Falkenberg

Göteborg

GÖTEBORG OCH BOHUS

ÄLVSBORG

Borås

SKARABORG

Vänern

JÖNKÖPING

Jönköping

ÖSTERGÖTLAND

Linköping

Norrköping

SÖDERMANLAND

Nyköping

STOCKHOLM

Uppsala

VÄSTMANLAND

Västerås

Eskilstuna

ÖREBRO

Örebro

KOPPARBERG

Falun

Borlänge

GÄVLEBORG

Gävle

Söderhamn

Hudiksvall

Sundsvall

VÄRMLAND

Karlstad

ÄLVSBORG

SVERIGE

HEDMARK

OPPLAND

Hamar

Lillehammer

OSLO

Drammen

BUSKERUD

TELEMARK

AUST-AGDER

VEST-AGDER

Kristiansand

Stavanger

HORDALAND

Bergen

SOGN OG FJORDANE

NORGE

Skagerrak

DENMARK

Ålborg

Århus

Randers

Viborg

Esbjerg

Odense

Fyn

Kolding

GERMANY

Flensburg

Kiel

Lübeck

Hamburg

Bremen

Bremerhaven

Wilhelmshaven

Oldenburg

Groningen

NETHERLANDS

Rostock

Schwerin

Wismar

Rügen

Stralsund

Greifswald

Anklam

GULF OF FINLAND

SEA

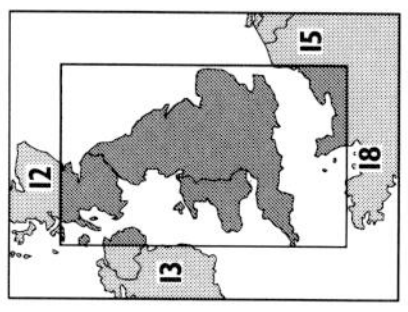

A   B   C   D

NORTH SEA

IRISH SEA

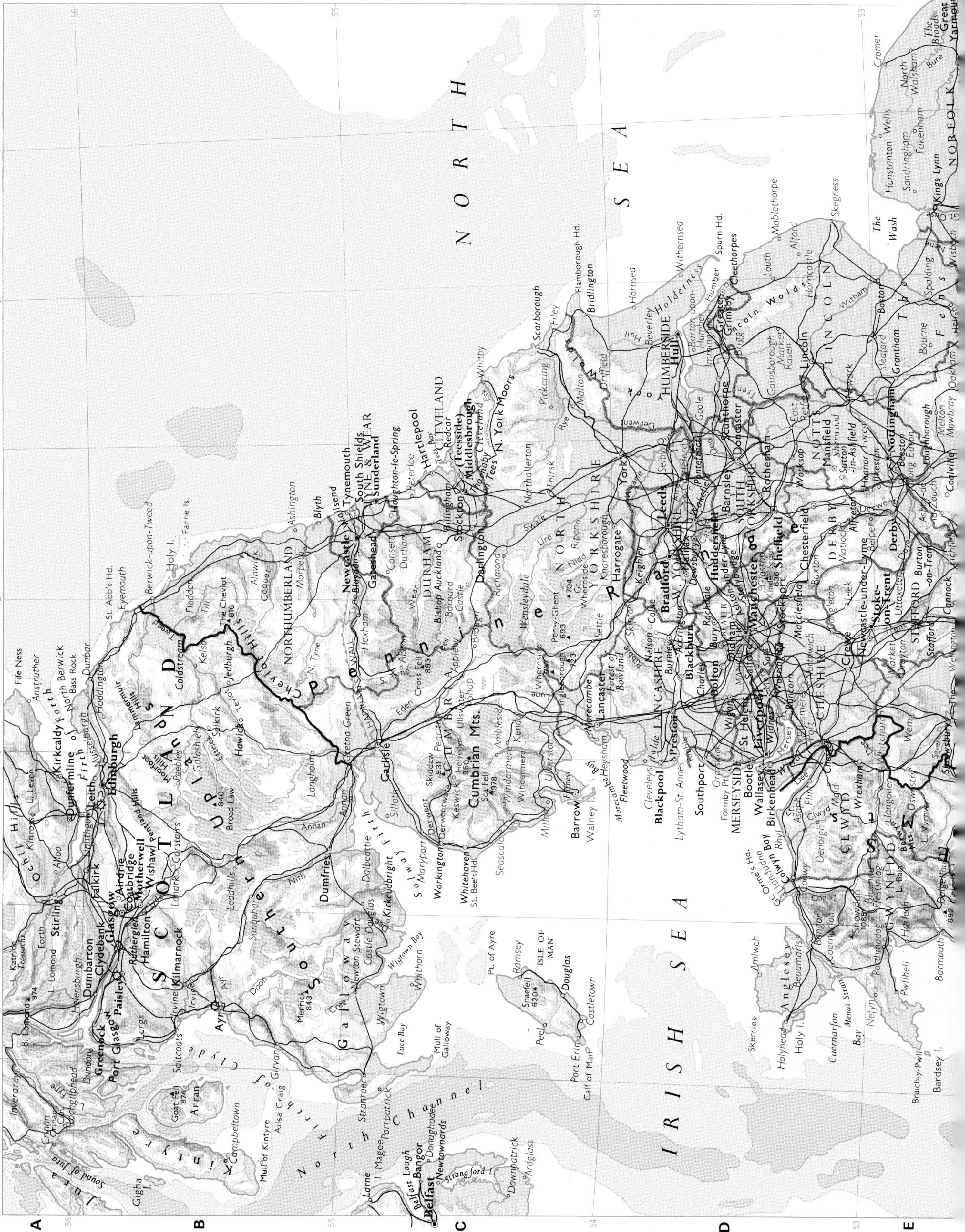

North Channel

SCOTLAND

Southern Uplands

Galloway

Cheviot Hills

NORTHUMBERLAND

Pennines

CUMBRIA

Cumbrian Mts.

LANCASHIRE

YORKSHIRE

N. YORK MOORS

CLEVELAND

DURHAM

TYNE & WEAR

N. YORKSHIRE

W. YORKSHIRE

S. YORKSHIRE

HUMBERSIDE

LINCOLN

LINCOLN WOLDS

The Fens

NORFOLK

DERBY

STAFFORD

CHESHIRE

MERSEYSIDE

CLWYD

GWYNEDD

Anglesey

Isle of Man

Fife Ness
Anstruther
North Berwick
Bass Rock
Edinburgh
Dunfermline
Kirkcaldy
Stirling
Falkirk
Glasgow
Greenock
Port Glasgow
Paisley
Dumbarton
Clydebank
Rutherglen
Hamilton
Motherwell
Wishaw
Coatbridge
Airdrie
Kilmarnock
Ayr
Irvine
Peebles
Galashiels
Selkirk
Hawick
Jedburgh
Kelso
Coldstream
Berwick-upon-Tweed
Eyemouth
St. Abb's Hd.
Holy I.
Farne Is.
Alnwick
Morpeth
Ashington
Blyth
Newcastle
Gateshead
South Shields
Tynemouth
Sunderland
Houghton-le-Spring
Peterlee
Hartlepool
Redcar
Middlesbrough
Stockton
Darlington
Bishop Auckland
Barnard Castle
Richmond
Northallerton
Thirsk
Whitby
Scarborough
Filey
Bridlington
Flamborough Hd.
Hornsea
Withernsea
Spurn Hd.
Cleethorpes
Grimsby
Hull
Beverley
Goole
Selby
York
Harrogate
Knaresborough
Leeds
Bradford
Keighley
Halifax
Huddersfield
Wakefield
Barnsley
Doncaster
Rotherham
Sheffield
Chesterfield
Worksop
Mansfield
Nottingham
Derby
Burton-on-Trent
Stoke-on-Trent
Newcastle-under-Lyme
Crewe
Macclesfield
Stockport
Manchester
Salford
Bolton
Bury
Rochdale
Oldham
Wigan
St. Helens
Warrington
Widnes
Runcorn
Liverpool
Bootle
Birkenhead
Wallasey
Southport
Blackpool
Lytham-St. Annes
Preston
Blackburn
Burnley
Nelson
Accrington
Morecambe
Lancaster
Fleetwood
Barrow
Whitehaven
Workington
Maryport
Carlisle
Penrith
Keswick
Windermere
Kendal
Ambleside
Dumfries
Annan
Kirkcudbright
Castle Douglas
Dalbeattie
Newton Stewart
Wigtown
Whithorn
Stranraer
Portpatrick
Douglas
Ramsey
Castletown
Peel
Port Erin
Holyhead
Caernarfon
Bangor
Beaumaris
Amlwch
Pwllheli
Barmouth
Harlech
Wrexham
Llangollen
Rhyl
Colwyn Bay
Llandudno
Conwy
Denbigh
Mold
Flint
Shrewsbury
Oswestry
Welshpool

Cross Fell 893
Scafell 978
Helvellyn
Skiddaw 931
Snaefell 620
Merrick 843
Goat Fell 874
The Cheviot 816
Peny-Ghent 693
Whernside 704

Solway Firth
Morecambe Bay
Cardigan Bay
The Wash
The Broads
Great Yarmouth
Cromer

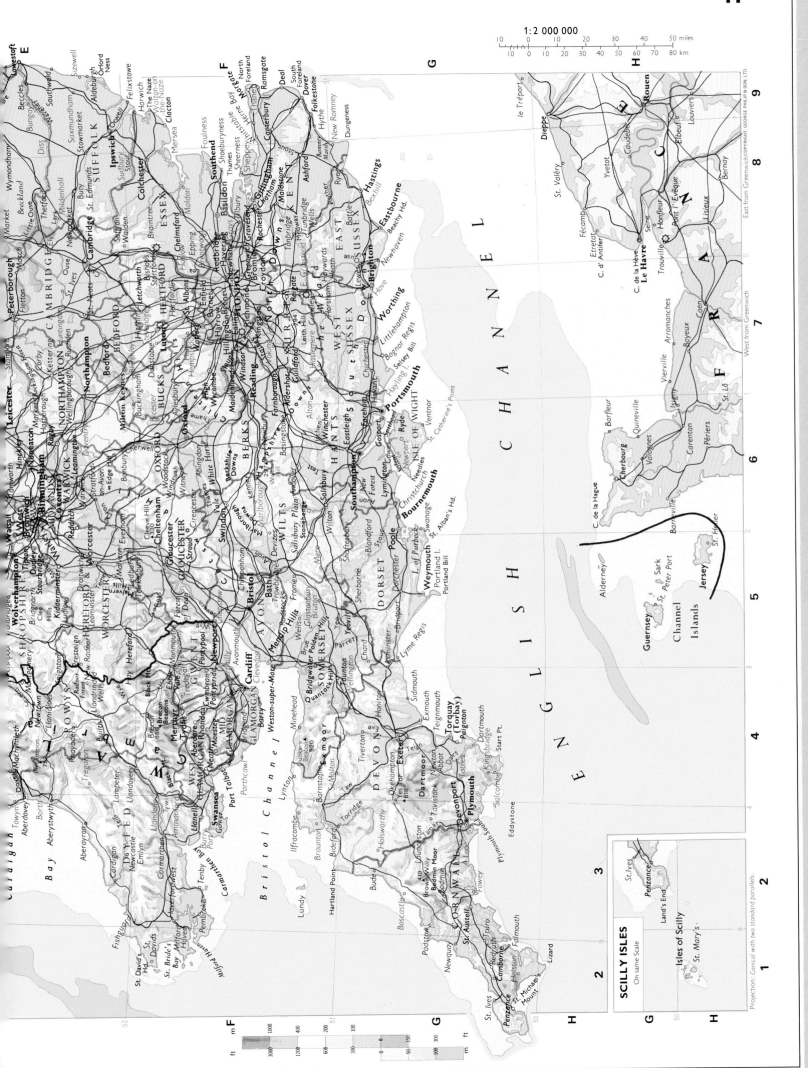

11

1 : 2 000 000

Projection : Conical with two standard parallels.

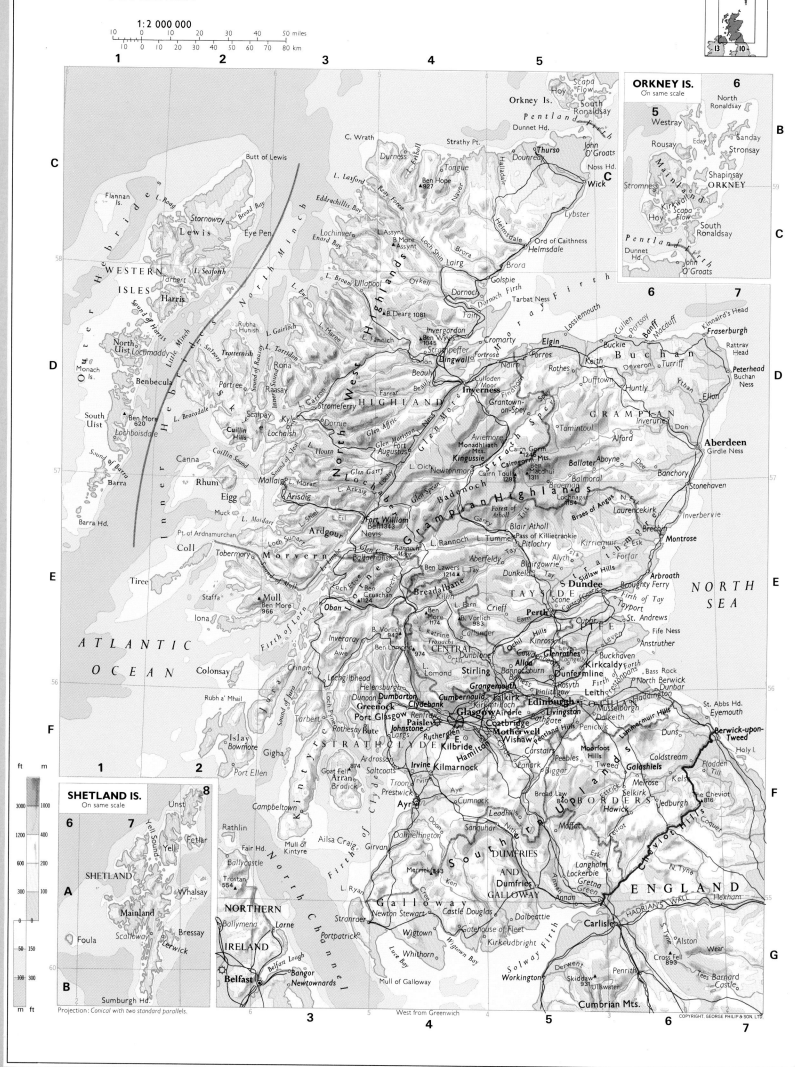

1:2 000 000

10   0   10   20   30   40   50 miles

10 0 10 20 30 40 50 60 70 80 km

ORKNEY IS.
On same scale

Hoy · Scapa Flow · South Ronaldsay
Orkney Is.
Westray
Rousay · Eday · Sanday · Stronsay
Mainland
Stromness · Shapinsay · ORKNEY
Hoy · Kirkwall · Scapa Flow
South Ronaldsay
Pentland Firth
Dunnet Hd. · John O'Groats

SHETLAND IS.
On same scale

Unst
Yell · Fetlar
Yell Sound
SHETLAND · Whalsay
Mainland · Bressay
Foula · Scalloway · Lerwick
Sumburgh Hd.

Scapa Flow · South Ronaldsay
Orkney Is. · Hoy
Pentland Firth · Dunnet Hd. · John O'Groats
C. Wrath · Strathy Pt. · Thurso · Dounreay · Halladale · Noss Hd. · Wick
Durness · L. Erboll · Tongue · Naver · Reay Forest · Ben Hope 927
L. Laxford · Lybster
Eddrachillis Bay · Helmsdale · Ord of Caithness
Butt of Lewis
Lochinver · Enard Bay · L. Assynt · B. More Assynt · Loch Shin · Lairg · Brora · Golspie
Flannan Is. · L. Roag · Broad Bay · Stornoway · Eye Pen.
Ullapool · Oykell · Dornoch · Helmsdale
Lewis
L. Seaforth · Tarbert · Harris · L. Broom · B. Dearg 1081 · Tain · Dornoch Firth · Tarbat Ness
WESTERN ISLES
L. Ewe · L. Fannich · Invergordon · Cromarty · Lossiemouth · Cullen · Portsoy · Banff · Macduff · Kinnaird's Head
North Uist · Lochmaddy · Rubha Hunish · L. Gairloch · Ben Wyvis 1045 · Strathpeffer · Dingwall · Conon · Nairn · Forres · Elgin · Buckie · Keith · Fraserburgh
Monach Is. · Trotternish · L. Maree · Beauly · Culloden Moor · Findhorn · Rothes · Deveron · Turriff · Rattray Head
Benbecula · Portree · Raasay · L. Torridon · Inverness · Grantown-on-Spey · Dufftown · Huntly · Ythan · Peterhead · Buchan Ness
South Uist · Scalpay · Rona · Stromeferry · Dornie · Glen Affric · Farrar · Aviemore · Cairn Gorm 1245 · Tomintoul · Alford · Inverurie · BUCHAN
Ben More 620 · Cuillin Hills · Kyle of Lochalsh · Glen Moriston · Fort Augustus · Monadhliath Mts. · Cairngorm Mts. · Ballater · Aboyne · GRAMPIAN · Aberdeen
Lochboisdale · L. Hourn · Kingussie · Newtonmore · Cairn Toul 1293 · B. Macdhui 1311 · Balmoral · Dee · Banchory · Girdle Ness
Barra · Canna · Rhum · L. Arkaig · Glen Garry · L. Oich · Braemar · Lochnagar 1154 · Stonehaven
Eigg · Mallaig · L. Morar · Glen Spean · Badenoch · Forest of Atholl · Braes of Angus · Laurencekirk · Inverbervie
Muck · L. Moidart · Arisaig · Fort William · Ben 1343 Nevis · Garry · Blair Atholl · Pass of Killiecrankie · Kirriemuir · Brechin · Montrose
Barra Hd. · Pt. of Ardnamurchan · Ardgour · Glen Coe · L. Rannoch · L. Tummel · Pitlochry · Forfar · Arbroath
Coll · Tobermory · MORVERN · Ballachulish · Rannoch Moor · Aberfeldy · Blairgowrie · Sidlaw Hills · NORTH SEA
Staffa · Mull · Ben More 966 · Ben Lawers 1214 · Tay · Dunkeld · Alyth · Forfar
Tiree · Iona · Loch Etive · Ben Cruachan 1124 · Killin · Breadalbane · TAYSIDE · Scone · Dundee · Broughty Ferry
ATLANTIC OCEAN · Oban · B. More 1174 · L. Earn · Crieff · Perth · Firth of Tay · Tayport · St. Andrews · Fife Ness · Anstruther
Colonsay · B. Vorlich 942 · L. Katrine · Trossachs · Ben Ledi 974 · CENTRAL · Dunblane · Cupar · FIFE · Leven
Inveraray · L. Awe · Ben Lomond · Callander · Kinross · Glenrothes · Buckhaven · Bass Rock
Crinan · Lochgilphead · L. Lomond · Stirling · Alloa · Kirkcaldy · North Berwick · Dunbar
Rubh' a' Mhail · Tarbert · Helensburgh · Dunoon · Dumbarton · Cumbernauld · Falkirk · Dunfermline · Rosyth · Leith · St. Abbs Hd. · Eyemouth
Islay · Bowmore · Gigha · Rothesay · Bute · Greenock · Port Glasgow · Clydebank · Grangemouth · Edinburgh · LOTHIAN · Haddington
Port Ellen · Johnstone · Paisley · GLASGOW · Airdrie · Livingston · Musselburgh · Dalkeith · Duns · Berwick-upon-Tweed
Goat Fell 874 · Saltcoats · Rutherglen · Coatbridge · Motherwell · Wishaw · Penicuik · Pentland Hills · Moorfoot Hills · Lammermuir Hills · Holy I.
Arran · Brodick · Ardrossan · Irvine · Kilmarnock · E. Kilbride · Hamilton · Carstairs · Peebles · Galashiels · Coldstream · Flodden · Till
Campbeltown · Troon · Prestwick · Lanark · Biggar · Tweed · Melrose · Selkirk · Kelso
Mull of Kintyre · Ayr · Cumnock · Leadhills · Broad Law 840 · BORDERS · Hawick · Jedburgh · The Cheviot 816
Rathlin · Fair Hd. · Ailsa Craig · Girvan · Doon · Sanquhar · Nith · Moffat · Ettrick · Teviot · Cheviot Hills · Coquet
Ballycastle · North Channel · Dalmellington · Leadhills · Esk · Langholm · N. Tyne · Hexham
Trostan 554 · Merrick 843 · DUMFRIES AND GALLOWAY · Lockerbie · ENGLAND
NORTHERN IRELAND · Ballymena · Larne · Portpatrick · Stranraer · Newton Stewart · Castle Douglas · Dumfries · Gretna Green · Carlisle · Alston
Belfast · Bangor · Newtownards · L. Ryan · Wigtown · Gatehouse of Fleet · Dalbeattie · Cross Fell 893 · Wear
Belfast Lough · Luce Bay · Whithorn · Kirkcudbright · Wigtown Bay · Workington · Derwent · Skiddaw 931 · Penrith · Tees · Barnard Castle
Mull of Galloway · Solway Firth · Ullswater · Cumbrian Mts.
GALLOWAY · SCOTLAND

HIGHLAND · NORTH WEST HIGHLANDS · Strath Spey · GRAMPIAN HIGHLANDS · Loch Ness · Loch Linnhe · Firth of Lorn · Sound of Jura · Kintyre · Firth of Clyde · Ayr · STRATHCLYDE · Moray Firth · Spey · Esk · Tay · Forth · South Uplands · Solway Firth · Little Minch · North Minch · Sound of Sleat · Inner Hebrides · Outer Hebrides

ft m
3000 1000
1200 400
600 200
300 100
0 0
50 150
100 300
m ft

Projection : Conical with two standard parallels.

West from Greenwich

COPYRIGHT. GEORGE PHILIP & SON. LTD.

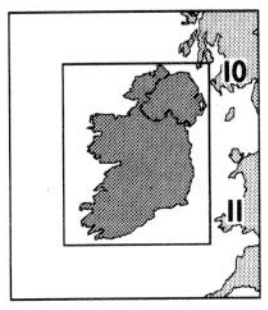

1:2 000 000

10   0   10   20   30   40   50 miles
10   0   10 20 30 40 50 60 70 80 km

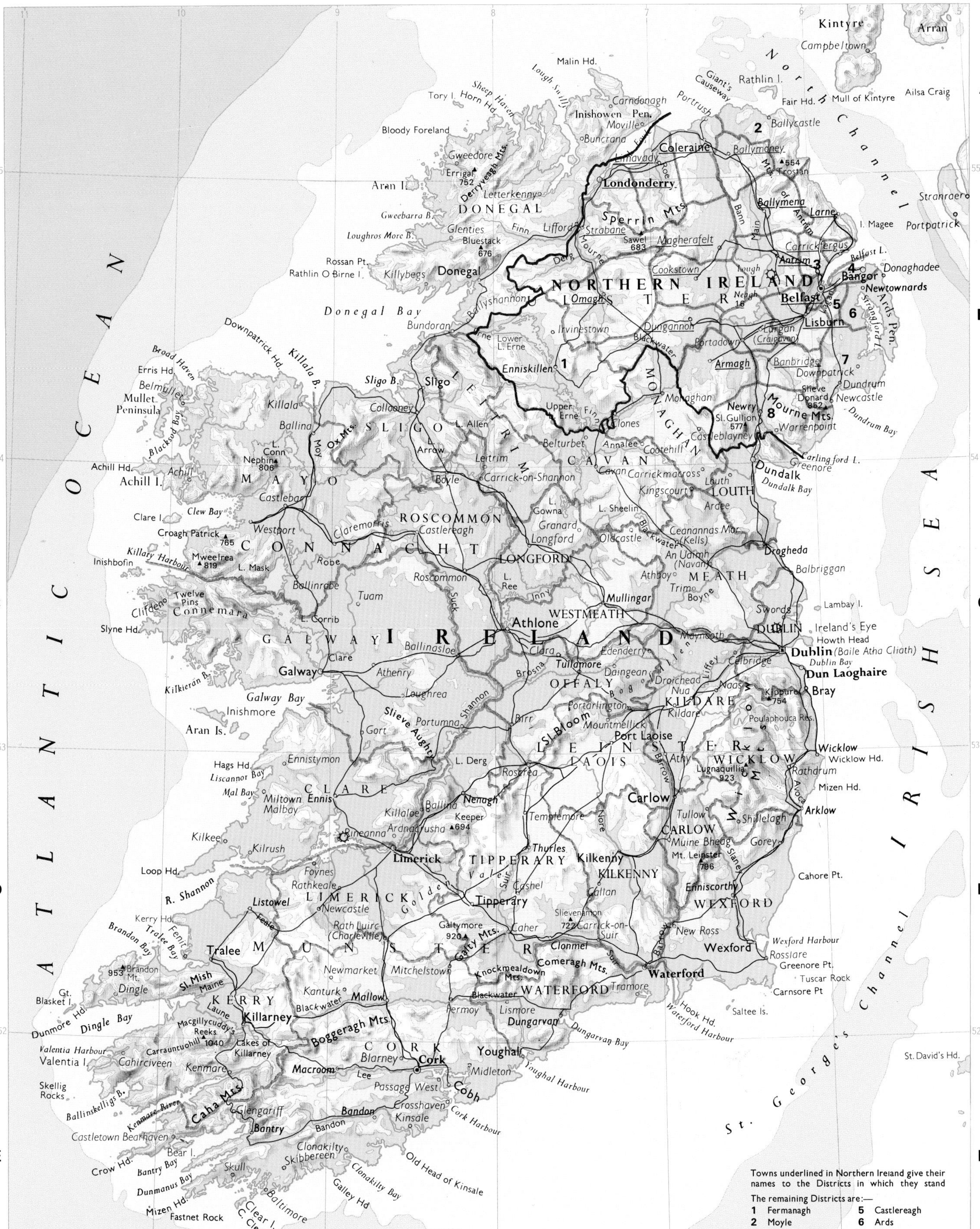

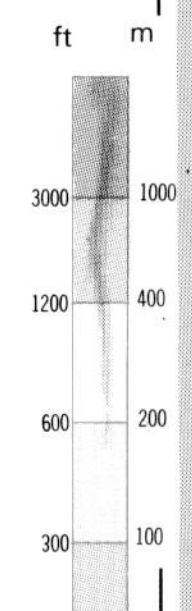

ATLANTIC OCEAN

NORTH CHANNEL

Kintyre
Arran
Campbeltown
Rathlin I.
Mull of Kintyre
Ailsa Craig
Stranraer
Portpatrick

Malin Hd.
Lough Swilly
Tory I. Horn Hd.
Sheep Haven
Carndonagh
Inishowen Pen.
Moville
Bunerana
Giant's Causeway
Portrush
Fair Hd.
Ballycastle

Bloody Foreland
Gweedore
Errigal 752
Derryveagh Mts.
Letterkenny
Coleraine
Limavady
Ballymoney
554 Trostan
Ballymena
Larne
I. Magee

Aran I.
Gweebarra B.
DONEGAL
Glenties
Bluestack 676
Lifford
Strabane
Sperrin Mts.
Sawel 683
Magherafelt
Carrickfergus
Antrim
Bangor
Newtownards
Ards Pen.

Loughros More B.
Rossan Pt.
Rathlin O Birne I.
Killybegs
Donegal
NORTHERN IRELAND
Omagh
Cookstown
Lough Neagh 18
Belfast
Lisburn
Lurgan (Craigavon)
Strangford

Downpatrick Hd.
Killala B.
Donegal Bay
Ballyshannon
Bundoran
Lower L. Erne
Irvinestown
Dungannon
Portadown
Banbridge
Downpatrick
Shieve Donard 852
Newcastle
Dundrum
Dundrum Bay

Broad Haven
Erris Hd.
Belmullet
Mullet Peninsula
Blacksod Bay
Sligo B.
Sligo
Collooney
L. Allen
Enniskillen
Upper L. Erne
Finn
Clones
Annalee
Cootehill
Armagh
Monaghan
Blackwater
Newry
St. Gullion 577
Warrenpoint
Mourne Mts.
Carlingford L.
Greenore

Achill Hd.
Achill
Achill I.
Killala
Ballina
Moy
Ox Mts.
SLIGO
LEITRIM
Leitrim
Arrow
Boyle
Carrick-on-Shannon
CAVAN
Cavan
Carrickmacross
Belturbet
Castleblayney
Dundalk
Dundalk Bay

Clare I.
Clew Bay
Croagh Patrick 765
MAYO
Castlebar
Westport
Claremorris
ROSCOMMON
Castlereagh
L. Gowna
Granard
Longford
Oldcastle
Ceanannus Mor (Kells)
Louth
Ardee
LOUTH

Inishbofin
Killary Harbour
Mweelrea 819
L. Mask
Ballinrobe
Tuam
CONNACHT
Roscommon
L. Ree
LONGFORD
Athboy
An Uaimh (Navan)
MEATH
Drogheda
Balbriggan

Twelve Pins
Slyne Hd.
Clifden
CONNEMARA
L. Corrib
GALWAY
IRELAND
Ballinasloe
Shannon
Trim
Boyne
Mullingar
WESTMEATH
Lambay I.
Swords
Ireland's Eye
Howth Head

Galway
Athenry
Loughrea
Clare
IRELAND
Athlone
Clara
Edenderry
Maynooth
DUBLIN
Celbridge
Dublin (Baile Átha Cliath)
Dublin Bay
Dun Laoghaire

Galway Bay
Inishmore
Aran Is.
Gort
Slieve Aughty
Portumna
Tullamore
Daingean
OFFALY
Bog
Droichead Nua
Naas
Kippure 754
Bray
Poulaphouca Res.

Hags Hd.
Liscannor Hd.
Birr
Sl. Bloom
Mountmellick
Portarlington
KILDARE
Kildare
WICKLOW
Wicklow
Wicklow Hd.

Lahinch
Ennistymon
CLARE
Ennis
Roscrea
Port Laoise
Athy
Barrow
Lugnaquillia 923
Rathdrum
Mizen Hd.

Mal Bay
Miltown Malbay
Killaloe
Nenagh
Templemore
LAOIS
LEINSTER
Tullow
Shillelagh
Arklow

Kilkee
Kilrush
Rineanna
Ardnacrusha
Keeper 694
Thurles
Carlow
CARLOW
Muine Bheag
Mt. Leinster 796
Gorey

Loop Hd.
Foynes
Limerick
TIPPERARY
Cashel
Kilkenny
KILKENNY
Callan
Enniscorthy
Cahore Pt.

R. Shannon
Rathkeale
LIMERICK
Newcastle
Tipperary
Caher
Clonmel
Carrick-on-Suir
WEXFORD

Listowel
Galtymore 920
Galty Mts.
Slievenamon 722
New Ross
Wexford
Wexford Harbour
Rosslare
Greenore Pt.
Tuscar Rock
Carnsore Pt.

Brandon Bay
Tralee
Fenit
KERRY
MUNSTER
Rathluirc (Charleville)
Newmarket
Mitchelstown
Knockmealdown Mts.
Comeragh Mts.
Waterford
Saltee I.

Brandon Bay
Brandon Mt. 953
St. Mish
Maine
Dingle
Killarney
Castleisland
Kanturk
Mallow
Fermoy
WATERFORD
Tramore

Gt. Blasket I.
Dunmore Hd.
Dingle Bay
Macgillycuddy's Reeks
Carrauntuohill 1040
Lakes of Killarney
Boggeragh Mts.
Blackwater
Lismore
Dungarvan
Dungarvan Bay

Valentia Harbour
Valentia I.
Cahirciveen
Kenmare
Macroom
CORK
Blarney
Lee
Cork
Midleton
Youghal
Youghal Harbour
Hook Hd.
Waterford Harbour

Skellig Rocks
Ballinskelligs B.
Kenmare River
Caha Mts.
Glengariff
Bantry
Bandon
Passage West
Crosshaven
Kinsale
Cork Harbour
Cobh

Crow Hd.
Bear I.
Bantry Bay
Dunmanus Bay
Mizen Hd.
Skull
Baltimore
Clear I.
C. Clear
Fastnet Rock
Clonakilty
Skibbereen
Clonakilty Bay
Galley Hd.
Old Head of Kinsale

IRISH SEA

St. George's Channel

St. David's Hd.

Towns underlined in Northern Ireland give their names to the Districts in which they stand
The remaining Districts are:—

1 Fermanagh	5 Castlereagh
2 Moyle	6 Ards
3 Newtownabbey	7 Down
4 North Down	8 Newry & Mourne

ft   m
3000   1000
1200   400
600   200
300   100
0   0
100   300
200   600
m   ft

1 : 4 000 000

20 0 20 40 60 miles
20 0 20 40 60 80 km

The DISTRICTS of Northern Ireland have been numbered and can be identified by reference to this table.

1 Londonderry	14 Craigavon
2 Limavady	15 Armagh
3 Coleraine	16 Newry & Mourne
4 Ballymoney	17 Banbridge
5 Moyle	18 Down
6 Larne	19 Lisburn
7 Ballymena	20 Antrim
8 Magherafelt	21 Newtownabbey
9 Cookstown	22 Carrickfergus
10 Strabane	23 North Down
11 Omagh	24 Ards
12 Fermanagh	25 Castlereagh
13 Dungannon	26 Belfast

ORKNEY

Kirkwall

HIGHLAND

SHETLAND

Lerwick

WESTERN ISLES

Stornoway

HIGHLAND

Inverness

GRAMPIAN

SCOTLAND

Aberdeen

TAYSIDE

Dundee

FIFE

Glenrothes

CENTRAL

Stirling

Edinburgh

LOTHIAN

Glasgow

STRATHCLYDE

North Channel

Newtown
St. Boswells

BORDERS

ATLANTIC

OCEAN

DUMFRIES
AND
GALLOWAY

Dumfries

NORTHUMBERLAND

Morpeth

Newcastle
TYNE AND
WEAR

Carlisle

Durham

DURHAM

CLEVELAND

Middlesbrough

CUMBRIA

Northallerton

Lifford

DONEGAL

Londonderry

Antrim

NORTHERN
IRELAND

Tyrone

Belfast

Down

NORTH
YORKSHIRE

ISLE OF
MAN

Douglas

HUMBERSIDE

Beverley

Sligo

LEITRIM

Monaghan

Fermanagh

MONAGHAN

Cavan

Dundalk

LANCASHIRE

Preston

WEST
YORKSHIRE

Wakefield

Carrick-on-
Shannon

CAVAN

LOUTH

GREATER
MANCHESTER

Barnsley

SOUTH
YORKSHIRE

SLIGO

MAYO

Castlebar

ROSCOMMON

Longford

LONGFORD

An tUaimh
(Navan)

MEATH

MERSEYSIDE

Liverpool

Manchester

ENGLAND

Lincoln

Roscommon

Mullingar

WESTMEATH

Chester

CHESHIRE

DERBYSHIRE

NOTTING-
HAM-
SHIRE

LINCOLNSHIRE

IRISH SEA

Caernarfon

Mold

Matlock

Nottingham

GALWAY

Galway

OFFALY

Tullamore

DUBLIN

Dublin

GWYNEDD

CLWYD

Stafford

STAFFORD-
SHIRE

Leicester

NORFOLK

Norwich

IRELAND

LAOIS

Port Laoise

KILDARE

Naas

Shrewsbury

SHROPSHIRE

LEICESTERSHIRE

CLARE

Ennis

WICKLOW

Wicklow

WEST
MIDLANDS

Birmingham

NORTH-
AMPTON-
SHIRE

CAMBRIDGE-
SHIRE

Cambridge

SUFFOLK

Ipswich

Limerick

TIPPERARY

Kilkenny

Carlow

CARLOW

WALES

WARWICK-
SHIRE

Warwick

Northampton

BEDFORD-
SHIRE

Bedford

LIMERICK

KILKENNY

Clonmel

WEXFORD

POWYS

HEREFORD
AND
WORCESTER

BUCK-
INGHAM-
SHIRE

Hertford

HERTFORD-
SHIRE

ESSEX

Chelmsford

Tralee

KERRY

WATERFORD

Waterford

Wexford

Llandrindod
Wells

Worcester

Gloucester

Oxford

OXFORDSHIRE

Aylesbury

GREATER
LONDON

CORK

Cork

DYFED

Carmarthen

GLOUCESTER-
SHIRE

GWENT

Cwmbran

BERKSHIRE

Reading

Kingston

Maidstone

St. George's Channel

WEST
GLAMORGAN

Swansea

MID
GLAMORGAN

SOUTH
GLAMORGAN

Cardiff

Bristol

AVON

Trowbridge

WILTSHIRE

SURREY

KENT

SOMERSET

Taunton

HAMPSHIRE

Winchester

WEST
SUSSEX

Chichester

EAST
SUSSEX

Lewes

CELTIC

SEA

DEVON

Exeter

DORSET

Dorchester

Newport

ISLE OF
WIGHT

CORNWALL

Truro

ENGLISH CHANNEL

FRANCE

NORTH

SEA

○ Norwich — Administrative headquarters
MERSEYSIDE — Metropolitan counties
Antrim — Former Northern Ireland counties

Projection: Conical with two standard parallels

West from Greenwich () East from Greenwich
COPYRIGHT. GEORGE PHILIP & SON. LTD.

1:2 500 000

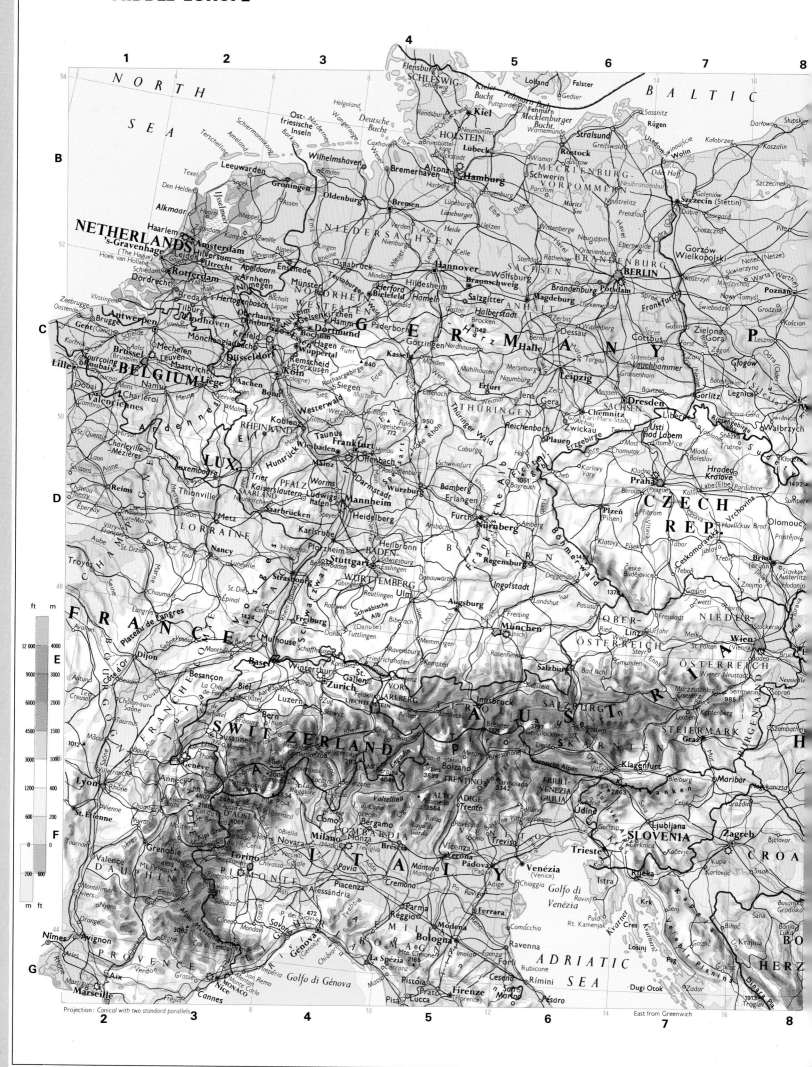

Projection : Conical with two standard parallels

East from Greenwich

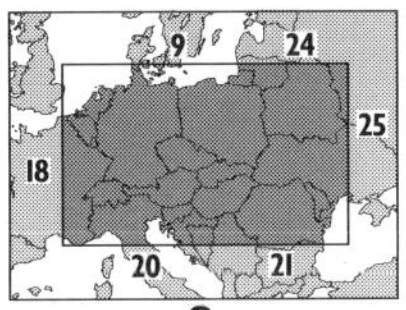

1:5 000 000

50   0   50   100 miles
50   0   50   100   150 km

9   10   11   12   13   14   15

A   54

SEA
Zelenogradsk

Zatoka Gdańska
Wejherowo
Sopot
Gdynia
Kaliningrad (Königsberg)
Pregolya
Chernyakhovsk
LITHUANIA
Vilnius
Alitus
Varena
Molodechno
Gorki
Borisov

Gdańsk (Danzig)
Elbląg
Braniewo
Malbork
Lyna
Kętrzyn
Gizycko
Suwałki
Augustów
Lida
Minsk
Mogilev
Krichev

Starogard
Chełmża
Kwidzyń
Ostróda
Olsztyn
Pojezierze Mazurskie
▲309
Grodno
Neman
Mosty
Novogrudok
Baranovichi
Bobruysk
B E L O R U S S I A

Bydgoszcz
Chełmno
Wąbrzeźno
Iława
Mława
Ostrołęka
Sokółka
▲238
Białystok
Volkavysk
Slonim
Shchara
Bereza

B   52

Noteć
Inowrocław
Toruń
Rypin
Lipno
Ciechanów
Ostrów Mazowiecka
Brańsk
Hajnówka
Luninets
Pripyat
Gomel (Homyel)

Gniezno
Włocławek
Płock
Wisła (Vistula)
Pułtusk
Wkra
Bug
Czeremcha
Zhabinka
Pinsk
Pripyat
Halmkovichi
Chernigov

Września
Warszawa (Warsaw)
Mińsk Mazowiecki
Siedlce
Biała Podlaska
Brest
Dubrovitsa
316
Uzh
Chernobyl

P O L A N D
Konin
Kutno
Łowicz
Pruszków
Żyrardów
Otwock
Łuków
Międzyrzec Podlaski
Włodawa
P o l e s y e
Sarny
Korosten

C   50

Krotoszyn
Kalisz
Zduńska Wola
Łódź
Skierniewice
Grójec
Pilica
Kozienice
Włodawa
Bug
Kovel
Styr
Goryn
Sluch
Novograd-Volynskiy
Radomyshl
Kiyev (Kiev)
Borispol

Ostrów Wielkopolski
Oleśnica
Wieluń
Piotrków Trybunalski
Końskie
Radom
Puławy
Chełm
Vladimir Volynskiy
Lutsk
Rovno
Korets

Wrocław
Breslau
Brzeg
Warta
Częstochowa
Jędrzejów
Kielce
Ostrowiec Świętokrzyski
Lublin
Zamość
Sokal
Dubno
Ostrog
Shepetovka
Zhitomir
Fastov
Belaya Tserkov

Opole
Tarnowskie Góry
Zawiercie
Pińczów
Sandomierz
Kraśnik
390▲
Radekhov
Brody
Kremenets
Berdichev
Kazatin

Nysa
Zabrze
Bytom
Sosnowiec
Katowice
Tarnobrzeg
San
Kamenka Bugskaya
U K R A I N E
Starokonstantinov

Gliwice
Chorzów
Tychy
Kraków
Tarnów
Dąbrowa Tarnowska
Rzeszów
Przeworsk
Jarosław
Gorodok
Lvov (Lviv)
Zolochev
Ternopol
Khmelnitskiy
384
Vinnitsa (Vinnytsya)

Racibórz
Opava
Ostrava
Bielsko-Biała
Wieliczka
Nowy Sącz
Jasło
Przemyśl
471
D   48

Frýdek Místek
Český Těšín
1725
Krosno
Sanok
Sambor
Dnestr
Buchach
Chortkov
Zhmerinka
Uman

Přerov
Zlin
Žilina
Ružomberok
2655
Zápádné Beskydy
Tatry
Vychodné Beskydy
Drohobych
Borislav
Stryi
Turka
Zaleshchiki
Kamenets-Podolskiy
Bug

Biele Karpaty
Nízke Tatry
Presov
Ivano-Frankovsk
1881
Nadvornaya
Kolomyya
Snyatyn
Khotin
Mogilev-Podolskiy
Pervomaysk

Kremnica
Košice
Uzhgorod
Per Yablonitse
931
Chernovtsy
Starozhinets
Yedintsy
Soroki
Kotovsk

S L O V A K   R E P.
Banská Bystrica
Zvolen
Slovenské Rudohorie
Banská Štiavnica
Lučenec
Sátoraljaújhely
Mukachevo
Beregovo
Khust
2061
Storozhinets
Dorohoi
M O L D A V I A

Nitra
Karpaty
Hron
N. Zámky
Miskolc
Tokaj
Eger
Nyiregyhaza
Satu Mare
Sighetu Marmatiei
2305
Pietrosul
Vatra-Dornei
Botoşani
429
Kishinev (Chişinău)
Tiraspol

Bratislava
Komárno
Vác
Gyöngyös
Mezőkövesd
Hajdúböszörmény
Carei
Baia Mare
2102
Pietrosu
Bistrita
Suceava
Iaşi
Bendery

Győr
Esztergom
Újpest
Hatvan
Jászberény
Debrecen
Someş
Dej
Bistrita
Piatra Neamţ
Roman
Vaslui
Odessa
E   46

Tatabánya
BUDAPEST
Cegléd
Szolnok
Karcag
Oradea
Cluj-Napoca
Turda
Reghin
Bacău
Bîrlad
Belgorod Dnestrovskiy

H U N G A R Y
Nagykőrös
Mezőtúr
Salonta
Negru
Aiud
Tirgu Mureş
Praid
Odorheiu Secuiesc
Miercurea
Piatra Neamţ
Tecuci

Veszprém
Székesfehérvár
Kecskemét
Kiskunfélegyháza
Gyula
Crişul Alb
Mtii Bihor
1848
T r a n s i l v a n i a
Sibiu
Mediaş
Sighişoara
Sfîntu Gheorghe
Focşani
Galaţi

Bakony
Dunaújváros
Kiskőrös
Kiskunhalas
Crişul Alb
Abrud
Brad
Alba-Iulia
Deva
Hunedoara
Simeria
Rîmnicu Sarat
Brăila
467

Kaposvár
Szekszárd
Szeged
Makó
Arad
Mureş
Lugoj
Simeria
Sibiu
Făgăraş
Carpaţii Meridionali
2535
Vf. Omu
2507
Buzău
Buzău
Sulina

Balaton
Batászék
Baja
Hódmezővásárhely
R O M A N I A
Caransebeş
Deva
Turnu Roşu
Petroşani
350
Braşov
Cîmpina
Ploieşti
F   44

Pécs
Mohács
Subotica
Timişoara
Banat
Reşiţa
2518
Peleaga
2509
Paring
Tîrgu-Jiu
Rîmnicu Vîlcea
Cîmpulung
Tîrgovişte
BLACK

Drava
Osijek
Novi Sad
Zrenjanin
Petrovgrad
Bela Crkva
Porta Orientali
Mehadia
Ile de Fier
Tîrgu-Jiu
Piteşti
Dîmboviţa
Argeş
Bucureşti (Bucharest)
Cernavodă
Constanţa

SNIA
Brčko
Bijeljina
Zemun
Beograd
Pančevo
Vojvodina
Dobreta-Turnu-Severin
Orşova
Jiu
V l a h i a
Ialomiţa
Călăraşi
Mangalia

EGOVINA
Tešanj
Tuzla
Sava
SERBIA
Smederevo
Požarevac
Y U G O S L A V I A
Negotin
Craiova
Slatina
Olteniţa
Silistra
SEA
G

Sarajevo
Travnik
Užice
Čačak
Kragujevac
Zaječar
Vidin
Lom
Dunărea (Danube)
Corabia
Turnu Măgurele
Vedea
Zimnicea
Giurgiu
Ruse (Ruschuk)
Dobrich

1346
Kraljevo
Morava
Timok
B U L G A R I A

18   9   20   10   22   11   24   12   26   13   28   14

COPYRIGHT. GEORGE PHILIP & SON. LTD.

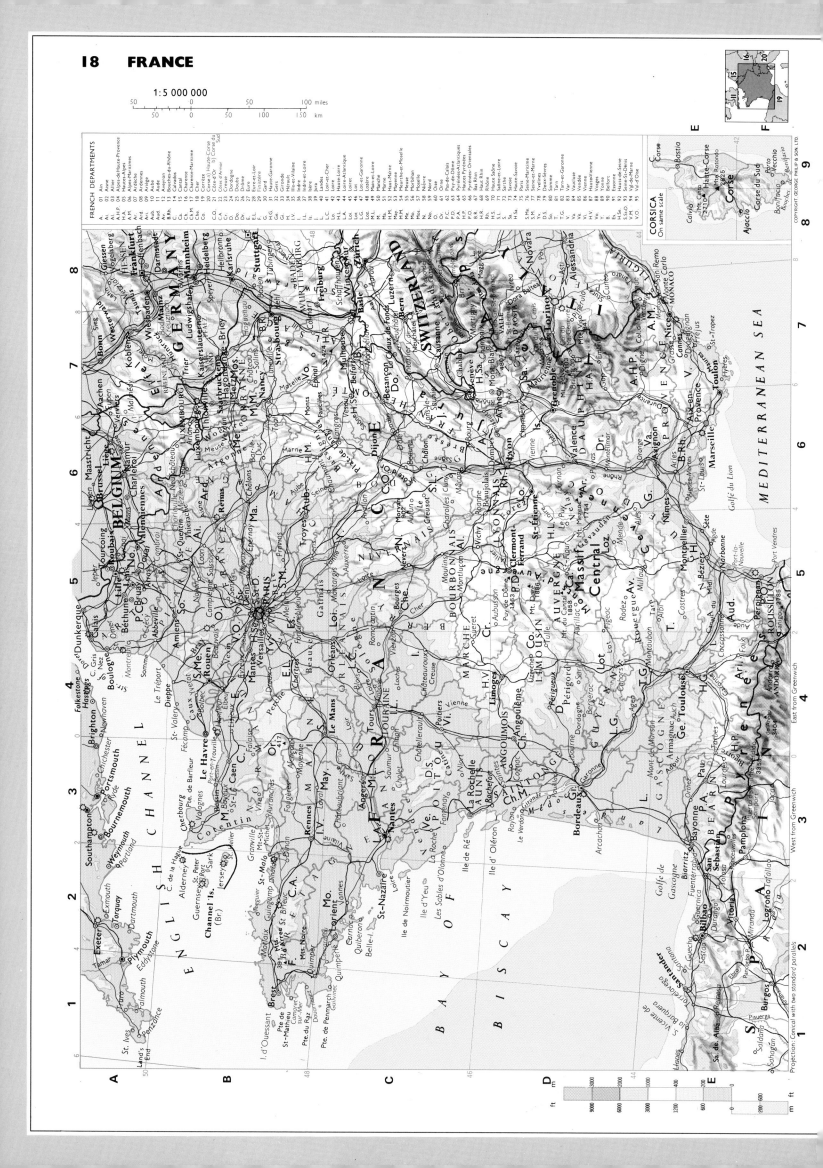

1:5 000 000

FRENCH DEPARTMENTS

Ai.	01 Ain
Ai.	02 Aisne
A.	03 Allier
A.H.P.	04 Alpes-de-Haute-Provence
H.A.	05 Hautes-Alpes
A.M.	06 Alpes-Maritimes
Ard.	07 Ardèche
Ard.	08 Ardennes
Ar.	09 Ariège
Aub.	10 Aube
Aud.	11 Aude
Av.	12 Aveyron
B.Rh.	13 Bouches-du-Rhône
C.	14 Calvados
Can.	15 Cantal
Cha.M	16 Charente-Maritime
Che.	17 Charente
Che.	18 Cher
C.O.	19 Corrèze
C.A.	20 a) Corse (Haute-Corse) du
	b) Corse du Sud
C.O.	21 Côte-d'Or
Do.	22 Côtes d'Armor
Cr.	23 Creuse
Do.	24 Dordogne
Do.	25 Doubs
Dr.	26 Drôme
E.	27 Eure
E.L.	28 Eure-et-Loir
F.	29 Finistère
Ga.	30 Gard
H.G.	31 Haute-Garonne
Ge.	32 Gers
Gi.	33 Gironde
H.V.	34 Hérault
I.V.	35 Ille-et-Vilaine
I.	36 Indre
I.L.	37 Indre-et-Loire
I.	38 Isère
Ju.	39 Jura
La.	40 Landes
L.C.	41 Loir-et-Cher
Lo.	42 Loire
H.L.	43 Haute-Loire
L.A.	44 Loire-Atlantique
Lo.	45 Loiret
Lo.	46 Lot
L.G.	47 Lot-et-Garonne
Lo.	48 Lozère
M.	49 Maine-et-Loire
Ma.	50 Manche
Ma.	51 Marne
H.M.	52 Haute-Marne
May.	53 Mayenne
M.M.	54 Meurthe-et-Moselle
Me.	55 Meuse
Mo.	56 Morbihan
Mo.	57 Moselle
Ni.	58 Nièvre
No.	59 Nord
O.	60 Oise
Or.	61 Orne
P.C.	62 Pas-de-Calais
P.D.	63 Puy-de-Dôme
P.A.	64 Pyrénées-Atlantiques
H.P.	65 Hautes-Pyrénées
P.O.	66 Pyrénées-Orientales
B.R.	67 Bas Rhin
H.R.	68 Haut Rhin
Rh.	69 Rhône
H.S.	70 Haute-Saône
S.L.	71 Saône-et-Loire
S.Me.	72 Sarthe
Sa.	73 Savoie
H.Sa.	74 Haute-Savoie
S.M.	75 Paris
S.M.	76 Seine-Maritime
Y.S.	77 Seine-et-Marne
Y.	78 Yvelines
Y.	79 Deux-Sèvres
S.	80 Somme
T.	81 Tarn
T.G.	82 Tarn-et-Garonne
Va.	83 Var
Va.	84 Vaucluse
V.	85 Vendée
V.	86 Vienne
H.V.	87 Haute-Vienne
Vo.	88 Vosges
Y.	89 Yonne
B.	90 Belfort
Esonne	91 Essonne
H.Se.	92 Hauts-de-Seine
S.S.o.D	93 Seine-St-Denis
V.M.	94 Val-de-Marne
V.d'O.	95 Val-d'Oise

CORSICA
On same scale

Corse

MEDITERRANEAN SEA

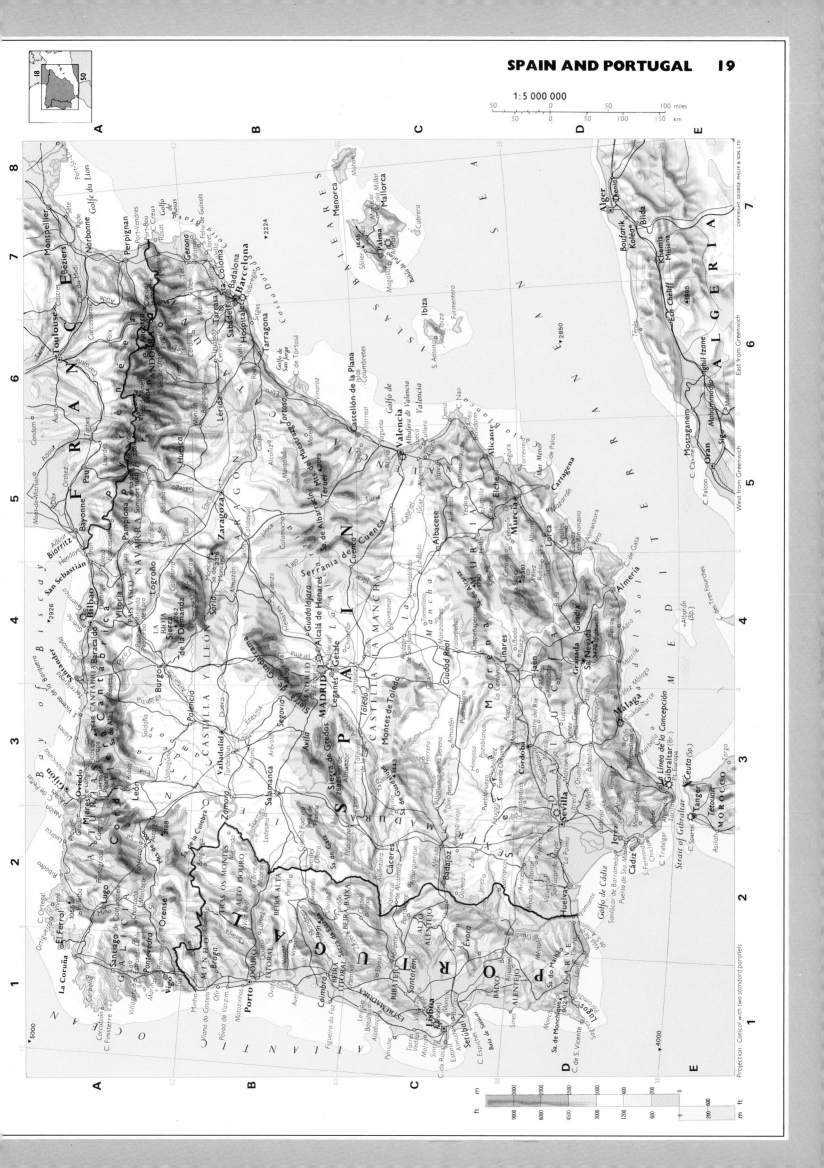

1:5 000 000

50   0   50   100 miles

50   0   50   100   150   km

COPYRIGHT GEORGE PHILIP & SON. LTD.

Projection: Conical with two standard parallels

Projection: Conical with two standard parallels

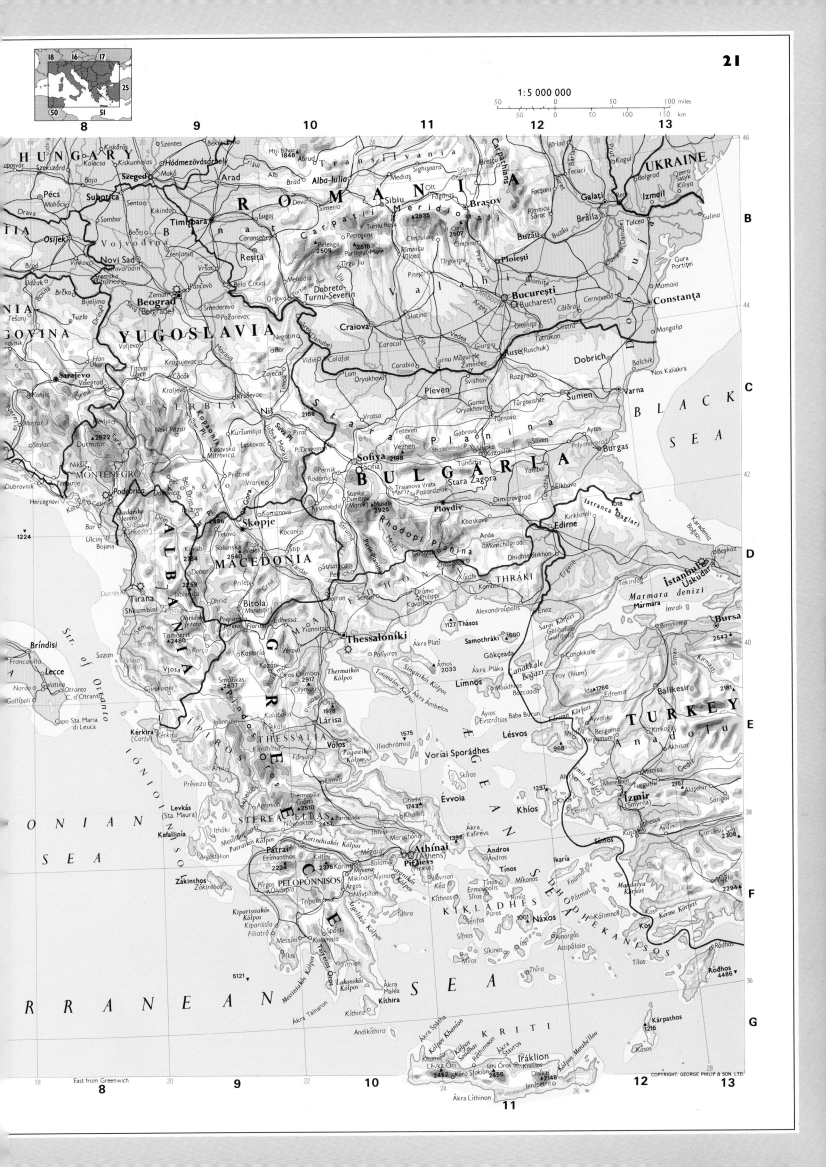

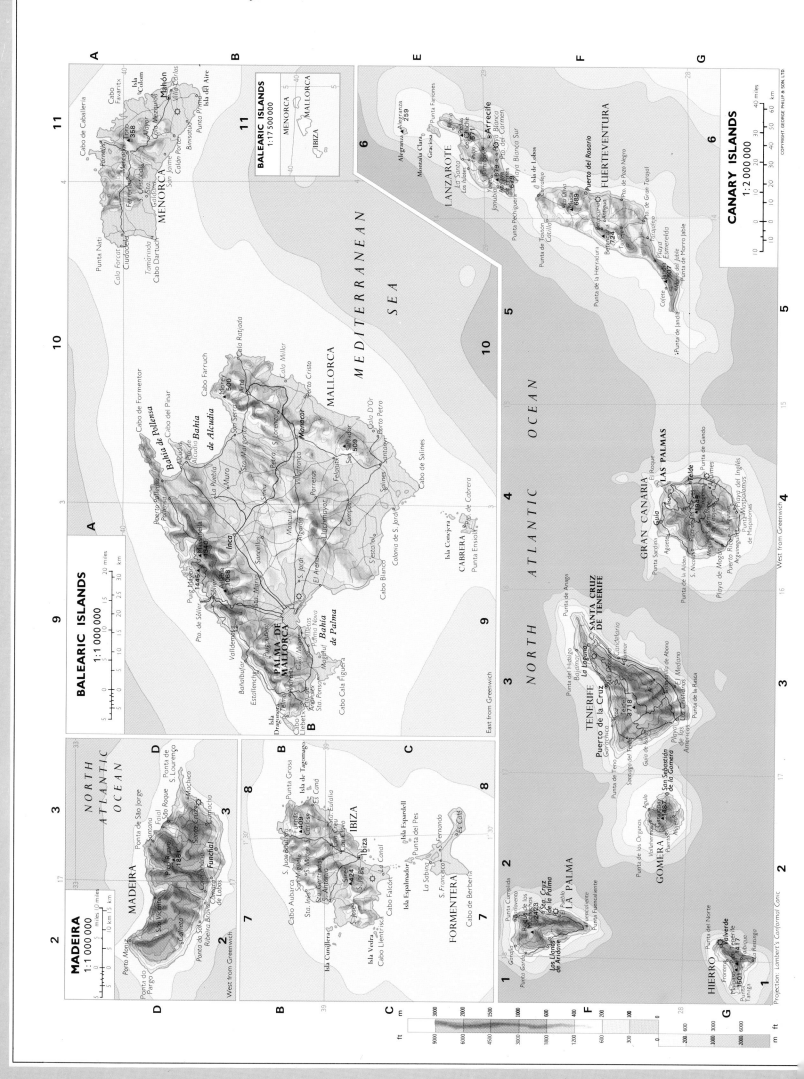

BALEARIC ISLANDS
1:1 000 000

BALEARIC ISLANDS
1:17 500 000

MENORCA
MALLORCA
IBIZA

MADEIRA
1:1 000 000

CANARY ISLANDS
1:2 000 000

MEDITERRANEAN SEA

MENORCA

Cabo de Caballeria
Punta Nati
Ciudadela
Ferrerías
Mercadal 358
Mahón
Cabo Favaritx
Isla Colom
Villa Carlos
Isla del Aire
Cala Forcat
Tamarinda
Cabo Dartuch
San Jaime
Colán Porter
Binisalem

MALLORCA

Cabo de Formentor
Cabo del Pinar
Cabo Farruch
Cala Ratjada
Cala Millor
Cala Bona
Porto Cristo
Cala D'Or
Porto Petro
Cabo de Salines
Cabo Blanco
Colonia de S. Jordi
Cabo Cala Figuera

Bahia de Pollensa
Bahia de Alcudia
Pollensa
Alcudia
Puerto de Pollensa
Puerto de Alcudia
La Puebla
Muro
Sta. Margarita
Artá 500
Son Serra
Son Servera
Manacor
Petra
Villafranca
Felanitx
San Salvador 509
Santany
S'Estanyol
S. Jordi
El Arenal
Lluchmayor
Campos
Porreras
Montuiri
Algaida
Sineu
Sancellas
Inca
La Puebla
Sa Pobla

PALMA DE MALLORCA
Bahia de Palma
Magaluf
Palma Nova
Illetas
Cala Mayor
El Terreno
Gènova
Puigpuñent
Esporlas
Estallenchs
Banalbufar
Valldemosa
Pto. de Sóller
Sóller
Masanella 1340
Puig Mayor 1445
Galatzó 1068
Isla Dragonera
Cabo Lliebetx
Sta. Ponsa
Pto. de Andraitx
Andraitx

CABRERA
Isla Conejera
Pto. de Cabrera
Punta Ensiolá
Cabo de Cabrera

MEDITERRANEAN SEA

NORTH ATLANTIC OCEAN

MADEIRA

Ponta de S. Lourenço
Ponta de São Jorge
Ponta do Sol
Ponta de Pargo
Porto Moniz
Porto do Seixal
Ribeira Brava
Machico
Camacha
Santana
Faial
São Roque
São Vicente
Santa Cruz
Funchal
Câmara de Lobos
Calheta
Pico Ruivo 1861
Caniço
Ponta de
Porto Santo

IBIZA

Cabo Aubarca
Punta Grosa
Isla de Tagomago
Es Caná
S. Juan Bautista
Sta. Inés
S. Miguel
Sta. Eulalia
S. Carlos
San Antonio Abad
S. Antonio
S. Agustín
S. Mateo
S. Gertrudis
S. Rafael
Can Cirèr
Ibiza
La Canal
Sarrá 424
S. José
S. Jorge
Isla Espardell
Punta del Pes
S. Fernando
Es Calò
Isla Vedra
Cabo Llentrisca
Cabo Falcón
Isla Espalmador
La Sabina
S. Francisco
Cabo de Berbería
S. Mola

FORMENTERA

LANZAROTE

Alegranza 259
Montaña Clara
Graciosa
Punta Fariones
La Santa
Arrecife
Haría
Puerto del Carmen
Tinajo
Yaiza
Playa Blanca
Pta. del Carmen
Los Islotes
Isla de Lobos
Corralejo

FUERTEVENTURA

Puerto del Rosario
La Oliva
Betancuria 689
Gran Tarajal
Pto. de Gran Tarajal
Antigua
Tuineje
Pájara
Cofete
Playa Esmeralda
Morro Jable
Punta de Jandía
Punta de Morro Jable
Punta de la Herradura

GRAN CANARIA

LAS PALMAS
El Roque
Telde
Teror
Arucas
Guía
Gáldar
Agaete
Maspalomas
Playa del Inglés
Punta Maspalomas
Puerto Rico
S. Nicolás
Punta Sardina
Agüimes
Punta de Gando
Punta de la Aldea
Aguaguinte

TENERIFE

SANTA CRUZ DE TENERIFE
La Laguna
Puerto de la Cruz
Teide 3718
Candelaria
Güímar
El Médano
Punta de Anaga
Punta del Hidalgo
Bajamar
Garachico
Buenavista
Punta de Teno
Guía de Isora
Playa de las Américas
Punta de la Rasca

GOMERA

San Sebastián de la Gomera
Garajonay 1487
Agulo
Vallehermoso
Hermigua
Valle Gran Rey
Puerto de Santiago
Punta de los Órganos
Alajeró

LA PALMA

Santa Cruz de la Palma
Roque de los Muchachos 2423
Punta Cumplida
Punta Gorda
Garafía
Los Llanos de Aridane
Fuencaliente
Punta Fuencaliente
El Paso

HIERRO

Valverde
Malpaso 1501
Frontera
Punta del Norte
Sabinosa
Tanajara
El Golfo

NORTH ATLANTIC OCEAN

West from Greenwich
East from Greenwich

Projection: Lambert's Conformal Conic

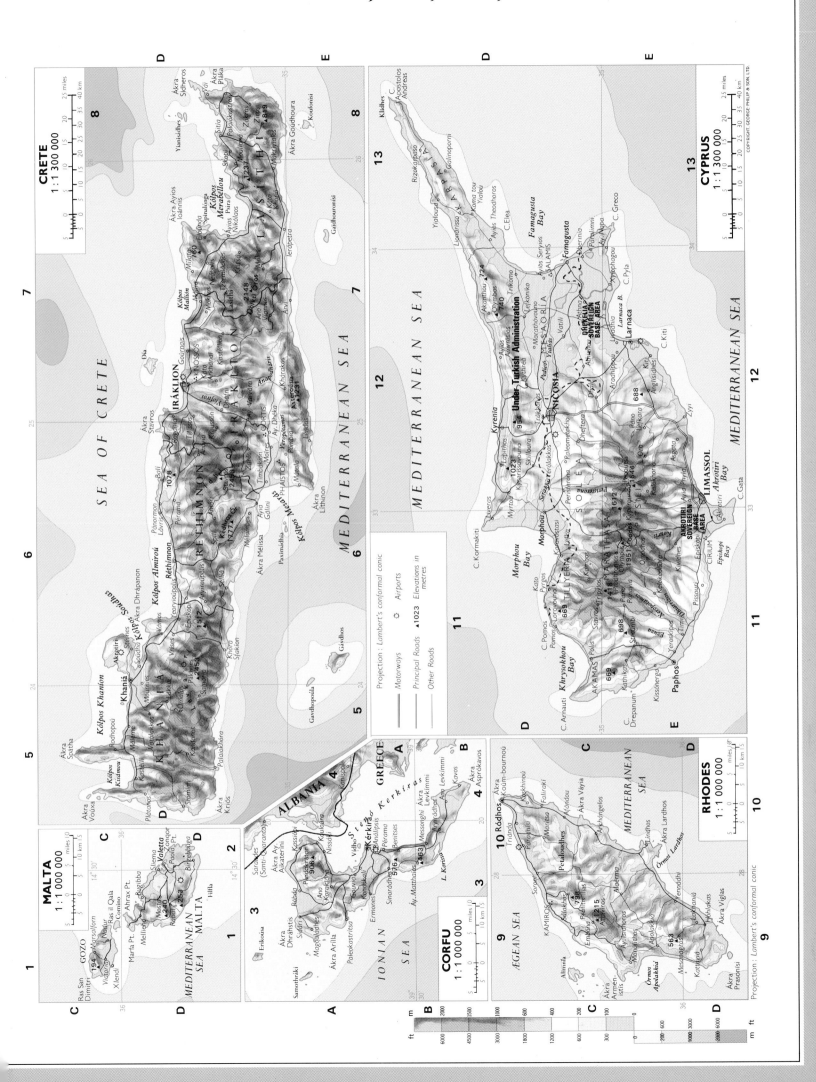

CRETE
1:1 300 000

25 miles
40 km

MALTA
1:1 000 000

CORFU
1:1 000 000

RHODES
1:1 000 000

CYPRUS
1:1 300 000

25 miles
40 km

Projection: Lambert's conformal conic

Airports
Elevations in
metres
▲1023

Motorways
Principal Roads
Other Roads

SEA OF CRETE

MEDITERRANEAN SEA

IONIAN SEA

AEGEAN SEA

GREECE

ALBANIA

IRÁKLION

Réthimnon

Khaniá

Kérkira

Ródhos

NICOSIA

Famagusta

Larnaca

LIMASSOL

Paphos

Under Turkish Administration

DHEKELIA SOVEREIGN BASE AREA

AKROTIRI SOVEREIGN BASE AREA

COPYRIGHT GEORGE PHILIP & SON LTD.

Projection: Lambert's conformal conic

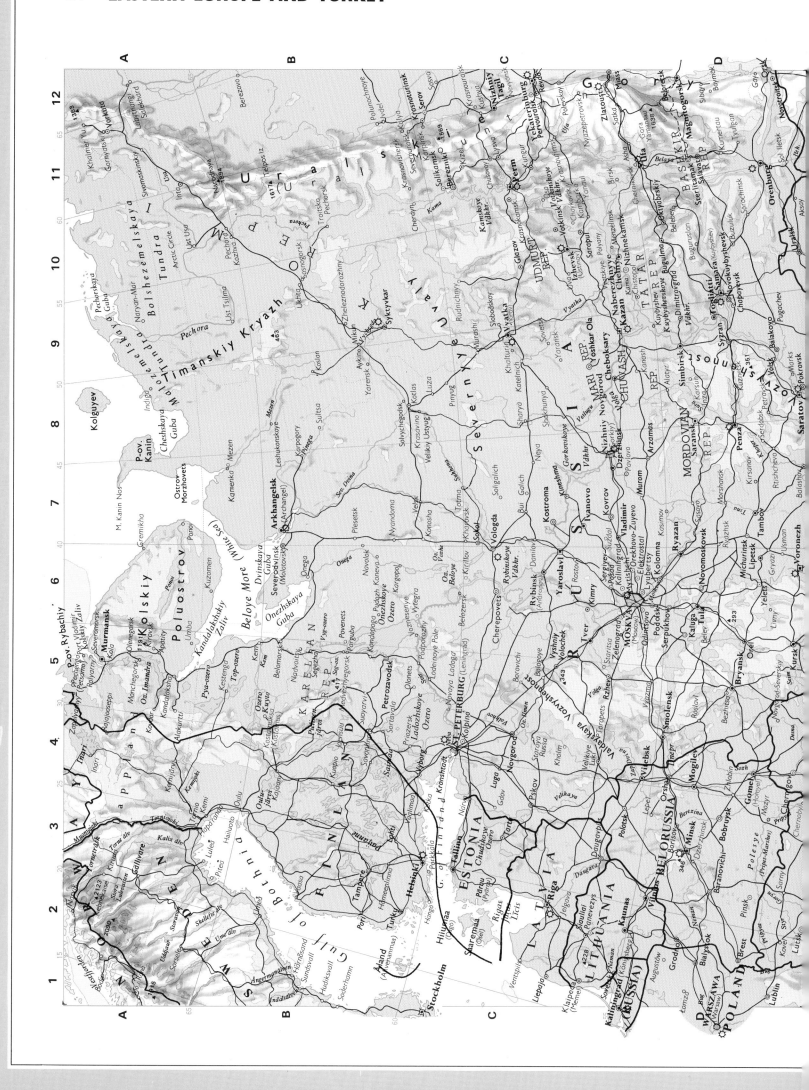

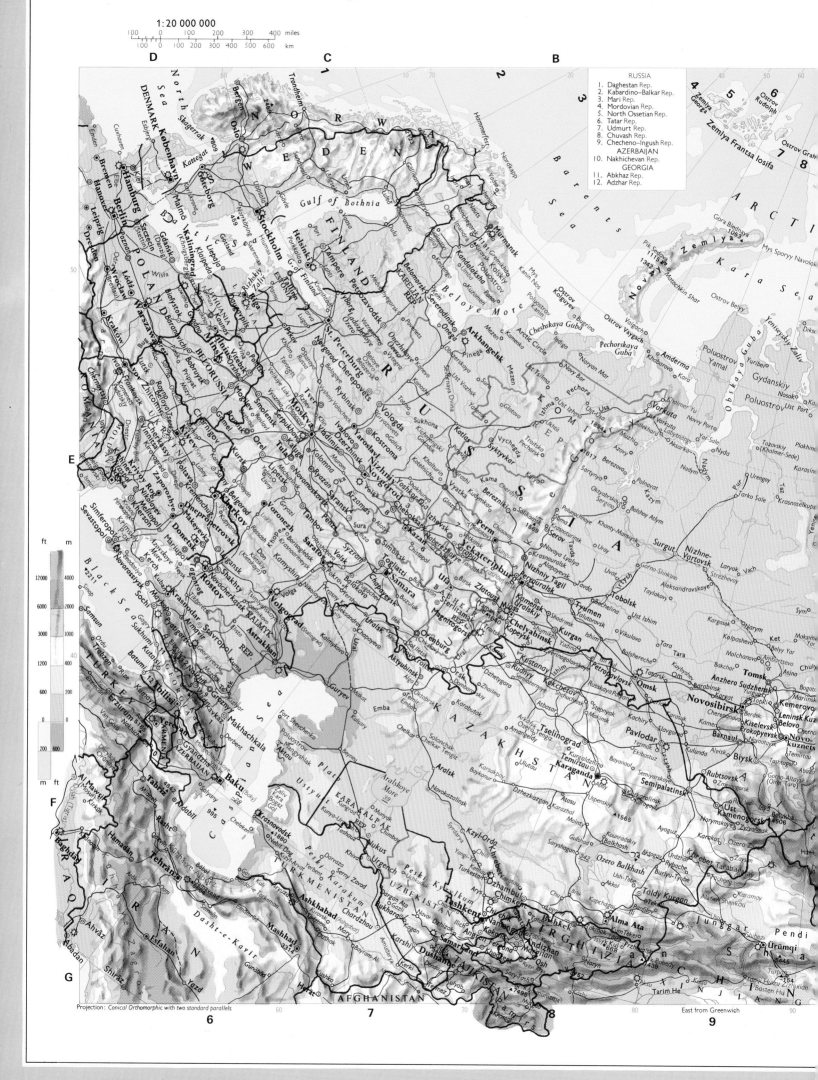

1:20 000 000

	RUSSIA
1.	Daghestan Rep.
2.	Kabardino–Balkar Rep.
3.	Mari Rep.
4.	Mordovian Rep.
5.	North Ossetian Rep.
6.	Tatar Rep.
7.	Udmurt Rep.
8.	Chuvash Rep.
9.	Checheno–Ingush Rep.
	AZERBAIJAN
10.	Nakhichevan Rep.
	GEORGIA
11.	Abkhaz Rep.
12.	Adzhar Rep.

Projection: Conical Orthomorphic with two standard parallels

East from Greenwich

1 : 50 000 000

250   0   250   500   750   1000 miles
250   0   500   1000   1500 km

PACIFIC OCEAN

ARCTIC OCEAN

INDIAN OCEAN

China

Plateau of Tibet

Himalaya

Kunlun Shan

Tien Shan

Altai

Plateau of Mongolia

West Siberian Plain

Central Siberian Plateau

Verkhoyansk Range

Stanovoy Ra.

Yablonovy Ra.

Sayan Mts.

Ural Mountains

Caspian Sea

Black Sea

Caucasus

Mesopotamia

Tigris   Euphrates

Arabia

Ar Rub' al Khali

Red Sea

Arabian Sea

Bay of Bengal

India

Deccan

Western Ghats

Eastern Ghats

Ceylon

Sumatra

Borneo

Malay Peninsula

Philippine Is.

Mindanao

Luzon

Celebes Sea

Sulu Sea

South China Sea

East China Sea

Yellow Sea

Sea of Japan

Honshu

Hokkaido

Korea

Formosa

Hainan

Mekong

Irrawaddy

Salween

Tarim Basin

Takla Makan

Turfan Basin

Gobi

Great Plain of China

Manchurian Plain

Great Khingan Mts.

Sikhote Alin Ra.

Amur

Lena

Yenisei

Angara

Ob

Irtysh

Turan Plain

Aral Sea

Syr Darya

Amu Darya

Plateau of Iran

Elburz Mts.

Zagros Mts.

The Gulf

G. of Oman

G. of Aden

Socotra

Somali Peninsula

Nile

Libyan Desert

Mediterranean Sea

Suez Canal

Cyprus

Anatolia

Taurus

Ararat

Dead Sea

Syrian Desert

Bosporus

Carpathians

Danube

Elbe

Odra

Vistula

Rhine

North Sea

Baltic Sea

North European Plain

Russian Central Uplands

Finland

Scandinavia

Iceland

Greenland

Svalbard

Novaya Zemlya

Kara Sea

Barents Sea

Laptev Sea

Taimyr Peninsula

New Siberian Is.

Wrangel I.

Bering Sea

Kamchatka Peninsula

Sea of Okhotsk

Sakhalin

Kuril Is.

Aleutian Is.

Adriatic Sea

British Isles

Maldive Is.

Laccadive Is.

Seychelles

Amirantes

Chagos Arch.

Andaman Is.

Nicobar Is.

Str. of Malacca

G. of Thailand

Chao Phraya

New Guinea

Australia

Arafura Sea

Timor

Flores

Bali

Java

Java Sea

Sunda Is.

Celebes

Ceram

Banda Sea

Moluccas

Halmahera

Palau Is.

Caroline Is.

Guam

Tropic of Cancer

Equator

Arctic Circle

Projection: Bonne

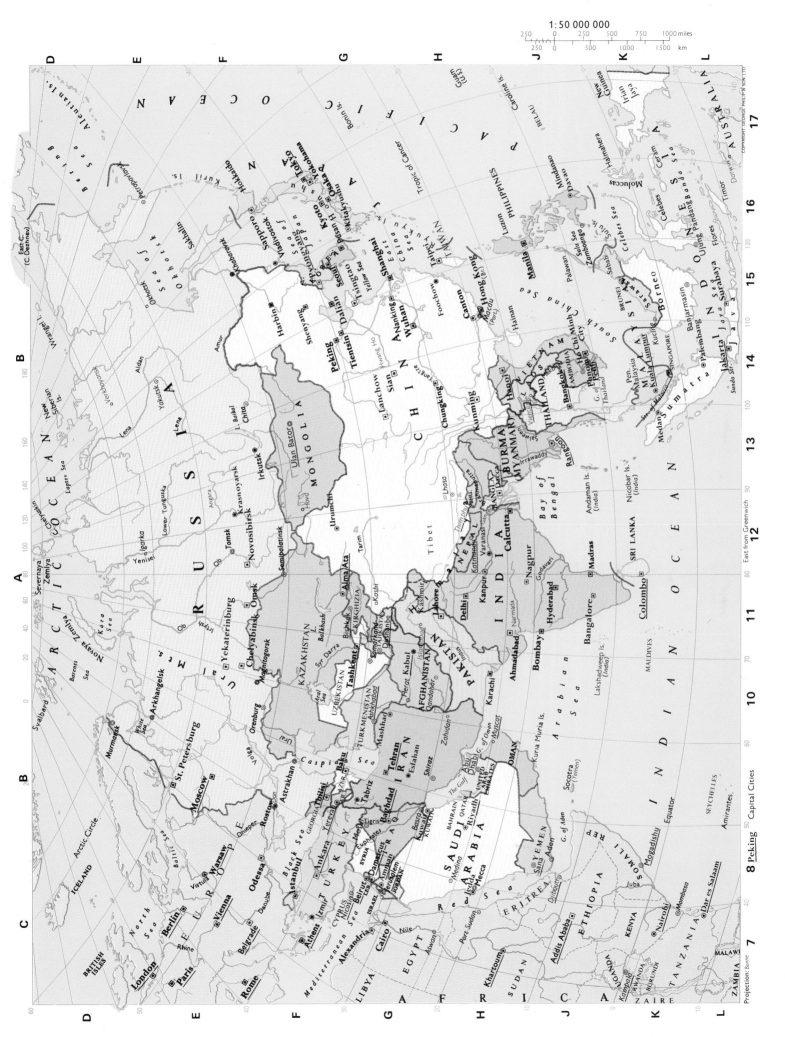

1:50 000 000

Projection: Bonne

East from Greenwich

8 **Peking** 50 Capital Cities

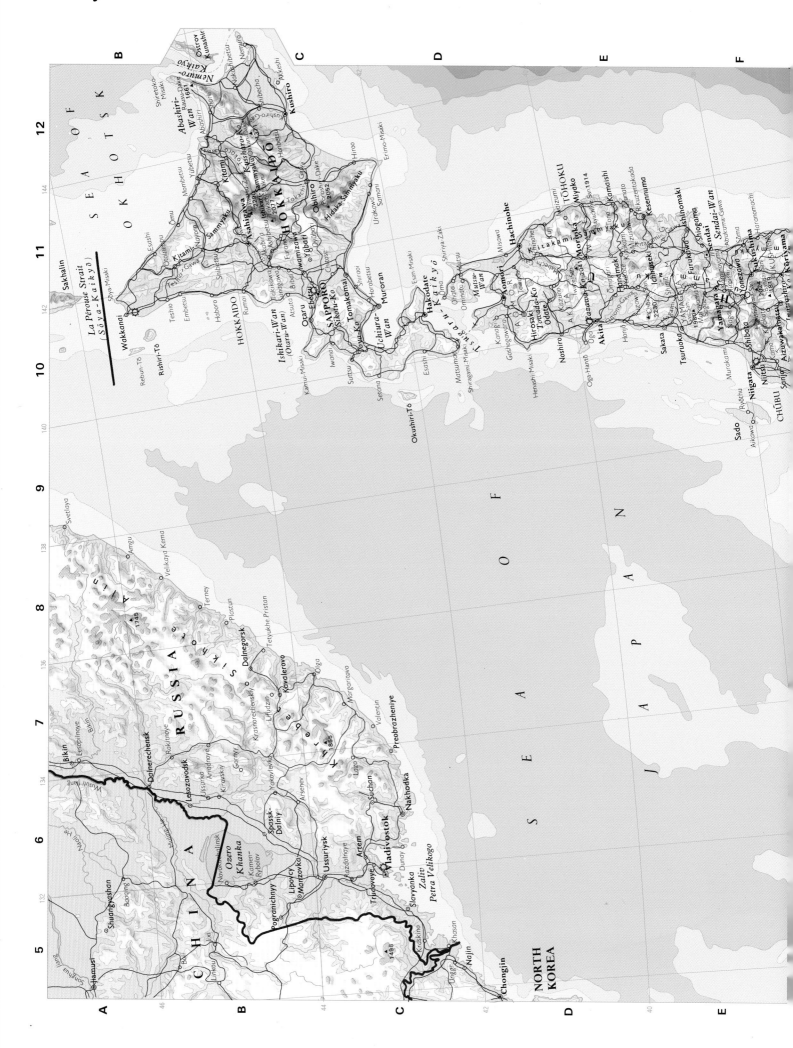

SEA OF OKHOTSK

HOKKAIDO

SEA OF JAPAN

RUSSIA

CHINA

NORTH KOREA

Sakhalin

La Pérouse Strait
(Sōya-Kaikyō)

Wakkanai

Nemuro-Kaikyō

Ostrov Kunashir

Abashiri-Wan

SAPPORO

Tsugaru-Kaikyō

Hakodate

TOHOKU

Hachinohe

Aomori

Akita

YAMAGATA

Sendai-Wan

Niigata

Sado

CHUBU

Vladivostok

Nakhodka

Zaliv Petra Velikogo

Ozero Khanka

Chongjin

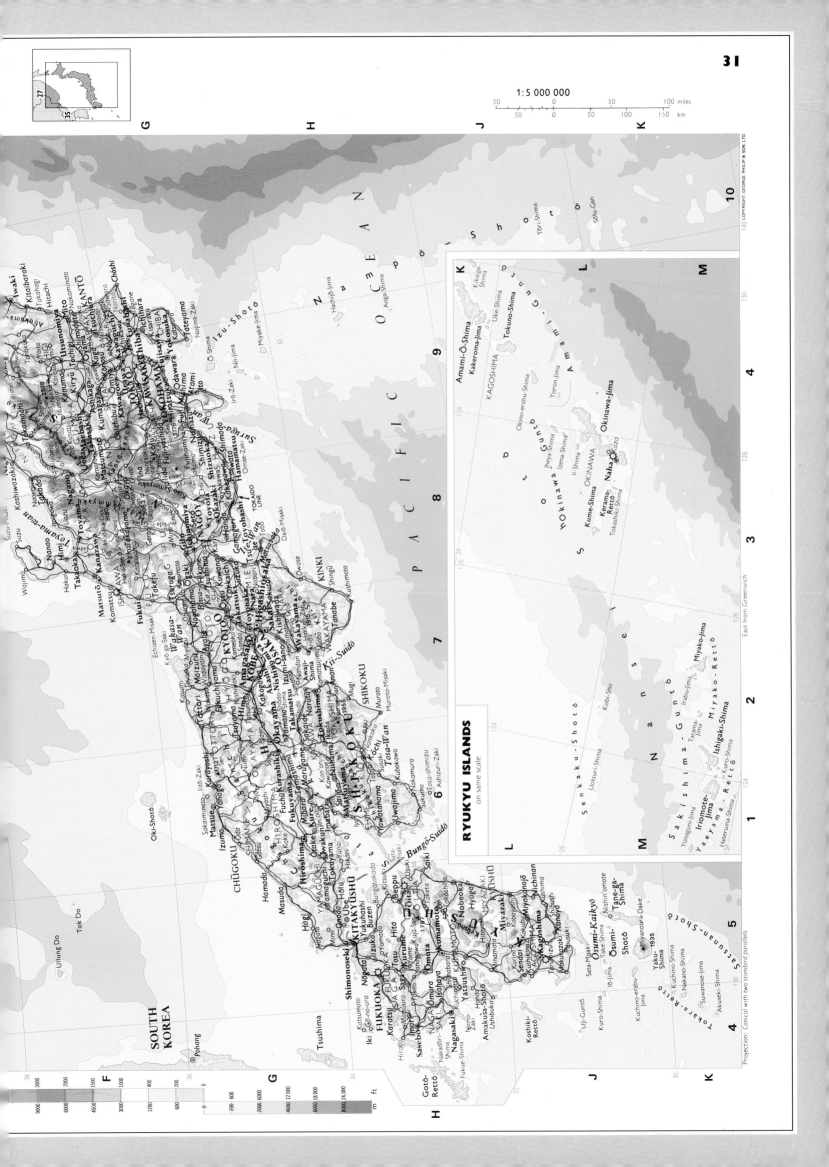

1:5 000 000

50   0   50   100 miles
50   0   50   100   150 km

RYUKYU ISLANDS
on same scale

PACIFIC OCEAN

SOUTH KOREA

East from Greenwich

Projection: Conical with two standard parallels

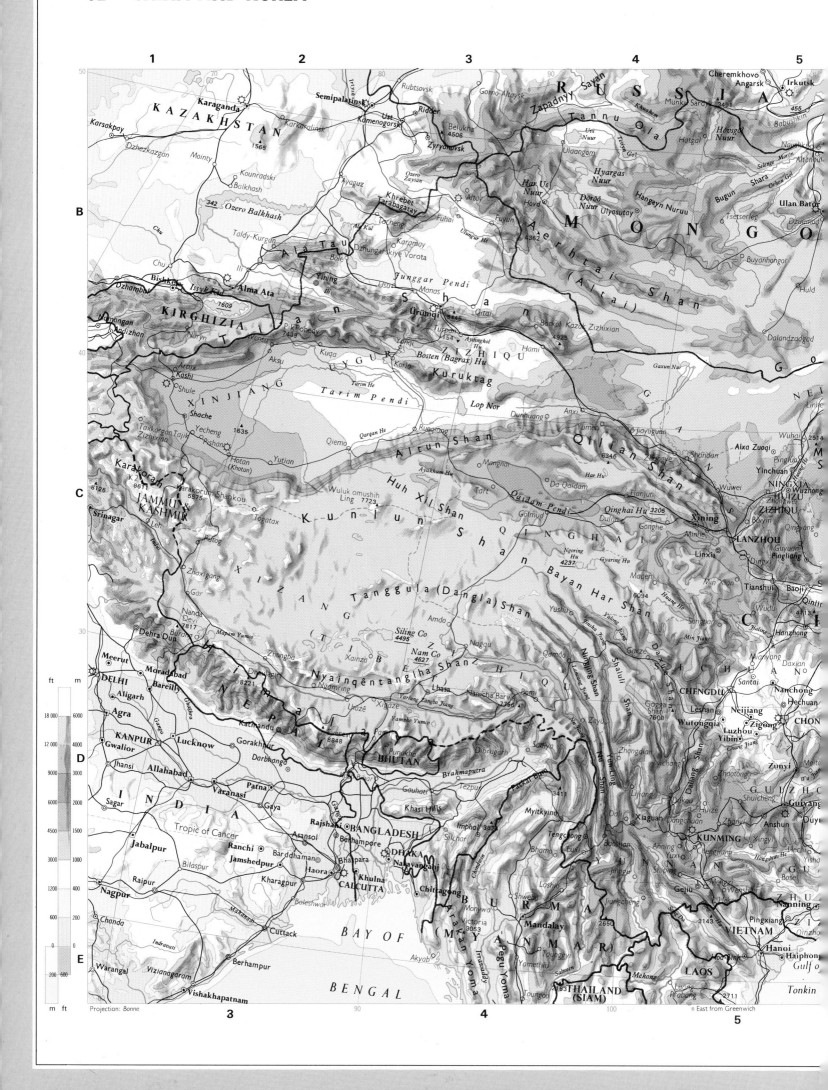

1:15 000 000

100   0   100   200   300   400 miles
100   0   100   200   300   400   500   600   km

27
40
38   37

6

7   8   9

B

Oz. Baykal
Ulan Ude
Chita
Nerchinsk
Sretensk
Bukachacha
Shimanovsk
Svobodny
Chegdomyn
Komsomolsk
Aleksandrovsk
C. Terpeniya
Sakhalin
Poronaysk
etrovsk
abaykalskiy
kita
Yablonovyy Khrebet
Olovyannaya
Borzya
Manzhouli
Nerchinsk
Blagoveshchensk
Aihui
Bureya
Birobidzhan
Obluchye
Khabarovsk
Traitskoye
Dolinsk
Yuzhno-Sakhalinsk
Kholmsk
Hentiyn
Nuruu
Dutulun Shan
Chaybalsan
Kerulen
Hailar
Oroqen Zizhiqi
Yichun
L. Bolon
Amur
Tartarskiy Proliv
Manzhouli
Saynshand
Hulun Nur
Buir Nur
Arxan
Solan
Nenjiang
Bei'an
Hegang
Jiamusi
Shuangyashan
Bikin
La Perouse Str.
Wakkanai
L I A
Butha Qi
Nong Jiang
Qiqihar
Anda
Suihua
HARBIN
Jixi
Mishan
Ozero Khanka
Ussuriysk
Artem
C. Erimo
Asahigawa
2290
Hokkaido
SAPPORO
Otaru
Kushiro
LIA
Dzamin Üüd
Erenhot
Abagnar Qi
Linxi
1949
Horqin Youyi Qianqi (Ulan Hot)
Baicheng
Shuangcheng
Fuyu
CHANGCHUN
Shuangliao
Jilin
Mudanjiang
Vladivostok
Nakhodka
Partizansk
Hakodate
Muroran
b   Si
MONGGOL
Saynshand
ZIZHIQU DA Hinggan Ling
Tao'an
Tongliao
Siping
Liaoyuan
Yanji
Hunchun
Chongjin
Tsugaru-kaikyo
Aomori
Hachinohe
Morioka
Hohhot
Jining
Zhangjiakou
Chengde
Jinzhou
Fuxin
Liaoyang
FUSHUN
SHENYANG
Benxi
Paektu 2744
NORTH
Akita
Sakata
Ishinomaki
Baotou
Datong
Xuanhua
Qinhuangdao
Chaoyang
Chifeng
Duolun
ANSHAN
Dandong
Yalu Jiang
Hungnam
Z
Sado
Niigata
Sendai
Koriyama
BEIJING
(Peking)
Baoding
2894
Tangshan
Liaodong Wan
Wŏnsan
JAPAN
Wajima
Utsunomiya
Us
mo
HEBEI
TIANJIN
TIANJIN SHI
Cangzhou
Korea Bay
P'YŎNGYANG
DALIAN
SEA OF
Toyama
Kanazawa
TOKYO
YOKOHAMA
Kawasaki
TAIYUAN
Shijiazhuang
Yangquan
Bo Hai
Yantai
Weihai
Kaesŏng
SŎUL
Inch'ŏn
SOUTH
Fuji 3776
Shizuoka
Yokosuka
GREAT WALL
Yuanping
Dezhou
Huang He
Ye Xian
Weifang
Taejŏn
TAEGU
NAGOYA
KYOTO
Hamamatsu
Fenyang
Yuci
JINAN
Handan
Tai'an
Tuyide Shan
QINGDAO
YELLOW
Masan
PUSAN
Okayama
KOBE
OSAKA
Sakai
Wakayama
an'an
Changzhi
Linfen
Anyang
Xinxiang
Jining
Zaozhuang
Lianyungang
SEA
Kwangju
1915
Hiroshima
Kure
Shikoku
Kochi
Matsuyama
Tongchuan
Luoyang
Kaifeng
Shangqiu
Huaibei
Qingjiang
Yancheng
Cheju Do
Shimonoseki
KITAKYUSHU
FUKUOKA
H
P
A
XI'AN
ZHENGZHOU
HENAN
Pingdingshan
Zhumadian
Hongze Hu
Qingjiang
1950
Sasebo
Kumamoto
anyang
Sanmenxia
Nanyang
N
Shangshui
Fuyang
Huainan
Bengbu
Yangzhou
Taizhou
Nagasaki
Kyushu
J
Han Shui
Xiangfan
ANHUI
Ma'anshan
Changzhou
Kagoshima
Ankang
Zhumadian
Huai He
NANJING
Wuxi
Suzhou
SHANGHAI
SHANGHAI SHI
Tanega-shima
aba Shan
engjie
Zhongxiang
Dable Shan
Hefei
Wuhu
Wuxing
Jiaxing
WUHAN
Anqing
Tongling
Huangshi
Hangzhou
Hangzhou Wan
Ningbo
Wanxian
Yichang
Shashi
Jiujiang
Tunxi
Shaoxing
EAST CHINA
Amami-ō-Shima
ING
Yueyang
Dongting Hu
Poyang Hu
Jingdezhen
Jinhua
Linhai
SEA
Changde
Yiyang
Nanchang
Shangrao
Qu Xian
Wenzhou
HUNAN
Changsha
JIANGXI
Jishan
2120
683
Xiangtan
Pingxiang
Fuzhou
Wuyi Shan
PACIFIC
Shaoyang
Ji'an
Nanping
Naha
Okinawa
Hongjiang
Hengyang
Sanming
RYŪKYŪ-rettō
Guilin
Ganzhou
Ruijin
FUJIAN
Longyan
Fuzhou
Quanzhou
Jilong
TAIBEI
Taizhong
Sakashima Gunto
Tropic of Cancer
NGXI
iuzhou
Xing'an
Nan Ling
Mei Xian
Shaoguan
Zhangzhou
Xiamen
Jiayi
Yu Shan 3997
Tainan
TAIWAN
GZU
Wuzhou
GUANGDONG
GUANGZHOU
Chao'an
Shantou
Pingdong
Gaoxiong
Formosa Strait
i Jiang
Zhaoqing
Huizhou
OCEAN
ethai
Jiangmen
Macau (Port.)
HONG KONG (Br.)
Batan Is.
Maoming
Xi Jiang
Foshan
eizhou
Bandao
Qiongzhou Haixia
Zhanjiang
Pratas
SOUTH CHINA
Hainan Dao
1879
HAINAN
Haikou
Yacheng
SEA
Babuyan Is.

110   120   130

6   7   8

E

COPYRIGHT. GEORGE PHILIP & SON LTD.

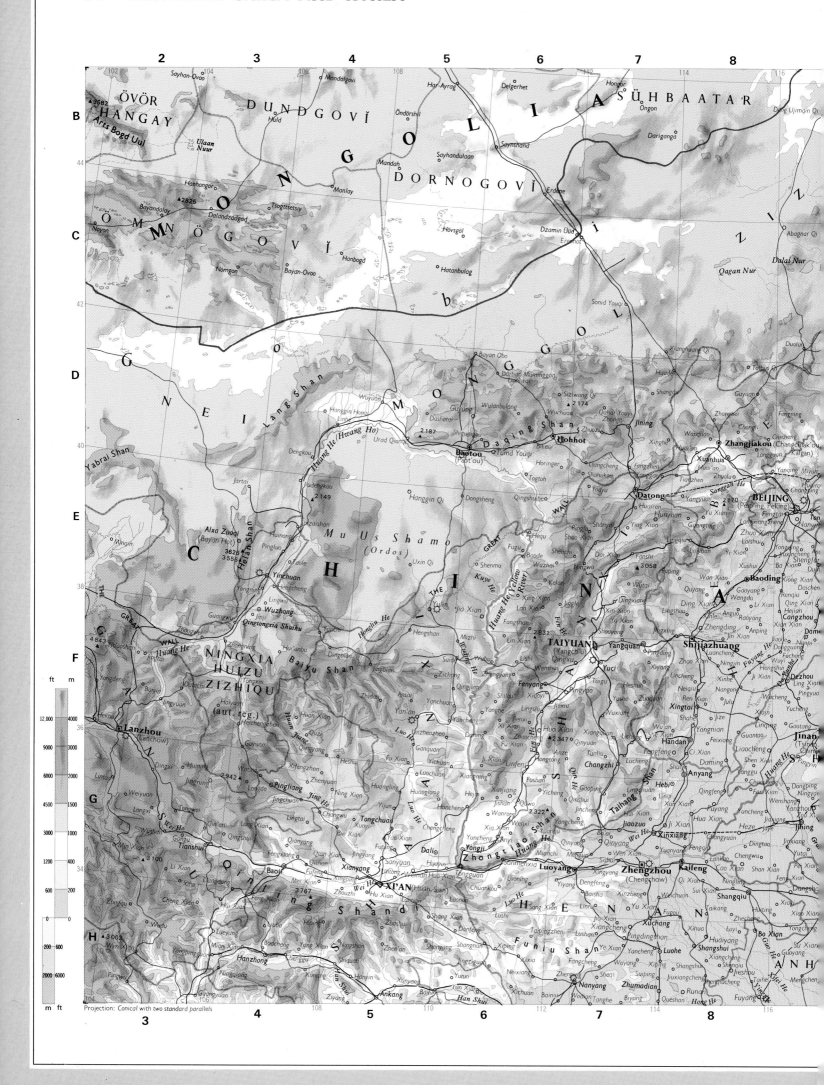

Projection: Conical with two standard parallels

1 : 6 000 000

50   0   50   100   150 miles
50   0   50   100   150   200 km

East from Greenwich

COPYRIGHT: GEORGE PHILIP & SON LTD.

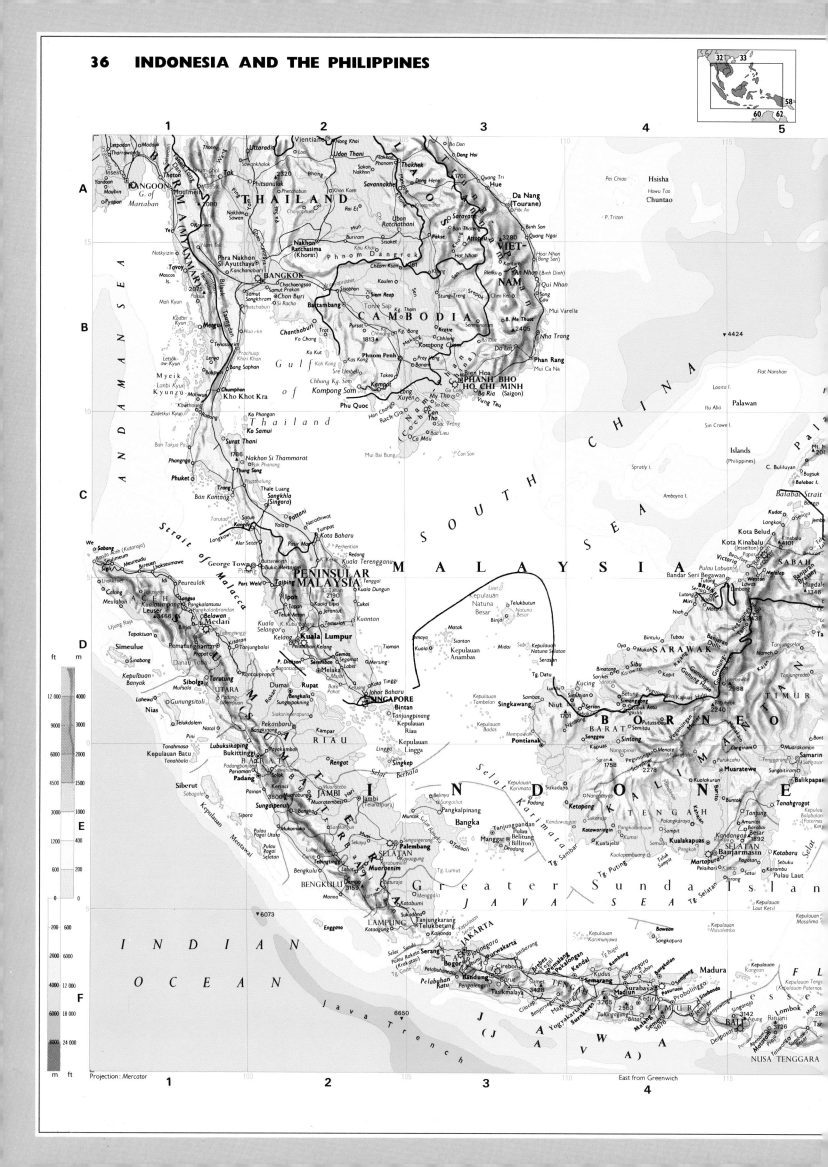

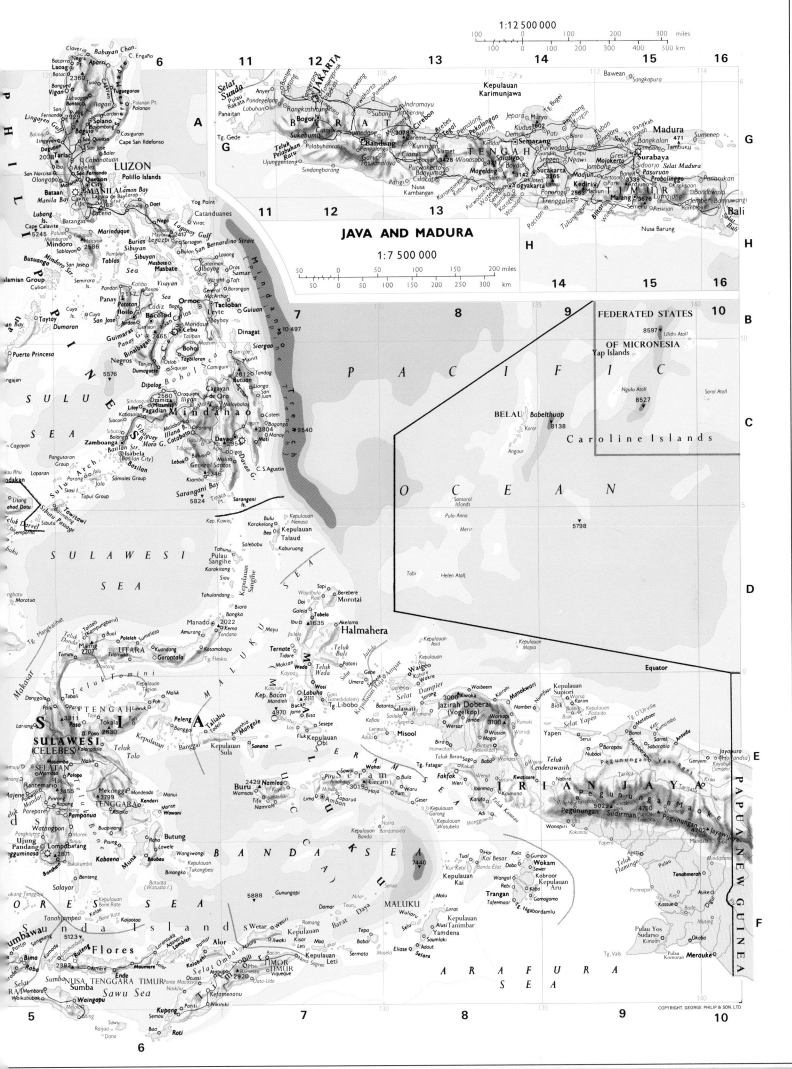

1:6 000 000

50 50 100 150 miles
50 0 50 100 150 200 km

MALAYSIA

SOUTH
CHINA

SEA

Gulf

of

Thailand

PENINSULAR
MALAYSIA

Strait of Malacca

BORNEO

Nha Trang
Phan Rang
Phan Thiet
Cao Nguyen 2287
Cu Lao Hon
Catwick Islands

Kompong Cham
PHNOM PENH
Phnom Penh
Kompong Som
Chuor Phnum Damrei
Phnum Kravanh
1813
1712

PHANH BHO
HO CHI MINH
(Saigon) Gia Dinh
Bien Hoa
My Tho
Vinh Long
Can Tho
Soc Trang
Mekong River Delta
Chau Phu Plain of Reeds
Rach Gia
Mekong
Quan Long (Ca Mau)
Mui Bai Bung
Con Son Islands
Hon Khoai

Koh Kong
Ko Kut
Ko Chang
Laem Ngop

Kepulauan Natuna
Telukbutun
Kepulauan Natuna Besar
Subi
Serasan
Panjang
Seraja
Binjai
P. Laut
P. Midai
Tanjong Datu
Kucing
SARAWAK

Kepulauan Anambas
Mubur
Matak
Jemaja
P. Santan
P. Arrabu
Penghibu
Kaju-ara

East from Greenwich

Kuala Terengganu
Kuala Dungun
Cukai
Kuantan
Pekan
P. Tioman
P. Babi Besar
P. Tinggi
P. Aur
P. Pemanggil

George Town
Butterworth
P. Pinang
Taiping
Cameron Highlands
2182
2130
Ipoh
Teluk Anson
Kuala Lumpur
Kelang
Seremban
Melaka
Bandar Penggaram
Bandar Maharani
Johor Baharu
SINGAPORE

Nakhon Si Thammarat
Hat Yai
Songkhla (Pattani)
Alor Setar
Pak Phanang
Surat Thani
Phuket
Ko Phuket
Ko Lanta Yai
Ko Tarutao
Langkawi

Kho Khot Kra
(Isthmus of Kra)
Prachuap Khiri Khan
Chumphon
1247

Thailand

Tebingtinggi
Rantauprapat
Pematangsiantar
Tanjungbalai
2009
Medan
Binjai
Langsa
Sibolga
SUMATERA

INDONESIA

Projection: Conical with two standard parallels

COPYRIGHT GEORGE PHILIP & SON LTD.

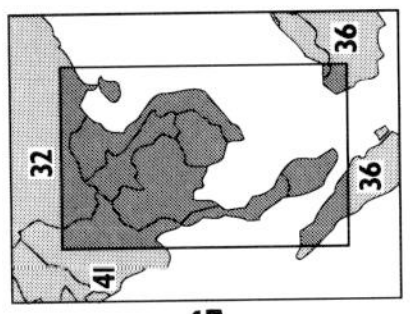

m ft
9000
6000
4500
3000
1500
600
200
0
m ft

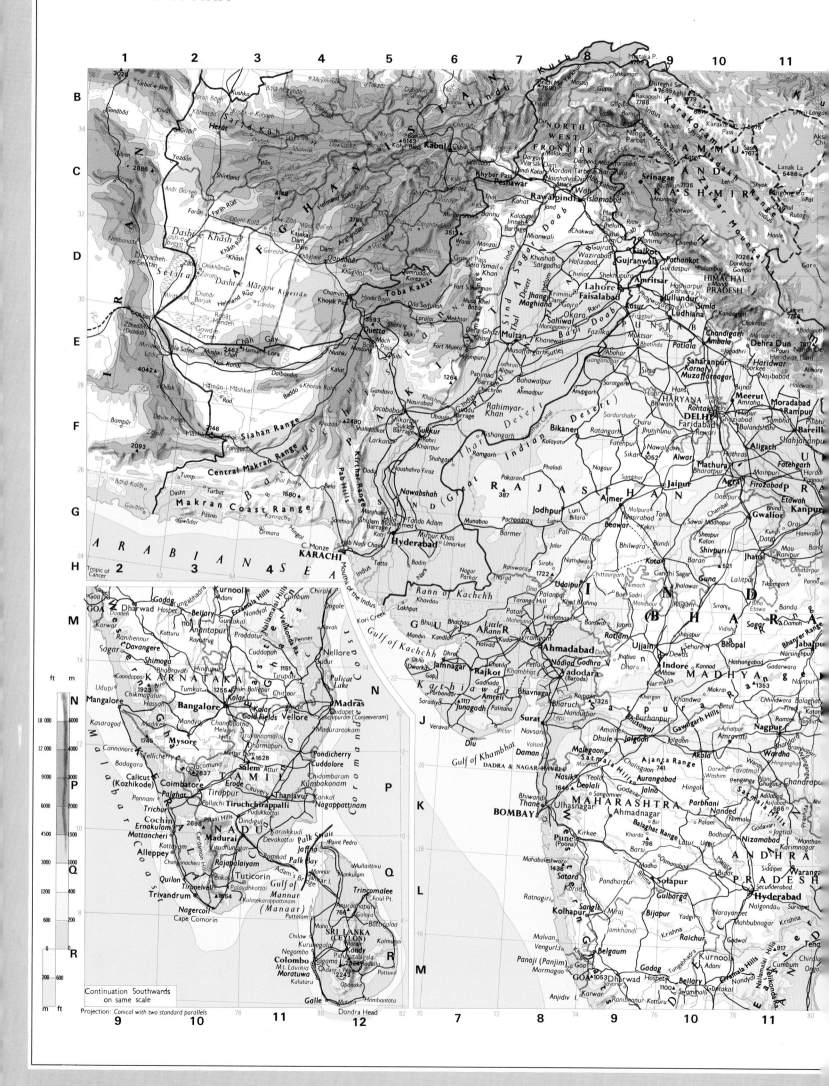

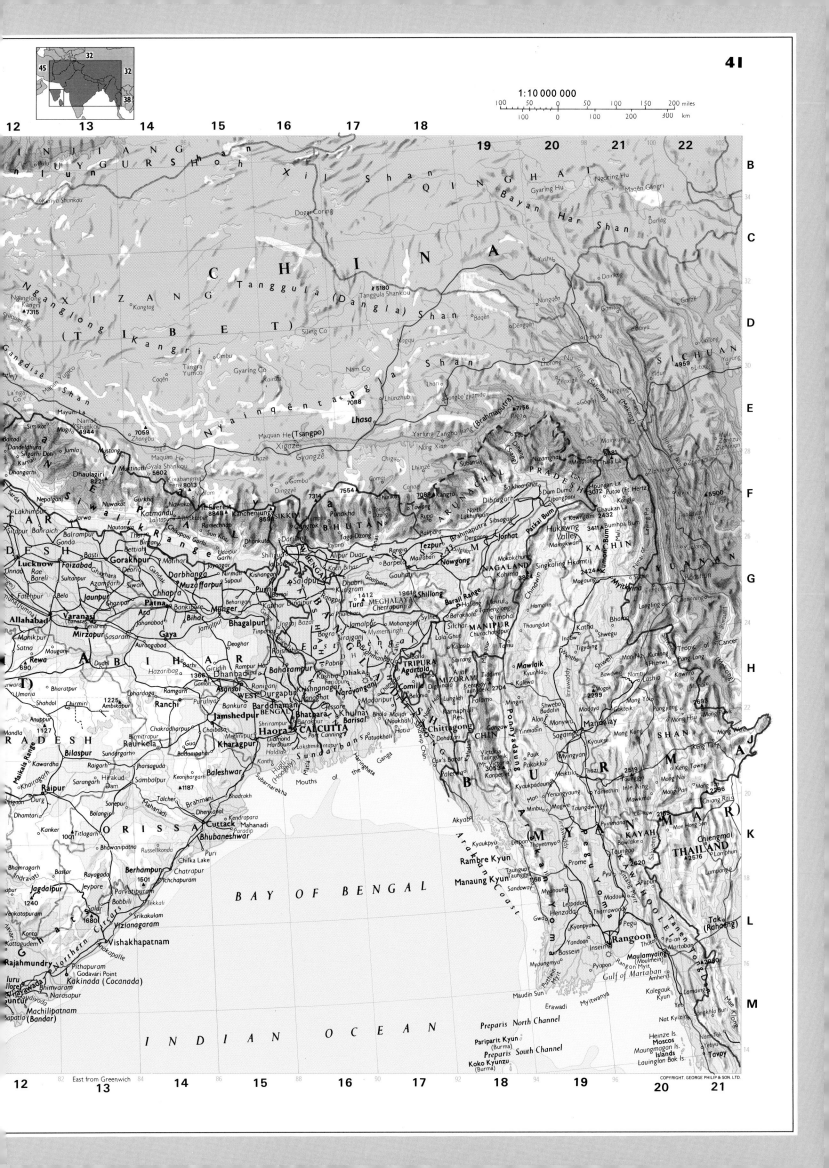

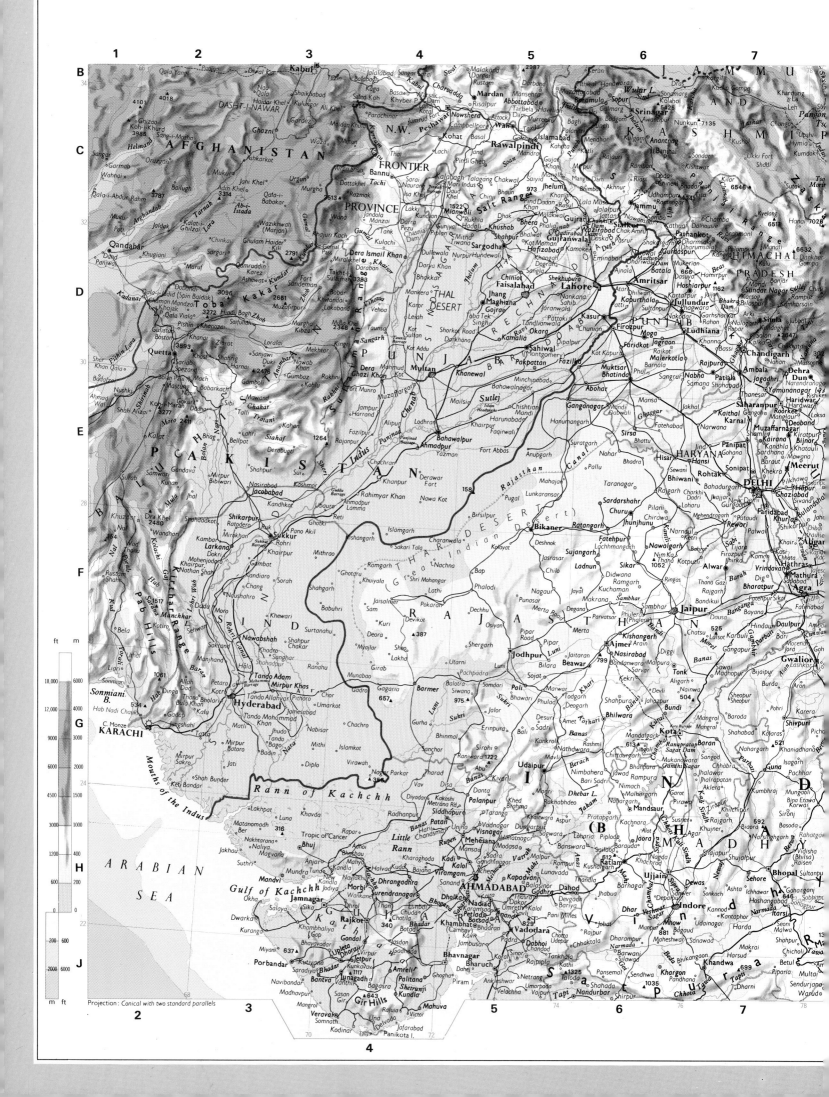

1:6 000 000

50   0   50   100 miles
50   0   50   100   150 km

JAMMU AND KASHMIR
On same scale as Main Map

East from Greenwich

COPYRIGHT. GEORGE PHILIP & SON. LTD.

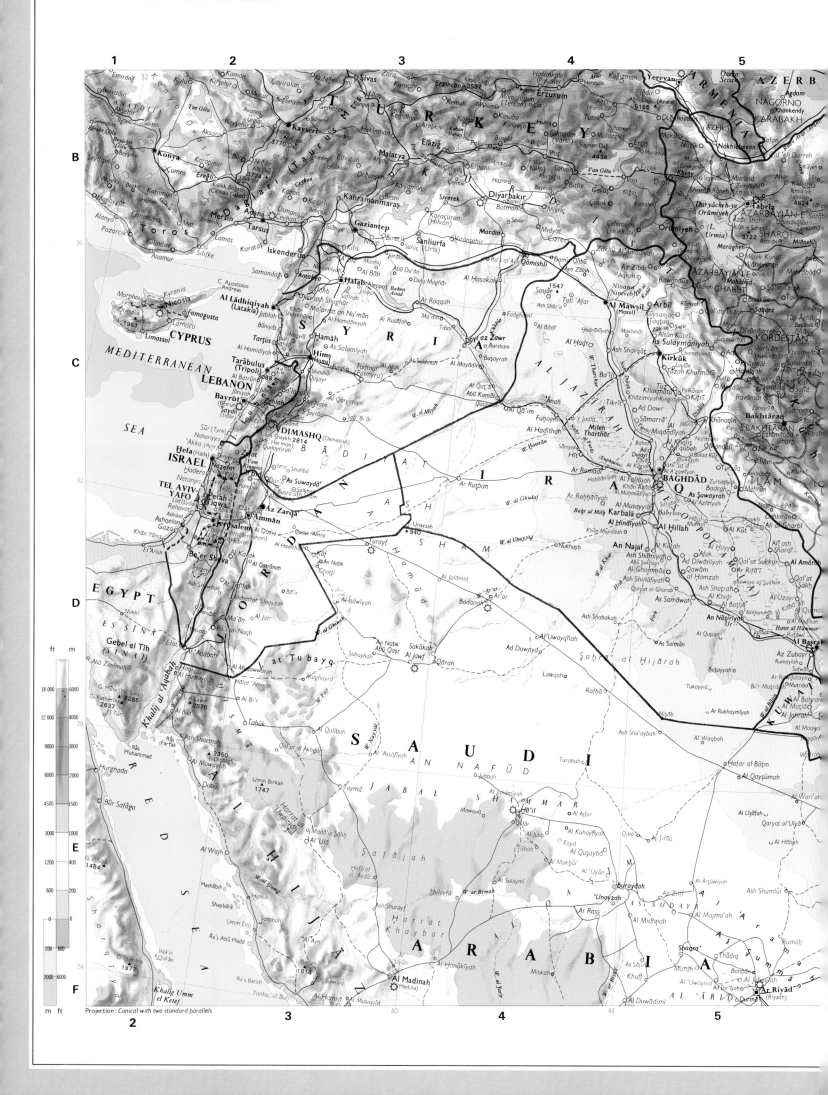

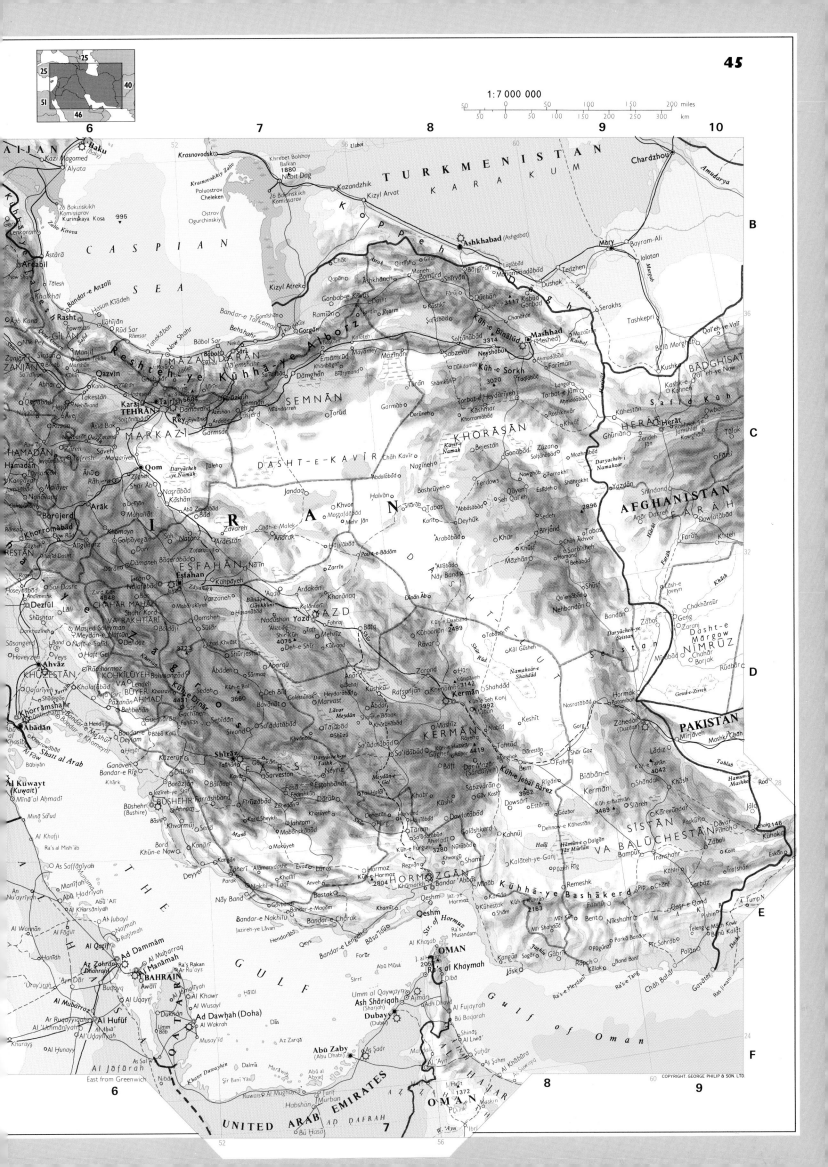

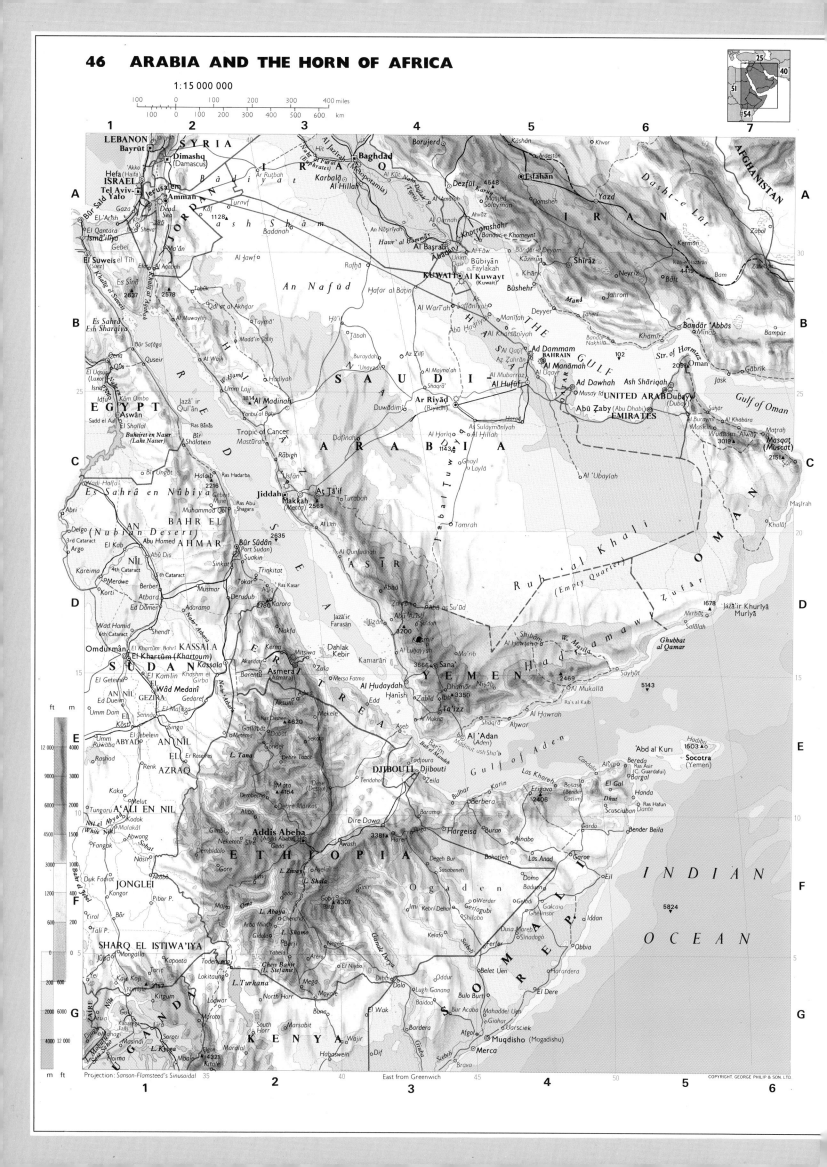

1:15 000 000

100   0   100   200   300   400 miles
100   0   100   200   300   400   500   600 km

1   2   3   4   5   6   7

LEBANON
Bayrūt
Dimashq
(Damascus)
Hefa (Haifa)
ISRAEL
Tel Aviv–Yafo
Jerusalem
Amman
SYRIA
Baghdad
IRAQ
Karbalā'
Al Hillah
Borujerd
Kāshān
Khvor
Ardestān
AFGHANISTAN

EGYPT
El Suweis (Suez)
An Nafūd
KUWAIT
Al Kuwayt (Kuwait)
IRAN
Eşfahān
Yazd
Kermān
Dasht-e Lut
Zābol

SAUDI
Al Madinah
Ar Riyad (Riyadh)
THE GULF
Ad Dammam
BAHRAIN
Al Manāmah
Ad Dawhah
Ash Shāriqah
UNITED ARAB
EMIRATES
Abū Zaby (Abu Dhabi)
Dubay (Dubai)
Str. of Hormuz
Gulf of Oman
Bandar 'Abbās

ARABIA
Makkah (Mecca)
Jiddah
At Ţā'if
Tropic of Cancer
Rub' al Khali
(Empty Quarter)
OMAN
Maşqaţ (Muscat)

RED SEA
'ASIR
Bür Sūdān
(Port Sudan)

SUDAN
Omdurmān
El Khartūm (Khartoum)
KASSALA
Asmera
ERITREA
YEMEN
Şan'ā'
Ta'izz
Al Hudaydah
Al Mukallā
Socotra
(Yemen)

Addis Abeha
(Addis Ababa)
ETHIOPIA
DJIBOUTI
Djibouti
Gulf of Aden
Al 'Adan (Aden)
Hargeisa
Berbera

SHARQ EL ISTIWA'IYA
INDIAN
OCEAN

SOMALI REP.
KENYA
L. Turkana
Muqdisho (Mogadishu)
Merca

Projection: Sanson-Flamsteed's Sinusoidal   East from Greenwich   COPYRIGHT: GEORGE PHILIP & SON, LTD.

1   2   3   4   5   6

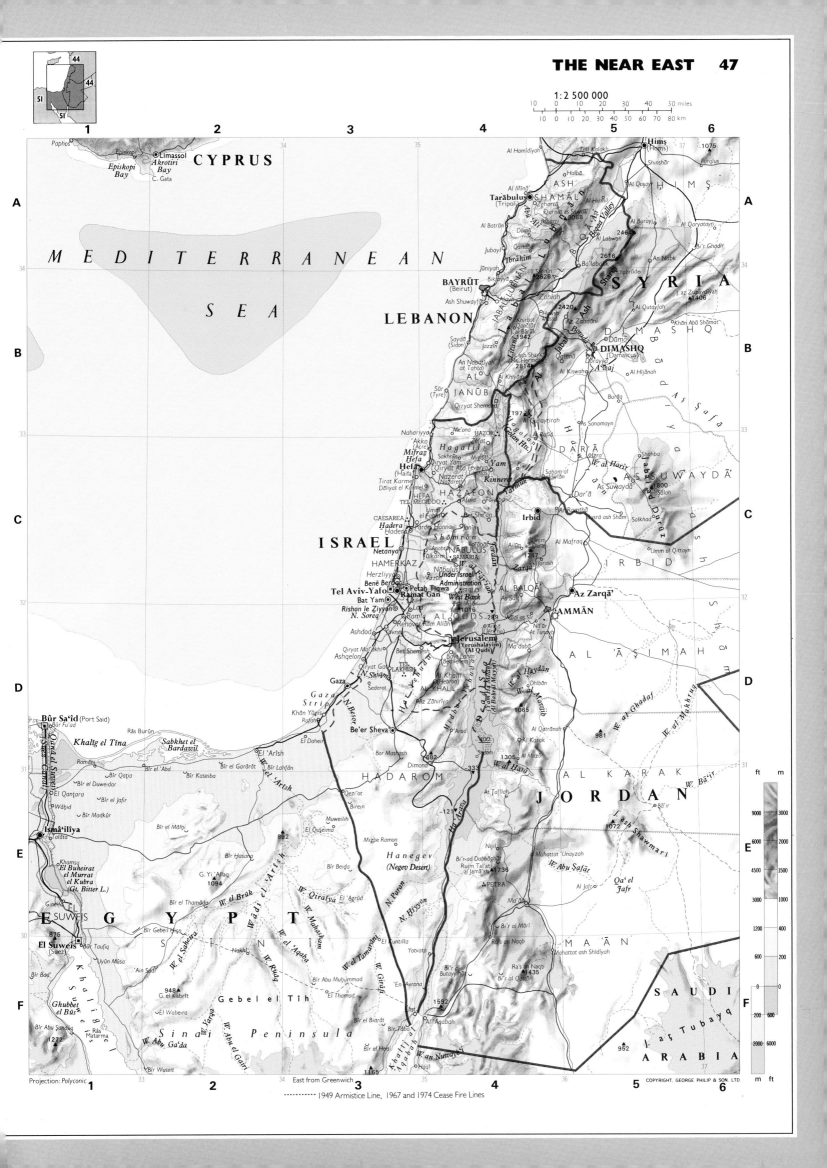

1 : 2 500 000

10   0   10   20   30   40   50 miles
10   0   10   20   30   40   50   60   70   80 km

CYPRUS

Paphos
Episkopi
Limassol
Akrotiri Bay
Episkopi Bay
C. Gata

M E D I T E R R A N E A N

S E A

Al Hamidiyah
Tall Kalakh
Ḥimṣ
(Homs)
1075
Farqlus
Halba
Al Qusayr
ASH
Azghartā
SHAMAL
Al Mīnā'
Ṭarābulus
(Tripoli)
Qurnat as Sawdā 3088
Al Ḥirmil
Ibrāhīm
Al Baṭrūn
Abu 'Alī
Dannā
Ba'labakk
2466
Al Labwah
Al Qaryatayn
Di'r Ghadīr

Jubayl
Jūniyah
2616
An Nabk
Ibrāhīm
2628
Sānīn
Bikfayyā
BAYRŪT
(Beirut)
Zaḥlah
Yubrūd
SYRIA
J. az Zubaydīyah
1406
Ash Shuwayfāt
2420
Az Zabdānī
Al Qutayfah
LEBANON
Jabal Barādā
Khān Abū Shāmāt
Saydā
(Sidon)
Chirbat
Qaṭanā
DIMASHQ
(Damascus)
Jazzīn
Ash Shaykh
2814
Dārayyā
A'ūaj
An Nabatīyah
at Taḥtā
Al Khiyām
Al Kiswah
Al Ḥijānah
AL
1974
Qunaytirah
Ayr al Ashmar
As Sanamayn
JANŪB
Ṣūr
(Tyre)
Naharīyya
HAZOR
Rāfiq
DAR'Ā
Shahba
Qiryat Shemona
'Akko
(Acre)
Me'ona
Golan Hts.
Zefat
'AS SUWAYDĀ'
Mifraẓ
Hefa
Hagalil
Sakhnīn
Migdal
Ẓra'
W. al Harīr
1800
Ḥefa
(Haifa)
Qiryat Yam
Yam
Dar'ā
As Suwaydā
Sālah
Tirat Karmel
Qiryat Ata
Teverya (Tiberias)
Sahm al
Dūruz
Daliyat el Karmel
Nazerat
Nazareth
Kinneret
Golan
HAZAFON
'Afula
Yarmūk
Busrā ash Shām
Salkhad
HEFA
Umm
'Afula
Bet She'an
Ailūn
TEL MEGIDDO
el Fahm
Irbid
Al Mafraq
IRBID
CAESAREA
Jenin
Umm al Qiṭṭayn
Hadera
Pardes Hanna
Shōmrōn
Ajlūn
Umm al Daraj
Ḥadera
2147
Jarash
ISRAEL
Anabta
NĀBULUS
Netanya
Ṭulkarm
SAMARIA
Zarqā'
HAMERKAZ
Nāpulus
Az Zil
Herzliyya
'Azzūn
Under Israeli
Administration
Benē Beraq
Peṭah Tiqwa
SHILO
Al BaLQā'
Tel Aviv-Yafo
Ramat Gan
1016
'AMMĀN
Bat Yam
Rishon le Ziyyon
West Bank
As Salt
Az Zarqā'
N. Soreq
Lod
Ar'
AL QUDS
Wadi as Sir
Na'ur
Rehovot
Ram Allāh
Ariḥa
At Tunayb
Ashdod
Yavne
Jericho
Qiryat Mal'akhi
Beth Jala
Ma'dabā
Ashqelon
Jerusalem
(Yerushalayim)
(Al Quds)
Bethlehem
AL 'ĀSIMAH
Bet Shemesh
Bayt Jāla
Qiryat Gat
TEL
W. al Ḥaydān
N. Shiqma
LAKHISH
Al Khalīl
Gaza
Sederot
(Hebron)
Dhībān
Az Ẓāhirīya
1065
Gaza Strip
N. Besor
Gaza Strip
Khān Yūnis
Yatta
Nahal Mē'arot
(Al Baḥr al Mayyit)
Rafaḥ
Arad
Al Qaṭrānah
El Daheir
Be'er Sheva
981
Bor Mashash
W. al Ḥasā
Dimona
682
Rās Burūn
Bûr Sa'îd (Port Said)
Bûr Fu'ad
Khalîg el Tîna
Sabkhet el Bardawîl
El 'Arîsh
333
Al Qatrānah
AL KARAK
Români
Bîr el 'Abd
Bîr Lahfân
Bîr Qaṭia
Bîr el Garârât
W. el 'Arîsh
W. Bâ'ir
El Qantara
Bîr Kaseiba
HADAROM
At Ṭafīlah
JORDAN
Bîr el Duweidar
Bîr el Jafir
Bîr Madkûr
Wâḥid
121
Bâ'ir
Ismâ'îlîya
Talâta
Qezi'ot
Nijil
1072
ash Shawmari
Birein
Muweilih
El Quseima
Mizpe Ramon
Khamsa
El Buheirat el Murrat el Kubra
(Gt. Bitter L.)
Bîr Hasana
Hanegev
(Negev Desert)
Bi'r ad Dabbāghiyah
Ruim Tal'at
al Jamā'in
4173
W. Abu Ṣafār
Qa' el Jafr
Gineifa
G. Yi 'Allaq
1094
Bîr Beiḍa
Nijil
EL SUWEIS
Bîr el Thamâda
W. el Brûk
W. Qiratya
El 'Agrûd
N. Paran
PETRA
Ma'ān
Al Jafr
E G Y P T
875
Bîr Gebeil Ḥisn
N. Hiyyon
Rās an Naqb
MA'ĀN
El Suweis
(Suez)
Bûr Taufîq
Bîr el Thamâda
Nakhl
El Thamad
Ra's an Naqb
1435
'Uyûn Mûsa
'Ain Sudr
W. el Sheira
W. el 'Aqaba
Yotvata
Mahattat ash Shīdīyah
Bîr Bad
Khamsa
W. Riaq
Bîr Abu Muhammad
SAUDI
G. el Kabrît
Gebel el Tîh
Bîr el Biârât
1592
Ghubbet el Bâs
948
'En Avrona
ARABIA
1272
G. el Kabrît
El Wabeira
Al 'Aqabah
952
Sinai Peninsula
Bîr el Biârât
Bîr Ṭâba
W. Abu el Gaîr
W. Yarga
W. Abu el Gairi
J. aṣ Ṭubayq
Bîr Wuselt
Bîr Abu Sandîq
W. an Nuweybi'
1165
Ṭaba
Naqb

ft   m
9000   3000
6000   2000
4500   1500
3000   1000
1200   400
600   200
0
200   600
2000   6000
m   ft

Projection: Polyconic

East from Greenwich

COPYRIGHT. GEORGE PHILIP & SON. LTD.

- - - - - - - - - 1949 Armistice Line, 1967 and 1974 Cease Fire Lines

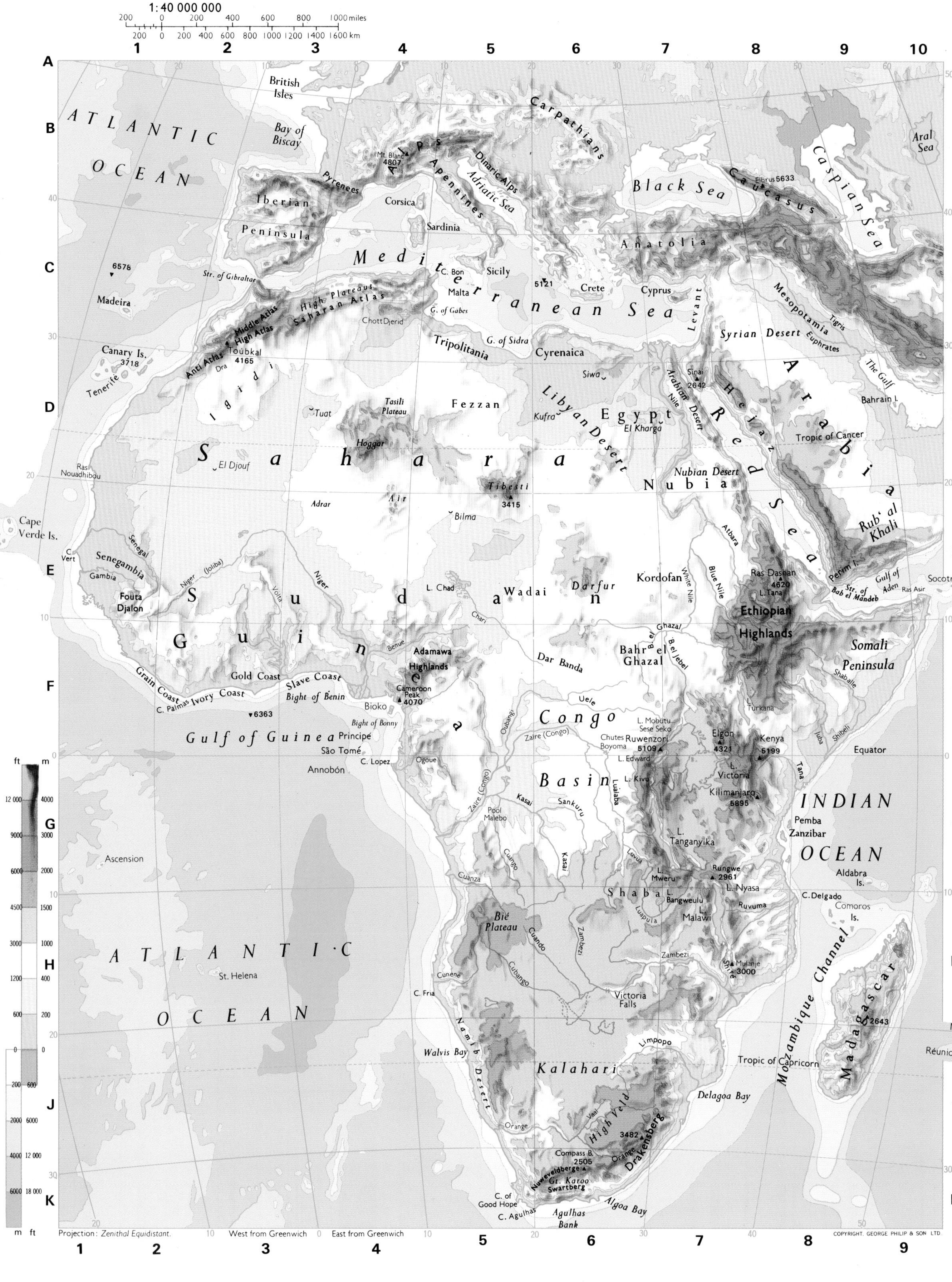

1 : 40 000 000

200 0 200 400 600 800 1000 miles
200 0 200 400 600 1000 1200 1400 1600 km

1   2   3   4   5   6   7   8   9   10

A
ATLANTIC

B
OCEAN
British Isles
Bay of Biscay
Carpathians
Alps
Mt. Blanc 4807
Dinaric Alps
Apennines
Adriatic Sea
Black Sea
Caucasus
Elbrus 5633
Caspian Sea
Aral Sea

C
Iberian
Pyrenees
Corsica
Sardinia
Mediterranean
C. Bon
Sicily
Malta
5121
Crete
Cyprus
Anatolia
Levant
Mesopotamia
Tigris
Euphrates
Str. of
The Gulf
6578
Peninsula
Str. of Gibraltar
High Plateau
Saharan Atlas
G. of Gabes
Sea
Syrian Desert
Madeira
Middle Atlas
High Atlas
Chott Djerid
G. of Sidra
Cyrenaica
Tripolitania
Bahrain I.

D
Canary Is.
3718
Anti Atlas
Toubkal 4165
Dra
Tuat
Tasili Plateau
Fezzan
Libyan Desert
Egypt
Sinai 2642
Arabian Desert
Nile
Red Sea
Hejaz
Arabia
Tropic of Cancer
Tenerife
I g i d i
S a h a r a
Hoggar
Kufra
El Kharga
Siwa
Rub' al Khali

E
Ras Nouadhibou
El Djouf
Adrar
Aïr
Tibesti 3415
Nubian Desert
Nubia
Attara
Ras Dashan 4620
L. Tana
Perim I.
Gulf of Aden
Ras Asir
Socotra
Cape Verde Is.
C. Vert
Senegal
Niger (Joliba)
Volta
Niger
L. Chad
Bilma
Wadai
Darfur
Kordofan
White Nile
Blue Nile
Str. of Bab el Mandeb
Senegambia
Gambia
Fouta Djalon
S u d a n
Chari
Benue
Dar Banda
Bahr el Ghazal
Ghazal
Ethiopian Highlands
Somali Peninsula

F
G u i n e a
Gold Coast
Grain Coast
C. Palmas
Ivory Coast
Slave Coast
Bight of Benin
Adamawa Highlands
Cameroon Peak 4070
Bioko
Uele
Oubangi
Uele
Congo
Bahr el Jebel
Shabelle
6363
Bight of Bonny
Gulf of Guinea
Príncipe
São Tomé
Zaire (Congo)
Ogoue
C. Lopez
Turkana
Juba

G
Annobón
Ascension
C. Lopez
Ogoue
Zaire (Congo)
Kasai
Pool Malebo
Basin
L. Mobutu Sese Seko
Chutes Boyoma
Ruwenzori 5109
L. Edward
L. Kivu
Lualaba
Elgon 4321
Kenya 5199
L. Victoria
Kilimanjaro 5895
INDIAN
Equator
St. Helena
Kasai
Sankuru
Pemba
Zanzibar
OCEAN

H
ATLANTIC
Cuanza
Cuango
Kasai
L. Tanganyika
Luvua
Aldabra Is.
C. Delgado
Comoros Is.
Bié Plateau
Mweru
Rungwe 2961
L. Nyasa
Ruvuma
Shaba
L. Bangweulu
Luapula
Malawi
Cunene
Cubango
Zambezi
Shire
Mulanje 3000
Mozambique Channel
Madagascar
2643

J
OCEAN
C. Fria
Cuando
Zambezi
Victoria Falls
Limpopo
Walvis Bay
Namib Desert
Kalahari
Victoria
Delagoa Bay
Tropic of Capricorn
Mauri
Réunion

K
Orange
High Veld
3482
Drakensberg
Delagoa Bay
Algoa Bay
C. of Good Hope
Compass B. 2505
Nuweveldberge
Gr. Karoo
Swartberg
Orange
C. Agulhas
Agulhas Bank

ft m
12 000 4000
9000 3000
6000 2000
4500 1500
3000 1000
1200 400
600 200
0 0
200 600
2000 6000
4000 12 000
6000 18 000
m ft

Projection: Zenithal Equidistant.   West from Greenwich 0 East from Greenwich

1 2 3 4 5 6 7 8 9

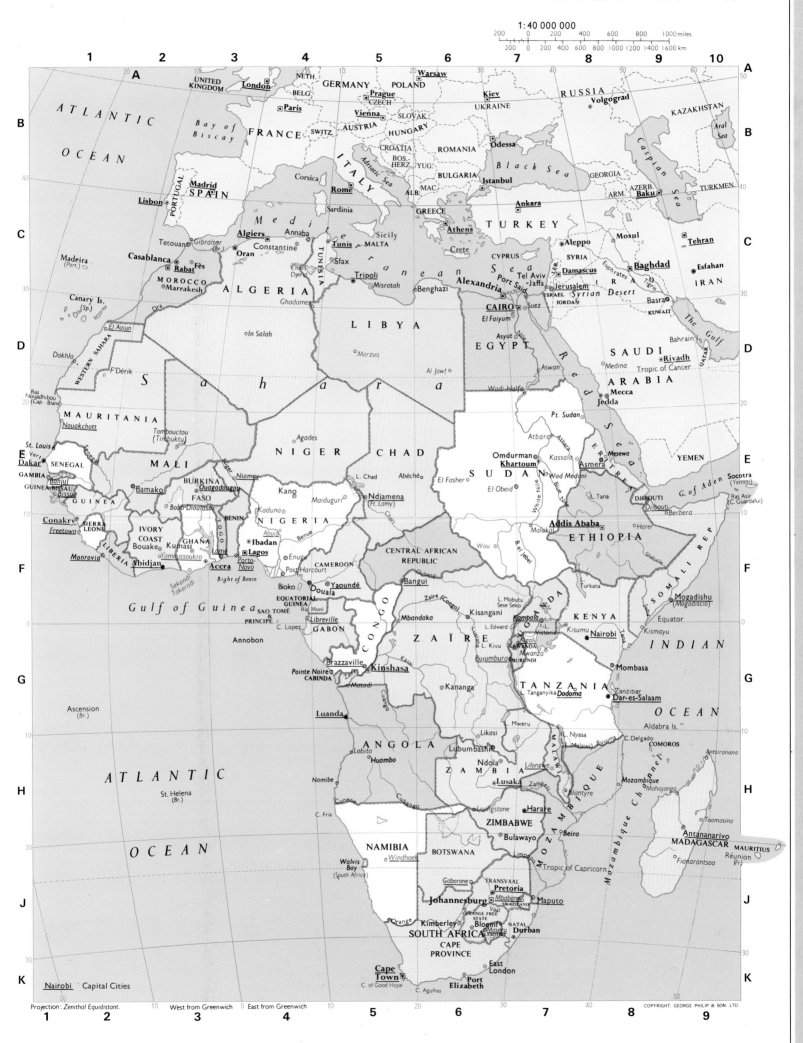

1:40 000 000

200 0 200 400 600 800 1000 miles
200 0 200 400 600 800 1000 1200 1400 1600 km

A 1   2   3   4   5   6   7   8   9   10 **A**

Projection : Zenithal Equidistant.   West from Greenwich   East from Greenwich   COPYRIGHT GEORGE PHILIP & SON. LTD.

<u>Nairobi</u>   Capital Cities

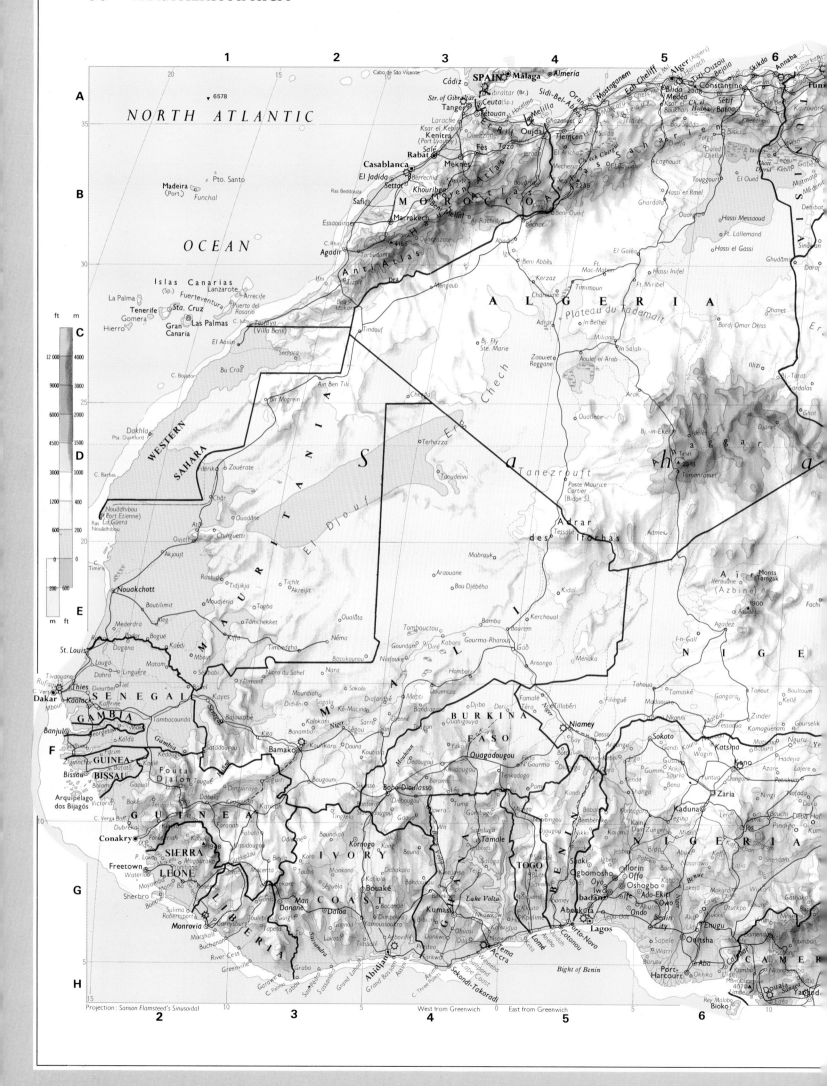

NORTH ATLANTIC

OCEAN

SPAIN

NORTH ATLANTIC
▾ 6578

Cabo de São Vicente

Cádiz · Málaga · Almería
Gibraltar (Br.) · Sidi-Bel-Abbès
Ceuta (Sp.) · Melilla · Oran · Mostaganem · Ech Cheliff
Str. of Gibraltar · Al Hoceima · Ghazaouet · Tlemcen · Blida · Médéa
Tánger · Tétouan · Oujda · Saïda · Bou Saâda
Larache · Ksar el Kebir · Taza · El Aricha · Laghouat
Kénitra (Port Lyautey) · Fès · Mecheria · Chott Ech Cherq · Djelfa
Salé · Meknès · Jerada · El Arícha
Rabat · · Touggourt · El Oued
Casablanca
El Jadida · Berrechid · MOROCCO
Settat · Khouribga · Beni Ounif · Ghardaïa · Hassi er Rmel
Safí · Ras Beddouza · Ar Rachidya · Hassi Messaoud
MOROCCO · Béchar · Ft. Lallemand
Marrakech · Beni Mellal · Hassi el Gassi
Essaouira · 4165 · Ouarzazate · El Goléa · Ghudâmi
Agadir · Taroudannt · Anti Atlas · Abadla · Igli · Ft. Mac-Mahon
Ifni · Tiznit · Mengoub · Beni Abbès · Kerzaz · Hassi Inifel
Dra · Bou Izakarn · ALGERIA · Ft. Miribel · Daraj
C. Rhir · Timimoun · Plateau du Tademaït
Islas Canarias (Sp.) · Lanzarote · Charouine · Ohanet
La Palma · Fuerteventura · Arrecife · Bj. Fly · Adrar · In Belbel
Tenerife · Sta. Cruz · Puerto del Rosario · Sté. Marie · Miliana · In Salah · Bordj Omar Driss
Gomera · Gran Canaria · Las Palmas · Zaouiet Reggane · Aoulef el Arab · Illizi
Hierro · C. Juby Tarfaya (Villa Bens) · Aoulef · Arak · Bj.-in-Eker · Bj.-Tarat
El Aaiún · Ain Ben Tili · Chech · Ouallene · Idelès · ardalas
Semara · Chesch · Tanezrouft · Djanet · Ghat
C. Bojador · Bu Craa · Bir Mogrein · Hoggar · 2918 · Tamanrasset
WESTERN SAHARA · Tessalit · Aïr
Dakhla · Pta. Durnfort · Frederik · Zouérate · Terhazza · Adrar des Iforhas · Admer
C. Barbas · MAURITANIA · El Djouf · Taoudenni · Poste Maurice Cartier (Bidon 5) · Monts Tamgak (Azbine)
Nouâdhibou (Port Étienne) · Chār · Ouadâne · Mabrouk · Iférouâne · 1900
Ras La Guéra · Atâr · Chinguetti · Araouane · Bou Djébéha · In-Gall · Agadez · Fachi
Timiris · Akjoujt · Rachid · Tidjikja · Akreijit · Kidal · Kerchoual
Nouakchott · Tichît · MALI · Bamba · Ménaka
Boutilimit · Moudjéria · Togba · Tâmchekket · Tombouctou · Bourem · Ansongo · NIGER
Mederdra · Aleg · Kiffa · Oualâta · Goundam · Diré · Gao · Tahoua · Tanout
St. Louis · Podor · Bogué · Kaédi · Néma · Bassikounou · Niafouké · Hombori · Tessaoua
Dagana · Magta Lahjar · Sélibabi · Nioro du Sahel · Nara · Douentza · Tillabéri · Filingué · Madaoua · Zinder · Gouré
Louga · Matam · Yélimane · Mopti · Djibo · Dori · Téra · Say · Niamey · Maradi · Katsina · Nguru
Tivaouane · Dahra · Linguère · Kayes · Didiéni · Sokolo · Diafarabé · Bandiagara · Kaya · Dosso · Sokoto · Kamoguenam
Rufisque · Thiès · Diourbel · Tiel · Nioro · BURKINA FASO · Birnin-Kebbi · Gusau · Kano
Dakar · Kaolack · SENEGAL · Bakel · Banamba · Ségou · Djenné · Ouahigouya · Gummi · Zaria
Mbour · Koungheul · Tambacounda · Kita · Koulikoro · Bobo-Dioulasso · Ouagadougou · Fada N'Gourma · Kandi · Kaduna
GAMBIA · Bafoulabé · MALI · Sikasso · Koutiala · Dédougou · Tenkodogo · NIGERIA
Banjul · Georgetown · Kolda · Satadougou · Bamako · Bougouni · Boromo · Nikki · Bida
GUINEA-BISSAU · Labé · Dabola · Kankan · Odienné · Korhogo · Kong · Tamale · Shaki · Ilorin · Jos
Bissau · Gaoual · Tougué · Dinguiraye · Siguiri · Bouna · Salaga · Oyo · Oshogbo · Makurdi
Arquipélago dos Bijagós · Boké · Telimélé · Kouroussa · Séguéla · IVORY · Bouaké · GHANA · TOGO · BENIN · Ogbomosho · Ife · Ado-Ekiti
GUINEA · Kindia · Beyla · Danané · Daloa · Yamoussoukro · Lake Volta · Ibadan · Abeokuta · Benin City · Onitsha
Conakry · Faranah · Man · COAST · Kumasi · Kpalime · Porto-Novo · Lagos · Enugu
SIERRA LEONE · Kenema · Tiassalé · Abengourou · Lomé · Cotonou · Warri
Freetown · Bo · LIBERIA · Sassandra · Abidjan · Accra · Sekondi-Takoradi · Port-Harcourt · Aba · CAMER
Monrovia · Buchanan · Grand Bassam · Cape Coast · Bight of Benin · Mt. Cameroun 4070 · Douala
Greenville · C. Palmas · C. Three Points · Bioko · Yaoundé

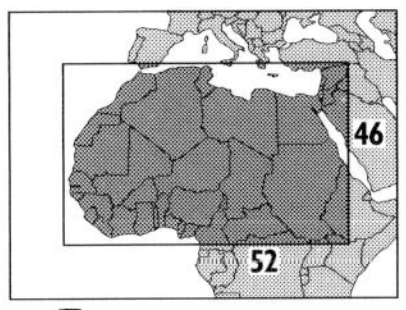

1:15 000 000

100   0   100   200   300   400 miles
100   0   200   300   400   500   600 km

46
52

MEDITERRANEAN SEA

MALTA

Sicily

Pantelleria (It.)
Lampedusa (It.)

Sousse
Monastir
Mahdia
Sfax
Golfe de Gabès
Île de Djerba
Zarzis
Ben Gardane

Tarâbulus (Tripoli)
Al Khums
Mişrâtah
Zlitan
Al Qaşbât
Ghoryân
Bani Walid
968
Mizdah
Jâdu
Nâlût
Zuwârah

Tarâbulus

Banghâzî (Benghazi)
Tûkrah
Barqah
Bananah
878
Suluq
Ajdâbiyah
Marsa Brega
Surt
Ra's Al-Unuf
Al 'Uqaylah
Khalîj Surt

L I B Y A

Hûn
Marâdah
Awjilah
Al 'Irq
Zillah
Al Jaghbûb

Cyrenaica

Sahrâ'

Adrî
Brach
Sabhah
1200
Awbârî   Fezzan
Tasawah   Marzûq   Tmassah
Wâw al Kabîr

Idehan Marzûq

Al Qaţrûn

Tropic of Cancer

Madama
Toumno
Wour
Djado
Chirfa
Bardai
3150
Emi Koussi
3415
Zouar
Gouro

Tibesti

Andye
Bilma

Borkou

Ounianga-Kébir
Ounianga-Serir
Depression du Mourdi

Faya-Largeau
Fada

Djourab

Ennedi

C H A D

Nguigmi
Rig-Rig
Zigey
Bol
Lac Tchad
Mao
Moussoro
Massakory
Massaguet
Massényé
Bokoro

Bahr el Ghazal (Soro)

Biltine
Arada
Kutum
Abéché
Adre
Oum Hadjer
Am Dam
Mongo
Abou-Deia
Melfi

Sarh
Ati
Bitkine
Haraze

D A R F Û R

Iriba
Kutum
El Fasher
Marrah
3088
Nyala

Massakory

Ndjamena (Ft. Lamy)
Kousseri
Dikwa
Marte

Chari

Bongor
Kélo
Léré
Pala

Baïbokoum
Bénoué
Moundou
Doba
Koumra
Goré
Bébédja

Massif de Adamaoua

Ngaoundéré

Garoua
Maroua

Bénoué

CENTRAL AFRICAN REPUBLIC

Bozoum
Bossangoa
Bouca
Batangafo
Kaga Bandoro
Bria
Ippy
Yalinga

Bakala

Fort-Sibut
Grimari
Bambari

Bossembélé
Boali
Bangui

Bimbo
Mbaïki

ZAÏRE (CONGO)

Bosobolo
Bondo
Libenge

EGYPT

El Iskandariya (Alexandria)
El Alamein
Marsa Matrûh
Sîdi Barrânî
Sallûm
Bardia
Tobruk
Ras Al Mîlh

El Qâhira (Cairo)
El Gîza
El Faiyûm
Beni Suef
El Minya
Mallawî
Asyût
Sohâg
Girga
Qinâ
Luxor (El Uqsur)
Qûş
Idfu
Aswân (Aswan High Dam)
El Shallâl

Qâra
Munkhafed el Qattâra (Qattâra Depression)
Siwa

El Bawîţi
Beni Mazâr
Es Sahrâ
Dairût
Manfalût
Akhmîm
Tahta

El Wâhât el-Dâkhla
El Qasr
Mût
El Wâhât el-Khârga
El Khârga
Bâris
Dunqul
Buheîret en Naser (Lake Nasser)

Wadi Halfa
El Wâhât el Selîma
1893
Ayn el 'Uwaynât

Es Sahrâ en Nûbiya
(Nubian Desert)

Laqiya Arba'in
Nukheila
Bir 'Atrun

Delgo
Abri
Kosha
Abu Hamed
Abû Dis
3rd Cataract
Argo
Dongola
Kareima
4th Cataract
Merowe
Korti
Atbara
Ed Debba
Berber
Ed Dâmer
Adarama

ESH SHAMALÎYA

AN NÎL

Haya Junction
Muhammad Qol
Ras Abu Shagara
Sinkat
Tokar
Trinkitat
Ras Kasar
Karora

BAHR EL AHMAR

Bûr Sûdân (Port Sudan)
2635
Suakin
Musmar
Derudeb

Wâd Hamid
Shendî
6th Cataract
Geili
Omdurmân
El Khartûm Bahrî (Khartoum)
El Khartûm
Kamlin
Wâd Medanî

S U D A N

Malha
Homrat esh Sheykh
Sodiri

SHAMÂL KORDOFAN

Umm Keddada
El Obeid
Umm Bel
Ed Dueim
En Nahud
Abû Zabad
El Odaiya
Dilling
Rashad

Kassala
Khashm el Girba
Gedaref
Gallabât
L. Tana
Metema

Kasala
GEZIRA
El Managil
Sennar
Singa
Er Roseires
Kurmuk

ERITREA
Nakfa
Keren
Akordat
Asmera
Massawa
Barentu
Adi Ugri

ETHIOPIA
Addis Abeba (Addis Ababa)
4620
Sekota
Dese
Debre Markos
Nekemte
Gimbi
Gambela
L. Tana
Mekele
Gonder

Dârfûr

JANUB DÂRFÛR
Idd al Ghanam
Rahad al Bardî
Buram
Abu Matariq

JANUB KORDOFAN
El Loqawa
Heiban
Talodi
Tungaru
Kodok

ABYAD AN NÎL
Renk
Melut
Kaka
Malakal
Kodok

EL AZRAQ
AN NÎL EL AZRAQ

Nil el Abyad (White Nile)
Bentiu
Nil el Azraq (Blue Nile)

BAHR EL GHAZAL
Kafia Kingi
Râga
Dem Zubeir
Wâw
Gogrial
Tonj
Meshra er Req

A'ALI EN NIL

EL BUHEIRAT
Rumbek
Yirol
Bôr

JONGLEI
Duk Fadiat
Kongor
Pibor P.
Akobo

GHARB EL ISTIWA'IYA
Tamburâ
Amadi
Maridi
Yambio
Yei

SHARQ EL ISTIWA'IYA
Juba
Torit
Kapoeta

KENYA
L. Turkana
Lokitaung
Todenyang
Mega

TURKEY
Antalya
Antalya Körfezi
İskenderun Körfezi
İskenderun
Antakya

SYRIA
Halab (Halab)
Al Mawşil (Mosul)
Nahr Dijla (Tigris)
Hamâh
Al Ladhiqiya
Hims
Tarabulus
Mesopotamia
Nahr al Furat

CYPRUS
Nicosia
Limassol

Ródhos
Iráklion
Karpathos
Kríti

LEBANON
Bayrût
Dimashq (Damascus)
Ar Rutbah

IRAQ
Bâdiyat ash Shâm

ISRAEL
Tel Aviv-Yafo
Haifa
Jerusalem (El Quds)
Gaza

JORDAN
Amman
Be'er Sheva
Ma'an
Al 'Aqabah
Tabûk

SAUDI ARABIA
Al Jawf
Al Muwaylih
Mada'in Salih
Taymâ'
An Nafûd
Al Wajh
Yanbu'al Bahr
Umm Lajj
Al Madînah (Medina)
Rabigh
Qasr
Râbigh
Jiddah
Makkah (Mecca)
At Ta'if
Ras Bânâs
Ras Shalatein

RED SEA

MEDINA

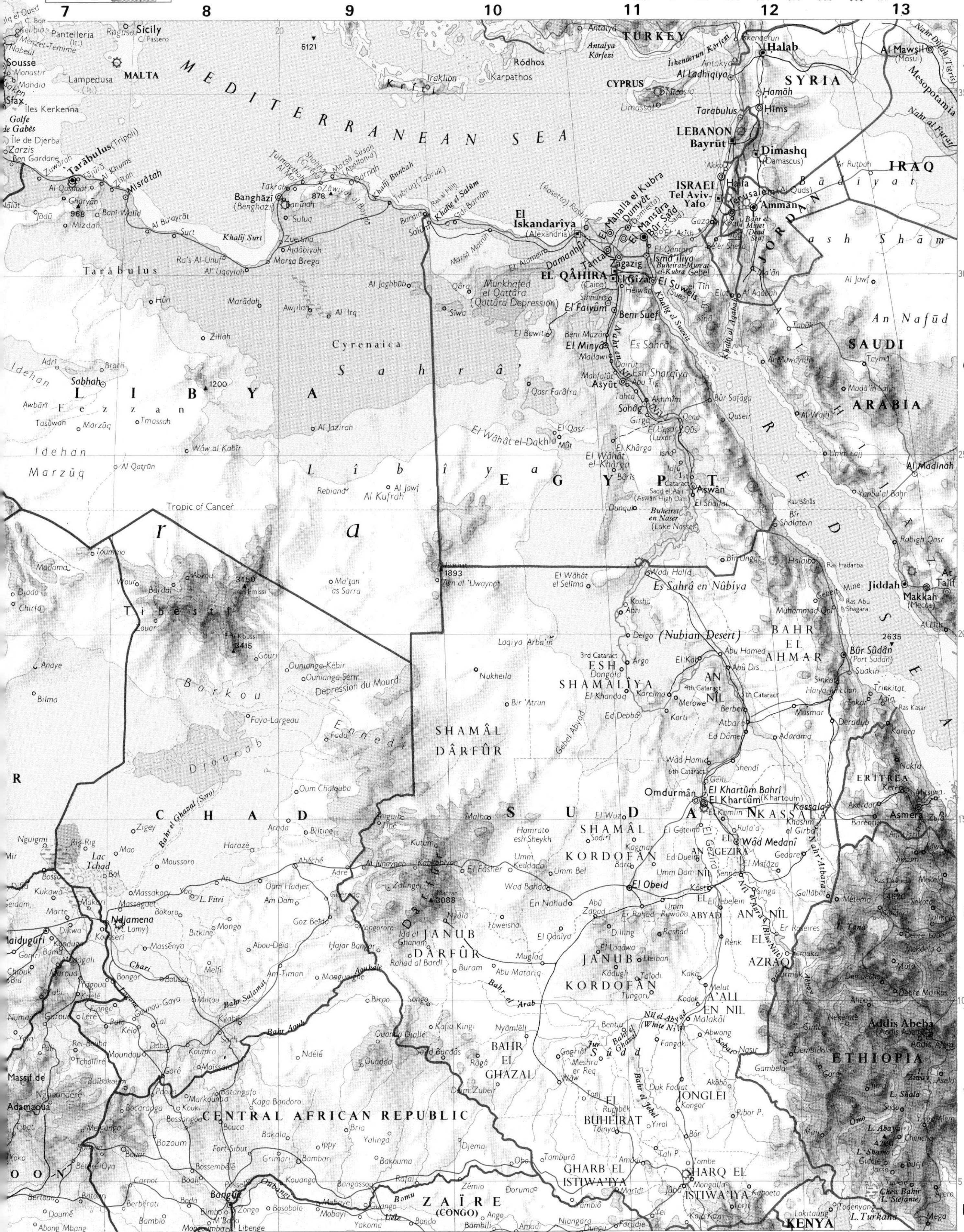

COPYRIGHT. GEORGE PHILIP & SON. LTD.

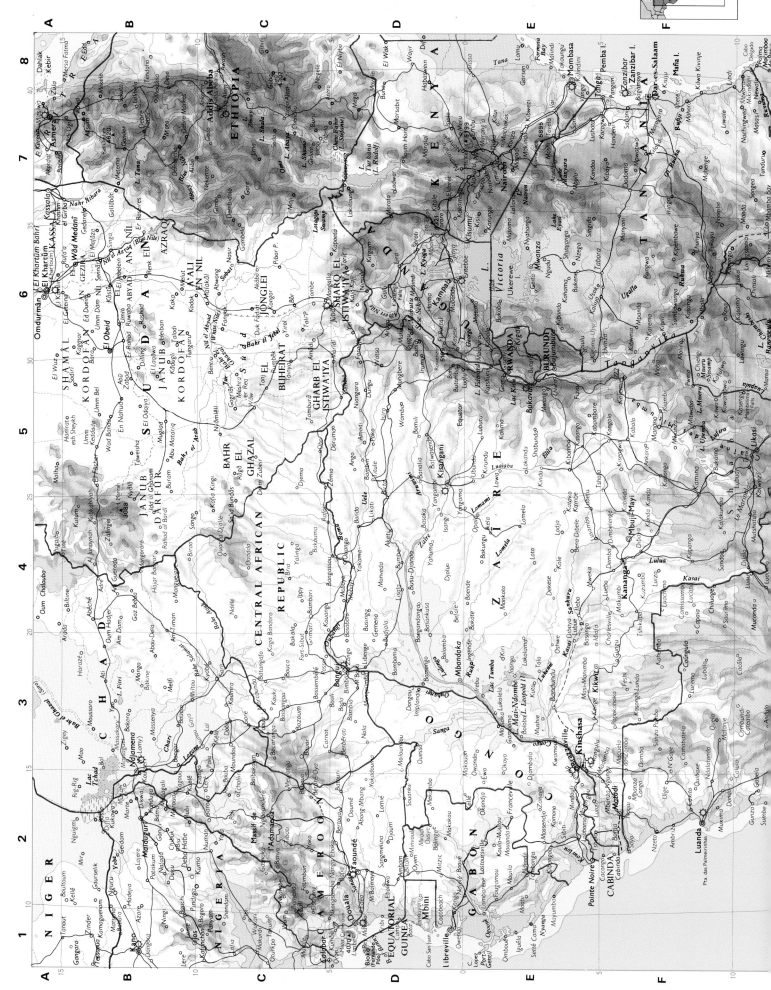

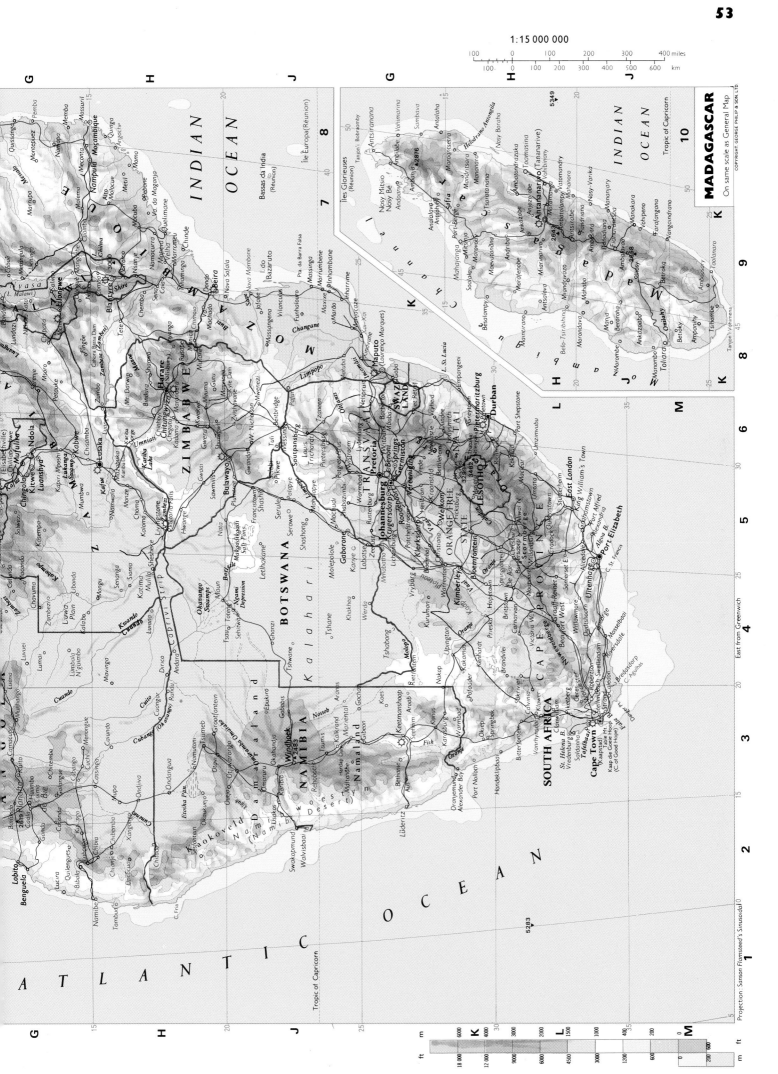

1:15 000 000

100   0   100   200   300   400 miles
100   0   100   200   300   400   500   600 km

MADAGASCAR
On same scale as General Map

COPYRIGHT GEORGE PHILIP & SON, LTD.

INDIAN OCEAN

ATLANTIC OCEAN

ANGOLA

NAMIBIA

BOTSWANA

Kalahari

ZAMBIA

ZIMBABWE

MOZAMBIQUE

SOUTH AFRICA

Cape Town

Port Elizabeth

Durban

LESOTHO

ORANGE FREE STATE

TRANSVAAL

Johannesburg

Pretoria

Bulawayo

Harare

Maputo

Windhoek

Gaborone

Lusaka

SWAZILAND

NATAL

CAPE PROVINCE

East from Greenwich

Tropic of Capricorn

Projection: Sanson Flamsteed's Sinusoidal

Tropic of Capricorn

Îles Glorieuses (Réunion)

Antananarivo (Tananarive)

m   ft
6000   18 000
4000   12 000
3000   9000
2000   6000
1500   4500
1000   3000
400   1200
200   600
0   0
ft   m

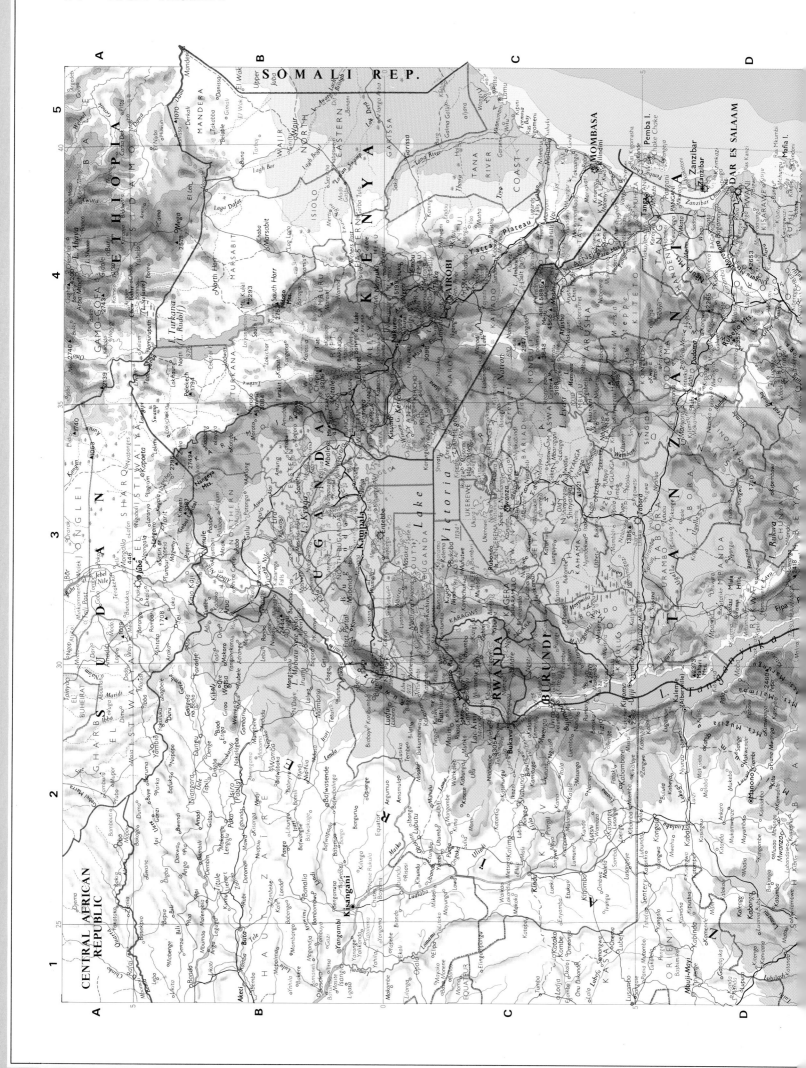

1:8 000 000

Projection: Lambert's Equivalent Azimuthal

East from Greenwich

COPYRIGHT GEORGE PHILIP & SON LTD.

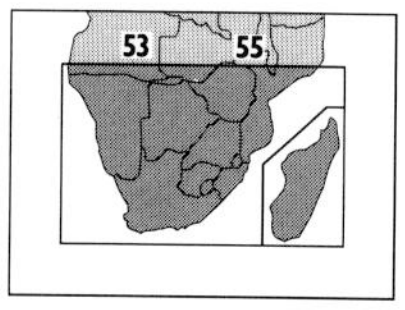

ANGOLA

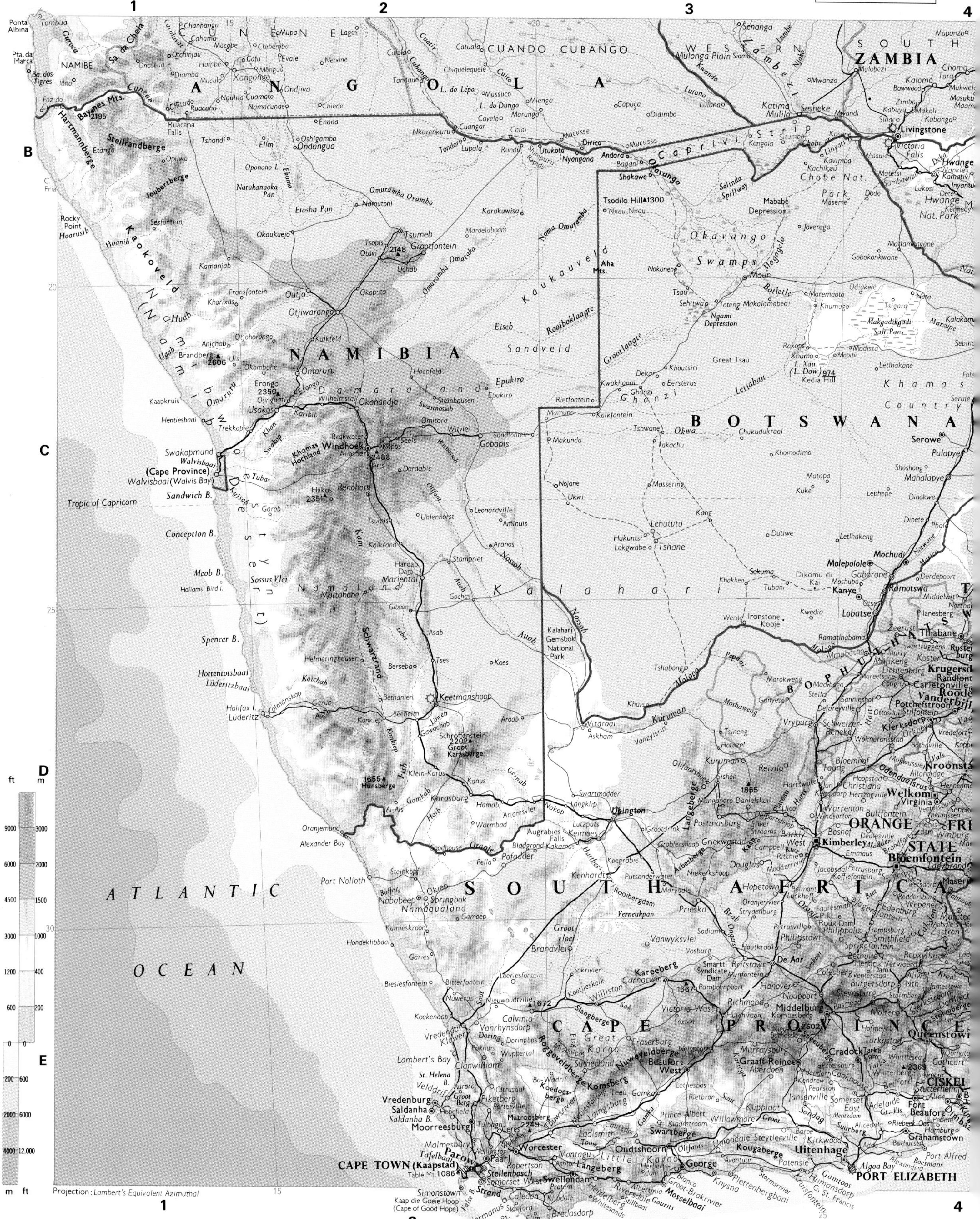

NAMIBIA

BOTSWANA

ZAMBIA

SOUTH AFRICA

CUANDO CUBANGO

WESTERN

Caprivi Strip

Okavango Swamps

Kalahari

Namib Desert

Kaokoveld

Etosha Pan

Windhoek

Swakopmund
Walvisbaai
(Cape Province)
Walvisbaai (Walvis Bay)

Tsumeb
Grootfontein
Otavi
Otjiwarongo
Outjo
Omaruru
Okahandja
Karibib
Rehoboth
Mariental
Keetmanshoop
Lüderitz
Karasburg

Tropic of Capricorn

Sandwich B.
Conception B.
Meob B.
Spencer B.

Sossus Vlei
Namaland

ATLANTIC

OCEAN

Port Nolloth
Alexander Bay
Oranjemund

Springbok
Namaqualand

CAPE PROVINCE

CAPE TOWN (Kaapstad)
Table Mt. 1086
Stellenbosch
Paarl
Strand
Somerset West
Swellendam

Saldanha
Saldanha B.
Vredenburg
Moorreesburg
Malmesbury

Worcester
Robertson
Montagu
George
Mosselbaai
Knysna
Oudtshoorn
Beaufort West
Calvinia
Carnarvon
Williston
Victoria West
Middelburg
Hanover
De Aar
Colesberg
Cradock
Graaff-Reinet
Somerset East
Grahamstown
Uitenhage
PORT ELIZABETH
Algoa Bay

CISKEI
Queenstown
Fort Beaufort

Kimberley
BLOEMFONTEIN
ORANGE FREE STATE
Welkom
Virginia
Kroonstad

BOPHUTHATSWANA
Kuruman
Upington
Vryburg
Klerksdorp
Potchefstroom
Krugersdorp

Gaborone
Kanye
Lobatse
Molepolole
Mochudi
Serowe
Palapye
Mahalapye

Ghanzi

Livingstone
Victoria Falls
Katima Mulilo
Sesheke

Chobe Nat. Park
Hwange Nat. Park

Projection: Lambert's Equivalent Azimuthal

Kaap die Goeie Hoop
(Cape of Good Hope)
C. Agulhas

ft	D m
9000	3000
6000	2000
4500	1500
3000	1000
1200	400
600	200
0	E 0
200	600
2000	6000
4000	12,000
m ft	

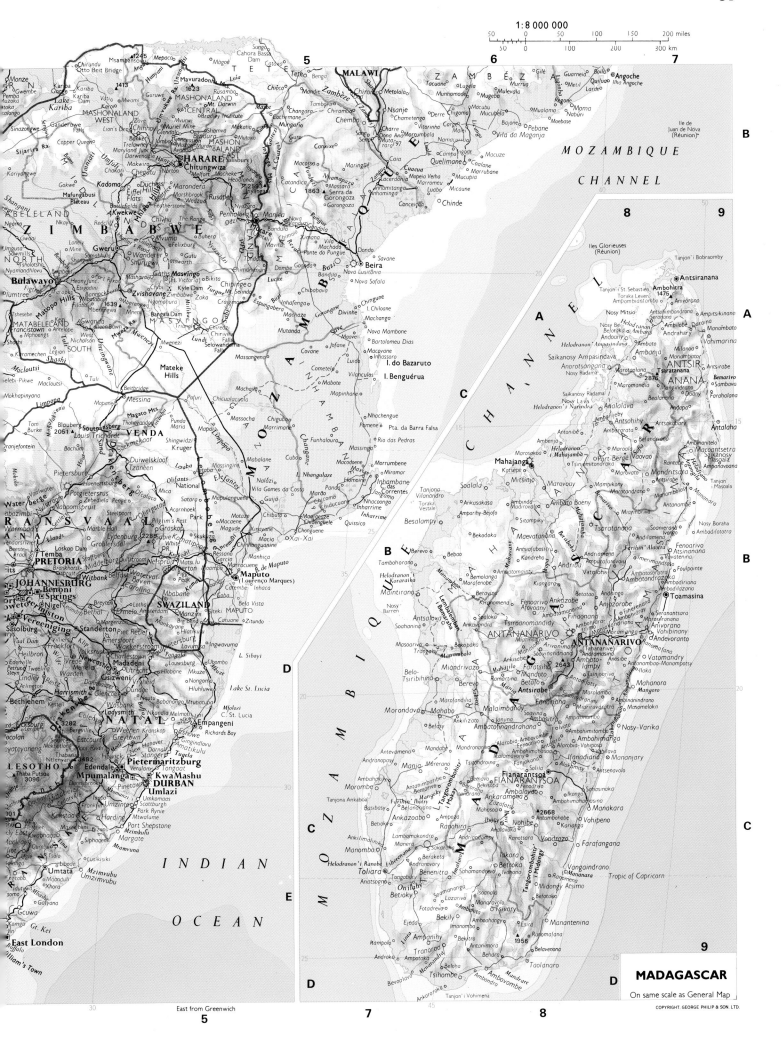

1:8 000 000

50   0   50   100   150   200 miles

0   100   200   300 km

5   6   7

MALAWI

ZAMBÉZIA

Ile de
Juan de Nova
(Réunion)

B

MOZAMBIQUE

CHANNEL

8   9

Iles Glorieuses
(Réunion)

Antsiranana

ZIMBABWE

MOZAMBIQUE

CHANNEL

Bulawayo

Beira

Nova Sofala

A

MADAGASCAR

Mahajanga

VENDA

Kruger

Antananarivo

ANTANANARIVO

Toamasina

PRETORIA

JOHANNESBURG
Springs

SWAZILAND

MAPUTO

Maputo
(Lourenço Marques)

Antsirabe

D

NATAL

Fianarantsoa

FIANARANTSOA

LESOTHO

Pietermaritzburg

DURBAN
Umlazi

INDIAN

C

Toliara

Tropic of Capricorn

Umtata

East London

OCEAN

MADAGASCAR

On same scale as General Map

East from Greenwich

5   7   8

COPYRIGHT. GEORGE PHILIP & SON. LTD.

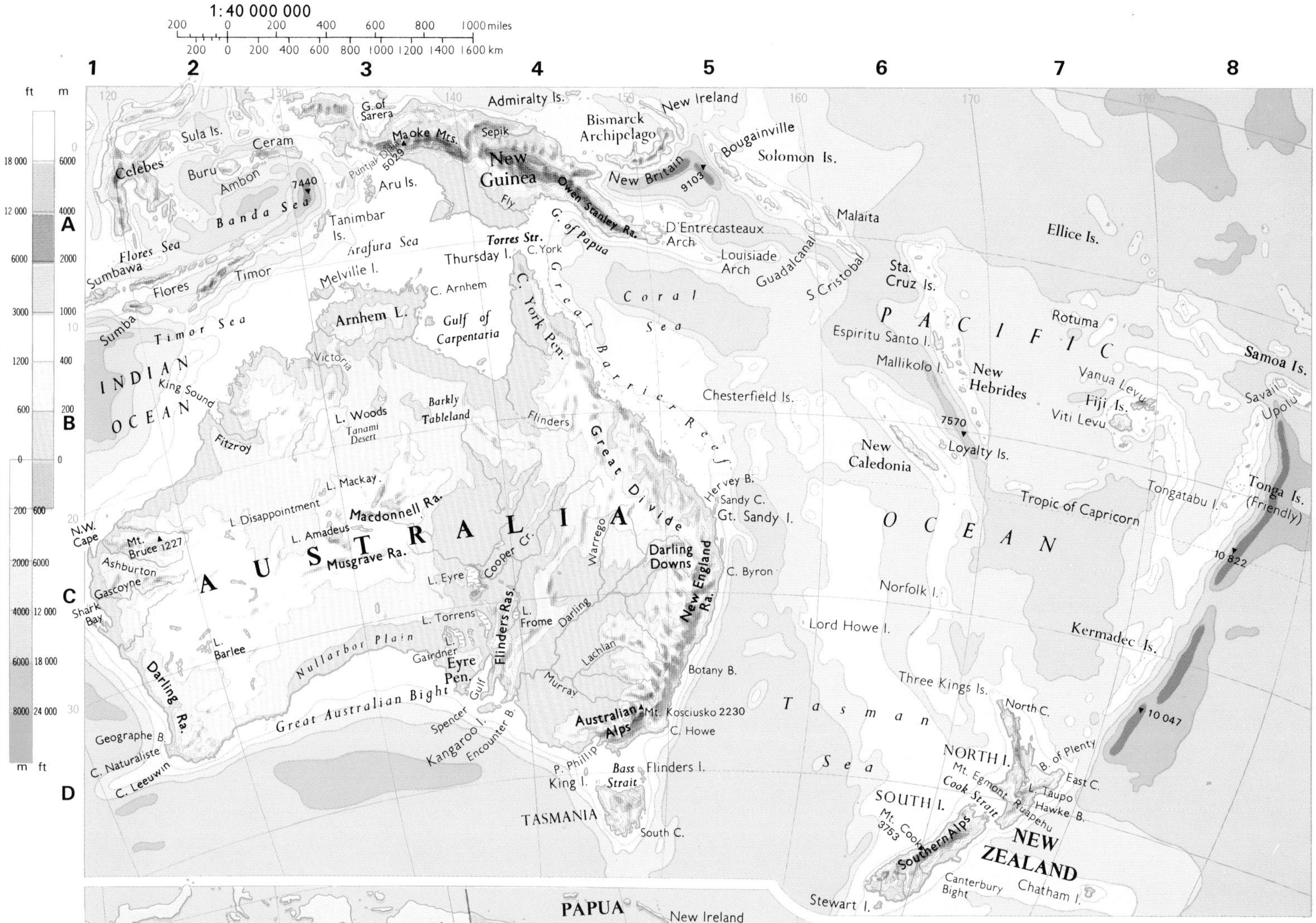

1:40 000 000

200 0 200 400 600 800 1000 miles
200 0 200 400 600 800 1000 1200 1400 1600 km

1 2 3 4 5 6 7 8

Top map (Physical)

ft m

18 000 6000
12 000 4000
6000 2000
3000 1000
1200 400
600 200
0 0
200 600
2000 6000
4000 12 000
6000 18 000
8000 24 000
m ft

G. of Sarera Admiralty Is. New Ireland
Maoke Mts. 5029 Sepik Bismarck Archipelago Bougainville Solomon Is.
Pumfar Dch. New Guinea Owen Stanley Ra. 9103 New Britain Malaita
Celebes Buru Ceram Ambon 7440 Aru Is. Fly G. of Papua D'Entrecasteaux Arch. S. Cristobal
Sula Is. Banda Sea Tanimbar Arafura Sea Torres Str. C. York Louisiade Arch. Guadalcanal Sta. Cruz Is.
Flores Sea Is. Thursday I. Great Barrier Reef Coral Sea Rotuma Ellice Is.
Sumbawa Flores Timor Melville I. C. Arnhem C. York Pen. Chesterfield Is. Espiritu Santo I. Samoa Is.
Sumba Timor Sea Arnhem L. Gulf of Carpentaria Mallikolo I. New Hebrides Vanua Levu Fiji Is. Savaii
INDIAN King Sound Victoria Great Divide 7570 Viti Levu Upolu
OCEAN L. Woods Tanami Desert Barkly Tableland Flinders New Caledonia Loyalty Is.
Fitzroy Macdonnell Ra. L. Mackay Hervey B. Tropic of Capricorn Tonga Is. (Friendly)
N.W. Cape Mt. Bruce 1227 L. Disappointment L. Amadeus Musgrave Ra. Cooper C. Warrego Gt. Sandy I. Sandy C. Norfolk I. Tongatabu I. 10 822
Ashburton AUSTRALIA Darling Downs C. Byron
Gascoyne L. Eyre New England Ra. PACIFIC OCEAN Lord Howe I. Kermadec Is.
Shark Bay L. Torrens Flinders Ras. Frome Darling Lachlan Botany B. Three Kings Is. North C. 10 047
Barlee Nullarbor Plain L. Gairdner Eyre Pen. Murray Tasman B. of Plenty East C.
Darling Ra. Great Australian Bight Gulf Spencer Australian Alps Mt. Kosciusko 2230 Sea NORTH I. Mt. Egmont Ruapehu Hawke B.
Geographe B. Kangaroo I. Encounter B. C. Howe SOUTH I. Mt. Cook 3753 Cook Strait Taupo
C. Naturaliste P. Philip Bass Flinders I. NEW ZEALAND Southern Alps Chatham I.
C. Leeuwin King I. Strait Canterbury Bight
TASMANIA South C. Stewart I.

Bottom map (Political)

IRIAN JAYA PAPUA New Ireland
Celebes Sula Is. Ceram NEW GUINEA Rabaul
Ujung Pandang Buru New Guinea Madang New Britain Choiseul SOLOMON IS. KIRIBATI
INDONESIA Banda Sea Aru Is. Lae Papua Ysabel Honiara Malaita Funafuti
Flores Sea Tanimbar Is. Torres Str. Port Moresby Guadalcanal S. Cristobal TUVALU
Sumbawa Timor Arafura Sea Fly Sta. Cruz Is. Tokelau Is. (N.Z.)
Sumba Flores Kupang Timor Sea Gulf of Carpentaria Coral Sea Rotuma Wallis and Futuna (Fr.) WESTERN SAMOA
Darwin Katherine Islands Territory Espiritu Santo VANUATU Upolu Savaii Apia
Wyndham NORTHERN Cooktown Chesterfield Is. (Fr.) Vila Vanua Levu AMER. SAMOA
Broome TERRITORY QUEENSLAND Cairns Townsville New Caledonia (Fr.) Loyalty Is. (Fr.) FIJI Viti Levu Suva TONGA (Friendly)
Mt. Isa Charters Towers Nouméa Tropic of Capricorn Nuku'alofa
WESTERN AUSTRALIA Alice Springs Rockhampton PACIFIC OCEAN
Dampier Longreach Norfolk I. (Aust.)
Onslow Charleville Toowoomba International Date Line
SOUTH Quilpie Cunnamulla Brisbane Lord Howe I. (Aust.) Kermadec Is. (N.Z.)
Oodnadatta L. Eyre Warwick
Wiluna AUSTRALIA Darling NEW SOUTH WALES Bourke
Laverton Broken Hill Hay Newcastle Tasman
Geraldton Kalgoorlie-Boulder P. Pirie Sydney Sea NORTH I.
Perth Great Australian Bight Mildura Murray Goulburn Canberra Auckland Hamilton
Fremantle Esperance Adelaide Bendigo VICTORIA New Plymouth Wellington Napier
Albany Ballarat Melbourne SOUTH I. Nelson
Geelong Bass Strait Greymouth NEW ZEALAND
King Launceston Christchurch Chatham
TASMANIA Hobart Invercargill Dunedin

Vila Capital Cities

East from Greenwich West from Greenwich COPYRIGHT GEORGE PHILIP & SON. LTD.

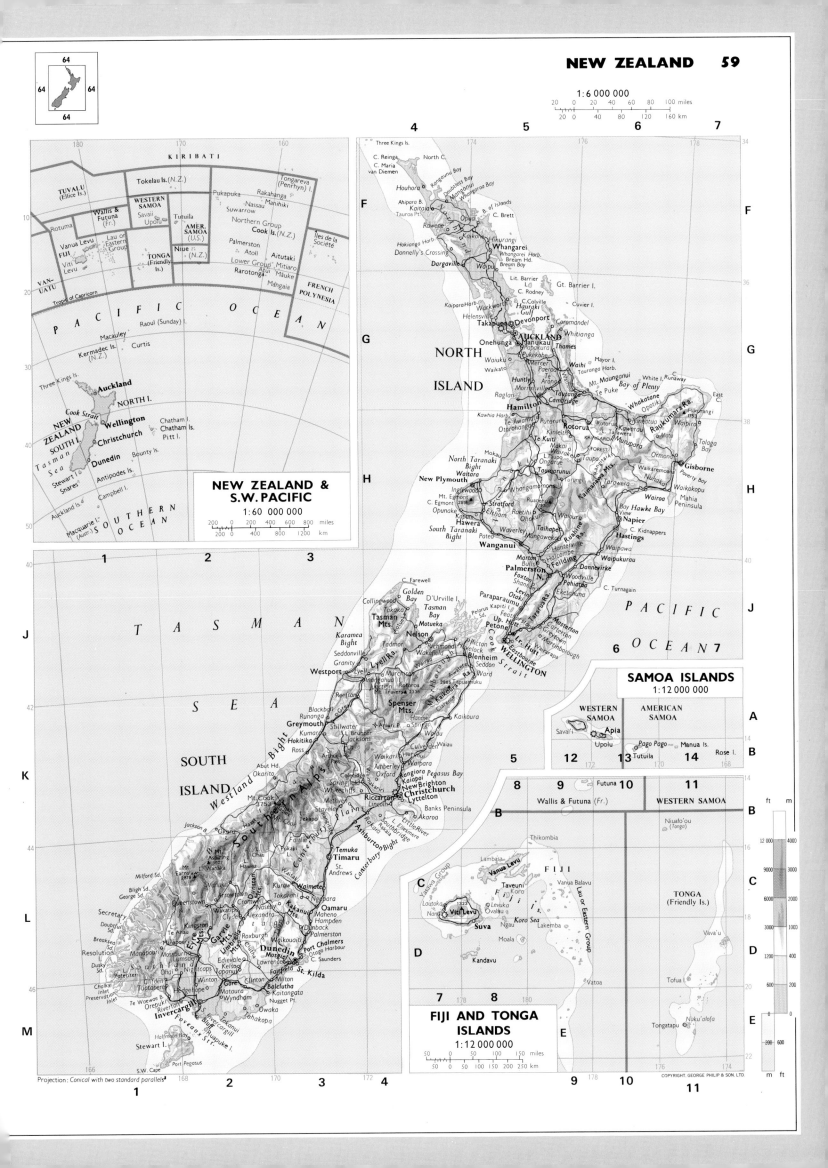

1:6 000 000

NEW ZEALAND &
S.W. PACIFIC
1:60 000 000

SAMOA ISLANDS
1:12 000 000

WESTERN SAMOA   AMERICAN SAMOA

FIJI AND TONGA
ISLANDS
1:12 000 000

Projection: Conical with two standard parallels

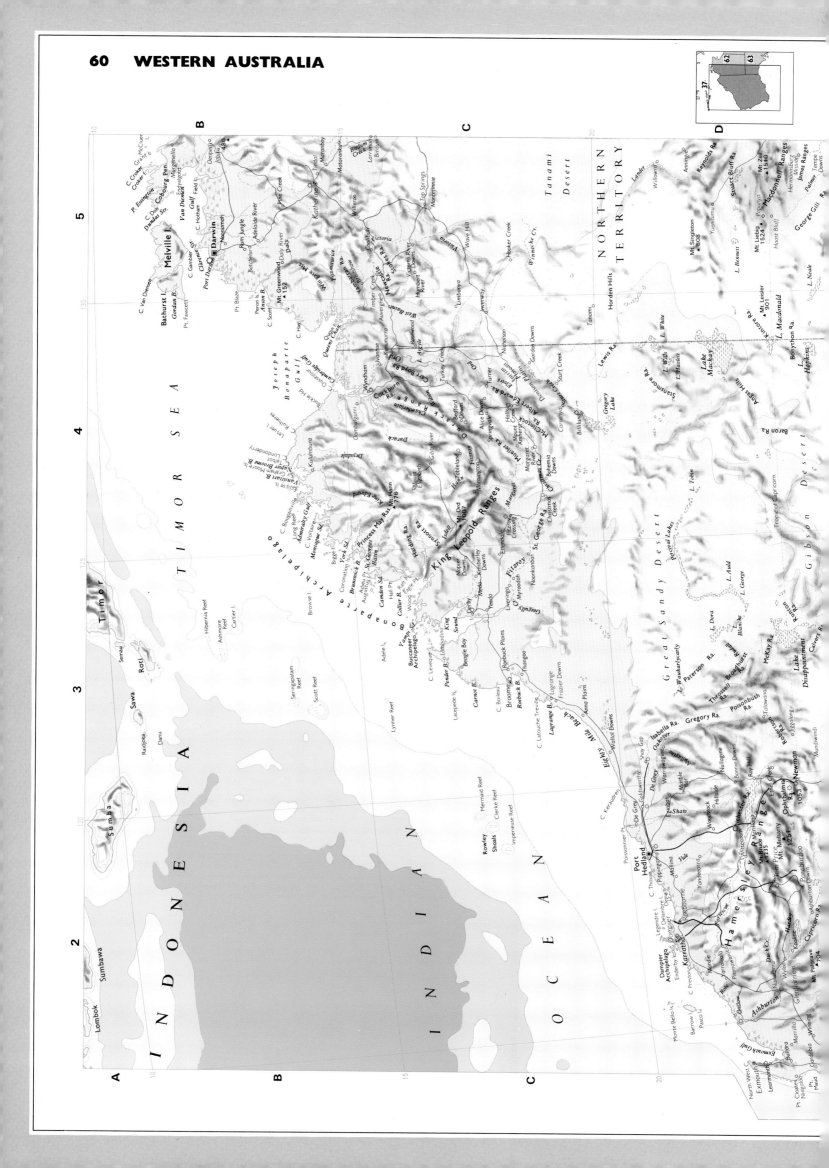

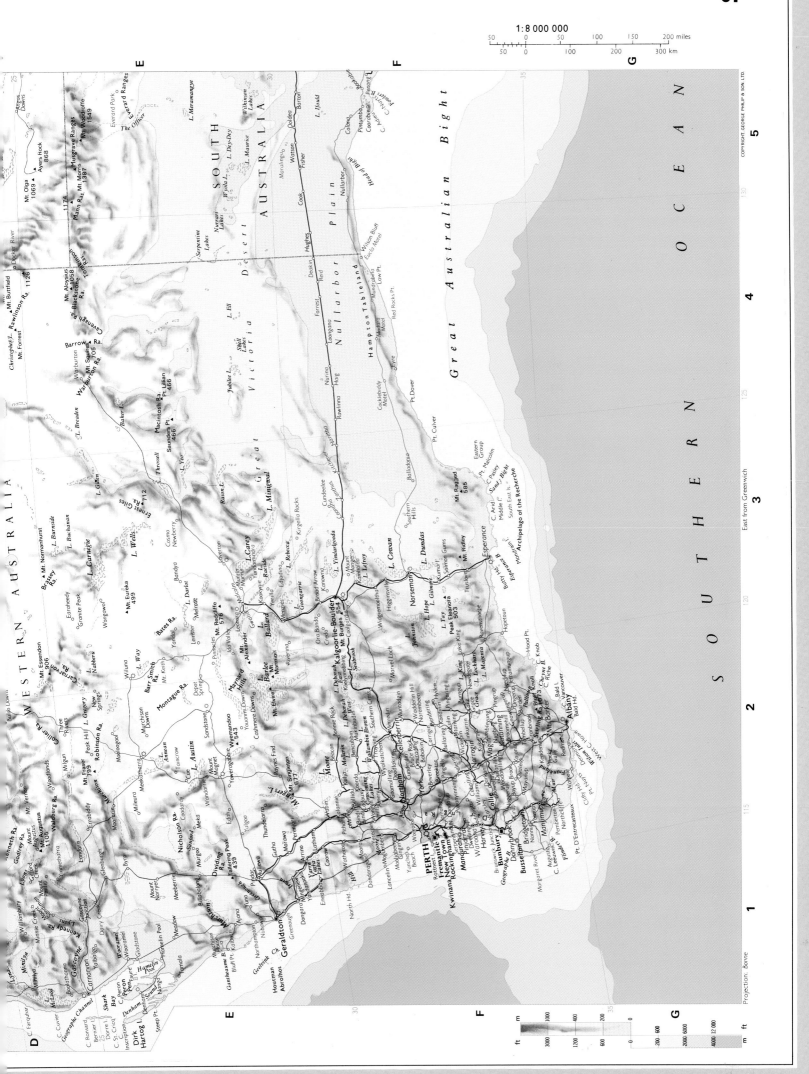

1:8 000 000

50   0   50   100   150   200 miles

50   0   50   100   200   300 km

WESTERN AUSTRALIA

SOUTH AUSTRALIA

Great Victoria Desert

Nullarbor Plain

Hampton Tableland

Great Australian Bight

SOUTHERN OCEAN

PERTH

Fremantle

Geraldton

Kalgoorlie-Boulder

Albany

Esperance

Norseman

Bunbury

Projection: Bonne

East from Greenwich

m ft
3000
1200
600
0
-200 -600
2000 6000 12 000
4000 12 000

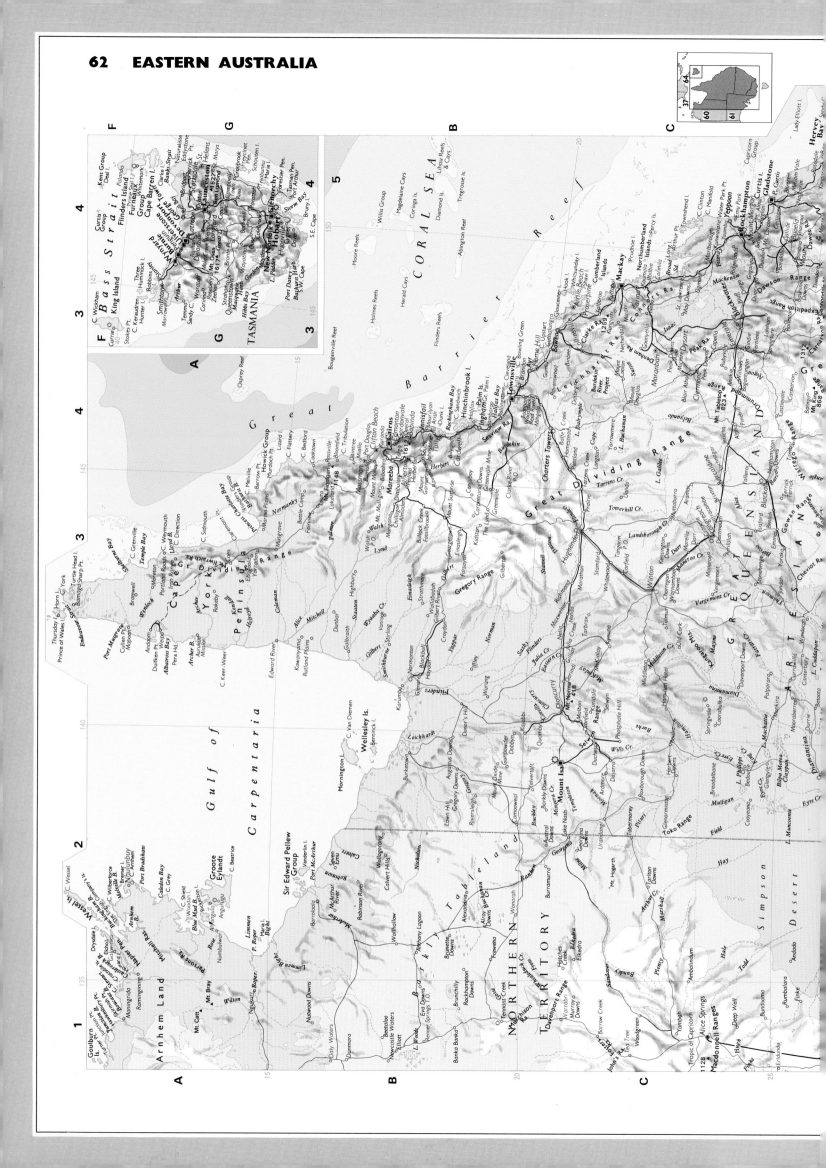

TASMANIA

Bass Strait

Kent Group
Deal I.
Flinders Island
Curtis Group
C. Wickham
F. Currie
Stokes Pt.
C. Keraudren
Hunter I.
Three Hummock I.
Robbins I.
Smithton
Stanley
Wynyard
Somerset
Burnie
Penguin
Ulverstone
Devonport
Latrobe
Deloraine
Westbury
Launceston
Longford
Ben Lomond
Naturaliste
Eddystone Pt.
Banks Strait
Clarke I.
St. Helens
St. Marys
Cranbrook
Freycinet Pen.
Schouten I.
Swansea
Ross
Tunbridge
Oatlands
Bothwell
Hamilton
New Norfolk
Glenorchy
Hobart
Huonville
Geeveston
Dover
L. Pedder
Port Davey
Bathurst Harb.
S.W. Cape
S.E. Cape
Storm Bay
Tasman Pen.
Port Arthur
Bruny I.
Hibbs Bay
Strahan
Queenstown
Zeehan
Rosebery
Mt. Ossa 1617
Waratah
Corinna
Temma
Sandy C.
Arthur
King Island
Naracoopa
Surrey Hills

Gulf of Carpentaria

CORAL SEA

Great Barrier Reef

Arnhem Land
Wessel Is.
C. Wessel
Goulburn Is.
Croker I.
C. Don
Cobourg Pen.
Port Bradshaw
Caledon Bay
Groote Eylandt
C. Grey
C. Shield
Blue Mud B.
C. Beatrice
Sir Edward Pellew Group
Vanderlin I.
Port McArthur
Mornington I.
Wellesley Is.
Bentinck I.
Sweers I.
C. Van Diemen
Cape York Peninsula
Thursday I.
Prince of Wales I.
Horn I.
Endeavour Str.
Turtle Head I.
Bamaga
C. York
Weipa
Albatross Bay
Aurukun Mission
Pera Hd.
C. Keer-Weer
Edward River
Mitchell
Coleman
Kowanyama
Rutland Plains
Staaten
Normanton
Karumba
Burketown
Doomadgee
Leichhardt
Nicholson
Wollogorang

NORTHERN TERRITORY

Barkly Tableland

Townsville
Cairns
Mossman
Mareeba
Atherton
Innisfail
Tully
Cardwell
Ingham
Halifax
Hinchinbrook I.
Palm Is.
Bowen
Ayr
Home Hill
Proserpine
Mackay
Sarina
Great Dividing Range
Charters Towers
Hughenden
Richmond
Julia Cr.
Cloncurry
Mount Isa
Duchess
Dajarra
Boulia
Winton
Longreach
Barcaldine
Blackall
Tropic of Capricorn

QUEENSLAND

Rockhampton
Gladstone
Yeppoon
Capricorn Group
Curtis I.
Hervey Bay
Lady Elliott I.
Bundaberg
Gin Gin

Macdonnell Ranges
Alice Springs
Simpson Desert

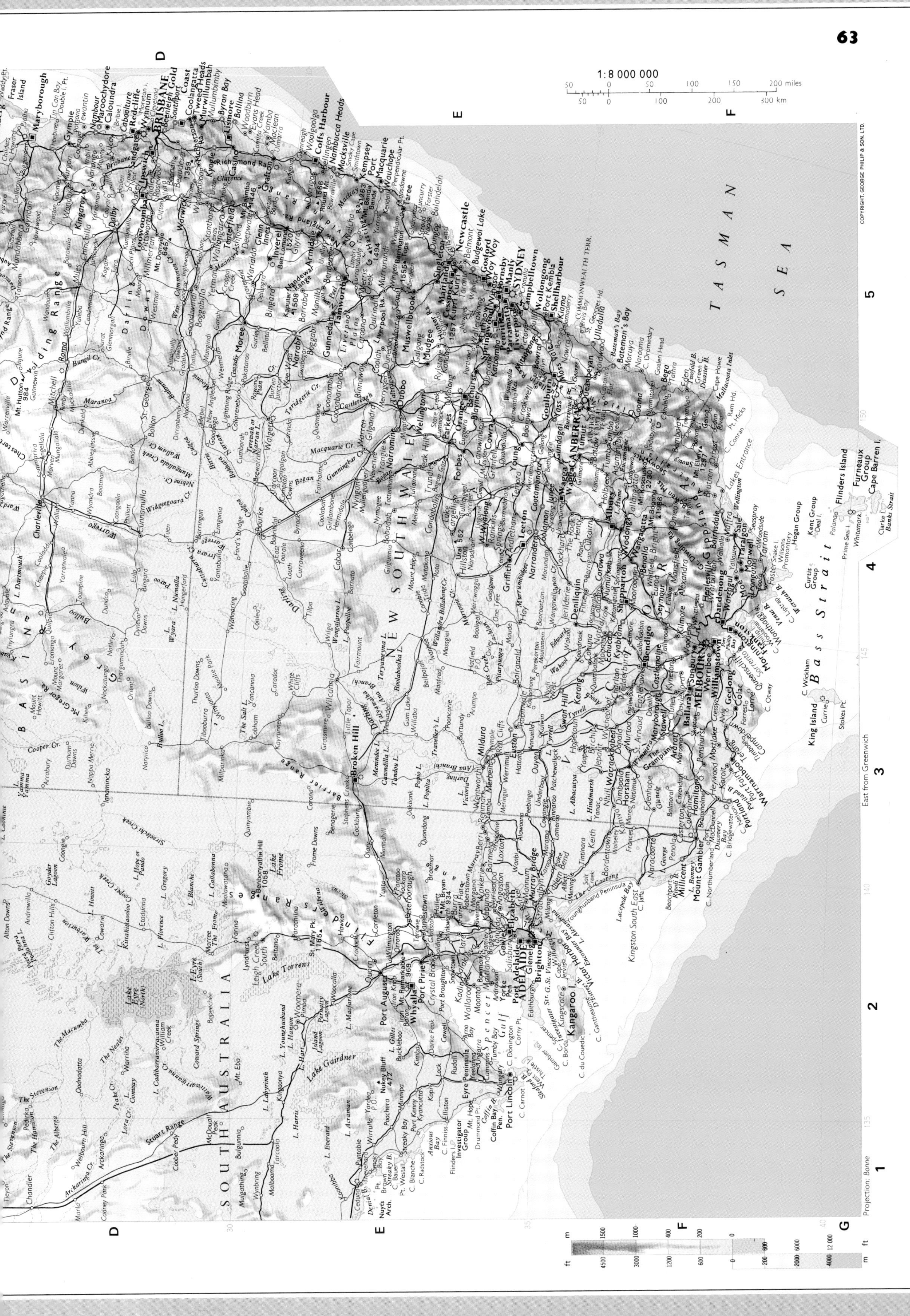

1:8 000 000

50 0 50 100 150 200 miles
50 0 100 200 300 km

TASMAN SEA

SOUTH AUSTRALIA

NEW SOUTH WALES

VICTORIA

QUEENSLAND

BRISBANE
SYDNEY
MELBOURNE
ADELAIDE
CANBERRA
Newcastle
Wollongong
Broken Hill
Geelong
Ballarat
Bendigo
Port Augusta
Whyalla
Port Pirie
Mount Gambier
Warrnambool
Coffs Harbour
Maryborough
Gympie
Toowoomba
Ipswich
Warwick
Dalby
Kingaroy
Armidale
Tamworth
Moree
Dubbo
Orange
Bathurst
Wagga Wagga
Albury
Wodonga
Griffith
Mildura
Gundagai
Goulburn
Queanbeyan
Cooma
Bega
Eden
Nambucca Heads
Macksville
Kempsey
Port Macquarie
Taree
Gosford
Lismore
Ballina
Byron Bay
Murwillumbah
Coolangatta
Southport
Redcliffe
Caboolture
Nambour
Caloundra
Gosford
Gunnedah
Narrabri
Inverell
Glen Innes
Tenterfield
Grafton
Casino
Forbes
Young
Cowra
Leeton
Narrandera
Deniliquin
Echuca
Shepparton
Wangaratta
Benalla
Sale
Traralgon
Morwell
Bairnsdale
Horsham
Ararat
Stawell
Colac
Portland
Hamilton
Murray Bridge
Victor Harbor
Elizabeth
Salisbury
Gawler
Kadina
Moonta
Wallaroo
Port Lincoln
Millicent
Bordertown

Barrier Range

Darling R.
Murray R.
Murrumbidgee R.
Lachlan R.

Lake Eyre North
Lake Eyre South
Lake Torrens
Lake Gairdner
Lake Frome
Lake Blanche
Lake Callabonna

Spencer Gulf
Gulf St. Vincent
Kangaroo I.
Investigator Group
Eyre Peninsula
Yorke Peninsula

Cooper Cr.
Warrego R.

Fraser Island

King Island
Flinders Island
Furneaux Group
Cape Barren I.
Bass Strait
Banks Strait

Charleville
Cunnamulla
Roma
Mitchell
St. George
Bourke
Cobar
Nyngan
Gilgandra
Coonamble
Walgett
Brewarrina
Wilcannia
Menindee

East from Greenwich

135 140 145 150

Projection: Bonne

COPYRIGHT. GEORGE PHILIP & SON, LTD.

m ft
1500 4500
1000 3000
400 1200
200 600
0 0
200 600
2000 6000
4000 12000

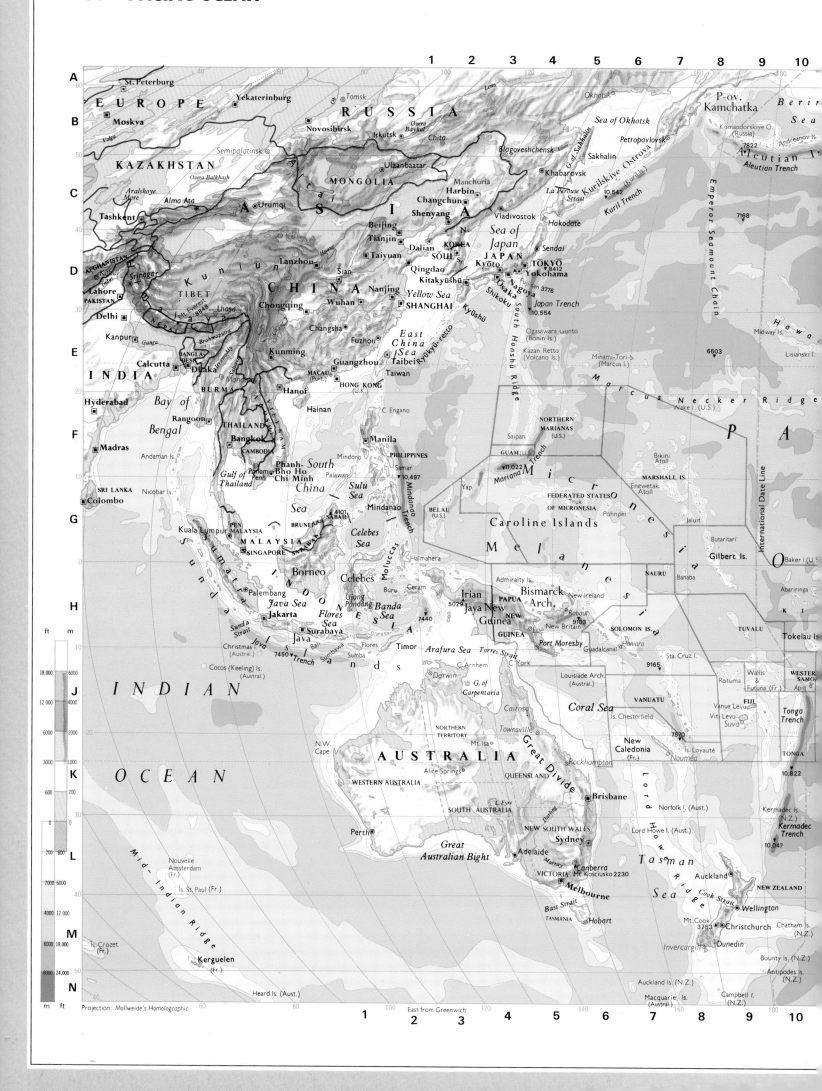

1:54 000 000

ALASKA
(U.S.)
6050
Gulf of Alaska
Bristol Bay
Prince of Wales I.
Queen Charlotte Is.
Prince Rupert
Kitimat
Juneau
GREENLAND
C. Farewell
U.K.

Hudson
Bay
CANADA
Edmonton
L. Winnipeg
NORTH AMERICA
Labrador
NORTH
Newfoundland

Vancouver
Vancouver I.
Victoria
Seattle
Calgary
Regina
Winnipeg
L. Superior
Montréal
Québec
St. Lawrence
Pr. Edward I.
Saint John

Portland
Boise
Snake
Missouri
Michigan
L. Huron
Ottawa
Toronto
L. Ontario
Buffalo
Boston
C. Sable
NEW YORK

C. Mendocino
Salt Lake
City
Denver
Kansas
City
St. Louis
CHICAGO
Detroit
Pittsburgh
Cincinnati
Philadelphia
Baltimore
Washington
Appalachian Mts.
ATLANTIC

San Francisco
4418
UNITED STATES
Oklahoma
Memphis
Atlanta
C. Hatteras

6741
Los Angeles
San Diego
Ciudad
Juárez
Dallas
Mississippi
Jacksonville
Bermuda (U.K.)

6225
Sierra Madre
M E X I C O
Houston
San Antonio
New
Orleans
Gulf of Mexico
Miami
OCEAN

Hawaiian Is.
(U.S.)
Honolulu
Oahu
Hawaii
Tropic of Cancer
Gulf of California
Is. Revilla Gigedo
(Mexico)
La Habana
CUBA
BAHAMAS
Florida
Strait
Yucatán Channel
West Indies
Hispaniola
DOM.
REP.
9200

Johnston I. (U.S.)
Guadalajara
Mérida
C. México
Puebla
5700
7680
JAMAICA
HAITI
Kingston
PUERTO
RICO
(U.S.)
Leeward
Is.

Acapulco
BELIZE
Caribbean Sea
BARBADOS
Windward
Is.
TRINIDAD &
TOBAGO

Palmyra Is. (U.S.)
I. Clipperton (Fr.)
GUATEMALA
Guatemala
HONDURAS
San Salvador
EL SALVADOR
NICARAGUA
Managua
CENTRAL
AMERICA
COSTA RICA
San José
Colón
PANAMA
Barranquilla
Maracaibo
Caracas
Orinoco
VENEZUELA

Teraina
Tabuaeran
Kiritimati
Jarvis I.
(U.S.)
I. del Coco
(Costa Rica)
Panama
Canal
Medellín
Bogotá
Cali
COLOMBIA

Duerbury I.
Phoenix Is.
Malden I.
Starbuck I.
Equator
Galápagos
(Ecuador)
Guayaquil
Quito
ECUADOR
Manaus
Amazonas

Îs. Marquises
C. Pariñas
BRAZIL
SOUTH

Tongareva
Penrhyn Is.
Manihiki
Suwarrow Is.
Vostok I.
Flint I.
Caroline I.
Trujillo
PERU
6369
Lima
AMERICA

Îs. de la
Société
Îs. Tuamotu
Cuzco
Arequipa
L. Titicaca
Hampu & Ancohum
6550

Cook
Islands
(N.Z.)
Manuae
Tahiti
FRENCH POLYNESIA
6866
Peru-
La Paz
BOLIVIA

Rarotonga
Iquique
Chile
PARAGUAY

Îs. Tubuai
(Îs. Australes)
Rapa
Pitcairn I. (U.K.)
Ducie I.
(U.K.)
Tropic of Capricorn
I. de Pascua
(Easter I.)
(Chile)
Sala-y-Gomez
(Chile)
San Félix (Chile)
San Ambrosio (Chile)
8050
Antofagasta
Trench
Asunción
Tucumán
Pto. Alegre

Arch. de Juan Fernández
(Chile)
Córdoba
Rosario
URUGUAY

Valparaíso
Santiago
Buenos Aires
Montevideo
Río de la Plata

Concepción
ARGENTINA
Chile Rise
SOUTH

6212
Falkland Is. (U.K.)
ATLANTIC
OCEAN

Punta Arenas
Str. of Magellan
Tierra del Fuego
C. Horn
South Georgia

PACIFIC
OCEAN
East Pacific Ridge
Pacific-Antarctic Ridge
Tuamotu Ridge
Seamount Chain
Austral
Patagonia

West from Greenwich

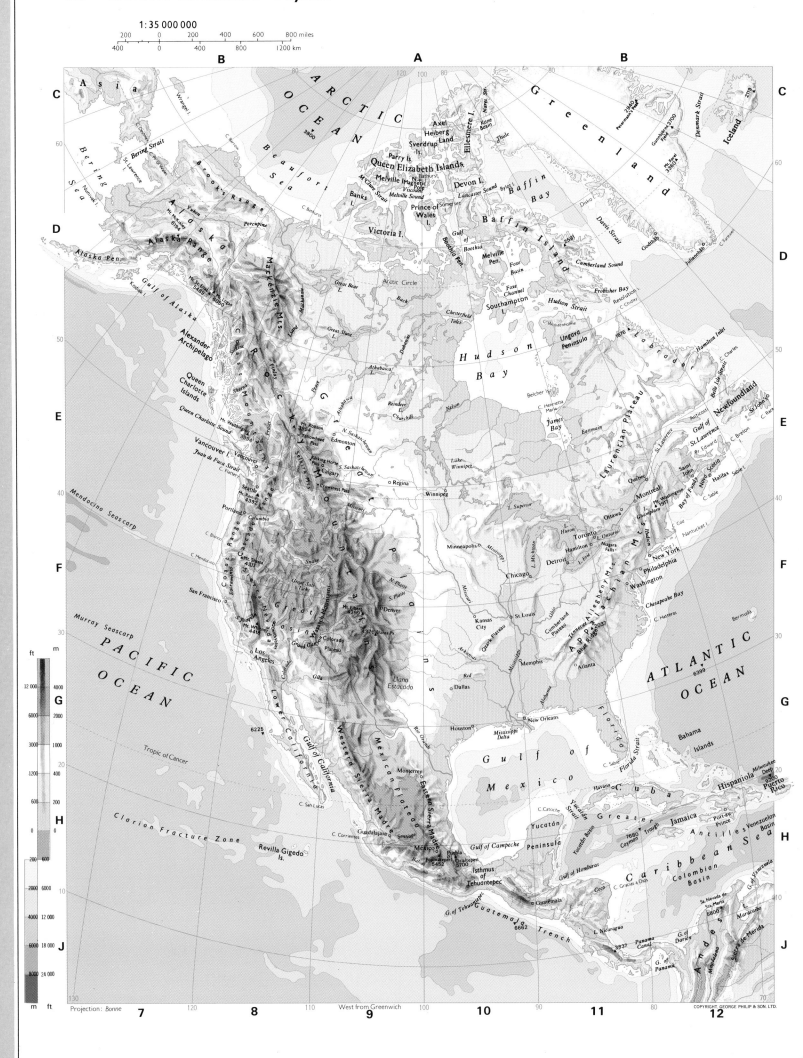

1:35 000 000

200   0   200   400   600   800 miles
400   0   400   800   1200 km

ft   m
12 000   4000
6000   2000
3000   1000
1200   400
600   200
0   0
200   600
2000   6000
4000   12 000
6000   18 000
8000   24 000
m   ft

Projection: Bonne

West from Greenwich

COPYRIGHT. GEORGE PHILIP & SON. LTD.

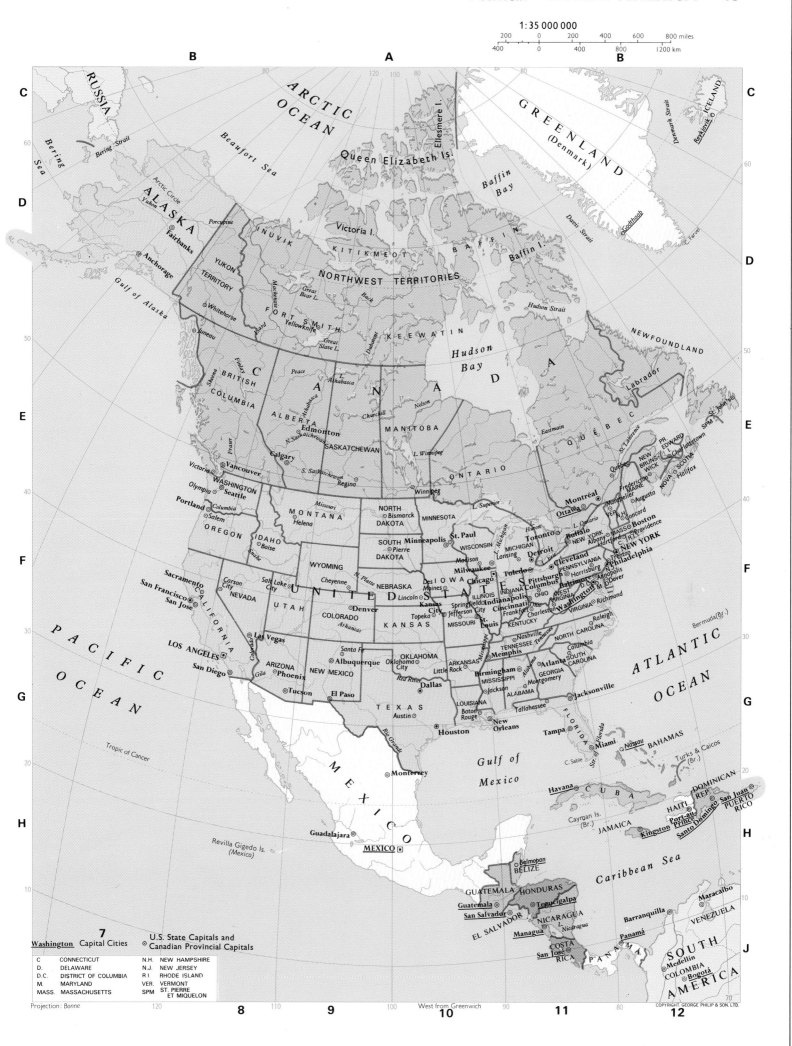

1:35 000 000

200   0   200   400   600   800 miles
400   0   400   800   1200 km

C   C

RUSSIA

ARCTIC OCEAN

GREENLAND (Denmark)

ICELAND
Reykjavik

Bering Strait
Beaufort Sea

Queen Elizabeth Is.
Ellesmere I.

Baffin Bay

Denmark Strait

Bering Sea

ALASKA
Yukon
Fairbanks
Anchorage
Arctic Circle
Porcupine

INUVIK
YUKON TERRITORY
Whitehorse
Juneau

VICTORIA I.

KITIKMEOT

NORTHWEST TERRITORIES

BAFFIN

Baffin I.

Godthåb

Gulf of Alaska

FORT SMITH
Great Bear L.
Yellowknife
Great Slave L.
Back

KEEWATIN

Hudson Strait

NEWFOUNDLAND

BRITISH COLUMBIA
Skeena
Finlay
Peace
ALBERTA
Edmonton
N. Saskatchewan
Calgary

CANADA

Athabasca
Churchill
Nelson

MANITOBA
SASKATCHEWAN
Regina
S. Saskatchewan

Hudson Bay

Eastmain

QUÉBEC

Labrador

St. John's
SPM

Fraser
Victoria
Vancouver
WASHINGTON
Seattle
Olympia
Portland
Salem
OREGON

L. Winnipeg
Winnipeg
ONTARIO

L. Superior

St. Lawrence
Québec
Montréal
Ottawa

NEW BRUNS WICK
Fredericton
MAINE
Augusta

PR. EDWARD
Charlottetown
NOVA SCOTIA
Halifax

Columbia
Missouri
MONTANA
Helena
IDAHO
Boise
Snake

NORTH DAKOTA
Bismarck
SOUTH DAKOTA
Pierre

MINNESOTA
Minneapolis St. Paul
WISCONSIN
Madison
Milwaukee

L. Michigan
MICHIGAN
Lansing
Detroit

L. Huron
Toronto
L. Ontario
Buffalo
NEW YORK
Albany
VER. N.H. Concord
Montpelier
Boston
MASS. Providence R.I.
Hartford CONN.

Sacramento
San Francisco
San Jose
Carson City
NEVADA
Salt Lake City
UTAH
WYOMING
Cheyenne

R. Platte
NEBRASKA
Lincoln
IOWA
Des Moines
Chicago
ILLINOIS
Springfield
Indianapolis
INDIANA
Cincinnati
OHIO
Columbus
Pittsburgh
PENNSYLVANIA
Harrisburg
Philadelphia
Trenton N.J.
NEW YORK

CALIFORNIA
Las Vegas
Colorado
Denver
COLORADO
KANSAS
Topeka
Kansas City
MISSOURI
St. Louis
Jefferson City
Frankfort
KENTUCKY
WEST VIRGINIA
Charleston
VIRGINIA
Richmond
Baltimore
Annapolis M. D.C.
Washington
Dover D.

Los Angeles
San Diego
ARIZONA
Phoenix
Tucson
Gila
Santa Fe
NEW MEXICO
Albuquerque
El Paso
Arkansas
OKLAHOMA
Oklahoma City
Red River
ARKANSAS
Little Rock
TENNESSEE
Nashville
Memphis
Mississippi
NORTH CAROLINA
Raleigh
Columbia
SOUTH CAROLINA

PACIFIC OCEAN

TEXAS
Dallas
Austin
Houston

LOUISIANA
Baton Rouge
New Orleans

MISSISSIPPI
Jackson
ALABAMA
Montgomery
Birmingham
GEORGIA
Atlanta
Alabama
Tallahassee
Jacksonville
FLORIDA
Tampa
Miami

ATLANTIC OCEAN

Bermuda (Br.)

Rio Grande
Monterrey

Gulf of Mexico
C. Sable
Str. of Florida

BAHAMAS
Nassau

Turks & Caicos (Br.)

Tropic of Cancer

MEXICO

Havana
CUBA
Cayman Is. (Br.)
JAMAICA
Kingston

HAITI
Port-au-Prince
DOMINICAN REP.
Santo Domingo
San Juan
PUERTO RICO

Guadalajara
Revilla Gigedo Is. (Mexico)
MEXICO

Caribbean Sea

Belmopan
BELIZE
GUATEMALA
HONDURAS
Tegucigalpa
Guatemala
San Salvador
EL SALVADOR
NICARAGUA
Managua
Nicaragua
Costa Rica
San José
PANAMÁ
Panamá

Maracaibo
VENEZUELA
Barranquilla
Medellín
COLOMBIA
Bogotá
SOUTH AMERICA

7
Washington Capital Cities

U.S. State Capitals and Canadian Provincial Capitals

C	CONNECTICUT	N.H.	NEW HAMPSHIRE
D.	DELAWARE	N.J.	NEW JERSEY
D.C.	DISTRICT OF COLUMBIA	R I	RHODE ISLAND
M.	MARYLAND	VER.	VERMONT
MASS.	MASSACHUSETTS	SPM	ST. PIERRE ET MIQUELON

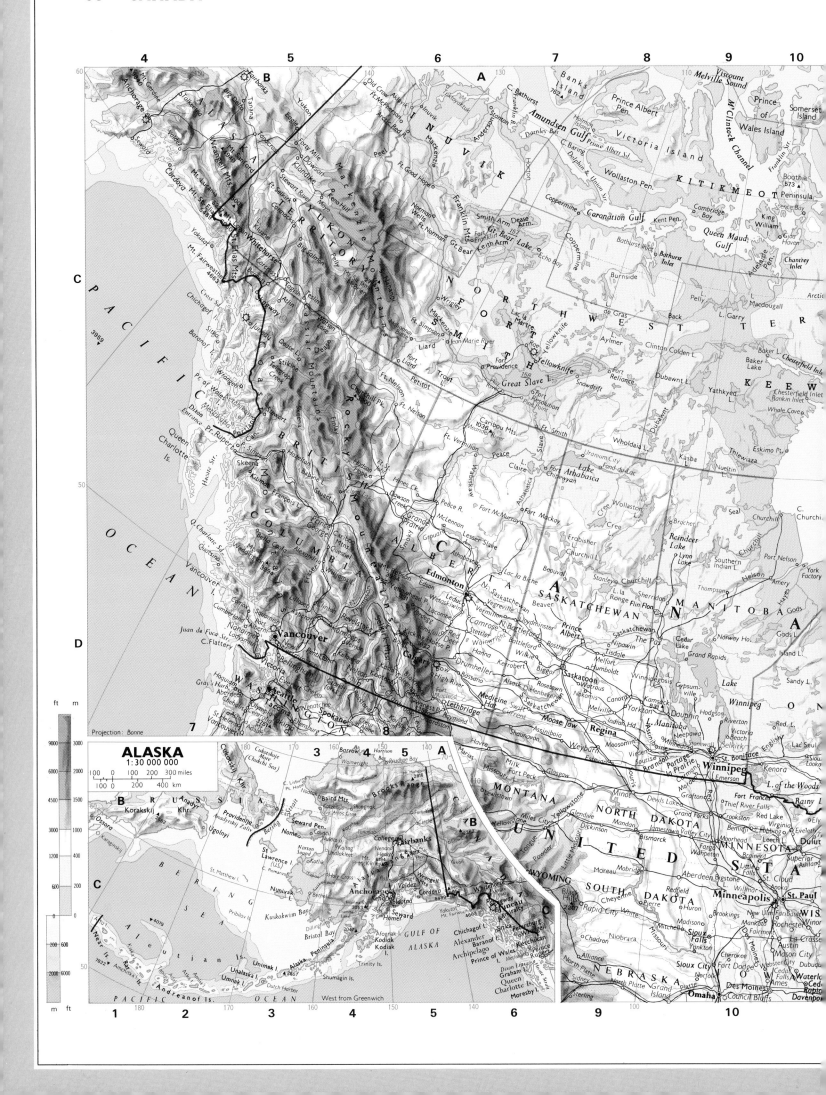

1:15 000 000

COPYRIGHT. GEORGE PHILIP & SON. LTD.

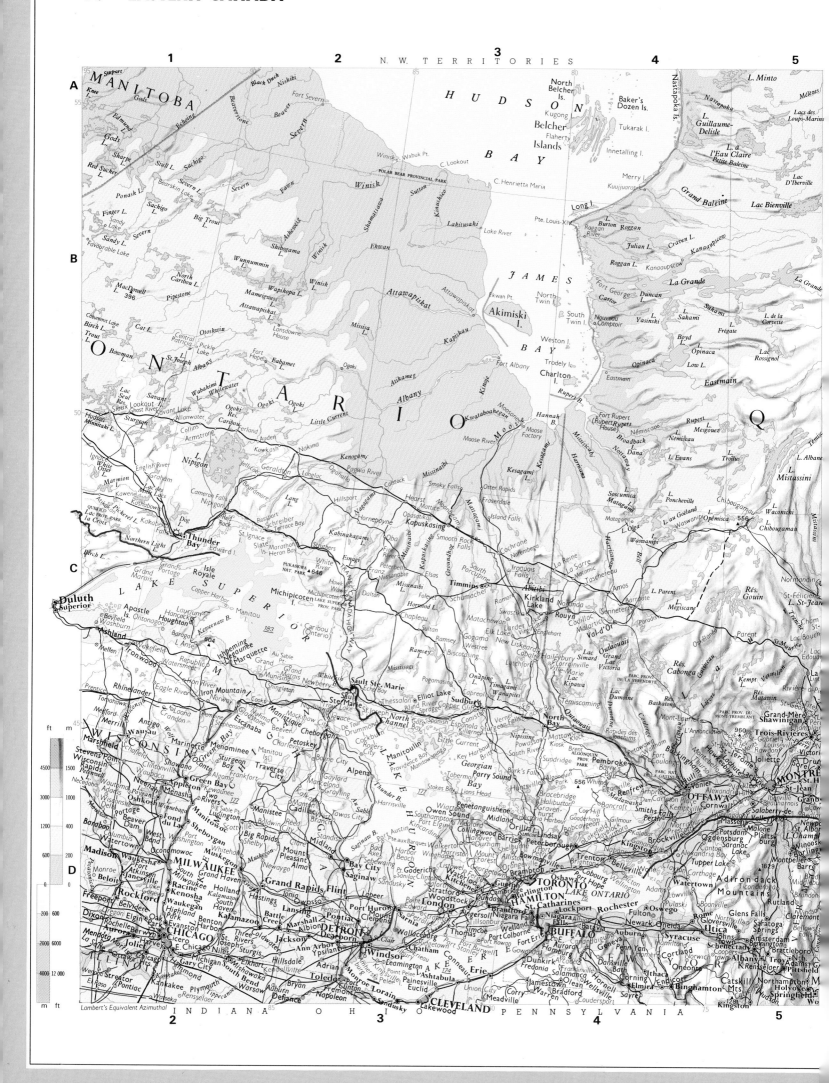

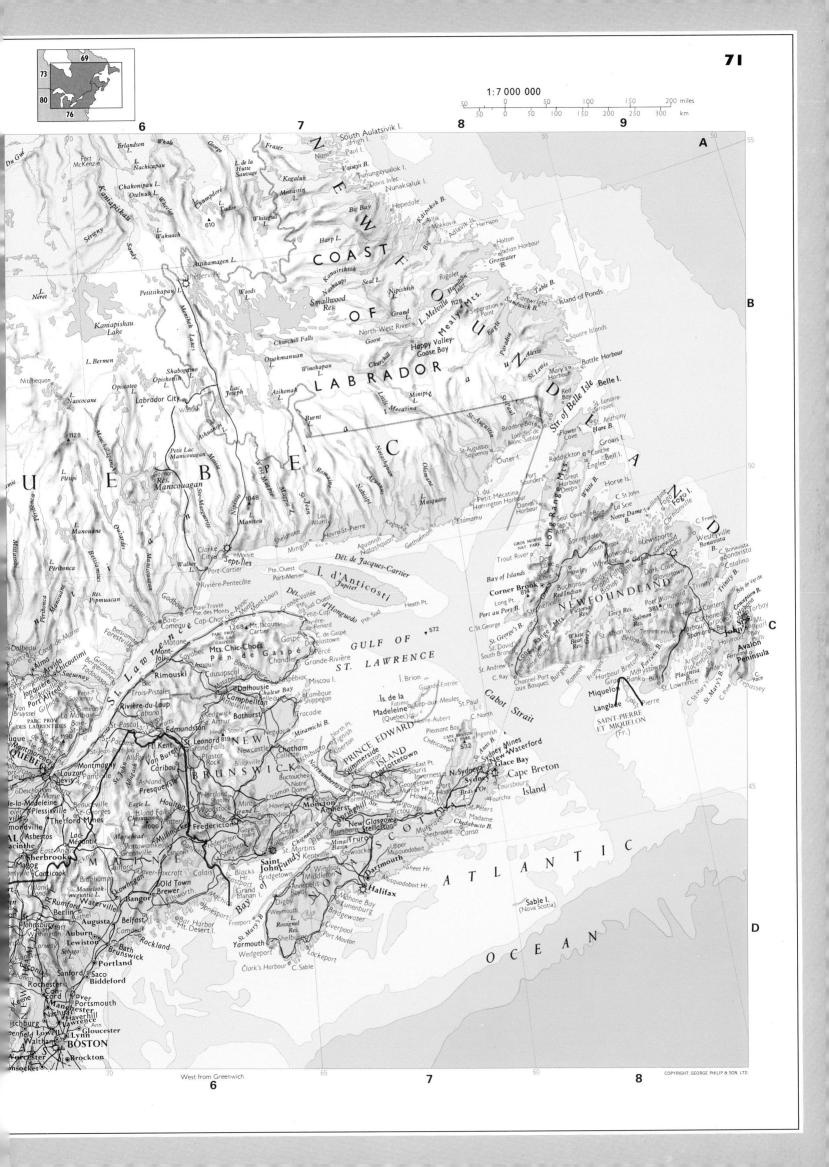

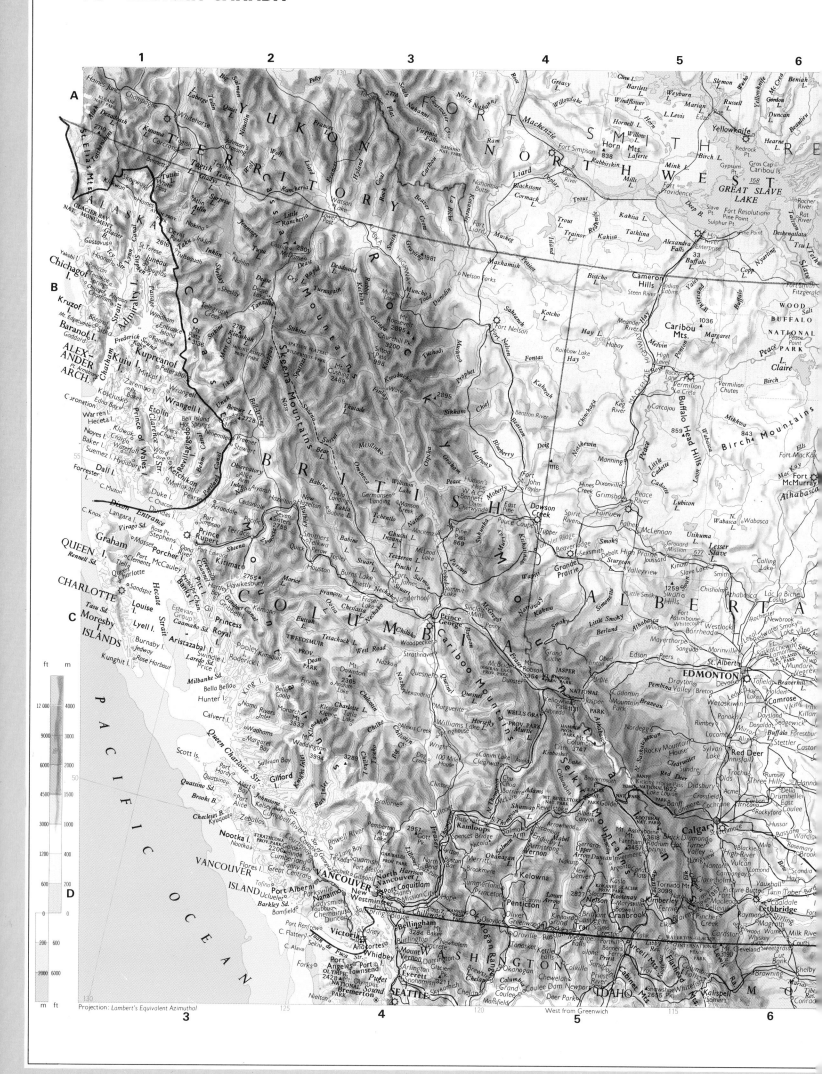

Projection: Lambert's Equivalent Azimuthal

West from Greenwich

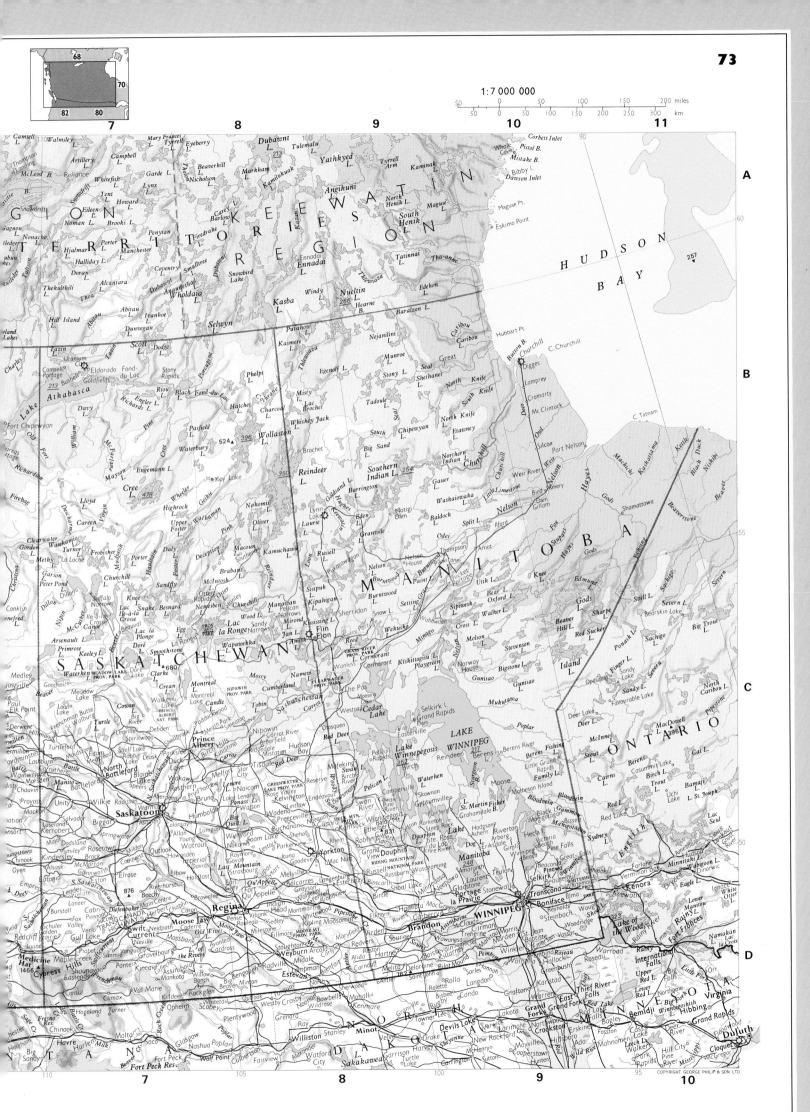

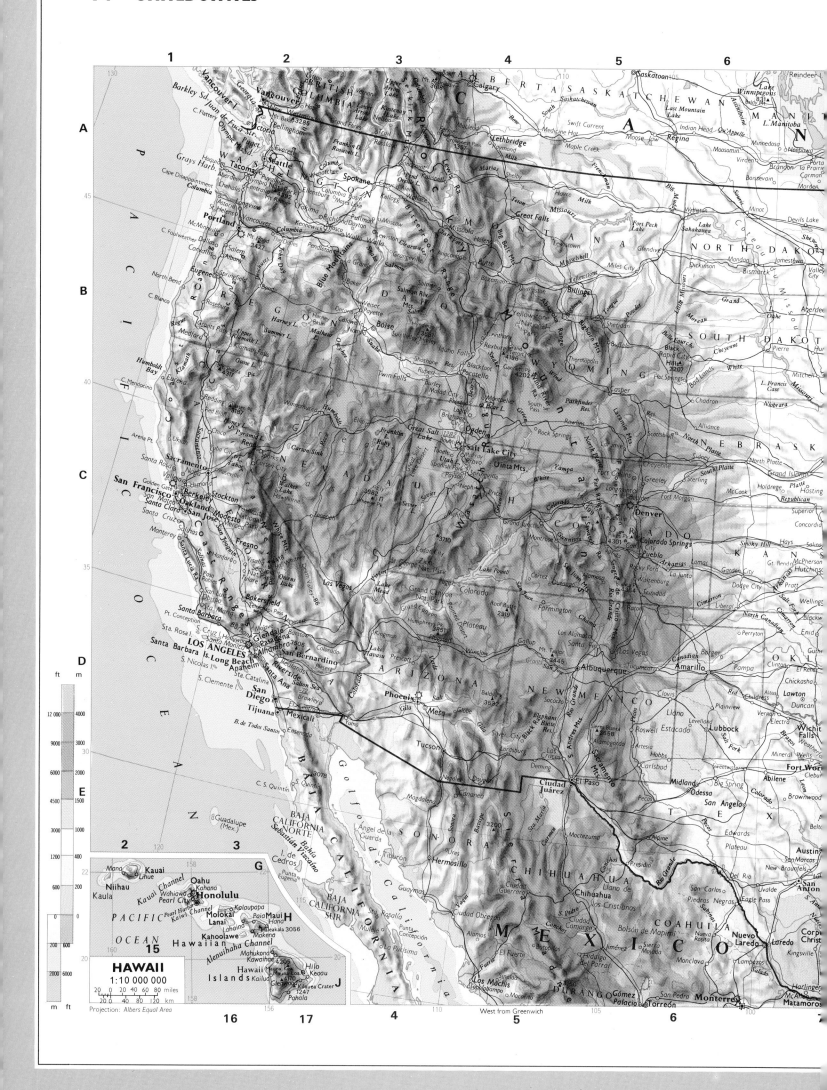

HAWAII
1:10 000 000

Projection: Albers Equal Area

West from Greenwich

LAKE SUPERIOR
ONTARIO
QUEBEC
CANADA
MICHIGAN
LAKE HURON
LAKE ONTARIO
LAKE ERIE
NEW YORK
PENNSYLVANIA
OHIO
INDIANA
KENTUCKY
WEST VIRGINIA
VIRGINIA
MARYLAND
NEW JERSEY
DELAWARE
VERMONT
NEW HAMPSHIRE
MASSACHUSETTS
CHICAGO
DETROIT
TORONTO
MONTREAL
BOSTON
NEW YORK
PHILADELPHIA
PITTSBURGH
CLEVELAND
BALTIMORE
WASHINGTON

1 : 6 000 000

50 0 50 100 150 miles
50 0 50 100 150 200 km

Continuation
Eastwards
On same scale

C A N A D A

M A I N E

NEW HAMPSHIRE

Bangor · Old Town · Brewer
Skowhegan · Waterville · Belfast
Augusta · Gardiner · Rockland
Auburn · Lewiston · Bath
Westbrook · Brunswick
Saco · S. Portland
Biddeford · Portland
Rochester
Dover · Portsmouth
Laconia · Haverhill · Newburyport

A T L A N T I C   O C E A N

BAHAMAS

Hope Town
Great Abaco I.
Little Abaco I. · Gt. Guana Cay
Grand Cays
Settlement Pt. · Grand
Freeport · Bahama I.

T E N N E S S E E

N O R T H   C A R O L I N A

S O U T H   C A R O L I N A

G E O R G I A

A L A B A M A

M I S S I S S I P P I

F L O R I D A

Nashville · Knoxville · Chattanooga
Asheville · Charlotte · Greensboro
Raleigh · Durham · Wilmington
Columbia · Charleston
Savannah · Augusta
Atlanta · Macon · Columbus
Montgomery · Birmingham · Mobile
Pensacola · Tallahassee
Jacksonville · Orlando · Tampa
St. Petersburg · Sarasota
West Palm Beach · Ft. Lauderdale
Hollywood · Miami · Hialeah · Homestead
Everglades Nat. Park
Key West

G U L F   O F   M E X I C O

EVERGLADES
NAT. PARK

BIG CYPRESS SWAMP

Okeechobee L.

West from Greenwich

Projection: Alber's Equal Area with two standard parallels

COPYRIGHT GEORGE PHILIP & SON Ltd

m ft
6000 · 4500 · 3000 · 2000 · 1500 · 1200 · 600 · 200 · 0 · 200-600 · 2000-6000 · 4000-12 000

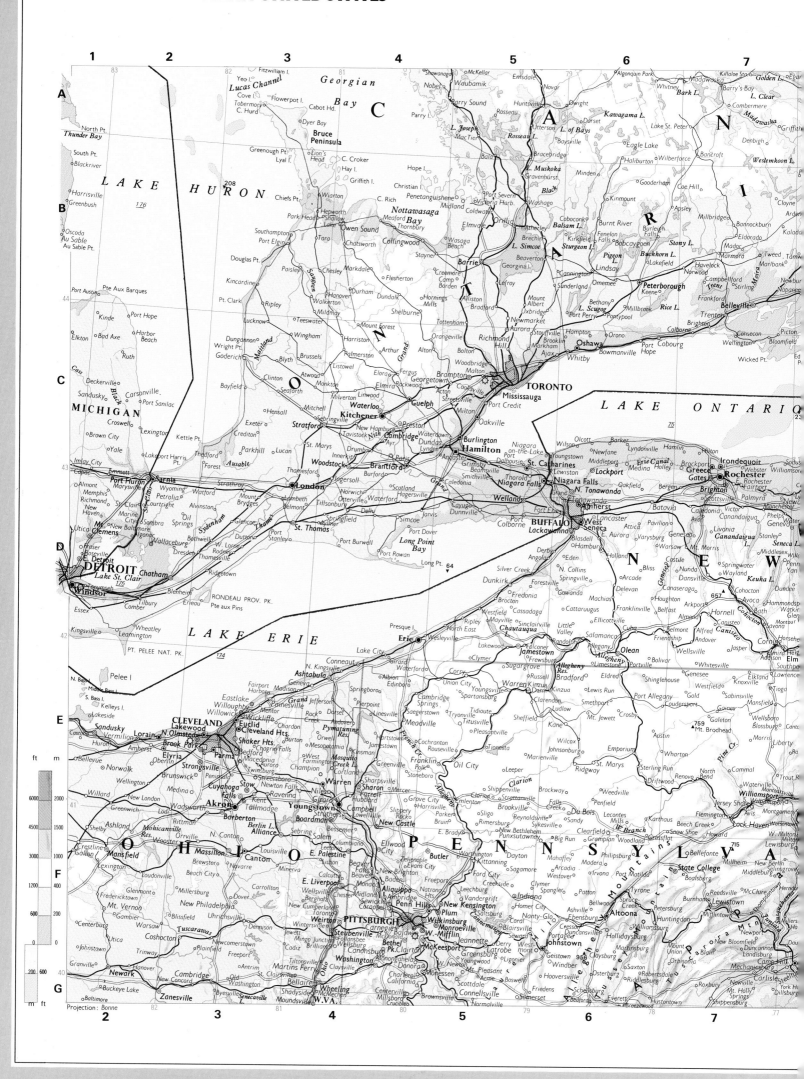

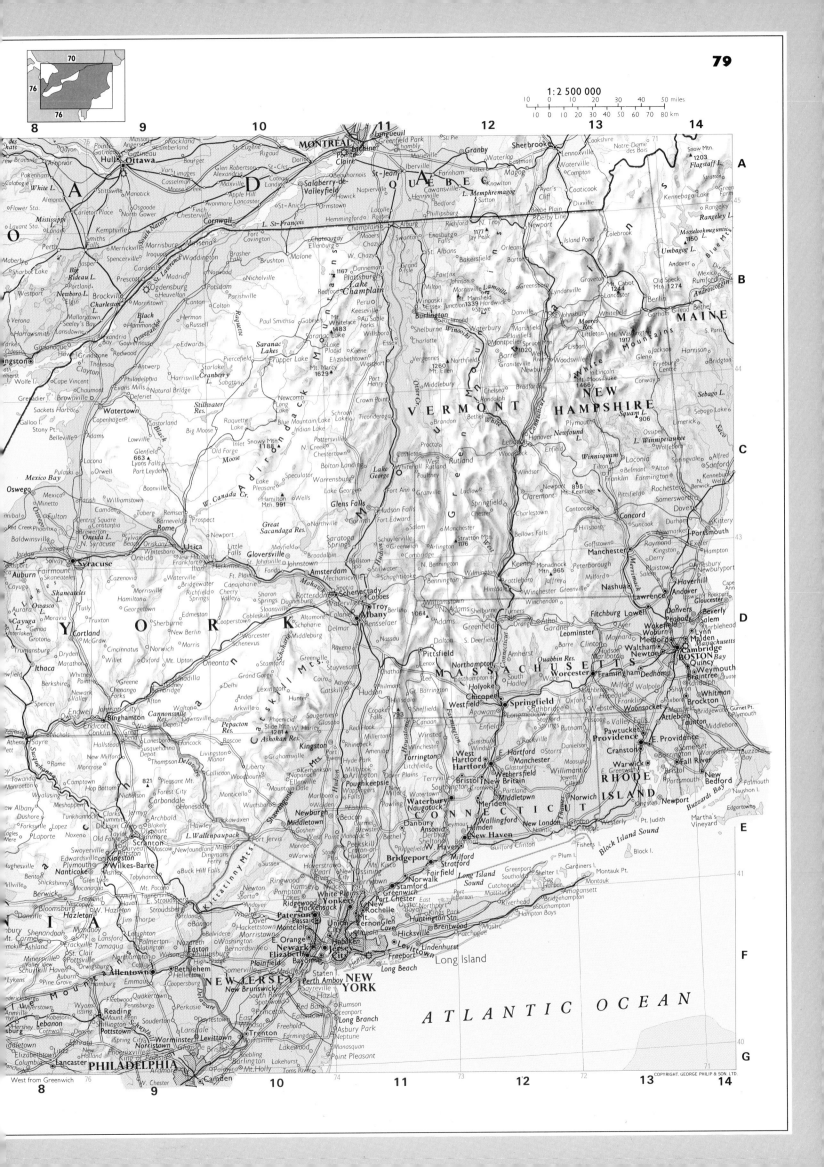

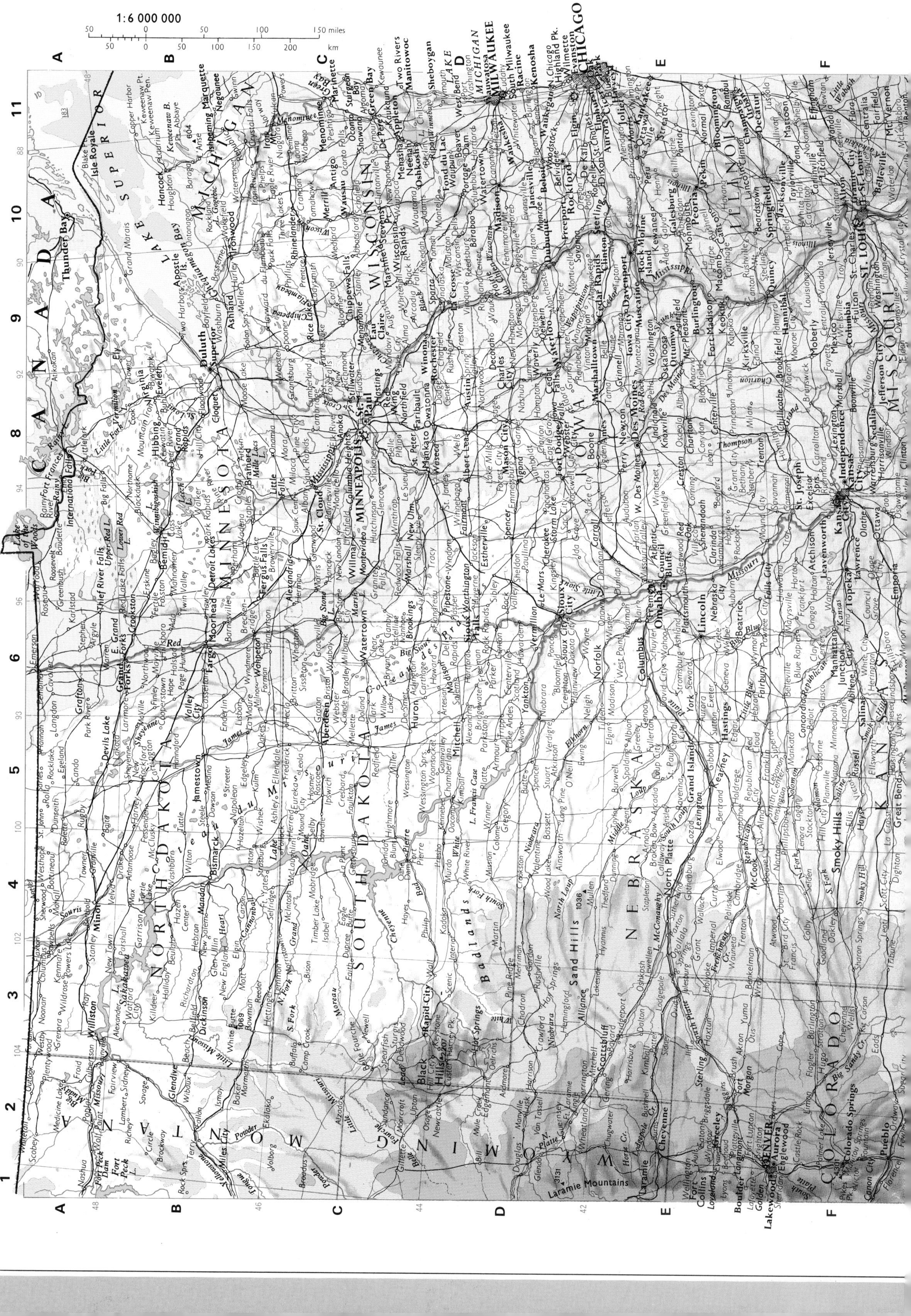

1:6 000 000

50   0   50   100   150 miles
50 0 50 100 150 200 km

CANADA

LAKE SUPERIOR

MICHIGAN

WISCONSIN

MILWAUKEE

CHICAGO

ILLINOIS

MINNESOTA

MINNEAPOLIS
St. Paul

Duluth

NORTH DAKOTA

SOUTH DAKOTA

Badlands

Black Hills

Rapid City

Sand Hills

NEBRASKA

IOWA

Des Moines

Sioux City

Omaha

Lincoln

MISSOURI

ST. LOUIS

Kansas City

KANSAS

Smoky Hills

MONTANA

WYOMING

Laramie Mountains

COLORADO

DENVER

Colorado Springs

Pueblo

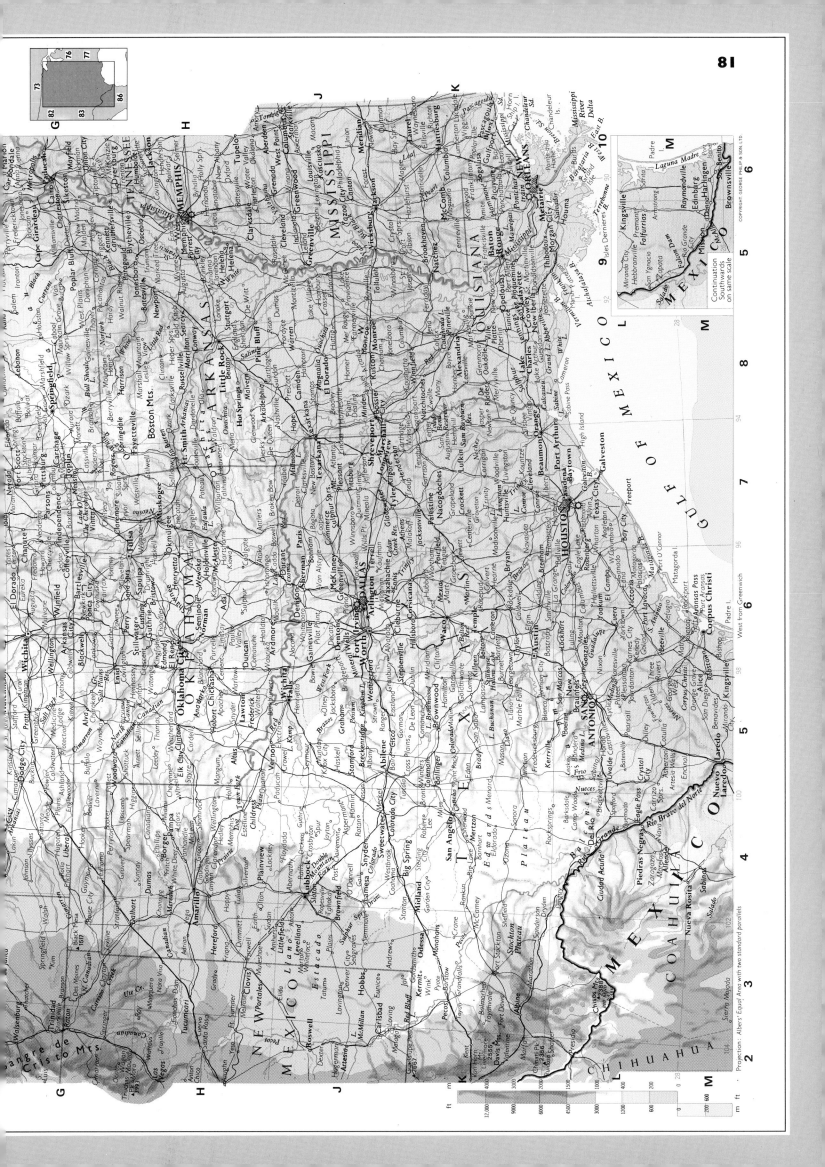

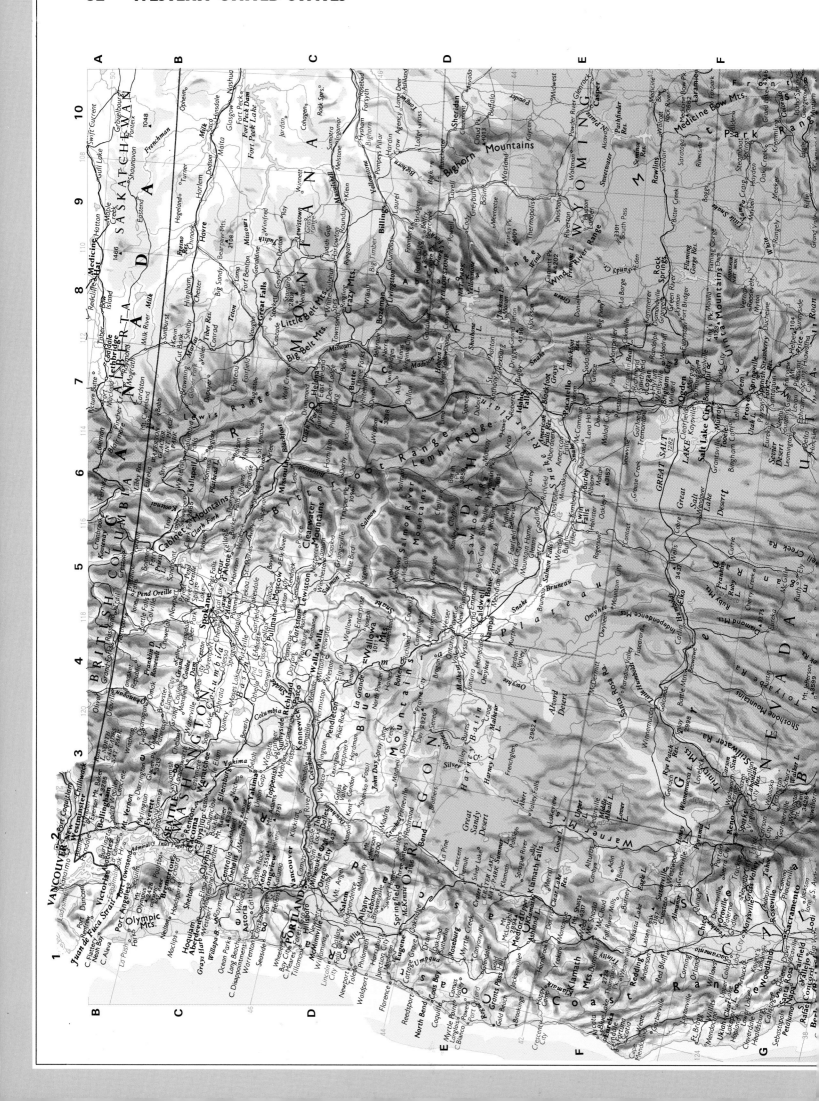

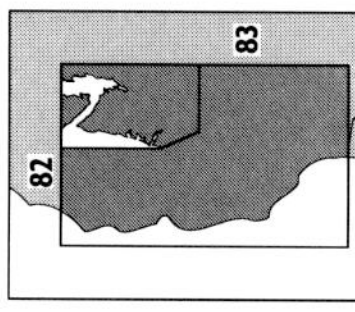

SEATTLE-PORTLAND REGION
On same scale

PACIFIC OCEAN

CANADA

Vancouver Island

Strait of Georgia

Juan de Fuca Strait

Olympic Mountains

WASHINGTON

SEATTLE
Tacoma
Lakewood Center
Olympia
Bremerton
Everett
Bellingham
Victoria
New Westminster
North Vancouver
VANCOUVER

Willapa Hills

PORTLAND
OREGON
Vancouver
Longview
Columbia

White Mts.

Inyo Mts.
Owens

SIERRA NEVADA

CALIFORNIA

Sacramento
San Francisco
Oakland
Berkeley
Richmond
Daly City
San Mateo
Redwood City
Palo Alto
Fremont
Hayward
San Leandro
SAN JOSE
Santa Clara
Sunnyvale
Campbell
Los Gatos
Saratoga
Santa Cruz
Monterey Bay
Pacific Grove
Monterey
Carmel-by-the-Sea
Seaside

Stockton
Modesto
Merced
Fresno
San Joaquin
Napa
Santa Rosa
Vallejo
Fairfield
Vacaville
Woodland
Davis
Chico
Marysville
Yuba City
Roseville
Concord
Walnut Creek
Antioch
Pittsburg
Livermore
Pleasanton
Mountain View
Watsonville
Salinas
Hollister
San Benito

San Joaquin

Diablo Range

Santa Lucia Range

Clear L.

Russian River

Sacramento Valley

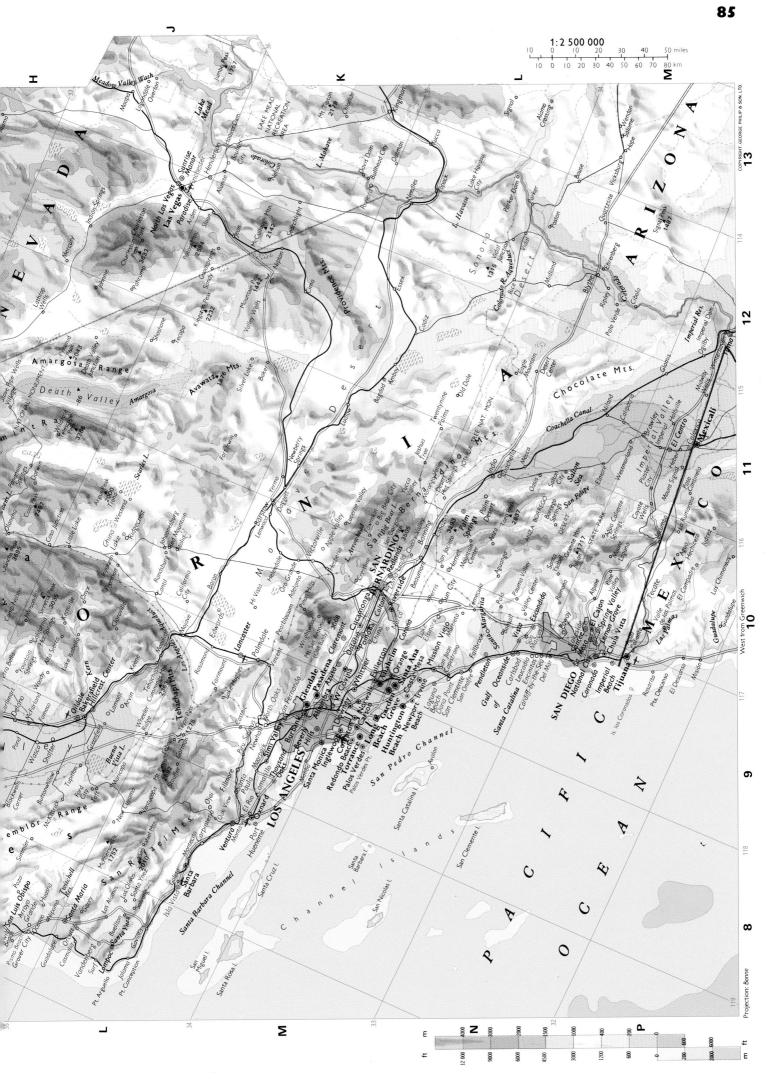

1 : 2 500 000

Projection: Bonne

COPYRIGHT GEORGE PHILIP & SON LTD.

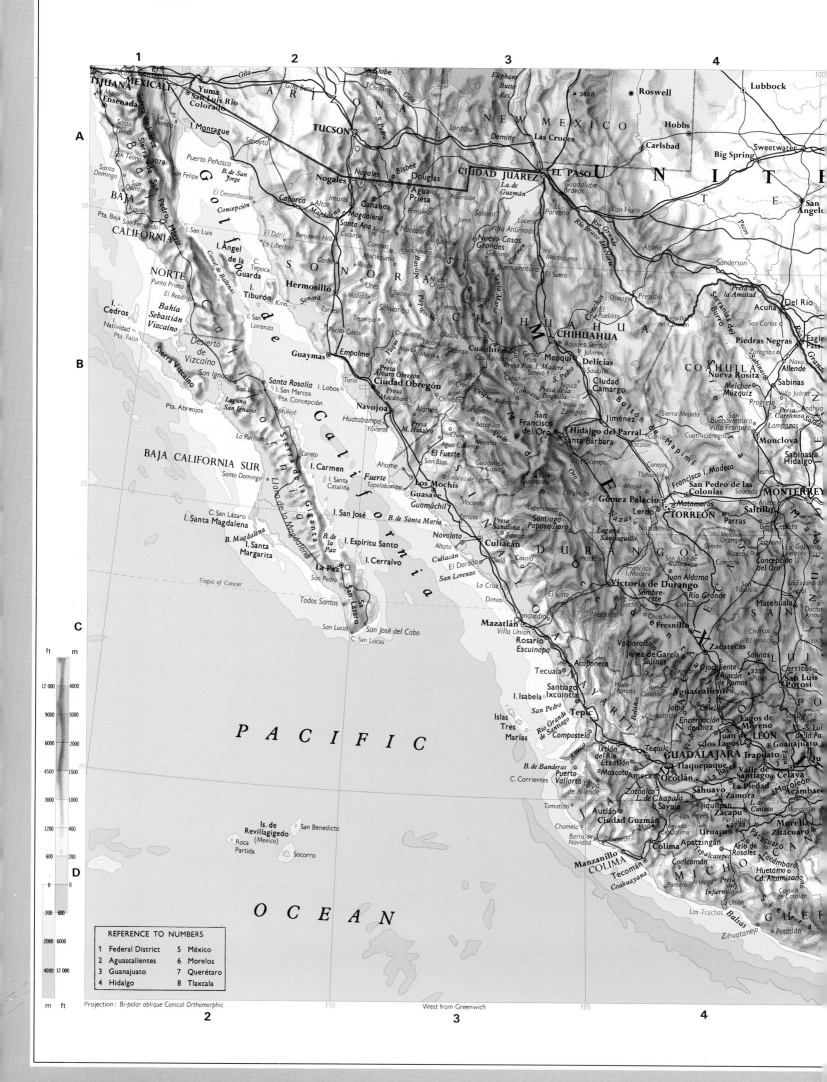

REFERENCE TO NUMBERS

1 Federal District   5 México
2 Aguascalientes   6 Morelos
3 Guanajuato   7 Querétaro
4 Hidalgo   8 Tlaxcala

Projection: Bi-polar oblique Conical Orthomorphic

West from Greenwich

PACIFIC

OCEAN

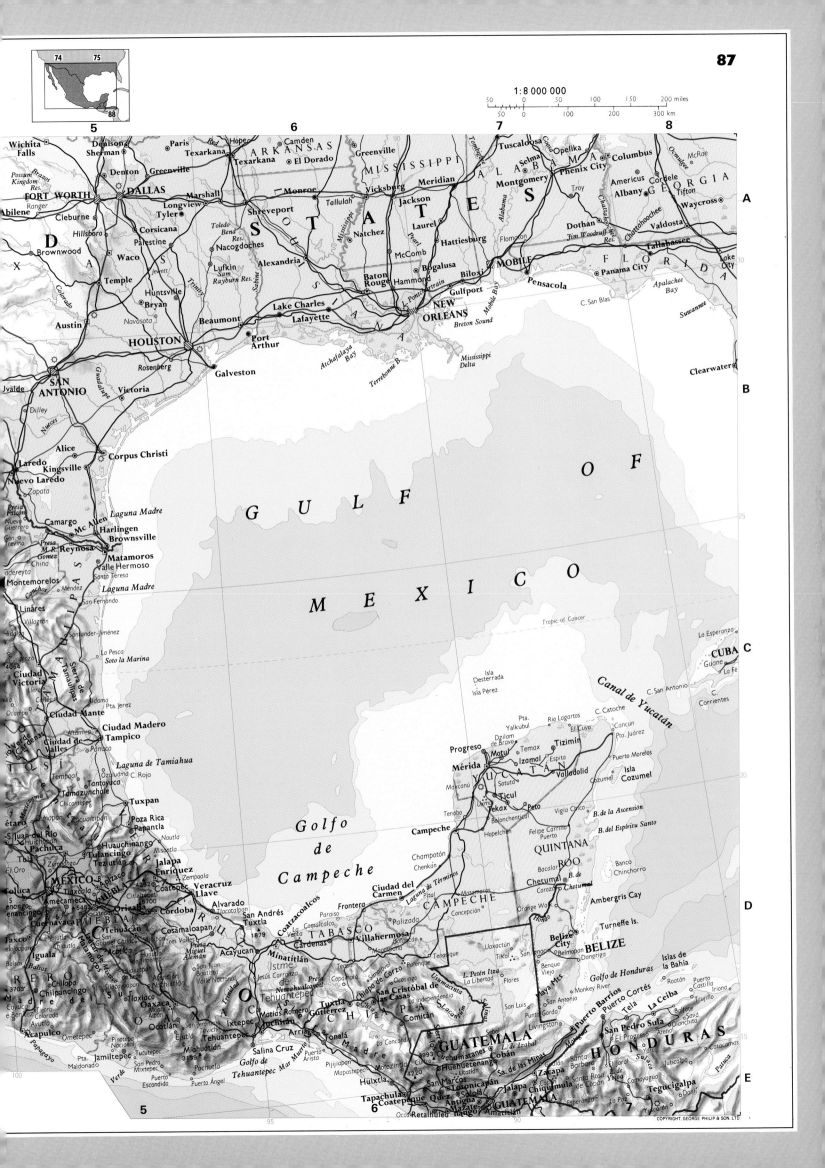

GULF OF MEXICO

U.S.A.
Fort Myers
Naples
C. Romano
C. Sable
Florida Bay
Dry Tortugas
Key West
Florida Keys
Everglades
Hialeah
MIAMI
West Palm Beach
Boca Raton
Fort Lauderdale
L. Okeechobee
The Everglades
Florida City

Little Abaco I.
West End
Normans Castle
Grand Bahama I.
Freeport
Hope Town
Great Abaco I.
Bimini Is.
Berry Is.
Nicolls Town
Eleuthera
Nassau
New Providence
Andros Town
Andros Island
BAH
Northwest Providence Channel
Northeast Providence Channel
Great Guana Cay
Great Exuma I.
Exuma I.
Jumentos Cays
Dunmore Town

(Havana) LA HABANA
LA MARIANAO
San Antonio de los Baños
Guanabacoa
Guanajay
Güines
Batabanó
Jagüey Grande
Matanzas
Cárdenas
Colón
Jovellanos
Sagua la Grande
Santa Clara
Caibarién
Placetas
Morón
Cayo Romano
Ciego de Ávila
Nuevitas
Puerto Manati
Puerto Padr
Gibara
Holguin
Bayamo
Palma Soriano
Manzanillo
SANTIAGO DE CUBA
Sierra Maestra
C. Cruz
Victoria de las Tunas
Camagüey
Florida
Sancti-Spíritus
Trinidad
Cienfuegos
Pinar del Río
Guane
La Fé
San Luis
Los Palacios
Bahía Honda
La Esperanza
Isla de la Juventud
Nueva Gerona
Corrientes
C. San Antonio
Pta. Yalkubul
Archipiélago de los Canarreos
Golfo de Guacanayabo
Golfo de Ana María
Canal de Yucatán
Straits of Florida
Santa Cruz del Norte
Canal Nicolás
Canal Viejo de Bahama
Cay Sal Bank
GREATER
Santaren Channel
GREAT BAHAMA BANK

CUBA

Progreso
Mérida
Motul
Temax
Izamal
Tizimín
Dzilam de Bravo
Río Lagartos
C. Catoche
Cancún
Pto. Juárez
Valladolid
Chichén Itzá
Cozumel
Isla Cozumel
Puerto Morelos
YUCATAN
Campeche
Champotón
QUINTANA ROO
CAMPECHE
Ciudad del Carmen
Laguna de Términos
Isla Desterrada
Isla Pérez

B. de la Ascensión
B. del Espíritu Santo
Banco Chinchorro
Ambergris Cay
BELIZE
Belize City
Turneffe Is.
Islas de la Bahía
Roatán
Swan Islands (U.S.A. & Honduras)

7680

Cayman Islands (Br.)
Georgetown
Grand Cayman
Little Cayman
Cayman Brac

JAMAICA
Montego Bay
Lucea
Falmouth
St. Ann's Bay
Port Maria
Port Antonio
Savanna la Mar
South Negril Pt.
Black River
Mandeville
May Pen
Spanish Town
KINGSTON
Morant
Pedro Cays (Jamaica)

GUATEMALA
Tikal
Flores
L. Petén Itzá
Comitán
Cobán
Puerto Barrios
Puerto Cortés
La Ceiba
Tela
San Pedro Sula
HONDURAS
Tegucigalpa
Golfo de Honduras
Monkey River
Maya Mts.
Punta Gorda
Livingston

Laguna Caratasca
Mosquitia
Puerto Lempira
C. Gracias á Dios
Puerto Cabo Gracias á Dios
Cayos Miskitos (Nicaragua)
Puerto Cabezas

CARIB

EL SALVADOR
SAN SALVADOR
Santa Ana
Ahuachapán
Usulután
San Miguel
Golfo de Fonseca
Chinandega
León
Corinto
NICARAGUA
MANAGUA
Masaya
Granada
Diriamba
Lago de Managua
Matagalpa
Estelí
Juigalpa
Boaco
Lago de Nicaragua
Isla de Ometepe
Rivas
San Juan del Sur
B. de Salinas
C. Sta. Elena
Golfo de Papagayo
Bluefields
El Bluff
Bahía de San Juan del Norte
San Juan del Norte
Río San Juan
San Carlos
Cord. de Yolaina
Rama
Prinzapolca
Siuna
Bonanza
Pta. Gorda
Pta. de Perlas
Islas del Maíz (Nicaragua, U.S.A.)
I. de Providencia (Colombia)
I. de San Andrés (Colombia)
Cayos Roncador (U.S.A. & Colombia)
Cayos de Albuquerque (Colombia)

COSTA RICA
Liberia
Santa Cruz
Nicoya
Puntarenas
Pen. de Nicoya
C. Blanco
Golfo de Nicoya
Alajuela
SAN JOSÉ
Cartago
Limón
Cord. Central
Cord. de Talamanca
Pen. de Osa
Golfo Dulce
Bahía de Coronado
Golfito
Puerto Quepos
Pta. Burica
Puerto Armuelles
David
CARTAGE

PANAMÁ
Colón
Gatún L.
Serranía de Tabasará
Golfo de los Mosquitos
Archipiélago de las Mulatas
Golfo del Darién
Arch. de las Perlas
San Miguel del Rey
Golfo de Panamá
Chitré
Las Tablas
Pen. de Azuero
Pta. Mala
I. de Coiba
I. de Cebaco
I. Jicarón
Is. de San Bernardo
G. de Morrosquill
Lorica
Cereté
Montería
G. de Urabá
Turbo

1:30 000 000

100 0 100 200 300 400 500 miles
100 0 200 400 600 800 km

1 2 3 4 5 6

ATLANTIC OCEAN

A
Sa. Nevada de Santa Marta
Barranquilla
5800
Maracaibo
G. of Darien
Maracaibo
Caracas
Margarita
Tobago I.
Trinidad
5994
Panama Canal

B
Gulf of Panamá
Medellín
Cali
Bogotá
Cord. de Mérida
Cordillera Occidental
Cordillera Central
Cordillera Oriental
Guaviare
Meta
Llanos
Orinoco
Guiana Highlands
2810 Roraima
Sierra Pacaraima
Georgetown
Serra de Tumucumaque
C. Orange
OCEAN

C. de San Francisco
Quito
Cotopaxi
Chimborazo 6267
5897
Guayaquil
G. of Guayaquil
Pta. Parinas
Pta. Aguja
Lobos Is.

C
Napo
Putumayo
Japurá
Negro
Casiquiare
Caquetá
Marañón
Ucayali
Juruá
Purus
Amazon
Madeira
Aripuaná
Roosevelt
Tapajós
Xingu
Teles Pires
Araguaia
Manaus
Equator
Marajó I.
Pará
Belém
Tocantins
Fortaleza
C. São Roque
Plateau of Borborema
C. Branco
Recife

PACIFIC
Lima
Chincha Is.
Huascarán 6768

D
Bolivian Plateau
L. Titicaca
Ancohuma & Illampu 6550
La Paz
L. Poopó
Madre de Dios
Guaporé
Mamoré
Beni
Plateau of Mato Grosso
Brasília
Belo Horizonte
Abrolhos Bank
São Francisco
Brazilian Highlands
Salvador
2890 Pico da Bandeira
Serra da Mantiqueira
Chile
Peru
Trench

E
Tropic of Capricorn
8050
S. Félix
S. Ambrosio
Atacama Desert
Ojos del Salado 6863
Tucumán
Salinas Grandes
Gran Chaco
Pilcomayo
Bermejo
Salado
Paraná
Paraguay
Asunción
Iguaçu Falls
Uruguay
São Paulo
Rio de Janeiro
C. Frio
Serra do Mar
Pôrto Alegre
Lagoa dos Patos
SOUTH
ATLANTIC
OCEAN

F
Arch. de Juan Fernández
Aconcagua 6960
Uspallata Pass
Santiago
Valparaíso
Sierra de Córdoba
Córdoba
L. Mar Chiquita
Rosario
Buenos Aires
La Plata
Montevideo
Rio de la Plata
Pta. Mogotes
Entre Rios
Colorado
Negro
Bahía Blanca
Andes
Pampa
Argentine Basin

G
Chile Rise
Chiloé I.
Chonos Archipelago
Taitao Peninsula
4058 S. Valentin
G. of Peñas
Wellington I.
Madre de Dios I.
Patagonia
G. of San Matias
Valdés Peninsula
G. of San Jorge
6212
OCEAN

H
Santa Ines I.
Magellan's Strait
Cockburn Chan.
Tierra del Fuego
Beagle Chan.
C. Horn
Staten I.
West Falkland
East Falkland
Falkland Islands

West from Greenwich

ft m
18 000 6000
12 000 4000
9000 3000
6000 2000
3000 1000
1200 400
600 200
0 0
200 600
2000 6000
4000 12 000
6000 18 000
8000 24 000
m ft

1:30 000 000

100   0   100 200 300 400 500 miles
100   0   200   400   600   800 km

NORTH ATLANTIC OCEAN

PACIFIC OCEAN

SOUTH ATLANTIC OCEAN

COSTA RICA
San José
PANAMA
Panamá
Golfo de Panamá
Colón
Golfo de Darién
Barranquilla
Cartagena
Maracaibo
Barquisimeto
Valencia
Caracas
Port of Spain
TRINIDAD AND TOBAGO
Cúcuta
San Cristóbal
Medellín
Bucaramanga
Bogotá
Cali
COLOMBIA
VENEZUELA
Orinoco
Ciudad Guayana
Georgetown
Paramaribo
GUYANA
SURINAM
FRENCH GUIANA
Cayenne
C. Orange
Esequibo
Corentijn
Maroni
C. de San Francisco
ECUADOR
Quito
Guayaquil
G. de Guayaquil
Iquitos
Napo
Putumayo
Caquetá
Negro
Branco
Orinoco
Japurá
Equator
Ilha de Marajó
Belém
São Luís
Amazonas (Amazon)
Manaus
Santarém
Fortaleza (Ceara)
Teresina
C. de São Roque
Natal
João Pessoa
Recife (Pernambuco)
Maceió
Aracaju
Salvador
Pta. Aguja
Chiclayo
Trujillo
Chimbote
PERU
Marañón
Ucayali
Juruá
Purus
Madeira
Tapajós
Xingu
Araguaia
Tocantins
Parnaíba
São Francisco
BRAZIL
Callao
Lima
Cuzco
Pôrto Velho
Madre de Dios
Guaporé
Mamoré
Titicaca
BOLIVIA
La Paz
Cochabamba
Santa Cruz
Sucre
Cuiabá
Brasília
Goiânia
Belo Horizonte
Vitória
Ribeirão Prêto
Juiz de Fora
Campos
Campinas
Niterói
RIO DE JANEIRO
Santos
SÃO PAULO
Londrina
Campo Grande
Arequipa
Iquique
Antofagasta
Tropic of Capricorn
Isla San Félix (Chile)
Isla San Ambrosio (Chile)
Salta
San Miguel de Tucumán
PARAGUAY
Asunción
Pilcomayo
Paraná
Paraguay
Uruguay
Curitiba
Resistencia
Corrientes
Pôrto Alegre
Lagoa dos Patos
Pelotas
Córdoba
San Juan
Santa Fe
Paraná
Rosario
URUGUAY
Mendoza
Viña del Mar
Valparaíso
Santiago
ARGENTINA
CHILE
Arch. de Juan Fernández (Chile)
Talca
Concepción
Valdivia
Puerto Montt
Salado
BUENOS AIRES
La Plata
Montevideo
Rio de la Plata
Mar del Plata
Bahía Blanca
Colorado
Negro
Viedma
Chubut
Golfo Comodoro Rivadavia
San Jorge
G. de Peñas
FALKLAND ISLANDS
West Falkland
Stanley
(U.K.)
East Falkland
Punta Arenas
Strait of Magellan
Tierra del Fuego
Cape Horn
West from Greenwich

Projection : Lambert's Equivalent Azimuthal

COPYRIGHT. GEORGE. PHILIP & SON. LTD.

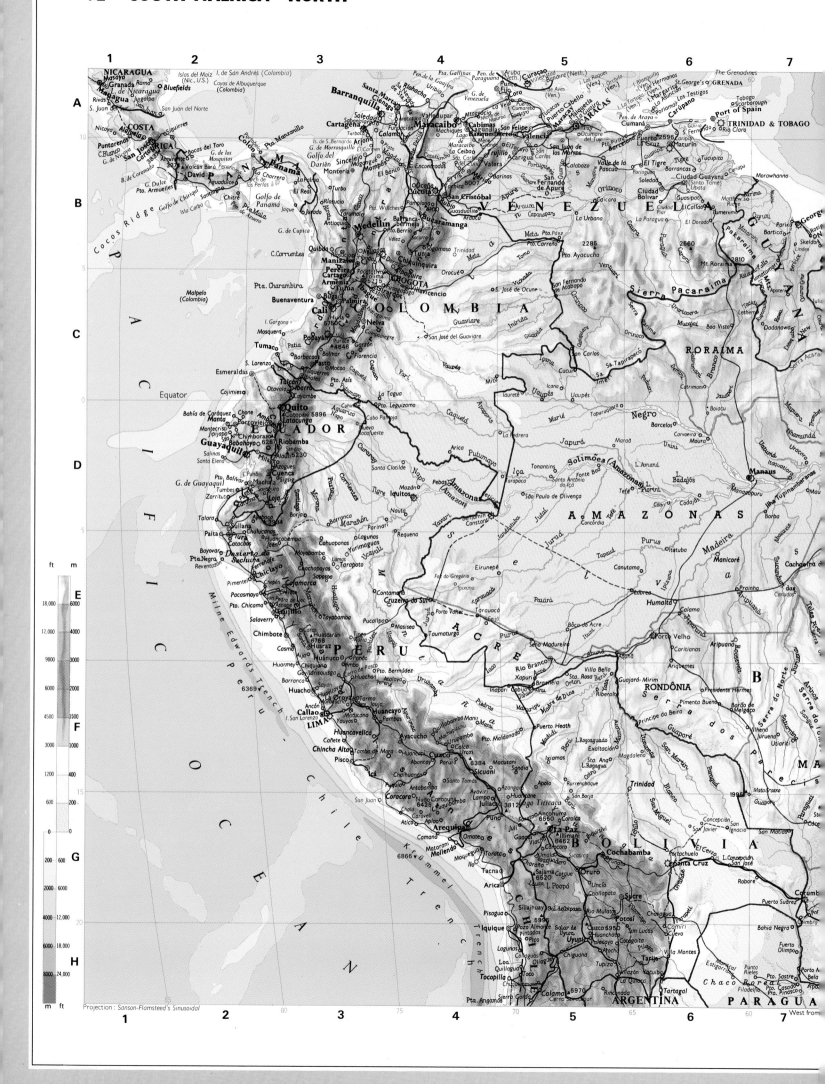

Projection : Sanson-Flamsteed's Sinusoidal

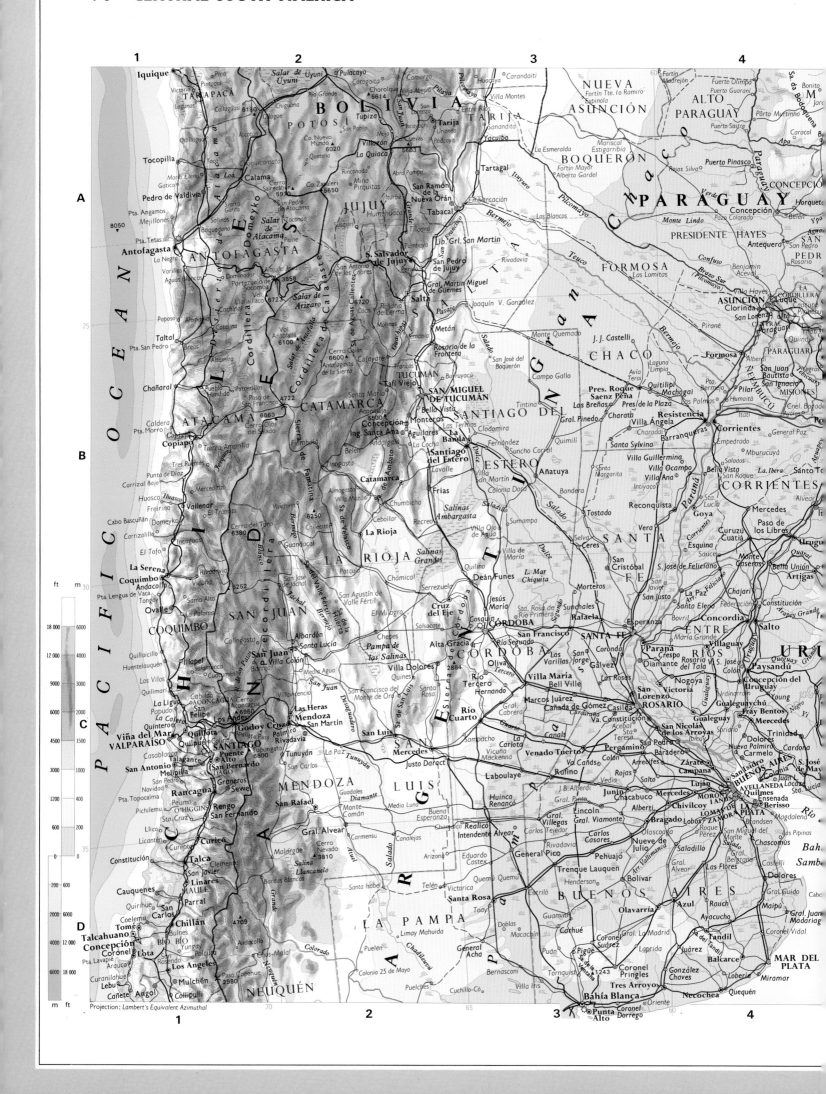

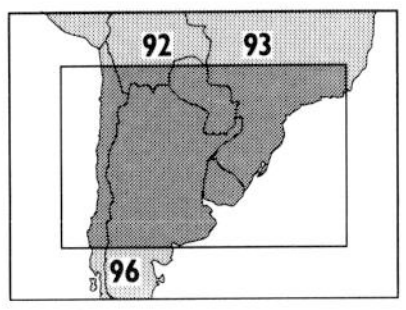

92 93 96

1:8 000 000

50 0 100 150 200 miles
50 0 100 200 300 km

5 **6** **7**

TO GROSSO
DO SUL

Nioaque
Guió Lopes
da Laguna
Maracaju
Dourados
Ponta Porã
Dourados
Rio Brilhante
Santo Anastácio
Três Lagoas
Andradina
Mirassol
S. José
do Rio Prêto
Olímpia
Batatais
Passos
Pardo
Xavantina
Mirandópolis
Araçatuba
Catanduva
Ribeirão Prêto
São José
do Rio Prêto
Oliveira
Cons.
Lafaiete
BELO
HORIZONTE
Lima
Itabirito
Vitória
Iatuari
Vila
Velha
Guarapari

Pedro Juan Caballero
Amambaí
Porto São José
Paraná
Presidente
Prudente
Panorama
Pres.
Epitácio
Adamantina
SÃO
PAULO
Penápolis
Birigui
Tupã
Lins
Novo
Horizonte
Jaboticabal
Mococa
Casa
Branca
Guaxupé
Campo Belo
Três
Pontas
Lavras
Barbacena
Cataguases
Ouro
Prêto
Ponte Nova
Carangola
Ubá
Muriaé
Alegre
Cachoeiro
de Itapemirim

Amambaí
Dourados
Martinópolis
Marília
Garça
Jaú
Rio Claro
Poços de
Caldas
Alfenas
Pouso
Alegre
Três
Corações
São
Lourenço
Juiz de Fora
São
João
del Rei
Leopoldina
Além Paraíba
Cambuci
Guarus
CAMPOS

Igatimí
Ponta Porã
Paranavaí
Rancharia
Paraguaçu
Paulista
Assis
Santa Cruz
do Rio Pardo
Bauru
Bariri
São
Carlos
Araraquara
da Boa Vista
Araras
Pinhal
Ouro Fino
Itajubá
2787
Volta
Redonda
Barra do Piraí
Petrópolis
Macaé
RIO DE JANEIRO
Nova Friburgo
Cabo de
São Tomé

Umuarama
Cruzeiro
do Oeste
Cianorte
Goio
Erê
Campo
Mourão
Londrina
Rolândia
Maringá
Apucarana
Arapongas
Cornélio
Procópio
Ourinhos
Avaré
Jacarèzinho
Piracicaba
CAMPINAS
Botucatu
Limeira
Americana
Mogi-Mirim
Itu
Jundiaí
Bragança
Taubaté
Jacareí
Cruzeiro
Guaratinguetá
S. J. dos Campos
Barra
Mansa
Nova Iguaçu
DUQUE DE CAXIAS
SÃO GONÇALO
NITERÓI
RIO DE JANEIRO
Cabo Frio
La. de Araruama

CANINDEYU
Ivaí
Nova
Esperança
Mandaguari
Tibagi
Jaguariaíva
Itapetininga
Tatuí
Sorocaba
SÃO PAULO
SANTO ANDRÉ
São Bernardo
del Campo
Santos
Tropic of Capricorn

Guaíra
Cruzeiro
do Oeste
Dourados
PARANÁ
BRAZIL
Prudentópolis
Castro
Ponta Grossa
Palmeira
CURITIBA
Antonina
Paranaguá
São Vicente
Guarujá
Itanhaém
Ilha de São Sebastião
Pta. do Boi

Piquiri
Sa. das Araras
Cascavel
Foz do Iguaçu
Guarapuava
Irati
Lapa
Iguape
Ilha Comprida
Ilha do Cardoso
Registro
Juquiá
Apiaí
Itararé
Itapeva

ALTO
PARANÁ
Iguaçu
Falls
Iguaçu
Laranjeiras
do Sul
União da
Vitória
Pto. União
Mafra
Rio Negro
Guaratuba
São Francisco do Sul
Joinvile

ITAPUÁ
Eldorado
São Pedro
Bernardo
de Irigoyen
Chopim
Palmas
Clevelândia
Caçador
Blumenau
Santa Cecília
Itajaí
Brusque

MISIONES
Uruguaí
Chapecó
Joaçaba
1340
SANTA CATARINA
Campos Novos
Rio do Sul
Ilha de Santa Catarina
Florianópolis

Encarnación
Obera
Montenegudo
Erechim
Lajes
1808
Ilha de Santa Catarina

Candelaria
Santa Rosa
Caràzinho
Passo Fundo
Vacaria
Tubarão
Laguna
Cabo Santa Marta Grande

Santo Ângelo
São Luís
Gonzaga
Cruz Alta
Criciúma
Araranguá

RIO GRANDE
Guaporé
Bento Gonçalves
Caxias do Sul

São Borja
Santa Maria
Santa Cruz
do Sul
Montenegro
Nôvo Hamburgo
Taquara
Osorio

DO SUL
Alegrete
Cachoeira do Sul
Rio Pardo
Canoas
São
Leopoldo
Viamão
PÔRTO ALEGRE

Rosário do Sul
São
Gabriel
Camaquã
Lagoa dos Patos
Mostardas

Santana do
Livramento
Dom Pedrito
Bagé
Camaquã
Canguçu

URUGUAY
Tacuarembó
Pelotas
Rio Grande

Melo
Jaguarão
Lagoa
Mirim
ATLANTIC

San Gregorio
Blanquillo
Rio Branco
Sta. Clara
de Olimar
Lagoa Mangueira

Sarandí del Yi
Treinta y Tres
Santa Vitória do Palmar
OCEAN

José Batlle
y Ordóñez
Lascano
Aigua
Castillos

Minas
Rocha
San Carlos
Maldonado

MONTEVIDEO

Plata

5304

West from Greenwich

A
B
C
D

25
30
35
40
45
50
55

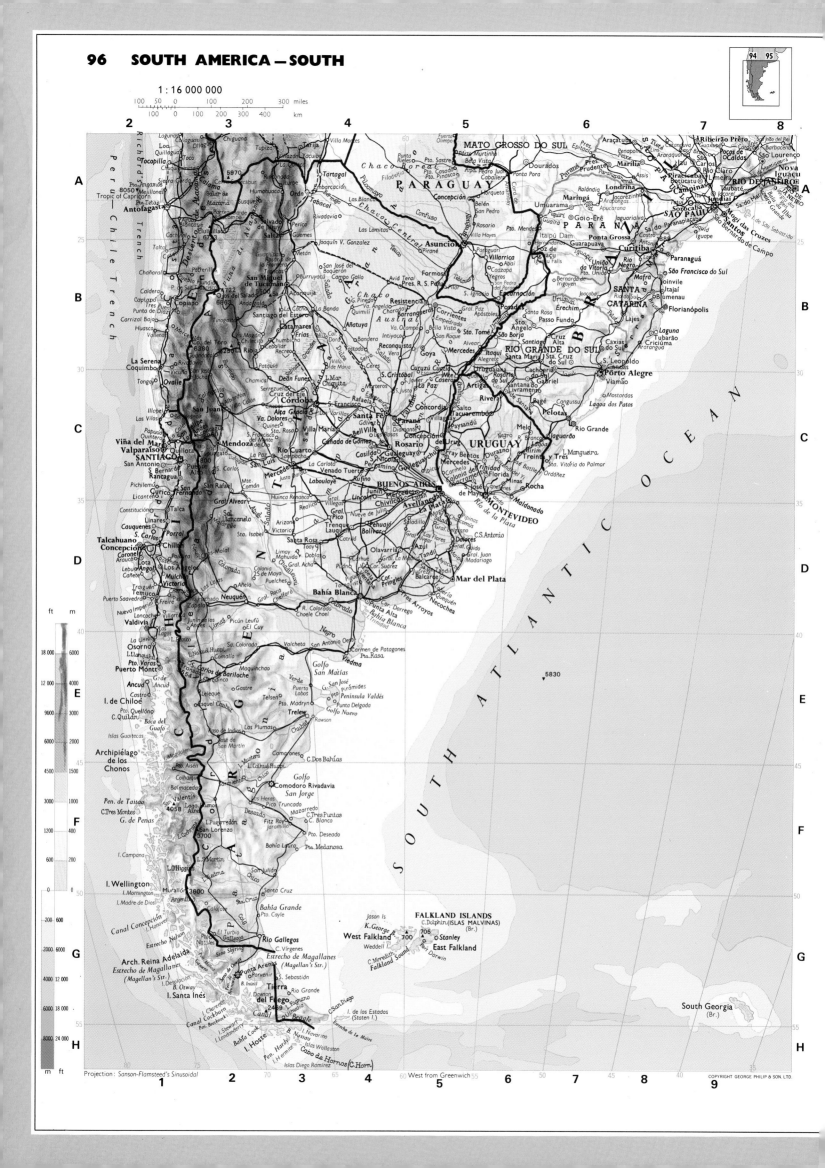

1 : 16 000 000

Projection: Sanson-Flamsteed's Sinusoidal

West from Greenwich

COPYRIGHT GEORGE PHILIP & SON LTD.

INDEX

The index contains the names of all the principal places and features shown on the World Maps. Each name is followed by an additional entry in italics giving the country or region within which it is located. The alphabetical order of names composed of two or more words is governed primarily by the first word and then by the second. This is an example of the rule:

Mīr Kūh, *Iran*	**45 E8**	26 22 N	58 55 E
Mīr Shahdād, *Iran*	**45 E8**	26 15 N	58 29 E
Miraj, *India*	**40 L9**	16 50 N	74 45 E
Miram Shah, *Pakistan*	**42 C4**	33 0 N	70 2 E
Miramar, *Mozam.*	**57 C6**	23 50 S	35 35 E

Physical features composed of a proper name (Erie) and a description (Lake) are positioned alphabetically by the proper name. The description is positioned after the proper name and is usually abbreviated:

Erie, L., *N. Amer.*	**78 D3**	42 15 N	81 0 W

Where a description forms part of a settlement or administrative name, however, it is always written in full and put in its true alphabetic position:

Mount Morris, *U.S.A.*	**78 D7**	42 44 N	77 52 W

Names beginning with M' and Mc are indexed as if they were spelled Mac. Names beginning St. are alphabetized under Saint, but Sankt, Sint, Sant', Santa and San are all spelled in full and are alphabetized accordingly. If the same place name occurs two or more times in the index and all are in the same country, each is followed by the name of the administrative subdivision in which it is located. The names are placed in the alphabetical order of the subdivisions. For example:

Jackson, *Ky., U.S.A.*	**76 G4**	37 33 N	83 23 W
Jackson, *Mich., U.S.A.*	**76 D3**	42 15 N	84 24 W
Jackson, *Minn., U.S.A.*	**80 D7**	43 37 N	95 1 W

The number in bold type which follows each name in the index refers to the number of the map page where that feature or place will be found. This is usually the largest scale at which the place or feature appears.

The letter and figure which are in bold type immediately after the page number give the grid square on the map page, within which the feature is situated. The letter represents the latitude and the figure the longitude.

In some cases the feature itself may fall within the specified square, while the name is outside. This is usually the case only with features which are larger than a grid square.

For a more precise location the geographical coordinates which follow the letter/figure references give the latitude and the longitude of each place. The first set of figures represent the latitude which is the distance north or south of the Equator measured as an angle at the centre of the Earth. The Equator is latitude 0°, the North Pole is 90°N, and the South Pole 90°S.

The second set of figures represent the longitude, which is the distance East or West of the prime meridian, which runs through Greenwich, England. Longitude is also measured as an angle at the centre of the earth and is given East or West of the prime meridian, from 0° to 180° in either direction.

The unit of measurement for latitude and longitude is the degree, which is subdivided into 60 minutes. Each index entry states the position of a place in degrees and minutes, a space being left between the degrees and the minutes.

The latitude is followed by N(orth) or S(outh) and the longitude by E(ast) or W(est).

Rivers are indexed to their mouths or confluences, and carry the symbol → after their names. A solid square ■ follows the name of a country while, an open square □ refers to a first order administrative area.

Abbreviations used in the index

A.C.T. — Australian Capital Territory
Afghan. — Afghanistan
Ala. — Alabama
Alta. — Alberta
Amer. — America(n)
Arch. — Archipelago
Ariz. — Arizona
Ark. — Arkansas
Atl. Oc. — Atlantic Ocean
B. — Baie, Bahía, Bay, Bucht, Bugt
B.C. — British Columbia
Bangla. — Bangladesh
Barr. — Barrage
Bos. & H. — Bosnia and Herzegovina
C. — Cabo, Cap, Cape, Coast
C.A.R. — Central African Republic
C. Prov. — Cape Province
Calif. — California
Cent. — Central
Chan. — Channel
Colo. — Colorado
Conn. — Connecticut
Cord. — Cordillera
Cr. — Creek
D.C. — District of Columbia
Del. — Delaware
Dep. — Dependency
Des. — Desert
Dist. — District
Dj. — Djebel
Domin. — Dominica
Dom. Rep. — Dominican Republic
E. — East
El Salv. — El Salvador

Eq. Guin. — Equatorial Guinea
Fla. — Florida
Falk. Is. — Falkland Is.
G. — Golfe, Golfo, Gulf, Guba, Gebel
Ga. — Georgia
Gt. — Great, Greater
Guinea-Biss. — Guinea-Bissau
H.K. — Hong Kong
H.P. — Himachal Pradesh
Hants. — Hampshire
Harb. — Harbor, Harbour
Hd. — Head
Hts. — Heights
I.(s). — Île, Ilha, Insel, Isla, Island, Isle
Ill. — Illinois
Ind. — Indiana
Ind. Oc. — Indian Ocean
Ivory C. — Ivory Coast
J. — Jabal, Jebel, Jazira
Junc. — Junction
K. — Kap, Kapp
Kans. — Kansas
Kep. — Kepulauan
Ky. — Kentucky
L. — Lac, Lacul, Lago, Lagoa, Lake, Limni, Loch, Lough
La. — Louisiana
Liech. — Liechtenstein
Lux. — Luxembourg
Mad. P. — Madhya Pradesh
Madag. — Madagascar
Man. — Manitoba
Mass. — Massachusetts
Md. — Maryland

Me. — Maine
Medit. S. — Mediterranean Sea
Mich. — Michigan
Minn. — Minnesota
Miss. — Mississippi
Mo. — Missouri
Mont. — Montana
Mozam. — Mozambique
Mt.(e). — Mont, Monte, Monti, Montaña, Mountain
N. — Nord, Norte, North, Northern, Nouveau
N.B. — New Brunswick
N.C. — North Carolina
N. Cal. — New Caledonia
N. Dak. — North Dakota
N.H. — New Hampshire
N.I. — North Island
N.J. — New Jersey
N. Mex. — New Mexico
N.S. — Nova Scotia
N.S.W. — New South Wales
N.W.T. — North West Territory
N.Y. — New York
N.Z. — New Zealand
Nebr. — Nebraska
Neths. — Netherlands
Nev. — Nevada
Nfld. — Newfoundland
Nic. — Nicaragua
O. — Oued, Ouadi
Occ. — Occidentale
O.F.S. — Orange Free State
Okla. — Oklahoma
Ont. — Ontario
Or. — Orientale

Oreg. — Oregon
Os. — Ostrov
Oz. — Ozero
P. — Pass, Passo, Pasul, Pulau
P.E.I. — Prince Edward Island
Pa. — Pennsylvania
Pac. Oc. — Pacific Ocean
Papua N.G. — Papua New Guinea
Pass. — Passage
Pen. — Peninsula, Péninsule
Phil. — Philippines
Pk. — Park, Peak
Plat. — Plateau
P-ov. — Poluostrov
Prov. — Province, Provincial
Pt. — Point
Pta. — Ponta, Punta
Pte. — Pointe
Qué. — Québec
Queens. — Queensland
R. — Rio, River
R.I. — Rhode Island
Ra.(s). — Range(s)
Raj. — Rajasthan
Reg. — Region
Rep. — Republic
Res. — Reserve, Reservoir
S. — San, South, Sea
Si. Arabia — Saudi Arabia
S.C. — South Carolina
S. Dak. — South Dakota
S.I. — South Island
S. Leone — Sierra Leone
Sa. — Serra, Sierra
Sask. — Saskatchewan
Scot. — Scotland

Sd. — Sound
Sev. — Severnaya
Sib. — Siberia
Sprs. — Springs
St. — Saint, Sankt, Sint
Sta. — Santa, Station
Ste. — Sainte
Sto. — Santo
Str. — Strait, Stretto
Switz. — Switzerland
Tas. — Tasmania
Tenn. — Tennessee
Tex. — Texas
Tg. — Tanjung
Trin. & Tob. — Trinidad & Tobago
U.A.E. — United Arab Emirates
U.K. — United Kingdom
U.S.A. — United States of America
Ut. P. — Uttar Pradesh
Va. — Virginia
Vdkhr. — Vodokhranilishche
Vf. — Vîrful
Vic. — Victoria
Vol. — Volcano
Vt. — Vermont
W. — Wadi, West
W. Va. — West Virginia
Wash. — Washington
Wis. — Wisconsin
Wlkp. — Wielkopolski
Wyo. — Wyoming
Yorks. — Yorkshire

A Coruña

A

A Coruña = La Coruña, Spain 19 A1 43 20N 8 25W
Aachen, Germany 16 C3 50 47N 6 4 E
Aalborg = Ålborg, Denmark 9 H10 57 2N 9 54 E
A'âli en Nîl □, Sudan 51 G11 9 30N 31 30 E
Aalsmeer, Neths. 15 B4 52 17N 4 43 E
Aalst, Belgium 15 D4 50 56N 4 2 E
Aalten, Neths. 15 C6 51 56N 6 35 E
Aarau, Switz. 16 E4 47 23N 8 4 E
Aare →, Switz. 16 E4 47 33N 8 14 E
Aarhus = Århus, Denmark 9 H11 56 8N 10 11 E
Aarschot, Belgium 15 D4 50 59N 4 49 E
Aba, Nigeria 50 G6 5 10N 7 19 E
Aba, Zaïre 54 B3 3 58N 30 17 E
Äbâdân, Iran 45 D6 30 22N 48 20 E
Äbâdeh, Iran 45 D7 31 8N 52 40 E
Abadla, Algeria 50 B4 31 2N 2 45W
Abaetetuba, Brazil 93 D9 1 40S 48 50W
Abagnar Qi, China 34 C9 43 52N 116 2 E
Abai, Paraguay 95 B4 25 58S 55 54W
Abakan, Russia 27 D10 53 40N 91 10 E
Abancay, Peru 92 F4 13 35S 72 55W
Abariringa, Kiribati 64 H10 2 50S 171 40W
Abarqū, Iran 45 D7 31 10N 53 20 E
Abashiri, Japan 30 B12 44 0N 144 15 E
Abashiri-Wan, Japan 30 B12 44 0N 144 30 E
Abay, Kazakhstan 26 E8 49 38N 72 53 E
Abaya, L., Ethiopia 51 G12 6 30N 37 50 E
Abaza, Russia 26 D10 52 39N 90 6 E
'Abbāsābād, Iran 45 C8 33 34N 58 23 E
Abbay = Nîl el Azraq →, Sudan 51 E11 15 38N 32 31 E
Abbaye, Pt., U.S.A. 76 B1 46 58N 88 8W
Abbeville, France 18 A4 50 6N 1 49 E
Abbeville, La., U.S.A. 81 K8 29 58N 92 8W
Abbeville, S.C., U.S.A. 77 H4 34 11N 82 23W
Abbieglassie, Australia 63 D4 27 15S 147 28 E
Abbot Ice Shelf, Antarctica 5 D16 73 0S 92 0W
Abbotsford, Canada 72 D4 49 5N 122 20W
Abbotsford, U.S.A. 80 C9 44 57N 90 19W
Abbottabad, Pakistan 42 B5 34 10N 73 15 E
Abd al Kūrī, Ind. Oc. 46 E5 12 5N 52 20 E
Äbdar, Iran 45 D7 30 16N 55 19 E
'Abdolābād, Iran 45 C8 34 12N 56 30 E
Abéché, Chad 51 F9 13 50N 20 35 E
Åbenrå, Denmark 9 J10 55 3N 9 25 E
Abeokuta, Nigeria 50 G5 7 3N 3 19 E
Aber, Uganda 54 B3 2 12N 32 25 E
Aberaeron, U.K. 11 E3 52 15N 4 16W
Aberayron = Aberaeron, U.K. 11 E3 52 15N 4 16W
Abercorn = Mbala, Zambia 55 D3 8 46S 31 24 E
Abercorn, Australia 63 D5 25 12S 151 5 E
Aberdare, U.K. 11 F4 51 43N 3 27W
Aberdare Ra., Kenya 54 C4 0 15S 36 50 E
Aberdeen, Australia 63 E5 32 9S 150 56 E
Aberdeen, Canada 73 C7 52 20N 106 8W
Aberdeen, S. Africa 56 E3 32 28S 24 2 E
Aberdeen, U.K. 12 D6 57 9N 2 6W
Aberdeen, Ala., U.S.A. 77 J1 33 49N 88 33W
Aberdeen, Idaho, U.S.A. 82 E7 42 57N 112 50W
Aberdeen, S. Dak., U.S.A. 80 C5 45 28N 98 29W
Aberdeen, Wash., U.S.A. 84 D3 46 59N 123 50W
Aberdovey = Aberdyfi, U.K. 11 E3 52 33N 4 3W
Aberdyfi, U.K. 11 E3 52 33N 4 3W
Aberfeldy, U.K. 12 E5 56 37N 3 50W
Abergavenny, U.K. 11 F4 51 49N 3 1W
Abernathy, U.S.A. 81 J4 33 50N 101 51W
Abert, L., U.S.A. 82 E3 42 38N 120 14W
Aberystwyth, U.K. 11 E3 52 25N 4 6W
Abhar, Iran 45 B6 36 9N 49 13 E
Abhayapuri, India 43 F14 26 24N 90 38 E
Abidjan, Ivory C. 50 G4 5 26N 3 58W
Abilene, Kans., U.S.A. 80 F6 38 55N 97 13W
Abilene, Tex., U.S.A. 81 J5 32 28N 99 43W
Abingdon, U.K. 11 F6 51 40N 1 17W
Abingdon, Ill., U.S.A. 80 E9 40 48N 90 24W
Abingdon, Va., U.S.A. 77 G5 36 43N 81 59W
Abington Reef, Australia 62 B4 18 0S 149 35 E
Abitau →, Canada 73 B7 59 53N 109 3W
Abitau L., Canada 73 A7 60 27N 107 15W
Abitibi L., Canada 70 C4 48 40N 79 40W
Abkhaz Republic □, Georgia 25 F7 43 0N 41 0 E
Abkit, Russia 27 C16 64 10N 157 10 E
Abminga, Australia 63 D1 26 8S 134 51 E
Abohar, India 42 D6 30 10N 74 10 E
Abomey, Benin 50 G5 7 10N 2 5 E
Abong-Mbang, Cameroon 52 D2 4 0N 13 8 E
Abou-Deïa, Chad 51 F8 11 20N 19 20 E
Aboyne, U.K. 12 D6 57 4N 2 48W
Abra Pampa, Argentina 94 A2 22 43S 65 42W
Abrantes, Portugal 19 C1 39 24N 8 7W
Abreojos, Pta., Mexico 86 B2 26 50N 113 40W
Abri, Sudan 51 D11 20 50N 30 27 E
Abrolhos, Banka, Brazil 93 G11 18 0S 38 0W
Abrud, Romania 17 E11 46 19N 23 5 E
Abruzzi □, Italy 20 C5 42 15N 14 0 E
Absaroka Range, U.S.A. 82 D9 44 45N 109 50W
Abū al Khaşīb, Iraq 45 D6 30 25N 48 0 E
Abū 'Alī, Si. Arabia 45 E6 27 20N 49 27 E
Abū 'Alī →, Lebanon 47 A4 34 25N 35 50 E
Abu 'Arīsh, Si. Arabia 46 D3 16 53N 42 48 E
Abu Dhabi = Abū Ẓāby, U.A.E. 45 E7 24 28N 54 22 E
Abū Dis, Sudan 51 E11 19 12N 33 38 E
Abū Du'ān, Syria 44 B3 36 25N 38 15 E
Abu el Gairi, W. →, Egypt 47 F2 29 35N 33 30 E
Abu Ga'da, W. →, Egypt 47 F1 29 15N 32 53 E
Abū Ḩadrīyah, Si. Arabia 45 E6 27 20N 48 58 E
Abu Hamed, Sudan 51 E11 19 32N 33 13 E
Abū Kamāl, Syria 44 C4 34 30N 41 0 E
Abū Madd, Ra's, Si. Arabia 44 E3 24 50N 37 7 E
Abu Matariq, Sudan 51 F10 10 59N 26 9 E
Abu Şafāt, W. →, Jordan 47 E5 30 24N 36 7 E
Abū Şukhayr, Iraq 44 D5 31 54N 44 30 E
Abu Tig, Egypt 51 C11 27 4N 31 15 E
Abû Zabad, Sudan 51 F10 12 25N 29 10 E
Abū Ẓāby, U.A.E. 45 E7 24 28N 54 22 E
Abū Zeydābād, Iran 45 C6 33 54N 51 45 E
Abuja, Nigeria 50 G6 9 16N 7 2 E
Abukuma-Gawa →, Japan 30 E10 38 6N 140 52 E
Abukuma-Sammyaku, Japan 30 F10 37 30N 140 45 E
Abunã, Brazil 92 E5 9 40S 65 20W
Abunã →, Brazil 92 E5 9 41S 65 20W
Aburo, Zaïre 54 B3 2 4N 30 53 E
Abut Hd., N.Z. 59 K3 43 7S 170 15 E
Abwong, Sudan 51 G11 9 2N 32 14 E
Acajutla, El Salv. 88 D2 13 36N 89 50W
Acámbaro, Mexico 86 C4 20 0N 100 40W
Acaponeta, Mexico 86 C3 22 30N 105 20W
Acapulco, Mexico 87 D5 16 51N 99 56W
Acarigua, Venezuela 92 B5 9 33N 69 12W
Acatlán, Mexico 87 D5 18 10N 98 3W
Acayucan, Mexico 87 D6 17 59N 94 58W
Accomac, U.S.A. 76 G8 37 43N 75 40W
Accra, Ghana 50 G4 5 35N 0 6W
Accrington, U.K. 10 D5 53 46N 2 22W
Acebal, Argentina 94 C3 33 20S 60 50W
Aceh □, Indonesia 36 D1 4 15N 97 30 E
Achalpur, India 40 J10 21 22N 77 32 E
Acheng, China 35 B14 45 30N 126 58 E
Acher, India 42 H5 23 10N 72 32 E
Achill, Ireland 13 C2 53 56N 9 55W
Achill Hd., Ireland 13 C1 53 59N 10 15W
Achill I., Ireland 13 C1 53 58N 10 5W
Achill Sd., Ireland 13 C2 53 53N 9 55W
Achinsk, Russia 27 D10 56 20N 90 20 E
Ackerman, U.S.A. 81 J10 33 19N 89 11W
Acklins I., Bahamas 89 B5 22 30N 74 0W
Acme, Canada 72 C6 51 33N 113 30W
Aconcagua □, Chile 94 C1 32 15S 70 30W
Aconcagua, Cerro, Argentina 94 C2 32 39S 70 0W
Aconquija, Mt., Argentina 94 B2 27 0S 66 0W
Açores, Is. dos = Azores, Atl. Oc. 2 C8 38 44N 29 0W
Acraman, L., Australia 63 E2 32 2S 135 23 E
Acre = 'Akko, Israel 47 C4 32 55N 35 4 E
Acre □, Brazil 92 E4 9 1S 71 0W
Acre →, Brazil 92 E5 8 45S 67 22W
Acton, Canada 78 C4 43 38N 80 3W
Ad Dammām, Si. Arabia 45 E6 26 20N 50 5 E
Ad Dawhah, Qatar 45 E6 25 15N 51 35 E
Ad Dawr, Iraq 44 C4 34 27N 43 47 E
Ad Dir'īyah, Si. Arabia 44 E5 24 44N 46 35 E
Ad Dīwānīyah, Iraq 44 D5 32 0N 45 0 E
Ad Dujayl, Iraq 44 C5 33 51N 44 14 E
Ad Durūz, J., Jordan 47 C5 32 35N 36 40 E
Ada, Minn., U.S.A. 80 B6 47 18N 96 31W
Ada, Okla., U.S.A. 81 H6 34 46N 96 41W
Adaja →, Spain 19 B3 41 32N 4 52W
Adamaoua, Massif de l', Cameroon 51 G7 7 20N 12 20 E
Adamawa Highlands = Adamaoua, Massif de l', Cameroon 51 G7 7 20N 12 20 E
Adamello, Mt., Italy 20 A4 46 10N 10 34 E
Adaminaby, Australia 63 F4 36 0S 148 45 E
Adams, Mass., U.S.A. 79 D11 42 38N 73 7W
Adams, N.Y., U.S.A. 79 C8 43 49N 76 1W
Adams, Wis., U.S.A. 80 D10 43 57N 89 49W
Adam's Bridge, Sri Lanka 40 Q11 9 15N 79 40 E
Adams L., Canada 72 C5 51 10N 119 40W
Adams Mt., U.S.A. 84 D5 46 12N 121 30W
Adam's Peak, Sri Lanka 40 R12 6 48N 80 30 E
Adana, Turkey 25 G6 37 0N 35 16 E
Adapazarı, Turkey 25 F5 40 48N 30 25 E
Adarama, Sudan 51 E11 17 10N 34 52 E
Adare, C., Antarctica 5 D11 71 0S 171 0 E
Adaut, Indonesia 37 F8 8 8S 131 7 E
Adavale, Australia 63 D3 25 52S 144 32 E
Adda →, Italy 20 B3 45 8N 9 53 E
Addis Ababa = Addis Abeba, Ethiopia 51 G12 9 2N 38 42 E
Addis Abeba, Ethiopia 51 G12 9 2N 38 42 E
Addis Alem, Ethiopia 51 G12 9 0N 38 17 E
Addison, U.S.A. 78 D7 42 1N 77 14W
Addo, S. Africa 56 E4 33 32S 25 45 E
Ädeh, Iran 44 B5 37 42N 45 11 E
Adel, U.S.A. 77 K4 31 8N 83 25W
Adelaide, Australia 63 E2 34 52S 138 30 E
Adelaide, Bahamas 88 A4 25 4N 77 31W
Adelaide, S. Africa 56 E4 32 42S 26 20 E
Adelaide I., Antarctica 5 C17 67 15S 68 30W
Adelaide Pen., Canada 68 B10 68 15N 97 30W
Adelaide River, Australia 60 B5 13 15S 131 7 E
Adelanto, U.S.A. 85 L9 34 35N 117 22W
Adele I., Australia 60 C3 15 32S 123 9 E
Adélie, Terre, Antarctica 5 C10 68 0S 140 0 E
Adélie Land = Adélie, Terre, Antarctica 5 C10 68 0S 140 0 E
Aden = Al 'Adan, Yemen 46 E4 12 45N 45 0 E
Aden, G. of, Asia 46 E4 12 30N 47 30 E
Adendorp, S. Africa 56 E3 32 15S 24 30 E
Adh Dhayd, U.A.E. 45 E7 25 17N 55 53 E
Adhoi, India 42 H4 23 26N 70 32 E
Adi, Indonesia 37 E8 4 15S 133 30 E
Adi Ugri, Eritrea 51 F12 14 58N 38 48 E
Adieu, C., Australia 61 F5 32 0S 132 10 E
Adieu Pt., Australia 60 C3 15 14S 124 35 E
Adige →, Italy 20 B5 45 9N 12 20 E
Adilabad, India 40 K11 19 33N 78 20 E
Adin, U.S.A. 82 F3 41 12N 120 57W
Adin Khel, Afghan. 40 C6 32 45N 68 5 E
Adirondack Mts., U.S.A. 79 C10 44 0N 74 0W
Adjumani, Uganda 54 B3 3 20N 31 50 E
Adlavik Is., Canada 71 B8 55 2N 57 45W
Admer, Algeria 50 D6 20 21N 5 27 E
Admiralty G., Australia 60 B4 14 20S 125 55 E
Admiralty I., U.S.A. 68 C6 57 30N 134 30W
Admiralty Inlet, U.S.A. 82 C2 48 8N 122 58W
Admiralty Is., Papua N. G. 64 H6 2 0S 147 0 E
Ado Ekiti, Nigeria 50 G6 7 38N 5 12 E
Adonara, Indonesia 37 F6 8 15S 123 5 E
Adoni, India 40 M10 15 33N 77 18 E
Adour →, France 18 E3 43 32N 1 32W
Adra, India 43 H12 23 30N 86 42 E
Adra, Spain 19 D4 36 43N 3 3W
Adrano, Italy 20 F6 37 40N 14 49 E
Adrar, Algeria 50 C4 27 51N 0 11W
Adré, Chad 51 F9 13 40N 22 20 E
Adrī, Libya 51 C7 27 32N 13 2 E
Adrian, Mich., U.S.A. 76 E3 41 54N 84 2W
Adrian, Tex., U.S.A. 81 H3 35 16N 102 40W
Adriatic Sea, Europe 20 C6 43 0N 16 0 E
Adua, Indonesia 37 E7 1 45S 129 50 E
Adwa, Ethiopia 51 F12 14 15N 38 52 E
Adzhar Republic □, Georgia 25 F7 41 30N 42 0 E
Ægean Sea, Europe 21 E11 38 30N 25 0 E
Æolian Is. = Eólie, Is., Italy 20 E6 38 30N 14 50 E
Aerhtai Shan, Mongolia 32 B4 46 40N 92 45 E
'Afak, Iraq 44 C5 32 4N 45 15 E
Afándou, Greece 23 C10 36 18N 28 12 E
Afars & Issas, Terr. of = Djibouti ■, Africa 46 E3 12 0N 43 0 E
Afghanistan ■, Asia 40 C4 33 0N 65 0 E
Afgoi, Somali Rep. 46 G3 2 7N 44 59 E
Afognak I., U.S.A. 68 C4 58 15N 152 30W
Africa 48 E6 10 0N 20 0 E
'Afrīn, Syria 44 B3 36 32N 36 50 E
Afton, U.S.A. 79 D9 42 14N 75 32W
Afuá, Brazil 93 D8 0 15S 50 20W
Afula, Israel 47 C4 32 37N 35 17 E
Afyonkarahisar, Turkey 25 G5 38 45N 30 33 E
Agadès = Agadez, Niger 50 E6 16 58N 7 59 E
Agadez, Niger 50 E6 16 58N 7 59 E
Agadir, Morocco 50 B3 30 28N 9 55W
Agaete, Canary Is. 22 F4 28 6N 15 43W
Agapa, Russia 27 B9 71 27N 89 15 E
Agar, India 42 H7 23 40N 76 2 E
Agartala, India 41 H17 23 50N 91 23 E
Agassiz, Canada 72 D4 49 14N 121 46W
Agats, Indonesia 37 F9 5 33S 138 0 E
Agboville, Ivory C. 50 G4 5 55N 4 15W
Agde, France 18 E5 43 19N 3 28 E
Agen, France 18 D4 44 12N 0 38 E
Āgh Kand, Iran 45 B6 37 15N 48 4 E
Aginskoye, Russia 27 D12 51 6N 114 32 E
Agra, India 42 F7 27 17N 77 58 E
Agri →, Italy 20 D7 40 13N 16 44 E
Ağrı Dağı, Turkey 25 G7 39 50N 44 15 E
Ağrı Karakose, Turkey 25 G7 39 44N 43 3 E
Agrigento, Italy 20 F5 37 19N 13 33 E
Agrinion, Greece 21 E9 38 37N 21 27 E
Agua Caliente, Baja Calif. N., Mexico 85 N10 32 29N 116 59W
Agua Caliente, Sinaloa, Mexico 86 B3 26 30N 108 20W
Agua Caliente Springs, U.S.A. 85 N10 32 56N 116 19W
Água Clara, Brazil 93 H8 20 25S 52 45W
Agua Hechicero, Mexico 85 N10 32 26N 116 14W
Agua Prieta, Mexico 86 A3 31 20N 109 32W
Aguadas, Colombia 92 B3 5 40N 75 38W
Aguadilla, Puerto Rico 89 C6 18 26N 67 10W
Aguadulce, Panama 88 E3 8 15N 80 32W
Aguanga, U.S.A. 85 M10 33 27N 116 51W
Aguanish, Canada 71 B7 50 14N 62 2W
Aguanus →, Canada 71 B7 50 13N 62 5W
Aguapey →, Argentina 94 B4 29 7S 56 36W
Aguaray Guazú →, Paraguay 94 A4 24 47S 57 19W
Aguarico →, Ecuador 92 D3 0 59S 75 11W
Aguas Blancas, Chile 94 A2 24 15S 69 55W
Aguas Calientes, Sierra de, Argentina 94 B2 25 26S 66 40W
Aguascalientes, Mexico 86 C4 21 53N 102 12W
Aguascalientes □, Mexico 86 C4 22 0N 102 20W
Aguilares, Argentina 94 B2 27 26S 65 35W
Aguilas, Spain 19 D5 37 23N 1 35W
Agüimes, Canary Is. 22 G4 27 58N 15 27W
Agulhas, C., S. Africa 56 E3 34 52S 20 0 E
Agulo, Canary Is. 22 F2 28 11N 17 12W
Agung, Indonesia 36 F5 8 20S 115 28 E
Agur, Uganda 54 B3 2 28N 32 55 E
Agusan →, Phil. 37 C7 9 0N 125 30 E
Aha Mts., Botswana 56 B3 19 45S 21 0 E
Ahaggar, Algeria 50 D6 23 0N 6 30 E
Ahar, Iran 44 B5 38 35N 47 0 E
Ahipara B., N.Z. 59 F4 35 5S 173 5 E
Ahiri, India 40 K12 19 30N 80 0 E
Ahmad Wal, Pakistan 42 E1 29 18N 65 58 E
Ahmadabad, India 42 H5 23 0N 72 40 E
Aḩmadābād, Khorāsān, Iran 45 C9 35 3N 60 50 E
Aḩmadābād, Khorāsān, Iran 45 C8 35 49N 59 42 E
Aḩmadī, Iran 45 E8 27 56N 56 42 E
Ahmadnagar, India 40 K9 19 7N 74 46 E
Ahmadpur, Pakistan 42 E4 29 12N 71 10 E
Ahmedabad = Ahmadabad, India 42 H5 23 0N 72 40 E
Ahmednagar = Ahmadnagar, India 40 K9 19 7N 74 46 E
Ahome, Mexico 86 B3 25 55N 109 11W
Ahram, Iran 45 D6 28 52N 51 16 E
Ahrax Pt., Malta 23 D1 35 59N 14 22 E
Āhū, Iran 45 C6 34 33N 50 2 E
Ahuachapán, El Salv. 88 D2 13 54N 89 52W
Ahvāz, Iran 45 D6 31 20N 48 40 E
Ahvenanmaa = Åland, Finland 9 F16 60 15N 20 0 E
Aḩwar, Yemen 46 E4 13 30N 46 40 E
Aichi □, Japan 31 G8 35 0N 137 15 E
Aigua, Uruguay 95 C5 34 13S 54 46W
Aigues-Mortes, France 18 E6 43 35N 4 12 E
Aihui, China 33 A7 50 10N 127 30 E
Aija, Peru 92 E3 9 50S 77 45W
Aikawa, Japan 30 E9 38 2N 138 15 E
Aiken, U.S.A. 77 J5 33 34N 81 43W
Aillik, Canada 71 A8 55 11N 59 18W
Ailsa Craig, U.K. 12 F3 55 15N 5 7W
'Ailūn, Jordan 47 C4 32 18N 35 47 E
Aim, Russia 27 D14 59 0N 133 55 E
Aimere, Indonesia 37 F6 8 45S 121 3 E
Aimogasta, Argentina 94 B2 28 33S 66 50W
Aimorés, Brazil 93 G10 19 30S 41 4W
Ain □, France 18 C6 46 5N 5 20 E
Aïn Beïda, Algeria 50 A6 35 50N 7 29 E
Aïn Ben Tili, Mauritania 50 C3 25 59N 9 27W
Aïn-Sefra, Algeria 50 B4 32 47N 0 37W
'Ain Sudr, Egypt 47 F2 29 50N 33 6 E
Ainabo, Somali Rep. 46 F4 9 0N 46 25 E
Ainsworth, U.S.A. 80 D5 42 33N 99 52W
Aïr, Niger 50 E6 18 30N 8 0 E
Air Hitam, Malaysia 39 M4 1 55N 103 11 E
Airdrie, U.K. 12 F5 55 53N 3 57W
Aire →, U.K. 10 D7 53 42N 0 55W
Aire, I. del, Spain 22 B11 39 48N 4 16 E
Airlie Beach, Australia 62 C4 20 16S 148 43 E
Aisne □, France 18 B5 49 42N 3 40 E
Aisne →, France 18 B5 49 26N 2 50 E
Aitkin, U.S.A. 80 B8 46 32N 93 42W
Aiud, Romania 17 E11 46 19N 23 44 E
Aix-en-Provence, France 18 E6 43 32N 5 27 E
Aix-la-Chapelle = Aachen, Germany 16 C3 50 47N 6 4 E
Aiyansh, Canada 72 B3 55 17N 129 2W
Aíyina, Greece 21 F10 37 45N 23 26 E
Aiyion, Greece 21 E10 38 15N 22 5 E
Aizawl, India 41 H18 23 40N 92 44 E
Aizuwakamatsu, Japan 30 F9 37 30N 139 56 E
Ajaccio, France 18 F8 41 55N 8 40 E
Ajalpan, Mexico 87 D5 18 22N 97 15W
Ajanta Ra., India 40 J9 20 28N 75 50 E
Ajari Rep. = Adzhar Republic □, Georgia 25 F7 41 30N 42 0 E
Ajax, Canada 78 C5 43 50N 79 1W
Ajdâbiyah, Libya 51 B9 30 54N 20 4 E
'Ajmān, U.A.E. 45 E7 25 25N 55 30 E
Ajmer, India 42 F6 26 28N 74 37 E
Ajo, U.S.A. 83 K7 32 22N 112 52W
Akabira, Japan 30 C11 43 33N 142 5 E
Akamas □, Cyprus 23 D11 35 3N 32 18 E
Akanthou, Cyprus 23 D12 35 22N 33 45 E
Akaroa, N.Z. 59 K4 43 49S 172 59 E
Akashi, Japan 31 G7 34 45N 134 58 E
Akelamo, Indonesia 37 D7 1 35N 129 40 E
Akershus fylke □, Norway 9 G11 60 0N 11 10 E
Aketi, Zaïre 52 D4 2 38N 23 47 E
Akhelóös →, Greece 21 E9 38 36N 21 14 E
Akhmîm, Egypt 51 C11 26 31N 31 47 E
Akhnur, India 43 C6 32 52N 74 45 E
Aki, Japan 31 H6 33 30N 133 54 E
Akimiski I., Canada 70 B3 52 50N 81 30W
Akita, Japan 30 E10 39 45N 140 7 E
Akita □, Japan 30 E10 39 40N 140 30 E
Akjoujt, Mauritania 50 E2 19 45N 14 15W
Akkeshi, Japan 30 C12 43 2N 144 51 E
'Akko, Israel 47 C4 32 55N 35 4 E
Akkol, Kazakhstan 26 E8 45 0N 75 39 E
Aklavik, Canada 68 B6 68 12N 135 0W
Akmolinsk = Tselinograd, Kazakhstan 26 D8 51 10N 71 30 E
Akō, Japan 31 G7 34 45N 134 24 E
Akobo →, Ethiopia 51 G11 7 48N 33 3 E
Akola, India 40 J10 20 42N 77 2 E
Akordat, Eritrea 51 E12 15 30N 37 40 E
Akpatok I., Canada 69 B13 60 25N 68 8W
Akranes, Iceland 8 D3 64 19N 22 5W
Akreïjit, Mauritania 50 E3 18 19N 9 11W
Akron, Colo., U.S.A. 80 E3 40 10N 103 13W
Akron, Ohio, U.S.A. 78 E3 41 5N 81 31W
Akrotiri, Cyprus 23 E11 34 36N 32 57 E
Akrotiri Bay, Cyprus 23 E12 34 35N 33 10 E
Aksai Chih, India 43 B8 35 15N 79 55 E
Aksarka, Russia 26 C7 66 31N 67 50 E
Aksay, Kazakhstan 24 D9 51 11N 53 0 E
Aksenovo Zilovskoye, Russia 27 D12 53 20N 117 40 E
Aksu, China 32 B3 41 5N 80 10 E
Aksum, Ethiopia 51 F12 14 5N 38 40 E
Aktogay, Kazakhstan 26 E8 46 57N 79 40 E
Aktyubinsk, Kazakhstan 25 D10 50 17N 57 10 E
Aku, Nigeria 50 G6 6 40N 7 18 E
Akure, Nigeria 50 G6 7 15N 5 5 E
Akureyri, Iceland 8 D4 65 40N 18 6W
Akuseki-Shima, Japan 31 K4 29 27N 129 37 E
Akyab = Sittwe, Burma 41 J18 20 18N 92 45 E
Al 'Adan, Yemen 46 E4 12 45N 45 0 E
Al Aḩsā, Si. Arabia 45 E6 25 50N 49 0 E
Al Ajfar, Si. Arabia 44 E4 27 26N 43 0 E
Al Amādīyah, Iraq 44 B4 37 5N 43 30 E
Al Amārah, Iraq 44 D5 31 55N 47 15 E
Al 'Aqabah, Jordan 47 F4 29 31N 35 0 E
Al Arak, Syria 44 C3 34 38N 38 35 E
Al 'Aramah, Si. Arabia 44 E5 25 30N 46 0 E
Al Arţāwīyah, Si. Arabia 44 E5 26 31N 45 20 E
Al 'Āşimah □, Jordan 47 D5 31 40N 36 30 E
Al' Assāfīyah, Si. Arabia 44 D3 28 17N 38 59 E
Al 'Ayn, Oman 45 E7 24 15N 55 45 E
Al 'Ayn, Si. Arabia 44 E3 25 4N 38 6 E
Al A'ẓamīyah, Iraq 44 C5 33 22N 44 22 E
Al 'Azīzīyah, Iraq 44 C5 32 54N 45 4 E
Al Bāb, Syria 44 B3 36 23N 37 29 E
Al Bad', Si. Arabia 44 D2 28 28N 35 1 E
Al Bādī, Iraq 44 C4 35 56N 41 32 E
Al Baḩrah, Kuwait 44 D5 29 40N 47 52 E
Al Balqā' □, Jordan 47 C4 32 5N 35 45 E
Al Bārūk, J., Lebanon 47 B4 33 39N 35 40 E
Al Başrah, Iraq 44 D5 30 30N 47 50 E
Al Baţḩā, Iraq 44 D5 31 6N 45 53 E
Al Batrūn, Lebanon 47 A4 34 15N 35 40 E
Al Bi'r, Si. Arabia 44 D3 28 51N 36 16 E
Al Bu'ayrāt, Libya 51 B8 31 24N 15 44 E
Al Burayj, Syria 47 A5 34 15N 36 46 E
Al Fallūjah, Iraq 44 C4 33 20N 43 55 E
Al Fāw, Iraq 45 D6 30 0N 48 30 E
Al Fujayrah, U.A.E. 45 E8 25 7N 56 18 E

Alvear

Al Ghadaf, W. →, Jordan 47 D5 31 26N 36 43 E
Al Ghammās, Iraq 44 D5 31 45N 44 37 E
Al Hābah, Si. Arabia 44 E5 27 10N 47 0 E
Al Ḥadīthah, Iraq 44 C4 34 0N 41 13 E
Al Ḥadīthah, Si. Arabia 44 D3 31 28N 37 8 E
Al Hājānah, Syria 47 B5 33 20N 36 33 E
Al Hāmad, Si. Arabia 44 D3 31 30N 39 30 E
Al Hamdānīyah, Syria 44 C3 35 25N 36 50 E
Al Hamidīyah, Syria 47 A4 34 42N 35 57 E
Al Ḥammār, Iraq 44 D5 30 57N 46 51 E
Al Ḥarīr, W. →, Syria 47 C4 32 44N 35 59 E
Al Ḥasā, W. →, Jordan 47 D4 31 4N 35 29 E
Al Ḥasakah, Syria 44 B4 36 35N 40 45 E
Al Ḥawrah, Yemen 46 E4 13 50N 47 35 E
Al Haydān, W. →, Jordan 47 D4 31 29N 35 34 E
Al Ḥayy, Iraq 44 C5 32 5N 46 5 E
Al Ḥijāz, Si. Arabia 46 B2 26 0N 37 30 E
Al Ḥillah, Iraq 44 C5 32 30N 44 25 E
Al Ḥillah, Si. Arabia 46 C4 23 35N 46 50 E
Al Hirmil, Lebanon 47 A5 34 26N 36 24 E
Al Hoceïma, Morocco 50 A4 35 8N 3 58W
Al Ḥudaydah, Yemen 46 E3 14 50N 43 0 E
Al Hufūf, Si. Arabia 45 E6 25 25N 49 45 E
Al Ḥumayḍah, Si. Arabia 44 D2 29 14N 34 56 E
Al Ḥunayy, Si. Arabia 45 E6 25 58N 48 45 E
Al Irq, Libya 51 C9 29 5N 21 35 E
Al Īsāwīyah, Si. Arabia 44 D3 30 43N 37 59 E
Al Ittihad = Madīnat ash Sha'b, Yemen 46 E3 12 50N 45 0 E
Al Jafr, Jordan 47 E5 30 18N 36 14 E
Al Jaghbūb, Libya 51 C9 29 42N 24 38 E
Al Jahrah, Kuwait 44 D5 29 25N 47 40 E
Al Jalāmīd, Si. Arabia 44 D3 31 20N 39 45 E
Al Jamalīyah, Qatar 45 E6 25 37N 51 5 E
Al Janūb □, Lebanon 47 B4 33 20N 35 20 E
Al Jawf, Libya 51 D9 24 10N 23 24 E
Al Jawf, Si. Arabia 44 D3 29 55N 39 40 E
Al Jazirah, Iraq 44 C5 33 30N 44 0 E
Al Jazirah, Libya 51 C9 26 10N 21 20 E
Al Jithāmīyah, Si. Arabia 44 E4 27 41N 41 43 E
Al Jubayl, Si. Arabia 45 E6 27 0N 49 50 E
Al Jubaylah, Si. Arabia 44 E5 24 55N 46 25 E
Al Jubb, Si. Arabia 44 E4 27 11N 42 17 E
Al Junaynah, Sudan 51 F9 13 27N 22 45 E
Al Kabā'ish, Iraq 44 D5 30 58N 47 0 E
Al Karak, Jordan 47 D4 31 11N 35 42 E
Al Karak □, Jordan 47 E5 31 0N 36 0 E
Al Kāzim Tyah, Iraq 44 C5 33 22N 44 12 E
Al Khalīl, Jordan 47 D4 31 32N 35 6 E
Al Khalīl □, Jordan 47 D4 31 35N 35 5 E
Al Khawr, Qatar 45 E6 25 41N 51 30 E
Al Khiḍr, Iraq 44 D5 31 12N 45 33 E
Al Khiyām, Lebanon 47 B4 33 20N 35 36 E
Al Kiswah, Syria 47 B5 33 23N 36 14 E
Al Kufrah, Libya 51 D9 24 17N 23 15 E
Al Kuhayfiyah, Si. Arabia 44 E4 27 12N 43 3 E
Al Kūt, Iraq 44 C5 32 30N 46 0 E
Al Kuwayt, Kuwait 44 D5 29 30N 48 0 E
Al Labwah, Lebanon 47 A5 34 11N 36 20 E
Al Lādhiqīyah, Syria 44 C2 35 30N 35 45 E
Al Liwā', Oman 45 E8 24 31N 56 36 E
Al Luḥayyah, Yemen 46 D3 15 45N 42 40 E
Al Madīnah, Iraq 44 D5 30 57N 47 16 E
Al Madīnah, Si. Arabia 46 C2 24 35N 39 52 E
Al-Mafraq, Jordan 47 C5 32 17N 36 14 E
Al Maḥmūdīyah, Iraq 44 C5 33 3N 44 21 E
Al Majma'ah, Si. Arabia 44 E5 25 57N 45 22 E
Al Makhruq, W. →, Jordan 47 D6 31 28N 37 0 E
Al Makḥūl, Si. Arabia 44 E4 26 37N 42 39 E
Al Manāmah, Bahrain 45 E6 26 10N 50 30 E
Al Maqwa', Kuwait 44 D5 29 10N 47 59 E
Al Marj, Libya 51 B9 32 25N 20 30 E
Al Maṭlā, Kuwait 44 D5 29 24N 47 40 E
Al Mawjib, W. →, Jordan 47 D4 31 28N 35 36 E
Al Mawṣil, Iraq 44 B4 36 15N 43 5 E
Al Mayādin, Syria 44 C4 35 1N 40 27 E
Al Mazār, Jordan 47 D4 31 4N 35 41 E
Al Midhnab, Si. Arabia 44 E5 25 50N 44 18 E
Al Minā', Lebanon 47 A4 34 24N 35 49 E
Al Miqdādīyah, Iraq 44 C5 34 0N 45 0 E
Al Mubarraz, Si. Arabia 45 E6 25 30N 49 40 E
Al Mughayrā', U.A.E. 45 E7 24 5N 53 32 E
Al Muḥarraq, Bahrain 45 E6 26 15N 50 40 E
Al Mukallā, Yemen 46 E4 14 33N 49 2 E
Al Mukhā, Yemen 46 E3 13 18N 43 15 E
Al Musayjīd, Si. Arabia 44 E3 24 5N 39 5 E
Al Musayyib, Iraq 44 C5 32 40N 44 25 E
Al Muwaylīḥ, Si. Arabia 44 E2 27 40N 35 30 E
Al Qā'im, Iraq 44 C4 34 21N 41 7 E
Al Qalībah, Si. Arabia 44 D3 28 24N 37 42 E
Al Qaryatayn, Syria 47 A6 34 12N 37 13 E
Al Qaṣabāt, Libya 51 B7 32 39N 14 1 E
Al Qaṭ'ā, Syria 44 C4 34 40N 40 48 E
Al Qaṭīf, Si. Arabia 45 E6 26 35N 50 0 E
Al Qatrānah, Jordan 47 D5 31 14N 36 26 E
Al Qaṭrūn, Libya 51 D8 24 56N 15 3 E
Al Qayṣūmah, Si. Arabia 44 D5 28 20N 46 7 E
Al Quds = Jerusalem, Israel 47 D4 31 47N 35 10 E
Al Quds □, Jordan 47 D4 31 50N 35 20 E
Al Qunaytirah, Syria 47 C4 32 55N 35 45 E
Al Qurnah, Iraq 44 D5 31 1N 47 25 E
Al Quṣayr, Iraq 44 D5 30 39N 45 50 E
Al Quṣayr, Syria 47 A5 34 31N 36 34 E
Al Quṭayfah, Syria 47 B5 33 44N 36 36 E
Al' Uḍaylīyah, Si. Arabia 45 E6 25 8N 49 18 E
Al 'Ulā, Si. Arabia 44 E3 26 35N 38 0 E
Al Uqaylah ash Sharqīgah, Libya 51 B8 30 12N 19 10 E
Al Uqayr, Si. Arabia 45 E6 25 40N 50 15 E
Al 'Uwaynid, Si. Arabia 44 E5 24 50N 46 0 E
Al' 'Uwayqīlah, Si. Arabia 44 D4 30 30N 42 10 E
Al 'Uyūn, Si. Arabia 44 E4 26 30N 43 50 E
Al 'Uyūn, Si. Arabia 44 E3 24 33N 39 35 E
Al Wajh, Si. Arabia 44 E3 26 10N 36 30 E
Al Wakrah, Qatar 45 E6 25 10N 51 40 E
Al Wannān, Si. Arabia 45 E6 26 55N 48 24 E
Al Waqbah, Si. Arabia 44 D5 28 48N 45 33 E
Al Wari'āh, Si. Arabia 44 E5 27 51N 47 25 E
Al Wusayl, Qatar 45 E6 25 29N 51 29 E
Ala Tau Shankou = Dzhungarskiye Vorota, Kazakhstan 32 B3 45 0N 82 0 E
Alabama □, U.S.A. 77 J2 33 0N 87 0W
Alabama →, U.S.A. 77 K2 31 8N 87 57W
Alaérma, Greece 23 C9 36 9N 27 57 E
Alagoa Grande, Brazil 93 E11 7 3S 35 35W
Alagoas □, Brazil 93 E11 9 0S 36 0W
Alagoinhas, Brazil 93 F11 12 7S 38 20W
Alajero, Canary Is. 22 F2 28 3N 17 13W
Alajuela, Costa Rica 88 D3 10 2N 84 8W
Alakamisy, Madag. 57 C8 21 19S 47 14 E
Alakurtti, Russia 24 A5 67 0N 30 30 E
Alameda, Calif., U.S.A. 84 H4 37 46N 122 15W
Alameda, N. Mex., U.S.A. 83 J10 35 11N 106 37W
Alamo, U.S.A. 85 J11 36 21N 115 10W
Alamo Crossing, U.S.A. 85 L13 34 16N 113 33W
Alamogordo, U.S.A. 83 K11 32 54N 105 57W
Alamos, Mexico 86 B3 27 0N 109 0W
Alamosa, U.S.A. 83 H11 37 28N 105 52W
Åland, Finland 9 F16 60 15N 20 0 E
Ålands hav, Sweden 9 G15 60 0N 19 30 E
Alandur, India 40 N12 13 0N 80 15 E
Alanya, Turkey 25 G5 36 38N 32 0 E
Alaotra, Farihin', Madag. 57 B8 17 30S 48 30 E
Alapayevsk, Russia 26 D7 57 52N 61 42 E
Alaşehir, Turkey 25 G4 38 23N 28 30 E
Alaska □, U.S.A. 68 B5 64 0N 154 0W
Alaska, G. of, Pac. Oc. 68 C5 58 0N 145 0W
Alaska Highway, Canada 72 B3 60 0N 130 0W
Alaska Peninsula, U.S.A. 68 C4 56 0N 159 0W
Alaska Range, U.S.A. 68 B4 62 50N 151 0W
Alatyr, Russia 24 D8 54 45N 46 35 E
Alausi, Ecuador 92 D3 2 0S 78 50W
Alava, C., U.S.A. 82 B1 48 10N 124 44W
Alawoona, Australia 63 E3 34 45S 140 30 E
'Alayh, Lebanon 47 B4 33 46N 35 33 E
Alayor, Spain 22 B11 39 57N 4 8 E
Alba, Italy 20 B3 44 41N 8 1 E
Alba Iulia, Romania 21 A10 46 8N 23 39 E
Albacete, Spain 19 C5 39 0N 1 50W
Albacutya, L., Australia 63 F3 35 45S 141 58 E
Albania ■, Europe 21 D9 41 0N 20 0 E
Albany, Australia 61 G2 35 1S 117 58 E
Albany, Ga., U.S.A. 77 K3 31 35N 84 10W
Albany, Minn., U.S.A. 80 C7 45 38N 94 34W
Albany, N.Y., U.S.A. 79 D11 42 39N 73 45W
Albany, Oreg., U.S.A. 82 D2 44 38N 123 6W
Albany, Tex., U.S.A. 81 J5 32 44N 99 18W
Albany →, Canada 70 B3 52 17N 81 31W
Albardón, Argentina 94 C2 31 20S 68 30W
Albarracín, Sierra de, Spain 19 B5 40 30N 1 30W
Albatross B., Australia 62 A3 12 45S 141 30 E
Albemarle, U.S.A. 77 H5 35 21N 80 11W
Albemarle Sd., U.S.A. 77 H7 36 5N 76 0W
Alberche →, Spain 19 C3 39 58N 4 46W
Alberdi, Paraguay 94 B4 26 14S 58 20W
Albert, L. = Mobutu Sese Seko, L., Africa 54 B3 1 30N 31 0 E
Albert, L., Australia 63 F2 35 30S 139 10 E
Albert Canyon, Canada 72 C5 51 8N 117 41W
Albert Edward Ra., Australia 60 C4 18 17S 127 57 E
Albert Lea, U.S.A. 80 D8 43 39N 93 22W
Albert Nile →, Uganda 54 B3 3 36N 32 2 E
Albert Town, Bahamas 89 B5 22 37N 74 33W
Alberta □, Canada 72 C6 54 40N 115 0W
Alberti, Argentina 94 D3 35 1S 60 16W
Albertinia, S. Africa 56 E3 34 11S 21 34 E
Alberton, Canada 71 C7 46 50N 64 0W
Albertville = Kalemie, Zaïre 54 D2 5 55S 29 9 E
Albi, France 18 E5 43 56N 2 9 E
Albia, U.S.A. 80 E8 41 2N 92 48W
Albina, Surinam 93 B8 5 37N 54 15W
Albina, Ponta, Angola 56 B1 15 52S 11 44 E
Albion, Idaho, U.S.A. 82 E7 42 25N 113 35W
Albion, Mich., U.S.A. 76 D3 42 15N 84 45W
Albion, Nebr., U.S.A. 80 E5 41 42N 98 0W
Albion, Pa., U.S.A. 78 E4 41 53N 80 22W
Ålborg, Denmark 9 H10 57 2N 9 54 E
Alborz, Reshteh-ye Kūhhā-ye, Iran 45 C7 36 0N 52 0 E
Albreda, Canada 72 C5 52 35N 119 10W
Albuquerque, U.S.A. 83 J10 35 5N 106 39W
Albuquerque, Cayos de, Caribbean 88 D3 12 10N 81 50W
Alburg, U.S.A. 79 B11 44 59N 73 18W
Alburquerque, Spain 19 C2 39 15N 6 59W
Albury, Australia 63 F4 36 3S 146 56 E
Alcalá de Henares, Spain 19 B4 40 28N 3 22W
Alcalá la Real, Spain 19 D4 37 27N 3 57W
Alcamo, Italy 20 F5 37 59N 12 55 E
Alcaníz, Spain 19 B5 41 2N 0 8W
Alcântara, Brazil 93 D10 2 20S 44 30W
Alcántara, Spain 19 C2 39 41N 6 57W
Alcantara L., Canada 73 A7 60 57N 108 9W
Alcaraz, Sierra de, Spain 19 C4 38 40N 2 20W
Alcaudete, Spain 19 D3 37 35N 4 5W
Alcázar de San Juan, Spain 19 C4 39 24N 3 12W
Alchevsk = Kommunarsk, Ukraine 25 E6 48 30N 38 45 E
Alcira, Spain 19 C5 39 9N 0 30W
Alcoa, U.S.A. 77 H4 35 48N 83 59W
Alcobaça, Portugal 19 C1 39 32N 8 58W
Alcova, U.S.A. 82 E10 42 34N 106 43W
Alcoy, Spain 19 C5 38 43N 0 30W
Alcudia, Spain 22 B10 39 51N 3 7 E
Alcudia, B. de, Spain 22 B10 39 47N 3 15 E
Aldabra Is., Seychelles 49 G8 9 22S 46 28 E
Aldama, Mexico 87 C5 23 0N 98 4W
Aldan, Russia 27 D13 58 40N 125 30 E
Aldan →, Russia 27 C13 63 28N 129 35 E
Aldea, Pta. de la, Canary Is. 22 G4 28 0N 15 50W
Aldeburgh, U.K. 11 E9 52 9N 1 35 E
Alder, U.S.A. 82 D7 45 19N 112 6W
Alder Pk., U.S.A. 84 K5 35 53N 121 22W
Alderney, Chan. Is. 11 H5 49 42N 2 12W
Aldershot, U.K. 11 F7 51 15N 0 43W
Aledo, U.S.A. 80 E9 41 12N 90 45W
Aleg, Mauritania 50 E2 17 3N 13 55W
Alegranza, Canary Is. 22 E6 29 23N 13 32W
Alegranza, I., Canary Is. 22 E6 29 23N 13 32W
Alegre, Brazil 95 A7 20 50S 41 30W
Alegrete, Brazil 95 B4 29 40S 56 0W
Aleisk, Russia 26 D9 52 40N 83 0 E
Aleksandrovsk-Sakhalinskiy, Russia 27 D15 50 50N 142 20 E
Aleksandrovskiy Zavod, Russia 27 D12 50 40N 117 50 E
Aleksandrovskoye, Russia 26 C8 60 35N 77 50 E
Além Paraíba, Brazil 95 A7 21 52S 42 41W
Alemania, Argentina 94 B2 25 40S 65 30W
Alemania, Chile 94 B2 25 10S 69 55W
Alençon, France 18 B4 48 27N 0 4 E
Alenuihaha Channel, U.S.A. 74 H17 20 30N 156 0W
Aleppo = Ḥalab, Syria 44 B3 36 10N 37 15 E
Aléria, France 20 C3 42 5N 9 26 E
Alert Bay, Canada 72 C3 50 30N 126 55W
Alès, France 18 D6 44 9N 4 5 E
Alessándria, Italy 20 B3 44 54N 8 37 E
Ålesund, Norway 8 E9 62 28N 6 12 E
Aleutian Is., Pac. Oc. 68 C2 52 0N 175 0W
Aleutian Trench, Pac. Oc. 64 B10 48 0N 180 0 E
Alexander, U.S.A. 80 B3 47 51N 103 39W
Alexander, Mt., Australia 61 E3 28 58S 120 16 E
Alexander Arch., U.S.A. 72 B2 56 0N 136 0W
Alexander B., S. Africa 56 D2 28 36S 16 33 E
Alexander Bay, S. Africa 56 D2 28 40S 16 30 E
Alexander City, U.S.A. 77 J3 32 56N 85 58W
Alexander I., Antarctica 5 C17 69 0S 70 0W
Alexandra, Australia 63 F4 37 8S 145 40 E
Alexandra, N.Z. 59 L2 45 14S 169 25 E
Alexandra Falls, Canada 72 A5 60 29N 116 18W
Alexandretta = İskenderun, Turkey 25 G6 36 32N 36 10 E
Alexandria = El Iskandarîya, Egypt 51 B10 31 0N 30 0 E
Alexandria, Australia 62 B2 19 5S 136 40 E
Alexandria, B.C., Canada 72 C4 52 35N 122 27W
Alexandria, Ont., Canada 70 C5 45 19N 74 38W
Alexandria, S. Africa 56 E4 33 38S 26 28 E
Alexandria, Ind., U.S.A. 76 E3 40 16N 85 41W
Alexandria, La., U.S.A. 81 K8 31 18N 92 27W
Alexandria, Minn., U.S.A. 80 C7 45 53N 95 22W
Alexandria, S. Dak., U.S.A. 80 D6 43 39N 97 47W
Alexandria, Va., U.S.A. 76 F7 38 48N 77 3W
Alexandria Bay, U.S.A. 79 B9 44 20N 75 55W
Alexandrina, L., Australia 63 F2 35 25S 139 10 E
Alexandroúpolis, Greece 21 D11 40 50N 25 54 E
Alexis →, Canada 71 B8 52 33N 56 8W
Alexis Creek, Canada 72 C4 52 10N 123 20W
Alfabia, Spain 22 B9 39 44N 2 44 E
Alfenas, Brazil 95 A6 21 20S 46 10W
Alford, U.K. 12 D6 57 13N 2 42W
Alfred, Maine, U.S.A. 79 C14 43 29N 70 43W
Alfred, N.Y., U.S.A. 78 D7 42 16N 77 48W
Alfreton, U.K. 10 D6 53 6N 1 22W
Alga, Kazakhstan 25 E10 49 53N 57 20 E
Algaida, Spain 22 B9 39 33N 2 53 E
Algarve, Portugal 19 D1 36 58N 8 20W
Algeciras, Spain 19 D3 36 9N 5 28W
Algemesí, Spain 19 C5 39 11N 0 27W
Alger, Algeria 50 A5 36 42N 3 8 E
Algeria ■, Africa 50 C5 28 30N 2 0 E
Alghero, Italy 20 D3 40 34N 8 20 E
Algiers = Alger, Algeria 50 A5 36 42N 3 8 E
Algoa B., S. Africa 56 E4 33 50S 25 45 E
Algoma, U.S.A. 76 C2 44 36N 87 26W
Algona, U.S.A. 80 D7 43 4N 94 14W
Algonac, U.S.A. 78 D2 42 37N 82 32W
Alhama de Murcia, Spain 19 D5 37 51N 1 25W
Alhambra, U.S.A. 85 L8 34 8N 118 6W
Alhucemas = Al Hoceïma, Morocco 50 A4 35 8N 3 58W
'Alī al Gharbī, Iraq 44 C5 32 30N 46 45 E
Alī ash Sharqī, Iraq 44 C5 32 7N 46 44 E
'Alī Khēl, Afghan. 42 C3 33 57N 69 43 E
Alī Shāh, Iran 44 B5 38 9N 45 50 E
Alīābād, Khorāsān, Iran 45 C8 32 30N 57 30 E
Alīābād, Kordestān, Iran 44 C5 35 4N 46 58 E
'Alīābād, Yazd, Iran 45 D7 31 41N 53 49 E
Aliákmon →, Greece 21 D10 40 30N 22 36 E
Alibo, Ethiopia 51 G12 9 52N 37 5 E
Alicante, Spain 19 C5 38 23N 0 30W
Alice, S. Africa 56 E4 32 48S 26 55 E
Alice, U.S.A. 81 M5 27 45N 98 5W
Alice →, Queens., Australia 62 C3 24 2S 144 50 E
Alice →, Queens., Australia 62 B3 15 35S 142 20 E
Alice Arm, Canada 72 B3 55 29N 129 31W
Alice Downs, Australia 60 C4 17 45S 127 56 E
Alice Springs, Australia 62 C1 23 40S 133 50 E
Alicedale, S. Africa 56 E4 33 15S 26 4 E
Aliceville, U.S.A. 77 J1 33 8N 88 9W
Alick Cr. →, Australia 62 C3 20 55S 142 20 E
Alida, Canada 73 D8 49 25N 101 55W
Aligarh, Raj., India 42 G7 25 55N 76 15 E
Aligarh, Ut. P., India 42 F8 27 55N 78 10 E
Alīgūdarz, Iran 45 C6 33 25N 49 45 E
Alimnía, Greece 23 C9 36 16N 27 43 E
Alingsås, Sweden 9 H12 57 56N 12 31 E
Alipur, Pakistan 42 E4 29 25N 70 55 E
Alipur Duar, India 41 F16 26 30N 89 35 E
Aliquippa, U.S.A. 78 F4 40 37N 80 15W
Aliwal North, S. Africa 56 E4 30 45S 26 45 E
Alix, Canada 72 C6 52 24N 113 11W
Aljustrel, Portugal 19 D1 37 55N 8 10W
Alkmaar, Neths. 15 B4 52 37N 4 45 E
All American Canal, U.S.A. 83 K6 32 45N 115 15W
Allah Dad, Pakistan 42 G2 25 38N 67 34 E
Allahabad, India 43 G9 25 25N 81 58 E
Allakh-Yun, Russia 27 C14 60 50N 137 5 E
Allan, Canada 73 C7 51 53N 106 4W
Allanmyo, Burma 41 K19 19 30N 95 17 E
Allanridge, S. Africa 56 D4 27 45S 26 40 E
Allanwater, Canada 70 B1 50 14N 90 10W
Allegan, U.S.A. 76 D3 42 32N 85 51W
Allegany, U.S.A. 78 D6 42 6N 78 30W
Allegheny →, U.S.A. 78 F5 40 27N 80 1W
Allegheny Plateau, U.S.A. 76 G6 38 0N 80 0W
Allegheny Reservoir, U.S.A. 78 E6 41 50N 79 0W
Allen, Bog of, Ireland 13 C4 53 15N 7 0W
Allen, L., Ireland 13 B3 54 12N 8 5W
Allende, Mexico 86 B4 28 20N 100 50W
Allentown, U.S.A. 79 F9 40 37N 75 29W
Alleppey, India 40 Q10 9 30N 76 28 E
Alliance, Nebr., U.S.A. 80 D3 42 6N 102 52W
Alliance, Ohio, U.S.A. 78 F3 40 55N 81 6W
Allier □, France 18 C5 46 25N 2 40 E
Allier →, France 18 C5 46 57N 3 4 E
Alliston, Canada 70 D4 44 9N 79 52W
Alloa, U.K. 12 E5 56 7N 3 49W
Allora, Australia 63 D5 28 2S 152 0 E
Alluitsup Paa = Sydprøven, Greenland 4 C5 60 30N 45 35W
Alma, Canada 71 C5 48 35N 71 40W
Alma, Ga., U.S.A. 77 K4 31 33N 82 28W
Alma, Kans., U.S.A. 80 F6 39 1N 96 17W
Alma, Mich., U.S.A. 76 D3 43 23N 84 39W
Alma, Nebr., U.S.A. 80 E5 40 6N 99 22W
Alma, Wis., U.S.A. 80 C9 44 20N 91 55W
Alma Ata, Kazakhstan 26 E8 43 15N 76 57 E
Almada, Portugal 19 C1 38 40N 9 9W
Almaden, Australia 62 B3 17 22S 144 40 E
Almadén, Spain 19 C3 38 49N 4 52W
Almanor, L., U.S.A. 82 F3 40 14N 121 9W
Almansa, Spain 19 C5 38 51N 1 5W
Almanzor, Pico de, Spain 19 B3 40 15N 5 18W
Almanzora →, Spain 19 D5 37 14N 1 46W
Almaty = Alma Ata, Kazakhstan 26 E8 43 15N 76 57 E
Almazán, Spain 19 B4 41 30N 2 30W
Almeirim, Brazil 93 D8 1 30S 52 34W
Almelo, Neths. 15 B6 52 22N 6 42 E
Almendralejo, Spain 19 C2 38 41N 6 26W
Almería, Spain 19 D4 36 52N 2 27W
Almirante, Panama 88 E3 9 10N 82 30W
Almiroú, Kólpos, Greece 23 D6 35 23N 24 20 E
Almont, U.S.A. 78 D1 42 55N 83 3W
Almonte, Canada 79 A8 45 14N 76 12W
Almora, India 43 E8 29 38N 79 40 E
Alnwick, U.K. 10 B6 55 25N 1 42W
Aloi, Uganda 54 B3 2 16N 33 10 E
Alon, Burma 41 H19 22 12N 95 5 E
Alor, Indonesia 37 F6 8 15S 124 30 E
Alor Setar, Malaysia 39 J3 6 7N 100 22 E
Aloysius, Mt., Australia 61 E4 26 0S 128 38 E
Alpaugh, U.S.A. 84 K7 35 53N 119 29W
Alpena, U.S.A. 76 C4 45 4N 83 27W
Alpes-de-Haute-Provence □, France 18 D7 44 8N 6 10 E
Alpes-Maritimes □, France 18 E7 43 55N 7 10 E
Alpha, Australia 62 C4 23 39S 146 37 E
Alpine, Ariz., U.S.A. 83 K9 33 51N 109 9W
Alpine, Calif., U.S.A. 85 N10 32 50N 116 46W
Alpine, Tex., U.S.A. 81 K3 30 22N 103 40W
Alps, Europe 16 E4 46 30N 9 30 E
Alroy Downs, Australia 62 B2 19 20S 136 5 E
Alsace, France 18 B7 48 15N 7 25 E
Alsask, Canada 73 C7 51 21N 109 59W
Alsásua, Spain 19 A4 42 54N 2 10W
Alsten, Norway 8 D12 65 58N 12 40 E
Alta, Norway 8 B17 69 57N 23 10 E
Alta Gracia, Argentina 94 C3 31 40S 64 30W
Alta Lake, Canada 72 C4 50 10N 123 0W
Alta Sierra, U.S.A. 85 K8 35 42N 118 33W
Altaelva →, Norway 8 B17 69 46N 23 45 E
Altafjorden, Norway 8 A17 70 5N 23 5 E
Altagracia, Venezuela 92 A4 10 45N 71 30W
Altai = Aerhtai Shan, Mongolia 32 B4 46 40N 92 45 E
Altai = Aerhtai Shan, Mongolia 32 B4 46 40N 92 45 E
Altamaha →, U.S.A. 77 K5 31 20N 81 20W
Altamira, Brazil 93 D8 3 12S 52 10W
Altamira, Chile 94 B2 25 47S 69 51W
Altamira, Mexico 87 C5 22 24N 97 55W
Altamont, U.S.A. 79 D10 42 43N 74 3W
Altanbulag, Mongolia 32 A5 50 16N 106 30 E
Altar, Mexico 86 A2 30 40N 111 50W
Altata, Mexico 86 C3 24 30N 108 0W
Altavista, U.S.A. 76 G6 37 6N 79 17W
Altay, China 32 B3 47 48N 88 10 E
Alto Adige = Trentino-Alto Adige □, Italy 20 A4 46 30N 11 0 E
Alto-Alentejo, Portugal 19 C2 39 0N 7 40W
Alto Araguaia, Brazil 93 G8 17 15S 53 20W
Alto Cuchumatanes = Cuchumatanes, Sierra de los, Guatemala 88 C1 15 35N 91 25W
Alto del Inca, Chile 94 A2 24 10S 68 10W
Alto Ligonha, Mozam. 55 F4 15 30S 38 11 E
Alto Molocue, Mozam. 55 F4 15 50S 37 35 E
Alto Paraguay □, Paraguay 94 A4 21 0S 58 30W
Alto Paraná □, Paraguay 95 B5 25 30S 54 50W
Alton, Canada 78 C4 43 54N 80 5W
Alton, U.S.A. 80 F9 38 53N 90 11W
Alton Downs, Australia 63 D2 26 7S 138 57 E
Altoona, U.S.A. 78 F6 40 31N 78 24W
Altūn Kūprī, Iraq 44 C5 35 45N 44 9 E
Altun Shan, China 32 C3 38 30N 88 0 E
Alturas, U.S.A. 82 F3 41 29N 120 32W
Altus, U.S.A. 81 H5 34 38N 99 20W
Alùla, Somali Rep. 46 E5 11 50N 50 45 E
Alunite, U.S.A. 85 K12 35 59N 114 55W
Alusi, Indonesia 37 F8 7 35S 131 40 E
Al'Uzayr, Iraq 44 D5 31 19N 47 25 E
Alva, U.S.A. 81 G5 36 48N 98 40W
Alvarado, Mexico 87 D5 18 40N 95 50W
Alvarado, U.S.A. 81 J6 32 24N 97 13W
Alvaro Obregón, Presa, Mexico 86 B3 27 55N 109 52W
Alvear, Argentina 94 B4 29 5S 56 30W

Alvesta

Alvesta, Sweden 9 H13 56 54N 14 35 E
Alvie, Australia 63 F3 38 14S 143 30 E
Alvin, U.S.A. 81 L7 29 26N 95 15W
Alvinston, Canada 78 D3 42 49N 81 52W
Älvkarleby, Sweden 9 F14 60 34N 17 26 E
Älvsborgs län □, Sweden 9 G12 58 30N 12 30 E
Älvsbyn, Sweden 8 D16 65 40N 21 0 E
Alwar, India 42 F7 27 38N 76 34 E
Alxa Zuoqi, China 34 E3 38 50N 105 40 E
Alyaskitovyy, Russia 27 C15 64 45N 141 30 E
Alyata, Azerbaijan 25 G8 39 58N 49 25 E
Alyth, U.K. 12 E5 56 38N 3 15W
Alzada, U.S.A. 80 C2 45 2N 104 25W
Am Dam, Chad 51 F9 12 40N 20 35 E
Am-Timan, Chad 51 F9 11 0N 20 10 E
Amadeus, L., Australia 61 D5 24 54S 131 0 E
Amâdi, Sudan 51 G11 5 29N 30 25 E
Amadi, Zaïre 54 B2 3 40N 26 40 E
Amadjuak, Canada 69 B12 64 0N 72 39W
Amadjuak L., Canada 69 B12 65 0N 71 8W
Amagasaki, Japan 31 G7 34 42N 135 20 E
Amakusa-Shotō, Japan 31 H5 32 15N 130 10 E
Amalner, India 40 J9 21 5N 75 5 E
Amambaí, Brazil 95 A4 23 5S 55 13W
Amambaí —►, Brazil 95 A5 23 22S 53 56W
Amambay □, Paraguay 95 A4 23 0S 56 0W
Amambay, Cordillera de, S. Amer. 95 A4 23 0S 55 45W
Amami-Guntō, Japan 31 L4 27 16N 129 21 E
Amami-Ō-Shima, Japan 31 L4 28 0N 129 0 E
Amanda Park, U.S.A. 84 C3 47 28N 123 55W
Amangeldy, Kazakhstan 26 D7 50 10N 65 10 E
Amapá, Brazil 93 C8 2 5N 50 50W
Amapá □, Brazil 93 C8 1 40N 52 0W
Amarante, Brazil 93 E10 6 14S 42 50W
Amaranth, Canada 73 C9 50 36N 98 43W
Amargosa, Brazil 93 F11 13 2S 39 36W
Amargosa —►, U.S.A. 85 J10 36 14N 116 51W
Amargosa Range, U.S.A. 85 J10 36 20N 116 45W
Amári, Greece 23 D6 35 13N 24 40 E
Amarillo, U.S.A. 81 H4 35 13N 101 50W
Amaro, Mt., Italy 20 C6 42 5N 14 6 E
Amarpur, India 43 G12 25 5N 87 0 E
Amatikulu, S. Africa 57 D5 29 3S 31 33 E
Amatitlán, Guatemala 88 D1 14 29N 90 38W
Amazon = Amazonas —►, S. Amer. 93 C9 0 5S 50 0W
Amazonas □, Brazil 92 E6 5 0S 65 0W
Amazonas —►, S. Amer. 93 C9 0 5S 50 0W
Ambahakily, Madag. 57 C7 21 36S 43 41 E
Ambala, India 42 D7 30 23N 76 56 E
Ambalavao, Madag. 57 C8 21 50S 46 56 E
Ambalindum, Australia 62 C2 23 23S 135 0 E
Ambam, Cameroon 52 D2 2 20N 11 15 E
Ambanja, Madag. 57 A8 13 40S 48 27 E
Ambarchik, Russia 27 C17 69 40N 162 20 E
Ambarijeby, Madag. 57 A8 14 56S 47 41 E
Ambaro, Helodranon', Madag. 57 A8 13 23S 48 38 E
Ambartsevo, Russia 26 D9 57 30N 83 52 E
Ambato, Ecuador 92 D3 1 5S 78 42W
Ambato, Sierra de, Argentina 94 B2 28 25S 66 10W
Ambato Boeny, Madag. 57 B8 16 28S 46 43 E
Ambatofinandrahana, Madag. 57 C8 20 33S 46 48 E
Ambatolampy, Madag. 57 B8 19 20S 47 35 E
Ambatondrazaka, Madag. 57 B8 17 55S 48 28 E
Ambatosoratra, Madag. 57 B8 17 37S 48 31 E
Ambenja, Madag. 57 B8 15 17S 46 58 E
Amberg, Germany 16 D5 49 25N 11 52 E
Ambergris Cay, Belize 87 D7 18 0N 88 0W
Amberley, N.Z. 59 K4 43 9S 172 44 E
Ambikapur, India 43 H10 23 15N 83 15 E
Ambilobé, Madag. 57 A8 13 10S 49 3 E
Ambinanindrano, Madag. 57 C8 20 5S 48 23 E
Ambleside, U.K. 10 C5 54 26N 2 58W
Ambo, Peru 92 F3 10 5S 76 10W
Ambodifototra, Madag. 57 B8 16 59S 49 52 E
Ambodilazana, Madag. 57 B8 18 6S 49 10 E
Ambohimahasoa, Madag. 57 C8 21 7S 47 13 E
Ambohimanga, Madag. 57 C8 20 52S 47 36 E
Ambohitra, Madag. 57 A8 12 30S 49 10 E
Ambon, Indonesia 37 E7 3 35S 128 20 E
Amboseli L., Kenya 54 C4 2 40S 37 10 E
Ambositra, Madag. 57 C8 20 31S 47 25 E
Ambovombé, Madag. 57 D8 25 11S 46 5 E
Amboy, U.S.A. 85 L11 34 33N 115 45W
Amboyna I., S. China Sea 36 C4 7 50N 112 50 E
Ambridge, U.S.A. 78 F4 40 36N 80 14W
Ambriz, Angola 52 F2 7 48S 13 8 E
Amby, Australia 63 D4 26 30S 148 11 E
Amchitka I., U.S.A. 68 C1 51 32N 179 0 E
Amderma, Russia 26 C7 69 45N 61 30 E
Ameca, Mexico 86 C4 20 30N 104 0W
Ameca —►, Mexico 86 C3 20 40N 105 15W
Amecameca, Mexico 87 D5 19 7N 98 46W
Ameland, Neths. 15 A5 53 27N 5 45 E
Amen, Russia 27 C18 68 45N 180 0 E
American Falls, U.S.A. 82 E7 42 47N 112 51W
American Falls Reservoir, U.S.A. 82 E7 42 47N 112 52W
American Highland, Antarctica 5 D6 73 0S 75 0 E
American Samoa ■, Pac. Oc. 59 B13 14 20S 170 40W
Americana, Brazil 95 A6 22 45S 47 20W
Americus, U.S.A. 77 J3 32 4N 84 14W
Amersfoort, Neths. 15 B5 52 9N 5 23 E
Amersfoort, S. Africa 57 D4 26 59S 29 53 E
Amery, Australia 61 F2 31 9S 117 5 E
Amery, Canada 73 B10 56 34N 94 3W
Amery Ice Shelf, Antarctica 5 C6 69 30S 72 0 E
Ames, U.S.A. 80 E8 42 2N 93 37W
Amesbury, U.S.A. 79 D14 42 51N 70 56W
Amga, Russia 27 C14 60 50N 132 0 E
Amga —►, Russia 27 C14 62 38N 134 32 E
Amgu, Russia 27 E14 45 45N 137 15 E
Amgun —►, Russia 27 D14 52 56N 139 38 E
Amherst, Burma 41 L20 16 2N 97 20 E
Amherst, Canada 71 C7 45 48N 64 8W
Amherst, Mass., U.S.A. 79 D12 42 23N 72 31W
Amherst, N.Y., U.S.A. 78 D6 42 59N 78 48W
Amherst, Ohio, U.S.A. 78 E2 41 24N 82 14W
Amherst, Tex., U.S.A. 81 H3 34 1N 102 25W
Amherst I., Canada 79 B8 44 8N 76 43W
Amherstburg, Canada 70 D3 42 6N 83 6W
Amiata, Mte., Italy 20 C4 42 54N 11 40 E
Amiens, France 18 B5 49 54N 2 16 E
Amīrābād, Iran 44 C5 33 20N 46 16 E
Amirante Is., Seychelles 3 E12 6 0S 53 0 E
Amisk L., Canada 73 C8 54 35N 102 15W
Amistad, Presa de la, Mexico 86 B4 29 24N 101 0W
Amite, U.S.A. 81 K9 30 44N 90 30W
Amlwch, U.K. 10 D3 53 24N 4 21W
'Ammān, Jordan 47 D4 31 57N 35 52 E
Ammanford, U.K. 11 F3 51 48N 4 4W
Ammassalik = Angmagssalik, Greenland 4 C6 65 40N 37 20W
Amnat Charoen, Thailand 38 E5 15 51N 104 38 E
Āmol, Iran 45 B7 36 23N 52 20 E
Amorgós, Greece 21 F11 36 50N 25 57 E
Amory, U.S.A. 77 J1 33 59N 88 29W
Amos, Canada 70 C4 48 35N 78 5W
Amoy = Xiamen, China 33 D6 24 25N 118 4 E
Ampang, Malaysia 39 L3 3 8N 101 45 E
Ampanihy, Madag. 57 C7 24 40S 44 45 E
Ampasindava, Helodranon', Madag. 57 A8 13 40S 48 15 E
Ampasindava, Saikanosy, Madag. 57 A8 13 42S 47 55 E
Ampenan, Indonesia 36 F5 8 35S 116 13 E
Ampotaka, Madag. 57 D7 25 3S 44 41 E
Ampoza, Madag. 57 C7 22 20S 44 44 E
Amqui, Canada 71 C6 48 28N 67 27W
Amravati, India 40 J10 20 55N 77 45 E
Amreli, India 42 J4 21 35N 71 17 E
Amritsar, India 42 D6 31 35N 74 57 E
Amroha, India 43 E8 28 53N 78 30 E
Amsterdam, Neths. 15 B4 52 23N 4 54 E
Amsterdam, U.S.A. 79 D10 42 56N 74 11W
Amsterdam, I., Ind. Oc. 3 F13 38 30S 77 30 E
Amudarya —►, Uzbekistan 26 E6 43 40N 59 0 E
Amundsen Gulf, Canada 68 A7 71 0N 124 0W
Amundsen Sea, Antarctica 5 D15 72 0S 115 0W
Amuntai, Indonesia 36 E5 2 28S 115 25 E
Amur —►, Russia 27 D15 52 56N 141 10 E
Amurang, Indonesia 37 D6 1 5N 124 40 E
Amuri Pass, N.Z. 59 K4 42 31S 172 11 E
Amursk, Russia 27 D14 50 14N 136 54 E
Amurzet, Russia 27 E14 47 50N 131 5 E
Amyderya = Amudarya —►, Uzbekistan 26 E6 43 40N 59 0 E
An Bien, Vietnam 39 H5 9 45N 105 0 E
An Hoa, Vietnam 38 E7 15 40N 108 5 E
An Khe, Vietnam 38 F7 13 57N 108 39 E
An Nabatīyah at Tahta, Lebanon 47 B4 33 23N 35 27 E
An Nabk, Si. Arabia 44 D3 31 20N 37 20 E
An Nabk, Syria 47 A5 34 2N 36 44 E
An Nabk Abū Qaşr, Si. Arabia 44 D3 30 21N 38 34 E
An Nafūd, Si. Arabia 44 D4 28 15N 41 0 E
An Najaf, Iraq 44 C5 32 3N 44 15 E
An Nāşirīyah, Iraq 44 D5 31 0N 46 15 E
An Nhon, Vietnam 38 F7 13 55N 109 7 E
An Nîl □, Sudan 51 E11 19 30N 33 0 E
An Nîl el Abyad □, Sudan 51 F11 14 0N 32 15 E
An Nîl el Azraq □, Sudan 51 F11 12 30N 34 30 E
An Nu'ayrīyah, Si. Arabia 45 E6 27 30N 48 30 E
An Nuwaybʻī, W. —►, Si. Arabia 47 F3 29 18N 34 57 E
An Thoi, Dao, Vietnam 39 H4 9 58N 104 0 E
An Uaimh, Ireland 13 C5 53 39N 6 40W
Anabar —►, Russia 27 B12 73 8N 113 36 E
'Anabtā, Jordan 47 C4 32 19N 35 7 E
Anaconda, U.S.A. 82 C7 46 8N 112 57W
Anacortes, U.S.A. 84 B4 48 30N 122 37W
Anadarko, U.S.A. 81 H5 35 4N 98 15W
Anadolu, Turkey 25 G5 39 0N 30 0 E
Anadyr, Russia 27 C18 64 35N 177 20 E
Anadyr —►, Russia 27 C18 64 55N 176 5 E
Anadyrskiy Zaliv, Russia 27 C19 64 0N 180 0 E
Anaga, Pta. de, Canary Is. 22 F3 28 34N 16 9W
'Ānah, Iraq 44 C4 34 25N 42 0 E
Anaheim, U.S.A. 85 M9 33 50N 117 55W
Anahim Lake, Canada 72 C3 52 28N 125 18W
Anáhuac, Mexico 86 B4 27 14N 100 9W
Anakapalle, India 41 L13 17 42N 83 6 E
Anakie, Australia 62 C4 23 32S 147 45 E
Analalava, Madag. 57 A8 14 35S 48 0 E
Análipsis, Greece 23 A3 39 36N 19 55 E
Anambar —►, Pakistan 42 D3 30 15N 68 50 E
Anambas, Kepulauan, Indonesia 36 D3 3 20N 106 30 E
Anambas Is. = Anambas, Kepulauan, Indonesia 36 D3 3 20N 106 30 E
Anamoose, U.S.A. 80 B4 47 53N 100 15W
Anamosa, U.S.A. 80 D9 42 7N 91 17W
Anamur, Turkey 25 G5 36 8N 32 58 E
Anan, Japan 31 H7 33 54N 134 40 E
Anand, India 42 H5 22 32N 72 59 E
Anantnag, India 43 C6 33 45N 75 10 E
Anapodháris —►, Greece 23 E7 34 59N 25 20 E
Anápolis, Brazil 93 G9 16 15S 48 50W
Anār, Iran 45 D7 30 55N 55 13 E
Anārak, Iran 45 C7 33 25N 53 40 E
Anatolia = Anadolu, Turkey 25 G5 39 0N 30 0 E
Anatone, U.S.A. 82 C5 46 8N 117 8W
Anatsogno, Madag. 57 C7 23 33S 43 46 E
Añatuya, Argentina 94 B3 28 20S 62 50W
Anaunethad L., Canada 73 A8 60 55N 104 25W
Anaye, Niger 51 E7 19 15N 12 50 E
Anbyŏn, N. Korea 35 E14 39 1N 127 35 E
Anchor Bay, U.S.A. 84 G3 38 48N 123 34W
Anchorage, U.S.A. 68 B5 61 13N 149 54W
Anci, China 34 E9 39 20N 116 40 E
Ancohuma, Nevada, Bolivia 92 G5 16 0S 68 50W
Ancón, Peru 92 F3 11 50S 77 10W
Ancona, Italy 20 C5 43 37N 13 30 E
Ancud, Chile 96 E2 42 0S 73 50W
Ancud, G. de, Chile 96 E2 42 0S 73 0W
Anda, China 33 B7 46 24N 125 19 E
Andacollo, Argentina 94 D1 37 10S 70 42W
Andacollo, Chile 94 C1 30 5S 71 10W
Andado, Australia 62 D2 25 25S 135 15 E
Andalgalá, Argentina 94 B2 27 40S 66 30W
Åndalsnes, Norway 8 E9 62 35N 7 43 E
Andalucía □, Spain 19 D3 37 35N 5 0W
Andalusia, U.S.A. 77 K2 31 18N 86 29W
Andalusia □ = Andalucía □, Spain 19 D3 37 35N 5 0W
Andaman Is., Ind. Oc. 28 H13 12 30N 92 30 E
Andaman Sea, Ind. Oc. 36 B1 13 0N 96 0 E
Andara, Namibia 56 B3 18 2S 21 9 E
Andenne, Belgium 15 D5 50 28N 5 5 E
Anderson, Calif., U.S.A. 82 F2 40 27N 122 18W
Anderson, Ind., U.S.A. 76 E3 40 10N 85 41W
Anderson, Mo., U.S.A. 81 G7 36 39N 94 27W
Anderson, S.C., U.S.A. 77 H4 34 31N 82 39W
Anderson —►, Canada 68 B7 69 42N 129 0W
Anderson, Mt., S. Africa 57 D5 25 5S 30 42 E
Andes = Andes, Cord. de los, S. Amer. 92 F4 20 0S 68 0W
Andes, S. Amer. 92 F3 10 0S 75 53W
Andes, Cord. de los, S. Amer. 92 F4 20 0S 68 0W
Andfjorden, Norway 8 B14 69 10N 16 20 E
Andhra Pradesh □, India 40 L11 18 0N 79 0 E
Andijon = Andizhan, Uzbekistan 26 E8 41 10N 72 15 E
Andikíthira, Greece 21 G10 35 52N 23 15 E
Andīmeshk, Iran 45 C6 32 27N 48 21 E
Andizhan, Uzbekistan 26 E8 41 10N 72 15 E
Andoany, Madag. 57 A8 13 25S 48 16 E
Andong, S. Korea 35 F15 36 40N 128 43 E
Andongwei, China 35 G10 35 6N 119 20 E
Andorra ■, Europe 19 A6 42 30N 1 30 E
Andorra La Vella, Andorra 19 A6 42 31N 1 32 E
Andover, U.K. 11 F6 51 13N 1 29W
Andover, Mass., U.S.A. 79 D13 42 40N 71 8W
Andover, N.Y., U.S.A. 78 D7 42 10N 77 48W
Andover, Ohio, U.S.A. 78 E4 41 36N 80 34W
Andrahary, Mt., Madag. 57 A8 13 37S 49 17 E
Andraitx, Spain 22 B9 39 39N 2 25 E
Andramasina, Madag. 57 B8 19 11S 47 35 E
Andranopasy, Madag. 57 C7 21 17S 43 44 E
Andreanof Is., U.S.A. 68 C2 52 0N 178 0W
Andrewilla, Australia 63 D2 26 31S 139 17 E
Andrews, S.C., U.S.A. 77 J6 33 27N 79 34W
Andrews, Tex., U.S.A. 81 J3 32 19N 102 33W
Ándria, Italy 20 D7 41 13N 16 17 E
Andriba, Madag. 57 B8 17 30S 46 58 E
Androka, Madag. 57 C7 24 58S 44 2 E
Andropov = Rybinsk, Russia 24 C6 58 5N 38 50 E
Ándros, Greece 21 F11 37 50N 24 57 E
Andros I., Bahamas 88 B4 24 30N 78 0W
Andros Town, Bahamas 88 B4 24 43N 77 47W
Andújar, Spain 19 C3 38 3N 4 5W
Andulo, Angola 52 G3 11 25S 16 45 E
Anegada I., Virgin Is. 89 C7 18 45N 64 20W
Anegada Passage, W. Indies 89 C7 18 15N 63 45W
Anécho, Togo 50 G5 6 12N 1 34 E
Aneto, Pico de, Spain 19 A6 42 37N 0 40 E
Ang Thong, Thailand 38 E3 14 35N 100 31 E
Angamos, Punta, Chile 94 A1 23 1S 70 32W
Angara —►, Russia 27 D10 58 5N 94 20 E
Angarsk, Russia 27 D11 52 30N 104 0 E
Angas Downs, Australia 61 E5 25 2S 132 14 E
Angas Hills, Australia 60 D4 23 0S 127 50 E
Angaston, Australia 63 E2 34 30S 139 8 E
Ånge, Sweden 8 E13 62 31N 15 35 E
Ángel de la Guarda, I., Mexico 86 B2 29 30N 113 30W
Angeles, Phil. 37 A6 15 9N 120 33 E
Ängelholm, Sweden 9 H12 56 15N 12 58 E
Angellala, Australia 63 D4 26 24S 146 54 E
Angels Camp, U.S.A. 83 G3 38 4N 120 32W
Ångermanälven —►, Sweden 8 E14 62 40N 18 0 E
Angers, Canada 79 A9 45 31N 75 29W
Angers, France 18 C3 47 30N 0 35W
Ängesån —►, Sweden 8 C17 66 50N 22 15 E
Angikuni L., Canada 73 A9 62 0N 100 0W
Angkor, Cambodia 38 F4 13 22N 103 50 E
Anglesey, U.K. 10 D3 53 17N 4 20W
Angleton, U.S.A. 81 L7 29 10N 95 26W
Anglisidhes, Cyprus 23 E12 34 51N 33 27 E
Angmagssalik, Greenland 4 C6 65 40N 37 20W
Ango, Zaïre 54 B2 4 10N 26 5 E
Angoche, Mozam. 55 F4 16 8S 39 55 E
Angoche, I., Mozam. 55 F4 16 20S 39 50 E
Angol, Chile 94 D1 37 56S 72 45W
Angola, Ind., U.S.A. 76 E3 41 38N 85 0W
Angola, N.Y., U.S.A. 78 D5 42 38N 79 2W
Angola ■, Africa 53 G3 12 0S 18 0 E
Angoon, U.S.A. 72 B2 57 30N 134 35W
Angoulême, France 18 D4 45 39N 0 10 E
Angoumois, France 18 D4 45 50N 0 25 E
Angra dos Reis, Brazil 95 A7 23 0S 44 10W
Angren, Uzbekistan 26 E8 41 1N 70 12 E
Angtassom, Cambodia 39 G5 11 1N 104 41 E
Angu, Zaïre 54 B1 3 25N 24 28 E
Anguang, China 35 B12 45 15N 123 45 E
Anguilla ■, W. Indies 89 C7 18 14N 63 5W
Anguo, China 34 E8 38 28N 115 15 E
Angurugu, Australia 62 A2 14 0S 136 25 E
Angus, Braes of, U.K. 12 E5 56 51N 3 10W
Anhanduí —►, Brazil 95 A5 21 46S 52 9W
Anholt, Denmark 9 H11 56 42N 11 33 E
Anhui □, China 33 C6 32 0N 117 0 E
Anhwei □ = Anhui □, China 33 C6 32 0N 117 0 E
Anichab, Namibia 56 C1 21 0S 14 46 E
Animas, U.S.A. 83 L9 31 57N 108 48W
Anivorano, Madag. 57 B8 18 44S 48 58 E
Anjar, India 42 H4 23 6N 70 10 E
Anjidiv I., India 40 M9 14 40N 74 10 E
Anjou, France 18 C3 47 20N 0 15W
Anjozorobe, Madag. 57 B8 18 22S 47 52 E
Anju, N. Korea 35 E13 39 36N 125 40 E
Anka, Nigeria 50 F6 12 13N 5 58 E
Ankaboa, Tanjona, Madag. 57 C7 21 58S 43 20 E
Ankang, China 34 H5 32 40N 109 1 E
Ankara, Turkey 25 G5 39 57N 32 54 E
Ankaramena, Madag. 57 C8 21 57S 46 39 E
Ankazoabo, Madag. 57 C7 22 18S 44 31 E
Ankazobe, Madag. 57 B8 18 20S 47 10 E
Ankisabe, Madag. 57 B8 19 17S 46 29 E
Ankoro, Zaïre 54 D2 6 45S 26 55 E
Anmyŏn-do, S. Korea 35 F14 36 25N 126 25 E
Ann, C., U.S.A. 79 D14 42 38N 70 35W
Ann Arbor, U.S.A. 76 D4 42 17N 83 45W
Anna, U.S.A. 81 G10 37 28N 89 15W
Anna Plains, Australia 60 C3 19 17S 121 37 E
Annaba, Algeria 50 A6 36 50N 7 46 E
Annalee —►, Ireland 13 B4 54 3N 7 15W
Annam = Trung-Phan, Vietnam 38 E7 16 0N 108 0 E
Annamitique, Chaîne, Asia 38 D6 17 0N 106 0 E
Annan, U.K. 12 G5 54 57N 3 17W
Annan —►, U.K. 12 G5 54 58N 3 18W
Annapolis, U.S.A. 76 F7 38 59N 76 30W
Annapolis Royal, Canada 71 D6 44 44N 65 32W
Annapurna, Nepal 43 E10 28 34N 83 50 E
Annean, L., Australia 61 E2 26 54S 118 14 E
Annecy, France 18 D7 45 55N 6 8 E
Anning, China 32 D5 24 55N 102 26 E
Anningie, Australia 60 D5 21 50S 133 7 E
Anniston, U.S.A. 77 J3 33 39N 85 50W
Annobón, Atl. Oc. 49 G4 1 25S 5 36 E
Annotto Bay, Jamaica 88 C4 18 17N 76 45W
Annuello, Australia 63 E3 34 53S 142 55 E
Annville, U.S.A. 79 F8 40 20N 76 31W
Áno Viánnos, Greece 23 D7 35 2N 25 21 E
Anoka, U.S.A. 80 C8 45 12N 93 23W
Anorotsangana, Madag. 57 A8 13 56S 47 55 E
Anóyia, Greece 23 D6 35 16N 24 52 E
Anping, Hebei, China 34 E8 38 15N 115 30 E
Anping, Liaoning, China 35 D12 41 5N 123 30 E
Anqing, China 33 C6 30 30N 117 3 E
Anqiu, China 35 F10 36 25N 119 10 E
Ansai, China 34 F5 36 50N 109 20 E
Ansbach, Germany 16 D5 49 17N 10 34 E
Anshan, China 33 B7 41 3N 122 58 E
Anshan, Liaoning, China 35 D12 41 5N 122 58 E
Anshun, China 32 D5 26 18N 105 57 E
Ansirabe, Madag. 57 B8 19 55S 47 2 E
Ansley, U.S.A. 80 E5 41 18N 99 23W
Anson, U.S.A. 81 J5 32 45N 99 54W
Anson B., Australia 60 B5 13 20S 130 6 E
Ansongo, Mali 50 E5 15 25N 0 35 E
Ansonia, U.S.A. 79 E11 41 21N 73 5W
Anstruther, U.K. 12 E6 56 14N 2 40W
Ansudu, Indonesia 37 E9 2 11S 139 22 E
Antabamba, Peru 92 F4 14 40S 73 0W
Antakya, Turkey 25 G6 36 14N 36 10 E
Antalaha, Madag. 57 A9 14 57S 50 20 E
Antalya, Turkey 25 G5 36 52N 30 45 E
Antalya Körfezi, Turkey 25 G5 36 15N 31 30 E
Antananarivo, Madag. 57 B8 18 55S 47 31 E
Antananarivo □, Madag. 57 B8 19 0S 47 0 E
Antanimbaribe, Madag. 57 C7 21 30S 44 48 E
Antarctic Pen., Antarctica 5 C18 67 0S 60 0W
Antarctica 5 E3 90 0S 0 0 E
Antelope, Zimbabwe 55 G2 21 2S 28 31 E
Antequera, Paraguay 94 A4 24 8S 57 7W
Antequera, Spain 19 D3 37 5N 4 33W
Antero, Mt., U.S.A. 83 G10 38 41N 106 15W
Anthony, Kans., U.S.A. 81 G5 37 9N 98 2W
Anthony, N. Mex., U.S.A. 83 K10 32 0N 106 36W
Anthony Lagoon, Australia 62 B2 18 0S 135 30 E
Anti Atlas, Morocco 50 C3 30 0N 8 30W
Anti-Lebanon = Ash Sharqi, Al Jabal, Lebanon 47 B5 33 40N 36 10 E
Anticosti, I. d', Canada 71 C7 49 30N 63 0W
Antigo, U.S.A. 80 C10 45 9N 89 9W
Antigonish, Canada 71 C7 45 38N 61 58W
Antigua, Canary Is. 22 F5 28 24N 14 1W
Antigua, Guatemala 88 D1 14 34N 90 41W
Antigua, W. Indies 89 C7 17 0N 61 50W
Antigua & Barbuda ■, W. Indies 89 C7 17 20N 61 48W
Antilla, Cuba 88 B4 20 40N 75 50W
Antimony, U.S.A. 83 G8 38 7N 112 0W
Antioch, U.S.A. 84 G5 38 1N 121 48W
Antioquia, Colombia 92 B3 6 40N 75 55W
Antipodes Is., Pac. Oc. 64 M9 49 45S 178 40 E
Antler, U.S.A. 80 A4 48 59N 101 17W
Antler —►, Canada 73 D8 49 8N 101 0W
Antlers, U.S.A. 81 H7 34 14N 95 37W
Antofagasta, Chile 94 A1 23 50S 70 30W
Antofagasta □, Chile 94 A2 24 0S 69 0W
Antofagasta de la Sierra, Argentina 94 B2 26 5S 67 20W
Antofalla, Argentina 94 B2 25 30S 68 5W
Antofalla, Salar de, Argentina 94 B2 25 40S 67 45W
Anton, U.S.A. 81 J3 33 49N 102 10W
Anton Chico, U.S.A. 83 J11 35 12N 105 9W
Antongila, Helodrano, Madag. 57 B8 15 30S 49 50 E
Antonibé, Madag. 57 B8 15 7S 47 24 E
Antonibé, Presqu'île d', Madag. 57 A8 14 55S 47 20 E
Antonina, Brazil 95 B6 25 26S 48 42W
Antonito, U.S.A. 83 H10 37 5N 106 0W
Antrim, U.K. 13 B5 54 43N 6 13W
Antrim □, U.K. 13 B5 54 55N 6 20W
Antrim, Mts. of, U.K. 13 B5 54 57N 6 8W
Antrim Plateau, Australia 60 C4 18 8S 128 20 E
Antsalova, Madag. 57 B7 18 40S 44 37 E

Ashland

Antsiranana, Madag. 57 A8 12 25S 49 20 E
Antsohihy, Madag. 57 A8 14 50S 47 59 E
Antsohimbondrona Seranana, Madag. 57 A8 13 7S 48 48 E
Antu, China 35 C15 42 30N 128 20 E
Antwerp = Antwerpen, Belgium 15 C4 51 13N 4 25 E
Antwerp, U.S.A. 79 B9 44 12N 75 37W
Antwerpen, Belgium 15 C4 51 13N 4 25 E
Antwerpen □, Belgium 15 C4 51 15N 4 40 E
Anupgarh, India 42 E5 29 10N 73 10 E
Anuradhapura, Sri Lanka 40 Q12 8 22N 80 28 E
Anveh, Iran 45 E7 27 23N 54 11 E
Anvers = Antwerpen, Belgium 15 C4 51 13N 4 25 E
Anvers I., Antarctica 5 C17 64 30S 63 40W
Anxi, China 32 B4 40 30N 95 43 E
Anxious B., Australia 63 E1 33 24S 134 45 E
Anyang, China 34 F8 36 5N 114 21 E
Anyi, China 34 G6 35 2N 111 2 E
Anza, U.S.A. 85 M10 33 35N 116 39W
Anze, China 34 F7 36 10N 112 12 E
Anzhero-Sudzhensk, Russia 26 D9 56 10N 86 0 E
Ánzio, Italy 20 D5 41 28N 12 37 E
Aoga-Shima, Japan 31 H9 32 28N 139 46 E
Aomori, Japan 30 D10 40 45N 140 45 E
Aomori □, Japan 30 D10 40 45N 140 40 E
Aonla, India 43 E8 28 16N 79 11 E
Aosta, Italy 20 B2 45 43N 7 20 E
Aoudéras, Niger 50 E6 17 45N 8 20 E
Aoulef el Arab, Algeria 50 C5 26 55N 1 2 E
Apa —►, S. Amer. 94 A4 22 6S 58 2W
Apache, U.S.A. 81 H5 34 54N 98 22W
Apalachee B., U.S.A. 77 L3 30 0N 84 0W
Apalachicola, U.S.A. 77 L3 29 43N 84 59W
Apalachicola —►, U.S.A. 77 L3 29 43N 84 58W
Apaporis —►, Colombia 92 D5 1 23S 69 25W
Aparri, Phil. 37 A6 18 22N 121 38 E
Apàtity, Russia 24 A5 67 34N 33 22 E
Apatzingán, Mexico 86 D4 19 0N 102 20W
Apeldoorn, Neths. 15 B5 52 13N 5 57 E
Apennines = Appennini, Italy 20 C4 44 0N 10 0 E
Apia, W. Samoa 59 A13 13 50S 171 50W
Apiacás, Serra dos, Brazil 92 E7 9 50S 57 0W
Apizaco, Mexico 87 D5 19 26N 98 9W
Aplao, Peru 92 G4 16 0S 72 40W
Apo, Mt., Phil. 37 C7 6 53N 125 14 E
Apolakkiá, Greece 23 C9 36 5N 27 48 E
Apolakkiá, Órmos, Greece 23 C9 36 5N 27 45 E
Apollonia = Marsá Susah, Libya 51 B9 32 52N 21 59 E
Apolo, Bolivia 92 F5 14 30S 68 30W
Apostle Is., U.S.A. 80 B9 47 0N 90 40W
Apóstoles, Argentina 95 B4 28 0S 56 0W
Apostolos Andreas, C., Cyprus 23 D13 35 42N 34 35 E
Apoteri, Guyana 92 C7 4 2N 58 32W
Appalachian Mts., U.S.A. 76 G6 38 0N 80 0W
Appennini, Italy 20 C4 44 0N 10 0 E
Apple Hill, Canada 79 A10 45 13N 74 46W
Apple Valley, U.S.A. 85 L9 34 32N 117 14W
Appleby, U.K. 10 C5 54 35N 2 29W
Appleton, U.S.A. 76 C1 44 16N 88 25W
Approuague, Fr. Guiana 93 C8 4 20N 52 0W
Apucarana, Brazil 95 A5 23 55S 51 33W
Apulia = Púglia □, Italy 20 D7 41 0N 16 30 E
Apure —►, Venezuela 92 B5 7 37N 66 25W
Apurimac —►, Peru 92 F4 12 17S 73 56W
Aqabah = Al 'Aqabah, Jordan 47 F4 29 31N 35 0 E
'Aqabah, Khalīj al, Red Sea 44 D2 28 15N 33 20 E
'Aqdā, Iran 45 C7 32 26N 53 37 E
Aqîq, Sudan 51 E12 18 14N 38 12 E
Aqmola = Tselinograd, Kazakhstan 26 D8 51 10N 71 30 E
Aqrah, Iraq 44 B4 36 46N 43 45 E
Aqtöbe = Aktyubinsk, Kazakhstan 25 D10 50 17N 57 10 E
Aquidauana, Brazil 93 H7 20 30S 55 50W
Aquiles Serdán, Mexico 86 B3 28 37N 105 54W
Aquin, Haiti 89 C5 18 16N 73 24W
Ar Rachidiya, Morocco 50 B4 31 58N 4 20W
Ar Rafid, Syria 47 C4 32 57N 35 52 E
Ar Raḩḩālīyah, Iraq 44 C4 32 44N 43 23 E
Ar Ramādī, Iraq 44 C4 33 25N 43 20 E
Ar Ramthā, Jordan 47 C5 32 34N 36 0 E
Ar Raqqah, Syria 44 C3 36 0N 38 55 E
Ar Rass, Si. Arabia 44 E4 25 50N 43 40 E
Ar Rifā'ī, Iraq 44 D5 31 50N 46 10 E
Ar Riyāḑ, Si. Arabia 46 C4 24 41N 46 42 E
Ar Ru'ays, Qatar 45 E6 26 8N 51 12 E
Ar Rukhaymīyah, Iraq 44 D5 29 22N 45 38 E
Ar Ruqayyidah, Si. Arabia 45 E6 25 21N 49 34 E
Ar Ruşāfah, Syria 44 C3 35 52N 36 53 E
Ar Ruţbah, Iraq 44 C4 33 0N 40 15 E
Ara, India 43 G11 25 35N 84 32 E
'Arab, Bahr el —►, Sudan 51 G10 9 0N 29 30 E
'Arabābād, Iran 45 C8 33 2N 57 41 E
Arabia, Asia 46 C4 25 0N 45 0 E
Arabian Desert = Es Sahrâ' Esh Sharqîya, Egypt 51 C11 27 30N 32 30 E
Arabian Gulf = Gulf, The, Asia 45 E6 27 0N 50 0 E
Arabian Sea, Ind. Oc. 29 H10 16 0N 65 0 E
Aracaju, Brazil 93 F11 10 55S 37 4W
Aracataca, Colombia 92 A4 10 38N 74 9W
Aracati, Brazil 93 D11 4 30S 37 44W
Araçatuba, Brazil 95 A5 21 10S 50 30W
Aracena, Spain 19 D2 37 53N 6 38W
Araçuaí, Brazil 93 G10 16 52S 42 4W
'Arad, Israel 47 D4 31 15N 35 12 E
Arad, Romania 17 E10 46 10N 21 20 E
Arada, Chad 51 F9 15 0N 20 20 E
Aradhippou, Cyprus 23 E12 34 57N 33 36 E
Arafura Sea, E. Indies 37 F8 9 0S 135 0 E
Aragón □, Spain 19 B5 41 25N 0 40W
Aragón —►, Spain 19 A5 42 13N 1 44W
Araguacema, Brazil 93 E9 8 50S 49 20W
Araguaia —►, Brazil 93 E9 5 21S 48 41W
Araguari, Brazil 93 G9 18 38S 48 11W
Araguari —►, Brazil 93 C9 1 15N 49 55W
Arak, Algeria 50 C5 25 20N 3 45 E
Arāk, Iran 45 C6 34 0N 49 40 E
Arakan Coast, Burma 41 K19 19 0N 94 0 E
Arakan Yoma, Burma 41 K19 20 0N 94 40 E
Araks = Aras, Rūd-e —►, Iran 44 B5 39 10N 47 10 E
Aral Sea = Aralskoye More, Asia 26 E7 44 30N 60 0 E
Aralsk, Kazakhstan 26 E7 46 50N 61 20 E
Aralskoye More, Asia 26 E7 44 30N 60 0 E
Aramac, Australia 62 C4 22 58S 145 14 E
Arambag, India 43 H12 22 53N 87 48 E
Aran I., Ireland 13 B3 55 0N 8 30W
Aran Is., Ireland 13 C2 53 5N 9 42W
Arandān, Iran 44 C5 35 23N 46 55 E
Aranjuez, Spain 19 B4 40 1N 3 40W
Aranos, Namibia 56 C2 24 9S 19 7 E
Aransas Pass, U.S.A. 81 M6 27 55N 97 9W
Araouane, Mali 50 E4 18 55N 3 30W
Arapahoe, U.S.A. 80 E5 40 18N 99 54W
Arapey Grande —►, Uruguay 94 C4 30 55S 57 49W
Arapiraca, Brazil 93 E11 9 45S 36 39W
Arapongas, Brazil 95 A5 23 29S 51 28W
Ar'ar, Si. Arabia 44 D4 30 59N 41 2 E
Araranguá, Brazil 95 B6 29 0S 49 30W
Araraquara, Brazil 93 H9 21 50S 48 0W
Ararás, Serra das, Brazil 95 B5 25 0S 53 10W
Ararat, Australia 63 F3 37 16S 143 0 E
Ararat, Mt. = Ağrı Dağı, Turkey 25 G7 39 50N 44 15 E
Araria, India 43 F12 26 9N 87 33 E
Araripe, Chapada do, Brazil 93 E11 7 20S 40 0W
Araruama, L. de, Brazil 95 A7 22 53S 42 12W
Aras, Rūd-e —►, Iran 44 B5 39 10N 47 10 E
Arauca, Colombia 92 B4 7 0N 70 40W
Arauca —►, Venezuela 92 B5 7 24N 66 35W
Arauco, Chile 94 D1 37 16S 73 25W
Arauco □, Chile 94 D1 37 40S 73 25W
Araxá, Brazil 93 G9 19 35S 46 55W
Araya, Pen. de, Venezuela 92 A6 10 40N 64 0W
Arbat, Iraq 44 C5 35 25N 45 35 E
Arbatax, Italy 20 E3 39 57N 9 42 E
Arbaza, Russia 27 D10 52 40N 92 30 E
Arbīl, Iraq 44 B5 36 15N 44 5 E
Arborfield, Canada 73 C8 53 6N 103 39W
Arborg, Canada 73 C9 50 54N 97 13W
Arbroath, U.K. 12 E6 56 34N 2 35W
Arbuckle, U.S.A. 84 F4 39 1N 122 3W
Arcachon, France 18 D3 44 40N 1 10W
Arcade, U.S.A. 78 D6 42 32N 78 25W
Arcadia, Fla., U.S.A. 77 M5 27 13N 81 52W
Arcadia, La., U.S.A. 81 J8 32 33N 92 55W
Arcadia, Nebr., U.S.A. 80 E5 41 25N 99 8W
Arcadia, Pa., U.S.A. 78 F6 40 47N 78 51W
Arcadia, Wis., U.S.A. 80 C9 44 15N 91 30W
Arcata, U.S.A. 82 F1 40 52N 124 5W
Archangel = Arkhangelsk, Russia 24 B7 64 40N 41 0 E
Archbald, U.S.A. 79 E9 41 30N 75 32W
Archer —►, Australia 62 A3 13 28S 141 41 E
Archer B., Australia 62 A3 13 20S 141 30 E
Archers Post, Kenya 54 B4 0 35N 37 35 E
Arcila = Asilah, Morocco 50 A3 35 29N 6 0W
Arckaringa, Australia 63 D1 27 56S 134 45 E
Arckaringa Cr. —►, Australia 63 D2 28 10S 135 22 E
Arco, U.S.A. 82 E7 43 38N 113 18W
Arcola, Canada 73 D8 49 40N 102 30W
Arcos, Spain 19 B4 41 12N 2 16W
Arcot, India 40 N11 12 53N 79 20 E
Arcoverde, Brazil 93 E11 8 25S 37 4W
Arctic Bay, Canada 69 A11 73 1N 85 7W
Arctic Ocean, Arctic 4 B18 78 0N 160 0W
Arctic Red River, Canada 68 B6 67 15N 134 0W
Arda —►, Bulgaria 21 D12 41 40N 26 29 E
Ardabīl, Iran 45 B6 38 15N 48 18 E
Ardakān = Sepīdān, Iran 45 D7 30 20N 52 5 E
Ardèche □, France 18 D6 44 42N 4 16 E
Ardee, Ireland 13 C5 53 51N 6 32W
Arden, Canada 78 B8 44 43N 76 56W
Arden, Calif., U.S.A. 84 G5 38 36N 121 33W
Arden, Nev., U.S.A. 85 J11 36 1N 115 14W
Ardenne, Belgium 15 E5 49 50N 5 5 E
Ardennes = Ardenne, Belgium 15 E5 49 50N 5 5 E
Ardennes □, France 18 B6 49 35N 4 40 E
Ardestān, Iran 45 C7 33 20N 52 25 E
Ardgour, U.K. 12 E3 56 45N 5 25W
Ardlethan, Australia 63 E4 34 22S 146 53 E
Ardmore, Australia 62 C2 21 39S 139 11 E
Ardmore, Okla., U.S.A. 81 H6 34 10N 97 8W
Ardmore, Pa., U.S.A. 79 G9 39 58N 75 18W
Ardmore, S. Dak., U.S.A. 80 D3 43 1N 103 40W
Ardnacrusha, Ireland 13 D3 52 43N 8 38W
Ardnamurchan, Pt. of, U.K. 12 E2 56 44N 6 14W
Ardrossan, Australia 63 E2 34 26S 137 53 E
Ardrossan, U.K. 12 F4 55 39N 4 50W
Ards □, U.K. 13 B6 54 35N 5 30W
Ards Pen., U.K. 13 B6 54 30N 5 25W
Arecibo, Puerto Rico 89 C6 18 29N 66 43W
Areia Branca, Brazil 93 D11 5 0S 37 0W
Arena, Pt., U.S.A. 84 G3 38 57N 123 44W
Arendal, Norway 9 G10 58 28N 8 46 E
Arequipa, Peru 92 G4 16 20S 71 30W
Arero, Ethiopia 51 H12 4 41N 38 50 E
Arévalo, Spain 19 B3 41 3N 4 43W
Arezzo, Italy 20 C4 43 28N 11 50 E
Argamakmur, Indonesia 36 E2 3 35S 102 0 E
Argentário, Mte., Italy 20 C4 42 23N 11 11 E
Argentia, Canada 71 C9 47 18N 53 58W
Argentina ■, S. Amer. 96 D3 35 0S 66 0W
Argentina Is., Antarctica 5 C17 66 0S 64 0W
Argentino, L., Argentina 96 G2 50 10S 73 0W
Argeş —►, Romania 17 F13 44 12N 26 14 E
Arghandab —►, Afghan. 42 D1 31 30N 64 15 E
Argo, Sudan 51 E11 19 28N 30 30 E
Argolikós Kólpos, Greece 21 F10 37 20N 22 52 E
Argonne, France 18 B6 49 10N 5 0 E
Árgos, Greece 21 F10 37 40N 22 43 E
Argostólion, Greece 21 E9 38 12N 20 33 E
Arguello, Pt., U.S.A. 85 L6 34 35N 120 39W
Arguineguín, Canary Is. 22 G4 27 46N 15 41W
Argun —►, Russia 27 D13 53 20N 121 28 E
Argungu, Nigeria 50 F5 12 40N 4 31 E
Argus Pk., U.S.A. 85 K9 35 52N 117 26W
Argyle, U.S.A. 80 A6 48 20N 96 49W
Argyle, L., Australia 60 C4 16 20S 128 40 E
Århus, Denmark 9 H11 56 8N 10 11 E
Ariadnoye, Russia 30 B7 45 8N 134 25 E
Ariamsvlei, Namibia 56 D2 28 9S 19 51 E
Arica, Chile 92 G4 18 32S 70 20W
Arica, Colombia 92 D4 2 0S 71 50W
Arico, Canary Is. 22 F3 28 9N 16 29W
Arid, C., Australia 61 F3 34 1S 123 10 E
Arida, Japan 31 G7 34 5N 135 8 E
Ariège □, France 18 E4 42 56N 1 30 E
Arīḩā, Syria 44 C3 35 49N 36 35 E
Arílla, Ákra, Greece 23 A3 39 43N 19 39 E
Arima, Trin. & Tob. 89 D7 10 38N 61 17W
Arinos —►, Brazil 92 F7 10 25S 58 20W
Ario de Rosales, Mexico 86 D4 19 12N 102 0W
Aripuanã, Brazil 92 E6 9 25S 60 30W
Aripuanã —►, Brazil 92 E6 5 7S 60 25W
Ariquemes, Brazil 92 E6 9 55S 63 6W
Arisaig, U.K. 12 E3 56 55N 5 50W
Aristazabal I., Canada 72 C3 52 40N 129 10W
Arivaca, U.S.A. 83 L8 31 37N 111 25W
Arivonimamo, Madag. 57 B8 19 1S 47 11 E
Arizaro, Salar de, Argentina 94 A2 24 40S 67 50W
Arizona, Argentina 94 D2 35 45S 65 25W
Arizona □, U.S.A. 83 J8 34 0N 112 0W
Arizpe, Mexico 86 A2 30 20N 110 11W
Arjeplog, Sweden 8 C15 66 3N 18 2 E
Arjona, Colombia 92 A3 10 14N 75 22W
Arjuno, Indonesia 37 G15 7 49S 112 34 E
Arka, Russia 27 C15 60 15N 142 0 E
Arkadelphia, U.S.A. 81 H8 34 7N 93 4W
Arkaig, L., U.K. 12 E3 56 58N 5 10W
Arkalyk, Kazakhstan 26 D7 50 13N 66 50 E
Arkansas □, U.S.A. 81 H8 35 0N 92 30W
Arkansas —►, U.S.A. 81 J9 33 47N 91 4W
Arkansas City, U.S.A. 81 G6 37 4N 97 2W
Arkhángelos, Greece 23 C10 36 13N 28 7 E
Arkhangelsk, Russia 24 B7 64 40N 41 0 E
Arklow, Ireland 13 D5 52 48N 6 10W
Arkticheskiy, Mys, Russia 27 A10 81 10N 95 0 E
Arlanzón —►, Spain 19 A3 42 3N 4 17W
Arlberg Pass, Austria 16 E5 47 9N 10 12 E
Arlee, U.S.A. 82 C6 47 10N 114 5W
Arles, France 18 E6 43 41N 4 40 E
Arlington, S. Africa 57 D4 28 1S 27 53 E
Arlington, Oreg., U.S.A. 82 D3 45 43N 120 12W
Arlington, S. Dak., U.S.A. 80 C6 44 22N 97 8W
Arlington, Va., U.S.A. 76 F7 38 53N 77 7W
Arlington, Wash., U.S.A. 84 B4 48 12N 122 8W
Arlon, Belgium 15 E5 49 42N 5 49 E
Armagh, U.K. 13 B5 54 22N 6 40W
Armagh □, U.K. 13 B5 54 18N 6 37W
Armagnac, France 18 E4 43 50N 0 10 E
Armavir, Russia 25 E7 45 2N 41 7 E
Armenia, Colombia 92 C3 4 35N 75 45W
Armenia ■, Asia 25 F7 40 20N 45 0 E
Armenistís, Ákra, Greece 23 C9 36 8N 27 42 E
Armidale, Australia 63 E5 30 30S 151 40 E
Armour, U.S.A. 80 D5 43 19N 98 21W
Armstrong, B.C., Canada 72 C5 50 25N 119 10W
Armstrong, Ont., Canada 70 B2 50 18N 89 4W
Armstrong, U.S.A. 81 M6 26 56N 97 47W
Armstrong —►, Australia 60 C5 16 35S 131 40 E
Arnarfjörður, Iceland 8 D2 65 48N 23 40W
Arnaud —►, Canada 69 B12 60 0N 70 0W
Arnauti, C., Cyprus 23 D11 35 6N 32 17 E
Arnes, Iceland 8 C3 66 1N 21 31W
Arnett, U.S.A. 81 G5 36 8N 99 46W
Arnhem, Neths. 15 C5 51 58N 5 55 E
Arnhem, C., Australia 62 A2 12 20S 137 30 E
Arnhem B., Australia 62 A2 12 20S 136 10 E
Arnhem Land, Australia 62 A1 13 10S 134 30 E
Arno —►, Italy 20 C4 43 41N 10 17 E
Arno Bay, Australia 63 E2 33 54S 136 34 E
Arnold, Calif., U.S.A. 84 G6 38 15N 120 20W
Arnold, Nebr., U.S.A. 80 E4 41 26N 100 12W
Arnot, Canada 73 B9 55 56N 96 41W
Arnøy, Norway 8 A16 70 9N 20 40 E
Arnprior, Canada 70 C4 45 26N 76 21W
Aroab, Namibia 56 D2 26 41S 19 39 E
Arrabury, Australia 63 D3 26 45S 141 0 E
Arrah = Ara, India 43 G11 25 35N 84 32 E
Arran, U.K. 12 F3 55 34N 5 12W
Arrandale, Canada 72 C3 54 57N 130 0W
Arras, France 18 A5 50 17N 2 46 E
Arrecife, Canary Is. 22 F6 28 57N 13 37W
Arrecifes, Argentina 94 C3 34 6S 60 9W
Arrée, Mts. d', France 18 B2 48 26N 3 55W
Arriaga, Chiapas, Mexico 87 D6 16 15N 93 52W
Arriaga, San Luis Potosí, Mexico 86 C4 21 55N 101 23W
Arrilalah P.O., Australia 62 C3 23 43S 143 54 E
Arrino, Australia 61 E2 29 30S 115 40 E
Arrow, L., Ireland 13 B3 54 3N 8 20W
Arrow Rock Res., U.S.A. 82 E6 43 45N 115 50W
Arrowhead, Canada 72 C5 50 40N 117 55W
Arrowhead, L., U.S.A. 85 L9 34 16N 117 10W
Arrowtown, N.Z. 59 L2 44 57S 168 50 E
Arroyo Grande, U.S.A. 85 K6 35 7N 120 35W
Ars, Iran 44 B5 37 9N 47 46 E
Arsenault L., Canada 73 B7 55 6N 108 32W
Arsenev, Russia 30 B6 44 10N 133 15 E
Árta, Greece 21 E9 39 8N 21 2 E
Artá, Spain 22 B10 39 41N 3 21 E
Arteaga, Mexico 86 D4 18 50N 102 20W
Artem, Russia 30 C6 43 22N 132 13 E
Artemovsk, Russia 27 D10 54 45N 93 35 E
Artesia = Mosomane, Botswana 56 C4 24 2S 26 19 E
Artesia, U.S.A. 81 J2 32 51N 104 24W
Artesia Wells, U.S.A. 81 L5 28 17N 99 17W
Artesian, U.S.A. 80 C6 44 1N 97 55W
Arthur —►, Australia 62 G3 41 2S 144 40 E
Arthur Cr. —►, Australia 62 C2 22 30S 136 25 E
Arthur Pt., Australia 62 C5 22 7S 150 3 E
Arthur's Pass, N.Z. 59 K3 42 54S 171 35 E
Arthur's Town, Bahamas 89 B4 24 38N 75 42W
Artigas, Uruguay 94 C4 30 20S 56 30W
Artillery L., Canada 73 A7 63 9N 107 52W
Artois, France 18 A5 50 20N 2 30 E
Artvin, Turkey 25 F7 41 14N 41 44 E
Aru, Kepulauan, Indonesia 37 F8 6 0S 134 30 E
Aru Is. = Aru, Kepulauan, Indonesia 37 F8 6 0S 134 30 E
Aru Meru □, Tanzania 54 C4 3 20S 36 50 E
Arua, Uganda 54 B3 3 1N 30 58 E
Aruanã, Brazil 93 F8 14 54S 51 10W
Aruba ■, W. Indies 89 D6 12 30N 70 0W
Arucas, Canary Is. 22 F4 28 7N 15 32W
Arumpo, Australia 63 E3 33 48S 142 55 E
Arun —►, Nepal 43 F12 26 55N 87 10 E
Arunachal Pradesh □, India 41 E19 28 0N 95 0 E
Arusha, Tanzania 54 C4 3 20S 36 40 E
Arusha □, Tanzania 54 C4 4 0S 36 30 E
Arusha Chini, Tanzania 54 C4 3 32S 37 20 E
Aruwimi —►, Zaïre 54 B1 1 13N 23 36 E
Arvada, U.S.A. 82 D10 44 39N 106 8W
Árvi, Greece 23 E7 34 59N 25 28 E
Arvida, Canada 71 C5 48 25N 71 14W
Arvidsjaur, Sweden 8 D15 65 35N 19 10 E
Arvika, Sweden 9 G12 59 40N 12 36 E
Arvin, U.S.A. 85 K8 35 12N 118 50W
Arxan, China 33 B6 47 11N 119 57 E
Aryirádhes, Greece 23 B3 39 27N 19 58 E
Aryiroúpolis, Greece 23 D6 35 17N 24 20 E
Arys, Kazakhstan 26 E7 42 26N 68 48 E
Arzamas, Russia 24 C7 55 27N 43 55 E
Arzew, Algeria 50 A4 35 50N 0 23W
Aş Şadr, U.A.E. 45 E7 24 40N 54 41 E
Aş Şafā, Syria 47 B6 33 10N 37 0 E
'As Saffānīyah, Si. Arabia 45 D6 28 5N 48 50 E
As Safīrah, Syria 44 B3 36 5N 37 21 E
Aş Şahm, Oman 45 E8 24 10N 56 53 E
As Sājir, Si. Arabia 44 E5 25 11N 44 36 E
As Salamīyah, Syria 44 C3 35 1N 37 2 E
As Salt, Jordan 47 C4 32 2N 35 43 E
As Sal'w'a, Qatar 45 E6 24 23N 50 50 E
As Samāwah, Iraq 44 D5 31 15N 45 15 E
As Sanamayn, Syria 47 B5 33 3N 36 10 E
As Sukhnah, Syria 44 C3 34 52N 38 52 E
As Sulaymānīyah, Iraq 44 C5 35 35N 45 29 E
As Sulaymī, Si. Arabia 44 E4 26 17N 41 21 E
As Summān, Si. Arabia 44 E5 25 0N 47 0 E
As Suwaydā', Syria 47 C5 32 40N 36 30 E
As Suwaydā' □, Syria 47 C5 32 45N 36 45 E
Aş Şuwayrah, Iraq 44 C5 32 55N 45 0 E
Asab, Namibia 56 D2 25 30S 18 0 E
Asahi-Gawa —►, Japan 31 G6 34 36N 133 58 E
Asahigawa, Japan 30 C11 43 46N 142 22 E
Asansol, India 43 H12 23 40N 87 1 E
Asbesberge, S. Africa 56 D3 29 0S 23 0 E
Asbestos, Canada 71 C5 45 47N 71 58W
Asbury Park, U.S.A. 79 F10 40 13N 74 1W
Ascensión, Mexico 86 A3 31 6N 107 59W
Ascensión, B. de la, Mexico 87 D7 19 50N 87 20W
Ascension I., Atl. Oc. 2 E9 8 0S 14 15W
Aschaffenburg, Germany 16 D4 49 58N 9 8 E
Áscoli Piceno, Italy 20 C5 42 51N 13 34 E
Ascope, Peru 92 E3 7 46S 79 8W
Ascotán, Chile 94 A2 21 45S 68 17W
Aseb, Eritrea 46 E3 13 0N 42 40 E
Asela, Ethiopia 51 G12 8 0N 39 0 E
Asgata, Cyprus 23 E12 34 46N 33 15 E
Ash Fork, U.S.A. 83 J7 35 13N 112 29W
Ash Grove, U.S.A. 81 G8 37 19N 93 35W
Ash Shamāl □, Lebanon 47 A5 34 25N 36 0 E
Ash Shāmīyah, Iraq 44 D5 31 55N 44 35 E
Ash Shāriqah, U.A.E. 45 E7 25 23N 55 26 E
Ash Sharmah, Si. Arabia 44 D2 28 1N 35 16 E
Ash Sharqāţ, Iraq 44 C4 35 27N 43 16 E
Ash Sharqi, Al Jabal, Lebanon 47 B5 33 40N 36 10 E
Ash Shaţrah, Iraq 44 D5 31 30N 46 10 E
Ash Shawbak, Jordan 44 D2 30 32N 35 34 E
Ash Shawmari, J., Jordan 47 E5 30 35N 36 35 E
Ash Shaykh, J., Lebanon 47 B4 33 25N 35 50 E
Ash Shināfīyah, Iraq 44 D5 31 35N 44 39 E
Ash Shu'aybah, Si. Arabia 44 E5 27 53N 44 43 E
Ash Shumlūl, Si. Arabia 44 E5 26 31N 47 20 E
Ash Shūr'a, Iraq 44 C4 35 58N 43 13 E
Ash Shurayf, Si. Arabia 44 E3 25 43N 39 14 E
Ash Shuwayfāt, Lebanon 47 B4 33 45N 35 30 E
Asha, Russia 24 D10 55 0N 57 16 E
Ashau, Vietnam 38 D6 16 6N 107 22 E
Ashburn, U.S.A. 77 K4 31 43N 83 39W
Ashburton, N.Z. 59 K3 43 53S 171 48 E
Ashburton —►, Australia 60 D1 21 40S 114 56 E
Ashburton Downs, Australia 60 D2 23 25S 117 4 E
Ashby de la Zouch, U.K. 10 E6 52 45N 1 29W
Ashcroft, Canada 72 C4 50 40N 121 20W
Ashdod, Israel 47 D3 31 49N 34 35 E
Asheboro, U.S.A. 77 H6 35 43N 79 49W
Asherton, U.S.A. 81 L5 28 27N 99 46W
Asheville, U.S.A. 77 H4 35 36N 82 33W
Asheweig —►, Canada 70 B2 54 17N 87 12W
Ashford, Australia 63 D5 29 15S 151 3 E
Ashford, U.K. 11 F8 51 8N 0 53 E
Ashford, U.S.A. 82 C2 46 46N 122 2W
Ashgabat = Ashkhabad, Turkmenistan 26 F6 38 0N 57 50 E
Ashibetsu, Japan 30 C11 43 31N 142 11 E
Ashikaga, Japan 31 F9 36 28N 139 29 E
Ashizuri-Zaki, Japan 31 H6 32 44N 133 0 E
Ashkarkot, Afghan. 42 C2 33 3N 67 58 E
Ashkhabad, Turkmenistan 26 F6 38 0N 57 50 E
Ashland, Kans., U.S.A. 81 G5 37 11N 99 46W
Ashland, Ky., U.S.A. 76 F4 38 28N 82 38W
Ashland, Maine, U.S.A. 71 C6 46 38N 68 24W

Ashland

Barakula

Bagé, *Brazil* **95 C5** 31 20S 54 15W
Bagenalstown = Muine Bheag, *Ireland* **13 D5** 52 42N 6 57W
Baggs, *U.S.A.* **82 F10** 41 2N 107 39W
Bagh, *Pakistan* **43 C5** 33 59N 73 45 E
Baghdād, *Iraq* **44 C5** 33 20N 44 30 E
Baghlān, *Afghan.* **40 A6** 36 12N 69 0 E
Bagley, *U.S.A.* **80 B7** 47 32N 95 24W
Bagotville, *Canada* **71 C5** 48 22N 70 54W
Baguio, *Phil.* **37 A6** 16 26N 120 34 E
Bahadurgarh, *India* **42 E7** 28 40N 76 57 E
Bahama, Canal Viejo de, *W. Indies* **88 B4** 22 10N 77 30W
Bahamas ■, *N. Amer.* **89 B5** 24 0N 75 0W
Baharampur, *India* **43 G13** 24 2N 88 27 E
Bahau, *Malaysia* **39 L4** 2 48N 102 26 E
Bahawalnagar, *Pakistan* **42 D5** 30 0N 73 15 E
Bahawalpur, *Pakistan* **42 E4** 29 24N 71 40 E
Baheri, *India* **43 E8** 28 45N 79 34 E
Bahi, *Tanzania* **54 D4** 5 58S 35 21 E
Bahi Swamp, *Tanzania* **54 D4** 6 10S 35 0 E
Bahía = Salvador, *Brazil* **93 F11** 13 0S 38 30W
Bahía □, *Brazil* **93 F10** 12 0S 42 0W
Bahía, Is. de la, *Honduras* **88 C2** 16 45N 86 15W
Bahía Blanca, *Argentina* **94 D3** 38 35S 62 13W
Bahía de Caráquez, *Ecuador* **92 D2** 0 40S 80 27W
Bahía Honda, *Cuba* **88 B3** 22 54N 83 10W
Bahía Laura, *Argentina* **96 F3** 48 10S 66 30W
Bahía Negra, *Paraguay* **92 H7** 20 5S 58 5W
Bahmanzād, *Iran* **45 D6** 31 15N 51 47 E
Bahr Aouk →, *C.A.R.* **52 C3** 8 40N 19 0 E
Bahr el Ahmar □, *Sudan* **51 E12** 20 0N 35 0 E
Bahr el Ghazâl □, *Sudan* **51 G10** 7 0N 28 0 E
Bahr Salamat →, *Chad* **51 G8** 9 20N 18 0 E
Bahraich, *India* **43 F9** 27 38N 81 37 E
Bahrain ■, *Asia* **45 E6** 26 0N 50 35 E
Bahret Assad, *Syria* **44 C3** 36 0N 38 15 E
Bahror, *India* **42 F7** 27 51N 76 20 E
Bāhū Kalāt, *Iran* **45 E9** 25 43N 61 25 E
Bai Bung, Mui, *Vietnam* **39 H5** 8 38N 104 44 E
Bai Duc, *Vietnam* **38 C5** 18 3N 105 49 E
Bai Thuong, *Vietnam* **38 C5** 19 54N 105 23 E
Baia Mare, *Romania* **17 E11** 47 40N 23 35 E
Baïbokoum, *Chad* **51 G8** 7 46N 15 43 E
Baicheng, *China* **35 B12** 45 38N 122 42 E
Baidoa, *Somali Rep.* **46 G3** 3 8N 43 30 E
Baie Comeau, *Canada* **71 C6** 49 12N 68 10W
Baie-St-Paul, *Canada* **71 C5** 47 28N 70 32W
Baie Trinité, *Canada* **71 C6** 49 25N 67 20W
Baie Verte, *Canada* **71 C8** 49 55N 56 12W
Baihe, *China* **34 H6** 32 50N 110 5 E
Ba'ījī, *Iraq* **44 C4** 35 0N 43 30 E
Baikal, L. = Baykal, Oz., *Russia* **27 D11** 53 0N 108 0 E
Baile Atha Cliath = Dublin, *Ireland* **13 C5** 53 20N 6 18W
Bailundo, *Angola* **53 G3** 12 10S 15 50 E
Bainbridge, *Ga., U.S.A.* **77 K3** 30 55N 84 35W
Bainbridge, *N.Y., U.S.A.* **79 D9** 42 18N 75 29W
Baing, *Indonesia* **37 F6** 10 14S 120 34 E
Bainiu, *China* **34 H7** 32 50N 112 15 E
Bainville, *U.S.A.* **80 A2** 48 8N 104 13W
Bā'ir, *Jordan* **47 E5** 30 45N 36 55 E
Baird, *U.S.A.* **81 J5** 32 24N 99 24W
Baird Mts., *U.S.A.* **68 B3** 67 0N 160 0W
Bairin Youqi, *China* **35 C10** 43 30N 118 35 E
Bairin Zuoqi, *China* **35 C10** 43 58N 119 15 E
Bairnsdale, *Australia* **63 F4** 37 48S 147 36 E
Baisha, *China* **34 G7** 34 20N 112 32 E
Baitadi, *Nepal* **43 E9** 29 35N 80 25 E
Baixo-Alentejo, *Portugal* **19 D1** 38 0N 8 30W
Baiyin, *China* **34 F3** 36 45N 104 14 E
Baiyu Shan, *China* **34 F4** 37 15N 107 30 E
Baj Baj, *India* **43 H13** 22 30N 88 5 E
Baja, *Hungary* **17 E9** 46 12N 18 59 E
Baja, Pta., *Mexico* **86 B1** 29 50N 116 0W
Baja California, *Mexico* **86 A1** 31 10N 115 12W
Bajamar, *Canary Is.* **22 F3** 28 33N 16 20W
Bajana, *India* **42 H4** 23 7N 71 49 E
Bājgīrān, *Iran* **45 B8** 37 36N 58 24 E
Bajimba, Mt., *Australia* **63 D5** 29 17S 152 6 E
Bajo Nuevo, *Caribbean* **88 C4** 15 40N 78 50W
Bajool, *Australia* **62 C5** 23 40S 150 35 E
Bakala, *C.A.R.* **51 G9** 6 15N 20 20 E
Bakchar, *Russia* **26 D9** 57 1N 82 5 E
Bakel, *Senegal* **50 F2** 14 56N 12 20W
Baker, *Calif., U.S.A.* **85 K10** 35 16N 116 4W
Baker, *Mont., U.S.A.* **80 B2** 46 22N 104 17W
Baker, *Oreg., U.S.A.* **82 D5** 44 47N 117 50W
Baker, L., *Canada* **68 B10** 64 0N 96 0W
Baker I., *Pac. Oc.* **64 G10** 0 10N 176 35W
Baker L., *Australia* **61 E4** 26 54S 126 5 E
Baker Lake, *Canada* **68 B10** 64 20N 96 3W
Baker Mt., *U.S.A.* **82 B3** 48 50N 121 49W
Bakers Creek, *Australia* **62 C4** 21 13S 149 7 E
Baker's Dozen Is., *Canada* **70 A4** 56 45N 78 45W
Bakersfield, *Calif., U.S.A.* **85 K7** 35 23N 119 1W
Bakersfield, *Vt., U.S.A.* **79 B12** 44 45N 72 48W
Bākhtarān, *Iran* **44 C5** 34 23N 47 0 E
Bākhtarān □, *Iran* **44 C5** 34 0N 46 30 E
Bakkafjörður, *Iceland* **8 C6** 66 2N 14 48W
Bakkagerði, *Iceland* **8 D7** 65 31N 13 49W
Bakony Forest = Bakony Hegyseg, *Hungary* **17 E8** 47 10N 17 30 E
Bakony Hegyseg, *Hungary* **17 E8** 47 10N 17 30 E
Bakouma, *C.A.R.* **51 G9** 5 40N 22 56 E
Baku, *Azerbaijan* **25 F8** 40 25N 49 45 E
Bakutis Coast, *Antarctica* **5 D15** 74 0S 120 0W
Baky = Baku, *Azerbaijan* **25 F8** 40 25N 49 45 E
Bala, *Canada* **78 A5** 45 1N 79 37W
Bala, L., *U.K.* **10 E4** 52 53N 3 38W
Balabac I., *Phil.* **36 C5** 8 0N 117 0 E
Balabac Str., *E. Indies* **36 C5** 7 53N 117 5 E
Balabagh, *Afghan.* **42 B4** 34 25N 70 12 E
Balabakk, *Lebanon* **47 A5** 34 0N 36 10 E
Balabalangan, Kepulauan, *Indonesia* **36 E5** 2 20S 117 30 E
Balad, *Iraq* **44 C5** 34 1N 44 9 E
Balad Rūz, *Iraq* **44 C5** 33 42N 45 5 E
Bālādeh, *Fārs, Iran* **45 D6** 29 17N 51 56 E
Baladeh, *Māzandaran, Iran* **45 B6** 36 12N 51 48 E
Balaghat, *India* **40 J12** 21 49N 80 12 E
Balaghat Ra., *India* **40 K10** 18 50N 76 30 E
Balaguer, *Spain* **19 B6** 41 50N 0 50 E
Balaklava, *Australia* **63 E2** 34 7S 138 22 E
Balaklava, *Ukraine* **25 F5** 44 30N 33 30 E
Balakovo, *Russia* **24 D8** 52 4N 47 55 E
Balancán, *Mexico* **87 D6** 17 48N 91 32W
Balashov, *Russia* **24 D7** 51 30N 43 10 E
Balasinor, *India* **42 H5** 22 57N 73 23 E
Balasore = Baleshwar, *India* **41 J15** 21 35N 87 3 E
Balaton, *Hungary* **17 E8** 46 50N 17 40 E
Balboa, *Panama* **88 E4** 8 57N 79 34W
Balbriggan, *Ireland* **13 C5** 53 35N 6 10W
Balcarce, *Argentina* **94 D4** 38 0S 58 10W
Balcarres, *Canada* **73 C8** 50 50N 103 35W
Balchik, *Bulgaria* **21 C13** 43 28N 28 11 E
Balclutha, *N.Z.* **59 M2** 46 15S 169 45 E
Bald Hd., *Australia* **61 G2** 35 6S 118 1 E
Bald I., *Australia* **61 F2** 34 57S 118 27 E
Bald Knob, *U.S.A.* **81 H9** 35 19N 91 34W
Baldock L., *Canada* **73 B9** 56 33N 97 57W
Baldwin, *Fla., U.S.A.* **77 K4** 30 18N 81 59W
Baldwin, *Mich., U.S.A.* **76 D3** 43 54N 85 51W
Baldwinsville, *U.S.A.* **79 C8** 43 10N 76 20W
Baldy Peak, *U.S.A.* **83 K9** 33 54N 109 34W
Baleares, Is., *Spain* **22 B10** 39 30N 3 0 E
Balearic Is. = Baleares, Is., *Spain* **22 B10** 39 30N 3 0 E
Baler, *Phil.* **37 A6** 15 46N 121 34 E
Baleshwar, *India* **41 J15** 21 35N 87 3 E
Balfate, *Honduras* **88 C2** 15 48N 86 25W
Balfe's Creek, *Australia* **62 C4** 20 12S 145 55 E
Balfour, *S. Africa* **57 D4** 26 38S 28 35 E
Bali, *Cameroon* **50 G7** 5 54N 10 0 E
Balí, *Greece* **23 D6** 35 25N 24 47 E
Bali, *Indonesia* **36 F5** 8 20S 115 0 E
Bali □, *Indonesia* **36 F5** 8 20S 115 0 E
Bali, Selat, *Indonesia* **37 H16** 8 18S 114 25 E
Balikeşir, *Turkey* **25 G4** 39 35N 27 58 E
Balikpapan, *Indonesia* **36 E5** 1 10S 116 55 E
Balimbing, *Phil.* **37 C5** 5 5N 119 58 E
Baling, *Malaysia* **39 K3** 5 41N 100 55 E
Balipara, *India* **41 F18** 26 50N 92 45 E
Baliza, *Brazil* **93 G8** 16 0S 52 20W
Balkan Mts. = Stara Planina, *Bulgaria* **21 C10** 43 15N 23 0 E
Balkan Peninsula, *Europe* **6 G10** 42 0N 23 0 E
Balkhash, *Kazakhstan* **26 E8** 46 50N 74 50 E
Balkhash, Ozero, *Kazakhstan* **26 E8** 46 0N 74 50 E
Balla, *Bangla.* **41 G17** 24 10N 91 35 E
Ballachulish, *U.K.* **12 E3** 56 40N 5 10W
Balladonia, *Australia* **61 F3** 32 27S 123 51 E
Ballarat, *Australia* **63 F3** 37 33S 143 50 E
Ballard, L., *Australia* **61 E3** 29 20S 120 40 E
Ballater, *U.K.* **12 D5** 57 2N 3 2W
Ballenas, Canal de, *Mexico* **86 B2** 29 10N 113 45W
Balleny Is., *Antarctica* **5 C11** 66 30S 163 0 E
Ballia, *India* **43 G11** 25 46N 84 12 E
Ballidu, *Australia* **61 F2** 30 35S 116 45 E
Ballina, *Australia* **63 D5** 28 50S 153 31 E
Ballina, *Mayo, Ireland* **13 B2** 54 7N 9 10W
Ballina, *Tipp., Ireland* **13 D3** 52 49N 8 27W
Ballinasloe, *Ireland* **13 C3** 53 20N 8 12W
Ballinger, *U.S.A.* **81 K5** 31 45N 99 57W
Ballinrobe, *Ireland* **13 C2** 53 36N 9 13W
Ballinskelligs B., *Ireland* **13 E1** 51 46N 10 11W
Ballycastle, *U.K.* **13 A5** 55 12N 6 15W
Ballymena, *U.K.* **13 B5** 54 53N 6 18W
Ballymena □, *U.K.* **13 B5** 54 53N 6 18W
Ballymoney, *U.K.* **13 A5** 55 5N 6 30W
Ballymoney □, *U.K.* **13 A5** 55 5N 6 23W
Ballyshannon, *Ireland* **13 B3** 54 30N 8 10W
Balmaceda, *Chile* **96 F2** 46 0S 71 50W
Balmoral, *Australia* **63 F3** 37 15S 141 48 E
Balmoral, *U.K.* **12 D5** 57 3N 3 13W
Balmorhea, *U.S.A.* **81 K3** 30 59N 103 45W
Balonne →, *Australia* **63 D4** 28 47S 147 56 E
Balqash Kol = Balkhash, Ozero, *Kazakhstan* **26 E8** 46 0N 74 50 E
Balrampur, *India* **43 F10** 27 30N 82 20 E
Balranald, *Australia* **63 E3** 34 38S 143 33 E
Balsas, *Mexico* **87 D5** 18 0N 99 40W
Balsas →, *Mexico* **86 D4** 17 55N 102 10W
Balston Spa, *U.S.A.* **79 D11** 43 0N 73 52W
Balta, *Ukraine* **25 E4** 48 2N 29 45 E
Balta, *U.S.A.* **80 A4** 48 10N 100 2W
Bălţi = Beltsy, *Moldavia* **25 E4** 47 48N 28 0 E
Baltic Sea, *Europe* **9 H15** 57 0N 19 0 E
Baltimore, *Ireland* **13 E2** 51 29N 9 22W
Baltimore, *U.S.A.* **76 F7** 39 17N 76 37W
Baltit, *Pakistan* **43 A6** 36 15N 74 40 E
Baluchistan □, *Pakistan* **40 F4** 27 30N 65 0 E
Balurghat, *India* **43 G13** 25 15N 88 44 E
Balygychan, *Russia* **27 C16** 63 56N 154 12 E
Bam, *Iran* **45 D8** 29 7N 58 14 E
Bama, *Nigeria* **51 F7** 11 33N 13 41 E
Bamako, *Mali* **50 F3** 12 34N 7 55W
Bamba, *Mali* **50 E4** 17 5N 1 24W
Bambari, *C.A.R.* **51 G9** 5 40N 20 35 E
Bambaroo, *Australia* **62 B4** 18 50S 146 10 E
Bamberg, *Germany* **16 D5** 49 54N 10 53 E
Bamberg, *U.S.A.* **77 J5** 33 18N 81 2W
Bambili, *Zaïre* **54 B2** 3 40N 26 0 E
Bamenda, *Cameroon* **50 G7** 5 57N 10 11 E
Bamfield, *Canada* **72 D3** 48 45N 125 10W
Bāmīān □, *Afghan.* **40 B5** 35 0N 67 0 E
Bamiancheng, *China* **35 C13** 43 15N 124 2 E
Bampūr, *Iran* **45 E9** 27 15N 60 21 E
Ban Aranyaprathet, *Thailand* **38 F4** 13 41N 102 30 E
Ban Ban, *Laos* **38 C4** 19 31N 103 30 E
Ban Bang Hin, *Thailand* **39 H2** 9 32N 98 35 E
Ban Chiang Klang, *Thailand* **38 C3** 19 25N 100 55 E
Ban Chik, *Laos* **38 D4** 17 15N 102 22 E
Ban Choho, *Thailand* **38 E4** 15 2N 102 9 E
Ban Dan Lan Hoi, *Thailand* **38 D2** 17 0N 99 35 E
Ban Don = Surat Thani, *Thailand* **39 H2** 9 6N 99 20 E
Ban Don, *Vietnam* **38 F6** 12 53N 107 48 E
Ban Don, Ao, *Thailand* **39 H2** 9 20N 99 25 E
Ban Dong, *Thailand* **38 C3** 19 30N 100 59 E
Ban Hong, *Thailand* **38 C2** 18 18N 98 50 E
Ban Kaeng, *Thailand* **38 D3** 17 29N 100 7 E
Ban Keun, *Laos* **38 C4** 18 22N 102 35 E
Ban Khai, *Thailand* **38 F3** 12 46N 101 18 E
Ban Kheun, *Laos* **38 B3** 20 13N 101 7 E
Ban Khlong Kua, *Thailand* **39 J3** 6 57N 100 8 E
Ban Khuan Mao, *Thailand* **39 J2** 7 50N 99 37 E
Ban Khun Yuam, *Thailand* **38 C1** 18 49N 97 57 E
Ban Ko Yai Chim, *Thailand* **39 G2** 11 17N 99 26 E
Ban Kok, *Thailand* **38 D4** 16 40N 103 40 E
Ban Laem, *Thailand* **38 F2** 13 13N 99 59 E
Ban Lao Ngam, *Laos* **38 E6** 15 28N 106 10 E
Ban Le Kathe, *Thailand* **38 E2** 15 49N 98 53 E
Ban Mae Chedi, *Thailand* **38 C2** 19 11N 99 31 E
Ban Mae Laeng, *Thailand* **38 B2** 20 1N 99 17 E
Ban Mae Sariang, *Thailand* **38 C1** 18 10N 97 56 E
Ban Mê Thuôt = Buon Me Thuot, *Vietnam* **38 F7** 12 40N 108 3 E
Ban Mi, *Thailand* **38 E3** 15 3N 100 32 E
Ban Muong Mo, *Laos* **38 C4** 19 4N 103 58 E
Ban Na Mo, *Laos* **38 D5** 17 7N 105 40 E
Ban Na San, *Thailand* **39 H2** 8 53N 99 52 E
Ban Na Tong, *Laos* **38 B3** 20 56N 101 47 E
Ban Nam Bac, *Laos* **38 B4** 20 38N 102 20 E
Ban Nam Ma, *Laos* **38 A3** 22 2N 101 37 E
Ban Ngang, *Laos* **38 E6** 15 59N 106 11 E
Ban Nong Bok, *Laos* **38 D5** 17 5N 104 48 E
Ban Nong Boua, *Laos* **38 E6** 15 40N 106 33 E
Ban Nong Pling, *Thailand* **38 E3** 15 40N 100 10 E
Ban Pak Chan, *Thailand* **39 G2** 10 32N 98 51 E
Ban Phai, *Thailand* **38 D4** 16 4N 102 44 E
Ban Pong, *Thailand* **38 F2** 13 50N 99 55 E
Ban Ron Phibun, *Thailand* **39 H2** 8 9N 99 51 E
Ban Sanam Chai, *Thailand* **39 J3** 7 33N 100 25 E
Ban Sangkha, *Thailand* **38 E4** 14 37N 103 52 E
Ban Tak, *Thailand* **38 D2** 17 2N 99 4 E
Ban Tako, *Thailand* **38 E4** 14 5N 102 40 E
Ban Tha Dua, *Thailand* **38 D2** 17 59N 98 39 E
Ban Tha Li, *Thailand* **38 D3** 17 37N 101 25 E
Ban Tha Nun, *Thailand* **39 H2** 8 12N 98 18 E
Ban Thahine, *Laos* **38 E5** 14 12N 105 33 E
Ban Xien Kok, *Laos* **38 B3** 20 54N 100 39 E
Ban Yen Nhan, *Vietnam* **38 B6** 20 57N 106 2 E
Banaba, *Kiribati* **64 H8** 0 45S 169 50 E
Bañalbufar, *Spain* **22 B9** 39 42N 2 31 E
Banalia, *Zaïre* **54 B2** 1 32N 25 5 E
Banam, *Cambodia* **39 G5** 11 20N 105 17 E
Banamba, *Mali* **50 F3** 13 29N 7 22W
Banana, *Australia* **62 C5** 24 28S 150 8 E
Bananal, I. do, *Brazil* **93 F8** 11 30S 50 30W
Banaras = Varanasi, *India* **43 G10** 25 22N 83 0 E
Banas →, *Gujarat, India* **42 H4** 23 45N 71 25 E
Banas →, *Mad. P., India* **43 G9** 24 15N 81 30 E
Banbān, *Si. Arabia* **44 E5** 25 1N 46 35 E
Banbridge, *U.K.* **13 B5** 54 21N 6 17W
Banbridge □, *U.K.* **13 B5** 54 21N 6 16W
Banbury, *U.K.* **11 E6** 52 4N 1 21W
Banchory, *U.K.* **12 D6** 57 3N 2 30W
Bancroft, *Canada* **70 C4** 45 3N 77 51W
Band Bonī, *Iran* **45 E8** 25 30N 59 33 E
Band Qīr, *Iran* **45 D6** 31 39N 48 53 E
Banda, *India* **43 G9** 25 30N 80 26 E
Banda, Kepulauan, *Indonesia* **37 E7** 4 37S 129 50 E
Banda Aceh, *Indonesia* **36 C1** 5 35N 95 20 E
Banda Banda, Mt., *Australia* **63 E5** 31 10S 152 28 E
Banda Elat, *Indonesia* **37 F8** 5 40S 133 5 E
Banda Is. = Banda, Kepulauan, *Indonesia* **37 E7** 4 37S 129 50 E
Banda Sea, *Indonesia* **37 F7** 6 0S 130 0 E
Bandai-San, *Japan* **30 F10** 37 36N 140 4 E
Bandān, *Iran* **45 D9** 31 23N 60 44 E
Bandanaira, *Indonesia* **37 E7** 4 32S 129 54 E
Bandanwara, *India* **42 F6** 26 9N 74 38 E
Bandar = Machilipatnam, *India* **41 L12** 16 12N 81 8 E
Bandar 'Abbās, *Iran* **45 E8** 27 15N 56 15 E
Bandar-e Anzalī, *Iran* **45 B6** 37 30N 49 30 E
Bandar-e Chārak, *Iran* **45 E7** 26 45N 54 20 E
Bandar-e Deylam, *Iran* **45 D6** 30 5N 50 10 E
Bandar-e Khomeyni, *Iran* **45 D6** 30 30N 49 5 E
Bandar-e Lengeh, *Iran* **45 E7** 26 35N 54 58 E
Bandar-e Maqām, *Iran* **45 E7** 26 56N 53 29 E
Bandar-e Ma'shur, *Iran* **45 D6** 30 35N 49 10 E
Bandar-e Nakhīlū, *Iran* **45 E7** 26 58N 53 30 E
Bandar-e Rīg, *Iran* **45 D6** 29 29N 50 38 E
Bandar-e Torkeman, *Iran* **45 B7** 37 0N 54 10 E
Bandar Maharani = Muar, *Malaysia* **39 L4** 2 3N 102 34 E
Bandar Penggaram = Batu Pahat, *Malaysia* **39 M4** 1 50N 102 56 E
Bandar Seri Begawan, *Brunei* **36 C4** 4 52N 115 0 E
Bandawe, *Malawi* **55 E3** 11 58S 34 5 E
Bandeira, Pico da, *Brazil* **95 A7** 20 26S 41 47W
Bandera, *Argentina* **94 B3** 28 55S 62 20W
Bandera, *U.S.A.* **81 L5** 29 44N 99 5W
Banderas, B. de, *Mexico* **86 C3** 20 40N 105 30W
Bandiagara, *Mali* **50 F4** 14 12N 3 29W
Bandırma, *Turkey* **25 F4** 40 20N 28 0 E
Bandon, *Ireland* **13 E3** 51 44N 8 45W
Bandon →, *Ireland* **13 E3** 51 40N 8 41W
Bandula, *Mozam.* **55 F3** 19 0S 33 7 E
Bandundu, *Zaïre* **52 E3** 3 15S 17 22 E
Bandung, *Indonesia* **37 G12** 6 54S 107 36 E
Bandya, *Australia* **61 E3** 27 40S 122 5 E
Bāneh, *Iran* **44 C5** 35 59N 45 53 E
Banes, *Cuba* **89 B4** 21 0N 75 42W
Banff, *Canada* **72 C5** 51 10N 115 34W
Banff, *U.K.* **12 D6** 57 40N 2 32W
Banff Nat. Park, *Canada* **72 C5** 51 30N 116 15W
Banfora, *Burkina Faso* **50 F4** 10 40N 4 40W
Bang Fai →, *Laos* **38 D5** 16 57N 104 45 E
Bang Hieng →, *Laos* **38 D5** 16 10N 105 10 E
Bang Krathum, *Thailand* **38 D3** 16 34N 100 18 E
Bang Lamung, *Thailand* **38 F3** 13 3N 100 56 E
Bang Mun Nak, *Thailand* **38 D3** 16 2N 100 23 E
Bang Pa In, *Thailand* **38 E3** 14 14N 100 35 E
Bang Rakam, *Thailand* **38 D3** 16 45N 100 7 E
Bang Saphan, *Thailand* **39 G2** 11 14N 99 28 E
Bangala Dam, *Zimbabwe* **55 G3** 21 7S 31 25 E
Bangalore, *India* **40 N10** 12 59N 77 40 E
Bangaon, *India* **43 H13** 23 0N 88 47 E
Bangassou, *C.A.R.* **52 D4** 4 55N 23 7 E
Banggai, Kepulauan, *Indonesia* **37 E6** 1 40S 123 30 E
Banggi, P., *Malaysia* **36 C5** 7 17N 117 12 E
Banghāzī, *Libya* **51 B9** 32 11N 20 3 E
Bangil, *Indonesia* **37 G15** 7 36S 112 50 E
Bangka, P., *Sulawesi, Indonesia* **37 D7** 1 50N 125 5 E
Bangka, P., *Sumatera, Indonesia* **36 E3** 2 0S 105 50 E
Bangka, Selat, *Indonesia* **36 E3** 2 30S 105 30 E
Bangkalan, *Indonesia* **37 G15** 7 2S 112 46 E
Bangkinang, *Indonesia* **36 D2** 0 18N 101 5 E
Bangko, *Indonesia* **36 E2** 2 5S 102 9 E
Bangkok, *Thailand* **38 F3** 13 45N 100 35 E
Bangladesh ■, *Asia* **41 H17** 24 0N 90 0 E
Bangong Co, *India* **43 B8** 35 50N 79 20 E
Bangor, *Down, U.K.* **13 B6** 54 40N 5 40W
Bangor, *Gwynedd, U.K.* **10 D3** 53 13N 4 9W
Bangor, *Maine, U.S.A.* **71 D6** 44 48N 68 46W
Bangor, *Pa., U.S.A.* **79 F9** 40 52N 75 13W
Bangued, *Phil.* **37 A6** 17 40N 120 37 E
Bangui, *C.A.R.* **52 D3** 4 23N 18 35 E
Banguru, *Zaïre* **54 B2** 0 30N 27 10 E
Bangweulu, L., *Zambia* **55 E3** 11 0S 30 0 E
Bangweulu Swamp, *Zambia* **55 E3** 11 20S 30 15 E
Bani, *Dom. Rep.* **89 C5** 18 16N 70 22W
Banī Sa'd, *Iraq* **44 C5** 33 34N 44 32 E
Banī Walīd, *Libya* **51 B7** 31 36N 13 53 E
Banihal Pass, *India* **43 C6** 33 30N 75 12 E
Banīnah, *Libya* **51 B9** 32 0N 20 12 E
Bāniyās, *Syria* **44 C3** 35 10N 36 0 E
Banja Luka, *Bos.-H.* **20 B7** 44 49N 17 11 E
Banjar, *Indonesia* **37 G13** 7 24S 108 30 E
Banjarmasin, *Indonesia* **36 E4** 3 20S 114 35 E
Banjarnegara, *Indonesia* **37 G13** 7 24S 109 42 E
Banjul, *Gambia* **50 F1** 13 28N 16 40W
Banka Banka, *Australia* **62 B1** 18 50S 134 0 E
Banket, *Zimbabwe* **55 F3** 17 27S 30 19 E
Bankipore, *India* **43 G11** 25 35N 85 10 E
Banks I., *B.C., Canada* **72 C3** 53 20N 130 0W
Banks I., *N.W.T., Canada* **68 A7** 73 15N 121 30W
Banks Pen., *N.Z.* **59 K4** 43 45S 173 15 E
Banks Str., *Australia* **62 G4** 40 40S 148 10 E
Bankura, *India* **43 H12** 23 11N 87 18 E
Bann →, *Down, U.K.* **13 B5** 54 30N 6 31W
Bann →, *L'derry., U.K.* **13 A5** 55 10N 6 34W
Bannang Sata, *Thailand* **39 J3** 6 16N 101 16 E
Banning, *U.S.A.* **85 M10** 33 56N 116 53W
Banningville = Bandundu, *Zaïre* **52 E3** 3 15S 17 22 E
Bannockburn, *Canada* **78 B7** 44 39N 77 33W
Bannockburn, *U.K.* **12 E5** 56 5N 3 55W
Bannockburn, *Zimbabwe* **55 G2** 20 17S 29 48 E
Bannu, *Pakistan* **40 C7** 33 0N 70 18 E
Banská Bystrica, *Slovakia* **17 D9** 48 46N 19 14 E
Banská Štiavnica, *Slovakia* **17 D9** 48 25N 18 55 E
Banswara, *India* **42 H6** 23 32N 74 24 E
Banten, *Indonesia* **37 G12** 6 5S 106 8 E
Bantry, *Ireland* **13 E2** 51 40N 9 28W
Bantry B., *Ireland* **13 E2** 51 35N 9 50W
Bantul, *Indonesia* **37 G14** 7 55S 110 19 E
Bantva, *India* **42 J4** 21 29N 70 12 E
Banu, *Afghan.* **40 B6** 35 35N 69 5 E
Banyak, Kepulauan, *Indonesia* **36 D1** 2 10N 97 10 E
Banyo, *Cameroon* **50 G7** 6 52N 11 45 E
Banyumas, *Indonesia* **37 G13** 7 32S 109 18 E
Banyuwangi, *Indonesia* **37 H16** 8 13S 114 21 E
Banzare Coast, *Antarctica* **5 C9** 68 0S 125 0 E
Banzyville = Mobayi, *Zaïre* **52 D4** 4 15N 21 8 E
Bao Ha, *Vietnam* **38 A5** 22 11N 104 21 E
Bao Lac, *Vietnam* **38 A5** 22 57N 105 40 E
Bao Loc, *Vietnam* **39 G6** 11 32N 107 48 E
Baocheng, *China* **34 H4** 33 12N 106 56 E
Baode, *China* **34 E6** 39 1N 111 5 E
Baodi, *China* **35 E9** 39 38N 117 20 E
Baoding, *China* **34 E8** 38 50N 115 28 E
Baoji, *China* **34 G4** 34 20N 107 5 E
Baoshan, *China* **32 D4** 25 10N 99 5 E
Baotou, *China* **34 D6** 40 32N 110 2 E
Baoying, *China* **35 H10** 33 17N 119 20 E
Bap, *India* **42 F5** 27 23N 72 18 E
Bapatla, *India* **41 M12** 15 55N 80 30 E
Bāqerābād, *Iran* **45 C6** 33 2N 51 58 E
Ba'qūbah, *Iraq* **44 C5** 33 45N 44 50 E
Baquedano, *Chile* **94 A2** 23 20S 69 52W
Bar, *Montenegro, Yug.* **21 C8** 42 8N 19 8 E
Bar Bigha, *India* **43 G11** 25 21N 85 47 E
Bar Harbor, *U.S.A.* **71 D6** 44 23N 68 13W
Bar-le-Duc, *France* **18 B6** 48 47N 5 10 E
Barabai, *Indonesia* **36 E5** 2 32S 115 34 E
Barabinsk, *Russia* **26 D8** 55 20N 78 20 E
Baraboo, *U.S.A.* **80 D10** 43 28N 89 45W
Baracaldo, *Spain* **19 A4** 43 18N 2 59W
Baracoa, *Cuba* **89 B5** 20 20N 74 30W
Baradero, *Argentina* **94 C4** 33 52S 59 29W
Baraga, *U.S.A.* **80 B10** 46 47N 88 30W
Barahona, *Dom. Rep.* **89 C5** 18 13N 71 7W
Barail Range, *India* **41 G18** 25 15N 93 20 E
Barakhola, *India* **41 G18** 25 0N 92 45 E
Barakot, *India* **43 J11** 21 33N 84 59 E
Barakpur, *India* **43 H13** 22 44N 88 30 E
Barakula, *Australia* **63 D5** 26 30S 150 33 E

Baralaba

Baralaba, Australia 62 C4 24 13S 149 50 E
Baralzon L., Canada 73 B9 60 0N 98 3W
Baramula, India 43 B6 34 15N 74 20 E
Baran, India 42 G7 25 9N 76 40 E
Baranavichy = Baranovichi, Belorussia 24 D4 53 10N 26 0 E
Baranof I., U.S.A. 72 B1 57 0N 135 0W
Baranovichi, Belorussia 24 D4 53 10N 26 0 E
Barão de Melgaço, Brazil 92 F6 11 50S 60 45W
Barapasi, Indonesia 37 E9 2 15S 137 5 E
Barasat, India 43 H13 22 46N 88 31 E
Barat Daya, Kepulauan, Indonesia 37 F7 7 30S 128 0 E
Barataria B., U.S.A. 81 L10 29 20N 89 55W
Baraut, India 42 E7 29 13N 77 7 E
Barbacena, Brazil 95 A7 21 15S 43 56W
Barbacoas, Colombia 92 C3 1 45N 78 0W
Barbados ■, W. Indies 89 D8 13 10N 59 30W
Barberton, S. Africa 57 D5 25 42S 31 2 E
Barberton, U.S.A. 78 E3 41 0N 81 39W
Barbourville, U.S.A. 77 G4 36 52N 83 53W
Barbuda, W. Indies 89 C7 17 30N 61 40W
Barcaldine, Australia 62 C4 23 43S 145 6 E
Barcelona, Spain 19 B7 41 21N 2 10 E
Barcelona, Venezuela 92 A6 10 10N 64 40W
Barcelos, Brazil 92 D6 1 0S 63 0W
Barcoo →, Australia 62 D3 25 30S 142 50 E
Bardai, Chad 51 D8 21 25N 17 0 E
Bardas Blancas, Argentina 94 D2 35 49S 69 45W
Barddhaman, India 43 H12 23 14N 87 39 E
Bardera, Somali Rep. 46 G3 2 20N 42 27 E
Bardia, Libya 51 B9 31 45N 25 5 E
Bardsey I., U.K. 10 E3 52 46N 4 47W
Bardstown, U.S.A. 76 G3 37 49N 85 28W
Bareilly, India 43 E8 28 22N 79 27 E
Barents Sea, Arctic 4 B9 73 0N 39 0 E
Barentu, Eritrea 51 E12 15 2N 37 35 E
Bargal, Somali Rep. 46 E5 11 25N 51 0 E
Bargara, Australia 62 C5 24 50S 152 25 E
Barguzin, Russia 27 D11 53 37N 109 37 E
Barh, India 43 G11 25 29N 85 46 E
Barhaj, India 43 F10 26 18N 83 44 E
Barhi, India 43 G11 24 15N 85 25 E
Bari, India 42 F7 26 39N 77 39 E
Bari, Italy 20 D7 41 6N 16 52 E
Bari Doab, Pakistan 42 D5 30 20N 73 0 E
Bariadi □, Tanzania 54 C3 2 45S 34 40 E
Barīm, Yemen 46 E3 12 39N 43 25 E
Barinas, Venezuela 92 B4 8 36N 70 15W
Baring, C., Canada 68 B8 70 0N 117 30W
Baringo, Kenya 54 B4 0 47N 36 16 E
Baringo □, Kenya 54 B4 0 55N 36 0 E
Baringo, L., Kenya 54 B4 0 47N 36 16 E
Bârîs, Egypt 51 D11 24 42N 30 31 E
Barisal, Bangla. 41 H17 22 45N 90 20 E
Barisan, Bukit, Indonesia 36 E2 3 30S 102 15 E
Barito →, Indonesia 36 E4 4 0S 114 50 E
Bark L., Canada 78 A7 45 27N 77 51W
Barker, U.S.A. 78 C6 43 20N 78 33W
Barkley Sound, Canada 72 D3 48 50N 125 10W
Barkly Downs, Australia 62 C2 20 30S 138 30 E
Barkly East, S. Africa 56 E4 30 58S 27 33 E
Barkly Tableland, Australia 62 B2 17 50S 136 40 E
Barkly West, S. Africa 56 D3 28 5S 24 31 E
Barkol, China 32 B4 43 37N 93 2 E
Barksdale, U.S.A. 81 L4 29 44N 100 2W
Barlee, L., Australia 61 E2 29 15S 119 30 E
Barlee, Mt., Australia 61 D4 24 38S 128 13 E
Barletta, Italy 20 D7 41 20N 16 17 E
Barlovento, Canary Is. 22 F2 28 48N 17 48W
Barlow L., Canada 73 A8 62 0N 103 0W
Barmedman, Australia 63 E4 34 9S 147 21 E
Barmer, India 42 G4 25 45N 71 20 E
Barmera, Australia 63 E3 34 15S 140 28 E
Barmouth, U.K. 10 E3 52 44N 4 3W
Barnagar, India 42 H6 23 7N 75 19 E
Barnard Castle, U.K. 10 C6 54 33N 1 55W
Barnato, Australia 63 E3 31 38S 145 0 E
Barnaul, Russia 26 D9 53 20N 83 40 E
Barnesville, U.S.A. 77 J3 33 3N 84 9W
Barnet, U.K. 11 F7 51 37N 0 15W
Barneveld, Neths. 15 B5 52 7N 5 36 E
Barneveld, U.S.A. 79 C9 43 16N 75 14W
Barngo, Australia 62 D4 25 3S 147 20 E
Barnhart, U.S.A. 81 K4 31 8N 101 10W
Barnsley, U.K. 10 D6 53 33N 1 29W
Barnstaple, U.K. 11 F3 51 5N 4 3W
Barnsville, U.S.A. 80 B6 46 43N 96 28W
Baro, Nigeria 50 G6 8 35N 6 18 E
Baroda = Vadodara, India 42 H5 22 20N 73 10 E
Baroda, India 42 G7 25 29N 76 35 E
Baroe, S. Africa 56 E3 33 13S 24 33 E
Baron Ra., Australia 60 D4 23 30S 127 45 E
Barpeta, India 41 F17 26 20N 91 10 E
Barques, Pt. Aux, U.S.A. 76 C4 44 4N 82 58W
Barquísimeto, Venezuela 92 A5 10 4N 69 19W
Barra, Brazil 93 F10 11 5S 43 10W
Barra, U.K. 12 E1 57 0N 7 30W
Barra, Sd. of, U.K. 12 D1 57 4N 7 25W
Barra de Navidad, Mexico 86 D4 19 12N 104 41W
Barra do Corda, Brazil 93 E9 5 30S 45 10W
Barra do Piraí, Brazil 95 A7 22 30S 43 50W
Barra Falsa, Pta. da, Mozam. 57 C6 22 58S 35 37 E
Barra Hd., U.K. 12 E1 56 47N 7 40W
Barra Mansa, Brazil 95 A7 22 35S 44 12W
Barraba, Australia 63 E5 30 21S 150 35 E
Barrackpur = Barakpur, India 43 H13 22 44N 88 30 E
Barranca, Lima, Peru 92 F3 10 45S 77 50W
Barranca, Loreto, Peru 92 D3 4 50S 76 50W
Barrancabermeja, Colombia 92 B4 7 0N 73 50W
Barrancas, Venezuela 92 B6 8 55N 62 5W
Barrancos, Portugal 19 C2 38 10N 6 58W
Barranqueras, Argentina 94 B4 27 30S 59 0W
Barranquilla, Colombia 92 A4 11 0N 74 50W
Barras, Brazil 93 D10 4 15S 42 18W
Barraute, Canada 70 C4 48 26N 77 38W
Barre, Mass., U.S.A. 75 B12 42 25N 72 6W
Barre, Vt., U.S.A. 79 B12 44 12N 72 30W
Barreal, Argentina 94 C2 31 33S 69 28W
Barreiras, Brazil 93 F10 12 8S 45 0W
Barreirinhas, Brazil 93 D10 2 30S 42 50W
Barreiro, Portugal 19 C1 38 40N 9 6W
Barreiros, Brazil 93 E11 8 49S 35 12W
Barren, Nosy, Madag. 57 B7 18 25S 43 40 E
Barretos, Brazil 93 H9 20 30S 48 35W
Barrhead, Canada 72 C6 54 10N 114 24W
Barrie, Canada 70 D4 44 24N 79 40W
Barrier Ra., Australia 63 E3 31 0S 141 30 E
Barrière, Canada 72 C4 51 12N 120 7W
Barrington, U.S.A. 79 E13 41 44N 71 18W
Barrington L., Canada 73 B8 56 55N 100 15W
Barrington Tops, Australia 63 E5 32 6S 151 28 E
Barringun, Australia 63 D4 29 1S 145 41 E
Barrow, U.S.A. 68 A4 71 18N 156 47W
Barrow →, Ireland 13 D4 52 10N 6 57W
Barrow, C., U.S.A. 66 B4 71 10N 156 20W
Barrow Creek, Australia 62 C1 21 30S 133 55 E
Barrow I., Australia 60 D2 20 45S 115 20 E
Barrow-in-Furness, U.K. 10 C4 54 8N 3 15W
Barrow Pt., Australia 62 A3 14 20S 144 40 E
Barrow Ra., Australia 61 E4 26 0S 127 40 E
Barrow Str., Canada 4 B3 74 20N 95 0W
Barry, U.K. 11 F4 51 23N 3 19W
Barry's Bay, Canada 70 C4 45 29N 77 41W
Barsat, Pakistan 43 A5 36 10N 72 45 E
Barsham, Syria 44 C4 35 21N 40 33 E
Barsi, India 40 K9 18 10N 75 50 E
Barsoi, India 41 G15 25 48N 87 57 E
Barstow, Calif., U.S.A. 85 L9 34 54N 117 1W
Barstow, Tex., U.S.A. 81 K3 31 28N 103 24W
Barthélemy, Col, Vietnam 38 C5 19 26N 104 6 E
Bartica, Guyana 92 B7 6 25N 58 40W
Bartlesville, U.S.A. 81 G7 36 45N 95 59W
Bartlett, Calif., U.S.A. 84 J8 36 29N 118 2W
Bartlett, Tex., U.S.A. 81 K6 30 48N 97 26W
Bartlett, L., Canada 72 A5 63 5N 118 20W
Bartolomeu Dias, Mozam. 55 G4 21 10S 35 8 E
Barton, Australia 61 F5 30 31S 132 39 E
Barton upon Humber, U.K. 10 D7 53 41N 0 27W
Bartow, U.S.A. 77 M5 27 54N 81 50W
Barú, Volcan, Panama 88 E3 8 55N 82 35W
Barumba, Zaïre 54 B1 1 3N 23 37 E
Barwani, India 42 H6 22 2N 74 57 E
Barysaw = Borisov, Belorussia 24 D4 54 17N 28 28 E
Barzān, Iraq 44 B5 36 55N 44 3 E
Bas-Rhin □, France 18 B7 48 40N 7 30 E
Bāsa'idū, Iran 45 E7 26 35N 55 20 E
Basal, Pakistan 42 C5 33 33N 72 13 E
Basankusa, Zaïre 52 D3 1 5N 19 50 E
Basawa, Afghan. 42 B4 34 15N 70 50 E
Bascuñán, C., Chile 94 B1 28 52S 71 35W
Basel, Switz. 16 E3 47 35N 7 35 E
Bāshī, Iran 45 D6 28 41N 51 4 E
Bashkir Republic □, Russia 24 D10 54 0N 57 0 E
Bashkortostan = Bashkir Republic □, Russia 24 D10 54 0N 57 0 E
Basilan, Phil. 37 C6 6 35N 122 0 E
Basilan Str., Phil. 37 C6 6 50N 122 0 E
Basildon, U.K. 11 F8 51 34N 0 29 E
Basilicata □, Italy 20 D7 40 30N 16 0 E
Basim = Washim, India 40 J10 20 3N 77 0 E
Basin, U.S.A. 82 D9 44 23N 108 2W
Basingstoke, U.K. 11 F6 51 15N 1 5W
Baskatong, Rés., Canada 70 C4 46 46N 75 50W
Basle = Basel, Switz. 16 E3 47 35N 7 35 E
Basoda, India 42 H7 23 52N 77 54 E
Basoka, Zaïre 54 B1 1 16N 23 40 E
Basongo, Zaïre 52 E4 4 15S 20 20 E
Basque Provinces = País Vasco □, Spain 19 A4 42 50N 2 45W
Basra = Al Başrah, Iraq 44 D5 30 30N 47 50 E
Bass Rock, U.K. 12 E6 56 5N 2 40W
Bass Str., Australia 62 F4 39 15S 146 30 E
Bassano, Canada 72 C6 50 48N 112 20W
Bassano del Grappa, Italy 20 B4 45 45N 11 45 E
Bassas da India, Ind. Oc. 53 J7 22 0S 39 0 E
Basse-Terre, Guadeloupe 89 C7 16 0N 61 44W
Bassein, Burma 41 L19 16 45N 94 30 E
Basseterre, St. Christopher-Nevis 89 C7 17 17N 62 43W
Bassett, Nebr., U.S.A. 80 D5 42 35N 99 32W
Bassett, Va., U.S.A. 77 G6 36 46N 79 59W
Bassi, India 42 D7 30 44N 76 21 E
Bassigny, France 18 C6 48 0N 5 10 E
Bassikounou, Mauritania 50 E3 15 55N 6 1W
Bastak, Iran 45 E7 27 15N 54 25 E
Baştām, Iran 45 B7 36 29N 55 4 E
Bastar, India 41 K12 19 15N 81 40 E
Basti, India 43 F10 26 52N 82 55 E
Bastia, France 18 E9 42 40N 9 30 E
Bastogne, Belgium 15 D5 50 1N 5 43 E
Bastrop, U.S.A. 81 K6 30 7N 97 19W
Bat Yam, Israel 47 C3 32 2N 34 44 E
Bata, Eq. Guin. 52 D1 1 57N 9 50 E
Bataan, Phil. 37 B6 14 40N 120 25 E
Batabanó, Cuba 88 B3 22 40N 82 20W
Batabanó, G. de, Cuba 88 B3 22 30N 82 30W
Batac, Phil. 37 A6 18 3N 120 34 E
Batagoy, Russia 27 C14 67 38N 134 38 E
Batalha, Portugal 19 C1 39 40N 8 50W
Batama, Zaïre 54 B2 0 58N 26 33 E
Batamay, Russia 27 C13 63 30N 129 15 E
Batang, Indonesia 37 G13 6 55S 109 45 E
Batangafo, C.A.R. 51 G8 7 25N 18 20 E
Batangas, Phil. 37 B6 13 35N 121 10 E
Batanta, Indonesia 37 E8 0 55S 130 40 E
Batatais, Brazil 95 A6 20 54S 47 37W
Batavia, U.S.A. 78 D6 43 0N 78 11W
Batchelor, Australia 60 B5 13 4S 131 1 E
Bateman's B., Australia 63 F5 35 40S 150 12 E
Batemans Bay, Australia 63 F5 35 44S 150 11 E
Bates Ra., Australia 61 E3 27 27S 121 5 E
Batesburg, U.S.A. 77 J5 33 54N 81 33W
Batesville, Ark., U.S.A. 81 H9 35 46N 91 39W
Batesville, Miss., U.S.A. 81 H10 34 19N 89 57W
Batesville, Tex., U.S.A. 81 L5 28 58N 99 37W
Bath, U.K. 11 F5 51 22N 2 22W
Bath, Maine, U.S.A. 71 D6 43 55N 69 49W
Bath, N.Y., U.S.A. 78 D7 42 20N 77 19W
Batheay, Cambodia 39 G5 11 59N 104 57 E
Bathgate, U.K. 12 F5 55 54N 3 38W
Bathurst = Banjul, Gambia 50 F1 13 28N 16 40W
Bathurst, Australia 63 E4 33 25S 149 31 E
Bathurst, Canada 71 C6 47 37N 65 43W
Bathurst, S. Africa 56 E4 33 30S 26 50 E
Bathurst, C., Canada 68 A7 70 34N 128 0W
Bathurst B., Australia 62 A3 14 16S 144 25 E
Bathurst Harb., Australia 62 G4 43 15S 146 10 E
Bathurst I., Australia 60 B5 11 30S 130 10 E
Bathurst I., Canada 4 B2 76 0N 100 30W
Bathurst Inlet, Canada 68 B9 66 50N 108 1W
Batlow, Australia 63 F4 35 31S 148 9 E
Batna, Algeria 50 A6 35 34N 6 15 E
Batoka, Zambia 55 F2 16 45S 27 15 E
Baton Rouge, U.S.A. 81 K9 30 27N 91 11W
Batong, Ko, Thailand 39 J2 6 32N 99 12 E
Batopilas, Mexico 86 B3 27 0N 107 45W
Batouri, Cameroon 52 D2 4 30N 14 25 E
Battambang, Cambodia 38 F4 13 7N 103 12 E
Batticaloa, Sri Lanka 40 R12 7 43N 81 45 E
Battle, U.K. 11 G8 50 55N 0 30 E
Battle →, Canada 73 C7 52 43N 108 15W
Battle Camp, Australia 62 B3 15 20S 144 40 E
Battle Creek, U.S.A. 76 D3 42 19N 85 11W
Battle Ground, U.S.A. 84 E4 45 47N 122 32W
Battle Harbour, Canada 71 B8 52 16N 55 35W
Battle Lake, U.S.A. 80 B7 46 17N 95 43W
Battle Mountain, U.S.A. 82 F5 40 38N 116 56W
Battlefields, Zimbabwe 55 F2 18 37S 29 47 E
Battleford, Canada 73 C7 52 45N 108 15W
Batu, Ethiopia 46 F2 6 55N 39 45 E
Batu, Kepulauan, Indonesia 36 E1 0 30S 98 25 E
Batu Caves, Malaysia 39 L3 3 15N 101 40 E
Batu Gajah, Malaysia 39 K3 4 28N 101 3 E
Batu Is. = Batu, Kepulauan, Indonesia 36 E1 0 30S 98 25 E
Batu Pahat, Malaysia 39 M4 1 50N 102 56 E
Batuata, Indonesia 37 F6 6 12S 122 42 E
Batumi, Georgia 25 F7 41 30N 41 30 E
Baturaja, Indonesia 36 E2 4 11S 104 15 E
Baturité, Brazil 93 D11 4 28S 38 45W
Bau, Malaysia 36 D4 1 25N 110 9 E
Baubau, Indonesia 37 F6 5 25S 122 38 E
Bauchi, Nigeria 50 F6 10 22N 9 48 E
Baudette, U.S.A. 80 A7 48 43N 94 36W
Bauer, C., Australia 63 E1 32 44S 134 4 E
Bauhinia Downs, Australia 62 C4 24 35S 149 18 E
Bauru, Brazil 95 A6 22 10S 49 0W
Baús, Brazil 93 G8 18 22S 52 47W
Bautzen, Germany 16 C7 51 11N 14 25 E
Bavānāt, Iran 45 D7 30 28N 53 27 E
Bavaria = Bayern □, Germany 16 D5 49 7N 11 30 E
Bavi Sadri, India 42 G6 24 28N 74 30 E
Bavispe →, Mexico 86 B3 29 30N 109 11W
Bawdwin, Burma 41 H20 23 5N 97 20 E
Bawean, Indonesia 36 F4 5 46S 112 35 E
Bawku, Ghana 50 F4 11 3N 0 19W
Bawlake, Burma 41 K20 19 11N 97 21 E
Baxley, U.S.A. 77 K4 31 47N 82 21W
Baxter Springs, U.S.A. 81 G7 37 2N 94 44W
Bay, L. de, Phil. 37 B6 14 20N 121 11 E
Bay Bulls, Canada 71 C9 47 19N 52 50W
Bay City, Mich., U.S.A. 76 D4 43 36N 83 54W
Bay City, Oreg., U.S.A. 82 D2 45 31N 123 53W
Bay City, Tex., U.S.A. 81 L7 28 59N 95 58W
Bay de Verde, Canada 71 C9 48 5N 52 54W
Bay Minette, U.S.A. 77 K2 30 53N 87 46W
Bay St. Louis, U.S.A. 81 K10 30 19N 89 20W
Bay Springs, U.S.A. 81 K10 31 59N 89 17W
Bay View, N.Z. 59 H6 39 25S 176 50 E
Baya, Zaïre 55 E2 11 53S 27 25 E
Bayamo, Cuba 88 B4 20 20N 76 40W
Bayamón, Puerto Rico 89 C6 18 24N 66 10W
Bayan Har Shan, China 32 C4 34 0N 98 0 E
Bayan Hot = Alxa Zuoqi, China 34 E3 38 50N 105 40 E
Bayan Obo, China 34 D5 41 52N 109 59 E
Bayan-Ovoo, Mongolia 34 C4 42 55N 106 5 E
Bayana, India 42 F7 26 55N 77 18 E
Bayanaul, Kazakhstan 26 D8 50 45N 75 45 E
Bayandalay, Mongolia 34 C2 43 30N 103 29 E
Bayanhongor, Mongolia 32 B5 46 8N 102 43 E
Bayard, U.S.A. 80 E3 41 45N 103 20W
Baybay, Phil. 37 B6 10 40N 124 55 E
Bayern □, Germany 16 D5 49 7N 11 30 E
Bayeux, France 18 B3 49 17N 0 42W
Bayfield, Canada 78 C3 43 34N 81 42W
Bayfield, U.S.A. 80 B9 46 49N 90 49W
Baykal, Oz., Russia 27 D11 53 0N 108 0 E
Baykit, Russia 27 C10 61 50N 95 50 E
Baykonur, Kazakhstan 26 E7 47 48N 65 50 E
Baymak, Russia 24 D10 52 36N 58 19 E
Baynes Mts., Namibia 56 B1 17 15S 13 0 E
Bayombong, Phil. 37 A6 16 30N 121 10 E
Bayonne, France 18 E3 43 30N 1 28W
Bayonne, U.S.A. 79 F10 40 40N 74 7W
Bayovar, Peru 92 E2 5 50S 81 0W
Bayram-Ali, Turkmenistan 26 F7 37 37N 62 10 E
Bayreuth, Germany 16 D5 49 56N 11 35 E
Bayrūt, Lebanon 47 B4 33 53N 35 31 E
Bayt Lahm, Jordan 47 D4 31 43N 35 12 E
Baytown, U.S.A. 81 L7 29 43N 94 59W
Baza, Spain 19 D4 37 30N 2 47W
Bazaruto, I. do, Mozam. 57 C6 21 40S 35 28 E
Bazmān, Kūh-e, Iran 45 D9 28 4N 60 1 E
Beach, U.S.A. 80 B3 46 58N 104 0W
Beach City, U.S.A. 78 F3 40 39N 81 35W
Beachport, Australia 63 F2 37 29S 140 0 E
Beachy Hd., U.K. 11 G8 50 44N 0 16 E
Beacon, Australia 61 F2 30 26S 117 52 E
Beacon, U.S.A. 79 E11 41 30N 73 58W
Beaconia, Canada 73 C9 50 25N 96 31W
Beagle, Canal, S. Amer. 96 G3 55 0S 68 30W
Beagle Bay, Australia 60 C3 16 58S 122 40 E
Bealanana, Madag. 57 A8 14 33S 48 44 E
Beamsville, Canada 78 C5 43 12N 79 28W
Bear →, U.S.A. 84 G5 38 56N 121 36W
Bear I., Ireland 13 E2 51 38N 9 50W
Bear L., B.C., Canada 72 B3 56 10N 126 52W
Bear L., Man., Canada 73 B9 55 8N 96 0W
Bear L., U.S.A. 82 E8 41 59N 111 21W
Bearcreek, U.S.A. 82 D9 45 11N 109 6W
Beardmore, Canada 70 C2 49 36N 87 57W
Beardmore Glacier, Antarctica 5 E11 84 30S 170 0 E
Beardstown, U.S.A. 80 F9 40 1N 90 26W
Béarn, France 18 E3 43 20N 0 30W
Bearpaw Mts., U.S.A. 82 B9 48 12N 109 30W
Bearskin Lake, Canada 70 B1 53 58N 91 2W
Beata, C., Dom. Rep. 89 C5 17 40N 71 30W
Beata, I., Dom. Rep. 89 C5 17 34N 71 31W
Beatrice, U.S.A. 80 E6 40 16N 96 45W
Beatrice, Zimbabwe 55 F3 18 15S 30 55 E
Beatrice, C., Australia 62 A2 14 20S 136 55 E
Beatton →, Canada 72 B4 56 15N 120 45W
Beatton River, Canada 72 B4 57 26N 121 20W
Beatty, U.S.A. 83 H5 36 54N 116 46W
Beauce, Plaine de la, France 18 B4 48 10N 1 45 E
Beauceville, Canada 71 C5 46 13N 70 46W
Beaudesert, Australia 63 D5 27 59S 153 0 E
Beaufort, Malaysia 36 C5 5 30N 115 40 E
Beaufort, N.C., U.S.A. 77 H7 34 43N 76 40W
Beaufort, S.C., U.S.A. 77 J5 32 26N 80 40W
Beaufort Sea, Arctic 4 B1 72 0N 140 0W
Beaufort West, S. Africa 56 E3 32 18S 22 36 E
Beauharnois, Canada 70 C5 45 20N 73 52W
Beaulieu →, Canada 72 A6 62 3N 113 11W
Beauly, U.K. 12 D4 57 29N 4 27W
Beauly →, U.K. 12 D4 57 26N 4 28W
Beaumaris, U.K. 10 D3 53 16N 4 7W
Beaumont, Calif., U.S.A. 85 M10 33 56N 116 58W
Beaumont, Tex., U.S.A. 81 K7 30 5N 94 6W
Beaune, France 18 C6 47 2N 4 50 E
Beauséjour, Canada 73 C9 50 5N 96 35W
Beauvais, France 18 B5 49 25N 2 8 E
Beauval, Canada 73 B7 55 9N 107 37W
Beaver, Alaska, U.S.A. 68 B5 66 22N 147 24W
Beaver, Okla., U.S.A. 81 G4 36 49N 100 31W
Beaver, Pa., U.S.A. 78 F4 40 42N 80 19W
Beaver, Utah, U.S.A. 83 G7 38 17N 112 38W
Beaver →, B.C., Canada 72 B4 59 52N 124 20W
Beaver →, Ont., Canada 70 A2 55 55N 87 48W
Beaver →, Sask., Canada 73 B7 55 26N 107 45W
Beaver City, U.S.A. 80 E5 40 8N 99 50W
Beaver Dam, U.S.A. 80 D10 43 28N 88 50W
Beaver Falls, U.S.A. 78 F4 40 46N 80 20W
Beaver Hill L., Canada 73 C10 54 5N 94 50W
Beaver I., U.S.A. 76 C3 45 40N 85 33W
Beaverhill L., Alta., Canada 72 C6 53 27N 112 32W
Beaverhill L., N.W.T., Canada 73 A8 63 2N 104 22W
Beaverlodge, Canada 72 B5 55 11N 119 29W
Beavermouth, Canada 72 C5 51 32N 117 23W
Beaverstone →, Canada 70 B2 54 59N 89 25W
Beaverton, Canada 78 B5 44 26N 79 9W
Beaverton, U.S.A. 84 E4 45 29N 122 48W
Beawar, India 42 F6 26 3N 74 18 E
Bebedouro, Brazil 95 A6 21 0S 48 25W
Beboa, Madag. 57 B7 17 22S 44 33 E
Beccles, U.K. 11 E9 52 27N 1 33 E
Bečej, Serbia 21 B9 45 36N 20 3 E
Béchar, Algeria 50 B4 31 38N 2 18W
Beckley, U.S.A. 76 G5 37 47N 81 11W
Bedford, Canada 70 C5 45 7N 72 59W
Bedford, S. Africa 56 E4 32 40S 26 10 E
Bedford, U.K. 11 E7 52 8N 0 29W
Bedford, Ind., U.S.A. 76 F2 38 52N 86 29W
Bedford, Iowa, U.S.A. 80 E7 40 40N 94 44W
Bedford, Ohio, U.S.A. 78 E3 41 23N 81 32W
Bedford, Pa., U.S.A. 78 F6 40 1N 78 30W
Bedford, Va., U.S.A. 76 G6 37 20N 79 31W
Bedford, C., Australia 62 B4 15 14S 145 21 E
Bedford Downs, Australia 60 C4 17 19S 127 20 E
Bedfordshire □, U.K. 11 E7 52 4N 0 28W
Bedourie, Australia 62 C2 24 30S 139 30 E
Beech Grove, U.S.A. 76 F2 39 44N 86 3W
Beechy, Canada 73 C7 50 53N 107 24W
Beenleigh, Australia 63 D5 27 43S 153 10 E
Be'er Menuha, Israel 44 D2 30 19N 35 8 E
Be'er Sheva', Israel 47 D3 31 15N 34 48 E
Beersheba = Be'er Sheva', Israel 47 D3 31 15N 34 48 E
Beeston, U.K. 10 E6 52 55N 1 11W
Beetaloo, Australia 62 B1 17 15S 133 50 E
Beeville, U.S.A. 81 L6 28 24N 97 45W
Befale, Zaïre 52 D4 0 25N 20 45 E
Befandriana, Madag. 57 C7 21 55S 44 0 E
Befotaka, Madag. 57 C8 23 49S 47 0 E
Bega, Australia 63 F4 36 41S 149 51 E
Begusarai, India 43 G12 25 24N 86 9 E
Behābād, Iran 45 C8 32 24N 59 47 E
Behara, Madag. 57 C8 24 55S 46 20 E
Behbehān, Iran 45 D6 30 30N 50 15 E
Behshahr, Iran 45 B7 36 45N 53 35 E
Bei Jiang →, China 33 D6 23 2N 112 58 E
Bei'an, China 33 B7 48 10N 126 20 E
Beihai, China 33 D5 21 28N 109 6 E
Beijing, China 34 E9 39 55N 116 20 E
Beijing □, China 34 E9 39 55N 116 20 E
Beilen, Neths. 15 B6 52 52N 6 27 E
Beilpajah, Australia 63 E3 32 54S 143 52 E
Beipiao, China 35 D11 41 52N 120 32 E
Beira, Mozam. 55 F3 19 50S 34 52 E
Beira-Alta, Portugal 19 B2 40 35N 7 35W
Beira-Baixa, Portugal 19 B2 40 2N 7 30W
Beira-Litoral, Portugal 19 B1 40 5N 8 30W
Beirut = Bayrūt, Lebanon 47 B4 33 53N 35 31 E
Beitaolaizhao, China 35 B13 44 58N 125 58 E
Beitbridge, Zimbabwe 55 G3 22 12S 30 0 E
Beizhen, Liaoning, China 35 D11 41 38N 121 54 E
Beizhen, Shandong, China 35 F10 37 20N 118 2 E

Big Lake

Beizhengzhen, China ... **35 B12** 44 31N 123 30 E
Beja, Portugal ... **19 C2** 38 2N 7 53W
Béja, Tunisia ... **50 A6** 36 43N 9 12 E
Bejaia, Algeria ... **50 A6** 36 42N 5 2 E
Bejestān, Iran ... **45 C8** 34 30N 58 5 E
Bekasi, Indonesia ... **37 G12** 6 14S 106 59 E
Békéscsaba, Hungary ... **17 E10** 46 40N 21 5 E
Bekily, Madag. ... **57 C8** 24 13S 45 19 E
Bekok, Malaysia ... **39 L4** 2 20N 103 7 E
Bela, India ... **43 G9** 25 50N 82 0 E
Bela, Pakistan ... **42 F2** 26 12N 66 20 E
Bela Crkva, Serbia, Yug. **21 B9** 44 55N 21 27 E
Bela Vista, Brazil ... **94 A4** 22 12S 56 20W
Bela Vista, Mozam. ... **57 D5** 26 10S 32 44 E
Belarus = Belorussia ■, Europe ... **24 D4** 53 30N 27 0 E
Belau ■, Pac. Oc. ... **64 G5** 7 30N 134 30 E
Belavenona, Madag. ... **57 C8** 24 50S 47 4 E
Belawan, Indonesia ... **36 D1** 3 33N 98 32 E
Belaya →, Russia ... **24 C9** 54 40N 56 0 E
Belaya Tserkov, Ukraine **25 E5** 49 45N 30 10 E
Belcher Is., Canada ... **69 C12** 56 15N 78 45W
Belden, U.S.A. ... **84 E5** 40 2N 121 17W
Belebey, Russia ... **24 D9** 54 7N 54 7 E
Belém, Brazil ... **93 D9** 1 20S 48 30W
Belén, Argentina ... **94 B2** 27 40S 67 5W
Belén, Paraguay ... **94 A4** 23 30S 57 6W
Belen, U.S.A. ... **83 J10** 34 40N 106 46W
Belet Uen, Somali Rep. **46 G4** 4 30N 45 5 E
Belev, Russia ... **24 D6** 53 50N 36 5 E
Belfair, U.S.A. ... **84 C4** 47 27N 122 50W
Belfast, S. Africa ... **57 D5** 25 42S 30 2 E
Belfast, U.K. ... **13 B6** 54 35N 5 56W
Belfast, Maine, U.S.A. ... **71 D6** 44 26N 69 1W
Belfast, N.Y., U.S.A. ... **78 D6** 42 21N 78 7W
Belfast □, U.K. ... **13 B6** 54 35N 5 56W
Belfast L., U.K. ... **13 B6** 54 40N 5 50W
Belfield, U.S.A. ... **80 B3** 46 53N 103 12W
Belfort, France ... **18 C7** 47 38N 6 50 E
Belfort, Territoire de □, France ... **18 C7** 47 40N 6 55 E
Belfry, U.S.A. ... **82 D9** 45 9N 109 1W
Belgaum, India ... **40 M9** 15 55N 74 35 E
Belgium ■, Europe ... **15 D5** 50 30N 5 0 E
Belgorod, Russia ... **25 D6** 50 35N 36 35 E
Belgorod-Dnestrovskiy, Ukraine ... **25 E5** 46 11N 30 23 E
Belgrade = Beograd, Serbia, Yug. ... **21 B9** 44 50N 20 37 E
Belgrade, U.S.A. ... **82 D8** 45 47N 111 11W
Belhaven, U.S.A. ... **77 H7** 35 33N 76 37W
Beli Drim →, Europe ... **21 C9** 42 6N 20 25 E
Belinga, Gabon ... **52 D2** 1 10N 13 2 E
Belinyu, Indonesia ... **36 E3** 1 35S 105 50 E
Beliton Is. = Belitung, Indonesia ... **36 E3** 3 10S 107 50 E
Belitung, Indonesia ... **36 E3** 3 10S 107 50 E
Belize ■, Cent. Amer. ... **87 D7** 17 0N 88 30W
Belize City, Belize ... **87 D7** 17 25N 88 0W
Belkovskiy, Ostrov, Russia ... **27 B14** 75 32N 135 44 E
Bell →, Canada ... **70 C4** 49 48N 77 38W
Bell Bay, Australia ... **62 G4** 41 6S 146 53 E
Bell I., Canada ... **71 B8** 50 46N 55 35W
Bell-Irving →, Canada ... **72 B3** 56 12N 129 5W
Bell Peninsula, Canada ... **69 B11** 63 50N 82 0W
Bell Ville, Argentina ... **94 C3** 32 40S 62 40W
Bella Bella, Canada ... **72 C3** 52 10N 128 10W
Bella Coola, Canada ... **72 C3** 52 25N 126 40W
Bella Unión, Uruguay ... **94 C4** 30 15S 57 40W
Bella Vista, Corrientes, Argentina ... **94 B4** 28 33S 59 0W
Bella Vista, Tucuman, Argentina ... **94 B2** 27 10S 65 25W
Bellaire, U.S.A. ... **78 F4** 40 1N 80 45W
Bellary, India ... **40 M10** 15 10N 76 56 E
Bellata, Australia ... **63 D4** 29 53S 149 46 E
Belle Fourche, U.S.A. ... **80 C3** 44 40N 103 51W
Belle Fourche →, U.S.A. **80 C3** 44 26N 102 18W
Belle Glade, U.S.A. ... **77 M5** 26 41N 80 40W
Belle-Ile, France ... **18 C2** 47 20N 3 10W
Belle Isle, Canada ... **71 B8** 51 57N 55 25W
Belle Isle, Str. of, Canada **71 B8** 51 30N 56 30W
Belle Plaine, Iowa, U.S.A. **80 E8** 41 54N 92 17W
Belle Plaine, Minn., U.S.A. ... **80 C8** 44 37N 93 46W
Belledune, Canada ... **71 C6** 47 55N 65 50W
Bellefontaine, U.S.A. ... **76 E4** 40 22N 83 46W
Bellefonte, U.S.A. ... **78 F7** 40 55N 77 47W
Belleoram, Canada ... **71 C8** 47 31N 55 25W
Belleville, Canada ... **70 D4** 44 10N 77 23W
Belleville, Ill., U.S.A. ... **80 F10** 38 31N 89 59W
Belleville, Kans., U.S.A. ... **80 F6** 39 50N 97 38W
Belleville, N.Y., U.S.A. ... **79 C8** 43 46N 76 10W
Bellevue, Canada ... **72 D6** 49 35N 114 22W
Bellevue, Idaho, U.S.A. ... **82 E6** 43 28N 114 16W
Bellevue, Ohio, U.S.A. ... **78 E2** 41 17N 82 51W
Bellevue, Wash., U.S.A. ... **84 C4** 47 37N 122 12W
Bellin, Canada ... **69 B13** 60 0N 70 0W
Bellingen, Australia ... **63 E5** 30 25S 152 50 E
Bellingham, U.S.A. ... **84 B4** 48 46N 122 29W
Bellingshausen Sea, Antarctica ... **5 C17** 66 0S 80 0W
Bellinzona, Switz. ... **16 E4** 46 11N 9 1 E
Bellows Falls, U.S.A. ... **79 C12** 43 8N 72 27W
Bellpat, Pakistan ... **42 E3** 29 0N 68 5 E
Belluno, Italy ... **20 A5** 46 8N 12 13 E
Bellville, U.S.A. ... **81 L6** 29 57N 96 15W
Bellwood, U.S.A. ... **78 F6** 40 36N 78 20W
Bélmez, Spain ... **19 C3** 38 17N 5 17W
Belmont, Australia ... **63 E5** 33 4S 151 42 E
Belmont, Canada ... **78 D3** 42 53N 81 5W
Belmont, S. Africa ... **56 D3** 29 28S 24 22 E
Belmont, U.S.A. ... **78 D6** 42 14N 78 2W
Belmonte, Brazil ... **93 G11** 16 0S 39 0W
Belmopan, Belize ... **87 D7** 17 18N 88 30W
Belmullet, Ireland ... **13 B2** 54 13N 9 58W
Belo Horizonte, Brazil ... **93 G10** 19 55S 43 56W
Belo-sur-Mer, Madag. ... **57 C7** 20 42S 44 0 E
Belo-Tsiribihina, Madag. **57 B7** 19 40S 44 30 E
Belogorsk, Russia ... **27 D13** 51 0N 128 20 E
Beloha, Madag. ... **57 D8** 25 10S 45 3 E
Beloit, Kans., U.S.A. ... **80 F5** 39 28N 98 6W
Beloit, Wis., U.S.A. ... **80 D10** 42 31N 89 2W
Belomorsk, Russia ... **24 B5** 64 35N 34 30 E
Belonia, India ... **41 H17** 23 15N 91 30 E
Beloretsk, Russia ... **24 D10** 53 58N 58 24 E
Belorussia ■, Europe ... **24 D4** 53 30N 27 0 E
Belovo, Russia ... **26 D9** 54 30N 86 0 E
Beloye, Oz., Russia ... **24 B6** 60 10N 37 35 E
Beloye More, Russia ... **24 A6** 66 30N 38 0 E
Belozersk, Russia ... **24 B6** 60 4N 37 30 E
Beltana, Australia ... **63 E2** 30 48S 138 25 E
Belterra, Brazil ... **93 D8** 2 45S 55 0W
Belton, S.C., U.S.A. ... **77 H4** 34 31N 82 30W
Belton, Tex., U.S.A. ... **81 K6** 31 3N 97 28W
Belton Res., U.S.A. ... **81 K6** 31 8N 97 32W
Beltsy, Moldavia ... **25 E4** 47 48N 28 0 E
Belturbet, Ireland ... **13 B4** 54 6N 7 28W
Belukha, Russia ... **26 E9** 49 50N 86 50 E
Beluran, Malaysia ... **36 C5** 5 48N 117 35 E
Belvidere, Ill., U.S.A. ... **80 D10** 42 15N 88 50W
Belvidere, N.J., U.S.A. ... **79 F9** 40 50N 75 5W
Belyando →, Australia ... **62 C4** 21 38S 146 50 E
Belyy, Ostrov, Russia ... **26 B8** 73 30N 71 0 E
Belyy Yar, Russia ... **26 D9** 58 26N 84 39 E
Belzoni, U.S.A. ... **81 J9** 33 11N 90 29W
Bemaraha, Lembalemban' i, Madag. ... **57 B7** 18 40S 44 45 E
Bemarivo, Madag. ... **57 C7** 21 45S 44 45 E
Bemarivo →, Madag. ... **57 B8** 15 27S 47 40 E
Bemavo, Madag. ... **57 C8** 21 33S 45 25 E
Bembéréke, Benin ... **50 F5** 10 11N 2 43 E
Bembesi, Zimbabwe ... **55 F2** 20 0S 28 58 E
Bembesi →, Zimbabwe ... **55 F2** 18 57S 27 47 E
Bemidji, U.S.A. ... **80 B7** 47 28N 94 53W
Ben, Iran ... **45 C6** 32 32N 50 45 E
Ben Cruachan, U.K. ... **12 E3** 56 26N 5 8W
Ben Dearg, U.K. ... **12 D4** 57 47N 4 58W
Ben Gardane, Tunisia ... **51 B7** 33 11N 11 11 E
Ben Hope, U.K. ... **12 C4** 58 24N 4 36W
Ben Lawers, U.K. ... **12 E4** 56 33N 4 13W
Ben Lomond, N.S.W., Australia ... **63 E5** 30 1S 151 43 E
Ben Lomond, Tas., Australia ... **62 G4** 41 38S 147 42 E
Ben Lomond, U.K. ... **12 E4** 56 12N 4 39W
Ben Luc, Vietnam ... **39 G6** 10 39N 106 29 E
Ben Macdhui, U.K. ... **12 D5** 57 4N 3 40W
Ben Mhor, U.K. ... **12 D1** 57 16N 7 21W
Ben More, Central, U.K. **12 E4** 56 23N 4 31W
Ben More, Strath., U.K. **12 E2** 56 26N 6 2W
Ben More Assynt, U.K. **12 C4** 58 7N 4 51W
Ben Nevis, U.K. ... **12 E4** 56 48N 4 58W
Ben Quang, Vietnam ... **38 D6** 17 3N 106 55 E
Ben Tre, Vietnam ... **39 G6** 10 3N 106 36 E
Ben Vorlich, U.K. ... **12 E4** 56 22N 4 15W
Ben Wyvis, U.K. ... **12 D4** 57 40N 4 35W
Bena, Nigeria ... **50 F6** 11 20N 5 50 E
Bena Dibele, Zaïre ... **52 E4** 4 4S 22 50 E
Benagerie, Australia ... **63 E3** 31 25S 140 22 E
Benalla, Australia ... **63 F4** 36 30S 146 0 E
Benambra, Mt., Australia **63 F4** 36 31S 147 34 E
Benares = Varanasi, India **43 G10** 25 22N 83 0 E
Benavides, U.S.A. ... **81 M5** 27 36N 98 25W
Benbecula, U.K. ... **12 D1** 57 26N 7 21W
Benbonyathe, Australia ... **63 E2** 30 25S 139 11 E
Bencubbin, Australia ... **61 F2** 30 48S 117 52 E
Bend, U.S.A. ... **82 D3** 44 4N 121 19W
Bender Beila, Somali Rep. ... **46 F5** 9 30N 50 48 E
Bendering, Australia ... **61 F2** 32 23S 118 18 E
Bendery, Moldavia ... **25 E4** 46 50N 29 30 E
Bendigo, Australia ... **63 F3** 36 40S 144 15 E
Benē Beraq, Israel ... **47 C3** 32 6N 34 51 E
Benenitra, Madag. ... **57 C8** 23 27S 45 5 E
Benevento, Italy ... **20 D6** 41 7N 14 45 E
Benga, Mozam. ... **55 F3** 16 11S 33 40 E
Bengal, Bay of, Ind. Oc. **41 K16** 15 0N 90 0 E
Bengbu, China ... **35 H9** 32 58N 117 20 E
Benghazi = Banghāzī, Libya ... **51 B9** 32 11N 20 3 E
Bengkalis, Indonesia ... **36 D2** 1 30N 102 10 E
Bengkulu, Indonesia ... **36 E2** 3 50S 102 12 E
Bengkulu □, Indonesia ... **36 E2** 3 48S 102 16 E
Bengough, Canada ... **73 D7** 49 25N 105 10W
Benguela, Angola ... **53 G2** 12 37S 13 25 E
Benguérua, I., Mozam. ... **57 C6** 21 58S 35 28 E
Beni, Zaïre ... **54 B2** 0 30N 29 27 E
Beni →, Bolivia ... **92 F5** 10 23S 65 24W
Beni Abbès, Algeria ... **50 B4** 30 5N 2 5W
Beni Mazâr, Egypt ... **51 C11** 28 32N 30 44 E
Beni Mellal, Morocco ... **50 B3** 32 21N 6 21W
Beni Ounif, Algeria ... **50 B4** 32 0N 1 10W
Beni Suef, Egypt ... **51 C11** 29 5N 31 6 E
Beniah L., Canada ... **72 A6** 63 23N 112 17W
Benicia, U.S.A. ... **84 G4** 38 3N 122 9W
Benidorm, Spain ... **19 C5** 38 33N 0 9W
Benin ■, Africa ... **50 G5** 10 0N 2 0 E
Benin, Bight of, W. Afr. **50 H5** 5 0N 3 0 E
Benin City, Nigeria ... **50 G6** 6 20N 5 31 E
Benitses, Greece ... **23 A3** 39 32N 19 55 E
Benjamin Aceval, Paraguay ... **94 A4** 24 58S 57 34W
Benjamin Constant, Brazil **92 D4** 4 40S 70 15W
Benjamin Hill, Mexico ... **86 A2** 30 10N 111 10W
Benkelman, U.S.A. ... **80 E4** 40 3N 101 32W
Benlidi, Australia ... **62 C3** 24 35S 144 50 E
Bennett, Canada ... **72 B2** 59 56N 134 53W
Bennett, L., Australia ... **60 D5** 22 50S 131 2 E
Bennett, Ostrov, Russia **27 B15** 76 21N 148 56 E
Bennettsville, U.S.A. ... **77 H6** 34 37N 79 41W
Bennington, U.S.A. ... **79 D11** 43 0N 71 55W
Benoni, S. Africa ... **57 D4** 26 11S 28 18 E
Benque Viejo, Belize ... **87 D7** 17 5N 89 8W
Benson, U.S.A. ... **83 L8** 31 58N 110 18W
Bent, Iran ... **45 E8** 26 20N 59 31 E
Benteng, Indonesia ... **37 F6** 6 10S 120 30 E
Bentinck I., Australia ... **62 B2** 17 3S 139 35 E
Bento Gonçalves, Brazil **95 B5** 29 10S 51 31W
Benton, Ark., U.S.A. ... **81 H8** 34 34N 92 35W
Benton, Calif., U.S.A. ... **84 H8** 37 48N 118 32W
Benton, Ill., U.S.A. ... **80 F10** 38 0N 88 55W
Benton Harbor, U.S.A. ... **76 D2** 42 6N 86 27W
Bentung, Malaysia ... **39 L3** 3 31N 101 55 E
Benue →, Nigeria ... **50 G6** 7 48N 6 46 E
Benxi, China ... **35 D12** 41 20N 123 48 E
Beo, Indonesia ... **37 D7** 4 25N 126 50 E
Beograd, Serbia, Yug. ... **21 B9** 44 50N 20 37 E
Beowawe, U.S.A. ... **82 F5** 40 35N 116 29W
Beppu, Japan ... **31 H5** 33 15N 131 30 E
Berau, Teluk, Indonesia ... **37 E8** 2 30S 132 30 E
Berber, Sudan ... **51 E11** 18 0N 34 0 E
Berbera, Somali Rep. ... **46 E4** 10 30N 45 2 E
Berbérati, C.A.R. ... **52 D3** 4 15N 15 40 E
Berberia, C. del, Spain ... **22 C7** 38 39N 1 24 E
Berbice →, Guyana ... **92 B7** 6 20N 57 32W
Berdichev, Ukraine ... **25 E4** 49 57N 28 30 E
Berdsk, Russia ... **26 D9** 54 47N 83 2 E
Berdyansk, Ukraine ... **25 E6** 46 45N 36 50 E
Berdychiv = Berdichev, Ukraine ... **25 E4** 49 57N 28 30 E
Berea, U.S.A. ... **76 G3** 37 34N 84 17W
Berebere, Indonesia ... **37 D7** 2 25N 128 45 E
Bereda, Somali Rep. ... **46 E5** 11 45N 51 0 E
Berekum, Ghana ... **50 G4** 7 29N 2 34W
Berens →, Canada ... **73 C9** 52 25N 97 2W
Berens I., Canada ... **73 C9** 52 18N 97 18W
Berens River, Canada ... **73 C9** 52 25N 97 0W
Berevo, Mahajanga, Madag. ... **57 B7** 17 14S 44 17 E
Berevo, Toliara, Madag. **57 B7** 19 44S 44 58 E
Berezina →, Belorussia **24 D5** 52 33N 30 14 E
Berezniki, Russia ... **26 D6** 59 24N 56 46 E
Berezovo, Russia ... **24 B11** 64 0N 65 0 E
Bérgamo, Italy ... **20 B3** 45 42N 9 40 E
Bergen, Neths. ... **15 B4** 52 40N 4 43 E
Bergen, Norway ... **9 F8** 60 23N 5 20 E
Bergen, U.S.A. ... **78 C7** 43 5N 77 57W
Bergen-op-Zoom, Neths. **15 C4** 51 28N 4 18 E
Bergerac, France ... **18 D4** 44 51N 0 30 E
Bergum, Neths. ... **15 A5** 53 13N 5 59 E
Bergville, S. Africa ... **57 D4** 28 52S 29 18 E
Berhala, Selat, Indonesia **36 E2** 1 0S 104 15 E
Berhampore = Baharampur, India ... **43 G13** 24 2N 88 27 E
Berhampur, India ... **41 K14** 19 15N 84 54 E
Bering Sea, Pac. Oc. ... **68 C1** 58 0N 171 0 E
Bering Strait, U.S.A. ... **68 B3** 65 30N 169 0W
Beringen, Belgium ... **15 C5** 51 3N 5 14 E
Beringovskiy, Russia ... **27 C18** 63 3N 179 19 E
Berisso, Argentina ... **94 C4** 34 56S 57 50W
Berja, Spain ... **19 D4** 36 50N 2 56W
Berkeley, U.K. ... **11 F5** 51 41N 2 28W
Berkeley, U.S.A. ... **84 H4** 37 52N 122 16W
Berkeley Springs, U.S.A. **76 F6** 39 38N 78 14W
Berkner I., Antarctica ... **5 D18** 79 30S 50 0W
Berkshire □, U.K. ... **11 F6** 51 30N 1 20W
Berland →, Canada ... **72 C5** 54 0N 116 50W
Berlin, Germany ... **16 B6** 52 32N 13 24 E
Berlin, Md., U.S.A. ... **76 F8** 38 20N 75 13W
Berlin, N.H., U.S.A. ... **79 B13** 44 28N 71 11W
Berlin, Wis., U.S.A. ... **76 D1** 43 58N 88 57W
Bermejo →, Formosa, Argentina ... **94 B4** 26 51S 58 23W
Bermejo →, San Juan, Argentina ... **94 C2** 32 30S 67 30W
Bermuda ■, Atl. Oc. ... **2 C6** 32 45N 65 0W
Bern, Switz. ... **16 E3** 46 57N 7 28 E
Bernado, U.S.A. ... **83 J10** 34 30N 106 53W
Bernalillo, U.S.A. ... **83 J10** 35 18N 106 33W
Bernardo de Irigoyen, Argentina ... **95 B5** 26 15S 53 40W
Bernardo O'Higgins □, Chile ... **94 C1** 34 15S 70 45W
Bernasconi, Argentina ... **94 D3** 37 55S 63 44W
Bernburg, Germany ... **16 C5** 51 40N 11 42 E
Berne = Bern, Switz. ... **16 E3** 46 57N 7 28 E
Bernier I., Australia ... **61 D1** 24 50S 113 12 E
Beroroha, Madag. ... **57 C8** 21 40S 45 10 E
Beroun, Czech. ... **16 D7** 49 57N 14 5 E
Berrechid, Morocco ... **50 B3** 33 18N 7 36W
Berri, Australia ... **63 E3** 34 14S 140 35 E
Berry, Australia ... **63 E5** 34 46S 150 43 E
Berry, France ... **18 C5** 46 50N 2 0 E
Berry Is., Bahamas ... **88 A4** 25 40N 77 50W
Berryessa L., U.S.A. ... **84 G4** 38 31N 122 6W
Berryville, U.S.A. ... **81 G8** 36 22N 93 34W
Berthold, U.S.A. ... **80 A4** 48 19N 101 44W
Berthoud, U.S.A. ... **80 E2** 40 19N 105 5W
Bertoua, Cameroon ... **52 D2** 4 30N 13 45 E
Bertrand, U.S.A. ... **80 E5** 40 32N 99 38W
Berufjörður, Iceland ... **8 D6** 64 48N 14 29W
Berwick, U.S.A. ... **79 E8** 41 3N 76 14W
Berwick-upon-Tweed, U.K. ... **10 B5** 55 47N 2 0W
Berwyn Mts., U.K. ... **10 E4** 52 54N 3 26W
Besal, Pakistan ... **43 B5** 35 4N 73 56 E
Besalampy, Madag. ... **57 B7** 16 43S 44 29 E
Besançon, France ... **18 C7** 47 15N 6 2 E
Besar, Indonesia ... **36 E5** 2 40S 116 0 E
Besnard L., Canada ... **73 B7** 55 25N 106 0W
Besor, N. →, Egypt ... **47 D3** 31 28N 34 22 E
Bessarabiya, Moldavia ... **17 E14** 47 0N 28 10 E
Bessemer, Ala., U.S.A. ... **77 J2** 33 24N 86 58W
Bessemer, Mich., U.S.A. **80 B9** 46 29N 90 3W
Bet She'an, Israel ... **47 C4** 32 30N 35 30 E
Bet Shemesh, Israel ... **47 D3** 31 44N 35 0 E
Betafo, Madag. ... **57 B8** 19 50S 46 51 E
Betancuria, Canary Is. ... **22 F5** 28 25N 14 3W
Bétaré Oya, Cameroon ... **52 C2** 5 40N 14 5 E
Bethal, S. Africa ... **57 D4** 26 27S 29 28 E
Bethanien, Namibia ... **56 D2** 26 31S 17 8 E
Bethany, S. Africa ... **56 D4** 29 34S 25 59 E
Bethany, U.S.A. ... **80 E7** 40 16N 94 2W
Bethel, Alaska, U.S.A. ... **68 B3** 60 48N 161 45W
Bethel, Vt., U.S.A. ... **79 C12** 43 50N 72 38W
Bethel Park, U.S.A. ... **78 F4** 40 20N 80 1W
Bethlehem = Bayt Lahm, Jordan ... **47 D4** 31 43N 35 12 E
Bethlehem, S. Africa ... **57 D4** 28 14S 28 18 E
Bethlehem, U.S.A. ... **79 F9** 40 37N 75 23W
Bethulie, S. Africa ... **56 E4** 30 30S 25 59 E
Béthune, France ... **18 A5** 50 30N 2 38 E
Bethungra, Australia ... **63 E4** 34 45S 147 51 E
Betioky, Madag. ... **57 C7** 23 48S 44 20 E
Betong, Thailand ... **39 K3** 5 45N 101 5 E
Betoota, Australia ... **62 D3** 25 45S 140 42 E
Betroka, Madag. ... **57 C8** 23 16S 46 0 E
Betsiamites, Canada ... **71 C6** 48 56N 68 40W
Betsiamites →, Canada **71 C6** 48 56N 68 38W
Betsiboka →, Madag. ... **57 B8** 16 3S 46 36 E
Betsjoeanaland, S. Africa **56 D3** 26 30S 22 30 E
Bettiah, India ... **43 F11** 26 48N 84 33 E
Betul, India ... **40 J10** 21 58N 77 59 E
Betung, Malaysia ... **36 D4** 1 24N 111 31 E
Beulah, U.S.A. ... **80 B4** 47 16N 101 47W
Beverley, Australia ... **61 F2** 32 9S 116 56 E
Beverley, U.K. ... **10 D7** 53 52N 0 26W
Beverly, Mass., U.S.A. ... **79 D14** 42 33N 70 53W
Beverly, Wash., U.S.A. ... **82 C4** 46 50N 119 56W
Beverly Hills, U.S.A. ... **85 L8** 34 4N 118 25W
Beverwijk, Neths. ... **15 B4** 52 28N 4 38 E
Beyānlū, Iran ... **44 C5** 36 0N 47 51 E
Beyla, Guinea ... **50 G3** 8 30N 8 38W
Beyneu, Kazakhstan ... **25 E10** 45 10N 55 3 E
Beypazan, Turkey ... **25 F5** 40 10N 31 56 E
Beyşehir Gölü, Turkey ... **25 G5** 37 40N 31 45 E
Bezhitsa, Russia ... **24 D5** 53 19N 34 17 E
Béziers, France ... **18 E5** 43 20N 3 12 E
Bezwada = Vijayawada, India ... **41 L12** 16 31N 80 39 E
Bhachau, India ... **40 H7** 23 20N 70 16 E
Bhadarwah, India ... **43 C6** 32 58N 75 46 E
Bhadrakh, India ... **41 J15** 21 10N 86 30 E
Bhadravati, India ... **40 N9** 13 49N 75 40 E
Bhagalpur, India ... **43 G12** 25 10N 87 0 E
Bhakkar, Pakistan ... **42 D4** 31 40N 71 5 E
Bhakra Dam, India ... **42 D7** 31 30N 76 45 E
Bhamo, Burma ... **41 G20** 24 15N 97 15 E
Bhandara, India ... **40 J11** 21 5N 79 42 E
Bhanrer Ra., India ... **42 H8** 23 40N 79 45 E
Bharat = India ■, Asia ... **40 K11** 20 0N 78 0 E
Bharatpur, India ... **42 F7** 27 15N 77 30 E
Bhatinda, India ... **42 D6** 30 15N 74 57 E
Bhatpara, India ... **43 H13** 22 50N 88 25 E
Bhaun, Pakistan ... **42 C5** 32 55N 72 40 E
Bhaunagar = Bhavnagar, India ... **42 J5** 21 45N 72 10 E
Bhavnagar, India ... **42 J5** 21 45N 72 10 E
Bhawanipatna, India ... **41 K12** 19 55N 80 10 E
Bhera, Pakistan ... **42 C5** 32 29N 72 57 E
Bhilsa = Vidisha, India ... **42 H7** 23 28N 77 53 E
Bhilwara, India ... **42 G6** 25 25N 74 38 E
Bhima →, India ... **40 L10** 16 25N 77 17 E
Bhimavaram, India ... **41 L12** 16 30N 81 30 E
Bhimbar, Pakistan ... **43 C6** 32 59N 74 3 E
Bhind, India ... **43 F8** 26 30N 78 46 E
Bhiwandi, India ... **40 K8** 19 20N 73 0 E
Bhiwani, India ... **42 E7** 28 50N 76 9 E
Bhola, Bangla. ... **41 H17** 22 45N 90 35 E
Bhopal, India ... **42 H7** 23 20N 77 30 E
Bhubaneshwar, India ... **41 J14** 20 15N 85 50 E
Bhuj, India ... **42 H3** 23 15N 69 49 E
Bhumiphol Dam = Phumiphon, Khuan, Thailand ... **38 D2** 17 15N 98 58 E
Bhusaval, India ... **40 J9** 21 3N 75 46 E
Bhutan ■, Asia ... **41 F17** 27 25N 90 30 E
Biafra, B. of = Bonny, Bight of, Africa ... **52 D1** 3 30N 9 20 E
Biak, Indonesia ... **37 E9** 1 10S 136 6 E
Biała Podlaska, Poland ... **17 B11** 52 4N 23 6 E
Białystok, Poland ... **17 B11** 53 10N 23 10 E
Bīārjmand, Iran ... **45 B7** 36 6N 55 53 E
Biaro, Indonesia ... **37 D7** 2 5N 125 26 E
Biarritz, France ... **18 E3** 43 29N 1 33W
Bibai, Japan ... **30 C10** 43 19N 141 52 E
Bibala, Angola ... **53 G2** 14 44S 13 24 E
Bibby I., Canada ... **73 A10** 61 55N 93 0W
Biberach, Germany ... **16 D4** 48 5N 9 49 E
Bibiani, Ghana ... **50 G4** 6 30N 2 8W
Biboohra, Australia ... **62 B4** 16 56S 145 25 E
Bibungwa, Zaïre ... **54 C2** 2 40S 28 15 E
Bic, Canada ... **71 C6** 48 20N 68 41W
Bickerton I., Australia ... **62 A2** 13 45S 136 10 E
Bicknell, Ind., U.S.A. ... **76 F2** 38 47N 87 19W
Bicknell, Utah, U.S.A. ... **83 G8** 38 20N 111 33W
Bida, Nigeria ... **50 G6** 9 3N 5 58 E
Bidar, India ... **40 L10** 17 55N 77 35 E
Biddeford, U.S.A. ... **71 D5** 43 30N 70 28W
Bideford, U.K. ... **11 F3** 51 1N 4 13W
Bidon 5 = Poste Maurice Cortier, Algeria ... **50 D5** 22 14N 1 2 E
Bidor, Malaysia ... **39 K3** 4 6N 101 15 E
Bié, Planalto de, Angola **53 G3** 12 0S 16 0 E
Bieber, U.S.A. ... **82 F3** 41 7N 121 8W
Biel, Switz. ... **16 E3** 47 8N 7 14 E
Bielé Karpaty, Europe ... **17 D8** 49 5N 18 0 E
Bielefeld, Germany ... **16 B4** 52 2N 8 31 E
Biella, Italy ... **20 B3** 45 33N 8 3 E
Bielsko-Biała, Poland ... **17 D9** 49 50N 19 2 E
Bien Hoa, Vietnam ... **39 G6** 10 57N 106 49 E
Bienfait, Canada ... **73 D8** 49 10N 102 50W
Bienne = Biel, Switz. ... **16 E3** 47 8N 7 14 E
Bienville, L., Canada ... **70 A5** 55 5N 72 40W
Biesiesfontein, S. Africa **56 E2** 30 57S 17 58 E
Big →, Canada ... **71 B8** 54 50N 58 55W
Big B., Canada ... **71 A7** 55 43N 60 35W
Big Bear City, U.S.A. ... **85 L10** 34 16N 116 51W
Big Bear Lake, U.S.A. ... **85 L10** 34 15N 116 56W
Big Beaver, Canada ... **73 D7** 49 10N 105 10W
Big Belt Mts., U.S.A. ... **82 C8** 46 30N 111 25W
Big Bend, Swaziland ... **57 D5** 26 50S 31 58 E
Big Bend National Park, U.S.A. ... **81 L3** 29 20N 103 5W
Big Black →, U.S.A. ... **81 J9** 32 3N 91 4W
Big Blue →, U.S.A. ... **80 F6** 39 35N 96 34W
Big Cr. →, Canada ... **72 C4** 51 42N 122 41W
Big Creek, U.S.A. ... **84 H7** 37 11N 119 14W
Big Cypress Swamp, U.S.A. ... **77 M5** 26 12N 81 10W
Big Falls, U.S.A. ... **80 A8** 48 12N 93 48W
Big Fork →, U.S.A. ... **80 A8** 48 31N 93 43W
Big Horn Mts. = Bighorn Mts., U.S.A. ... **82 D10** 44 30N 107 30W
Big Lake, U.S.A. ... **81 K4** 31 12N 101 28W

Big Moose

Big Moose, U.S.A. 79 C10 43 49N 74 58W
Big Muddy Cr. →, U.S.A. 80 A2 48 8N 104 36W
Big Pine, U.S.A. 83 H4 37 10N 118 17W
Big Piney, U.S.A. 82 E8 42 32N 110 7W
Big Quill L., Canada 73 C8 51 55N 104 50W
Big Rapids, U.S.A. 76 D3 43 42N 85 29W
Big River, Canada 73 C7 53 50N 107 0W
Big Run, U.S.A. 78 F6 40 57N 78 55W
Big Sable Pt., U.S.A. 76 C2 44 3N 86 1W
Big Sand L., Canada 73 B9 57 45N 99 45W
Big Sandy, U.S.A. 82 B8 48 11N 110 7W
Big Sandy Cr. →, U.S.A. 80 F3 38 7N 102 29W
Big Sioux →, U.S.A. 80 D6 42 29N 96 27W
Big Spring, U.S.A. 81 J4 32 15N 101 28W
Big Springs, U.S.A. 80 E3 41 4N 102 5W
Big Stone City, U.S.A. 80 C6 45 18N 96 28W
Big Stone Gap, U.S.A. 77 G4 36 52N 82 47W
Big Stone L., U.S.A. 80 C6 45 30N 96 35W
Big Sur, U.S.A. 84 J5 36 15N 121 48W
Big Timber, U.S.A. 82 D9 45 50N 109 57W
Big Trout L., Canada 70 B1 53 40N 90 0W
Bigfork, U.S.A. 82 B6 48 4N 114 4W
Biggar, Canada 73 C7 52 4N 108 0W
Biggar, U.K. 12 F5 55 38N 3 31W
Bigge I., Australia 60 B4 14 35S 125 10 E
Biggenden, Australia 63 D5 25 31S 152 4 E
Biggs, U.S.A. 84 F5 39 25N 121 43W
Bighorn, U.S.A. 82 C10 46 10N 107 27W
Bighorn →, U.S.A. 82 C10 46 10N 107 28W
Bighorn Mts., U.S.A. 82 D10 44 30N 107 30W
Bigorre, France 18 E4 43 10N 0 5 E
Bigstone L., Canada 73 C9 53 42N 95 44W
Bigwa, Tanzania 54 D4 7 10S 39 10 E
Bihać, Bos.-H. 20 B6 44 49N 15 57 E
Bihar, India 43 G11 25 5N 85 40 E
Bihar □, India 43 G11 25 0N 86 0 E
Biharamulo, Tanzania 54 C3 2 25S 31 25 E
Biharamulo □, Tanzania 54 C3 2 30S 31 20 E
Bijagós, Arquipélago dos, Guinea-Biss. 50 F1 11 15N 16 10W
Bijaipur, India 42 F7 26 2N 77 20 E
Bijapur, Karnataka, India 40 L9 16 50N 75 55 E
Bijapur, Mad. P., India 41 K12 18 50N 80 50 E
Bījār, Iran 44 C5 35 52N 47 35 E
Bijeljina, Bos.-H. 21 B8 44 46N 19 17 E
Bijnor, India 42 E8 29 27N 78 11 E
Bikaner, India 42 E5 28 2N 73 18 E
Bikapur, India 43 F10 26 30N 82 7 E
Bikeqi, China 34 D6 40 43N 111 20 E
Bikfayyā, Lebanon 47 B4 33 55N 35 41 E
Bikin, Russia 27 E14 46 50N 134 20 E
Bikin →, Russia 30 A7 46 51N 134 2 E
Bikini Atoll, Pac. Oc. 64 F8 12 0N 167 30 E
Bila Tserkva = Belaya Tserkov, Ukraine 25 E5 49 45N 30 10 E
Bilara, India 42 F5 26 14N 73 53 E
Bilaspur, Mad. P., India 43 H10 22 2N 82 15 E
Bilaspur, Punjab, India 42 D7 31 19N 76 50 E
Bilauk Taungdan, Thailand 38 F2 13 0N 99 0 E
Bilbao, Spain 19 A4 43 16N 2 56W
Bilbo = Bilbao, Spain 19 A4 43 16N 2 56W
Bíldudalur, Iceland 8 D2 65 41N 23 36W
Bilecik, Turkey 25 F5 40 5N 30 5 E
Bilibino, Russia 27 C17 68 3N 166 20 E
Bilibiza, Mozam. 55 E5 12 30S 40 20 E
Bilir, Russia 27 C14 65 40N 131 20 E
Bill, U.S.A. 80 D2 43 14N 105 16W
Billabalong, Australia 61 E2 27 25S 115 49 E
Billiluna, Australia 60 C4 19 37S 127 41 E
Billingham, U.K. 10 C6 54 36N 1 18W
Billings, U.S.A. 82 D9 45 47N 108 30W
Billiton Is. = Belitung, Indonesia 36 E3 3 10S 107 50 E
Bilma, Niger 51 E7 18 50N 13 30 E
Biloela, Australia 62 C5 24 24S 150 31 E
Biloxi, U.S.A. 81 K10 30 24N 88 53W
Bilpa Morea Claypan, Australia 62 D2 25 0S 140 0 E
Biltine, Chad 51 F9 14 40N 20 50 E
Bilyana, Australia 62 B4 18 5S 145 50 E
Bima, Indonesia 37 F5 8 22S 118 49 E
Bimbo, C.A.R. 52 D3 4 15N 18 33 E
Bimini Is., Bahamas 88 A4 25 42N 79 25W
Bin Xian, Heilongjiang, China 35 B14 45 42N 127 32 E
Bin Xian, Shaanxi, China 34 G5 35 2N 108 4 E
Bina-Etawah, India 42 G8 24 13N 78 14 E
Bināb, Iran 45 B6 36 35N 48 41 E
Binalbagan, Phil. 37 B6 10 12N 122 50 E
Binalong, Australia 63 E4 34 40S 148 39 E
Bīnālūd, Kūh-e, Iran 45 B8 36 30N 58 30 E
Binatang, Malaysia 36 D4 2 10N 111 40 E
Binbee, Australia 62 C4 20 19S 147 56 E
Binche, Belgium 15 D4 50 26N 4 10 E
Binda, Australia 63 D4 27 52S 147 21 E
Bindle, Australia 63 D4 27 40S 148 45 E
Bindura, Zimbabwe 55 F3 17 18S 31 18 E
Bingara, N.S.W., Australia 63 D5 29 52S 150 36 E
Bingara, Queens., Australia 63 D3 28 10S 144 37 E
Bingham, U.S.A. 71 C6 45 3N 69 53W
Bingham Canyon, U.S.A. 82 F7 40 32N 112 9W
Binghamton, U.S.A. 79 D9 42 6N 75 55W
Binh Dinh = An Nhon, Vietnam 38 F7 13 55N 109 7 E
Binh Khe, Vietnam 38 F7 13 57N 108 51 E
Binh Son, Vietnam 38 E7 15 20N 108 40 E
Binhai, China 35 G10 34 2N 119 49 E
Binisatua, Spain 22 B11 39 50N 4 11 E
Binjai, Indonesia 36 D1 3 20N 98 30 E
Binnaway, Australia 63 E4 31 28S 149 24 E
Binongko, Indonesia 37 F6 5 55S 123 55 E
Binscarth, Canada 73 C8 50 37N 101 17W
Bintan, Indonesia 36 D2 1 0N 104 0 E
Bintulu, Malaysia 36 D4 3 10N 113 0 E
Bintuni, Indonesia 37 E8 2 7S 133 32 E
Binzert = Bizerte, Tunisia 50 A6 37 15N 9 50 E
Bío Bío □, Chile 94 D1 37 35S 72 0W
Bioko, Eq. Guin. 50 H6 3 30N 8 40 E
Bir, India 40 K9 19 0N 75 54 E
Bîr Abu Muḥammad, Egypt 47 F3 29 44N 34 14 E
Bi'r ad Dabbāghāt, Jordan 47 E4 30 26N 35 32 E
Bi'r al Butayyihāt, Jordan 47 F4 29 47N 35 20 E
Bi'r al Mārī, Jordan 47 E4 30 4N 35 33 E
Bi'r al Qattār, Jordan 47 F4 29 47N 35 32 E
Bir Autrun, Sudan 51 E10 18 15N 26 40 E
Bîr Beiḍa, Egypt 47 E3 30 25N 34 29 E
Bîr el 'Abd, Egypt 47 D2 31 2N 33 0 E
Bîr el Biarât, Egypt 47 F3 29 30N 34 43 E
Bîr el Duweidar, Egypt 47 E1 30 56N 32 32 E
Bîr el Garârât, Egypt 47 D2 31 3N 33 34 E
Bîr el Heisi, Egypt 47 F3 29 22N 34 36 E
Bîr el Jafir, Egypt 47 E1 30 50N 32 41 E
Bîr el Mâlḥi, Egypt 47 E2 30 38N 33 19 E
Bîr el Thamâda, Egypt 47 E2 30 12N 33 27 E
Bîr Gebeil Ḥiṣn, Egypt 47 E2 30 2N 33 18 E
Bi'r Ghadīr, Syria 47 A6 34 6N 37 3 E
Bîr Ḥasana, Egypt 47 E2 30 29N 33 46 E
Bi'r Jadīd, Iraq 44 C4 34 1N 42 54 E
Bîr Kaseiba, Egypt 47 E2 31 0N 33 17 E
Bîr Lahfân, Egypt 47 D2 31 0N 33 51 E
Bîr Madkûr, Egypt 47 E1 30 44N 32 33 E
Bir Mogrein, Mauritania 50 C2 25 10N 11 25W
Bi'r Muṭribah, Kuwait 44 D5 29 54N 47 17 E
Bîr Qaṭia, Egypt 47 E1 30 58N 32 45 E
Bîr Ungât, Egypt 51 D11 22 8N 33 48 E
Bira, Indonesia 37 E8 2 3S 132 2 E
Birao, C.A.R. 51 F9 10 20N 22 47 E
Birawa, Zaïre 54 C2 2 20S 28 48 E
Birch Hills, Canada 73 C7 52 59N 105 25W
Birch I., Canada 73 C9 52 26N 99 54W
Birch L., N.W.T., Canada 72 A5 62 4N 116 33W
Birch L., Ont., Canada 70 B1 51 23N 92 18W
Birch L., U.S.A. 70 C1 47 45N 91 51W
Birch Mts., Canada 72 B6 57 30N 113 10W
Birch River, Canada 73 C8 52 24N 101 6W
Birchip, Australia 63 F3 35 56S 142 55 E
Bird, Canada 73 B10 56 30N 94 13W
Bird City, U.S.A. 80 F4 39 45N 101 32W
Bird I. = Aves, I. de, W. Indies 89 C7 15 45N 63 55W
Bird I., S. Africa 56 E2 32 3S 18 17 E
Birdlip, U.K. 11 F5 51 50N 2 7W
Birdsville, Australia 62 D2 25 51S 139 20 E
Birdum, Australia 60 C5 15 39S 133 13 E
Birein, Israel 47 E3 30 50N 34 28 E
Bireuen, Indonesia 36 C1 5 14N 96 39 E
Birigui, Brazil 95 A5 21 18S 50 16W
Birkenhead, U.K. 10 D4 53 24N 3 1W
Bîrlad, Romania 17 E13 46 15N 27 38 E
Birmingham, U.K. 11 E6 52 30N 1 55W
Birmingham, U.S.A. 77 J2 33 31N 86 48W
Birmitrapur, India 41 H14 22 24N 84 46 E
Birni Nkonni, Niger 50 F6 13 55N 5 15 E
Birnin Kebbi, Nigeria 50 F5 12 32N 4 12 E
Birobidzhan, Russia 27 E14 48 50N 132 50 E
Birr, Ireland 13 C4 53 7N 7 55W
Birrie →, Australia 63 D4 29 43S 146 37 E
Birsilpur, India 42 E5 28 11N 72 15 E
Birsk, Russia 24 C10 55 25N 55 30 E
Birtle, Canada 73 C8 50 30N 101 5W
Birur, India 40 N9 13 30N 75 55 E
Birzebugga, Malta 23 D2 35 49N 14 32 E
Bisa, Indonesia 37 E7 1 15S 127 28 E
Bisalpur, India 43 E8 28 14N 79 48 E
Bisbee, U.S.A. 83 L9 31 27N 109 55W
Biscay, B. of, Atl. Oc. 18 F4 45 0N 2 0W
Biscayne B., U.S.A. 77 N5 25 40N 80 12W
Biscoe Bay, Antarctica 5 D13 77 0S 152 0W
Biscoe Is., Antarctica 5 C17 66 0S 67 0W
Biscostasing, Canada 70 C3 47 18N 82 9W
Bishkek, Kirghizia 26 E8 42 54N 74 46 E
Bishnupur, India 43 H12 23 8N 87 20 E
Bisho, S. Africa 57 E4 32 50S 27 23 E
Bishop, Calif., U.S.A. 83 H4 37 22N 118 24W
Bishop, Tex., U.S.A. 81 M6 27 35N 97 48W
Bishop Auckland, U.K. 10 C6 54 40N 1 40W
Bishop's Falls, Canada 71 C8 49 2N 55 30W
Bishop's Stortford, U.K. 11 F8 51 52N 0 11 E
Bisina, L., Uganda 54 B3 1 38N 33 56 E
Biskra, Algeria 50 B6 34 50N 5 44 E
Bislig, Phil. 37 C7 8 15N 126 27 E
Bismarck, U.S.A. 80 B4 46 48N 100 47W
Bismarck Arch., Papua N. G. 64 H6 2 30S 150 0 E
Biso, Uganda 54 B3 1 44N 31 26 E
Bison, U.S.A. 80 C3 45 31N 102 28W
Bīsotūn, Iran 44 C5 34 23N 47 26 E
Bispfors, Sweden 8 E14 63 1N 16 37 E
Bissagos = Bijagós, Arquipélago dos, Guinea-Biss. 50 F1 11 15N 16 10W
Bissau, Guinea-Biss. 50 F1 11 45N 15 45W
Bissett, Canada 73 C9 51 2N 95 41W
Bistcho L., Canada 72 B5 59 45N 118 50W
Bistriţa, Romania 17 E12 47 9N 24 35 E
Bistriţa →, Romania 17 E13 46 30N 26 57 E
Biswan, India 43 F9 27 29N 81 2 E
Bitam, Gabon 52 D2 2 5N 11 25 E
Bitkine, Chad 51 F8 11 59N 18 13 E
Bitlis, Turkey 25 G7 38 20N 42 3 E
Bitola, Macedonia 21 D9 41 5N 21 10 E
Bitolj = Bitola, Macedonia 21 D9 41 5N 21 10 E
Bitter Creek, U.S.A. 82 F9 41 33N 108 33W
Bitter L. = Buheirat-Murrat-el-Kubra, Egypt 51 B11 30 15N 32 40 E
Bitterfontein, S. Africa 56 E2 31 1S 18 32 E
Bitterroot →, U.S.A. 82 C6 46 52N 114 7W
Bitterroot Range, U.S.A. 82 D6 46 0N 114 20W
Bitterwater, U.S.A. 84 J6 36 23N 121 0W
Biu, Nigeria 51 F7 10 40N 12 3 E
Biwa-Ko, Japan 31 G8 35 15N 136 10 E
Biwabik, U.S.A. 80 B8 47 32N 92 21W
Biyang, China 34 H7 32 38N 113 21 E
Biysk, Russia 26 D9 52 40N 85 0 E
Bizana, S. Africa 57 E4 30 50S 29 52 E
Bizen, Japan 31 G7 34 43N 134 8 E
Bizerte, Tunisia 50 A6 37 15N 9 50 E
Bjargtangar, Iceland 8 D1 65 30N 24 30W
Bjelovar, Croatia 20 B7 45 56N 16 49 E
Bjørnøya, Arctic 4 B8 74 30N 19 0 E
Black = Da →, Vietnam 38 B5 21 15N 105 20 E
Black →, Canada 78 B5 44 42N 79 19W
Black →, Ark., U.S.A. 81 H9 35 38N 91 20W
Black →, N.Y., U.S.A. 79 C8 43 59N 76 4W
Black →, Wis., U.S.A. 80 D9 43 57N 91 22W
Black Diamond, Canada 72 C6 50 45N 114 14W
Black Forest = Schwarzwald, Germany 16 E4 48 0N 8 0 E
Black Hills, U.S.A. 80 C3 44 0N 103 45W
Black I., Canada 73 C9 51 12N 96 30W
Black L., Canada 73 B7 59 12N 105 15W
Black L., U.S.A. 76 C3 45 28N 84 16W
Black Mesa, U.S.A. 81 G3 36 58N 102 58W
Black Mt. = Mynydd Du, U.K. 11 F4 51 45N 3 45W
Black Mts., U.K. 11 F4 51 52N 3 5W
Black Range, U.S.A. 83 K10 33 15N 107 50W
Black River, Jamaica 88 C4 18 0N 77 50W
Black River Falls, U.S.A. 80 C9 44 18N 90 51W
Black Sea, Europe 25 F6 43 30N 35 0 E
Black Volta →, Africa 50 G4 8 41N 1 33W
Black Warrior →, U.S.A. 77 J2 32 32N 87 51W
Blackall, Australia 62 C4 24 25S 145 45 E
Blackball, N.Z. 59 K3 42 22S 171 26 E
Blackbull, Australia 62 B3 17 55S 141 45 E
Blackburn, U.K. 10 D5 53 44N 2 30W
Blackduck, U.S.A. 80 B7 47 44N 94 33W
Blackfoot, U.S.A. 82 E7 43 11N 112 21W
Blackfoot →, U.S.A. 82 C7 46 52N 113 53W
Blackfoot River Reservoir, U.S.A. 82 E8 43 0N 111 43W
Blackie, Canada 72 C6 50 36N 113 37W
Blackpool, U.K. 10 D4 53 48N 3 3W
Blackriver, U.S.A. 78 B1 44 46N 83 17W
Blacks Harbour, Canada 71 C6 45 3N 66 49W
Blacksburg, U.S.A. 76 G5 37 14N 80 25W
Blacksod B., Ireland 13 B2 54 6N 10 0W
Blackstone, U.S.A. 76 G6 37 4N 78 0W
Blackstone →, Canada 72 A4 61 5N 122 55W
Blackstone Ra., Australia 61 E4 26 0S 128 30 E
Blackville, Canada 71 C6 46 44N 65 50W
Blackwater, Australia 62 C4 23 35S 148 53 E
Blackwater →, Ireland 13 E4 51 55N 7 50W
Blackwater →, U.K. 13 B5 54 31N 6 35W
Blackwater Cr. →, Australia 63 D3 25 56S 144 30 E
Blackwell, U.S.A. 81 G6 36 48N 97 17W
Blackwells Corner, U.S.A. 85 K7 35 37N 119 47W
Blaenau Ffestiniog, U.K. 10 E4 52 59N 3 57W
Blagodarnoye, Russia 25 E7 45 7N 43 37 E
Blagoveshchensk, Russia 27 D13 50 20N 127 30 E
Blaine, U.S.A. 84 B4 48 59N 122 45W
Blaine Lake, Canada 73 C7 52 51N 106 52W
Blair, U.S.A. 80 E6 41 33N 96 8W
Blair Athol, Australia 62 C4 22 42S 147 31 E
Blair Atholl, U.K. 12 E5 56 46N 3 50W
Blairgowrie, U.K. 12 E5 56 36N 3 20W
Blairmore, Canada 72 D6 49 40N 114 25W
Blairsden, U.S.A. 84 F6 39 47N 120 37W
Blairsville, U.S.A. 78 F5 40 26N 79 16W
Blake Pt., U.S.A. 80 A10 48 11N 88 25W
Blakely, U.S.A. 77 K3 31 23N 84 56W
Blanc, Mont, Alps 18 D7 45 48N 6 50 E
Blanca, B., Argentina 96 D4 39 10S 61 30W
Blanca Peak, U.S.A. 83 H11 37 35N 105 29W
Blanchard, U.S.A. 81 H6 35 8N 97 39W
Blanche, C., Australia 63 E1 33 1S 134 9 E
Blanche, L., S. Austral., Australia 63 D2 29 15S 139 40 E
Blanche, L., W. Austral., Australia 60 D3 22 25S 123 17 E
Blanco, S. Africa 56 E3 33 55S 22 23 E
Blanco, U.S.A. 81 K5 30 6N 98 25W
Blanco →, Argentina 94 C2 30 20S 68 42W
Blanco, C., U.S.A. 82 E1 42 51N 124 34W
Blanco, C., Costa Rica 88 E2 9 34N 85 8W
Blanco, C., Spain 22 B9 39 21N 2 51 E
Blanda →, Iceland 8 D4 65 20N 19 40W
Blandford Forum, U.K. 11 G5 50 52N 2 10W
Blanding, U.S.A. 83 H9 37 37N 109 29W
Blankenberge, Belgium 15 C3 51 20N 3 9 E
Blanquillo, Uruguay 95 C4 32 53S 55 37W
Blantyre, Malawi 55 F4 15 45S 35 0 E
Blarney, Ireland 13 E3 51 57N 8 35W
Blåvands Huk, Denmark 9 J10 55 33N 8 4 E
Blaydon, U.K. 10 C6 54 56N 1 47W
Blayney, Australia 63 E4 33 32S 149 14 E
Blaze, Pt., Australia 60 B5 12 56S 130 11 E
Blednaya, Gora, Russia 26 B7 76 20N 65 0 E
Bleiburg, Austria 16 E7 46 35N 14 49 E
Blekinge län □, Sweden 9 H13 56 20N 15 20 E
Blenheim, Canada 78 D2 42 20N 82 0W
Blenheim, N.Z. 59 J4 41 38S 173 57 E
Bletchley, U.K. 11 F7 51 59N 0 44W
Blida, Algeria 50 A5 36 30N 2 49 E
Bligh Sound, N.Z. 59 L1 44 47S 167 32 E
Blind River, Canada 70 C3 46 10N 82 58W
Blitar, Indonesia 37 H15 8 5S 112 11 E
Blitta, Togo 50 G5 8 23N 1 6 E
Block I., U.S.A. 79 E13 41 11N 71 35W
Block Island Sd., U.S.A. 79 E13 41 15N 71 40W
Blodgett Iceberg Tongue, Antarctica 5 C9 66 8S 130 35 E
Bloemfontein, S. Africa 56 D4 29 6S 26 7 E
Bloemhof, S. Africa 56 D4 27 38S 25 32 E
Blois, France 18 C4 47 35N 1 20 E
Blönduós, Iceland 8 D3 65 40N 20 12W
Bloodvein →, Canada 73 C9 51 47N 96 43W
Bloody Foreland, Ireland 13 A3 55 10N 8 18W
Bloomer, U.S.A. 80 C9 45 6N 91 29W
Bloomfield, Australia 62 B4 15 56S 145 22 E
Bloomfield, Canada 78 C7 43 59N 77 14W
Bloomfield, Iowa, U.S.A. 80 E8 40 45N 92 25W
Bloomfield, N. Mex., U.S.A. 83 H10 36 43N 107 59W
Bloomfield, Nebr., U.S.A. 80 D6 42 36N 97 39W
Bloomington, Ill., U.S.A. 80 E10 40 28N 89 0W
Bloomington, Ind., U.S.A. 76 F2 39 10N 86 32W
Bloomsburg, U.S.A. 79 F8 41 0N 76 27W
Blora, Indonesia 37 G14 6 57S 111 25 E
Blossburg, U.S.A. 78 E7 41 41N 77 4W
Blouberg, S. Africa 57 C4 23 8S 28 59 E
Blountstown, U.S.A. 77 K3 30 27N 85 3W
Blue Island, U.S.A. 76 E2 41 40N 87 40W
Blue Lake, U.S.A. 82 F2 40 53N 123 59W
Blue Mesa Reservoir, U.S.A. 83 G10 38 28N 107 20W
Blue Mts., Oreg., U.S.A. 82 D4 45 15N 119 0W
Blue Mts., Pa., U.S.A. 79 F8 40 30N 76 30W
Blue Mud B., Australia 62 A2 13 30S 136 0 E
Blue Nile = An Nîl el Azraq □, Sudan 51 F11 12 30N 34 30 E
Blue Nile = Nîl el Azraq →, Sudan 51 E11 15 38N 32 31 E
Blue Rapids, U.S.A. 80 F6 39 41N 96 39W
Blue Ridge Mts., U.S.A. 77 G5 36 30N 80 15W
Blue Stack Mts., Ireland 13 B3 54 46N 8 5W
Blueberry →, Canada 72 B4 56 45N 120 49W
Bluefield, U.S.A. 76 G5 37 15N 81 17W
Bluefields, Nic. 88 D3 12 20N 83 50W
Bluff, Australia 62 C4 23 35S 149 4 E
Bluff, N.Z. 59 M2 46 37S 168 20 E
Bluff, U.S.A. 83 H9 37 17N 109 33W
Bluff Knoll, Australia 61 F2 34 24S 118 15 E
Bluff Pt., Australia 61 E1 27 50S 114 5 E
Bluffton, U.S.A. 76 E3 40 44N 85 11W
Blumenau, Brazil 95 B6 27 0S 49 0W
Blunt, U.S.A. 80 C4 44 31N 99 59W
Bly, U.S.A. 82 E3 42 24N 121 3W
Blyth, Canada 78 C3 43 44N 81 26W
Blyth, U.K. 10 B6 55 8N 1 32W
Blyth Bridge, U.K. 10 E5 52 58N 2 4W
Blythe, U.S.A. 85 M12 33 37N 114 36W
Bo, S. Leone 50 G2 7 55N 11 50W
Bo Duc, Vietnam 39 G6 11 58N 106 50 E
Bo Hai, China 35 E10 39 0N 119 0 E
Bo Xian, China 34 H8 33 55N 115 41 E
Boa Vista, Brazil 92 C6 2 48N 60 30W
Boaco, Nic. 88 D2 12 29N 85 35W
Bo'ai, China 34 G7 35 10N 113 3 E
Boardman, U.S.A. 78 E4 41 2N 80 40W
Boatman, Australia 63 D4 27 16S 146 55 E
Bobadah, Australia 63 E4 32 19S 146 41 E
Bobbili, India 41 K13 18 35N 83 30 E
Bobcaygeon, Canada 70 D4 44 33N 78 33W
Bobo-Dioulasso, Burkina Faso 50 F4 11 8N 4 13W
Bóbr →, Poland 16 B7 52 4N 15 4 E
Bobraomby, Tanjon' i, Madag. 57 A8 12 40S 49 10 E
Bobruysk, Belorussia 24 D4 53 10N 29 15 E
Bôca do Acre, Brazil 92 E5 8 50S 67 27W
Boca Raton, U.S.A. 77 M5 26 21N 80 5W
Bocaiúva, Brazil 93 G10 17 7S 43 49W
Bocanda, Ivory C. 50 G4 7 5N 4 31W
Bocaranga, C.A.R. 51 G8 7 0N 15 35 E
Bocas del Toro, Panama 88 E3 9 15N 82 20W
Bocholt, Germany 16 C3 51 50N 6 35 E
Bochum, Germany 16 C3 51 28N 7 12 E
Bocoyna, Mexico 86 B3 27 52N 107 35W
Boda, C.A.R. 52 D3 4 19N 17 26 E
Bodaybo, Russia 27 D12 57 50N 114 0 E
Boddington, Australia 61 F2 32 50S 116 30 E
Bodega Bay, U.S.A. 84 G3 38 20N 123 3W
Boden, Sweden 8 D16 65 50N 21 42 E
Bodensee, Europe 16 E4 47 35N 9 25 E
Bodhan, India 40 K10 18 40N 77 44 E
Bodmin, U.K. 11 G3 50 28N 4 44W
Bodmin Moor, U.K. 11 G3 50 33N 4 36W
Bodrog →, Hungary 17 D10 48 15N 21 35 E
Boegoebergdam, S. Africa 56 D3 29 7S 22 9 E
Boende, Zaïre 52 E4 0 24S 21 12 E
Boerne, U.S.A. 81 L5 29 47N 98 44W
Boffa, Guinea 50 F2 10 16N 14 3W
Bogalusa, U.S.A. 81 K10 30 47N 89 52W
Bogan Gate, Australia 63 E4 33 7S 147 49 E
Bogantungan, Australia 62 C4 23 41S 147 17 E
Bogata, U.S.A. 81 J7 33 28N 95 13W
Boggabilla, Australia 63 D5 28 36S 150 24 E
Boggabri, Australia 63 E5 30 45S 150 5 E
Boggeragh Mts., Ireland 13 D3 52 2N 8 55W
Bognor Regis, U.K. 11 G7 50 47N 0 40W
Bogo, Phil. 37 B6 11 3N 124 0 E
Bogong, Mt., Australia 63 F4 36 47S 147 17 E
Bogor, Indonesia 37 G12 6 36S 106 48 E
Bogorodskoye, Russia 27 D15 52 22N 140 30 E
Bogotá, Colombia 92 C4 4 34N 74 0W
Bogotol, Russia 26 D9 56 15N 89 50 E
Bogra, Bangla. 41 G16 24 51N 89 22 E
Boguchany, Russia 27 D10 58 40N 97 30 E
Bogué, Mauritania 50 E2 16 45N 14 10W
Bohemia Downs, Australia 60 C4 18 53S 126 14 E
Bohemian Forest = Böhmerwald, Germany 16 D6 49 30N 12 40 E
Bohena Cr. →, Australia 63 E4 30 17S 149 42 E
Böhmerwald, Germany 16 D6 49 30N 12 40 E
Bohol, Phil. 37 C6 9 50N 124 10 E
Bohol Sea, Phil. 37 C6 9 0N 124 0 E
Bohotleh, Somali Rep. 46 F4 8 20N 46 25 E
Boi, Pta. de, Brazil 95 A6 23 55S 45 15W
Boileau, C., Australia 60 C3 17 40S 122 7 E
Boise, U.S.A. 82 E5 43 37N 116 13W
Boise City, U.S.A. 81 G3 36 44N 102 31W
Boissevain, Canada 73 D8 49 15N 100 5W
Bojador C., W. Sahara 50 C2 26 0N 14 30W
Bojana →, Albania 21 D8 41 52N 19 22 E
Bojnūrd, Iran 45 B8 37 30N 57 20 E
Bojonegoro, Indonesia 37 G14 7 11S 111 54 E
Boké, Guinea 50 F2 10 56N 14 17W
Bokhara →, Australia 63 D4 29 55S 146 42 E
Boknafjorden, Norway 9 G8 59 14N 5 40 E
Bokoro, Chad 51 F8 12 25N 17 14 E
Bokote, Zaïre 52 E4 0 12S 21 8 E
Bokungu, Zaïre 52 E4 0 35S 22 50 E
Bol, Chad 51 F7 13 30N 14 40 E
Bolama, Guinea-Biss. 50 F1 11 30N 15 30W
Bolan Pass, Pakistan 40 E5 29 50N 67 20 E
Bolaños →, Mexico 86 C4 21 14N 104 8W
Bolbec, France 18 B4 49 30N 0 30 E
Boldājī, Iran 45 D6 31 56N 51 3 E

Bridgton

Bridgton, *U.S.A.* 79 B14 44 3N 70 42W
Bridgwater, *U.K.* 11 F4 51 7N 3 0W
Bridlington, *U.K.* 10 C7 54 6N 0 11W
Bridport, *Australia* 62 G4 40 59S 147 23 E
Bridport, *U.K.* 11 G5 50 43N 2 45W
Brie, Plaine de la, *France* 18 B5 48 35N 3 10 E
Brig, *Switz.* 16 E3 46 18N 7 59 E
Brigg, *U.K.* 10 D7 53 33N 0 30W
Briggsdale, *U.S.A.* 80 E2 40 38N 104 20W
Brigham City, *U.S.A.* 82 F7 41 31N 112 1W
Bright, *Australia* 63 F4 36 42S 146 56 E
Brighton, *Australia* 63 F2 35 5S 138 30 E
Brighton, *Canada* 70 D4 44 2N 77 44W
Brighton, *U.K.* 11 G7 50 50N 0 9W
Brighton, *U.S.A.* 80 F2 39 59N 104 49W
Brilliant, *Canada* 72 D5 49 19N 117 38W
Brilliant, *U.S.A.* 78 F4 40 15N 80 39W
Bríndisi, *Italy* 21 D7 40 39N 17 55 E
Brinkley, *U.S.A.* 81 H9 34 53N 91 12W
Brinkworth, *Australia* 63 E2 33 42S 138 26 E
Brinnon, *U.S.A.* 84 C4 47 41N 122 54W
Brion, I., *Canada* 71 C7 47 46N 61 26W
Brisbane, *Australia* 63 D5 27 25S 153 2 E
Brisbane →, *Australia* 63 D5 27 24S 153 9 E
Bristol, *U.K.* 11 F5 51 26N 2 35W
Bristol, *Conn., U.S.A.* 79 E12 41 40N 72 57W
Bristol, *Pa., U.S.A.* 79 F10 40 6N 74 51W
Bristol, *R.I., U.S.A.* 79 E13 41 40N 71 16W
Bristol, *S. Dak., U.S.A.* 80 C6 45 21N 97 45W
Bristol, *Tenn., U.S.A.* 77 G4 36 36N 82 11W
Bristol B., *U.S.A.* 68 C4 58 0N 160 0W
Bristol Channel, *U.K.* 11 F3 51 18N 4 30W
Bristol I., *Antarctica* 5 B1 58 45S 28 0W
Bristol L., *U.S.A.* 83 J5 34 23N 116 50W
Bristow, *U.S.A.* 81 H6 35 50N 96 23W
British Columbia □, *Canada* 72 C3 55 0N 125 15W
British Guiana = Guyana ■, *S. Amer.* 92 C7 5 0N 59 0W
British Honduras = Belize ■, *Cent. Amer.* 87 D7 17 0N 88 30W
British Isles, *Europe* 6 E5 54 0N 4 0W
Brits, *S. Africa* 57 D4 25 37S 27 48 E
Britstown, *S. Africa* 56 E3 30 37S 23 30 E
Britt, *Canada* 70 C3 45 46N 80 34W
Brittany = Bretagne, *France* 18 B2 48 10N 3 0W
Britton, *U.S.A.* 80 C6 45 48N 97 45W
Brixton, *Australia* 62 C3 23 32S 144 57 E
Brlik, *Kazakhstan* 26 E8 43 40N 73 49 E
Brno, *Czech.* 16 D8 49 10N 16 35 E
Broad →, *U.S.A.* 77 J5 34 1N 81 4W
Broad Arrow, *Australia* 61 F3 30 23S 121 15 E
Broad B., *U.K.* 12 C2 58 14N 6 16W
Broad Haven, *Ireland* 13 B2 54 20N 9 55W
Broad Law, *U.K.* 12 F5 55 30N 3 22W
Broad Sd., *Australia* 62 C4 22 0S 149 45 E
Broadhurst Ra., *Australia* 60 D3 22 30S 122 30 E
Broads, The, *U.K.* 10 E9 52 45N 1 30 E
Broadus, *U.S.A.* 80 C2 45 27N 105 25W
Broadview, *Canada* 73 C8 50 22N 102 35W
Brochet, *Canada* 73 B8 57 53N 101 40W
Brochet, L., *Canada* 73 B8 58 36N 101 35W
Brock, *Canada* 73 C7 51 26N 108 43W
Brocken, *Germany* 16 C5 51 48N 10 40 E
Brockport, *U.S.A.* 78 C7 43 13N 77 56W
Brockton, *U.S.A.* 79 D13 42 5N 71 1W
Brockville, *Canada* 70 D4 44 35N 75 41W
Brockway, *Mont., U.S.A.* 80 B2 47 18N 105 45W
Brockway, *Pa., U.S.A.* 78 E6 41 15N 78 47W
Brocton, *U.S.A.* 78 D5 42 23N 79 26W
Brodeur Pen., *Canada* 69 A11 72 30N 88 10W
Brodick, *U.K.* 12 F3 55 34N 5 9W
Brogan, *U.S.A.* 82 D5 44 15N 117 31W
Broken Bow, *Nebr., U.S.A.* 80 E5 41 24N 99 38W
Broken Bow, *Okla., U.S.A.* 81 H7 34 2N 94 44W
Broken Hill = Kabwe, *Zambia* 55 E2 14 30S 28 29 E
Broken Hill, *Australia* 63 E3 31 58S 141 29 E
Bromfield, *U.K.* 11 E5 52 25N 2 45W
Bromley, *U.K.* 11 F8 51 20N 0 5 E
Brønderslev, *Denmark* 9 H10 57 16N 9 57 E
Bronkhorstspruit, *S. Africa* 57 D4 25 46S 28 45 E
Bronte, *U.S.A.* 81 K4 31 53N 100 18W
Bronte Park, *Australia* 62 G4 42 8S 146 30 E
Brook Park, *U.S.A.* 78 E4 41 24N 80 51W
Brookfield, *U.S.A.* 80 F8 39 47N 93 4W
Brookhaven, *U.S.A.* 81 K9 31 35N 90 26W
Brookings, *Oreg., U.S.A.* 82 E1 42 3N 124 17W
Brookings, *S. Dak., U.S.A.* 80 C6 44 19N 96 48W
Brooklin, *Canada* 78 C6 43 55N 78 55W
Brookmere, *Canada* 72 D4 49 52N 120 53W
Brooks, *Canada* 72 C6 50 35N 111 55W
Brooks B., *Canada* 72 C3 50 15N 127 55W
Brooks L., *Canada* 73 A7 61 55N 106 35W
Brooks Ra., *U.S.A.* 68 B5 68 40N 147 0W
Brooksville, *U.S.A.* 77 L4 28 33N 82 23W
Brookville, *U.S.A.* 76 F3 39 25N 85 1W
Brooloo, *Australia* 63 D5 26 30S 152 43 E
Broom, L., *U.K.* 12 D3 57 55N 5 15W
Broome, *Australia* 60 C3 18 0S 122 15 E
Broomehill, *Australia* 61 F2 33 51S 117 39 E
Brora, *U.K.* 12 C5 58 3N 3 50W
Brora →, *U.K.* 12 C5 58 4N 3 52W
Brosna →, *Ireland* 13 C4 53 8N 8 0W
Brothers, *U.S.A.* 82 E3 43 49N 120 36W
Brough, *U.K.* 10 C5 54 32N 2 19W
Broughton Island, *Canada* 69 B13 67 33N 63 0W
Broughty Ferry, *U.K.* 12 E6 56 29N 2 50W
Brouwershaven, *Neths.* 15 C3 51 45N 3 55 E
Browerville, *U.S.A.* 80 B7 46 5N 94 52W
Brown, Pt., *Australia* 63 E1 32 32S 133 50 E
Brown Willy, *U.K.* 11 G3 50 35N 4 34W
Brownfield, *U.S.A.* 81 J3 33 11N 102 17W
Browning, *U.S.A.* 82 B7 48 34N 113 1W
Brownlee, *Canada* 73 C7 50 43N 106 1W
Brownsville, *Oreg., U.S.A.* 82 D2 44 24N 122 59W
Brownsville, *Tenn., U.S.A.* 81 H10 35 36N 89 16W
Brownsville, *Tex., U.S.A.* 81 N6 25 54N 97 30W
Brownwood, *U.S.A.* 81 K5 31 43N 98 59W
Brownwood, L., *U.S.A.* 81 K5 31 51N 98 35W
Browse I., *Australia* 60 B3 14 7S 123 33 E
Bruas, *Malaysia* 39 K3 4 30N 100 47 E
Bruay-en-Artois, *France* 18 A5 50 29N 2 33 E
Bruce, Mt., *Australia* 60 D2 22 37S 118 8 E
Bruce Pen., *Canada* 78 A3 45 0N 81 30W
Bruce Rock, *Australia* 61 F2 31 52S 118 8 E
Bruck an der Leitha, *Austria* 16 D8 48 1N 16 47 E
Brue →, *U.K.* 11 F5 51 10N 2 59W
Bruges = Brugge, *Belgium* 15 C3 51 13N 3 13 E
Brugge, *Belgium* 15 C3 51 13N 3 13 E
Brûlé, *Canada* 72 C5 53 15N 117 58W
Brumado, *Brazil* 93 F10 14 14S 41 40W
Brunchilly, *Australia* 62 B1 18 50S 134 30 E
Brundidge, *U.S.A.* 77 K3 31 43N 85 49W
Bruneau, *U.S.A.* 82 E6 42 53N 115 48W
Bruneau →, *U.S.A.* 82 E6 42 56N 115 57W
Brunei = Bandar Seri Begawan, *Brunei* 36 C4 4 52N 115 0 E
Brunei ■, *Asia* 36 D4 4 50N 115 0 E
Brunette Downs, *Australia* 62 B2 18 40S 135 55 E
Brunner, L., *N.Z.* 59 K3 42 37S 171 27 E
Bruno, *Canada* 73 C7 52 20N 105 30W
Brunsbüttel, *Germany* 16 B4 53 52N 9 13 E
Brunswick = Braunschweig, *Germany* 16 B5 52 17N 10 28 E
Brunswick, *Ga., U.S.A.* 77 K5 31 10N 81 30W
Brunswick, *Maine, U.S.A.* 71 D6 43 55N 69 58W
Brunswick, *Md., U.S.A.* 76 F7 39 19N 77 38W
Brunswick, *Mo., U.S.A.* 80 F8 39 26N 93 8W
Brunswick, *Ohio, U.S.A.* 78 E3 41 14N 81 51W
Brunswick, Pen. de, *Chile* 96 G2 53 30S 71 30W
Brunswick B., *Australia* 60 C3 15 15S 124 50 E
Brunswick Junction, *Australia* 61 F2 33 15S 115 50 E
Bruny I., *Australia* 62 G4 43 20S 147 15 E
Brus Laguna, *Honduras* 88 C3 15 47N 84 35W
Brush, *U.S.A.* 80 E3 40 15N 103 37W
Brushton, *U.S.A.* 79 B10 44 50N 74 31W
Brusque, *Brazil* 95 B6 27 5S 49 0W
Brussel, *Belgium* 15 D4 50 51N 4 21 E
Brussels = Brussel, *Belgium* 15 D4 50 51N 4 21 E
Brussels, *Canada* 78 C3 43 44N 81 15W
Bruthen, *Australia* 63 F4 37 42S 147 50 E
Bruxelles = Brussel, *Belgium* 15 D4 50 51N 4 21 E
Bryan, *Ohio, U.S.A.* 76 E3 41 28N 84 33W
Bryan, *Tex., U.S.A.* 81 K6 30 40N 96 22W
Bryan, Mt., *Australia* 63 E2 33 30S 139 0 E
Bryansk, *Russia* 24 D5 53 13N 34 25 E
Bryant, *U.S.A.* 80 C6 44 35N 97 28W
Bryne, *Norway* 9 G8 58 44N 5 38 E
Bryson City, *U.S.A.* 77 H4 35 26N 83 27W
Bsharri, *Lebanon* 47 A5 34 15N 36 0 E
Bū Baqarah, *U.A.E.* 45 E8 25 35N 56 25 E
Bu Craa, *W. Sahara* 50 C2 26 45N 12 50W
Bū Ḩasā, *U.A.E.* 45 F7 23 30N 53 20 E
Bua Yai, *Thailand* 38 E4 15 33N 102 26 E
Buapinang, *Indonesia* 37 E6 4 40S 121 30 E
Buayan, *Phil.* 37 C7 6 3N 125 6 E
Bubanza, *Burundi* 54 C2 3 6S 29 23 E
Būbiyān, *Kuwait* 45 D6 29 45N 48 15 E
Bucaramanga, *Colombia* 92 B4 7 0N 73 0W
Buccaneer Arch., *Australia* 60 C3 16 7S 123 20 E
Buchan, *U.K.* 12 D6 57 32N 2 8W
Buchan Ness, *U.K.* 12 D7 57 29N 1 48W
Buchanan, *Canada* 73 C8 51 40N 102 45W
Buchanan, *Liberia* 50 G2 5 57N 10 2W
Buchanan, L., *Queens., Australia* 62 C4 21 35S 145 52 E
Buchanan, L., *W. Austral., Australia* 61 E3 25 33S 123 2 E
Buchanan, L., *U.S.A.* 81 K5 30 45N 98 25W
Buchanan Cr. →, *Australia* 62 B2 19 13S 136 33 E
Buchans, *Canada* 71 C8 48 50N 56 52W
Bucharest = Bucureşti, *Romania* 21 B12 44 27N 26 10 E
Buchon, Pt., *U.S.A.* 84 K6 35 15N 120 54W
Buckeye, *U.S.A.* 83 K7 33 22N 112 35W
Buckhannon, *U.S.A.* 76 F5 39 0N 80 8W
Buckhaven, *U.K.* 12 E5 56 10N 3 2W
Buckie, *U.K.* 12 D6 57 40N 2 58W
Buckingham, *Canada* 70 C4 45 37N 75 24W
Buckingham, *U.K.* 11 F7 52 0N 0 59W
Buckingham B., *Australia* 62 A2 12 10S 135 40 E
Buckinghamshire □, *U.K.* 11 F7 51 50N 0 55W
Buckle Hd., *Australia* 60 B4 14 26S 127 52 E
Buckleboo, *Australia* 63 E2 32 54S 136 12 E
Buckley, *U.S.A.* 82 C2 47 10N 122 2W
Buckley →, *Australia* 62 C2 20 10S 138 49 E
Bucklin, *U.S.A.* 81 G5 37 33N 99 38W
Bucks L., *U.S.A.* 84 F5 39 54N 121 12W
Buctouche, *Canada* 71 C7 46 30N 64 45W
Bucureşti, *Romania* 21 B12 44 27N 26 10 E
Bucyrus, *U.S.A.* 76 E4 40 48N 82 59W
Budalin, *Burma* 41 H19 22 20N 95 10 E
Budapest, *Hungary* 17 E9 47 29N 19 5 E
Budaun, *India* 43 E8 28 5N 79 10 E
Budd Coast, *Antarctica* 5 C8 68 0S 112 0 E
Bude, *U.K.* 11 G3 50 49N 4 33W
Budennovsk, *Russia* 25 F7 44 50N 44 10 E
Budge Budge = Baj Baj, *India* 43 H13 22 30N 88 5 E
Budgewoi, *Australia* 63 E5 33 13S 151 34 E
Búðareyri, *Iceland* 8 D6 65 2N 14 13W
Búðir, *Iceland* 8 D2 64 49N 23 23W
Budjala, *Zaïre* 52 D3 2 50N 19 40 E
Buellton, *U.S.A.* 85 L6 34 37N 120 12W
Buena Vista, *Colo., U.S.A.* 83 G10 38 51N 106 8W
Buena Vista, *Va., U.S.A.* 76 G6 37 44N 79 21W
Buena Vista L., *U.S.A.* 85 K7 35 12N 119 18W
Buenaventura, *Colombia* 92 C3 3 53N 77 4W
Buenaventura, *Mexico* 86 B3 29 50N 107 30W
Buenos Aires, *Argentina* 94 C4 34 30S 58 20W
Buenos Aires, *Costa Rica* 88 E3 9 10N 83 20W
Buenos Aires □, *Argentina* 94 D4 36 30S 60 0W
Buenos Aires, L., *Chile* 96 F2 46 35S 72 30W
Buffalo, *Mo., U.S.A.* 81 G8 37 39N 93 6W
Buffalo, *N.Y., U.S.A.* 78 D6 42 53N 78 53W
Buffalo, *Okla., U.S.A.* 81 G5 36 50N 99 38W
Buffalo, *S. Dak., U.S.A.* 80 C3 45 35N 103 33W
Buffalo, *Wyo., U.S.A.* 82 D10 44 21N 106 42W
Buffalo →, *Canada* 72 A5 60 5N 115 5W
Buffalo Head Hills, *Canada* 72 B5 57 25N 115 55W
Buffalo L., *Canada* 72 C6 52 27N 112 54W
Buffalo Narrows, *Canada* 73 B7 55 51N 108 29W
Buffels →, *S. Africa* 56 D2 29 36S 17 3 E
Buford, *U.S.A.* 77 H4 34 10N 84 0W
Bug →, *Poland* 17 B10 52 31N 21 5 E
Bug →, *Ukraine* 25 E5 46 59N 31 58 E
Buga, *Colombia* 92 C3 4 0N 76 15W
Buganda, *Uganda* 54 C3 0 0 31 30 E
Buganga, *Uganda* 54 C3 0 3S 32 0 E
Bugel, Tanjung, *Indonesia* 36 F4 6 26S 111 3 E
Bugibba, *Malta* 23 D1 35 57N 14 25 E
Bugsuk, *Phil.* 36 C5 8 15N 117 15 E
Bugulma, *Russia* 24 D9 54 33N 52 48 E
Bugun Shara, *Mongolia* 32 B5 49 0N 104 0 E
Bugun Shara, *Mongolia* 32 B5 49 0N 104 0 E
Buguruslan, *Russia* 24 D9 53 39N 52 26 E
Buheirat-Murrat-el-Kubra, *Egypt* 51 B11 30 15N 32 40 E
Buhl, *Idaho, U.S.A.* 82 E6 42 36N 114 46W
Buhl, *Minn., U.S.A.* 80 B8 47 30N 92 46W
Buick, *U.S.A.* 81 G9 37 38N 91 2W
Builth Wells, *U.K.* 11 E4 52 10N 3 26W
Buir Nur, *Mongolia* 33 B6 47 50N 117 42 E
Bujumbura, *Burundi* 54 C2 3 16S 29 18 E
Bukachacha, *Russia* 27 D12 52 55N 116 50 E
Bukama, *Zaïre* 55 D2 9 10S 25 50 E
Bukavu, *Zaïre* 54 C2 2 20S 28 52 E
Bukene, *Tanzania* 54 C3 4 15S 32 48 E
Bukhara, *Uzbekistan* 26 F7 39 48N 64 25 E
Bukhoro = Bukhara, *Uzbekistan* 26 F7 39 48N 64 25 E
Bukima, *Tanzania* 54 C3 1 50S 33 25 E
Bukit Mertajam, *Malaysia* 39 K3 5 22N 100 28 E
Bukittinggi, *Indonesia* 36 E2 0 20S 100 20 E
Bukoba, *Tanzania* 54 C3 1 20S 31 49 E
Bukoba □, *Tanzania* 54 C3 1 30S 32 0 E
Bukuya, *Uganda* 54 B3 0 40N 31 52 E
Bula, *Indonesia* 37 E8 3 6S 130 30 E
Bulahdelah, *Australia* 63 E5 32 23S 152 13 E
Bulan, *Phil.* 37 B6 12 40N 123 52 E
Bulandshahr, *India* 42 E7 28 28N 77 51 E
Bulawayo, *Zimbabwe* 55 G2 20 7S 28 32 E
Bulgaria ■, *Europe* 21 C11 42 35N 25 30 E
Bulgroo, *Australia* 63 D3 25 47S 143 58 E
Bulgunnia, *Australia* 63 E1 30 10S 134 53 E
Bulhar, *Somali Rep.* 46 E3 10 25N 44 30 E
Buli, Teluk, *Indonesia* 37 D7 1 5N 128 25 E
Buliluyan, C., *Phil.* 36 C5 8 20N 117 15 E
Bulkley →, *Canada* 72 B3 55 15N 127 40W
Bull Shoals L., *U.S.A.* 81 G8 36 22N 92 35W
Bullara, *Australia* 60 D1 22 40S 114 3 E
Bullaring, *Australia* 61 F2 32 30S 117 45 E
Bulli, *Australia* 63 E5 34 15S 150 57 E
Bullock Creek, *Australia* 62 B3 17 43S 144 31 E
Bulloo →, *Australia* 63 D3 28 43S 142 30 E
Bulloo Downs, *Queens., Australia* 63 D3 28 31S 142 57 E
Bulloo Downs, *W. Austral., Australia* 60 D2 24 0S 119 32 E
Bulloo L., *Australia* 63 D3 28 43S 142 25 E
Bulls, *N.Z.* 59 J5 40 10S 175 24 E
Bulnes, *Chile* 94 D1 36 42S 72 19W
Bulo Burti, *Somali Rep.* 46 G4 3 50N 45 33 E
Bulsar = Valsad, *India* 40 J8 20 40N 72 58 E
Bultfontein, *S. Africa* 56 D4 28 18S 26 10 E
Bulu Karakelong, *Indonesia* 37 D7 4 35N 126 50 E
Bulukumba, *Indonesia* 37 F6 5 33S 120 11 E
Bulun, *Russia* 27 B13 70 37N 127 30 E
Bumba, *Zaïre* 52 D4 2 13N 22 30 E
Bumbiri I., *Tanzania* 54 C3 1 40S 31 55 E
Bumhpa Bum, *Burma* 41 F20 26 51N 97 14 E
Bumi →, *Zimbabwe* 55 F2 17 0S 28 20 E
Buna, *Kenya* 54 B4 2 58N 39 30 E
Bunazi, *Tanzania* 54 C3 1 3S 31 23 E
Bunbah, Khalīj, *Libya* 51 B9 32 20N 23 15 E
Bunbury, *Australia* 61 F2 33 20S 115 35 E
Buncrana, *Ireland* 13 A4 55 8N 7 28W
Bundaberg, *Australia* 63 C5 24 54S 152 22 E
Bundey →, *Australia* 62 C2 21 46S 135 37 E
Bundi, *India* 42 G6 25 30N 75 35 E
Bundooma, *Australia* 62 C1 24 54S 134 16 E
Bundoran, *Ireland* 13 B3 54 24N 8 17W
Bung Kan, *Thailand* 38 C4 18 23N 103 37 E
Bungatakada, *Japan* 31 H5 33 35N 131 25 E
Bungil Cr. →, *Australia* 63 D4 27 5S 149 5 E
Bungo-Suidō, *Japan* 31 H6 33 0N 132 15 E
Bungoma, *Kenya* 54 B3 0 34N 34 34 E
Bungu, *Tanzania* 54 D4 7 35S 39 0 E
Bunia, *Zaïre* 54 B3 1 35N 30 20 E
Bunji, *Pakistan* 43 B6 35 45N 74 40 E
Bunkie, *U.S.A.* 81 K8 30 57N 92 11W
Bunnell, *U.S.A.* 77 L5 29 28N 81 16W
Buntok, *Indonesia* 36 E4 1 40S 114 58 E
Bunyu, *Indonesia* 36 D5 3 35N 117 50 E
Buol, *Indonesia* 37 D6 1 15N 121 32 E
Buon Brieng, *Vietnam* 38 F7 13 9N 108 12 E
Buon Me Thuot, *Vietnam* 38 F7 12 40N 108 3 E
Buong Long, *Cambodia* 38 F6 13 44N 106 59 E
Buorkhaya, Mys, *Russia* 27 B14 71 50N 132 40 E
Buqayq, *Si. Arabia* 45 E6 26 0N 49 45 E
Bur Acaba, *Somali Rep.* 46 G3 3 12N 44 20 E
Bûr Safâga, *Egypt* 51 C11 26 43N 33 57 E
Bûr Sa'îd, *Egypt* 51 B11 31 16N 32 18 E
Bûr Sûdân, *Sudan* 51 E12 19 32N 37 9 E
Bura, *Kenya* 54 C4 1 4S 39 58 E
Burao, *Somali Rep.* 46 F4 9 32N 45 32 E
Burāq, *Syria* 47 B5 33 11N 36 29 E
Buras, *U.S.A.* 81 L10 29 22N 89 32W
Buraydah, *Si. Arabia* 44 E5 26 20N 44 8 E
Burbank, *U.S.A.* 85 L8 34 11N 118 19W
Burcher, *Australia* 63 E4 33 30S 147 16 E
Burdekin →, *Australia* 62 B4 19 38S 147 25 E
Burdett, *Canada* 72 D6 49 50N 111 32W
Burdur, *Turkey* 25 G5 37 45N 30 17 E
Burdwan = Barddhaman, *India* 43 H12 23 14N 87 39 E
Bure →, *U.K.* 10 E9 52 38N 1 45 E
Bureya →, *Russia* 27 E13 49 27N 129 30 E
Burford, *Canada* 78 C4 43 7N 80 27W
Burgas, *Bulgaria* 21 C12 42 33N 27 29 E
Burgenland □, *Austria* 16 E8 47 20N 16 20 E
Burgeo, *Canada* 71 C8 47 37N 57 38W
Burgersdorp, *S. Africa* 56 E4 31 0S 26 20 E
Burges, Mt., *Australia* 61 F3 30 50S 121 5 E
Burgos, *Spain* 19 A4 42 21N 3 41W
Burgsvik, *Sweden* 9 H15 57 3N 18 19 E
Burgundy = Bourgogne, *France* 18 C6 47 0N 4 50 E
Burhanpur, *India* 40 J10 21 18N 76 14 E
Burias, *Phil.* 37 B6 12 55N 123 5 E
Burica, Pta., *Costa Rica* 88 E3 8 3N 82 51W
Burigi, L., *Tanzania* 54 C3 2 2S 31 22 E
Burin, *Canada* 71 C8 47 1N 55 14W
Buriram, *Thailand* 38 E4 15 0N 103 0 E
Burj Sāfitā, *Syria* 44 C3 34 48N 36 7 E
Burji, *Ethiopia* 51 G12 5 29N 37 51 E
Burkburnett, *U.S.A.* 81 H5 34 6N 98 34W
Burke, *U.S.A.* 82 C6 47 31N 115 49W
Burke →, *Australia* 62 C2 23 12S 139 33 E
Burketown, *Australia* 62 B2 17 45S 139 33 E
Burkina Faso ■, *Africa* 50 F4 12 0N 1 0W
Burk's Falls, *Canada* 70 C4 45 37N 79 24W
Burley, *U.S.A.* 82 E7 42 32N 113 48W
Burlingame, *U.S.A.* 84 H4 37 35N 122 21W
Burlington, *Canada* 78 C5 43 18N 79 45W
Burlington, *Colo., U.S.A.* 80 F3 39 18N 102 16W
Burlington, *Iowa, U.S.A.* 80 E9 40 49N 91 14W
Burlington, *Kans., U.S.A.* 80 F7 38 12N 95 45W
Burlington, *N.C., U.S.A.* 77 G6 36 6N 79 26W
Burlington, *N.J., U.S.A.* 79 F10 40 4N 74 51W
Burlington, *Vt., U.S.A.* 79 B11 44 29N 73 12W
Burlington, *Wash., U.S.A.* 84 B4 48 28N 122 20W
Burlington, *Wis., U.S.A.* 76 D1 42 41N 88 17W
Burlyu-Tyube, *Kazakhstan* 26 E8 46 30N 79 10 E
Burma ■, *Asia* 41 J20 21 0N 96 30 E
Burnaby I., *Canada* 72 C2 52 25N 131 19W
Burnet, *U.S.A.* 81 K5 30 45N 98 14W
Burney, *U.S.A.* 82 F3 40 53N 121 40W
Burngup, *Australia* 61 F2 33 2S 118 42 E
Burnham, *U.S.A.* 78 F7 40 38N 77 34W
Burnie, *Australia* 62 G4 41 4S 145 56 E
Burnley, *U.K.* 10 D5 53 47N 2 15W
Burns, *Oreg., U.S.A.* 82 E4 43 35N 119 3W
Burns, *Wyo., U.S.A.* 80 E2 41 12N 104 21W
Burns Lake, *Canada* 72 C3 54 20N 125 45W
Burnside →, *Canada* 68 B9 66 51N 108 4W
Burnside, L., *Australia* 61 E3 25 22S 123 0 E
Burnt River, *Canada* 78 B6 44 41N 78 42W
Burntwood →, *Canada* 73 B9 56 8N 96 34W
Burntwood L., *Canada* 73 B8 55 22N 100 26W
Burqān, *Kuwait* 44 D5 29 0N 47 57 E
Burra, *Australia* 63 E2 33 40S 138 55 E
Burramurra, *Australia* 62 C2 20 25S 137 15 E
Burren Junction, *Australia* 63 E4 30 7S 148 59 E
Burrendong Dam, *Australia* 63 E4 32 39S 149 6 E
Burrinjuck Res., *Australia* 63 F4 35 0S 148 36 E
Burro, Serranías del, *Mexico* 86 B4 29 0N 102 0W
Burruyacú, *Argentina* 94 B3 26 30S 64 40W
Burry Port, *U.K.* 11 F3 51 41N 4 17W
Bursa, *Turkey* 25 F4 40 15N 29 5 E
Burstall, *Canada* 73 C7 50 39N 109 54W
Burton L., *Canada* 70 B4 54 45N 78 20W
Burton upon Trent, *U.K.* 10 E6 52 48N 1 39W
Burtundy, *Australia* 63 E3 33 45S 142 15 E
Buru, *Indonesia* 37 E7 3 30S 126 30 E
Burûn, Râs, *Egypt* 47 D2 31 14N 33 7 E
Burundi ■, *Africa* 54 C3 3 15S 30 0 E
Bururi, *Burundi* 54 C2 3 57S 29 37 E
Burutu, *Nigeria* 50 G6 5 20N 5 29 E
Burwell, *U.S.A.* 80 E5 41 47N 99 8W
Bury, *U.K.* 10 D5 53 36N 2 19W
Bury St. Edmunds, *U.K.* 11 E8 52 15N 0 42 E
Buryat Republic □, *Russia* 27 D11 53 0N 110 0 E
Busango Swamp, *Zambia* 55 E2 14 15S 25 45 E
Buşayrah, *Syria* 44 C4 35 9N 40 26 E
Buşayyah, *Iraq* 44 D5 30 0N 46 10 E
Būshehr, *Iran* 45 D6 28 55N 50 55 E
Būshehr □, *Iran* 45 D6 28 20N 51 45 E
Bushell, *Canada* 73 B7 59 31N 108 45W
Bushenyi, *Uganda* 54 C3 0 35S 30 10 E
Bushire = Būshehr, *Iran* 45 D6 28 55N 50 55 E
Bushnell, *Ill., U.S.A.* 80 E9 40 33N 90 31W
Bushnell, *Nebr., U.S.A.* 80 E3 41 14N 103 54W
Busia □, *Kenya* 54 B3 0 25N 34 6 E
Businga, *Zaïre* 52 D4 3 16N 20 59 E
Buskerud fylke □, *Norway* 9 F10 60 13N 9 0 E
Busoga □, *Uganda* 54 B3 0 5N 33 30 E
Busra ash Shām, *Syria* 47 C5 32 30N 36 25 E
Busselton, *Australia* 61 F2 33 42S 115 15 E
Bussum, *Neths.* 15 B5 52 16N 5 10 E
Busto Arsizio, *Italy* 20 B3 45 40N 8 50 E
Busu-Djanoa, *Zaïre* 52 D4 1 43N 21 23 E
Busuanga, *Phil.* 37 B5 12 10N 120 0 E
Buta, *Zaïre* 54 B1 2 50N 24 53 E
Butare, *Rwanda* 54 C2 2 31S 29 52 E
Butaritari, *Kiribati* 64 G9 3 30N 174 0 E
Bute, *U.K.* 12 F3 55 48N 5 2W
Bute Inlet, *Canada* 72 C4 50 40N 124 53W
Butemba, *Uganda* 54 B3 1 9N 31 37 E
Butembo, *Zaïre* 54 B2 0 9N 29 18 E
Butha Qi, *China* 33 B7 48 0N 122 32 E
Butiaba, *Uganda* 54 B3 1 50N 31 20 E
Butler, *Mo., U.S.A.* 80 F7 38 16N 94 20W
Butler, *Pa., U.S.A.* 78 F5 40 52N 79 54W
Butte, *Mont., U.S.A.* 82 C7 46 0N 112 32W

Canton L.

Butte, *Nebr., U.S.A.* **80 D5** 42 58N 98 51W
Butte Creek →, *U.S.A.* **84 F5** 39 12N 121 56W
Butterworth = Gcuwa, *S. Africa* **57 E4** 32 20S 28 11 E
Butterworth, *Malaysia* **39 K3** 5 24N 100 23 E
Buttfield, Mt., *Australia* **61 D4** 24 45S 128 9 E
Button B., *Canada* **73 B10** 58 45N 94 23W
Buttonwillow, *U.S.A.* **85 K7** 35 24N 119 28W
Butty Hd., *Australia* **61 F3** 33 54S 121 39 E
Butuan, *Phil.* **37 C7** 8 57N 125 33 E
Butung, *Indonesia* **37 E6** 5 0S 122 45 E
Buturlinovka, *Russia* **25 D7** 50 50N 40 35 E
Buxar, *India* **43 G10** 25 34N 83 58 E
Buxton, *S. Africa* **56 D3** 27 38S 24 42 E
Buxton, *U.K.* **10 D6** 53 16N 1 54W
Buy, *Russia* **24 C7** 58 28N 41 28 E
Buyaga, *Russia* **27 D13** 59 50N 127 0 E
Buzău, *Romania* **21 B12** 45 10N 26 50 E
Buzău →, *Romania* **17 F13** 45 26N 27 44 E
Buzen, *Japan* **31 H5** 33 35N 131 5 E
Buzi →, *Mozam.* **55 F3** 19 50S 34 43 E
Buzuluk, *Russia* **24 D9** 52 48N 52 12 E
Buzzards Bay, *U.S.A.* **79 E14** 41 45N 70 37W
Bwana Mkubwe, *Zaïre* **55 E2** 13 8S 28 38 E
Bydgoszcz, *Poland* **17 B8** 53 10N 18 0 E
Byelarus = Belorussia ■, *Europe* **24 D4** 53 30N 27 0 E
Byelorussia = Belorussia ■, *Europe* **24 D4** 53 30N 27 0 E
Byers, *U.S.A.* **80 F2** 39 43N 104 14W
Byesville, *U.S.A.* **78 G3** 39 58N 81 32W
Byhalia, *U.S.A.* **81 H10** 34 52N 89 41W
Bylas, *U.S.A.* **83 K8** 33 8N 110 7W
Bylot I., *Canada* **69 A12** 73 13N 78 34W
Byrd, C., *Antarctica* **5 C17** 69 38S 76 7W
Byro, *Australia* **61 E2** 26 5S 116 11 E
Byrock, *Australia* **63 E4** 30 40S 146 27 E
Byron Bay, *Australia* **63 D5** 28 43S 153 37 E
Byrranga, Gory, *Russia* **27 B11** 75 0N 100 0 E
Byrranga Mts. = Byrranga, Gory, *Russia* **27 B11** 75 0N 100 0 E
Byske, *Sweden* **8 D16** 64 57N 21 11 E
Byske älv →, *Sweden* **8 D16** 64 57N 21 13 E
Bytom, *Poland* **17 C9** 50 25N 18 54 E
Byumba, *Rwanda* **54 C3** 1 35S 30 4 E

C

C.I.S. = Commonwealth of Independent States ■, *Eurasia* **27 D11** 60 0N 100 0 E
Ca →, *Vietnam* **38 C5** 18 45N 105 45 E
Ca Mau = Quan Long, *Vietnam* **39 H5** 9 7N 105 8 E
Ca Mau, Mui = Bai Bung, Mui, *Vietnam* **39 H5** 8 38N 104 44 E
Ca Na, *Vietnam* **39 G7** 11 20N 108 54 E
Caacupé, *Paraguay* **94 B4** 25 23S 57 5W
Caála, *Angola* **53 G3** 12 46S 15 30 E
Caamano Sd., *Canada* **72 C3** 52 55N 129 25W
Caazapá, *Paraguay* **94 B4** 26 8S 56 19W
Caazapá □, *Paraguay* **95 B4** 26 10S 56 0W
Caballeria, C. de, *Spain* **22 A11** 40 5N 4 5 E
Cabanatuan, *Phil.* **37 A6** 15 30N 120 58 E
Cabano, *Canada* **71 C6** 47 40N 68 56W
Cabazon, *U.S.A.* **85 M10** 33 55N 116 47W
Cabedelo, *Brazil* **93 E12** 7 0S 34 50W
Cabildo, *Chile* **94 C1** 32 30S 71 5W
Cabimas, *Venezuela* **92 A4** 10 23N 71 25W
Cabinda, *Angola* **52 F2** 5 33S 12 11 E
Cabinda □, *Angola* **52 F2** 5 0S 12 30 E
Cabinet Mts., *U.S.A.* **82 C6** 48 0N 115 30W
Cabo Blanco, *Argentina* **96 F3** 47 15S 65 47W
Cabo Frio, *Brazil* **95 A7** 22 51S 42 3W
Cabo Pantoja, *Peru* **92 D3** 1 0S 75 10W
Cabonga, Réservoir, *Canada* **70 C4** 47 20N 76 40W
Cabool, *U.S.A.* **81 G8** 37 7N 92 6W
Caboolture, *Australia* **63 D5** 27 5S 152 58 E
Cabora Bassa Dam = Cahora Bassa Dam, *Mozam.* **55 F3** 15 20S 32 50 E
Caborca, *Mexico* **86 A2** 30 40N 112 10W
Cabot, Mt., *U.S.A.* **79 B13** 44 30N 71 25W
Cabot Str., *Canada* **71 C8** 47 15N 59 40W
Cabrera, I., *Spain* **22 B9** 39 8N 2 57 E
Cabri, *Canada* **73 C7** 50 35N 108 25W
Cabriel →, *Spain* **19 C5** 39 14N 1 3W
Čačak, *Serbia, Yug.* **21 C9** 43 54N 20 20 E
Cáceres, *Brazil* **92 G7** 16 5S 57 40W
Cáceres, *Spain* **19 C2** 39 26N 6 23W
Cache Bay, *Canada* **70 C4** 46 22N 80 0W
Cache Cr. →, *U.S.A.* **84 G5** 38 42N 121 42W
Cachi, *Argentina* **94 B2** 25 5S 66 10W
Cachimbo, Serra do, *Brazil* **93 E7** 9 30S 55 30W
Cachoeira, *Brazil* **93 F11** 12 30S 39 0W
Cachoeira de Itapemirim, *Brazil* **95 A7** 20 51S 41 7W
Cachoeira do Sul, *Brazil* **95 C5** 30 3S 52 53W
Cacólo, *Angola* **52 G3** 10 9S 19 21 E
Caconda, *Angola* **53 G3** 13 48S 15 8 E
Cacongo, *Angola* **52 F2** 5 11S 12 5 E
Caddo, *U.S.A.* **81 H6** 34 7N 96 16W
Cadell Cr. →, *Australia* **62 C3** 22 35S 141 51 E
Cader Idris, *U.K.* **10 E4** 52 43N 3 56W
Cadibarrawirracanna, L., *Australia* **63 D2** 28 52S 135 27 E
Cadillac, *Canada* **70 C4** 48 14N 78 23W
Cadillac, *U.S.A.* **76 C3** 44 15N 85 24W
Cadiz, *Phil.* **37 B6** 10 57N 123 15 E
Cádiz, *Spain* **19 D2** 36 30N 6 20W
Cadiz, *U.S.A.* **78 F4** 40 22N 81 0W
Cádiz, G. de, *Spain* **19 D2** 36 40N 7 0W
Cadney Park, *Australia* **63 D1** 27 55S 134 3 E
Cadomin, *Canada* **72 C5** 53 2N 117 20W
Cadotte →, *Canada* **72 B5** 56 43N 117 10W
Cadoux, *Australia* **61 F2** 30 46S 117 7 E
Caen, *France* **18 B3** 49 10N 0 22W
Caernarfon, *U.K.* **10 D3** 53 8N 4 17W
Caernarfon B., *U.K.* **10 D3** 53 4N 4 40W
Caernarvon = Caernarfon, *U.K.* **10 D3** 53 8N 4 17W
Caerphilly, *U.K.* **11 F4** 51 34N 3 13W
Caesarea, *Israel* **47 C3** 32 30N 34 53 E
Caeté, *Brazil* **93 G10** 19 55S 43 40W
Caetité, *Brazil* **93 F10** 13 50S 42 32W
Cafayate, *Argentina* **94 B2** 26 2S 66 0W
Cafu, *Angola* **56 B2** 16 30S 15 8 E
Cagayan →, *Phil.* **37 A6** 18 25N 121 42 E
Cagayan de Oro, *Phil.* **37 C6** 8 30N 124 40 E
Cágliari, *Italy* **20 E3** 39 15N 9 6 E
Cágliari, G. di, *Italy* **20 E3** 39 8N 9 10 E
Caguas, *Puerto Rico* **89 C6** 18 14N 66 2W
Caha Mts., *Ireland* **13 E2** 51 45N 9 40W
Cahama, *Angola* **56 B1** 16 17S 14 19 E
Caher, *Ireland* **13 D4** 52 23N 7 56W
Cahersiveen, *Ireland* **13 E1** 51 57N 10 13W
Cahora Bassa Dam, *Mozam.* **55 F3** 15 20S 32 50 E
Cahore Pt., *Ireland* **13 D5** 52 34N 6 11W
Cahors, *France* **18 D4** 44 27N 1 27 E
Cahuapanas, *Peru* **92 E3** 5 15S 77 0W
Cai Bau, Dao, *Vietnam* **38 B6** 21 10N 107 27 E
Cai Nuoc, *Vietnam* **39 H5** 8 56N 105 1 E
Caia, *Mozam.* **55 F4** 17 51S 35 24 E
Caianda, *Angola* **55 E1** 11 2S 23 31 E
Caibarién, *Cuba* **88 B4** 22 30N 79 30W
Caicara, *Venezuela* **92 B5** 7 38N 66 10W
Caicó, *Brazil* **93 E11** 6 20S 37 0W
Caicos Is., *W. Indies* **89 B5** 21 40N 71 40W
Caicos Passage, *W. Indies* **89 B5** 22 45N 72 45W
Caird Coast, *Antarctica* **5 D1** 75 0S 25 0W
Cairn Gorm, *U.K.* **12 D5** 57 7N 3 40W
Cairn Toul, *U.K.* **12 D5** 57 3N 3 44W
Cairngorm Mts., *U.K.* **12 D5** 57 6N 3 42W
Cairns, *Australia* **62 B4** 16 57S 145 45 E
Cairo = El Qâhira, *Egypt* **51 B11** 30 1N 31 14 E
Cairo, *Ga., U.S.A.* **77 K3** 30 52N 84 13W
Cairo, *Ill., U.S.A.* **81 G10** 37 0N 89 11W
Caithness, Ord of, *U.K.* **12 C5** 58 9N 3 37W
Caiundo, *Angola* **53 H3** 15 50S 17 28 E
Caiza, *Bolivia* **92 H5** 20 2S 65 40W
Cajamarca, *Peru* **92 E3** 7 5S 78 28W
Cajàzeiras, *Brazil* **93 E11** 6 52S 38 30W
Cala d'Or, *Spain* **22 B10** 39 23N 3 14 E
Cala Figuera, C., *Spain* **22 B9** 39 27N 2 31 E
Cala Forcat, *Spain* **22 A10** 40 0N 3 47 E
Cala Mayor, *Spain* **22 B9** 39 33N 2 37 E
Cala Mezquida, *Spain* **22 B11** 39 55N 4 16 E
Cala Millor, *Spain* **22 B10** 39 35N 3 22 E
Cala Ratjada, *Spain* **22 B10** 39 43N 3 27 E
Calabar, *Nigeria* **50 H6** 4 57N 8 20 E
Calábria □, *Italy* **20 E7** 39 24N 16 30 E
Calafate, *Argentina* **96 G2** 50 19S 72 15W
Calahorra, *Spain* **19 A5** 42 18N 1 59W
Calais, *France* **18 A4** 50 57N 1 56 E
Calais, *U.S.A.* **71 C6** 45 11N 67 17W
Calalaste, Cord. de, *Argentina* **94 B2** 25 0S 67 0W
Calama, *Brazil* **92 E6** 8 0S 62 50W
Calama, *Chile* **94 A2** 22 30S 68 55W
Calamar, *Bolívar, Colombia* **92 A4** 10 15N 74 55W
Calamar, *Vaupés, Colombia* **92 C4** 1 58N 72 32W
Calamian Group, *Phil.* **37 B5** 11 50N 119 55 E
Calamocha, *Spain* **19 B5** 40 50N 1 17W
Calán Porter, *Spain* **22 B11** 39 52N 4 8 E
Calang, *Indonesia* **36 D1** 4 37N 95 37 E
Calapan, *Phil.* **37 B6** 13 25N 121 7 E
Calatayud, *Spain* **19 B5** 41 20N 1 40W
Calauag, *Phil.* **37 B6** 13 55N 122 15 E
Calavite, C., *Phil.* **37 B6** 13 26N 120 20 E
Calbayog, *Phil.* **37 B6** 12 4N 124 38 E
Calca, *Peru* **92 F4** 13 22S 72 0W
Calcasieu L., *U.S.A.* **81 L8** 29 55N 93 18W
Calcutta, *India* **43 H13** 22 36N 88 24 E
Calder →, *U.K.* **10 D6** 53 44N 1 21W
Caldera, *Chile* **94 B1** 27 5S 70 55W
Caldwell, *Idaho, U.S.A.* **82 E5** 43 40N 116 41W
Caldwell, *Kans., U.S.A.* **81 G6** 37 2N 97 37W
Caldwell, *Tex., U.S.A.* **81 K6** 30 32N 96 42W
Caledon, *S. Africa* **56 E2** 34 14S 19 26 E
Caledon →, *S. Africa* **56 E4** 30 31S 26 5 E
Caledon B., *Australia* **62 A2** 12 45S 137 0 E
Caledonia, *Canada* **78 C5** 43 7N 79 58W
Caledonia, *U.S.A.* **78 D7** 42 58N 77 51W
Calella, *Spain* **19 B7** 41 37N 2 40 E
Calemba, *Angola* **56 B2** 16 0S 15 44 E
Calexico, *U.S.A.* **85 N11** 32 40N 115 30W
Calf of Man, *U.K.* **10 C3** 54 3N 4 49W
Calgary, *Canada* **72 C6** 51 0N 114 10W
Calheta, *Madeira* **22 D2** 32 44N 17 11W
Calhoun, *U.S.A.* **77 H3** 34 30N 84 57W
Cali, *Colombia* **92 C3** 3 25N 76 35W
Calicut, *India* **40 P9** 11 15N 75 43 E
Caliente, *U.S.A.* **83 H6** 37 37N 114 31W
California, *Mo., U.S.A.* **80 F8** 38 38N 92 34W
California, *Pa., U.S.A.* **78 F5** 40 4N 79 54W
California □, *U.S.A.* **83 H4** 37 30N 119 30W
California, Baja, *Mexico* **86 A1** 32 10N 115 12W
California, Baja, T.N. □, *Mexico* **86 B2** 30 0N 115 0W
California, Baja, T.S. □, *Mexico* **86 B2** 25 50N 111 50W
California, G. de, *Mexico* **86 B2** 27 0N 111 0W
California City, *U.S.A.* **85 K9** 35 10N 117 55W
California Hot Springs, *U.S.A.* **85 K8** 35 51N 118 41W
Calingasta, *Argentina* **94 C2** 31 15S 69 30W
Calipatria, *U.S.A.* **85 M11** 33 8N 115 31W
Calistoga, *U.S.A.* **84 G4** 38 35N 122 35W
Calitzdorp, *S. Africa* **56 E3** 33 33S 21 42 E
Callabonna, L., *Australia* **63 D3** 29 40S 140 5 E
Callan, *Ireland* **13 D4** 52 33N 7 25W
Callander, *U.K.* **12 E4** 56 15N 4 14W
Callao, *Peru* **92 F3** 12 0S 77 0W
Callaway, *U.S.A.* **80 E5** 41 18N 99 56W
Calles, *Mexico* **87 C5** 23 2N 98 42W
Callide, *Australia* **62 C5** 24 18S 150 28 E
Calling Lake, *Canada* **72 B6** 55 15N 113 12W
Calliope, *Australia* **62 C5** 24 0S 151 16 E
Calola, *Angola* **56 B2** 16 25S 17 48 E
Caloundra, *Australia* **63 D5** 26 45S 153 10 E
Calpella, *U.S.A.* **84 F3** 39 14N 123 12W
Calpine, *U.S.A.* **84 F6** 39 40N 120 27W
Calstock, *Canada* **70 C3** 49 47N 84 9W
Caltagirone, *Italy* **20 F6** 37 13N 14 30 E
Caltanissetta, *Italy* **20 F6** 37 30N 14 3 E
Calulo, *Angola* **52 G2** 10 1S 14 56 E
Calumet, *U.S.A.* **76 B1** 47 14N 88 27W
Calunda, *Angola* **53 G4** 12 7S 23 36 E
Calvados □, *France* **18 B3** 49 5N 0 15W
Calvert, *U.S.A.* **81 K6** 30 59N 96 40W
Calvert →, *Australia* **62 B2** 16 17S 137 44 E
Calvert Hills, *Australia* **62 B2** 17 15S 137 20 E
Calvert I., *Canada* **72 C3** 51 30N 128 0W
Calvert Ra., *Australia* **60 D3** 24 0S 122 30 E
Calvi, *France* **18 E8** 42 34N 8 45 E
Calvillo, *Mexico* **86 C4** 21 51N 102 43W
Calvinia, *S. Africa* **56 E2** 31 28S 19 45 E
Calwa, *U.S.A.* **84 J7** 36 42N 119 46W
Cam →, *U.K.* **11 E8** 52 21N 0 16 E
Cam Lam, *Vietnam* **39 G7** 11 54N 109 10 E
Cam Pha, *Vietnam* **38 B6** 21 7N 107 18 E
Cam Ranh, *Vietnam* **39 G7** 11 54N 109 12 E
Cam Xuyen, *Vietnam* **38 C6** 18 15N 106 0 E
Camabatela, *Angola* **52 F3** 8 20S 15 26 E
Camacha, *Madeira* **22 D3** 32 41N 16 49W
Camacho, *Mexico* **86 C4** 24 25N 102 18W
Camacupa, *Angola* **53 G3** 11 58S 17 22 E
Camagüey, *Cuba* **88 B4** 21 20N 78 0W
Camaná, *Peru* **92 G4** 16 30S 72 50W
Camanche Reservoir, *U.S.A.* **84 G6** 38 14N 121 1W
Camaquã →, *Brazil* **95 C5** 31 17S 51 47W
Câmara de Lobos, *Madeira* **22 D3** 32 39N 16 59W
Camaret, *France* **18 B1** 48 16N 4 37W
Camargo, *Bolivia* **92 H5** 20 38S 65 15W
Camarillo, *U.S.A.* **85 L7** 34 13N 119 2W
Camarón, C., *Honduras* **88 C2** 16 0N 85 5W
Camarones, *Argentina* **96 E3** 44 50S 65 40W
Camas, *U.S.A.* **84 E4** 45 35N 122 24W
Camas Valley, *U.S.A.* **82 E2** 43 2N 123 40W
Cambará, *Brazil* **95 A5** 23 2S 50 5W
Cambay = Khambhat, *India* **42 H5** 22 23N 72 33 E
Cambay, G. of = Khambat, G. of, *India* **42 J5** 20 45N 72 30 E
Cambodia ■, *Asia* **38 F5** 12 15N 105 0 E
Camborne, *U.K.* **11 G2** 50 13N 5 18W
Cambrai, *France* **18 A5** 50 11N 3 14 E
Cambria, *U.S.A.* **83 J3** 35 34N 121 5W
Cambrian Mts., *U.K.* **11 E4** 52 25N 3 52W
Cambridge, *Canada* **70 D3** 43 23N 80 15W
Cambridge, *Jamaica* **88 C4** 18 18N 77 54W
Cambridge, *N.Z.* **59 G5** 37 54S 175 29 E
Cambridge, *U.K.* **11 E8** 52 13N 0 8 E
Cambridge, *Idaho, U.S.A.* **82 D5** 44 34N 116 41W
Cambridge, *Mass., U.S.A.* **79 D13** 42 22N 71 6W
Cambridge, *Md., U.S.A.* **76 F7** 38 34N 76 5W
Cambridge, *Minn., U.S.A.* **80 C8** 45 34N 93 13W
Cambridge, *N.Y., U.S.A.* **79 C11** 43 2N 73 22W
Cambridge, *Nebr., U.S.A.* **80 E4** 40 17N 100 10W
Cambridge, *Ohio, U.S.A.* **78 F3** 40 2N 81 35W
Cambridge Bay, *Canada* **68 B9** 69 10N 105 0W
Cambridge G., *Australia* **60 B4** 14 55S 128 15 E
Cambridge Springs, *U.S.A.* **78 E4** 41 48N 80 4W
Cambridgeshire □, *U.K.* **11 E8** 52 12N 0 7 E
Cambuci, *Brazil* **95 A7** 21 35S 41 55W
Cambundi-Catembo, *Angola* **52 G3** 10 10S 17 35 E
Camden, *Ala., U.S.A.* **77 K2** 31 59N 87 17W
Camden, *Ark., U.S.A.* **81 J8** 33 35N 92 50W
Camden, *Maine, U.S.A.* **71 D6** 44 13N 69 4W
Camden, *N.J., U.S.A.* **79 G9** 39 56N 75 7W
Camden, *S.C., U.S.A.* **77 H5** 34 16N 80 36W
Camden Sd., *Australia* **60 C3** 15 27S 124 25 E
Camdenton, *U.S.A.* **81 F8** 38 1N 92 45W
Cameron, *Ariz., U.S.A.* **83 J8** 35 53N 111 25W
Cameron, *La., U.S.A.* **81 L8** 29 48N 93 20W
Cameron, *Mo., U.S.A.* **80 F7** 39 44N 94 14W
Cameron, *Tex., U.S.A.* **81 K6** 30 51N 96 59W
Cameron Falls, *Canada* **70 C2** 49 8N 88 19W
Cameron Highlands, *Malaysia* **39 K3** 4 27N 101 22 E
Cameron Hills, *Canada* **72 B5** 59 48N 118 0W
Cameroon ■, *Africa* **51 G7** 6 0N 12 30 E
Cameroun, Mt., *Cameroon* **50 H6** 4 13N 9 10 E
Cametá, *Brazil* **93 D9** 2 12S 49 30W
Caminha, *Portugal* **19 B1** 41 50N 8 50W
Camino, *U.S.A.* **84 G6** 38 44N 120 41W
Camira Creek, *Australia* **63 D5** 29 15S 152 58 E
Camissombo, *Angola* **52 F4** 8 7S 20 38 E
Cammal, *U.S.A.* **78 E7** 41 24N 77 28W
Camocim, *Brazil* **93 D10** 2 55S 40 50W
Camooweal, *Australia* **62 B2** 19 56S 138 7 E
Camopi →, *Fr. Guiana* **93 C8** 3 10N 52 20W
Camp Crook, *U.S.A.* **80 C3** 45 33N 103 59W
Camp Nelson, *U.S.A.* **85 J8** 36 8N 118 39W
Camp Wood, *U.S.A.* **81 L4** 29 40N 100 1W
Campana, *Argentina* **94 C4** 34 10S 58 55W
Campana, I., *Chile* **96 F1** 48 20S 75 20W
Campanário, *Madeira* **22 D2** 32 39N 17 2W
Campania □, *Italy* **20 D6** 40 50N 14 45 E
Campbell, *S. Africa* **56 D3** 28 48S 23 44 E
Campbell, *Calif., U.S.A.* **84 H5** 37 17N 121 57W
Campbell, *Ohio, U.S.A.* **78 E4** 41 5N 80 37W
Campbell I., *Pac. Oc.* **64 N8** 52 30S 169 0 E
Campbell L., *Canada* **73 A7** 63 14N 106 55W
Campbell River, *Canada* **72 C3** 50 5N 125 20W
Campbell Town, *Australia* **62 G4** 41 52S 147 30 E
Campbellford, *Canada* **78 B7** 44 18N 77 48W
Campbellpur, *Pakistan* **42 C5** 33 46N 72 26 E
Campbellsville, *U.S.A.* **76 G3** 37 21N 85 20W
Campbellton, *Canada* **71 C6** 47 57N 66 43W
Campbelltown, *Australia* **63 E5** 34 4S 150 49 E
Campbeltown, *U.K.* **12 F3** 55 25N 5 36W
Campeche, *Mexico* **87 D6** 19 50N 90 32W
Campeche □, *Mexico* **87 D6** 19 50N 90 32W
Campeche, B. de, *Mexico* **87 D6** 19 30N 93 0W
Camperdown, *Australia* **63 F3** 38 14S 143 9 E
Camperville, *Canada* **73 C8** 51 59N 100 9W
Campina Grande, *Brazil* **93 E11** 7 20S 35 47W
Campinas, *Brazil* **95 A6** 22 50S 47 0W
Campo, *Cameroon* **52 D1** 2 22N 9 50 E
Campo Belo, *Brazil* **93 H9** 20 52S 45 16W
Campo Formoso, *Brazil* **93 F10** 10 30S 40 20W
Campo Grande, *Brazil* **93 H8** 20 25S 54 40W
Campo Maíor, *Brazil* **93 D10** 4 50S 42 12W
Campo Mourão, *Brazil* **95 A5** 24 3S 52 22W
Campoalegre, *Colombia* **92 C3** 2 41N 75 20W
Campobasso, *Italy* **20 D6** 41 34N 14 40 E
Campos, *Brazil* **95 A7** 21 50S 41 20W
Campos Belos, *Brazil* **93 F9** 13 10S 47 3W
Campos del Puerto, *Spain* **22 B10** 39 26N 3 1 E
Campos Novos, *Brazil* **95 B5** 27 21S 51 50W
Camptonville, *U.S.A.* **84 F5** 39 27N 121 3W
Campuya →, *Peru* **92 D4** 1 40S 73 30W
Camrose, *Canada* **72 C6** 53 0N 112 50W
Camsell Portage, *Canada* **73 B7** 59 37N 109 15W
Can Clavo, *Spain* **22 C7** 38 57N 1 27 E
Can Creu, *Spain* **22 C7** 38 58N 1 28 E
Can Gio, *Vietnam* **39 G6** 10 25N 106 58 E
Can Tho, *Vietnam* **39 G5** 10 2N 105 46 E
Canaan, *U.S.A.* **79 D11** 42 2N 73 20W
Canada ■, *N. Amer.* **68 C10** 60 0N 100 0W
Cañada de Gómez, *Argentina* **94 C3** 32 40S 61 30W
Canadian, *U.S.A.* **81 H4** 35 55N 100 23W
Canadian →, *U.S.A.* **81 H7** 35 28N 95 3W
Canadian Shield, *Canada* **69 C10** 53 0N 75 0W
Çanakkale, *Turkey* **25 F4** 40 8N 26 24 E
Çanakkale Boğazı, *Turkey* **21 D12** 40 17N 26 32 E
Canal Flats, *Canada* **72 C5** 50 10N 115 48W
Canalejas, *Argentina* **94 D2** 35 15S 66 34W
Canals, *Argentina* **94 C3** 33 35S 62 53W
Canandaigua, *U.S.A.* **78 D7** 42 54N 77 17W
Cananea, *Mexico* **86 A2** 31 0N 110 20W
Canarias, Is., *Atl. Oc.* **22 F4** 28 30N 16 0W
Canarreos, Arch. de los, *Cuba* **88 B3** 21 35N 81 40W
Canary Is. = Canarias, Is., *Atl. Oc.* **22 F4** 28 30N 16 0W
Canatlán, *Mexico* **86 C4** 24 31N 104 47W
Canaveral, C., *U.S.A.* **77 L5** 28 27N 80 32W
Canavieiras, *Brazil* **93 G11** 15 39S 39 0W
Canbelego, *Australia* **63 E4** 31 32S 146 18 E
Canberra, *Australia* **63 F4** 35 15S 149 8 E
Canby, *Calif., U.S.A.* **82 F3** 41 27N 120 52W
Canby, *Minn., U.S.A.* **80 C6** 44 43N 96 16W
Canby, *Oreg., U.S.A.* **84 E4** 45 16N 122 42W
Cancún, *Mexico* **87 C7** 21 8N 86 44W
Candala, *Somali Rep.* **46 E4** 11 30N 49 58 E
Candelaria, *Argentina* **95 B4** 27 29S 55 44W
Candelaria, *Canary Is.* **22 F3** 28 22N 16 22W
Candelo, *Australia* **63 F4** 36 47S 149 43 E
Candia = Iráklion, *Greece* **23 D7** 35 20N 25 12 E
Candle L., *Canada* **73 C7** 53 50N 105 18W
Candlemas I., *Antarctica* **5 B1** 57 3S 26 40W
Cando, *U.S.A.* **80 A5** 48 32N 99 12W
Canea = Khaniá, *Greece* **23 D6** 35 30N 24 4 E
Canelones, *Uruguay* **95 C4** 34 32S 56 17W
Cañete, *Chile* **94 D1** 37 50S 73 30W
Cañete, *Peru* **92 F3** 13 8S 76 30W
Cangas, *Spain* **19 A1** 42 16N 8 47W
Canguaretama, *Brazil* **93 E11** 6 20S 35 5W
Canguçu, *Brazil* **95 C5** 31 22S 52 43W
Cangzhou, *China* **34 E9** 38 19N 116 52 E
Canim Lake, *Canada* **72 C4** 51 47N 120 54W
Canindeyu □, *Paraguay* **95 A4** 24 10S 55 0W
Canipaan, *Phil.* **36 C5** 8 33N 117 15 E
Canisteo, *U.S.A.* **78 D7** 42 16N 77 36W
Canisteo →, *U.S.A.* **78 D7** 42 7N 77 8W
Cañitas, *Mexico* **86 C4** 23 36N 102 43W
Çankırı, *Turkey* **25 F5** 40 40N 33 37 E
Cankuzo, *Burundi* **54 C3** 3 10S 30 31 E
Canmore, *Canada* **72 C5** 51 7N 115 18W
Cann River, *Australia* **63 F4** 37 35S 149 7 E
Canna, *U.K.* **12 D2** 57 3N 6 33W
Cannanore, *India* **40 P9** 11 53N 75 27 E
Cannes, *France* **18 E7** 43 32N 7 1 E
Canning Town = Port Canning, *India* **43 H13** 22 23N 88 40 E
Cannington, *Canada* **78 B5** 44 20N 79 2W
Cannock, *U.K.* **10 E5** 52 42N 2 2W
Cannon Ball →, *U.S.A.* **80 B4** 46 20N 100 38W
Cannondale Mt., *Australia* **62 D4** 25 13S 148 57 E
Canoas, *Brazil* **95 B5** 29 56S 51 11W
Canoe L., *Canada* **73 B7** 55 10N 108 15W
Canon City, *U.S.A.* **80 F2** 38 27N 105 14W
Canora, *Canada* **73 C8** 51 40N 102 30W
Canowindra, *Australia* **63 E4** 33 35S 148 38 E
Canso, *Canada* **71 C7** 45 20N 61 0W
Cantabria □, *Spain* **19 A4** 43 10N 4 0W
Cantabrian Mts. = Cantábrica, Cordillera, *Spain* **19 A3** 43 0N 5 10W
Cantábrica, Cordillera, *Spain* **19 A3** 43 0N 5 10W
Cantal □, *France* **18 D5** 45 5N 2 45 E
Canterbury, *Australia* **62 D3** 25 23S 141 53 E
Canterbury, *U.K.* **11 F9** 51 17N 1 5 E
Canterbury □, *N.Z.* **59 K3** 43 45S 171 19 E
Canterbury Bight, *N.Z.* **59 L3** 44 16S 171 55 E
Canterbury Plains, *N.Z.* **59 K3** 43 55S 171 22 E
Cantil, *U.S.A.* **85 K9** 35 18N 117 58W
Canton = Guangzhou, *Guangdong, China* **33 D6** 23 5N 113 10 E
Canton = Guangzhou, *Guangdong, China* **33 D6** 23 5N 113 10 E
Canton, *Ga., U.S.A.* **77 H3** 34 14N 84 29W
Canton, *Ill., U.S.A.* **80 E9** 40 33N 90 2W
Canton, *Miss., U.S.A.* **81 J9** 32 37N 90 2W
Canton, *Mo., U.S.A.* **80 E9** 40 8N 91 32W
Canton, *N.Y., U.S.A.* **79 B9** 44 36N 75 10W
Canton, *Ohio, U.S.A.* **78 F3** 40 48N 81 23W
Canton, *Okla., U.S.A.* **81 G5** 36 3N 98 35W
Canton, *S. Dak., U.S.A.* **80 D6** 43 18N 96 35W
Canton L., *U.S.A.* **81 G5** 36 6N 98 35W

Canudos

Canudos, *Brazil* **92 E7** 7 13S 58 5W
Canutama, *Brazil* **92 E6** 6 30S 64 20W
Canutillo, *U.S.A.* **83 L10** 31 55N 106 36W
Canyon, *Tex., U.S.A.* **81 H4** 34 59N 101 55W
Canyon, *Wyo., U.S.A.* **82 D8** 44 43N 110 36W
Canyonlands National Park, *U.S.A.* **83 G9** 38 15N 110 0W
Canyonville, *U.S.A.* **82 E2** 42 56N 123 17W
Cao Bang, *Vietnam* **38 A6** 22 40N 106 15 E
Cao He →, *China* **35 D13** 40 10N 124 32 E
Cao Lanh, *Vietnam* **39 G5** 10 27N 105 38 E
Cao Xian, *China* **34 G8** 34 50N 115 35 E
Cap-aux-Meules, *Canada* **71 C7** 47 23N 61 52W
Cap-Chat, *Canada* **71 C6** 49 6N 66 40W
Cap-de-la-Madeleine, *Canada* **70 C5** 46 22N 72 31W
Cap-Haïtien, *Haiti* **89 C5** 19 40N 72 20W
Cap St.-Jacques = Vung Tau, *Vietnam* **39 G6** 10 21N 107 4 E
Capa, *Vietnam* **38 A4** 22 21N 103 50 E
Capaia, *Angola* **52 F4** 8 27S 20 13 E
Capanaparo →, *Venezuela* **92 B5** 7 1N 67 7W
Cape →, *Australia* **62 C4** 20 59S 146 51 E
Cape Barren I., *Australia* **62 G4** 40 25S 148 15 E
Cape Breton Highlands Nat. Park, *Canada* **71 C7** 46 50N 60 40W
Cape Breton I., *Canada* **71 C7** 46 0N 60 30W
Cape Charles, *U.S.A.* **76 G8** 37 16N 76 1W
Cape Coast, *Ghana* **50 G4** 5 5N 1 15W
Cape Dorset, *Canada* **69 B12** 64 14N 76 32W
Cape Dyer, *Canada* **69 B13** 66 30N 61 22W
Cape Fear →, *U.S.A.* **77 H6** 33 53N 78 1W
Cape Girardeau, *U.S.A.* **81 G10** 37 19N 89 32W
Cape Jervis, *Australia* **63 F2** 35 40S 138 5 E
Cape May, *U.S.A.* **76 F8** 38 56N 74 56W
Cape May Point, *U.S.A.* **75 C12** 38 56N 74 58W
Cape Province □, *S. Africa* **56 E3** 32 0S 23 0 E
Cape Tormentine, *Canada* **71 C7** 46 8N 63 47W
Cape Town, *S. Africa* **56 E2** 33 55S 18 22 E
Cape Verde Is. ■, *Atl. Oc.* **2 D8** 17 10N 25 20W
Cape Vincent, *U.S.A.* **79 B8** 44 8N 76 20W
Cape York Peninsula, *Australia* **62 A3** 12 0S 142 30 E
Capela, *Brazil* **93 F11** 10 30S 37 0W
Capella, *Australia* **62 C4** 23 2S 148 1 E
Capim →, *Brazil* **93 D9** 1 40S 47 47W
Capitan, *U.S.A.* **83 K11** 33 35N 105 35W
Capitola, *U.S.A.* **84 J5** 36 59N 121 57W
Capoche →, *Mozam.* **55 F3** 15 35S 33 0 E
Capraia, *Italy* **20 C3** 43 2N 9 50 E
Capreol, *Canada* **70 C3** 46 43N 80 56W
Caprera, *Italy* **20 D3** 41 12N 9 28 E
Capri, *Italy* **20 D6** 40 34N 14 15 E
Capricorn Group, *Australia* **62 C5** 23 30S 151 55 E
Capricorn Ra., *Australia* **60 D2** 23 20S 116 50 E
Caprivi Strip, *Namibia* **56 B3** 18 0S 23 0 E
Captainganj, *India* **43 F10** 26 55N 83 45 E
Captain's Flat, *Australia* **63 F4** 35 35S 149 27 E
Caquetá →, *Colombia* **92 D5** 1 15S 69 15W
Caracal, *Romania* **17 F12** 44 8N 24 22 E
Caracas, *Venezuela* **92 A5** 10 30N 66 55W
Caracol, *Brazil* **93 E10** 9 15S 43 22W
Caradoc, *Australia* **63 E3** 30 35S 143 5 E
Carajás, Serra dos, *Brazil* **93 E8** 6 0S 51 30W
Carangola, *Brazil* **95 A7** 20 44S 42 5W
Carani, *Australia* **61 F2** 30 57S 116 28 E
Caransebeş, *Romania* **17 F11** 45 28N 22 18 E
Caratasca, L., *Honduras* **88 C3** 15 20N 83 40W
Caratinga, *Brazil* **93 G10** 19 50S 42 10W
Caraúbas, *Brazil* **93 E11** 5 43S 37 33W
Caravaca, *Spain* **19 C5** 38 8N 1 52W
Caravelas, *Brazil* **93 G11** 17 45S 39 15W
Caraveli, *Peru* **92 G4** 15 45S 73 25W
Caràzinho, *Brazil* **95 B5** 28 16S 52 46W
Carballo, *Spain* **19 A1** 43 13N 8 41W
Carberry, *Canada* **73 D9** 49 50N 99 25W
Carbó, *Mexico* **86 B2** 29 42N 110 58W
Carbon, *Canada* **72 C6** 51 30N 113 9W
Carbonara, C., *Italy* **20 E3** 39 8N 9 30 E
Carbondale, *Colo., U.S.A.* **82 G10** 39 24N 107 13W
Carbondale, *Ill., U.S.A.* **81 G10** 37 44N 89 13W
Carbondale, *Pa., U.S.A.* **79 E9** 41 35N 75 30W
Carbonear, *Canada* **71 C9** 47 42N 53 13W
Carbonia, *Italy* **20 E3** 39 10N 8 30 E
Carcajou, *Canada* **72 B5** 57 47N 117 6W
Carcasse, C., *Haiti* **89 C5** 18 30N 74 28W
Carcassonne, *France* **18 E5** 43 13N 2 20 E
Carcross, *Canada* **68 B6** 60 13N 134 45W
Cardabia, *Australia* **60 D1** 23 2S 113 48 E
Cardamon Hills, *India* **40 Q10** 9 30N 77 15 E
Cárdenas, *Cuba* **88 B3** 23 0N 81 30W
Cárdenas, *San Luis Potosí, Mexico* **87 C5** 22 0N 99 41W
Cárdenas, *Tabasco, Mexico* **87 D6** 17 59N 93 21W
Cardiff, *U.K.* **11 F4** 51 28N 3 11W
Cardiff-by-the-Sea, *U.S.A.* **85 M9** 33 1N 117 17W
Cardigan, *U.K.* **11 E3** 52 6N 4 41W
Cardigan B., *U.K.* **11 E3** 52 30N 4 30W
Cardinal, *Canada* **79 B9** 44 47N 75 23W
Cardona, *Spain* **19 B6** 41 56N 1 40 E
Cardona, *Uruguay* **94 C4** 33 53S 57 18W
Cardross, *Canada* **73 D7** 49 50N 105 40W
Cardston, *Canada* **72 D6** 49 15N 113 20W
Cardwell, *Australia* **62 B4** 18 14S 146 2 E
Careen L., *Canada* **73 B7** 57 0N 108 11W
Carei, *Romania* **17 E11** 47 40N 22 29 E
Careme, *Indonesia* **37 G13** 6 55S 108 27 E
Carey, *Idaho, U.S.A.* **82 E7** 43 19N 113 57W
Carey, *Ohio, U.S.A.* **76 E4** 40 57N 83 23W
Carey, L., *Australia* **61 E3** 29 0S 122 15 E
Carey L., *Canada* **73 A8** 62 12N 102 55W
Careysburg, *Liberia* **50 G2** 6 34N 10 30W
Carhué, *Argentina* **94 D3** 37 10S 62 50W
Cariacica, *Brazil* **93 H10** 20 16S 40 25W
Caribbean Sea, *W. Indies* **89 C5** 15 0N 75 0W
Cariboo Mts., *Canada* **72 C4** 53 0N 121 0W
Caribou, *U.S.A.* **71 C6** 46 52N 68 1W
Caribou →, *Man., Canada* **73 B10** 59 20N 94 44W
Caribou →, *N.W.T., Canada* **72 A3** 61 27N 125 45W
Caribou I., *Canada* **70 C2** 47 22N 85 49W
Caribou Is., *Canada* **72 A6** 61 55N 113 15W
Caribou L., *Man., Canada* **73 B9** 59 21N 96 10W
Caribou L., *Ont., Canada* **70 B2** 50 25N 89 5W
Caribou Mts., *Canada* **72 B5** 59 12N 115 40W
Carichic, *Mexico* **86 B3** 27 56N 107 3W
Carillo, *Mexico* **86 B4** 26 50N 103 55W
Carinda, *Australia* **63 E4** 30 28S 147 41 E
Carinhanha, *Brazil* **93 F10** 14 15S 44 46W
Carinthia □ = Kärnten □, *Austria* **16 E6** 46 52N 13 30 E
Caripito, *Venezuela* **92 A6** 10 8N 63 6W
Caritianas, *Brazil* **92 E6** 9 20S 63 6W
Carleton Place, *Canada* **70 C4** 45 8N 76 9W
Carletonville, *S. Africa* **56 D4** 26 23S 27 22 E
Carlin, *U.S.A.* **82 F5** 40 43N 116 7W
Carlingford, L., *Ireland* **13 B5** 54 2N 6 5W
Carlinville, *U.S.A.* **80 F10** 39 17N 89 53W
Carlisle, *U.K.* **10 C5** 54 54N 2 55W
Carlisle, *U.S.A.* **78 F7** 40 12N 77 12W
Carlos Casares, *Argentina* **94 D3** 35 32S 61 20W
Carlos Tejedor, *Argentina* **94 D3** 35 25S 62 25W
Carlow, *Ireland* **13 D5** 52 50N 6 58W
Carlow □, *Ireland* **13 D5** 52 43N 6 50W
Carlsbad, *Calif., U.S.A.* **85 M9** 33 10N 117 21W
Carlsbad, *N. Mex., U.S.A.* **81 J2** 32 25N 104 14W
Carlyle, *Canada* **73 D8** 49 40N 102 20W
Carlyle, *U.S.A.* **80 F10** 38 37N 89 22W
Carmacks, *Canada* **68 B6** 62 5N 136 16W
Carman, *Canada* **73 D9** 49 30N 98 0W
Carmangay, *Canada* **72 C6** 50 10N 113 10W
Carmanville, *Canada* **71 C9** 49 23N 54 19W
Carmarthen, *U.K.* **11 F3** 51 52N 4 20W
Carmarthen B., *U.K.* **11 F3** 51 40N 4 30W
Carmel, *U.S.A.* **79 E11** 41 26N 73 41W
Carmel-by-the-Sea, *U.S.A.* **83 H3** 36 33N 121 55W
Carmel Valley, *U.S.A.* **84 J5** 36 29N 121 43W
Carmelo, *Uruguay* **94 C4** 34 0S 58 20W
Carmen, *Colombia* **92 B3** 9 43N 75 8W
Carmen, *Paraguay* **95 B4** 27 13S 56 12W
Carmen →, *Mexico* **86 A3** 30 42N 106 29W
Carmen, I., *Mexico* **86 B2** 26 0N 111 20W
Carmen de Patagones, *Argentina* **96 E4** 40 50S 63 0W
Carmensa, *Argentina* **94 D2** 35 15S 67 40W
Carmi, *U.S.A.* **76 F1** 38 5N 88 10W
Carmichael, *U.S.A.* **84 G5** 38 38N 121 19W
Carmila, *Australia* **62 C4** 21 55S 149 24 E
Carmona, *Spain* **19 D3** 37 28N 5 42W
Carnarvon, *Queens., Australia* **62 C4** 24 48S 147 45 E
Carnarvon, *W. Austral., Australia* **61 D1** 24 51S 113 42 E
Carnarvon, *S. Africa* **56 E3** 30 56S 22 8 E
Carnarvon Ra., *Queens., Australia* **62 D4** 25 15S 148 30 E
Carnarvon Ra., *W. Austral., Australia* **61 E3** 25 20S 120 45 E
Carnation, *U.S.A.* **84 C5** 47 39N 121 55W
Carndonagh, *Ireland* **13 A4** 55 15N 7 16W
Carnduff, *Canada* **73 D8** 49 10N 101 50W
Carnegie, *U.S.A.* **78 F4** 40 24N 80 5W
Carnegie, L., *Australia* **61 E3** 26 5S 122 30 E
Carnic Alps = Karnische Alpen, *Europe* **20 A5** 46 36N 13 0 E
Carniche Alpi = Karnische Alpen, *Europe* **20 A5** 46 36N 13 0 E
Carnot, *C.A.R.* **52 D3** 4 59N 15 56 E
Carnot, C., *Australia* **63 E2** 34 57S 135 38 E
Carnot B., *Australia* **60 C3** 17 20S 122 15 E
Carnsore Pt., *Ireland* **13 D5** 52 10N 6 20W
Caro, *U.S.A.* **76 D4** 43 29N 83 24W
Carol City, *U.S.A.* **77 N5** 25 56N 80 16W
Carolina, *Brazil* **93 E9** 7 10S 47 30W
Carolina, *Puerto Rico* **89 C6** 18 23N 65 58W
Carolina, *S. Africa* **57 D5** 26 5S 30 6 E
Caroline I., *Kiribati* **65 H12** 9 15S 150 3W
Caroline Is., *Pac. Oc.* **64 G6** 8 0N 150 0 E
Caron, *Canada* **73 C7** 50 30N 105 50W
Caroni →, *Venezuela* **92 B6** 8 21N 62 43W
Caroona, *Australia* **63 E5** 31 24S 150 26 E
Carpathians, *Europe* **17 D10** 49 30N 21 0 E
Carpaţii Meridionali, *Romania* **17 F12** 45 30N 25 0 E
Carpentaria, G. of, *Australia* **62 A2** 14 0S 139 0 E
Carpentaria Downs, *Australia* **62 B3** 18 44S 144 20 E
Carpinteria, *U.S.A.* **85 L7** 34 24N 119 31W
Carpolac = Morea, *Australia* **63 F3** 36 45S 141 18 E
Carr Boyd Ra., *Australia* **60 C4** 16 15S 128 35 E
Carrabelle, *U.S.A.* **77 L3** 29 51N 84 40W
Carranya, *Australia* **60 C4** 19 14S 127 46 E
Carrara, *Italy* **20 B4** 44 5N 10 7 E
Carrauntoohill, *Ireland* **13 E2** 52 0N 9 49W
Carrick-on-Shannon, *Ireland* **13 C3** 53 57N 8 7W
Carrick-on-Suir, *Ireland* **13 D4** 52 22N 7 30W
Carrickfergus, *U.K.* **13 B6** 54 43N 5 50W
Carrickfergus □, *U.K.* **13 B6** 54 43N 5 49W
Carrickmacross, *Ireland* **13 C5** 53 58N 6 43W
Carrieton, *Australia* **63 E2** 32 25S 138 31 E
Carrington, *U.S.A.* **80 B5** 47 27N 99 8W
Carrizal Bajo, *Chile* **94 B1** 28 5S 71 20W
Carrizalillo, *Chile* **94 B1** 29 5S 71 30W
Carrizo Cr. →, *U.S.A.* **81 G3** 36 55N 103 55W
Carrizo Springs, *U.S.A.* **81 L5** 28 31N 99 52W
Carrizozo, *U.S.A.* **83 K11** 33 38N 105 53W
Carroll, *U.S.A.* **80 D7** 42 4N 94 52W
Carrollton, *Ga., U.S.A.* **77 J3** 33 35N 85 5W
Carrollton, *Ill., U.S.A.* **80 F9** 39 18N 90 24W
Carrollton, *Ky., U.S.A.* **76 F3** 38 41N 85 11W
Carrollton, *Mo., U.S.A.* **80 F8** 39 22N 93 30W
Carrollton, *Ohio, U.S.A.* **78 F3** 40 34N 81 5W
Carron →, *U.K.* **12 D3** 57 30N 5 30W
Carron, L., *U.K.* **12 D3** 57 22N 5 35W
Carrot →, *Canada* **73 C8** 53 50N 101 17W
Carrot River, *Canada* **73 C8** 53 17N 103 35W
Carruthers, *Canada* **73 C7** 52 52N 109 16W
Carse of Gowrie, *U.K.* **12 E5** 56 30N 3 10W
Carson, *Calif., U.S.A.* **85 M8** 33 48N 118 17W
Carson, *N. Dak., U.S.A.* **80 B4** 46 25N 101 34W
Carson →, *U.S.A.* **84 F8** 39 45N 118 40W
Carson City, *U.S.A.* **84 F7** 39 10N 119 46W
Carson Sink, *U.S.A.* **82 G4** 39 50N 118 25W
Carstairs, *U.K.* **12 F5** 55 42N 3 41W
Cartagena, *Colombia* **92 A3** 10 25N 75 33W
Cartagena, *Spain* **19 D5** 37 38N 0 59W
Cartago, *Colombia* **92 C3** 4 45N 75 55W
Cartago, *Costa Rica* **88 E3** 9 50N 83 55W
Cartersville, *U.S.A.* **77 H3** 34 10N 84 48W
Carterton, *N.Z.* **59 J5** 41 2S 175 31 E
Carthage, *Ark., U.S.A.* **81 H8** 34 4N 92 33W
Carthage, *Ill., U.S.A.* **80 E9** 40 25N 91 8W
Carthage, *Mo., U.S.A.* **81 G7** 37 11N 94 19W
Carthage, *S. Dak., U.S.A.* **80 C6** 44 10N 97 43W
Carthage, *Tex., U.S.A.* **81 J7** 32 9N 94 20W
Cartier I., *Australia* **60 B3** 12 31S 123 29 E
Cartwright, *Canada* **71 B8** 53 41N 56 58W
Caruaru, *Brazil* **93 E11** 8 15S 35 55W
Carúpano, *Venezuela* **92 A6** 10 39N 63 15W
Caruthersville, *U.S.A.* **81 G10** 36 11N 89 39W
Carvoeiro, *Brazil* **92 D6** 1 30S 61 59W
Casa Grande, *U.S.A.* **83 K8** 32 53N 111 45W
Casablanca, *Chile* **94 C1** 33 20S 71 25W
Casablanca, *Morocco* **50 B3** 33 36N 7 36W
Casale Monferrato, *Italy* **20 B3** 45 8N 8 28 E
Casas Grandes, *Mexico* **86 A3** 30 22N 108 0W
Cascade, *Idaho, U.S.A.* **82 D5** 44 31N 116 2W
Cascade, *Mont., U.S.A.* **82 C8** 47 16N 111 42W
Cascade Locks, *U.S.A.* **84 E5** 45 40N 121 54W
Cascade Ra., *U.S.A.* **84 D5** 47 0N 121 30W
Cascavel, *Brazil* **95 A5** 24 57S 53 28W
Caserta, *Italy* **20 D6** 41 5N 14 20 E
Cashel, *Ireland* **13 D4** 52 31N 7 53W
Cashmere, *U.S.A.* **82 C3** 47 31N 120 28W
Cashmere Downs, *Australia* **61 E2** 28 57S 119 35 E
Casiguran, *Phil.* **37 A6** 16 22N 122 7 E
Casilda, *Argentina* **94 C3** 33 10S 61 10W
Casino, *Australia* **63 D5** 28 52S 153 3 E
Casiquiare →, *Venezuela* **92 C5** 2 1N 67 7W
Caslan, *Canada* **72 C6** 54 38N 112 31W
Casma, *Peru* **92 E3** 9 30S 78 20W
Casmalia, *U.S.A.* **85 L6** 34 50N 120 32W
Caspe, *Spain* **19 B5** 41 14N 0 1W
Casper, *U.S.A.* **82 E10** 42 51N 106 19W
Caspian Sea, *Asia* **25 F9** 43 0N 50 0 E
Cass City, *U.S.A.* **76 D4** 43 36N 83 11W
Cass Lake, *U.S.A.* **80 B7** 47 23N 94 37W
Casselman, *Canada* **79 A9** 45 19N 75 5W
Casselton, *U.S.A.* **80 B6** 46 54N 97 13W
Cassiar, *Canada* **72 B3** 59 16N 129 40W
Cassiar Mts., *Canada* **72 B2** 59 30N 130 30W
Cassinga, *Angola* **53 H3** 15 5S 16 4 E
Cassville, *U.S.A.* **81 G8** 36 41N 93 52W
Castaic, *U.S.A.* **85 L8** 34 30N 118 38W
Castellammare del Golfo, *Italy* **20 E5** 38 2N 12 53 E
Castellammare di Stábia, *Italy* **20 D6** 40 47N 14 29 E
Castelli, *Argentina* **94 D4** 36 7S 57 47W
Castellón de la Plana, *Spain* **19 C5** 39 58N 0 3W
Castelo, *Brazil* **95 A7** 20 33S 41 14W
Castelo Branco, *Portugal* **19 C2** 39 50N 7 31W
Castelvetrano, *Italy* **20 F5** 37 40N 12 46 E
Casterton, *Australia* **63 F3** 37 30S 141 30 E
Castilla La Mancha □, *Spain* **19 C4** 39 30N 3 30W
Castilla La Nueva = Castilla La Mancha □, *Spain* **19 C4** 39 30N 3 30W
Castilla La Vieja = Castilla y Leon □, *Spain* **19 B3** 42 0N 5 0W
Castilla y Leon □, *Spain* **19 B3** 42 0N 5 0W
Castillos, *Uruguay* **95 C5** 34 12S 53 52W
Castle Dale, *U.S.A.* **82 G8** 39 13N 111 1W
Castle Douglas, *U.K.* **12 G5** 54 57N 3 57W
Castle Rock, *Colo., U.S.A.* **80 F2** 39 22N 104 51W
Castle Rock, *Wash., U.S.A.* **84 D4** 46 17N 122 54W
Castlebar, *Ireland* **13 C2** 53 52N 9 17W
Castleblaney, *Ireland* **13 B5** 54 7N 6 44W
Castlegar, *Canada* **72 D5** 49 20N 117 40W
Castlemaine, *Australia* **63 F3** 37 2S 144 12 E
Castlereagh, *Ireland* **13 C3** 53 47N 8 30W
Castlereagh □, *U.K.* **13 B6** 54 33N 5 53W
Castlereagh →, *Australia* **63 E4** 30 12S 147 32 E
Castlereagh B., *Australia* **62 A2** 12 10S 135 10 E
Castletown, *I. of Man* **10 C3** 54 4N 4 40W
Castletown Bearhaven, *Ireland* **13 E2** 51 40N 9 54W
Castlevale, *Australia* **62 C4** 24 30S 146 48 E
Castor, *Canada* **72 C6** 52 15N 111 50W
Castres, *France* **18 E5** 43 37N 2 13 E
Castries, *St. Lucia* **89 D7** 14 2N 60 58W
Castro, *Brazil* **95 A5** 24 45S 50 0W
Castro, *Chile* **96 E2** 42 30S 73 50W
Castro Alves, *Brazil* **93 F11** 12 46S 39 33W
Castro del Río, *Spain* **19 D3** 37 41N 4 29W
Castroville, *Calif., U.S.A.* **84 J5** 36 46N 121 45W
Castroville, *Tex., U.S.A.* **81 L5** 29 21N 98 53W
Casummit Lake, *Canada* **70 B1** 51 29N 92 22W
Cat Ba, Dao, *Vietnam* **38 B6** 20 50N 107 0 E
Cat I., *Bahamas* **89 B4** 24 30N 75 30W
Cat I., *U.S.A.* **81 K10** 30 14N 89 6W
Cat L., *Canada* **70 B1** 51 40N 91 50W
Catacamas, *Honduras* **88 D2** 14 54N 85 56W
Catacáos, *Peru* **92 E2** 5 20S 80 45W
Cataguases, *Brazil* **95 A7** 21 23S 42 39W
Catahoula L., *U.S.A.* **81 K8** 31 31N 92 7W
Catalão, *Brazil* **93 G9** 18 10S 47 57W
Catalina, *Canada* **71 C9** 48 31N 53 4W
Catalonia = Cataluña □, *Spain* **19 B6** 41 40N 1 15 E
Cataluña □, *Spain* **19 B6** 41 40N 1 15 E
Catamarca, *Argentina* **94 B2** 28 30S 65 50W
Catamarca □, *Argentina* **94 B2** 27 0S 65 50W
Catanduanes, *Phil.* **37 B6** 13 50N 124 20 E
Catanduva, *Brazil* **95 A6** 21 5S 48 58W
Catánia, *Italy* **20 F6** 37 31N 15 4 E
Catanzaro, *Italy* **20 E7** 38 54N 16 38 E
Catarman, *Phil.* **37 B6** 12 28N 124 35 E
Cateel, *Phil.* **37 C7** 7 47N 126 24 E
Cathcart, *S. Africa* **56 E4** 32 18S 27 10 E
Cathlamet, *U.S.A.* **84 D3** 46 12N 123 23W
Catlettsburg, *U.S.A.* **76 F4** 38 25N 82 36W
Catoche, C., *Mexico* **87 C7** 21 40N 87 8W
Catrimani, *Brazil* **92 C6** 0 27N 61 41W
Catskill, *U.S.A.* **79 D11** 42 14N 73 52W
Catskill Mts., *U.S.A.* **79 D10** 42 10N 74 25W
Catt, Mt., *Australia* **62 A1** 13 49S 134 23 E
Cattaraugus, *U.S.A.* **78 D6** 42 22N 78 52W
Catuala, *Angola* **56 B2** 16 25S 19 2 E
Catur, *Mozam.* **55 E4** 13 45S 35 30 E
Catwick Is., *Vietnam* **39 G7** 10 0N 109 0 E
Cauca →, *Colombia* **92 B4** 8 54N 74 28W
Caucaia, *Brazil* **93 D11** 3 40S 38 35W
Caucasus = Bolshoi Kavkas, *Asia* **25 F7** 42 50N 44 0 E
Caúngula, *Angola* **52 F3** 8 26S 18 38 E
Cauquenes, *Chile* **94 D1** 36 0S 72 22W
Caura →, *Venezuela* **92 B6** 7 38N 64 53W
Cauresi →, *Mozam.* **55 F3** 17 8S 33 0 E
Causapscal, *Canada* **71 C6** 48 19N 67 12W
Cauvery →, *India* **40 P11** 11 9N 78 52 E
Caux, Pays de, *France* **18 B4** 49 38N 0 35 E
Cavalier, *U.S.A.* **80 A6** 48 48N 97 37W
Cavan, *Ireland* **13 C4** 54 0N 7 22W
Cavan □, *Ireland* **13 C4** 53 58N 7 10W
Cave City, *U.S.A.* **76 G3** 37 8N 85 58W
Cavenagh Ra., *Australia* **61 E4** 26 12S 127 55 E
Cavendish, *Australia* **63 F3** 37 31S 142 2 E
Caviana, I., *Brazil* **93 C8** 0 10N 50 10W
Cavite, *Phil.* **37 B6** 14 29N 120 55 E
Cawndilla L., *Australia* **63 E3** 32 30S 142 15 E
Cawnpore = Kanpur, *India* **43 F9** 26 28N 80 20 E
Caxias, *Brazil* **93 D10** 4 55S 43 20W
Caxias do Sul, *Brazil* **95 B5** 29 10S 51 10W
Caxito, *Angola* **52 F2** 8 30S 13 30 E
Cay Sal Bank, *Bahamas* **88 B3** 23 45N 80 0W
Cayambe, *Ecuador* **92 C3** 0 3N 78 8W
Cayenne, *Fr. Guiana* **93 B8** 5 5N 52 18W
Cayman Brac, *Cayman Is.* **88 C4** 19 43N 79 49W
Cayman Is. ■, *W. Indies* **88 C3** 19 40N 80 30W
Cayo Romano, *Cuba* **89 B4** 22 0N 78 0W
Cayuga, *Canada* **78 D5** 42 59N 79 50W
Cayuga, *U.S.A.* **79 D8** 42 54N 76 44W
Cayuga L., *U.S.A.* **79 D8** 42 41N 76 41W
Cazombo, *Angola* **53 G4** 11 54S 22 56 E
Ceanannus Mor, *Ireland* **13 C5** 53 42N 6 53W
Ceará = Fortaleza, *Brazil* **93 D11** 3 45S 38 35W
Ceará □, *Brazil* **93 E11** 5 0S 40 0W
Ceará Mirim, *Brazil* **93 E11** 5 38S 35 25W
Cebaco, I. de, *Panama* **88 E3** 7 33N 81 9W
Cebollar, *Argentina* **94 B2** 29 10S 66 35W
Cebu, *Phil.* **37 B6** 10 18N 123 54 E
Cecil Plains, *Australia* **63 D5** 27 30S 151 11 E
Cedar →, *U.S.A.* **80 E9** 41 17N 91 21W
Cedar City, *U.S.A.* **83 H7** 37 41N 113 4W
Cedar Creek Reservoir, *U.S.A.* **81 J6** 32 11N 96 4W
Cedar Falls, *Iowa, U.S.A.* **80 D8** 42 32N 92 27W
Cedar Falls, *Wash., U.S.A.* **84 C5** 47 25N 121 45W
Cedar Key, *U.S.A.* **77 L4** 29 8N 83 2W
Cedar L., *Canada* **73 C8** 53 10N 100 0W
Cedar Rapids, *U.S.A.* **80 E9** 41 59N 91 40W
Cedartown, *U.S.A.* **77 H3** 34 1N 85 15W
Cedarvale, *Canada* **72 B3** 55 1N 128 22W
Cedarville, *S. Africa* **57 E4** 30 23S 29 3 E
Cedarville, *U.S.A.* **82 F3** 41 32N 120 10W
Cedral, *Mexico* **86 C4** 23 50N 100 42W
Cedro, *Brazil* **93 E11** 6 34S 39 3W
Cedros, I. de, *Mexico* **86 B1** 28 10N 115 20W
Ceduna, *Australia* **63 E1** 32 7S 133 46 E
Cefalù, *Italy* **20 E6** 38 3N 14 1 E
Cegléd, *Hungary* **17 E9** 47 11N 19 47 E
Cehegín, *Spain* **19 C5** 38 6N 1 48W
Celaya, *Mexico* **86 C4** 20 31N 100 37W
Celbridge, *Ireland* **13 C5** 53 20N 6 33W
Celebes = Sulawesi □, *Indonesia* **37 E6** 2 0S 120 0 E
Celebes Sea = Sulawesi Sea, *Indonesia* **37 D6** 3 0N 123 0 E
Celina, *U.S.A.* **76 E3** 40 33N 84 35W
Celje, *Slovenia* **20 A6** 46 16N 15 18 E
Celle, *Germany* **16 B5** 52 37N 10 4 E
Cement, *U.S.A.* **81 H5** 34 56N 98 8W
Center, *N. Dak., U.S.A.* **80 B4** 47 7N 101 18W
Center, *Tex., U.S.A.* **81 K7** 31 48N 94 11W
Centerfield, *U.S.A.* **83 G8** 39 8N 111 49W
Centerville, *Calif., U.S.A.* **84 J7** 36 44N 119 30W
Centerville, *Iowa, U.S.A.* **80 E8** 40 44N 92 52W
Centerville, *Pa., U.S.A.* **78 F5** 40 3N 79 59W
Centerville, *S. Dak., U.S.A.* **80 D6** 43 7N 96 58W
Centerville, *Tenn., U.S.A.* **77 H2** 35 47N 87 28W
Centerville, *Tex., U.S.A.* **81 K7** 31 16N 95 59W
Central, *U.S.A.* **83 K9** 32 47N 108 9W
Central □, *Kenya* **54 C4** 0 30S 37 30 E
Central □, *Malawi* **55 E3** 13 30S 33 30 E
Central □, *U.K.* **12 E4** 56 10N 4 30W
Central □, *Zambia* **55 E2** 14 25S 28 50 E
Central, Cordillera, *Colombia* **92 C4** 5 0N 75 0W
Central, Cordillera, *Costa Rica* **88 D3** 10 10N 84 5W
Central, Cordillera, *Dom. Rep.* **89 C5** 19 15N 71 0W
Central African Rep. ■, *Africa* **51 G9** 7 0N 20 0 E
Central City, *Ky., U.S.A.* **76 G2** 37 18N 87 7W
Central City, *Nebr., U.S.A.* **80 E5** 41 7N 98 0W

Chicago Heights

Central I., Kenya 54 B4 3 30N 36 0 E
Central Makran Range, Pakistan 40 F4 26 30N 64 15 E
Central Patricia, Canada 70 B1 51 30N 90 9W
Central Russian Uplands, Europe 6 E13 54 0N 36 0 E
Central Siberian Plateau, Russia 28 C14 65 0N 105 0 E
Centralia, Ill., U.S.A. 80 F10 38 32N 89 8W
Centralia, Mo., U.S.A. 80 F8 39 13N 92 8W
Centralia, Wash., U.S.A. 84 D4 46 43N 122 58W
Centreville, Ala., U.S.A. 77 J2 32 57N 87 8W
Centreville, Miss., U.S.A. 81 K9 31 5N 91 4W
Cephalonia = Kefallinía, Greece 21 E9 38 20N 20 30 E
Cepu, Indonesia 37 G14 7 9S 111 35 E
Ceram = Seram, Indonesia 37 E7 3 10S 129 0 E
Ceram Sea = Seram Sea, Indonesia 37 E7 2 30S 128 30 E
Ceres, Argentina 94 B3 29 55S 61 55W
Ceres, S. Africa 56 E2 33 21S 19 18 E
Ceres, U.S.A. 84 H6 37 35N 120 57W
Cerignola, Italy 20 D6 41 17N 15 53 E
Cerigo = Kíthira, Greece 21 F11 36 9N 23 12 E
Cerknica, Slovenia 20 B6 45 48N 14 21 E
Cernavodă, Romania 17 F14 44 22N 28 3 E
Cerralvo, I., Mexico 86 C3 24 20N 109 45W
Cerritos, Mexico 86 C4 22 27N 100 20W
Cervera, Spain 19 B6 41 40N 1 16 E
Cervera del Río Alhama, Spain 19 A5 42 2N 1 58W
Cesena, Italy 20 B5 44 9N 12 14 E
České Budějovice, Czech. 16 D7 48 55N 14 25 E
Ceskomoravská Vrchovina, Czech. 16 D7 49 30N 15 40 E
Český Těšin, Czech. 17 D9 49 45N 18 39 E
Cessnock, Australia 63 E5 32 50S 151 21 E
Cetinje, Montenegro, Yug. 21 C8 42 23N 18 59 E
Ceuta, Morocco 50 A3 35 52N 5 18W
Cévennes, France 18 D5 44 10N 3 50 E
Ceyhan →, Turkey 25 G6 36 38N 35 40 E
Ceylon = Sri Lanka ■, Asia 40 R12 7 30N 80 50 E
Cha-am, Thailand 38 F2 12 48N 99 58 E
Chablais, France 18 C7 46 20N 6 36 E
Chacabuco, Argentina 94 C3 34 40S 60 27W
Chachapoyas, Peru 92 E3 6 15S 77 50W
Chachoengsao, Thailand 38 F3 13 42N 101 5 E
Chachran, Pakistan 40 E7 28 55N 70 30 E
Chachro, Pakistan 42 G4 25 5N 70 15 E
Chaco □, Argentina 94 B3 26 30S 61 0W
Chaco □, Paraguay 94 B3 26 0S 60 0W
Chad ■, Africa 51 E8 15 0N 17 15 E
Chad, L. = Tchad, L., Chad 51 F7 13 30N 14 30 E
Chadan, Russia 27 D10 51 17N 91 35 E
Chadileuvú →, Argentina 94 D2 37 46S 66 0W
Chadiza, Zambia 55 E3 14 45S 32 27 E
Chadron, U.S.A. 80 D3 42 50N 103 0W
Chae Hom, Thailand 38 C2 18 43N 99 35 E
Chaem →, Thailand 38 C2 18 11N 98 38 E
Chaeryŏng, N. Korea 35 E13 38 24N 125 36 E
Chagai Hills, Afghan. 40 E3 29 30N 63 0 E
Chagda, Russia 27 D14 58 45N 130 38 E
Chagos Arch., Ind. Oc. 28 K11 6 0S 72 0 E
Chāh Ākhvor, Iran 45 C8 32 41N 59 40 E
Chāh Bahār, Iran 45 E9 25 20N 60 40 E
Chāh-e-Malek, Iran 45 D8 28 35N 59 7 E
Chāh Kavīr, Iran 45 D7 31 45N 54 52 E
Chahar Burjak, Afghan. 40 D3 30 15N 62 0 E
Chaibasa, India 41 H14 22 42N 85 49 E
Chainat, Thailand 38 E3 15 11N 100 8 E
Chaiya, Thailand 39 H2 9 23N 99 14 E
Chaj Doab, Pakistan 42 C5 32 15N 73 0 E
Chajari, Argentina 94 C4 30 42S 58 0W
Chake Chake, Tanzania 54 D4 5 15S 39 45 E
Chakhānsūr, Afghan. 40 D3 31 10N 62 0 E
Chakonipau, L., Canada 71 A6 56 18N 68 30W
Chakradharpur, India 43 H11 22 45N 85 40 E
Chakwal, Pakistan 42 C5 32 56N 72 53 E
Chala, Peru 92 G4 15 48S 74 20W
Chalchihuites, Mexico 86 C4 23 29N 103 53W
Chalcis = Khalkís, Greece 21 E10 38 27N 23 42 E
Chaleur B., Canada 71 C6 47 55N 65 30W
Chalfant, U.S.A. 84 H8 37 32N 118 21W
Chalhuanca, Peru 92 F4 14 15S 73 15W
Chalisgaon, India 40 J9 20 30N 75 10 E
Chalky Inlet, N.Z. 59 M1 46 3S 166 31 E
Challapata, Bolivia 92 G5 18 53S 66 50W
Challis, U.S.A. 82 D6 44 30N 114 14W
Chalna, India 43 H13 22 36N 89 35 E
Châlons-sur-Marne, France 18 B6 48 58N 4 20 E
Chalyaphum, Thailand 38 E4 15 48N 102 2 E
Cham, Cu Lao, Vietnam 38 E7 15 57N 108 30 E
Chama, U.S.A. 83 H10 36 54N 106 35W
Chaman, Pakistan 40 D5 30 58N 66 25 E
Chamba, India 42 C7 32 35N 76 10 E
Chamba, Tanzania 55 E4 11 37S 37 0 E
Chambal →, India 43 F8 26 29N 79 15 E
Chamberlain, U.S.A. 80 D5 43 49N 99 20W
Chamberlain →, Australia 60 C4 15 30S 127 54 E
Chambers, U.S.A. 83 J9 35 11N 109 26W
Chambersburg, U.S.A. 76 F7 39 56N 77 40W
Chambéry, France 18 D6 45 34N 5 55 E
Chambly, Canada 79 A11 45 27N 73 17W
Chambord, Canada 71 C5 48 25N 72 6W
Chamchamal, Iraq 44 C5 35 32N 44 50 E
Chamela, Mexico 86 D3 19 32N 105 5W
Chamical, Argentina 94 C2 30 22S 66 27W
Chamkar Luong, Cambodia 39 G4 11 0N 103 45 E
Chamonix-Mont-Blanc, France 18 D7 45 55N 6 51 E
Champa, India 43 H10 22 2N 82 43 E
Champagne, Canada 72 A1 60 49N 136 30W
Champagne, Plaine de, France 18 B6 49 0N 4 30 E
Champaign, U.S.A. 76 E1 40 7N 88 15W
Champassak, Laos 38 E5 14 53N 105 52 E
Champlain, Canada 76 B9 46 27N 72 24W
Champlain, U.S.A. 79 B11 44 59N 73 27W
Champlain, L., U.S.A. 79 B11 44 40N 73 20W
Champotón, Mexico 87 D6 19 20N 90 50W
Chana, Thailand 39 J3 6 55N 100 44 E
Chañaral, Chile 94 B1 26 23S 70 40W
Chanārān, Iran 45 B8 36 39N 59 6 E
Chanasma, India 42 H5 23 44N 72 5 E
Chandannagar, India 43 H13 22 52N 88 24 E
Chandausi, India 43 E8 28 27N 78 49 E
Chandeleur Is., U.S.A. 81 L10 29 55N 88 57W
Chandeleur Sd., U.S.A. 81 L10 29 55N 89 0W
Chandigarh, India 42 D7 30 43N 76 47 E
Chandler, Australia 63 D1 27 0S 133 19 E
Chandler, Canada 71 C7 48 18N 64 46W
Chandler, Ariz., U.S.A. 83 K8 33 18N 111 50W
Chandler, Okla., U.S.A. 81 H6 35 42N 96 53W
Chandpur, Bangla. 41 H17 23 8N 90 45 E
Chandpur, India 42 E8 29 8N 78 19 E
Chandrapur, India 40 K11 19 57N 79 25 E
Chānf, Iran 45 E9 26 38N 60 29 E
Chang, Pakistan 42 F3 26 59N 68 30 E
Chang, Ko, Thailand 39 F4 12 0N 102 23 E
Ch'ang Chiang = Chang Jiang →, China 33 C7 31 48N 121 10 E
Chang Jiang →, China 33 C7 31 48N 121 10 E
Changa, India 43 C7 33 53N 77 35 E
Changanacheri, India 40 Q10 9 25N 76 31 E
Changane →, Mozam. 57 C5 24 30S 33 30 E
Changbai, China 35 D15 41 25N 128 5 E
Changbai Shan, China 35 C15 42 20N 129 0 E
Changchiak'ou = Zhangjiakou, China 34 D8 40 48N 114 55 E
Ch'angchou = Changzhou, China 33 C6 31 47N 119 58 E
Changchun, China 35 C13 43 57N 125 17 E
Changchunling, China 35 B13 45 18N 125 27 E
Changde, China 33 D6 29 4N 111 35 E
Changdo-ri, N. Korea 35 E14 38 30N 127 40 E
Changhai = Shanghai, China 33 C7 31 15N 121 26 E
Changhŭng, S. Korea 35 G14 34 41N 126 52 E
Changhŭngni, N. Korea 35 D15 40 24N 128 19 E
Changjiang, China 38 C7 19 20N 108 55 E
Changjin, N. Korea 35 D14 40 23N 127 15 E
Changjin-chŏsuji, N. Korea 35 D14 40 30N 127 15 E
Changli, China 35 E10 39 40N 119 13 E
Changling, China 35 B12 44 20N 123 58 E
Changlun, Malaysia 39 J3 6 25N 100 26 E
Changping, China 34 D9 40 14N 116 12 E
Changsha, China 33 D6 28 12N 113 0 E
Changwu, China 34 G4 35 10N 107 45 E
Changyi, China 35 F10 36 40N 119 30 E
Changyŏn, N. Korea 35 E13 38 15N 125 6 E
Changyuan, China 34 G8 35 15N 114 42 E
Changzhi, China 34 F7 36 10N 113 6 E
Changzhou, China 33 C6 31 47N 119 58 E
Chanhanga, Angola 56 B1 16 0S 14 8 E
Channapatna, India 40 N10 12 40N 77 15 E
Channel Is., U.K. 11 H5 49 30N 2 40W
Channel Is., U.S.A. 85 M7 33 40N 119 15W
Channel-Port aux Basques, Canada 71 C8 47 30N 59 9W
Channing, Mich., U.S.A. 76 B1 46 9N 88 5W
Channing, Tex., U.S.A. 81 H3 35 41N 102 20W
Chantada, Spain 19 A2 42 36N 7 46W
Chanthaburi, Thailand 38 F4 12 38N 102 12 E
Chantrey Inlet, Canada 68 B10 67 48N 96 20W
Chanute, U.S.A. 81 G7 37 41N 95 27W
Chao Phraya →, Thailand 38 F3 13 32N 100 36 E
Chao Phraya Lowlands, Thailand 38 E3 15 30N 100 0 E
Chao'an, China 33 D6 23 42N 116 32 E
Chaocheng, China 34 F8 36 4N 115 37 E
Chaoyang, China 35 D11 41 35N 120 22 E
Chapala, Mozam. 55 F4 15 50S 37 35 E
Chapala, L. de, Mexico 86 C4 20 10N 103 20W
Chapayevo, Kazakhstan 25 D9 50 25N 51 10 E
Chapayevsk, Russia 24 D8 53 0N 49 40 E
Chapecó, Brazil 95 B5 27 14S 52 41W
Chapel Hill, U.S.A. 77 H6 35 55N 79 4W
Chapleau, Canada 70 C3 47 50N 83 24W
Chaplin, Canada 73 C7 50 28N 106 40W
Chapra = Chhapra, India 43 G11 25 48N 84 44 E
Châr, Mauritania 50 D2 21 32N 12 45W
Chara, Russia 27 D12 56 54N 118 20 E
Charadai, Argentina 94 B4 27 35S 59 55W
Charagua, Bolivia 92 G6 19 45S 63 10W
Charaña, Bolivia 92 G5 17 30S 69 25W
Charata, Argentina 94 B3 27 13S 61 14W
Charcas, Mexico 86 C4 23 10N 101 20W
Charcoal L., Canada 73 B8 58 49N 102 22W
Chard, U.K. 11 G5 50 52N 2 59W
Chardara, Kazakhstan 26 E7 41 16N 67 59 E
Chardon, U.S.A. 78 E3 41 35N 81 12W
Chardzhou, Turkmenistan 26 F7 39 6N 63 34 E
Charente □, France 18 D4 45 50N 0 16 E
Charente-Maritime □, France 18 D3 45 45N 0 45W
Chari →, Chad 51 F7 12 58N 14 31 E
Chārīkār, Afghan. 40 B6 35 0N 69 10 E
Chariton →, U.S.A. 80 F8 39 19N 92 58W
Chärjew = Chardzhou, Turkmenistan 26 F7 39 6N 63 34 E
Charkhari, India 43 G8 25 24N 79 45 E
Charkhi Dadri, India 42 E7 28 37N 76 17 E
Charleroi, Belgium 15 D4 50 24N 4 27 E
Charleroi, U.S.A. 78 F5 40 9N 79 57W
Charles, C., U.S.A. 76 G8 37 7N 75 58W
Charles City, U.S.A. 80 D8 43 4N 92 41W
Charles L., Canada 73 B6 59 50N 110 33W
Charles Town, U.S.A. 76 F7 39 17N 77 52W
Charleston, Ill., U.S.A. 76 F1 39 30N 88 10W
Charleston, Miss., U.S.A. 81 H9 34 1N 90 4W
Charleston, Mo., U.S.A. 81 G10 36 55N 89 21W
Charleston, S.C., U.S.A. 77 J6 32 46N 79 56W
Charleston, W. Va., U.S.A. 76 F5 38 21N 81 38W
Charleston Peak, U.S.A. 85 J11 36 16N 115 42W
Charlestown, S. Africa 57 D4 27 26S 29 53 E
Charlestown, U.S.A. 76 F3 38 27N 85 40W
Charlesville, Zaïre 52 F4 5 27S 20 59 E
Charleville = Rath Luirc, Ireland 13 D3 52 21N 8 40W
Charleville, Australia 63 D4 26 24S 146 15 E
Charleville-Mézières, France 18 B6 49 44N 4 40 E
Charlevoix, U.S.A. 76 C3 45 19N 85 16W
Charlotte, Mich., U.S.A. 76 D3 42 34N 84 50W
Charlotte, N.C., U.S.A. 77 H5 35 13N 80 51W
Charlotte Amalie, Virgin Is. 89 C7 18 21N 64 56W
Charlotte Harbor, U.S.A. 77 M4 26 50N 82 10W
Charlottesville, U.S.A. 76 F6 38 2N 78 30W
Charlottetown, Canada 71 C7 46 14N 63 8W
Charlton, Australia 63 F3 36 16S 143 24 E
Charlton, U.S.A. 80 E8 40 59N 93 20W
Charlton I., Canada 70 B4 52 0N 79 20W
Charny, Canada 71 C5 46 43N 71 15W
Charolles, France 18 C6 46 27N 4 16 E
Charouine, Algeria 50 C4 29 0N 0 15W
Charre, Mozam. 55 F4 17 13S 35 10 E
Charsadda, Pakistan 42 B4 34 7N 71 45 E
Charters Towers, Australia 62 C4 20 5S 146 13 E
Chartres, France 18 B4 48 29N 1 30 E
Chascomús, Argentina 94 D4 35 30S 58 0W
Chasefu, Zambia 55 E3 11 55S 33 8 E
Chasovnya-Uchurskaya, Russia 27 D14 57 15N 132 50 E
Chāt, Iran 45 B7 37 59N 55 16 E
Château-Salins, France 18 B7 48 50N 6 30 E
Châteaubriant, France 18 C3 47 43N 1 23W
Châteauroux, France 18 C4 46 50N 1 40 E
Châtellerault, France 18 C4 46 50N 0 30 E
Chatfield, U.S.A. 80 D9 43 51N 92 11W
Chatham, N.B., Canada 71 C6 47 2N 65 28W
Chatham, Ont., Canada 70 D3 42 24N 82 11W
Chatham, U.K. 11 F8 51 22N 0 32 E
Chatham, La., U.S.A. 81 J8 32 18N 92 27W
Chatham, N.Y., U.S.A. 79 D11 42 21N 73 36W
Chatham Is., Pac. Oc. 64 M10 44 0S 176 40W
Chatham Str., U.S.A. 72 B2 57 0N 134 40W
Chatmohar, Bangla. 43 G13 24 15N 89 15 E
Chatra, India 43 G11 24 12N 84 56 E
Chatrapur, India 41 K14 19 22N 85 2 E
Chats, L. des, Canada 79 A8 45 30N 76 20W
Chatsworth, Canada 78 B4 44 27N 80 54W
Chatsworth, Zimbabwe 55 F3 19 38S 31 13 E
Chattahoochee →, U.S.A. 77 K3 30 54N 84 57W
Chattanooga, U.S.A. 77 H3 35 3N 85 19W
Chaturat, Thailand 38 E3 15 40N 101 51 E
Chau Doc, Vietnam 39 G5 10 42N 105 7 E
Chauk, Burma 41 J19 20 53N 94 49 E
Chaukan La, Burma 41 F20 27 0N 97 15 E
Chaumont, France 18 B6 48 7N 5 8 E
Chaumont, U.S.A. 79 B8 44 4N 76 8W
Chautauqua L., U.S.A. 78 D5 42 10N 79 24W
Chauvin, Canada 73 C6 52 45N 110 10W
Chaves, Brazil 93 D9 0 15S 49 55W
Chaves, Portugal 19 B2 41 45N 7 32W
Chavuma, Zambia 53 G4 13 4S 22 40 E
Chawang, Thailand 39 H2 8 25N 99 30 E
Chaykovskiy, Russia 24 C9 56 47N 54 9 E
Chazy, U.S.A. 79 B11 44 53N 73 26W
Cheb, Czech. 16 C6 50 9N 12 28 E
Cheboksary, Russia 24 C8 56 8N 47 12 E
Cheboygan, U.S.A. 76 C3 45 39N 84 29W
Chech, Erg, Africa 50 D4 25 0N 2 15W
Checheno-Ingush Republic □, Russia 25 F8 43 30N 45 29 E
Chechon, S. Korea 35 F15 37 8N 128 12 E
Checleset B., Canada 72 C3 50 5N 127 35W
Checotah, U.S.A. 81 H7 35 28N 95 31W
Chedabucto B., Canada 71 C7 45 25N 61 8W
Cheduba I., Burma 41 K18 18 45N 93 40 E
Cheepie, Australia 63 D4 26 33S 145 1 E
Chegdomyn, Russia 27 D14 51 7N 133 1 E
Chegga, Mauritania 50 C3 25 27N 5 40W
Chegutu, Zimbabwe 55 F3 18 10S 30 14 E
Chehalis, U.S.A. 84 D4 46 40N 122 58W
Cheju Do, S. Korea 35 H14 33 29N 126 34 E
Chekiang = Zhejiang □, China 33 D7 29 0N 120 0 E
Chela, Sa. da, Angola 56 B1 16 20S 13 20 E
Chelan, U.S.A. 82 C4 47 51N 120 1W
Chelan, L., U.S.A. 82 C3 48 11N 120 30W
Cheleken, Turkmenistan 25 G9 39 26N 53 7 E
Chelforó, Argentina 96 D3 39 0S 66 33W
Chelkar, Kazakhstan 26 E6 47 48N 59 39 E
Chelkar Tengiz, Solonchak, Kazakhstan 26 E7 48 0N 62 30 E
Chełm, Poland 17 C11 51 8N 23 30 E
Chełmno, Poland 17 B9 53 20N 18 30 E
Chelmsford, U.K. 11 F8 51 44N 0 29 E
Chelmsford Dam, S. Africa 57 D4 27 55S 29 59 E
Chełmża, Poland 17 B9 53 10N 18 39 E
Chelsea, Okla., U.S.A. 81 G7 36 32N 95 26W
Chelsea, Vt., U.S.A. 79 C12 43 59N 72 27W
Cheltenham, U.K. 11 F5 51 55N 2 5W
Chelyabinsk, Russia 26 D7 55 10N 61 24 E
Chelyuskin, C., Russia 28 B14 77 30N 103 0 E
Chemainus, Canada 72 D4 48 55N 123 42W
Chemnitz, Germany 16 C6 50 50N 12 55 E
Chemult, U.S.A. 82 E3 43 14N 121 47W
Chen, Gora, Russia 27 C15 65 16N 141 50 E
Chenab →, Pakistan 42 D4 30 23N 71 2 E
Chenango Forks, U.S.A. 79 D9 42 15N 75 51W
Chencha, Ethiopia 51 G12 6 15N 37 32 E
Cheney, U.S.A. 82 C5 47 30N 117 35W
Cheng Xian, China 34 H3 33 43N 105 42 E
Chengcheng, China 34 G5 35 8N 109 56 E
Chengchou = Zhengzhou, China 34 G7 34 45N 113 34 E
Chengde, China 35 D9 40 59N 117 58 E
Chengdu, China 32 C5 30 38N 104 2 E
Chenggu, China 34 H4 33 10N 107 21 E
Chengjiang, China 32 D5 24 39N 103 0 E
Ch'engtu = Chengdu, China 32 C5 30 38N 104 2 E
Chengwu, China 34 G8 34 58N 115 50 E
Chengyang, China 35 F11 36 18N 120 21 E
Chenjiagang, China 35 G10 34 23N 119 47 E
Chenkán, Mexico 87 D6 19 8N 90 58W
Cheo Reo, Vietnam 38 F7 13 25N 108 28 E
Cheom Ksan, Cambodia 38 E5 14 13N 104 56 E
Chepén, Peru 92 E3 7 15S 79 23W
Chepes, Argentina 94 C2 31 20S 66 35W
Chepo, Panama 88 E4 9 10N 79 6W
Cheptulil, Mt., Kenya 54 B4 1 25N 35 35 E
Chequamegon B., U.S.A. 80 B9 46 40N 90 30W
Cher □, France 18 C5 47 10N 2 30 E
Cher →, France 18 C4 47 21N 0 29 E
Cheraw, U.S.A. 77 H6 34 42N 79 53W
Cherbourg, France 18 B3 49 39N 1 40W
Cherchell, Algeria 50 A5 36 35N 2 12 E
Cherdyn, Russia 24 B10 60 24N 56 29 E
Cheremkhovo, Russia 27 D11 53 8N 103 1 E
Cherepanovo, Russia 26 D9 54 15N 83 30 E
Cherepovets, Russia 24 C6 59 5N 37 55 E
Chergui, Chott ech, Algeria 50 B5 34 21N 0 25 E
Cherkassy, Ukraine 25 E5 49 27N 32 4 E
Cherkasy = Cherkassy, Ukraine 25 E5 49 27N 32 4 E
Cherlak, Russia 26 D8 54 15N 74 55 E
Chernigov, Ukraine 24 D5 51 28N 31 20 E
Chernihiv = Chernigov, Ukraine 24 D5 51 28N 31 20 E
Chernikovsk, Russia 24 D10 54 48N 56 8 E
Chernivtsi = Chernovtsy, Ukraine 25 E4 48 15N 25 52 E
Chernogorsk, Russia 27 D10 53 49N 91 18 E
Chernovtsy, Ukraine 25 E4 48 15N 25 52 E
Chernoye, Russia 27 B9 70 30N 89 10 E
Chernyakhovsk, Russia 26 D3 54 36N 21 48 E
Chernyshovskiy, Russia 27 C12 63 0N 112 30 E
Cherokee, Iowa, U.S.A. 80 D7 42 45N 95 33W
Cherokee, Okla., U.S.A. 81 G5 36 45N 98 21W
Cherokees, Lake O' The, U.S.A. 81 G7 36 28N 95 2W
Cherquenco, Chile 96 D2 38 35S 72 0W
Cherrapunji, India 41 G17 25 17N 91 47 E
Cherry Creek, U.S.A. 82 G6 39 54N 114 53W
Cherry Valley, U.S.A. 85 M10 33 59N 116 57W
Cherryvale, U.S.A. 81 G7 37 16N 95 33W
Cherskiy, Russia 27 C17 68 45N 161 18 E
Cherskogo Khrebet, Russia 27 C15 65 0N 143 0 E
Cherwell →, U.K. 11 F6 51 46N 1 18W
Chesapeake, U.S.A. 76 G7 36 50N 76 17W
Chesapeake B., U.S.A. 76 F7 38 0N 76 10W
Cheshire □, U.K. 10 D5 53 14N 2 30W
Cheshskaya Guba, Russia 24 A8 67 20N 47 0 E
Cheslatta L., Canada 72 C3 53 49N 125 20W
Chesley, Canada 78 B3 44 17N 81 5W
Chester, U.K. 10 D5 53 12N 2 53W
Chester, Calif., U.S.A. 82 F3 40 19N 121 14W
Chester, Ill., U.S.A. 81 G10 37 55N 89 49W
Chester, Mont., U.S.A. 82 B8 48 31N 110 58W
Chester, Pa., U.S.A. 76 F8 39 51N 75 22W
Chester, S.C., U.S.A. 77 H5 34 43N 81 12W
Chesterfield, U.K. 10 D6 53 14N 1 26W
Chesterfield, Is., N. Cal. 64 J7 19 52S 158 15 E
Chesterfield Inlet, Canada 68 B10 63 30N 90 45W
Chesterton Ra., Australia 63 D4 25 30S 147 27 E
Chesterville, Canada 79 A9 45 6N 75 14W
Chesuncook L., U.S.A. 71 C6 46 0N 69 21W
Chéticamp, Canada 71 C7 46 37N 60 59W
Chetumal, B. de, Mexico 87 D7 18 40N 88 10W
Chetwynd, Canada 72 B4 55 45N 121 36W
Cheviot, The, U.K. 10 B5 55 29N 2 8W
Cheviot Hills, U.K. 10 B5 55 20N 2 30W
Cheviot Ra., Australia 62 D3 25 20S 143 45 E
Chew Bahir, Ethiopia 51 H12 4 40N 36 50 E
Chewelah, U.S.A. 82 B5 48 17N 117 43W
Cheyenne, Okla., U.S.A. 81 H5 35 37N 99 40W
Cheyenne, Wyo., U.S.A. 80 E2 41 8N 104 49W
Cheyenne →, U.S.A. 80 C4 44 41N 101 18W
Cheyenne Wells, U.S.A. 80 F3 38 49N 102 21W
Cheyne B., Australia 61 F2 34 35S 118 50 E
Chhabra, India 42 G7 24 40N 76 54 E
Chhapra, India 43 G11 25 48N 84 44 E
Chhata, India 42 F7 27 42N 77 30 E
Chhatarpur, India 43 G8 24 55N 79 35 E
Chhep, Cambodia 38 F5 13 45N 105 24 E
Chhindwara, India 43 H8 22 2N 78 59 E
Chhlong, Cambodia 39 F5 12 15N 105 58 E
Chhuk, Cambodia 39 G5 10 46N 104 28 E
Chi →, Thailand 38 E5 15 11N 104 43 E
Chiai, Taiwan 33 D7 23 29N 120 25 E
Chiamis, Indonesia 37 G13 7 20S 108 21 E
Chiamussu = Jiamusi, China 33 B8 46 40N 130 26 E
Chiang Dao, Thailand 38 C2 19 22N 98 58 E
Chiang Kham, Thailand 38 C3 19 32N 100 18 E
Chiang Khan, Thailand 38 D3 17 52N 101 36 E
Chiang Khong, Thailand 38 B3 20 17N 100 24 E
Chiang Mai, Thailand 38 C2 18 47N 98 59 E
Chiang Saen, Thailand 38 B3 20 16N 100 5 E
Chiange, Angola 53 H2 15 35S 13 40 E
Chiapa →, Mexico 87 D6 16 42N 93 0W
Chiapa de Corzo, Mexico 87 D6 16 42N 93 0W
Chiapas □, Mexico 87 D6 17 0N 92 45W
Chiautla, Mexico 87 D5 18 18N 98 34W
Chiba, Japan 31 G10 35 30N 140 7 E
Chibabava, Mozam. 57 C5 20 17S 33 35 E
Chibatu, Indonesia 37 G12 7 6S 107 59 E
Chibemba, Cunene, Angola 53 H2 15 48S 14 8 E
Chibemba, Huila, Angola 56 B2 16 20S 15 20 E
Chibia, Angola 53 H2 15 10S 13 42 E
Chibougamau, Canada 70 C5 49 56N 74 24W
Chibougamau L., Canada 70 C5 49 50N 74 20W
Chibuk, Nigeria 51 F7 10 52N 12 50 E
Chic-Chocs, Mts., Canada 71 C6 48 55N 66 0W
Chicacole = Srikakulam, India 41 K13 18 14N 83 58 E
Chicago, U.S.A. 76 E2 41 53N 87 38W
Chicago Heights, U.S.A. 76 E2 41 30N 87 38W

Chichagof I.

Chichagof I., U.S.A. 72 B1 57 30N 135 30W
Chicheng, China 34 D8 40 55N 115 55 E
Chichester, U.K. 11 G7 50 50N 0 47W
Chichibu, Japan 31 F9 36 5N 139 10 E
Ch'ich'ihaerh = Qiqihar, China 27 E13 47 26N 124 0 E
Chickasha, U.S.A. 81 H5 35 3N 97 58W
Chiclana de la Frontera, Spain 19 D2 36 26N 6 9W
Chiclayo, Peru 92 E3 6 42S 79 50W
Chico, U.S.A. 84 F5 39 44N 121 50W
Chico →, Chubut, Argentina 96 E3 44 0S 67 0W
Chico →, Santa Cruz, Argentina 96 G3 50 0S 68 30W
Chicomo, Mozam. 57 C5 24 31S 34 6 E
Chicontepec, Mexico 87 C5 20 58N 98 10W
Chicopee, U.S.A. 79 D12 42 9N 72 37W
Chicoutimi, Canada 71 C5 48 28N 71 5W
Chicualacuala, Mozam. 57 C5 22 6S 31 42 E
Chidambaram, India 40 P11 11 20N 79 45 E
Chidenguele, Mozam. 57 C5 24 55S 34 11 E
Chidley, C., Canada 69 B13 60 23N 64 26W
Chiede, Angola 56 B2 17 15S 16 22 E
Chiefs Pt., Canada 78 B3 44 41N 81 18W
Chiem Hoa, Vietnam 38 A5 22 12N 105 17 E
Chiengi, Zambia 55 D2 8 45S 29 10 E
Chiengmai = Chiang Mai, Thailand 38 C2 18 47N 98 59 E
Chiese →, Italy 20 B4 45 8N 10 25 E
Chieti, Italy 20 C6 42 22N 14 10 E
Chifeng, China 35 C10 42 18N 118 58 E
Chignecto B., Canada 71 C7 45 30N 64 40W
Chiguana, Bolivia 94 A2 21 0S 67 58W
Chiha-ri, N. Korea 35 E14 38 40N 126 30 E
Chihli, G. of = Bo Hai, China 35 E10 39 0N 119 0 E
Chihuahua, Mexico 86 B3 28 40N 106 3W
Chihuahua □, Mexico 86 B3 28 40N 106 3W
Chiili, Kazakhstan 26 E7 44 20N 66 15 E
Chik Bollapur, India 40 N10 13 25N 77 45 E
Chikmagalur, India 40 N9 13 15N 75 45 E
Chikwawa, Malawi 55 F3 16 2S 34 50 E
Chilac, Mexico 87 D5 18 20N 97 24W
Chilako →, Canada 72 C4 53 53N 122 57W
Chilam Chavki, Pakistan 43 B6 35 5N 75 5 E
Chilanga, Zambia 55 F2 15 33S 28 16 E
Chilapa, Mexico 87 D5 17 40N 99 11W
Chilas, Pakistan 43 B6 35 25N 74 5 E
Chilaw, Sri Lanka 40 R11 7 30N 79 50 E
Chilcotin →, Canada 72 C4 51 44N 122 23W
Childers, Australia 63 D5 25 15S 152 17 E
Childress, U.S.A. 81 H4 34 25N 100 13W
Chile ■, S. Amer. 96 D2 35 0S 72 0W
Chile Rise, Pac. Oc. 65 L18 38 0S 92 0W
Chilecito, Argentina 94 B2 29 10S 67 30W
Chilete, Peru 92 E3 7 10S 78 50W
Chililabombwe, Zambia 55 E2 12 18S 27 43 E
Chilin = Jilin, China 35 C14 43 44N 126 30 E
Chilka L., India 41 K14 19 40N 85 25 E
Chilko →, Canada 72 C4 52 0N 123 40W
Chilko, L., Canada 72 C4 51 20N 124 10W
Chillagoe, Australia 62 B3 17 7S 144 33 E
Chillán, Chile 94 D1 36 40S 72 10W
Chillicothe, Ill., U.S.A. 80 E10 40 55N 89 29W
Chillicothe, Mo., U.S.A. 80 F8 39 48N 93 33W
Chillicothe, Ohio, U.S.A. 76 F4 39 20N 82 59W
Chilliwack, Canada 72 D4 49 10N 121 54W
Chilo, India 42 F5 27 25N 73 32 E
Chiloane, I., Mozam. 57 C5 20 40S 34 55 E
Chiloé, I. de, Chile 96 E2 42 30S 73 50W
Chilpancingo, Mexico 87 D5 17 30N 99 30W
Chiltern Hills, U.K. 11 F7 51 44N 0 42W
Chilton, U.S.A. 76 C1 44 2N 88 10W
Chiluage, Angola 52 F4 9 30S 21 50 E
Chilubi, Zambia 55 E2 11 5S 29 58 E
Chilubula, Zambia 55 E3 10 14S 30 51 E
Chilumba, Malawi 55 E3 10 28S 34 12 E
Chilung, Taiwan 33 D7 25 3N 121 45 E
Chilwa, L., Malawi 55 F4 15 15S 35 40 E
Chimaltitán, Mexico 86 C4 21 46N 103 50W
Chimán, Panama 88 E4 8 45N 78 40W
Chimay, Belgium 15 D4 50 3N 4 20 E
Chimbay, Uzbekistan 26 E6 42 57N 59 47 E
Chimborazo, Ecuador 92 D3 1 29S 78 55W
Chimbote, Peru 92 E3 9 0S 78 35W
Chimkent, Kazakhstan 26 E7 42 18N 69 36 E
Chimoio, Mozam. 55 F3 19 4S 33 30 E
Chimpembe, Zambia 55 D2 9 31S 29 33 E
Chin □, Burma 41 J18 22 0N 93 0 E
Chin Ling Shan = Qinling Shandi, China 34 H5 33 50N 108 10 E
China, Mexico 87 B5 25 40N 99 20W
China ■, Asia 34 E3 30 0N 110 0 E
China Lake, U.S.A. 85 K9 35 44N 117 37W
Chinan = Jinan, China 34 F9 36 38N 117 1 E
Chinandega, Nic. 88 D2 12 35N 87 12W
Chinati Peak, U.S.A. 81 K2 29 57N 104 29W
Chincha Alta, Peru 92 F3 13 25S 76 7W
Chinchilla, Australia 63 D5 26 45S 150 38 E
Chinchón, Spain 19 B4 40 9N 3 26W
Chinchorro, Banco, Mexico 87 D7 18 35N 87 20W
Chinchou = Jinzhou, China 35 D11 41 5N 121 3 E
Chincoteague, U.S.A. 76 G8 37 56N 75 23W
Chinde, Mozam. 55 F4 18 35S 36 30 E
Chindo, S. Korea 35 G14 34 28N 126 15 E
Chindwin →, Burma 41 J19 21 26N 95 15 E
Chineni, India 43 C6 33 2N 75 15 E
Chinga, Mozam. 55 F4 15 13S 38 35 E
Chingola, Zambia 55 E2 12 31S 27 53 E
Chingole, Malawi 55 E3 13 4S 34 17 E
Ch'ingtao = Qingdao, China 35 F11 36 5N 120 20 E
Chinguetti, Mauritania 50 D2 20 25N 12 24W
Chingune, Mozam. 57 C5 20 33S 34 58 E
Chinhae, S. Korea 35 G15 35 9N 128 47 E
Chinhanguanine, Mozam. 57 D5 25 21S 32 30 E
Chinhoyi, Zimbabwe 55 F3 17 20S 30 8 E
Chiniot, Pakistan 42 D5 31 45N 73 0 E
Chínipas, Mexico 86 B3 27 22N 108 32W
Chinju, S. Korea 35 G15 35 12N 128 2 E
Chinle, U.S.A. 83 H9 36 9N 109 33W
Chinnampo, N. Korea 35 E13 38 52N 125 10 E
Chino, Japan 31 G9 35 59N 138 9 E
Chino, U.S.A. 85 L9 34 1N 117 41W
Chino Valley, U.S.A. 83 J7 34 45N 112 27W
Chinon, France 18 C4 47 10N 0 15 E
Chinook, Canada 73 C6 51 28N 110 59W
Chinook, U.S.A. 82 B9 48 35N 109 14W
Chinsali, Zambia 55 E3 10 30S 32 2 E
Chióggia, Italy 20 B5 45 13N 12 15 E
Chíos = Khíos, Greece 21 E12 38 27N 26 9 E
Chipata, Zambia 55 E3 13 38S 32 28 E
Chipewyan L., Canada 73 B9 58 0N 98 27W
Chipinge, Zimbabwe 55 G3 20 13S 32 28 E
Chipley, U.S.A. 77 K3 30 47N 85 32W
Chipman, Canada 71 C6 46 6N 65 53W
Chipoka, Malawi 55 E3 13 57S 34 28 E
Chippenham, U.K. 11 F5 51 27N 2 7W
Chippewa →, U.S.A. 80 C8 44 25N 92 5W
Chippewa Falls, U.S.A. 80 C9 44 56N 91 24W
Chiquián, Peru 92 F3 10 10S 77 0W
Chiquimula, Guatemala 88 D2 14 51N 89 37W
Chiquinquira, Colombia 92 B4 5 37N 73 50W
Chirala, India 40 M12 15 50N 80 26 E
Chiramba, Mozam. 55 F3 16 55S 34 39 E
Chirawa, India 42 E6 28 14N 75 42 E
Chirchik, Uzbekistan 26 E7 41 29N 69 35 E
Chiricahua Peak, U.S.A. 83 L9 31 51N 109 18W
Chiriquí, G. de, Panama 88 E3 8 0N 82 10W
Chiriquí, L. de, Panama 88 E3 9 10N 82 0W
Chirivira Falls, Zimbabwe 55 G3 21 10S 32 12 E
Chirmiri, India 41 H13 23 15N 82 20 E
Chiromo, Malawi 53 H7 16 30S 35 7 E
Chirripó Grande, Cerro, Costa Rica 88 E3 9 29N 83 29W
Chisamba, Zambia 55 E2 14 55S 28 20 E
Chisapani Garhi, Nepal 41 F14 27 30N 84 2 E
Chisholm, Canada 72 C6 54 55N 114 10W
Chishtian Mandi, Pakistan 42 E5 29 50N 72 55 E
Chisimba Falls, Zambia 55 E3 10 12S 30 56 E
Chişinău = Kishinev, Moldavia 25 E4 47 0N 28 50 E
Chisos Mts., U.S.A. 81 L3 29 5N 103 15W
Chistopol, Russia 24 C9 55 25N 50 38 E
Chita, Russia 27 D12 52 0N 113 35 E
Chitado, Angola 53 H2 17 10S 14 8 E
Chitembo, Angola 53 G3 13 30S 16 50 E
Chitipa, Malawi 55 D3 9 41S 33 19 E
Chitose, Japan 30 C10 42 49N 141 39 E
Chitral, Pakistan 40 B7 35 50N 71 56 E
Chitré, Panama 88 E3 7 59N 80 27W
Chittagong, Bangla. 41 H17 22 19N 91 48 E
Chittagong □, Bangla. 41 G17 24 5N 91 0 E
Chittaurgarh, India 42 G6 24 52N 74 38 E
Chittoor, India 40 N11 13 15N 79 5 E
Chitungwiza, Zimbabwe 55 F3 18 0S 31 6 E
Chiusi, Italy 20 C4 43 1N 11 58 E
Chivasso, Italy 20 B2 45 10N 7 52 E
Chivhu, Zimbabwe 55 F3 19 2S 30 52 E
Chivilcoy, Argentina 94 C4 34 55S 60 0W
Chiwanda, Tanzania 55 E3 11 23S 34 55 E
Chizera, Zambia 55 E1 13 10S 25 0 E
Chkalov = Orenburg, Russia 24 D10 51 45N 55 6 E
Chloride, U.S.A. 85 K12 35 25N 114 12W
Cho Bo, Vietnam 38 B5 20 46N 105 10 E
Cho-do, N. Korea 35 E13 38 30N 124 40 E
Cho Phuoc Hai, Vietnam 39 G6 10 26N 107 18 E
Choba, Kenya 54 B4 2 30N 38 5 E
Chobe National Park, Botswana 56 B3 18 0S 25 0 E
Chochiwŏn, S. Korea 35 F14 36 37N 127 18 E
Choctawhatchee B., U.S.A. 75 D9 30 20N 86 20W
Choele Choel, Argentina 96 D3 39 11S 65 40W
Choix, Mexico 86 B3 26 40N 108 23W
Chojnice, Poland 17 B8 53 42N 17 32 E
Chōkai-San, Japan 30 E10 39 6N 140 3 E
Chokurdakh, Russia 27 B15 70 38N 147 55 E
Cholame, U.S.A. 84 K6 35 44N 120 18W
Cholet, France 18 C3 47 4N 0 52W
Choluteca, Honduras 88 D2 13 20N 87 14W
Choluteca →, Honduras 88 D2 13 0N 87 20W
Chom Bung, Thailand 38 F2 13 37N 99 36 E
Chom Thong, Thailand 38 C2 18 25N 98 41 E
Choma, Zambia 55 F2 16 48S 26 59 E
Chomun, India 42 F6 27 15N 75 40 E
Chomutov, Czech. 16 C6 50 28N 13 23 E
Chon Buri, Thailand 38 F3 13 21N 101 1 E
Chon Thanh, Vietnam 39 G6 11 24N 106 36 E
Chonan, S. Korea 35 F14 36 48N 127 9 E
Chone, Ecuador 92 D2 0 40S 80 0W
Chong Kai, Cambodia 38 F4 13 57N 103 35 E
Chong Mek, Thailand 38 E5 15 10N 105 27 E
Chŏngdo, S. Korea 35 G15 35 38N 128 42 E
Chŏngha, S. Korea 35 F15 36 12N 129 21 E
Chongjin, N. Korea 35 D15 41 47N 129 50 E
Chŏngju, N. Korea 35 E13 39 40N 125 5 E
Chŏngju, S. Korea 35 F14 36 39N 127 27 E
Chongli, China 34 D8 40 58N 115 15 E
Chongqing, China 32 D5 29 35N 106 25 E
Chŏngŭp, S. Korea 35 G14 35 35N 126 50 E
Chŏnju, S. Korea 35 G14 35 50N 127 4 E
Chonos, Arch. de los, Chile 96 F2 45 0S 75 0W
Chopim →, Brazil 95 B5 25 35S 53 5W
Chorbat La, India 43 B7 34 42N 76 37 E
Chorley, U.K. 10 D5 53 39N 2 39W
Chorolque, Cerro, Bolivia 94 A2 20 59S 66 5W
Chorregon, Australia 62 C3 22 40S 143 32 E
Chŏrwŏn, S. Korea 35 E14 38 15N 127 10 E
Chorzów, Poland 17 C9 50 18N 18 57 E
Chos-Malal, Argentina 94 D1 37 20S 70 15W
Chosan, N. Korea 35 D13 40 50N 125 47 E
Choszczno, Poland 16 B7 53 7N 15 25 E
Choteau, U.S.A. 82 C7 47 49N 112 11W
Chotila, India 42 H4 22 23N 71 15 E
Chowchilla, U.S.A. 83 H3 37 7N 120 16W
Choybalsan, Mongolia 33 B6 48 4N 114 30 E
Christchurch, N.Z. 59 K4 43 33S 172 47 E
Christchurch, U.K. 11 G6 50 44N 1 45W
Christian I., Canada 78 B4 44 50N 80 12W
Christiana, S. Africa 56 D4 27 52S 25 8 E
Christiansted, Virgin Is. 89 C7 17 45N 64 42W
Christie B., Canada 73 A6 62 32N 111 10W
Christina →, Canada 73 B6 56 40N 111 3W
Christmas Cr. →, Australia 60 C4 18 29S 125 23 E
Christmas Creek, Australia 60 C4 18 29S 125 23 E
Christmas I. = Kiritimati, Kiribati 65 G12 1 58N 157 27W
Christmas I., Ind. Oc. 64 J2 10 30S 105 40 E
Christopher L., Australia 61 D4 24 49S 127 42 E
Chtimba, Malawi 55 E3 10 35S 34 13 E
Chu, Kazakhstan 26 E8 43 36N 73 42 E
Chu →, Vietnam 38 C5 19 53N 105 45 E
Chu Chua, Canada 72 C4 51 22N 120 10W
Chu Lai, Vietnam 38 E7 15 28N 108 45 E
Ch'uanchou = Quanzhou, China 33 D6 24 55N 118 34 E
Chuankou, China 34 G6 34 20N 110 59 E
Chūbu □, Japan 31 F8 36 45N 137 30 E
Chubut →, Argentina 96 E3 43 20S 65 5W
Chuchi L., Canada 72 B4 55 12N 124 30W
Chudskoye, Oz., Estonia 24 C4 58 13N 27 30 E
Chūgoku □, Japan 31 G6 35 0N 133 0 E
Chūgoku-Sanchi, Japan 31 G6 35 0N 133 0 E
Chugwater, U.S.A. 80 E2 41 46N 104 50W
Chukotskiy Khrebet, Russia 27 C18 68 0N 175 0 E
Chukotskoye More, Russia 27 C19 68 0N 175 0W
Chula Vista, U.S.A. 85 N9 32 39N 117 5W
Chulman, Russia 27 D13 56 52N 124 52 E
Chulucanas, Peru 92 E2 5 8S 80 10W
Chulym →, Russia 26 D9 57 43N 83 51 E
Chum Phae, Thailand 38 D4 16 40N 102 6 E
Chum Saeng, Thailand 38 E3 15 55N 100 15 E
Chumar, India 43 C8 32 40N 78 35 E
Chumbicha, Argentina 94 B2 29 0S 66 10W
Chumikan, Russia 27 D14 54 40N 135 10 E
Chumphon, Thailand 39 G2 10 35N 99 14 E
Chumuare, Mozam. 55 E3 14 31S 31 50 E
Chumunjin, S. Korea 35 F15 37 55N 128 54 E
Chuna →, Russia 27 D10 57 47N 94 37 E
Chunchŏn, S. Korea 35 F14 37 58N 127 44 E
Chunchura, India 43 H13 22 53N 88 27 E
Chunga, Zambia 55 F2 15 0S 26 2 E
Chunggang-ŭp, N. Korea 35 D14 41 48N 126 48 E
Chunghwa, N. Korea 35 E13 38 52N 125 47 E
Chungju, S. Korea 35 F14 36 58N 127 58 E
Chungking = Chongqing, China 32 D5 29 35N 106 25 E
Chungmu, S. Korea 35 G15 34 50N 128 20 E
Chungt'iaoshan = Zhongtiao Shan, China 34 G6 35 0N 111 10 E
Chunian, Pakistan 42 D6 30 57N 74 0 E
Chunya, Tanzania 55 D3 8 30S 33 27 E
Chunya □, Tanzania 54 D3 7 48S 33 0 E
Chunyang, China 35 C15 43 38N 129 23 E
Chuquibamba, Peru 92 G4 15 47S 72 44W
Chuquicamata, Chile 94 A2 22 15S 69 0W
Chur, Switz. 16 E4 46 52N 9 32 E
Churachandpur, India 41 G18 24 20N 93 40 E
Churchill, Canada 73 B10 58 47N 94 11W
Churchill →, Man., Canada 73 B10 58 47N 94 12W
Churchill →, Nfld., Canada 71 B7 53 19N 60 10W
Churchill, C., Canada 73 B10 58 46N 93 12W
Churchill Falls, Canada 71 B7 53 36N 64 19W
Churchill L., Canada 73 B7 55 55N 108 20W
Churchill Pk., Canada 72 B3 58 10N 125 10W
Churu, India 42 E6 28 20N 74 50 E
Chushal, India 43 C8 33 40N 78 40 E
Chusovoy, Russia 24 C10 58 15N 57 40 E
Chuuronjang, N. Korea 35 D15 41 35N 129 40 E
Chuvash Republic □, Russia 24 C8 55 30N 47 0 E
Chuwārtah, Iraq 44 C5 35 43N 45 34 E
Ci Xian, China 34 F8 36 20N 114 25 E
Cianjur, Indonesia 37 G12 6 49S 107 8 E
Cibadok, Indonesia 37 G12 6 53S 106 47 E
Cibatu, Indonesia 37 G12 7 8S 107 59 E
Cibola, U.S.A. 85 M12 33 17N 114 42W
Cicero, U.S.A. 76 E2 41 48N 87 48W
Ciechanów, Poland 17 B10 52 52N 20 38 E
Ciego de Avila, Cuba 88 B4 21 50N 78 50W
Ciénaga, Colombia 92 A4 11 1N 74 15W
Cienfuegos, Cuba 88 B3 22 10N 80 30W
Cieszyn, Poland 17 D9 49 45N 18 35 E
Cieza, Spain 19 C5 38 17N 1 23W
Cihuatlán, Mexico 86 D4 19 14N 104 35W
Cijulang, Indonesia 37 G13 7 42S 108 27 E
Cikajang, Indonesia 37 G12 7 25S 107 48 E
Cikampek, Indonesia 37 G12 6 23S 107 28 E
Cilacap, Indonesia 37 G13 7 43S 109 0 E
Cima, U.S.A. 85 K11 35 14N 115 30W
Cimahi, Indonesia 37 G12 6 53S 107 33 E
Cimarron, Kans., U.S.A. 81 G4 37 48N 100 21W
Cimarron, N. Mex., U.S.A. 81 G2 36 31N 104 55W
Cimarron →, U.S.A. 81 G6 36 10N 96 17W
Cimone, Mte., Italy 20 B4 44 10N 10 40 E
Cimpina, Romania 17 F12 45 10N 25 45 E
Cimpulung, Romania 17 F12 45 17N 25 3 E
Cinca →, Spain 19 B6 41 26N 0 21 E
Cincinnati, U.S.A. 76 F3 39 6N 84 31W
Ciney, Belgium 15 D5 50 18N 5 5 E
Cinto, Mte., France 18 E8 42 24N 8 54 E
Circle, Alaska, U.S.A. 68 B5 65 50N 144 4W
Circle, Mont., U.S.A. 80 B2 47 25N 105 35W
Circleville, Ohio, U.S.A. 76 F4 39 36N 82 57W
Circleville, Utah, U.S.A. 83 G7 38 10N 112 16W
Cirebon, Indonesia 37 G13 6 45S 108 32 E
Cirencester, U.K. 11 F6 51 43N 1 59W
Cirium, Cyprus 23 E11 34 40N 32 53 E
Cisco, U.S.A. 81 J5 32 23N 98 59W
Ciskei □, S. Africa 57 E4 33 0S 27 0 E
Citlaltépetl, Mexico 87 D5 19 0N 97 20W
Citrus Heights, U.S.A. 84 G5 38 42N 121 17W
Citrusdal, S. Africa 56 E2 32 35S 19 0 E
Ciudad Altamirano, Mexico 86 D4 18 20N 100 40W
Ciudad Bolívar, Venezuela 92 B6 8 5N 63 36W
Ciudad Camargo, Mexico 86 B3 27 41N 105 10W
Ciudad Chetumal, Mexico 87 D7 18 30N 88 20W
Ciudad de Valles, Mexico 87 C5 22 0N 99 0W
Ciudad del Carmen, Mexico 87 D6 18 38N 91 50W
Ciudad Delicias = Delicias, Mexico 86 B3 28 10N 105 30W
Ciudad Guayana, Venezuela 92 B6 8 0N 62 30W
Ciudad Guerrero, Mexico 86 B3 28 33N 107 28W
Ciudad Guzmán, Mexico 86 D4 19 40N 103 30W
Ciudad Juárez, Mexico 86 A3 31 40N 106 28W
Ciudad Madero, Mexico 87 C5 22 19N 97 50W
Ciudad Mante, Mexico 87 C5 22 50N 99 0W
Ciudad Obregón, Mexico 86 B3 27 28N 109 59W
Ciudad Real, Spain 19 C4 38 59N 3 55W
Ciudad Rodrigo, Spain 19 B2 40 35N 6 32W
Ciudad Trujillo = Santo Domingo, Dom. Rep. 89 C6 18 30N 69 59W
Ciudad Victoria, Mexico 87 C5 23 41N 99 9W
Ciudadela, Spain 22 B10 40 0N 3 50 E
Civitanova Marche, Italy 20 C5 43 18N 13 41 E
Civitavécchia, Italy 20 C4 42 6N 11 46 E
Cizre, Turkey 25 G7 37 19N 42 10 E
Clacton-on-Sea, U.K. 11 F9 51 47N 1 10 E
Claire, L., Canada 72 B6 58 35N 112 5W
Clairemont, U.S.A. 81 J4 33 9N 100 44W
Clairton, U.S.A. 78 F5 40 18N 79 53W
Clallam Bay, U.S.A. 84 B2 48 15N 124 16W
Clanton, U.S.A. 77 J2 32 51N 86 38W
Clanwilliam, S. Africa 56 E2 32 11S 18 52 E
Clara, Ireland 13 C4 53 20N 7 38W
Clara →, Australia 62 B3 19 8S 142 30 E
Claraville, U.S.A. 85 K8 35 24N 118 20W
Clare, Australia 63 E2 33 50S 138 37 E
Clare, U.S.A. 76 D3 43 49N 84 46W
Clare □, Ireland 13 D3 52 45N 9 0W
Clare →, Ireland 13 C2 53 22N 9 5W
Clare I., Ireland 13 C2 53 48N 10 0W
Claremont, Calif., U.S.A. 85 L9 34 6N 117 43W
Claremont, N.H., U.S.A. 79 C12 43 23N 72 20W
Claremont Pt., Australia 62 A3 14 1S 143 41 E
Claremore, U.S.A. 81 G7 36 19N 95 36W
Claremorris, Ireland 13 C3 53 45N 9 0W
Clarence →, Australia 63 D5 29 25S 153 22 E
Clarence →, N.Z. 59 K4 42 10S 173 56 E
Clarence, I., Chile 96 G2 54 0S 72 0W
Clarence I., Antarctica 5 C18 61 10S 54 0W
Clarence Str., Australia 60 B5 12 0S 131 0 E
Clarence Str., U.S.A. 72 B2 55 40N 132 10W
Clarence Town, Bahamas 89 B5 23 6N 74 59W
Clarendon, Ark., U.S.A. 81 H9 34 42N 91 19W
Clarendon, Tex., U.S.A. 81 H4 34 56N 100 53W
Clarenville, Canada 71 C9 48 10N 54 1W
Claresholm, Canada 72 C6 50 0N 113 33W
Clarie Coast, Antarctica 5 C9 68 0S 135 0 E
Clarinda, U.S.A. 80 E7 40 44N 95 2W
Clarion, Iowa, U.S.A. 80 D8 42 44N 93 44W
Clarion, Pa., U.S.A. 78 E5 41 13N 79 23W
Clarion →, U.S.A. 78 E5 41 7N 79 41W
Clark, U.S.A. 80 C6 44 53N 97 44W
Clark, Pt., Canada 78 B3 44 4N 81 45W
Clark Fork, U.S.A. 82 B5 48 9N 116 11W
Clark Fork →, U.S.A. 82 B5 48 9N 116 15W
Clark Hill Res., U.S.A. 77 J4 33 45N 82 20W
Clarkdale, U.S.A. 83 J7 34 46N 112 3W
Clarke City, Canada 71 B6 50 12N 66 38W
Clarke I., Australia 62 G4 40 32S 148 10 E
Clarke L., Canada 73 C7 54 24N 106 54W
Clarke Ra., Australia 62 C4 20 40S 148 30 E
Clark's Fork →, U.S.A. 82 D9 45 39N 108 43W
Clark's Harbour, Canada 71 D6 43 25N 65 38W
Clarks Summit, U.S.A. 79 E9 41 30N 75 42W
Clarksburg, U.S.A. 76 F5 39 17N 80 30W
Clarksdale, U.S.A. 81 H9 34 12N 90 35W
Clarkston, U.S.A. 82 C5 46 25N 117 3W
Clarksville, Ark., U.S.A. 81 H8 35 28N 93 28W
Clarksville, Tenn., U.S.A. 77 G2 36 32N 87 21W
Clarksville, Tex., U.S.A. 81 J7 33 37N 95 3W
Clatskanie, U.S.A. 84 D3 46 6N 123 12W
Claude, U.S.A. 81 H4 35 7N 101 22W
Claveria, Phil. 37 A6 18 37N 121 4 E
Clay, U.S.A. 84 G5 38 17N 121 10W
Clay Center, U.S.A. 80 F6 39 23N 97 8W
Claypool, U.S.A. 83 K8 33 25N 110 51W
Claysville, U.S.A. 78 F4 40 7N 80 25W
Clayton, Idaho, U.S.A. 82 D6 44 16N 114 24W
Clayton, N. Mex., U.S.A. 81 G3 36 27N 103 11W
Cle Elum, U.S.A. 82 C3 47 12N 120 56W
Clear, C., Ireland 13 E2 51 26N 9 30W
Clear I., Ireland 13 E2 51 26N 9 30W
Clear L., U.S.A. 84 F4 39 2N 122 47W
Clear Lake, S. Dak., U.S.A. 80 C6 44 45N 96 41W
Clear Lake, Wash., U.S.A. 82 B2 48 27N 122 15W
Clear Lake Reservoir, U.S.A. 82 F3 41 56N 121 5W
Clearfield, Pa., U.S.A. 76 E6 41 2N 78 27W
Clearfield, Utah, U.S.A. 82 F7 41 7N 112 2W
Clearlake Highlands, U.S.A. 84 G4 38 57N 122 38W
Clearmont, U.S.A. 82 D10 44 38N 106 23W
Clearwater, Canada 72 C4 51 38N 120 2W
Clearwater, U.S.A. 77 M4 27 58N 82 48W
Clearwater →, Alta., Canada 72 C6 52 22N 114 57W
Clearwater →, Alta., Canada 73 B6 56 44N 111 23W
Clearwater Cr. →, Canada 72 A3 61 36N 125 30W
Clearwater Mts., U.S.A. 82 C6 46 5N 115 20W
Clearwater Prov. Park, Canada 73 C8 54 0N 101 0W
Cleburne, U.S.A. 81 J6 32 21N 97 23W
Cleethorpes, U.K. 10 D7 53 33N 0 2W
Cleeve Hill, U.K. 11 F6 51 54N 2 0W
Clerke Reef, Australia 60 C2 17 22S 119 20 E
Clermont, Australia 62 C4 22 49S 147 39 E

Cooper Cr.

Clermont-Ferrand, France 18 D5 45 46N 3 4 E
Clervaux, Lux. 15 D6 50 4N 6 2 E
Cleveland, Australia 63 D5 27 30S 153 15 E
Cleveland, Miss., U.S.A. 81 J9 33 45N 90 43W
Cleveland, Ohio, U.S.A. 78 E3 41 30N 81 42W
Cleveland, Okla., U.S.A. 81 G6 36 19N 96 28W
Cleveland, Tenn., U.S.A. 77 H3 35 10N 84 53W
Cleveland, Tex., U.S.A. 81 K7 30 21N 95 5W
Cleveland □, U.K. 10 C9 54 35N 1 8 E
Cleveland, C., Australia 62 B4 19 11S 147 1 E
Cleveland Heights, U.S.A. 78 E3 41 30N 81 34W
Clevelândia, Brazil 95 B5 26 24S 52 23W
Clew B., Ireland 13 C2 53 54N 9 50W
Clewiston, U.S.A. 77 M5 26 45N 80 56W
Clifden, Ireland 13 C1 53 30N 10 2W
Clifden, N.Z. 59 M1 46 1S 167 42 E
Cliffdell, U.S.A. 84 D5 46 56N 121 5W
Clifton, Australia 63 D5 27 59S 151 53 E
Clifton, Ariz., U.S.A. 83 K9 33 3N 109 18W
Clifton, Tex., U.S.A. 81 K6 31 47N 97 35W
Clifton Beach, Australia 62 B4 16 46S 145 39 E
Clifton Forge, U.S.A. 76 G6 37 49N 79 50W
Clifton Hills, Australia 63 D2 27 1S 138 54 E
Climax, Canada 73 D7 49 10N 108 20W
Clinch →, U.S.A. 77 H3 35 53N 84 29W
Clingmans Dome, U.S.A. 77 H4 35 34N 83 30W
Clint, U.S.A. 83 L10 31 35N 106 14W
Clinton, B.C., Canada 72 C4 51 6N 121 35W
Clinton, Ont., Canada 70 D3 43 37N 81 32W
Clinton, N.Z. 59 M2 46 12S 169 23 E
Clinton, Ark., U.S.A. 81 H8 35 36N 92 28W
Clinton, Ill., U.S.A. 80 E10 40 9N 88 57W
Clinton, Ind., U.S.A. 76 F2 39 40N 87 24W
Clinton, Iowa, U.S.A. 80 E9 41 51N 90 12W
Clinton, Mass., U.S.A. 79 D13 42 25N 71 41W
Clinton, Mo., U.S.A. 80 F8 38 22N 93 46W
Clinton, N.C., U.S.A. 77 H6 35 0N 78 22W
Clinton, Okla., U.S.A. 81 H5 35 31N 98 58W
Clinton, S.C., U.S.A. 77 H5 34 29N 81 53W
Clinton, Tenn., U.S.A. 77 G3 36 6N 84 8W
Clinton, Wash., U.S.A. 84 C4 47 59N 122 21W
Clinton C., Australia 62 C5 22 30S 150 45 E
Clinton Colden L., Canada 68 B9 63 58N 107 27W
Clintonville, U.S.A. 80 C10 44 37N 88 46W
Clipperton, I., Pac. Oc. 65 F17 10 18N 109 13W
Clive L., Canada 72 A5 63 13N 118 54W
Cloates, Pt., Australia 60 D1 22 43S 113 40 E
Clocolan, S. Africa 57 D4 28 55S 27 34 E
Clodomira, Argentina 94 B3 27 35S 64 14W
Clonakilty, Ireland 13 E3 51 37N 8 53W
Clonakilty B., Ireland 13 E3 51 33N 8 50W
Cloncurry, Australia 62 C3 20 40S 140 28 E
Cloncurry →, Australia 62 B3 18 37S 140 40 E
Clones, Ireland 13 B4 54 10N 7 13W
Clonmel, Ireland 13 D4 52 22N 7 42W
Cloquet, U.S.A. 80 B8 46 43N 92 28W
Clorinda, Argentina 94 B4 25 16S 57 45W
Cloud Peak, U.S.A. 82 D10 44 23N 107 11W
Cloudcroft, U.S.A. 83 K11 32 58N 105 45W
Cloverdale, U.S.A. 84 G4 38 48N 123 1W
Clovis, Calif., U.S.A. 83 H4 36 49N 119 42W
Clovis, N. Mex., U.S.A. 81 H3 34 24N 103 12W
Cluj-Napoca, Romania 17 E11 46 47N 23 38 E
Clunes, Australia 63 F3 37 20S 143 45 E
Cluny, France 18 C6 46 26N 4 38 E
Clutha →, N.Z. 59 M2 46 20S 169 49 E
Clwyd □, U.K. 10 D4 53 5N 3 20W
Clwyd →, U.K. 10 D4 53 20N 3 30W
Clyde, N.Z. 59 L2 45 12S 169 20 E
Clyde, U.S.A. 78 C8 43 5N 76 52W
Clyde →, U.K. 12 F4 55 56N 4 29W
Clyde, Firth of, U.K. 12 F4 55 20N 5 0W
Clyde River, Canada 69 A13 70 30N 68 30W
Clydebank, U.K. 12 F4 55 54N 4 25W
Clymer, U.S.A. 78 D5 40 40N 79 1W
Coachella, U.S.A. 85 M10 33 41N 116 10W
Coachella Canal, U.S.A. 85 N12 32 43N 114 57W
Coahoma, U.S.A. 81 J4 32 18N 101 18W
Coahuayana →, Mexico 86 D4 18 41N 103 45W
Coahuayutla, Mexico 86 D4 18 19N 101 42W
Coahuila □, Mexico 86 B4 27 0N 103 0W
Coal →, Canada 72 B3 59 39N 126 57W
Coalane, Mozam. 55 F4 17 48S 37 2 E
Coalcomán, Mexico 86 D4 18 40N 103 10W
Coaldale, Canada 72 D6 49 45N 112 35W
Coalgate, U.S.A. 81 H6 34 32N 96 13W
Coalinga, U.S.A. 83 H3 36 9N 120 21W
Coalville, U.K. 10 E6 52 43N 1 21W
Coalville, U.S.A. 82 F8 40 55N 111 24W
Coari, Brazil 92 D6 4 8S 63 7W
Coast □, Kenya 54 C4 2 40S 39 45 E
Coast Mts., Canada 72 C3 55 0N 129 20W
Coast Ranges, U.S.A. 84 E4 41 0N 123 0W
Coatbridge, U.K. 12 F4 55 52N 4 2W
Coatepec, Mexico 87 D5 19 27N 96 58W
Coatepeque, Guatemala 88 D1 14 46N 91 55W
Coatesville, U.S.A. 76 F8 39 59N 75 50W
Coaticook, Canada 71 C5 45 10N 71 46W
Coats I., Canada 69 B11 62 30N 83 0W
Coats Land, Antarctica 5 D1 77 0S 25 0W
Coatzacoalcos, Mexico 87 D6 18 7N 94 25W
Cobalt, Canada 70 C4 47 25N 79 42W
Cobán, Guatemala 88 C1 15 30N 90 21W
Cobar, Australia 63 E4 31 27S 145 48 E
Cóbh, Ireland 13 E3 51 50N 8 18W
Cobham, Australia 63 E3 30 18S 142 7 E
Cobija, Bolivia 92 F5 11 0S 68 50W
Cobleskill, U.S.A. 79 D10 42 41N 74 29W
Coboconk, Canada 78 B6 44 39N 78 48W
Cobourg, Canada 70 D4 43 58N 78 10W
Cobourg Pen., Australia 60 B5 11 20S 132 15 E
Cobram, Australia 63 F4 35 54S 145 40 E
Cobre, U.S.A. 82 F6 41 7N 114 24W
Cóbué, Mozam. 55 E3 12 0S 34 58 E
Coburg, Germany 16 C5 50 15N 10 58 E
Cocanada = Kakinada, India 41 L13 16 57N 82 11 E
Cochabamba, Bolivia 92 G5 17 26S 66 10W
Cochemane, Mozam. 55 F3 17 0S 32 54 E
Cochin, India 40 Q10 9 59N 76 22 E
Cochin China = Nam-Phan, Vietnam 39 G6 10 30N 106 0 E
Cochise, U.S.A. 83 K9 32 7N 109 55W
Cochran, U.S.A. 77 J4 32 23N 83 21W
Cochrane, Alta., Canada 72 C6 51 11N 114 30W
Cochrane, Ont., Canada 70 C3 49 0N 81 0W
Cochrane →, Canada 73 B8 59 0N 103 40W
Cochrane, L., Chile 96 F2 47 10S 72 0W
Cockburn, Australia 63 E3 32 5S 141 0 E
Cockburn, Canal, Chile 96 G2 54 30S 72 0W
Cockburn I., Canada 70 C3 45 55N 83 22W
Cockburn Ra., Australia 60 C4 15 46S 128 0 E
Cocklebiddy Motel, Australia 61 F4 32 0S 126 3 E
Coco →, Cent. Amer. 88 D3 15 0N 83 8W
Cocoa, U.S.A. 77 L5 28 21N 80 44W
Cocobeach, Gabon 52 D1 0 59N 9 34 E
Cocos, I. del, Pac. Oc. 65 G19 5 25N 87 55W
Cocos Is., Ind. Oc. 64 J1 12 10S 96 55 E
Cod, C., U.S.A. 75 B13 42 5N 70 10W
Codajás, Brazil 92 D6 3 55S 62 0W
Coderre, Canada 73 C7 50 11N 106 31W
Codó, Brazil 93 D10 4 30S 43 55W
Cody, U.S.A. 82 D9 44 32N 109 3W
Coe Hill, Canada 70 D4 44 52N 77 50W
Coelemu, Chile 94 D1 36 30S 72 48W
Coen, Australia 62 A3 13 52S 143 12 E
Cœur d'Alene, U.S.A. 82 C5 47 45N 116 51W
Cœur d'Alene L., U.S.A. 82 C5 47 32N 116 48W
Coevorden, Neths. 15 B6 52 40N 6 44 E
Cofete, Canary Is. 22 F5 28 6N 14 23W
Coffeyville, U.S.A. 81 G7 37 2N 95 37W
Coffin B., Australia 63 E2 34 38S 135 28 E
Coffin Bay Peninsula, Australia 63 E2 34 32S 135 15 E
Coffs Harbour, Australia 63 E5 30 16S 153 5 E
Coghinas →, Italy 20 D3 40 55N 8 48 E
Cognac, France 18 D3 45 41N 0 20W
Cohagen, U.S.A. 82 C10 47 3N 106 37W
Cohoes, U.S.A. 79 D11 42 46N 73 42W
Cohuna, Australia 63 F3 35 45S 144 15 E
Coiba, I., Panama 88 E3 7 30N 81 40W
Coig →, Argentina 96 G3 51 0S 69 10W
Coihaique, Chile 96 F2 45 30S 71 45W
Coimbatore, India 40 P10 11 2N 76 59 E
Coimbra, Brazil 92 G7 19 55S 57 48W
Coimbra, Portugal 19 B1 40 15N 8 27W
Coín, Spain 19 D3 36 40N 4 48W
Cojimies, Ecuador 92 C2 0 20N 80 0W
Cojutepequé, El Salv. 88 D2 13 41N 88 54W
Cokeville, U.S.A. 82 E8 42 5N 110 57W
Colac, Australia 63 F3 38 21S 143 35 E
Colatina, Brazil 93 G10 19 32S 40 37W
Colbeck, C., Antarctica 5 D17 77 6S 157 48W
Colbinabbin, Australia 63 F3 36 38S 144 48 E
Colborne, Canada 78 B7 44 0N 77 53W
Colby, U.S.A. 80 F4 39 24N 101 3W
Colchagua □, Chile 94 C1 34 30S 71 0W
Colchester, U.K. 11 F8 51 54N 0 55 E
Coldstream, U.K. 12 F6 55 39N 2 14W
Coldwater, Canada 78 B5 44 42N 79 40W
Coldwater, U.S.A. 81 G5 37 16N 99 20W
Colebrook, Australia 62 G4 42 31S 147 21 E
Colebrook, U.S.A. 79 B13 44 54N 71 30W
Coleman, Canada 72 D6 49 40N 114 30W
Coleman, U.S.A. 81 K5 31 50N 99 26W
Coleman →, Australia 62 B3 15 6S 141 38 E
Colenso, S. Africa 57 D4 28 44S 29 50 E
Coleraine, Australia 63 F3 37 36S 141 40 E
Coleraine, U.K. 13 A5 55 8N 6 40W
Coleraine □, U.K. 13 A5 55 8N 6 40W
Coleridge, L., N.Z. 59 K3 43 17S 171 30 E
Colesberg, S. Africa 56 E4 30 45S 25 5 E
Coleville, U.S.A. 84 G7 38 34N 119 30W
Colfax, Calif., U.S.A. 84 F6 39 6N 120 57W
Colfax, La., U.S.A. 81 K8 31 31N 92 42W
Colfax, Wash., U.S.A. 82 C5 46 53N 117 22W
Colhué Huapi, L., Argentina 96 F3 45 30S 69 0W
Coligny, S. Africa 57 D4 26 17S 26 15 E
Colima, Mexico 86 D4 19 10N 103 40W
Colima □, Mexico 86 D4 19 10N 103 40W
Colima, Nevado de, Mexico 86 D4 19 35N 103 45W
Colina, Chile 94 C1 33 13S 70 45W
Colinas, Brazil 93 E10 6 0S 44 10W
Coll, U.K. 12 E2 56 40N 6 35W
Collaguasi, Chile 94 A2 21 5S 68 45W
Collarenebri, Australia 63 D4 29 33S 148 34 E
Collbran, U.S.A. 83 G10 39 14N 107 58W
Colleen Bawn, Zimbabwe 55 G2 21 0S 29 12 E
College Park, U.S.A. 77 J3 33 40N 84 27W
Collette, Canada 71 C6 46 40N 65 30W
Collie, Australia 61 F2 33 22S 116 8 E
Collier B., Australia 60 C3 16 10S 124 15 E
Collier Ra., Australia 60 D2 24 45S 119 10 E
Collingwood, Canada 78 B4 44 29N 80 13W
Collingwood, N.Z. 59 J4 40 41S 172 40 E
Collins, Canada 70 B2 50 17N 89 27W
Collinsville, Australia 62 C4 20 30S 147 56 E
Collipulli, Chile 94 D1 37 55S 72 30W
Collooney, Ireland 13 B3 54 11N 8 28W
Colmar, France 18 B7 48 5N 7 20 E
Colne, U.K. 10 D5 53 51N 2 11W
Colo →, Australia 63 E5 33 25S 150 52 E
Cologne = Köln, Germany 16 C3 50 56N 6 58 E
Colom, I., Spain 22 B11 39 58N 4 16 E
Coloma, U.S.A. 84 G6 38 48N 120 53W
Colomb-Béchar = Béchar, Algeria 50 B4 31 38N 2 18W
Colômbia, Brazil 93 H9 20 10S 48 40W
Colombia ■, S. Amer. 92 C4 3 45N 73 0W
Colombo, Sri Lanka 40 R11 6 56N 79 58 E
Colome, U.S.A. 80 D5 43 16N 99 43W
Colón, Argentina 94 C4 32 12S 58 10W
Colón, Cuba 88 B3 22 42N 80 54W
Colón, Panama 88 E4 9 20N 79 54W
Colona, Australia 61 F5 31 38S 132 4 E
Colonia, Uruguay 94 C4 34 25S 57 50W
Colonia de San Jordi, Spain 22 B9 39 19N 2 59 E
Colonia Dora, Argentina 94 B3 28 34S 62 59W
Colonial Heights, U.S.A. 76 G7 37 15N 77 25W
Colonsay, Canada 73 C7 51 59N 105 52W
Colonsay, U.K. 12 E2 56 4N 6 12W
Colorado □, U.S.A. 83 G10 39 30N 105 30W
Colorado →, Argentina 96 D4 39 50S 62 8W
Colorado →, N. Amer. 83 L6 31 45N 114 40W
Colorado →, U.S.A. 81 L7 28 36N 95 59W
Colorado City, U.S.A. 81 J4 32 24N 100 52W
Colorado Desert, U.S.A. 74 D3 34 20N 116 0W
Colorado Plateau, U.S.A. 83 H8 37 0N 111 0W
Colorado River Aqueduct, U.S.A. 85 L12 34 17N 114 10W
Colorado Springs, U.S.A. 80 F2 38 50N 104 49W
Colotlán, Mexico 86 C4 22 6N 103 16W
Colton, Calif., U.S.A. 85 L9 34 4N 117 20W
Colton, N.Y., U.S.A. 79 B10 44 33N 74 56W
Colton, Wash., U.S.A. 82 C5 46 34N 117 8W
Columbia, La., U.S.A. 81 J8 32 6N 92 5W
Columbia, Miss., U.S.A. 81 K10 31 15N 89 50W
Columbia, Mo., U.S.A. 80 F8 38 57N 92 20W
Columbia, Pa., U.S.A. 79 F8 40 2N 76 30W
Columbia, S.C., U.S.A. 77 H5 34 0N 81 2W
Columbia, Tenn., U.S.A. 77 H2 35 37N 87 2W
Columbia →, U.S.A. 82 C1 46 15N 124 5W
Columbia, C., Canada 4 A4 83 0N 70 0W
Columbia, District of □, U.S.A. 76 F7 38 55N 77 0W
Columbia, Mt., Canada 72 C5 52 8N 117 20W
Columbia Basin, U.S.A. 82 C4 46 45N 119 5W
Columbia Falls, U.S.A. 82 B6 48 23N 114 11W
Columbia Heights, U.S.A. 80 C8 45 3N 93 15W
Columbiana, U.S.A. 78 F4 40 53N 80 42W
Columbretes, Is., Spain 19 C6 39 50N 0 50 E
Columbus, Ga., U.S.A. 77 J3 32 28N 84 59W
Columbus, Ind., U.S.A. 76 F3 39 13N 85 55W
Columbus, Kans., U.S.A. 81 G7 37 10N 94 50W
Columbus, Miss., U.S.A. 77 J1 33 30N 88 25W
Columbus, Mont., U.S.A. 82 D9 45 38N 109 15W
Columbus, N. Dak., U.S.A. 80 A3 48 54N 102 47W
Columbus, N. Mex., U.S.A. 83 L10 31 50N 107 38W
Columbus, Nebr., U.S.A. 80 E6 41 26N 97 22W
Columbus, Ohio, U.S.A. 76 F4 39 58N 83 0W
Columbus, Tex., U.S.A. 81 L6 29 42N 96 33W
Columbus, Wis., U.S.A. 80 D10 43 21N 89 1W
Colusa, U.S.A. 84 F4 39 13N 122 1W
Colville, U.S.A. 82 B5 48 33N 117 54W
Colville →, U.S.A. 68 A4 70 25N 150 30W
Colville, C., N.Z. 59 G5 36 29S 175 21 E
Colwyn Bay, U.K. 10 D4 53 17N 3 44W
Comácchio, Italy 20 B5 44 41N 12 10 E
Comalcalco, Mexico 87 D6 18 16N 93 13W
Comallo, Argentina 96 E2 41 0S 70 5W
Comanche, Okla., U.S.A. 81 H6 34 22N 97 58W
Comanche, Tex., U.S.A. 81 K5 31 54N 98 36W
Comayagua, Honduras 88 D2 14 25N 87 37W
Combahee →, U.S.A. 77 J5 32 30N 80 31W
Comber, Canada 78 D2 42 14N 82 33W
Comblain-au-Pont, Belgium 15 D5 50 29N 5 35 E
Comeragh Mts., Ireland 13 D4 52 17N 7 35W
Comet, Australia 62 C4 23 36S 148 38 E
Comilla, Bangla. 41 H17 23 28N 91 10 E
Comino, Malta 23 C1 36 2N 14 20 E
Comitán, Mexico 87 D6 16 18N 92 9W
Commerce, Ga., U.S.A. 77 H4 34 12N 83 28W
Commerce, Tex., U.S.A. 81 J7 33 15N 95 54W
Committee B., Canada 69 B11 68 30N 86 30W
Commonwealth B., Antarctica 5 C10 67 0S 144 0 E
Commonwealth of Independent States ■, Eurasia 27 D11 60 0N 100 0 E
Commoron Cr. →, Australia 63 D5 28 22S 150 8 E
Communism Pk. = Kommunizma, Pik, Tajikistan 26 F8 39 0N 72 2 E
Como, Italy 20 B3 45 48N 9 5 E
Como, L. di, Italy 20 A3 46 5N 9 17 E
Comodoro Rivadavia, Argentina 96 F3 45 50S 67 40W
Comorin, C., India 40 Q10 8 3N 77 40 E
Comoro Is. = Comoros ■, Ind. Oc. 49 H8 12 10S 44 15 E
Comoros ■, Ind. Oc. 49 H8 12 10S 44 15 E
Comox, Canada 72 D4 49 42N 124 55W
Compiègne, France 18 B5 49 24N 2 50 E
Compostela, Mexico 86 C4 21 15N 104 53W
Comprida, I., Brazil 95 A6 24 50S 47 42W
Compton, U.S.A. 85 M8 33 54N 118 13W
Compton Downs, Australia 63 E4 30 28S 146 30 E
Con Cuong, Vietnam 38 C5 19 2N 104 54 E
Con Son, Is., Vietnam 39 H6 8 41N 106 37 E
Conakry, Guinea 50 G2 9 29N 13 49W
Conara Junction, Australia 62 G4 41 50S 147 26 E
Concarneau, France 18 C2 47 52N 3 56W
Conceição, Mozam. 55 F4 18 47S 36 7 E
Conceição da Barra, Brazil 93 G11 18 35S 39 45W
Conceição do Araguaia, Brazil 93 E9 8 0S 49 2W
Concepción, Argentina 94 B2 27 20S 65 35W
Concepción, Bolivia 92 G6 16 15S 62 8W
Concepción, Chile 94 D1 36 50S 73 0W
Concepción, Mexico 87 D6 18 15N 90 5W
Concepción, Paraguay 94 A4 23 22S 57 26W
Concepción □, Chile 94 D1 37 0S 72 30W
Concepción →, Mexico 86 A2 30 32N 113 2W
Concepción, L., Bolivia 92 G6 17 20S 61 20W
Concepción, Punta, Mexico 86 B2 26 55N 111 59W
Concepción del Oro, Mexico 86 C4 24 40N 101 30W
Concepción del Uruguay, Argentina 94 C4 32 35S 58 20W
Conception, Pt., U.S.A. 85 L6 34 27N 120 28W
Conception B., Namibia 56 C1 23 55S 14 22 E
Conception I., Bahamas 89 B4 23 52N 75 9W
Concession, Zimbabwe 55 F3 17 27S 30 56 E
Conchas Dam, U.S.A. 81 H2 35 22N 104 11W
Conche, Canada 71 B8 50 55N 55 58W
Concho, U.S.A. 83 J9 34 28N 109 36W
Concho →, U.S.A. 81 K5 31 34N 99 43W
Conchos →, Chihuahua, Mexico 86 B4 29 32N 105 0W
Conchos →, Tamaulipas, Mexico 87 B5 25 9N 98 35W
Concord, Calif., U.S.A. 84 H4 37 59N 122 2W
Concord, N.C., U.S.A. 77 H5 35 25N 80 35W
Concord, N.H., U.S.A. 79 C13 43 12N 71 32W
Concordia, Argentina 94 C4 31 20S 58 2W
Concórdia, Brazil 92 D5 4 36S 66 36W
Concordia, Mexico 86 C3 23 18N 106 2W
Concordia, U.S.A. 80 F6 39 34N 97 40W
Concrete, U.S.A. 82 B3 48 32N 121 45W
Condamine, Australia 63 D5 26 56S 150 9 E
Conde, U.S.A. 80 C5 45 9N 98 6W
Condeúba, Brazil 93 F10 14 52S 42 0W
Condobolin, Australia 63 E4 33 4S 147 6 E
Condon, U.S.A. 82 D3 45 14N 120 11W
Conejera, I., Spain 22 B9 39 11N 2 58 E
Conejos, Mexico 86 B4 26 14N 103 53W
Confuso →, Paraguay 94 B4 25 9S 57 34W
Congleton, U.K. 10 D5 53 10N 2 12W
Congo = Zaïre →, Africa 52 F2 6 4S 12 24 E
Congo (Kinshasa) = Zaïre ■, Africa 52 E4 3 0S 23 0 E
Congo ■, Africa 52 E3 1 0S 16 0 E
Congo Basin, Africa 48 G6 0 10S 24 30 E
Congonhas, Brazil 95 A7 20 30S 43 52W
Congress, U.S.A. 83 J7 34 9N 112 51W
Coniston, Canada 70 C3 46 29N 80 51W
Conjeeveram = Kanchipuram, India 40 N11 12 52N 79 45 E
Conjuboy, Australia 62 B3 18 35S 144 35 E
Conklin, Canada 73 B6 55 38N 111 5W
Conlea, Australia 63 E3 30 7S 144 35 E
Conn, L., Ireland 13 B2 54 3N 9 15W
Connacht, Ireland 13 C3 53 23N 8 40W
Conneaut, U.S.A. 78 E4 41 57N 80 34W
Connecticut □, U.S.A. 79 E12 41 30N 72 45W
Connecticut →, U.S.A. 79 E12 41 16N 72 20W
Connell, U.S.A. 82 C4 46 40N 118 52W
Connellsville, U.S.A. 78 F5 40 1N 79 35W
Connemara, Ireland 13 C2 53 29N 9 45W
Connemaugh →, U.S.A. 78 F5 40 28N 79 19W
Connersville, U.S.A. 76 F3 39 39N 85 8W
Connors Ra., Australia 62 C4 21 40S 149 10 E
Conoble, Australia 63 E3 32 55S 144 33 E
Cononaco →, Ecuador 92 D3 1 32S 75 35W
Cononbridge, U.K. 12 D4 57 32N 4 30W
Conquest, Canada 73 C7 51 32N 107 14W
Conrad, U.S.A. 82 B8 48 10N 111 57W
Conran, C., Australia 63 F4 37 49S 148 44 E
Conroe, U.S.A. 81 K7 30 19N 95 27W
Conselheiro Lafaiete, Brazil 95 A7 20 40S 43 48W
Consort, Canada 73 C6 52 1N 110 46W
Constance = Konstanz, Germany 16 E4 47 39N 9 10 E
Constance, L. = Bodensee, Europe 16 E4 47 35N 9 25 E
Constanța, Romania 21 B13 44 14N 28 38 E
Constantina, Spain 19 D3 37 51N 5 40W
Constantine, Algeria 50 A6 36 25N 6 42 E
Constitución, Chile 94 D1 35 20S 72 30W
Constitución, Uruguay 94 C4 31 0S 57 50W
Consul, Canada 73 D7 49 20N 109 30W
Contact, U.S.A. 82 F6 41 46N 114 45W
Contai, India 43 J12 21 54N 87 46 E
Contamana, Peru 92 E4 7 19S 74 55W
Contas →, Brazil 93 F11 14 17S 39 1W
Contoocook, U.S.A. 79 C13 43 13N 71 45W
Contra Costa, Mozam. 57 D5 25 9S 33 30 E
Conway = Conwy, U.K. 10 D4 53 17N 3 50W
Conway = Conwy →, U.K. 10 D4 53 18N 3 50W
Conway, Ark., U.S.A. 81 H8 35 5N 92 26W
Conway, N.H., U.S.A. 79 C13 43 59N 71 7W
Conway, S.C., U.S.A. 77 J6 33 51N 79 3W
Conway, L., Australia 63 D2 28 17S 135 35 E
Conwy, U.K. 10 D4 53 17N 3 50W
Conwy →, U.K. 10 D4 53 18N 3 50W
Coober Pedy, Australia 63 D1 29 1S 134 43 E
Cooch Behar = Koch Bihar, India 41 F16 26 22N 89 29 E
Coodardy, Australia 61 E2 27 15S 117 39 E
Cook, Australia 61 F5 30 37S 130 25 E
Cook, U.S.A. 80 B8 47 49N 92 39W
Cook, B., Chile 96 H3 55 10S 70 0W
Cook, Mt., N.Z. 59 K3 43 36S 170 9 E
Cook Inlet, U.S.A. 68 C4 60 0N 152 0W
Cook Is., Pac. Oc. 65 J11 17 0S 160 0W
Cook Strait, N.Z. 59 J5 41 15S 174 29 E
Cookeville, U.S.A. 77 G3 36 10N 85 30W
Cookhouse, S. Africa 56 E4 32 44S 25 47 E
Cookshire, Canada 79 A13 45 25N 71 38W
Cookstown, U.K. 13 B5 54 40N 6 43W
Cookstown □, U.K. 13 B5 54 40N 6 43W
Cooksville, Canada 78 C5 43 36N 79 35W
Cooktown, Australia 62 B4 15 30S 145 16 E
Coolabah, Australia 63 E4 31 1S 146 15 E
Cooladdi, Australia 63 D4 26 37S 145 23 E
Coolah, Australia 63 E4 31 48S 149 41 E
Coolamon, Australia 63 E4 34 46S 147 8 E
Coolangatta, Australia 63 D5 28 11S 153 29 E
Coolgardie, Australia 61 F3 30 55S 121 8 E
Coolibah, Australia 60 C5 15 33S 130 56 E
Coolidge, U.S.A. 83 K8 32 59N 111 31W
Coolidge Dam, U.S.A. 83 K8 33 0N 110 20W
Cooma, Australia 63 F4 36 12S 149 8 E
Coonabarabran, Australia 63 E4 31 14S 149 18 E
Coonamble, Australia 63 E4 30 56S 148 27 E
Coonana, Australia 61 F3 31 0S 123 0 E
Coondapoor, India 40 N9 13 42N 74 40 E
Coongie, Australia 63 D3 27 9S 140 8 E
Coongoola, Australia 63 D4 27 43S 145 51 E
Cooninie, L., Australia 63 D2 26 4S 139 59 E
Cooper, U.S.A. 81 J7 33 23N 95 42W
Cooper →, U.S.A. 77 J6 32 50N 79 56W
Cooper Cr. →, Australia 63 D2 28 29S 137 46 E

Cooperstown

Cooperstown, *N. Dak., U.S.A.* **80 B5** 47 27N 98 8W
Cooperstown, *N.Y., U.S.A.* **79 D10** 42 42N 74 56W
Coorabie, *Australia* **61 F5** 31 54S 132 18 E
Coorabulka, *Australia* **62 C3** 23 41S 140 20 E
Coorow, *Australia* **61 E2** 29 53S 116 2 E
Cooroy, *Australia* **63 D5** 26 22S 152 54 E
Coos Bay, *U.S.A.* **82 E1** 43 22N 124 13W
Cootamundra, *Australia* **63 E4** 34 36S 148 1 E
Cootehill, *Ireland* **13 B4** 54 5N 7 5W
Cooyar, *Australia* **63 D5** 26 59S 151 51 E
Cooyeana, *Australia* **62 C2** 24 29S 138 45 E
Copahue Paso, *Argentina* **94 D1** 37 49S 71 8W
Copainalá, *Mexico* **87 D6** 17 8N 93 11W
Copán, *Honduras* **88 D2** 14 50N 89 9W
Cope, *U.S.A.* **80 F3** 39 40N 102 51W
Copenhagen = København, *Denmark* **9 J12** 55 41N 12 34 E
Copiapó, *Chile* **94 B1** 27 30S 70 20W
Copiapó →, *Chile* **94 B1** 27 19S 70 56W
Copley, *Australia* **63 E2** 30 36S 138 26 E
Copp L., *Canada* **72 A6** 60 14N 114 40W
Copper Center, *U.S.A.* **68 B5** 61 58N 145 18W
Copper Cliff, *Canada* **70 C3** 46 28N 81 4W
Copper Harbor, *U.S.A.* **76 B2** 47 28N 87 53W
Copper Queen, *Zimbabwe* **55 F2** 17 29S 29 18 E
Copperbelt □, *Zambia* **55 E2** 13 15S 27 30 E
Coppermine, *Canada* **68 B8** 67 50N 115 5W
Coppermine →, *Canada* **68 B8** 67 49N 116 4W
Copperopolis, *U.S.A.* **84 H6** 37 58N 120 38W
Coquet →, *U.K.* **10 B6** 55 18N 1 45W
Coquilhatville = Mbandaka, *Zaïre* **52 D3** 0 1N 18 18 E
Coquille, *U.S.A.* **82 E1** 43 11N 124 11W
Coquimbo, *Chile* **94 B1** 30 0S 71 20W
Coquimbo □, *Chile* **94 C1** 31 0S 71 0W
Corabia, *Romania* **17 G12** 43 48N 24 30 E
Coracora, *Peru* **92 G4** 15 5S 73 45W
Coral Gables, *U.S.A.* **77 N5** 25 45N 80 16W
Coral Harbour, *Canada* **69 B11** 64 8N 83 10W
Coral Sea, *Pac. Oc.* **64 J7** 15 0S 150 0 E
Coraopolis, *U.S.A.* **78 F4** 40 31N 80 10W
Corato, *Italy* **20 D7** 41 12N 16 22 E
Corbin, *U.S.A.* **76 G3** 36 57N 84 6W
Corby, *U.K.* **11 E7** 52 29N 0 41W
Corby Glen, *U.K.* **11 E7** 52 49N 0 31W
Corcoran, *U.S.A.* **83 H4** 36 6N 119 33W
Corcubión, *Spain* **19 A1** 42 56N 9 12W
Cordele, *U.S.A.* **77 K4** 31 58N 83 47W
Cordell, *U.S.A.* **81 H5** 35 17N 98 59W
Córdoba, *Argentina* **94 C3** 31 20S 64 10W
Córdoba, *Mexico* **87 D5** 18 50N 97 0W
Córdoba, *Spain* **19 D3** 37 50N 4 50W
Córdoba □, *Argentina* **94 C3** 31 22S 64 15W
Córdoba, Sierra de, *Argentina* **94 C3** 31 10S 64 25W
Cordon, *Phil.* **37 A6** 16 42N 121 32 E
Cordova, *Ala., U.S.A.* **77 J2** 33 46N 87 11W
Cordova, *Alaska, U.S.A.* **68 B5** 60 33N 145 45W
Corella →, *Australia* **62 B3** 19 34S 140 47 E
Corfield, *Australia* **62 C3** 21 40S 143 21 E
Corfu = Kérkira, *Greece* **23 A3** 39 38N 19 50 E
Corfu, Str of, *Greece* **23 A4** 39 34N 20 0 E
Corigliano Cálabro, *Italy* **20 E7** 39 36N 16 31 E
Coringa Is., *Australia* **62 B4** 16 58S 149 58 E
Corinna, *Australia* **62 G4** 41 35S 145 10 E
Corinth = Kórinthos, *Greece* **21 F10** 37 56N 22 55 E
Corinth, *Miss., U.S.A.* **77 H1** 34 56N 88 31W
Corinth, *N.Y., U.S.A.* **79 C11** 43 15N 73 49W
Corinth, G. of = Korinthiakós Kólpos, *Greece* **21 E10** 38 16N 22 30 E
Corinto, *Brazil* **93 G10** 18 20S 44 30W
Corinto, *Nic.* **88 D2** 12 30N 87 10W
Cork, *Ireland* **13 E3** 51 54N 8 30W
Cork □, *Ireland* **13 E3** 51 50N 8 50W
Cork Harbour, *Ireland* **13 E3** 51 46N 8 16W
Cormack L., *Canada* **72 A4** 60 56N 121 37W
Cormorant, *Canada* **73 C8** 54 14N 100 35W
Cormorant L., *Canada* **73 C8** 54 15N 100 50W
Corn Is. = Maiz, Is. del, *Nic.* **88 D3** 12 15N 83 4W
Cornélio Procópio, *Brazil* **95 A5** 23 7S 50 40W
Cornell, *U.S.A.* **80 C9** 45 10N 91 9W
Corner Brook, *Canada* **71 C8** 48 57N 57 58W
Corning, *Ark., U.S.A.* **81 G9** 36 25N 90 35W
Corning, *Calif., U.S.A.* **82 G2** 39 56N 122 11W
Corning, *Iowa, U.S.A.* **80 E7** 40 59N 94 44W
Corning, *N.Y., U.S.A.* **78 D7** 42 9N 77 3W
Cornwall, *Canada* **70 C5** 45 2N 74 44W
Cornwall □, *U.K.* **11 G3** 50 26N 4 40W
Corny Pt., *Australia* **63 E2** 34 55S 137 0 E
Coro, *Venezuela* **92 A5** 11 25N 69 41W
Coroatá, *Brazil* **93 D10** 4 8S 44 0W
Corocoro, *Bolivia* **92 G5** 17 15S 68 28W
Coroico, *Bolivia* **92 G5** 16 0S 67 50W
Coromandel, *N.Z.* **59 G5** 36 45S 175 31 E
Coromandel Coast, *India* **40 N12** 12 30N 81 0 E
Corona, *Australia* **63 E3** 31 16S 141 24 E
Corona, *Calif., U.S.A.* **85 M9** 33 53N 117 34W
Corona, *N. Mex., U.S.A.* **83 J11** 34 15N 105 36W
Coronado, *U.S.A.* **85 N9** 32 41N 117 11W
Coronado, B. de, *Costa Rica* **88 E3** 9 0N 83 40W
Coronados, Is. los, *U.S.A.* **85 N9** 32 25N 117 15W
Coronation, *Canada* **72 C6** 52 5N 111 27W
Coronation Gulf, *Canada* **68 B8** 68 25N 110 0W
Coronation I., *Antarctica* **5 C18** 60 45S 46 0W
Coronation I., *U.S.A.* **72 B2** 55 52N 134 20W
Coronation Is., *Australia* **60 B3** 14 57S 124 55 E
Coronda, *Argentina* **94 C3** 31 58S 60 56W
Coronel, *Chile* **94 D1** 37 0S 73 10W
Coronel Bogado, *Paraguay* **94 B4** 27 11S 56 18W
Coronel Dorrego, *Argentina* **94 D3** 38 40S 61 10W
Coronel Oviedo, *Paraguay* **94 B4** 25 24S 56 30W
Coronel Pringles, *Argentina* **94 D3** 38 0S 61 30W
Coronel Suárez, *Argentina* **94 D3** 37 30S 61 52W
Coronel Vidal, *Argentina* **94 D4** 37 28S 57 45W
Corowa, *Australia* **63 F4** 35 58S 146 21 E
Corozal, *Belize* **87 D7** 18 23N 88 23W
Corpus, *Argentina* **95 B4** 27 10S 55 30W
Corpus Christi, *U.S.A.* **81 M6** 27 47N 97 24W
Corpus Christi, L., *U.S.A.* **81 L6** 28 2N 97 52W
Corque, *Bolivia* **92 G5** 18 20S 67 41W
Corralejo, *Canary Is.* **22 F6** 28 43N 13 53W
Correntes, C. das, *Mozam.* **57 C6** 24 6S 35 34 E
Corrèze □, *France* **18 D4** 45 20N 1 45 E
Corrib, L., *Ireland* **13 C2** 53 5N 9 10W
Corrientes, *Argentina* **94 B4** 27 30S 58 45W
Corrientes □, *Argentina* **94 B4** 28 0S 57 0W
Corrientes →, *Argentina* **94 C4** 30 42S 59 38W
Corrientes →, *Peru* **92 D4** 3 43S 74 35W
Corrientes, C., *Colombia* **92 B3** 5 30N 77 34W
Corrientes, C., *Cuba* **88 B3** 21 43N 84 30W
Corrientes, C., *Mexico* **86 C3** 20 25N 105 42W
Corrigan, *U.S.A.* **81 K7** 31 0N 94 52W
Corrigin, *Australia* **61 F2** 32 20S 117 53 E
Corry, *U.S.A.* **78 E5** 41 55N 79 39W
Corse, *France* **18 F9** 42 0N 9 0 E
Corse, C., *France* **20 C3** 43 1N 9 25 E
Corse-du-Sud □, *France* **18 F9** 41 45N 9 0 E
Corsica = Corse, *France* **18 F9** 42 0N 9 0 E
Corsicana, *U.S.A.* **81 J6** 32 6N 96 28W
Cortez, *U.S.A.* **83 H9** 37 21N 108 35W
Cortland, *U.S.A.* **79 D8** 42 36N 76 11W
Cortona, *Italy* **20 C4** 43 16N 12 0 E
Çorum, *Turkey* **25 F5** 40 30N 34 57 E
Corumbá, *Brazil* **92 G7** 19 0S 57 30W
Corumbá de Goiás, *Brazil* **93 G9** 16 0S 48 50W
Corunna = La Coruña, *Spain* **19 A1** 43 20N 8 25W
Corvallis, *U.S.A.* **82 D2** 44 34N 123 16W
Corvette, L. de la, *Canada* **70 B5** 53 25N 74 3W
Corydon, *U.S.A.* **80 E8** 40 46N 93 19W
Cosalá, *Mexico* **86 C3** 24 28N 106 40W
Cosamaloapan, *Mexico* **87 D5** 18 23N 95 50W
Cosenza, *Italy* **20 E7** 39 17N 16 14 E
Coshocton, *U.S.A.* **78 F3** 40 16N 81 51W
Cosmo Newberry, *Australia* **61 E3** 28 0S 122 54 E
Coso Junction, *U.S.A.* **85 J9** 36 3N 117 57W
Coso Pk., *U.S.A.* **85 J9** 36 13N 117 44W
Cosquín, *Argentina* **94 C3** 31 15S 64 30W
Costa Blanca, *Spain* **19 C5** 38 25N 0 10W
Costa Brava, *Spain* **19 B7** 41 30N 3 0 E
Costa del Sol, *Spain* **19 D3** 36 30N 4 30W
Costa Dorada, *Spain* **19 B6** 40 45N 1 15 E
Costa Mesa, *U.S.A.* **85 M9** 33 38N 117 55W
Costa Rica ■, *Cent. Amer.* **88 D3** 10 0N 84 0W
Costilla, *U.S.A.* **83 H11** 36 59N 105 32W
Cosumnes →, *U.S.A.* **84 G5** 38 16N 121 26W
Cotabato, *Phil.* **37 C6** 7 14N 124 15 E
Cotagaita, *Bolivia* **94 A2** 20 45S 65 40W
Côte-d'Ivoire ■ = Ivory Coast ■, *Africa* **50 G3** 7 30N 5 0W
Côte-d'Or □, *France* **18 C6** 47 30N 4 50 E
Coteau des Prairies, *U.S.A.* **80 C6** 45 20N 97 50W
Coteau du Missouri, *U.S.A.* **80 B4** 47 0N 100 0W
Coteau Landing, *Canada* **79 A10** 45 15N 74 13W
Cotentin, *France* **18 B3** 49 15N 1 30W
Côtes-d'Armor □, *France* **18 B2** 48 25N 2 40W
Côtes-du-Nord = Côtes-d'Armor □, *France* **18 B2** 48 25N 2 40W
Cotillo, *Canary Is.* **22 F5** 28 41N 14 1W
Cotonou, *Benin* **50 G5** 6 20N 2 25 E
Cotopaxi, *Ecuador* **92 D3** 0 40S 78 30W
Cotswold Hills, *U.K.* **11 F5** 51 42N 2 10W
Cottage Grove, *U.S.A.* **82 E2** 43 48N 123 3W
Cottbus, *Germany* **16 C7** 51 44N 14 20 E
Cottingham, *U.K.* **10 C5** 53 47N 0 23W
Cottonwood, *U.S.A.* **83 J7** 34 45N 112 1W
Cotulla, *U.S.A.* **81 L5** 28 26N 99 14W
Coudersport, *U.S.A.* **78 E6** 41 46N 78 1W
Couedic, C. du, *Australia* **63 F2** 36 5S 136 40 E
Coulee City, *U.S.A.* **82 C4** 47 37N 119 17W
Coulman I., *Antarctica* **5 D11** 73 35S 170 0 E
Coulonge →, *Canada* **70 C4** 45 52N 76 46W
Coulterville, *U.S.A.* **84 H6** 37 43N 120 12W
Council, *Alaska, U.S.A.* **68 B3** 64 55N 163 45W
Council, *Idaho, U.S.A.* **82 D5** 44 44N 116 26W
Council Bluffs, *U.S.A.* **80 E7** 41 16N 95 52W
Council Grove, *U.S.A.* **80 F6** 38 40N 96 29W
Coupeville, *U.S.A.* **84 B4** 48 13N 122 41W
Courantyne →, *S. Amer.* **92 B7** 5 55N 57 5W
Courtenay, *Canada* **72 D3** 49 45N 125 0W
Courtland, *U.S.A.* **84 G5** 38 20N 121 34W
Courtrai = Kortrijk, *Belgium* **15 D3** 50 50N 3 17 E
Courtright, *Canada* **78 D2** 42 49N 82 28W
Coushatta, *U.S.A.* **81 J8** 32 1N 93 21W
Coutts, *Canada* **72 D6** 49 0N 111 57W
Coventry, *U.K.* **11 E6** 52 25N 1 31W
Coventry L., *Canada* **73 A7** 61 15N 106 15W
Covilhã, *Portugal* **19 B2** 40 17N 7 31W
Covington, *Ga., U.S.A.* **77 J4** 33 36N 83 51W
Covington, *Ky., U.S.A.* **76 F3** 39 5N 84 31W
Covington, *Okla., U.S.A.* **81 G6** 36 18N 97 35W
Covington, *Tenn., U.S.A.* **81 H10** 35 34N 89 39W
Cowal, L., *Australia* **63 E4** 33 40S 147 25 E
Cowan, *Canada* **73 C8** 52 5N 100 45W
Cowan, L., *Australia* **61 F3** 31 45S 121 45 E
Cowan L., *Canada* **73 C7** 54 0N 107 15W
Cowangie, *Australia* **63 F3** 35 12S 141 26 E
Cowansville, *Canada* **79 A12** 45 14N 72 46W
Cowarie, *Australia* **63 D2** 27 45S 138 15 E
Cowcowing Lakes, *Australia* **61 F2** 30 55S 117 20 E
Cowdenbeath, *U.K.* **12 E5** 56 7N 3 20W
Cowell, *Australia* **63 E2** 33 39S 136 56 E
Cowes, *U.K.* **11 G6** 50 45N 1 18W
Cowlitz →, *U.S.A.* **84 D4** 46 6N 122 55W
Cowra, *Australia* **63 E4** 33 49S 148 42 E
Coxilha Grande, *Brazil* **95 B5** 28 18S 51 30W
Coxim, *Brazil* **93 G8** 18 30S 54 55W
Cox's Bazar, *Bangla.* **41 J17** 21 26N 91 59 E
Cox's Cove, *Canada* **71 C8** 49 7N 58 5W
Coyame, *Mexico* **86 B3** 29 28N 105 6W
Coyote Wells, *U.S.A.* **85 N11** 32 44N 115 58W
Coyuca de Benítez, *Mexico* **87 D4** 17 1N 100 8W
Coyuca de Catalan, *Mexico* **86 D4** 18 18N 100 41W
Cozad, *U.S.A.* **80 E5** 40 52N 99 59W
Cozumel, *Mexico* **87 C7** 20 31N 86 55W
Cozumel, I. de, *Mexico* **87 C7** 20 30N 86 40W
Craboon, *Australia* **63 E4** 32 3S 149 30 E
Cracow = Kraków, *Poland* **17 C9** 50 4N 19 57 E
Cracow, *Australia* **63 D5** 25 17S 150 17 E
Cradock, *S. Africa* **56 E4** 32 8S 25 36 E
Craig, *Alaska, U.S.A.* **72 B2** 55 29N 133 9W
Craig, *Colo., U.S.A.* **82 F10** 40 31N 107 33W
Craigavon = Lurgan, *U.K.* **13 B5** 54 28N 6 20W
Craigmore, *Zimbabwe* **55 G3** 20 28S 32 50 E
Craiova, *Romania* **21 B10** 44 21N 23 48 E
Cramsie, *Australia* **62 C3** 23 20S 144 15 E
Cranberry Portage, *Canada* **73 C8** 54 35N 101 23W
Cranbrook, *Tas., Australia* **62 G4** 42 0S 148 5 E
Cranbrook, *W. Austral., Australia* **61 F2** 34 18S 117 33 E
Cranbrook, *Canada* **72 D5** 49 30N 115 46W
Crandon, *U.S.A.* **80 C10** 45 34N 88 54W
Crane, *Oreg., U.S.A.* **82 E4** 43 25N 118 35W
Crane, *Tex., U.S.A.* **81 K3** 31 24N 102 21W
Cranston, *U.S.A.* **79 E13** 41 47N 71 26W
Crater L., *U.S.A.* **82 E2** 42 56N 122 6W
Crateús, *Brazil* **93 E10** 5 10S 40 39W
Crato, *Brazil* **93 E11** 7 10S 39 25W
Crawford, *U.S.A.* **80 D3** 42 41N 103 25W
Crawfordsville, *U.S.A.* **76 E2** 40 2N 86 54W
Crawley, *U.K.* **11 F7** 51 7N 0 10W
Crazy Mts., *U.S.A.* **82 C8** 46 12N 110 20W
Crean L., *Canada* **73 C7** 54 5N 106 9W
Crécy-en-Ponthieu, *France* **18 A4** 50 15N 1 53 E
Crediton, *Canada* **78 C3** 43 17N 81 33W
Credo, *Australia* **61 F3** 30 28S 120 45 E
Cree →, *Canada* **73 B7** 58 57N 105 47W
Cree →, *U.K.* **12 G4** 54 51N 4 24W
Cree L., *Canada* **73 B7** 57 30N 106 30W
Creede, *U.S.A.* **83 H10** 37 51N 106 56W
Creel, *Mexico* **86 B3** 27 45N 107 38W
Creighton, *U.S.A.* **80 D6** 42 28N 97 54W
Cremona, *Italy* **20 B4** 45 8N 10 2 E
Cres, *Croatia* **20 B6** 44 58N 14 25 E
Cresbard, *U.S.A.* **80 C5** 45 10N 98 57W
Crescent, *Okla., U.S.A.* **81 H6** 35 57N 97 36W
Crescent, *Oreg., U.S.A.* **82 E3** 43 28N 121 42W
Crescent City, *U.S.A.* **82 F1** 41 45N 124 12W
Crespo, *Argentina* **94 C3** 32 2S 60 19W
Cressy, *Australia* **63 F3** 38 2S 143 40 E
Crested Butte, *U.S.A.* **83 G10** 38 52N 106 59W
Crestline, *Calif., U.S.A.* **85 L9** 34 14N 117 18W
Crestline, *Ohio, U.S.A.* **78 F2** 40 47N 82 44W
Creston, *Canada* **72 D5** 49 10N 116 31W
Creston, *Calif., U.S.A.* **84 K6** 35 32N 120 33W
Creston, *Iowa, U.S.A.* **80 E7** 41 4N 94 22W
Creston, *Wash., U.S.A.* **82 C4** 47 46N 118 31W
Crestview, *Calif., U.S.A.* **84 H8** 37 46N 118 58W
Crestview, *Fla., U.S.A.* **77 K2** 30 46N 86 34W
Crete = Kríti, *Greece* **23 D7** 35 15N 25 0 E
Crete, *U.S.A.* **80 E6** 40 38N 96 58W
Creus, C., *Spain* **19 A7** 42 20N 3 19 E
Creuse □, *France* **18 C5** 46 10N 2 0 E
Creuse →, *France* **18 C4** 47 0N 0 34 E
Crewe, *U.K.* **10 D5** 53 6N 2 28W
Criciúma, *Brazil* **95 B6** 28 40S 49 23W
Crieff, *U.K.* **12 E5** 56 22N 3 50W
Crimea = Krymskiy Poluostrov, *Ukraine* **25 E5** 45 0N 34 0 E
Crişul Alb →, *Romania* **17 E10** 46 42N 21 17 E
Crişul Negru →, *Romania* **17 E10** 46 42N 21 16 E
Crna Gora = Montenegro □, *Montenegro, Yug.* **21 C8** 42 40N 19 20 E
Crna Gora, *Serbia* **21 C9** 42 10N 21 30 E
Crna Reka →, *Macedonia* **21 D9** 41 33N 21 59 E
Croaghpatrick, *Ireland* **13 C2** 53 46N 9 40W
Croatia ■, *Europe* **20 B7** 45 20N 16 0 E
Crocker, Banjaran, *Malaysia* **36 C5** 5 40N 116 30 E
Crockett, *U.S.A.* **81 K7** 31 19N 95 27W
Crocodile = Krokodil →, *Mozam.* **57 D5** 25 14S 32 18 E
Crocodile Is., *Australia* **62 A1** 12 3S 134 58 E
Croix, L. La, *Canada* **70 C1** 48 20N 92 15W
Croker, C., *Australia* **60 B5** 10 58S 132 35 E
Croker I., *Australia* **60 B5** 11 12S 132 32 E
Cromarty, *Canada* **73 B10** 58 3N 94 9W
Cromarty, *U.K.* **12 D4** 57 40N 4 2W
Cromer, *U.K.* **10 E9** 52 56N 1 18 E
Cromwell, *N.Z.* **59 L2** 45 3S 169 14 E
Cronulla, *Australia* **63 E5** 34 3S 151 8 E
Crooked →, *Canada* **72 C4** 54 50N 122 54W
Crooked →, *U.S.A.* **82 D3** 44 32N 121 16W
Crooked I., *Bahamas* **89 B5** 22 50N 74 10W
Crooked Island Passage, *Bahamas* **89 B5** 23 0N 74 30W
Crookston, *Minn., U.S.A.* **80 B6** 47 47N 96 37W
Crookston, *Nebr., U.S.A.* **80 D4** 42 56N 100 45W
Crooksville, *U.S.A.* **76 F4** 39 46N 82 6W
Crookwell, *Australia* **63 E4** 34 28S 149 24 E
Crosby, *Minn., U.S.A.* **80 B8** 46 29N 93 58W
Crosby, *N. Dak., U.S.A.* **73 D8** 48 55N 103 18W
Crosby, *Pa., U.S.A.* **78 E6** 41 45N 78 23W
Crosbyton, *U.S.A.* **81 J4** 33 40N 101 14W
Cross City, *U.S.A.* **77 L4** 29 38N 83 7W
Cross Fell, *U.K.* **10 C5** 54 44N 2 29W
Cross L., *Canada* **73 C9** 54 45N 97 30W
Cross Plains, *U.S.A.* **81 J5** 32 8N 99 11W
Cross Sound, *U.S.A.* **68 C6** 58 0N 135 0W
Crossett, *U.S.A.* **81 J9** 33 8N 91 58W
Crossfield, *Canada* **72 C6** 51 25N 114 0W
Crosshaven, *Ireland* **13 E3** 51 48N 8 19W
Croton-on-Hudson, *U.S.A.* **79 E11** 41 12N 73 55W
Crotone, *Italy* **20 E7** 39 5N 17 6 E
Crow →, *Canada* **72 B4** 59 41N 124 20W
Crow Agency, *U.S.A.* **82 D10** 45 36N 107 28W
Crow Hd., *Ireland* **13 E1** 51 34N 10 9W
Crowell, *U.S.A.* **81 J5** 33 59N 99 43W
Crowley, *U.S.A.* **81 K8** 30 13N 92 22W
Crowley, L., *U.S.A.* **84 H8** 37 35N 118 42W
Crown Point, *U.S.A.* **76 E2** 41 25N 87 22W
Crows Landing, *U.S.A.* **84 H5** 37 23N 121 6W
Crows Nest, *Australia* **63 D5** 27 16S 152 4 E
Crowsnest Pass, *Canada* **72 D6** 49 40N 114 40W
Croydon, *Australia* **62 B3** 18 13S 142 14 E
Croydon, *U.K.* **11 F7** 51 18N 0 5W
Crozet Is., *Ind. Oc.* **3 G12** 46 27S 52 0 E
Cruz, C., *Cuba* **88 C4** 19 50N 77 50W
Cruz Alta, *Brazil* **95 B5** 28 45S 53 40W
Cruz del Eje, *Argentina* **94 C3** 30 45S 64 50W
Cruzeiro, *Brazil* **95 A7** 22 33S 45 0W
Cruzeiro do Oeste, *Brazil* **95 A5** 23 46S 53 4W
Cruzeiro do Sul, *Brazil* **92 E4** 7 35S 72 35W
Cry L., *Canada* **72 B3** 58 45N 129 0W
Crystal Bay, *U.S.A.* **84 F7** 39 15N 120 0W
Crystal Brook, *Australia* **63 E2** 33 21S 138 12 E
Crystal City, *Mo., U.S.A.* **80 F9** 38 13N 90 23W
Crystal City, *Tex., U.S.A.* **81 L5** 28 41N 99 50W
Crystal Falls, *U.S.A.* **76 B1** 46 5N 88 20W
Crystal River, *U.S.A.* **77 L4** 28 54N 82 35W
Crystal Springs, *U.S.A.* **81 K9** 31 59N 90 21W
Csongrád, *Hungary* **17 E10** 46 43N 20 12 E
Cu Lao Hon, *Vietnam* **39 G7** 10 54N 108 18 E
Cua Rao, *Vietnam* **38 C5** 19 16N 104 27 E
Cuácua →, *Mozam.* **55 F4** 17 54S 37 0 E
Cuamato, *Angola* **56 B2** 17 2S 15 7 E
Cuamba, *Mozam.* **55 E4** 14 45S 36 22 E
Cuando →, *Angola* **53 H4** 17 30S 23 15 E
Cuando Cubango □, *Angola* **56 B3** 16 25S 20 0 E
Cuangar, *Angola* **56 B2** 17 36S 18 39 E
Cuanza →, *Angola* **48 G5** 9 2S 13 30 E
Cuarto →, *Argentina* **94 C3** 33 25S 63 2W
Cuatrociénegas, *Mexico* **86 B4** 26 59N 102 5W
Cuauhtémoc, *Mexico* **86 B3** 28 25N 106 52W
Cuba, *N. Mex., U.S.A.* **83 J10** 36 1N 107 4W
Cuba, *N.Y., U.S.A.* **78 D6** 42 13N 78 17W
Cuba ■, *W. Indies* **88 B4** 22 0N 79 0W
Cuballing, *Australia* **61 F2** 32 50S 117 10 E
Cubango →, *Africa* **56 B3** 18 50S 22 25 E
Cucamonga, *U.S.A.* **85 L9** 34 10N 117 30W
Cuchi, *Angola* **53 G3** 14 37S 16 58 E
Cuchumatanes, Sierra de los, *Guatemala* **88 C1** 15 35N 91 25W
Cucurpe, *Mexico* **86 A2** 30 20N 110 43W
Cúcuta, *Colombia* **92 B4** 7 54N 72 31W
Cudahy, *U.S.A.* **76 D2** 42 58N 87 52W
Cuddalore, *India* **40 P11** 11 46N 79 45 E
Cuddapah, *India* **40 M11** 14 30N 78 47 E
Cuddapan, L., *Australia* **62 D3** 25 45S 141 26 E
Cudgewa, *Australia* **63 F4** 36 10S 147 42 E
Cue, *Australia* **61 E2** 27 25S 117 54 E
Cuenca, *Ecuador* **92 D3** 2 50S 79 9W
Cuenca, *Spain* **19 B4** 40 5N 2 10W
Cuenca, Serranía de, *Spain* **19 C5** 39 55N 1 50W
Cuernavaca, *Mexico* **87 D5** 18 50N 99 20W
Cuero, *U.S.A.* **81 L6** 29 6N 97 17W
Cuervo, *U.S.A.* **81 H2** 35 2N 104 25W
Cuevas del Almanzora, *Spain* **19 D5** 37 18N 1 58W
Cuevo, *Bolivia* **92 H6** 20 15S 63 30W
Cuiabá, *Brazil* **93 G7** 15 30S 56 0W
Cuiabá →, *Brazil* **93 G7** 17 5S 56 36W
Cuilco, *Guatemala* **88 C1** 15 24N 91 58W
Cuillin Hills, *U.K.* **12 D2** 57 14N 6 15W
Cuillin Sd., *U.K.* **12 D2** 57 4N 6 20W
Cuima, *Angola* **53 G3** 13 25S 15 45 E
Cuito →, *Angola* **56 B3** 18 1S 20 48 E
Cuitzeo, L. de, *Mexico* **86 D4** 19 55N 101 5W
Cukai, *Malaysia* **39 K4** 4 13N 103 25 E
Culbertson, *U.S.A.* **80 A2** 48 9N 104 31W
Culcairn, *Australia* **63 F4** 35 41S 147 3 E
Culebra, Sierra de la, *Spain* **19 B2** 41 55N 6 20W
Culgoa →, *Australia* **63 D4** 29 56S 146 20 E
Culiacán, *Mexico* **86 C3** 24 50N 107 23W
Culiacán →, *Mexico* **86 C3** 24 30N 107 42W
Culion, *Phil.* **37 B6** 11 54N 120 1 E
Cullarin Ra., *Australia* **63 E4** 34 30S 149 30 E
Cullen, *U.K.* **12 D6** 57 45N 2 50W
Cullen Pt., *Australia* **62 A3** 11 57S 141 54 E
Cullera, *Spain* **19 C5** 39 9N 0 17W
Cullman, *U.S.A.* **77 H2** 34 11N 86 51W
Culloden Moor, *U.K.* **12 D4** 57 29N 4 7W
Culpeper, *U.S.A.* **76 F7** 38 30N 78 0W
Culuene →, *Brazil* **93 F8** 12 56S 52 51W
Culver, Pt., *Australia* **61 F3** 32 54S 124 43 E
Culverden, *N.Z.* **59 K4** 42 47S 172 49 E
Cumaná, *Venezuela* **92 A6** 10 30N 64 5W
Cumberland, *Canada* **72 D3** 49 40N 125 0W
Cumberland, *Md., U.S.A.* **76 F6** 39 39N 78 46W
Cumberland, *Wis., U.S.A.* **80 C8** 45 32N 92 1W
Cumberland →, *U.S.A.* **77 G2** 36 15N 87 0W
Cumberland I., *U.S.A.* **77 K5** 30 50N 81 25W
Cumberland Is., *Australia* **62 C4** 20 35S 149 10 E
Cumberland L., *Canada* **73 C8** 54 3N 102 18W
Cumberland Pen., *Canada* **69 B13** 67 0N 64 0W
Cumberland Plateau, *U.S.A.* **77 H3** 36 0N 85 0W
Cumberland Sd., *Canada* **69 B13** 65 30N 66 0W
Cumborah, *Australia* **63 D4** 29 40S 147 45 E
Cumbria □, *U.K.* **10 C5** 54 35N 2 55W
Cumbrian Mts., *U.K.* **10 C4** 54 30N 3 0W
Cumbum, *India* **40 M11** 15 40N 79 10 E
Cummings Mt., *U.S.A.* **85 K8** 35 2N 118 34W
Cummins, *Australia* **63 E2** 34 16S 135 43 E
Cumnock, *Australia* **63 E4** 32 59S 148 46 E
Cumnock, *U.K.* **12 F4** 55 27N 4 18W
Cumpas, *Mexico* **86 A3** 30 0N 109 48W

Delaware

Cumplida, Pta., Canary Is. 22 F2 28 50N 17 48W
Cuncumén, Chile 94 C1 31 53S 70 38W
Cundeelee, Australia 61 F3 30 43S 123 26 E
Cunderdin, Australia 61 F2 31 37S 117 12 E
Cunene →, Angola 56 B1 17 20S 11 50 E
Cúneo, Italy 20 B2 44 23N 7 31 E
Cunillera, I., Spain 22 C7 38 59N 1 13 E
Cunnamulla, Australia 63 D4 28 2S 145 38 E
Cupar, Canada 73 C8 50 57N 104 10W
Cupar, U.K. 12 E5 56 20N 3 3W
Cupica, G. de, Colombia 92 B3 6 25N 77 30W
Curaçao, Neth. Ant. 89 D6 12 10N 69 0W
Curanilahue, Chile 94 D1 37 29S 73 28W
Curaray →, Peru 92 D4 2 20S 74 5W
Curepto, Chile 94 D1 35 8S 72 1W
Curiapo, Venezuela 92 B6 8 33N 61 5W
Curicó, Chile 94 C1 34 55S 71 20W
Curicó □, Chile 94 C1 34 50S 71 15W
Curitiba, Brazil 95 B6 25 20S 49 10W
Currabubula, Australia 63 E5 31 16S 150 44 E
Currais Novos, Brazil 93 E11 6 13S 36 30W
Curralinho, Brazil 93 D9 1 45S 49 46W
Currant, U.S.A. 82 G6 38 51N 115 32W
Curraweena, Australia 63 E4 30 47S 145 54 E
Currawilla, Australia 62 D3 25 10S 141 20 E
Current →, U.S.A. 81 G9 36 15N 90 55W
Currie, Australia 62 F3 39 56S 143 53 E
Currie, U.S.A. 82 F6 40 16N 114 45W
Currie, Mt., S. Africa 57 E4 30 29S 29 21 E
Currituck Sd., U.S.A. 77 G8 36 20N 75 52W
Curtis, U.S.A. 80 E4 40 38N 100 31W
Curtis Group, Australia 62 F4 39 30S 146 37 E
Curtis I., Australia 62 C5 23 35S 151 10 E
Curuápanema →, Brazil 93 D7 2 25S 55 2W
Curuçá, Brazil 93 D9 0 43S 47 50W
Curuguaty, Paraguay 95 A4 24 31S 55 42W
Çürüksu Çayi →, Turkey 25 G4 37 27N 27 11 E
Curup, Indonesia 36 E2 4 26S 102 13 E
Cururupu, Brazil 93 D10 1 50S 44 50W
Curuzú Cuatiá, Argentina 94 B4 29 50S 58 5W
Cushing, U.S.A. 81 H6 35 59N 96 46W
Cushing, Mt., Canada 72 B3 57 35N 126 57W
Cusihuiriáchic, Mexico 86 B3 28 10N 106 50W
Custer, U.S.A. 80 D3 43 46N 103 36W
Cut Bank, U.S.A. 82 B7 48 38N 112 20W
Cuthbert, U.S.A. 77 K3 31 46N 84 48W
Cutler, U.S.A. 84 J7 36 31N 119 17W
Cuttaburra →, Australia 63 D3 29 43S 144 22 E
Cuttack, India 41 J14 20 25N 85 57 E
Cuvier, C., Australia 61 D1 23 14S 113 22 E
Cuvier I., N.Z. 59 G5 36 27S 175 50 E
Cuxhaven, Germany 16 B5 53 51N 8 41 E
Cuyahoga Falls, U.S.A. 78 E3 41 8N 81 29W
Cuyo, Phil. 37 B6 10 50N 121 5 E
Cuzco, Bolivia 92 H5 20 0S 66 50W
Cuzco, Peru 92 F4 13 32S 72 0W
Cwmbran, U.K. 11 F4 51 39N 3 2W
Cyangugu, Rwanda 54 C2 2 29S 28 54 E
Cyclades = Kikládhes, Greece 21 F11 37 20N 24 30 E
Cygnet, Australia 62 G4 43 8S 147 1 E
Cynthiana, U.S.A. 76 F3 38 23N 84 18W
Cypress Hills, Canada 73 D7 49 40N 109 30W
Cyprus ■, Asia 23 E12 35 0N 33 0 E
Cyrenaica, Libya 51 C9 27 0N 23 0 E
Cyrene = Shaḥḥāt, Libya 51 B9 32 48N 21 54 E
Czar, Canada 73 C6 52 27N 110 50W
Czech Rep. ■, Europe 16 D7 50 0N 15 0 E
Czeremcha, Poland 17 B11 52 31N 23 21 E
Częstochowa, Poland 17 C9 50 49N 19 7 E

D

Da →, Vietnam 38 B5 21 15N 105 20 E
Da Hinggan Ling, China 33 B7 48 0N 121 0 E
Da Lat, Vietnam 39 G7 11 56N 108 25 E
Da Nang, Vietnam 38 D7 16 4N 108 13 E
Da Qaidam, China 32 C4 37 50N 95 15 E
Da Yunhe →, China 35 G11 34 25N 120 5 E
Da'an, China 35 B13 45 30N 124 7 E
Daba Shan, China 33 C5 32 0N 109 0 E
Dabakala, Ivory C. 50 G4 8 15N 4 20W
Dabhoi, India 42 H5 22 10N 73 20 E
Dąbie, Poland 16 B7 53 27N 14 45 E
Dabo, Indonesia 36 E2 0 30S 104 33 E
Dabola, Guinea 50 F2 10 50N 11 5W
Daboya, Ghana 50 G4 9 30N 1 20W
Dabrowa Tarnówska, Poland 17 C10 50 10N 20 59 E
Dabung, Malaysia 39 K4 5 23N 102 1 E
Dacca = Dhaka, Bangla. 43 H14 23 43N 90 26 E
Dacca = Dhaka □, Bangla. 43 G14 24 25N 90 25 E
Dadanawa, Guyana 92 C7 2 50N 59 30W
Dade City, U.S.A. 77 L4 28 22N 82 11W
Dadra and Nagar Haveli □, India 40 J8 20 5N 73 0 E
Dadri = Charkhi Dadri, India 42 E7 28 37N 76 17 E
Dadu, Pakistan 42 F2 26 45N 67 45 E
Daet, Phil. 37 B6 14 2N 122 55 E
Dagana, Senegal 50 E1 16 30N 15 35W
Daggett, U.S.A. 85 L10 34 52N 116 52W
Daghestan Republic □, Russia 25 F8 42 30N 47 0 E
Dagö = Hiiumaa, Estonia 24 C3 58 50N 22 45 E
Dagu, China 35 E9 38 59N 117 40 E
Dagupan, Phil. 37 A6 16 3N 120 20 E
Dahlak Kebir, Eritrea 46 D3 15 50N 40 10 E
Dahlonega, U.S.A. 77 H4 34 32N 83 59W
Dahod, India 42 H6 22 50N 74 15 E
Dahomey = Benin ■, Africa 50 G5 10 0N 2 0 E
Dahra, Senegal 50 E1 15 22N 15 30W
Dai Hao, Vietnam 38 C6 18 1N 106 25 E
Dai-Sen, Japan 31 G6 35 22N 133 32 E
Dai Xian, China 34 E7 39 4N 112 58 E
Daicheng, China 34 E9 38 42N 116 38 E
Daingean, Ireland 13 C4 53 18N 7 15W
Daintree, Australia 62 B4 16 20S 145 20 E
Daiō-Misaki, Japan 31 G8 34 15N 136 45 E
Dairût, Egypt 51 C11 27 34N 30 43 E
Daisetsu-Zan, Japan 30 C11 43 30N 142 57 E
Dajarra, Australia 62 C2 21 42S 139 30 E
Dak Dam, Cambodia 38 F6 12 20N 107 21 E
Dak Nhe, Vietnam 38 E6 15 28N 107 48 E
Dak Pek, Vietnam 38 E6 15 4N 107 44 E
Dak Song, Vietnam 39 F6 12 19N 107 35 E
Dak Sui, Vietnam 38 E6 14 55N 107 43 E
Dakar, Senegal 50 F1 14 34N 17 29W
Dakhla, W. Sahara 50 D1 23 50N 15 53W
Dakhla, El Wâhât el-, Egypt 51 C10 25 30N 28 50 E
Dakhovskaya, Russia 25 F7 44 13N 40 13 E
Dakor, India 42 H5 22 45N 73 11 E
Dakota City, U.S.A. 80 D6 42 25N 96 25W
Đakovica, Serbia 21 C9 42 22N 20 26 E
Dalachi, China 34 F3 36 48N 105 0 E
Dalai Nur, China 34 C9 43 20N 116 45 E
Dālakī, Iran 45 D6 29 26N 51 17 E
Dalälven, Sweden 9 F14 60 12N 16 43 E
Dalandzadgad, Mongolia 34 C3 43 27N 104 30 E
Dalarö, Sweden 9 G15 59 8N 18 24 E
Dālbandīn, Pakistan 40 E4 29 0N 64 23 E
Dalbeattie, U.K. 12 G5 54 55N 3 50W
Dalby, Australia 63 D5 27 10S 151 17 E
Dalgān, Iran 45 E8 27 31N 59 19 E
Dalhart, U.S.A. 81 G3 36 4N 102 31W
Dalhousie, Canada 71 C6 48 5N 66 26W
Dalhousie, India 42 C6 32 38N 75 58 E
Dali, Shaanxi, China 34 G5 34 48N 109 58 E
Dali, Yunnan, China 32 D5 25 40N 100 10 E
Dalian, China 35 E11 38 50N 121 40 E
Daliang Shan, China 32 D5 28 0N 102 45 E
Daling He →, China 35 D11 40 55N 121 40 E
Dāliyat el Karmel, Israel 47 C4 32 43N 35 2 E
Dalkeith, U.K. 12 F5 55 54N 3 5W
Dall I., U.S.A. 72 C2 54 59N 133 25W
Dallarnil, Australia 63 D5 25 19S 152 2 E
Dallas, Oreg., U.S.A. 82 D2 44 55N 123 19W
Dallas, Tex., U.S.A. 81 J6 32 47N 96 49W
Dalmacija = Dalmatia □, Croatia 20 C7 43 20N 17 0 E
Dalmatia □, Croatia 20 C7 43 20N 17 0 E
Dalmellington, U.K. 12 F4 55 20N 4 25W
Dalnegorsk, Russia 27 E14 44 32N 135 33 E
Dalnerechensk, Russia 27 E14 45 50N 133 40 E
Daloa, Ivory C. 50 G3 7 0N 6 30W
Daltenganj, India 43 G11 24 0N 84 4 E
Dalton, Canada 70 C3 48 11N 84 1W
Dalton, Ga., U.S.A. 77 H3 34 46N 84 58W
Dalton, Mass., U.S.A. 79 D11 42 28N 73 11W
Dalton, Nebr., U.S.A. 80 E3 41 25N 102 58W
Dalton Iceberg Tongue, Antarctica 5 C9 66 15S 121 30 E
Dalvík, Iceland 8 D4 65 58N 18 32W
Daly →, Australia 60 B5 13 35S 130 19 E
Daly City, U.S.A. 84 H4 37 42N 122 28W
Daly L., Canada 73 B7 56 32N 105 39W
Daly Waters, Australia 62 B1 16 15S 133 24 E
Dam Doi, Vietnam 39 H5 8 50N 105 12 E
Dam Ha, Vietnam 38 B6 21 21N 107 36 E
Daman, India 40 J8 20 25N 72 57 E
Dāmaneh, Iran 45 C6 33 1N 50 29 E
Damanhûr, Egypt 51 B11 31 0N 30 30 E
Damanzhuang, China 34 E9 38 5N 116 35 E
Damar, Indonesia 37 F7 7 7S 128 40 E
Damaraland, Namibia 56 C2 21 0S 17 0 E
Damascus = Dimashq, Syria 47 B5 33 30N 36 18 E
Damāvand, Iran 45 C7 35 47N 52 0 E
Damāvand, Qolleh-ye, Iran 45 C7 35 56N 52 10 E
Damba, Angola 52 F3 6 44S 15 20 E
Dame Marie, Haiti 89 C5 18 36N 74 26W
Dāmghān, Iran 45 B7 36 10N 54 17 E
Damietta = Dumyât, Egypt 51 B11 31 24N 31 48 E
Daming, China 34 F8 36 15N 115 6 E
Damīr Qābū, Syria 44 B4 36 58N 41 51 E
Dammam = Ad Dammām, Si. Arabia 45 E6 26 20N 50 5 E
Damodar →, India 43 H12 23 17N 87 35 E
Damoh, India 43 H8 23 50N 79 28 E
Dampier, Australia 60 D2 20 41S 116 42 E
Dampier, Selat, Indonesia 37 E8 0 40S 131 0 E
Dampier Arch., Australia 60 D2 20 38S 116 32 E
Damrei, Chuor Phnum, Cambodia 39 G4 11 30N 103 0 E
Dana, Indonesia 37 F6 11 0S 122 52 E
Dana, L., Canada 70 B4 50 53N 77 20W
Dana, Mt., U.S.A. 84 H7 37 54N 119 12W
Danbury, U.S.A. 79 E11 41 24N 73 28W
Danby L., U.S.A. 83 J6 34 13N 115 5W
Dand, Afghan. 42 D1 31 28N 65 32 E
Dandaragan, Australia 61 F2 30 40S 115 40 E
Dandeldhura, Nepal 43 E9 29 20N 80 35 E
Dandeli, India 40 M9 15 5N 74 30 E
Dandenong, Australia 63 F4 38 0S 145 15 E
Dandong, China 35 D13 40 10N 124 20 E
Danfeng, China 34 H6 33 45N 110 25 E
Danforth, U.S.A. 71 C6 45 40N 67 52W
Danger Is. = Pukapuka, Cook Is. 65 J11 10 53S 165 49W
Danger Pt., S. Africa 56 E2 34 40S 19 17 E
Dangora, Nigeria 50 F6 11 30N 8 7 E
Dangrek, Phnom, Thailand 38 E5 14 15N 105 0 E
Dangriga, Belize 87 D7 17 0N 88 13W
Dangshan, China 34 G9 34 27N 116 22 E
Daniel, U.S.A. 82 E8 42 52N 110 4W
Daniel's Harbour, Canada 71 B8 50 13N 57 35W
Danielskuil, S. Africa 56 D3 28 11S 23 33 E
Danielson, U.S.A. 79 E13 41 48N 71 53W
Danilov, Russia 24 C7 58 16N 40 13 E
Daning, China 34 F6 36 28N 110 45 E
Danissa, Kenya 54 B5 3 15N 40 58 E
Dankhar Gompa, India 40 C11 32 10N 78 10 E
Danlí, Honduras 88 D2 14 4N 86 35W
Dannemora, Sweden 9 F14 60 12N 17 51 E
Dannemora, U.S.A. 79 B11 44 43N 73 44W
Dannevirke, N.Z. 59 J6 40 12S 176 8 E
Dannhauser, S. Africa 57 D5 28 0S 30 3 E
Dansville, U.S.A. 78 D7 42 34N 77 42W
Dantan, India 43 J12 21 57N 87 20 E
Dante, Somali Rep. 46 E5 10 25N 51 16 E
Danube →, Europe 21 B13 45 20N 29 40 E
Danvers, U.S.A. 79 D14 42 34N 70 56W
Danville, Ill., U.S.A. 76 E2 40 8N 87 37W
Danville, Ky., U.S.A. 76 G3 37 39N 84 46W
Danville, Va., U.S.A. 77 G6 36 36N 79 23W
Danzig = Gdańsk, Poland 17 A9 54 22N 18 40 E
Dao, Phil. 37 B6 10 30N 121 57 E
Daoud = Aïn Beïda, Algeria 50 A6 35 50N 7 29 E
Daqing Shan, China 34 D6 40 40N 111 0 E
Dar es Salaam, Tanzania 54 D4 6 50S 39 12 E
Dar Mazār, Iran 45 D8 29 14N 57 20 E
Dar'ā, Syria 47 C5 32 36N 36 7 E
Dar'ā □, Syria 47 C5 32 55N 36 10 E
Dārāb, Iran 45 D7 28 50N 54 30 E
Daraj, Libya 50 B7 30 10N 10 28 E
Dārān, Iran 45 C6 32 59N 50 24 E
Dārayyā, Syria 47 B5 33 28N 36 15 E
Darband, Pakistan 42 B5 34 20N 72 50 E
Darband, Kūh-e, Iran 45 D8 31 34N 57 8 E
Darbhanga, India 43 F11 26 15N 85 55 E
Darby, U.S.A. 82 C6 46 1N 114 11W
Dardanelle, Ark., U.S.A. 81 H8 35 13N 93 9W
Dardanelle, Calif., U.S.A. 84 G7 38 20N 119 50W
Dardanelles = Çanakkale Boğazı, Turkey 21 D12 40 17N 26 32 E
Dārestān, Iran 45 D8 29 9N 58 42 E
Dârfûr, Sudan 51 F9 13 40N 24 0 E
Dargai, Pakistan 42 B4 34 25N 71 55 E
Dargan Ata, Uzbekistan 26 E7 40 29N 62 10 E
Dargaville, N.Z. 59 F4 35 57S 173 52 E
Darhan Muminggan Lianheqi, China 34 D6 41 40N 110 28 E
Darién, G. del, Colombia 92 B3 9 0N 77 0W
Dariganga, Mongolia 34 B7 45 21N 113 45 E
Darjeeling = Darjiling, India 43 F13 27 3N 88 18 E
Darjiling, India 43 F13 27 3N 88 18 E
Dark Cove, Canada 71 C9 48 47N 54 13W
Darkan, Australia 61 F2 33 20S 116 43 E
Darkhazīneh, Iran 45 D6 31 54N 48 39 E
Darkot Pass, Pakistan 43 A5 36 45N 73 26 E
Darling →, Australia 63 E3 34 4S 141 54 E
Darling Downs, Australia 63 D5 27 30S 150 30 E
Darling Ra., Australia 61 F2 32 30S 116 0 E
Darlington, U.K. 10 C6 54 33N 1 33W
Darlington, S.C., U.S.A. 77 H6 34 18N 79 52W
Darlington, Wis., U.S.A. 80 D9 42 41N 90 7W
Darlot, L., Australia 61 E3 27 48S 121 35 E
Darłowo, Poland 16 A8 54 25N 16 25 E
Darmstadt, Germany 16 D4 49 51N 8 40 E
Darnah, Libya 51 B9 32 40N 22 35 E
Darnall, S. Africa 57 D5 29 23S 31 18 E
Darnley, C., Antarctica 5 C6 68 0S 69 0 E
Darnley B., Canada 68 B7 69 30N 123 30W
Darr, Australia 62 C3 23 13S 144 7 E
Darr →, Australia 62 C3 23 39S 143 50 E
Darrington, U.S.A. 82 B3 48 15N 121 36W
Dart →, U.K. 11 G4 50 24N 3 36W
Dart, C., Antarctica 5 D14 73 6S 126 20W
Dartmoor, U.K. 11 G4 50 36N 4 0W
Dartmouth, Australia 62 C3 23 31S 144 44 E
Dartmouth, Canada 71 D7 44 40N 63 30W
Dartmouth, U.K. 11 G4 50 21N 3 35W
Dartmouth, L., Australia 63 D4 26 4S 145 18 E
Dartuch, C., Spain 22 B10 39 55N 3 49 E
Darvaza, Turkmenistan 26 E6 40 11N 58 24 E
Darvel, Teluk, Malaysia 37 D5 4 50N 118 20 E
Darwha, India 40 J10 20 15N 77 45 E
Darwin, Australia 60 B5 12 25S 130 51 E
Darwin, U.S.A. 85 J9 36 15N 117 35W
Darwin River, Australia 60 B5 12 50S 130 58 E
Daryoi Amu = Amudarya →, Uzbekistan 26 E6 43 40N 59 0 E
Dās, U.A.E. 45 E7 25 20N 53 30 E
Dashetai, China 34 D5 41 0N 109 5 E
Dasht, Iran 45 B8 37 17N 56 7 E
Dasht →, Pakistan 40 G2 25 10N 61 40 E
Dasht-e Mārgow, Afghan. 40 D3 30 40N 62 30 E
Dasht-i-Nawar, Afghan. 42 C3 33 52N 68 0 E
Daska, Pakistan 42 C6 32 20N 74 20 E
Dasseneiland, S. Africa 56 E2 33 25S 18 3 E
Datia, India 43 G8 25 39N 78 27 E
Datong, China 34 D7 40 6N 113 18 E
Datu, Tanjung, Indonesia 36 D3 2 5N 109 39 E
Datu Piang, Phil. 37 C6 7 2N 124 30 E
Daugava →, Latvia 24 C3 57 4N 24 3 E
Daugavpils, Latvia 24 C4 55 53N 26 32 E
Daulpur, India 42 F7 26 45N 77 59 E
Dauphin, Canada 73 C8 51 9N 100 5W
Dauphin I., U.S.A. 77 K1 30 15N 88 11W
Dauphin L., Canada 73 C9 51 20N 99 45W
Dauphiné, France 18 D6 45 15N 5 25 E
Dausa, India 42 F7 26 52N 76 20 E
Davangere, India 40 M9 14 25N 75 55 E
Davao, Phil. 37 C7 7 0N 125 40 E
Davao, G. of, Phil. 37 C7 6 30N 125 48 E
Dāvar Panāh, Iran 45 E9 27 25N 62 15 E
Davenport, Calif., U.S.A. 84 H4 37 1N 122 12W
Davenport, Iowa, U.S.A. 80 E9 41 32N 90 35W
Davenport, Wash., U.S.A. 82 C4 47 39N 118 9W
Davenport Downs, Australia 62 C3 24 8S 141 7 E
Davenport Ra., Australia 62 C1 20 28S 134 0 E
David, Panama 88 E3 8 30N 82 30W
David City, U.S.A. 80 E6 41 15N 97 8W
Davidson, Canada 73 C7 51 16N 105 59W
Davis, U.S.A. 84 G5 38 33N 121 44W
Davis Dam, U.S.A. 85 K12 35 11N 114 34W
Davis Inlet, Canada 71 A7 55 50N 60 59W
Davis Mts., U.S.A. 81 K2 30 50N 103 55W
Davis Sea, Antarctica 5 C7 66 0S 92 0 E
Davis Str., N. Amer. 69 B14 65 0N 58 0W
Davos, Switz. 16 E4 46 48N 9 49 E
Davy L., Canada 73 B7 58 53N 108 18W
Dawes Ra., Australia 62 C5 24 40S 150 40 E
Dawson, Canada 68 B6 64 10N 139 30W
Dawson, Ga., U.S.A. 77 K3 31 46N 84 27W
Dawson, N. Dak., U.S.A. 80 B5 46 52N 99 45W
Dawson, I., Chile 96 G2 53 50S 70 50W
Dawson Creek, Canada 72 B4 55 45N 120 15W
Dawson Inlet, Canada 73 A10 61 50N 93 25W
Dawson Ra., Australia 62 C4 24 30S 149 48 E
Daxian, China 32 C5 31 15N 107 23 E
Daxindian, China 35 F11 37 30N 120 50 E
Daxinggou, China 35 C15 43 25N 129 40 E
Daxue Shan, China 32 C5 30 30N 101 30 E
Daylesford, Australia 63 F3 37 21S 144 9 E
Dayr az Zawr, Syria 44 C4 35 20N 40 5 E
Daysland, Canada 72 C6 52 50N 112 20W
Dayton, Nev., U.S.A. 84 F7 39 14N 119 36W
Dayton, Ohio, U.S.A. 76 F3 39 45N 84 12W
Dayton, Pa., U.S.A. 78 F5 40 53N 79 15W
Dayton, Tenn., U.S.A. 77 H3 35 30N 85 1W
Dayton, Wash., U.S.A. 82 C4 46 19N 117 59W
Daytona Beach, U.S.A. 77 L5 29 13N 81 1W
Dayville, U.S.A. 82 D4 44 28N 119 32W
De Aar, S. Africa 56 E3 30 39S 24 0 E
De Funiak Springs, U.S.A. 77 K2 30 43N 86 7W
De Grey, Australia 60 D2 20 12S 119 12 E
De Grey →, Australia 60 D2 20 12S 119 13 E
De Kalb, U.S.A. 80 E10 41 56N 88 46W
De Land, U.S.A. 77 L5 29 2N 81 18W
De Leon, U.S.A. 81 J5 32 7N 98 32W
De Pere, U.S.A. 76 C1 44 27N 88 4W
De Queen, U.S.A. 81 H7 34 2N 94 21W
De Quincy, U.S.A. 81 K8 30 27N 93 26W
De Ridder, U.S.A. 81 K8 30 51N 93 17W
De Smet, U.S.A. 80 C6 44 23N 97 33W
De Soto, U.S.A. 80 F9 38 8N 90 34W
De Tour Village, U.S.A. 76 C4 46 0N 83 56W
De Witt, U.S.A. 81 H9 34 18N 91 20W
Dead Sea, Asia 47 D4 31 30N 35 30 E
Deadwood, U.S.A. 80 C3 44 23N 103 44W
Deadwood L., Canada 72 B3 59 10N 128 30W
Deakin, Australia 61 F4 30 46S 128 58 E
Deal, U.K. 11 F9 51 13N 1 25 E
Deal I., Australia 62 F4 39 30S 147 20 E
Dealesville, S. Africa 56 D4 28 41S 25 44 E
Dean, Forest of, U.K. 11 F5 51 50N 2 35W
Deán Funes, Argentina 94 C3 30 20S 64 20W
Dearborn, U.S.A. 70 D3 42 19N 83 11W
Dease →, Canada 72 B3 59 56N 128 32W
Dease L., Canada 72 B2 58 40N 130 5W
Dease Lake, Canada 72 B2 58 25N 130 6W
Death Valley, U.S.A. 85 J10 36 15N 116 50W
Death Valley Junction, U.S.A. 85 J10 36 20N 116 25W
Death Valley National Monument, U.S.A. 85 J10 36 45N 117 15W
Deba Habe, Nigeria 50 F7 10 14N 11 20 E
Debar, Macedonia 21 D9 41 31N 20 30 E
Debden, Canada 73 C7 53 30N 106 50W
Debolt, Canada 72 B5 55 12N 118 1W
Deborah East, L., Australia 61 F2 30 45S 119 0 E
Deborah West, L., Australia 61 F2 30 45S 118 50 E
Debre Markos, Ethiopia 51 F12 10 20N 37 40 E
Debre Tabor, Ethiopia 51 F12 11 50N 38 26 E
Debrecen, Hungary 17 E10 47 33N 21 42 E
Decatur, Ala., U.S.A. 77 H2 34 36N 86 59W
Decatur, Ga., U.S.A. 77 J3 33 47N 84 18W
Decatur, Ill., U.S.A. 80 F10 39 51N 88 57W
Decatur, Ind., U.S.A. 76 E3 40 50N 84 56W
Decatur, Tex., U.S.A. 81 J6 33 14N 97 35W
Deccan, India 40 M10 18 0N 79 0 E
Deception L., Canada 73 B8 56 33N 104 13W
Deckerville, U.S.A. 78 C2 43 32N 82 44W
Decorah, U.S.A. 80 D9 43 18N 91 48W
Dedéagach = Alexandroúpolis, Greece 21 D11 40 50N 25 54 E
Dedham, U.S.A. 79 D13 42 15N 71 10W
Dédougou, Burkina Faso 50 F4 12 30N 3 25W
Dedza, Malawi 55 E3 14 20S 34 20 E
Dee →, Clwyd, U.K. 10 D4 53 15N 3 7W
Dee →, Gramp., U.K. 12 D6 57 4N 2 7W
Deep B., Canada 72 A5 61 15N 116 35W
Deep Well, Australia 62 C1 24 20S 134 0 E
Deepwater, Australia 63 D5 29 25S 151 51 E
Deer →, Canada 73 B10 58 23N 94 13W
Deer Lake, Nfld., Canada 71 C8 49 11N 57 27W
Deer Lake, Ont., Canada 73 C10 52 36N 94 20W
Deer Lodge, U.S.A. 82 C7 46 24N 112 44W
Deer Park, U.S.A. 82 C5 47 57N 117 28W
Deer River, U.S.A. 80 B8 47 20N 93 48W
Deeral, Australia 62 B4 17 14S 145 55 E
Deerdepoort, S. Africa 56 C4 24 37S 26 27 E
Deferiet, U.S.A. 79 B9 44 2N 75 41W
Defiance, U.S.A. 76 E3 41 17N 84 22W
Degeh Bur, Ethiopia 46 F3 8 11N 43 31 E
Deggendorf, Germany 16 D6 48 49N 12 59 E
Deh Bīd, Iran 45 D7 30 39N 53 11 E
Deh-e Shīr, Iran 45 D7 31 29N 53 45 E
Dehaj, Iran 45 D7 30 42N 54 53 E
Dehdez, Iran 45 D6 31 43N 50 17 E
Dehestān, Iran 45 D7 28 30N 55 35 E
Dehgolān, Iran 44 C5 35 17N 47 25 E
Dehi Titan, Afghan. 40 C3 33 45N 63 50 E
Dehibat, Tunisia 50 B7 32 0N 10 47 E
Dehlorān, Iran 44 C5 32 41N 47 16 E
Dehnow-e Kūhestān, Iran 45 E8 27 58N 58 32 E
Dehra Dun, India 42 D8 30 20N 78 4 E
Dehri, India 43 G11 24 50N 84 15 E
Dehui, China 35 B13 44 30N 125 40 E
Deinze, Belgium 15 D3 50 59N 3 32 E
Dej, Romania 17 E11 47 10N 23 52 E
Dekese, Zaïre 52 E4 3 24S 21 24 E
Del Mar, U.S.A. 85 N9 32 58N 117 16W
Del Norte, U.S.A. 83 H10 37 41N 106 21W
Del Rio, U.S.A. 81 L4 29 22N 100 54W
Delano, U.S.A. 85 K7 35 46N 119 15W
Delareyville, S. Africa 56 D4 26 41S 25 26 E
Delavan, U.S.A. 80 D10 42 38N 88 39W
Delaware, U.S.A. 76 E4 40 18N 83 4W
Delaware □, U.S.A. 76 F8 39 0N 75 20W
Delaware →, U.S.A. 76 F8 39 15N 75 20W

Delaware B.

Delaware B., U.S.A. 75 C12 39 0N 75 10W
Delegate, Australia 63 F4 37 4S 148 56 E
Delft, Neths. 15 B4 52 1N 4 22 E
Delfzijl, Neths. 15 A6 53 20N 6 55 E
Delgado, C., Mozam. 55 E5 10 45S 40 40 E
Delgerhet, Mongolia 34 B6 45 50N 110 30 E
Delgo, Sudan 51 D11 20 6N 30 40 E
Delhi, Canada 78 D4 42 51N 80 30W
Delhi, India 42 E7 28 38N 77 17 E
Delhi, U.S.A. 79 D10 42 17N 74 55W
Delia, Canada 72 C6 51 38N 112 23W
Delice ~►, Turkey 25 G5 39 45N 34 15 E
Delicias, Mexico 86 B3 28 10N 105 30W
Delījān, Iran 45 C6 33 59N 50 40 E
Dell City, U.S.A. 83 L11 31 56N 105 12W
Dell Rapids, U.S.A. 80 D6 43 50N 96 43W
Delmar, U.S.A. 79 D11 42 37N 73 47W
Delmiro Gouveia, Brazil 93 E11 9 24S 38 6W
Delong, Ostrova, Russia 27 B15 76 40N 149 20 E
Deloraine, Australia 62 G4 41 30S 146 40 E
Deloraine, Canada 73 D8 49 15N 100 29W
Delphi, U.S.A. 76 E2 40 36N 86 41W
Delphos, U.S.A. 76 E3 40 51N 84 21W
Delportshoop, S. Africa 56 D3 28 22S 24 20 E
Delray Beach, U.S.A. 77 M5 26 28N 80 4W
Delta, Colo., U.S.A. 83 G9 38 44N 108 4W
Delta, Utah, U.S.A. 82 G7 39 21N 112 35W
Delungra, Australia 63 D5 29 39S 150 51 E
Demanda, Sierra de la, Spain 19 A4 42 15N 3 0W
Demavand = Damāvand, Iran 45 C7 35 47N 52 0 E
Demba, Zaïre 52 F4 5 28S 22 15 E
Dembecha, Ethiopia 51 F12 10 32N 37 30 E
Dembia, Zaïre 54 B2 3 33N 25 48 E
Dembidolo, Ethiopia 51 G11 8 34N 34 50 E
Demer ~►, Belgium 15 D4 50 57N 4 42 E
Deming, N. Mex., U.S.A. 83 K10 32 16N 107 46W
Deming, Wash., U.S.A. 84 B4 48 50N 122 13W
Demini ~►, Brazil 92 D6 0 46S 62 56W
Demopolis, U.S.A. 77 J2 32 31N 87 50W
Dempo, Indonesia 36 E2 4 2S 103 15 E
Den Burg, Neths. 15 A4 53 3N 4 47 E
Den Chai, Thailand 38 D3 17 59N 100 4 E
Den Haag = 's-Gravenhage, Neths. 15 B4 52 7N 4 17 E
Den Helder, Neths. 15 B4 52 57N 4 45 E
Den Oever, Neths. 15 B5 52 56N 5 2 E
Denain, France 18 D3 50 20N 3 22 E
Denair, U.S.A. 84 H6 37 32N 120 48W
Denau, Uzbekistan 26 F7 38 16N 67 54 E
Denbigh, U.K. 10 D4 53 12N 3 26W
Dendang, Indonesia 36 E3 3 7S 107 56 E
Dendermonde, Belgium 15 C4 51 2N 4 5 E
Dengfeng, China 34 G7 34 25N 113 2 E
Dengkou, China 34 D4 40 18N 106 55 E
Denham, Australia 61 E1 25 56S 113 31 E
Denham Ra., Australia 62 C4 21 55S 147 46 E
Denham Sd., Australia 61 E1 25 45S 113 15 E
Denia, Spain 19 C6 38 49N 0 8 E
Denial B., Australia 63 E1 32 14S 133 32 E
Deniliquin, Australia 63 F3 35 30S 144 58 E
Denison, Iowa, U.S.A. 80 D7 42 1N 95 21W
Denison, Tex., U.S.A. 81 J6 33 45N 96 33W
Denison Plains, Australia 60 C4 18 35S 128 0 E
Denizli, Turkey 25 G4 37 42N 29 2 E
Denman Glacier, Antarctica 5 C7 66 45S 99 25 E
Denmark, Australia 61 F2 34 59S 117 25 E
Denmark ■, Europe 9 J10 55 30N 9 0 E
Denmark Str., Atl. Oc. 4 C6 66 0N 30 0W
Dennison, U.S.A. 78 F3 40 24N 81 19W
Denpasar, Indonesia 36 F5 8 45S 115 14 E
Denton, Mont., U.S.A. 82 C9 47 19N 109 57W
Denton, Tex., U.S.A. 81 J6 33 13N 97 8W
D'Entrecasteaux, Pt., Australia 61 F2 34 50S 115 57 E
Denver, U.S.A. 80 F2 39 44N 104 59W
Denver City, U.S.A. 81 J3 32 58N 102 50W
Deoband, India 42 E7 29 42N 77 43 E
Deoghar, India 43 G12 24 30N 86 42 E
Deolali, India 40 K8 19 58N 73 50 E
Deoli = Devli, India 42 G6 25 50N 75 20 E
Deoria, India 43 F10 26 31N 83 48 E
Deosai Mts., Pakistan 43 B6 35 40N 75 0 E
Deping, China 35 F9 37 25N 116 58 E
Deposit, U.S.A. 79 D9 42 4N 75 25W
Depot Springs, Australia 61 E3 27 55S 120 3 E
Deputatskiy, Russia 27 C14 69 18N 139 54 E
Dera Ghazi Khan, Pakistan 42 D4 30 5N 70 43 E
Dera Ismail Khan, Pakistan 42 D4 31 50N 70 50 E
Derbent, Russia 25 F8 42 5N 48 15 E
Derby, Australia 60 C3 17 18S 123 38 E
Derby, U.K. 10 E6 52 55N 1 28W
Derby, Conn., U.S.A. 79 E11 41 19N 73 5W
Derby, N.Y., U.S.A. 78 D6 42 41N 78 58W
Derbyshire □, U.K. 10 E6 52 55N 1 28W
Derg ~►, U.K. 13 B4 54 42N 7 26W
Derg, L., Ireland 13 D3 53 0N 8 20W
Dergaon, India 41 F19 26 45N 94 0 E
Dernieres, Isles, U.S.A. 81 L9 29 2N 90 50W
Derry = Londonderry, U.K. 13 B4 55 0N 7 23W
Derryveagh Mts., Ireland 13 B3 55 0N 8 4W
Derudub, Sudan 51 E12 17 31N 36 7 E
Derwent, Canada 73 C6 53 41N 110 58W
Derwent ~►, Derby, U.K. 10 E6 52 53N 1 17W
Derwent ~►, N. Yorks., U.K. 10 D7 53 45N 0 57W
Derwent Water, U.K. 10 C4 54 35N 3 9W
Des Moines, Iowa, U.S.A. 80 E8 41 35N 93 37W
Des Moines, N. Mex., U.S.A. 81 G3 36 46N 103 50W
Des Moines ~►, U.S.A. 80 E9 40 23N 91 25W
Desaguadero ~►, Argentina 94 C2 34 30S 66 46W
Desaguadero ~►, Bolivia 92 G5 16 35S 69 5W
Descanso, Pta., Mexico 85 N9 32 21N 117 3W
Deschaillons, Canada 71 C5 46 32N 72 7W
Descharme ~►, Canada 73 B7 56 51N 109 13W
Deschutes ~►, U.S.A. 82 D3 45 38N 120 55W
Dese, Ethiopia 46 E2 11 5N 39 40 E
Desert Center, U.S.A. 85 M11 33 43N 115 24W
Desert Hot Springs, U.S.A. 85 M10 33 58N 116 30W
Désirade, I., Guadeloupe 89 C7 16 18N 61 3W
Deskenatlata L., Canada 72 A6 60 55N 112 3W
Desna ~►, Ukraine 24 D5 50 33N 30 32 E
Desolación, I., Chile 96 G2 53 0S 74 0W
Despeñaperros, Paso, Spain 19 C4 38 24N 3 30W
Dessau, Germany 16 C6 51 49N 12 15 E
Dessye = Dese, Ethiopia 46 E2 11 5N 39 40 E
D'Estrees B., Australia 63 F2 35 55S 137 45 E
Desuri, India 42 G5 25 18N 73 35 E
Det Udom, Thailand 38 E5 14 54N 105 5 E
Dete, Zimbabwe 55 F2 18 38S 26 50 E
Detmold, Germany 16 C4 51 55N 8 50 E
Detour, Pt., U.S.A. 76 C2 45 40N 86 40W
Detroit, Mich., U.S.A. 75 B10 42 20N 83 3W
Detroit, Tex., U.S.A. 81 J7 33 40N 95 16W
Detroit Lakes, U.S.A. 80 B7 46 49N 95 51W
Deurne, Belgium 15 C4 51 12N 4 24 E
Deurne, Neths. 15 C5 51 27N 5 49 E
Deutsche Bucht, Germany 16 A4 54 15N 8 0 E
Deux-Sèvres □, France 18 C3 46 35N 0 20W
Deva, Romania 17 F11 45 53N 22 55 E
Devakottai, India 40 Q11 9 55N 78 45 E
Devaprayag, India 43 D8 30 13N 78 35 E
Deventer, Neths. 15 B6 52 15N 6 10 E
Deveron ~►, U.K. 12 D6 57 40N 2 31W
Devgadh Bariya, India 42 H5 22 40N 73 55 E
Devils Den, U.S.A. 84 K7 35 46N 119 58W
Devils Lake, U.S.A. 80 A5 48 7N 98 52W
Devils Paw, Canada 72 B2 58 47N 134 0W
Devizes, U.K. 11 F6 51 21N 2 0W
Devli, India 42 G6 25 50N 75 20 E
Devon, Canada 72 C6 53 24N 113 44W
Devon □, U.K. 11 G4 50 50N 3 40W
Devon I., Canada 4 B3 75 10N 85 0W
Devonport, Australia 62 G4 41 10S 146 22 E
Devonport, N.Z. 59 G5 36 49S 174 49 E
Devonport, U.K. 11 G3 50 23N 4 11W
Dewas, India 42 H7 22 59N 76 3 E
Dewetsdorp, S. Africa 56 D4 29 33S 26 39 E
Dewsbury, U.K. 10 D6 53 42N 1 38W
Dexter, Mo., U.S.A. 81 G9 36 48N 89 57W
Dexter, N. Mex., U.S.A. 81 J2 33 12N 104 22W
Dey-Dey, L., Australia 61 E5 29 12S 131 4 E
Deyhūk, Iran 45 C8 33 15N 57 30 E
Deyyer, Iran 45 E6 27 55N 51 55 E
Dezadeash L., Canada 72 A1 60 28N 136 58W
Dezfūl, Iran 45 C6 32 20N 48 30 E
Dezhneva, Mys, Russia 27 C19 66 5N 169 40W
Dezhou, China 34 F9 37 26N 116 18 E
Dháfni, Greece 23 D7 35 13N 25 3 E
Dhahiriya = Az̧ Z̧āhirīyah, Jordan 47 D3 31 25N 34 58 E
Dhahran = Az̧ Z̧ahrān, Si. Arabia 45 E6 26 10N 50 7 E
Dhaka, Bangla. 43 H14 23 43N 90 26 E
Dhaka □, Bangla. 43 G14 24 25N 90 25 E
Dhali, Cyprus 23 D12 35 1N 33 25 E
Dhamar, Yemen 46 E3 14 30N 44 20 E
Dhampur, India 43 E8 29 19N 78 33 E
Dhamtari, India 41 J12 20 42N 81 35 E
Dhanbad, India 43 H12 23 50N 86 30 E
Dhangarhi, Nepal 41 E12 28 55N 80 40 E
Dhankuta, Nepal 43 F12 26 55N 87 40 E
Dhar, India 42 H6 22 35N 75 26 E
Dharampur, India 42 H6 22 13N 75 18 E
Dharamsala = Dharmsala, India 42 C7 32 16N 76 23 E
Dharmapuri, India 40 N11 12 10N 78 10 E
Dharmsala, India 42 C7 32 16N 76 23 E
Dharwad, India 40 M9 15 22N 75 15 E
Dharwar, India 40 M9 15 43N 75 1 E
Dhaulagiri, Nepal 43 E10 28 39N 83 28 E
Dhebar, L., India 42 G6 24 10N 74 0 E
Dheftera, Cyprus 23 D12 35 5N 33 16 E
Dhenkanal, India 41 J14 20 45N 85 35 E
Dherinia, Cyprus 23 D12 35 3N 33 57 E
Dhiarrizos ~►, Cyprus 23 E11 34 41N 32 34 E
Dhībān, Jordan 47 D4 31 30N 35 46 E
Dhidhimótikhon, Greece 21 D12 41 22N 26 29 E
Dhíkti Óros, Greece 23 D7 35 8N 25 22 E
Dhírfis, Greece 21 E10 38 40N 23 54 E
Dhodhekánisos, Greece 21 F12 36 35N 27 0 E
Dholka, India 42 H5 22 44N 72 29 E
Dhoraji, India 42 J4 21 45N 70 37 E
Dhráhstis, Ákra, Greece 23 A3 39 48N 19 40 E
Dhrangadhra, India 42 H4 22 59N 71 31 E
Dhrápanon, Ákra, Greece 23 D6 35 28N 24 14 E
Dhrol, India 42 H4 22 33N 70 25 E
Dhuburi, India 41 F16 26 2N 89 59 E
Dhule, India 40 J9 20 58N 74 50 E
Dhut ~►, Somali Rep. 46 E5 10 30N 50 0 E
Di Linh, Vietnam 39 G7 11 35N 108 4 E
Di Linh, Cao Nguyen, Vietnam 39 G7 11 30N 108 0 E
Día, Greece 23 D7 35 28N 25 14 E
Diablo, Mt., U.S.A. 84 H5 37 53N 121 56W
Diablo Range, U.S.A. 84 J5 37 20N 121 25W
Diafarabé, Mali 50 F4 14 9N 4 57W
Diamante, Argentina 94 C3 32 5S 60 40W
Diamante ~►, Argentina 94 C2 34 30S 66 46W
Diamantina, Brazil 93 G10 18 17S 43 40W
Diamantina ~►, Australia 63 D2 26 45S 139 10 E
Diamantino, Brazil 93 F7 14 30S 56 30W
Diamond Harbour, India 43 H13 22 11N 88 14 E
Diamond Is., Australia 62 B5 17 25S 151 5 E
Diamond Mts., U.S.A. 82 G6 39 50N 115 30W
Diamond Springs, U.S.A. 84 G6 38 42N 120 49W
Diamondville, U.S.A. 82 F8 41 47N 110 32W
Diapaga, Burkina Faso 50 F5 12 5N 1 46 E
Dībā, Oman 45 E8 25 45N 56 16 E
Dibaya, Zaïre 52 F4 6 30S 22 57 E
Dibaya-Lubue, Zaïre 52 E3 4 12S 19 54 E
Dibbi, Ethiopia 46 G3 4 10N 41 52 E
Dibete, Botswana 56 C4 23 45S 26 32 E
Dibrugarh, India 41 F19 27 29N 94 55 E
Dickinson, U.S.A. 80 B3 46 53N 102 47W
Dickson, U.S.A. 77 G2 36 5N 87 23W
Dickson City, U.S.A. 79 E9 41 29N 75 40W
Didiéni, Mali 50 F3 13 53N 8 6W
Didsbury, Canada 72 C6 51 35N 114 10W
Didwana, India 42 F6 27 23N 74 36 E
Diébougou, Burkina Faso 50 F4 11 0N 3 15W
Diefenbaker L., Canada 73 C7 51 0N 106 55W
Diego Garcia, Ind. Oc. 3 E13 7 50S 72 50 E
Diekirch, Lux. 15 E6 49 52N 6 10 E
Dien Ban, Vietnam 38 E7 15 53N 108 16 E
Dien Bien, Vietnam 38 B4 21 20N 103 0 E
Dien Khanh, Vietnam 39 F7 12 15N 109 6 E
Dieppe, France 18 B4 49 54N 1 4 E
Dieren, Neths. 15 B6 52 3N 6 6 E
Dierks, U.S.A. 81 H7 34 7N 94 1W
Diest, Belgium 15 D5 50 58N 5 4 E
Differdange, Lux. 15 E5 49 31N 5 54 E
Dig, India 42 F7 27 28N 77 20 E
Digba, Zaïre 54 B2 4 25N 25 48 E
Digby, Canada 71 D6 44 38N 65 50W
Digges, Canada 73 B10 58 40N 94 0W
Digges Is., Canada 69 B12 62 40N 77 50W
Dighinala, Bangla. 41 H18 23 15N 92 5 E
Dighton, U.S.A. 80 F4 38 29N 100 28W
Digne, France 18 D7 44 5N 6 12 E
Digos, Phil. 37 C7 6 45N 125 20 E
Digranes, Iceland 8 C6 66 4N 14 44W
Digul ~►, Indonesia 37 F9 7 7S 138 42 E
Dihang ~►, India 41 F19 27 48N 95 30 E
Dīhōk, Iraq 44 B3 36 55N 38 57 E
Dijlah, Nahr ~►, Asia 44 D5 31 0N 47 25 E
Dijon, France 18 C6 47 20N 5 3 E
Dikomu di Kai, Botswana 56 C3 24 58S 24 36 E
Diksmuide, Belgium 15 C2 51 2N 2 52 E
Dikson, Russia 26 B9 73 40N 80 5 E
Dikwa, Nigeria 51 F7 12 4N 13 30 E
Dili, Indonesia 37 F7 8 39S 125 34 E
Dilley, U.S.A. 81 L5 28 40N 99 10W
Dilling, Sudan 51 F10 12 3N 29 35 E
Dillingham, U.S.A. 68 C4 59 3N 158 28W
Dillon, Canada 73 B7 55 56N 108 35W
Dillon, Mont., U.S.A. 82 D7 45 13N 112 38W
Dillon, S.C., U.S.A. 77 H6 34 25N 79 22W
Dillon ~►, Canada 73 B7 55 56N 108 56W
Dilolo, Zaïre 52 G4 10 28S 22 18 E
Dilston, Australia 62 G4 41 22S 147 10 E
Dimas, Mexico 86 C3 23 43N 106 47W
Dimashq, Syria 47 B5 33 30N 36 18 E
Dimashq □, Syria 47 B5 33 30N 36 30 E
Dimbaza, S. Africa 57 E4 32 50S 27 14 E
Dimbokro, Ivory C. 50 G4 6 45N 4 46W
Dimboola, Australia 63 F3 36 28S 142 7 E
Dîmboviţa ~►, Romania 17 F13 44 5N 26 35 E
Dimbulah, Australia 62 B4 17 8S 145 4 E
Dimitrovgrad, Bulgaria 21 C11 42 5N 25 35 E
Dimitrovgrad, Russia 24 D8 54 14N 49 39 E
Dimmitt, U.S.A. 81 H3 34 33N 102 19W
Dimona, Israel 47 D4 31 2N 35 1 E
Dinagat, Phil. 37 B7 10 10N 125 40 E
Dinajpur, Bangla. 41 G16 25 33N 88 43 E
Dinan, France 18 B2 48 28N 2 2W
Dīnān Āb, Iran 45 C8 32 4N 56 49 E
Dinant, Belgium 15 D4 50 16N 4 55 E
Dinapur, India 43 G11 25 38N 85 5 E
Dinara Planina, Croatia 20 B6 44 0N 16 30 E
Dinard, France 18 B2 48 38N 2 6W
Dinaric Alps = Dinara Planina, Croatia 20 B6 44 0N 16 30 E
Dindigul, India 40 P11 10 25N 78 0 E
Ding Xian, China 34 E8 38 30N 114 59 E
Dingbian, China 34 F4 37 35N 107 32 E
Dingle, Ireland 13 D1 52 9N 10 17W
Dingle B., Ireland 13 D1 52 3N 10 20W
Dingmans Ferry, U.S.A. 79 E10 41 13N 74 55W
Dingo, Australia 62 C4 23 38S 149 19 E
Dingtao, China 34 G8 35 5N 115 35 E
Dinguiraye, Guinea 50 F2 11 18N 10 49W
Dingwall, U.K. 12 D4 57 36N 4 26W
Dingxi, China 34 G3 35 30N 104 33 E
Dingxiang, China 34 E7 38 30N 112 58 E
Dinh, Mui, Vietnam 39 G7 11 22N 109 1 E
Dinh Lap, Vietnam 38 B6 21 33N 107 6 E
Dinokwe, Botswana 56 C4 23 29S 26 37 E
Dinosaur National Monument, U.S.A. 82 F9 40 30N 108 45W
Dinuba, U.S.A. 83 H4 36 32N 119 23W
Diourbel, Senegal 50 F1 14 39N 16 12W
Diplo, Pakistan 42 G3 24 35N 69 35 E
Dipolog, Phil. 37 C6 8 36N 123 20 E
Dir, Pakistan 40 B7 35 8N 71 59 E
Diré, Mali 50 E4 16 20N 3 25W
Dire Dawa, Ethiopia 46 F3 9 35N 41 45 E
Diriamba, Nic. 88 D2 11 51N 86 19W
Dirico, Angola 53 H4 17 50S 20 42 E
Dirk Hartog I., Australia 61 E1 25 50S 113 5 E
Dirranbandi, Australia 63 D4 28 33S 148 17 E
Disa, India 42 G5 24 18N 72 10 E
Disappointment, C., U.S.A. 82 C1 46 18N 124 5W
Disappointment, L., Australia 60 D3 23 20S 122 40 E
Disaster B., Australia 63 F4 37 15S 149 58 E
Discovery B., Australia 63 F3 38 10S 140 40 E
Disko, Greenland 4 C5 69 45N 53 30W
Disko Bugt, Greenland 4 C5 69 10N 52 0W
Disteghil Sar, Pakistan 43 A6 36 20N 75 12 E
Distrito Federal □, Brazil 93 G9 15 45S 47 45W
Diu, India 42 J4 20 45N 70 58 E
Dīvāndarreh, Iran 44 C5 35 55N 47 2 E
Divide, U.S.A. 82 D7 45 45N 112 45W
Dividing Ra., Australia 61 E2 27 45S 116 0 E
Divinópolis, Brazil 93 H10 20 10S 44 54W
Divnoye, Russia 25 E7 45 55N 43 21 E
Dīwāl Kol, Afghan. 42 B2 34 23N 67 52 E
Dixie Mt., U.S.A. 84 F6 39 55N 120 16W
Dixon, Calif., U.S.A. 84 G5 38 27N 121 49W
Dixon, Ill., U.S.A. 80 E10 41 50N 89 29W
Dixon, Mont., U.S.A. 82 C6 47 19N 114 19W
Dixon, N. Mex., U.S.A. 83 H11 36 12N 105 53W
Dixon Entrance, U.S.A. 72 C2 54 30N 132 0W
Dixonville, Canada 72 B5 56 32N 117 40W
Diyarbakır, Turkey 25 G7 37 55N 40 18 E
Djado, Niger 51 D7 21 4N 12 14 E
Djakarta = Jakarta, Indonesia 37 G12 6 9S 106 49 E
Djamba, Angola 56 B1 16 45S 13 58 E
Djambala, Congo 52 E2 2 32S 14 30 E
Djanet, Algeria 50 D6 24 35N 9 32 E
Djawa = Jawa, Indonesia 37 G14 7 0S 110 0 E
Djelfa, Algeria 50 B5 34 40N 3 15 E
Djema, C.A.R. 54 A2 6 3N 25 15 E
Djenné, Mali 50 F4 14 0N 4 30W
Djerba, I. de, Tunisia 51 B7 33 50N 10 48 E
Djerid, Chott, Tunisia 50 B6 33 42N 8 30 E
Djibo, Burkina Faso 50 F4 14 9N 1 35W
Djibouti, Djibouti 46 E3 11 30N 43 5 E
Djibouti ■, Africa 46 E3 12 0N 43 0 E
Djolu, Zaïre 52 D4 0 35N 22 5 E
Djougou, Benin 50 G5 9 40N 1 45 E
Djoum, Cameroon 52 D2 2 41N 12 35 E
Djourab, Chad 51 E8 16 40N 18 50 E
Djugu, Zaïre 54 B3 1 55N 30 35 E
Djúpivogur, Iceland 8 D6 64 39N 14 17W
Dmitriya Lapteva, Proliv, Russia 27 B15 73 0N 140 0 E
Dnepr ~►, Ukraine 25 E5 46 30N 32 18 E
Dneprodzerzhinsk, Ukraine 25 E5 48 32N 34 37 E
Dnepropetrovsk, Ukraine 25 E5 48 30N 35 0 E
Dnestr ~►, Europe 25 E5 46 18N 30 17 E
Dnestrovski = Belgorod, Russia 25 D6 50 35N 36 35 E
Dnieper = Dnepr ~►, Ukraine 25 E5 46 30N 32 18 E
Dniester = Dnestr ~►, Europe 25 E5 46 18N 30 17 E
Dnipro = Dnepr ~►, Ukraine 25 E5 46 30N 32 18 E
Dniprodzerzhynsk = Dneprodzerzhinsk, Ukraine 25 E5 48 32N 34 37 E
Dnipropetrovsk = Dnepropetrovsk, Ukraine 25 E5 48 30N 35 0 E
Dnyapro = Dnepr ~►, Ukraine 25 E5 46 30N 32 18 E
Doan Hung, Vietnam 38 B5 21 30N 105 10 E
Doba, Chad 51 G8 8 40N 16 50 E
Dobbyn, Australia 62 B3 19 44S 140 2 E
Doberai, Jazirah, Indonesia 37 E8 1 25S 133 0 E
Doblas, Argentina 94 D3 37 5S 64 0W
Dobo, Indonesia 37 F8 5 45S 134 15 E
Dobreta-Turnu-Severin, Romania 21 B10 44 39N 22 41 E
Dobruja, Romania 17 F14 44 30N 28 15 E
Doc, Mui, Vietnam 38 D6 17 58N 106 30 E
Doda, India 43 C6 33 10N 75 34 E
Dodecanese = Dhodhekánisos, Greece 21 F12 36 35N 27 0 E
Dodge Center, U.S.A. 80 C8 44 2N 92 52W
Dodge City, U.S.A. 81 G5 37 45N 100 1W
Dodge L., Canada 73 B7 59 50N 105 36W
Dodgeville, U.S.A. 80 D9 42 58N 90 8W
Dodoma, Tanzania 54 D4 6 8S 35 45 E
Dodoma □, Tanzania 54 D4 6 0S 36 0 E
Dodsland, Canada 73 C7 51 50N 108 45W
Dodson, U.S.A. 82 B9 48 24N 108 15W
Doetinchem, Neths. 15 C6 51 59N 6 18 E
Dog Creek, Canada 72 C4 51 35N 122 14W
Dog L., Man., Canada 73 C9 51 2N 98 31W
Dog L., Ont., Canada 70 C2 48 48N 89 30W
Dogger Bank, N. Sea 6 E6 54 50N 2 0 E
Dogi, Afghan. 40 C3 32 20N 62 50 E
Dogran, Pakistan 42 D5 31 48N 73 35 E
Doha = Ad Dawhah, Qatar 45 E6 25 15N 51 35 E
Dohazari, Bangla. 41 H18 22 10N 92 5 E
Doi, Indonesia 37 D7 2 14N 127 49 E
Doi Luang, Thailand 38 C3 18 30N 101 0 E
Doi Saket, Thailand 38 C2 18 52N 99 9 E
Doig ~►, Canada 72 B4 56 25N 120 40W
Dois Irmãos, Sa., Brazil 93 E10 9 0S 42 30W
Dokka, Norway 9 F11 60 49N 10 7 E
Dokkum, Neths. 15 A5 53 20N 5 59 E
Dokri, Pakistan 42 F3 27 25N 68 7 E
Doland, U.S.A. 80 C5 44 54N 98 6W
Dolbeau, Canada 71 C5 48 53N 72 18W
Dole, France 18 C6 47 7N 5 31 E
Dolgellau, U.K. 10 E4 52 44N 3 53W
Dolgelley = Dolgellau, U.K. 10 E4 52 44N 3 53W
Dollart, Neths. 15 A7 53 20N 7 10 E
Dolomites = Dolomiti, Italy 20 A4 46 30N 11 40 E
Dolomiti, Italy 20 A4 46 30N 11 40 E
Dolores, Argentina 94 D4 36 20S 57 40W
Dolores, Uruguay 94 C4 33 34S 58 15W
Dolores, U.S.A. 83 H9 37 28N 108 30W
Dolores ~►, U.S.A. 83 G9 38 49N 109 17W
Dolphin, C., Falk. Is. 96 G5 51 10S 59 0W
Dolphin and Union Str., Canada 68 B8 69 5N 114 45W
Dom Pedrito, Brazil 95 C5 31 0S 54 40W
Domasi, Malawi 55 F4 15 15S 35 22 E
Dombarovskiy, Russia 26 D6 50 46N 59 32 E
Dombås, Norway 9 E10 62 4N 9 8 E
Dombes, France 18 C6 45 58N 5 0 E
Domburg, Neths. 15 C3 51 34N 3 30 E
Domeyko, Chile 94 B1 29 0S 71 0W
Domeyko, Cordillera, Chile 94 A2 24 30S 69 0W
Dominador, Chile 94 A2 24 21S 69 20W
Dominica ■, W. Indies 89 C7 15 20N 61 20W
Dominica Passage, W. Indies 89 C7 15 10N 61 20W
Dominican Rep. ■, W. Indies 89 C5 19 0N 70 30W
Domo, Ethiopia 46 F4 7 50N 47 10 E
Domodóssola, Italy 20 A3 46 6N 8 19 E
Domville, Mt., Australia 63 D5 28 1S 151 15 E
Don ~►, Russia 25 E6 47 4N 39 18 E
Don ~►, Gramp., U.K. 12 D6 57 14N 2 5W

East C.

Don →, S. Yorks., U.K. 10 D7 53 41N 0 51W
Don, C., Australia 60 B5 11 18S 131 46 E
Don Benito, Spain 19 C3 38 53N 5 51W
Don Duong, Vietnam 39 G7 11 51N 108 35 E
Don Martín, Presa de, Mexico 86 B4 27 30N 100 50W
Dona Ana = Nhamaabué, Mozam. 55 F4 17 25S 35 5 E
Donaghadee, U.K. 13 B6 54 38N 5 32W
Donald, Australia 63 F3 36 23S 143 0 E
Donalda, Canada 72 C6 52 35N 112 34W
Donaldsonville, U.S.A. 81 K9 30 6N 90 59W
Donalsonville, U.S.A. 77 K3 31 3N 84 53W
Donau →, Austria 16 D8 48 10N 17 0 E
Donauwörth, Germany 16 D5 48 42N 10 47 E
Doncaster, U.K. 10 D6 53 31N 1 9W
Dondo, Angola 52 F2 9 45S 14 25 E
Dondo, Mozam. 55 F3 19 33S 34 46 E
Dondo, Teluk, Indonesia 37 D6 0 29N 120 30 E
Dondra Head, Sri Lanka 40 S12 5 55N 80 40 E
Donegal, Ireland 13 B3 54 39N 8 8W
Donegal □, Ireland 13 B4 54 53N 8 0W
Donegal B., Ireland 13 B3 54 30N 8 35W
Donets →, Russia 25 E7 47 33N 40 55 E
Donetsk, Ukraine 25 E6 48 0N 37 45 E
Dong Ba Thin, Vietnam 39 F7 12 8N 109 13 E
Dong Dang, Vietnam 38 B6 21 54N 106 42 E
Dong Giam, Vietnam 38 C5 19 25N 105 31 E
Dong Ha, Vietnam 38 D6 16 55N 107 8 E
Dong Hene, Laos 38 D5 16 40N 105 18 E
Dong Hoi, Vietnam 38 D6 17 29N 106 36 E
Dong Khe, Vietnam 38 A6 22 26N 106 27 E
Dong Ujimqin Qi, China 34 B9 45 32N 116 55 E
Dong Van, Vietnam 38 A5 23 16N 105 22 E
Dong Xoai, Vietnam 39 G6 11 32N 106 55 E
Dongara, Australia 61 E1 29 14S 114 57 E
Dongbei, China 35 D13 42 0N 125 0 E
Dongchuan, China 32 D5 26 8N 103 1 E
Dongfang, China 38 C7 18 50N 108 33 E
Dongfeng, China 35 C13 42 40N 125 34 E
Donggala, Indonesia 37 E5 0 30S 119 40 E
Donggou, China 35 E13 39 52N 124 10 E
Dongguang, China 34 F9 37 50N 116 30 E
Dongjingcheng, China 35 B15 44 5N 129 10 E
Dongning, China 35 B16 44 2N 131 5 E
Dongola, Sudan 51 E11 19 9N 30 22 E
Dongou, Congo 52 D3 2 0N 18 5 E
Dongping, China 34 G9 35 55N 116 20 E
Dongsheng, China 34 E6 39 50N 110 0 E
Dongtai, China 35 H11 32 51N 120 21 E
Dongting Hu, China 33 D6 29 18N 112 45 E
Donington, C., Australia 63 E2 34 45S 136 0 E
Doniphan, U.S.A. 81 G9 36 37N 90 50W
Dønna, Norway 8 C12 66 6N 12 30 E
Donna, U.S.A. 81 M5 26 9N 98 4W
Donnaconna, Canada 71 C5 46 41N 71 41W
Donnelly's Crossing, N.Z. 59 F4 35 42S 173 38 E
Donnybrook, Australia 61 F2 33 34S 115 48 E
Donnybrook, S. Africa 57 D4 29 59S 29 48 E
Donora, U.S.A. 78 F5 40 11N 79 52W
Donor's Hill, Australia 62 B3 18 42S 140 33 E
Donostia = San Sebastián, Spain 19 A5 43 17N 1 58W
Doon →, U.K. 12 F4 55 26N 4 41W
Dora, L., Australia 60 D3 22 0S 123 0 E
Dora Báltea →, Italy 20 B3 45 11N 8 5 E
Doran L., Canada 73 A7 61 13N 108 6W
Dorchester, U.K. 11 G5 50 42N 2 28W
Dorchester, C., Canada 69 B12 65 27N 77 27W
Dordogne □, France 18 D4 45 5N 0 40 E
Dordogne →, France 18 D3 45 2N 0 36W
Dordrecht, Neths. 15 C4 51 48N 4 39 E
Dordrecht, S. Africa 56 E4 31 20S 27 3 E
Dore, Mts., France 18 D5 45 32N 2 50 E
Doré L., Canada 73 C7 54 46N 107 17W
Doré Lake, Canada 73 C7 54 38N 107 36W
Dori, Burkina Faso 50 F4 14 3N 0 2W
Doring →, S. Africa 56 E2 31 54S 18 39 E
Doringbos, S. Africa 56 E2 31 59S 19 16 E
Dorion, Canada 70 C5 45 23N 74 3W
Dornoch, U.K. 12 D4 57 52N 4 5W
Dornoch Firth, U.K. 12 D4 57 52N 4 0W
Dornogovĭ □, Mongolia 34 B6 44 0N 110 0 E
Dorohoi, Romania 17 E13 47 56N 26 30 E
Döröö Nuur, Mongolia 32 B4 48 0N 93 0 E
Dorr, Iran 45 C6 33 17N 50 38 E
Dorre I., Australia 61 E1 25 13S 113 12 E
Dorrigo, Australia 63 E5 30 20S 152 44 E
Dorris, U.S.A. 82 F3 41 58N 121 55W
Dorset, Canada 78 A6 45 14N 78 54W
Dorset, U.S.A. 78 E4 41 4N 80 40W
Dorset □, U.K. 11 G5 50 48N 2 25W
Dortmund, Germany 16 C3 51 32N 7 28 E
Doruma, Zaïre 54 B2 4 42N 27 33 E
Dorūneh, Iran 45 C8 35 10N 57 18 E
Dos Bahías, C., Argentina 96 E3 44 58S 65 32W
Dos Palos, U.S.A. 84 J6 36 59N 120 37W
Dosso, Niger 50 F5 13 0N 3 13 E
Dothan, U.S.A. 77 K3 31 13N 85 24W
Doty, U.S.A. 84 D3 46 38N 123 17W
Douai, France 18 A5 50 21N 3 4 E
Douala, Cameroon 50 H6 4 0N 9 45 E
Douarnenez, France 18 B1 48 6N 4 21W
Double Island Pt., Australia 63 D5 25 56S 153 11 E
Doubs □, France 18 C7 47 10N 6 20 E
Doubs →, France 18 C6 46 53N 5 1 E
Doubtful Sd., N.Z. 59 L1 45 20S 166 49 E
Doubtless B., N.Z. 59 F4 34 55S 173 26 E
Douentza, Mali 50 F4 14 58N 2 48W
Douglas, S. Africa 56 D3 29 4S 23 46 E
Douglas, U.K. 10 C3 54 9N 4 29W
Douglas, Alaska, U.S.A. 72 B2 58 17N 134 24W
Douglas, Ariz., U.S.A. 83 L9 31 21N 109 33W
Douglas, Ga., U.S.A. 77 K4 31 31N 82 51W
Douglas, Wyo., U.S.A. 80 D2 42 45N 105 24W
Douglastown, Canada 71 C7 48 46N 64 24W
Douglasville, U.S.A. 77 J3 33 45N 84 45W
Doumé, Cameroon 52 D2 4 15N 13 25 E
Dounreay, U.K. 12 C5 58 34N 3 44W
Dourados, Brazil 95 A5 22 9S 54 50W
Dourados →, Brazil 95 A5 21 58S 54 18W
Douro →, Europe 19 B1 41 8N 8 40W
Douro Litoral, Portugal 19 B1 41 10N 8 20W
Dove →, U.K. 10 E6 52 51N 1 36W
Dove Creek, U.S.A. 83 H9 37 46N 108 54W
Dover, Australia 62 G4 43 18S 147 2 E
Dover, U.K. 11 F9 51 7N 1 19 E
Dover, Del., U.S.A. 76 F8 39 10N 75 32W
Dover, N.H., U.S.A. 79 C14 43 12N 70 56W
Dover, N.J., U.S.A. 79 F10 40 53N 74 34W
Dover, Ohio, U.S.A. 78 F3 40 32N 81 29W
Dover, Pt., Australia 61 F4 32 32S 125 32 E
Dover, Str. of, Europe 18 A4 51 0N 1 30 E
Dover-Foxcroft, U.S.A. 71 C6 45 11N 69 13W
Dover Plains, U.S.A. 79 E11 41 43N 73 35W
Dovey = Dyfi →, U.K. 11 E4 52 32N 4 0W
Dovrefjell, Norway 8 E10 62 15N 9 33 E
Dow Rūd, Iran 45 C6 33 28N 49 4 E
Dowa, Malawi 55 E3 13 38S 33 58 E
Dowagiac, U.S.A. 76 E2 41 59N 86 6W
Dowgha'i, Iran 45 B8 36 54N 58 32 E
Dowlatābād, Iran 45 D8 28 20N 56 40 E
Down □, U.K. 13 B6 54 20N 5 47W
Downey, Calif., U.S.A. 85 M8 33 56N 118 7W
Downey, Idaho, U.S.A. 82 E7 42 26N 112 7W
Downham Market, U.K. 11 E8 52 36N 0 22 E
Downieville, U.S.A. 84 F6 39 34N 120 50W
Downpatrick, U.K. 13 B6 54 20N 5 43W
Downpatrick Hd., Ireland 13 B2 54 20N 9 21W
Dowsārī, Iran 45 D8 28 25N 57 59 E
Doyle, U.S.A. 84 E6 40 2N 120 6W
Doylestown, U.S.A. 79 F9 40 21N 75 10W
Draa, Oued →, Morocco 50 C2 28 40N 11 10W
Drachten, Neths. 15 A6 53 7N 6 5 E
Dragoman, Prokhod, Bulgaria 21 C10 42 58N 22 53 E
Dragonera, I., Spain 22 B9 39 35N 2 19 E
Draguignan, France 18 E7 43 32N 6 27 E
Drain, U.S.A. 82 E2 43 40N 123 19W
Drake, Australia 63 D5 28 55S 152 25 E
Drake, U.S.A. 80 B4 47 55N 100 23W
Drake Passage, S. Ocean 5 B17 58 0S 68 0W
Drakensberg, S. Africa 57 E4 31 0S 28 0 E
Dráma, Greece 21 D11 41 9N 24 10 E
Drammen, Norway 9 G11 59 42N 10 12 E
Drangajökull, Iceland 8 C2 66 9N 22 15W
Dras, India 43 B6 34 25N 75 48 E
Drau = Drava →, Croatia 21 B8 45 33N 18 55 E
Drava →, Croatia 21 B8 45 33N 18 55 E
Drayton Valley, Canada 72 C6 53 12N 114 58W
Drenthe □, Neths. 15 B6 52 52N 6 40 E
Drepanum, C., Cyprus 23 E11 34 54N 32 19 E
Dresden, Canada 78 D2 42 35N 82 11W
Dresden, Germany 16 C6 51 2N 13 45 E
Dreux, France 18 B4 48 44N 1 23 E
Driffield = Great Driffield, U.K. 10 C7 54 0N 0 25W
Driftwood, U.S.A. 78 E6 41 20N 78 8W
Driggs, U.S.A. 82 E8 43 44N 111 6W
Drina →, Bos.-H. 21 B8 44 53N 19 21 E
Drøbak, Norway 9 G11 59 39N 10 39 E
Drogheda, Ireland 13 C5 53 45N 6 20W
Drogobych, Ukraine 25 E3 49 20N 23 30 E
Drohobych = Drogobych, Ukraine 25 E3 49 20N 23 30 E
Droichead Nua, Ireland 13 C5 53 11N 6 50W
Droitwich, U.K. 11 E5 52 16N 2 10W
Drôme □, France 18 D6 44 38N 5 15 E
Dromedary, C., Australia 63 F5 36 17S 150 10 E
Dronfield, Australia 62 C3 21 12S 140 3 E
Drumbo, Canada 78 C4 43 16N 80 35W
Drumheller, Canada 72 C6 51 25N 112 40W
Drummond, U.S.A. 82 C7 46 40N 113 9W
Drummond I., U.S.A. 70 C3 46 1N 83 39W
Drummond Pt., Australia 63 E2 34 9S 135 16 E
Drummond Ra., Australia 62 C4 23 45S 147 10 E
Drummondville, Canada 70 C5 45 55N 72 25W
Drumright, U.S.A. 81 H6 35 59N 96 36W
Druzhina, Russia 27 C15 68 14N 145 18 E
Dry Tortugas, U.S.A. 88 B3 24 38N 82 55W
Dryden, Canada 73 D10 49 47N 92 50W
Dryden, U.S.A. 81 K3 30 3N 102 7W
Drygalski I., Antarctica 5 C7 66 0S 92 0 E
Drysdale →, Australia 60 B4 13 59S 126 51 E
Drysdale I., Australia 62 A2 11 41S 136 0 E
Dschang, Cameroon 50 G7 5 32N 10 3 E
Du Bois, U.S.A. 78 E6 41 8N 78 46W
Du Quoin, U.S.A. 80 G10 38 1N 89 14W
Duanesburg, U.S.A. 79 D10 42 45N 74 11W
Duaringa, Australia 62 C4 23 42S 149 42 E
Ḑubā, Si. Arabia 44 E2 27 10N 35 40 E
Dubai = Dubayy, U.A.E. 45 E7 25 18N 55 20 E
Dubawnt →, Canada 73 A8 64 33N 100 6W
Dubawnt, L., Canada 73 A8 63 4N 101 42W
Dubayy, U.A.E. 45 E7 25 18N 55 20 E
Dubbo, Australia 63 E4 32 11S 148 35 E
Dubele, Zaïre 54 B2 2 56N 29 35 E
Dublin, Ireland 13 C5 53 20N 6 18W
Dublin, Ga., U.S.A. 77 J4 32 32N 82 54W
Dublin, Tex., U.S.A. 81 J5 32 5N 98 21W
Dublin □, Ireland 13 C5 53 24N 6 20W
Dublin B., Ireland 13 C5 53 18N 6 5W
Dubois, U.S.A. 82 D7 44 10N 112 14W
Dubovka, Russia 25 E7 49 5N 44 50 E
Dubrajpur, India 43 H12 23 48N 87 25 E
Dubréka, Guinea 50 G2 9 46N 13 31W
Dubrovnik, Croatia 21 C8 42 39N 18 6 E
Dubrovskoye, Russia 27 D12 58 55N 111 10 E
Dubuque, U.S.A. 80 D9 42 30N 90 41W
Duchesne, U.S.A. 82 F8 40 10N 110 24W
Duchess, Australia 62 C2 21 20S 139 50 E
Ducie I., Pac. Oc. 65 K15 24 40S 124 48W
Duck Cr. →, Australia 60 D2 22 37S 116 53 E
Duck Lake, Canada 73 C7 52 50N 106 16W
Duck Mountain Prov. Park, Canada 73 C8 51 45N 101 0W
Duckwall, Mt., U.S.A. 84 H6 37 58N 120 7W
Dudhi, India 41 G13 24 15N 83 10 E
Dudinka, Russia 27 C9 69 30N 86 13 E
Dudley, U.K. 11 E5 52 30N 2 5W
Duero = Douro →, Europe 19 B1 41 8N 8 40W
Dufftown, U.K. 12 D5 57 26N 3 9W
Dugi Otok, Croatia 20 B6 44 0N 15 3 E
Duifken Pt., Australia 62 A3 12 33S 141 38 E
Duisburg, Germany 16 C3 51 27N 6 42 E
Duiwelskloof, S. Africa 57 C5 23 42S 30 10 E
Dūkdamīn, Iran 45 C8 35 59N 57 43 E
Duke I., U.S.A. 72 C2 54 50N 131 20W
Dukhān, Qatar 45 E6 25 25N 50 50 E
Duki, Pakistan 40 D6 30 14N 68 25 E
Duku, Nigeria 50 F7 10 43N 10 43 E
Dulce →, Argentina 94 C3 30 32S 62 33W
Dulce, G., Costa Rica 88 E3 8 40N 83 20W
Dulf, Iraq 44 C5 35 7N 45 51 E
Dulit, Banjaran, Malaysia 36 D4 3 15N 114 30 E
Duliu, China 34 E9 39 2N 116 55 E
Dullewala, Pakistan 42 D4 31 50N 71 25 E
Dulq Maghār, Syria 44 B3 36 22N 38 39 E
Dululu, Australia 62 C5 23 48S 150 15 E
Duluth, U.S.A. 80 B8 46 47N 92 6W
Dum Dum, India 43 H13 22 39N 88 33 E
Dum Duma, India 41 F19 27 40N 95 40 E
Dum Hadjer, Chad 51 F8 13 18N 19 41 E
Dūmā, Lebanon 47 A4 34 12N 35 50 E
Dūmā, Syria 47 B5 33 34N 36 24 E
Dumaguete, Phil. 37 C6 9 17N 123 15 E
Dumai, Indonesia 36 D2 1 35N 101 28 E
Dumaran, Phil. 37 B5 10 33N 119 50 E
Dumas, Ark., U.S.A. 81 J9 33 53N 91 29W
Dumas, Tex., U.S.A. 81 H4 35 52N 101 58W
Dumbarton, U.K. 12 F4 55 58N 4 35W
Dumbleyung, Australia 61 F2 33 17S 117 42 E
Dumfries, U.K. 12 F5 55 4N 3 37W
Dumfries & Galloway □, U.K. 12 F5 55 5N 4 0W
Dumka, India 43 G12 24 12N 87 15 E
Dumoine →, Canada 70 C4 46 13N 77 51W
Dumoine L., Canada 70 C4 46 55N 77 55W
Dumraon, India 43 G11 25 33N 84 8 E
Dumyât, Egypt 51 B11 31 24N 31 48 E
Dun Laoghaire, Ireland 13 C5 53 17N 6 9W
Dunaföldvár, Hungary 17 E9 46 50N 18 57 E
Dunărea →, Romania 17 F14 45 20N 29 40 E
Dunay, Russia 30 C6 42 52N 132 22 E
Dunback, N.Z. 59 L3 45 23S 170 36 E
Dunbar, Australia 62 B3 16 0S 142 22 E
Dunbar, U.K. 12 E6 56 0N 2 32W
Dunblane, U.K. 12 E5 56 10N 3 58W
Duncan, Canada 72 D4 48 45N 123 40W
Duncan, Ariz., U.S.A. 83 K9 32 43N 109 6W
Duncan, Okla., U.S.A. 81 H6 34 30N 97 57W
Duncan, L., Canada 70 B4 53 29N 77 58W
Duncan L., Canada 72 A6 62 51N 113 58W
Duncan Town, Bahamas 88 B4 22 15N 75 45W
Duncannon, U.S.A. 78 F7 40 23N 77 2W
Dundalk, Canada 78 B4 44 10N 80 24W
Dundalk, Ireland 13 B5 54 1N 6 25W
Dundalk Bay, Ireland 13 C5 53 55N 6 15W
Dundas, Canada 70 D4 43 17N 79 59W
Dundas, L., Australia 61 F3 32 35S 121 50 E
Dundas I., Canada 72 C2 54 30N 130 50W
Dundas Str., Australia 60 B5 11 15S 131 35 E
Dundee, S. Africa 57 D5 28 11S 30 15 E
Dundee, U.K. 12 E6 56 29N 3 0W
Dundgovĭ □, Mongolia 34 B4 45 10N 106 0 E
Dundoo, Australia 63 D3 27 40S 144 37 E
Dundrum, U.K. 13 B6 54 17N 5 50W
Dundrum B., U.K. 13 B6 54 12N 5 40W
Dundwara, India 43 F8 27 48N 79 9 E
Dunedin, N.Z. 59 L3 45 50S 170 33 E
Dunedin, U.S.A. 77 L4 28 1N 82 47W
Dunedin →, Canada 72 B4 59 30N 124 5W
Dunfermline, U.K. 12 E5 56 5N 3 28W
Dungannon, Canada 78 C3 43 51N 81 36W
Dungannon, U.K. 13 B5 54 30N 6 47W
Dungannon □, U.K. 13 B5 54 30N 6 55W
Dungarpur, India 42 H5 23 52N 73 45 E
Dungarvan, Ireland 13 D4 52 6N 7 40W
Dungarvan Bay, Ireland 13 D4 52 5N 7 35W
Dungeness, U.K. 11 G8 50 54N 0 59 E
Dungo, L. do, Angola 56 B2 17 15S 19 0 E
Dungog, Australia 63 E5 32 22S 151 46 E
Dungu, Zaïre 54 B2 3 40N 28 32 E
Dunhua, China 35 C15 43 20N 128 14 E
Dunhuang, China 32 B4 40 8N 94 36 E
Dunk I., Australia 62 B4 17 59S 146 29 E
Dunkeld, U.K. 12 E5 56 34N 3 36W
Dunkerque, France 18 A5 51 2N 2 20 E
Dunkery Beacon, U.K. 11 F4 51 15N 3 37W
Dunkirk = Dunkerque, France 18 A5 51 2N 2 20 E
Dunkirk, U.S.A. 78 D5 42 29N 79 20W
Dunkwa, Ghana 50 G4 6 0N 1 47W
Dunlap, U.S.A. 80 E7 41 51N 95 36W
Dúnleary = Dun Laoghaire, Ireland 13 C5 53 17N 6 9W
Dunmanus B., Ireland 13 E2 51 31N 9 50W
Dunmara, Australia 62 B1 16 42S 133 25 E
Dunmore, U.S.A. 79 E9 41 25N 75 38W
Dunmore Hd., Ireland 13 D1 52 10N 10 35W
Dunmore Town, Bahamas 88 A4 25 30N 76 39W
Dunn, U.S.A. 77 H6 35 19N 78 37W
Dunnellon, U.S.A. 77 L4 29 3N 82 28W
Dunnet Hd., U.K. 12 C5 58 38N 3 22W
Dunning, U.S.A. 80 E4 41 50N 100 6W
Dunnville, Canada 78 D5 42 54N 79 36W
Dunolly, Australia 63 F3 36 51S 143 44 E
Dunoon, U.K. 12 F4 55 57N 4 56W
Dunqul, Egypt 51 D11 23 26N 31 37 E
Duns, U.K. 12 F6 55 47N 2 20W
Dunseith, U.S.A. 80 A4 48 50N 100 3W
Dunsmuir, U.S.A. 82 F2 41 13N 122 16W
Dunstable, U.K. 11 F7 51 53N 0 31W
Dunstan Mts., N.Z. 59 L2 44 53S 169 35 E
Dunster, Canada 72 C5 53 8N 119 50W
Dunvegan L., Canada 73 A7 60 8N 107 10W
Duolun, China 34 C9 42 12N 116 28 E
Duong Dong, Vietnam 39 G4 10 13N 103 58 E
Dupree, U.S.A. 80 C4 45 4N 101 35W
Dupuyer, U.S.A. 82 B7 48 13N 112 30W
Duque de Caxias, Brazil 95 A7 22 45S 43 19W
Durack →, Australia 60 C4 15 33S 127 52 E
Durack Ra., Australia 60 C4 16 50S 127 40 E
Durance →, France 18 E6 43 55N 4 45 E
Durand, U.S.A. 76 D4 42 55N 83 59W
Durango = Victoria de Durango, Mexico 86 C4 24 3N 104 39W
Durango, Spain 19 A4 43 13N 2 40W
Durango, U.S.A. 83 H10 37 16N 107 53W
Durango □, Mexico 86 C4 25 0N 105 0W
Duranillin, Australia 61 F2 33 30S 116 45 E
Durant, U.S.A. 81 J6 33 59N 96 25W
Durazno, Uruguay 94 C4 33 25S 56 31W
Durazzo = Durrësi, Albania 21 D8 41 19N 19 28 E
Durban, S. Africa 57 D5 29 49S 31 1 E
Durg, India 41 J12 21 15N 81 22 E
Durgapur, India 43 H12 23 30N 87 20 E
Durham, Canada 78 B4 44 10N 80 49W
Durham, U.K. 10 C6 54 47N 1 34W
Durham, Calif., U.S.A. 84 F5 39 39N 121 48W
Durham, N.C., U.S.A. 77 G6 35 59N 78 54W
Durham □, U.K. 10 C6 54 42N 1 45W
Durham Downs, Australia 63 D4 26 6S 149 5 E
Durmitor, Montenegro, Yug. 21 C8 43 10N 19 0 E
Durness, U.K. 12 C4 58 34N 4 45W
Durrësi, Albania 21 D8 41 19N 19 28 E
Durrie, Australia 62 D3 25 40S 140 15 E
Duru, Zaïre 54 B2 4 14N 28 50 E
D'Urville, Tanjung, Indonesia 37 E9 1 28S 137 54 E
D'Urville I., N.Z. 59 J4 40 50S 173 55 E
Duryea, U.S.A. 79 E9 41 20N 75 45W
Dusa Mareb, Somali Rep. 46 F4 5 30N 46 15 E
Dushak, Turkmenistan 26 F7 37 13N 60 1 E
Dushanbe, Tajikistan 26 F7 38 33N 68 48 E
Dusky Sd., N.Z. 59 L1 45 47S 166 30 E
Dussejour, C., Australia 60 B4 14 45S 128 13 E
Düsseldorf, Germany 16 C3 51 15N 6 46 E
Dutch Harbor, U.S.A. 68 C3 53 53N 166 32W
Dutlwe, Botswana 56 C3 23 58S 23 46 E
Dutton, Canada 78 D3 42 39N 81 30W
Dutton →, Australia 62 C3 20 44S 143 10 E
Duyun, China 32 D5 26 18N 107 29 E
Duzdab = Zāhedān, Iran 45 D9 29 30N 60 50 E
Dvina, Sev. →, Russia 24 B7 64 32N 40 30 E
Dvinsk = Daugavpils, Latvia 24 C4 55 53N 26 32 E
Dvinskaya Guba, Russia 24 B6 65 0N 39 0 E
Dwarka, India 42 H3 22 18N 69 8 E
Dwellingup, Australia 61 F2 32 43S 116 4 E
Dwight, Canada 78 A5 45 20N 79 1W
Dwight, U.S.A. 76 E1 41 5N 88 26W
Dyer, C., Canada 69 B13 66 40N 61 0W
Dyer Plateau, Antarctica 5 D17 70 45S 65 30W
Dyersburg, U.S.A. 81 G10 36 3N 89 23W
Dyfed □, U.K. 11 E3 52 0N 4 30W
Dyfi →, U.K. 11 E4 52 32N 4 0W
Dynevor Downs, Australia 63 D3 28 10S 144 20 E
Dysart, Canada 73 C8 50 57N 104 2W
Dzamin Üüd, Mongolia 34 C6 43 50N 111 58 E
Dzerzhinsk, Belorussia 24 D4 53 40N 27 1 E
Dzerzhinsk, Russia 24 C7 56 14N 43 30 E
Dzhalinda, Russia 27 D13 53 26N 124 0 E
Dzhambul, Kazakhstan 26 E8 42 54N 71 22 E
Dzhankoi, Ukraine 25 E5 45 40N 34 20 E
Dzhardzhan, Russia 27 C13 68 10N 124 10 E
Dzhelinde, Russia 27 C12 70 0N 114 20 E
Dzhetygara, Kazakhstan 26 D7 52 11N 61 12 E
Dzhezkazgan, Kazakhstan 26 E7 47 44N 67 40 E
Dzhikimde, Russia 27 D13 59 1N 121 47 E
Dzhizak, Uzbekistan 26 E7 40 6N 67 50 E
Dzhugdzur, Khrebet, Russia 27 D14 57 30N 138 0 E
Dzhungarskiye Vorota, Kazakhstan 32 B3 45 0N 82 0 E
Dzilam de Bravo, Mexico 87 C7 21 24N 88 53W
Dzungaria = Junggar Pendi, China 32 B3 44 30N 86 0 E
Dzungarian Gates = Dzhungarskiye Vorota, Kazakhstan 32 B3 45 0N 82 0 E
Dzuumod, Mongolia 32 B5 47 45N 106 58 E

E

Eabamet, L., Canada 70 B2 51 30N 87 46W
Eads, U.S.A. 80 F3 38 29N 102 47W
Eagle, U.S.A. 82 G10 39 39N 106 50W
Eagle →, Canada 71 B8 53 36N 57 26W
Eagle Butte, U.S.A. 80 C4 45 0N 101 10W
Eagle Grove, U.S.A. 80 D8 42 40N 93 54W
Eagle L., Calif., U.S.A. 82 F3 40 39N 120 45W
Eagle L., Maine, U.S.A. 71 C6 46 20N 69 22W
Eagle Lake, U.S.A. 81 L6 29 35N 96 20W
Eagle Mountain, U.S.A. 85 M11 33 49N 115 27W
Eagle Nest, U.S.A. 83 H11 36 33N 105 16W
Eagle Pass, U.S.A. 81 L4 28 43N 100 30W
Eagle Pk., U.S.A. 84 G7 38 10N 119 25W
Eagle Pt., Australia 60 C3 16 11S 124 23 E
Eagle River, U.S.A. 80 C10 45 55N 89 15W
Ealing, U.K. 11 F7 51 30N 0 19W
Earaheedy, Australia 61 E3 25 34S 121 29 E
Earl Grey, Canada 73 C8 50 57N 104 43W
Earle, U.S.A. 81 H9 35 16N 90 28W
Earlimart, U.S.A. 85 K7 35 53N 119 16W
Earn →, U.K. 12 E5 56 20N 3 19W
Earn, L., U.K. 12 E4 56 23N 4 14W
Earnslaw, Mt., N.Z. 59 L2 44 32S 168 27 E
Earth, U.S.A. 81 H3 34 14N 102 24W
Easley, U.S.A. 77 H4 34 50N 82 36W
East Angus, Canada 71 C5 45 30N 71 40W
East Aurora, U.S.A. 78 D6 42 46N 78 37W
East B., U.S.A. 81 L10 29 0N 89 15W
East Bengal, Bangla. 41 G17 24 0N 90 0 E
East Beskids = Vychodné Beskydy, Europe 17 D10 49 20N 22 0 E
East Brady, U.S.A. 78 F5 40 59N 79 36W
East C., N.Z. 59 G7 37 42S 178 35 E

East Chicago

East Chicago, U.S.A. ... 76 E2 41 38N 87 27W
East China Sea, Asia ... 33 C7 30 5N 126 0 E
East Coulee, Canada ... 72 C6 51 23N 112 27W
East Falkland, Falk. Is. .. 96 G5 51 30S 58 30W
East Grand Forks, U.S.A. 80 B6 47 56N 97 1W
East Greenwich, U.S.A. . 79 E13 41 40N 71 27W
East Hartford, U.S.A. ... 79 E12 41 46N 72 39W
East Helena, U.S.A. 82 C8 46 35N 111 56W
East Indies, Asia .... 37 E6 0 0 120 0 E
East Jordan, U.S.A. 76 C3 45 10N 85 7W
East Lansing, U.S.A. 76 D3 42 44N 84 29W
East Liverpool, U.S.A. .. 78 F4 40 37N 80 35W
East London, S. Africa .. 57 E4 33 0S 27 55 E
East Main = Eastmain, Canada 70 B4 52 10N 78 30W
East Orange, U.S.A. 79 F10 40 46N 74 13W
East Pacific Ridge, Pac. Oc. 65 J17 15 0S 110 0W
East Pakistan = Bangladesh ■, Asia .. 41 H17 24 0N 90 0 E
East Palestine, U.S.A. .. 78 F4 40 50N 80 33W
East Pine, Canada 72 B4 55 48N 120 12W
East Pt., Canada 71 C7 46 27N 61 58W
East Point, U.S.A. 77 J3 33 41N 84 27W
East Providence, U.S.A. . 79 E13 41 49N 71 23W
East Retford, U.K. 10 D7 53 19N 0 55W
East St. Louis, U.S.A. ... 80 F9 38 37N 90 9W
East Schelde —► = Oosterschelde, Neths. 15 C4 51 33N 4 0 E
East Siberian Sea, Russia 27 B17 73 0N 160 0 E
East Stroudsburg, U.S.A. 79 E9 41 1N 75 11W
East Sussex □, U.K. .... 11 G8 51 0N 0 20 E
East Tawas, U.S.A. 76 C4 44 17N 83 29W
East Toorale, Australia . 63 E4 30 27S 145 28 E
East Walker —►, U.S.A. . 84 G7 38 52N 119 10W
Eastbourne, N.Z. .... 59 J5 41 19S 174 55 E
Eastbourne, U.K. ...... 11 G8 50 46N 0 18 E
Eastend, Canada 73 D7 49 32N 108 50W
Easter Dal = Österdalälven —►, Sweden .... 9 F12 61 30N 13 45 E
Easter Islands = Pascua, I. de, Pac. Oc. 65 K17 27 0S 109 0W
Eastern □, Kenya ...... 54 B4 0 0 38 30 E
Eastern □, Uganda ..... 54 B3 1 50N 33 45 E
Eastern Cr. —►, Australia 62 C3 20 40S 141 35 E
Eastern Ghats, India .... 40 N11 14 0N 78 50 E
Eastern Group = Lau Group, Fiji 59 C9 17 0S 178 30W
Eastern Group, Australia 61 F3 33 30S 124 30 E
Easterville, Canada 73 C9 53 8N 99 49W
Easthampton, U.S.A. ... 79 D12 42 16N 72 40W
Eastland, U.S.A. 81 J5 32 24N 98 49W
Eastleigh, U.K. 11 G6 50 58N 1 21W
Eastmain, Canada 70 B4 52 10N 78 30W
Eastmain —►, Canada .. 70 B4 52 27N 78 26W
Eastman, Canada 79 A12 45 18N 72 19W
Eastman, U.S.A. 77 J4 32 12N 83 11W
Easton, Md., U.S.A. 76 F7 38 47N 76 5W
Easton, Pa., U.S.A. 79 F9 40 41N 75 13W
Easton, Wash., U.S.A. .. 84 C5 47 14N 121 11W
Eastport, U.S.A. 71 D6 44 56N 67 0W
Eastsound, U.S.A. 84 B4 48 42N 122 55W
Eaton, U.S.A. 80 E2 40 32N 104 42W
Eatonia, Canada 73 C7 51 13N 109 25W
Eatonton, U.S.A. 77 J4 33 20N 83 23W
Eatontown, U.S.A. 79 F10 40 19N 74 4W
Eatonville, U.S.A. 84 D4 46 52N 122 16W
Eau Claire, U.S.A. 80 C9 44 49N 91 30W
Ebagoola, Australia .... 62 A3 14 15S 143 12 E
Ebbw Vale, U.K. 11 F4 51 47N 3 12W
Ebeltoft, Denmark ..... 9 H11 56 12N 10 41 E
Ebensburg, U.S.A. 78 F6 40 29N 78 44W
Eberswalde, Germany .. 16 B6 52 49N 13 50 E
Ebetsu, Japan .... 30 C10 43 7N 141 34 E
Eboli, Italy .... 20 D6 40 39N 15 2 E
Ebolowa, Cameroon .... 52 D2 2 55N 11 10 E
Ebro —►, Spain .... 19 B6 40 43N 0 54 E
Ech Cheliff, Algeria ..... 50 A5 36 10N 1 20 E
Echigo-Sammyaku, Japan 31 F9 36 50N 139 50 E
Echizen-Misaki, Japan .. 31 G7 35 59N 135 57 E
Echo Bay, N.W.T., Canada 68 B8 66 5N 117 55W
Echo Bay, Ont., Canada . 70 C3 46 29N 84 4W
Echoing —►, Canada ... 73 B10 55 51N 92 5W
Echternach, Lux. .... 15 E6 49 49N 6 25 E
Echuca, Australia ...... 63 F3 36 10S 144 20 E
Ecija, Spain 19 D3 37 30N 5 10W
Eclipse Is., Australia .... 60 B4 13 54S 126 19 E
Ecuador ■, S. Amer. ... 92 D3 2 0S 78 0W
Ed Dâmer, Sudan ...... 51 E11 17 27N 34 0 E
Ed Debba, Sudan ...... 51 E11 18 0N 30 51 E
Ed Dueim, Sudan ...... 51 F11 14 0N 32 10 E
Edah, Australia .... 61 E2 28 16S 117 10 E
Edam, Canada 73 C7 53 11N 108 46W
Edam, Neths. .... 15 B5 52 31N 5 3 E
Eday, U.K. 12 B6 59 11N 2 47W
Edd, Eritrea .... 46 E3 14 0N 41 38 E
Eddrachillis B., U.K. 12 C3 58 16N 5 10W
Eddystone, U.K. 11 G3 50 11N 4 16W
Eddystone Pt., Australia 62 G4 40 59S 148 20 E
Ede, Neths. .... 15 B5 52 4N 5 40 E
Édea, Cameroon .... 50 H7 3 51N 10 9 E
Edehon L., Canada 73 A9 60 25N 97 15W
Eden, Australia .... 63 F4 37 3S 149 55 E
Eden, N.C., U.S.A. 77 G6 36 29N 79 53W
Eden, N.Y., U.S.A. 78 D6 42 39N 78 55W
Eden, Tex., U.S.A. 81 K5 31 13N 99 51W
Eden, Wyo., U.S.A. 82 E9 42 3N 109 26W
Eden —►, U.K. 10 C4 54 57N 3 2W
Eden L., Canada 73 B8 56 38N 100 15W
Edenburg, S. Africa .... 56 D4 29 43S 25 58 E
Edendale, S. Africa ..... 57 D5 29 39S 30 18 E
Edenderry, Ireland 13 C4 53 21N 7 3W
Edenton, U.S.A. 77 G7 36 4N 76 39W
Edenville, S. Africa ..... 57 D4 27 37S 27 34 E
Edgar, U.S.A. 80 E6 40 22N 97 58W
Edgartown, U.S.A. 79 E14 41 23N 70 31W
Edge Hill, U.K. 11 E6 52 7N 1 28W
Edgefield, U.S.A. 77 J5 33 47N 81 56W
Edgeley, U.S.A. 80 B5 46 22N 98 43W
Edgemont, U.S.A. 80 D3 43 18N 103 50W
Edgeøya, Svalbard ..... 4 B9 77 45N 22 30 E
Edhessa, Greece .... 21 D10 40 48N 22 5 E
Edievale, N.Z. .... 59 L2 45 49S 169 22 E
Edina, U.S.A. 80 E8 40 10N 92 11W
Edinburg, U.S.A. 81 M5 26 18N 98 10W
Edinburgh, U.K. 12 F5 55 57N 3 12W
Edirne, Turkey .... 21 D12 41 40N 26 34 E
Edison, U.S.A. 84 B4 48 33N 122 27W
Edithburgh, Australia ... 63 F2 35 5S 137 43 E
Edjudina, Australia ..... 61 E3 29 48S 122 23 E
Edmeston, U.S.A. 79 D9 42 42N 75 15W
Edmond, U.S.A. 81 H6 35 39N 97 29W
Edmonds, U.S.A. 84 C4 47 49N 122 23W
Edmonton, Australia .... 62 B4 17 2S 145 46 E
Edmonton, Canada 72 C6 53 30N 113 30W
Edmund L., Canada 73 C10 54 45N 93 17W
Edmundston, Canada .. 71 C6 47 23N 68 20W
Edna, U.S.A. 81 L6 28 59N 96 39W
Edna Bay, U.S.A. 72 B2 55 55N 133 40W
Edson, Canada 72 C5 53 35N 116 28W
Eduardo Castex, Argentina 94 D3 35 50S 64 18W
Edward —►, Australia .. 63 F3 35 5S 143 30 E
Edward, L., Africa ...... 54 C2 0 25S 29 40 E
Edward I., Canada 70 C2 48 22N 88 37W
Edward River, Australia . 62 A3 14 59S 141 26 E
Edward VII Land, Antarctica 5 E13 80 0S 150 0W
Edwards, U.S.A. 85 L9 34 55N 117 51W
Edwards Plateau, U.S.A. 81 K4 30 45N 101 20W
Edwardsville, U.S.A. ... 79 E9 41 15N 75 56W
Edzo, Canada 72 A5 62 49N 116 4W
Eekloo, Belgium .... 15 C3 51 11N 3 33 E
Effingham, U.S.A. 76 F1 39 7N 88 33W
Égadi, Ísole, Italy ...... 20 F5 37 55N 12 16 E
Eganville, Canada 70 C4 45 32N 77 5W
Egeland, U.S.A. 80 A5 48 38N 99 6W
Egenolf L., Canada 73 B9 59 3N 100 0W
Eger = Cheb, Czech. ... 16 C6 50 9N 12 28 E
Eger, Hungary .... 17 E10 47 53N 20 27 E
Egersund, Norway ...... 9 G9 58 26N 6 1 E
Egg L., Canada 73 B7 55 5N 105 30W
Eginbah, Australia ..... 60 D2 20 53S 119 47 E
Egmont, C., N.Z. .... 59 H4 39 16S 173 45 E
Egmont, Mt., N.Z. ...... 59 H5 39 17S 174 5 E
Eğridir, Turkey .... 25 G5 37 52N 30 51 E
Eğridir Gölü, Turkey .... 25 G5 37 53N 30 50 E
Egvekinot, Russia 27 C19 66 19N 179 50W
Egypt ■, Africa .... 51 C11 28 0N 31 0 E
Ehime □, Japan .... 31 H6 33 30N 132 40 E
Ehrenberg, U.S.A. 85 M12 33 36N 114 31W
Eidsvold, Australia ..... 63 D5 25 25S 151 12 E
Eidsvoll, Norway .... 9 F11 60 19N 11 14 E
Eifel, Germany .... 16 C3 50 10N 6 45 E
Eiffel Flats, Zimbabwe .. 55 F3 18 20S 30 0 E
Eigg, U.K. 12 E2 56 54N 6 10W
Eighty Mile Beach, Australia .... 60 C3 19 30S 120 40 E
Eil, Somali Rep. .... 46 F4 8 0N 49 50 E
Eil, L., U.K. 12 E3 56 50N 5 15W
Eildon, L., Australia .... 63 F4 37 10S 146 0 E
Eileen L., Canada 73 A7 62 16N 107 37W
Einasleigh, Australia ... 62 B3 18 32S 144 5 E
Einasleigh —►, Australia 62 B3 17 30S 142 17 E
Eindhoven, Neths. ..... 15 C5 51 26N 5 28 E
Eire ■ = Ireland ■, Europe 13 D4 53 0N 8 0W
Eiríksjökull, Iceland 8 D3 64 46N 20 24W
Eirunepé, Brazil 92 E5 6 35S 69 53W
Eisenach, Germany .... 16 C5 50 58N 10 18 E
Eisenerz, Austria ...... 16 E7 47 32N 14 54 E
Eivissa = Ibiza, Spain .. 22 C7 38 54N 1 26 E
Ejutla, Mexico 87 D5 16 34N 96 44W
Ekalaka, U.S.A. 80 C2 45 53N 104 33W
Eketahuna, N.Z. .... 59 J5 40 38S 175 43 E
Ekibastuz, Kazakhstan .. 26 D8 51 50N 75 10 E
Ekimchan, Russia ...... 27 D14 53 0N 133 0 E
Ekoli, Zaïre .... 54 C1 0 23S 24 13 E
Ekwan —►, Canada 70 B3 53 12N 82 15W
Ekwan Pt., Canada 70 B3 53 16N 82 7W
El Aaiún, W. Sahara 50 C2 27 9N 13 12W
El 'Agrûd, Egypt .... 47 E3 30 14N 34 24 E
El Alamein, Egypt ..... 51 B10 30 48N 28 58 E
El 'Aqaba, W. —►, Egypt 47 E2 30 7N 33 54 E
El Arenal, Spain .... 22 B9 39 30N 2 45 E
El Aricha, Algeria 50 B4 34 13N 1 10W
El Arīhā, Jordan ...... 47 D4 31 52N 35 27 E
El Arish, Australia ..... 62 B4 17 35S 146 1 E
El 'Arîsh, Egypt .... 47 D2 31 8N 33 50 E
El 'Arîsh, W. —►, Egypt . 47 D2 31 8N 33 47 E
El Asnam = Ech Cheliff, Algeria .... 50 A5 36 10N 1 20 E
El Bawiti, Egypt .... 51 C10 28 25N 28 45 E
El Bayadh, Algeria ..... 50 B5 33 40N 1 1 E
El Bluff, Nic. 88 D3 11 59N 83 40W
El Brûk, W. —►, Egypt . 47 E2 30 15N 33 50 E
El Buheirat □, Sudan ... 51 G10 7 0N 30 0 E
El Cajon, U.S.A. 85 N10 32 48N 116 58W
El Callao, Venezuela ... 92 B6 7 18N 61 50W
El Campo, U.S.A. 81 L6 29 12N 96 16W
El Centro, U.S.A. 85 N11 32 48N 115 34W
El Cerro, Bolivia 92 G6 17 30S 61 40W
El Compadre, Mexico ... 85 N10 32 20N 116 14W
El Cuy, Argentina 96 D3 39 55S 68 25W
El Cuyo, Mexico 87 C7 21 30N 87 40W
El Daheir, Egypt ...... 47 D3 31 13N 34 10 E
El Dere, Somali Rep. ... 46 G4 3 50N 47 8 E
El Descanso, Mexico ... 85 N10 32 12N 116 58W
El Desemboque, Mexico 86 A2 30 30N 112 57W
El Diviso, Colombia 92 C3 1 22N 78 14W
El Djouf, Mauritania 50 E3 20 0N 9 0W
El Dorado, Ark., U.S.A. . 81 J8 33 12N 92 40W
El Dorado, Kans., U.S.A. 81 G6 37 49N 96 52W
El Dorado, Venezuela .. 92 B6 6 55N 61 37W
El Escorial, Spain 19 B3 40 35N 4 7W
El Faiyûm, Egypt ...... 51 C11 29 19N 30 50 E
El Fâsher, Sudan ...... 51 F10 13 33N 25 26 E
El Ferrol, Spain 19 A1 43 29N 8 15W
El Fuerte, Mexico 86 B3 26 30N 108 40W
El Gal, Somali Rep. .... 46 E5 10 58N 50 20 E
El Geteina, Sudan ..... 51 F11 14 50N 32 27 E
El Gezira □, Sudan ..... 51 F11 15 0N 33 0 E
El Gîza, Egypt .... 51 C11 30 0N 31 10 E
El Goléa, Algeria ...... 50 B5 30 30N 2 50 E
El Harrach, Algeria ..... 50 A5 36 45N 3 5 E
El Iskandarîya, Egypt ... 51 B10 31 0N 30 0 E
El Jadida, Morocco 50 B3 33 11N 8 17W
El Jebelein, Sudan ..... 51 F11 12 40N 32 55 E
El Kab, Sudan .... 51 E11 19 27N 32 46 E
El Kabrît, G., Egypt .... 47 F2 29 42N 33 16 E
El Kala, Algeria .... 50 A6 36 50N 8 30 E
El Kamlin, Sudan ...... 51 E11 15 3N 33 11 E
El Kef, Tunisia .... 50 A6 36 12N 8 47 E
El Khandaq, Sudan ..... 51 E11 18 30N 30 30 E
El Khârga, Egypt ...... 51 C11 25 30N 30 33 E
El Khartûm, Sudan ..... 51 E11 15 31N 32 35 E
El Khartûm Bahrî, Sudan 51 E11 15 40N 32 31 E
El Kuntilla, Egypt ...... 47 E3 30 1N 34 45 E
El Laqâwa, Sudan ..... 51 F10 11 25N 29 1 E
El Mafâza, Sudan ...... 51 F11 13 38N 34 30 E
El Mahalla el Kubra, Egypt .... 51 B11 31 0N 31 0 E
El Mansûra, Egypt ..... 51 B11 31 0N 31 19 E
El Medano, Canary Is. .. 22 F3 28 3N 16 32W
El Milagro, Argentina ... 94 C2 30 59S 65 59W
El Minyâ, Egypt .... 51 C11 28 7N 30 33 E
El Obeid, Sudan .... 51 F11 13 8N 30 10 E
El Odaiya, Sudan ...... 51 F10 12 8N 28 12 E
El Oro, Mexico 87 D4 19 48N 100 8W
El Oued, Algeria .... 50 B6 33 20N 6 58 E
El Palmito, Presa, Mexico 86 B3 25 40N 105 30W
El Paso, U.S.A. 83 L10 31 45N 106 29W
El Paso Robles, U.S.A. .. 84 K6 35 38N 120 41W
El Portal, U.S.A. 83 H4 37 41N 119 47W
El Porvenir, Mexico 86 A3 31 15N 105 51W
El Progreso, Honduras .. 88 C2 15 26N 87 51W
El Pueblito, Mexico 86 B3 29 3N 105 4W
El Pueblo, Canary Is. ... 22 F2 28 36N 17 47W
El Qâhira, Egypt .... 51 B11 30 1N 31 14 E
El Qantara, Egypt ..... 47 E1 30 51N 32 20 E
El Qasr, Egypt .... 51 C10 25 44N 28 42 E
El Quseima, Egypt ..... 47 E3 30 40N 34 15 E
El Reno, U.S.A. 81 H6 35 32N 97 57W
El Rio, U.S.A. 85 L7 34 14N 119 10W
El Roque, Pta., Canary Is. 22 F4 28 10N 15 25W
El Rosarito, Mexico 86 B2 28 38N 114 4W
El Saheira, W. —►, Egypt 47 E2 30 5N 33 25 E
El Salto, Mexico 86 C3 23 47N 105 22W
El Salvador ■, Cent. Amer. 88 D2 13 50N 89 0W
El Sauce, Nic. 88 D2 13 0N 86 40W
El Shallal, Egypt ...... 51 D11 24 0N 32 53 E
El Suweis, Egypt ...... 51 C11 29 58N 32 31 E
El Tamarâni, W. —►, Egypt .... 47 E3 30 7N 34 43 E
El Thamad, Egypt ...... 47 F3 29 40N 34 28 E
El Tigre, Venezuela 92 B6 8 44N 64 15W
El Tîh, G., Egypt ...... 47 F2 29 40N 33 50 E
El Tîna, Khalîg, Egypt .. 47 D1 31 10N 32 40 E
El Tocuyo, Venezuela .. 92 B5 9 47N 69 48W
El Tofo, Chile 94 B1 29 22S 71 18W
El Tránsito, Chile 94 B1 28 52S 70 17W
El Turbio, Argentina 96 G2 51 45S 72 5W
El Uqsur, Egypt .... 51 C11 25 41N 32 38 E
El Venado, Mexico 86 C4 22 56N 101 10W
El Vigía, Venezuela 92 B4 8 38N 71 39W
El Wabeira, Egypt ..... 47 F2 29 34N 33 6 E
El Wak, Kenya .... 54 B5 2 49N 40 56 E
El Wuz, Sudan .... 51 E11 15 5N 30 7 E
Elandsvlei, S. Africa ... 56 E2 32 19S 19 31 E
Elat, Israel .... 47 F3 29 30N 34 56 E
Elâzığ, Turkey .... 25 G6 38 37N 39 14 E
Elba, Italy .... 20 C4 42 48N 10 15 E
Elba, U.S.A. 77 K2 31 25N 86 4W
Elbasani, Albania ...... 21 D9 41 9N 20 9 E
Elbe, U.S.A. 84 D4 46 45N 122 10W
Elbe —►, Europe .... 16 B4 53 50N 9 0 E
Elbert, Mt., U.S.A. 83 G10 39 7N 106 27W
Elberta, U.S.A. 76 C2 44 37N 86 14W
Elberton, U.S.A. 77 H4 34 7N 82 52W
Elbeuf, France .... 18 B4 49 17N 1 2 E
Elbing = Elbląg, Poland 17 A9 54 10N 19 25 E
Elbląg, Poland .... 17 A9 54 10N 19 25 E
Elbow, Canada 73 C7 51 7N 106 35W
Elbrus, Asia .... 25 F7 43 21N 42 30 E
Elburg, Neths. .... 15 B5 52 26N 5 50 E
Elburz Mts. = Alborz, Reshteh-ye Kūhhā-ye, Iran .... 45 C7 36 0N 52 0 E
Elche, Spain 19 C5 38 15N 0 42W
Elcho I., Australia ..... 62 A2 11 55S 135 45 E
Eldon, Mo., U.S.A. 80 F8 38 21N 92 35W
Eldon, Wash., U.S.A. ... 84 C3 47 33N 123 3W
Eldora, U.S.A. 80 D8 42 22N 93 5W
Eldorado, Argentina 95 B5 26 28S 54 43W
Eldorado, Canada 73 B7 59 35N 108 30W
Eldorado, Mexico 86 C3 24 20N 107 22W
Eldorado, Ill., U.S.A. ... 76 G1 37 49N 88 26W
Eldorado, Tex., U.S.A. .. 81 K4 30 52N 100 36W
Eldorado Springs, U.S.A. 81 G8 37 52N 94 1W
Eldoret, Kenya .... 54 B4 0 30N 35 17 E
Eldred, U.S.A. 78 E6 41 58N 78 23W
Elea, C., Cyprus ...... 23 D13 35 19N 34 4 E
Electra, U.S.A. 81 H5 34 2N 98 55W
Elefantes —►, Mozam. .. 57 C5 24 10S 32 40 E
Elektrostal, Russia ..... 24 C6 55 41N 38 32 E
Elephant Butte Reservoir, U.S.A. 83 K10 33 9N 107 11W
Elephant I., Antarctica .. 5 C18 61 0S 55 0W
Eleuthera, Bahamas 88 A4 25 0N 76 20W
Elgeyo-Marakwet □, Kenya .... 54 B4 0 45N 35 30 E
Elgin, N.B., Canada 71 C6 45 48N 65 10W
Elgin, Ont., Canada 79 B8 44 36N 76 13W
Elgin, U.K. 12 D5 57 39N 3 20W
Elgin, Ill., U.S.A. 76 D1 42 2N 88 17W
Elgin, N. Dak., U.S.A. .. 80 B4 46 24N 101 51W
Elgin, Nebr., U.S.A. 80 E5 41 59N 98 5W
Elgin, Nev., U.S.A. 83 H6 37 21N 114 32W
Elgin, Oreg., U.S.A. 82 D5 45 34N 117 55W
Elgin, Tex., U.S.A. 81 K6 30 21N 97 22W
Elgon, Mt., Africa ...... 54 B3 1 10N 34 30 E
Eliase, Indonesia ...... 37 F8 8 21S 130 48 E
Elida, U.S.A. 81 J3 33 57N 103 39W
Elim, S. Africa .... 56 E2 34 35S 19 45 E
Elisabethville = Lubumbashi, Zaïre ... 55 E2 11 40S 27 28 E
Elista, Russia .... 25 E7 46 16N 44 14 E
Elizabeth, Australia .... 63 E2 34 42S 138 41 E
Elizabeth, U.S.A. 79 F10 40 40N 74 13W
Elizabeth City, U.S.A. ... 77 G7 36 18N 76 14W
Elizabethton, U.S.A. ... 77 G4 36 21N 82 13W
Elizabethtown, Ky., U.S.A. 76 G3 37 42N 85 52W
Elizabethtown, N.Y., U.S.A. 79 B11 44 13N 73 36W
Elizabethtown, Pa., U.S.A. 79 F8 40 9N 76 36W
Elk City, U.S.A. 81 H5 35 25N 99 25W
Elk Creek, U.S.A. 84 F4 39 36N 122 32W
Elk Grove, U.S.A. 84 G5 38 25N 121 22W
Elk Island Nat. Park, Canada 72 C6 53 35N 112 59W
Elk Lake, Canada 70 C3 47 40N 80 25W
Elk Point, Canada 73 C6 53 54N 110 55W
Elk River, Idaho, U.S.A. . 82 C5 46 47N 116 11W
Elk River, Minn., U.S.A. . 80 C8 45 18N 93 35W
Elkedra, Australia ..... 62 C2 21 9S 135 33 E
Elkedra —►, Australia .. 62 C2 21 8S 136 22 E
Elkhart, Ind., U.S.A. ... 76 E3 41 41N 85 58W
Elkhart, Kans., U.S.A. .. 81 G4 37 0N 101 54W
Elkhorn, Canada 73 D8 49 59N 101 14W
Elkhorn —►, U.S.A. ... 80 E6 41 8N 96 19W
Elkhovo, Bulgaria ..... 21 C12 42 10N 26 40 E
Elkin, U.S.A. 77 G5 36 15N 80 51W
Elkins, U.S.A. 76 F6 38 55N 79 51W
Elko, Canada 72 D5 49 20N 115 10W
Elko, U.S.A. 82 F6 40 50N 115 46W
Ell, L., Australia ...... 61 E4 29 13S 127 46 E
Ellef Ringnes I., Canada . 4 B2 78 30N 102 2W
Ellendale, Australia .... 60 C3 17 56S 124 48 E
Ellendale, U.S.A. 80 B5 46 0N 98 32W
Ellensburg, U.S.A. 82 C3 46 59N 120 34W
Ellenville, U.S.A. 79 E10 41 43N 74 24W
Ellery, Mt., Australia ... 63 F4 37 28S 148 47 E
Ellesmere, L., N.Z. .... 59 M4 47 47S 172 28 E
Ellesmere I., Canada ... 4 B4 79 30N 80 0W
Ellesmere Port, U.K. ... 10 D5 53 17N 2 55W
Ellice Is. = Tuvalu ■, Pac. Oc. .... 64 H9 8 0S 178 0 E
Ellinwood, U.S.A. 80 F5 38 21N 98 35W
Elliot, Australia .... 62 B1 17 33S 133 32 E
Elliot, S. Africa .... 57 E4 31 22S 27 48 E
Elliot Lake, Canada 70 C3 46 25N 82 35W
Elliotdale = Xhora, S. Africa .... 57 E4 31 55S 28 38 E
Ellis, U.S.A. 80 F5 38 56N 99 34W
Elliston, Australia ..... 63 E1 33 39S 134 53 E
Ellisville, U.S.A. 81 K10 31 36N 89 12W
Ellon, U.K. 12 D6 57 21N 2 5W
Ellore = Eluru, India ... 41 L12 16 48N 81 8 E
Ells —►, Canada 72 B6 57 18N 111 40W
Ellsworth, U.S.A. 80 F5 38 44N 98 14W
Ellsworth Land, Antarctica 5 D16 76 0S 89 0W
Ellsworth Mts., Antarctica 5 D16 78 30S 85 0W
Ellwood City, U.S.A. ... 78 F4 40 52N 80 17W
Elma, Canada 73 D9 49 52N 95 55W
Elma, U.S.A. 84 D3 47 0N 123 25W
Elmalı, Turkey .... 25 G4 36 44N 29 56 E
Elmhurst, U.S.A. 76 E2 41 53N 87 56W
Elmira, Canada 78 C4 43 36N 80 33W
Elmira, U.S.A. 78 D8 42 6N 76 48W
Elmore, Australia ..... 63 F3 36 30S 144 37 E
Elmore, U.S.A. 85 M11 33 7N 115 49W
Elmvale, Canada 78 B5 44 35N 79 52W
Elora, Canada 78 C4 43 41N 80 26W
Eloúnda, Greece ...... 23 D7 35 16N 25 42 E
Eloy, U.S.A. 83 K8 32 45N 111 33W
Elrose, Canada 73 C7 51 12N 108 0W
Elsas, Canada 70 C3 48 32N 82 55W
Elsie, U.S.A. 84 E3 45 52N 123 36W
Elsinore = Helsingør, Denmark .... 9 H12 56 2N 12 35 E
Elsinore, U.S.A. 83 G7 38 41N 112 9W
Eltham, N.Z. .... 59 H5 39 26S 174 19 E
Eluru, India .... 41 L12 16 48N 81 8 E
Elvas, Portugal 19 C2 38 50N 7 10W
Elverum, Norway ...... 9 F11 60 53N 11 34 E
Elvire —►, Australia ... 60 C4 17 51S 128 11 E
Elwood, Ind., U.S.A. ... 76 E3 40 17N 85 50W
Elwood, Nebr., U.S.A. .. 80 E5 40 36N 99 52W
Elx = Elche, Spain 19 C5 38 15N 0 42W
Ely, U.K. .... 11 E8 52 24N 0 16 E
Ely, Minn., U.S.A. 80 B9 47 55N 91 51W
Ely, Nev., U.S.A. 82 G6 39 15N 114 54W
Elyria, U.S.A. 78 E2 41 22N 82 7W
Emāmrūd, Iran .... 45 B7 36 30N 55 0 E
Emba, Kazakhstan ..... 26 E6 48 50N 58 8 E
Emba —►, Kazakhstan . 25 E9 46 38N 53 14 E
Embarcación, Argentina 94 A3 23 10S 64 0W
Embarras Portage, Canada 73 B6 58 27N 111 28W
Embetsu, Japan ...... 30 B10 44 44N 141 47 E
Embóna, Greece ...... 23 C9 36 13N 27 51 E
Embrun, France ...... 18 D7 44 34N 6 30 E
Embu, Kenya .... 54 C4 0 32S 37 38 E
Embu □, Kenya ...... 54 C4 0 30S 37 35 E
Emden, Germany ..... 16 B3 53 22N 7 12 E
Emerald, Australia .... 62 C4 23 32S 148 10 E
Emerson, Canada 73 D9 49 0N 97 10W
Emery, U.S.A. 83 G8 38 55N 111 15W
Emi Koussi, Chad ..... 51 E8 19 45N 18 55 E
Emilia-Romagna □, Italy 20 B4 44 33N 10 40 E
Eminabad, Pakistan ... 42 C6 32 2N 74 8 E
Emlenton, U.S.A. 78 E5 41 11N 79 43W
Emmeloord, Neths. .... 15 B5 52 44N 5 46 E
Emmen, Neths. .... 15 B6 52 48N 6 57 E
Emmet, Australia ..... 62 C3 24 45S 144 30 E
Emmetsburg, U.S.A. ... 80 D7 43 7N 94 41W
Emmett, U.S.A. 82 E5 43 52N 116 30W
Empalme, Mexico 86 B2 28 1N 110 49W
Empangeni, S. Africa .. 57 D5 28 50S 31 52 E
Empedrado, Argentina . 94 B4 28 0S 58 46W
Emperor Seamount Chain, Pac. Oc. ...... 64 D9 40 0N 170 0 E
Emporia, Kans., U.S.A. . 80 F6 38 25N 96 11W
Emporia, Va., U.S.A. ... 77 G7 36 42N 77 32W

Fannich, L.

Emporium, U.S.A. **78 E6** 41 31N 78 14W
Empress, Canada **73 C6** 50 57N 110 0W
Empty Quarter = Rub' al Khali, Si. Arabia **46 D4** 18 0N 48 0 E
Ems →, Germany **16 B3** 53 22N 7 15 E
Emsdale, Canada **78 A5** 45 32N 79 19W
Emu, China **35 C15** 43 40N 128 6 E
Emu Park, Australia **62 C5** 23 13S 150 50 E
'En 'Avrona, Israel **47 F3** 29 43N 35 0 E
En Nahud, Sudan **51 F10** 12 45N 28 25 E
Ena, Japan **31 G8** 35 25N 137 25 E
Enana, Namibia **56 B2** 17 30S 16 23 E
Enaratoli, Indonesia **37 E9** 3 55S 136 21 E
Enard B., U.K. **12 C3** 58 5N 5 20W
Enare = Inarijärvi, Finland **8 B19** 69 0N 28 0 E
Encantadas, Serra, Brazil **95 C5** 30 40S 53 0W
Encanto, C., Phil. **37 A6** 15 45N 121 38 E
Encarnación, Paraguay **95 B4** 27 15S 55 50W
Encarnación de Diaz, Mexico **86 C4** 21 30N 102 13W
Encinal, U.S.A. **81 L5** 28 2N 99 21W
Encinitas, U.S.A. **85 M9** 33 3N 117 17W
Encino, U.S.A. **83 J11** 34 39N 105 28W
Encounter B., Australia **63 F2** 35 45S 138 45 E
Ende, Indonesia **37 F6** 8 45S 121 40 E
Endeavour, Canada **73 C8** 52 10N 102 39W
Endeavour Str., Australia **62 A3** 10 45S 142 0 E
Enderbury I., Kiribati **64 H10** 3 8S 171 5W
Enderby, Canada **72 C5** 50 35N 119 10W
Enderby I., Australia **60 D2** 20 35S 116 30 E
Enderby Land, Antarctica **5 C5** 66 0S 53 0 E
Enderlin, U.S.A. **80 B6** 46 38N 97 36W
Endicott, N.Y., U.S.A. **79 D8** 42 6N 76 4W
Endicott, Wash., U.S.A. **82 C5** 46 56N 117 41W
Endyalgout I., Australia **60 B5** 11 40S 132 35 E
Enewetak Atoll, Pac. Oc. **64 F8** 11 30N 162 15 E
Enez, Turkey **21 D12** 40 45N 26 5 E
Enfield, U.K. **11 F7** 51 39N 0 4W
Engadin, Switz. **16 E5** 46 45N 10 10 E
Engaño, C., Dom. Rep. **89 C6** 18 30N 68 20W
Engaño, C., Phil. **37 A6** 18 35N 122 23 E
Engcobo, S. Africa **57 E4** 31 37S 28 0 E
Engels = Pokrovsk, Russia **24 D8** 51 28N 46 6 E
Engemann L., Canada **73 B7** 58 0N 106 55W
Enggano, Indonesia **36 F2** 5 20S 102 40 E
Enghien, Belgium **15 D4** 50 37N 4 2 E
Engkilili, Malaysia **36 D4** 1 3N 111 42 E
England, U.S.A. **81 H9** 34 33N 91 58W
England □, U.K. **7 E5** 53 0N 2 0W
Englee, Canada **71 B8** 50 45N 56 5W
Englehart, Canada **70 C4** 47 49N 79 52W
Engler L., Canada **73 B7** 59 8N 106 52W
Englewood, Colo., U.S.A. **80 F2** 39 39N 104 59W
Englewood, Kans., U.S.A. **81 G5** 37 2N 99 59W
English →, Canada **73 C10** 50 35N 93 30W
English Bazar = Ingraj Bazar, India **43 G13** 24 58N 88 10 E
English Channel, Europe **11 G6** 50 0N 2 0W
English River, Canada **70 C1** 49 14N 91 0W
Enid, U.S.A. **81 G6** 36 24N 97 53W
Enkhuizen, Neths. **15 B5** 52 42N 5 17 E
Enna, Italy **20 F6** 37 34N 14 15 E
Ennadai, Canada **73 A8** 61 8N 100 53W
Ennadai L., Canada **73 A8** 61 0N 101 0W
Ennedi, Chad **51 E9** 17 15N 22 0 E
Enngonia, Australia **63 D4** 29 21S 145 50 E
Ennis, Ireland **13 D3** 52 51N 8 59W
Ennis, Mont., U.S.A. **82 D8** 45 21N 111 44W
Ennis, Tex., U.S.A. **81 J6** 32 20N 96 38W
Enniscorthy, Ireland **13 D5** 52 30N 6 35W
Enniskillen, U.K. **13 B4** 54 20N 7 40W
Ennistimon, Ireland **13 D2** 52 56N 9 18W
Enns →, Austria **16 D7** 48 14N 14 32 E
Enontekiö, Finland **8 B17** 68 23N 23 37 E
Enriquillo, L., Dom. Rep. **89 C5** 18 20N 72 5W
Enschede, Neths. **15 B6** 52 13N 6 53 E
Ensenada, Argentina **94 C4** 34 55S 57 55W
Ensenada, Mexico **86 A1** 31 50N 116 50W
Ensiola, Pta., Spain **22 B9** 39 7N 2 55 E
Entebbe, Uganda **54 B3** 0 4N 32 28 E
Enterprise, Canada **72 A5** 60 47N 115 45W
Enterprise, Oreg., U.S.A. **82 D5** 45 25N 117 17W
Enterprise, Utah, U.S.A. **83 H7** 37 34N 113 43W
Entre Ríos, Bolivia **94 A3** 21 30S 64 25W
Entre Ríos □, Argentina **94 C4** 30 30S 58 30W
Enugu, Nigeria **50 G6** 6 20N 7 30 E
Enugu Ezike, Nigeria **50 G6** 7 0N 7 29 E
Enumclaw, U.S.A. **84 C5** 47 12N 121 59W
Eólie, Is., Italy **20 E6** 38 30N 14 50 E
Epe, Neths. **15 B5** 52 21N 5 59 E
Épernay, France **18 B5** 49 3N 3 56 E
Ephraim, U.S.A. **82 G8** 39 22N 111 35W
Ephrata, U.S.A. **82 C4** 47 19N 119 33W
Épinal, France **18 B7** 48 10N 6 27 E
Episkopi, Cyprus **23 E11** 34 40N 32 54 E
Episkopí, Greece **23 D6** 35 20N 24 20 E
Episkopi Bay, Cyprus **23 E11** 34 35N 32 50 E
Epping, U.K. **11 F8** 51 42N 0 8 E
Epukiro, Namibia **56 C2** 21 40S 19 9 E
Equatorial Guinea ■, Africa **52 D1** 2 0N 8 0 E
Er Rahad, Sudan **51 F11** 12 45N 30 32 E
Er Rif, Morocco **50 A4** 35 1N 4 1W
Er Roseires, Sudan **51 F11** 11 55N 34 30 E
Erāwadī Myit = Irrawaddy →, Burma **41 M19** 15 50N 95 6 E
Erbil = Arbīl, Iraq **44 B5** 36 15N 44 5 E
Ercha, Russia **27 C15** 69 45N 147 20 E
Erciyaş Dağı, Turkey **25 G6** 38 30N 35 30 E
Erdao Jiang →, China **35 C14** 43 0N 127 0 E
Erdene, Mongolia **34 B6** 44 13N 111 10 E
Erebus, Mt., Antarctica **5 D11** 77 35S 167 0 E
Erechim, Brazil **95 B5** 27 35S 52 15W
Ereğli, Konya, Turkey **25 G5** 37 31N 34 4 E
Ereğli, Zonguldak, Turkey **25 F5** 41 15N 31 24 E
Erenhot, China **34 C7** 43 48N 112 2 E
Eresma →, Spain **19 B3** 41 26N 4 45W
Erewadi Myitwanya, Burma **41 M19** 15 30N 95 0 E
Erfenisdam, S. Africa **56 D4** 28 30S 26 50 E
Erfurt, Germany **16 C5** 50 58N 11 2 E
Erğene →, Turkey **21 D12** 41 1N 26 22 E
Ergeni Vozvyshennost, Russia **25 E7** 47 0N 44 0 E
Eriboll, L., U.K. **12 C4** 58 28N 4 41W
Érice, Italy **20 E5** 38 4N 12 34 E
Erie, U.S.A. **78 D4** 42 8N 80 5W
Erie, L., N. Amer. **78 D3** 42 15N 81 0W
Erie Canal, U.S.A. **78 C6** 43 5N 78 43W
Erieau, Canada **78 D3** 42 16N 81 57W
Erigavo, Somali Rep. **46 E4** 10 35N 47 20 E
Erikoúsa, Greece **23 A3** 39 53N 19 34 E
Eriksdale, Canada **73 C9** 50 52N 98 7W
Erímanthos, Greece **21 F9** 37 57N 21 50 E
Erimo-misaki, Japan **30 D11** 41 50N 143 15 E
Eritrea ■, Africa **51 F12** 14 0N 38 30 E
Erlangen, Germany **16 D5** 49 35N 11 2 E
Erldunda, Australia **62 D1** 25 14S 133 12 E
Ermelo, Neths. **15 B5** 52 18N 5 35 E
Ermelo, S. Africa **57 D4** 26 31S 29 59 E
Ermones, Greece **23 A3** 39 37N 19 46 E
Ermoúpolis = Síros, Greece **21 F11** 37 28N 24 57 E
Ernakulam = Cochin, India **40 Q10** 9 59N 76 22 E
Erne →, Ireland **13 B3** 54 30N 8 16W
Erne, Lower L., U.K. **13 B4** 54 26N 7 46W
Erne, Upper L., U.K. **13 B4** 54 14N 7 22W
Ernest Giles Ra., Australia **61 E3** 27 0S 123 45 E
Erode, India **40 P10** 11 24N 77 45 E
Eromanga, Australia **63 D3** 26 40S 143 11 E
Erongo, Namibia **56 C2** 21 39S 15 58 E
Errabiddy, Australia **61 E2** 25 25S 117 5 E
Erramala Hills, India **40 M11** 15 30N 78 15 E
Errigal, Ireland **13 A3** 55 2N 8 8W
Erris Hd., Ireland **13 B1** 54 19N 10 0W
Erskine, U.S.A. **80 B7** 47 40N 96 0W
Ertis = Irtysh →, Russia **26 C7** 61 4N 68 52 E
Erwin, U.S.A. **77 G4** 36 9N 82 25W
Erzgebirge, Germany **16 C6** 50 25N 13 0 E
Erzin, Russia **27 D10** 50 15N 95 10 E
Erzincan, Turkey **25 G6** 39 46N 39 30 E
Erzurum, Turkey **25 G7** 39 57N 41 15 E
Es Caló, Spain **22 C8** 38 40N 1 30 E
Es Caná, Spain **22 B8** 39 2N 1 36 E
Es Sahrâ' Esh Sharqîya, Egypt **51 C11** 27 30N 32 30 E
Es Sînâ', Egypt **51 C11** 29 0N 34 0 E
Esambo, Zaïre **54 C1** 3 48S 23 30 E
Esan-Misaki, Japan **30 D10** 41 40N 141 10 E
Esashi, Hokkaidō, Japan **30 B11** 44 56N 142 35 E
Esashi, Hokkaidō, Japan **30 D10** 41 52N 140 7 E
Esbjerg, Denmark **9 J10** 55 29N 8 29 E
Escalante, U.S.A. **83 H8** 37 47N 111 36W
Escalante →, U.S.A. **83 H8** 37 24N 110 57W
Escalón, Mexico **86 B4** 26 46N 104 20W
Escambia →, U.S.A. **77 K2** 30 32N 87 11W
Escanaba, U.S.A. **76 C2** 45 45N 87 4W
Esch-sur-Alzette, Lux. **15 E5** 49 32N 6 0 E
Escondido, U.S.A. **85 M9** 33 7N 117 5W
Escuinapa, Mexico **86 C3** 22 50N 105 50W
Escuintla, Guatemala **88 D1** 14 20N 90 48W
Eşfahān, Iran **45 C6** 33 0N 51 30 E
Esfideh, Iran **45 C8** 33 39N 59 46 E
Esh Sham = Dimashq, Syria **47 B5** 33 30N 36 18 E
Esh Shamâlîya □, Sudan **51 E10** 19 0N 29 0 E
Eshowe, S. Africa **57 D5** 28 50S 31 30 E
Esil = Ishim →, Russia **26 D8** 57 45N 71 10 E
Esk →, Dumf. & Gall., U.K. **12 G5** 54 58N 3 4W
Esk →, N. Yorks., U.K. **10 C7** 54 27N 0 36W
Eskifjörður, Iceland **8 D7** 65 3N 13 55W
Eskilstuna, Sweden **9 G14** 59 22N 16 32 E
Eskimo Pt., Canada **73 A10** 61 10N 94 15W
Eskişehir, Turkey **25 G5** 39 50N 30 35 E
Esla →, Spain **19 B2** 41 29N 6 3W
Eslāmābād-e Gharb, Iran **44 C5** 34 10N 46 30 E
Esmeraldas, Ecuador **92 C3** 1 0N 79 40W
Espalmador, I., Spain **22 C7** 38 47N 1 26 E
Espanola, Canada **70 C3** 46 15N 81 46W
Espardell, I. del, Spain **22 C7** 38 48N 1 29 E
Esparta, Costa Rica **88 E3** 9 59N 84 40W
Esperance, Australia **61 F3** 33 45S 121 55 E
Esperance B., Australia **61 F3** 33 48S 121 55 E
Esperanza, Argentina **94 C3** 31 29S 61 3W
Espichel, C., Portugal **19 C1** 38 22N 9 16W
Espigão, Serra do, Brazil **95 B5** 26 35S 50 30W
Espinal, Colombia **92 C4** 4 9N 74 53W
Espinazo, Sierra del = Espinhaço, Serra do, Brazil **93 G10** 17 30S 43 30W
Espinhaço, Serra do, Brazil **93 G10** 17 30S 43 30W
Espinilho, Serra do, Brazil **95 B5** 28 30S 55 0W
Espírito Santo □, Brazil **93 G10** 20 0S 40 45W
Espíritu Santo, B. del, Mexico **87 D7** 19 15N 87 0W
Espíritu Santo, I., Mexico **86 C2** 24 30N 110 23W
Espita, Mexico **87 C7** 21 1N 88 19W
Espungabera, Mozam. **57 C5** 20 29S 32 45 E
Esquel, Argentina **96 E2** 42 55S 71 20W
Esquina, Argentina **94 B4** 30 0S 59 30W
Essaouira, Morocco **50 B3** 31 32N 9 42W
Essebie, Zaïre **54 B3** 2 58N 30 40 E
Essen, Belgium **15 C4** 51 28N 4 28 E
Essen, Germany **16 C3** 51 28N 6 59 E
Essendon, Mt., Australia **61 E3** 25 0S 120 29 E
Essequibo →, Guyana **92 B7** 6 50N 58 30W
Essex, Canada **78 D2** 42 10N 82 49W
Essex, Calif., U.S.A. **85 L11** 34 44N 115 15W
Essex, N.Y., U.S.A. **79 B11** 44 19N 73 21W
Essex □, U.K. **11 F8** 51 48N 0 30 E
Esslingen, Germany **16 D4** 48 43N 9 19 E
Essonne □, France **18 B5** 48 30N 2 20 E
Estados, I. de Los, Argentina **96 G4** 54 40S 64 30W
Eşţahbānāt, Iran **45 D7** 29 8N 54 4 E
Estallenchs, Spain **22 B9** 39 39N 2 29 E
Estância, Brazil **93 F11** 11 16S 37 26W
Estancia, U.S.A. **83 J10** 34 46N 106 4W
Estārm, Iran **45 D8** 28 21N 58 21 E
Estcourt, S. Africa **57 D4** 29 0S 29 53 E
Estelí, Nic. **88 D2** 13 9N 86 22W
Estelline, S. Dak., U.S.A. **80 C6** 44 35N 96 54W
Estelline, Tex., U.S.A. **81 H4** 34 33N 100 26W
Esterhazy, Canada **73 C8** 50 37N 102 5W
Estevan, Canada **73 D8** 49 10N 102 59W
Estevan Group, Canada **72 C3** 53 3N 129 38W
Estherville, U.S.A. **80 D7** 43 24N 94 50W
Eston, Canada **73 C7** 51 8N 108 40W
Estonia ■, Europe **24 C4** 58 30N 25 30 E
Estoril, Portugal **19 C1** 38 42N 9 23W
Estrêla, Serra da, Portugal **19 B2** 40 10N 7 45W
Estremadura, Portugal **19 C1** 39 0N 9 0W
Estrondo, Serra do, Brazil **93 E9** 7 20S 48 0W
Esztergom, Hungary **17 E9** 47 47N 18 44 E
Etadunna, Australia **63 D2** 28 43S 138 38 E
Etah, India **43 F8** 27 35N 78 40 E
Etamamu, Canada **71 B8** 50 18N 59 59W
Etanga, Namibia **56 B1** 17 55S 13 0 E
Etawah, India **43 F8** 26 48N 79 6 E
Etawah →, U.S.A. **77 H3** 34 20N 84 15W
Etawney L., Canada **73 B9** 57 50N 96 50W
Ethel, U.S.A. **84 D4** 46 32N 122 46W
Ethel Creek, Australia **60 D3** 22 55S 120 11 E
Ethelbert, Canada **73 C8** 51 32N 100 25W
Ethiopia ■, Africa **46 F3** 8 0N 40 0 E
Ethiopian Highlands, Ethiopia **48 E7** 10 0N 37 0 E
Etive, L., U.K. **12 E3** 56 30N 5 12W
Etna, Italy **20 F6** 37 45N 15 0 E
Etoile, Zaïre **55 E2** 11 33S 27 30 E
Etolin I., U.S.A. **72 B2** 56 5N 132 20W
Etosha Pan, Namibia **56 B2** 18 40S 16 30 E
Etowah, U.S.A. **77 H3** 35 20N 84 32W
Ettrick Water →, U.K. **12 F6** 55 31N 2 55W
Etuku, Zaïre **54 C2** 3 42S 25 45 E
Etzatlán, Mexico **86 C4** 20 48N 104 5W
Euboea = Évvoia, Greece **21 E11** 38 30N 24 0 E
Eucla Motel, Australia **61 F4** 31 41S 128 52 E
Euclid, U.S.A. **78 E3** 41 34N 81 32W
Eucumbene, L., Australia **63 F4** 36 2S 148 40 E
Eudora, U.S.A. **81 J9** 33 7N 91 16W
Eufaula, Ala., U.S.A. **77 K3** 31 54N 85 9W
Eufaula, Okla., U.S.A. **81 H7** 35 17N 95 35W
Eufaula L., U.S.A. **81 H7** 35 18N 95 21W
Eugene, U.S.A. **82 E2** 44 5N 123 4W
Eugowra, Australia **63 E4** 33 22S 148 24 E
Eulo, Australia **63 D4** 28 10S 145 3 E
Eunice, La., U.S.A. **81 K8** 30 30N 92 25W
Eunice, N. Mex., U.S.A. **81 J3** 32 26N 103 10W
Eupen, Belgium **15 D6** 50 37N 6 3 E
Euphrates = Furāt, Nahr al →, Asia **44 D5** 31 0N 47 25 E
Eure □, France **18 B4** 49 10N 1 0 E
Eure-et-Loir □, France **18 B4** 48 22N 1 30 E
Eureka, Canada **4 B3** 80 0N 85 56W
Eureka, Calif., U.S.A. **82 F1** 40 47N 124 9W
Eureka, Kans., U.S.A. **81 G6** 37 49N 96 17W
Eureka, Mont., U.S.A. **82 B6** 48 53N 115 3W
Eureka, Nev., U.S.A. **82 G5** 39 31N 115 58W
Eureka, S. Dak., U.S.A. **80 C5** 45 46N 99 38W
Eureka, Utah, U.S.A. **82 G7** 39 58N 112 7W
Eureka, Mt., Australia **61 E3** 26 35S 121 35 E
Euroa, Australia **63 F4** 36 44S 145 35 E
Europa, I., Ind. Oc. **53 J8** 22 20S 40 22 E
Europa, Picos de, Spain **19 A3** 43 10N 4 49W
Europa, Pta. de, Gib. **19 D3** 36 3N 5 21W
Europa Pt. = Europa, Pta. de, Gib. **19 D3** 36 3N 5 21W
Europe **6 F10** 50 0N 20 0 E
Europoort, Neths. **15 C4** 51 57N 4 10 E
Eustis, U.S.A. **77 L5** 28 51N 81 41W
Eutsuk L., Canada **72 C3** 53 20N 126 45W
Eva Downs, Australia **62 B1** 18 1S 134 52 E
Evale, Angola **56 B2** 16 33S 15 44 E
Evans, U.S.A. **80 E2** 40 23N 104 41W
Evans Head, Australia **63 D5** 29 7S 153 27 E
Evans L., Canada **70 B4** 50 50N 77 0W
Evans Mills, U.S.A. **79 B9** 44 6N 75 48W
Evanston, Ill., U.S.A. **76 D2** 42 3N 87 41W
Evanston, Wyo., U.S.A. **82 F8** 41 16N 110 58W
Evansville, Ind., U.S.A. **76 F2** 37 58N 87 35W
Evansville, Wis., U.S.A. **80 D10** 42 47N 89 18W
Evaz, Iran **45 E7** 27 46N 53 59 E
Eveleth, U.S.A. **80 B8** 47 28N 92 32W
Evensk, Russia **27 C16** 62 12N 159 30 E
Everard, L., Australia **63 E1** 31 30S 135 0 E
Everard Park, Australia **61 E5** 27 1S 132 43 E
Everard Ras., Australia **61 E5** 27 5S 132 28 E
Everest, Mt., Nepal **43 E12** 28 5N 86 58 E
Everett, Pa., U.S.A. **78 F6** 40 1N 78 23W
Everett, Wash., U.S.A. **84 C4** 47 59N 122 12W
Everglades, The, U.S.A. **77 N5** 25 50N 81 0W
Everglades City, U.S.A. **77 N5** 25 52N 81 23W
Everglades National Park, U.S.A. **77 N5** 25 30N 81 0W
Evergreen, U.S.A. **77 K2** 31 26N 86 57W
Everson, U.S.A. **82 B2** 48 57N 122 22W
Evesham, U.K. **11 E6** 52 6N 1 57W
Evinayong, Eq. Guin. **52 D2** 1 26N 10 35 E
Évora, Portugal **19 C2** 38 33N 7 57W
Evowghlī, Iran **44 B5** 38 43N 45 13 E
Évreux, France **18 B4** 49 3N 1 8 E
Évvoia, Greece **21 E11** 38 30N 24 0 E
Ewe, L., U.K. **12 D3** 57 49N 5 38W
Ewing, U.S.A. **80 D5** 42 16N 98 21W
Ewo, Congo **52 E2** 0 48S 14 45 E
Exaltación, Bolivia **92 F5** 13 10S 65 20W
Excelsior Springs, U.S.A. **80 F7** 39 20N 94 13W
Exe →, U.K. **11 G4** 50 38N 3 27W
Exeter, Canada **78 C3** 43 21N 81 29W
Exeter, U.K. **11 G4** 50 43N 3 31W
Exeter, Calif., U.S.A. **83 H4** 36 18N 119 9W
Exeter, N.H., U.S.A. **79 D14** 42 59N 70 57W
Exeter, Nebr., U.S.A. **80 E6** 40 39N 97 27W
Exmoor, U.K. **11 F4** 51 10N 3 59W
Exmouth, Australia **60 D1** 21 54S 114 10 E
Exmouth, U.K. **11 G4** 50 37N 3 26W
Exmouth G., Australia **60 D1** 22 15S 114 15 E
Expedition Ra., Australia **62 C4** 24 30S 149 12 E
Extremadura □, Spain **19 C2** 39 30N 6 5W
Exuma Sound, Bahamas **88 B4** 24 30N 76 20W
Eyasi, L., Tanzania **54 C4** 3 30S 35 0 E
Eyeberry L., Canada **73 A8** 63 8N 104 43W
Eyemouth, U.K. **12 F6** 55 53N 2 5W
Eyjafjörður, Iceland **8 C4** 66 15N 18 30W
Eyrarbakki, Iceland **8 E3** 63 52N 21 9W
Eyre, Australia **61 F4** 32 15S 126 18 E
Eyre (North), L., Australia **63 D2** 28 30S 137 20 E
Eyre (South), L., Australia **63 D2** 29 18S 137 25 E
Eyre Cr. →, Australia **63 D2** 26 40S 139 0 E
Eyre Mts., N.Z. **59 L2** 45 25S 168 25 E
Eyre Pen., Australia **63 E2** 33 30S 136 17 E
Eyvānkī, Iran **45 C6** 35 24N 51 56 E
Ezouza →, Cyprus **23 E11** 34 44N 32 27 E

F

Fabens, U.S.A. **83 L10** 31 30N 106 10W
Fabriano, Italy **20 C5** 43 20N 12 52 E
Facatativá, Colombia **92 C4** 4 49N 74 22W
Fachi, Niger **50 E7** 18 6N 11 34 E
Fada, Chad **51 E9** 17 13N 21 34 E
Fada-n-Gourma, Burkina Faso **50 F5** 12 10N 0 30 E
Faddeyevskiy, Ostrov, Russia **27 B15** 76 0N 144 0 E
Fadghāmī, Syria **44 C4** 35 53N 40 52 E
Faenza, Italy **20 B4** 44 17N 11 53 E
Făgăras, Romania **17 F12** 45 48N 24 58 E
Fagernes, Norway **9 F10** 60 59N 9 14 E
Fagersta, Sweden **9 F13** 60 1N 15 46 E
Fagnano, L., Argentina **96 G3** 54 30S 68 0W
Fahlīān, Iran **45 D6** 30 11N 51 28 E
Fahraj, Kermān, Iran **45 D8** 29 0N 59 0 E
Fahraj, Yazd, Iran **45 D7** 31 46N 54 36 E
Faial, Madeira **22 D3** 32 47N 16 53W
Fair Hd., U.K. **13 A5** 55 14N 6 10W
Fair Oaks, U.S.A. **84 G5** 38 39N 121 16W
Fairbank, U.S.A. **83 L8** 31 43N 110 11W
Fairbanks, U.S.A. **68 B5** 64 51N 147 43W
Fairbury, U.S.A. **80 E6** 40 8N 97 11W
Fairfax, U.S.A. **81 G6** 36 34N 96 42W
Fairfield, Ala., U.S.A. **77 J2** 33 29N 86 55W
Fairfield, Calif., U.S.A. **84 G4** 38 15N 122 3W
Fairfield, Conn., U.S.A. **79 E11** 41 9N 73 16W
Fairfield, Idaho, U.S.A. **82 E6** 43 21N 114 44W
Fairfield, Ill., U.S.A. **76 F1** 38 23N 88 22W
Fairfield, Iowa, U.S.A. **80 E9** 40 56N 91 57W
Fairfield, Mont., U.S.A. **82 C8** 47 37N 111 59W
Fairfield, Tex., U.S.A. **81 K7** 31 44N 96 10W
Fairford, Canada **73 C9** 51 37N 98 38W
Fairhope, U.S.A. **77 K2** 30 31N 87 54W
Fairlie, N.Z. **59 L3** 44 5S 170 49 E
Fairmead, U.S.A. **84 H6** 37 5N 120 10W
Fairmont, Minn., U.S.A. **80 D7** 43 39N 94 28W
Fairmont, W. Va., U.S.A. **76 F5** 39 29N 80 9W
Fairmount, U.S.A. **85 L8** 34 45N 118 26W
Fairplay, U.S.A. **83 G11** 39 15N 106 2W
Fairport, U.S.A. **78 C7** 43 6N 77 27W
Fairport Harbor, U.S.A. **78 E3** 41 45N 81 17W
Fairview, Australia **62 B3** 15 31S 144 17 E
Fairview, Canada **72 B5** 56 5N 118 25W
Fairview, Mont., U.S.A. **80 B2** 47 51N 104 3W
Fairview, Okla., U.S.A. **81 G5** 36 16N 98 29W
Fairview, Utah, U.S.A. **82 G8** 39 50N 111 0W
Fairweather, Mt., U.S.A. **68 C6** 58 55N 137 32W
Faisalabad, Pakistan **42 D5** 31 30N 73 5 E
Faith, U.S.A. **80 C3** 45 2N 102 2W
Faizabad, India **43 F10** 26 45N 82 10 E
Fajardo, Puerto Rico **89 C6** 18 20N 65 39W
Fakfak, Indonesia **37 E8** 3 0S 132 15 E
Faku, China **35 C12** 42 32N 123 21 E
Falaise, France **18 B3** 48 54N 0 12W
Falaise, Mui, Vietnam **38 C5** 19 6N 105 45 E
Falam, Burma **41 H18** 23 0N 93 45 E
Falcón, C., Spain **22 C7** 38 50N 1 23 E
Falcon Dam, U.S.A. **81 M5** 26 50N 99 20W
Falconer, U.S.A. **78 D5** 42 7N 79 13W
Falfurrias, U.S.A. **81 M5** 27 14N 98 9W
Falher, Canada **72 B5** 55 44N 117 15W
Falirakí, Greece **23 C10** 36 22N 28 12 E
Falkenberg, Sweden **9 H12** 56 54N 12 30 E
Falkirk, U.K. **12 F5** 56 0N 3 47W
Falkland Is. ■, Atl. Oc. **96 G5** 51 30S 59 0W
Falkland Is. Dependency □, Atl. Oc. **5 B1** 57 0S 40 0W
Falkland Sd., Falk. Is. **96 G5** 52 0S 60 0W
Falköping, Sweden **9 G12** 58 12N 13 33 E
Fall River, U.S.A. **79 E13** 41 43N 71 10W
Fall River Mills, U.S.A. **82 F3** 41 3N 121 26W
Fallbrook, U.S.A. **83 K5** 33 25N 117 12W
Fallbrook, Calif., U.S.A. **85 M9** 33 23N 117 15W
Fallon, Mont., U.S.A. **80 B2** 46 50N 105 8W
Fallon, Nev., U.S.A. **82 G4** 39 28N 118 47W
Falls City, Nebr., U.S.A. **80 E7** 40 3N 95 36W
Falls City, Oreg., U.S.A. **82 D2** 44 52N 123 26W
Falls Creek, U.S.A. **78 E6** 41 9N 78 48W
Falmouth, Jamaica **88 C4** 18 30N 77 40W
Falmouth, U.K. **11 G2** 50 9N 5 5W
Falmouth, U.S.A. **76 F3** 38 41N 84 20W
False B., S. Africa **56 E2** 34 15S 18 40 E
Falso, C., Honduras **88 C3** 15 12N 83 21W
Falster, Denmark **9 J11** 54 45N 11 55 E
Falsterbo, Sweden **9 J12** 55 23N 12 50 E
Falun, Sweden **9 F13** 60 37N 15 37 E
Famagusta, Cyprus **23 D12** 35 8N 33 55 E
Famagusta Bay, Cyprus **23 D13** 35 15N 34 0 E
Famatina, Sierra de, Argentina **94 B2** 27 30S 68 0W
Family L., Canada **73 C9** 51 54N 95 27W
Famoso, U.S.A. **85 K7** 35 37N 119 12W
Fan Xian, China **34 G8** 35 55N 115 38 E
Fandriana, Madag. **57 C8** 20 14S 47 21 E
Fang, Thailand **38 C2** 19 55N 99 13 E
Fangcheng, China **34 H7** 33 18N 112 59 E
Fangshan, China **34 E6** 38 3N 111 25 E
Fangzi, China **35 F10** 36 33N 119 10 E
Fenjiatun, China **35 C13** 43 40N 125 15 E
Fannich, L., U.K. **12 D4** 57 40N 5 0W

G

Gangtok

Gangtok, *India* **41 F16** 27 20N 88 37 E
Gangu, *China* **34 G3** 34 40N 105 15 E
Gangyao, *China* **35 B14** 44 12N 126 37 E
Gani, *Indonesia* **37 E7** 0 48S 128 14 E
Ganj, *India* **43 F8** 27 45N 78 57 E
Gannett Peak, *U.S.A.* **82 E9** 43 11N 109 39W
Gannvalley, *U.S.A.* **80 C5** 44 2N 98 59W
Ganquan, *China* **34 F5** 36 20N 109 20 E
Gansu □, *China* **34 G3** 36 0N 104 0 E
Ganta, *Liberia* **50 G3** 7 15N 8 59W
Gantheaume, C., *Australia* **63 F2** 36 4S 137 32 E
Gantheaume B., *Australia* **61 E1** 27 40S 114 10 E
Ganyem, *Indonesia* **37 E10** 2 46S 140 12 E
Ganyu, *China* **35 G10** 34 50N 119 8 E
Ganzhou, *China* **33 D6** 25 51N 114 56 E
Gaomi, *China* **35 F10** 36 20N 119 42 E
Gaoping, *China* **34 G7** 35 45N 112 55 E
Gaotang, *China* **34 F9** 36 50N 116 15 E
Gaoua, *Burkina Faso* **50 F4** 10 20N 3 8W
Gaoual, *Guinea* **50 F2** 11 45N 13 25W
Gaoxiong, *Taiwan* **33 D7** 22 38N 120 18 E
Gaoyang, *China* **34 E8** 38 40N 115 45 E
Gaoyou Hu, *China* **35 H10** 32 45N 119 20 E
Gaoyuan, *China* **35 F9** 37 8N 117 58 E
Gap, *France* **18 D7** 44 33N 6 5 E
Gar, *China* **32 C2** 32 10N 79 58 E
Garabogazkol Aylagy = Kara Bogaz Gol, Zaliv, *Turkmenistan* **25 F9** 41 0N 53 30 E
Garachico, *Canary Is.* **22 F3** 28 22N 16 46W
Garachiné, *Panama* **88 E4** 8 0N 78 12W
Garafia, *Canary Is.* **22 F2** 28 48N 17 57W
Garajonay, *Canary Is.* **22 F2** 28 7N 17 14W
Garanhuns, *Brazil* **93 E11** 8 50S 36 30W
Garawe, *Liberia* **50 H3** 4 35N 8 0W
Garba Tula, *Kenya* **54 B4** 0 30N 38 32 E
Garber, *U.S.A.* **81 G6** 36 26N 97 35W
Garberville, *U.S.A.* **82 F2** 40 6N 123 48W
Gard, *Somali Rep.* **46 F4** 9 30N 49 6 E
Gard □, *France* **18 D6** 44 2N 4 10 E
Garda, L. di, *Italy* **20 B4** 45 40N 10 40 E
Garde L., *Canada* **73 A7** 62 50N 106 13W
Garden City, *Kans., U.S.A.* **81 G4** 37 58N 100 53W
Garden City, *Tex., U.S.A.* **81 K4** 31 52N 101 29W
Garden Grove, *U.S.A.* **85 M9** 33 47N 117 55W
Gardēz, *Afghan.* **42 C3** 33 37N 69 9 E
Gardiner, *U.S.A.* **82 D8** 45 2N 110 22W
Gardiners I., *U.S.A.* **79 E12** 41 6N 72 6W
Gardner, *U.S.A.* **79 D13** 42 34N 71 59W
Gardner Canal, *Canada* **72 C3** 53 27N 128 8W
Gardnerville, *U.S.A.* **84 G7** 38 56N 119 45W
Garey, *U.S.A.* **85 L6** 34 53N 120 19W
Garfield, *U.S.A.* **82 C5** 47 1N 117 9W
Gargano, Mte., *Italy* **20 D6** 41 43N 15 43 E
Garhshankar, *India* **42 D7** 31 13N 76 11 E
Garibaldi Prov. Park, *Canada* **72 D4** 49 50N 122 40W
Garies, *S. Africa* **56 E2** 30 32S 17 59 E
Garigliano →, *Italy* **20 D5** 41 13N 13 44 E
Garissa, *Kenya* **54 C4** 0 25S 39 40 E
Garissa □, *Kenya* **54 C5** 0 20S 40 0 E
Garland, *U.S.A.* **82 F7** 41 47N 112 10W
Garm, *Tajikistan* **26 F8** 39 0N 70 20 E
Garmāb, *Iran* **45 C8** 35 25N 56 45 E
Garmsār, *Iran* **45 C7** 35 20N 52 25 E
Garner, *U.S.A.* **80 D8** 43 6N 93 36W
Garnett, *U.S.A.* **80 F7** 38 17N 95 14W
Garo Hills, *India* **43 G14** 25 30N 90 30 E
Garoe, *Somali Rep.* **46 F4** 8 25N 48 33 E
Garonne →, *France* **18 D3** 45 2N 0 36W
Garoua, *Cameroon* **51 G7** 9 19N 13 21 E
Garrison, *Mont., U.S.A.* **82 C7** 46 31N 112 49W
Garrison, *N. Dak., U.S.A.* **80 B4** 47 40N 101 25W
Garrison, *Tex., U.S.A.* **81 K7** 31 49N 94 30W
Garrison Res. = Sakakawea, L., *U.S.A.* **80 B3** 47 30N 101 25W
Garry →, *U.K.* **12 E5** 56 47N 3 47W
Garry, L., *Canada* **68 B9** 65 58N 100 18W
Garsen, *Kenya* **54 C5** 2 20S 40 5 E
Garson L., *Canada* **73 B6** 56 19N 110 2W
Garub, *Namibia* **56 D2** 26 37S 16 0 E
Garut, *Indonesia* **37 G12** 7 14S 107 53 E
Garvie Mts., *N.Z.* **59 L2** 45 30S 168 50 E
Garwa = Garoua, *Cameroon* **51 G7** 9 19N 13 21 E
Garwa, *India* **43 G10** 24 11N 83 47 E
Gary, *U.S.A.* **76 E2** 41 36N 87 20W
Garzê, *China* **32 C5** 31 38N 100 1 E
Garzón, *Colombia* **92 C3** 2 10N 75 40W
Gas-San, *Japan* **30 E10** 38 32N 140 1 E
Gasan Kuli, *Turkmenistan* **26 F6** 37 40N 54 20 E
Gascogne, *France* **18 E4** 43 45N 0 20 E
Gascogne, G. de, *Europe* **18 E3** 44 0N 2 0W
Gascony = Gascogne, *France* **18 E4** 43 45N 0 20 E
Gascoyne →, *Australia* **61 D1** 24 52S 113 37 E
Gascoyne Junc. T.O., *Australia* **61 E2** 25 2S 115 17 E
Gashaka, *Nigeria* **50 G7** 7 20N 11 29 E
Gasherbrum, *Pakistan* **43 B7** 35 40N 76 40 E
Gaspé, *Canada* **71 C7** 48 52N 64 30W
Gaspé, C. de, *Canada* **71 C7** 48 48N 64 7W
Gaspé, Pén. de, *Canada* **71 C6** 48 45N 65 40W
Gaspésie, Parc Prov. de la, *Canada* **71 C6** 48 55N 65 50W
Gassaway, *U.S.A.* **76 F5** 38 41N 80 47W
Gasteiz = Vitoria, *Spain* **19 A4** 42 50N 2 41W
Gastonia, *U.S.A.* **77 H5** 35 16N 81 11W
Gastre, *Argentina* **96 E3** 42 20S 69 15W
Gata, C., *Cyprus* **23 E12** 34 34N 33 2 E
Gata, C. de, *Spain* **19 D4** 36 41N 2 13W
Gata, Sierra de, *Spain* **19 B2** 40 20N 6 45W
Gataga →, *Canada* **72 B3** 58 35N 126 59W
Gates, *U.S.A.* **78 C7** 43 9N 77 42W
Gateshead, *U.K.* **10 C6** 54 57N 1 37W
Gatesville, *U.S.A.* **81 K6** 31 26N 97 45W
Gaths, *Zimbabwe* **55 G3** 20 2S 30 32 E
Gatico, *Chile* **94 A1** 22 29S 70 20W
Gâtinais, *France* **18 B5** 48 5N 2 40 E
Gatineau →, *Canada* **70 C4** 45 27N 75 42W
Gatineau, Parc de la, *Canada* **70 C4** 45 40N 76 0W
Gatun, L., *Panama* **88 E4** 9 7N 79 56W
Gatyana, *S. Africa* **57 E4** 32 16S 28 31 E
Gau, *Fiji* **59 D8** 18 2S 179 18 E
Gauer L., *Canada* **73 B9** 57 0N 97 50W
Gauhati, *India* **43 F14** 26 10N 91 45 E
Gaula →, *Norway* **8 E11** 63 21N 10 14 E
Gausta, *Norway* **9 G10** 59 50N 8 37 E
Gāv Koshī, *Iran* **45 D8** 28 38N 57 12 E
Gāvakān, *Iran* **45 D7** 29 37N 53 10 E
Gavāter, *Iran* **45 E9** 25 10N 61 31 E
Gāvbandī, *Iran* **45 E7** 27 12N 53 4 E
Gavdhopoúla, *Greece* **23 E6** 34 56N 24 0 E
Gávdhos, *Greece* **23 E6** 34 50N 24 5 E
Gaviota, *U.S.A.* **85 L6** 34 29N 120 13W
Gävleborgs län □, *Sweden* **9 F14** 61 30N 16 15 E
Gawachab, *Namibia* **56 D2** 27 4S 17 55 E
Gawilgarh Hills, *India* **40 J10** 21 15N 76 45 E
Gawler, *Australia* **63 E2** 34 30S 138 42 E
Gaxun Nur, *China* **32 B5** 42 22N 100 30 E
Gay, *Russia* **24 D10** 51 27N 58 27 E
Gaya, *India* **43 G11** 24 47N 85 4 E
Gaya, *Niger* **50 F5** 11 52N 3 28 E
Gaylord, *U.S.A.* **76 C3** 45 2N 84 41W
Gayndah, *Australia* **63 D5** 25 35S 151 32 E
Gaza, *Egypt* **47 D3** 31 30N 34 28 E
Gaza □, *Mozam.* **57 C5** 23 10S 32 45 E
Gaza Strip, *Egypt* **47 D3** 31 29N 34 25 E
Gāzbor, *Iran* **45 D8** 28 5N 58 51 E
Gazi, *Zaïre* **54 B1** 1 3N 24 30 E
Gaziantep, *Turkey* **25 G6** 37 6N 37 23 E
Gazli, *Uzbekistan* **26 E7** 40 14N 63 24 E
Gcuwa, *S. Africa* **57 E4** 32 20S 28 11 E
Gdańsk, *Poland* **17 A9** 54 22N 18 40 E
Gdańska, Zatoka, *Poland* **17 A9** 54 30N 19 20 E
Gdov, *Russia* **24 C4** 58 48N 27 55 E
Gdynia, *Poland* **17 A9** 54 35N 18 33 E
Gebe, *Indonesia* **37 D7** 0 5N 129 25 E
Gebeit Mine, *Sudan* **51 D12** 21 3N 36 29 E
Gedaref, *Sudan* **51 F12** 14 2N 35 28 E
Gede, Tanjung, *Indonesia* **36 F3** 6 46S 105 12 E
Gedser, *Denmark* **9 J11** 54 35N 11 55 E
Geegully Cr. →, *Australia* **60 C3** 18 32S 123 41 E
Geelong, *Australia* **63 F3** 38 10S 144 22 E
Geelvink Chan., *Australia* **61 E1** 28 30S 114 0 E
Geidam, *Nigeria* **51 F7** 12 57N 11 57 E
Geikie →, *Canada* **73 B8** 57 45N 103 52W
Geili, *Sudan* **51 E11** 16 1N 32 37 E
Geita, *Tanzania* **54 C3** 2 48S 32 12 E
Geita □, *Tanzania* **54 C3** 2 50S 32 10 E
Gejiu, *China* **32 D5** 23 20N 103 10 E
Gela, *Italy* **20 F6** 37 6N 14 18 E
Geladi, *Ethiopia* **46 F4** 6 59N 46 30 E
Gelderland □, *Neths.* **15 B6** 52 5N 6 10 E
Geldermalsen, *Neths.* **15 C5** 51 53N 5 17 E
Geldrop, *Neths.* **15 C5** 51 25N 5 32 E
Geleen, *Neths.* **15 D5** 50 57N 5 49 E
Gelehun, *S. Leone* **50 G2** 8 20N 11 40W
Gelibolu, *Turkey* **21 D12** 40 28N 26 43 E
Gelsenkirchen, *Germany* **16 C3** 51 30N 7 5 E
Gemas, *Malaysia* **39 L4** 2 37N 102 36 E
Gembloux, *Belgium* **15 D4** 50 34N 4 43 E
Gemena, *Zaïre* **52 D3** 3 13N 19 48 E
Gendringen, *Neths.* **15 C6** 51 52N 6 21 E
General Acha, *Argentina* **94 D3** 37 20S 64 38W
General Alvear, *Buenos Aires, Argentina* **94 D3** 36 0S 60 0W
General Alvear, *Mendoza, Argentina* **94 D2** 35 0S 67 40W
General Artigas, *Paraguay* **94 B4** 26 52S 56 16W
General Belgrano, *Argentina* **94 D4** 36 35S 58 47W
General Cabrera, *Argentina* **94 C3** 32 53S 63 52W
General Cepeda, *Mexico* **86 B4** 25 23N 101 27W
General Guido, *Argentina* **94 D4** 36 40S 57 50W
General Juan Madariaga, *Argentina* **94 D4** 37 0S 57 0W
General La Madrid, *Argentina* **94 D3** 37 17S 61 20W
General MacArthur, *Phil.* **37 B7** 11 18N 125 28 E
General Martin Miguel de Güemes, *Argentina* **94 A3** 24 50S 65 0W
General Paz, *Argentina* **94 B4** 27 45S 57 36W
General Pico, *Argentina* **94 D3** 35 45S 63 50W
General Pinedo, *Argentina* **94 B3** 27 15S 61 20W
General Pinto, *Argentina* **94 C3** 34 45S 61 50W
General Santos, *Phil.* **37 C7** 6 5N 125 14 E
General Trevino, *Mexico* **87 B5** 26 14N 99 29W
General Trías, *Mexico* **86 B3** 28 21N 106 22W
General Viamonte, *Argentina* **94 D3** 35 1S 61 3W
General Villegas, *Argentina* **94 D3** 35 5S 63 0W
Genesee, *Idaho, U.S.A.* **82 C5** 46 33N 116 56W
Genesee, *Pa., U.S.A.* **78 E7** 41 59N 77 54W
Genesee →, *U.S.A.* **78 C7** 43 16N 77 36W
Geneseo, *Ill., U.S.A.* **80 E9** 41 27N 90 9W
Geneseo, *Kans., U.S.A.* **80 F5** 38 31N 98 10W
Geneseo, *N.Y., U.S.A.* **78 D7** 42 48N 77 49W
Geneva = Genève, *Switz.* **16 E3** 46 12N 6 9 E
Geneva, *Ala., U.S.A.* **77 K3** 31 2N 85 52W
Geneva, *N.Y., U.S.A.* **78 D7** 42 52N 76 59W
Geneva, *Nebr., U.S.A.* **80 E6** 40 32N 97 36W
Geneva, *Ohio, U.S.A.* **78 E4** 41 48N 80 57W
Geneva, L. = Léman, Lac, *Switz.* **16 E3** 46 26N 6 30 E
Geneva, L., *U.S.A.* **76 D1** 42 38N 88 30W
Genève, *Switz.* **16 E3** 46 12N 6 9 E
Genil →, *Spain* **19 D3** 37 42N 5 19W
Genk, *Belgium* **15 D5** 50 58N 5 32 E
Gennargentu, Mti. del, *Italy* **20 D3** 40 0N 9 10 E
Gennep, *Neths.* **15 C5** 51 41N 5 59 E
Genoa = Génova, *Italy* **20 B3** 44 24N 8 56 E
Genoa, *Australia* **63 F4** 37 29S 149 35 E
Genoa, *N.Y., U.S.A.* **79 D8** 42 40N 76 32W
Genoa, *Nebr., U.S.A.* **80 E6** 41 27N 97 44W
Genoa, *Nev., U.S.A.* **84 F7** 39 2N 119 50W
Génova, *Italy* **20 B3** 44 24N 8 56 E
Génova, G. di, *Italy* **20 B3** 44 0N 9 0 E
Gent, *Belgium* **15 C3** 51 2N 3 42 E
Geographe B., *Australia* **61 F2** 33 30S 115 15 E
Geographe Chan., *Australia* **61 D1** 24 30S 113 0 E
Georga, Zemlya, *Russia* **26 A5** 80 30N 49 0 E
George, *S. Africa* **56 E3** 33 58S 22 29 E
George →, *Canada* **71 A6** 58 49N 66 10W
George, L., *N.S.W., Australia* **63 F4** 35 10S 149 25 E
George, L., *S. Austral., Australia* **63 F3** 37 25S 140 0 E
George, L., *W. Austral., Australia* **60 D3** 22 45S 123 40 E
George, L., *Uganda* **54 B3** 0 5N 30 10 E
George, L., *Fla., U.S.A.* **77 L5** 29 17N 81 36W
George, L., *N.Y., U.S.A.* **79 C11** 43 37N 73 33W
George Gill Ra., *Australia* **60 D5** 24 22S 131 45 E
George River = Port Nouveau-Québec, *Canada* **69 C13** 58 30N 65 59W
George Sound, *N.Z.* **59 L1** 44 52S 167 25 E
George Town, *Bahamas* **88 B4** 23 33N 75 47W
George Town, *Malaysia* **39 K3** 5 25N 100 15 E
George V Land, *Antarctica* **5 C10** 69 0S 148 0 E
George VI Sound, *Antarctica* **5 D17** 71 0S 68 0W
George West, *U.S.A.* **81 L5** 28 20N 98 7W
Georgetown, *Australia* **62 B3** 18 17S 143 33 E
Georgetown, *Ont., Canada* **70 D4** 43 40N 79 56W
Georgetown, *P.E.I., Canada* **71 C7** 46 13N 62 24W
Georgetown, *Cayman Is.* **88 C3** 19 20N 81 24W
Georgetown, *Gambia* **50 F2** 13 30N 14 47W
Georgetown, *Guyana* **92 B7** 6 50N 58 12W
Georgetown, *Calif., U.S.A.* **84 G6** 38 54N 120 50W
Georgetown, *Colo., U.S.A.* **82 G11** 39 42N 105 42W
Georgetown, *Ky., U.S.A.* **76 F3** 38 13N 84 33W
Georgetown, *S.C., U.S.A.* **77 J6** 33 23N 79 17W
Georgetown, *Tex., U.S.A.* **81 K6** 30 38N 97 41W
Georgia □, *U.S.A.* **77 J4** 32 50N 83 15W
Georgia ■, *Asia* **25 F7** 42 0N 43 0 E
Georgia, Str. of, *Canada* **72 D4** 49 25N 124 0W
Georgian B., *Canada* **70 C3** 45 15N 81 0W
Georgievsk, *Russia* **25 F7** 44 12N 43 28 E
Georgina →, *Australia* **62 C2** 23 30S 139 47 E
Georgina Downs, *Australia* **62 C2** 21 10S 137 40 E
Georgiu-Dezh = Liski, *Russia* **25 D6** 51 3N 39 30 E
Gera, *Germany* **16 C6** 50 53N 12 11 E
Geraardsbergen, *Belgium* **15 D3** 50 45N 3 53 E
Geral, Serra, *Brazil* **95 B6** 26 25S 50 0W
Geral de Goiás, Serra, *Brazil* **93 F9** 12 0S 46 0W
Geraldine, *U.S.A.* **82 C8** 47 36N 110 16W
Geraldton, *Australia* **61 E1** 28 48S 114 32 E
Geraldton, *Canada* **70 C2** 49 44N 86 59W
Gereshk, *Afghan.* **40 D4** 31 47N 64 35 E
Gerik, *Malaysia* **39 K3** 5 50N 101 15 E
Gering, *U.S.A.* **80 E3** 41 50N 103 40W
Gerlach, *U.S.A.* **82 F4** 40 39N 119 21W
Gerlogubi, *Ethiopia* **46 F4** 6 53N 45 3 E
Germansen Landing, *Canada* **72 B4** 55 43N 124 40W
Germany ■, *Europe* **16 C5** 51 0N 10 0 E
Germiston, *S. Africa* **57 D4** 26 15S 28 10 E
Gero, *Japan* **31 G8** 35 48N 137 14 E
Gerona, *Spain* **19 B7** 41 58N 2 46 E
Gerrard, *Canada* **72 C5** 50 30N 117 17W
Gers □, *France* **18 E4** 43 35N 0 30 E
Geser, *Indonesia* **37 E8** 3 50S 130 54 E
Gethsémani, *Canada* **71 B7** 50 13N 60 40W
Gettysburg, *Pa., U.S.A.* **76 F7** 39 50N 77 14W
Gettysburg, *S. Dak., U.S.A.* **80 C5** 45 1N 99 57W
Getz Ice Shelf, *Antarctica* **5 D14** 75 0S 130 0W
Gévaudan, *France* **18 D5** 44 40N 3 40 E
Geyser, *U.S.A.* **82 C8** 47 16N 110 30W
Geyserville, *U.S.A.* **84 G4** 38 42N 122 54W
Geysir, *Iceland* **8 D3** 64 19N 20 18W
Ghaghara →, *India* **43 G11** 25 45N 84 40 E
Ghana ■, *W. Afr.* **50 G4** 8 0N 1 0W
Ghansor, *India* **43 H9** 22 39N 80 1 E
Ghanzi, *Botswana* **56 C3** 21 50S 21 34 E
Ghanzi □, *Botswana* **56 C3** 21 50S 21 45 E
Gharb el Istiwa'iya □, *Sudan* **51 G10** 5 0N 30 0 E
Ghardaïa, *Algeria* **50 B5** 32 20N 3 37 E
Gharyān, *Libya* **51 B7** 32 10N 13 0 E
Ghat, *Libya* **50 D7** 24 59N 10 11 E
Ghatal, *India* **43 H12** 22 40N 87 46 E
Ghatampur, *India* **43 F9** 26 8N 80 13 E
Ghaţţī, *Si. Arabia* **44 D3** 31 16N 37 31 E
Ghawdex = Gozo, *Malta* **23 C1** 36 3N 14 13 E
Ghazal, Bahr el →, *Chad* **51 F8** 13 0N 15 47 E
Ghazâl, Bahr el →, *Sudan* **51 G11** 9 31N 30 25 E
Ghazaouet, *Algeria* **50 A4** 35 8N 1 50W
Ghaziabad, *India* **42 E7** 28 42N 77 26 E
Ghazipur, *India* **43 G10** 25 38N 83 35 E
Ghaznī, *Afghan.* **42 C3** 33 30N 68 28 E
Ghaznī □, *Afghan.* **40 C6** 32 10N 68 20 E
Ghèlinsor, *Somali Rep.* **46 F4** 6 28N 46 39 E
Ghent = Gent, *Belgium* **15 C3** 51 2N 3 42 E
Ghizao, *Afghan.* **42 C1** 33 20N 65 44 E
Ghizar →, *Pakistan* **43 A5** 36 15N 73 43 E
Ghogha, *India* **42 J5** 21 40N 72 20 E
Ghotaru, *India* **42 F4** 27 20N 70 1 E
Ghotki, *Pakistan* **42 E3** 28 5N 69 21 E
Ghowr □, *Afghan.* **40 C4** 34 0N 64 20 E
Ghudaf, W. al →, *Iraq* **44 C4** 32 56N 43 30 E
Ghughri, *India* **43 H9** 22 39N 80 41 E
Ghugus, *India* **40 K11** 19 58N 79 12 E
Ghulam Mohammad Barrage, *Pakistan* **42 G3** 25 30N 68 20 E
Ghūriān, *Afghan.* **40 B2** 34 17N 61 25 E
Gia Dinh, *Vietnam* **39 G6** 10 49N 106 42 E
Gia Lai = Pleiku, *Vietnam* **38 F7** 13 57N 108 0 E
Gia Nghia, *Vietnam* **39 G6** 11 58N 107 42 E
Gia Ngoc, *Vietnam* **38 E7** 14 50N 108 58 E
Gia Vuc, *Vietnam* **38 E7** 14 42N 108 34 E
Gian, *Phil.* **37 C7** 5 45N 125 20 E
Giant Forest, *U.S.A.* **84 J8** 36 36N 118 43W
Giants Causeway, *U.K.* **13 A5** 55 15N 6 30W
Giarabub = Al Jaghbūb, *Libya* **51 C9** 29 42N 24 38 E
Giarre, *Italy* **20 F6** 37 44N 15 10 E
Gibara, *Cuba* **88 B4** 21 9N 76 11W
Gibb River, *Australia* **60 C4** 16 26S 126 26 E
Gibbon, *U.S.A.* **80 E5** 40 45N 98 51W
Gibraltar ■, *Europe* **19 D3** 36 7N 5 22W
Gibraltar, Str. of, *Medit. S.* **19 E3** 35 55N 5 40W
Gibson Desert, *Australia* **60 D4** 24 0S 126 0 E
Gibsons, *Canada* **72 D4** 49 24N 123 32W
Gibsonville, *U.S.A.* **84 F6** 39 46N 120 54W
Giddings, *U.S.A.* **81 K6** 30 11N 96 56W
Gidole, *Ethiopia* **51 G12** 5 40N 37 25 E
Giessen, *Germany* **16 C4** 50 34N 8 40 E
Gīfān, *Iran* **45 B8** 37 54N 57 28 E
Gifford Creek, *Australia* **60 D2** 24 3S 116 16 E
Gifu, *Japan* **31 G8** 35 30N 136 45 E
Gifu □, *Japan* **31 G8** 35 40N 137 0 E
Giganta, Sa. de la, *Mexico* **86 B2** 25 30N 111 30W
Gigha, *U.K.* **12 F3** 55 42N 5 45W
Gijón, *Spain* **19 A3** 43 32N 5 42W
Gil I., *Canada* **72 C3** 53 12N 129 15W
Gila →, *U.S.A.* **83 K6** 32 43N 114 33W
Gila Bend, *U.S.A.* **83 K7** 32 57N 112 43W
Gila Bend Mts., *U.S.A.* **83 K7** 33 10N 113 0W
Gīlān □, *Iran* **45 B6** 37 0N 50 0 E
Gilbert →, *Australia* **62 B3** 16 35S 141 15 E
Gilbert Is., *Kiribati* **64 G9** 1 0N 176 0 E
Gilbert Plains, *Canada* **73 C8** 51 9N 100 28W
Gilbert River, *Australia* **62 B3** 18 9S 142 52 E
Gilberton, *Australia* **62 B3** 19 16S 143 35 E
Gilford I., *Canada* **72 C3** 50 40N 126 30W
Gilgandra, *Australia* **63 E4** 31 43S 148 39 E
Gilgil, *Kenya* **54 C4** 0 30S 36 20 E
Gilgit, *India* **43 B6** 35 50N 74 15 E
Gilgit →, *Pakistan* **43 B6** 35 44N 74 37 E
Gillam, *Canada* **73 B10** 56 20N 94 40W
Gillen, L., *Australia* **61 E3** 26 11S 124 38 E
Gilles, L., *Australia* **63 E2** 32 50S 136 45 E
Gillette, *U.S.A.* **80 C2** 44 18N 105 30W
Gilliat, *Australia* **62 C3** 20 40S 141 28 E
Gillingham, *U.K.* **11 F8** 51 23N 0 34 E
Gilmer, *U.S.A.* **81 J7** 32 44N 94 57W
Gilmore, *Australia* **63 F4** 35 20S 148 12 E
Gilmore, L., *Australia* **61 F3** 32 29S 121 37 E
Gilmour, *Canada* **70 D4** 44 48N 77 37W
Gilroy, *U.S.A.* **83 H3** 37 1N 121 34W
Gimbi, *Ethiopia* **51 G12** 9 3N 35 42 E
Gimli, *Canada* **73 C9** 50 40N 97 0W
Gin Gin, *Australia* **63 D5** 25 0S 151 58 E
Gindie, *Australia* **62 C4** 23 44S 148 8 E
Gingin, *Australia* **61 F2** 31 22S 115 54 E
Ginir, *Ethiopia* **46 F3** 7 6N 40 40 E
Giohar, *Somali Rep.* **46 G4** 2 48N 45 30 E
Gióna, Óros, *Greece* **21 E10** 38 38N 22 14 E
Gir Hills, *India* **42 J4** 21 0N 71 0 E
Girab, *India* **42 F4** 26 2N 70 38 E
Girâfi, W. →, *Egypt* **47 F3** 29 58N 34 39 E
Girard, *Kans., U.S.A.* **81 G7** 37 31N 94 51W
Girard, *Ohio, U.S.A.* **78 E4** 41 9N 80 42W
Girard, *Pa., U.S.A.* **78 D4** 42 0N 80 19W
Girardot, *Colombia* **92 C4** 4 18N 74 48W
Girdle Ness, *U.K.* **12 D6** 57 9N 2 2W
Giresun, *Turkey* **25 F6** 40 55N 38 30 E
Girga, *Egypt* **51 C11** 26 17N 31 55 E
Giridih, *India* **43 G12** 24 10N 86 21 E
Girilambone, *Australia* **63 E4** 31 16S 146 57 E
Girona = Gerona, *Spain* **19 B7** 41 58N 2 46 E
Gironde □, *France* **18 D3** 44 45N 0 30W
Gironde →, *France* **18 D3** 45 32N 1 7W
Giru, *Australia* **62 B4** 19 30S 147 5 E
Girvan, *U.K.* **12 F4** 55 15N 4 50W
Gisborne, *N.Z.* **59 H7** 38 39S 178 5 E
Gisenyi, *Rwanda* **54 C2** 1 41S 29 15 E
Gitega, *Burundi* **54 C2** 3 26S 29 56 E
Giuba →, *Somali Rep.* **46 G3** 1 30N 42 35 E
Giurgiu, *Romania* **17 G12** 43 52N 25 57 E
Giza = El Gîza, *Egypt* **51 C11** 30 0N 31 10 E
Gizhiga, *Russia* **27 C17** 62 3N 160 30 E
Gizhiginskaya Guba, *Russia* **27 C16** 61 0N 158 0 E
Giżycko, *Poland* **17 A10** 54 2N 21 48 E
Gjirokastra, *Albania* **21 D9** 40 7N 20 10 E
Gjoa Haven, *Canada* **68 B10** 68 20N 96 8W
Gjøvik, *Norway* **9 F11** 60 47N 10 43 E
Glace Bay, *Canada* **71 C8** 46 11N 59 58W
Glacier Bay, *U.S.A.* **72 B1** 58 40N 136 0W
Glacier Nat. Park, *Canada* **72 C5** 51 15N 117 30W
Glacier Park, *U.S.A.* **82 B7** 48 30N 113 18W
Glacier Peak, *U.S.A.* **82 B3** 48 7N 121 7W
Gladewater, *U.S.A.* **81 J7** 32 33N 94 56W
Gladstone, *Queens., Australia* **62 C5** 23 52S 151 16 E
Gladstone, *S. Austral., Australia* **63 E2** 33 15S 138 22 E
Gladstone, *W. Austral., Australia* **61 E1** 25 57S 114 17 E
Gladstone, *Canada* **73 C9** 50 13N 98 57W
Gladstone, *U.S.A.* **76 C2** 45 51N 87 1W
Gladwin, *U.S.A.* **76 D3** 43 59N 84 29W
Gladys L., *Canada* **72 B2** 59 50N 133 0W
Gláma, *Iceland* **8 D2** 65 48N 23 0W
Glåma →, *Norway* **9 G11** 59 12N 10 57 E
Glamis, *U.S.A.* **85 N11** 32 55N 115 5W
Glasco, *Kans., U.S.A.* **80 F6** 39 22N 97 50W
Glasco, *N.Y., U.S.A.* **79 D11** 42 3N 73 57W
Glasgow, *U.K.* **12 F4** 55 52N 4 14W
Glasgow, *Ky., U.S.A.* **76 G3** 37 0N 85 55W
Glasgow, *Mont., U.S.A.* **82 B10** 48 12N 106 38W

Hab Nadi Chauki, Pakistan 42 G2 25 0N 66 50 E
Habaswein, Kenya 54 B4 1 2N 39 30 E
Habay, Canada 72 B5 58 50N 118 44W
Habbānīyah, Iraq 44 C4 33 17N 43 29 E
Haboro, Japan 30 B10 44 22N 141 42 E
Hachijō-Jima, Japan 31 H9 33 5N 139 45 E
Hachinohe, Japan 30 D10 40 30N 141 29 E
Hachiōji, Japan 31 G9 35 40N 139 20 E
Hachŏn, N. Korea 35 D15 41 29N 129 2 E
Hackensack, U.S.A. 79 F10 40 53N 74 3W
Hadali, Pakistan 42 C5 32 16N 72 11 E
Hadarba, Ras, Sudan 51 D12 22 4N 36 51 E
Hadarom □, Israel 47 E3 31 0N 35 0 E
Haddington, U.K. 12 F6 55 57N 2 48W
Hadejia, Nigeria 50 F7 12 30N 10 5 E
Haden, Australia 63 D5 27 13S 151 54 E
Hadera, Israel 47 C3 32 27N 34 55 E
Hadera, N. →, Israel 47 C3 32 28N 34 52 E
Hadhramaut = Ḥaḍramawt, Yemen 46 D4 15 30N 49 30 E
Hadong, S. Korea 35 G14 35 5N 127 44 E
Ḥaḍramawt, Yemen 46 D4 15 30N 49 30 E
Ḥadrānīyah, Iraq 44 C4 35 38N 43 14 E
Hadrian's Wall, U.K. 10 C5 55 0N 2 30W
Haeju, N. Korea 35 E13 38 3N 125 45 E
Haenam, S. Korea 35 G14 34 34N 126 35 E
Haerhpin = Harbin, China 35 B14 45 48N 126 40 E
Hafar al Bāṭin, Si. Arabia 44 D5 28 25N 46 0 E
Ḥafirat al 'Aydā, Si. Arabia 44 E3 26 26N 39 12 E
Hafizabad, Pakistan 42 C5 32 5N 73 40 E
Haflong, India 41 G18 25 10N 93 5 E
Hafnarfjörður, Iceland 8 D3 64 4N 21 57W
Hafun, Ras, Somali Rep. 46 E5 10 29N 51 30 E
Hagalil, Israel 47 C4 32 53N 35 18 E
Hagen, Germany 16 C3 51 21N 7 29 E
Hagerman, U.S.A. 81 J2 33 7N 104 20W
Hagerstown, U.S.A. 76 F7 39 39N 77 43W
Hagfors, Sweden 9 F12 60 3N 13 45 E
Hagi, Iceland 8 D2 65 28N 23 25W
Hagi, Japan 31 G5 34 30N 131 22 E
Hagolan, Syria 47 B4 33 0N 35 45 E
Hagondange-Briey, France 18 B7 49 16N 6 11 E
Hags Hd., Ireland 13 D2 52 57N 9 30W
Hague, C. de la, France 18 B3 49 44N 1 56W
Hague, The = 's-Gravenhage, Neths. 15 B4 52 7N 4 17 E
Haguenau, France 18 B7 48 49N 7 47 E
Hai □, Tanzania 54 C4 3 10S 37 10 E
Hai Duong, Vietnam 38 B6 20 56N 106 19 E
Haicheng, China 35 D12 40 50N 122 45 E
Haidar Khel, Afghan. 42 C3 33 58N 68 38 E
Haifa = Hefa, Israel 47 C3 32 46N 35 0 E
Haig, Australia 61 F4 30 55S 126 10 E
Haikou, China 33 D6 20 1N 110 16 E
Ḥā'il, Si. Arabia 44 E4 27 28N 41 45 E
Hailar, China 33 B6 49 10N 119 38 E
Hailey, U.S.A. 82 E6 43 31N 114 19W
Haileybury, Canada 70 C4 47 30N 79 38W
Hailin, China 35 B15 44 37N 129 30 E
Hailong, China 35 C13 42 32N 125 40 E
Hailuoto, Finland 8 D18 65 3N 24 45 E
Hainan □, China 33 E5 19 0N 109 30 E
Hainaut □, Belgium 15 D4 50 30N 4 0 E
Haines, U.S.A. 82 D5 44 55N 117 56W
Haines City, U.S.A. 77 L5 28 7N 81 38W
Haines Junction, Canada 72 A1 60 45N 137 30W
Haiphong, Vietnam 32 D5 20 47N 106 41 E
Haiti ■, W. Indies 89 C5 19 0N 72 30W
Haiya Junction, Sudan 51 E12 18 20N 36 21 E
Haiyang, China 35 F11 36 47N 121 9 E
Haiyuan, China 34 F3 36 35N 105 52 E
Haizhou, China 35 G10 34 37N 119 7 E
Haizhou Wan, China 35 G10 34 50N 119 20 E
Haja, Indonesia 37 E7 3 19S 129 37 E
Hajar Bangar, Sudan 51 F9 10 40N 22 45 E
Hajdúböszörmény, Hungary 17 E10 47 40N 21 30 E
Hajipur, India 43 G11 25 45N 85 13 E
Hājjī Muḥsin, Iraq 44 C5 32 35N 45 29 E
Ḥājjīābād, Eṣfahan, Iran 45 C7 33 41N 54 50 E
Ḥājjīābād, Hormozgān, Iran 45 D7 28 19N 55 55 E
Hajnówka, Poland 17 B11 52 47N 23 35 E
Hakansson, Mts., Zaïre 55 D2 8 40S 25 45 E
Hakken-Zan, Japan 31 G7 34 10N 135 54 E
Hakodate, Japan 30 D10 41 45N 140 44 E
Haku-San, Japan 31 F8 36 9N 136 46 E
Hakui, Japan 31 F8 36 53N 136 47 E
Hala, Pakistan 40 G6 25 43N 68 20 E
Ḥalab, Syria 44 B3 36 10N 37 15 E
Ḥalabjah, Iraq 44 C5 35 10N 45 58 E
Halaib, Sudan 51 D12 22 12N 36 30 E
Hālat' Ammār, Si. Arabia 44 D3 29 10N 36 4 E
Halbā, Lebanon 47 A5 34 34N 36 6 E
Halberstadt, Germany 16 C5 51 53N 11 2 E
Halcombe, N.Z. 59 J5 40 8S 175 30 E
Halcon, Mt., Phil. 37 B6 13 0N 121 30 E
Halden, Norway 9 G11 59 9N 11 23 E
Haldia, India 41 H16 22 5N 88 3 E
Haldwani, India 43 E8 29 31N 79 30 E
Hale →, Australia 62 C2 24 56S 135 53 E
Haleakala Crater, U.S.A. 74 H16 20 43N 156 16W
Haleyville, U.S.A. 77 H2 34 14N 87 37W
Halfway →, Canada 72 B4 56 12N 121 32W
Haliburton, Canada 70 C4 45 3N 78 30W
Halifax, Australia 62 B4 18 32S 146 22 E
Halifax, Canada 71 D7 44 38N 63 35W
Halifax, U.K. 10 D6 53 43N 1 51W
Halifax B., Australia 62 B4 18 50S 147 0 E
Halifax I., Namibia 56 D2 26 38S 15 4 E
Halīl →, Iran 45 E8 27 40N 58 30 E
Hall Beach, Canada 69 B11 68 46N 81 12W
Hall Pt., Australia 60 C3 15 40S 124 23 E
Hallands län □, Sweden 9 H12 56 50N 12 50 E
Halle, Belgium 15 D4 50 44N 4 13 E
Halle, Germany 16 C5 51 29N 12 0 E
Hällefors, Sweden 9 G13 59 47N 14 31 E
Hallett, Australia 63 E2 33 25S 138 55 E
Hallettsville, U.S.A. 81 L6 29 27N 96 57W
Halliday, U.S.A. 80 B3 47 21N 102 20W
Halliday L., Canada 73 A7 61 21N 108 56W
Hallim, S. Korea 35 H14 33 24N 126 15 E
Hallingdal →, Norway 9 F10 60 34N 9 12 E
Hällnäs, Sweden 8 D15 64 19N 19 36 E
Hallock, U.S.A. 73 D9 48 47N 96 57W
Halls Creek, Australia 60 C4 18 16S 127 38 E
Hallstead, U.S.A. 79 E9 41 58N 75 45W
Halmahera, Indonesia 37 D7 0 40N 128 0 E
Halmstad, Sweden 9 H12 56 41N 12 52 E
Halq el Oued, Tunisia 51 A7 36 53N 10 18 E
Hals, Denmark 9 H11 56 59N 10 18 E
Hälsingborg = Helsingborg, Sweden 9 H12 56 3N 12 42 E
Halstad, U.S.A. 80 B6 47 21N 96 50W
Halul, Qatar 45 E7 25 40N 52 40 E
Ḥalvān, Iran 45 C8 33 57N 56 15 E
Ham Tan, Vietnam 39 G6 10 40N 107 45 E
Ham Yen, Vietnam 38 A5 22 4N 105 3 E
Hamab, Namibia 56 D2 28 7S 19 16 E
Hamada, Japan 31 G6 34 56N 132 4 E
Hamadān, Iran 45 C6 34 52N 48 32 E
Hamadān □, Iran 45 C6 35 0N 49 0 E
Ḥamāh, Syria 44 C3 35 5N 36 40 E
Hamamatsu, Japan 31 G8 34 45N 137 45 E
Hamar, Norway 9 F11 60 48N 11 7 E
Hamarøy, Norway 8 B13 68 5N 15 38 E
Hambantota, Sri Lanka 40 R12 6 10N 81 10 E
Hamber Prov. Park, Canada 72 C5 52 20N 118 0W
Hamburg, Germany 16 B4 53 32N 9 59 E
Hamburg, Ark., U.S.A. 81 J9 33 14N 91 48W
Hamburg, Iowa, U.S.A. 80 E7 40 36N 95 39W
Hamburg, N.Y., U.S.A. 78 D6 42 43N 78 50W
Hamburg, Pa., U.S.A. 79 F9 40 33N 75 59W
Ḥamḍ, W. al →, Si. Arabia 44 E3 24 55N 36 20 E
Hamden, U.S.A. 79 E12 41 23N 72 54W
Hame □ = Hämeen lääni □, Finland 9 F18 61 30N 24 0 E
Hämeen lääni □, Finland 9 F18 61 30N 24 0 E
Hämeenlinna, Finland 9 F18 61 0N 24 28 E
Hamelin Pool, Australia 61 E1 26 22S 114 20 E
Hameln, Germany 16 B4 52 7N 9 24 E
Hamerkaz □, Israel 47 C3 32 15N 34 55 E
Hamersley Ra., Australia 60 D2 22 0S 117 45 E
Hamhung, N. Korea 35 E14 39 54N 127 30 E
Hami, China 32 B4 42 55N 93 25 E
Hamilton, Australia 63 F3 37 45S 142 2 E
Hamilton, Canada 70 D4 43 15N 79 50W
Hamilton, N.Z. 59 G5 37 47S 175 19 E
Hamilton, U.K. 12 F4 55 47N 4 2W
Hamilton, Mo., U.S.A. 80 F8 39 45N 93 59W
Hamilton, Mont., U.S.A. 82 C6 46 15N 114 10W
Hamilton, N.Y., U.S.A. 79 D9 42 50N 75 33W
Hamilton, Ohio, U.S.A. 76 F3 39 24N 84 34W
Hamilton, Tex., U.S.A. 81 K5 31 42N 98 7W
Hamilton →, Australia 62 C2 23 30S 139 47 E
Hamilton City, U.S.A. 84 F4 39 45N 122 1W
Hamilton Hotel, Australia 62 C3 22 45S 140 40 E
Hamilton Inlet, Canada 71 B8 54 0N 57 30W
Hamiota, Canada 73 C8 50 11N 100 38W
Hamlet, U.S.A. 77 H6 34 53N 79 42W
Hamley Bridge, Australia 63 E2 34 17S 138 35 E
Hamlin = Hameln, Germany 16 B4 52 7N 9 24 E
Hamlin, N.Y., U.S.A. 78 C7 43 17N 77 55W
Hamlin, Tex., U.S.A. 81 J4 32 53N 100 8W
Hamm, Germany 16 C3 51 40N 7 49 E
Hammerfest, Norway 8 A17 70 39N 23 41 E
Hammond, Ind., U.S.A. 76 E2 41 38N 87 30W
Hammond, La., U.S.A. 81 K9 30 30N 90 28W
Hammonton, U.S.A. 76 F8 39 39N 74 48W
Hampden, N.Z. 59 L3 45 18S 170 50 E
Hampshire □, U.K. 11 F6 51 3N 1 20W
Hampshire Downs, U.K. 11 F6 51 10N 1 10W
Hampton, Ark., U.S.A. 81 J8 33 32N 92 28W
Hampton, Iowa, U.S.A. 80 D8 42 45N 93 13W
Hampton, N.H., U.S.A. 79 D14 42 57N 70 50W
Hampton, S.C., U.S.A. 77 J5 32 52N 81 7W
Hampton, Va., U.S.A. 76 G7 37 2N 76 21W
Hampton Tableland, Australia 61 F4 32 0S 127 0 E
Hamrat esh Sheykh, Sudan 51 F10 14 38N 27 55 E
Hamyang, S. Korea 35 G14 35 32N 127 42 E
Hana, U.S.A. 74 H17 20 45N 155 59W
Hanak, Si. Arabia 44 E3 25 32N 37 0 E
Hanamaki, Japan 30 E10 39 23N 141 7 E
Hanang, Tanzania 54 C4 4 30S 35 25 E
Hanau, Germany 16 C4 50 8N 8 56 E
Hanbogd, Mongolia 34 C4 43 11N 107 10 E
Hancheng, China 34 G6 35 31N 110 25 E
Hancock, Mich., U.S.A. 80 B10 47 8N 88 35W
Hancock, Minn., U.S.A. 80 C7 45 30N 95 48W
Hancock, N.Y., U.S.A. 79 E9 41 57N 75 17W
Handa, Japan 31 G8 34 53N 136 55 E
Handa, Somali Rep. 46 E5 10 37N 51 2 E
Handan, China 34 F8 36 35N 114 28 E
Handeni, Tanzania 54 D4 5 25S 38 2 E
Handeni □, Tanzania 54 D4 5 30S 38 0 E
Handwara, India 43 B6 34 21N 74 20 E
Hanegev, Israel 47 E3 30 50N 35 0 E
Haney, Canada 72 D4 49 12N 122 40W
Hanford, U.S.A. 83 H4 36 20N 119 39W
Hang Chat, Thailand 38 C2 18 20N 99 21 E
Hang Dong, Thailand 38 C2 18 41N 98 55 E
Hangang →, S. Korea 35 F14 37 50N 126 30 E
Hangayn Nuruu, Mongolia 32 B4 47 30N 99 0 E
Hangchou = Hangzhou, China 33 C7 30 18N 120 11 E
Hanggin Houqi, China 34 D4 40 58N 107 4 E
Hanggin Qi, China 34 E5 39 52N 108 50 E
Hangö, Finland 9 G17 59 50N 22 57 E
Hangu, China 35 E9 39 18N 117 53 E
Hangzhou, China 33 C7 30 18N 120 11 E
Hangzhou Wan, China 33 C7 30 15N 120 45 E
Hanhongor, Mongolia 34 C3 43 55N 104 28 E
Ḥanīdh, Si. Arabia 45 E6 26 35N 48 38 E
Ḥanīsh, Yemen 46 E3 13 45N 42 46 E
Hankinson, U.S.A. 80 B6 46 4N 96 54W
Hanko = Hangö, Finland 9 G17 59 50N 22 57 E
Hanko, Finland 9 G17 59 50N 22 57 E
Hanksville, U.S.A. 83 G8 38 22N 110 43W
Hanle, India 43 C8 32 42N 79 4 E
Hanmer Springs, N.Z. 59 K4 42 32S 172 50 E
Hann →, Australia 60 C4 17 26S 126 17 E
Hann, Mt., Australia 60 C4 15 45S 126 0 E
Hanna, Canada 72 C6 51 40N 111 54W
Hannaford, U.S.A. 80 B5 47 19N 98 11W
Hannah, U.S.A. 80 A5 48 58N 98 42W
Hannah B., Canada 70 B4 51 40N 80 0W
Hannibal, U.S.A. 80 F9 39 42N 91 22W
Hannover, Germany 16 B4 52 23N 9 43 E
Hanoi, Vietnam 32 D5 21 5N 105 55 E
Hanover = Hannover, Germany 16 B4 52 23N 9 43 E
Hanover, Canada 78 B3 44 9N 81 2W
Hanover, S. Africa 56 E3 31 4S 24 29 E
Hanover, N.H., U.S.A. 79 C12 43 42N 72 17W
Hanover, Ohio, U.S.A. 78 F2 40 4N 82 16W
Hanover, Pa., U.S.A. 76 F7 39 48N 76 59W
Hanover, I., Chile 96 G2 51 0S 74 50W
Hansi, India 42 E6 29 10N 75 57 E
Hanson, L., Australia 63 E2 31 0S 136 15 E
Hanzhong, China 34 H4 33 10N 107 1 E
Hanzhuang, China 35 G9 34 33N 117 23 E
Haora, India 43 H13 22 37N 88 20 E
Haparanda, Sweden 8 D18 65 52N 24 8 E
Happy, U.S.A. 81 H4 34 45N 101 52W
Happy Camp, U.S.A. 82 F2 41 48N 123 23W
Happy Valley-Goose Bay, Canada 71 B7 53 15N 60 20W
Hapsu, N. Korea 35 D15 41 13N 128 51 E
Hapur, India 42 E7 28 45N 77 45 E
Ḥaql, Si. Arabia 47 F3 29 10N 34 58 E
Har, Indonesia 37 F8 5 16S 133 14 E
Har-Ayrag, Mongolia 34 B5 45 47N 109 16 E
Har Hu, China 32 C4 38 20N 97 38 E
Har Us Nuur, Mongolia 32 B4 48 0N 92 0 E
Har Yehuda, Israel 47 D3 31 35N 34 57 E
Ḥaraḍ, Si. Arabia 46 C4 24 22N 49 0 E
Haranomachi, Japan 30 F10 37 38N 140 58 E
Harardera, Somali Rep. 46 G4 4 33N 47 38 E
Harare, Zimbabwe 55 F3 17 43S 31 2 E
Harazé, Chad 51 F8 14 20N 19 12 E
Harbin, China 35 B14 45 48N 126 40 E
Harbor Beach, U.S.A. 76 D4 43 51N 82 39W
Harbor Springs, U.S.A. 76 C3 45 26N 85 0W
Harbour Breton, Canada 71 C8 47 29N 55 50W
Harbour Grace, Canada 71 C9 47 40N 53 22W
Harburg, Germany 16 B4 53 27N 9 58 E
Harda, India 42 H7 22 27N 77 5 E
Hardangerfjorden, Norway 9 F8 60 15N 6 0 E
Hardap Dam, Namibia 56 C2 24 32S 17 50 E
Hardenberg, Neths. 15 B6 52 34N 6 37 E
Harderwijk, Neths. 15 B5 52 21N 5 38 E
Hardey →, Australia 60 D2 22 45S 116 8 E
Hardin, U.S.A. 82 D10 45 44N 107 37W
Harding, S. Africa 57 E4 30 35S 29 55 E
Harding Ra., Australia 60 C3 16 17S 124 55 E
Hardisty, Canada 72 C6 52 40N 111 18W
Hardman, U.S.A. 82 D4 45 10N 119 41W
Hardoi, India 43 F9 27 26N 80 6 E
Hardwar = Haridwar, India 42 E8 29 58N 78 9 E
Hardwick, U.S.A. 79 B12 44 30N 72 22W
Hardy, U.S.A. 81 G9 36 19N 91 29W
Hardy, Pen., Chile 96 H3 55 30S 68 20W
Hare B., Canada 71 B8 51 15N 55 45W
Harer, Ethiopia 46 F3 9 20N 42 8 E
Hargeisa, Somali Rep. 46 F3 9 30N 44 2 E
Hargshamn, Sweden 9 F15 60 12N 18 30 E
Hari →, Indonesia 36 E2 1 16S 104 5 E
Haria, Canary Is. 22 E6 29 8N 13 32W
Haridwar, India 42 E8 29 58N 78 9 E
Haringhata →, Bangla. 41 J16 22 0N 89 58 E
Harīrūd →, Asia 40 A2 37 24N 60 38 E
Harlan, Iowa, U.S.A. 80 E7 41 39N 95 19W
Harlan, Ky., U.S.A. 77 G4 36 51N 83 19W
Harlech, U.K. 10 E3 52 52N 4 7W
Harlem, U.S.A. 82 B9 48 32N 108 47W
Harlingen, Neths. 15 A5 53 11N 5 25 E
Harlingen, U.S.A. 81 M6 26 12N 97 42W
Harlowton, U.S.A. 82 C9 46 26N 109 50W
Harney Basin, U.S.A. 82 E4 43 30N 119 0W
Harney L., U.S.A. 82 E4 43 14N 119 8W
Harney Peak, U.S.A. 80 D3 43 52N 103 32W
Härnösand, Sweden 8 E14 62 38N 17 55 E
Harp L., Canada 71 A7 55 5N 61 50W
Harrand, Pakistan 42 E4 29 28N 70 3 E
Harriman, U.S.A. 77 H3 35 56N 84 33W
Harrington Harbour, Canada 71 B8 50 31N 59 30W
Harris, U.K. 12 D2 57 50N 6 55W
Harris, Sd. of, U.K. 12 D1 57 44N 7 6W
Harris L., Australia 63 E2 31 10S 135 10 E
Harrisburg, Ill., U.S.A. 81 G10 37 44N 88 32W
Harrisburg, Nebr., U.S.A. 80 E3 41 33N 103 44W
Harrisburg, Oreg., U.S.A. 82 D2 44 16N 123 10W
Harrisburg, Pa., U.S.A. 78 F8 40 16N 76 53W
Harrismith, S. Africa 57 D4 28 15S 29 8 E
Harrison, Ark., U.S.A. 81 G8 36 14N 93 7W
Harrison, Idaho, U.S.A. 82 C5 47 27N 116 47W
Harrison, Nebr., U.S.A. 80 D3 42 41N 103 53W
Harrison, C., Canada 71 B8 54 55N 57 55W
Harrison Bay, U.S.A. 68 A4 70 40N 151 0W
Harrison L., Canada 72 D4 49 33N 121 50W
Harrisonburg, U.S.A. 76 F6 38 27N 78 52W
Harrisonville, U.S.A. 80 F7 38 39N 94 21W
Harriston, Canada 70 D3 43 57N 80 53W
Harrisville, U.S.A. 78 B1 44 39N 83 17W
Harrogate, U.K. 10 D6 53 59N 1 32W
Harrow, U.K. 11 F7 51 35N 0 15W
Harsīn, Iran 44 C5 34 18N 47 33 E
Harstad, Norway 8 B14 68 48N 16 30 E
Hart, U.S.A. 76 D2 43 42N 86 22W
Hart, L., Australia 63 E2 31 10S 136 25 E
Hartbees →, S. Africa 56 D3 28 45S 20 32 E
Hartford, Conn., U.S.A. 79 E12 41 46N 72 41W
Hartford, Ky., U.S.A. 76 G2 37 27N 86 55W
Hartford, S. Dak., U.S.A. 80 D6 43 38N 96 57W
Hartford, Wis., U.S.A. 80 D10 43 19N 88 22W
Hartford City, U.S.A. 76 E3 40 27N 85 22W
Hartland, Canada 71 C6 46 20N 67 32W
Hartland Pt., U.K. 11 F3 51 2N 4 32W
Hartlepool, U.K. 10 C6 54 42N 1 11W
Hartley Bay, Canada 72 C3 53 25N 129 15W
Hartmannberge, Namibia 56 B1 17 0S 13 0 E
Hartney, Canada 73 D8 49 30N 100 35W
Harts →, S. Africa 56 D3 28 24S 24 17 E
Hartselle, U.S.A. 77 H2 34 27N 86 56W
Hartshorne, U.S.A. 81 H7 34 51N 95 34W
Hartsville, U.S.A. 77 H5 34 23N 80 4W
Hartwell, U.S.A. 77 H4 34 21N 82 56W
Harunabad, Pakistan 42 E5 29 35N 73 8 E
Harvand, Iran 45 D7 28 25N 55 43 E
Harvey, Australia 61 F2 33 5S 115 54 E
Harvey, Ill., U.S.A. 76 E2 41 36N 87 50W
Harvey, N. Dak., U.S.A. 80 B5 47 47N 99 56W
Harwich, U.K. 11 F9 51 56N 1 18 E
Haryana □, India 42 E7 29 0N 76 10 E
Harz, Germany 16 C5 51 40N 10 40 E
Hasan Kīādeh, Iran 45 B6 37 24N 49 58 E
Ḥasanābād, Iran 45 C7 32 8N 52 44 E
Hasanpur, India 42 E8 28 43N 78 17 E
Hashimoto, Japan 31 G7 34 19N 135 37 E
Hashtjerd, Iran 45 C6 35 52N 50 40 E
Haskell, Okla., U.S.A. 81 H7 35 50N 95 40W
Haskell, Tex., U.S.A. 81 J5 33 10N 99 44W
Hasselt, Belgium 15 D5 50 56N 5 21 E
Hassi Inifel, Algeria 50 C5 29 50N 3 41 E
Hassi Messaoud, Algeria 50 B6 31 43N 6 8 E
Hastings, N.Z. 59 H6 39 39S 176 52 E
Hastings, U.K. 11 G8 50 51N 0 36 E
Hastings, Mich., U.S.A. 76 D3 42 39N 85 17W
Hastings, Minn., U.S.A. 80 C8 44 44N 92 51W
Hastings, Nebr., U.S.A. 80 E5 40 35N 98 23W
Hastings Ra., Australia 63 E5 31 15S 152 14 E
Hat Yai, Thailand 39 J3 7 1N 100 27 E
Hatanbulag, Mongolia 34 C5 43 8N 109 5 E
Hatay = Antalya, Turkey 25 G5 36 52N 30 45 E
Hatch, U.S.A. 83 K10 32 40N 107 9W
Hatches Creek, Australia 62 C2 20 56S 135 12 E
Hatchet L., Canada 73 B8 58 36N 103 40W
Hateruma-Shima, Japan 31 M1 24 3N 123 47 E
Hatfield P.O., Australia 63 E3 33 54S 143 49 E
Hatgal, Mongolia 32 A5 50 26N 100 9 E
Hathras, India 42 F8 27 36N 78 6 E
Hatia, Bangla. 41 H17 22 30N 91 5 E
Hato Mayor, Dom. Rep. 89 C6 18 46N 69 15W
Hattah, Australia 63 E3 34 48S 142 17 E
Hatteras, C., U.S.A. 77 H8 35 14N 75 32W
Hattiesburg, U.S.A. 81 K10 31 20N 89 17W
Hatvan, Hungary 17 E9 47 40N 19 45 E
Hau Bon = Cheo Reo, Vietnam 38 F7 13 25N 108 28 E
Hau Duc, Vietnam 38 E7 15 20N 108 13 E
Haugesund, Norway 9 G8 59 23N 5 13 E
Haultain →, Canada 73 B7 55 51N 106 46W
Hauraki G., N.Z. 59 G5 36 35S 175 5 E
Haut Atlas, Morocco 50 B3 32 30N 5 0W
Haut-Rhin □, France 18 C7 48 0N 7 15 E
Haut Zaïre □, Zaïre 54 B2 2 20N 26 0 E
Haute-Corse □, France 18 E9 42 30N 9 30 E
Haute-Garonne □, France 18 E4 43 30N 1 30 E
Haute-Loire □, France 18 D5 45 5N 3 50 E
Haute-Marne □, France 18 B6 48 10N 5 20 E
Haute-Saône □, France 18 C7 47 45N 6 10 E
Haute-Savoie □, France 18 C7 46 0N 6 20 E
Haute-Vienne □, France 18 D4 45 50N 1 10 E
Hauterive, Canada 71 C6 49 10N 68 16W
Hautes-Alpes □, France 18 D7 44 42N 6 20 E
Hautes Fagnes = Hohe Venn, Belgium 15 D6 50 30N 6 5 E
Hautes-Pyrénées □, France 18 E4 43 0N 0 10 E
Hauts-de-Seine □, France 18 B5 48 52N 2 15 E
Hauts Plateaux, Algeria 50 A5 35 0N 1 0 E
Havana = La Habana, Cuba 88 B3 23 8N 82 22W
Havana, U.S.A. 80 E9 40 18N 90 4W
Havant, U.K. 11 G7 50 51N 0 59W
Havasu, L., U.S.A. 85 L12 34 18N 114 28W
Havel →, Germany 16 B6 52 40N 12 1 E
Havelange, Belgium 15 D5 50 23N 5 15 E
Havelian, Pakistan 42 B5 34 2N 73 10 E
Havelock, N.B., Canada 71 C6 46 2N 65 24W
Havelock, Ont., Canada 70 D4 44 26N 77 53W
Havelock, N.Z. 59 J4 41 17S 173 48 E
Haverfordwest, U.K. 11 F3 51 48N 4 59W
Haverhill, U.S.A. 79 D13 42 47N 71 5W
Havering, U.K. 11 F8 51 33N 0 20 E
Haverstraw, U.S.A. 79 E11 41 12N 73 58W
Havlíčkův Brod, Czech. 16 D7 49 36N 15 33 E
Havre, U.S.A. 82 B9 48 33N 109 41W
Havre-Aubert, Canada 71 C7 47 12N 61 56W
Havre-St.-Pierre, Canada 71 B7 50 18N 63 33W
Haw →, U.S.A. 77 H6 35 36N 79 3W
Hawaii □, U.S.A. 74 H16 19 30N 156 30W
Hawaii I., Pac. Oc. 74 J17 20 0N 155 0W
Hawaiian Is., Pac. Oc. 74 H17 20 30N 156 0W
Hawaiian Ridge, Pac. Oc. 65 E11 24 0N 165 0W
Hawarden, Canada 73 C7 51 25N 106 36W
Hawarden, U.S.A. 80 D6 43 0N 96 29W
Hawea, L., N.Z. 59 L2 44 28S 169 19 E
Hawera, N.Z. 59 H5 39 35S 174 19 E
Hawick, U.K. 12 F6 55 25N 2 48W
Hawk Junction, Canada 70 C3 48 5N 84 38W
Hawke B., N.Z. 59 H6 39 25S 177 20 E
Hawker, Australia 63 E2 31 59S 138 22 E
Hawkesbury, Canada 70 C5 45 37N 74 37W
Hawkesbury I., Canada 72 C3 53 37N 129 3W
Hawkesbury Pt., Australia 62 A1 11 55S 134 5 E
Hawkinsville, U.S.A. 77 J4 32 17N 83 28W
Hawkwood, Australia 63 D5 25 45S 150 50 E
Hawley, U.S.A. 80 B6 46 53N 96 19W
Hawrān, Syria 47 C5 32 45N 36 15 E
Hawsh Müssá, Lebanon 47 B4 33 45N 35 55 E
Hawthorne, U.S.A. 82 G4 38 32N 118 38W
Haxtun, U.S.A. 80 E3 40 39N 102 38W
Hay, Australia 63 E3 34 30S 144 51 E
Hay →, Australia 62 C2 24 50S 138 0 E
Hay →, Canada 72 A5 60 50N 116 26W

Hay, C., Australia 60 B4 14 5S 129 29 E
Hay L., Canada 72 B5 58 50N 118 50W
Hay Lakes, Canada 72 C6 53 12N 113 2W
Hay-on-Wye, U.K. 11 E4 52 4N 3 9W
Hay River, Canada 72 A5 60 51N 115 44W
Hay Springs, U.S.A. 80 D3 42 41N 102 41W
Hayachine-San, Japan 30 E10 39 34N 141 29 E
Hayden, Ariz., U.S.A. 83 K8 33 0N 110 47W
Hayden, Colo., U.S.A. 82 F10 40 30N 107 16W
Haydon, Australia 62 B3 18 0S 141 30 E
Hayes, U.S.A. 80 C4 44 23N 101 1W
Hayes →, Canada 73 B10 57 3N 92 12W
Haynesville, U.S.A. 81 J8 32 58N 93 8W
Hays, Canada 72 C6 50 6N 111 48W
Hays, U.S.A. 80 F5 38 53N 99 20W
Hayward, Calif., U.S.A. 84 H4 37 40N 122 5W
Hayward, Wis., U.S.A. 80 B9 46 1N 91 29W
Haywards Heath, U.K. 11 F7 51 1N 0 6W
Hazafon □, Israel 47 C4 32 40N 35 20 E
Hazarām, Kūh-e, Iran 45 D8 29 30N 57 18 E
Hazard, U.S.A. 76 G4 37 15N 83 12W
Hazaribag, India 43 H11 23 58N 85 26 E
Hazaribag Road, India 43 G11 24 12N 85 57 E
Hazelton, Canada 72 B3 55 20N 127 42W
Hazelton, U.S.A. 80 B4 46 29N 100 17W
Hazen, N. Dak., U.S.A. 80 B4 47 18N 101 38W
Hazen, Nev., U.S.A. 82 G4 39 34N 119 3W
Hazlehurst, Ga., U.S.A. 77 K4 31 52N 82 36W
Hazlehurst, Miss., U.S.A. 81 K9 31 52N 90 24W
Hazleton, U.S.A. 79 F9 40 57N 75 59W
Hazlett, L., Australia 60 D4 21 30S 128 48 E
Hazor, Israel 47 B4 33 2N 35 32 E
Head of Bight, Australia 61 F5 31 30S 131 25 E
Headlands, Zimbabwe 55 F3 18 15S 32 2 E
Healdsburg, U.S.A. 84 G4 38 37N 122 52W
Healdton, U.S.A. 81 H6 34 14N 97 29W
Healesville, Australia 63 F4 37 35S 145 30 E
Heanor, U.K. 10 D6 53 1N 1 20W
Heard I., Ind. Oc. 3 G13 53 0S 74 0 E
Hearne, U.S.A. 81 K6 30 53N 96 36W
Hearne B., Canada 73 A9 60 10N 99 10W
Hearne L., Canada 72 A6 62 20N 113 10W
Hearst, Canada 70 C3 49 40N 83 41W
Heart →, U.S.A. 80 B4 46 46N 100 50W
Heart's Content, Canada 71 C9 47 54N 53 27W
Heath Pt., Canada 71 C7 49 8N 61 40W
Heath Steele, Canada 71 C6 47 17N 66 5W
Heavener, U.S.A. 81 H7 34 53N 94 36W
Hebbronville, U.S.A. 81 M5 27 18N 98 41W
Hebei □, China 34 E9 39 0N 116 0 E
Hebel, Australia 63 D4 28 58S 147 47 E
Heber, U.S.A. 85 N11 32 44N 115 32W
Heber Springs, U.S.A. 81 H9 35 30N 92 2W
Hebert, Canada 73 C7 50 30N 107 10W
Hebgen L., U.S.A. 82 D8 44 52N 111 20W
Hebi, China 34 G8 35 57N 114 7 E
Hebrides, U.K. 12 D1 57 30N 7 0W
Hebron = Al Khalīl, Jordan 47 D4 31 32N 35 6 E
Hebron, Canada 69 C13 58 5N 62 30W
Hebron, N. Dak., U.S.A. 80 B3 46 54N 102 3W
Hebron, Nebr., U.S.A. 80 E6 40 10N 97 35W
Hecate Str., Canada 72 C2 53 10N 130 30W
Hechi, China 32 D5 24 40N 108 2 E
Hechuan, China 32 C5 30 2N 106 12 E
Hecla, U.S.A. 80 C5 45 53N 98 9W
Hecla I., Canada 73 C9 51 10N 96 43W
Hede, Sweden 8 E12 62 23N 13 30 E
Hedemora, Sweden 9 F13 60 18N 15 58 E
Hedley, U.S.A. 81 H4 34 52N 100 39W
Heemstede, Neths. 15 B4 52 22N 4 37 E
Heerde, Neths. 15 B6 52 24N 6 2 E
Heerenveen, Neths. 15 B5 52 57N 5 55 E
Heerlen, Neths. 15 D5 50 55N 5 58 E
Hefa, Israel 47 C3 32 46N 35 0 E
Hefa □, Israel 47 C4 32 40N 35 0 E
Hefei, China 33 C6 31 52N 117 18 E
Hegang, China 33 B8 47 20N 130 19 E
Heichengzhen, China 34 F4 36 24N 106 3 E
Heidelberg, Germany 16 D4 49 23N 8 41 E
Heidelberg, C. Prov., S. Africa 56 E3 34 6S 20 59 E
Heidelberg, Trans., S. Africa 57 D4 26 30S 28 23 E
Heilbron, S. Africa 57 D4 27 16S 27 59 E
Heilbronn, Germany 16 D4 49 8N 9 13 E
Heilongjiang □, China 35 B14 48 0N 126 0 E
Heilunkiang = Heilongjiang □, China 35 B14 48 0N 126 0 E
Heinola, Finland 9 F19 61 13N 26 2 E
Heinze Is., Burma 41 M20 14 25N 97 45 E
Heishan, China 35 D12 41 40N 122 5 E
Heishui, China 35 C10 42 8N 119 30 E
Hejaz = Al Ḥijāz, Si. Arabia 46 B2 26 0N 37 30 E
Hejian, China 34 E9 38 25N 116 5 E
Hejin, China 34 G6 35 35N 110 42 E
Hekla, Iceland 8 E4 63 56N 19 35W
Hekou, Gansu, China 34 F2 36 10N 103 28 E
Hekou, Yunnan, China 32 D5 22 30N 103 59 E
Helan Shan, China 34 E3 38 30N 105 55 E
Helena, Ark., U.S.A. 81 H9 34 32N 90 36W
Helena, Mont., U.S.A. 82 C7 46 36N 112 2W
Helendale, U.S.A. 85 L9 34 44N 117 19W
Helensburgh, U.K. 12 E4 56 0N 4 44W
Helensville, N.Z. 59 G5 36 41S 174 29 E
Helgoland, Germany 16 A3 54 10N 7 51 E
Heligoland = Helgoland, Germany 16 A3 54 10N 7 51 E
Heligoland B. = Deutsche Bucht, Germany 16 A4 54 15N 8 0 E
Hellendoorn, Neths. 15 B6 52 24N 6 27 E
Hellevoetsluis, Neths. 15 C4 51 50N 4 8 E
Hellín, Spain 19 C5 38 31N 1 40W
Helmand □, Afghan. 40 D4 31 20N 64 0 E
Helmand →, Afghan. 40 D2 31 12N 61 34 E
Helmond, Neths. 15 C5 51 29N 5 41 E
Helmsdale, U.K. 12 C5 58 7N 3 40W
Helong, China 35 C15 42 40N 129 0 E
Helper, U.S.A. 82 G8 39 41N 110 51W
Helsingborg, Sweden 9 H12 56 3N 12 42 E
Helsingfors, Finland 9 F18 60 15N 25 3 E
Helsingør, Denmark 9 H12 56 2N 12 35 E
Helsinki, Finland 9 F18 60 15N 25 3 E
Helston, U.K. 11 G2 50 7N 5 17W
Helvellyn, U.K. 10 C4 54 31N 3 1W
Helwân, Egypt 51 C11 29 50N 31 20 E
Hemet, U.S.A. 85 M10 33 45N 116 58W
Hemingford, U.S.A. 80 D3 42 19N 103 4W
Hemphill, U.S.A. 81 K8 31 20N 93 51W
Hempstead, U.S.A. 81 K6 30 6N 96 5W
Hemse, Sweden 9 H15 57 15N 18 22 E
Henan □, China 34 G8 34 0N 114 0 E
Henares →, Spain 19 B4 40 24N 3 30W
Henashi-Misaki, Japan 30 D9 40 37N 139 51 E
Henderson, Argentina 94 D3 36 18S 61 43W
Henderson, Ky., U.S.A. 76 G2 37 50N 87 35W
Henderson, N.C., U.S.A. 77 G6 36 20N 78 25W
Henderson, Nev., U.S.A. 85 J12 36 2N 114 59W
Henderson, Tenn., U.S.A. 77 H1 35 26N 88 38W
Henderson, Tex., U.S.A. 81 J7 32 9N 94 48W
Hendersonville, U.S.A. 77 H4 35 19N 82 28W
Hendījān, Iran 45 D6 30 14N 49 43 E
Hendon, Australia 63 D5 28 5S 151 50 E
Hengcheng, China 34 E4 38 18N 106 28 E
Hengdaohezi, China 35 B15 44 52N 129 0 E
Hengelo, Neths. 15 B6 52 3N 6 19 E
Hengshan, China 34 F5 37 58N 109 5 E
Hengshui, China 34 F8 37 41N 115 40 E
Hengyang, China 33 D6 26 52N 112 33 E
Henlopen, C., U.S.A. 76 F8 38 48N 75 6W
Hennenman, S. Africa 56 D4 27 59S 27 1 E
Hennessey, U.S.A. 81 G6 36 6N 97 54W
Henrietta, U.S.A. 81 J5 33 49N 98 12W
Henrietta, Ostrov, Russia 27 B16 77 6N 156 30 E
Henrietta Maria C., Canada 70 A3 55 9N 82 20W
Henry, U.S.A. 80 E10 41 7N 89 22W
Henryetta, U.S.A. 81 H6 35 27N 95 59W
Hensall, Canada 78 C3 43 26N 81 30W
Hentiyn Nuruu, Mongolia 33 B5 48 30N 108 30 E
Henty, Australia 63 F4 35 30S 147 0 E
Henzada, Burma 41 L19 17 38N 95 26 E
Heppner, U.S.A. 82 D4 45 21N 119 33W
Hepworth, Canada 78 B3 44 37N 81 9W
Hequ, China 34 E6 39 20N 111 15 E
Héraðsflói, Iceland 8 D6 65 42N 14 12W
Héraðsvötn →, Iceland 8 D4 65 45N 19 25W
Herald Cays, Australia 62 B4 16 58S 149 9 E
Herāt, Afghan. 40 B3 34 20N 62 7 E
Herāt □, Afghan. 40 B3 35 0N 62 0 E
Hérault □, France 18 E5 43 34N 3 15 E
Herbert →, Australia 62 B4 18 31S 146 17 E
Herbert Downs, Australia 62 C2 23 7S 139 9 E
Herberton, Australia 62 B4 17 20S 145 25 E
Hercegnovi, Montenegro, Yug. 21 C8 42 30N 18 33 E
Herðubreið, Iceland 8 D5 65 11N 16 21W
Hereford, U.K. 11 E5 52 4N 2 42W
Hereford, U.S.A. 81 H3 34 49N 102 24W
Hereford and Worcester □, U.K. 11 E5 52 10N 2 30W
Herentals, Belgium 15 C4 51 12N 4 51 E
Herford, Germany 16 B4 52 7N 8 40 E
Herington, U.S.A. 80 F6 38 40N 96 57W
Herjehogna, Norway 9 F12 61 43N 12 7 E
Herkimer, U.S.A. 79 D10 43 0N 74 59W
Herlong, U.S.A. 84 E6 40 8N 120 8W
Herman, U.S.A. 80 C6 45 49N 96 9W
Hermann, U.S.A. 80 F9 38 42N 91 27W
Hermannsburg Mission, Australia 60 D5 23 57S 132 45 E
Hermanus, S. Africa 56 E2 34 27S 19 12 E
Hermidale, Australia 63 E4 31 30S 146 42 E
Hermiston, U.S.A. 82 D4 45 51N 119 17W
Hermitage, N.Z. 59 K3 43 44S 170 5 E
Hermite, I., Chile 96 H3 55 50S 68 0W
Hermon, Mt. = Ash Shaykh, J., Lebanon 47 B4 33 25N 35 50 E
Hermosillo, Mexico 86 B2 29 10N 111 0W
Hernád →, Hungary 17 E10 47 56N 21 8 E
Hernandarias, Paraguay 95 B5 25 20S 54 40W
Hernandez, U.S.A. 84 J6 36 24N 120 46W
Hernando, Argentina 94 C3 32 28S 63 40W
Hernando, U.S.A. 81 H10 34 50N 90 0W
Herne, Germany 15 C7 51 33N 7 12 E
Herne Bay, U.K. 11 F9 51 22N 1 8 E
Herning, Denmark 9 H10 56 8N 8 58 E
Heroica = Caborca, Mexico 86 A2 30 40N 112 10W
Heroica Nogales = Nogales, Mexico 86 A2 31 20N 110 56W
Heron Bay, Canada 70 C2 48 40N 86 25W
Herradura, Pta. de la, Canary Is. 22 F5 28 26N 14 8W
Herreid, U.S.A. 80 C4 45 50N 100 4W
Herrera, Spain 19 D3 37 26N 4 55W
Herrick, Australia 62 G4 41 5S 147 55 E
Herrin, U.S.A. 81 G10 37 48N 89 2W
Hersonissos, Greece 23 D7 35 18N 25 22 E
Herstal, Belgium 15 D5 50 40N 5 38 E
Hertford, U.K. 11 F7 51 47N 0 4W
Hertfordshire □, U.K. 11 F7 51 51N 0 5W
's-Hertogenbosch, Neths. 15 C5 51 42N 5 17 E
Hertzogville, S. Africa 56 D4 28 9S 25 30 E
Herzliyya, Israel 47 C3 32 10N 34 50 E
Ḥeşār, Fārs, Iran 45 D6 29 52N 50 16 E
Ḥeşār, Markazī, Iran 45 C6 35 49N 49 24 E
Heshui, China 34 G5 36 0N 108 0 E
Heshun, China 34 F7 37 22N 113 32 E
Hesperia, U.S.A. 85 L9 34 25N 117 18W
Hesse = Hessen □, Germany 16 C4 50 40N 9 20 E
Hessen □, Germany 16 C4 50 40N 9 20 E
Hetch Hetchy Aqueduct, U.S.A. 84 H5 37 29N 122 19W
Hettinger, U.S.A. 80 C3 46 0N 102 42W
Hewett, C., Canada 69 A13 70 16N 67 45W
Hexham, U.K. 10 C5 54 58N 2 7W
Hexigten Qi, China 35 C9 43 18N 117 30 E
Hexrivier, S. Africa 56 E2 33 30S 19 35 E
Ḥeydarābād, Iran 45 D7 30 33N 55 38 E
Heyfield, Australia 63 F4 37 59S 146 47 E
Heysham, U.K. 10 C5 54 5N 2 53W
Heywood, Australia 63 F3 38 8S 141 37 E
Heze, China 34 G8 35 14N 115 20 E
Hi Vista, U.S.A. 85 L9 34 45N 117 46W
Hialeah, U.S.A. 77 N5 25 50N 80 17W
Hiawatha, Kans., U.S.A. 80 F7 39 51N 95 32W
Hiawatha, Utah, U.S.A. 82 G8 39 29N 111 1W
Hibbing, U.S.A. 80 B8 47 25N 92 56W
Hibbs B., Australia 62 G4 42 35S 145 15 E
Hibernia Reef, Australia 60 B3 12 0S 123 23 E
Hickory, U.S.A. 77 H5 35 44N 81 21W
Hicks, Pt., Australia 63 F4 37 49S 149 17 E
Hicksville, U.S.A. 79 F11 40 46N 73 32W
Hida-Gawa →, Japan 31 G8 35 26N 137 3 E
Hida-Sammyaku, Japan 31 F8 36 30N 137 40 E
Hidaka-Sammyaku, Japan 30 C11 42 35N 142 45 E
Hidalgo, Mexico 87 C5 24 15N 99 26W
Hidalgo □, Mexico 87 C5 20 30N 99 10W
Hidalgo, Presa M., Mexico 86 B3 26 30N 108 35W
Hidalgo, Pta. del, Canary Is. 22 F3 28 33N 16 19W
Hidalgo del Parral, Mexico 86 B3 26 58N 105 40W
Hierro, Canary Is. 22 G1 27 44N 18 0W
Higashiajima-San, Japan 30 F10 37 40N 140 10 E
Higashiōsaka, Japan 31 G7 34 40N 135 37 E
Higgins, U.S.A. 81 G4 36 7N 100 2W
Higgins Corner, U.S.A. 84 F5 39 2N 121 5W
Higginsville, Australia 61 F3 31 42S 121 38 E
High Atlas = Haut Atlas, Morocco 50 B3 32 30N 5 0W
High I., Canada 71 A7 56 40N 61 10W
High Island, U.S.A. 81 L7 29 34N 94 24W
High Level, Canada 72 B5 58 31N 117 8W
High Point, U.S.A. 77 H6 35 57N 80 0W
High Prairie, Canada 72 B5 55 30N 116 30W
High River, Canada 72 C6 50 30N 113 50W
High Springs, U.S.A. 77 L4 29 50N 82 36W
High Tatra = Tatry, Slovakia 17 D10 49 20N 20 0 E
High Wycombe, U.K. 11 F7 51 37N 0 45W
Highbury, Australia 62 B3 16 25S 143 9 E
Highland □, U.K. 12 D4 57 30N 5 0W
Highland Park, U.S.A. 76 D2 42 11N 87 48W
Highmore, U.S.A. 80 C5 44 31N 99 27W
Highrock L., Canada 73 B7 57 5N 105 32W
Higüay, Dom. Rep. 89 C6 18 37N 68 42W
Hiiumaa, Estonia 24 C3 58 50N 22 45 E
Ḥijāz □, Si. Arabia 46 C2 24 0N 40 0 E
Hijo = Tagum, Phil. 37 C7 7 33N 125 53 E
Hikari, Japan 31 H5 33 58N 131 58 E
Hiko, U.S.A. 83 H6 37 32N 115 14W
Hikone, Japan 31 G8 35 15N 136 10 E
Hikurangi, N.Z. 59 F5 35 36S 174 17 E
Hikurangi, Mt., N.Z. 59 H6 38 21S 176 52 E
Hildesheim, Germany 16 B4 52 9N 9 55 E
Hill →, Australia 61 F2 30 23S 115 3 E
Hill City, Idaho, U.S.A. 82 E6 43 18N 115 3W
Hill City, Kans., U.S.A. 80 F5 39 22N 99 51W
Hill City, Minn., U.S.A. 80 B8 46 59N 93 36W
Hill City, S. Dak., U.S.A. 80 D3 43 56N 103 35W
Hill Island L., Canada 73 A7 60 30N 109 50W
Hillcrest Center, U.S.A. 85 K8 35 23N 118 57W
Hillegom, Neths. 15 B4 52 18N 4 35 E
Hillingdon, U.K. 11 F7 51 33N 0 29W
Hillman, U.S.A. 76 C4 45 4N 83 54W
Hillmond, Canada 73 C7 53 26N 109 41W
Hillsboro, Kans., U.S.A. 80 F6 38 21N 97 12W
Hillsboro, N. Dak., U.S.A. 80 B6 47 26N 97 3W
Hillsboro, N.H., U.S.A. 79 C13 43 7N 71 54W
Hillsboro, N. Mex., U.S.A. 83 K10 32 55N 107 34W
Hillsboro, Oreg., U.S.A. 84 E4 45 31N 122 59W
Hillsboro, Tex., U.S.A. 81 J6 32 1N 97 8W
Hillsborough, Grenada 89 D7 12 28N 61 28W
Hillsdale, Mich., U.S.A. 76 E3 41 56N 84 38W
Hillsdale, N.Y., U.S.A. 79 D11 42 11N 73 30W
Hillside, Australia 60 D2 21 45S 119 23 E
Hillsport, Canada 70 C2 49 27N 85 34W
Hillston, Australia 63 E4 33 30S 145 31 E
Hilo, U.S.A. 74 J17 19 44N 155 5W
Hilton, U.S.A. 78 C7 43 17N 77 48W
Hilversum, Neths. 15 B5 52 14N 5 10 E
Himachal Pradesh □, India 42 D7 31 30N 77 0 E
Himalaya, Asia 43 E11 29 0N 84 0 E
Himatnagar, India 40 H8 23 37N 72 57 E
Himeji, Japan 31 G7 34 50N 134 40 E
Himi, Japan 31 F8 36 50N 136 55 E
Ḥimş, Syria 47 A5 34 40N 36 45 E
Ḥimş □, Syria 47 A5 34 30N 37 0 E
Hinche, Haiti 89 C5 19 9N 72 1W
Hinchinbrook I., Australia 62 B4 18 20S 146 15 E
Hinckley, U.K. 11 E6 52 33N 1 21W
Hinckley, U.S.A. 82 G7 39 20N 112 40W
Hindaun, India 42 F7 26 44N 77 5 E
Hindmarsh, L., Australia 63 F3 36 5S 141 55 E
Hindu Bagh, Pakistan 42 D2 30 56N 67 50 E
Hindu Kush, Asia 40 B7 36 0N 71 0 E
Hindubagh, Pakistan 40 D5 30 56N 67 57 E
Hindupur, India 40 N10 13 49N 77 32 E
Hines Creek, Canada 72 B5 56 20N 118 40W
Hinganghat, India 40 J11 20 30N 78 52 E
Hingham, U.S.A. 82 B8 48 33N 110 25W
Hingoli, India 40 K10 19 41N 77 15 E
Hinna = Imi, Ethiopia 46 F3 6 28N 42 10 E
Hinsdale, U.S.A. 82 B10 48 24N 107 5W
Hinton, Canada 72 C5 53 26N 117 34W
Hinton, U.S.A. 76 G5 37 40N 80 54W
Hippolytushoef, Neths. 15 B4 52 54N 4 58 E
Hirado, Japan 31 H4 33 22N 129 33 E
Hirakud Dam, India 41 J13 21 32N 83 45 E
Hiratsuka, Japan 31 G9 35 19N 139 21 E
Hiroo, Japan 30 C11 42 17N 143 19 E
Hirosaki, Japan 30 D10 40 34N 140 28 E
Hiroshima, Japan 31 G6 34 24N 132 30 E
Hiroshima □, Japan 31 G6 34 50N 133 0 E
Hisar, India 42 E6 29 12N 75 45 E
Hisb →, Iraq 44 D5 31 45N 44 17 E
Ḥismá, Si. Arabia 44 D3 28 30N 36 0 E
Hispaniola, W. Indies 89 C5 19 0N 71 0W
Hīt, Iraq 44 C4 33 38N 42 49 E
Hita, Japan 31 H5 33 20N 130 58 E
Hitachi, Japan 31 F10 36 36N 140 39 E
Hitchin, U.K. 11 F7 51 57N 0 16W
Hitoyoshi, Japan 31 H5 32 13N 130 45 E
Hitra, Norway 8 E10 63 30N 8 45 E
Ḥiyyon, N. →, Israel 47 E4 30 25N 35 10 E
Hjalmar L., Canada 73 A7 61 33N 109 25W
Hjälmaren, Sweden 9 G13 59 18N 15 40 E
Hjørring, Denmark 9 H10 57 29N 9 59 E
Hluhluwe, S. Africa 57 D5 28 1S 32 15 E
Ho, Ghana 50 G5 6 37N 0 27 E
Ho Chi Minh City = Phanh Bho Ho Chi Minh, Vietnam 39 G6 10 58N 106 40 E
Ho Thuong, Vietnam 38 C5 19 32N 105 48 E
Hoa Binh, Vietnam 38 B5 20 50N 105 20 E
Hoa Da, Vietnam 39 G7 11 16N 108 40 E
Hoa Hiep, Vietnam 39 G5 11 34N 105 51 E
Hoai Nhon, Vietnam 38 E7 14 28N 109 1 E
Hoare B., Canada 69 B13 65 17N 62 30W
Hobart, Australia 62 G4 42 50S 147 21 E
Hobart, U.S.A. 81 H5 35 1N 99 6W
Hobbs, U.S.A. 81 J3 32 42N 103 8W
Hobbs Coast, Antarctica 5 D14 74 50S 131 0W
Hoboken, Belgium 15 C4 51 11N 4 21 E
Hoboken, U.S.A. 79 F10 40 45N 74 4W
Hobro, Denmark 9 H10 56 39N 9 46 E
Hoburgen, Sweden 9 H15 56 55N 18 7 E
Hodaka-Dake, Japan 31 F8 36 17N 137 39 E
Hodgson, Canada 73 C9 51 13N 97 36W
Hódmezövásárhely, Hungary 17 E10 46 28N 20 22 E
Hodna, Chott el, Algeria 50 A5 35 30N 5 0 E
Hodonín, Czech. 16 D8 48 50N 17 10 E
Hoeamdong, N. Korea 35 C16 42 30N 130 16 E
Hoek van Holland, Neths. 15 C4 52 0N 4 7 E
Hoengsŏng, S. Korea 35 F14 37 29N 127 59 E
Hoeryong, N. Korea 35 C15 42 30N 129 45 E
Hoeyang, N. Korea 35 E14 38 43N 127 36 E
Hof, Germany 16 C5 50 18N 11 55 E
Hof, Iceland 8 D6 64 33N 14 40W
Höfðakaupstaður, Iceland 8 D3 65 50N 20 19W
Hofmeyr, S. Africa 56 E4 31 39S 25 50 E
Hofsjökull, Iceland 8 D4 64 49N 18 48W
Hofsós, Iceland 8 D4 65 53N 19 26W
Hōfu, Japan 31 G5 34 3N 131 34 E
Hogan Group, Australia 62 F4 39 13S 147 1 E
Hogansville, U.S.A. 77 J3 33 10N 84 55W
Hogeland, U.S.A. 82 B9 48 51N 108 40W
Hoggar = Ahaggar, Algeria 50 D6 23 0N 6 30 E
Hogsty Reef, Bahamas 89 B5 21 41N 73 48W
Hoh →, U.S.A. 84 C2 47 45N 124 29W
Hohe Rhön, Germany 16 C4 50 24N 9 58 E
Hohe Venn, Belgium 15 D6 50 30N 6 5 E
Hohenwald, U.S.A. 77 H2 35 33N 87 33W
Hohhot, China 34 D6 40 52N 111 40 E
Hóhlakas, Greece 23 D9 35 57N 27 53 E
Hoi An, Vietnam 38 E7 15 30N 108 19 E
Hoi Xuan, Vietnam 38 B5 20 25N 105 9 E
Hoisington, U.S.A. 80 F5 38 31N 98 47W
Hōjō, Japan 31 H6 33 58N 132 46 E
Hokianga Harbour, N.Z. 59 F4 35 31S 173 22 E
Hokitika, N.Z. 59 K3 42 42S 171 0 E
Hokkaidō □, Japan 30 C11 43 30N 143 0 E
Holbrook, Australia 63 F4 35 42S 147 18 E
Holbrook, U.S.A. 83 J8 34 54N 110 10W
Holden, Canada 72 C6 53 13N 112 11W
Holden, U.S.A. 82 G7 39 6N 112 16W
Holdenville, U.S.A. 81 H6 35 5N 96 24W
Holderness, U.K. 10 D7 53 45N 0 5W
Holdfast, Canada 73 C7 50 58N 105 25W
Holdrege, U.S.A. 80 E5 40 26N 99 23W
Holguín, Cuba 88 B4 20 50N 76 20W
Hollams Bird I., Namibia 56 C1 24 40S 14 30 E
Holland, U.S.A. 76 D2 42 47N 86 7W
Hollandia = Jayapura, Indonesia 37 E10 2 28S 140 38 E
Hollidaysburg, U.S.A. 78 F6 40 26N 78 24W
Hollis, U.S.A. 81 H5 34 41N 99 55W
Hollister, Calif., U.S.A. 83 H3 36 51N 121 24W
Hollister, Idaho, U.S.A. 82 E6 42 21N 114 35W
Holly, U.S.A. 80 F3 38 3N 102 7W
Holly Hill, U.S.A. 77 L5 29 16N 81 3W
Holly Springs, U.S.A. 81 H10 34 46N 89 27W
Hollywood, Calif., U.S.A. 83 J4 34 7N 118 25W
Hollywood, Fla., U.S.A. 77 N5 26 1N 80 9W
Holman Island, Canada 68 A8 70 42N 117 41W
Hólmavík, Iceland 8 D3 65 42N 21 40W
Holmes Reefs, Australia 62 B4 16 27S 148 0 E
Holmsund, Sweden 8 E16 63 41N 20 20 E
Holroyd →, Australia 62 A3 14 10S 141 36 E
Holstebro, Denmark 9 H10 56 22N 8 37 E
Holsworthy, U.K. 11 G3 50 48N 4 21W
Holt, Iceland 8 E4 63 33N 19 48W
Holton, Canada 71 B8 54 31N 57 12W
Holton, U.S.A. 80 F7 39 28N 95 44W
Holtville, U.S.A. 85 N11 32 49N 115 23W
Holwerd, Neths. 15 A5 53 22N 5 54 E
Holy Cross, U.S.A. 68 B4 62 12N 159 46W
Holy I., Gwynedd, U.K. 10 D3 53 17N 4 37W
Holy I., Northumb., U.K. 10 B6 55 42N 1 48W
Holyhead, U.K. 10 D3 53 18N 4 38W
Holyoke, Colo., U.S.A. 80 E3 40 35N 102 18W
Holyoke, Mass., U.S.A. 79 D12 42 12N 72 37W
Holyrood, Canada 71 C9 47 27N 53 8W
Homa Bay, Kenya 54 C3 0 36S 34 30 E
Homa Bay □, Kenya 54 C3 0 50S 34 30 E
Homalin, Burma 41 G19 24 55N 95 0 E
Homand, Iran 45 C8 32 28N 59 37 E
Hombori, Mali 50 E4 15 20N 1 38W
Home B., Canada 69 B13 68 40N 67 10W
Home Hill, Australia 62 B4 19 43S 147 25 E
Homedale, U.S.A. 82 E5 43 37N 116 56W
Homer, Alaska, U.S.A. 68 C4 59 39N 151 33W
Homer, La., U.S.A. 81 J8 32 48N 93 4W
Homestead, Australia 62 C4 20 20S 145 40 E
Homestead, Fla., U.S.A. 77 N5 25 28N 80 29W
Homestead, Oreg., U.S.A. 82 D5 45 2N 116 51W
Homewood, U.S.A. 84 F6 39 4N 120 8W
Hominy, U.S.A. 81 G6 36 25N 96 24W
Homoine, Mozam. 57 C6 23 55S 35 8 E
Homs = Ḥimş, Syria 47 A5 34 40N 36 45 E

Ichchapuram

Homyel = Gomel, *Belorussia* **24 D5** 52 28N 31 0 E
Hon Chong, *Vietnam* **39 G5** 10 25N 104 30 E
Hon Me, *Vietnam* **38 C5** 19 23N 105 56 E
Hon Quan, *Vietnam* **39 G6** 11 40N 106 50 E
Honan = Henan □, *China* **34 G8** 34 0N 114 0 E
Honbetsu, *Japan* **30 C11** 43 7N 143 37 E
Honcut, *U.S.A.* **84 F5** 39 20N 121 32W
Honda, *Colombia* **92 B4** 5 12N 74 45W
Hondeklipbaai, *S. Africa* **56 E2** 30 19S 17 17 E
Hondo, *Japan* **31 H5** 32 27N 130 12 E
Hondo, *U.S.A.* **81 L5** 29 21N 99 9W
Hondo →, *Belize* **87 D7** 18 25N 88 21W
Honduras ■, *Cent. Amer.* **88 D2** 14 40N 86 30W
Honduras, G. de, *Caribbean* **88 C2** 16 50N 87 0W
Hønefoss, *Norway* **9 F11** 60 10N 10 18 E
Honesdale, *U.S.A.* **79 E9** 41 34N 75 16W
Honey L., *U.S.A.* **84 E6** 40 15N 120 19W
Honfleur, *France* **18 B4** 49 25N 0 13 E
Hong Gai, *Vietnam* **38 B6** 20 57N 107 5 E
Hong He →, *China* **34 H8** 32 25N 115 35 E
Hong Kong ■, *Asia* **33 D6** 22 11N 114 14 E
Hongchŏn, *S. Korea* **35 F14** 37 44N 127 53 E
Hongha →, *Vietnam* **32 D5** 22 0N 104 0 E
Hongjiang, *China* **33 D5** 27 7N 109 59 E
Hongliu He →, *China* **34 F5** 38 0N 109 50 E
Hongor, *Mongolia* **34 B7** 45 45N 112 50 E
Hongsa, *Laos* **38 C3** 19 43N 101 20 E
Hongshui He →, *China* **33 D5** 23 48N 109 30 E
Hongsŏng, *S. Korea* **35 F14** 36 37N 126 38 E
Hongtong, *China* **34 F6** 36 16N 111 40 E
Honguedo, Détroit d', *Canada* **71 C7** 49 15N 64 0W
Hongwon, *N. Korea* **35 E14** 40 0N 127 56 E
Hongze Hu, *China* **35 H10** 33 15N 118 35 E
Honiara, *Solomon Is.* **64 H7** 9 27S 159 57 E
Honiton, *U.K.* **11 G4** 50 48N 3 11W
Honjō, *Japan* **30 E10** 39 23N 140 3 E
Honolulu, *U.S.A.* **74 H16** 21 19N 157 52W
Honshū, *Japan* **31 G9** 36 0N 138 0 E
Hood, Mt., *U.S.A.* **82 D3** 45 23N 121 42W
Hood, Pt., *Australia* **61 F2** 34 23S 119 34 E
Hood River, *U.S.A.* **82 D3** 45 43N 121 31W
Hoodsport, *U.S.A.* **84 C3** 47 24N 123 9W
Hoogeveen, *Neths.* **15 B6** 52 44N 6 28 E
Hoogezand, *Neths.* **15 A6** 53 11N 6 45 E
Hooghly → = Hughli →, *India* **43 J13** 21 56N 88 4 E
Hooghly-Chinsura = Chunchura, *India* **43 H13** 22 53N 88 27 E
Hook Hd., *Ireland* **13 D5** 52 8N 6 57W
Hook I., *Australia* **62 C4** 20 4S 149 0 E
Hook of Holland = Hoek van Holland, *Neths.* **15 C4** 52 0N 4 7 E
Hooker, *U.S.A.* **81 G4** 36 52N 101 13W
Hooker Creek, *Australia* **60 C5** 18 23S 130 38 E
Hoopeston, *U.S.A.* **76 E2** 40 28N 87 40W
Hoopstad, *S. Africa* **56 D4** 27 50S 25 55 E
Hoorn, *Neths.* **15 B5** 52 38N 5 4 E
Hoover Dam, *U.S.A.* **85 K12** 36 1N 114 44W
Hooversville, *U.S.A.* **78 F6** 40 9N 78 55W
Hop Bottom, *U.S.A.* **79 E9** 41 42N 75 46W
Hope, *Canada* **72 D4** 49 25N 121 25W
Hope, *Ariz., U.S.A.* **85 M13** 33 43N 113 42W
Hope, *Ark., U.S.A.* **81 J8** 33 40N 93 36W
Hope, *N. Dak., U.S.A.* **80 B6** 47 19N 97 43W
Hope, L., *Australia* **63 D2** 28 24S 139 18 E
Hope Pt., *U.S.A.* **68 B3** 68 20N 166 50W
Hope Town, *Bahamas* **88 A4** 26 35N 76 57W
Hopedale, *Canada* **71 A7** 55 28N 60 13W
Hopefield, *S. Africa* **56 E2** 33 3S 18 22 E
Hopei = Hebei □, *China* **34 E9** 39 0N 116 0 E
Hopelchén, *Mexico* **87 D7** 19 46N 89 50W
Hopetoun, *Vic., Australia* **63 F3** 35 42S 142 22 E
Hopetoun, *W. Austral., Australia* **61 F3** 33 57S 120 7 E
Hopetown, *S. Africa* **56 D3** 29 34S 24 3 E
Hopkins, *U.S.A.* **80 E7** 40 33N 94 49W
Hopkins, L., *Australia* **60 D4** 24 15S 128 35 E
Hopkinsville, *U.S.A.* **77 G2** 36 52N 87 29W
Hopland, *U.S.A.* **84 G3** 38 58N 123 7W
Hoquiam, *U.S.A.* **84 D3** 46 59N 123 53W
Hordaland fylke □, *Norway* **9 F9** 60 25N 6 15 E
Horden Hills, *Australia* **60 D5** 20 15S 130 0 E
Horinger, *China* **34 D6** 40 28N 111 48 E
Horlick Mts., *Antarctica* **5 E15** 84 0S 102 0W
Horlivka = Gorlovka, *Ukraine* **25 E6** 48 19N 38 5 E
Hormoz, *Iran* **45 E7** 27 35N 55 0 E
Hormoz, Jaz. ye, *Iran* **45 E8** 27 8N 56 28 E
Hormuz Str. of, *The Gulf* **45 E8** 26 30N 56 30 E
Horn, *Austria* **16 D7** 48 39N 15 40 E
Horn, *Ísafjarðarsýsla, Iceland* **8 C2** 66 28N 22 28W
Horn, *Suður-Múlasýsla, Iceland* **8 D7** 65 10N 13 31W
Horn →, *Canada* **72 A5** 61 30N 118 1W
Horn, Cape = Hornos, C. de, *Chile* **96 H3** 55 50S 67 30W
Horn Head, *Ireland* **13 A3** 55 13N 8 0W
Horn I., *Australia* **62 A3** 10 37S 142 17 E
Horn I., *U.S.A.* **77 K1** 30 14N 88 39W
Horn Mts., *Canada* **72 A5** 62 15N 119 15W
Hornavan, *Sweden* **8 C14** 66 15N 17 30 E
Hornbeck, *U.S.A.* **81 K8** 31 20N 93 24W
Hornbrook, *U.S.A.* **82 F2** 41 55N 122 33W
Horncastle, *U.K.* **10 D7** 53 13N 0 8W
Hornell, *U.S.A.* **78 D7** 42 20N 77 40W
Hornell L., *Canada* **72 A5** 62 20N 119 25W
Hornepayne, *Canada* **70 C3** 49 14N 84 48W
Hornitos, *U.S.A.* **84 H6** 37 30N 120 14W
Hornos, C. de, *Chile* **96 H3** 55 50S 67 30W
Hornsby, *Australia* **63 E5** 33 42S 151 2 E
Hornsea, *U.K.* **10 D7** 53 55N 0 10W
Horobetsu, *Japan* **30 C10** 42 24N 141 6 E
Horqin Youyi Qianqi, *China* **35 A12** 46 5N 122 3 E
Horqueta, *Paraguay* **94 A4** 23 15S 56 55W
Horse Creek, *U.S.A.* **80 E3** 41 57N 105 10W
Horse Is., *Canada* **71 B8** 50 15N 55 50W
Horsefly L., *Canada* **72 C4** 52 25N 121 0W
Horsens, *Denmark* **9 J10** 55 52N 9 51 E
Horsham, *Australia* **63 F3** 36 44S 142 13 E
Horsham, *U.K.* **11 F7** 51 4N 0 20W
Horten, *Norway* **9 G11** 59 25N 10 32 E
Horton, *U.S.A.* **80 F7** 39 40N 95 32W
Horton →, *Canada* **68 B7** 69 56N 126 52W
Horwood, L., *Canada* **70 C3** 48 5N 82 20W
Hose, Gunung-Gunung, *Malaysia* **36 D4** 2 5N 114 6 E
Ḥoseynābād, *Khuzestān, Iran* **45 C6** 32 45N 48 20 E
Ḥoseynābād, *Kordestān, Iran* **44 C5** 35 33N 47 8 E
Hoshangabad, *India* **42 H7** 22 45N 77 45 E
Hoshiarpur, *India* **42 D6** 31 30N 75 58 E
Hosmer, *U.S.A.* **80 C5** 45 34N 99 28W
Hospet, *India* **40 M10** 15 15N 76 20 E
Hospitalet de Llobregat, *Spain* **19 B7** 41 21N 2 6 E
Hoste, I., *Chile* **96 H3** 55 0S 69 0W
Hot, *Thailand* **38 C2** 18 8N 98 29 E
Hot Creek Range, *U.S.A.* **82 G5** 38 40N 116 20W
Hot Springs, *Ark., U.S.A.* **81 H8** 34 31N 93 3W
Hot Springs, *S. Dak., U.S.A.* **80 D3** 43 26N 103 29W
Hotagen, *Sweden* **8 E13** 63 50N 14 30 E
Hotan, *China* **32 C2** 37 25N 79 55 E
Hotazel, *S. Africa* **56 D3** 27 17S 22 58 E
Hotchkiss, *U.S.A.* **83 G10** 38 48N 107 43W
Hotham, C., *Australia* **60 B5** 12 2S 131 18 E
Hoting, *Sweden* **8 D14** 64 8N 16 15 E
Hotte, Massif de la, *Haiti* **89 C5** 18 30N 73 45W
Hottentotsbaai, *Namibia* **56 D1** 26 8S 14 59 E
Houck, *U.S.A.* **83 J9** 35 20N 109 10W
Houei Sai, *Laos* **38 B3** 20 18N 100 26 E
Houffalize, *Belgium* **15 D5** 50 8N 5 48 E
Houghton, *U.S.A.* **80 B10** 47 7N 88 34W
Houghton L., *U.S.A.* **76 C3** 44 21N 84 44W
Houghton-le-Spring, *U.K.* **10 C6** 54 51N 1 28W
Houhora Heads, *N.Z.* **59 F4** 34 49S 173 9 E
Houlton, *U.S.A.* **71 C6** 46 8N 67 51W
Houma, *U.S.A.* **81 L9** 29 36N 90 43W
Houston, *Canada* **72 C3** 54 25N 126 39W
Houston, *Mo., U.S.A.* **81 G9** 37 22N 91 58W
Houston, *Tex., U.S.A.* **81 L7** 29 46N 95 22W
Houtman Abrolhos, *Australia* **61 E1** 28 43S 113 48 E
Hovd, *Mongolia* **32 B4** 48 2N 91 37 E
Hove, *U.K.* **11 G7** 50 50N 0 10W
Hoveyzeh, *Iran* **45 D6** 31 27N 48 4 E
Hövsgöl, *Mongolia* **34 C5** 43 37N 109 39 E
Hövsgöl Nuur, *Mongolia* **32 A5** 51 0N 100 30 E
Howard, *Australia* **63 D5** 25 16S 152 32 E
Howard, *Kans., U.S.A.* **81 G6** 37 28N 96 16W
Howard, *Pa., U.S.A.* **78 E7** 41 1N 77 40W
Howard, *S. Dak., U.S.A.* **80 C6** 44 1N 97 32W
Howard I., *Australia* **62 A2** 12 10S 135 24 E
Howard L., *Canada* **73 A7** 62 15N 105 57W
Howe, *U.S.A.* **82 E7** 43 48N 113 0W
Howe, C., *Australia* **63 F5** 37 30S 150 0 E
Howell, *U.S.A.* **76 D4** 42 36N 83 56W
Howick, *Canada* **79 A11** 45 11N 73 51W
Howick, *S. Africa* **57 D5** 29 28S 30 14 E
Howick Group, *Australia* **62 A4** 14 20S 145 30 E
Howitt, L., *Australia* **63 D2** 27 40S 138 40 E
Howley, *Canada* **71 C8** 49 12N 57 2W
Howrah = Haora, *India* **43 H13** 22 37N 88 20 E
Howth Hd., *Ireland* **13 C5** 53 21N 6 3W
Hoy, *U.K.* **12 C5** 58 50N 3 15W
Høyanger, *Norway* **9 F9** 61 13N 6 4 E
Hpungan Pass, *Burma* **41 F20** 27 30N 96 55 E
Hradec Králové, *Czech.* **16 C7** 50 15N 15 50 E
Hrodna = Grodno, *Belorussia* **24 D3** 53 42N 23 52 E
Hron →, *Slovakia* **17 E9** 47 49N 18 45 E
Hrvatska = Croatia ■, *Europe* **20 B7** 45 20N 16 0 E
Hsenwi, *Burma* **41 H20** 23 22N 97 55 E
Hsiamen = Xiamen, *China* **33 D6** 24 25N 118 4 E
Hsian = Xi'an, *China* **34 G5** 34 15N 109 0 E
Hsinhailien = Lianyungang, *China* **35 G10** 34 40N 119 11 E
Hsisha Chuntao, *Pac. Oc.* **36 A4** 15 50N 112 0 E
Hsüchou = Xuzhou, *China* **35 G9** 34 18N 117 10 E
Hu Xian, *China* **34 G5** 34 8N 108 42 E
Hua Hin, *Thailand* **38 F2** 12 34N 99 58 E
Hua Xian, *Henan, China* **34 G8** 35 30N 114 30 E
Hua Xian, *Shaanxi, China* **34 G5** 34 30N 109 48 E
Huachinera, *Mexico* **86 A3** 30 9N 108 55W
Huacho, *Peru* **92 F3** 11 10S 77 35W
Huachón, *Peru* **92 F3** 10 35S 76 0W
Huade, *China* **34 D7** 41 55N 113 59 E
Huadian, *China* **35 C14** 43 0N 126 40 E
Huai He →, *China* **33 C6** 33 0N 118 30 E
Huai Yot, *Thailand* **39 J2** 7 45N 99 37 E
Huai'an, *Hebei, China* **34 D8** 40 30N 114 20 E
Huai'an, *Jiangsu, China* **35 H10** 33 30N 119 10 E
Huaide, *China* **35 C13** 43 30N 124 40 E
Huaidezhen, *China* **35 C13** 43 48N 124 50 E
Huainan, *China* **33 C6** 32 38N 116 58 E
Huairen, *China* **34 E7** 39 48N 113 20 E
Huairou, *China* **34 D9** 40 20N 116 35 E
Huaiyang, *China* **34 H8** 33 40N 114 52 E
Huaiyuan, *China* **35 H9** 32 55N 117 10 E
Huajianzi, *China* **35 D13** 41 23N 125 20 E
Huajuapan de Leon, *Mexico* **87 D5** 17 50N 97 48W
Hualapai Peak, *U.S.A.* **83 J7** 35 5N 113 54W
Huallaga →, *Peru* **92 E3** 5 15S 75 30W
Huambo, *Angola* **53 G3** 12 42S 15 54 E
Huan Jiang →, *China* **34 G5** 34 28N 109 0 E
Huan Xian, *China* **34 F4** 36 33N 107 7 E
Huancabamba, *Peru* **92 E3** 5 10S 79 15W
Huancane, *Peru* **92 G5** 15 10S 69 44W
Huancapi, *Peru* **92 F4** 13 40S 74 0W
Huancavelica, *Peru* **92 F3** 12 50S 75 5W
Huancayo, *Peru* **92 F3** 12 5S 75 12W
Huang Hai = Yellow Sea, *China* **35 G12** 35 0N 123 0 E
Huang He →, *China* **35 F10** 37 55N 118 50 E
Huang Xian, *China* **35 F11** 37 38N 120 30 E
Huangling, *China* **34 G5** 35 34N 109 15 E
Huanglong, *China* **34 G5** 35 30N 109 59 E
Huangshi, *China* **33 C6** 30 10N 115 3 E
Huangsongdian, *China* **35 C14** 43 45N 127 25 E
Huantai, *China* **35 F9** 36 58N 117 56 E
Huánuco, *Peru* **92 E3** 9 55S 76 15W
Huaraz, *Peru* **92 E3** 9 30S 77 32W
Huarmey, *Peru* **92 F3** 10 5S 78 5W
Huascarán, *Peru* **92 E3** 9 8S 77 36W
Huasco, *Chile* **94 B1** 28 30S 71 15W
Huasco →, *Chile* **94 B1** 28 27S 71 13W
Huasna, *U.S.A.* **85 K6** 35 6N 120 24W
Huatabampo, *Mexico* **86 B3** 26 50N 109 50W
Huauchinango, *Mexico* **87 C5** 20 11N 98 3W
Huautla de Jiménez, *Mexico* **87 D5** 18 8N 96 51W
Huay Namota, *Mexico* **86 C4** 21 56N 104 30W
Huayin, *China* **34 G6** 34 35N 110 5 E
Huayllay, *Peru* **92 F3** 11 3S 76 21W
Hubbard, *U.S.A.* **81 K6** 31 51N 96 48W
Hubbart Pt., *Canada* **73 B10** 59 21N 94 41W
Hubei □, *China* **33 C6** 31 0N 112 0 E
Hubli-Dharwad = Dharwad, *India* **40 M9** 15 22N 75 15 E
Huchang, *N. Korea* **35 D14** 41 25N 127 2 E
Huddersfield, *U.K.* **10 D6** 53 38N 1 49W
Hudiksvall, *Sweden* **9 F14** 61 43N 17 10 E
Hudson, *Canada* **73 C10** 50 6N 92 9W
Hudson, *Mass., U.S.A.* **79 D13** 42 23N 71 34W
Hudson, *Mich., U.S.A.* **76 E3** 41 51N 84 21W
Hudson, *N.Y., U.S.A.* **79 D11** 42 15N 73 46W
Hudson, *Wis., U.S.A.* **80 C8** 44 58N 92 45W
Hudson, *Wyo., U.S.A.* **82 E9** 42 54N 108 35W
Hudson →, *U.S.A.* **79 F10** 40 42N 74 2W
Hudson Bay, *N.W.T., Canada* **69 C11** 60 0N 86 0W
Hudson Bay, *Sask., Canada* **73 C8** 52 51N 102 23W
Hudson Falls, *U.S.A.* **79 C11** 43 18N 73 35W
Hudson Mts., *Antarctica* **5 D16** 74 32S 99 20W
Hudson Str., *Canada* **69 B13** 62 0N 70 0W
Hudson's Hope, *Canada* **72 B4** 56 0N 121 54W
Hue, *Vietnam* **38 D6** 16 30N 107 35 E
Huehuetenango, *Guatemala* **88 C1** 15 20N 91 28W
Huejúcar, *Mexico* **86 C4** 22 21N 103 13W
Huelva, *Spain* **19 D2** 37 18N 6 57W
Huentelauquén, *Chile* **94 C1** 31 38S 71 33W
Huerta, Sa. de la, *Argentina* **94 C2** 31 10S 67 30W
Huesca, *Spain* **19 A5** 42 8N 0 25W
Huetamo, *Mexico* **86 D4** 18 36N 100 54W
Hugh →, *Australia* **62 D1** 25 1S 134 1 E
Hughenden, *Australia* **62 C3** 20 52S 144 10 E
Hughes, *Australia* **61 F4** 30 42S 129 31 E
Hughli →, *India* **43 J13** 21 56N 88 4 E
Hugo, *U.S.A.* **80 F3** 39 8N 103 28W
Hugoton, *U.S.A.* **81 G4** 37 11N 101 21W
Hui Xian, *Gansu, China* **34 H4** 33 50N 106 4 E
Hui Xian, *Henan, China* **34 G7** 35 27N 113 12 E
Hui'anbu, *China* **34 F4** 37 28N 106 38 E
Huichapán, *Mexico* **87 C5** 20 24N 99 40W
Huifa He →, *China* **35 C14** 43 0N 127 50 E
Huila, Nevado del, *Colombia* **92 C3** 3 0N 76 0W
Huimin, *China* **35 F9** 37 27N 117 28 E
Huinan, *China* **35 C14** 42 40N 126 2 E
Huinca Renancó, *Argentina* **94 C3** 34 51S 64 22W
Huining, *China* **34 G3** 35 38N 105 0 E
Huinong, *China* **34 E4** 39 5N 106 35 E
Huiting, *China* **34 G9** 34 5N 116 5 E
Huixtla, *Mexico* **87 D6** 15 9N 92 28W
Huize, *China* **32 D5** 26 24N 103 15 E
Hukawng Valley, *Burma* **41 F20** 26 30N 96 30 E
Hukuntsi, *Botswana* **56 C3** 23 58S 21 45 E
Ḥulayfā', *Si. Arabia* **44 E4** 25 58N 40 45 E
Huld, *Mongolia* **34 B3** 45 5N 105 30 E
Hulin He →, *China* **35 B12** 45 0N 122 10 E
Hull = Kingston upon Hull, *U.K.* **10 D7** 53 45N 0 20W
Hull, *Canada* **70 C4** 45 25N 75 44W
Hull →, *U.K.* **10 D7** 53 43N 0 25W
Hulst, *Neths.* **15 C4** 51 17N 4 2 E
Hulun Nur, *China* **33 B6** 49 0N 117 30 E
Humahuaca, *Argentina* **94 A2** 23 10S 65 25W
Humaitá, *Brazil* **92 E6** 7 35S 63 1W
Humaitá, *Paraguay* **94 B4** 27 2S 58 31W
Humansdorp, *S. Africa* **56 E3** 34 2S 24 46 E
Humbe, *Angola* **56 B1** 16 40S 14 55 E
Humber →, *U.K.* **10 D7** 53 40N 0 10W
Humberside □, *U.K.* **10 D7** 53 50N 0 30W
Humbert River, *Australia* **60 C5** 16 30S 130 45 E
Humble, *U.S.A.* **81 L8** 29 59N 93 18W
Humboldt, *Canada* **73 C7** 52 15N 105 9W
Humboldt, *Iowa, U.S.A.* **80 D7** 42 44N 94 13W
Humboldt, *Tenn., U.S.A.* **81 H10** 35 50N 88 55W
Humboldt →, *U.S.A.* **82 F4** 39 59N 118 36W
Humboldt Gletscher, *Greenland* **4 B4** 79 30N 62 0W
Hume, *U.S.A.* **84 J8** 36 48N 118 54W
Hume, L., *Australia* **63 F4** 36 0S 147 5 E
Humphreys, Mt., *U.S.A.* **84 H8** 37 17N 118 40W
Humphreys Peak, *U.S.A.* **83 J8** 35 21N 111 41W
Humptulips, *U.S.A.* **84 C3** 47 14N 123 57W
Hūn, *Libya* **51 C8** 29 2N 16 0 E
Hun Jiang →, *China* **35 D13** 40 50N 125 38 E
Húnaflói, *Iceland* **8 D3** 65 50N 20 50W
Hunan □, *China* **33 D6** 27 30N 112 0 E
Hunchun, *China* **35 C16** 42 52N 130 28 E
Hundred Mile House, *Canada* **72 C4** 51 38N 121 18W
Hunedoara, *Romania* **17 F11** 45 40N 22 50 E
Hung Yen, *Vietnam* **38 B6** 20 39N 106 4 E
Hungary ■, *Europe* **17 E9** 47 20N 19 20 E
Hungary, Plain of, *Europe* **6 F9** 47 0N 20 0 E
Hungerford, *Australia* **63 D3** 28 58S 144 24 E
Hŭngnam, *N. Korea* **35 E14** 39 49N 127 45 E
Hunsberge, *Namibia* **56 D2** 27 45S 17 12 E
Hunsrück, *Germany* **16 D3** 49 30N 7 0 E
Hunstanton, *U.K.* **10 E8** 52 57N 0 30 E
Hunter, *N. Dak., U.S.A.* **80 B6** 47 12N 97 13W
Hunter, *N.Y., U.S.A.* **79 D10** 42 13N 74 13W
Hunter I., *Australia* **62 G3** 40 30S 144 45 E
Hunter I., *Canada* **72 C3** 51 55N 128 0W
Hunter Ra., *Australia* **63 E5** 32 45S 150 15 E
Hunters Road, *Zimbabwe* **55 F2** 19 9S 29 49 E
Hunterville, *N.Z.* **59 H5** 39 56S 175 35 E
Huntingburg, *U.S.A.* **76 F2** 38 18N 86 57W
Huntingdon, *Canada* **70 C5** 45 6N 74 10W
Huntingdon, *U.K.* **11 E7** 52 20N 0 11W
Huntingdon, *U.S.A.* **78 F6** 40 30N 78 1W
Huntington, *Ind., U.S.A.* **76 E3** 40 53N 85 30W
Huntington, *N.Y., U.S.A.* **79 F11** 40 52N 73 26W
Huntington, *Oreg., U.S.A.* **82 D5** 44 21N 117 16W
Huntington, *Utah, U.S.A.* **82 G8** 39 20N 110 58W
Huntington, *W. Va., U.S.A.* **76 F4** 38 25N 82 27W
Huntington Beach, *U.S.A.* **85 M8** 33 40N 118 5W
Huntington Park, *U.S.A.* **83 K4** 33 58N 118 15W
Huntly, *N.Z.* **59 G5** 37 34S 175 11 E
Huntly, *U.K.* **12 D6** 57 27N 2 48W
Huntsville, *Canada* **70 C4** 45 20N 79 14W
Huntsville, *Ala., U.S.A.* **77 H2** 34 44N 86 35W
Huntsville, *Tex., U.S.A.* **81 K7** 30 43N 95 33W
Hunyani →, *Zimbabwe* **55 F3** 15 57S 30 39 E
Hunyuan, *China* **34 E7** 39 42N 113 42 E
Hunza →, *India* **43 B6** 35 54N 74 20 E
Huo Xian, *China* **34 F6** 36 36N 111 42 E
Huong Hoa, *Vietnam* **38 D6** 16 37N 106 45 E
Huong Khe, *Vietnam* **38 C5** 18 13N 105 41 E
Huonville, *Australia* **62 G4** 43 0S 147 5 E
Hupeh = Hubei □, *China* **33 C6** 31 0N 112 0 E
Ḥūr, *Iran* **45 D8** 30 50N 57 7 E
Hure Qi, *China* **35 C11** 42 45N 121 45 E
Hurley, *N. Mex., U.S.A.* **83 K9** 32 42N 108 8W
Hurley, *Wis., U.S.A.* **80 B9** 46 27N 90 11W
Huron, *Calif., U.S.A.* **84 J6** 36 12N 120 6W
Huron, *Ohio, U.S.A.* **78 E2** 41 24N 82 33W
Huron, *S. Dak., U.S.A.* **80 C5** 44 22N 98 13W
Huron, L., *U.S.A.* **78 B2** 44 30N 82 40W
Hurricane, *U.S.A.* **83 H7** 37 11N 113 17W
Hurunui →, *N.Z.* **59 K4** 42 54S 173 18 E
Húsavík, *Iceland* **8 C5** 66 3N 17 21W
Huskvarna, *Sweden* **9 H13** 57 47N 14 15 E
Hussar, *Canada* **72 C6** 51 3N 112 41W
Hutchinson, *Kans., U.S.A.* **81 F6** 38 5N 97 56W
Hutchinson, *Minn., U.S.A.* **80 C7** 44 54N 94 22W
Huttig, *U.S.A.* **81 J8** 33 2N 92 11W
Hutton, Mt., *Australia* **63 D4** 25 51S 148 20 E
Huy, *Belgium* **15 D5** 50 31N 5 15 E
Hvammur, *Iceland* **8 D3** 65 13N 21 49W
Hvar, *Croatia* **20 C7** 43 11N 16 28 E
Hvítá, *Iceland* **8 D3** 64 40N 21 5W
Hvítá →, *Iceland* **8 D3** 64 30N 21 58W
Hvítárvatn, *Iceland* **8 D4** 64 37N 19 50W
Hwachon-chosuji, *S. Korea* **35 E14** 38 5N 127 50 E
Hwang Ho = Huang He →, *China* **35 F10** 37 55N 118 50 E
Hwange, *Zimbabwe* **55 F2** 18 18S 26 30 E
Hwange Nat. Park, *Zimbabwe* **56 B4** 19 0S 26 30 E
Hyannis, *U.S.A.* **80 E4** 42 0N 101 46W
Hyargas Nuur, *Mongolia* **32 B4** 49 0N 93 0 E
Hyden, *Australia* **61 F2** 32 24S 118 53 E
Hyderabad, *India* **40 L11** 17 22N 78 29 E
Hyderabad, *Pakistan* **42 G3** 25 23N 68 24 E
Hyères, *France* **18 E7** 43 8N 6 9 E
Hyesan, *N. Korea* **35 D15** 41 20N 128 10 E
Hyland →, *Canada* **72 B3** 59 52N 128 12W
Hymia, *India* **43 C8** 33 40N 78 2 E
Hyndman Peak, *U.S.A.* **82 E6** 43 45N 114 8W
Hyōgo □, *Japan* **31 G7** 35 15N 134 50 E
Hyrum, *U.S.A.* **82 F8** 41 38N 111 51W
Hysham, *U.S.A.* **82 C10** 46 18N 107 14W
Hythe, *U.K.* **11 F9** 51 4N 1 5 E
Hyūga, *Japan* **31 H5** 32 25N 131 35 E
Hyvinge = Hyvinkää, *Finland* **9 F18** 60 38N 24 50 E
Hyvinkää, *Finland* **9 F18** 60 38N 24 50 E

I

I-n-Gall, *Niger* **50 E6** 16 51N 7 1 E
Iaco →, *Brazil* **92 E5** 9 3S 68 34W
Iakora, *Madag.* **57 C8** 23 6S 46 40 E
Ialomiţa →, *Romania* **21 B12** 44 42N 27 51 E
Iaşi, *Romania* **17 E13** 47 10N 27 40 E
Iba, *Phil.* **37 A6** 15 22N 120 0 E
Ibadan, *Nigeria* **50 G5** 7 22N 3 58 E
Ibagué, *Colombia* **92 C3** 4 20N 75 20W
Ibar →, *Serbia, Yug.* **21 C9** 43 43N 20 45 E
Ibaraki □, *Japan* **31 F10** 36 10N 140 10 E
Ibarra, *Ecuador* **92 C3** 0 21N 78 7W
Ibembo, *Zaïre* **54 B1** 2 35N 23 35 E
Ibera, L., *Argentina* **94 B4** 28 30S 57 9W
Iberian Peninsula, *Europe* **6 G5** 40 0N 5 0W
Iberville, *Canada* **70 C5** 45 19N 73 17W
Iberville, Lac d', *Canada* **70 A5** 55 55N 73 15W
Ibi, *Nigeria* **50 G6** 8 15N 9 44 E
Ibiá, *Brazil* **93 G9** 19 30S 46 30W
Ibicuy, *Argentina* **94 C4** 33 55S 59 10W
Ibioapaba, Sa. da, *Brazil* **93 D10** 4 0S 41 30W
Ibiza, *Spain* **22 C7** 38 54N 1 26 E
Ibo, *Mozam.* **55 E5** 12 22S 40 40 E
Ibonma, *Indonesia* **37 E8** 3 29S 133 31 E
Ibotirama, *Brazil* **93 F10** 12 13S 43 12W
Ibrāhīm →, *Lebanon* **47 A4** 34 4N 35 38 E
Ibu, *Indonesia* **37 D7** 1 35N 127 33 E
Ibusuki, *Japan* **31 J5** 31 12N 130 40 E
Icá, *Peru* **92 F3** 14 0S 75 48W
Içá →, *Brazil* **92 D5** 2 55S 67 58W
Içana, *Brazil* **92 C5** 0 21N 67 19W
Içel = Mersin, *Turkey* **25 G5** 36 51N 34 36 E
Iceland ■, *Europe* **8 D4** 64 45N 19 0W
Icha, *Russia* **27 D16** 55 30N 156 0 E
Ich'ang = Yichang, *China* **33 C6** 30 40N 111 20 E
Ichchapuram, *India* **41 K14** 19 10N 84 40 E

Ichihara

Ichihara, Japan 31 G10 35 28N 140 5 E
Ichikawa, Japan 31 G9 35 44N 139 55 E
Ichilo →, Bolivia 92 G6 15 57S 64 50W
Ichinohe, Japan 30 D10 40 13N 141 17 E
Ichinomiya, Japan 31 G8 35 18N 136 48 E
Ichinoseki, Japan 30 E10 38 55N 141 8 E
Ichŏn, S. Korea 35 F14 37 17N 127 27 E
Icod, Canary Is. 22 F3 28 22N 16 43W
Icy Str., U.S.A. 72 B1 58 20N 135 30W
Ida Grove, U.S.A. 80 D7 42 21N 95 28W
Ida Valley, Australia 61 E3 28 42S 120 29 E
Idabel, U.S.A. 81 J7 33 54N 94 50W
Idaho □, U.S.A. 82 D6 45 0N 115 0W
Idaho City, U.S.A. 82 E6 43 50N 115 50W
Idaho Falls, U.S.A. 82 E7 43 30N 112 2W
Idaho Springs, U.S.A. 82 G11 39 45N 105 31W
Idd el Ghanam, Sudan 51 F9 11 30N 24 19 E
Iddan, Somali Rep. 46 F4 6 10N 48 55 E
Idehan, Libya 51 C7 27 10N 11 30 E
Idehan Marzūq, Libya 51 D7 24 50N 13 51 E
Idelès, Algeria 50 D6 23 50N 5 53 E
Idfû, Egypt 51 D11 24 55N 32 49 E
Ídhi Óros, Greece 23 D6 35 15N 24 45 E
Ídhra, Greece 21 F10 37 20N 23 28 E
Idi, Indonesia 36 C1 5 2N 97 37 E
Idiofa, Zaïre 52 E3 4 55S 19 42 E
Idlib, Syria 44 C3 35 55N 36 36 E
Idria, U.S.A. 84 J6 36 25N 120 41W
Idutywa, S. Africa 57 E4 32 8S 28 18 E
Ieper, Belgium 15 D2 50 51N 2 53 E
Ierápetra, Greece 23 E7 35 0N 25 44 E
Ierzu, Italy 20 E3 39 48N 9 32 E
'Ifâl, W. al →, Si. Arabia 44 D2 28 7N 35 3 E
Ifanadiana, Madag. 57 C8 21 19S 47 39 E
Ife, Nigeria 50 G5 7 30N 4 31 E
Iffley, Australia 62 B3 18 53S 141 12 E
Ifni, Morocco 50 C2 29 29N 10 12W
Iforas, Adrar des, Mali 50 E5 19 40N 1 40 E
Ifould, L., Australia 61 F5 30 52S 132 6 E
Iganga, Uganda 54 B3 0 37N 33 28 E
Igarapava, Brazil 93 H9 20 3S 47 47W
Igarapé Açu, Brazil 93 D9 1 4S 47 33W
Igarka, Russia 26 C9 67 30N 86 33 E
Igatimi, Paraguay 95 A4 24 5S 55 40W
Igbetti, Nigeria 50 G5 8 44N 4 8 E
Iggesund, Sweden 9 F14 61 39N 17 10 E
Iglésias, Italy 20 E3 39 19N 8 27 E
Igli, Algeria 50 B4 30 25N 2 19W
Igloolik, Canada 69 B11 69 20N 81 49W
Ignace, Canada 70 C1 49 30N 91 40W
Iguaçu →, Brazil 95 B5 25 36S 54 36W
Iguaçu, Cat. del, Brazil 95 B5 25 41S 54 26W
Iguaçu Falls = Iguaçu, Cat. del, Brazil 95 B5 25 41S 54 26W
Iguala, Mexico 87 D5 18 20N 99 40W
Igualada, Spain 19 B6 41 37N 1 37 E
Iguassu = Iguaçu →, Brazil 95 B5 25 36S 54 36W
Iguatu, Brazil 93 E11 6 20S 39 18W
Iguéla, Gabon 52 E1 2 0S 9 16 E
Igunga □, Tanzania 54 C3 4 20S 33 45 E
Iheya-Shima, Japan 31 L3 27 4N 127 58 E
Ihosy, Madag. 57 C8 22 24S 46 8 E
Ihotry, L., Madag. 57 C7 21 56S 43 41 E
Ii, Finland 8 D18 65 19N 25 22 E
Ii-Shima, Japan 31 L3 26 43N 127 47 E
Iida, Japan 31 G8 35 35N 137 50 E
Iijoki →, Finland 8 D18 65 20N 25 20 E
Iisalmi, Finland 8 E19 63 32N 27 10 E
Iiyama, Japan 31 F9 36 51N 138 22 E
Iizuka, Japan 31 H5 33 38N 130 42 E
Ijebu-Ode, Nigeria 50 G5 6 47N 3 58 E
IJmuiden, Neths. 15 B4 52 28N 4 35 E
IJssel →, Neths. 15 B5 52 35N 5 50 E
IJsselmeer, Neths. 15 B5 52 45N 5 20 E
Ijuí →, Brazil 95 B4 27 58S 55 20W
Ikaría, Greece 21 F12 37 35N 26 10 E
Ikeda, Japan 31 G6 34 1N 133 48 E
Ikela, Zaïre 52 E4 1 6S 23 6 E
Iki, Japan 31 H4 33 45N 129 42 E
Ikimba L., Tanzania 54 C3 1 30S 31 20 E
Ikopa →, Madag. 57 B8 16 45S 46 40 E
Ikungu, Tanzania 54 C3 1 33S 33 42 E
Ilagan, Phil. 37 A6 17 7N 121 53 E
Īlām, Iran 44 C5 33 0N 46 0 E
Ilam, Nepal 43 F12 26 58N 87 58 E
Ilanskiy, Russia 27 D10 56 14N 96 3 E
Ilbilbie, Australia 62 C4 21 45S 149 20 E
Ile-à-la-Crosse, Canada 73 B7 55 27N 107 53W
Ile-à-la-Crosse, Lac, Canada 73 B7 55 40N 107 45W
Île-de-France, France 18 B5 49 0N 2 20 E
Ilebo, Zaïre 52 E4 4 17S 20 55 E
Ileje □, Tanzania 55 D3 9 30S 33 25 E
Ilek, Russia 26 D6 51 32N 53 21 E
Ilek →, Russia 24 D9 51 30N 53 22 E
Ilford, Canada 73 B9 56 4N 95 35W
Ilfracombe, Australia 62 C3 23 30S 144 30 E
Ilfracombe, U.K. 11 F3 51 13N 4 8W
Ilhéus, Brazil 93 F11 14 49S 39 2W
Ili →, Kazakhstan 26 E8 45 53N 77 10 E
Ilich, Kazakhstan 26 E7 40 50N 68 27 E
Iliff, U.S.A. 80 E3 40 45N 103 4W
Iligan, Phil. 37 C6 8 12N 124 13 E
Iliodhrómia, Greece 21 E10 39 12N 23 50 E
Ilion, U.S.A. 79 D9 43 1N 75 2W
Ilkeston, U.K. 10 E6 52 59N 1 19W
Illampu = Ancohuma, Nevada, Bolivia 92 G5 16 0S 68 50W
Illana B., Phil. 37 C6 7 35N 123 45 E
Illapel, Chile 94 C1 32 0S 71 10W
Ille-et-Vilaine □, France 18 B3 48 10N 1 30W
Iller →, Germany 16 D4 48 23N 9 58 E
Illetas, Spain 22 B9 39 32N 2 35 E
Illimani, Bolivia 92 G5 16 30S 67 50W
Illinois □, U.S.A. 75 C9 40 15N 89 30W
Illinois →, U.S.A. 75 C8 38 58N 90 28W
Ilmen, Oz., Russia 24 C5 58 15N 31 10 E
Ilo, Peru 92 G4 17 40S 71 20W
Iloilo, Phil. 37 B6 10 45N 122 33 E
Ilorin, Nigeria 50 G5 8 30N 4 35 E
Ilwaco, U.S.A. 84 D2 46 19N 124 3W
Ilwaki, Indonesia 37 F7 7 55S 126 30 E
Imabari, Japan 31 G6 34 4N 133 0 E
Imaloto →, Madag. 57 C8 23 27S 45 13 E
Imandra, Oz., Russia 24 A5 67 30N 33 0 E
Imari, Japan 31 H4 33 15N 129 52 E
Imbler, U.S.A. 82 D5 45 28N 117 58W
imeni 26 Bakinskikh Komissarov, Azerbaijan 25 G8 39 19N 49 12 E
imeni 26 Bakinskikh Komissarov, Turkmenistan 25 G9 39 22N 54 10 E
Imeni Poliny Osipenko, Russia 27 D14 52 30N 136 29 E
Imeri, Serra, Brazil 92 C5 0 50N 65 25W
Imerimandroso, Madag. 57 B8 17 26S 48 35 E
Imi, Ethiopia 46 F3 6 28N 42 10 E
Imlay, U.S.A. 82 F4 40 40N 118 9W
Imlay City, U.S.A. 78 C1 43 2N 83 5W
Immingham, U.K. 10 D7 53 37N 0 12W
Immokalee, U.S.A. 77 M5 26 25N 81 25W
Imola, Italy 20 B4 44 20N 11 42 E
Imperatriz, Brazil 93 E9 5 30S 47 29W
Impéria, Italy 20 C3 43 52N 8 3 E
Imperial, Canada 73 C7 51 21N 105 28W
Imperial, Calif., U.S.A. 85 N11 32 51N 115 34W
Imperial, Nebr., U.S.A. 80 E4 40 31N 101 39W
Imperial Beach, U.S.A. 85 N9 32 35N 117 8W
Imperial Dam, U.S.A. 85 N12 32 55N 114 25W
Imperial Reservoir, U.S.A. 85 N12 32 53N 114 28W
Imperial Valley, U.S.A. 85 N11 33 0N 115 30W
Imperieuse Reef, Australia 60 C2 17 36S 118 50 E
Impfondo, Congo 52 D3 1 40N 18 0 E
Imphal, India 41 G18 24 48N 93 56 E
Imuruan B., Phil. 37 B5 10 40N 119 10 E
In Belbel, Algeria 50 C5 27 55N 1 12 E
In Salah, Algeria 50 C5 27 10N 2 32 E
Ina, Japan 31 G8 35 50N 137 55 E
Inangahua Junction, N.Z. 59 J3 41 52S 171 59 E
Inanwatan, Indonesia 37 E8 2 10S 132 14 E
Iñapari, Peru 92 F5 11 0S 69 40W
Inari, Finland 8 B19 68 54N 27 5 E
Inarijärvi, Finland 8 B19 69 0N 28 0 E
Inawashiro-Ko, Japan 30 F10 37 29N 140 6 E
Inca, Spain 22 B9 39 43N 2 54 E
Incaguasi, Chile 94 B1 29 12S 71 5W
İnce-Burnu, Turkey 25 F5 42 7N 34 56 E
Inchon, S. Korea 35 F14 37 27N 126 40 E
Incomáti →, Mozam. 57 D5 25 46S 32 43 E
Indalsälven →, Sweden 8 E14 62 36N 17 30 E
Indaw, Burma 41 G20 24 15N 96 5 E
Independence, Calif., U.S.A. 83 H4 36 48N 118 12W
Independence, Iowa, U.S.A. 80 D9 42 28N 91 54W
Independence, Kans., U.S.A. 81 G7 37 14N 95 42W
Independence, Mo., U.S.A. 80 F7 39 6N 94 25W
Independence, Oreg., U.S.A. 82 D2 44 51N 123 11W
Independence Fjord, Greenland 4 A6 82 10N 29 0W
Independence Mts., U.S.A. 82 F5 41 20N 116 0W
Index, U.S.A. 84 C5 47 50N 121 33W
India ■, Asia 40 K11 20 0N 78 0 E
Indian →, U.S.A. 77 M5 27 59N 80 34W
Indian Cabins, Canada 72 B5 59 52N 117 40W
Indian Harbour, Canada 71 B8 54 27N 57 13W
Indian Head, Canada 73 C8 50 30N 103 41W
Indian Ocean 28 K11 5 0S 75 0 E
Indian Springs, U.S.A. 85 J11 36 35N 115 40W
Indiana, U.S.A. 78 F5 40 37N 79 9W
Indiana □, U.S.A. 76 E3 40 0N 86 0W
Indianapolis, U.S.A. 76 F2 39 46N 86 9W
Indianola, Iowa, U.S.A. 80 E8 41 22N 93 34W
Indianola, Miss., U.S.A. 81 J9 33 27N 90 39W
Indiga, Russia 24 A8 67 50N 48 50 E
Indigirka →, Russia 27 B15 70 48N 148 54 E
Indio, U.S.A. 85 M10 33 43N 116 13W
Indonesia ■, Asia 36 F5 5 0S 115 0 E
Indore, India 42 H6 22 42N 75 53 E
Indramayu, Indonesia 37 G13 6 20S 108 19 E
Indravati →, India 41 K12 19 20N 80 20 E
Indre □, France 18 C4 46 50N 1 39 E
Indre-et-Loire □, France 18 C4 47 20N 0 40 E
Indus →, Pakistan 42 G2 24 20N 67 47 E
Indus, Mouth of the, Pakistan 42 H2 24 0N 68 0 E
İnebolu, Turkey 25 F5 41 55N 33 40 E
Infiernillo, Presa del, Mexico 86 D4 18 9N 102 0W
Ingende, Zaïre 52 E3 0 12S 18 57 E
Ingenio, Canary Is. 22 G4 27 55N 15 26W
Ingenio Santa Ana, Argentina 94 B2 27 25S 65 40W
Ingersoll, Canada 78 C4 43 4N 80 55W
Ingham, Australia 62 B4 18 43S 146 10 E
Ingleborough, U.K. 10 C5 54 11N 2 23W
Inglewood, Queens., Australia 63 D5 28 25S 151 2 E
Inglewood, Vic., Australia 63 F3 36 29S 143 53 E
Inglewood, N.Z. 59 H5 39 9S 174 14 E
Inglewood, U.S.A. 85 M8 33 58N 118 21W
Ingólfshöfði, Iceland 8 E5 63 48N 16 39W
Ingolstadt, Germany 16 D5 48 45N 11 26 E
Ingomar, U.S.A. 82 C10 46 35N 107 23W
Ingonish, Canada 71 C7 46 42N 60 18W
Ingraj Bazar, India 43 G13 24 58N 88 10 E
Ingrid Christensen Coast, Antarctica 5 C6 69 30S 76 0 E
Ingulec, Ukraine 25 E5 47 42N 33 14 E
Ingwavuma, S. Africa 57 D5 27 9S 31 59 E
Inhaca, I., Mozam. 57 D5 26 1S 32 57 E
Inhafenga, Mozam. 57 C5 20 36S 33 53 E
Inhambane, Mozam. 57 C6 23 54S 35 30 E
Inhambane □, Mozam. 57 C5 22 30S 34 20 E
Inhaminga, Mozam. 55 F4 18 26S 35 0 E
Inharrime, Mozam. 57 C6 24 30S 35 0 E
Inharrime →, Mozam. 57 C6 24 30S 35 0 E
Ining = Yining, China 26 E9 43 58N 81 10 E
Inírida →, Colombia 92 C5 3 55N 67 52W
Inishbofin, Ireland 13 C1 53 35N 10 12W
Inishmore, Ireland 13 C2 53 8N 9 45W
Inishowen, Ireland 13 A4 55 14N 7 15W
Injune, Australia 63 D4 25 53S 148 32 E
Inklin, Canada 72 B2 58 56N 133 5W
Inklin →, Canada 72 B2 58 50N 133 10W
Inkom, U.S.A. 82 E7 42 48N 112 15W
Inle L., Burma 41 J20 20 30N 96 58 E
Inn →, Austria 16 D6 48 35N 13 28 E
Innamincka, Australia 63 D3 27 44S 140 46 E
Inner Hebrides, U.K. 12 D2 57 0N 6 30W
Inner Mongolia = Nei Monggol Zizhiqu □, China 34 C6 42 0N 112 0 E
Inner Sound, U.K. 12 D3 57 30N 5 55W
Innerkip, Canada 78 C4 43 13N 80 42W
Innetalling I., Canada 70 A4 56 0N 79 0W
Innisfail, Australia 62 B4 17 33S 146 5 E
Innisfail, Canada 72 C6 52 0N 113 57W
In'no-shima, Japan 31 G6 34 19N 133 10 E
Innsbruck, Austria 16 E5 47 16N 11 23 E
Inny →, Ireland 13 C4 53 30N 7 50W
Inongo, Zaïre 52 E3 1 55S 18 30 E
Inoucdjouac, Canada 69 C12 58 25N 78 15W
Inowrocław, Poland 17 B9 52 50N 18 12 E
Inpundong, N. Korea 35 D14 41 25N 126 34 E
Inquisivi, Bolivia 92 G5 16 50S 67 10W
Inscription, C., Australia 61 E1 25 29S 112 59 E
Insein, Burma 41 L20 16 50N 96 5 E
Inta, Russia 24 A11 66 5N 60 8 E
Intendente Alvear, Argentina 94 D3 35 12S 63 32W
Interior, U.S.A. 80 D4 43 44N 101 59W
Interlaken, Switz. 16 E3 46 41N 7 50 E
International Falls, U.S.A. 80 A8 48 36N 93 25W
Intiyaco, Argentina 94 B3 28 43S 60 5W
Inútil, B., Chile 96 G2 53 30S 70 15W
Inuvik, Canada 68 B6 68 16N 133 40W
Inveraray, U.K. 12 E3 56 13N 5 5W
Inverbervie, U.K. 12 E6 56 50N 2 17W
Invercargill, N.Z. 59 M2 46 24S 168 24 E
Inverell, Australia 63 D5 29 45S 151 8 E
Invergordon, U.K. 12 D4 57 41N 4 10W
Invermere, Canada 72 C5 50 30N 116 2W
Inverness, Canada 71 C7 46 15N 61 19W
Inverness, U.K. 12 D4 57 29N 4 12W
Inverness, U.S.A. 77 L4 28 50N 82 20W
Inverurie, U.K. 12 D6 57 15N 2 21W
Inverway, Australia 60 C4 17 50S 129 38 E
Investigator Group, Australia 63 E1 34 45S 134 20 E
Investigator Str., Australia 63 F2 35 30S 137 0 E
Inya, Russia 26 D9 50 28N 86 37 E
Inyanga, Zimbabwe 55 F3 18 12S 32 40 E
Inyangani, Zimbabwe 55 F3 18 5S 32 50 E
Inyantue, Zimbabwe 55 F2 18 30S 26 40 E
Inyo Mts., U.S.A. 83 H5 36 40N 118 0W
Inyokern, U.S.A. 85 K9 35 39N 117 49W
Inza, Russia 24 D8 53 55N 46 25 E
Iō-Jima, Japan 31 J5 30 48N 130 18 E
Iola, U.S.A. 81 G7 37 55N 95 24W
Iona, U.K. 12 E2 56 20N 6 25W
Ione, Calif., U.S.A. 84 G6 38 21N 120 56W
Ione, Wash., U.S.A. 82 B5 48 45N 117 25W
Ionia, U.S.A. 76 D3 42 59N 85 4W
Ionian Is. = Iónioi Nísoi, Greece 21 E9 38 40N 20 0 E
Ionian Sea, Europe 21 E7 37 30N 17 30 E
Iónioi Nísoi, Greece 21 E9 38 40N 20 0 E
Íos, Greece 21 F11 36 41N 25 20 E
Iowa □, U.S.A. 80 D8 42 18N 93 30W
Iowa City, U.S.A. 80 E9 41 40N 91 32W
Iowa Falls, U.S.A. 80 D8 42 31N 93 16W
Ipala, Tanzania 54 C3 4 30S 32 52 E
Ipameri, Brazil 93 G9 17 44S 48 9W
Ipatinga, Brazil 93 G10 19 32S 42 30W
Ipiales, Colombia 92 C3 0 50N 77 37W
Ipin = Yibin, China 32 D5 28 45N 104 32 E
Ípiros □, Greece 21 E9 39 30N 20 30 E
Ipixuna, Brazil 92 E4 7 0S 71 40W
Ipoh, Malaysia 39 K3 4 35N 101 5 E
Ippy, C.A.R. 51 G9 6 5N 21 7 E
Ipswich, Australia 63 D5 27 35S 152 40 E
Ipswich, U.K. 11 E9 52 4N 1 9 E
Ipswich, Mass., U.S.A. 79 D14 42 41N 70 50W
Ipswich, S. Dak., U.S.A. 80 C5 45 27N 99 2W
Ipu, Brazil 93 D10 4 23S 40 44W
Iqaluit, Canada 69 B13 63 44N 68 31W
Iquique, Chile 92 H4 20 19S 70 5W
Iquitos, Peru 92 D4 3 45S 73 10W
Irabu-Jima, Japan 31 M2 24 50N 125 10 E
Iracoubo, Fr. Guiana 93 B8 5 30N 53 10W
Īrafshān, Iran 45 E9 26 42N 61 56 E
Iráklion, Greece 23 D7 35 20N 25 12 E
Iráklion □, Greece 23 D7 35 10N 25 10 E
Irala, Paraguay 95 B5 25 55S 54 35W
Iramba □, Tanzania 54 C3 4 30S 34 30 E
Iran ■, Asia 45 C7 33 0N 53 0 E
Iran, Gunung-Gunung, Malaysia 36 D4 2 20N 114 50 E
Iran Ra. = Iran, Gunung-Gunung, Malaysia 36 D4 2 20N 114 50 E
Īrānshahr, Iran 45 E9 27 15N 60 40 E
Irapuato, Mexico 86 C4 20 40N 101 30W
Iraq ■, Asia 44 C5 33 0N 44 0 E
Irati, Brazil 95 B5 25 25S 50 38W
Irbid, Jordan 47 C4 32 35N 35 48 E
Irbid □, Jordan 47 C5 32 15N 36 35 E
Irebu, Zaïre 52 E3 0 40S 17 46 E
Ireland ■, Europe 13 D4 53 0N 8 0W
Ireland's Eye, Ireland 13 C5 53 25N 6 4W
Iret, Russia 27 C16 60 3N 154 20 E
Irhyangdong, N. Korea 35 D15 41 15N 129 30 E
Iri, S. Korea 35 G14 35 59N 127 0 E
Irian Jaya □, Indonesia 37 E9 4 0S 137 0 E
Iringa, Tanzania 54 D4 7 48S 35 43 E
Iringa □, Tanzania 54 D4 7 48S 35 43 E
Iriomote-Jima, Japan 31 M1 24 19N 123 48 E
Iriona, Honduras 88 C2 15 57N 85 11W
Iriri →, Brazil 93 D8 3 52S 52 37W
Irish Republic ■, Europe 13 D4 53 0N 8 0W
Irish Sea, Europe 10 D3 54 0N 5 0W
Irkineyeva, Russia 27 D10 58 30N 96 49 E
Irkutsk, Russia 27 D11 52 18N 104 20 E
Irma, Canada 73 C6 52 55N 111 14W
Irō-Zaki, Japan 31 G9 34 36N 138 51 E
Iron Baron, Australia 63 E2 32 58S 137 11 E
Iron Gate = Portile de Fier, Europe 17 F11 44 42N 22 30 E
Iron Knob, Australia 63 E2 32 46S 137 8 E
Iron Mountain, U.S.A. 76 C1 45 49N 88 4W
Iron Ra., Australia 62 A3 12 46S 143 16 E
Iron River, U.S.A. 80 B10 46 6N 88 39W
Ironbridge, U.K. 11 E5 52 38N 2 29W
Irondequoit, U.S.A. 78 C7 43 13N 77 35W
Ironstone Kopje, Botswana 56 D3 25 17S 24 5 E
Ironton, Mo., U.S.A. 81 G9 37 36N 90 38W
Ironton, Ohio, U.S.A. 76 F4 38 32N 82 41W
Ironwood, U.S.A. 80 B9 46 27N 90 9W
Iroquois Falls, Canada 70 C3 48 46N 80 41W
Irrara Cr. →, Australia 63 D4 29 35S 145 31 E
Irrawaddy □, Burma 41 L19 17 0N 95 0 E
Irrawaddy →, Burma 41 M19 15 50N 95 6 E
Irtysh →, Russia 26 C7 61 4N 68 52 E
Irumu, Zaïre 54 B2 1 32N 29 53 E
Irún, Spain 19 A5 43 20N 1 52W
Irunea = Pamplona, Spain 19 A5 42 48N 1 38W
Irvine, Canada 73 D6 49 57N 110 16W
Irvine, U.K. 12 F4 55 37N 4 40W
Irvine, Calif., U.S.A. 85 M9 33 41N 117 46W
Irvine, Ky., U.S.A. 76 G4 37 42N 83 58W
Irvinestown, U.K. 13 B4 54 28N 7 38W
Irving, U.S.A. 81 J6 32 49N 96 56W
Irvona, U.S.A. 78 F6 40 46N 78 33W
Irwin →, Australia 61 E1 29 15S 114 54 E
Irymple, Australia 63 E3 34 14S 142 8 E
Isaac →, Australia 62 C4 22 55S 149 20 E
Isabel, U.S.A. 80 C4 45 24N 101 26W
Isabela, I., Mexico 86 C3 21 51N 105 55W
Isabella, Phil. 37 C6 6 40N 122 10 E
Isabella, Cord., Nic. 88 D2 13 30N 85 25W
Isabella Ra., Australia 60 D3 21 0S 121 4 E
Ísafjarðardjúp, Iceland 8 C2 66 10N 23 0W
Ísafjörður, Iceland 8 C2 66 5N 23 9W
Isagarh, India 42 G7 24 48N 77 51 E
Isahaya, Japan 31 H5 32 52N 130 2 E
Isaka, Tanzania 54 C3 3 56S 32 59 E
Isangi, Zaïre 52 D4 0 52N 24 10 E
Isar →, Germany 16 D6 48 49N 12 58 E
Ischia, Italy 20 D5 40 45N 13 51 E
Isdell →, Australia 60 C3 16 27S 124 51 E
Ise, Japan 31 G8 34 25N 136 45 E
Ise-Wan, Japan 31 G8 34 43N 136 43 E
Iseramagazi, Tanzania 54 C3 4 37S 32 10 E
Isère □, France 18 D6 45 15N 5 40 E
Isère →, France 18 D6 44 59N 4 51 E
Ishigaki-Shima, Japan 31 M2 24 20N 124 10 E
Ishikari-Gawa →, Japan 30 C10 43 15N 141 23 E
Ishikari-Sammyaku, Japan 30 C11 43 30N 143 0 E
Ishikari-Wan, Japan 30 C10 43 25N 141 1 E
Ishikawa □, Japan 31 F8 36 30N 136 30 E
Ishim, Russia 26 D7 56 10N 69 30 E
Ishim →, Russia 26 D8 57 45N 71 10 E
Ishinomaki, Japan 30 E10 38 32N 141 20 E
Ishioka, Japan 31 F10 36 11N 140 16 E
Ishkuman, Pakistan 43 A5 36 30N 73 50 E
Ishpeming, U.S.A. 76 B2 46 29N 87 40W
Isil Kul, Russia 26 D8 54 55N 71 16 E
Isiolo, Kenya 54 B4 0 24N 37 33 E
Isiolo □, Kenya 54 B4 2 30N 37 30 E
Isipingo Beach, S. Africa 57 E5 30 0S 30 57 E
Isiro, Zaïre 54 B2 2 53N 27 40 E
Isisford, Australia 62 C3 24 15S 144 21 E
İskenderun, Turkey 25 G6 36 32N 36 10 E
İskenderun Körfezi, Turkey 25 G6 36 40N 35 50 E
Iskut →, Canada 72 B2 56 45N 131 49W
Isla →, U.K. 12 E5 56 32N 3 20W
Isla Vista, U.S.A. 85 L7 34 25N 119 53W
Islamabad, Pakistan 42 C5 33 40N 73 10 E
Islamkot, Pakistan 42 G4 24 42N 70 13 E
Island →, Canada 72 A4 60 25N 121 12W
Island Falls, Canada 70 C3 49 35N 81 20W
Island Falls, U.S.A. 71 C6 46 1N 68 16W
Island L., Canada 73 C10 53 47N 94 25W
Island Lagoon, Australia 63 E2 31 30S 136 40 E
Island Pond, U.S.A. 79 B13 44 49N 71 53W
Islands, B. of, Canada 71 C8 49 11N 58 15W
Islay, U.K. 12 F2 55 46N 6 10W
Isle aux Morts, Canada 71 C8 47 35N 59 0W
Isle of Wight □, U.K. 11 G6 50 40N 1 20W
Isle Royale, U.S.A. 80 A10 48 0N 88 54W
Isleta, U.S.A. 83 J10 34 55N 106 42W
Isleton, U.S.A. 84 G5 38 10N 121 37W
Ismail, Ukraine 25 E4 45 22N 28 46 E
Ismâ'ilîya, Egypt 51 B11 30 37N 32 18 E
Ismay, U.S.A. 80 B2 46 30N 104 48W
Isna, Egypt 51 C11 25 17N 32 30 E
Isogstalo, India 43 B8 34 15N 78 46 E
Isparta, Turkey 25 G5 37 47N 30 30 E
Íspica, Italy 20 F6 36 47N 14 53 E
Israel ■, Asia 47 D3 32 0N 34 50 E
Issyk-Kul, Ozero, Kirghizia 26 E8 42 25N 77 15 E
Istaihah, U.A.E. 45 F7 23 19N 54 4 E
İstanbul, Turkey 21 D13 41 0N 29 0 E
Istokpoga, L., U.S.A. 77 M5 27 23N 81 17W
Istra, Croatia 20 B5 45 10N 14 0 E
Istria = Istra, Croatia 20 B5 45 10N 14 0 E
Itá, Paraguay 94 B4 25 29S 57 21W
Itaberaba, Brazil 93 F10 12 32S 40 18W
Itabaiana, Brazil 93 E11 7 18S 35 19W
Itabira, Brazil 93 G10 19 37S 43 13W
Itabirito, Brazil 95 A7 20 15S 43 48W
Itabuna, Brazil 93 F11 14 48S 39 16W
Itaipu Dam, Brazil 95 B5 25 30S 54 30W
Itaituba, Brazil 93 D7 4 10S 55 50W
Itajaí, Brazil 95 B6 27 50S 48 39W
Itajubá, Brazil 95 A6 22 24S 45 30W
Itaka, Tanzania 55 D3 8 50S 32 49 E

Jiuxincheng

Jiuxincheng, China 34 E8 39 17N 115 59 E
Jixi, China 35 B16 45 20N 130 50 E
Jiyang, China 35 F9 37 0N 117 12 E
Jīzān, Si. Arabia 46 D3 17 0N 42 20 E
Jize, China 34 F8 36 54N 114 56 E
Jizō-Zaki, Japan 31 G6 35 34N 133 20 E
Joaçaba, Brazil 95 B5 27 5S 51 31W
João Pessoa, Brazil 93 E12 7 10S 34 52W
Joaquín V. González, Argentina 94 B3 25 10S 64 0W
Jodhpur, India 42 F5 26 23N 73 8 E
Joensuu, Finland 24 B4 62 37N 29 49 E
Jofane, Mozam. 57 C5 21 15S 34 18 E
Joggins, Canada 71 C7 45 42N 64 27W
Jogjakarta = Yogyakarta, Indonesia 37 G14 7 49S 110 22 E
Johannesburg, S. Africa 57 D4 26 10S 28 2 E
Johannesburg, U.S.A. 85 K9 35 22N 117 38W
John Day, U.S.A. 82 D4 44 25N 118 57W
John Day →, U.S.A. 82 D3 45 44N 120 39W
John H. Kerr Reservoir, U.S.A. 77 G6 36 36N 78 18W
John o' Groats, U.K. 12 C5 58 39N 3 3W
Johnnie, U.S.A. 85 J10 36 25N 116 5W
John's Ra., Australia 62 C1 21 55S 133 23 E
Johnson, U.S.A. 81 G4 37 34N 101 45W
Johnson City, N.Y., U.S.A. 79 D9 42 7N 75 58W
Johnson City, Tenn., U.S.A. 77 G4 36 19N 82 21W
Johnson City, Tex., U.S.A. 81 K5 30 17N 98 25W
Johnsonburg, U.S.A. 78 E6 41 29N 78 41W
Johnsondale, U.S.A. 85 K8 35 58N 118 32W
Johnson's Crossing, Canada 72 A2 60 29N 133 18W
Johnston, L., Australia 61 F3 32 25S 120 30 E
Johnston Falls = Mambilima Falls, Zambia 55 E2 10 31S 28 45 E
Johnston I., Pac. Oc. 65 F11 17 10N 169 8W
Johnstone Str., Canada 72 C3 50 28N 126 0W
Johnstown, N.Y., U.S.A. 79 C10 43 0N 74 22W
Johnstown, Pa., U.S.A. 78 F6 40 20N 78 55W
Johor Baharu, Malaysia 39 M4 1 28N 103 46 E
Joinvile, Brazil 95 B6 26 15S 48 55W
Joinville I., Antarctica 5 C18 65 0S 55 30W
Jojutla, Mexico 87 D5 18 37N 99 11W
Jokkmokk, Sweden 8 C15 66 35N 19 50 E
Jökulsá á Bru →, Iceland 8 D6 65 40N 14 16W
Jökulsá á Fjöllum →, Iceland 8 C5 66 10N 16 30W
Jolfā, Āzarbājān-e Sharqī, Iran 44 B5 38 57N 45 38 E
Jolfā, Eşfahan, Iran 45 C6 32 58N 51 37 E
Joliet, U.S.A. 76 E1 41 32N 88 5W
Joliette, Canada 70 C5 46 3N 73 24W
Jolo, Phil. 37 C6 6 0N 121 0 E
Jolon, U.S.A. 84 K5 35 58N 121 9W
Jombang, Indonesia 37 G15 7 33S 112 14 E
Jome, Indonesia 37 E7 1 16S 127 30 E
Jones Sound, Canada 4 B3 76 0N 85 0W
Jonesboro, Ark., U.S.A. 81 H9 35 50N 90 42W
Jonesboro, Ill., U.S.A. 81 G10 37 27N 89 16W
Jonesboro, La., U.S.A. 81 J8 32 15N 92 43W
Jonesport, U.S.A. 71 D6 44 32N 67 37W
Jonglei □, Sudan 51 G11 7 30N 32 30 E
Jönköping, Sweden 9 H13 57 45N 14 10 E
Jönköpings län □, Sweden 9 H13 57 30N 14 30 E
Jonquière, Canada 71 C5 48 27N 71 14W
Joplin, U.S.A. 81 G7 37 6N 94 31W
Jordan, U.S.A. 82 C10 47 19N 106 55W
Jordan ■, Asia 47 E5 31 0N 36 0 E
Jordan →, Asia 47 D4 31 48N 35 32 E
Jordan Valley, U.S.A. 82 E5 42 59N 117 3W
Jorhat, India 41 F19 26 45N 94 12 E
Jörn, Sweden 8 D16 65 4N 20 1 E
Jorong, Indonesia 36 E4 3 58S 114 56 E
Jorquera →, Chile 94 B2 28 3S 69 58W
Jos, Nigeria 50 G6 9 53N 8 51 E
José Batlle y Ordóñez, Uruguay 95 C4 33 20S 55 10W
Joseph, U.S.A. 82 D5 45 21N 117 14W
Joseph, L., Nfld., Canada 71 B6 52 45N 65 18W
Joseph, L., Ont., Canada 78 A5 45 10N 79 44W
Joseph Bonaparte G., Australia 60 B4 14 35S 128 50 E
Joseph City, U.S.A. 83 J8 34 57N 110 20W
Joshua Tree, U.S.A. 85 L10 34 8N 116 19W
Joshua Tree National Monument, U.S.A. 85 M10 33 55N 116 0W
Jostedal, Norway 9 F9 61 35N 7 15 E
Jotunheimen, Norway 9 F10 61 35N 8 25 E
Jourdanton, U.S.A. 81 L5 28 55N 98 33W
Joussard, Canada 72 B5 55 22N 115 50W
Jovellanos, Cuba 88 B3 22 40N 81 10W
Ju Xian, China 35 F10 36 35N 118 20 E
Juan Aldama, Mexico 86 C4 24 20N 103 23W
Juan Bautista Alberdi, Argentina 94 C3 34 26S 61 48W
Juan de Fuca Str., Canada 84 B2 48 15N 124 0W
Juan de Nova, Ind. Oc. 57 B7 17 3S 43 45 E
Juan Fernández, Arch. de, Pac. Oc. 65 L20 33 50S 80 0W
Juan José Castelli, Argentina 94 B3 25 27S 60 57W
Juan L. Lacaze, Uruguay 94 C4 34 26S 57 25W
Juárez, Argentina 94 D4 37 40S 59 43W
Juárez, Mexico 85 N11 32 20N 115 57W
Juárez, Sierra de, Mexico 86 A1 32 0N 116 0W
Juàzeiro, Brazil 93 E10 9 30S 40 30W
Juàzeiro do Norte, Brazil 93 E11 7 10S 39 18W
Jubayl, Lebanon 47 A4 34 5N 35 39 E
Jubbah, Si. Arabia 44 D4 28 2N 40 56 E
Jubbulpore = Jabalpur, India 43 H8 23 9N 79 58 E
Jubilee L., Australia 61 E4 29 0S 126 50 E
Juby, C., Morocco 50 C2 28 0N 12 59W
Júcar →, Spain 19 C5 39 5N 0 10W
Júcaro, Cuba 88 B4 21 37N 78 51W
Juchitán, Mexico 87 D5 16 27N 95 5W
Judaea = Har Yehuda, Israel 47 D3 31 35N 34 57 E
Judith →, U.S.A. 82 C9 47 44N 109 39W
Judith, Pt., U.S.A. 79 E13 41 22N 71 29W
Judith Gap, U.S.A. 82 C9 46 41N 109 45W
Jugoslavia = Yugoslavia ■, Europe 21 C8 44 0N 20 0 E
Juigalpa, Nic. 88 D2 12 6N 85 26W
Juiz de Fora, Brazil 95 A7 21 43S 43 19W
Jujuy □, Argentina 94 A2 23 20S 65 40W
Julesburg, U.S.A. 80 E3 40 59N 102 16W
Juli, Peru 92 G5 16 10S 69 25W
Julia Cr. →, Australia 62 C3 20 0S 141 11 E
Julia Creek, Australia 62 C3 20 39S 141 44 E
Juliaca, Peru 92 G4 15 25S 70 10W
Julian, U.S.A. 85 M10 33 4N 116 38W
Julianehåb, Greenland 4 C5 60 43N 46 0W
Julimes, Mexico 86 B3 28 25N 105 27W
Jullundur, India 42 D6 31 20N 75 40 E
Julu, China 34 F8 37 15N 115 2 E
Jumbo, Zimbabwe 55 F3 17 30S 30 58 E
Jumbo Pk., U.S.A. 85 J12 36 12N 114 11W
Jumentos Cays, Bahamas 89 B4 23 0N 75 40W
Jumet, Belgium 15 D4 50 27N 4 25 E
Jumilla, Spain 19 C5 38 28N 1 19W
Jumla, Nepal 43 E10 29 15N 82 13 E
Jumna = Yamuna →, India 43 G9 25 30N 81 53 E
Junagadh, India 42 J4 21 30N 70 30 E
Junction, Tex., U.S.A. 81 K5 30 29N 99 46W
Junction, Utah, U.S.A. 83 G7 38 14N 112 13W
Junction B., Australia 62 A1 11 52S 133 55 E
Junction City, Kans., U.S.A. 80 F6 39 2N 96 50W
Junction City, Oreg., U.S.A. 82 D2 44 13N 123 12W
Junction Pt., Australia 62 A1 11 45S 133 50 E
Jundah, Australia 62 C3 24 46S 143 2 E
Jundiaí, Brazil 95 A6 24 30S 47 0W
Juneau, U.S.A. 68 C6 58 18N 134 25W
Junee, Australia 63 E4 34 53S 147 35 E
Junggar Pendi, China 32 B3 44 30N 86 0 E
Jungshahi, Pakistan 42 G2 24 52N 67 44 E
Juniata →, U.S.A. 78 F7 40 30N 77 40W
Junín, Argentina 94 C3 34 33S 60 57W
Junín de los Andes, Argentina 96 D2 39 45S 71 0W
Jūniyah, Lebanon 47 B4 33 59N 35 38 E
Juntura, U.S.A. 82 E4 43 45N 118 5W
Jupiter →, Canada 71 C7 49 29N 63 37W
Jur, Nahr el →, Sudan 51 G10 8 45N 29 15 E
Jura = Jura, Mts. du, Europe 18 C7 46 40N 6 5 E
Jura = Schwäbische Alb, Germany 16 D4 48 30N 9 30 E
Jura, U.K. 12 F3 56 0N 5 50W
Jura □, France 18 C6 46 47N 5 45 E
Jura, Mts. du, Europe 18 C7 46 40N 6 5 E
Jura, Sd. of, U.K. 12 F3 55 57N 5 45W
Jurado, Colombia 92 B3 7 7N 77 46W
Juruá →, Brazil 92 D5 2 37S 65 44W
Juruena →, Brazil 92 E7 7 20S 58 3W
Juruti, Brazil 93 D7 2 9S 56 4W
Justo Daract, Argentina 94 C2 33 52S 65 12W
Juticalpa, Honduras 88 D2 14 40N 86 12W
Jutland = Jylland, Denmark 9 H10 56 25N 9 30 E
Juventud, I. de la, Cuba 88 B3 21 40N 82 40W
Juwain, Afghan. 40 D2 31 45N 61 30 E
Jūy Zar, Iran 44 C5 33 50N 46 18 E
Juye, China 34 G9 35 22N 116 5 E
Jylland, Denmark 9 H10 56 25N 9 30 E
Jyväskylä, Finland 8 E18 62 14N 25 50 E

K

K2, Mt., Pakistan 43 B7 35 58N 76 32 E
Kaap Plateau, S. Africa 56 D3 28 30S 24 0 E
Kaapkruis, Namibia 56 C1 21 55S 13 57 E
Kaapstad = Cape Town, S. Africa 56 E2 33 55S 18 22 E
Kabaena, Indonesia 37 F6 5 15S 122 0 E
Kabala, S. Leone 50 G2 9 38N 11 37W
Kabale, Uganda 54 C3 1 15S 30 0 E
Kabalo, Zaïre 54 D2 6 0S 27 0 E
Kabambare, Zaïre 54 C2 4 41S 27 39 E
Kabango, Zaïre 55 D2 8 35S 28 30 E
Kabanjahe, Indonesia 36 D1 3 6N 98 30 E
Kabara, Mali 50 E4 16 40N 2 50W
Kabardino-Balkar Republic □, Russia 25 F7 43 30N 43 30 E
Kabare, Indonesia 37 E8 0 4S 130 58 E
Kabarega Falls, Uganda 54 B3 2 15N 31 30 E
Kabasalan, Phil. 37 C6 7 47N 122 44 E
Kabba, Nigeria 50 G6 7 50N 6 3 E
Kabin Buri, Thailand 38 F3 13 57N 101 43 E
Kabinakagami L., Canada 70 C3 48 54N 84 25W
Kabīr, Zab al →, Iraq 44 C4 36 0N 43 0 E
Kabkabīyah, Sudan 51 F9 13 50N 24 0 E
Kabompo, Zambia 55 E1 13 36S 24 14 E
Kabompo →, Zambia 53 G4 14 10S 23 11 E
Kabondo, Zaïre 55 D2 8 58S 25 40 E
Kabongo, Zaïre 54 D2 7 22S 25 33 E
Kabra, Australia 62 C5 23 25S 150 25 E
Kabūd Gonbad, Iran 45 B8 37 5N 59 45 E
Kābul, Afghan. 42 B3 34 28N 69 11 E
Kābul □, Afghan. 40 B6 34 30N 69 0 E
Kabul →, Pakistan 42 C5 33 55N 72 14 E
Kabunga, Zaïre 54 C2 1 38S 28 3 E
Kaburuang, Indonesia 37 D7 3 50N 126 30 E
Kabwe, Zambia 55 E2 14 30S 28 29 E
Kachchh, Gulf of, India 42 H3 22 50N 69 15 E
Kachchh, Rann of, India 42 G4 24 0N 70 0 E
Kachebera, Zambia 55 E3 13 50S 32 50 E
Kachin □, Burma 41 F20 26 0N 97 30 E
Kachira, L., Uganda 54 C3 0 40S 31 7 E
Kachiry, Kazakhstan 26 D8 53 10N 75 50 E
Kachot, Cambodia 39 G4 11 30N 103 3 E
Kackar, Turkey 25 F7 40 45N 41 10 E
Kadan Kyun, Burma 36 B1 12 30N 98 20 E
Kadanai →, Afghan. 42 D1 31 22N 65 45 E
Kadi, India 42 H5 23 18N 72 23 E
Kadina, Australia 63 E2 33 55S 137 43 E
Kadiyevka, Ukraine 25 E6 48 35N 38 40 E
Kadoka, U.S.A. 80 D4 43 50N 101 31W
Kadoma, Zimbabwe 55 F2 18 20S 29 52 E
Kâdugli, Sudan 51 F10 11 0N 29 45 E
Kaduna, Nigeria 50 F6 10 30N 7 21 E
Kaédi, Mauritania 50 E2 16 9N 13 28W
Kaelé, Cameroon 51 F7 10 7N 14 27 E
Kaeng Khoï, Thailand 38 E3 14 35N 101 0 E
Kaesŏng, N. Korea 35 F14 37 58N 126 35 E
Kāf, Si. Arabia 44 D3 31 25N 37 29 E
Kafakumba, Zaïre 52 F4 9 38S 23 46 E
Kafan, Armenia 25 G8 39 18N 46 15 E
Kafanchan, Nigeria 50 G6 9 40N 8 20 E
Kaffrine, Senegal 50 F1 14 8N 15 36W
Kafia Kingi, Sudan 51 G9 9 20N 24 25 E
Kafinda, Zambia 55 E3 12 32S 30 20 E
Kafirévs, Ákra, Greece 21 E11 38 9N 24 38 E
Kafue, Zambia 55 F2 15 46S 28 9 E
Kafue →, Zambia 53 H5 15 30S 29 0 E
Kafue Flats, Zambia 55 F2 15 40S 27 25 E
Kafue Nat. Park, Zambia 55 F2 15 0S 25 30 E
Kafulwe, Zambia 55 D2 9 0S 29 1 E
Kaga, Afghan. 42 B4 34 14N 70 10 E
Kaga Bandoro, C.A.R. 51 G8 7 0N 19 10 E
Kagan, Uzbekistan 26 F7 39 43N 64 33 E
Kagawa □, Japan 31 G6 34 15N 134 0 E
Kagera □, Tanzania 54 C3 2 0S 31 30 E
Kagera →, Uganda 54 C3 0 57S 31 47 E
Kagoshima, Japan 31 J5 31 35N 130 33 E
Kagoshima □, Japan 31 J5 31 30N 130 30 E
Kahak, Iran 45 B6 36 6N 49 46 E
Kahama, Tanzania 54 C3 4 8S 32 30 E
Kahama □, Tanzania 54 C3 3 50S 32 0 E
Kahang, Malaysia 39 L4 2 12N 103 32 E
Kahayan →, Indonesia 36 E4 3 40S 114 0 E
Kahe, Tanzania 54 C4 3 30S 37 25 E
Kahemba, Zaïre 52 F3 7 18S 18 55 E
Kahniah →, Canada 72 B4 58 15N 120 55W
Kahnūj, Iran 45 E8 27 55N 57 40 E
Kahoka, U.S.A. 80 E9 40 25N 91 44W
Kahoolawe, U.S.A. 74 H16 20 33N 156 37W
Kahramanmaraş, Turkey 25 G6 37 37N 36 53 E
Kahuta, Pakistan 42 C5 33 35N 73 24 E
Kai, Kepulauan, Indonesia 37 F8 5 55S 132 45 E
Kai Besar, Indonesia 37 F8 5 35S 133 0 E
Kai Is. = Kai, Kepulauan, Indonesia 37 F8 5 55S 132 45 E
Kai-Ketil, Indonesia 37 F8 5 45S 132 40 E
Kaiama, Nigeria 50 G5 9 36N 4 1 E
Kaiapoi, N.Z. 59 K4 43 24S 172 40 E
Kaieteur Falls, Guyana 92 B7 5 1N 59 10W
Kaifeng, China 34 G8 34 48N 114 21 E
Kaikohe, N.Z. 59 F4 35 25S 173 49 E
Kaikoura, N.Z. 59 K4 42 25S 173 43 E
Kaikoura Ra., N.Z. 59 J4 41 59S 173 41 E
Kailu, China 35 C11 43 38N 121 18 E
Kailua Kona, U.S.A. 74 J17 19 39N 155 59W
Kaimana, Indonesia 37 E8 3 39S 133 45 E
Kaimanawa Mts., N.Z. 59 H5 39 15S 175 56 E
Kaimganj, India 43 F8 27 33N 79 24 E
Kaimur Hills, India 43 G9 24 30N 82 0 E
Kaingaroa Forest, N.Z. 59 H6 38 24S 176 30 E
Kainji Res., Nigeria 50 F5 10 1N 4 40 E
Kaipara Harbour, N.Z. 59 G5 36 25S 174 14 E
Kaipokok B., Canada 71 B8 54 54N 59 47W
Kairana, India 42 E7 29 24N 77 15 E
Kaironi, Indonesia 37 E8 0 47S 133 40 E
Kairouan, Tunisia 50 A7 35 45N 10 5 E
Kaiserslautern, Germany 16 D3 49 30N 7 43 E
Kaitaia, N.Z. 59 F4 35 8S 173 17 E
Kaitangata, N.Z. 59 M2 46 17S 169 51 E
Kaithal, India 42 E7 29 48N 76 26 E
Kaitu →, Pakistan 42 C4 33 10N 70 30 E
Kaiwi Channel, U.S.A. 74 H16 21 15N 157 30W
Kaiyuan, China 35 C13 42 28N 124 1 E
Kajaani, Finland 8 D19 64 17N 27 46 E
Kajabbi, Australia 62 B3 20 0S 140 1 E
Kajana = Kajaani, Finland 8 D19 64 17N 27 46 E
Kajang, Malaysia 39 L3 2 59N 101 48 E
Kajiado, Kenya 54 C4 1 53S 36 48 E
Kajiado □, Kenya 54 C4 2 0S 36 30 E
Kajo Kaji, Sudan 51 H11 3 58N 31 40 E
Kaka, Sudan 51 F11 10 38N 32 10 E
Kakabeka Falls, Canada 70 C2 48 24N 89 37W
Kakamas, S. Africa 56 D3 28 45S 20 33 E
Kakamega, Kenya 54 B3 0 20N 34 46 E
Kakamega □, Kenya 54 B3 0 20N 34 46 E
Kakanui Mts., N.Z. 59 L3 45 10S 170 30 E
Kake, Japan 31 G6 34 36N 132 19 E
Kakegawa, Japan 31 G9 34 45N 138 1 E
Kakeroma-Jima, Japan 31 K4 28 8N 129 14 E
Kakhovka, Ukraine 25 E5 46 40N 33 15 E
Kakhovskoye Vdkhr., Ukraine 25 E5 47 5N 34 0 E
Kakinada, India 41 L13 16 57N 82 11 E
Kakisa →, Canada 72 A5 61 3N 118 10W
Kakisa L., Canada 72 A5 60 56N 117 43W
Kakogawa, Japan 31 G7 34 46N 134 51 E
Kakwa →, Canada 72 C5 54 37N 118 28W
Kāl Gūsheh, Iran 45 D8 30 59N 58 12 E
Kal Safid, Iran 44 C5 34 52N 47 23 E
Kalabagh, Pakistan 42 C4 33 0N 71 28 E
Kalabahi, Indonesia 37 F6 8 13S 124 31 E
Kalabáka, Greece 21 E9 39 42N 21 39 E
Kalabo, Zambia 53 G4 14 58S 22 40 E
Kalach, Russia 25 D7 50 22N 41 0 E
Kaladan →, Burma 41 J18 20 20N 93 5 E
Kaladar, Canada 78 B7 44 37N 77 5W
Kalahari, Africa 56 C3 24 0S 21 30 E
Kalahari Gemsbok Nat. Park, S. Africa 56 D3 25 30S 20 30 E
Kālak, Iran 45 E8 25 29N 59 22 E
Kalakamati, Botswana 57 C4 20 40S 27 25 E
Kalakan, Russia 27 D12 55 15N 116 45 E
Kalakh, Syria 44 C3 34 55N 36 10 E
K'alak'unlun Shank'ou, Pakistan 43 B7 35 33N 77 46 E
Kalam, Pakistan 43 B5 35 34N 72 30 E
Kalama, U.S.A. 84 E4 46 1N 122 51W
Kalama, Zaïre 54 C2 2 52S 28 35 E
Kalamata, Greece 21 F10 37 3N 22 10 E
Kalamazoo, U.S.A. 76 D3 42 17N 85 35W
Kalamazoo →, U.S.A. 76 D2 42 40N 86 10W
Kalambo Falls, Tanzania 55 D3 8 37S 31 35 E
Kalannie, Australia 61 F2 30 22S 117 5 E
Kalāntarī, Iran 45 C7 32 10N 54 8 E
Kalao, Indonesia 37 F6 7 21S 121 0 E
Kalaotoa, Indonesia 37 F6 7 20S 121 50 E
Kalasin, Thailand 38 D4 16 26N 103 30 E
Kalat, Pakistan 40 E5 29 8N 66 31 E
Kalāteh, Iran 45 B7 36 33N 55 41 E
Kalāteh-ye-Ganj, Iran 45 E8 27 31N 57 55 E
Kalbarri, Australia 61 E1 27 40S 114 10 E
Kalegauk Kyun, Burma 41 M20 15 33N 97 35 E
Kalehe, Zaïre 54 C2 2 6S 28 50 E
Kalema, Tanzania 54 C3 1 12S 31 55 E
Kalemie, Zaïre 54 D2 5 55S 29 9 E
Kalewa, Burma 41 H19 23 10N 94 15 E
Kálfafellsstaður, Iceland 8 D6 64 11N 15 53W
Kalgan = Zhangjiakou, China 34 D8 40 48N 114 55 E
Kalgoorlie-Boulder, Australia 61 F3 30 40S 121 22 E
Kaliakra, Nos, Bulgaria 21 C13 43 21N 28 30 E
Kalianda, Indonesia 36 F3 5 50S 105 45 E
Kalibo, Phil. 37 B6 11 43N 122 22 E
Kaliganj, Bangla. 43 H13 22 25N 89 8 E
Kalima, Zaïre 54 C2 2 33S 26 32 E
Kalimantan, Indonesia 36 E4 0 0 114 0 E
Kalimantan Barat □, Indonesia 36 E4 0 0 110 30 E
Kalimantan Selatan □, Indonesia 36 E5 2 30S 115 30 E
Kalimantan Tengah □, Indonesia 36 E4 2 0S 113 30 E
Kalimantan Timur □, Indonesia 36 D5 1 30N 116 30 E
Kálimnos, Greece 21 F12 37 0N 27 0 E
Kalimpong, India 43 F13 27 4N 88 35 E
Kalinin = Tver, Russia 24 C6 56 55N 35 55 E
Kaliningrad, Russia 24 C6 55 58N 37 54 E
Kaliningrad, Russia 24 D3 54 42N 20 32 E
Kaliro, Uganda 54 B3 0 56N 33 30 E
Kalispell, U.S.A. 82 B6 48 12N 114 19W
Kalisz, Poland 17 C9 51 45N 18 8 E
Kaliua, Tanzania 54 D3 5 5S 31 48 E
Kalix →, Sweden 8 D17 65 50N 23 11 E
Kalka, India 42 D7 30 46N 76 57 E
Kalkaska, U.S.A. 76 C3 44 44N 85 11W
Kalkfeld, Namibia 56 C2 20 57S 16 14 E
Kalkfontein, Botswana 56 C3 22 4S 20 57 E
Kalkrand, Namibia 56 C2 24 1S 17 35 E
Kallsjön, Sweden 8 E12 63 38N 13 0 E
Kalmar, Sweden 9 H14 56 40N 16 20 E
Kalmyk Republic □, Russia 25 E8 46 5N 46 1 E
Kalmykovo, Kazakhstan 25 E9 49 0N 51 47 E
Kalna, India 43 H13 23 13N 88 25 E
Kalocsa, Hungary 17 E9 46 32N 19 0 E
Kalokhorio, Cyprus 23 E12 34 51N 33 2 E
Kaloko, Zaïre 54 D2 6 47S 25 48 E
Kalol, Gujarat, India 42 H5 22 37N 73 31 E
Kalol, Gujarat, India 42 H5 23 15N 72 33 E
Kalomo, Zambia 55 F2 17 0S 26 30 E
Kalpi, India 43 F8 26 8N 79 47 E
Kalu, Pakistan 42 G2 25 5N 67 39 E
Kaluga, Russia 24 D6 54 35N 36 10 E
Kalulushi, Zambia 55 E2 12 50S 28 3 E
Kalundborg, Denmark 9 J11 55 41N 11 5 E
Kalutara, Sri Lanka 40 R11 6 35N 80 0 E
Kalya, Russia 24 B10 60 15N 59 59 E
Kama, Zaïre 54 C2 3 30S 27 5 E
Kama →, Russia 24 C9 55 45N 52 0 E
Kamachumu, Tanzania 54 C3 1 37S 31 37 E
Kamaishi, Japan 30 E10 39 16N 141 53 E
Kamalia, Pakistan 42 D5 30 44N 72 42 E
Kamapanda, Zambia 55 E1 12 5S 24 0 E
Kamaran, Yemen 46 D3 15 21N 42 35 E
Kamativi, Zimbabwe 55 F2 18 15S 27 27 E
Kambalda, Australia 61 F3 31 10S 121 37 E
Kambar, Pakistan 42 F3 27 37N 68 1 E
Kambarka, Russia 24 C9 56 15N 54 11 E
Kambolé, Zambia 55 D3 8 47S 30 48 E
Kambos, Cyprus 23 D11 35 2N 32 44 E
Kambove, Zaïre 55 E2 10 51S 26 33 E
Kamchatka, P-ov., Russia 27 D16 57 0N 160 0 E
Kamchatka Pen. = Kamchatka, P-ov., Russia 27 D16 57 0N 160 0 E
Kamen, Russia 26 D9 53 50N 81 30 E
Kamen-Rybolov, Russia 30 B6 44 46N 132 2 E
Kamenets-Podolskiy, Ukraine 25 E4 48 45N 26 40 E
Kamenjak, Rt., Croatia 20 B5 44 47N 13 55 E
Kamenka, Russia 24 A7 65 58N 44 0 E
Kamensk Uralskiy, Russia 26 D7 56 25N 62 2 E
Kamenskoye, Russia 27 C17 62 45N 165 30 E
Kameoka, Japan 31 G7 35 0N 135 35 E
Kamiah, U.S.A. 82 C5 46 14N 116 2W
Kamieskroon, S. Africa 56 E2 30 9S 17 56 E
Kamilukuak, L., Canada 73 A8 62 22N 101 40W
Kamina, Zaïre 55 D1 8 45S 25 0 E
Kaminak L., Canada 73 A9 62 10N 95 0W
Kaminoyama, Japan 30 E10 38 9N 140 17 E
Kamiros, Greece 23 C9 36 20N 27 56 E
Kamituga, Zaïre 54 C2 3 2S 28 10 E
Kamloops, Canada 72 C4 50 40N 120 20W
Kamo, Japan 30 F9 37 39N 139 3 E
Kamoke, Pakistan 42 C6 32 4N 74 4 E
Kampala, Uganda 54 B3 0 20N 32 30 E
Kampar, Malaysia 39 K3 4 18N 101 9 E
Kampar →, Indonesia 36 D2 0 30N 103 8 E
Kampen, Neths. 15 B5 52 33N 5 53 E
Kamphaeng Phet, Thailand 38 D2 16 28N 99 30 E
Kampolombo, L., Zambia 55 E2 11 37S 29 42 E
Kampong To, Thailand 39 J3 6 3N 101 13 E

Kelloselkä

Kampot, Cambodia 39 G5 10 36N 104 10 E
Kampuchea = Cambodia ■, Asia 38 F5 12 15N 105 0 E
Kampung →, Indonesia 37 F9 5 44S 138 24 E
Kampung Air Putih, Malaysia 39 K4 4 15N 103 10 E
Kampung Jerangau, Malaysia 39 K4 4 50N 103 10 E
Kampung Raja, Malaysia 39 K4 5 45N 102 35 E
Kampungbaru = Tolitoli, Indonesia 37 D6 1 5N 120 50 E
Kamrau, Teluk, Indonesia 37 E8 3 30S 133 36 E
Kamsack, Canada 73 C8 51 34N 101 54W
Kamskoye Vdkhr., Russia 24 C10 58 0N 56 0 E
Kamuchawie L., Canada 73 B8 56 18N 101 59W
Kamui-Misaki, Japan 30 C10 43 20N 140 21 E
Kamyanets-Podilskyy = Kamenets-Podolskiy, Ukraine 25 E4 48 45N 26 40 E
Kāmyārān, Iran 44 C5 34 47N 46 56 E
Kamyshin, Russia 25 D8 50 10N 45 24 E
Kanaaupscow, Canada 70 B4 54 2N 76 30W
Kanab, U.S.A. 83 H7 37 3N 112 32W
Kanab →, U.S.A. 83 H7 36 24N 112 38W
Kanagi, Japan 30 D10 40 54N 140 27 E
Kanairiktok →, Canada 71 A7 55 2N 60 18W
Kananga, Zaïre 52 F4 5 55S 22 18 E
Kanarraville, U.S.A. 83 H7 37 32N 113 11W
Kanash, Russia 24 C8 55 30N 47 32 E
Kanaskat, U.S.A. 84 C5 47 19N 121 54W
Kanawha →, U.S.A. 76 F4 38 50N 82 9W
Kanazawa, Japan 31 F8 36 30N 136 38 E
Kanchanaburi, Thailand 38 E2 14 2N 99 31 E
Kanchenjunga, Nepal 43 F13 27 50N 88 10 E
Kanchipuram, India 40 N11 12 52N 79 45 E
Kanda Kanda, Zaïre 52 F4 6 52S 23 48 E
Kandahar = Qandahār, Afghan. 40 D4 31 32N 65 30 E
Kandalaksha, Russia 24 A5 67 9N 32 30 E
Kandalakshkiy Zaliv, Russia 24 A5 66 0N 35 0 E
Kandalu, Afghan. 40 E3 29 55N 63 20 E
Kandangan, Indonesia 36 E5 2 50S 115 20 E
Kandanos, Greece 23 D5 35 20N 23 45 E
Kandhkot, Pakistan 42 E3 28 16N 69 8 E
Kandhla, India 42 E7 29 18N 77 19 E
Kandi, Benin 50 F5 11 7N 2 55 E
Kandi, India 43 H13 23 58N 88 5 E
Kandla, India 42 H4 23 0N 70 10 E
Kandos, Australia 63 E4 32 45S 149 58 E
Kandy, Sri Lanka 40 R12 7 18N 80 43 E
Kane, U.S.A. 78 E6 41 40N 78 49W
Kane Basin, Greenland 4 B4 79 1N 70 0W
Kangān, Fārs, Iran 45 E7 27 50N 52 3 E
Kangān, Hormozgān, Iran 45 E8 25 48N 57 28 E
Kangar, Malaysia 39 J3 6 27N 100 12 E
Kangaroo I., Australia 63 F2 35 45S 137 0 E
Kangavar, Iran 45 C6 34 40N 48 0 E
Kāngdong, N. Korea 35 E14 39 9N 126 5 E
Kangean, Kepulauan, Indonesia 36 F5 6 55S 115 23 E
Kangean Is. = Kangean, Kepulauan, Indonesia 36 F5 6 55S 115 23 E
Kanggye, N. Korea 35 D14 41 0N 126 35 E
Kanggyŏng, S. Korea 35 F14 36 10N 127 0 E
Kanghwa, S. Korea 35 F14 37 45N 126 30 E
Kangnŭng, S. Korea 35 F15 37 45N 128 54 E
Kango, Gabon 52 D2 0 11N 10 5 E
Kangping, China 35 C12 42 43N 123 18 E
Kangto, India 41 F18 27 50N 92 35 E
Kaniama, Zaïre 54 D1 7 30S 24 12 E
Kaniapiskau →, Canada 71 A6 56 40N 69 30W
Kaniapiskau L., Canada 71 B6 54 10N 69 55W
Kanin, P-ov., Russia 24 A8 68 0N 45 0 E
Kanin Nos, Mys, Russia 24 A7 68 45N 43 20 E
Kanin Pen. = Kanin, P-ov., Russia 24 A8 68 0N 45 0 E
Kaniva, Australia 63 F3 36 22S 141 18 E
Kanjut Sar, Pakistan 43 A6 36 7N 75 25 E
Kankakee, U.S.A. 76 E2 41 7N 87 52W
Kankakee →, U.S.A. 76 E1 41 23N 88 15W
Kankan, Guinea 50 F3 10 23N 9 15W
Kanker, India 41 J12 20 10N 81 40 E
Kankunskiy, Russia 27 D13 57 37N 126 8 E
Kannapolis, U.S.A. 77 H5 35 30N 80 37W
Kannauj, India 43 F8 27 3N 79 56 E
Kannod, India 40 H10 22 45N 76 40 E
Kano, Nigeria 50 F6 12 2N 8 30 E
Kan'onji, Japan 31 G6 34 7N 133 39 E
Kanowit, Malaysia 36 D4 2 14N 112 20 E
Kanowna, Australia 61 F3 30 32S 121 31 E
Kanoya, Japan 31 J5 31 25N 130 50 E
Kanpetlet, Burma 41 J18 21 10N 93 59 E
Kanpur, India 43 F9 26 28N 80 20 E
Kansas □, U.S.A. 80 F6 38 30N 99 0W
Kansas →, U.S.A. 80 F7 39 7N 94 37W
Kansas City, Kans., U.S.A. 80 F7 39 7N 94 38W
Kansas City, Mo., U.S.A. 80 F7 39 6N 94 35W
Kansenia, Zaïre 55 E2 10 20S 26 0 E
Kansk, Russia 27 D10 56 20N 95 37 E
Kansŏng, S. Korea 35 E15 38 24N 128 30 E
Kansu = Gansu □, China 34 G3 36 0N 104 0 E
Kantang, Thailand 39 J2 7 25N 99 31 E
Kantharalak, Thailand 38 E5 14 39N 104 39 E
Kantō □, Japan 31 F9 36 15N 139 30 E
Kantō-Sanchi, Japan 31 G9 35 59N 138 50 E
Kanturk, Ireland 13 D3 52 10N 8 55W
Kanuma, Japan 31 F9 36 34N 139 42 E
Kanus, Namibia 56 D2 27 50S 18 39 E
Kanye, Botswana 56 C4 24 55S 25 28 E
Kanzenze, Zaïre 55 E2 10 30S 25 12 E
Kanzi, Ras, Tanzania 54 D4 7 1S 39 33 E
Kaohsiung = Gaoxiong, Taiwan 33 D7 22 38N 120 18 E
Kaohsiung, Taiwan 33 D7 22 35N 120 16 E
Kaokoveld, Namibia 56 B1 19 15S 14 30 E
Kaolack, Senegal 50 F1 14 5N 16 8W
Kaoshan, China 35 B13 44 38N 124 50 E
Kapadvanj, India 42 H5 23 5N 73 0 E
Kapanga, Zaïre 52 F4 8 30S 22 40 E
Kapchagai, Kazakhstan 26 E8 43 51N 77 14 E
Kapela, Croatia 20 B6 44 40N 15 40 E
Kapema, Zaïre 55 E2 10 45S 28 22 E
Kapfenberg, Austria 16 E7 47 26N 15 18 E
Kapiri Mposhi, Zambia 55 E2 13 59S 28 43 E
Kapiskau →, Canada 70 B3 52 47N 81 55W
Kapit, Malaysia 36 D4 2 0N 112 55 E
Kapiti I., N.Z. 59 J5 40 50S 174 56 E
Kapoe, Thailand 39 H2 9 34N 98 32 E
Kapoeta, Sudan 51 H11 4 50N 33 35 E
Kaposvár, Hungary 17 E8 46 25N 17 47 E
Kapowsin, U.S.A. 84 D4 46 59N 122 13W
Kapps, Namibia 56 C2 22 32S 17 18 E
Kapsan, N. Korea 35 D15 41 4N 128 19 E
Kapuas →, Indonesia 36 E3 0 25S 109 20 E
Kapuas Hulu, Pegunungan, Malaysia 36 D4 1 30N 113 30 E
Kapuas Hulu Ra. = Kapuas Hulu, Pegunungan, Malaysia 36 D4 1 30N 113 30 E
Kapulo, Zaïre 55 D2 8 18S 29 15 E
Kapunda, Australia 63 E2 34 20S 138 56 E
Kapuni, N.Z. 59 H5 39 29S 174 8 E
Kapurthala, India 42 D6 31 23N 75 25 E
Kapuskasing, Canada 70 C3 49 25N 82 30W
Kapuskasing →, Canada 70 C3 49 49N 82 0W
Kaputar, Australia 63 E5 30 15S 150 10 E
Kaputir, Kenya 54 B4 2 5N 35 28 E
Kara, Russia 26 C7 69 10N 65 0 E
Kara Bogaz Gol, Zaliv, Turkmenistan 25 F9 41 0N 53 30 E
Kara Kalpak Republic □, Uzbekistan 26 E6 43 0N 58 0 E
Kara Kum = Karakum, Peski, Turkmenistan 26 F6 39 30N 60 0 E
Kara Sea, Russia 26 B7 75 0N 70 0 E
Karabutak, Kazakhstan 26 E7 49 59N 60 14 E
Karachi, Pakistan 42 G2 24 53N 67 0 E
Karad, India 40 L9 17 15N 74 10 E
Karadeniz Boğazı, Turkey 21 D13 41 10N 29 10 E
Karaganda, Kazakhstan 26 E8 49 50N 73 10 E
Karagayly, Kazakhstan 26 E8 49 26N 76 0 E
Karaginskiy, Ostrov, Russia 27 D17 58 45N 164 0 E
Karagiye Depression, Kazakhstan 25 F9 43 27N 51 45 E
Karagwe □, Tanzania 54 C3 2 0S 31 0 E
Karaikal, India 40 P11 10 59N 79 50 E
Karaikkudi, India 40 P11 10 5N 78 45 E
Karaj, Iran 45 C6 35 48N 51 0 E
Karak, Malaysia 39 L4 3 25N 102 2 E
Karakas, Kazakhstan 26 E9 48 20N 83 30 E
Karakitang, Indonesia 37 D7 3 14N 125 28 E
Karaklis, Armenia 25 F7 40 48N 44 30 E
Karakoram Pass, Pakistan 43 B7 35 33N 77 50 E
Karakoram Ra., Pakistan 43 B7 35 30N 77 0 E
Karakum, Peski, Turkmenistan 26 F6 39 30N 60 0 E
Karalon, Russia 27 D12 57 5N 115 50 E
Karaman, Turkey 25 G5 37 14N 33 13 E
Karamay, China 32 B3 45 30N 84 58 E
Karambu, Indonesia 36 E5 3 53S 116 6 E
Karamea Bight, N.Z. 59 J3 41 22S 171 40 E
Karamoja □, Uganda 54 B3 3 0N 34 15 E
Karamsad, India 42 H5 22 35N 72 50 E
Karand, Iran 44 C5 34 16N 46 15 E
Karanganyar, Indonesia 37 G13 7 38S 109 37 E
Karasburg, Namibia 56 D2 28 0S 18 44 E
Karasino, Russia 26 C9 66 50N 86 50 E
Karasjok, Norway 8 B18 69 27N 25 30 E
Karasuk, Russia 26 D8 53 44N 78 2 E
Karasuyama, Japan 31 F10 36 39N 140 9 E
Karatau, Kazakhstan 26 E8 43 10N 70 28 E
Karatau, Khrebet, Kazakhstan 26 E7 43 30N 69 30 E
Karauli, India 42 F7 26 30N 77 4 E
Karavostasi, Cyprus 23 D11 35 8N 32 50 E
Karawang, Indonesia 37 G12 6 30S 107 15 E
Karawanken, Europe 20 A6 46 30N 14 40 E
Karazhal, Kazakhstan 26 E8 48 2N 70 49 E
Karbalā, Iraq 44 C5 32 36N 44 3 E
Karcag, Hungary 17 E10 47 19N 20 57 E
Karcha →, Pakistan 43 B7 34 45N 76 10 E
Karda, Russia 27 D11 55 0N 103 16 E
Kardhítsa, Greece 21 E9 39 23N 21 54 E
Kareeberge, S. Africa 56 E3 30 59S 21 50 E
Karelian Republic □, Russia 24 A5 65 30N 32 30 E
Karevāndar, Iran 45 E9 27 53N 60 44 E
Kargasok, Russia 26 D9 59 3N 80 53 E
Kargat, Russia 26 D9 55 10N 80 15 E
Kargil, India 43 B7 34 32N 76 12 E
Kargopol, Russia 24 B6 61 30N 38 58 E
Karīān, Iran 45 E8 26 57N 57 14 E
Kariba, Zimbabwe 55 F2 16 28S 28 50 E
Kariba, L., Zimbabwe 55 F2 16 40S 28 25 E
Kariba Dam, Zimbabwe 55 F2 16 30S 28 35 E
Kariba Gorge, Zambia 55 F2 16 30S 28 50 E
Karibib, Namibia 56 C2 22 0S 15 56 E
Karimata, Kepulauan, Indonesia 36 E3 1 25S 109 0 E
Karimata, Selat, Indonesia 36 E3 2 0S 108 40 E
Karimata Is. = Karimata, Kepulauan, Indonesia 36 E3 1 25S 109 0 E
Karimnagar, India 40 K11 18 26N 79 10 E
Karimunjawa, Kepulauan, Indonesia 36 F4 5 50S 110 30 E
Karin, Somali Rep. 46 E4 10 50N 45 52 E
Karit, Iran 45 C8 33 29N 56 55 E
Kariya, Japan 31 G8 34 58N 137 1 E
Karkaralinsk, Kazakhstan 26 E8 49 26N 75 30 E
Karkinitskiy Zaliv, Ukraine 25 E5 45 56N 33 0 E
Karl-Marx-Stadt = Chemnitz, Germany 16 C6 50 50N 12 55 E
Karlovac, Croatia 20 B6 45 31N 15 36 E
Karlovy Vary, Czech. 16 C6 50 13N 12 51 E
Karlsbad = Karlovy Vary, Czech. 16 C6 50 13N 12 51 E
Karlsborg, Sweden 9 G13 58 33N 14 33 E
Karlshamn, Sweden 9 H13 56 10N 14 51 E
Karlskoga, Sweden 9 G13 59 22N 14 33 E
Karlskrona, Sweden 9 H13 56 10N 15 35 E
Karlsruhe, Germany 16 D4 49 3N 8 23 E
Karlstad, Sweden 9 G12 59 23N 13 30 E
Karlstad, U.S.A. 80 A6 48 35N 96 31W
Karnal, India 42 E7 29 42N 77 2 E
Karnali →, Nepal 43 E9 28 45N 81 16 E
Karnaphuli Res., Bangla. 41 H18 22 40N 92 20 E
Karnataka □, India 40 N10 13 15N 77 0 E
Karnes City, U.S.A. 81 L6 28 53N 97 54W
Karnische Alpen, Europe 20 A5 46 36N 13 0 E
Kärnten □, Austria 16 E6 46 52N 13 30 E
Karoi, Zimbabwe 55 F2 16 48S 29 45 E
Karonga, Malawi 55 D3 9 57S 33 55 E
Karoonda, Australia 63 F2 35 1S 139 59 E
Karora, Sudan 51 E12 17 44N 38 15 E
Karpasia □, Cyprus 23 D13 35 32N 34 15 E
Kárpathos, Greece 21 G12 35 37N 27 10 E
Karpinsk, Russia 24 C11 59 45N 60 1 E
Karpogory, Russia 24 B7 63 59N 44 27 E
Kars, Turkey 25 F7 40 40N 43 5 E
Karsakpay, Kazakhstan 26 E7 47 55N 66 40 E
Karshi, Uzbekistan 26 F7 38 53N 65 48 E
Karsiyang, India 43 F13 26 56N 88 18 E
Karsun, Russia 24 D8 54 14N 46 57 E
Kartaly, Russia 26 D7 53 3N 60 40 E
Kartapur, India 42 D6 31 27N 75 32 E
Karthaus, U.S.A. 78 E6 41 8N 78 9W
Karufa, Indonesia 37 E8 3 50S 133 20 E
Karumba, Australia 62 B3 17 31S 140 50 E
Karumo, Tanzania 54 C3 2 25S 32 50 E
Karumwa, Tanzania 54 C3 3 12S 32 38 E
Karungu, Kenya 54 C3 0 50S 34 10 E
Karwar, India 40 M9 14 55N 74 13 E
Karwi, India 43 G9 25 12N 80 57 E
Kasache, Malawi 55 E3 13 25S 34 20 E
Kasai →, Zaïre 52 E3 3 30S 16 10 E
Kasai Oriental □, Zaïre 54 C1 5 0S 24 30 E
Kasaji, Zaïre 55 E1 10 25S 23 27 E
Kasama, Zambia 55 E3 10 16S 31 9 E
Kasan-dong, N. Korea 35 D14 41 18N 126 55 E
Kasane, Namibia 56 B3 17 34S 24 50 E
Kasanga, Tanzania 55 D3 8 30S 31 10 E
Kasangulu, Zaïre 52 E3 4 33S 15 15 E
Kasaragod, India 40 N9 12 30N 74 58 E
Kasba L., Canada 73 A8 60 20N 102 10W
Kāseh Garān, Iran 44 C5 34 5N 46 2 E
Kasempa, Zambia 55 E2 13 30S 25 44 E
Kasenga, Zaïre 55 E2 10 20S 28 45 E
Kasese, Uganda 54 B3 0 13N 30 3 E
Kasewa, Zambia 55 E2 14 28S 28 53 E
Kasganj, India 43 F8 27 48N 78 42 E
Kashabowie, Canada 70 C1 48 40N 90 26W
Kāshān, Iran 45 C6 34 5N 51 30 E
Kashi, China 32 C2 39 30N 76 2 E
Kashimbo, Zaïre 55 E2 11 12S 26 19 E
Kashipur, India 43 E8 29 15N 79 0 E
Kashiwazaki, Japan 31 F9 37 22N 138 33 E
Kashk-e Kohneh, Afghan. 40 B3 34 55N 62 30 E
Kāshmar, Iran 45 C8 35 16N 58 26 E
Kashmir, Asia 43 C7 34 0N 76 0 E
Kashmor, Pakistan 42 E3 28 28N 69 32 E
Kashun Noerh = Gaxun Nur, China 32 B5 42 22N 100 30 E
Kasimov, Russia 24 D7 54 55N 41 20 E
Kasinge, Zaïre 54 D2 6 15S 26 58 E
Kasiruta, Indonesia 37 E7 0 25S 127 12 E
Kaskaskia →, U.S.A. 80 G10 37 58N 89 57W
Kaskattama →, Canada 73 B10 57 3N 90 4W
Kaskinen, Finland 8 E16 62 22N 21 15 E
Kaskö, Finland 8 E16 62 22N 21 15 E
Kaslo, Canada 72 D5 49 55N 116 55W
Kasmere L., Canada 73 B8 59 34N 101 10W
Kasongo, Zaïre 54 C2 4 30S 26 33 E
Kasongo Lunda, Zaïre 52 F3 6 35S 16 49 E
Kásos, Greece 21 G12 35 20N 26 55 E
Kassala, Sudan 51 E12 15 30N 36 0 E
Kassalâ □, Sudan 51 E12 15 20N 36 26 E
Kassel, Germany 16 C4 51 19N 9 32 E
Kassiópi, Greece 23 A3 39 48N 19 53 E
Kassue, Indonesia 37 F9 6 58S 139 21 E
Kastamonu, Turkey 25 F5 41 25N 33 43 E
Kastélli, Greece 23 D5 35 29N 23 38 E
Kastéllion, Greece 23 D7 35 12N 25 20 E
Kastoría, Greece 21 D9 40 30N 21 19 E
Kasulu, Tanzania 54 C3 4 37S 30 5 E
Kasulu □, Tanzania 54 C3 4 37S 30 5 E
Kasumi, Japan 31 G7 35 38N 134 38 E
Kasungu, Malawi 55 E3 13 0S 33 29 E
Kasur, Pakistan 42 D6 31 5N 74 25 E
Kata, Russia 27 D11 58 46N 102 40 E
Kataba, Zambia 55 F2 16 5S 25 10 E
Katako Kombe, Zaïre 54 C1 3 25S 24 20 E
Katale, Tanzania 54 C3 4 52S 31 7 E
Katamatite, Australia 63 F4 36 6S 145 41 E
Katanda, Kivu, Zaïre 54 C2 0 55S 29 21 E
Katanda, Shaba, Zaïre 54 D1 7 52S 24 13 E
Katangi, India 40 J11 21 56N 79 50 E
Katangli, Russia 27 D15 51 42N 143 14 E
Katavi Swamp, Tanzania 54 D3 6 50S 31 10 E
Katha, Burma 41 G20 24 10N 96 30 E
Katherine, Australia 60 B5 14 27S 132 20 E
Kathiawar, India 42 H4 22 20N 71 0 E
Kathikas, Cyprus 23 E11 34 55N 32 25 E
Katihar, India 43 G12 25 34N 87 36 E
Katima Mulilo, Zambia 56 B3 17 28S 24 13 E
Katimbira, Malawi 55 E3 12 40S 34 0 E
Katingan = Mendawai →, Indonesia 36 E4 3 30S 113 0 E
Katiola, Ivory C. 50 G3 8 10N 5 10W
Katmandu, Nepal 43 F11 27 45N 85 20 E
Káto Arkhánai, Greece 23 D7 35 15N 25 10 E
Káto Khorió, Greece 23 D7 35 3N 25 47 E
Kato Pyrgos, Cyprus 23 D11 35 11N 32 41 E
Katompe, Zaïre 54 D2 6 2S 26 23 E
Katonga →, Uganda 54 B3 0 34N 31 50 E
Katoomba, Australia 63 E5 33 41S 150 19 E
Katowice, Poland 17 C9 50 17N 19 5 E
Katrine, L., U.K. 12 E4 56 15N 4 30W
Katrineholm, Sweden 9 G14 59 9N 16 12 E
Katsepe, Madag. 57 B8 15 45S 46 15 E
Katsina, Nigeria 50 F6 13 0N 7 32 E
Katsumoto, Japan 31 H4 33 51N 129 42 E
Katsuura, Japan 31 G10 35 10N 140 20 E
Katsuyama, Japan 31 F8 36 3N 136 30 E
Kattaviá, Greece 23 D9 35 57N 27 46 E
Kattegatt, Denmark 9 H11 57 0N 11 20 E
Katumba, Zaïre 54 D2 7 40S 25 17 E
Katungu, Kenya 54 C5 2 55S 40 3 E
Katwa, India 43 H13 23 30N 88 5 E
Katwijk-aan-Zee, Neths. 15 B4 52 12N 4 24 E
Kauai, U.S.A. 74 H15 22 3N 159 30W
Kauai Channel, U.S.A. 74 H15 21 45N 158 50W
Kaufman, U.S.A. 81 J6 32 35N 96 19W
Kaukauna, U.S.A. 76 C1 44 17N 88 17W
Kaukauveld, Namibia 56 C3 20 0S 20 15 E
Kaukonen, Finland 8 C18 67 31N 24 53 E
Kauliranta, Finland 8 C17 66 27N 23 41 E
Kaunas, Lithuania 24 D3 54 54N 23 54 E
Kaura Namoda, Nigeria 50 F6 12 37N 6 33 E
Kautokeino, Norway 8 B17 69 0N 23 4 E
Kavacha, Russia 27 C17 60 16N 169 51 E
Kavalerovo, Russia 30 B7 44 15N 135 4 E
Kavali, India 40 M12 14 55N 80 1 E
Kavála, Greece 21 D11 40 57N 24 28 E
Kavār, Iran 45 D7 29 11N 52 44 E
Kavos, Greece 23 B4 39 23N 20 3 E
Kaw, Fr. Guiana 93 C8 4 30N 52 15W
Kawagama L., Canada 78 A6 45 18N 78 45W
Kawagoe, Japan 31 G9 35 55N 139 29 E
Kawaguchi, Japan 31 G9 35 52N 139 45 E
Kawaihae, U.S.A. 74 H17 20 3N 155 50W
Kawambwa, Zambia 55 D2 9 48S 29 3 E
Kawanoe, Japan 31 G6 34 1N 133 34 E
Kawardha, India 43 J9 22 0N 81 17 E
Kawasaki, Japan 31 G9 35 35N 139 42 E
Kawene, Canada 70 C1 48 45N 91 15W
Kawerau, N.Z. 59 H6 38 7S 176 42 E
Kawhia Harbour, N.Z. 59 H5 38 5S 174 51 E
Kawio, Kepulauan, Indonesia 37 D7 4 30N 125 30 E
Kawnro, Burma 41 H21 22 48N 99 8 E
Kawthoolei = Kawthule □, Burma 41 L20 18 0N 97 30 E
Kawthule □, Burma 41 L20 18 0N 97 30 E
Kaya, Burkina Faso 50 F4 13 4N 1 10W
Kayah □, Burma 41 K20 19 15N 97 15 E
Kayan →, Indonesia 36 D5 2 55N 117 35 E
Kaycee, U.S.A. 82 E10 43 43N 106 38W
Kayeli, Indonesia 37 E7 3 20S 127 10 E
Kayenta, U.S.A. 83 H8 36 44N 110 15W
Kayes, Mali 50 F2 14 25N 11 30W
Kayoa, Indonesia 37 D7 0 1N 127 28 E
Kayomba, Zambia 55 E1 13 11S 24 2 E
Kayrunnera, Australia 63 E3 30 40S 142 30 E
Kayseri, Turkey 25 G6 38 45N 35 30 E
Kaysville, U.S.A. 82 F8 41 2N 111 56W
Kayuagung, Indonesia 36 E2 3 24S 104 50 E
Kazachinskoye, Russia 27 D11 56 16N 107 36 E
Kazachye, Russia 27 B14 70 52N 135 58 E
Kazakhstan ■, Asia 26 E7 50 0N 70 0 E
Kazan, Russia 24 C8 55 48N 49 3 E
Kazan-Rettō, Pac. Oc. 64 E6 25 0N 141 0 E
Kazanlúk, Bulgaria 21 C11 42 38N 25 20 E
Kāzerūn, Iran 45 D6 29 38N 51 40 E
Kazumba, Zaïre 52 F4 6 25S 22 5 E
Kazuno, Japan 30 D10 40 10N 140 45 E
Kazym →, Russia 26 C7 63 54N 65 50 E
Ké-Macina, Mali 50 F3 13 58N 5 22W
Kéa, Greece 21 F11 37 35N 24 22 E
Keams Canyon, U.S.A. 83 J8 35 49N 110 12W
Kearney, U.S.A. 80 E5 40 42N 99 5W
Keban, Turkey 25 G6 38 50N 38 50 E
Kebnekaise, Sweden 8 C15 67 53N 18 33 E
Kebri Dehar, Ethiopia 46 F3 6 45N 44 17 E
Kebumen, Indonesia 37 G13 7 42S 109 40 E
Kechika →, Canada 72 B3 59 41N 127 12W
Kecskemét, Hungary 17 E9 46 57N 19 42 E
Kedgwick, Canada 71 C6 47 40N 67 20W
Kédhros Óros, Greece 23 D6 35 11N 24 37 E
Kedia Hill, Botswana 56 C3 21 28S 24 37 E
Kediri, Indonesia 37 G15 7 51S 112 1 E
Kédougou, Senegal 50 F2 12 35N 12 10W
Keeler, U.S.A. 84 J9 36 29N 117 52W
Keeley L., Canada 73 C7 54 54N 108 8W
Keeling Is. = Cocos Is., Ind. Oc. 64 J1 12 10S 96 55 E
Keelung = Chilung, Taiwan 33 D7 25 3N 121 45 E
Keene, Calif., U.S.A. 85 K8 35 13N 118 33W
Keene, N.H., U.S.A. 79 D12 42 56N 72 17W
Keeper Hill, Ireland 13 D3 52 46N 8 17W
Keer-Weer, C., Australia 62 A3 14 0S 141 32 E
Keeseville, U.S.A. 79 B11 44 29N 73 30W
Keetmanshoop, Namibia 56 D2 26 35S 18 8 E
Keewatin, U.S.A. 80 B8 47 24N 93 5W
Keewatin □, Canada 73 A9 63 20N 95 0W
Keewatin →, Canada 73 B8 56 29N 100 46W
Kefallinía, Greece 21 E9 38 20N 20 30 E
Kefamenanu, Indonesia 37 F6 9 28S 124 29 E
Keffi, Nigeria 50 G6 8 55N 7 43 E
Keflavík, Iceland 8 D2 64 2N 22 35W
Keg River, Canada 72 B5 57 54N 117 55W
Kegaska, Canada 71 B7 50 9N 61 18W
Keighley, U.K. 10 D6 53 52N 1 54W
Keimoes, S. Africa 56 D3 28 41S 20 59 E
Keith, Australia 63 F3 36 6S 140 20 E
Keith, U.K. 12 D6 57 33N 2 58W
Keith Arm, Canada 68 B7 64 20N 122 15W
Kejser Franz Joseph Fjord = Kong Franz Joseph Fd., Greenland 4 B6 73 30N 24 30W
Kekri, India 42 G6 26 0N 75 10 E
Kël, Russia 27 C13 69 30N 124 10 E
Kelan, China 34 E6 38 43N 111 31 E
Kelang, Malaysia 39 L3 3 2N 101 26 E
Kelantan →, Malaysia 39 J4 6 13N 102 14 E
Kelibia, Tunisia 51 A7 36 50N 11 3 E
Kellé, Congo 52 E2 0 8S 14 38 E
Keller, U.S.A. 82 B4 48 5N 118 41W
Kellerberrin, Australia 61 F2 31 36S 117 38 E
Kellett, C., Canada 4 B1 72 0N 126 0W
Kelleys I., U.S.A. 78 E2 41 36N 82 42W
Kellogg, U.S.A. 82 C5 47 32N 116 7W
Kelloselkä, Finland 8 C20 66 56N 28 53 E

Kells

132

Kontum

Kingfisher, *U.S.A.* **81 H6** 35 52N 97 56W
Kingirbān, *Iraq* **44 C5** 34 40N 44 54 E
Kingman, *Ariz., U.S.A.* **85 K12** 35 12N 114 4W
Kingman, *Kans., U.S.A.* **81 G5** 37 39N 98 7W
Kingoonya, *Australia* **63 E2** 30 55S 135 19 E
Kings →, *U.S.A.* **83 H4** 36 3N 119 50W
Kings Canyon National Park, *U.S.A.* **83 H4** 36 50N 118 40W
King's Lynn, *U.K.* **10 E8** 52 45N 0 25 E
Kings Mountain, *U.S.A.* **77 H5** 35 15N 81 20W
King's Peak, *U.S.A.* **82 F8** 40 46N 110 27W
Kingsbridge, *U.K.* **11 G4** 50 17N 3 46W
Kingsburg, *U.S.A.* **83 H4** 36 31N 119 33W
Kingscote, *Australia* **63 F2** 35 40S 137 38 E
Kingscourt, *Ireland* **13 C5** 53 55N 6 48W
Kingsley, *U.S.A.* **80 D7** 42 35N 95 58W
Kingsport, *U.S.A.* **77 G4** 36 33N 82 33W
Kingston, *Canada* **70 D4** 44 14N 76 30W
Kingston, *Jamaica* **88 C4** 18 0N 76 50W
Kingston, *N.Z.* **59 L2** 45 20S 168 43 E
Kingston, *N.Y., U.S.A.* **79 E10** 41 56N 73 59W
Kingston, *Pa., U.S.A.* **79 E9** 41 16N 75 54W
Kingston, *R.I., U.S.A.* **79 E13** 41 29N 71 30W
Kingston Pk., *U.S.A.* **85 K11** 35 45N 115 54W
Kingston South East, *Australia* **63 F2** 36 51S 139 55 E
Kingston upon Hull, *U.K.* **10 D7** 53 45N 0 20W
Kingston-upon-Thames, *U.K.* **11 F7** 51 23N 0 20W
Kingstown, *St. Vincent* **89 D7** 13 10N 61 10W
Kingstree, *U.S.A.* **77 J6** 33 40N 79 50W
Kingsville, *Canada* **70 D3** 42 2N 82 45W
Kingsville, *U.S.A.* **81 M6** 27 31N 97 52W
Kingussie, *U.K.* **12 D4** 57 5N 4 2W
Kinistino, *Canada* **73 C7** 52 57N 105 2W
Kinkala, *Congo* **52 E2** 4 18S 14 49 E
Kinki □, *Japan* **31 H8** 33 45N 136 0 E
Kinleith, *N.Z.* **59 H5** 38 20S 175 56 E
Kinmount, *Canada* **78 B6** 44 48N 78 45W
Kinnaird, *Canada* **72 D5** 49 17N 117 39W
Kinnairds Hd., *U.K.* **12 D7** 57 40N 2 0W
Kinnarodden, *Norway* **6 A11** 71 8N 27 40 E
Kino, *Mexico* **86 B2** 28 45N 111 59W
Kinoje →, *Canada* **70 B3** 52 8N 81 25W
Kinomoto, *Japan* **31 G8** 35 30N 136 13 E
Kinoni, *Uganda* **54 C3** 0 41S 30 28 E
Kinross, *U.K.* **12 E5** 56 13N 3 25W
Kinsale, *Ireland* **13 E3** 51 42N 8 31W
Kinsale, Old Hd. of, *Ireland* **13 E3** 51 37N 8 32W
Kinsha = Chang Jiang →, *China* **33 C7** 31 48N 121 10 E
Kinshasa, *Zaïre* **52 E3** 4 20S 15 15 E
Kinsley, *U.S.A.* **81 G5** 37 55N 99 25W
Kinston, *U.S.A.* **77 H7** 35 16N 77 35W
Kintampo, *Ghana* **50 G4** 8 5N 1 41W
Kintap, *Indonesia* **36 E5** 3 51S 115 13 E
Kintore Ra., *Australia* **60 D4** 23 15S 128 47 E
Kintyre, *U.K.* **12 F3** 55 30N 5 35W
Kintyre, Mull of, *U.K.* **12 F3** 55 17N 5 55W
Kinushseo →, *Canada* **70 A3** 55 15N 83 45W
Kinuso, *Canada* **72 B5** 55 20N 115 25W
Kinyangiri, *Tanzania* **54 C3** 4 25S 34 37 E
Kinzua, *U.S.A.* **78 E6** 41 52N 78 58W
Kinzua Dam, *U.S.A.* **78 E5** 41 53N 79 0W
Kiosk, *Canada* **70 C4** 46 6N 78 53W
Kiowa, *Kans., U.S.A.* **81 G5** 37 1N 98 29W
Kiowa, *Okla., U.S.A.* **81 H7** 34 43N 95 54W
Kipahigan L., *Canada* **73 B8** 55 20N 101 55W
Kipanga, *Tanzania* **54 D4** 6 15S 35 20 E
Kiparissía, *Greece* **21 F9** 37 15N 21 40 E
Kiparissiakós Kólpos, *Greece* **21 F9** 37 25N 21 25 E
Kipembawe, *Tanzania* **54 D3** 7 38S 33 27 E
Kipengere Ra., *Tanzania* **55 D3** 9 12S 34 15 E
Kipili, *Tanzania* **54 D3** 7 28S 30 32 E
Kipini, *Kenya* **54 C5** 2 30S 40 32 E
Kipling, *Canada* **73 C8** 50 6N 102 38W
Kippure, *Ireland* **13 C5** 53 11N 6 23W
Kipushi, *Zaïre* **55 E2** 11 48S 27 12 E
Kiratpur, *India* **42 E8** 29 32N 78 12 E
Kirensk, *Russia* **27 D11** 57 50N 107 55 E
Kirgella Rocks, *Australia* **61 F3** 30 5S 122 50 E
Kirghizia ■, *Asia* **26 E8** 42 0N 75 0 E
Kirghizstan = Kirghizia ■, *Asia* **26 E8** 42 0N 75 0 E
Kirgizia = Kirghizia ■, *Asia* **26 E8** 42 0N 75 0 E
Kirgiziya Steppe, *Kazakhstan* **25 D10** 50 0N 55 0 E
Kiri, *Zaïre* **52 E3** 1 29S 19 0 E
Kiribati ■, *Pac. Oc.* **64 H10** 5 0S 180 0 E
Kırıkkale, *Turkey* **25 G5** 39 51N 33 32 E
Kirillov, *Russia* **24 C6** 59 51N 38 14 E
Kirin = Jilin, *China* **35 C14** 43 44N 126 30 E
Kirin = Jilin □, *China* **35 C13** 44 0N 127 0 E
Kiritimati, *Kiribati* **65 G12** 1 58N 157 27W
Kirkcaldy, *U.K.* **12 E5** 56 7N 3 10W
Kirkcudbright, *U.K.* **12 G4** 54 50N 4 3W
Kirkee, *India* **40 K8** 18 34N 73 56 E
Kirkenes, *Norway* **8 B21** 69 40N 30 5 E
Kirkintilloch, *U.K.* **12 F4** 55 57N 4 10W
Kirkjubæjarklaustur, *Iceland* **8 E4** 63 47N 18 4W
Kirkland, *U.S.A.* **83 J7** 34 25N 112 43W
Kirkland Lake, *Canada* **70 C3** 48 9N 80 2W
Kırklareli, *Turkey* **21 D12** 41 44N 27 15 E
Kirksville, *U.S.A.* **80 E8** 40 12N 92 35W
Kirkūk, *Iraq* **44 C5** 35 30N 44 21 E
Kirkwall, *U.K.* **12 C6** 58 59N 2 59W
Kirkwood, *S. Africa* **56 E4** 33 22S 25 15 E
Kirov = Vyatka, *Russia* **26 D5** 58 35N 49 40 E
Kirovabad = Gyandzha, *Azerbaijan* **25 F8** 40 45N 46 20 E
Kirovakan = Karaklis, *Armenia* **25 F7** 40 48N 44 30 E
Kirovograd, *Ukraine* **25 E5** 48 35N 32 20 E
Kirovohrad = Kirovograd, *Ukraine* **25 E5** 48 35N 32 20 E
Kirovsk, *Russia* **24 A5** 67 48N 33 50 E
Kirovsk, *Turkmenistan* **26 F7** 37 42N 60 23 E
Kirovskiy, *Russia* **27 D16** 54 27N 155 42 E
Kirovskiy, *Russia* **30 B6** 45 7N 133 30 E
Kirriemuir, *U.K.* **12 E6** 56 41N 2 58W
Kirsanov, *Russia* **24 D7** 52 35N 42 40 E
Kırşehir, *Turkey* **25 G5** 39 14N 34 5 E
Kirstonia, *S. Africa* **56 D3** 25 30S 23 45 E
Kirthar Range, *Pakistan* **42 F2** 27 0N 67 0 E
Kiruna, *Sweden* **8 C16** 67 52N 20 15 E
Kirundu, *Zaïre* **54 C2** 0 50S 25 35 E
Kirup, *Australia* **61 F2** 33 40S 115 50 E
Kiryū, *Japan* **31 F9** 36 24N 139 20 E
Kisaga, *Tanzania* **54 C3** 4 30S 34 23 E
Kisalaya, *Nic.* **88 D3** 14 40N 84 3W
Kisámou, Kólpos, *Greece* **23 D5** 35 30N 23 38 E
Kisanga, *Zaïre* **54 B2** 2 30N 26 35 E
Kisangani, *Zaïre* **54 B2** 0 35N 25 15 E
Kisar, *Indonesia* **37 F7** 8 5S 127 10 E
Kisaran, *Indonesia* **36 D1** 3 0N 99 37 E
Kisarawe, *Tanzania* **54 D4** 6 53S 39 0 E
Kisarawe □, *Tanzania* **54 D4** 7 3S 39 0 E
Kisarazu, *Japan* **31 G9** 35 23N 139 55 E
Kiselevsk, *Russia* **26 D9** 54 5N 86 39 E
Kishanganga →, *Pakistan* **43 B5** 34 18N 73 28 E
Kishanganj, *India* **43 F13** 26 3N 88 14 E
Kishangarh, *India* **42 F4** 27 50N 70 30 E
Kishinev, *Moldavia* **25 E4** 47 0N 28 50 E
Kishiwada, *Japan* **31 G7** 34 28N 135 22 E
Kishtwar, *India* **43 C6** 33 20N 75 48 E
Kisii, *Kenya* **54 C3** 0 40S 34 45 E
Kisii □, *Kenya* **54 C3** 0 40S 34 45 E
Kisiju, *Tanzania* **54 D4** 7 23S 39 19 E
Kisizi, *Uganda* **54 C2** 1 0S 29 58 E
Kiska I., *U.S.A.* **68 C1** 51 59N 177 30 E
Kiskatinaw →, *Canada* **72 B4** 56 8N 120 10W
Kiskittogisu L., *Canada* **73 C9** 54 13N 98 20W
Kiskörös, *Hungary* **17 E9** 46 37N 19 20 E
Kiskunfélegyháza, *Hungary* **17 E9** 46 42N 19 53 E
Kiskunhalas, *Hungary* **17 E9** 46 28N 19 37 E
Kislovodsk, *Russia* **25 F7** 43 50N 42 45 E
Kiso-Gawa →, *Japan* **31 G8** 35 20N 136 45 E
Kiso-Sammyaku, *Japan* **31 G8** 35 45N 137 45 E
Kisofukushima, *Japan* **31 G8** 35 52N 137 43 E
Kisoro, *Uganda* **54 C2** 1 17S 29 48 E
Kissidougou, *Guinea* **50 G2** 9 5N 10 5W
Kissimmee, *U.S.A.* **77 L5** 28 18N 81 24W
Kissimmee →, *U.S.A.* **77 M5** 27 9N 80 52W
Kississing L., *Canada* **73 B8** 55 10N 101 20W
Kissónerga, *Cyprus* **23 E11** 34 49N 32 24 E
Kisumu, *Kenya* **54 C3** 0 3S 34 45 E
Kiswani, *Tanzania* **54 C4** 4 5S 37 57 E
Kiswere, *Tanzania* **55 D4** 9 27S 39 30 E
Kit Carson, *U.S.A.* **80 F3** 38 46N 102 48W
Kita, *Mali* **50 F3** 13 5N 9 25W
Kitab, *Uzbekistan* **26 F7** 39 7N 66 52 E
Kitaibaraki, *Japan* **31 F10** 36 50N 140 45 E
Kitakami, *Japan* **30 E10** 39 20N 141 10 E
Kitakami-Gawa →, *Japan* **30 E10** 38 25N 141 19 E
Kitakami-Sammyaku, *Japan* **30 E10** 39 30N 141 30 E
Kitakata, *Japan* **30 F9** 37 39N 139 52 E
Kitakyūshū, *Japan* **31 H5** 33 50N 130 50 E
Kitale, *Kenya* **54 B4** 1 0N 35 0 E
Kitami, *Japan* **30 C11** 43 48N 143 54 E
Kitami-Sammyaku, *Japan* **30 B11** 44 22N 142 43 E
Kitangiri, L., *Tanzania* **54 C3** 4 5S 34 20 E
Kitaya, *Tanzania* **55 E5** 10 38S 40 8 E
Kitchener, *Australia* **61 F3** 30 55S 124 8 E
Kitchener, *Canada* **70 D3** 43 27N 80 29W
Kitega = Gitega, *Burundi* **54 C2** 3 26S 29 56 E
Kitengo, *Zaïre* **54 D1** 7 26S 24 8 E
Kiteto □, *Tanzania* **54 C4** 5 0S 37 0 E
Kitgum, *Uganda* **54 B3** 3 17N 32 52 E
Kíthira, *Greece* **21 F10** 36 9N 23 12 E
Kíthnos, *Greece* **21 F11** 37 26N 24 27 E
Kiti, *Cyprus* **23 E12** 34 50N 33 34 E
Kiti, C., *Cyprus* **23 E12** 34 48N 33 36 E
Kitikmeot □, *Canada* **68 A9** 70 0N 110 0W
Kitimat, *Canada* **72 C3** 54 3N 128 38W
Kitinen →, *Finland* **8 C19** 67 34N 26 40 E
Kitsuki, *Japan* **31 H5** 33 25N 131 37 E
Kittakittaooloo, L., *Australia* **63 D2** 28 3S 138 14 E
Kittanning, *U.S.A.* **78 F5** 40 49N 79 31W
Kittatinny Mts., *U.S.A.* **79 E10** 41 0N 75 0W
Kittery, *U.S.A.* **77 D10** 43 5N 70 45W
Kitui, *Kenya* **54 C4** 1 17S 38 0 E
Kitui □, *Kenya* **54 C4** 1 30S 38 25 E
Kitwe, *Zambia* **55 E2** 12 54S 28 13 E
Kivalo, *Finland* **8 C19** 66 18N 26 0 E
Kivarli, *India* **42 G5** 24 33N 72 46 E
Kividhes, *Cyprus* **23 E11** 34 46N 32 51 E
Kivu □, *Zaïre* **54 C2** 3 10S 27 0 E
Kivu, L., *Zaïre* **54 C2** 1 48S 29 0 E
Kiyev, *Ukraine* **25 D5** 50 30N 30 28 E
Kiyevskoye Vdkhr., *Ukraine* **25 D5** 51 0N 30 25 E
Kizel, *Russia* **24 C10** 59 3N 57 40 E
Kiziguru, *Rwanda* **54 C3** 1 46S 30 23 E
Kızıl Irmak →, *Turkey* **25 F6** 41 44N 35 58 E
Kizil Jilga, *India* **43 B8** 35 26N 78 50 E
Kizimkazi, *Tanzania* **54 D4** 6 28S 39 30 E
Kizlyar, *Russia* **25 F8** 43 51N 46 40 E
Kizyl-Arvat, *Turkmenistan* **26 F6** 38 58N 56 15 E
Kladno, *Czech.* **16 C7** 50 10N 14 7 E
Klaeng, *Thailand* **38 F3** 12 47N 101 39 E
Klagenfurt, *Austria* **16 E7** 46 38N 14 20 E
Klaipéda, *Lithuania* **24 C3** 55 43N 21 10 E
Klamath →, *U.S.A.* **82 F1** 41 33N 124 5W
Klamath Falls, *U.S.A.* **82 E3** 42 13N 121 46W
Klamath Mts., *U.S.A.* **82 F2** 41 20N 123 0W
Klappan →, *Canada* **72 B3** 58 0N 129 43W
Klarälven →, *Sweden* **9 G12** 59 23N 13 32 E
Klaten, *Indonesia* **37 G14** 7 43S 110 36 E
Klatovy, *Czech.* **16 D6** 49 23N 13 18 E
Klawer, *S. Africa* **56 E2** 31 44S 18 36 E
Klawock, *U.S.A.* **72 B2** 55 33N 133 6W
Kleena Kleene, *Canada* **72 C4** 52 0N 124 59W
Klein, *U.S.A.* **82 C9** 46 24N 108 33W
Klein-Karas, *Namibia* **56 D2** 27 33S 18 7 E
Klerksdorp, *S. Africa* **56 D4** 26 53S 26 38 E
Kletskiy, *Russia* **26 E5** 49 20N 43 0 E
Klickitat, *U.S.A.* **82 D3** 45 49N 121 9W
Klickitat →, *U.S.A.* **84 E5** 45 42N 121 17W
Klidhes, *Cyprus* **23 D13** 35 42N 34 36 E
Klin, *Russia* **26 D4** 56 20N 36 48 E
Klinaklini →, *Canada* **72 C3** 51 21N 125 40W
Klipdale, *S. Africa* **56 E2** 34 19S 19 57 E
Klipplaat, *S. Africa* **56 E3** 33 1S 24 22 E
Kłodzko, *Poland* **16 C8** 50 28N 16 38 E
Klondike, *Canada* **68 B6** 64 0N 139 26W
Klouto, *Togo* **50 G5** 6 57N 0 44 E
Kluane L., *Canada* **68 B6** 61 15N 138 40W
Klyuchevskaya, Gora, *Russia* **27 D17** 55 50N 160 30 E
Knaresborough, *U.K.* **10 C6** 54 1N 1 29W
Knee L., *Man., Canada* **73 B10** 55 3N 94 45W
Knee L., *Sask., Canada* **73 B7** 55 51N 107 0W
Knight Inlet, *Canada* **72 C3** 50 45N 125 40W
Knighton, *U.K.* **11 E4** 52 21N 3 2W
Knights Ferry, *U.S.A.* **84 H6** 37 50N 120 40W
Knights Landing, *U.S.A.* **84 G5** 38 48N 121 43W
Knob, C., *Australia* **61 F2** 34 32S 119 16 E
Knockmealdown Mts., *Ireland* **13 D4** 52 16N 8 0W
Knokke, *Belgium* **15 C3** 51 20N 3 17 E
Knossós, *Greece* **23 D7** 35 16N 25 10 E
Knox, *U.S.A.* **76 E2** 41 18N 86 37W
Knox, C., *Canada* **72 C2** 54 11N 133 5W
Knox City, *U.S.A.* **81 J5** 33 25N 99 49W
Knox Coast, *Antarctica* **5 C8** 66 30S 108 0 E
Knoxville, *Iowa, U.S.A.* **80 E8** 41 19N 93 6W
Knoxville, *Tenn., U.S.A.* **77 H4** 35 58N 83 55W
Knysna, *S. Africa* **56 E3** 34 2S 23 2 E
Ko Kha, *Thailand* **38 C2** 18 11N 99 24 E
Ko Tao, *Thailand* **39 G2** 10 6N 99 48 E
Koartac, *Canada* **69 B13** 60 55N 69 40W
Koba, *Aru, Indonesia* **37 F8** 6 37S 134 37 E
Koba, *Bangka, Indonesia* **36 E3** 2 26S 106 14 E
Kobarid, *Slovenia* **20 A5** 46 15N 13 30 E
Kobayashi, *Japan* **31 J5** 31 56N 130 59 E
Kobdo = Hovd, *Mongolia* **32 B4** 48 2N 91 37 E
Kōbe, *Japan* **31 G7** 34 45N 135 10 E
København, *Denmark* **9 J12** 55 41N 12 34 E
Kōbi-Sho, *Japan* **31 M1** 25 56N 123 41 E
Koblenz, *Germany* **16 C3** 50 21N 7 36 E
Kobroor, Kepulauan, *Indonesia* **37 F8** 6 10S 134 30 E
Kocaeli = İzmit, *Turkey* **25 F4** 40 45N 29 50 E
Kočani, *Macedonia* **21 D10** 41 55N 22 25 E
Kočevje, *Slovenia* **20 B6** 45 39N 14 50 E
Koch Bihar, *India* **41 F16** 26 22N 89 29 E
Kochang, *S. Korea* **35 G14** 35 41N 127 55 E
Kochas, *India* **43 G10** 25 15N 83 56 E
Kocheya, *Russia* **27 D13** 52 32N 120 42 E
Kōchi, *Japan* **31 H6** 33 30N 133 35 E
Kōchi □, *Japan* **31 H6** 33 40N 133 30 E
Kochiu = Gejiu, *China* **32 D5** 23 20N 103 10 E
Kodiak, *U.S.A.* **68 C4** 57 47N 152 24W
Kodiak I., *U.S.A.* **68 C4** 57 30N 152 45W
Kodinar, *India* **42 J4** 20 46N 70 46 E
Koes, *Namibia* **56 D2** 26 0S 19 15 E
Koffiefontein, *S. Africa* **56 D4** 29 30S 25 0 E
Kofiau, *Indonesia* **37 E7** 1 11S 129 50 E
Koforidua, *Ghana* **50 G4** 6 3N 0 17W
Kōfu, *Japan* **31 G9** 35 40N 138 30 E
Koga, *Japan* **31 F9** 36 11N 139 43 E
Kogaluk →, *Canada* **71 A7** 56 12N 61 44W
Kogan, *Australia* **63 D5** 27 2S 150 40 E
Koh-i-Bābā, *Afghan.* **40 B5** 34 30N 67 0 E
Koh-i-Khurd, *Afghan.* **42 C1** 33 30N 65 59 E
Kohat, *Pakistan* **42 C4** 33 40N 71 29 E
Kohima, *India* **41 G19** 25 35N 94 10 E
Kohkīlūyeh va Būyer Aḥmadi □, *Iran* **45 D6** 31 30N 50 30 E
Kohler Ra., *Antarctica* **5 D15** 77 0S 110 0W
Koin-dong, *N. Korea* **35 D14** 40 28N 126 18 E
Kojŏ, *N. Korea* **35 E14** 38 58N 127 58 E
Kojonup, *Australia* **61 F2** 33 48S 117 10 E
Kojūr, *Iran* **45 B6** 36 23N 51 43 E
Kokand, *Uzbekistan* **26 E8** 40 30N 70 57 E
Kokanee Glacier Prov. Park, *Canada* **72 D5** 49 47N 117 10W
Kokas, *Indonesia* **37 E8** 2 42S 132 26 E
Kokchetav, *Kazakhstan* **26 D7** 53 20N 69 25 E
Kokemäenjoki, *Finland* **9 F16** 61 32N 21 44 E
Kokkola, *Finland* **8 E17** 63 50N 23 8 E
Koko Kyunzu, *Burma* **41 M18** 14 10N 93 25 E
Kokomo, *U.S.A.* **76 E2** 40 29N 86 8W
Kokonau, *Indonesia* **37 E9** 4 43S 136 26 E
Koksan, *N. Korea* **35 E14** 38 46N 126 40 E
Koksoak →, *Canada* **69 C13** 58 30N 68 10W
Kokstad, *S. Africa* **57 E4** 30 32S 29 29 E
Kokubu, *Japan* **31 J5** 31 44N 130 46 E
Kokuora, *Russia* **27 B15** 71 35N 144 50 E
Kola, *Indonesia* **37 F8** 5 35S 134 30 E
Kola, *Russia* **24 A5** 68 45N 33 8 E
Kola Pen. = Kolskiy Poluostrov, *Russia* **24 A6** 67 30N 38 0 E
Kolahoi, *India* **43 B6** 34 12N 75 22 E
Kolaka, *Indonesia* **37 E6** 4 3S 121 46 E
Kolar, *India* **40 N11** 13 12N 78 15 E
Kolar Gold Fields, *India* **40 N11** 12 58N 78 16 E
Kolari, *Finland* **8 C17** 67 20N 23 48 E
Kolayat, *India* **40 F8** 27 50N 72 50 E
Kolchugino = Leninsk-Kuznetskiy, *Russia* **26 D9** 54 44N 86 10 E
Kolda, *Senegal* **50 F2** 12 55N 14 57W
Kolding, *Denmark* **9 J10** 55 30N 9 29 E
Kole, *Zaïre* **52 E4** 3 16S 22 42 E
Kolepom = Yos Sudarso, Pulau, *Indonesia* **37 F9** 8 0S 138 30 E
Kolguyev, Ostrov, *Russia* **24 A8** 69 20N 48 30 E
Kolhapur, *India* **40 L9** 16 43N 74 15 E
Kolín, *Czech.* **16 C7** 50 2N 15 9 E
Kolmanskog, *Namibia* **56 D2** 26 45S 15 14 E
Köln, *Germany* **16 C3** 50 56N 6 58 E
Koło, *Poland* **17 B9** 52 14N 18 40 E
Kołobrzeg, *Poland* **16 A7** 54 10N 15 35 E
Kolokani, *Mali* **50 F3** 13 35N 7 45W
Kolomna, *Russia* **24 C6** 55 8N 38 45 E
Kolomyya, *Ukraine* **25 E4** 48 31N 25 2 E
Kolonodale, *Indonesia* **37 E6** 2 3S 121 25 E
Kolosib, *India* **41 G18** 24 15N 92 45 E
Kolpashevo, *Russia* **26 D9** 58 20N 83 5 E
Kolpino, *Russia* **24 C5** 59 44N 30 39 E
Kolskiy Poluostrov, *Russia* **24 A6** 67 30N 38 0 E
Kolskiy Zaliv, *Russia* **24 A5** 69 23N 34 0 E
Kolwezi, *Zaïre* **55 E2** 10 40S 25 25 E
Kolyma →, *Russia* **27 C17** 69 30N 161 0 E
Kolymskoye, Okhotsko, *Russia* **27 C16** 63 0N 157 0 E
Komandorskiye Is. = Komandorskiye Ostrova, *Russia* **27 D17** 55 0N 167 0 E
Komandorskiye Ostrova, *Russia* **27 D17** 55 0N 167 0 E
Komárno, *Slovakia* **17 E9** 47 49N 18 5 E
Komatipoort, *S. Africa* **57 D5** 25 25S 31 55 E
Komatou Yialou, *Cyprus* **23 D13** 35 25N 34 8 E
Komatsu, *Japan* **31 F8** 36 25N 136 30 E
Komatsujima, *Japan* **31 H7** 34 0N 134 35 E
Komi Republic □, *Russia* **24 B10** 64 0N 55 0 E
Kommunarsk, *Ukraine* **25 E6** 48 30N 38 45 E
Kommunizma, Pik, *Tajikistan* **26 F8** 39 0N 72 2 E
Komodo, *Indonesia* **37 F5** 8 37S 119 20 E
Komono, *Congo* **52 E2** 3 10S 13 20 E
Komoran, Pulau, *Indonesia* **37 F9** 8 18S 138 45 E
Komoro, *Japan* **31 F9** 36 19N 138 26 E
Komotini, *Greece* **21 D11** 41 9N 25 26 E
Kompasberg, *S. Africa* **56 E3** 31 45S 24 32 E
Kompong Bang, *Cambodia* **39 F5** 12 24N 104 40 E
Kompong Cham, *Cambodia* **39 F5** 12 0N 105 30 E
Kompong Chhnang, *Cambodia* **39 F5** 12 20N 104 35 E
Kompong Chikreng, *Cambodia* **38 F5** 13 5N 104 18 E
Kompong Kleang, *Cambodia* **38 F5** 13 6N 104 8 E
Kompong Luong, *Cambodia* **39 G5** 11 49N 104 48 E
Kompong Pranak, *Cambodia* **38 F5** 13 35N 104 55 E
Kompong Som, *Cambodia* **39 G4** 10 38N 103 30 E
Kompong Som, Chhung, *Cambodia* **39 G4** 10 50N 103 32 E
Kompong Speu, *Cambodia* **39 G5** 11 26N 104 32 E
Kompong Sralao, *Cambodia* **38 E5** 14 5N 105 46 E
Kompong Thom, *Cambodia* **38 F5** 12 35N 104 51 E
Kompong Trabeck, *Cambodia* **38 F5** 13 6N 105 14 E
Kompong Trabeck, *Cambodia* **39 G5** 11 9N 105 28 E
Kompong Trach, *Cambodia* **39 G5** 11 25N 105 48 E
Kompong Tralach, *Cambodia* **39 G5** 11 54N 104 47 E
Komsberg, *S. Africa* **56 E3** 32 40S 20 45 E
Komsomolets, Ostrov, *Russia* **27 A10** 80 30N 95 0 E
Komsomolsk, *Russia* **27 D14** 50 30N 137 0 E
Konarhá □, *Afghan.* **40 B7** 35 30N 71 3 E
Konārī, *Iran* **45 D6** 28 13N 51 36 E
Konawa, *U.S.A.* **81 H6** 34 58N 96 45W
Konch, *India* **43 G8** 26 0N 79 10 E
Kondakovo, *Russia* **27 C16** 69 36N 152 0 E
Konde, *Tanzania* **54 C4** 4 57S 39 45 E
Kondinin, *Australia* **61 F2** 32 34S 118 8 E
Kondoa, *Tanzania* **54 C4** 4 55S 35 50 E
Kondoa □, *Tanzania* **54 D4** 5 0S 36 0 E
Kondókali, *Greece* **23 A3** 39 38N 19 51 E
Kondopaga, *Russia* **24 B5** 62 12N 34 17 E
Kondratyevo, *Russia* **27 D10** 57 22N 98 15 E
Konduga, *Nigeria* **51 F7** 11 35N 13 26 E
Konevo, *Russia* **24 B6** 62 8N 39 20 E
Kong, *Ivory C.* **50 G4** 8 54N 4 36W
Kong →, *Cambodia* **38 F5** 13 32N 105 58 E
Kong, Koh, *Cambodia* **39 G4** 11 20N 103 0 E
Kong Christian IX.s Land, *Greenland* **4 C6** 68 0N 36 0W
Kong Christian X.s Land, *Greenland* **4 B6** 74 0N 29 0W
Kong Franz Joseph Fd., *Greenland* **4 B6** 73 30N 24 30W
Kong Frederik IX.s Land, *Greenland* **4 C5** 67 0N 52 0W
Kong Frederik VI.s Kyst, *Greenland* **4 C5** 63 0N 43 0W
Kong Frederik VIII.s Land, *Greenland* **4 B6** 78 30N 26 0W
Kong Oscar Fjord, *Greenland* **4 B6** 72 20N 24 0W
Kongju, *S. Korea* **35 F14** 36 30N 127 0 E
Konglu, *Burma* **41 F20** 27 13N 97 57 E
Kongolo, *Kasai Or., Zaïre* **54 D1** 5 26S 24 49 E
Kongolo, *Shaba, Zaïre* **54 D2** 5 22S 27 0 E
Kongor, *Sudan* **51 G11** 7 1N 31 27 E
Kongsberg, *Norway* **9 G10** 59 39N 9 39 E
Kongsvinger, *Norway* **9 F12** 60 12N 12 2 E
Kongwa, *Tanzania* **54 D4** 6 11S 36 26 E
Koni, *Zaïre* **55 E2** 10 40S 27 11 E
Koni, Mts., *Zaïre* **55 E2** 10 36S 27 10 E
Königsberg = Kaliningrad, *Russia* **24 D3** 54 42N 20 32 E
Konin, *Poland* **17 B9** 52 12N 18 15 E
Konjic, *Bos.-H.* **21 C7** 43 42N 17 58 E
Konkiep, *Namibia* **56 D2** 26 49S 17 15 E
Konosha, *Russia* **24 B7** 61 0N 40 5 E
Kōnosu, *Japan* **31 F9** 36 3N 139 31 E
Konotop, *Ukraine* **25 D5** 51 12N 33 7 E
Końskie, *Poland* **17 C10** 51 15N 20 23 E
Konstanz, *Germany* **16 E4** 47 39N 9 10 E
Kont, *Iran* **45 E9** 26 55N 61 50 E
Kontagora, *Nigeria* **50 F6** 10 23N 5 27 E
Kontum, *Vietnam* **38 E7** 14 24N 108 0 E

Lankao

Kwamouth, Zaïre 52 E3 3 9S 16 12 E
Kwando ~►, Africa 56 B3 18 27S 23 32 E
Kwangdaeri, N. Korea 35 D14 40 31N 127 32 E
Kwangju, S. Korea 35 G14 35 9N 126 54 E
Kwangsi-Chuang = Guangxi Zhuangzu Zizhiqu □, China 33 D5 24 0N 109 0 E
Kwangtung = Guangdong □, China 33 D6 23 0N 113 0 E
Kwataboahegan ~►, Canada 70 B3 51 9N 80 50W
Kwatisore, Indonesia 37 E8 3 18S 134 50 E
Kweichow = Guizhou □, China 32 D5 27 0N 107 0 E
Kwekwe, Zimbabwe 55 F2 18 58S 29 48 E
Kwimba □, Tanzania 54 C3 3 0S 33 0 E
Kwinana New Town, Australia 61 F2 32 15S 115 47 E
Kwoka, Indonesia 37 E8 0 31S 132 27 E
Kyabé, Chad 51 G8 9 30N 19 0 E
Kyabra Cr. ~►, Australia 63 D3 25 36S 142 55 E
Kyabram, Australia 63 F4 36 19S 145 4 E
Kyaikto, Burma 38 D1 17 20N 97 3 E
Kyakhta, Russia 27 D11 50 30N 106 25 E
Kyancutta, Australia 63 E2 33 8S 135 33 E
Kyangin, Burma 41 K19 18 20N 95 20 E
Kyaukpadaung, Burma 41 J19 20 52N 95 8 E
Kyaukpyu, Burma 41 K18 19 28N 93 30 E
Kyaukse, Burma 41 J20 21 36N 96 10 E
Kyburz, U.S.A. 84 G6 38 47N 120 18W
Kycen, Russia 27 D11 51 45N 101 45 E
Kyenjojo, Uganda 54 B3 0 40N 30 37 E
Kyle Dam, Zimbabwe 55 G3 20 15S 31 0 E
Kyle of Lochalsh, U.K. 12 D3 57 17N 5 43W
Kyneton, Australia 63 F3 37 10S 144 29 E
Kynuna, Australia 62 C3 21 37S 141 55 E
Kyō-ga-Saki, Japan 31 G7 35 45N 135 15 E
Kyoga, L., Uganda 54 B3 1 35N 33 0 E
Kyogle, Australia 63 D5 28 40S 153 0 E
Kyongju, S. Korea 35 G15 35 51N 129 14 E
Kyongpyaw, Burma 41 L19 17 12N 95 10 E
Kyŏngsŏng, N. Korea 35 D15 41 35N 129 36 E
Kyōto, Japan 31 G7 35 0N 135 45 E
Kyōto □, Japan 31 G7 35 15N 135 45 E
Kyparissovouno, Cyprus 23 D12 35 19N 33 10 E
Kyperounda, Cyprus 23 E11 34 56N 32 58 E
Kyrenia, Cyprus 23 D12 35 20N 33 20 E
Kyrgystan = Kirghizia ■, Asia 26 E8 42 0N 75 0 E
Kystatyam, Russia 27 C13 67 20N 123 10 E
Kytal Ktakh, Russia 27 C13 65 30N 123 40 E
Kythréa, Cyprus 23 D12 35 15N 33 29 E
Kyulyunken, Russia 27 C14 64 10N 137 5 E
Kyunhla, Burma 41 H19 23 25N 95 15 E
Kyuquot, Canada 72 C3 50 3N 127 25W
Kyūshū, Japan 31 H5 33 0N 131 0 E
Kyūshū □, Japan 31 H5 33 0N 131 0 E
Kyūshū-Sanchi, Japan 31 H5 32 35N 131 17 E
Kyustendil, Bulgaria 21 C10 42 16N 22 41 E
Kyusyur, Russia 27 B13 70 19N 127 30 E
Kywong, Australia 63 E4 34 58S 146 44 E
Kyyiv = Kiyev, Ukraine 25 D5 50 30N 30 28 E
Kyzyl, Russia 27 D10 51 50N 94 30 E
Kyzyl-Kiya, Kirghizia 26 E8 40 16N 72 8 E
Kyzylkum, Peski, Uzbekistan 26 E7 42 30N 65 0 E
Kzyl-Orda, Kazakhstan 26 E7 44 48N 65 28 E

L

La Asunción, Venezuela 92 A6 11 2N 63 53W
La Banda, Argentina 94 B3 27 45S 64 10W
La Barca, Mexico 86 C4 20 20N 102 40W
La Barge, U.S.A. 82 E8 42 16N 110 12W
La Belle, U.S.A. 77 M5 26 46N 81 26W
La Biche ~►, Canada 72 B4 59 57N 123 50W
La Bomba, Mexico 86 A1 31 53N 115 2W
La Calera, Chile 94 C1 32 50S 71 10W
La Canal, Spain 22 C7 38 51N 1 23 E
La Carlota, Argentina 94 C3 33 30S 63 20W
La Carolina, Spain 19 C4 38 17N 3 38W
La Ceiba, Honduras 88 C2 15 40N 86 50W
La Chaux de Fonds, Switz. 16 E3 47 7N 6 50 E
La Cocha, Argentina 94 B2 27 50S 65 40W
La Concordia, Mexico 87 D6 16 8N 92 38W
La Conner, U.S.A. 82 B2 48 23N 122 30W
La Coruña, Spain 19 A1 43 20N 8 25W
La Crete, Canada 72 B5 58 11N 116 24W
La Crosse, Kans., U.S.A. 80 F5 38 32N 99 18W
La Crosse, Wis., U.S.A. 80 D9 43 48N 91 15W
La Cruz, Costa Rica 88 D2 11 4N 85 39W
La Cruz, Mexico 86 C3 23 55N 106 54W
La Dorada, Colombia 92 B4 5 30N 74 40W
La Escondida, Mexico 86 C5 24 6N 99 55W
La Esmeralda, Paraguay 94 A3 22 16S 62 33W
La Esperanza, Cuba 88 B3 22 46N 83 44W
La Esperanza, Honduras 88 D2 14 15N 88 10W
La Estrada, Spain 19 A1 42 43N 8 27W
La Fayette, U.S.A. 77 H3 34 42N 85 17W
La Fé, Cuba 88 B3 22 2N 84 15W
La Follette, U.S.A. 77 G3 36 23N 84 7W
La Grande, U.S.A. 82 D4 45 20N 118 5W
La Grange, Calif., U.S.A. 84 H6 37 42N 120 27W
La Grange, Ga., U.S.A. 77 J3 33 2N 85 2W
La Grange, Ky., U.S.A. 76 F3 38 25N 85 23W
La Grange, Tex., U.S.A. 81 L6 29 54N 96 52W
La Guaira, Venezuela 92 A5 10 36N 66 56W
La Güera, Mauritania 50 D1 20 51N 17 0W
La Habana, Cuba 88 B3 23 8N 82 22W
La Harpe, U.S.A. 80 E9 40 35N 90 58W
La Independencia, Mexico 87 D6 16 31N 91 47W
La Isabela, Dom. Rep. 89 C5 19 58N 71 2W
La Jara, U.S.A. 83 H11 37 16N 105 58W
La Junta, U.S.A. 81 F3 37 59N 103 33W
La Laguna, Canary Is. 22 F3 28 28N 16 18W
La Libertad, Guatemala 88 C1 16 47N 90 7W
La Libertad, Mexico 86 B2 29 55N 112 41W
La Ligua, Chile 94 C1 32 30S 71 16W
La Línea de la Concepción, Spain 19 D3 36 15N 5 23W
La Loche, Canada 73 B7 56 29N 109 26W
La Louvière, Belgium 15 D4 50 27N 4 10 E
La Malbaie, Canada 71 C5 47 40N 70 10W
La Mancha, Spain 19 C4 39 10N 2 54W
La Mesa, Calif., U.S.A. 85 N9 32 46N 117 3W
La Mesa, N. Mex., U.S.A. 83 K10 32 7N 106 42W
La Misión, Mexico 86 A1 32 5N 116 50W
La Moure, U.S.A. 80 B5 46 21N 98 18W
La Negra, Chile 94 A1 23 46S 70 18W
La Oliva, Canary Is. 22 F6 28 36N 13 57W
La Orotava, Canary Is. 22 F3 28 22N 16 31W
La Palma, Canary Is. 22 F2 28 40N 17 50W
La Palma, Panama 88 E4 8 15N 78 0W
La Palma, Spain 19 D2 37 21N 6 38W
La Paloma, Chile 94 C1 30 35S 71 0W
La Pampa □, Argentina 94 D2 36 50S 66 0W
La Paragua, Venezuela 92 B6 6 50N 63 20W
La Paz, Entre Ríos, Argentina 94 C4 30 50S 59 45W
La Paz, San Luis, Argentina 94 C2 33 30S 67 20W
La Paz, Bolivia 92 G5 16 20S 68 10W
La Paz, Honduras 88 D2 14 20N 87 47W
La Paz, Mexico 86 C2 24 10N 110 20W
La Paz Centro, Nic. 88 D2 12 20N 86 41W
La Pedrera, Colombia 92 D5 1 18S 69 43W
La Perouse Str., Asia 30 B11 45 40N 142 0 E
La Pesca, Mexico 87 C5 23 46N 97 47W
La Piedad, Mexico 86 C4 20 20N 102 1W
La Pine, U.S.A. 82 E3 43 40N 121 30W
La Plant, U.S.A. 80 C4 45 9N 100 39W
La Plata, Argentina 94 D4 35 0S 57 55W
La Porte, U.S.A. 76 E2 41 36N 86 43W
La Purísima, Mexico 86 B2 26 10N 112 4W
La Push, U.S.A. 84 C2 47 55N 124 38W
La Quiaca, Argentina 94 A2 22 5S 65 35W
La Reine, Canada 70 C4 48 50N 79 30W
La Restinga, Canary Is. 22 G2 27 38N 17 59W
La Rioja, Argentina 94 B2 29 20S 67 0W
La Rioja □, Argentina 94 B2 29 30S 67 0W
La Rioja □, Spain 19 A4 42 20N 2 20W
La Robla, Spain 19 A3 42 50N 5 41W
La Rochelle, France 18 C3 46 10N 1 9W
La Roda, Spain 19 C4 39 13N 2 15W
La Romana, Dom. Rep. 89 C6 18 27N 68 57W
La Ronge, Canada 73 B7 55 5N 105 20W
La Rumorosa, Mexico 85 N10 32 33N 116 4W
La Sabina, Spain 22 C7 38 44N 1 25 E
La Sagra, Spain 19 D4 37 57N 2 35W
La Salle, U.S.A. 80 E10 41 20N 89 6W
La Santa, Canary Is. 22 E6 29 5N 13 40W
La Sarre, Canada 70 C4 48 45N 79 15W
La Scie, Canada 71 C8 49 57N 55 36W
La Selva Beach, U.S.A. 84 J5 36 56N 121 51W
La Serena, Chile 94 B1 29 55S 71 10W
La Spézia, Italy 20 B3 44 8N 9 50 E
La Tortuga, Venezuela 89 D6 11 0N 65 22W
La Tuque, Canada 70 C5 47 30N 72 50W
La Unión, Chile 96 E2 40 10S 73 0W
La Unión, El Salv. 88 D2 13 20N 87 50W
La Unión, Mexico 86 D4 17 58N 101 49W
La Urbana, Venezuela 92 B5 7 8N 66 56W
La Vega, Dom. Rep. 89 C5 19 20N 70 30W
La Venta, Mexico 87 D6 18 8N 94 3W
La Ventura, Mexico 86 C4 24 38N 100 54W
Labe = Elbe ~►, Europe 16 B4 53 50N 9 0 E
Labé, Guinea 50 F2 11 24N 12 16W
Laberge, L., Canada 72 A1 61 11N 135 12W
Labis, Malaysia 39 L4 2 22N 103 2 E
Laboulaye, Argentina 94 C3 34 10S 63 30W
Labrador, Coast of □, Canada 71 B7 53 20N 61 0W
Labrador City, Canada 71 B6 52 57N 66 55W
Lábrea, Brazil 92 E6 7 15S 64 51W
Labuan, Pulau, Malaysia 36 C5 5 21N 115 13 E
Labuha, Indonesia 37 E7 0 30S 127 30 E
Labuhan, Indonesia 37 G11 6 22S 105 50 E
Labuhanbajo, Indonesia 37 F6 8 28S 120 1 E
Labuk, Telok, Malaysia 36 C5 6 10N 117 50 E
Labyrinth, L., Australia 63 E2 30 40S 135 11 E
Labytnangi, Russia 24 A12 66 39N 66 21 E
Lac Allard, Canada 71 B7 50 33N 63 24W
Lac Bouchette, Canada 71 C5 48 16N 72 11W
Lac du Flambeau, U.S.A. 80 B10 45 58N 89 53W
Lac Édouard, Canada 70 C5 47 40N 72 16W
Lac La Biche, Canada 72 C6 54 45N 111 58W
Lac la Martre, Canada 68 B8 63 8N 117 16W
Lac-Mégantic, Canada 71 C5 45 35N 70 53W
Lac Seul, Res., Canada 70 B1 50 25N 92 30W
Lac Thien, Vietnam 38 F7 12 25N 108 11 E
Lacantúm ~►, Mexico 87 D6 16 36N 90 40W
Laccadive Is. = Lakshadweep Is., Ind. Oc. 28 J11 10 0N 72 30 E
Lacepede B., Australia 63 F2 36 40S 139 40 E
Lacepede Is., Australia 60 C3 16 55S 122 0 E
Lacerdónia, Mozam. 55 F4 18 3S 35 35 E
Lacey, U.S.A. 84 C4 47 7N 122 49W
Lachhmangarh, India 42 F6 27 50N 75 4 E
Lachi, Pakistan 42 C4 33 25N 71 20 E
Lachine, Canada 70 C5 45 30N 73 40W
Lachlan ~►, Australia 63 E3 34 22S 143 55 E
Lachute, Canada 70 C5 45 39N 74 21W
Lackawanna, U.S.A. 78 D6 42 50N 78 50W
Lacolle, Canada 79 A11 45 5N 73 22W
Lacombe, Canada 72 C6 52 30N 113 44W
Lacona, U.S.A. 79 C8 43 39N 76 10W
Laconia, U.S.A. 79 C13 43 32N 71 28W
Lacrosse, U.S.A. 82 C5 46 51N 117 58W
Ladakh Ra., India 43 B8 34 0N 78 0 E
Ladismith, S. Africa 56 E3 33 28S 21 15 E
Lādīz, Iran 45 D9 28 55N 61 15 E
Ladnun, India 42 F6 27 38N 74 25 E
Ladoga, L. = Ladozhskoye Ozero, Russia 24 B5 61 15N 30 30 E
Ladozhskoye Ozero, Russia 24 B5 61 15N 30 30 E
Lady Grey, S. Africa 56 E4 30 43S 27 13 E
Ladybrand, S. Africa 56 D4 29 9S 27 29 E
Ladysmith, Canada 72 D4 49 0N 123 49W
Ladysmith, S. Africa 57 D4 28 32S 29 46 E
Ladysmith, U.S.A. 80 C9 45 28N 91 12W
Lae, Papua N. G. 64 H6 6 40S 147 2 E
Laem Ngop, Thailand 39 F4 12 10N 102 26 E
Laem Pho, Thailand 39 J3 6 55N 101 19 E
Læsø, Denmark 9 H11 57 15N 10 53 E
Lafayette, Colo., U.S.A. 80 F2 39 58N 105 12W
Lafayette, Ind., U.S.A. 76 E2 40 25N 86 54W
Lafayette, La., U.S.A. 81 K9 30 14N 92 1W
Lafayette, Tenn., U.S.A. 77 G3 36 31N 86 2W
Laferte ~►, Canada 72 A5 61 53N 117 44W
Lafia, Nigeria 50 G6 8 30N 8 34 E
Lafleche, Canada 73 D7 49 45N 106 40W
Lagan ~►, U.K. 13 B6 54 35N 5 55W
Lagarfljót ~►, Iceland 8 D6 65 40N 14 18W
Lågen ~►, Norway 9 F11 61 8N 10 25 E
Laghouat, Algeria 50 B5 33 50N 2 59 E
Lagonoy Gulf, Phil. 37 B6 13 50N 123 50 E
Lagos, Nigeria 50 G5 6 25N 3 27 E
Lagos, Portugal 19 D1 37 5N 8 41W
Lagos de Moreno, Mexico 86 C4 21 21N 101 55W
Lagrange, Australia 60 C3 18 45S 121 43 E
Lagrange B., Australia 60 C3 18 38S 121 42 E
Laguna, Brazil 95 B6 28 30S 48 50W
Laguna, U.S.A. 83 J10 35 2N 107 25W
Laguna Beach, U.S.A. 85 M9 33 33N 117 47W
Laguna Limpia, Argentina 94 B4 26 32S 59 45W
Laguna Madre, U.S.A. 87 B5 27 0N 97 20W
Lagunas, Chile 94 A2 21 0S 69 45W
Lagunas, Peru 92 E3 5 10S 75 35W
Lahad Datu, Malaysia 37 C5 5 0N 118 20 E
Lahan Sai, Thailand 38 E4 14 25N 102 52 E
Lahanam, Laos 38 D5 16 16N 105 16 E
Laharpur, India 43 F9 27 43N 80 56 E
Lahat, Indonesia 36 E2 3 45S 103 30 E
Lahewa, Indonesia 36 D1 1 22N 97 12 E
Lahijan, Iran 45 B6 37 10N 50 6 E
Lahn ~►, Germany 16 C3 50 17N 7 38 E
Laholm, Sweden 9 H12 56 30N 13 2 E
Lahontan Reservoir, U.S.A. 82 G4 39 28N 119 4W
Lahore, Pakistan 42 D6 31 32N 74 22 E
Lahti, Finland 9 F18 60 58N 25 40 E
Lahtis = Lahti, Finland 9 F18 60 58N 25 40 E
Laï, Chad 51 G8 9 25N 16 18 E
Lai Chau, Vietnam 38 A4 22 5N 103 3 E
Laidley, Australia 63 D5 27 39S 152 20 E
Laikipia □, Kenya 54 B4 0 30N 36 30 E
Laingsburg, S. Africa 56 E3 33 9S 20 52 E
Lairg, U.K. 12 C4 58 1N 4 24W
Laishui, China 34 E8 39 23N 115 45 E
Laiwu, China 35 F9 36 15N 117 40 E
Laixi, China 35 F11 36 50N 120 31 E
Laiyang, China 35 F11 36 59N 120 45 E
Laiyuan, China 34 E8 39 20N 114 40 E
Laizhou Wan, China 35 F10 37 30N 119 30 E
Laja ~►, Mexico 86 C4 20 55N 100 46W
Lajere, Nigeria 50 F7 12 10N 11 25 E
Lajes, Brazil 95 B5 27 48S 50 20W
Lak Sao, Laos 38 C5 18 11N 104 59 E
Lakaband, Pakistan 42 D3 31 2N 69 15 E
Lakar, Indonesia 37 F7 8 15S 128 17 E
Lake Alpine, U.S.A. 84 G7 38 29N 120 0W
Lake Andes, U.S.A. 80 D5 43 9N 98 32W
Lake Anse, U.S.A. 76 B1 46 42N 88 25W
Lake Arthur, U.S.A. 81 K8 30 5N 92 41W
Lake Cargelligo, Australia 63 E4 33 15S 146 22 E
Lake Charles, U.S.A. 81 K8 30 14N 93 13W
Lake City, Colo., U.S.A. 83 G10 38 2N 107 19W
Lake City, Fla., U.S.A. 77 K4 30 11N 82 38W
Lake City, Iowa, U.S.A. 80 D7 42 16N 94 44W
Lake City, Mich., U.S.A. 76 C3 44 20N 85 13W
Lake City, Minn., U.S.A. 80 C8 44 27N 92 16W
Lake City, Pa., U.S.A. 78 D4 42 1N 80 21W
Lake City, S.C., U.S.A. 77 J6 33 52N 79 45W
Lake George, U.S.A. 79 C11 43 26N 73 43W
Lake Grace, Australia 61 F2 33 7S 118 28 E
Lake Harbour, Canada 69 B13 62 50N 69 50W
Lake Havasu City, U.S.A. 85 L12 34 27N 114 22W
Lake Hughes, U.S.A. 85 L8 34 41N 118 26W
Lake Isabella, U.S.A. 85 K8 35 38N 118 28W
Lake King, Australia 61 F2 33 5S 119 45 E
Lake Lenore, Canada 73 C8 52 24N 104 59W
Lake Louise, Canada 72 C5 51 30N 116 10W
Lake Mead National Recreation Area, U.S.A. 85 K12 36 15N 114 30W
Lake Mills, U.S.A. 80 D8 43 25N 93 32W
Lake Nash, Australia 62 C2 20 57S 138 0 E
Lake Providence, U.S.A. 81 J9 32 48N 91 10W
Lake River, Canada 70 B3 54 30N 82 31W
Lake Superior Prov. Park, Canada 70 C3 47 45N 84 45W
Lake Village, U.S.A. 81 J9 33 20N 91 17W
Lake Wales, U.S.A. 77 M5 27 54N 81 35W
Lake Worth, U.S.A. 77 M5 26 37N 80 3W
Lakefield, Canada 70 D4 44 25N 78 16W
Lakeland, Australia 62 B3 15 49S 144 57 E
Lakeland, U.S.A. 77 L5 28 3N 81 57W
Lakemba, Fiji 59 D9 18 13S 178 47W
Lakeport, U.S.A. 84 F4 39 3N 122 55W
Lakes Entrance, Australia 63 F4 37 50S 148 0 E
Lakeside, Ariz., U.S.A. 83 J9 34 9N 109 58W
Lakeside, Calif., U.S.A. 85 N10 32 52N 116 55W
Lakeside, Nebr., U.S.A. 80 D3 42 3N 102 26W
Lakeview, U.S.A. 82 E3 42 11N 120 21W
Lakewood, Colo., U.S.A. 80 F2 39 44N 105 5W
Lakewood, N.J., U.S.A. 79 F10 40 6N 74 13W
Lakewood, Ohio, U.S.A. 78 E3 41 29N 81 48W
Lakewood Center, U.S.A. 84 C4 47 11N 122 32W
Lakhaniá, Greece 23 D9 35 58N 27 54 E
Lakhonpheng, Laos 38 E5 15 54N 105 34 E
Lakhpat, India 42 H3 23 48N 68 47 E
Laki, Iceland 8 D4 64 4N 18 14W
Lakin, U.S.A. 81 G4 37 57N 101 15W
Lakitusaki ~►, Canada 70 B3 54 21N 82 25W
Lákkoi, Greece 23 D5 35 24N 23 57 E
Lakonikós Kólpos, Greece 21 F10 36 40N 22 40 E
Lakota, Ivory C. 50 G3 5 50N 5 30W
Lakota, U.S.A. 80 A5 48 2N 98 21W
Laksefjorden, Norway 8 A19 70 45N 26 50 E
Lakselv, Norway 8 A18 70 2N 24 56 E
Lakshadweep Is., Ind. Oc. 28 J11 10 0N 72 30 E
Lakshmikantapur, India 43 H13 22 5N 88 20 E
Lala Ghat, India 41 G18 24 30N 92 40 E
Lala Musa, Pakistan 42 C5 32 40N 73 57 E
Lalago, Tanzania 54 C3 3 28S 33 58 E
Lalapanzi, Zimbabwe 55 F3 19 20S 30 15 E
Lalganj, India 43 G11 25 52N 85 13 E
Lalibela, Ethiopia 51 F12 12 2N 39 2 E
Lalin, China 35 B14 45 12N 127 0 E
Lalín, Spain 19 A1 42 40N 8 5W
Lalin He ~►, China 35 B13 45 32N 125 40 E
Lalitapur = Patan, Nepal 41 F14 27 40N 85 20 E
Lalitpur, India 43 G8 24 42N 78 28 E
Lam, Vietnam 38 B6 21 21N 106 31 E
Lam Pao Res., Thailand 38 D4 16 50N 103 15 E
Lamaing, Burma 41 M20 15 25N 97 53 E
Lamar, Colo., U.S.A. 80 F3 38 5N 102 37W
Lamar, Mo., U.S.A. 81 G7 37 30N 94 16W
Lamas, Peru 92 E3 6 28S 76 31W
Lambaréné, Gabon 52 E2 0 41S 10 12 E
Lambasa, Fiji 59 C8 16 30S 179 10 E
Lambay I., Ireland 13 C5 53 30N 6 0W
Lambert, U.S.A. 80 B2 47 41N 104 37W
Lambert Glacier, Antarctica 5 D6 71 0S 70 0 E
Lamberts Bay, S. Africa 56 E2 32 5S 18 17 E
Lame, Nigeria 50 F6 10 30N 9 20 E
Lame Deer, U.S.A. 82 D10 45 37N 106 40W
Lamego, Portugal 19 B2 41 5N 7 52W
Lamèque, Canada 71 C7 47 45N 64 38W
Lameroo, Australia 63 F3 35 19S 140 33 E
Lamesa, U.S.A. 81 J4 32 44N 101 58W
Lamía, Greece 21 E10 38 55N 22 26 E
Lammermuir Hills, U.K. 12 F6 55 50N 2 40W
Lamon Bay, Phil. 37 B6 14 30N 122 20 E
Lamont, Canada 72 C6 53 46N 112 50W
Lamont, U.S.A. 85 K8 35 15N 118 55W
Lampa, Peru 92 G4 15 22S 70 22W
Lampang, Thailand 38 C2 18 16N 99 32 E
Lampasas, U.S.A. 81 K5 31 4N 98 11W
Lampazos de Naranjo, Mexico 86 B4 27 2N 100 32W
Lampedusa, Medit. S. 20 G5 35 36N 12 40 E
Lampeter, U.K. 11 E3 52 6N 4 6W
Lampman, Canada 73 D8 49 25N 102 50W
Lamprey, Canada 73 B10 58 33N 94 8W
Lampung □, Indonesia 36 F2 5 30S 104 30 E
Lamu, Kenya 54 C5 2 16S 40 55 E
Lamu □, Kenya 54 C5 2 0S 40 45 E
Lamy, U.S.A. 83 J11 35 29N 105 53W
Lan Xian, China 34 E6 38 15N 111 35 E
Lanai I., U.S.A. 74 H16 20 50N 156 55W
Lanak La, India 43 B8 34 27N 79 32 E
Lanak'o Shank'ou = Lanak La, India 43 B8 34 27N 79 32 E
Lanao, L., Phil. 37 C6 7 52N 124 15 E
Lanark, Canada 79 A8 45 1N 76 22W
Lanark, U.K. 12 F5 55 40N 3 48W
Lancang Jiang ~►, China 32 D5 21 40N 101 10 E
Lancashire □, U.K. 10 D5 53 40N 2 30W
Lancaster, Canada 79 A10 45 10N 74 30W
Lancaster, U.K. 10 C5 54 3N 2 48W
Lancaster, Calif., U.S.A. 85 L8 34 42N 118 8W
Lancaster, Ky., U.S.A. 76 G3 37 37N 84 35W
Lancaster, N.H., U.S.A. 79 B13 44 29N 71 34W
Lancaster, N.Y., U.S.A. 78 D6 42 54N 78 40W
Lancaster, Pa., U.S.A. 79 F8 40 2N 76 19W
Lancaster, S.C., U.S.A. 77 H5 34 43N 80 46W
Lancaster, Wis., U.S.A. 80 D9 42 51N 90 43W
Lancaster Sd., Canada 69 A11 74 13N 84 0W
Lancer, Canada 73 C7 50 48N 108 53W
Lanchow = Lanzhou, China 34 F2 36 1N 103 52 E
Lanciano, Italy 20 C6 42 15N 14 22 E
Lancun, China 35 F11 36 25N 120 10 E
Landeck, Austria 16 E5 47 9N 10 34 E
Landen, Belgium 15 D5 50 45N 5 3 E
Lander, U.S.A. 82 E9 42 50N 108 44W
Lander ~►, Australia 60 D5 22 0S 132 0 E
Landes, France 18 E3 44 0N 1 0W
Landes □, France 18 E3 43 57N 0 48W
Landi Kotal, Pakistan 42 B4 34 7N 71 6 E
Landor, Australia 61 E2 25 10S 116 54 E
Land's End, U.K. 11 G2 50 4N 5 43W
Landsborough Cr. ~►, Australia 62 C3 22 28S 144 35 E
Landshut, Germany 16 D6 48 31N 12 10 E
Landskrona, Sweden 9 J12 55 53N 12 50 E
Lanesboro, U.S.A. 79 E9 41 57N 75 34W
Lanett, U.S.A. 77 J3 32 52N 85 12W
Lang Bay, Canada 72 D4 49 45N 124 21W
Lang Qua, Vietnam 38 A5 22 16N 104 27 E
Lang Shan, China 34 D4 41 0N 106 30 E
Lang Son, Vietnam 38 B6 21 52N 106 42 E
Lang Suan, Thailand 39 H2 9 57N 99 4 E
La'nga Co, China 41 D12 30 45N 81 15 E
Langar, Iran 45 C9 35 23N 60 25 E
Langara I., Canada 72 C2 54 14N 133 1W
Langdon, U.S.A. 80 A5 48 45N 98 22W
Langeberg, S. Africa 56 E3 33 55S 21 0 E
Langeberge, S. Africa 56 D3 28 15S 22 33 E
Langenburg, Canada 73 C8 50 51N 101 43W
Langholm, U.K. 12 F6 55 9N 2 59W
Langjökull, Iceland 8 D3 64 39N 20 12W
Langkawi, P., Malaysia 39 J2 6 25N 99 45 E
Langklip, S. Africa 56 D3 28 12S 20 20 E
Langkon, Malaysia 36 C5 6 30N 116 40 E
Langlade, St- P. & M. 71 C8 46 50N 56 20W
Langlois, U.S.A. 82 E1 42 56N 124 27W
Langøya, Norway 8 B13 68 45N 14 50 E
Langres, France 18 C6 47 52N 5 20 E
Langres, Plateau de, France 18 C6 47 45N 5 3 E
Langsa, Indonesia 36 D1 4 30N 97 57 E
Langtry, U.S.A. 81 L4 29 49N 101 34W
Langu, Thailand 39 J2 6 53N 99 47 E
Languedoc, France 18 E5 43 58N 3 55 E
Langxiangzhen, China 34 E9 39 43N 116 8 E
Lanigan, Canada 73 C7 51 51N 105 2W
Lankao, China 34 G8 34 48N 114 50 E

Long Pt.

Libode, S. Africa 57 E4 31 33S 29 2 E
Libonda, Zambia 53 G4 14 28S 23 12 E
Libourne, France 18 D3 44 55N 0 14W
Libramont, Belgium 15 E5 49 55N 5 23 E
Libreville, Gabon 52 D1 0 25N 9 26 E
Libya ■, N. Afr. 51 C8 27 0N 17 0 E
Libyan Desert = Lîbîya, Sahrâ', Africa 51 C9 25 0N 25 0 E
Licantén, Chile 94 D1 35 55S 72 0W
Licata, Italy 20 F5 37 6N 13 55 E
Licheng, China 34 F7 36 28N 113 20 E
Lichfield, U.K. 10 E6 52 40N 1 50W
Lichinga, Mozam. 55 E4 13 13S 35 11 E
Lichtenburg, S. Africa 56 D4 26 8S 26 8 E
Lida, U.S.A. 83 H5 37 28N 117 30W
Lidköping, Sweden 9 G12 58 31N 13 14 E
Liebig, Mt., Australia 60 D5 23 18S 131 22 E
Liechtenstein ■, Europe 16 E4 47 8N 9 35 E
Liège, Belgium 15 D5 50 38N 5 35 E
Liège □, Belgium 15 D5 50 32N 5 35 E
Liegnitz = Legnica, Poland 16 C8 51 12N 16 10 E
Lienart, Zaïre 54 B2 3 3N 25 31 E
Lienyünchiangshih = Lianyungang, China 35 G10 34 40N 119 11 E
Lienz, Austria 16 E6 46 50N 12 46 E
Liepāja, Latvia 24 C3 56 30N 21 0 E
Lier, Belgium 15 C4 51 7N 4 34 E
Lièvre →, Canada 70 C4 45 31N 75 26W
Liffey →, Ireland 13 C5 53 21N 6 20W
Lifford, Ireland 13 B4 54 50N 7 29W
Lifudzin, Russia 30 B7 44 21N 134 58 E
Lightning Ridge, Australia 63 D4 29 22S 148 0 E
Liguria □, Italy 20 B3 44 30N 9 0 E
Ligurian Sea, Italy 20 C3 43 20N 9 0 E
Lihou Reefs and Cays, Australia 62 B5 17 25S 151 40 E
Lihue, U.S.A. 74 H15 21 59N 159 23W
Lijiang, China 32 D5 26 55N 100 20 E
Likasi, Zaïre 55 E2 10 55S 26 48 E
Likati, Zaïre 52 D4 3 20N 24 0 E
Likoma I., Malawi 55 E3 12 3S 34 45 E
Likumburu, Tanzania 55 D4 9 43S 35 8 E
Lille, France 18 A5 50 38N 3 3 E
Lille Bælt, Denmark 9 J10 55 20N 9 45 E
Lillehammer, Norway 9 F11 61 8N 10 30 E
Lillesand, Norway 9 G10 58 15N 8 23 E
Lilleshall, U.K. 11 E5 52 45N 2 22W
Lillestrøm, Norway 9 G11 59 58N 11 5 E
Lillian Point, Mt., Australia 61 E4 27 40S 126 6 E
Lillooet →, Canada 72 D4 49 15N 121 57W
Lilongwe, Malawi 55 E3 14 0S 33 48 E
Liloy, Phil. 37 C6 8 4N 122 39 E
Lima, Indonesia 37 E7 3 37S 128 4 E
Lima, Peru 92 F3 12 0S 77 0W
Lima, Mont., U.S.A. 82 D7 44 38N 112 36W
Lima, Ohio, U.S.A. 76 E3 40 44N 84 6W
Limages, Canada 79 A9 45 20N 75 16W
Limassol, Cyprus 23 E12 34 42N 33 1 E
Limavady, U.K. 13 A5 55 3N 6 58W
Limavady □, U.K. 13 B5 55 0N 6 55W
Limay →, Argentina 96 D3 39 0S 68 0W
Limay Mahuida, Argentina 94 D2 37 10S 66 45W
Limbang, Brunei 36 D5 4 42N 115 6 E
Limbdi, India 42 H4 22 34N 71 51 E
Limbe, Cameroon 50 H6 4 1N 9 10 E
Limbri, Australia 63 E5 31 3S 151 5 E
Limbunya, Australia 60 C4 17 14S 129 50 E
Limburg □, Belgium 15 C5 51 2N 5 25 E
Limburg □, Neths. 15 C5 51 20N 5 55 E
Limeira, Brazil 95 A6 22 35S 47 28W
Limerick, Ireland 13 D3 52 40N 8 38W
Limerick □, Ireland 13 D3 52 30N 8 50W
Limestone, U.S.A. 78 D6 42 2N 78 38W
Limestone →, Canada 73 B10 56 31N 94 7W
Limfjorden, Denmark 9 H10 56 55N 9 0 E
Limmen Bight, Australia 62 A2 14 40S 135 35 E
Limmen Bight →, Australia 62 B2 15 7S 135 44 E
Límnos, Greece 21 E11 39 50N 25 5 E
Limoeiro do Norte, Brazil 93 E11 5 5S 38 0W
Limoges, France 18 D4 45 50N 1 15 E
Limón, Costa Rica 88 D3 10 0N 83 2W
Limon, U.S.A. 80 F3 39 16N 103 41W
Limousin, France 18 D4 45 30N 1 30 E
Limpopo →, Africa 57 D5 25 5S 33 30 E
Limuru, Kenya 54 C4 1 2S 36 35 E
Lin Xian, China 34 F6 37 57N 110 58 E
Linares, Chile 94 D1 35 50S 71 40W
Linares, Mexico 87 C5 24 50N 99 40W
Linares, Spain 19 C4 38 10N 3 40W
Linares □, Chile 94 D1 36 0S 71 0W
Lincheng, China 34 F8 37 25N 114 30 E
Lincoln, Argentina 94 C3 34 55S 61 30W
Lincoln, N.Z. 59 K4 43 38S 172 30 E
Lincoln, U.K. 10 D7 53 14N 0 32W
Lincoln, Calif., U.S.A. 84 G5 38 54N 121 17W
Lincoln, Ill., U.S.A. 80 E10 40 9N 89 22W
Lincoln, Kans., U.S.A. 80 F5 39 3N 98 9W
Lincoln, Maine, U.S.A. 71 C6 45 22N 68 30W
Lincoln, N.H., U.S.A. 79 B13 44 3N 71 40W
Lincoln, N. Mex., U.S.A. 83 K11 33 30N 105 23W
Lincoln, Nebr., U.S.A. 80 E6 40 49N 96 41W
Lincoln Hav = Lincoln Sea, Arctic 4 A5 84 0N 55 0W
Lincoln Sea, Arctic 4 A5 84 0N 55 0W
Lincoln Wolds, U.K. 10 D7 53 20N 0 5W
Lincolnshire □, U.K. 10 D7 53 14N 0 32W
Lincolnton, U.S.A. 77 H5 35 29N 81 16W
Lind, U.S.A. 82 C4 46 58N 118 37W
Linda, U.S.A. 84 F5 39 8N 121 34W
Linden, Guyana 92 B7 6 0N 58 10W
Linden, Calif., U.S.A. 84 G5 38 1N 121 5W
Linden, Tex., U.S.A. 81 J7 33 1N 94 22W
Lindenhurst, U.S.A. 79 F11 40 41N 73 23W
Lindesnes, Norway 6 D7 57 58N 7 3 E
Líndhos, Greece 23 C10 36 6N 28 4 E
Lindi, Tanzania 55 D4 9 58S 39 38 E
Lindi □, Tanzania 55 D4 9 40S 38 30 E
Lindi →, Zaïre 54 B2 0 33N 25 5 E
Lindsay, Canada 70 D4 44 22N 78 43W
Lindsay, Calif., U.S.A. 83 H4 36 12N 119 5W
Lindsay, Okla., U.S.A. 81 H6 34 50N 97 38W
Lindsborg, U.S.A. 80 F6 38 35N 97 40W
Linfen, China 34 F6 36 3N 111 30 E
Ling Xian, China 34 F9 37 22N 116 30 E
Lingao, China 38 C7 19 56N 109 42 E
Lingayen, Phil. 37 A6 16 1N 120 14 E
Lingayen G., Phil. 37 A6 16 10N 120 15 E
Lingbi, China 35 H9 33 33N 117 33 E
Lingchuan, China 34 G7 35 45N 113 12 E
Lingen, Germany 16 B3 52 32N 7 21 E
Lingga, Indonesia 36 E2 0 12S 104 37 E
Lingga, Kepulauan, Indonesia 36 E2 0 10S 104 30 E
Lingga Arch. = Lingga, Kepulauan, Indonesia 36 E2 0 10S 104 30 E
Lingle, U.S.A. 80 D2 42 8N 104 21W
Lingqiu, China 34 E8 39 28N 114 22 E
Lingshi, China 34 F6 36 48N 111 48 E
Lingshou, China 34 E8 38 20N 114 20 E
Lingshui, China 38 C8 18 27N 110 0 E
Lingtai, China 34 G4 35 0N 107 40 E
Linguéré, Senegal 50 E1 15 25N 15 5W
Lingwu, China 34 E4 38 6N 106 20 E
Lingyuan, China 35 D10 41 10N 119 15 E
Linh Cam, Vietnam 38 C5 18 31N 105 31 E
Linhai, China 33 D7 28 50N 121 8 E
Linhares, Brazil 93 G10 19 25S 40 4W
Linhe, China 34 D4 40 48N 107 20 E
Linjiang, China 35 D14 41 50N 127 0 E
Linköping, Sweden 9 G13 58 28N 15 36 E
Linkou, China 35 B16 45 15N 130 18 E
Linlithgow, U.K. 12 F5 55 58N 3 38W
Linnhe, L., U.K. 12 E3 56 36N 5 25W
Linqi, China 34 G7 35 45N 113 52 E
Linqing, China 34 F8 36 50N 115 42 E
Linqu, China 35 F10 36 25N 118 30 E
Linru, China 34 G7 34 11N 112 52 E
Lins, Brazil 95 A6 21 40S 49 44W
Lintao, China 34 G2 35 18N 103 52 E
Lintlaw, Canada 73 C8 52 4N 103 14W
Linton, Canada 71 C5 47 15N 72 16W
Linton, Ind., U.S.A. 76 F2 39 2N 87 10W
Linton, N. Dak., U.S.A. 80 B4 46 16N 100 14W
Lintong, China 34 G5 34 20N 109 10 E
Linville, Australia 63 D5 26 50S 152 11 E
Linwood, Canada 78 C4 43 35N 80 43W
Linxi, China 35 C10 43 36N 118 2 E
Linxia, China 32 C5 35 36N 103 10 E
Linyanti →, Africa 56 B4 17 50S 25 5 E
Linyi, China 35 G10 35 5N 118 21 E
Linz, Austria 16 D7 48 18N 14 18 E
Linzhenzhen, China 34 F5 36 30N 109 59 E
Linzi, China 35 F10 36 50N 118 20 E
Lion, G. du, France 18 E6 43 10N 4 0 E
Lionárisso, Cyprus 23 D13 35 28N 34 8 E
Lions, G. of = Lion, G. du, France 18 E6 43 10N 4 0 E
Lion's Den, Zimbabwe 55 F3 17 15S 30 5 E
Lion's Head, Canada 70 D3 44 58N 81 15W
Lipa, Phil. 37 B6 13 57N 121 10 E
Lipali, Mozam. 55 F4 15 50S 35 50 E
Lípari, Is., Italy 20 E6 38 30N 14 50 E
Lipetsk, Russia 24 D6 52 37N 39 35 E
Lipovcy Manzovka, Russia 30 B6 44 12N 132 26 E
Lippe →, Germany 16 C3 51 39N 6 38 E
Lipscomb, U.S.A. 81 G4 36 14N 100 16W
Liptrap C., Australia 63 F4 38 50S 145 55 E
Lira, Uganda 54 B3 2 17N 32 57 E
Liria, Spain 19 C5 39 37N 0 35W
Lisala, Zaïre 52 D4 2 12N 21 38 E
Lisboa, Portugal 19 C1 38 42N 9 10W
Lisbon = Lisboa, Portugal 19 C1 38 42N 9 10W
Lisbon, N. Dak., U.S.A. 80 B6 46 27N 97 41W
Lisbon, N.H., U.S.A. 79 B13 44 13N 71 55W
Lisbon, Ohio, U.S.A. 78 F4 40 46N 80 46W
Lisburn, U.K. 13 B5 54 30N 6 9W
Lisburne, C., U.S.A. 68 B3 68 53N 166 13W
Liscannor, B., Ireland 13 D2 52 57N 9 24W
Lishi, China 34 F6 37 31N 111 8 E
Lishu, China 35 C13 43 20N 124 18 E
Lisianski I., Pac. Oc. 64 E10 26 2N 174 0W
Lisichansk, Ukraine 25 E6 48 55N 38 30 E
Lisieux, France 18 B4 49 10N 0 12 E
Liski, Russia 25 D6 51 3N 39 30 E
Lismore, Australia 63 D5 28 44S 153 21 E
Lismore, Ireland 13 D4 52 8N 7 58W
Lisse, Neths. 15 B4 52 16N 4 33 E
Lista, Norway 9 G9 58 7N 6 39 E
Lista, Sweden 9 G14 59 19N 16 16 E
Lister, Mt., Antarctica 5 D11 78 0S 162 0 E
Liston, Australia 63 D5 28 39S 152 6 E
Listowel, Canada 70 D3 43 44N 80 58W
Listowel, Ireland 13 D2 52 27N 9 30W
Litang, Malaysia 37 C5 5 27N 118 31 E
Litani →, Lebanon 47 B4 33 20N 35 15 E
Litchfield, Calif., U.S.A. 84 E6 40 24N 120 23W
Litchfield, Conn., U.S.A. 79 E11 41 45N 73 11W
Litchfield, Ill., U.S.A. 80 F10 39 11N 89 39W
Litchfield, Minn., U.S.A. 80 C7 45 8N 94 32W
Lithgow, Australia 63 E5 33 25S 150 8 E
Líthinon, Ákra, Greece 23 E6 34 55N 24 44 E
Lithuania ■, Europe 24 C3 55 30N 24 0 E
Litoměřice, Czech. 16 C7 50 33N 14 10 E
Little Abaco I., Bahamas 88 A4 26 50N 77 30W
Little Barrier I., N.Z. 59 G5 36 12S 175 8 E
Little Belt Mts., U.S.A. 82 C8 46 40N 110 45W
Little Blue →, U.S.A. 80 F6 39 42N 96 41W
Little Bushman Land, S. Africa 56 D2 29 10S 18 10 E
Little Cadotte →, Canada 72 B5 56 41N 117 6W
Little Cayman, I., Cayman Is. 88 C3 19 41N 80 3W
Little Churchill →, Canada 73 B9 57 30N 95 22W
Little Colorado →, U.S.A. 83 H8 36 12N 111 48W
Little Current, Canada 70 C3 45 55N 82 0W
Little Current →, Canada 70 B3 50 57N 84 36W
Little Falls, Minn., U.S.A. 80 C7 45 59N 94 22W
Little Falls, N.Y., U.S.A. 79 C10 43 3N 74 51W
Little Fork →, U.S.A. 80 A8 48 31N 93 35W
Little Grand Rapids, Canada 73 C9 52 0N 95 29W
Little Humboldt →, U.S.A. 82 F5 41 1N 117 43W
Little Inagua I., Bahamas 89 B5 21 40N 73 50W
Little Karoo, S. Africa 56 E3 33 45S 21 0 E
Little Lake, U.S.A. 85 K9 35 56N 117 55W
Little Laut Is. = Laut Ketil, Kepulauan, Indonesia 36 E5 4 45S 115 40 E
Little Minch, U.K. 12 D2 57 35N 6 45W
Little Missouri →, U.S.A. 80 B3 47 36N 102 25W
Little Namaqualand, S. Africa 56 D2 29 0S 17 9 E
Little Ouse →, U.K. 11 E8 52 25N 0 50 E
Little Rann, India 42 H4 23 25N 71 25 E
Little Red →, U.S.A. 81 H9 35 11N 91 27W
Little River, N.Z. 59 K4 43 45S 172 49 E
Little Rock, U.S.A. 81 H8 34 45N 92 17W
Little Ruaha →, Tanzania 54 D4 7 57S 37 53 E
Little Sable Pt., U.S.A. 76 D2 43 38N 86 33W
Little Sioux →, U.S.A. 80 E6 41 48N 96 4W
Little Smoky →, Canada 72 C5 54 44N 117 11W
Little Snake →, U.S.A. 82 F9 40 27N 108 26W
Little Valley, U.S.A. 78 D6 42 15N 78 48W
Little Wabash →, U.S.A. 76 G1 37 55N 88 5W
Littlefield, U.S.A. 81 J3 33 55N 102 20W
Littlefork, U.S.A. 80 A8 48 24N 93 34W
Littlehampton, U.K. 11 G7 50 48N 0 32W
Littleton, U.S.A. 79 B13 44 18N 71 46W
Liu He →, China 35 D11 40 55N 121 35 E
Liuba, China 34 H4 33 38N 106 55 E
Liugou, China 35 D10 40 57N 118 15 E
Liuhe, China 35 C13 42 17N 125 43 E
Liukang Tenggaja, Indonesia 37 F5 6 45S 118 50 E
Liuli, Tanzania 55 E3 11 3S 34 38 E
Liuwa Plain, Zambia 53 G4 14 20S 22 30 E
Liuzhou, China 33 D5 24 22N 109 22 E
Liuzhuang, China 35 H11 33 12N 120 18 E
Livadhia, Cyprus 23 E12 34 57N 33 38 E
Live Oak, Calif., U.S.A. 84 F5 39 17N 121 40W
Live Oak, Fla., U.S.A. 77 K4 30 18N 82 59W
Liveras, Cyprus 23 D11 35 23N 32 57 E
Liveringa, Australia 60 C3 18 3S 124 10 E
Livermore, U.S.A. 84 H5 37 41N 121 47W
Livermore, Mt., U.S.A. 81 K2 30 38N 104 11W
Liverpool, Australia 63 E5 33 54S 150 58 E
Liverpool, Canada 71 D7 44 5N 64 41W
Liverpool, U.K. 10 D5 53 25N 3 0W
Liverpool Plains, Australia 63 E5 31 15S 150 15 E
Liverpool Ra., Australia 63 E5 31 50S 150 30 E
Livingston, Guatemala 88 C2 15 50N 88 50W
Livingston, Calif., U.S.A. 84 H6 37 23N 120 43W
Livingston, Mont., U.S.A. 82 D8 45 40N 110 34W
Livingston, Tex., U.S.A. 81 K7 30 43N 94 56W
Livingstone, Zambia 55 F2 17 46S 25 52 E
Livingstone Mts., Tanzania 55 D3 9 40S 34 20 E
Livingstonia, Malawi 55 E3 10 38S 34 5 E
Livny, Russia 24 D6 52 30N 37 30 E
Livonia, U.S.A. 76 D4 42 23N 83 23W
Livorno, Italy 20 C4 43 32N 10 18 E
Livramento, Brazil 95 C4 30 55S 55 30W
Liwale, Tanzania 55 D4 9 48S 37 58 E
Liwale □, Tanzania 55 D4 9 0S 38 0 E
Lizard I., Australia 62 A4 14 42S 145 30 E
Lizard Pt., U.K. 11 H2 49 57N 5 11W
Ljubljana, Slovenia 20 A6 46 4N 14 33 E
Ljungan →, Sweden 8 E14 62 18N 17 23 E
Ljungby, Sweden 9 H12 56 49N 13 55 E
Ljusdal, Sweden 9 F14 61 46N 16 3 E
Ljusnan = Ljungan →, Sweden 8 E14 62 18N 17 23 E
Ljusnan →, Sweden 9 F14 61 12N 17 8 E
Ljusne, Sweden 9 F14 61 13N 17 7 E
Llancanelo, Salina, Argentina 94 D2 35 40S 69 8W
Llandeilo, U.K. 11 F3 51 53N 4 3W
Llandovery, U.K. 11 F4 51 59N 3 49W
Llandrindod Wells, U.K. 11 E4 52 15N 3 23W
Llandudno, U.K. 10 D4 53 19N 3 51W
Llanelli, U.K. 11 F3 51 41N 4 11W
Llanes, Spain 19 A3 43 25N 4 50W
Llangollen, U.K. 10 E4 52 58N 3 10W
Llanidloes, U.K. 11 E4 52 28N 3 31W
Llano, U.S.A. 81 K5 30 45N 98 41W
Llano →, U.S.A. 81 K5 30 39N 98 26W
Llano Estacado, U.S.A. 81 J3 33 30N 103 0W
Llanos, S. Amer. 90 B2 5 0N 71 35W
Llebetx, C., Spain 22 B9 39 33N 2 18 E
Lleida = Lérida, Spain 19 B6 41 37N 0 39 E
Llentrisca, C., Spain 22 C7 38 52N 1 15 E
Llera, Mexico 87 C5 23 19N 99 1W
Llico, Chile 94 C1 34 46S 72 5W
Llobregat →, Spain 19 B7 41 19N 2 9 E
Lloret de Mar, Spain 19 B7 41 41N 2 53 E
Lloyd B., Australia 62 A3 12 45S 143 27 E
Lloyd L., Canada 73 B7 57 22N 108 57W
Lloydminster, Canada 73 C6 53 17N 110 0W
Lluchmayor, Spain 22 B9 39 29N 2 53 E
Llullaillaco, Volcán, S. Amer. 94 A2 24 43S 68 30W
Lo →, Vietnam 38 B5 21 18N 105 25 E
Loa, U.S.A. 83 G8 38 24N 111 39W
Loa →, Chile 94 A1 21 26S 70 41W
Lobatse, Botswana 56 D4 25 12S 25 40 E
Lobería, Argentina 94 D4 38 10S 58 40W
Lobito, Angola 53 G2 12 18S 13 35 E
Lobos, Argentina 94 D4 35 10S 59 0W
Lobos, I., Mexico 86 B2 27 15N 110 30W
Lobos, Is., Peru 90 C1 6 57S 80 45W
Lobos, I. de, Canary Is. 22 F6 28 45N 13 50W
Loc Binh, Vietnam 38 B6 21 46N 106 54 E
Loc Ninh, Vietnam 39 G6 11 50N 106 34 E
Locarno, Switz. 16 E4 46 10N 8 47 E
Lochaber, U.K. 12 E4 56 55N 5 0W
Lochcarron, U.K. 12 D3 57 25N 5 30W
Lochem, Neths. 15 B6 52 9N 6 26 E
Loches, France 18 C4 47 7N 1 0 E
Lochgelly, U.K. 12 E5 56 7N 3 18W
Lochgilphead, U.K. 12 E3 56 2N 5 37W
Lochinver, U.K. 12 C3 58 9N 5 15W
Lochnagar, Australia 62 C4 23 33S 145 38 E
Lochnagar, U.K. 12 E5 56 57N 3 14W
Lochy →, U.K. 12 E3 56 52N 5 3W
Lock, Australia 63 E2 33 34S 135 46 E
Lock Haven, U.S.A. 78 E7 41 8N 77 28W
Lockeford, U.S.A. 84 G5 38 10N 121 9W
Lockeport, Canada 71 D6 43 47N 65 4W
Lockerbie, U.K. 12 F5 55 7N 3 21W
Lockhart, U.S.A. 81 L6 29 53N 97 40W
Lockhart, L., Australia 61 F2 33 15S 119 3 E
Lockney, U.S.A. 81 H4 34 7N 101 27W
Lockport, U.S.A. 78 C6 43 10N 78 42W
Lod, Israel 47 D3 31 57N 34 54 E
Lodeinoye Pole, Russia 24 B5 60 44N 33 33 E
Lodge Grass, U.S.A. 82 D10 45 19N 107 22W
Lodgepole, U.S.A. 80 E3 41 9N 102 38W
Lodgepole Cr. →, U.S.A. 80 E2 41 20N 104 30W
Lodhran, Pakistan 42 E4 29 32N 71 30 E
Lodi, U.S.A. 84 G5 38 8N 121 16W
Lodja, Zaïre 54 C1 3 30S 23 23 E
Lodwar, Kenya 54 B4 3 10N 35 40 E
Łódź, Poland 17 C9 51 45N 19 27 E
Loei, Thailand 38 D3 17 29N 101 35 E
Loengo, Zaïre 54 C2 4 48S 26 30 E
Loeriesfontein, S. Africa 56 E2 31 0S 19 26 E
Lofoten, Norway 8 B13 68 30N 15 0 E
Logan, Kans., U.S.A. 80 F5 39 40N 99 34W
Logan, Ohio, U.S.A. 76 F4 39 32N 82 25W
Logan, Utah, U.S.A. 82 F8 41 44N 111 50W
Logan, W. Va., U.S.A. 76 G5 37 51N 81 59W
Logan, Mt., Canada 68 B5 60 31N 140 22W
Logan Pass, U.S.A. 72 D6 48 41N 113 44W
Logandale, U.S.A. 85 J12 36 36N 114 29W
Logansport, Ind., U.S.A. 76 E2 40 45N 86 22W
Logansport, La., U.S.A. 81 K8 31 58N 94 0W
Logone →, Chad 51 F8 12 6N 15 2 E
Logroño, Spain 19 A4 42 28N 2 27W
Lohardaga, India 43 H11 23 27N 84 45 E
Loi-kaw, Burma 41 K20 19 40N 97 17 E
Loimaa, Finland 9 F17 60 50N 23 5 E
Loir →, France 18 C3 47 33N 0 32W
Loir-et-Cher □, France 18 C4 47 40N 1 20 E
Loire □, France 18 D6 45 40N 4 5 E
Loire →, France 18 C2 47 16N 2 10W
Loire-Atlantique □, France 18 C3 47 25N 1 40W
Loiret □, France 18 C5 47 55N 2 30 E
Loja, Ecuador 92 D3 3 59S 79 16W
Loja, Spain 19 D3 37 10N 4 10W
Loji, Indonesia 37 E7 1 38S 127 28 E
Lokandu, Zaïre 54 C2 2 30S 25 45 E
Lokeren, Belgium 15 C3 51 6N 3 59 E
Lokichokio, Kenya 54 B3 4 19N 34 13 E
Lokitaung, Kenya 54 B4 4 12N 35 48 E
Lokka, Finland 8 C19 67 55N 27 35 E
Løkken Verk, Norway 8 E10 63 7N 9 43 E
Lokoja, Nigeria 50 G6 7 47N 6 45 E
Lokolama, Zaïre 52 E3 2 35S 19 50 E
Lola, Mt., U.S.A. 84 F6 39 26N 120 22W
Loliondo, Tanzania 54 C4 2 2S 35 39 E
Lolland, Denmark 9 J11 54 45N 11 30 E
Lolo, U.S.A. 82 C6 46 45N 114 5W
Lom, Bulgaria 21 C10 43 48N 23 12 E
Lom Kao, Thailand 38 D3 16 53N 101 14 E
Lom Sak, Thailand 38 D3 16 47N 101 15 E
Loma, U.S.A. 82 C8 47 56N 110 30W
Loma Linda, U.S.A. 85 L9 34 3N 117 16W
Lomami →, Zaïre 54 B1 0 46N 24 16 E
Lomas de Zamóra, Argentina 94 C4 34 45S 58 25W
Lombadina, Australia 60 C3 16 31S 122 54 E
Lombardia □, Italy 20 B3 45 35N 9 45 E
Lombardy = Lombardia □, Italy 20 B3 45 35N 9 45 E
Lomblen, Indonesia 37 F6 8 30S 123 32 E
Lombok, Indonesia 36 F5 8 45S 116 30 E
Lomé, Togo 50 G5 6 9N 1 20 E
Lomela, Zaïre 52 E4 2 19S 23 15 E
Lomela →, Zaïre 52 E4 0 15S 20 40 E
Lometa, U.S.A. 81 K5 31 13N 98 24W
Lomié, Cameroon 52 D2 3 13N 13 38 E
Lomond, Canada 72 C6 50 24N 112 36W
Lomond, L., U.K. 12 E4 56 8N 4 38W
Lomphat, Cambodia 38 F6 13 30N 106 59 E
Lompobatang, Indonesia 37 F5 5 24S 119 56 E
Lompoc, U.S.A. 85 L6 34 38N 120 28W
Łomża, Poland 17 B11 53 10N 22 2 E
Loncoche, Chile 96 D2 39 20S 72 50W
Londa, India 40 M9 15 30N 74 30 E
Londiani, Kenya 54 C4 0 10S 35 33 E
London, Canada 70 D3 42 59N 81 15W
London, U.K. 11 F7 51 30N 0 5W
London, Ky., U.S.A. 76 G3 37 8N 84 5W
London, Ohio, U.S.A. 76 F4 39 53N 83 27W
London, Greater □, U.K. 11 F7 51 30N 0 5W
Londonderry, U.K. 13 B4 55 0N 7 23W
Londonderry □, U.K. 13 B4 55 0N 7 20W
Londonderry, C., Australia 60 B4 13 45S 126 55 E
Londonderry, I., Chile 96 H2 55 0S 71 0W
Londrina, Brazil 95 A5 23 18S 51 10W
Lone Pine, U.S.A. 83 H4 36 36N 118 4W
Long Beach, Calif., U.S.A. 85 M8 33 47N 118 11W
Long Beach, N.Y., U.S.A. 79 F11 40 35N 73 39W
Long Beach, Wash., U.S.A. 84 D2 46 21N 124 3W
Long Branch, U.S.A. 79 F11 40 18N 74 0W
Long Creek, U.S.A. 82 D4 44 43N 119 6W
Long Eaton, U.K. 10 E6 52 54N 1 16W
Long I., Australia 62 C4 22 8S 149 53 E
Long I., Bahamas 89 B4 23 20N 75 10W
Long I., U.S.A. 79 F11 40 45N 73 30W
Long Island Sd., U.S.A. 79 E12 41 10N 73 0W
Long L., Canada 70 C2 49 30N 86 50W
Long Lake, U.S.A. 79 C10 43 58N 74 25W
Long Pine, U.S.A. 80 D5 42 32N 99 42W
Long Pt., Nfld., Canada 71 C8 48 47N 58 46W

Malita

McDouall Peak, Australia 63 D1 29 51S 134 55 E
Macdougall L., Canada 68 B10 66 0N 98 27W
MacDowell L., Canada 70 B1 52 15N 92 45W
Macduff, U.K. 12 D6 57 40N 2 30W
Macedonia = Makedhonía □, Greece 21 D10 40 39N 22 0 E
Macedonia ■, Europe 21 D9 41 53N 21 40 E
Maceió, Brazil 93 E11 9 40S 35 41W
Macenta, Guinea 50 G3 8 35N 9 32W
Macerata, Italy 20 C5 43 19N 13 28 E
McFarland, U.S.A. 85 K7 35 41N 119 14W
McFarlane →, Canada 73 B7 59 12N 107 58W
Macfarlane, L., Australia 63 E2 32 0S 136 40 E
McGehee, U.S.A. 81 J9 33 38N 91 24W
McGill, U.S.A. 82 G6 39 23N 114 47W
Macgillycuddy's Reeks, Ireland 13 D2 52 2N 9 45W
MacGregor, Canada 73 D9 49 57N 98 48W
McGregor, U.S.A. 80 D9 43 1N 91 11W
McGregor →, Canada 72 B4 55 10N 122 0W
McGregor Ra., Australia 63 D3 27 0S 142 45 E
Mach, Pakistan 40 E5 29 50N 67 20 E
Māch Kowr, Iran 45 E9 25 48N 61 28 E
Machado = Jiparaná →, Brazil 92 E6 8 3S 62 52W
Machagai, Argentina 94 B3 26 56S 60 2W
Machakos, Kenya 54 C4 1 30S 37 15 E
Machakos □, Kenya 54 C4 1 30S 37 15 E
Machala, Ecuador 92 D3 3 20S 79 57W
Machanga, Mozam. 57 C6 20 59S 35 0 E
Machattie, L., Australia 62 C2 24 50S 139 48 E
Machava, Mozam. 57 D5 25 54S 32 28 E
Machece, Mozam. 55 F4 19 15S 35 32 E
Machevna, Russia 27 C18 61 20N 172 20 E
Machias, U.S.A. 71 D6 44 43N 67 28W
Machichi →, Canada 73 B10 57 3N 92 6W
Machico, Madeira 22 D3 32 43N 16 44W
Machilipatnam, India 41 L12 16 12N 81 8 E
Machiques, Venezuela 92 A4 10 4N 72 34W
Machupicchu, Peru 92 F4 13 8S 72 30W
Machynlleth, U.K. 11 E4 52 36N 3 51W
McIlwraith Ra., Australia 62 A3 13 50S 143 20 E
McIntosh, U.S.A. 80 C4 45 55N 101 21W
McIntosh L., Canada 73 B8 55 45N 105 0W
Macintosh Ra., Australia 61 E4 27 39S 125 32 E
Macintyre →, Australia 63 D5 28 37S 150 47 E
Mackay, Australia 62 C4 21 8S 149 11 E
Mackay, U.S.A. 82 E7 43 55N 113 37W
MacKay →, Canada 72 B6 57 10N 111 38W
Mackay, L., Australia 60 D4 22 30S 129 0 E
McKay Ra., Australia 60 D3 23 0S 122 30 E
McKeesport, U.S.A. 78 F5 40 21N 79 52W
McKenna, U.S.A. 84 D4 46 56N 122 33W
Mackenzie, Canada 72 B4 55 20N 123 5W
McKenzie, U.S.A. 77 G1 36 8N 88 31W
Mackenzie →, Australia 62 C4 23 38S 149 46 E
Mackenzie →, Canada 68 B6 69 10N 134 20W
McKenzie →, U.S.A. 82 D2 44 7N 123 6W
Mackenzie Bay, Canada 4 B1 69 0N 137 30W
Mackenzie City = Linden, Guyana 92 B7 6 0N 58 10W
Mackenzie Highway, Canada 72 B5 58 0N 117 15W
Mackenzie Mts., Canada 68 B6 64 0N 130 0W
Mackinaw City, U.S.A. 76 C3 45 47N 84 44W
McKinlay, Australia 62 C3 21 16S 141 18 E
McKinlay →, Australia 62 C3 20 50S 141 28 E
McKinley, Mt., U.S.A. 68 B4 63 4N 151 0W
McKinley Sea, Arctic 4 A7 84 0N 10 0W
McKinney, U.S.A. 81 J6 33 12N 96 37W
Mackinnon Road, Kenya 54 C4 3 40S 39 1 E
Macksville, Australia 63 E5 30 40S 152 56 E
McLaughlin, U.S.A. 80 C4 45 49N 100 49W
Maclean, Australia 63 D5 29 26S 153 16 E
McLean, U.S.A. 81 H4 35 14N 100 36W
McLeansboro, U.S.A. 80 F10 38 6N 88 32W
Maclear, S. Africa 57 E4 31 2S 28 23 E
Macleay →, Australia 63 E5 30 56S 153 0 E
McLennan, Canada 72 B5 55 42N 116 50W
MacLeod, B., Canada 73 A7 62 53N 110 0W
McLeod, L., Australia 61 D1 24 9S 113 47 E
MacLeod Lake, Canada 72 C4 54 58N 123 0W
M'Clintock Chan., Canada 68 A9 72 0N 102 0W
McLoughlin, Mt., U.S.A. 82 E2 42 27N 122 19W
McLure, Canada 72 C4 51 2N 120 13W
McMechen, U.S.A. 78 G4 39 57N 80 44W
McMillan, L., U.S.A. 81 J2 32 36N 104 21W
McMinnville, Oreg., U.S.A. 82 D2 45 13N 123 12W
McMinnville, Tenn., U.S.A. 77 H3 35 41N 85 46W
McMorran, Canada 73 C7 51 19N 108 42W
McMurdo Sd., Antarctica 5 D11 77 0S 170 0 E
McMurray = Fort McMurray, Canada 72 B6 56 44N 111 7W
McMurray, U.S.A. 84 B4 48 19N 122 14W
McNary, U.S.A. 83 J9 34 4N 109 51W
MacNutt, Canada 73 C8 51 5N 101 36W
Macodoene, Mozam. 57 C6 23 32S 35 5 E
Macomb, U.S.A. 80 E9 40 27N 90 40W
Mâcon, France 18 C6 46 19N 4 50 E
Macon, Ga., U.S.A. 77 J4 32 51N 83 38W
Macon, Miss., U.S.A. 77 J1 33 7N 88 34W
Macon, Mo., U.S.A. 80 F8 39 44N 92 28W
Macondo, Angola 53 G4 12 37S 23 46 E
Macossa, Mozam. 55 F3 17 55S 33 56 E
Macoun L., Canada 73 B8 56 32N 103 40W
Macovane, Mozam. 57 C6 21 30S 35 2 E
McPherson, U.S.A. 80 F6 38 22N 97 40W
McPherson Pk., U.S.A. 85 L7 34 53N 119 53W
McPherson Ra., Australia 63 D5 28 15S 153 15 E
Macquarie Harbour, Australia 62 G4 42 15S 145 23 E
Macquarie Is., Pac. Oc. 64 N7 54 36S 158 55 E
MacRobertson Land, Antarctica 5 D6 71 0S 64 0 E
Macroom, Ireland 13 E3 51 54N 8 57W
Macroy, Australia 60 D2 20 53S 118 2 E
MacTier, Canada 78 A5 45 9N 79 46W
Macubela, Mozam. 55 F4 16 53S 37 49 E
Macuiza, Mozam. 55 F3 18 7S 34 29 E
Macuse, Mozam. 55 F4 17 45S 37 10 E
Macuspana, Mexico 87 D6 17 46N 92 36W
Macusse, Angola 56 B3 17 48S 20 23 E
McVille, U.S.A. 80 B5 47 46N 98 11W
Madadeni, S. Africa 57 D5 27 43S 30 3 E
Madagali, Nigeria 51 F7 10 56N 13 33 E
Madagascar ■, Africa 57 C8 20 0S 47 0 E
Madā'in Sālih, Si. Arabia 44 E3 26 46N 37 57 E
Madama, Niger 51 D7 22 0N 13 40 E
Madame I., Canada 71 C7 45 30N 60 58W
Madaoua, Niger 50 F6 14 5N 6 27 E
Madaripur, Bangla. 41 H17 23 19N 90 15 E
Madauk, Burma 41 L20 17 56N 96 52 E
Madawaska, Canada 78 A7 45 30N 78 0W
Madawaska →, Canada 78 A8 45 27N 76 21W
Madaya, Burma 41 H20 22 12N 96 10 E
Madeira, Atl. Oc. 22 D3 32 50N 17 0W
Madeira →, Brazil 92 D7 3 22S 58 45W
Madeleine, Is. de la, Canada 71 C7 47 30N 61 40W
Madera, U.S.A. 83 H3 36 57N 120 3W
Madha, India 40 L9 18 0N 75 30 E
Madhubani, India 43 F12 26 21N 86 7 E
Madhya Pradesh □, India 42 J7 21 50N 78 0 E
Madikeri, India 40 N9 12 30N 75 45 E
Madill, U.S.A. 81 H6 34 6N 96 46W
Madimba, Zaïre 52 E3 4 58S 15 5 E
Ma'din, Syria 44 C3 35 45N 39 36 E
Madīnat ash Sha'b, Yemen 46 E3 12 50N 45 0 E
Madingou, Congo 52 E2 4 10S 13 33 E
Madirovalo, Madag. 57 B8 16 26S 46 32 E
Madison, Calif., U.S.A. 84 G5 38 41N 121 59W
Madison, Fla., U.S.A. 77 K4 30 28N 83 25W
Madison, Ind., U.S.A. 76 F3 38 44N 85 23W
Madison, Nebr., U.S.A. 80 E6 41 50N 97 27W
Madison, Ohio, U.S.A. 78 E3 41 46N 81 3W
Madison, S. Dak., U.S.A. 80 D6 44 0N 97 7W
Madison, Wis., U.S.A. 80 D10 43 4N 89 24W
Madison →, U.S.A. 82 D8 45 56N 111 31W
Madisonville, Ky., U.S.A. 76 G2 37 20N 87 30W
Madisonville, Tex., U.S.A. 81 K7 30 57N 95 55W
Madista, Botswana 56 C4 21 15S 25 6 E
Madiun, Indonesia 37 G14 7 38S 111 32 E
Madley, U.K. 11 E5 52 3N 2 51W
Madras = Tamil Nadu □, India 40 P10 11 0N 77 0 E
Madras, India 40 N12 13 8N 80 19 E
Madras, U.S.A. 82 D3 44 38N 121 8W
Madre, L., Mexico 87 B5 25 0N 97 30W
Madre, Laguna, U.S.A. 81 M6 27 0N 97 30W
Madre, Sierra, Phil. 37 A6 17 0N 122 0 E
Madre de Dios →, Bolivia 92 F5 10 59S 66 8W
Madre de Dios, I., Chile 96 G1 50 20S 75 10W
Madre del Sur, Sierra, Mexico 87 D5 17 30N 100 0W
Madre Occidental, Sierra, Mexico 86 B3 27 0N 107 0W
Madre Oriental, Sierra, Mexico 86 C4 25 0N 100 0W
Madri, India 42 G5 24 16N 73 32 E
Madrid, Spain 19 B4 40 25N 3 45W
Madura, Selat, Indonesia 37 G15 7 30S 113 20 E
Madura Motel, Australia 61 F4 31 55S 127 0 E
Madurai, India 40 Q11 9 55N 78 10 E
Madurantakam, India 40 N11 12 30N 79 50 E
Mae Chan, Thailand 38 B2 20 9N 99 52 E
Mae Hong Son, Thailand 38 C2 19 16N 98 1 E
Mae Khlong →, Thailand 38 F3 13 24N 100 0 E
Mae Phrik, Thailand 38 D2 17 27N 99 7 E
Mae Ramat, Thailand 38 D2 16 58N 98 31 E
Mae Rim, Thailand 38 C2 18 54N 98 57 E
Mae Sot, Thailand 38 D2 16 43N 98 34 E
Mae Suai, Thailand 38 C2 19 39N 99 33 E
Mae Tha, Thailand 38 C2 18 28N 99 8 E
Maebashi, Japan 31 F9 36 24N 139 4 E
Maesteg, U.K. 11 F4 51 36N 3 40W
Maestra, Sierra, Cuba 88 B4 20 15N 77 0W
Maestrazgo, Mts. del, Spain 19 B5 40 30N 0 25W
Maevatanana, Madag. 57 B8 16 56S 46 49 E
Mafeking = Mafikeng, S. Africa 56 D4 25 50S 25 38 E
Mafeking, Canada 73 C8 52 40N 101 10W
Mafeteng, Lesotho 56 D4 29 51S 27 15 E
Maffra, Australia 63 F4 37 53S 146 58 E
Mafia I., Tanzania 54 D4 7 45S 39 50 E
Mafikeng, S. Africa 56 D4 25 50S 25 38 E
Mafra, Brazil 95 B6 26 10S 49 55W
Mafra, Portugal 19 C1 38 55N 9 20W
Mafungabusi Plateau, Zimbabwe 55 F2 18 30S 29 8 E
Magadan, Russia 27 D16 59 38N 150 50 E
Magadi, Kenya 54 C4 1 54S 36 19 E
Magadi, L., Kenya 54 C4 1 54S 36 19 E
Magaliesburg, S. Africa 57 D4 26 0S 27 32 E
Magallanes, Estrecho de, Chile 96 G2 52 30S 75 0W
Magangué, Colombia 92 B4 9 14N 74 45W
Magburaka, S. Leone 50 G2 8 47N 12 0W
Magdalen Is. = Madeleine, Is. de la, Canada 71 C7 47 30N 61 40W
Magdalena, Argentina 94 D4 35 5S 57 30W
Magdalena, Bolivia 92 F6 13 13S 63 57W
Magdalena, Malaysia 36 D5 4 25N 117 55 E
Magdalena, Mexico 86 A2 30 50N 112 0W
Magdalena, U.S.A. 83 J10 34 7N 107 15W
Magdalena →, Colombia 92 A4 11 6N 74 51W
Magdalena →, Mexico 86 A2 30 40N 112 25W
Magdalena, B., Mexico 86 C2 24 30N 112 10W
Magdalena, Llano de la, Mexico 86 C2 25 0N 111 30W
Magdeburg, Germany 16 B5 52 8N 11 36 E
Magdelaine Cays, Australia 62 B5 16 33S 150 18 E
Magee, U.S.A. 81 K10 31 52N 89 44W
Magee, I., U.K. 13 B6 54 48N 5 44W
Magelang, Indonesia 37 G14 7 29S 110 13 E
Magellan's Str. = Magallanes, Estrecho de, Chile 96 G2 52 30S 75 0W
Magenta, L., Australia 61 F2 33 30S 119 2 E
Maggiore, L., Italy 20 A3 46 0N 8 35 E
Magherafelt, U.K. 13 B5 54 44N 6 37W
Magnetic Pole (North) = North Magnetic Pole, Canada 4 B2 77 58N 102 8W
Magnetic Pole (South) = South Magnetic Pole, Antarctica 5 C9 64 8S 138 8 E
Magnitogorsk, Russia 24 D10 53 27N 59 4 E
Magnolia, Ark., U.S.A. 81 J8 33 16N 93 14W
Magnolia, Miss., U.S.A. 81 K9 31 9N 90 28W
Magog, Canada 71 C5 45 18N 72 9W
Magoro, Uganda 54 B3 1 45N 34 12 E
Magosa = Famagusta, Cyprus 23 D12 35 8N 33 55 E
Magouládhes, Greece 23 A3 39 45N 19 42 E
Magoye, Zambia 55 F2 16 1S 27 30 E
Magpie L., Canada 71 B7 51 0N 64 41W
Magrath, Canada 72 D6 49 25N 112 50W
Magu □, Tanzania 54 C3 2 31S 33 28 E
Maguarinho, C., Brazil 93 D9 0 15S 48 30W
Maguse L., Canada 73 A9 61 40N 95 10W
Maguse Pt., Canada 73 A10 61 20N 93 50W
Magwe, Burma 41 J19 20 10N 95 0 E
Maha Sarakham, Thailand 38 D4 16 12N 103 16 E
Mahābād, Iran 44 B5 36 50N 45 45 E
Mahabharat Lekh, Nepal 43 E9 28 30N 82 0 E
Mahabo, Madag. 57 C7 20 23S 44 40 E
Mahadeo Hills, India 42 H8 22 20N 78 30 E
Mahagi, Zaïre 54 B3 2 20N 31 0 E
Mahajamba →, Madag. 57 B8 15 33S 47 8 E
Mahajamba, Helodranon' i, Madag. 57 B8 15 24S 47 5 E
Mahajan, India 42 E5 28 48N 73 56 E
Mahajanga, Madag. 57 B8 15 40S 46 25 E
Mahajanga □, Madag. 57 B8 17 0S 47 0 E
Mahajilo →, Madag. 57 B8 19 42S 45 22 E
Mahakam →, Indonesia 36 E5 0 35S 117 17 E
Mahalapye, Botswana 56 C4 23 1S 26 51 E
Mahallāt, Iran 45 C6 33 55N 50 30 E
Māhān, Iran 45 D8 30 5N 57 18 E
Mahanadi →, India 41 J15 20 20N 86 25 E
Mahanoro, Madag. 57 B8 19 54S 48 48 E
Mahanoy City, U.S.A. 79 F8 40 49N 76 9W
Maharashtra □, India 40 J9 20 30N 75 30 E
Mahari Mts., Tanzania 54 D2 6 20S 30 0 E
Mahasham, W. →, Egypt 47 E3 30 15N 34 10 E
Mahasolo, Madag. 57 B8 19 7S 46 22 E
Mahattat ash Shīdīyah, Jordan 47 F4 29 55N 35 55 E
Mahattat 'Unayzah, Jordan 47 E4 30 30N 35 47 E
Mahaxay, Laos 38 D5 17 22N 105 12 E
Mahbubnagar, India 40 L10 16 45N 77 59 E
Mahḍah, Oman 45 E7 24 24N 55 59 E
Mahdia, Tunisia 51 A7 35 28N 11 0 E
Mahe, India 43 C8 33 10N 78 32 E
Mahenge, Tanzania 55 D4 8 45S 36 41 E
Maheno, N.Z. 59 L3 45 10S 170 50 E
Mahesana, India 42 H5 23 39N 72 26 E
Mahia Pen., N.Z. 59 H6 39 9S 177 55 E
Mahilyow = Mogilev, Belorussia 24 D5 53 55N 30 18 E
Mahmud Kot, Pakistan 42 D4 30 16N 71 0 E
Mahnomen, U.S.A. 80 B7 47 19N 95 58W
Mahoba, India 43 G8 25 15N 79 55 E
Mahón, Spain 22 B11 39 53N 4 16 E
Mahone Bay, Canada 71 D7 44 30N 64 20W
Mai-Ndombe, L., Zaïre 52 E3 2 0S 18 20 E
Mai-Sai, Thailand 38 B2 20 20N 99 55 E
Maicurú →, Brazil 93 D8 2 14S 54 17W
Maidan Khula, Afghan. 42 C3 33 36N 69 50 E
Maidenhead, U.K. 11 F7 51 31N 0 42W
Maidstone, Canada 73 C7 53 5N 109 20W
Maidstone, U.K. 11 F8 51 16N 0 31 E
Maiduguri, Nigeria 51 F7 12 0N 13 20 E
Maijdi, Bangla. 41 H17 22 48N 91 10 E
Maikala Ra., India 41 J12 22 0N 81 0 E
Mailsi, Pakistan 42 E5 29 48N 72 15 E
Main →, Germany 16 D4 50 0N 8 18 E
Main →, U.K. 13 B5 54 49N 6 20W
Main Centre, Canada 73 C7 50 35N 107 21W
Maine, France 18 C3 47 55N 0 25W
Maine □, U.S.A. 71 C6 45 20N 69 0W
Maine →, Ireland 13 D2 52 10N 9 40W
Maine-et-Loire □, France 18 C3 47 31N 0 30W
Maingkwan, Burma 41 F20 26 15N 96 37 E
Mainit, L., Phil. 37 C7 9 31N 125 30 E
Mainland, Orkney, U.K. 12 C5 59 0N 3 10W
Mainland, Shet., U.K. 12 A7 60 15N 1 22W
Mainpuri, India 43 F8 27 18N 79 4 E
Maintirano, Madag. 57 B7 18 3S 44 1 E
Mainz, Germany 16 D4 50 0N 8 17 E
Maipú, Argentina 94 D4 36 52S 57 50W
Maiquetía, Venezuela 92 A5 10 36N 66 57W
Mairabari, India 41 F18 26 30N 92 22 E
Maisí, Cuba 89 B5 20 17N 74 9W
Maisi, Pta. de, Cuba 89 B5 20 10N 74 10W
Maitland, N.S.W., Australia 63 E5 32 33S 151 36 E
Maitland, S. Austral., Australia 63 E2 34 23S 137 40 E
Maitland →, Canada 78 C3 43 45N 81 43W
Maiz, Is. del, Nic. 88 D3 12 15N 83 4W
Maizuru, Japan 31 G7 35 25N 135 22 E
Majalengka, Indonesia 37 G13 6 50S 108 13 E
Majene, Indonesia 37 E5 3 38S 118 57 E
Maji, Ethiopia 51 G12 6 12N 35 30 E
Major, Canada 73 C7 51 52N 109 37W
Majorca = Mallorca, Spain 22 B10 39 30N 3 0 E
Maka, Senegal 50 F2 13 40N 14 10W
Makale, Indonesia 37 E5 3 6S 119 51 E
Makamba, Burundi 54 C2 4 8S 29 49 E
Makari, Cameroon 52 B2 12 35N 14 28 E
Makarikari = Makgadikgadi Salt Pans, Botswana 56 C4 20 40S 25 45 E
Makarovo, Russia 27 D11 57 40N 107 45 E
Makasar = Ujung Pandang, Indonesia 37 F5 5 10S 119 20 E
Makasar, Selat, Indonesia 37 E5 1 0S 118 20 E
Makasar, Str. of = Makasar, Selat, Indonesia 37 E5 1 0S 118 20 E
Makat, Kazakhstan 25 E9 47 39N 53 19 E
Makedhonía □, Greece 21 D10 40 39N 22 0 E
Makedonija = Macedonia ■, Europe 21 D9 41 53N 21 40 E
Makena, U.S.A. 74 H16 20 39N 156 27W
Makeni, S. Leone 50 G2 8 55N 12 5W
Makeyevka, Ukraine 25 E6 48 0N 38 0 E
Makgadikgadi Salt Pans, Botswana 56 C4 20 40S 25 45 E
Makhachkala, Russia 25 F8 43 0N 47 30 E
Makhmūr, Iraq 44 C4 35 46N 43 35 E
Makian, Indonesia 37 D7 0 20N 127 20 E
Makindu, Kenya 54 C4 2 18S 37 50 E
Makinsk, Kazakhstan 26 D8 52 37N 70 26 E
Makiyivka = Makeyevka, Ukraine 25 E6 48 0N 38 0 E
Makkah, Si. Arabia 46 C2 21 30N 39 54 E
Makkovik, Canada 71 A8 55 10N 59 10W
Makó, Hungary 17 E10 46 14N 20 33 E
Makokou, Gabon 52 D2 0 40N 12 50 E
Makongo, Zaïre 54 B2 3 25N 26 17 E
Makoro, Zaïre 54 B2 3 10N 29 59 E
Makoua, Congo 52 E3 0 5S 15 50 E
Makrai, India 40 H10 22 2N 77 0 E
Makran Coast Range, Pakistan 40 G4 25 40N 64 0 E
Makrana, India 42 F6 27 2N 74 46 E
Makriyialos, Greece 23 D7 35 2N 25 59 E
Maksimkin Yar, Russia 26 D9 58 42N 86 50 E
Mākū, Iran 44 B5 39 15N 44 31 E
Makumbi, Zaïre 52 F4 5 50S 20 43 E
Makunda, Botswana 56 C3 22 30S 20 7 E
Makurazaki, Japan 31 J5 31 15N 130 20 E
Makurdi, Nigeria 50 G6 7 43N 8 35 E
Makūyeh, Iran 45 D7 28 7N 53 9 E
Makwassie, S. Africa 56 D4 27 17S 26 0 E
Mal B., Ireland 13 D2 52 50N 9 30W
Mala, Pta., Panama 88 E3 7 28N 80 2W
Malabang, Phil. 37 C6 7 36N 124 3 E
Malabar Coast, India 40 P9 11 0N 75 0 E
Malabo = Rey Malabo, Eq. Guin. 50 H6 3 45N 8 50 E
Malacca, Str. of, Indonesia 39 L3 3 0N 101 0 E
Malad City, U.S.A. 82 E7 42 12N 112 15W
Málaga, Spain 19 D3 36 43N 4 23W
Malaga, U.S.A. 81 J2 32 14N 104 4W
Málaga □, Spain 19 D3 36 38N 4 58W
Malagarasi, Tanzania 54 D3 5 5S 30 50 E
Malagarasi →, Tanzania 54 D2 5 12S 29 47 E
Malaimbandy, Madag. 57 C8 20 20S 45 36 E
Malakâl, Sudan 51 G11 9 33N 31 40 E
Malakand, Pakistan 42 B4 34 40N 71 55 E
Malakoff, U.S.A. 81 J7 32 10N 96 1W
Malamyzh, Russia 27 E14 49 50N 136 50 E
Malang, Indonesia 37 G15 7 59S 112 45 E
Malanje, Angola 52 F3 9 36S 16 17 E
Mälaren, Sweden 9 G14 59 30N 17 10 E
Malargüe, Argentina 94 D2 35 32S 69 30W
Malartic, Canada 70 C4 48 9N 78 9W
Malatya, Turkey 25 G6 38 25N 38 20 E
Malawi ■, Africa 55 E3 11 55S 34 0 E
Malawi, L., Africa 55 E3 12 30S 34 30 E
Malay Pen., Asia 39 J3 7 25N 100 0 E
Malaybalay, Phil. 37 C7 8 5N 125 7 E
Malāyer, Iran 45 C6 34 19N 48 51 E
Malaysia ■, Asia 36 D4 5 0N 110 0 E
Malazgirt, Turkey 25 G7 39 10N 42 33 E
Malbon, Australia 62 C3 21 5S 140 17 E
Malbooma, Australia 63 E1 30 41S 134 11 E
Malbork, Poland 17 A9 54 3N 19 1 E
Malcolm, Australia 61 E3 28 51S 121 25 E
Malcolm, Pt., Australia 61 F3 33 48S 123 45 E
Maldegem, Belgium 15 C3 51 14N 3 26 E
Malden, Mass., U.S.A. 79 D13 42 26N 71 4W
Malden, Mo., U.S.A. 81 G10 36 34N 89 57W
Malden I., Kiribati 65 H12 4 3S 155 1W
Maldives ■, Ind. Oc. 29 J11 5 0N 73 0 E
Maldonado, Uruguay 95 C5 34 59S 55 0W
Maldonado, Punta, Mexico 87 D5 16 19N 98 35W
Malé Karpaty, Slovakia 16 D8 48 30N 17 20 E
Maléa, Ákra, Greece 21 F10 36 28N 23 7 E
Malebo, Pool, Africa 48 G5 4 17S 15 20 E
Malegaon, India 40 J9 20 30N 74 38 E
Malei, Mozam. 55 F4 17 12S 36 58 E
Malek Kandī, Iran 44 B5 37 9N 46 6 E
Malela, Zaïre 54 C2 4 22S 26 8 E
Malema, Mozam. 55 E4 14 57S 37 20 E
Máleme, Greece 23 D5 35 31N 23 49 E
Malerkotla, India 42 D6 30 32N 75 58 E
Máles, Greece 23 D7 35 6N 25 35 E
Malgomaj, Sweden 8 D14 64 40N 16 30 E
Malha, Sudan 51 E10 15 8N 25 10 E
Malhão, Sa. do, Portugal 19 D1 37 25N 8 0W
Malheur →, U.S.A. 82 D5 44 4N 116 59W
Malheur L., U.S.A. 82 E4 43 20N 118 48W
Mali ■, Africa 50 E4 17 0N 3 0W
Mali →, Burma 41 G20 25 40N 97 40 E
Malibu, U.S.A. 85 L8 34 2N 118 41W
Malik, Indonesia 37 E6 0 39S 123 16 E
Malili, Indonesia 37 E6 2 42S 121 6 E
Malimba, Mts., Zaïre 54 D2 7 30S 29 30 E
Malin Hd., Ireland 13 A4 55 18N 7 24W
Malindi, Kenya 54 C5 3 12S 40 5 E
Malines = Mechelen, Belgium 15 C4 51 2N 4 29 E
Maling, Indonesia 37 D6 1 0N 121 0 E
Malinyi, Tanzania 55 D4 8 56S 36 0 E
Malita, Phil. 37 C7 6 19N 125 39 E

Medicine Bow Ra.

Markazī □, Iran .... 45 C6 35 0N 49 30 E
Markdale, Canada 78 B4 44 19N 80 39W
Marked Tree, U.S.A. 81 H9 35 32N 90 25W
Marken, Neths. .... 15 B5 52 26N 5 12 E
Market Drayton, U.K. ... 10 E5 52 55N 2 30W
Market Harborough, U.K. 11 E7 52 29N 0 55W
Markham, Canada 78 C5 43 52N 79 16W
Markham, Mt., Antarctica 5 E11 83 0S 164 0 E
Markham L., Canada ... 73 A8 62 30N 102 35W
Markleeville, U.S.A. 84 G7 38 42N 119 47W
Markovo, Russia .... 27 C17 64 40N 169 40 E
Marks, Russia .... 24 D8 51 45N 46 50 E
Marksville, U.S.A. 81 K8 31 8N 92 4W
Marla, Australia .... 63 D1 27 19S 133 33 E
Marlboro, U.S.A. 79 D13 42 19N 71 33W
Marlborough, Australia . 62 C4 22 46S 149 52 E
Marlborough Downs, U.K. 11 F6 51 25N 1 55W
Marlin, U.S.A. 81 K6 31 18N 96 54W
Marlow, U.S.A. 81 H6 34 39N 97 58W
Marmagao, India ...... 40 M8 15 25N 73 56 E
Marmara, Turkey ...... 21 D12 40 35N 27 38 E
Marmara, Sea of = Marmara Denizi, Turkey 21 D13 40 45N 28 15 E
Marmara Denizi, Turkey 21 D13 40 45N 28 15 E
Marmarth, U.S.A. 80 B3 46 18N 103 54W
Marmion, Mt., Australia 61 E2 29 16S 119 50 E
Marmion L., Canada ... 70 C1 48 55N 91 20W
Marmolada, Mte., Italy . 20 A4 46 25N 11 55 E
Marmora, Canada 70 D4 44 28N 77 41W
Marne □, France ...... 18 B6 48 50N 4 10 E
Marne →, France ..... 18 B5 48 48N 2 24 E
Maroala, Madag. ...... 57 B8 15 23S 47 59 E
Maroantsetra, Madag. .. 57 B8 15 26S 49 44 E
Maromandia, Madag. .. 57 A8 14 13S 48 5 E
Marondera, Zimbabwe . 55 F3 18 5S 31 42 E
Maroni →, Fr. Guiana . 93 B8 5 30N 54 0W
Maroochydore, Australia 63 D5 26 29S 153 5 E
Maroona, Australia .... 63 F3 37 27S 142 54 E
Marosakoa, Madag. .... 57 B8 15 26S 46 38 E
Maroua, Cameroon .... 51 F7 10 40N 14 20 E
Marovoay, Madag. ..... 57 B8 16 6S 46 39 E
Marquard, S. Africa .... 56 D4 28 40S 27 28 E
Marquesas Is. = Marquises, Is., Pac. Oc. 65 H14 9 30S 140 0W
Marquette, U.S.A. 76 B2 46 33N 87 24W
Marquises, Is., Pac. Oc. . 65 H14 9 30S 140 0W
Marracuene, Mozam. ... 57 D5 25 45S 32 35 E
Marrakech, Morocco ... 50 B3 31 9N 8 0W
Marrawah, Australia ... 62 G3 40 55S 144 42 E
Marree, Australia ...... 63 D2 29 39S 138 1 E
Marrilla, Australia ..... 60 D1 22 31S 114 25 E
Marrimane, Mozam. .... 57 C5 22 58S 33 34 E
Marromeu, Mozam. .... 57 B6 18 15S 36 25 E
Marrowie Cr. →, Australia .... 63 E4 33 23S 145 40 E
Marrubane, Mozam. .... 55 F4 18 0S 37 0 E
Marrupa, Mozam. ...... 55 E4 13 8S 37 30 E
Marsá Matrûh, Egypt ... 51 B10 31 19N 27 9 E
Marsá Susah, Libya .... 51 B9 32 52N 21 59 E
Marsabit, Kenya .... 54 B4 2 18N 38 0 E
Marsabit □, Kenya ..... 54 B4 2 45N 37 45 E
Marsala, Italy .... 20 F5 37 48N 12 25 E
Marsalforn, Malta ...... 23 C1 36 4N 14 15 E
Marsden, Australia ..... 63 E4 33 47S 147 32 E
Marseille, France ...... 18 E6 43 18N 5 23 E
Marseilles = Marseille, France .... 18 E6 43 18N 5 23 E
Marsh I., U.S.A. 81 L9 29 34N 91 53W
Marsh L., U.S.A. 80 C6 45 5N 96 0W
Marshall, Liberia 50 G2 6 8N 10 22W
Marshall, Ark., U.S.A. .. 81 H8 35 55N 92 38W
Marshall, Mich., U.S.A. . 76 D3 42 16N 84 58W
Marshall, Minn., U.S.A. . 80 C7 44 25N 95 45W
Marshall, Mo., U.S.A. ... 80 F8 39 7N 93 12W
Marshall, Tex., U.S.A. .. 81 J7 32 33N 94 23W
Marshall →, Australia . 62 C2 22 59S 136 59 E
Marshall Is. ■, Pac. Oc. . 64 G9 9 0N 171 0 E
Marshalltown, U.S.A. ... 80 D8 42 3N 92 55W
Marshfield, Mo., U.S.A. . 81 G8 37 15N 92 54W
Marshfield, Wis., U.S.A. 80 C9 44 40N 90 10W
Marshūn, Iran .... 45 B6 36 19N 49 23 E
Marstrand, Sweden .... 9 H11 57 53N 11 35 E
Mart, U.S.A. 81 K6 31 33N 96 50W
Martaban, Burma ...... 41 L20 16 30N 97 35 E
Martaban, G. of, Burma . 41 L20 16 5N 96 30 E
Martapura, Kalimantan, Indonesia .... 36 E4 3 22S 114 47 E
Martapura, Sumatera, Indonesia .... 36 E2 4 19S 104 22 E
Marte, Nigeria .... 51 F7 12 23N 13 46 E
Martelange, Belgium ... 15 E5 49 49N 5 43 E
Martha's Vineyard, U.S.A. 79 E14 41 25N 70 38W
Martin, S. Dak., U.S.A. .. 80 D4 43 11N 101 44W
Martin, Tenn., U.S.A. ... 81 G10 36 21N 88 51W
Martin L., U.S.A. 77 J3 32 41N 85 55W
Martinborough, N.Z. ... 59 J5 41 14S 175 29 E
Martinez, U.S.A. 84 G4 38 1N 122 8W
Martinique ■, W. Indies 89 D7 14 40N 61 0W
Martinique Passage, W. Indies 89 C7 15 15N 61 0W
Martinópolis, Brazil 95 A5 22 11S 51 12W
Martins Ferry, U.S.A. ... 78 F4 40 6N 80 44W
Martinsburg, Pa., U.S.A. 78 F6 40 19N 78 20W
Martinsburg, W. Va., U.S.A. 76 F7 39 27N 77 58W
Martinsville, Ind., U.S.A. 76 F2 39 26N 86 25W
Martinsville, Va., U.S.A. . 77 G6 36 41N 79 52W
Marton, N.Z. .... 59 J5 40 4S 175 23 E
Martos, Spain 19 D4 37 44N 3 58W
Marudi, Malaysia ...... 36 D4 4 11N 114 19 E
Ma'ruf, Afghan. .... 40 D5 31 30N 67 6 E
Marugame, Japan ..... 31 G6 34 15N 133 40 E
Marulan, Australia ..... 63 E5 34 43S 150 3 E
Marunga, Angola ...... 56 B3 17 28S 20 2 E
Marungu, Mts., Zaïre ... 54 D2 7 30S 30 0 E
Marvast, Iran .... 45 D7 30 30N 54 15 E
Marwar, India .... 42 G5 25 43N 73 45 E
Mary, Turkmenistan .... 26 F7 37 40N 61 50 E
Mary Frances L., Canada 73 A7 63 19N 106 13W
Mary Kathleen, Australia 62 C2 20 44S 139 48 E
Maryborough = Port Laoise, Ireland 13 C4 53 2N 7 20W
Maryborough, Queens., Australia .... 63 D5 25 31S 152 37 E
Maryborough, Vic., Australia .... 63 F3 37 0S 143 44 E
Maryfield, Canada 73 D8 49 50N 101 35W
Maryland □, U.S.A. 76 F7 39 0N 76 30W
Maryland Junction, Zimbabwe .... 55 F3 17 45S 30 31 E
Maryport, U.K. 10 C4 54 43N 3 30W
Mary's Harbour, Canada 71 B8 52 18N 55 51W
Marystown, Canada 71 C8 47 10N 55 10W
Marysvale, U.S.A. 83 G7 38 27N 112 14W
Marysville, Canada 72 D5 49 35N 116 0W
Marysville, Calif., U.S.A. 84 F5 39 9N 121 35W
Marysville, Kans., U.S.A. 80 F6 39 51N 96 39W
Marysville, Mich., U.S.A. 78 D2 42 54N 82 29W
Marysville, Ohio, U.S.A. 76 E4 40 14N 83 22W
Marysville, Wash., U.S.A. 84 B4 48 3N 122 11W
Maryvale, Australia .... 63 D5 28 4S 152 12 E
Maryville, U.S.A. 77 H4 35 46N 83 58W
Marzūq, Libya .... 51 C7 25 53N 13 57 E
Masahunga, Tanzania .. 54 C3 2 6S 33 18 E
Masai, Malaysia .... 39 M4 1 29N 103 55 E
Masai Steppe, Tanzania 54 C4 4 30S 36 30 E
Masaka, Uganda .... 54 C3 0 21S 31 45 E
Masalembo, Kepulauan, Indonesia .... 36 F4 5 35S 114 30 E
Masalima, Kepulauan, Indonesia .... 36 F5 5 4S 117 5 E
Masamba, Indonesia ... 37 E6 2 30S 120 15 E
Masan, S. Korea .... 35 G15 35 11N 128 32 E
Masasi, Tanzania ...... 55 E4 10 45S 38 52 E
Masasi □, Tanzania .... 55 E4 10 45S 38 50 E
Masaya, Nic. 88 D2 12 0N 86 7W
Masbate, Phil. .... 37 B6 12 21N 123 36 E
Mascara, Algeria ...... 50 A5 35 26N 0 6 E
Mascota, Mexico 86 C4 20 30N 104 50W
Masela, Indonesia ..... 37 F7 8 9S 129 51 E
Maseru, Lesotho .... 56 D4 29 18S 27 30 E
Mashaba, Zimbabwe ... 55 G3 20 2S 30 29 E
Mashābih, Si. Arabia ... 44 E3 25 35N 36 30 E
Masherbrum, Pakistan .. 43 B7 35 38N 76 18 E
Mashhad, Iran .... 45 B8 36 20N 59 35 E
Mashīz, Iran .... 45 D8 29 56N 56 37 E
Mashkel, Hamun-i-, Pakistan .... 40 E3 28 30N 63 0 E
Mashki Chāh, Pakistan .. 40 E3 29 5N 62 30 E
Mashonaland Central □, Zimbabwe .... 57 B5 17 30S 31 0 E
Mashonaland East □, Zimbabwe .... 57 B5 18 0S 32 0 E
Mashonaland West □, Zimbabwe .... 57 B4 17 30S 29 30 E
Masi, Norway .... 8 B17 69 26N 23 40 E
Masi Manimba, Zaïre ... 52 E3 4 40S 17 54 E
Masindi, Uganda ...... 54 B3 1 40N 31 43 E
Masindi Port, Uganda .. 54 B3 1 43N 32 2 E
Masisea, Peru 92 E4 8 35S 74 22W
Masisi, Zaïre .... 54 C2 1 23S 28 49 E
Masjed Soleyman, Iran . 45 D6 31 55N 49 18 E
Mask, L., Ireland 13 C2 53 36N 9 24W
Masoala, Tanjon' i, Madag. .... 57 B9 15 59S 50 13 E
Masoarivo, Madag. .... 57 B7 19 3S 44 19 E
Masohi, Indonesia ..... 37 E7 3 20S 128 55 E
Masomeloka, Madag. ... 57 C8 20 17S 48 37 E
Mason, Nev., U.S.A. 84 G7 38 56N 119 8W
Mason, Tex., U.S.A. 81 K5 30 45N 99 14W
Mason City, U.S.A. 80 D8 43 9N 93 12W
Maspalomas, Canary Is. 22 G4 27 46N 15 35W
Maspalomas, Pta., Canary Is. 22 G4 27 43N 15 36W
Masqat, Oman .... 46 C6 23 37N 58 36 E
Massa, Italy .... 20 B4 44 2N 10 7 E
Massachusetts □, U.S.A. 79 D12 42 30N 72 0W
Massachusetts B., U.S.A. 79 D14 42 20N 70 50W
Massaguet, Chad ...... 51 F8 12 28N 15 26 E
Massakory, Chad ...... 51 F8 13 0N 15 49 E
Massanella, Spain ..... 22 B9 39 48N 2 51 E
Massangena, Mozam. ... 57 C5 21 34S 33 0 E
Massawa = Mitsiwa, Eritrea .... 51 E12 15 35N 39 25 E
Massena, U.S.A. 79 B10 44 56N 74 54W
Massénya, Chad .... 51 F8 11 21N 16 9 E
Masset, Canada 72 C2 54 2N 132 10W
Massif Central, France . 18 D5 44 55N 3 0 E
Massillon, U.S.A. 78 F3 40 48N 81 32W
Massinga, Mozam. ..... 57 C6 23 15S 35 22 E
Masson, Canada 79 A9 45 32N 75 25W
Masson I., Antarctica ... 5 C7 66 10S 93 20 E
Masterton, N.Z. .... 59 J5 40 56S 175 39 E
Mastuj, Pakistan ...... 43 A5 36 20N 72 36 E
Mastung, Pakistan ..... 40 E5 29 50N 66 56 E
Masuda, Japan .... 31 G5 34 40N 131 51 E
Masvingo, Zimbabwe .. 55 G3 20 8S 30 49 E
Masvingo □, Zimbabwe . 55 G3 21 0S 31 30 E
Maswa □, Tanzania .... 54 C3 3 30S 34 0 E
Maşyāf, Syria .... 44 C3 35 4N 36 20 E
Matabeleland North □, Zimbabwe .... 55 F2 19 0S 28 0 E
Matabeleland South □, Zimbabwe .... 55 G2 21 0S 29 0 E
Mataboor, Indonesia ... 37 E9 1 41S 138 3 E
Matachewan, Canada .. 70 C3 47 56N 80 39W
Matadi, Zaïre .... 52 F2 5 52S 13 31 E
Matagalpa, Nic. 88 D2 13 0N 85 58W
Matagami, Canada 70 C4 49 45N 77 34W
Matagami, L., Canada .. 70 C4 49 50N 77 40W
Matagorda, U.S.A. 81 L7 28 42N 95 58W
Matagorda B., U.S.A. ... 81 L6 28 40N 96 0W
Matagorda I., U.S.A. ... 81 L6 28 15N 96 30W
Matak, P., Indonesia ... 39 L6 3 18N 106 16 E
Matakana, Australia .... 63 E4 32 59S 145 54 E
Mátala, Greece .... 23 E6 34 59N 24 45 E
Matam, Senegal 50 E2 15 34N 13 17W
Matamoros, Campeche, Mexico 87 D6 18 50N 90 50W
Matamoros, Coahuila, Mexico 86 B4 25 33N 103 15W
Matamoros, Puebla, Mexico 87 D5 18 2N 98 17W
Matamoros, Tamaulipas, Mexico 87 B5 25 50N 97 30W
Ma'ţan as Sarra, Libya .. 51 D9 21 45N 22 0 E
Matandu →, Tanzania . 55 D3 8 45S 34 19 E
Matane, Canada 71 C6 48 50N 67 33W
Matanzas, Cuba 88 B3 23 0N 81 40W
Matapan, C. = Taínaron, Ákra, Greece .... 21 F10 36 22N 22 27 E
Matapédia, Canada 71 C6 48 0N 66 59W
Matara, Sri Lanka ...... 40 S12 5 58N 80 30 E
Mataram, Indonesia .... 36 F5 8 41S 116 10 E
Matarani, Peru 92 G4 17 0S 72 10W
Mataranka, Australia ... 60 B5 14 55S 133 4 E
Matarma, Râs, Egypt ... 47 E1 30 27N 32 44 E
Matatiele, S. Africa .... 57 E4 30 20S 28 49 E
Mataura, N.Z. .... 59 M2 46 11S 168 51 E
Matehuala, Mexico 86 C4 23 40N 100 40W
Mateke Hills, Zimbabwe 55 G3 21 48S 31 0 E
Matera, Italy .... 20 D7 40 40N 16 37 E
Matetsi, Zimbabwe .... 55 F2 18 12S 26 0 E
Matheson Island, Canada 73 C9 51 45N 96 56W
Mathis, U.S.A. 81 L6 28 6N 97 50W
Mathura, India .... 42 F7 27 30N 77 40 E
Mati, Phil. .... 37 C7 6 55N 126 15 E
Matías Romero, Mexico 87 D5 16 53N 95 2W
Matibane, Mozam. ..... 55 E5 14 49S 40 45 E
Matima, Botswana ..... 56 C3 20 15S 24 26 E
Matiri Ra., N.Z. .... 59 J4 41 38S 172 20 E
Matlock, U.K. 10 D6 53 8N 1 32W
Matmata, Tunisia ..... 50 B6 33 37N 9 59 E
Mato Grosso □, Brazil .. 93 F8 14 0S 55 0W
Mato Grosso, Planalto do, Brazil 93 G8 15 0S 55 0W
Mato Grosso do Sul □, Brazil 93 G8 18 0S 55 0W
Matochkin Shar, Russia . 26 B6 73 10N 56 40 E
Matopo Hills, Zimbabwe 55 G2 20 36S 28 20 E
Matopos, Zimbabwe ... 55 G2 20 20S 28 29 E
Matosinhos, Portugal .. 19 B1 41 11N 8 42W
Matsue, Japan .... 31 G6 35 25N 133 10 E
Matsumae, Japan ..... 30 D10 41 26N 140 7 E
Matsumoto, Japan ..... 31 F9 36 15N 138 0 E
Matsusaka, Japan ..... 31 G8 34 34N 136 32 E
Matsuura, Japan ...... 31 H4 33 20N 129 49 E
Matsuyama, Japan .... 31 H6 33 45N 132 45 E
Mattagami →, Canada . 70 B3 50 43N 81 29W
Mattancheri, India .... 40 Q10 9 50N 76 15 E
Mattawa, Canada 70 C4 46 20N 78 45W
Mattawamkeag, U.S.A. . 71 C6 45 32N 68 21W
Matterhorn, Switz. ..... 16 F3 45 58N 7 39 E
Matthew Town, Bahamas 89 B5 20 57N 73 40W
Matthew's Ridge, Guyana 92 B6 7 37N 60 10W
Mattice, Canada 70 C3 49 40N 83 20W
Mattituck, U.S.A. 79 F12 40 59N 72 32W
Matuba, Mozam. ...... 57 C5 24 28S 32 49 E
Matucana, Peru 92 F3 11 55S 76 25W
Matun, Afghan. .... 42 C3 33 22N 69 58 E
Maturín, Venezuela 92 B6 9 45N 63 11W
Mau, India .... 43 G10 25 56N 83 33 E
Mau Escarpment, Kenya 54 C4 0 40S 36 0 E
Mau Ranipur, India .... 43 G8 25 16N 79 8 E
Maud, Pt., Australia .... 60 D1 23 6S 113 45 E
Maude, Australia ...... 63 E3 34 29S 144 18 E
Maudin Sun, Burma .... 41 M19 16 0N 94 30 E
Maués, Brazil 92 D7 3 20S 57 45W
Mauganj, India .... 41 G12 24 50N 81 55 E
Maui, U.S.A. 74 H16 20 48N 156 20W
Maulamyaing = Moulmein, Burma ..... 41 L20 16 30N 97 40 E
Maule □, Chile 94 D1 36 5S 72 30W
Maumee, U.S.A. 76 E4 41 34N 83 39W
Maumee →, U.S.A. 76 E4 41 42N 83 28W
Maumere, Indonesia ... 37 F6 8 38S 122 13 E
Maun, Botswana .... 56 B3 20 0S 23 26 E
Mauna Kea, U.S.A. 74 J17 19 50N 155 28W
Mauna Loa, U.S.A. 74 J17 19 30N 155 35W
Maungmagan Kyunzu, Burma .... 41 M20 14 0N 97 48 E
Maupin, U.S.A. 82 D3 45 11N 121 5W
Maurepas, L., U.S.A. ... 81 K9 30 15N 90 30W
Maures, France .... 18 E7 43 15N 6 15 E
Maurice, L., Australia .. 61 E5 29 30S 131 0 E
Mauritania ■, Africa ... 50 D3 20 50N 10 0W
Mauritius ■, Ind. Oc. ... 49 J9 20 0S 57 0 E
Mauston, U.S.A. 80 D9 43 48N 90 5W
Mavinga, Angola ...... 53 H4 15 50S 20 21 E
Mavli, India .... 42 G5 24 45N 73 55 E
Mavuradonha Mts., Zimbabwe .... 55 F3 16 30S 31 30 E
Mawa, Zaïre .... 54 B2 2 45N 26 40 E
Mawana, India .... 42 E7 29 6N 77 58 E
Mawand, Pakistan ..... 42 E3 29 33N 68 38 E
Mawk Mai, Burma ..... 41 J20 20 14N 97 37 E
Mawlaik, Burma .... 41 H19 23 40N 94 26 E
Mawqaq, Si. Arabia .... 44 E4 27 25N 41 8 E
Mawson Coast, Antarctica .... 5 C6 68 30S 63 0 E
Max, U.S.A. 80 B4 47 49N 101 18W
Maxcanú, Mexico 87 C6 20 40N 92 0W
Maxesibeni, S. Africa .. 57 E4 30 49S 29 23 E
Maxhamish L., Canada . 72 B4 59 50N 123 17W
Maxixe, Mozam. .... 57 C6 23 54S 35 17 E
Maxville, Canada 79 A10 45 17N 74 51W
Maxwell, U.S.A. 84 F4 39 17N 122 11W
Maxwelton, Australia ... 62 C3 20 43S 142 41 E
May Downs, Australia .. 62 C4 22 38S 148 55 E
May Pen, Jamaica 88 C4 17 58N 77 15W
Maya →, Russia ...... 27 D14 60 28N 134 28 E
Maya Mts., Belize 87 D7 16 30N 89 0W
Mayaguana, Bahamas .. 89 B5 22 30N 72 44W
Mayagüez, Puerto Rico . 89 C6 18 12N 67 9W
Mayāmey, Iran .... 45 B7 36 24N 55 42 E
Mayari, Cuba 89 B4 20 40N 75 41W
Maybell, U.S.A. 82 F9 40 31N 108 5W
Maydān, Iraq .... 44 C5 34 55N 45 37 E
Maydena, Australia .... 62 G4 42 45S 146 30 E
Mayenne, France 18 B3 48 20N 0 38W
Mayenne □, France 18 B3 48 10N 0 40W
Mayer, U.S.A. 83 J7 34 24N 112 14W
Mayerthorpe, Canada .. 72 C5 53 57N 115 8W
Mayfield, U.S.A. 77 G1 36 44N 88 38W
Mayhill, U.S.A. 83 K11 32 53N 105 29W
Maykop, Russia .... 25 F7 44 35N 40 10 E
Maymyo, Burma .... 38 A1 22 2N 96 28 E
Maynard, U.S.A. 84 C4 47 59N 122 55W
Maynard Hills, Australia 61 E2 28 28S 119 49 E
Mayne →, Australia ... 62 C3 23 40S 141 55 E
Maynooth, Ireland 13 C5 53 22N 6 38W
Mayo, Canada 68 B6 63 38N 135 57W
Mayo □, Ireland 13 C2 53 47N 9 7W
Mayo L., Canada 68 B6 63 45N 135 0W
Mayon Volcano, Phil. ... 37 B6 13 15N 123 41 E
Mayor I., N.Z. .... 59 G6 37 16S 176 17 E
Mayson L., Canada 73 B7 57 55N 107 10W
Maysville, U.S.A. 76 F4 38 39N 83 46W
Mayu, Indonesia ...... 37 D7 1 30N 126 30 E
Mayville, N. Dak., U.S.A. 80 B6 47 30N 97 20W
Mayville, N.Y., U.S.A. .. 78 D5 42 15N 79 30W
Mayya, Russia .... 27 C14 61 44N 130 18 E
Mazabuka, Zambia ..... 55 F2 15 52S 27 44 E
Mazagán = El Jadida, Morocco 50 B3 33 11N 8 17W
Mazagão, Brazil 93 D8 0 7S 51 16W
Mazán, Peru 92 D4 3 30S 73 0W
Māzandarān □, Iran .... 45 B7 36 30N 52 0 E
Mazapil, Mexico 86 C4 24 38N 101 34W
Mazarredo, Argentina .. 96 F3 47 10S 66 50W
Mazarrón, Spain 19 D5 37 38N 1 19W
Mazaruni →, Guyana .. 92 B7 6 25N 58 35W
Mazatán, Mexico 86 B2 29 0N 110 8W
Mazatenango, Guatemala 88 D1 14 35N 91 30W
Mazatlán, Mexico 86 C3 23 10N 106 30W
Māzhān, Iran .... 45 C8 32 30N 59 0 E
Mazīnān, Iran .... 45 B8 36 19N 56 56 E
Mazoe, Mozam. .... 55 F3 16 42S 33 7 E
Mazoe →, Mozam. ..... 55 F3 16 20S 33 30 E
Mazowe, Zimbabwe .... 55 F3 17 28S 30 58 E
Mazurian Lakes = Mazurski, Pojezierze, Poland .... 17 B10 53 50N 21 0 E
Mazurski, Pojezierze, Poland .... 17 B10 53 50N 21 0 E
Mbabane, Swaziland ... 57 D5 26 18S 31 6 E
Mbaïki, C.A.R. .... 52 D3 3 53N 18 1 E
Mbala, Zambia .... 55 D3 8 46S 31 24 E
Mbale, Uganda .... 54 B3 1 8N 34 12 E
Mbalmayo, Cameroon .. 52 D2 3 33N 11 33 E
Mbamba Bay, Tanzania . 55 E3 11 13S 34 49 E
Mbandaka, Zaïre ...... 52 D3 0 1N 18 18 E
Mbanza Congo, Angola . 52 F2 6 18S 14 16 E
Mbanza Ngungu, Zaïre . 52 F2 5 12S 14 53 E
Mbarara, Uganda ...... 54 C3 0 35S 30 40 E
Mbashe →, S. Africa .. 57 E4 32 15S 28 54 E
Mbenkuru →, Tanzania 55 D4 9 25S 39 50 E
Mberengwa, Zimbabwe 55 G2 20 29S 29 57 E
Mberengwa, Mt., Zimbabwe .... 55 G2 20 37S 29 55 E
Mbesuma, Zambia ..... 55 D3 10 0S 32 2 E
Mbeya, Tanzania ...... 55 D3 8 54S 33 29 E
Mbeya □, Tanzania .... 54 D3 8 15S 33 30 E
Mbinga, Tanzania ..... 55 E4 10 50S 35 0 E
Mbinga □, Tanzania ... 55 E3 10 50S 35 0 E
Mbini □, Eq. Guin. ..... 52 D2 1 30N 10 0 E
Mbour, Senegal 50 F1 14 22N 16 54W
Mbout, Mauritania 50 E2 16 1N 12 38W
Mbozi □, Tanzania ..... 55 D3 9 0S 32 50 E
Mbuji-Mayi, Zaïre ..... 54 D1 6 9S 23 40 E
Mbulu, Tanzania ...... 54 C4 3 45S 35 30 E
Mbulu □, Tanzania .... 54 C4 3 52S 35 33 E
Mburucuyá, Argentina . 94 B4 28 1S 58 14W
Mchinja, Tanzania ..... 55 D4 9 44S 39 45 E
Mchinji, Malawi .... 55 E3 13 47S 32 58 E
Mead, L., U.S.A. 85 J12 36 1N 114 44W
Meade, U.S.A. 81 G4 37 17N 100 20W
Meadow, Australia ..... 61 E1 26 35S 114 40 E
Meadow Lake, Canada . 73 C7 54 10N 108 26W
Meadow Lake Prov. Park, Canada 73 C7 54 27N 109 0W
Meadow Valley Wash →, U.S.A. 85 J12 36 40N 114 34W
Meadville, U.S.A. 78 E4 41 39N 80 9W
Meaford, Canada 70 D3 44 36N 80 35W
Mealy Mts., Canada 71 B8 53 10N 58 0W
Meander River, Canada . 72 B5 59 2N 117 42W
Meares, C., U.S.A. 82 D2 45 37N 124 0W
Mearim →, Brazil 93 D10 3 4S 44 35W
Meath □, Ireland 13 C5 53 32N 6 40W
Meath Park, Canada ... 73 C7 53 27N 105 22W
Meaux, France .... 18 B5 48 58N 2 50 E
Mebechi-Gawa →, Japan .... 30 D10 40 31N 141 31 E
Mecanhelas, Mozam. ... 55 F4 15 12S 35 54 E
Mecca = Makkah, Si. Arabia .... 46 C2 21 30N 39 54 E
Mecca, U.S.A. 85 M10 33 34N 116 5W
Mechanicsburg, U.S.A. . 78 F8 40 13N 77 1W
Mechanicville, U.S.A. .. 79 D11 42 54N 73 41W
Mechelen, Belgium ..... 15 C4 51 2N 4 29 E
Mecheria, Algeria 50 B4 33 35N 0 18W
Mecklenburger Bucht, Germany .... 16 A5 54 20N 11 40 E
Meconta, Mozam. ..... 55 E4 14 59S 39 50 E
Meda, Australia .... 60 C3 17 22S 123 59 E
Medan, Indonesia ..... 36 D1 3 40N 98 38 E
Medanosa, Pta., Argentina 96 F3 48 8S 66 0W
Medéa, Algeria .... 50 A5 36 12N 2 50 E
Medellín, Colombia 92 B3 6 15N 75 35W
Medemblik, Neths. ..... 15 B5 52 46N 5 8 E
Mederdra, Mauritania .. 50 E1 17 0N 15 38W
Medford, Mass., U.S.A. . 79 D13 42 25N 71 7W
Medford, Oreg., U.S.A. . 82 E2 42 19N 122 52W
Medford, Wis., U.S.A. .. 80 C9 45 9N 90 20W
Media Agua, Argentina . 94 C2 31 58S 68 25W
Media Luna, Argentina . 94 C2 34 45S 66 44W
Mediaş, Romania ...... 17 E12 46 9N 24 22 E
Medical Lake, U.S.A. ... 82 C5 47 34N 117 41W
Medicine Bow, U.S.A. .. 82 F10 41 54N 106 12W
Medicine Bow Pk., U.S.A. 82 F10 41 21N 106 19W
Medicine Bow Ra., U.S.A. 82 F10 41 10N 106 25W

Medicine Hat

Medicine Hat, *Canada* **73 D6** 50 0N 110 45W
Medicine Lake, *U.S.A.* **80 A2** 48 30N 104 30W
Medicine Lodge, *U.S.A.* **81 G5** 37 17N 98 35W
Medina = Al Madīnah, *Si. Arabia* **46 C2** 24 35N 39 52 E
Medina, *N. Dak., U.S.A.* **80 B5** 46 54N 99 18W
Medina, *N.Y., U.S.A.* **78 C6** 43 13N 78 23W
Medina, *Ohio, U.S.A.* **78 E3** 41 8N 81 52W
Medina →, *U.S.A.* **81 L5** 29 16N 98 29W
Medina del Campo, *Spain* **19 B3** 41 18N 4 55W
Medina L., *U.S.A.* **81 L5** 29 32N 98 56W
Medina-Sidonia, *Spain* **19 D3** 36 28N 5 57W
Medinipur, *India* **43 H12** 22 25N 87 21 E
Mediterranean Sea, *Europe* **48 C5** 35 0N 15 0 E
Medley, *Canada* **73 C6** 54 25N 110 16W
Médoc, *France* **18 D3** 45 10N 0 50W
Medstead, *Canada* **73 C7** 53 19N 108 5W
Medveditsa →, *Russia* **25 E7** 49 35N 42 41 E
Medvezhi, Ostrava, *Russia* **27 B17** 71 0N 161 0 E
Medvezhyegorsk, *Russia* **24 B5** 63 0N 34 25 E
Medway →, *U.K.* **11 F8** 51 28N 0 45 E
Meeberrie, *Australia* **61 E2** 26 57S 115 51 E
Meekatharra, *Australia* **61 E2** 26 32S 118 29 E
Meeker, *U.S.A.* **82 F10** 40 2N 107 55W
Meerut, *India* **42 E7** 29 1N 77 42 E
Meeteetse, *U.S.A.* **82 D9** 44 9N 108 52W
Mega, *Ethiopia* **51 H12** 3 57N 38 19 E
Mégara, *Greece* **21 F10** 37 58N 23 22 E
Meghalaya □, *India* **41 G17** 25 50N 91 0 E
Mégiscane, L., *Canada* **70 C4** 48 35N 75 55W
Mehadia, *Romania* **17 F11** 44 56N 22 23 E
Mehndawal, *India* **43 F10** 26 58N 83 5 E
Mehr Jān, *Iran* **45 C7** 33 50N 55 6 E
Mehrābād, *Iran* **44 B5** 36 53N 47 55 E
Mehrān, *Iran* **44 C5** 33 7N 46 10 E
Mehrīz, *Iran* **45 D7** 31 35N 54 28 E
Mei Xian, *Guangdong, China* **33 D6** 24 16N 116 6 E
Mei Xian, *Shaanxi, China* **34 G4** 34 18N 107 55 E
Meiganga, *Cameroon* **52 C2** 6 30N 14 25 E
Meiktila, *Burma* **41 J19** 20 53N 95 54 E
Meissen, *Germany* **16 C6** 51 10N 13 29 E
Mejillones, *Chile* **94 A1** 23 10S 70 30W
Meka, *Australia* **61 E2** 27 25S 116 48 E
Mékambo, *Gabon* **52 D2** 1 2N 13 50 E
Mekdela, *Ethiopia* **51 F12** 11 24N 39 10 E
Mekhtar, *Pakistan* **40 D6** 30 30N 69 15 E
Meknès, *Morocco* **50 B3** 33 57N 5 33W
Mekong →, *Asia* **39 H6** 9 30N 106 15 E
Mekongga, *Indonesia* **37 E6** 3 39S 121 15 E
Melagiri Hills, *India* **40 N10** 12 20N 77 30 E
Melaka, *Malaysia* **39 L4** 2 15N 102 15 E
Melalap, *Malaysia* **36 C5** 5 10N 116 5 E
Mélambes, *Greece* **23 D6** 35 8N 24 40 E
Melanesia, *Pac. Oc.* **64 H7** 4 0S 155 0 E
Melbourne, *Australia* **63 F3** 37 50S 145 0 E
Melbourne, *U.S.A.* **77 L5** 28 5N 80 37W
Melchor Múzquiz, *Mexico* **86 B4** 27 50N 101 30W
Melchor Ocampo, *Mexico* **86 C4** 24 52N 101 40W
Mélèzes →, *Canada* **69 C12** 57 30N 71 0W
Melfi, *Chad* **51 F8** 11 0N 17 59 E
Melfort, *Canada* **73 C8** 52 50N 104 37W
Melfort, *Zimbabwe* **55 F3** 18 0S 31 25 E
Melilla, *Morocco* **50 A4** 35 21N 2 57W
Melipilla, *Chile* **94 C1** 33 42S 71 15W
Mélissa, Ákra, *Greece* **23 D6** 35 6N 24 33 E
Melita, *Canada* **73 D8** 49 15N 101 0W
Melitopol, *Ukraine* **25 E6** 46 50N 35 22 E
Melk, *Austria* **16 D7** 48 13N 15 20 E
Mellansel, *Sweden* **8 E15** 63 25N 18 17 E
Mellen, *U.S.A.* **80 B9** 46 20N 90 40W
Mellerud, *Sweden* **9 G12** 58 41N 12 28 E
Mellette, *U.S.A.* **80 C5** 45 9N 98 30W
Mellieha, *Malta* **23 D1** 35 57N 14 21 E
Melo, *Uruguay* **95 C5** 32 20S 54 10W
Melolo, *Indonesia* **37 F6** 9 53S 120 40 E
Melouprey, *Cambodia* **38 F5** 13 48N 105 16 E
Melrose, *N.S.W., Australia* **63 E4** 32 42S 146 57 E
Melrose, *W. Austral., Australia* **61 E3** 27 50S 121 15 E
Melrose, *U.K.* **12 F6** 55 35N 2 44W
Melrose, *U.S.A.* **81 H3** 34 26N 103 38W
Melstone, *U.S.A.* **82 C10** 46 36N 107 52W
Melton Mowbray, *U.K.* **10 E7** 52 46N 0 52W
Melun, *France* **18 B5** 48 32N 2 39 E
Melut, *Sudan* **51 F11** 10 30N 32 13 E
Melville, *Canada* **73 C8** 50 55N 102 50W
Melville, C., *Australia* **62 A3** 14 11S 144 30 E
Melville, L., *Canada* **71 B8** 53 30N 60 0W
Melville B., *Australia* **62 A2** 12 0S 136 45 E
Melville I., *Australia* **60 B5** 11 30S 131 0 E
Melville I., *Canada* **4 B2** 75 30N 112 0W
Melville Pen., *Canada* **69 B11** 68 0N 84 0W
Melvin →, *Canada* **72 B5** 59 11N 117 31W
Memba, *Mozam.* **55 E5** 14 11S 40 30 E
Memboro, *Indonesia* **37 F5** 9 30S 119 30 E
Memel = Klaipėda, *Lithuania* **24 C3** 55 43N 21 10 E
Memel, *S. Africa* **57 D4** 27 38S 29 36 E
Memmingen, *Germany* **16 E5** 47 59N 10 12 E
Mempawah, *Indonesia* **36 D3** 0 30N 109 5 E
Memphis, *Tenn., U.S.A.* **81 H10** 35 8N 90 3W
Memphis, *Tex., U.S.A.* **81 H4** 34 44N 100 33W
Mena, *U.S.A.* **81 H7** 34 35N 94 15W
Menai Strait, *U.K.* **10 D3** 53 14N 4 10W
Ménaka, *Mali* **50 E5** 15 59N 2 18 E
Menan = Chao Phraya →, *Thailand* **38 F3** 13 32N 100 36 E
Menarandra →, *Madag.* **57 D7** 25 17S 44 30 E
Menard, *U.S.A.* **81 K5** 30 55N 99 47W
Menasha, *U.S.A.* **76 C1** 44 13N 88 26W
Menate, *Indonesia* **36 E4** 0 12S 113 3 E
Mendawai →, *Indonesia* **36 E4** 3 30S 113 0 E
Mende, *France* **18 D5** 44 31N 3 30 E
Mendez, *Mexico* **87 B5** 25 7N 98 34W
Mendhar, *India* **43 C6** 33 35N 74 10 E
Mendip Hills, *U.K.* **11 F5** 51 17N 2 40W
Mendocino, *U.S.A.* **82 G2** 39 19N 123 48W
Mendocino, C., *U.S.A.* **82 F1** 40 26N 124 25W
Mendota, *Calif., U.S.A.* **83 H3** 36 45N 120 23W
Mendota, *Ill., U.S.A.* **80 E10** 41 33N 89 7W
Mendoza, *Argentina* **94 C2** 32 50S 68 52W
Mendoza □, *Argentina* **94 C2** 33 0S 69 0W
Mene Grande, *Venezuela* **92 B4** 9 49N 70 56W
Menen, *Belgium* **15 D3** 50 47N 3 7 E
Menfi, *Italy* **20 F5** 37 36N 12 57 E
Menggala, *Indonesia* **36 E3** 4 30S 105 15 E
Mengjin, *China* **34 G7** 34 55N 112 45 E
Mengyin, *China* **35 G9** 35 40N 117 58 E
Mengzi, *China* **32 D5** 23 20N 103 22 E
Menihek L., *Canada* **71 B6** 54 0N 67 0W
Menin = Menen, *Belgium* **15 D3** 50 47N 3 7 E
Menindee, *Australia* **63 E3** 32 20S 142 25 E
Menindee L., *Australia* **63 E3** 32 20S 142 25 E
Meningie, *Australia* **63 F2** 35 50S 139 18 E
Menlo Park, *U.S.A.* **84 H4** 37 27N 122 12W
Menominee, *U.S.A.* **76 C2** 45 6N 87 37W
Menominee →, *U.S.A.* **76 C2** 45 6N 87 36W
Menomonie, *U.S.A.* **80 C9** 44 53N 91 55W
Menongue, *Angola* **53 G3** 14 48S 17 52 E
Menorca, *Spain* **22 B11** 40 0N 4 0 E
Mentakab, *Malaysia* **39 L4** 3 29N 102 21 E
Mentawai, Kepulauan, *Indonesia* **36 E1** 2 0S 99 0 E
Menton, *France* **18 E7** 43 50N 7 29 E
Mentor, *U.S.A.* **78 E3** 41 40N 81 21W
Mentz Dam, *S. Africa* **56 E4** 33 10S 25 9 E
Menzelinsk, *Russia* **24 C9** 55 53N 53 1 E
Menzies, *Australia* **61 E3** 29 40S 121 2 E
Me'ona, *Israel* **47 B4** 33 1N 35 15 E
Meoqui, *Mexico* **86 B3** 28 17N 105 29W
Mepaco, *Mozam.* **55 F3** 15 57S 30 48 E
Meppel, *Neths.* **15 B6** 52 42N 6 12 E
Mer Rouge, *U.S.A.* **81 J9** 32 47N 91 48W
Merabéllou, Kólpos, *Greece* **23 D7** 35 10N 25 50 E
Meramangye, L., *Australia* **61 E5** 28 25S 132 13 E
Meran = Merano, *Italy* **20 A4** 46 40N 11 10 E
Merano, *Italy* **20 A4** 46 40N 11 10 E
Merauke, *Indonesia* **37 F10** 8 29S 140 24 E
Merbabu, *Indonesia* **37 G14** 7 30S 110 40 E
Merbein, *Australia* **63 E3** 34 10S 142 2 E
Merca, *Somali Rep.* **46 G3** 1 48N 44 50 E
Mercadal, *Spain* **22 B11** 39 59N 4 5 E
Merced, *U.S.A.* **83 H3** 37 18N 120 29W
Merced Pk., *U.S.A.* **84 H7** 37 36N 119 24W
Mercedes, *Buenos Aires, Argentina* **94 C4** 34 40S 59 30W
Mercedes, *Corrientes, Argentina* **94 B4** 29 10S 58 5W
Mercedes, *San Luis, Argentina* **94 C2** 33 40S 65 21W
Mercedes, *Uruguay* **94 C4** 33 12S 58 0W
Merceditas, *Chile* **94 B1** 28 20S 70 35W
Mercer, *N.Z.* **59 G5** 37 16S 175 5 E
Mercer, *U.S.A.* **78 E4** 41 14N 80 15W
Mercury, *U.S.A.* **85 J11** 36 40N 115 58W
Mercy C., *Canada* **69 B13** 65 0N 63 30W
Meredith, C., *Falk. Is.* **96 G4** 52 15S 60 40W
Meredith, L., *U.S.A.* **81 H4** 35 43N 101 33W
Merga = Nukheila, *Sudan* **51 E10** 19 1N 26 21 E
Mergui Arch. = Myeik Kyunzu, *Burma* **39 G1** 11 30N 97 30 E
Mérida, *Mexico* **87 C7** 20 9N 89 40W
Mérida, *Spain* **19 C2** 38 55N 6 25W
Mérida, *Venezuela* **92 B4** 8 24N 71 8W
Mérida, Cord. de, *Venezuela* **90 B2** 9 0N 71 0W
Meriden, *U.S.A.* **79 E12** 41 32N 72 48W
Meridian, *Calif., U.S.A.* **84 F5** 39 9N 121 55W
Meridian, *Idaho, U.S.A.* **82 E5** 43 37N 116 24W
Meridian, *Miss., U.S.A.* **77 J1** 32 22N 88 42W
Meridian, *Tex., U.S.A.* **81 K6** 31 56N 97 39W
Meriruma, *Brazil* **93 C8** 1 15N 54 50W
Merkel, *U.S.A.* **81 J4** 32 28N 100 1W
Merksem, *Belgium* **15 C4** 51 16N 4 25 E
Mermaid Reef, *Australia* **60 C2** 17 6S 119 36 E
Merowe, *Sudan* **51 E11** 18 29N 31 46 E
Merredin, *Australia* **61 F2** 31 28S 118 18 E
Merrick, *U.K.* **12 F4** 55 8N 4 30W
Merrickville, *Canada* **79 B9** 44 55N 75 50W
Merrill, *Oreg., U.S.A.* **82 E3** 42 1N 121 36W
Merrill, *Wis., U.S.A.* **80 C10** 45 11N 89 41W
Merriman, *U.S.A.* **80 D4** 42 55N 101 42W
Merritt, *Canada* **72 C4** 50 10N 120 45W
Merriwa, *Australia* **63 E5** 32 6S 150 22 E
Merriwagga, *Australia* **63 E4** 33 47S 145 43 E
Merry I., *Canada* **70 A4** 55 29N 77 31W
Merrygoen, *Australia* **63 E4** 31 51S 149 12 E
Merryville, *U.S.A.* **81 K8** 30 45N 93 33W
Mersa Fatma, *Eritrea* **46 E3** 14 57N 40 17 E
Mersch, *Lux.* **15 E6** 49 44N 6 7 E
Merseburg, *Germany* **16 C5** 51 20N 12 0 E
Mersey →, *U.K.* **10 D5** 53 20N 2 56W
Merseyside □, *U.K.* **10 D5** 53 25N 2 55W
Mersin, *Turkey* **25 G5** 36 51N 34 36 E
Mersing, *Malaysia* **39 L4** 2 25N 103 50 E
Merta, *India* **42 F6** 26 39N 74 4 E
Merthyr Tydfil, *U.K.* **11 F4** 51 45N 3 23W
Mertzon, *U.S.A.* **81 K4** 31 16N 100 49W
Meru, *Kenya* **54 B4** 0 3N 37 40 E
Meru, *Tanzania* **54 C4** 3 15S 36 46 E
Meru □, *Kenya* **54 B4** 0 3N 37 46 E
Mesa, *U.S.A.* **83 K8** 33 25N 111 50W
Mesanagrós, *Greece* **23 C9** 36 1N 27 49 E
Mesaoría □, *Cyprus* **23 D12** 35 12N 33 14 E
Mesarás, Kólpos, *Greece* **23 D6** 35 6N 24 47 E
Mesgouez, L., *Canada* **70 B4** 51 20N 75 0W
Meshed = Mashhad, *Iran* **45 B8** 36 20N 59 35 E
Meshoppen, *U.S.A.* **79 E8** 41 36N 76 3W
Meshra er Req, *Sudan* **51 G10** 8 25N 29 18 E
Mesick, *U.S.A.* **76 C3** 44 24N 85 43W
Mesilinka →, *Canada* **72 B4** 56 6N 124 30W
Mesilla, *U.S.A.* **83 K10** 32 16N 106 48W
Mesolóngion, *Greece* **21 E9** 38 21N 21 28 E
Mesopotamia = Al Jazirah, *Iraq* **44 C5** 33 30N 44 0 E
Mesquite, *U.S.A.* **83 H6** 36 47N 114 6W
Mess Cr. →, *Canada* **72 B2** 57 55N 131 14W
Messalo →, *Mozam.* **55 E4** 12 25S 39 15 E
Messina, *Italy* **20 E6** 38 10N 15 32 E
Messina, *S. Africa* **57 C5** 22 20S 30 5 E
Messina, Str. di, *Italy* **20 E6** 38 5N 15 35 E
Messíni, *Greece* **21 F10** 37 4N 22 1 E
Messiniakós Kólpos, *Greece* **21 F10** 36 45N 22 5 E
Messonghi, *Greece* **23 B3** 39 29N 19 56 E
Mesta →, *Bulgaria* **21 D11** 41 30N 24 12 E
Meta →, *S. Amer.* **92 B5** 6 12N 67 28W
Metairie, *U.S.A.* **81 L9** 29 58N 90 10W
Metaline Falls, *U.S.A.* **82 B5** 48 52N 117 22W
Metán, *Argentina* **94 B3** 25 30S 65 0W
Metangula, *Mozam.* **55 E3** 12 40S 34 50 E
Metema, *Ethiopia* **51 F12** 12 56N 36 13 E
Metengobalame, *Mozam.* **55 E3** 14 49S 34 30 E
Methven, *N.Z.* **59 K3** 43 38S 171 40 E
Methy L., *Canada* **73 B7** 56 28N 109 30W
Metil, *Mozam.* **55 F4** 16 24S 39 0 E
Metlakatla, *U.S.A.* **72 B2** 55 8N 131 35W
Metropolis, *U.S.A.* **81 G10** 37 9N 88 44W
Mettur Dam, *India* **40 P10** 11 45N 77 45 E
Metz, *France* **18 B7** 49 8N 6 10 E
Meulaboh, *Indonesia* **36 D1** 4 11N 96 3 E
Meureudu, *Indonesia* **36 C1** 5 19N 96 10 E
Meurthe-et-Moselle □, *France* **18 B7** 48 52N 6 0 E
Meuse □, *France* **18 B6** 49 8N 5 25 E
Meuse →, *Europe* **18 A6** 50 45N 5 41 E
Mexborough, *U.K.* **10 D6** 53 29N 1 18W
Mexia, *U.S.A.* **81 K6** 31 41N 96 29W
Mexiana, I., *Brazil* **93 C9** 0 0 49 30W
Mexicali, *Mexico* **86 A1** 32 40N 115 30W
México, *Mexico* **87 D5** 19 20N 99 10W
Mexico, *Maine, U.S.A.* **79 B14** 44 34N 70 33W
Mexico, *Mo., U.S.A.* **80 F9** 39 10N 91 53W
México □, *Mexico* **86 D5** 19 20N 99 10W
Mexico ■, *Cent. Amer.* **86 C4** 25 0N 105 0W
Mexico, G. of, *Cent. Amer.* **87 C7** 25 0N 90 0W
Meymaneh, *Afghan.* **40 B4** 35 53N 64 38 E
Mezen, *Russia* **24 A7** 65 50N 44 20 E
Mezen →, *Russia* **24 A7** 66 11N 43 59 E
Mezőkövesd, *Hungary* **17 E10** 47 49N 20 35 E
Mezőtúr, *Hungary* **17 E10** 46 58N 20 41 E
Mezquital, *Mexico* **86 C4** 23 29N 104 23W
Mgeta, *Tanzania* **55 D4** 8 22S 36 6 E
Mhlaba Hills, *Zimbabwe* **55 F3** 18 30S 30 30 E
Mhow, *India* **42 H6** 22 33N 75 50 E
Miahuatlán, *Mexico* **87 D5** 16 21N 96 36W
Miallo, *Australia* **62 B4** 16 28S 145 22 E
Miami, *Ariz., U.S.A.* **83 K8** 33 24N 110 52W
Miami, *Fla., U.S.A.* **77 N5** 25 47N 80 11W
Miami, *Tex., U.S.A.* **81 H4** 35 42N 100 38W
Miami →, *U.S.A.* **76 F3** 39 20N 84 40W
Miami Beach, *U.S.A.* **77 N5** 25 47N 80 8W
Miamisburg, *U.S.A.* **76 F3** 39 38N 84 17W
Mian Xian, *China* **34 H4** 33 10N 106 32 E
Mianchi, *China* **34 G6** 34 48N 111 48 E
Mīāndowāb, *Iran* **44 B5** 37 0N 46 5 E
Miandrivazo, *Madag.* **57 B8** 19 31S 45 29 E
Mīāneh, *Iran* **44 B5** 37 30N 47 40 E
Mianwali, *Pakistan* **42 C4** 32 38N 71 28 E
Miarinarivo, *Madag.* **57 B8** 18 57S 46 55 E
Miass, *Russia* **24 D11** 54 59N 60 6 E
Michigan □, *U.S.A.* **76 C3** 44 0N 85 0W
Michigan, L., *U.S.A.* **76 C2** 44 0N 87 0W
Michigan City, *U.S.A.* **76 E2** 41 43N 86 54W
Michikamau L., *Canada* **71 B7** 54 20N 63 10W
Michipicoten, *Canada* **70 C3** 47 55N 84 55W
Michipicoten I., *Canada* **70 C2** 47 40N 85 40W
Michoacan □, *Mexico* **86 D4** 19 0N 102 0W
Michurinsk, *Russia* **24 D7** 52 58N 40 27 E
Miclere, *Australia* **62 C4** 22 34S 147 32 E
Mico, Pta., *Nic.* **88 D3** 12 0N 83 30W
Micronesia, Federated States of ■, *Pac. Oc.* **64 G7** 9 0N 150 0 E
Mid Glamorgan □, *U.K.* **11 F4** 51 40N 3 25W
Midai, P., *Indonesia* **39 L6** 3 0N 107 47 E
Midale, *Canada* **73 D8** 49 25N 103 20W
Middelburg, *Neths.* **15 C3** 51 30N 3 36 E
Middelburg, *C. Prov., S. Africa* **56 E3** 31 30S 25 0 E
Middelburg, *Trans., S. Africa* **57 D4** 25 49S 29 28 E
Middelwit, *S. Africa* **56 C4** 24 51S 27 3 E
Middle Alkali L., *U.S.A.* **82 F3** 41 27N 120 5W
Middle Fork Feather →, *U.S.A.* **84 F5** 38 33N 121 30W
Middle I., *Australia* **61 F3** 34 6S 123 11 E
Middle Loup →, *U.S.A.* **80 E5** 41 17N 98 24W
Middleboro, *U.S.A.* **79 E14** 41 54N 70 55W
Middleburg, *N.Y., U.S.A.* **79 D10** 42 36N 74 20W
Middleburg, *Pa., U.S.A.* **78 F7** 40 47N 77 3W
Middlebury, *U.S.A.* **79 B11** 44 1N 73 10W
Middleport, *U.S.A.* **76 F4** 39 0N 82 3W
Middlesboro, *U.S.A.* **77 G4** 36 36N 83 43W
Middlesboro, *Ky., U.S.A.* **75 C10** 36 36N 83 43W
Middlesbrough, *U.K.* **10 C6** 54 35N 1 14W
Middlesex, *Belize* **88 C2** 17 2N 88 31W
Middlesex, *U.S.A.* **79 F10** 40 36N 74 30W
Middleton, *Australia* **62 C3** 22 22S 141 32 E
Middleton, *Canada* **71 D6** 44 57N 65 4W
Middletown, *Calif., U.S.A.* **84 G4** 38 45N 122 37W
Middletown, *Conn., U.S.A.* **79 E12** 41 34N 72 39W
Middletown, *N.Y., U.S.A.* **79 E10** 41 27N 74 25W
Middletown, *Ohio, U.S.A.* **76 F3** 39 31N 84 24W
Middletown, *Pa., U.S.A.* **79 F8** 40 12N 76 44W
Midi, Canal du →, *France* **18 E4** 43 45N 1 21 E
Midland, *Canada* **70 D4** 44 45N 79 50W
Midland, *Calif., U.S.A.* **85 M12** 33 52N 114 48W
Midland, *Mich., U.S.A.* **76 D3** 43 37N 84 14W
Midland, *Pa., U.S.A.* **78 F4** 40 39N 80 27W
Midland, *Tex., U.S.A.* **81 K3** 32 0N 102 3W
Midlands □, *Zimbabwe* **55 F2** 19 40S 29 0 E
Midleton, *Ireland* **13 E3** 51 52N 8 12W
Midlothian, *U.S.A.* **81 J6** 32 30N 97 0W
Midongy, Tangorombohitr' i, *Madag.* **57 C8** 23 30S 47 0 E
Midongy Atsimo, *Madag.* **57 C8** 23 35S 47 1 E
Midway Is., *Pac. Oc.* **64 E10** 28 13N 177 22W
Midway Wells, *U.S.A.* **85 N11** 32 41N 115 7W
Midwest, *U.S.A.* **75 B9** 42 0N 90 0W
Midwest, *Wyo., U.S.A.* **82 E10** 43 25N 106 16W
Mie □, *Japan* **31 G8** 34 30N 136 10 E
Międzychód, *Poland* **16 B7** 52 35N 15 53 E
Międzyrzec Podlaski, *Poland* **17 C11** 51 58N 22 45 E
Mienga, *Angola* **56 B2** 17 12S 19 48 E
Miercurea Ciuc, *Romania* **17 E12** 46 21N 25 48 E
Mieres, *Spain* **19 A3** 43 18N 5 48W
Mifflintown, *U.S.A.* **78 F7** 40 34N 77 24W
Mifraz Hefa, *Israel* **47 C4** 32 52N 35 0 E
Migdāl, *Israel* **47 C4** 32 51N 35 30 E
Miguel Alemán, Presa, *Mexico* **87 D5** 18 15N 96 40W
Miguel Alves, *Brazil* **93 D10** 4 11S 42 55W
Mihara, *Japan* **31 G6** 34 24N 133 5 E
Mikese, *Tanzania* **54 D4** 6 48S 37 55 E
Mikínai, *Greece* **21 F10** 37 43N 22 46 E
Mikkeli, *Finland* **9 F19** 61 43N 27 15 E
Mikkeli □ = Mikkelin lääni □, *Finland* **8 F20** 61 56N 28 0 E
Mikkelin lääni □, *Finland* **8 F20** 61 56N 28 0 E
Mikkwa →, *Canada* **72 B6** 58 25N 114 46W
Míkonos, *Greece* **21 F11** 37 30N 25 25 E
Mikumi, *Tanzania* **54 D4** 7 26S 37 0 E
Mikun, *Russia* **24 B9** 62 20N 50 0 E
Milaca, *U.S.A.* **80 C8** 45 45N 93 39W
Milagro, *Ecuador* **92 D3** 2 11S 79 36W
Milan = Milano, *Italy* **20 B3** 45 28N 9 10 E
Milan, *Mo., U.S.A.* **80 E8** 40 12N 93 7W
Milan, *Tenn., U.S.A.* **77 H1** 35 55N 88 46W
Milang, *Australia* **63 E2** 32 2S 139 10 E
Milange, *Mozam.* **55 F4** 16 3S 35 45 E
Milano, *Italy* **20 B3** 45 28N 9 10 E
Milâs, *Turkey* **25 G4** 37 20N 27 50 E
Milatos, *Greece* **23 D7** 35 18N 25 34 E
Milazzo, *Italy* **20 E6** 38 13N 15 13 E
Milbank, *U.S.A.* **80 C6** 45 13N 96 38W
Milden, *Canada* **73 C7** 51 29N 107 32W
Mildmay, *Canada* **78 B3** 44 3N 81 7W
Mildura, *Australia* **63 E3** 34 13S 142 9 E
Mileh Tharthār, *Iraq* **44 C4** 34 0N 43 15 E
Miles, *Australia* **63 D5** 26 40S 150 9 E
Miles, *U.S.A.* **81 K4** 31 36N 100 11W
Miles City, *U.S.A.* **80 B2** 46 25N 105 51W
Milestone, *Canada* **73 D8** 49 59N 104 31W
Mileura, *Australia* **61 E2** 26 22S 117 20 E
Milford, *Calif., U.S.A.* **84 E6** 40 10N 120 22W
Milford, *Conn., U.S.A.* **79 E11** 41 14N 73 3W
Milford, *Del., U.S.A.* **76 F8** 38 55N 75 26W
Milford, *Mass., U.S.A.* **79 D13** 42 8N 71 31W
Milford, *Pa., U.S.A.* **79 E10** 41 19N 74 48W
Milford, *Utah, U.S.A.* **83 G7** 38 24N 113 1W
Milford Haven, *U.K.* **11 F2** 51 43N 5 2W
Milford Sd., *N.Z.* **59 L1** 44 41S 167 47 E
Milgun, *Australia* **61 D2** 24 56S 118 18 E
Milh, Bahr al, *Iraq* **44 C4** 32 40N 43 35 E
Miliana, *Algeria* **50 C5** 27 20N 2 32 E
Miling, *Australia* **61 F2** 30 30S 116 17 E
Milk →, *U.S.A.* **82 B10** 48 4N 106 19W
Milk River, *Canada* **72 D6** 49 10N 112 5W
Mill City, *U.S.A.* **82 D2** 44 45N 122 29W
Mill I., *Antarctica* **5 C8** 66 0S 101 30 E
Mill Valley, *U.S.A.* **84 H4** 37 54N 122 32W
Millbridge, *Canada* **78 B7** 44 41N 77 36W
Millbrook, *Canada* **78 B6** 44 10N 78 29W
Mille Lacs, L. des, *Canada* **70 C1** 48 45N 90 35W
Mille Lacs L., *U.S.A.* **80 B8** 46 15N 93 39W
Milledgeville, *U.S.A.* **77 J4** 33 5N 83 14W
Millen, *U.S.A.* **77 J5** 32 48N 81 57W
Miller, *U.S.A.* **80 C5** 44 31N 98 59W
Millersburg, *Ohio, U.S.A.* **78 F3** 40 33N 81 55W
Millersburg, *Pa., U.S.A.* **78 F8** 40 32N 76 58W
Millerton, *U.S.A.* **79 E11** 41 57N 73 31W
Millerton L., *U.S.A.* **84 J7** 37 1N 119 41W
Millicent, *Australia* **63 F3** 37 34S 140 21 E
Millinocket, *U.S.A.* **71 C6** 45 39N 68 43W
Millmerran, *Australia* **63 D5** 27 53S 151 16 E
Mills L., *Canada* **72 A5** 61 30N 118 20W
Millsboro, *U.S.A.* **78 G4** 40 0N 80 0W
Milltown Malbay, *Ireland* **13 D2** 52 51N 9 25W
Millville, *U.S.A.* **76 F8** 39 24N 75 2W
Millwood L., *U.S.A.* **81 J8** 33 42N 93 58W
Milne →, *Australia* **62 C2** 21 10S 137 33 E
Milne Inlet, *Canada* **69 A11** 72 30N 80 0W
Milnor, *U.S.A.* **80 B6** 46 16N 97 27W
Milo, *Canada* **72 C6** 50 34N 112 53W
Mílos, *Greece* **21 F11** 36 44N 24 25 E
Milparinka P.O., *Australia* **63 D3** 29 46S 141 57 E
Milton, *Canada* **78 C5** 43 33N 79 53W
Milton, *N.Z.* **59 M2** 46 7S 169 59 E
Milton, *U.K.* **12 D4** 57 18N 4 32W
Milton, *Calif., U.S.A.* **84 G6** 38 3N 120 51W
Milton, *Fla., U.S.A.* **77 K2** 30 38N 87 3W
Milton, *Pa., U.S.A.* **78 F8** 41 1N 76 51W
Milton-Freewater, *U.S.A.* **82 D4** 45 56N 118 23W
Milton Keynes, *U.K.* **11 E7** 52 3N 0 42W
Miltou, *Chad* **51 F8** 10 14N 17 26 E
Milverton, *Canada* **78 C4** 43 34N 80 55W
Milwaukee, *U.S.A.* **76 D2** 43 2N 87 55W
Milwaukee Deep, *Atl. Oc.* **89 C6** 19 50N 68 0W
Milwaukie, *U.S.A.* **84 E4** 45 27N 122 38W
Min Chiang →, *China* **33 D6** 26 0N 119 35 E
Min Jiang →, *China* **32 D5** 28 45N 104 40 E
Min Xian, *China* **34 G3** 34 25N 104 5 E
Mina, *U.S.A.* **83 G4** 38 24N 118 7W
Mina Pirquitas, *Argentina* **94 A2** 22 40S 66 30W
Mīnā Su'ud, *Si. Arabia* **45 D6** 28 45N 48 28 E
Mīnā'al Aḥmadī, *Kuwait* **45 D6** 29 5N 48 10 E
Mīnāb, *Iran* **45 E8** 27 10N 57 1 E
Minago →, *Canada* **73 C9** 54 33N 98 59W
Minaki, *Canada* **73 D10** 49 59N 94 40W
Minamata, *Japan* **31 H5** 32 10N 130 30 E
Minami-Tori-Shima, *Pac. Oc.* **64 E7** 24 0N 153 45 E
Minas, *Uruguay* **95 C4** 34 20S 55 10W
Minas, Sierra de las, *Guatemala* **88 C2** 15 9N 89 31W

Monterey B.

Minas Basin, Canada 71 C7 45 20N 64 12W
Minas de Rio Tinto, Spain 19 D2 37 42N 6 35W
Minas Gerais □, Brazil 93 G9 18 50S 46 0W
Minatitlán, Mexico 87 D6 17 58N 94 35W
Minbu, Burma 41 J19 20 10N 94 52 E
Mindanao, Phil. 37 C6 8 0N 125 0 E
Mindanao Sea = Bohol Sea, Phil. 37 C6 9 0N 124 0 E
Mindanao Trench, Pac. Oc. 37 B7 12 0N 126 6 E
Minden, Canada 78 B6 44 55N 78 43W
Minden, Germany 16 B4 52 18N 8 45 E
Minden, La., U.S.A. 81 J8 32 37N 93 17W
Minden, Nev., U.S.A. 84 G7 38 57N 119 46W
Mindiptana, Indonesia 37 F10 5 55S 140 22 E
Mindoro, Phil. 37 B6 13 0N 121 0 E
Mindoro Str., Phil. 37 B6 12 30N 120 30 E
Mindouli, Congo 52 E2 4 12S 14 28 E
Mine, Japan 31 G5 34 12N 131 7 E
Minehead, U.K. 11 F4 51 12N 3 29W
Mineola, U.S.A. 81 J7 32 40N 95 29W
Mineral King, U.S.A. 84 J8 36 27N 118 36W
Mineral Wells, U.S.A. 81 J5 32 48N 98 7W
Minersville, Pa., U.S.A. 79 F8 40 41N 76 16W
Minersville, Utah, U.S.A. 83 G7 38 13N 112 56W
Minerva, U.S.A. 78 F3 40 44N 81 6W
Minetto, U.S.A. 79 C8 43 24N 76 28W
Mingan, Canada 71 B7 50 20N 64 0W
Mingechaurskoye Vdkhr., Azerbaijan 25 F8 40 56N 47 20 E
Mingela, Australia 62 B4 19 52S 146 38 E
Mingenew, Australia 61 E2 29 12S 115 21 E
Mingera Cr. →, Australia 62 C2 20 38S 137 45 E
Mingin, Burma 41 H19 22 50N 94 30 E
Mingt'iehkaitafan = Mintaka Pass, Pakistan 43 A6 37 0N 74 58 E
Mingyuegue, China 35 C15 43 2N 128 50 E
Minho, Portugal 19 B1 41 25N 8 20W
Minho →, Spain 19 B1 41 58N 8 40W
Minidoka, U.S.A. 82 E7 42 45N 113 29W
Minigwal, L., Australia 61 E3 29 31S 123 14 E
Minilya, Australia 61 D1 23 55S 114 0 E
Minilya →, Australia 61 D1 23 45S 114 0 E
Minipi, L., Canada 71 B7 52 25N 60 45W
Mink L., Canada 72 A5 61 54N 117 40W
Minna, Nigeria 50 G6 9 37N 6 30 E
Minneapolis, Kans., U.S.A. 80 F6 39 8N 97 42W
Minneapolis, Minn., U.S.A. 80 C8 44 59N 93 16W
Minnedosa, Canada 73 C9 50 14N 99 50W
Minnesota □, U.S.A. 80 B7 46 0N 94 15W
Minnie Creek, Australia 61 D2 24 3S 115 42 E
Minnipa, Australia 63 E2 32 51S 135 9 E
Minnitaki L., Canada 70 C1 49 57N 92 10W
Mino, Japan 31 G8 35 32N 136 55 E
Miño →, Spain 19 B1 41 52N 8 40W
Minorca = Menorca, Spain 22 B11 40 0N 4 0 E
Minore, Australia 63 E4 32 14S 148 27 E
Minot, U.S.A. 80 A4 48 14N 101 18W
Minqin, China 34 E2 38 38N 103 20 E
Minsk, Belorussia 24 D4 53 52N 27 30 E
Mińsk Mazowiecki, Poland 17 B10 52 10N 21 33 E
Mintaka Pass, Pakistan 43 A6 37 0N 74 58 E
Minto, U.S.A. 68 B5 64 53N 149 11W
Minton, Canada 73 D8 49 10N 104 35W
Minturn, U.S.A. 82 G10 39 35N 106 26W
Minusinsk, Russia 27 D10 53 50N 91 20 E
Minutang, India 41 E20 28 15N 96 30 E
Minvoul, Gabon 52 D2 2 9N 12 8 E
Mir, Niger 51 F7 14 5N 11 59 E
Mīr Kūh, Iran 45 E8 26 22N 58 55 E
Mīr Shahdād, Iran 45 E8 26 15N 58 29 E
Mira por vos Cay, Bahamas 89 B5 22 9N 74 30W
Miraj, India 40 L9 16 50N 74 45 E
Miram Shah, Pakistan 42 C4 33 0N 70 2 E
Miramar, Argentina 94 D4 38 15S 57 50W
Miramar, Mozam. 57 C6 23 50S 35 35 E
Miramichi B., Canada 71 C7 47 15N 65 0W
Miranda, Brazil 93 H7 20 10S 56 15W
Miranda de Ebro, Spain 19 A4 42 41N 2 57W
Mirando City, U.S.A. 81 M5 27 26N 99 0W
Mirandópolis, Brazil 95 A5 21 9S 51 6W
Mirango, Malawi 55 E3 13 32S 34 58 E
Mirani, Australia 62 C4 21 9S 148 53 E
Mirassol, Brazil 95 A6 20 46S 49 28W
Mirbāţ, Oman 46 D5 17 0N 54 45 E
Miri, Malaysia 36 D4 4 23N 113 59 E
Miriam Vale, Australia 62 C5 24 20S 151 33 E
Mirim, L., S. Amer. 95 C5 32 45S 52 50W
Mirnyy, Russia 27 C12 62 33N 113 53 E
Mirond L., Canada 73 B8 55 6N 102 47W
Mirpur, Pakistan 43 C5 33 32N 73 56 E
Mirpur Bibiwari, Pakistan 42 E2 28 33N 67 44 E
Mirpur Khas, Pakistan 42 G3 25 30N 69 0 E
Mirpur Sakro, Pakistan 42 G2 24 33N 67 41 E
Mirror, Canada 72 C6 52 30N 113 7W
Miryang, S. Korea 35 G15 35 31N 128 44 E
Mirzapur, India 43 G10 25 10N 82 34 E
Mirzapur-cum-Vindhyachal = Mirzapur, India 43 G10 25 10N 82 34 E
Misantla, Mexico 87 D5 19 56N 96 50W
Misawa, Japan 30 D10 40 41N 141 24 E
Miscou I., Canada 71 C7 47 57N 64 31W
Mish'āb, Ra'as al, Si. Arabia 45 D6 28 15N 48 43 E
Mishan, China 33 B8 45 37N 131 48 E
Mishawaka, U.S.A. 76 E2 41 40N 86 11W
Mishima, Japan 31 G9 35 10N 138 52 E
Misión, Mexico 85 N10 32 6N 116 53W
Misiones □, Argentina 95 B5 27 0S 55 0W
Misiones □, Paraguay 94 B4 27 0S 56 0W
Miskah, Si. Arabia 44 E4 24 49N 42 56 E
Miskitos, Cayos, Nic. 88 D3 14 26N 82 50W
Miskolc, Hungary 17 D10 48 7N 20 50 E
Misoke, Zaïre 54 C2 0 42S 28 2 E
Misool, Indonesia 37 E8 1 52S 130 10 E
Misrātah, Libya 51 B8 32 24N 15 3 E
Missanabie, Canada 70 C3 48 20N 84 6W
Missinaibi →, Canada 70 B3 50 43N 81 29W
Missinaibi L., Canada 70 C3 48 23N 83 40W
Mission, S. Dak., U.S.A. 80 D4 43 18N 100 39W
Mission, Tex., U.S.A. 81 M5 26 13N 98 20W
Mission City, Canada 72 D4 49 10N 122 15W
Mission Viejo, U.S.A. 85 M9 33 36N 117 40W
Missisa L., Canada 70 B2 52 20N 85 7W
Mississagi →, Canada 70 C3 46 15N 83 9W
Mississippi □, U.S.A. 81 J10 33 0N 90 0W
Mississippi →, U.S.A. 81 L10 29 9N 89 15W
Mississippi L., Canada 79 A8 45 5N 76 10W
Mississippi River Delta, U.S.A. 81 L9 29 10N 89 15W
Mississippi Sd., U.S.A. 81 K10 30 20N 89 0W
Missoula, U.S.A. 82 C6 46 52N 114 1W
Missouri □, U.S.A. 80 F8 38 25N 92 30W
Missouri →, U.S.A. 80 F9 38 49N 90 7W
Missouri Valley, U.S.A. 80 E7 41 34N 95 53W
Mist, U.S.A. 84 E3 45 59N 123 15W
Mistake B., Canada 73 A10 62 8N 93 0W
Mistassini →, Canada 71 C5 48 42N 72 20W
Mistassini L., Canada 70 B5 51 0N 73 30W
Mistastin L., Canada 71 A7 55 57N 63 20W
Mistatim, Canada 73 C8 52 52N 103 22W
Mistretta, Italy 20 F6 37 56N 14 20 E
Misty L., Canada 73 B8 58 53N 101 40W
Misurata = Misrātah, Libya 51 B8 32 24N 15 3 E
Mitchell, Australia 63 D4 26 29S 147 58 E
Mitchell, Canada 78 C3 43 28N 81 12W
Mitchell, Ind., U.S.A. 76 F2 38 44N 86 28W
Mitchell, Nebr., U.S.A. 80 E3 41 57N 103 49W
Mitchell, Oreg., U.S.A. 82 D3 44 34N 120 9W
Mitchell, S. Dak., U.S.A. 80 D5 43 43N 98 2W
Mitchell →, Australia 62 B3 15 12S 141 35 E
Mitchell, Mt., U.S.A. 77 H4 35 46N 82 16W
Mitchell Ras., Australia 62 A2 12 49S 135 36 E
Mitchelstown, Ireland 13 D3 52 16N 8 18W
Mitha Tiwana, Pakistan 42 C5 32 13N 72 6 E
Mito, Japan 31 F10 36 20N 140 30 E
Mitsinjo, Madag. 57 B8 16 1S 45 52 E
Mitsiwa, Eritrea 51 E12 15 35N 39 25 E
Mitsukaidō, Japan 31 F9 36 1N 139 59 E
Mittagong, Australia 63 E5 34 28S 150 29 E
Mitú, Colombia 92 C4 1 8N 70 3W
Mitumba, Tanzania 54 D3 7 8S 31 2 E
Mitumba, Chaîne des, Zaïre 54 D2 7 0S 27 30 E
Mitumba Mts. = Mitumba, Chaîne des, Zaïre 54 D2 7 0S 27 30 E
Mitwaba, Zaïre 55 D2 8 2S 27 17 E
Mityana, Uganda 54 B3 0 23N 32 2 E
Mitzic, Gabon 52 D2 0 45N 11 40 E
Mixteco →, Mexico 87 D5 18 11N 98 30W
Miyagi □, Japan 30 E10 38 15N 140 45 E
Miyah, W. el →, Syria 44 C3 34 44N 39 57 E
Miyake-Jima, Japan 31 G9 34 5N 139 30 E
Miyako, Japan 30 E10 39 40N 141 59 E
Miyako-Jima, Japan 31 M2 24 45N 125 20 E
Miyako-Rettō, Japan 31 M2 24 24N 125 0 E
Miyakonojō, Japan 31 J5 31 40N 131 5 E
Miyanoura-Dake, Japan 31 J5 30 20N 130 31 E
Miyazaki, Japan 31 J5 31 56N 131 30 E
Miyazaki □, Japan 31 H5 32 30N 131 30 E
Miyazu, Japan 31 G7 35 35N 135 10 E
Miyet, Bahr el = Dead Sea, Asia 47 D4 31 30N 35 30 E
Miyoshi, Japan 31 G6 34 48N 132 51 E
Miyun, China 34 D9 40 28N 116 50 E
Miyun Shuiku, China 35 D9 40 30N 117 0 E
Mizamis = Ozamiz, Phil. 37 C6 8 15N 123 50 E
Mizdah, Libya 51 B7 31 30N 13 0 E
Mizen Hd., Cork, Ireland 13 E2 51 27N 9 50W
Mizen Hd., Wick., Ireland 13 D5 52 52N 6 4W
Mizhi, China 34 F6 37 47N 110 12 E
Mizoram □, India 41 H18 23 30N 92 40 E
Mizpe Ramon, Israel 47 E3 30 34N 34 49 E
Mizusawa, Japan 30 E10 39 8N 141 8 E
Mjölby, Sweden 9 G13 58 20N 15 10 E
Mjøsa, Norway 9 F11 60 48N 11 0 E
Mkata, Tanzania 54 D4 5 45S 38 20 E
Mkokotoni, Tanzania 54 D4 5 55S 39 15 E
Mkomazi, Tanzania 54 C4 4 40S 38 7 E
Mkomazi →, S. Africa 57 E5 30 12S 30 50 E
Mkulwe, Tanzania 55 D3 8 37S 32 20 E
Mkumbi, Ras, Tanzania 54 D4 7 38S 39 55 E
Mkushi, Zambia 55 E2 14 25S 29 15 E
Mkushi River, Zambia 55 E2 13 32S 29 45 E
Mkuze, S. Africa 57 D5 27 10S 32 0 E
Mkuze →, S. Africa 57 D5 27 45S 32 30 E
Mladá Boleslav, Czech. 16 C7 50 27N 14 53 E
Mlala Hills, Tanzania 54 D3 6 50S 31 40 E
Mlange, Malawi 55 F4 16 2S 35 33 E
Mława, Poland 17 B10 53 9N 20 25 E
Mmabatho, S. Africa 56 D4 25 49S 25 30 E
Mo i Rana, Norway 8 C13 66 15N 14 7 E
Moa, Indonesia 37 F7 8 0S 128 0 E
Moab, U.S.A. 83 G9 38 35N 109 33W
Moabi, Gabon 52 E2 2 24S 10 59 E
Moala, Fiji 59 D8 18 36S 179 53 E
Moalie Park, Australia 63 D3 29 42S 143 3 E
Moba, Zaïre 54 D2 7 0S 29 48 E
Mobārakābād, Iran 45 D7 28 24N 53 20 E
Mobārakīyeh, Iran 45 C6 35 5N 51 47 E
Mobaye, C.A.R. 52 D4 4 25N 21 5 E
Mobayi, Zaïre 52 D4 4 15N 21 8 E
Moberly, U.S.A. 80 F8 39 25N 92 26W
Moberly →, Canada 72 B4 56 12N 120 55W
Mobile, U.S.A. 77 K1 30 41N 88 3W
Mobile B., U.S.A. 77 K2 30 30N 88 0W
Mobridge, U.S.A. 80 C4 45 32N 100 26W
Mobutu Sese Seko, L., Africa 54 B3 1 30N 31 0 E
Moc Chau, Vietnam 38 B5 20 50N 104 38 E
Moc Hoa, Vietnam 39 G5 10 46N 105 56 E
Mocabe Kasari, Zaïre 55 D2 9 58S 26 12 E
Moçambique, Mozam. 55 F5 15 3S 40 42 E
Moçâmedes = Namibe, Angola 53 H2 15 7S 12 11 E
Mochudi, Botswana 56 C4 24 27S 26 7 E
Mocimboa da Praia, Mozam. 55 E5 11 25S 40 20 E
Moclips, U.S.A. 84 C2 47 14N 124 13W
Mocoa, Colombia 92 C3 1 7N 76 35W
Mococa, Brazil 95 A6 21 28S 47 0W
Mocorito, Mexico 86 B3 25 30N 107 53W
Moctezuma, Mexico 86 B3 29 50N 109 0W
Moctezuma →, Mexico 87 C5 21 59N 98 34W
Mocuba, Mozam. 55 F4 16 54S 36 57 E
Mocúzari, Presa, Mexico 86 B3 27 10N 109 10W
Modane, France 18 D7 45 12N 6 40 E
Modasa, India 42 H5 23 30N 73 21 E
Modder →, S. Africa 56 D3 29 2S 24 37 E
Modderrivier, S. Africa 56 D3 29 2S 24 38 E
Módena, Italy 20 B4 44 39N 10 55 E
Modena, U.S.A. 83 H7 37 48N 113 56W
Modesto, U.S.A. 83 H3 37 39N 121 0W
Módica, Italy 20 F6 36 52N 14 45 E
Moe, Australia 63 F4 38 12S 146 19 E
Moebase, Mozam. 55 F4 17 3S 38 41 E
Moengo, Surinam 93 B8 5 45N 54 20W
Moffat, U.K. 12 F5 55 20N 3 27W
Moga, India 42 D6 30 48N 75 8 E
Mogadishu = Muqdisho, Somali Rep. 46 G4 2 2N 45 25 E
Mogador = Essaouira, Morocco 50 B3 31 32N 9 42W
Mogalakwena →, S. Africa 57 C4 22 38S 28 40 E
Mogami →, Japan 30 E10 38 45N 140 0 E
Mogán, Canary Is. 22 G4 27 53N 15 43W
Mogaung, Burma 41 G20 25 20N 97 0 E
Mogi das Cruzes, Brazil 95 A6 23 31S 46 11W
Mogi-Guaçu →, Brazil 95 A6 20 53S 48 10W
Mogi-Mirim, Brazil 95 A6 22 29S 47 0W
Mogilev, Belorussia 24 D5 53 55N 30 18 E
Mogilev-Podolskiy, Moldavia 25 E4 48 20N 27 40 E
Mogincual, Mozam. 55 F5 15 35S 40 25 E
Mogocha, Russia 27 D12 53 40N 119 50 E
Mogoi, Indonesia 37 E8 1 55S 133 10 E
Mogok, Burma 41 H20 23 0N 96 40 E
Mogumber, Australia 61 F2 31 2S 116 3 E
Mohács, Hungary 17 F9 45 58N 18 41 E
Mohales Hoek, Lesotho 56 E4 30 7S 27 26 E
Mohall, U.S.A. 80 A4 48 46N 101 31W
Moḩammadābād, Iran 45 B8 37 52N 59 5 E
Mohave, L., U.S.A. 85 K12 35 12N 114 34W
Mohawk →, U.S.A. 79 D11 42 47N 73 41W
Mohoro, Tanzania 54 D4 8 6S 39 8 E
Moidart, L., U.K. 12 E3 56 47N 5 40W
Mointy, Kazakhstan 26 E8 47 10N 73 18 E
Moíres, Greece 23 D6 35 4N 24 56 E
Moisie, Canada 71 B6 50 12N 66 1W
Moisie →, Canada 71 B6 50 14N 66 5W
Moïssala, Chad 51 G8 8 21N 17 46 E
Mojave, U.S.A. 85 K8 35 3N 118 10W
Mojave Desert, U.S.A. 85 L10 35 0N 116 30W
Mojo, Bolivia 94 A2 21 48S 65 33W
Mojokerto, Indonesia 37 G15 7 28S 112 26 E
Mokai, N.Z. 59 H5 38 32S 175 56 E
Mokambo, Zaïre 55 E2 12 25S 28 20 E
Mokameh, India 43 G11 25 24N 85 55 E
Mokelumne →, U.S.A. 84 G5 38 13N 121 28W
Mokelumne Hill, U.S.A. 84 G6 38 18N 120 43W
Mokhós, Greece 23 D7 35 16N 25 27 E
Mokhotlong, Lesotho 57 D4 29 22S 29 2 E
Mokokchung, India 41 F19 26 15N 94 30 E
Mol, Belgium 15 C5 51 11N 5 5 E
Molchanovo, Russia 26 D9 57 40N 83 50 E
Mold, U.K. 10 D4 53 10N 3 10W
Moldavia ■, Europe 25 E4 47 0N 28 0 E
Molde, Norway 8 E9 62 45N 7 9 E
Moldova ■ = Moldavia ■, Europe 25 E4 47 0N 28 0 E
Molepolole, Botswana 56 C4 24 28S 25 28 E
Molfetta, Italy 20 D7 41 12N 16 35 E
Moline, U.S.A. 80 E9 41 30N 90 31W
Molinos, Argentina 94 B2 25 28S 66 15W
Moliro, Zaïre 54 D3 8 12S 30 30 E
Molise □, Italy 20 D6 41 45N 14 30 E
Mollahat, Bangla. 43 H13 22 56N 89 48 E
Mollendo, Peru 92 G4 17 0S 72 0W
Mollerin, L., Australia 61 F2 30 30S 117 35 E
Mölndal, Sweden 9 H12 57 40N 12 3 E
Molokai, U.S.A. 74 H16 21 8N 157 0W
Molong, Australia 63 E4 33 5S 148 54 E
Molopo →, Africa 56 D3 27 30S 20 13 E
Molotov = Perm, Russia 24 C10 58 0N 56 10 E
Moloundou, Cameroon 52 D3 2 8N 15 15 E
Molson L., Canada 73 C9 54 22N 96 40W
Molteno, S. Africa 56 E4 31 22S 26 22 E
Molu, Indonesia 37 F8 6 45S 131 40 E
Molucca Sea = Maluku Sea, Indonesia 37 E6 2 0S 124 0 E
Moluccas = Maluku, Indonesia 37 E7 1 0S 127 0 E
Moma, Mozam. 55 F4 16 47S 39 4 E
Moma, Zaïre 54 C1 1 35S 23 52 E
Mombasa, Kenya 54 C4 4 2S 39 43 E
Mombetsu, Japan 30 B11 44 21N 143 22 E
Momi, Zaïre 54 C2 1 42S 27 0 E
Mompós, Colombia 92 B4 9 14N 74 26W
Møn, Denmark 9 J12 54 57N 12 15 E
Mon →, Burma 41 J19 20 25N 94 30 E
Mona, Canal de la, W. Indies 89 C6 18 30N 67 45W
Mona, Isla, Puerto Rico 89 C6 18 5N 67 54W
Mona, Pta., Costa Rica 88 E3 9 37N 82 36W
Monach Is., U.K. 12 D1 57 32N 7 40W
Monaco ■, Europe 18 E7 43 46N 7 23 E
Monadhliath Mts., U.K. 12 D4 57 10N 4 4W
Monaghan, Ireland 13 B5 54 15N 6 58W
Monaghan □, Ireland 13 B5 54 15N 7 0W
Monahans, U.S.A. 81 K3 31 36N 102 54W
Monapo, Mozam. 55 E5 14 56S 40 19 E
Monarch Mt., Canada 72 C3 51 55N 125 57W
Monastir = Bitola, Macedonia 21 D9 41 5N 21 10 E
Monastir, Tunisia 51 A7 35 50N 10 49 E
Moncayo, Sierra del, Spain 19 B5 41 48N 1 50W
Mönchengladbach, Germany 16 C3 51 12N 6 23 E
Monchique, Portugal 19 D1 37 19N 8 38W
Monchique, Sa. de, Portugal 19 D1 37 18N 8 39W
Monclova, Mexico 86 B4 26 50N 101 30W
Moncton, Canada 71 C7 46 7N 64 51W
Mondego →, Portugal 19 B1 40 9N 8 52W
Mondeodo, Indonesia 37 E6 3 34S 122 9 E
Mondovì, Italy 20 B2 44 23N 7 49 E
Mondovi, U.S.A. 80 C9 44 34N 91 40W
Mondrain I., Australia 61 F3 34 9S 122 14 E
Monduli □, Tanzania 54 C4 3 0S 36 0 E
Monessen, U.S.A. 78 F5 40 9N 79 54W
Monett, U.S.A. 81 G8 36 55N 93 55W
Monforte, Portugal 19 C2 39 6N 7 25W
Mong Hsu, Burma 41 J21 21 54N 98 30 E
Mong Kung, Burma 41 J20 21 35N 97 35 E
Mong Nai, Burma 41 J20 20 32N 97 46 E
Mong Pawk, Burma 41 H21 22 4N 99 16 E
Mong Ton, Burma 41 J21 20 17N 98 45 E
Mong Wa, Burma 41 J22 21 26N 100 27 E
Mong Yai, Burma 41 H21 22 21N 98 3 E
Mongalla, Sudan 51 G11 5 8N 31 42 E
Mongers, L., Australia 61 E2 29 25S 117 5 E
Monghyr = Munger, India 43 G12 25 23N 86 30 E
Mongo, Chad 51 F8 12 14N 18 43 E
Mongolia ■, Asia 27 E10 47 0N 103 0 E
Mongororo, Chad 51 F9 12 3N 22 26 E
Mongu, Zambia 53 H4 15 16S 23 12 E
Mônguа, Angola 56 B2 16 43S 15 20 E
Monkey Bay, Malawi 55 E4 14 7S 35 1 E
Monkey River, Belize 87 D7 16 22N 88 29W
Monkira, Australia 62 C3 24 46S 140 30 E
Monkoto, Zaïre 52 E4 1 38S 20 35 E
Monmouth, U.K. 11 F5 51 48N 2 43W
Monmouth, U.S.A. 80 E9 40 55N 90 39W
Mono L., U.S.A. 83 H4 38 1N 119 1W
Monolith, U.S.A. 85 K8 35 7N 118 22W
Monólithos, Greece 23 C9 36 7N 27 45 E
Monongahela, U.S.A. 78 F5 40 12N 79 56W
Monópoli, Italy 20 D7 40 57N 17 18 E
Monqoumba, C.A.R. 52 D3 3 33N 18 40 E
Monroe, Ga., U.S.A. 77 J4 33 47N 83 43W
Monroe, La., U.S.A. 81 J8 32 30N 92 7W
Monroe, Mich., U.S.A. 76 E4 41 55N 83 24W
Monroe, N.C., U.S.A. 77 H5 34 59N 80 33W
Monroe, N.Y., U.S.A. 79 E10 41 20N 74 11W
Monroe, Utah, U.S.A. 83 G7 38 38N 112 7W
Monroe, Wash., U.S.A. 84 C5 47 51N 121 58W
Monroe, Wis., U.S.A. 80 D10 42 36N 89 38W
Monroe City, U.S.A. 80 F9 39 39N 91 44W
Monroeville, Ala., U.S.A. 77 K2 31 31N 87 20W
Monroeville, Pa., U.S.A. 78 F5 40 26N 79 45W
Monrovia, Liberia 50 G2 6 18N 10 47W
Monrovia, U.S.A. 83 J4 34 7N 118 1W
Mons, Belgium 15 D3 50 27N 3 58 E
Monse, Indonesia 37 E6 4 0S 123 10 E
Mont-de-Marsan, France 18 E3 43 54N 0 31W
Mont-Joli, Canada 71 C6 48 37N 68 10W
Mont-Laurier, Canada 70 C4 46 35N 75 30W
Mont-St.-Michel, Le = Le Mont-St.-Michel, France 18 B3 48 40N 1 30W
Mont Tremblant Prov. Park, Canada 70 C5 46 30N 74 30W
Montagu, S. Africa 56 E3 33 45S 20 8 E
Montagu I., Antarctica 5 B1 58 25S 26 20W
Montague, Canada 71 C7 46 10N 62 39W
Montague, U.S.A. 82 F2 41 44N 122 32W
Montague, I., Mexico 86 A2 31 40N 114 56W
Montague Ra., Australia 61 E2 27 15S 119 30 E
Montague Sd., Australia 60 B4 14 28S 125 20 E
Montalbán, Spain 19 B5 40 50N 0 45W
Montalvo, U.S.A. 85 L7 34 15N 119 12W
Montaña, Peru 92 E4 6 0S 73 0W
Montana □, U.S.A. 82 C9 47 0N 110 0W
Montaña Clara, I., Canary Is. 22 E6 29 17N 13 33W
Montargis, France 18 C5 47 59N 2 43 E
Montauban, France 18 D4 44 2N 1 21 E
Montauk, U.S.A. 79 E13 41 3N 71 57W
Montauk Pt., U.S.A. 79 E13 41 4N 71 52W
Montbéliard, France 18 C7 47 31N 6 48 E
Montclair, U.S.A. 79 F10 40 49N 74 13W
Monte Albán, Mexico 87 D5 17 2N 96 45W
Monte Alegre, Brazil 93 D8 2 0S 54 0W
Monte Azul, Brazil 93 G10 15 9S 42 53W
Monte Bello Is., Australia 60 D2 20 30S 115 45 E
Monte-Carlo, Monaco 18 E7 43 46N 7 23 E
Monte Caseros, Argentina 94 C4 30 10S 57 50W
Monte Comán, Argentina 94 C2 34 40S 67 53W
Monte Cristi, Dom. Rep. 89 C5 19 52N 71 39W
Monte Lindo →, Paraguay 94 A4 23 56S 57 12W
Monte Quemado, Argentina 94 B3 25 53S 62 41W
Monte Rio, U.S.A. 84 G4 38 28N 123 0W
Monte Sant' Ángelo, Italy 20 D6 41 42N 15 59 E
Monte Santu, C. di, Italy 20 D3 40 5N 9 42 E
Monte Vista, U.S.A. 83 H10 37 35N 106 9W
Monteagudo, Argentina 95 B5 27 14S 54 8W
Montebello, Canada 70 C5 45 40N 74 55W
Montecito, U.S.A. 85 L7 34 26N 119 40W
Montecristi, Ecuador 92 D2 1 0S 80 40W
Montego Bay, Jamaica 88 C4 18 30N 78 0W
Montejinnie, Australia 60 C5 16 40S 131 38 E
Montélimar, France 18 D6 44 33N 4 45 E
Montello, U.S.A. 80 D10 43 48N 89 20W
Montemorelos, Mexico 87 B5 25 11N 99 42W
Montenegro, Brazil 95 B5 29 39S 51 29W
Montenegro □, Montenegro 21 C8 42 40N 19 20 E
Montepuez, Mozam. 55 E4 13 8S 39 28 E
Montepuez →, Mozam. 55 E5 12 32S 40 27 E
Monterey, U.S.A. 83 H3 36 37N 121 55W
Monterey B., U.S.A. 84 J5 36 45N 122 0W

Nakuru

Nakuru □, *Kenya* **54 C4** 0 15S 35 5 E
Nakuru, L., *Kenya* **54 C4** 0 23S 36 5 E
Nakusp, *Canada* **72 C5** 50 20N 117 45W
Nal →, *Pakistan* **42 G1** 25 20N 65 30 E
Nalchik, *Russia* **25 F7** 43 30N 43 33 E
Nalgonda, *India* **40 L11** 17 6N 79 15 E
Nalhati, *India* **43 G12** 24 17N 87 52 E
Nallamalai Hills, *India* **40 M11** 15 30N 78 50 E
Nalón →, *Spain* **19 A2** 43 32N 6 4W
Nālūt, *Libya* **51 B7** 31 54N 11 0 E
Nam Can, *Vietnam* **39 H5** 8 46N 104 59 E
Nam Co, *China* **32 C4** 30 30N 90 45 E
Nam Dinh, *Vietnam* **38 B6** 20 25N 106 5 E
Nam Du, Hon, *Vietnam* **39 H5** 9 41N 104 21 E
Nam Ngum Dam, *Laos* **38 C4** 18 35N 102 34 E
Nam-Phan, *Vietnam* **39 G6** 10 30N 106 0 E
Nam Phong, *Thailand* **38 D4** 16 42N 102 52 E
Nam Tha, *Laos* **38 B3** 20 58N 101 30 E
Nam Tok, *Thailand* **38 E2** 14 21N 99 4 E
Namacunde, *Angola* **56 B2** 17 18S 15 50 E
Namacurra, *Mozam.* **57 B6** 17 30S 36 50 E
Namak, Daryācheh-ye, *Iran* **45 C7** 34 30N 52 0 E
Namak, Kavir-e, *Iran* **45 C8** 34 30N 57 30 E
Namaland, *Namibia* **56 C2** 24 30S 17 0 E
Namangan, *Uzbekistan* **26 E8** 41 0N 71 40 E
Namapa, *Mozam.* **55 E4** 13 43S 39 50 E
Namaqualand, *S. Africa* **56 D2** 30 0S 17 25 E
Namasagali, *Uganda* **54 B3** 1 2N 33 0 E
Namber, *Indonesia* **37 E8** 1 2S 134 49 E
Nambour, *Australia* **63 D5** 26 32S 152 58 E
Nambucca Heads, *Australia* **63 E5** 30 37S 153 0 E
Namcha Barwa, *China* **32 D4** 29 40N 95 10 E
Namche Bazar, *Nepal* **43 F12** 27 51N 86 47 E
Namchonjŏm, *N. Korea* **35 E14** 38 15N 126 26 E
Namecunda, *Mozam.* **55 E4** 14 54S 37 37 E
Nameh, *Indonesia* **36 D5** 2 34N 116 21 E
Nameponda, *Mozam.* **55 F4** 15 50S 39 50 E
Nametil, *Mozam.* **55 F4** 15 40S 39 21 E
Namew L., *Canada* **73 C8** 54 14N 101 56W
Namib Desert = Namibwoestyn, *Namibia* **56 C2** 22 30S 15 0 E
Namibe, *Angola* **53 H2** 15 7S 12 11 E
Namibe □, *Angola* **56 B1** 16 35S 12 30 E
Namibia ■, *Africa* **56 C2** 22 0S 18 9 E
Namibwoestyn, *Namibia* **56 C2** 22 30S 15 0 E
Namlea, *Indonesia* **37 E7** 3 18S 127 5 E
Namoi →, *Australia* **63 E4** 30 12S 149 30 E
Nampa, *U.S.A.* **82 E5** 43 34N 116 34W
Nampō-Shotō, *Japan* **31 J10** 32 0N 140 0 E
Nampula, *Mozam.* **55 F4** 15 6S 39 15 E
Namrole, *Indonesia* **37 E7** 3 46S 126 46 E
Namse Shankou, *China* **41 E13** 30 0N 82 25 E
Namsen →, *Norway* **8 D11** 64 27N 11 42 E
Namsos, *Norway* **8 D11** 64 29N 11 30 E
Namtay, *Russia* **27 C13** 62 43N 129 37 E
Namtu, *Burma* **41 H20** 23 5N 97 28 E
Namtumbo, *Tanzania* **55 E4** 10 30S 36 4 E
Namu, *Canada* **72 C3** 51 52N 127 50W
Namur, *Belgium* **15 D4** 50 27N 4 52 E
Namur □, *Belgium* **15 D4** 50 17N 5 0 E
Namutoni, *Namibia* **56 B2** 18 49S 16 55 E
Namwala, *Zambia* **55 F2** 15 44S 26 30 E
Namwŏn, *S. Korea* **35 G14** 35 23N 127 23 E
Nan, *Thailand* **38 C3** 18 48N 100 46 E
Nan →, *Thailand* **38 E3** 15 42N 100 9 E
Nanaimo, *Canada* **72 D4** 49 10N 124 0W
Nanam, *N. Korea* **35 D15** 41 44N 129 40 E
Nanango, *Australia* **63 D5** 26 40S 152 0 E
Nanao, *Japan* **31 F8** 37 0N 137 0 E
Nanchang, *China* **33 D6** 28 42N 115 55 E
Nanching = Nanjing, *China* **33 C6** 32 2N 118 47 E
Nanchong, *China* **32 C5** 30 43N 106 2 E
Nancy, *France* **18 B7** 48 42N 6 12 E
Nanda Devi, *India* **43 D8** 30 23N 79 59 E
Nandan, *Japan* **31 G7** 34 10N 134 42 E
Nanded, *India* **40 K10** 19 10N 77 20 E
Nandewar Ra., *Australia* **63 E5** 30 15S 150 35 E
Nandi, *Fiji* **59 C7** 17 42S 177 20 E
Nandi □, *Kenya* **54 B4** 0 15N 35 0 E
Nandurbar, *India* **40 J9** 21 20N 74 15 E
Nandyal, *India* **40 M11** 15 30N 78 30 E
Nanga, *Australia* **61 E1** 26 7S 113 45 E
Nanga-Eboko, *Cameroon* **52 D2** 4 41N 12 22 E
Nanga Parbat, *Pakistan* **43 B6** 35 10N 74 35 E
Nangade, *Mozam.* **55 E4** 11 5S 39 36 E
Nangapinoh, *Indonesia* **36 E4** 0 20S 111 44 E
Nangarhár □, *Afghan.* **40 B7** 34 20N 70 0 E
Nangatayap, *Indonesia* **36 E4** 1 32S 110 34 E
Nangeya Mts., *Uganda* **54 B3** 3 30N 33 30 E
Nangong, *China* **34 F8** 37 23N 115 22 E
Nanhuang, *China* **35 F11** 36 58N 121 48 E
Nanjeko, *Zambia* **55 F1** 15 31S 23 30 E
Nanjing, *China* **33 C6** 32 2N 118 47 E
Nanjirinji, *Tanzania* **55 D4** 9 41S 39 5 E
Nankana Sahib, *Pakistan* **42 D5** 31 27N 73 38 E
Nanking = Nanjing, *Jiangsu, China* **33 C6** 32 2N 118 47 E
Nanking = Nanjing, *Jiangsu, China* **33 C6** 32 2N 118 47 E
Nankoku, *Japan* **31 H6** 33 39N 133 44 E
Nanning, *China* **32 D5** 22 48N 108 20 E
Nannup, *Australia* **61 F2** 33 59S 115 48 E
Nanpara, *India* **43 F9** 27 52N 81 33 E
Nanpi, *China* **34 E9** 38 2N 116 45 E
Nanping, *China* **33 D6** 26 38N 118 10 E
Nanripe, *Mozam.* **55 E4** 13 52S 38 52 E
Nansei-Shotō = Ryūkyū-rettō, *Japan* **31 M2** 26 0N 126 0 E
Nansen Sd., *Canada* **4 A3** 81 0N 91 0W
Nansio, *Tanzania* **54 C3** 2 3S 33 4 E
Nantes, *France* **18 C3** 47 12N 1 33W
Nanticoke, *U.S.A.* **79 E8** 41 12N 76 0W
Nanton, *Canada* **72 C6** 50 21N 113 46W
Nantong, *China* **33 C7** 32 1N 120 52 E
Nantucket I., *U.S.A.* **66 E12** 41 16N 70 5W
Nanuque, *Brazil* **93 G10** 17 50S 40 21W
Nanutarra, *Australia* **60 D2** 22 32S 115 30 E
Nanyang, *China* **34 H7** 33 11N 112 30 E
Nanyuan, *China* **34 E9** 39 44N 116 22 E
Nanyuki, *Kenya* **54 B4** 0 2N 37 4 E
Náo, C. de la, *Spain* **19 C6** 38 44N 0 14 E
Naococane L., *Canada* **71 B5** 52 50N 70 45W
Naoetsu, *Japan* **31 F9** 37 12N 138 10 E
Napa, *U.S.A.* **84 G4** 38 18N 122 17W
Napa →, *U.S.A.* **84 G4** 38 10N 122 19W
Napanee, *Canada* **70 D4** 44 15N 77 0W
Napanoch, *U.S.A.* **79 E10** 41 44N 74 22W
Nape, *Laos* **38 C5** 18 18N 105 6 E
Nape Pass = Keo Neua, Deo, *Vietnam* **38 C5** 18 23N 105 10 E
Napier, *N.Z.* **59 H6** 39 30S 176 56 E
Napier Broome B., *Australia* **60 B4** 14 2S 126 37 E
Napier Downs, *Australia* **60 C3** 17 11S 124 36 E
Napier Pen., *Australia* **62 A2** 12 4S 135 43 E
Naples = Nápoli, *Italy* **20 D6** 40 50N 14 17 E
Naples, *U.S.A.* **77 M5** 26 8N 81 48W
Napo →, *Peru* **92 D4** 3 20S 72 40W
Napoleon, *N. Dak., U.S.A.* **80 B5** 46 30N 99 46W
Napoleon, *Ohio, U.S.A.* **76 E3** 41 23N 84 8W
Nápoli, *Italy* **20 D6** 40 50N 14 17 E
Napopo, *Zaïre* **54 B2** 4 15N 28 0 E
Nappa Merrie, *Australia* **63 D3** 27 36S 141 7 E
Naqqāsh, *Iran* **45 C6** 35 40N 49 6 E
Nara, *Japan* **31 G7** 34 40N 135 49 E
Nara, *Mali* **50 E3** 15 10N 7 20W
Nara □, *Japan* **31 G8** 34 30N 136 0 E
Nara Canal, *Pakistan* **42 G3** 24 30N 69 20 E
Nara Visa, *U.S.A.* **81 H3** 35 37N 103 6W
Naracoorte, *Australia* **63 F3** 36 58S 140 45 E
Naradhan, *Australia* **63 E4** 33 34S 146 17 E
Narasapur, *India* **41 L12** 16 26N 81 40 E
Narathiwat, *Thailand* **39 J3** 6 30N 101 48 E
Narayanganj, *Bangla.* **41 H17** 23 40N 90 33 E
Narayanpet, *India* **40 L10** 16 45N 77 30 E
Narbonne, *France* **18 E5** 43 11N 3 0 E
Nardīn, *Iran* **45 B7** 37 3N 55 59 E
Nardò, *Italy* **21 D8** 40 10N 18 0 E
Narembeen, *Australia* **61 F2** 32 7S 118 24 E
Nares Str., *Arctic* **66 B13** 80 0N 70 0W
Naretha, *Australia* **61 F3** 31 0S 124 45 E
Nari →, *Pakistan* **42 E2** 28 0N 67 40 E
Narin, *Afghan.* **40 A6** 36 5N 69 0 E
Narindra, Helodranon' i, *Madag.* **57 A8** 14 55S 47 30 E
Narita, *Japan* **31 G10** 35 47N 140 19 E
Narmada →, *India* **42 J5** 21 38N 72 36 E
Narnaul, *India* **42 E7** 28 5N 76 11 E
Narodnaya, *Russia* **24 A10** 65 5N 59 58 E
Narok, *Kenya* **54 C4** 1 55S 35 52 E
Narok □, *Kenya* **54 C4** 1 20S 36 30 E
Narooma, *Australia* **63 F5** 36 14S 150 4 E
Narowal, *Pakistan* **42 C6** 32 6N 74 52 E
Narrabri, *Australia* **63 E4** 30 19S 149 46 E
Narran →, *Australia* **63 D4** 28 37S 148 12 E
Narrandera, *Australia* **63 E4** 34 42S 146 31 E
Narraway →, *Canada* **72 B5** 55 44N 119 55W
Narrogin, *Australia* **61 F2** 32 58S 117 14 E
Narromine, *Australia* **63 E4** 32 12S 148 12 E
Narsimhapur, *India* **43 H8** 22 54N 79 14 E
Naruto, *Japan* **31 G7** 34 11N 134 37 E
Narva, *Estonia* **24 C4** 59 23N 28 12 E
Narvik, *Norway* **8 B14** 68 28N 17 26 E
Narwana, *India* **42 E7** 29 39N 76 6 E
Naryan-Mar, *Russia* **24 A9** 68 0N 53 0 E
Naryilco, *Australia* **63 D3** 28 37S 141 53 E
Narym, *Russia* **26 D9** 59 0N 81 30 E
Narymskoye, *Kazakhstan* **26 E9** 49 10N 84 15 E
Naryn, *Kirghizia* **26 E8** 41 26N 75 58 E
Nasa, *Norway* **8 C13** 66 29N 15 23 E
Nasarawa, *Nigeria* **50 G6** 8 32N 7 41 E
Naseby, *N.Z.* **59 L3** 45 1S 170 10 E
Naselle, *U.S.A.* **84 D3** 46 22N 123 49W
Naser, Buheirat en, *Egypt* **51 D11** 23 0N 32 30 E
Nashua, *Iowa, U.S.A.* **80 D8** 42 57N 92 32W
Nashua, *Mont., U.S.A.* **82 B10** 48 8N 106 22W
Nashua, *N.H., U.S.A.* **79 D13** 42 45N 71 28W
Nashville, *Ark., U.S.A.* **81 J8** 33 57N 93 51W
Nashville, *Ga., U.S.A.* **77 K4** 31 12N 83 15W
Nashville, *Tenn., U.S.A.* **77 G2** 36 10N 86 47W
Nasik, *India* **40 K8** 19 58N 73 50 E
Nasirabad, *India* **42 F6** 26 15N 74 45 E
Naskaupi →, *Canada* **71 B7** 53 47N 60 51W
Naṣriān-e Pā'īn, *Iran* **44 C5** 32 52N 46 52 E
Nass →, *Canada* **72 B3** 55 0N 129 40W
Nassau, *Bahamas* **88 A4** 25 5N 77 20W
Nassau, *U.S.A.* **79 D11** 42 31N 73 37W
Nassau, B., *Chile* **96 H3** 55 20S 68 0W
Nasser, L. = Naser, Buheirat en, *Egypt* **51 D11** 23 0N 32 30 E
Nässjö, *Sweden* **9 H13** 57 39N 14 42 E
Nat Kyizin, *Burma* **41 M20** 14 57N 97 59 E
Nata, *Botswana* **56 C4** 20 12S 26 12 E
Natagaima, *Colombia* **92 C3** 3 37N 75 6W
Natal, *Brazil* **93 E11** 5 47S 35 13W
Natal, *Canada* **72 D6** 49 43N 114 51W
Natal, *Indonesia* **36 D1** 0 35N 99 7 E
Natal □, *S. Africa* **57 D5** 28 30S 30 30 E
Naṭanz, *Iran* **45 C6** 33 30N 51 55 E
Natashquan, *Canada* **71 B7** 50 14N 61 46W
Natashquan →, *Canada* **71 B7** 50 7N 61 50W
Natchez, *U.S.A.* **81 K9** 31 34N 91 24W
Natchitoches, *U.S.A.* **81 K8** 31 46N 93 5W
Nathalia, *Australia* **63 F4** 36 1S 145 13 E
Nathdwara, *India* **42 G5** 24 55N 73 50 E
Nati, Pta., *Spain* **22 A10** 40 3N 3 50 E
Natimuk, *Australia* **63 F3** 36 42S 142 0 E
Nation →, *Canada* **72 B4** 55 30N 123 32W
National City, *U.S.A.* **85 N9** 32 41N 117 6W
Natitingou, *Benin* **50 F5** 10 20N 1 26 E
Natividad, I., *Mexico* **86 B1** 27 50N 115 10W
Natoma, *U.S.A.* **80 F5** 39 11N 99 2W
Natron, L., *Tanzania* **54 C4** 2 20S 36 0 E
Natrona Heights, *U.S.A.* **78 F5** 40 37N 79 44W
Natuna Besar, Kepulauan, *Indonesia* **39 L7** 4 0N 108 15 E
Natuna Is. = Natuna Besar, Kepulauan, *Indonesia* **39 L7** 4 0N 108 15 E
Natuna Selatan, Kepulauan, *Indonesia* **39 L7** 2 45N 109 0 E
Natural Bridge, *U.S.A.* **79 B9** 44 5N 75 30W
Naturaliste, C., *Australia* **62 G4** 40 50S 148 15 E
Nau Qala, *Afghan.* **42 B3** 34 5N 68 5 E
Naubinway, *U.S.A.* **70 C2** 46 6N 85 27W
Naugatuck, *U.S.A.* **79 E11** 41 30N 73 3W
Naumburg, *Germany* **16 C5** 51 10N 11 48 E
Nā'ūr at Tunayb, *Jordan* **47 D4** 31 48N 35 57 E
Nauru ■, *Pac. Oc.* **64 H8** 1 0S 166 0 E
Naushahra = Nowshera, *Pakistan* **40 B8** 34 0N 72 0 E
Nauta, *Peru* **92 D4** 4 31S 73 35W
Nautanwa, *India* **41 F13** 27 20N 83 25 E
Nautla, *Mexico* **87 C5** 20 20N 96 50W
Nava, *Mexico* **86 B4** 28 25N 100 46W
Navadwip, *India* **43 H13** 23 34N 88 20 E
Navajo Reservoir, *U.S.A.* **83 H10** 36 48N 107 36W
Navalcarnero, *Spain* **19 B3** 40 17N 4 5W
Navan = An Uaimh, *Ireland* **13 C5** 53 39N 6 40W
Navarino, I., *Chile* **96 H3** 55 0S 67 40W
Navarra □, *Spain* **19 A5** 42 40N 1 40W
Navarre, *U.S.A.* **78 F3** 40 43N 81 31W
Navarro →, *U.S.A.* **84 F3** 39 11N 123 45W
Navasota, *U.S.A.* **81 K6** 30 23N 96 5W
Navassa, *W. Indies* **89 C4** 18 30N 75 0W
Naver →, *U.K.* **12 C4** 58 34N 4 15W
Navidad, *Chile* **94 C1** 33 57S 71 50W
Navoi, *Uzbekistan* **26 E7** 40 9N 65 22 E
Navojoa, *Mexico* **86 B3** 27 0N 109 30W
Navolato, *Mexico* **86 C3** 24 47N 107 42W
Navolok, *Russia* **24 B6** 62 33N 39 57 E
Návpaktos, *Greece* **21 E9** 38 23N 21 50 E
Návplion, *Greece* **21 F10** 37 33N 22 50 E
Navsari, *India* **40 J8** 20 57N 72 59 E
Nawa Kot, *Pakistan* **42 E4** 28 21N 71 24 E
Nawabganj, *Ut. P., India* **43 F9** 26 56N 81 14 E
Nawabganj, *Ut. P., India* **43 E8** 28 32N 79 40 E
Nawabshah, *Pakistan* **42 F3** 26 15N 68 25 E
Nawada, *India* **43 G11** 24 50N 85 33 E
Nawakot, *Nepal* **43 F11** 27 55N 85 10 E
Nawalgarh, *India* **42 F6** 27 50N 75 15 E
Nawanshahr, *India* **43 C6** 32 33N 74 48 E
Náxos, *Greece* **21 F11** 37 8N 25 25 E
Nāy Band, *Iran* **45 E7** 27 20N 52 40 E
Nayakhan, *Russia* **27 C16** 61 56N 159 0 E
Nayarit □, *Mexico* **86 C4** 22 0N 105 0W
Nayoro, *Japan* **30 B11** 44 21N 142 28 E
Nayyāl, W. →, *Si. Arabia* **44 D3** 28 35N 39 4 E
Nazareth = Naẓerat, *Israel* **47 C4** 32 42N 35 17 E
Nazas, *Mexico* **86 B4** 25 10N 104 6W
Nazas →, *Mexico* **86 B4** 25 35N 103 25W
Naze, The, *U.K.* **11 F9** 51 53N 1 19 E
Naẓerat, *Israel* **47 C4** 32 42N 35 17 E
Nāzīk, *Iran* **44 B5** 39 1N 45 4 E
Nazir Hat, *Bangla.* **41 H17** 22 35N 91 49 E
Nazko, *Canada* **72 C4** 53 1N 123 37W
Nazko →, *Canada* **72 C4** 53 7N 123 34W
Nchanga, *Zambia* **55 E2** 12 30S 27 49 E
Ncheu, *Malawi* **55 E3** 14 50S 34 47 E
Ndala, *Tanzania* **54 C3** 4 45S 33 15 E
Ndalatando, *Angola* **52 F2** 9 12S 14 48 E
Ndareda, *Tanzania* **54 C4** 4 12S 35 30 E
Ndélé, *C.A.R.* **51 G9** 8 25N 20 36 E
Ndendé, *Gabon* **52 E2** 2 22S 11 23 E
Ndjamena, *Chad* **51 F7** 12 10N 14 59 E
Ndjolé, *Gabon* **52 E2** 0 10S 10 45 E
Ndola, *Zambia* **55 E2** 13 0S 28 34 E
Ndoto Mts., *Kenya* **54 B4** 2 0N 37 0 E
Nduguti, *Tanzania* **54 C3** 4 18S 34 41 E
Neagh, Lough, *U.K.* **13 B5** 54 35N 6 25W
Neah Bay, *U.S.A.* **84 B2** 48 22N 124 37W
Neale, L., *Australia* **60 D5** 24 15S 130 0 E
Neápolis, *Greece* **23 D7** 35 15N 25 37 E
Near Is., *U.S.A.* **68 C1** 53 0N 172 0 E
Neath, *U.K.* **11 F4** 51 39N 3 49W
Nebine Cr. →, *Australia* **63 D4** 29 27S 146 56 E
Nebit Dag, *Turkmenistan* **25 G9** 39 30N 54 22 E
Nebraska □, *U.S.A.* **80 E5** 41 30N 99 30W
Nebraska City, *U.S.A.* **80 E7** 40 41N 95 52W
Nébrodi, Monti, *Italy* **20 F6** 37 55N 14 50 E
Necedah, *U.S.A.* **80 C9** 44 2N 90 4W
Nechako →, *Canada* **72 C4** 53 30N 122 44W
Neches →, *U.S.A.* **81 L8** 29 58N 93 51W
Neckar →, *Germany* **16 D4** 49 31N 8 26 E
Necochea, *Argentina* **94 D4** 38 30S 58 50W
Needles, *U.S.A.* **85 L12** 34 51N 114 37W
Needles, The, *U.K.* **11 G6** 50 39N 1 35W
Ñeembucú □, *Paraguay* **94 B4** 27 0S 58 0W
Neemuch = Nimach, *India* **42 G6** 24 30N 74 56 E
Neenah, *U.S.A.* **76 C1** 44 11N 88 28W
Neepawa, *Canada* **73 C9** 50 15N 99 30W
Neft-chala = imeni 26 Bakinskikh Komissarov, *Azerbaijan* **25 G8** 39 19N 49 12 E
Nefta, *Tunisia* **50 B6** 33 53N 7 50 E
Neftyannyye Kamni, *Azerbaijan* **25 F9** 40 20N 50 55 E
Negapatam = Nagappattinam, *India* **40 P11** 10 46N 79 51 E
Negaunee, *U.S.A.* **76 B2** 46 30N 87 36W
Negele, *Ethiopia* **46 F2** 5 20N 39 36 E
Negev Desert = Hanegev, *Israel* **47 E3** 30 50N 35 0 E
Negoiul, Vf., *Romania* **17 F12** 45 38N 24 35 E
Negombo, *Sri Lanka* **40 R11** 7 12N 79 50 E
Negotin, *Serbia, Yug.* **21 B10** 44 16N 22 37 E
Negra Pt., *Phil.* **37 A6** 18 40N 120 50 E
Negrais, C. = Maudin Sun, *Burma* **41 M19** 16 0N 94 30 E
Negro →, *Argentina* **96 E4** 41 2S 62 47W
Negro →, *Brazil* **92 D6** 3 0S 60 0W
Negro →, *Uruguay* **95 C4** 33 24S 58 22W
Negros, *Phil.* **37 C6** 9 30N 122 40 E
Nehalem →, *U.S.A.* **84 E3** 45 40N 123 56W
Nehāvand, *Iran* **45 C6** 35 56N 49 31 E
Nehbandān, *Iran* **45 D9** 31 35N 60 5 E
Nei Monggol Zizhiqu □, *China* **34 C6** 42 0N 112 0 E
Neidpath, *Canada* **73 C7** 50 12N 107 20W
Neihart, *U.S.A.* **82 C8** 47 0N 110 44W
Neijiang, *China* **32 D5** 29 35N 104 55 E
Neilton, *U.S.A.* **82 C2** 47 25N 123 53W
Neiqiu, *China* **34 F8** 37 15N 114 30 E
Neiva, *Colombia* **92 C3** 2 56N 75 18W
Neixiang, *China* **34 H6** 33 10N 111 52 E
Nejanilini L., *Canada* **73 B9** 59 33N 97 48W
Nekā, *Iran* **45 B7** 36 39N 53 19 E
Nekemte, *Ethiopia* **51 G12** 9 4N 36 30 E
Neksø, *Denmark* **9 J13** 55 4N 15 8 E
Nelia, *Australia* **62 C3** 20 39S 142 12 E
Neligh, *U.S.A.* **80 D5** 42 8N 98 2W
Nelkan, *Russia* **27 D14** 57 40N 136 4 E
Nellore, *India* **40 M11** 14 27N 79 59 E
Nelma, *Russia* **27 E14** 47 39N 139 0 E
Nelson, *Canada* **72 D5** 49 30N 117 20W
Nelson, *N.Z.* **59 J4** 41 18S 173 16 E
Nelson, *U.K.* **10 D5** 53 50N 2 14W
Nelson, *U.S.A.* **83 J7** 35 31N 113 19W
Nelson →, *Canada* **73 C9** 54 33N 98 2W
Nelson, C., *Australia* **63 F3** 38 26S 141 32 E
Nelson, Estrecho, *Chile* **96 G2** 51 30S 75 0W
Nelson Forks, *Canada* **72 B4** 59 30N 124 0W
Nelson House, *Canada* **73 B9** 55 47N 98 51W
Nelson L., *Canada* **73 B8** 55 48N 100 7W
Nelspoort, *S. Africa* **56 E3** 32 7S 23 0 E
Nelspruit, *S. Africa* **57 D5** 25 29S 30 59 E
Néma, *Mauritania* **50 E3** 16 40N 7 15W
Neman →, *Lithuania* **24 C3** 55 25N 21 10 E
Nemeiben L., *Canada* **73 B7** 55 20N 105 20W
Nemunas = Neman →, *Lithuania* **24 C3** 55 25N 21 10 E
Nemuro, *Japan* **30 C12** 43 20N 145 35 E
Nemuro-Kaikyō, *Japan* **30 C12** 43 30N 145 30 E
Nemuy, *Russia* **27 D14** 55 40N 136 9 E
Nen Jiang →, *China* **35 B13** 45 28N 124 30 E
Nenagh, *Ireland* **13 D3** 52 52N 8 11W
Nenana, *U.S.A.* **68 B5** 64 34N 149 5W
Nenasi, *Malaysia* **39 L4** 3 9N 103 23 E
Nene →, *U.K.* **10 E8** 52 38N 0 13 E
Nenjiang, *China* **33 B7** 49 10N 125 10 E
Neno, *Malawi* **55 F3** 15 25S 34 40 E
Nenusa, Kepulauan, *Indonesia* **37 D7** 4 45N 127 1 E
Neodesha, *U.S.A.* **81 G7** 37 25N 95 41W
Neosho, *U.S.A.* **81 G7** 36 52N 94 22W
Neosho →, *U.S.A.* **81 H7** 36 48N 95 18W
Nepal ■, *Asia* **43 F11** 28 0N 84 30 E
Nepalganj, *Nepal* **43 E9** 28 5N 81 40 E
Nephi, *U.S.A.* **82 G8** 39 43N 111 50W
Nephin, *Ireland* **13 B2** 54 1N 9 21W
Neptune, *U.S.A.* **79 F10** 40 13N 74 2W
Nerchinsk, *Russia* **27 D12** 52 0N 116 39 E
Nerchinskiy Zavod, *Russia* **27 D12** 51 20N 119 40 E
Néret L., *Canada* **71 B5** 54 45N 70 44W
Neretva →, *Croatia* **21 C7** 43 1N 17 27 E
Nerva, *Spain* **19 D2** 37 42N 6 30W
Nes, *Iceland* **8 D5** 65 53N 17 24W
Neskaupstaður, *Iceland* **8 D7** 65 9N 13 42W
Ness, L., *U.K.* **12 D4** 57 15N 4 30W
Nesttun, *Norway* **9 F8** 60 19N 5 21 E
Netanya, *Israel* **47 C3** 32 20N 34 51 E
Nète →, *Belgium* **15 C4** 51 7N 4 14 E
Netherdale, *Australia* **62 C4** 21 10S 148 33 E
Netherlands ■, *Europe* **15 C5** 52 0N 5 30 E
Netherlands Antilles ■, *W. Indies* **92 A5** 12 15N 69 0W
Nettilling L., *Canada* **69 B12** 66 30N 71 0W
Netzahualcoyotl, Presa, *Mexico* **87 D6** 17 10N 93 30W
Neubrandenburg, *Germany* **16 B6** 53 33N 13 17 E
Neuchâtel, *Switz.* **16 E3** 47 0N 6 55 E
Neuchâtel, Lac de, *Switz.* **16 E3** 46 53N 6 50 E
Neufchâteau, *Belgium* **15 E5** 49 50N 5 25 E
Neumünster, *Germany* **16 A4** 54 4N 9 58 E
Neunkirchen, *Germany* **16 D3** 49 23N 7 12 E
Neuquén, *Argentina* **96 D3** 38 55S 68 0W
Neuquén □, *Argentina* **94 D2** 38 0S 69 50W
Neuruppin, *Germany* **16 B6** 52 56N 12 48 E
Neuse →, *U.S.A.* **77 H7** 35 5N 76 29W
Neusiedler See, *Austria* **16 E8** 47 50N 16 47 E
Neuss, *Germany* **15 C6** 51 12N 6 39 E
Neustrelitz, *Germany* **16 B6** 53 22N 13 4 E
Neva →, *Russia* **24 C5** 59 50N 30 30 E
Nevada, *U.S.A.* **81 G7** 37 51N 94 22W
Nevada □, *U.S.A.* **82 G5** 39 0N 117 0W
Nevada, Sierra, *Spain* **19 D4** 37 3N 3 15W
Nevada, Sierra, *U.S.A.* **82 G3** 39 0N 120 30W
Nevada City, *U.S.A.* **84 F6** 39 16N 121 1W
Nevado, Cerro, *Argentina* **94 D2** 35 30S 68 32W
Nevanka, *Russia* **27 D10** 56 31N 98 55 E
Nevers, *France* **18 C5** 47 0N 3 9 E
Nevertire, *Australia* **63 E4** 31 50S 147 44 E
Neville, *Canada* **73 D7** 49 58N 107 39W
Nevinnomyssk, *Russia* **25 F7** 44 40N 42 0 E
Nevis, *W. Indies* **89 C7** 17 0N 62 30W
Nevyansk, *Russia* **24 C11** 57 30N 60 13 E
New Albany, *Ind., U.S.A.* **76 F3** 38 18N 85 49W
New Albany, *Miss., U.S.A.* **81 H10** 34 29N 89 0W
New Albany, *Pa., U.S.A.* **79 E8** 41 36N 76 27W
New Amsterdam, *Guyana* **92 B7** 6 15N 57 36W
New Angledool, *Australia* **63 D4** 29 5S 147 55 E
New Bedford, *U.S.A.* **79 E14** 41 38N 70 56W
New Bern, *U.S.A.* **77 H7** 35 7N 77 3W
New Bethlehem, *U.S.A.* **78 E5** 41 0N 79 20W
New Bloomfield, *U.S.A.* **78 F7** 40 25N 77 11W
New Boston, *U.S.A.* **81 J7** 33 28N 94 25W
New Braunfels, *U.S.A.* **81 L5** 29 42N 98 8W
New Brighton, *N.Z.* **59 K4** 43 29S 172 43 E
New Brighton, *U.S.A.* **78 F4** 40 42N 80 19W
New Britain, *Papua N. G.* **64 H7** 5 50S 150 20 E
New Britain, *U.S.A.* **79 E12** 41 40N 72 47W
New Brunswick, *U.S.A.* **79 F10** 40 30N 74 27W
New Brunswick □, *Canada* **71 C6** 46 50N 66 30W

Normanhurst, Mt.

New Caledonia, *Pac. Oc.* **64 K8** 21 0S 165 0 E
New Castile = Castilla La Mancha □, *Spain* **19 C4** 39 30N 3 30W
New Castle, *Ind., U.S.A.* **76 F3** 39 55N 85 22W
New Castle, *Pa., U.S.A.* **78 E4** 41 0N 80 21W
New City, *U.S.A.* **79 E11** 41 9N 73 59W
New Cumberland, *U.S.A.* **78 F4** 40 30N 80 36W
New Cuyama, *U.S.A.* **85 L7** 34 57N 119 38W
New Delhi, *India* **42 E7** 28 37N 77 13 E
New Denver, *Canada* **72 D5** 50 0N 117 25W
New Don Pedro Reservoir, *U.S.A.* **84 H6** 37 43N 120 24W
New England, *U.S.A.* **80 B3** 46 32N 102 52W
New England Ra., *Australia* **63 E5** 30 20S 151 45 E
New Forest, *U.K.* **11 G6** 50 53N 1 40W
New Glasgow, *Canada* **71 C7** 45 35N 62 36W
New Guinea, *Oceania* **64 H5** 4 0S 136 0 E
New Hamburg, *Canada* **78 C4** 43 23N 80 42W
New Hampshire □, *U.S.A.* **79 C13** 44 0N 71 30W
New Hampton, *U.S.A.* **80 D8** 43 3N 92 19W
New Hanover, *S. Africa* **57 D5** 29 22S 30 31 E
New Haven, *Conn., U.S.A.* **79 E12** 41 18N 72 55W
New Haven, *Mich., U.S.A.* **78 D2** 42 44N 82 48W
New Hazelton, *Canada* **72 B3** 55 20N 127 30W
New Hebrides = Vanuatu ■, *Pac. Oc.* **64 J8** 15 0S 168 0 E
New Iberia, *U.S.A.* **81 K9** 30 1N 91 49W
New Ireland, *Papua N. G.* **64 H7** 3 20S 151 50 E
New Jersey □, *U.S.A.* **79 F10** 40 0N 74 30W
New Kensington, *U.S.A.* **78 F5** 40 34N 79 46W
New Lexington, *U.S.A.* **76 F4** 39 43N 82 13W
New Liskeard, *Canada* **70 C4** 47 31N 79 41W
New London, *Conn., U.S.A.* **79 E12** 41 22N 72 6W
New London, *Minn., U.S.A.* **80 C7** 45 18N 94 56W
New London, *Ohio, U.S.A.* **78 E2** 41 5N 82 24W
New London, *Wis., U.S.A.* **80 C10** 44 23N 88 45W
New Madrid, *U.S.A.* **81 G10** 36 36N 89 32W
New Meadows, *U.S.A.* **82 D5** 44 58N 116 18W
New Melones L., *U.S.A.* **84 H6** 37 57N 120 31W
New Mexico □, *U.S.A.* **83 J10** 34 30N 106 0W
New Milford, *Conn., U.S.A.* **79 E11** 41 35N 73 25W
New Milford, *Pa., U.S.A.* **79 E9** 41 52N 75 44W
New Norcia, *Australia* **61 F2** 30 57S 116 13 E
New Norfolk, *Australia* **62 G4** 42 46S 147 2 E
New Orleans, *U.S.A.* **81 K9** 29 58N 90 4W
New Philadelphia, *U.S.A.* **78 F3** 40 30N 81 27W
New Plymouth, *N.Z.* **59 H5** 39 4S 174 5 E
New Plymouth, *U.S.A.* **82 E5** 43 58N 116 49W
New Providence, *Bahamas* **88 A4** 25 25N 78 35W
New Radnor, *U.K.* **11 E4** 52 15N 3 10W
New Richmond, *U.S.A.* **80 C8** 45 7N 92 32W
New Roads, *U.S.A.* **81 K9** 30 42N 91 26W
New Rochelle, *U.S.A.* **79 F11** 40 55N 73 47W
New Rockford, *U.S.A.* **80 B5** 47 41N 99 8W
New Ross, *Ireland* **13 D5** 52 24N 6 58W
New Salem, *U.S.A.* **80 B4** 46 51N 101 25W
New Scone, *U.K.* **12 E5** 56 25N 3 26W
New Siberian I. = Novaya Sibir, Ostrov, *Russia* **27 B16** 75 10N 150 0 E
New Siberian Is. = Novosibirskiye Ostrova, *Russia* **27 B15** 75 0N 142 0 E
New Smyrna Beach, *U.S.A.* **77 L5** 29 1N 80 56W
New South Wales □, *Australia* **63 E4** 33 0S 146 0 E
New Springs, *Australia* **61 E3** 25 49S 120 1 E
New Town, *U.S.A.* **80 A3** 47 59N 102 30W
New Ulm, *U.S.A.* **80 C7** 44 19N 94 28W
New Waterford, *Canada* **71 C7** 46 13N 60 4W
New Westminster, *Canada* **72 D4** 49 13N 122 55W
New York □, *U.S.A.* **79 D9** 43 0N 75 0W
New York City, *U.S.A.* **79 F11** 40 45N 74 0W
New Zealand ■, *Oceania* **59 J5** 40 0S 176 0 E
Newala, *Tanzania* **55 E4** 10 58S 39 18 E
Newala □, *Tanzania* **55 E4** 10 46S 39 20 E
Newark, *Del., U.S.A.* **76 F8** 39 41N 75 46W
Newark, *N.J., U.S.A.* **79 F10** 40 44N 74 10W
Newark, *N.Y., U.S.A.* **78 C7** 43 3N 77 6W
Newark, *Ohio, U.S.A.* **78 F2** 40 3N 82 24W
Newark-on-Trent, *U.K.* **10 D7** 53 6N 0 48W
Newaygo, *U.S.A.* **76 D3** 43 25N 85 48W
Newberg, *U.S.A.* **82 D2** 45 18N 122 58W
Newberry, *Mich., U.S.A.* **76 B3** 46 21N 85 30W
Newberry, *S.C., U.S.A.* **77 H5** 34 17N 81 37W
Newberry Springs, *U.S.A.* **85 L10** 34 50N 116 41W
Newbrook, *Canada* **72 C6** 54 24N 112 57W
Newburgh, *U.S.A.* **79 E10** 41 30N 74 1W
Newbury, *U.K.* **11 F6** 51 24N 1 19W
Newbury, *U.S.A.* **79 B12** 43 19N 72 3W
Newburyport, *U.S.A.* **79 D14** 42 49N 70 53W
Newcastle, *Australia* **63 E5** 33 0S 151 46 E
Newcastle, *Canada* **71 C6** 47 1N 65 38W
Newcastle, *S. Africa* **57 D4** 27 45S 29 58 E
Newcastle, *U.K.* **13 B6** 54 13N 5 54W
Newcastle, *Calif., U.S.A.* **84 G5** 38 53N 121 8W
Newcastle, *Wyo., U.S.A.* **80 D2** 43 50N 104 11W
Newcastle Emlyn, *U.K.* **11 E3** 52 2N 4 29W
Newcastle Ra., *Australia* **60 C5** 15 45S 130 15 E
Newcastle-under-Lyme, *U.K.* **10 D5** 53 2N 2 15W
Newcastle-upon-Tyne, *U.K.* **10 C6** 54 59N 1 37W
Newcastle Waters, *Australia* **62 B1** 17 30S 133 28 E
Newdegate, *Australia* **61 F2** 33 6S 119 0 E
Newell, *U.S.A.* **80 C3** 44 43N 103 25W
Newfoundland □, *Canada* **71 B8** 53 0N 58 0W
Newhalem, *U.S.A.* **72 D4** 48 40N 121 15W
Newhall, *U.S.A.* **85 L8** 34 23N 118 32W
Newham, *U.K.* **11 F8** 51 31N 0 2 E
Newhaven, *U.K.* **11 G8** 50 47N 0 4 E
Newkirk, *U.S.A.* **81 G6** 36 53N 97 3W
Newman, *Australia* **60 D2** 23 18S 119 45 E
Newman, *U.S.A.* **84 H5** 37 19N 121 1W
Newmarket, *Canada* **78 B5** 44 3N 79 28W
Newmarket, *Ireland* **13 D3** 52 13N 9 0W
Newmarket, *U.K.* **11 E8** 52 15N 0 23 E
Newmarket, *U.S.A.* **79 C14** 43 5N 70 56W
Newnan, *U.S.A.* **77 J3** 33 23N 84 48W
Newport, *Gwent, U.K.* **11 F5** 51 35N 3 0W
Newport, *I. of W., U.K.* **11 G6** 50 42N 1 18W
Newport, *Shrops., U.K.* **11 E5** 52 47N 2 22W
Newport, *Ark., U.S.A.* **81 H9** 35 37N 91 16W
Newport, *Ky., U.S.A.* **76 F3** 39 5N 84 30W
Newport, *N.H., U.S.A.* **79 C12** 43 22N 72 10W
Newport, *Oreg., U.S.A.* **82 D1** 44 39N 124 3W
Newport, *Pa., U.S.A.* **78 F7** 40 29N 77 8W
Newport, *R.I., U.S.A.* **79 E13** 41 29N 71 19W
Newport, *Tenn., U.S.A.* **77 H4** 35 58N 83 11W
Newport, *Vt., U.S.A.* **79 B12** 44 56N 72 13W
Newport, *Wash., U.S.A.* **82 B5** 48 11N 117 3W
Newport Beach, *U.S.A.* **85 M9** 33 37N 117 56W
Newport News, *U.S.A.* **76 G7** 36 59N 76 25W
Newquay, *U.K.* **11 G2** 50 24N 5 6W
Newry, *U.K.* **13 B5** 54 10N 6 20W
Newry & Mourne □, *U.K.* **13 B5** 54 10N 6 15W
Newton, *Iowa, U.S.A.* **80 E8** 41 42N 93 3W
Newton, *Mass., U.S.A.* **79 D13** 42 21N 71 12W
Newton, *Miss., U.S.A.* **81 J10** 32 19N 89 10W
Newton, *N.C., U.S.A.* **77 H5** 35 40N 81 13W
Newton, *N.J., U.S.A.* **79 E10** 41 3N 74 45W
Newton, *Tex., U.S.A.* **81 K8** 30 51N 93 46W
Newton Abbot, *U.K.* **11 G4** 50 32N 3 37W
Newton Boyd, *Australia* **63 D5** 29 45S 152 16 E
Newton Stewart, *U.K.* **12 G4** 54 57N 4 30W
Newtonmore, *U.K.* **12 D4** 57 4N 4 7W
Newtown, *U.K.* **11 E4** 52 31N 3 19W
Newtownabbey □, *U.K.* **13 B6** 54 45N 6 0W
Newtownards, *U.K.* **13 B6** 54 37N 5 40W
Newville, *U.S.A.* **78 F7** 40 10N 77 24W
Neya, *Russia* **24 C7** 58 21N 43 49 E
Neyrīz, *Iran* **45 D7** 29 15N 54 19 E
Neyshābūr, *Iran* **45 B8** 36 10N 58 50 E
Nezhin, *Ukraine* **25 D5** 51 5N 31 55 E
Nezperce, *U.S.A.* **82 C5** 46 14N 116 14W
Ngabang, *Indonesia* **36 D3** 0 23N 109 55 E
Ngabordamlu, Tanjung, *Indonesia* **37 F8** 6 56S 134 11 E
Ngami Depression, *Botswana* **56 C3** 20 30S 22 46 E
Ngamo, *Zimbabwe* **55 F2** 19 3S 27 32 E
Nganglong Kangri, *China* **41 C12** 33 0N 81 0 E
Nganjuk, *Indonesia* **37 G14** 7 32S 111 55 E
Ngao, *Thailand* **38 C2** 18 46N 99 59 E
Ngaoundéré, *Cameroon* **52 C2** 7 15N 13 35 E
Ngapara, *N.Z.* **59 L3** 44 57S 170 46 E
Ngara, *Tanzania* **54 C3** 2 29S 30 40 E
Ngara □, *Tanzania* **54 C3** 2 29S 30 40 E
Ngawi, *Indonesia* **37 G14** 7 24S 111 26 E
Nghia Lo, *Vietnam* **38 B5** 21 33N 104 28 E
Ngoma, *Malawi* **55 E3** 13 8S 33 45 E
Ngomahura, *Zimbabwe* **55 G3** 20 26S 30 43 E
Ngomba, *Tanzania* **55 D3** 8 20S 32 53 E
Ngoring Hu, *China* **32 C4** 34 55N 97 5 E
Ngorongoro, *Tanzania* **54 C4** 3 11S 35 32 E
Ngozi, *Burundi* **54 C2** 2 54S 29 50 E
Ngudu, *Tanzania* **54 C3** 2 58S 33 25 E
Nguigmi, *Niger* **51 F7** 14 20N 13 20 E
Ngukurr, *Australia* **62 A1** 14 44S 134 44 E
Ngunga, *Tanzania* **54 C3** 3 37S 33 37 E
Nguru, *Nigeria* **50 F7** 12 56N 10 29 E
Nguru Mts., *Tanzania* **54 D4** 6 0S 37 30 E
Nguyen Binh, *Vietnam* **38 A5** 22 39N 105 56 E
Nha Trang, *Vietnam* **39 F7** 12 16N 109 10 E
Nhacoongo, *Mozam.* **57 C6** 24 18S 35 14 E
Nhamaabué, *Mozam.* **55 F4** 17 25S 35 5 E
Nhangutazi, L., *Mozam.* **57 C5** 24 0S 34 30 E
Nhill, *Australia* **63 F3** 36 18S 141 40 E
Nho Quan, *Vietnam* **38 B5** 20 18N 105 45 E
Nhulunbuy, *Australia* **62 A2** 12 10S 137 20 E
Nia-nia, *Zaïre* **54 B2** 1 30N 27 40 E
Niafounké, *Mali* **50 E4** 16 0N 4 5W
Niagara, *U.S.A.* **76 C1** 45 45N 88 0W
Niagara Falls, *Canada* **70 D4** 43 7N 79 5W
Niagara Falls, *U.S.A.* **78 C6** 43 5N 79 4W
Niagara-on-the-Lake, *Canada* **78 C5** 43 15N 79 4W
Niah, *Malaysia* **36 D4** 3 58N 113 46 E
Niamey, *Niger* **50 F5** 13 27N 2 6 E
Niangara, *Zaïre* **54 B2** 3 42N 27 50 E
Nias, *Indonesia* **36 D1** 1 0N 97 30 E
Niassa □, *Mozam.* **55 E4** 13 30S 36 0 E
Nicaragua ■, *Cent. Amer.* **88 D2** 11 40N 85 30W
Nicaragua, L. de, *Nic.* **88 D2** 12 0N 85 30W
Nicastro, *Italy* **20 E7** 38 58N 16 18 E
Nice, *France* **18 E7** 43 42N 7 14 E
Niceville, *U.S.A.* **77 K2** 30 31N 86 30W
Nichinan, *Japan* **31 J5** 31 38N 131 23 E
Nicholás, Canal, *W. Indies* **88 B3** 23 30N 80 5W
Nicholasville, *U.S.A.* **76 G3** 37 53N 84 34W
Nichols, *U.S.A.* **79 D8** 42 1N 76 22W
Nicholson, *Australia* **60 C4** 18 2S 128 54 E
Nicholson, *U.S.A.* **79 E9** 41 37N 75 47W
Nicholson →, *Australia* **62 B2** 17 31S 139 36 E
Nicholson Ra., *Australia* **61 E2** 27 15S 116 45 E
Nicobar Is., *Ind. Oc.* **28 J13** 9 0N 93 0 E
Nicola, *Canada* **72 C4** 50 12N 120 40W
Nicolet, *Canada* **70 C5** 46 17N 72 35W
Nicolls Town, *Bahamas* **88 A4** 25 8N 78 0W
Nicosia, *Cyprus* **23 D12** 35 10N 33 25 E
Nicoya, *Costa Rica* **88 D2** 10 9N 85 27W
Nicoya, G. de, *Costa Rica* **88 E3** 10 0N 85 0W
Nicoya, Pen. de, *Costa Rica* **88 E2** 9 45N 85 40W
Nidd →, *U.K.* **10 C6** 54 1N 1 32W
Niekerkshoop, *S. Africa* **56 D3** 29 19S 22 51 E
Niemba, *Zaïre* **54 D2** 5 58S 28 24 E
Niemen = Neman →, *Lithuania* **24 C3** 55 25N 21 10 E
Nienburg, *Germany* **16 B4** 52 38N 9 15 E
Nieu Bethesda, *S. Africa* **56 E3** 31 51S 24 34 E
Nieuw Amsterdam, *Surinam* **93 B7** 5 53N 55 5W
Nieuw Nickerie, *Surinam* **93 B7** 6 0N 56 59W
Nieuwoudtville, *S. Africa* **56 E2** 31 23S 19 7 E
Nieuwpoort, *Belgium* **15 C2** 51 8N 2 45 E
Nieves, Pico de las, *Canary Is.* **22 G4** 27 57N 15 35W
Nièvre □, *France* **18 C5** 47 10N 3 40 E
Niğde, *Turkey* **25 G5** 37 58N 34 40 E
Nigel, *S. Africa* **57 D4** 26 27S 28 25 E
Niger ■, *W. Afr.* **50 E6** 17 30N 10 0 E
Niger →, *W. Afr.* **50 G6** 5 33N 6 33 E
Nigeria ■, *W. Afr.* **50 G6** 8 30N 8 0 E
Nightcaps, *N.Z.* **59 L2** 45 57S 168 2 E
Nihtaur, *India* **43 E8** 29 20N 78 23 E
Nii-Jima, *Japan* **31 G9** 34 20N 139 15 E
Niigata, *Japan* **30 F9** 37 58N 139 0 E
Niigata □, *Japan* **31 F9** 37 15N 138 45 E
Niihama, *Japan* **31 H6** 33 55N 133 16 E
Niihau, *U.S.A.* **74 H14** 21 54N 160 9W
Niimi, *Japan* **31 G6** 34 59N 133 28 E
Niitsu, *Japan* **30 F9** 37 48N 139 7 E
Nijil, *Jordan* **47 E4** 30 32N 35 33 E
Nijkerk, *Neths.* **15 B5** 52 13N 5 30 E
Nijmegen, *Neths.* **15 C5** 51 50N 5 52 E
Nijverdal, *Neths.* **15 B6** 52 22N 6 28 E
Nīk Pey, *Iran* **45 B6** 36 50N 48 10 E
Nikel, *Russia* **8 B21** 69 24N 30 12 E
Nikiniki, *Indonesia* **37 F6** 9 49S 124 30 E
Nikki, *Benin* **50 G5** 9 58N 3 12 E
Nikkō, *Japan* **31 F9** 36 45N 139 35 E
Nikolayev, *Ukraine* **25 E5** 46 58N 32 0 E
Nikolayevsk, *Russia* **25 D8** 50 0N 45 35 E
Nikolayevsk-na-Amur, *Russia* **27 D15** 53 8N 140 44 E
Nikolskoye, *Russia* **27 D17** 55 12N 166 0 E
Nikopol, *Ukraine* **25 E5** 47 35N 34 25 E
Nīkshahr, *Iran* **45 E9** 26 15N 60 10 E
Nîl, Nahr en →, *Africa* **51 B11** 30 10N 31 6 E
Nîl el Abyad →, *Sudan* **51 E11** 15 38N 32 31 E
Nîl el Azraq →, *Sudan* **51 E11** 15 38N 32 31 E
Niland, *U.S.A.* **85 M11** 33 14N 115 31W
Nile = Nîl, Nahr en →, *Africa* **51 B11** 30 10N 31 6 E
Nile □, *Uganda* **54 B3** 2 0N 31 30 E
Niles, *U.S.A.* **78 E4** 41 11N 80 46W
Nimach, *India* **42 G6** 24 30N 74 56 E
Nimbahera, *India* **42 G6** 24 37N 74 45 E
Nîmes, *France* **18 E6** 43 50N 4 23 E
Nimmitabel, *Australia* **63 F4** 36 29S 149 15 E
Nimneryskiy, *Russia* **27 D13** 57 50N 125 10 E
Nimule, *Sudan* **52 D6** 3 32N 32 3 E
Nīnawá, *Iraq* **44 B4** 36 25N 43 10 E
Nindigully, *Australia* **63 D4** 28 21S 148 50 E
Ninemile, *U.S.A.* **72 B2** 56 0N 130 7W
Nineveh = Nīnawá, *Iraq* **44 B4** 36 25N 43 10 E
Ning Xian, *China* **34 G4** 35 30N 107 58 E
Ningaloo, *Australia* **60 D1** 22 41S 113 41 E
Ning'an, *China* **35 B15** 44 22N 129 20 E
Ningbo, *China* **33 D7** 29 51N 121 28 E
Ningcheng, *China* **35 D10** 41 32N 119 53 E
Ningjin, *China* **34 F8** 37 35N 114 57 E
Ningjing Shan, *China* **32 C4** 30 0N 98 20 E
Ningling, *China* **34 G8** 34 25N 115 22 E
Ningpo = Ningbo, *China* **33 D7** 29 51N 121 28 E
Ningqiang, *China* **34 H4** 32 47N 106 15 E
Ningshan, *China* **34 H5** 33 21N 108 21 E
Ningsia Hui A.R. = Ningxia Huizu Zizhiqu □, *China* **34 E3** 38 0N 106 0 E
Ningwu, *China* **34 E7** 39 0N 112 18 E
Ningxia Huizu Zizhiqu □, *China* **34 E3** 38 0N 106 0 E
Ningyang, *China* **34 G9** 35 47N 116 45 E
Ninh Binh, *Vietnam* **38 B5** 20 15N 105 55 E
Ninh Giang, *Vietnam* **38 B6** 20 44N 106 24 E
Ninh Hoa, *Vietnam* **38 F7** 12 30N 109 7 E
Ninh Ma, *Vietnam* **38 F7** 12 48N 109 21 E
Ninove, *Belgium* **15 D4** 50 51N 4 2 E
Nioaque, *Brazil* **95 A4** 21 5S 55 50W
Niobrara, *U.S.A.* **80 D6** 42 45N 98 2W
Niobrara →, *U.S.A.* **80 D6** 42 46N 98 3W
Nioro du Sahel, *Mali* **50 E3** 15 15N 9 30W
Niort, *France* **18 C3** 46 19N 0 29W
Nipawin, *Canada* **73 C8** 53 20N 104 0W
Nipawin Prov. Park, *Canada* **73 C8** 54 0N 104 37W
Nipigon, *Canada* **70 C2** 49 0N 88 17W
Nipigon, L., *Canada* **70 C2** 49 50N 88 30W
Nipin →, *Canada* **73 B7** 55 46N 108 35W
Nipishish L., *Canada* **71 B7** 54 12N 60 45W
Nipissing L., *Canada* **70 C4** 46 20N 80 0W
Nipomo, *U.S.A.* **85 K6** 35 3N 120 29W
Nipton, *U.S.A.* **85 K11** 35 28N 115 16W
Niquelândia, *Brazil* **93 F9** 14 33S 48 23W
Nīr, *Iran* **44 B5** 38 2N 47 59 E
Nirasaki, *Japan* **31 G9** 35 42N 138 27 E
Nirmal, *India* **40 K11** 19 3N 78 20 E
Nirmali, *India* **43 F12** 26 20N 86 35 E
Niš, *Serbia, Yug.* **21 C9** 43 19N 21 58 E
Nişāb, *Yemen* **46 E4** 14 25N 46 29 E
Nishinomiya, *Japan* **31 G7** 34 45N 135 20 E
Nishin'omote, *Japan* **31 J5** 30 43N 130 59 E
Nishiwaki, *Japan* **31 G7** 34 59N 134 58 E
Niskibi →, *Canada* **70 A2** 56 29N 88 9W
Nisqually →, *U.S.A.* **84 C4** 47 6N 122 42W
Nissáki, *Greece* **23 A3** 39 43N 19 52 E
Nisutlin →, *Canada* **72 A2** 60 14N 132 34W
Nitchequon, *Canada* **71 B5** 53 10N 70 58W
Niterói, *Brazil* **95 A7** 22 52S 43 0W
Nith →, *U.K.* **12 F5** 55 20N 3 5W
Nitra, *Slovakia* **17 D9** 48 19N 18 4 E
Nitra →, *Slovakia* **17 E9** 47 46N 18 10 E
Niuafo'ou, *Tonga* **59 B11** 15 30S 175 58W
Niue, *Cook Is.* **65 J11** 19 2S 169 54W
Niut, *Indonesia* **36 D4** 0 55N 110 6 E
Niuzhuang, *China* **35 D12** 40 58N 122 28 E
Nivelles, *Belgium* **15 D4** 50 35N 4 20 E
Nivernais, *France* **18 C5** 47 15N 3 30 E
Nixon, *U.S.A.* **81 L6** 29 16N 97 46W
Nizamabad, *India* **40 K11** 18 45N 78 7 E
Nizamghat, *India* **41 E19** 28 20N 95 45 E
Nizhne Kolymsk, *Russia* **27 C17** 68 34N 160 55 E
Nizhne-Vartovsk, *Russia* **26 C8** 60 56N 76 38 E
Nizhneangarsk, *Russia* **27 D11** 55 47N 109 30 E
Nizhnekamsk, *Russia* **24 C9** 55 38N 51 49 E
Nizhneudinsk, *Russia* **27 D10** 54 54N 99 3 E
Nizhneyansk, *Russia* **27 B14** 71 26N 136 4 E
Nizhniy Novgorod, *Russia* **24 C7** 56 20N 44 0 E
Nizhniy Tagil, *Russia* **24 C10** 57 55N 59 57 E
Nizké Tatry, *Slovakia* **17 D9** 48 55N 19 30 E
Njakwa, *Malawi* **55 E3** 11 1S 33 56 E
Njanji, *Zambia* **55 E3** 14 25S 31 46 E
Njinjo, *Tanzania* **55 D4** 8 48S 38 54 E
Njombe, *Tanzania* **55 D3** 9 20S 34 50 E
Njombe □, *Tanzania* **55 D3** 9 20S 34 49 E
Njombe →, *Tanzania* **54 D4** 6 56S 35 6 E
Nkambe, *Cameroon* **50 G7** 6 35N 10 40 E
Nkana, *Zambia* **55 E2** 12 50S 28 8 E
Nkawkaw, *Ghana* **50 G4** 6 36N 0 49W
Nkayi, *Zimbabwe* **55 F2** 19 41S 29 20 E
Nkhata Bay, *Malawi* **52 G6** 11 33S 34 16 E
Nkhota Kota, *Malawi* **55 E3** 12 56S 34 15 E
Nkongsamba, *Cameroon* **50 H6** 4 55N 9 55 E
Nkurenkuru, *Namibia* **56 B2** 17 42S 18 32 E
Nmai →, *Burma* **41 G20** 25 30N 97 25 E
Noakhali = Maijdi, *Bangla.* **41 H17** 22 48N 91 10 E
Noatak, *U.S.A.* **68 B3** 67 34N 162 58W
Nobel, *Canada* **78 A4** 45 25N 80 6W
Nobeoka, *Japan* **31 H5** 32 36N 131 41 E
Noblesville, *U.S.A.* **76 E3** 40 3N 86 1W
Nocera Inferiore, *Italy* **20 D6** 40 45N 14 37 E
Nockatunga, *Australia* **63 D3** 27 42S 142 42 E
Nocona, *U.S.A.* **81 J6** 33 47N 97 44W
Noda, *Japan* **31 G9** 35 56N 139 52 E
Noel, *U.S.A.* **81 G7** 36 33N 94 29W
Nogales, *Mexico* **86 A2** 31 20N 110 56W
Nogales, *U.S.A.* **83 L8** 31 20N 110 56W
Nōgata, *Japan* **31 H5** 33 48N 130 44 E
Noggerup, *Australia* **61 F2** 33 32S 116 5 E
Noginsk, *Russia* **27 C10** 64 30N 90 50 E
Nogoa →, *Australia* **62 C4** 23 40S 147 55 E
Nogoyá, *Argentina* **94 C4** 32 24S 59 48W
Nohar, *India* **42 E6** 29 11N 74 49 E
Noirmoutier, I. de, *France* **18 C2** 46 58N 2 10W
Nojane, *Botswana* **56 C3** 23 15S 20 14 E
Nojima-Zaki, *Japan* **31 G9** 34 54N 139 53 E
Nok Kundi, *Pakistan* **40 E3** 28 50N 62 45 E
Nokaneng, *Botswana* **56 B3** 19 40S 22 17 E
Nokhtuysk, *Russia* **27 C12** 60 0N 117 45 E
Nokomis, *Canada* **73 C8** 51 35N 105 0W
Nokomis L., *Canada* **73 B8** 57 0N 103 0W
Nola, *C.A.R.* **52 D3** 3 35N 16 4 E
Noma Omuramba →, *Namibia* **56 B3** 18 52S 20 53 E
Noman L., *Canada* **73 A7** 62 15N 108 55W
Nombre de Dios, *Panama* **88 E4** 9 34N 79 28W
Nome, *U.S.A.* **68 B3** 64 30N 165 25W
Nomo-Zaki, *Japan* **31 H4** 32 35N 129 44 E
Nonacho L., *Canada* **73 A7** 61 42N 109 40W
Nonda, *Australia* **62 C3** 20 40S 142 28 E
Nong Chang, *Thailand* **38 E2** 15 23N 99 51 E
Nong Het, *Laos* **38 C4** 19 29N 103 59 E
Nong Khai, *Thailand* **38 D4** 17 50N 102 46 E
Nong'an, *China* **35 B13** 44 25N 125 5 E
Nongoma, *S. Africa* **57 D5** 27 58S 31 35 E
Nonoava, *Mexico* **86 B3** 27 28N 106 44W
Nonthaburi, *Thailand* **38 F3** 13 51N 100 34 E
Noonamah, *Australia* **60 B5** 12 40S 131 4 E
Noonan, *U.S.A.* **80 A3** 48 54N 103 1W
Noondoo, *Australia* **63 D4** 28 35S 148 30 E
Noonkanbah, *Australia* **60 C3** 18 30S 124 50 E
Noord Brabant □, *Neths.* **15 C5** 51 40N 5 0 E
Noord Holland □, *Neths.* **15 B4** 52 30N 4 45 E
Noordbeveland, *Neths.* **15 C3** 51 35N 3 50 E
Noordoostpolder, *Neths.* **15 B5** 52 45N 5 45 E
Noordwijk aan Zee, *Neths.* **15 B4** 52 14N 4 26 E
Nootka, *Canada* **72 D3** 49 38N 126 38W
Nootka I., *Canada* **72 D3** 49 32N 126 42W
Nóqui, *Angola* **52 F2** 5 55S 13 30 E
Noranda, *Canada* **70 C4** 48 20N 79 0W
Norco, *U.S.A.* **85 M9** 33 56N 117 33W
Nord □, *France* **18 A5** 50 15N 3 30 E
Nord-Ostsee Kanal, *Germany* **16 A4** 54 15N 9 40 E
Nord-Trøndelag fylke □, *Norway* **8 D12** 64 20N 12 10 E
Nordaustlandet, *Svalbard* **4 B9** 79 14N 23 0 E
Nordegg, *Canada* **72 C5** 52 29N 116 5W
Nordhausen, *Germany* **16 C5** 51 29N 10 47 E
Nordkapp, *Norway* **8 A18** 71 10N 25 44 E
Nordkapp, *Svalbard* **4 A9** 80 31N 20 0 E
Nordkinn = Kinnarodden, *Norway* **6 A11** 71 8N 27 40 E
Nordland fylke □, *Norway* **8 D12** 65 40N 13 0 E
Nordrhein-Westfalen □, *Germany* **16 C3** 51 45N 7 30 E
Nordvik, *Russia* **27 B12** 74 2N 111 32 E
Norembega, *Canada* **70 C3** 48 59N 80 43W
Norfolk, *Nebr., U.S.A.* **80 D6** 42 2N 97 25W
Norfolk, *Va., U.S.A.* **76 G7** 36 51N 76 17W
Norfolk □, *U.K.* **10 E9** 52 39N 1 0 E
Norfolk Broads, *U.K.* **10 E9** 52 30N 1 15 E
Norfolk I., *Pac. Oc.* **64 K8** 28 58S 168 3 E
Norfork Res., *U.S.A.* **81 G8** 36 13N 92 15W
Norilsk, *Russia* **27 C9** 69 20N 88 6 E
Norley, *Australia* **63 D3** 27 45S 143 48 E
Norma, Mt., *Australia* **62 C3** 20 55S 140 42 E
Normal, *U.S.A.* **80 E10** 40 31N 88 59W
Norman, *U.S.A.* **81 H6** 35 13N 97 26W
Norman →, *Australia* **62 B3** 19 18S 141 51 E
Norman Wells, *Canada* **68 B7** 65 17N 126 51W
Normanby →, *Australia* **62 A3** 14 23S 144 10 E
Normandie, *France* **18 B4** 48 45N 0 10 E
Normandin, *Canada* **70 C5** 48 49N 72 31W
Normandy = Normandie, *France* **18 B4** 48 45N 0 10 E
Normanhurst, Mt., *Australia* **61 E3** 25 4S 122 30 E

Oromocto

Obskaya Guba, Russia 26 C8 69 0N 73 0 E
Obuasi, Ghana 50 G4 6 17N 1 40W
Ocala, U.S.A. 77 L4 29 11N 82 8W
Ocampo, Mexico 86 B3 28 9N 108 24W
Ocaña, Spain 19 C4 39 55N 3 30W
Oconomowoc, U.S.A. 80 D10 43 7N 88 30W
Ocate, U.S.A. 81 G2 36 11N 105 3W
Occidental, Cordillera, Colombia 92 C3 5 0N 76 0W
Ocean City, N.J., U.S.A. 76 F8 39 17N 74 35W
Ocean City, Wash., U.S.A. 84 C2 47 4N 124 10W
Ocean I. = Banaba, Kiribati 64 H8 0 45S 169 50 E
Ocean Park, U.S.A. 84 D2 46 30N 124 3W
Oceano, U.S.A. 85 K6 35 6N 120 37W
Oceanport, U.S.A. 79 F10 40 19N 74 3W
Oceanside, U.S.A. 85 M9 33 12N 117 23W
Ochil Hills, U.K. 12 E5 56 14N 3 40W
Ochre River, Canada 73 C9 51 4N 99 47W
Ocilla, U.S.A. 77 K4 31 36N 83 15W
Ocmulgee →, U.S.A. 77 K4 31 58N 82 33W
Oconee →, U.S.A. 77 K4 31 58N 82 33W
Oconto, U.S.A. 76 C2 44 53N 87 52W
Oconto Falls, U.S.A. 76 C1 44 52N 88 9W
Ocosingo, Mexico 87 D6 17 10N 92 15W
Ocotal, Nic. 88 D2 13 41N 86 31W
Ocotlán, Mexico 86 C4 20 21N 102 42W
Octave, U.S.A. 83 J7 34 10N 112 43W
Ocumare del Tuy, Venezuela 92 A5 10 7N 66 46W
Ōda, Japan 31 G6 35 11N 132 30 E
Ódáðahraun, Iceland 8 D5 65 5N 17 0W
Odate, Japan 30 D10 40 16N 140 34 E
Odawara, Japan 31 G9 35 20N 139 6 E
Odda, Norway 9 F9 60 3N 6 35 E
Oddur, Somali Rep. 46 G3 4 11N 43 52 E
Odei →, Canada 73 B9 56 6N 96 54W
Odendaalsrus, S. Africa 56 D4 27 48S 26 45 E
Odense, Denmark 9 J11 55 22N 10 23 E
Oder →, Germany 16 B7 53 33N 14 38 E
Odesa = Odessa, Ukraine 25 E5 46 30N 30 45 E
Odessa, Canada 79 B8 44 17N 76 43W
Odessa, Ukraine 25 E5 46 30N 30 45 E
Odessa, Tex., U.S.A. 81 K3 31 52N 102 23W
Odessa, Wash., U.S.A. 82 C4 47 20N 118 41W
Odiakwe, Botswana 56 C4 20 12S 25 17 E
Odienné, Ivory C. 50 G3 9 30N 7 34W
Odintsovo, Russia 24 C6 55 39N 37 15 E
O'Donnell, U.S.A. 81 J4 32 58N 101 50W
Odorheiu Secuiesc, Romania 17 E12 46 21N 25 21 E
Odra →, Poland 16 B7 53 33N 14 38 E
Odžak, Bos.-H. 21 B8 45 3N 18 18 E
Odzi, Zimbabwe 57 B5 19 0S 32 20 E
Oeiras, Brazil 93 E10 7 0S 42 8W
Oelrichs, U.S.A. 80 D3 43 11N 103 14W
Oelwein, U.S.A. 80 D9 42 41N 91 55W
Oenpelli, Australia 60 B5 12 20S 133 4 E
Ofanto →, Italy 20 D7 41 22N 16 13 E
Offa, Nigeria 50 G5 8 13N 4 42 E
Offaly □, Ireland 13 C4 53 15N 7 30W
Offenbach, Germany 16 C4 50 6N 8 46 E
Ofotfjorden, Norway 8 B14 68 27N 16 40 E
Ōfunato, Japan 30 E10 39 4N 141 43 E
Oga, Japan 30 E9 39 55N 139 50 E
Oga-Hantō, Japan 30 E9 39 58N 139 47 E
Ogahalla, Canada 70 B2 50 6N 85 51W
Ōgaki, Japan 31 G8 35 21N 136 37 E
Ogallala, U.S.A. 80 E4 41 8N 101 43W
Ogasawara Gunto, Pac. Oc. 64 E6 27 0N 142 0 E
Ogbomosho, Nigeria 50 G5 8 1N 4 11 E
Ogden, Iowa, U.S.A. 80 D8 42 2N 94 2W
Ogden, Utah, U.S.A. 82 F7 41 13N 111 58W
Ogdensburg, U.S.A. 79 B9 44 42N 75 30W
Ogeechee →, U.S.A. 77 K5 31 50N 81 3W
Ogilby, U.S.A. 85 N12 32 49N 114 50W
Oglio →, Italy 20 B4 45 2N 10 39 E
Ogmore, Australia 62 C4 22 37S 149 35 E
Ogoki →, Canada 70 B2 51 38N 85 57W
Ogoki L., Canada 70 B2 50 50N 87 10W
Ogoki Res., Canada 70 B2 50 45N 88 15W
Ogooué →, Gabon 52 E1 1 0S 9 0 E
Ogowe = Ogooué →, Gabon 52 E1 1 0S 9 0 E
Ohai, N.Z. 59 L2 45 55S 168 0 E
Ohakune, N.Z. 59 H5 39 24S 175 24 E
Ohanet, Algeria 50 C6 28 44N 8 46 E
Ohata, Japan 30 D10 41 24N 141 10 E
Ohau, L., N.Z. 59 L2 44 15S 169 53 E
Ohey, Belgium 15 D5 50 26N 5 8 E
Ohio □, U.S.A. 76 E3 40 15N 82 45W
Ohio →, U.S.A. 76 G1 36 59N 89 8W
Ohre →, Czech. 16 C7 50 30N 14 10 E
Ohridsko, Jezero, Macedonia 21 D9 41 8N 20 52 E
Ohrigstad, S. Africa 57 C5 24 39S 30 36 E
Oikou, China 35 E9 38 35N 117 42 E
Oil City, U.S.A. 78 E5 41 26N 79 42W
Oildale, U.S.A. 85 K7 35 25N 119 1W
Oise □, France 18 B5 49 28N 2 30 E
Ōita, Japan 31 H5 33 14N 131 36 E
Ōita □, Japan 31 H5 33 15N 131 30 E
Oiticica, Brazil 93 E10 5 3S 41 5W
Ojai, U.S.A. 85 L7 34 27N 119 15W
Ojinaga, Mexico 86 B4 29 34N 104 25W
Ojiya, Japan 31 F9 37 18N 138 48 E
Ojos del Salado, Cerro, Argentina 94 B2 27 0S 68 40W
Oka →, Russia 26 D5 56 20N 43 59 E
Okaba, Indonesia 37 F9 8 6S 139 42 E
Okahandja, Namibia 56 C2 22 0S 16 59 E
Okahukura, N.Z. 59 H5 38 48S 175 14 E
Okanagan L., Canada 72 C5 50 0N 119 30W
Okandja, Gabon 52 E2 0 35S 13 45 E
Okanogan, U.S.A. 82 B4 48 22N 119 35W
Okanogan →, U.S.A. 82 B4 48 6N 119 44W
Okaputa, Namibia 56 C2 20 5S 17 0 E
Okara, Pakistan 42 D5 30 50N 73 31 E
Okarito, N.Z. 59 K3 43 15S 170 9 E
Okaukuejo, Namibia 56 B2 19 10S 16 0 E
Okavango Swamps, Botswana 56 B3 18 45S 22 45 E
Okaya, Japan 31 F9 36 5N 138 10 E
Okayama, Japan 31 G6 34 40N 133 54 E
Okayama □, Japan 31 G6 35 0N 133 50 E
Okazaki, Japan 31 G8 34 57N 137 10 E
Okeechobee, U.S.A. 77 M5 27 15N 80 50W
Okeechobee, L., U.S.A. 77 M5 27 0N 80 50W
Okefenokee Swamp, U.S.A. 77 K4 30 40N 82 20W
Okehampton, U.K. 11 G3 50 44N 4 1W
Okha, Russia 27 D15 53 40N 143 0 E
Okhotsk, Russia 27 D15 59 20N 143 10 E
Okhotsk, Sea of, Asia 27 D15 55 0N 145 0 E
Okhotskiy Perevoz, Russia 27 C14 61 52N 135 35 E
Okhotsko Kolymskoye, Russia 27 C16 63 0N 157 0 E
Oki-Shotō, Japan 31 F6 36 5N 133 15 E
Okiep, S. Africa 56 D2 29 39S 17 53 E
Okinawa □, Japan 31 L3 26 40N 128 0 E
Okinawa-Guntō, Japan 31 L3 26 40N 128 0 E
Okinawa-Jima, Japan 31 L4 26 32N 128 0 E
Okino-erabu-Shima, Japan 31 L4 27 21N 128 33 E
Oklahoma □, U.S.A. 81 H6 35 20N 97 30W
Oklahoma City, U.S.A. 81 H6 35 30N 97 30W
Okmulgee, U.S.A. 81 H7 35 37N 95 58W
Okolo, Uganda 54 B3 2 37N 31 8 E
Okolona, U.S.A. 81 H10 34 0N 88 45W
Okrika, Nigeria 50 H6 4 40N 7 10 E
Oktabrsk, Kazakhstan 25 E10 49 28N 57 25 E
Oktyabrskiy, Russia 24 D9 54 28N 53 28 E
Oktyabrskoy Revolyutsii, Os., Russia 27 B10 79 30N 97 0 E
Oktyabrskoye, Russia 26 C7 62 28N 66 3 E
Okuru, N.Z. 59 K2 43 55S 168 55 E
Okushiri-Tō, Japan 30 C9 42 15N 139 30 E
Okwa →, Botswana 56 C3 22 30S 23 0 E
Ola, U.S.A. 81 H8 35 2N 93 13W
Ólafsfjörður, Iceland 8 C4 66 4N 18 39W
Ólafsvík, Iceland 8 D2 64 53N 23 43W
Olancha, U.S.A. 85 J8 36 17N 118 1W
Olancha Pk., U.S.A. 85 J8 36 15N 118 7W
Olanchito, Honduras 88 C2 15 30N 86 30W
Öland, Sweden 9 H14 56 45N 16 38 E
Olary, Australia 63 E3 32 18S 140 19 E
Olascoaga, Argentina 94 D3 35 15S 60 39W
Olathe, U.S.A. 80 F7 38 53N 94 49W
Olavarría, Argentina 94 D3 36 55S 60 20W
Ólbia, Italy 20 D3 40 55N 9 30 E
Old Bahama Chan. = Bahama, Canal Viejo de, W. Indies 88 B4 22 10N 77 30W
Old Baldy Pk. = San Antonio, Mt., U.S.A. 85 L9 34 17N 117 38W
Old Castile = Castilla y Leon □, Spain 19 B3 42 0N 5 0W
Old Castle, Ireland 13 C4 53 46N 7 10W
Old Cork, Australia 62 C3 22 57S 141 52 E
Old Crow, Canada 68 B6 67 30N 139 55W
Old Dale, U.S.A. 85 L11 34 8N 115 47W
Old Fletton, U.K. 11 E7 52 34N 0 13W
Old Forge, N.Y., U.S.A. 79 C10 43 43N 74 58W
Old Forge, Pa., U.S.A. 79 E9 41 22N 75 45W
Old Fort →, Canada 73 B6 58 36N 110 24W
Old Shinyanga, Tanzania 54 C3 3 33S 33 27 E
Old Speck Mt., U.S.A. 79 B14 44 34N 70 57W
Old Town, U.S.A. 71 D6 44 56N 68 39W
Old Wives L., Canada 73 C7 50 5N 106 0W
Oldbury, U.K. 11 F5 51 38N 2 30W
Oldeani, Tanzania 54 C4 3 22S 35 35 E
Oldenburg, Germany 16 B4 53 10N 8 10 E
Oldenzaal, Neths. 15 B6 52 19N 6 53 E
Oldham, U.K. 10 D5 53 33N 2 8W
Oldman →, Canada 72 D6 49 57N 111 42W
Olds, Canada 72 C6 51 50N 114 10W
Olean, U.S.A. 78 D6 42 5N 78 26W
Olekma →, Russia 27 C13 60 22N 120 42 E
Olekminsk, Russia 27 C13 60 25N 120 30 E
Olema, U.S.A. 84 G4 38 3N 122 47W
Olenegorsk, Russia 24 A5 68 9N 33 18 E
Olenek, Russia 27 C12 68 28N 112 18 E
Olenek →, Russia 27 B13 73 0N 120 10 E
Oléron, I. d', France 18 D3 45 55N 1 15W
Oleśnica, Poland 17 C8 51 13N 17 22 E
Olga, Russia 27 E14 43 50N 135 14 E
Olga, L., Canada 70 C4 49 47N 77 15W
Olga, Mt., Australia 61 E5 25 20S 130 50 E
Olifants →, Africa 57 C5 23 57S 31 58 E
Olifantshoek, S. Africa 56 D3 27 57S 22 42 E
Ólimbos, Óros, Greece 21 D10 40 6N 22 23 E
Olímpia, Brazil 95 A6 20 44S 48 54W
Olinda, Brazil 93 E12 8 1S 34 51W
Olivia, Argentina 94 C3 32 0S 63 38W
Olivehurst, U.S.A. 84 F5 39 6N 121 34W
Oliveira, Brazil 93 H10 20 39S 44 50W
Olivenza, Spain 19 C2 38 41N 7 9W
Oliver, Canada 72 D5 49 13N 119 37W
Oliver L., Canada 73 B8 56 56N 103 22W
Ollagüe, Chile 94 A2 21 15S 68 10W
Olney, Ill., U.S.A. 76 F1 38 44N 88 5W
Olney, Tex., U.S.A. 81 J5 33 22N 98 45W
Olomane →, Canada 71 B7 50 14N 60 37W
Olomouc, Czech. 16 D8 49 38N 17 12 E
Olonets, Russia 24 B5 61 10N 33 0 E
Olongapo, Phil. 37 B6 14 50N 120 18 E
Olovo, Bos.-H. 21 B8 44 8N 18 35 E
Olovo, Yugoslavia 21 B8 44 3N 19 0 E
Olovyannaya, Russia 27 D12 50 58N 115 35 E
Oloy →, Russia 27 C16 66 29N 159 29 E
Olsztyn, Poland 17 B10 53 48N 20 29 E
Olt →, Romania 21 C11 43 43N 24 51 E
Olteniţa, Romania 17 F13 44 7N 26 42 E
Olton, U.S.A. 81 H3 34 11N 102 8W
Olymbos, Cyprus 23 D12 35 21N 33 45 E
Olympia, Greece 21 F9 37 39N 21 39 E
Olympia, U.S.A. 84 D4 47 3N 122 53W
Olympic Mts., U.S.A. 84 C3 47 55N 123 45W
Olympic Nat. Park, U.S.A. 84 C3 47 48N 123 30W
Olympus, Cyprus 23 E11 34 56N 32 52 E
Olympus, Mt. = Ólimbos, Óros, Greece 21 D10 40 6N 22 23 E
Olympus, Mt., U.S.A. 84 C3 47 48N 123 43W
Olyphant, U.S.A. 79 E9 41 27N 75 36W
Om →, Russia 26 D8 54 59N 73 22 E
Om Koi, Thailand 38 D2 17 48N 98 22 E
Ōma, Japan 30 D10 41 45N 141 5 E
Ōmachi, Japan 31 F8 36 30N 137 50 E
Omae-Zaki, Japan 31 G9 34 36N 138 14 E
Ōmagari, Japan 30 E10 39 27N 140 29 E
Omagh, U.K. 13 B4 54 36N 7 20W
Omagh □, U.K. 13 B4 54 35N 7 15W
Omaha, U.S.A. 80 E7 41 17N 95 58W
Omak, U.S.A. 82 B4 48 25N 119 31W
Omalos, Greece 23 D5 35 19N 23 55 E
Oman ■, Asia 46 C6 23 0N 58 0 E
Oman, G. of, Asia 45 E8 24 30N 58 30 E
Omaruru, Namibia 56 C2 21 26S 16 0 E
Omaruru →, Namibia 56 C1 22 7S 14 15 E
Omate, Peru 92 G4 16 45S 71 0W
Ombai, Selat, Indonesia 37 F6 8 30S 124 50 E
Omboué, Gabon 52 E1 1 35S 9 15 E
Ombrone →, Italy 20 C4 42 39N 11 0 E
Omdurmân, Sudan 51 E11 15 40N 32 28 E
Omeonga, Zaïre 54 C1 3 40S 24 22 E
Ometepe, I. de, Nic. 88 D2 11 32N 85 35W
Ometepec, Mexico 87 D5 16 39N 98 23W
Ominato, Japan 30 D10 41 17N 141 10 E
Omineca →, Canada 72 B4 56 3N 124 16W
Omitara, Namibia 56 C2 22 16S 18 2 E
Ōmiya, Japan 31 G9 35 54N 139 38 E
Ommen, Neths. 15 B6 52 31N 6 26 E
Ömnögovĭ □, Mongolia 34 C3 43 15N 104 0 E
Omo →, Ethiopia 51 G12 6 25N 36 10 E
Omodhos, Cyprus 23 E11 34 51N 32 48 E
Omolon →, Russia 27 C16 68 42N 158 36 E
Omono-Gawa →, Japan 30 E10 39 46N 140 3 E
Omsk, Russia 26 D8 55 0N 73 12 E
Omsukchan, Russia 27 C16 62 32N 155 48 E
Ōmu, Japan 30 B11 44 34N 142 58 E
Omul, Vf., Romania 17 F12 45 27N 25 29 E
Ōmura, Japan 31 H4 32 56N 129 57 E
Omuramba Omatako →, Namibia 53 H4 17 45S 20 25 E
Ōmuta, Japan 31 H5 33 5N 130 26 E
Onaga, U.S.A. 80 F6 39 29N 96 10W
Onalaska, U.S.A. 80 D9 43 53N 91 14W
Onamia, U.S.A. 80 B8 46 4N 93 40W
Onancock, U.S.A. 76 G8 37 43N 75 45W
Onang, Indonesia 37 E5 3 2S 118 49 E
Onaping L., Canada 70 C3 47 3N 81 30W
Onavas, Mexico 86 B3 28 28N 109 30W
Onawa, U.S.A. 80 D6 42 2N 96 6W
Onaway, U.S.A. 76 C3 45 21N 84 14W
Oncócua, Angola 56 B1 16 30S 13 25 E
Onda, Spain 19 C5 39 55N 0 17W
Ondaejin, N. Korea 35 D15 41 34N 129 40 E
Ondangua, Namibia 56 B2 17 57S 16 4 E
Ondjiva, Angola 56 B2 16 48S 15 50 E
Ondo, Nigeria 50 G5 7 4N 4 47 E
Öndörshil, Mongolia 34 B5 45 13N 108 5 E
Öndverðarnes, Iceland 8 D1 64 52N 24 0W
Onega, Russia 24 B6 64 0N 38 10 E
Onega →, Russia 24 B6 63 58N 37 55 E
Onega, G. of = Onezhskaya Guba, Russia 24 B6 64 30N 37 0 E
Onega, L. = Onezhskoye Ozero, Russia 24 B6 62 0N 35 30 E
Onehunga, N.Z. 59 G5 36 55S 174 48 E
Oneida, U.S.A. 79 C9 43 6N 75 39W
Oneida L., U.S.A. 79 C9 43 12N 75 54W
O'Neill, U.S.A. 80 D5 42 27N 98 39W
Onekotan, Ostrov, Russia 27 E16 49 25N 154 45 E
Onema, Zaïre 54 C1 4 35S 24 30 E
Oneonta, Ala., U.S.A. 77 J2 33 57N 86 28W
Oneonta, N.Y., U.S.A. 79 D9 42 27N 75 4W
Onezhskaya Guba, Russia 24 B6 64 30N 37 0 E
Onezhskoye Ozero, Russia 24 B6 62 0N 35 30 E
Ongarue, N.Z. 59 H5 38 42S 175 19 E
Ongerup, Australia 61 F2 33 58S 118 28 E
Ongjin, N. Korea 35 F13 37 56N 125 21 E
Ongkharak, Thailand 38 E3 14 8N 101 1 E
Ongniud Qi, China 35 C10 43 0N 118 38 E
Ongoka, Zaïre 54 C2 1 20S 26 0 E
Ongole, India 40 M12 15 33N 80 2 E
Ongon, Mongolia 34 B7 45 41N 113 5 E
Onguren, Russia 27 D11 53 38N 107 36 E
Onida, U.S.A. 80 C4 44 42N 100 4W
Onilahy →, Madag. 57 C7 23 34S 43 45 E
Onitsha, Nigeria 50 G6 6 6N 6 42 E
Onoda, Japan 31 G5 34 2N 131 25 E
Onpyŏng-ni, S. Korea 35 H14 33 25N 126 55 E
Onslow, Australia 60 D2 21 40S 115 12 E
Onslow B., U.S.A. 77 H7 34 20N 77 15W
Onstwedde, Neths. 15 A7 53 2N 7 4 E
Ontake-San, Japan 31 G8 35 53N 137 29 E
Ontario, Calif., U.S.A. 85 L9 34 4N 117 39W
Ontario, Oreg., U.S.A. 82 D5 44 2N 116 58W
Ontario □, Canada 70 B2 48 0N 83 0W
Ontario, L., U.S.A. 70 D4 43 20N 78 0W
Ontonagon, U.S.A. 80 B10 46 52N 89 19W
Onyx, U.S.A. 85 K8 35 41N 118 14W
Oodnadatta, Australia 63 D2 27 33S 135 30 E
Ooldea, Australia 61 F5 30 27S 131 50 E
Oombulgurri, Australia 60 C4 15 15S 127 45 E
Oona River, Canada 72 C2 53 57N 130 16W
Oorindi, Australia 62 C3 20 40S 141 1 E
Oost-Vlaanderen □, Belgium 15 C3 51 5N 3 50 E
Oostende, Belgium 15 C2 51 15N 2 54 E
Oosterhout, Neths. 15 C4 51 39N 4 47 E
Oosterschelde, Neths. 15 C4 51 33N 4 0 E
Ootacamund, India 40 P10 11 30N 76 44 E
Ootsa L., Canada 72 C3 53 50N 126 2W
Opala, Russia 27 D16 51 58N 156 30 E
Opala, Zaïre 54 C1 0 40S 24 20 E
Opanake, Sri Lanka 40 R12 6 35N 80 40 E
Opasatika, Canada 70 C3 49 30N 82 50W
Opasquia, Canada 73 C10 53 16N 93 34W
Opava, Czech. 17 D8 49 57N 17 58 E
Opelousas, U.S.A. 81 K8 30 32N 92 5W
Opémisca, L., Canada 70 C5 49 56N 74 52W
Opheim, U.S.A. 82 B10 48 51N 106 24W
Ophthalmia Ra., Australia 60 D2 23 15S 119 30 E
Opinaca →, Canada 70 B4 52 15N 78 2W
Opinaca L., Canada 70 B4 52 39N 76 20W
Opiskotish, L., Canada 71 B6 53 10N 67 50W
Opole, Poland 17 C8 50 42N 17 58 E
Oporto = Porto, Portugal 19 B1 41 8N 8 40W
Opotiki, N.Z. 59 H6 38 1S 177 19 E
Opp, U.S.A. 77 K2 31 17N 86 16W
Oppland fylke □, Norway 9 F10 61 15N 9 40 E
Opua, N.Z. 59 F5 35 19S 174 9 E
Opunake, N.Z. 59 H4 39 26S 173 52 E
Ora, Cyprus 23 E12 34 51N 33 12 E
Ora Banda, Australia 61 F3 30 20S 121 0 E
Oracle, U.S.A. 83 K8 32 37N 110 46W
Oradea, Romania 17 E10 47 2N 21 58 E
Öræfajökull, Iceland 8 D5 64 2N 16 39W
Orai, India 43 G8 25 58N 79 30 E
Oral = Ural →, Kazakhstan 25 E9 47 0N 51 48 E
Oral = Uralsk, Kazakhstan 24 D9 51 20N 51 20 E
Oran, Algeria 50 A4 35 45N 0 39W
Oran, Argentina 94 A3 23 10S 64 20W
Orange = Oranje →, S. Africa 56 D2 28 41S 16 28 E
Orange, Australia 63 E4 33 15S 149 7 E
Orange, France 18 D6 44 8N 4 47 E
Orange, Calif., U.S.A. 85 M9 33 47N 117 51W
Orange, Mass., U.S.A. 79 D12 42 35N 72 19W
Orange, Tex., U.S.A. 81 K8 30 6N 93 44W
Orange, Va., U.S.A. 76 F6 38 15N 78 7W
Orange, C., Brazil 93 C8 4 20N 51 30W
Orange Cove, U.S.A. 84 J7 36 38N 119 19W
Orange Free State □, S. Africa 56 D4 28 30S 27 0 E
Orange Grove, U.S.A. 81 M6 27 58N 97 56W
Orange Walk, Belize 87 D7 18 6N 88 33W
Orangeburg, U.S.A. 77 J5 33 30N 80 52W
Orangeville, Canada 70 D3 43 55N 80 5W
Oranienburg, Germany 16 B6 52 45N 13 15 E
Oranje →, S. Africa 56 D2 28 41S 16 28 E
Oranje Vrystaat = Orange Free State □, S. Africa 56 D4 28 30S 27 0 E
Oranjemund, Namibia 56 D2 28 38S 16 29 E
Oranjerivier, S. Africa 56 D3 29 40S 24 12 E
Oras, Phil. 37 B7 12 9N 125 28 E
Oraşul Stalin = Braşov, Romania 21 B11 45 38N 25 35 E
Orbetello, Italy 20 C4 42 26N 11 11 E
Orbost, Australia 63 F4 37 40S 148 29 E
Orchila, I., Venezuela 92 A5 11 48N 66 10W
Orcutt, U.S.A. 85 L6 34 52N 120 27W
Ord →, Australia 60 C4 15 33S 128 15 E
Ord, Mt., Australia 60 C4 17 20S 125 34 E
Orderville, U.S.A. 83 H7 37 17N 112 38W
Ordos = Mu Us Shamo, Nei Mongol Zizhiqu, China 34 E5 39 0N 109 0 E

Ordos = Mu Us Shamo, Nei Mongol Zizhiqu, China 34 E5 39 0N 109 0 E

Ordway, U.S.A. 80 F3 38 13N 103 46W
Ordzhonikidze = Vladikavkaz, Russia 25 F7 43 0N 44 35 E
Ore, Zaïre 54 B2 3 17N 29 30 E
Ore Mts. = Erzgebirge, Germany 16 C6 50 25N 13 0 E
Örebro, Sweden 9 G13 59 20N 15 18 E
Örebro län □, Sweden 9 G13 59 27N 15 0 E
Oregon, U.S.A. 80 D10 42 1N 89 20W
Oregon □, U.S.A. 82 E3 44 0N 121 0W
Oregon City, U.S.A. 84 E4 45 21N 122 36W
Orekhovo-Zuyevo, Russia 24 C6 55 50N 38 55 E
Orel, Russia 24 D6 52 57N 36 3 E
Orem, U.S.A. 82 F8 40 19N 111 42W
Orenburg, Russia 24 D10 51 45N 55 6 E
Orense, Spain 19 A2 42 19N 7 55W
Orepuki, N.Z. 59 M1 46 19S 167 46 E
Øresund, Europe 9 H12 55 45N 12 45 E
Orford Ness, U.K. 11 E9 52 6N 1 31 E
Organos, Pta. de los, Canary Is. 22 F2 28 12N 17 17W
Orhon Gol →, Mongolia 32 A5 50 21N 106 0 E
Orient, Australia 63 D3 28 7S 142 50 E
Oriental, Cordillera, Colombia 92 B4 6 0N 73 0W
Oriente, Argentina 94 D3 38 44S 60 37W
Orihuela, Spain 19 C5 38 7N 0 55W
Orinoco →, Venezuela 92 B6 9 15N 61 30W
Orissa □, India 41 K14 20 0N 84 0 E
Oristano, Italy 20 E3 39 54N 8 35 E
Oristano, G. di, Italy 20 E3 39 50N 8 22 E
Orizaba, Mexico 87 D5 18 50N 97 10W
Orkanger, Norway 8 E10 63 18N 9 52 E
Orkla →, Norway 8 E10 63 18N 9 51 E
Orkney, S. Africa 56 D4 26 58S 26 40 E
Orkney □, U.K. 12 C6 59 0N 3 0W
Orkney Is., U.K. 12 C6 59 0N 3 0W
Orland, U.S.A. 84 F4 39 45N 122 12W
Orlando, U.S.A. 77 L5 28 33N 81 23W
Orléanais, France 18 C4 48 0N 2 0 E
Orléans, France 18 C4 47 54N 1 52 E
Orleans, U.S.A. 79 B12 44 49N 72 12W
Orléans, I. d', Canada 71 C5 46 54N 70 58W
Ormara, Pakistan 40 G4 25 16N 64 33 E
Ormoc, Phil. 37 B6 11 0N 124 37 E
Ormond, N.Z. 59 H6 38 33S 177 56 E
Ormond Beach, U.S.A. 77 L5 29 17N 81 3W
Ormstown, Canada 79 A11 45 8N 74 0W
Orne □, France 18 B4 48 40N 0 5 E
Örnsköldsvik, Sweden 8 E15 63 17N 18 40 E
Oro, N. Korea 35 D14 40 1N 127 27 E
Oro →, Mexico 86 B3 25 35N 105 2W
Oro Grande, U.S.A. 85 L9 34 36N 117 20W
Orocué, Colombia 92 C4 4 48N 71 20W
Orogrande, U.S.A. 83 K10 32 24N 106 5W
Oromocto, Canada 71 C6 45 54N 66 29W

Orono

Orono, Canada 78 C6 43 59N 78 37W
Oroqen Zizhiqi, China 33 A7 50 34N 123 43 E
Oroquieta, Phil. 37 C6 8 32N 123 44 E
Orós, Brazil 93 E11 6 15S 38 55W
Orotukan, Russia 27 C16 62 16N 151 42 E
Oroville, Calif., U.S.A. 84 F5 39 31N 121 33W
Oroville, Wash., U.S.A. 82 B4 48 56N 119 26W
Oroville, L., U.S.A. 84 F5 39 33N 121 29W
Orroroo, Australia 63 E2 32 43S 138 38 E
Orrville, U.S.A. 78 F3 40 50N 81 46W
Orsha, Belorussia 24 D5 54 30N 30 25 E
Orsk, Russia 24 D10 51 12N 58 34 E
Orşova, Romania 17 F11 44 41N 22 25 E
Ortegal, C., Spain 19 A2 43 43N 7 52W
Orthez, France 18 E3 43 29N 0 48W
Ortigueira, Spain 19 A2 43 40N 7 50W
Orting, U.S.A. 84 C4 47 6N 122 12W
Ortles, Italy 20 A4 46 31N 10 33 E
Ortón →, Bolivia 92 F5 10 50S 67 0W
Ortona, Italy 20 C6 42 21N 14 24 E
Orūmīyeh, Iran 44 B5 37 40N 45 0 E
Orūmīyeh, Daryācheh-ye, Iran 44 B5 37 50N 45 30 E
Oruro, Bolivia 92 G5 18 0S 67 9W
Oruzgān □, Afghan. 40 C5 33 30N 66 0 E
Orvieto, Italy 20 C5 42 43N 12 8 E
Orwell, U.S.A. 78 E4 41 32N 80 52W
Orwell →, U.K. 11 E9 52 2N 1 12 E
Oryakhovo, Bulgaria 21 C10 43 40N 23 57 E
Osa, Russia 24 C10 57 17N 55 26 E
Osa, Pen. de, Costa Rica 88 E3 8 0N 84 0W
Osage, Iowa, U.S.A. 80 D8 43 17N 92 49W
Osage, Wyo., U.S.A. 80 D2 43 59N 104 25W
Osage →, U.S.A. 80 F9 38 35N 91 57W
Osage City, U.S.A. 80 F7 38 38N 95 50W
Ōsaka, Japan 31 G7 34 40N 135 30 E
Osan, S. Korea 35 F14 37 11N 127 4 E
Osawatomie, U.S.A. 80 F7 38 31N 94 57W
Osborne, U.S.A. 80 F5 39 26N 98 42W
Osceola, Ark., U.S.A. 81 H10 35 42N 89 58W
Osceola, Iowa, U.S.A. 80 E8 41 2N 93 46W
Oscoda, U.S.A. 78 B1 44 26N 83 20W
Ösel = Saaremaa, Estonia 24 C3 58 30N 22 30 E
Osh, Kirghizia 26 E8 40 37N 72 49 E
Oshawa, Canada 70 D4 43 50N 78 50W
Oshkosh, Nebr., U.S.A. 80 E3 41 24N 102 21W
Oshkosh, Wis., U.S.A. 80 C10 44 1N 88 33W
Oshnovīyeh, Iran 44 B5 37 2N 45 6 E
Oshogbo, Nigeria 50 G5 7 48N 4 37 E
Oshtorīnān, Iran 45 C6 34 1N 48 38 E
Oshwe, Zaïre 52 E3 3 25S 19 28 E
Osijek, Croatia 21 B8 45 34N 18 41 E
Osipenko = Berdyansk, Ukraine 25 E6 46 45N 36 50 E
Osizweni, S. Africa 57 D5 27 49S 30 7 E
Oskaloosa, U.S.A. 80 E8 41 18N 92 39W
Oskarshamn, Sweden 9 H14 57 15N 16 27 E
Oskélanéo, Canada 70 C4 48 5N 75 15W
Öskemen = Ust-Kamenogorsk, Kazakhstan 26 E9 50 0N 82 36 E
Oslo, Norway 9 G11 59 55N 10 45 E
Oslob, Phil. 37 C6 9 31N 123 26 E
Oslofjorden, Norway 9 G11 59 20N 10 35 E
Osmanabad, India 40 K10 18 5N 76 10 E
Osmaniye, Turkey 25 G6 37 5N 36 10 E
Osnabrück, Germany 16 B4 52 16N 8 2 E
Osorio, Brazil 95 B5 29 53S 50 17W
Osorno, Chile 96 E2 40 25S 73 0W
Osoyoos, Canada 72 D5 49 0N 119 30W
Ospika →, Canada 72 B4 56 20N 124 0W
Osprey Reef, Australia 62 A4 13 52S 146 36 E
Oss, Neths. 15 C5 51 46N 5 32 E
Ossa, Mt., Australia 62 G4 41 52S 146 3 E
Óssa, Oros, Greece 21 E10 39 47N 22 42 E
Ossabaw I., U.S.A. 77 K5 31 50N 81 5W
Ossining, U.S.A. 79 E11 41 10N 73 55W
Ossipee, U.S.A. 79 C13 43 41N 71 7W
Ossokmanuan L., Canada 71 B7 53 25N 65 0W
Ossora, Russia 27 D17 59 20N 163 13 E
Ostend = Oostende, Belgium 15 C2 51 15N 2 54 E
Österdalälven →, Sweden 9 F12 61 30N 13 45 E
Östergötlands län □, Sweden 9 G13 58 35N 15 45 E
Östersund, Sweden 8 E13 63 10N 14 38 E
Østfold fylke □, Norway 9 G11 59 25N 11 25 E
Ostfriesische Inseln, Germany 16 B3 53 45N 7 15 E
Ostrava, Czech. 17 D9 49 51N 18 18 E
Ostróda, Poland 17 B9 53 42N 19 58 E
Ostrołęka, Poland 17 B10 53 4N 21 32 E
Ostrów Mazowiecka, Poland 17 B10 52 50N 21 51 E
Ostrów Wielkopolski, Poland 17 C8 51 36N 17 44 E
Ostrowiec-Świętokrzyski, Poland 17 C10 50 55N 21 22 E
Ōsumi-Kaikyō, Japan 31 J5 30 55N 131 0 E
Ōsumi-Shotō, Japan 31 J5 30 30N 130 0 E
Osuna, Spain 19 D3 37 14N 5 8W
Oswego, U.S.A. 79 C8 43 27N 76 31W
Oswestry, U.K. 10 E4 52 52N 3 3W
Otago □, N.Z. 59 L2 45 15S 170 0 E
Otago Harbour, N.Z. 59 L3 45 47S 170 42 E
Ōtake, Japan 31 G6 34 12N 132 13 E
Otaki, N.Z. 59 J5 40 45S 175 10 E
Otaru, Japan 30 C10 43 10N 141 0 E
Otaru-Wan = Ishikari-Wan, Japan 30 C10 43 25N 141 1 E
Otavalo, Ecuador 92 C3 0 13N 78 20W
Otavi, Namibia 56 B2 19 40S 17 24 E
Otchinjau, Angola 56 B1 16 30S 13 56 E
Othello, U.S.A. 82 C4 46 50N 119 10W
Otira Gorge, N.Z. 59 K3 42 53S 171 33 E
Otis, U.S.A. 80 E3 40 9N 102 58W
Otjiwarongo, Namibia 56 C2 20 30S 16 33 E
Otoineppu, Japan 30 B11 44 44N 142 16 E
Otorohanga, N.Z. 59 H5 38 12S 175 14 E
Otoskwin →, Canada 70 B2 52 13N 88 6W
Otosquen, Canada 73 C8 53 17N 102 1W
Otranto, Italy 21 D8 40 9N 18 28 E
Otranto, C. d', Italy 21 D8 40 7N 18 30 E
Otranto, Str. of, Italy 21 D8 40 15N 18 40 E
Otse, S. Africa 56 D4 25 2S 25 45 E
Ōtsu, Japan 31 G7 35 0N 135 50 E
Ōtsuki, Japan 31 G9 35 36N 138 57 E
Ottawa = Outaouais →, Canada 70 C5 45 27N 74 8W
Ottawa, Canada 70 C4 45 27N 75 42W
Ottawa, Ill., U.S.A. 80 E10 41 21N 88 51W
Ottawa, Kans., U.S.A. 80 F7 38 37N 95 16W
Ottawa Is., Canada 69 C11 59 35N 80 10W
Otter L., Canada 73 B8 55 35N 104 39W
Otter Rapids, Ont., Canada 70 B3 50 11N 81 39W
Otter Rapids, Sask., Canada 73 B8 55 38N 104 44W
Otterville, Canada 78 D4 42 55N 80 36W
Otto Beit Bridge, Zimbabwe 55 F2 15 59S 28 56 E
Ottosdal, S. Africa 56 D4 26 46S 25 59 E
Ottoshoop, S. Africa 56 D4 25 45S 25 58 E
Ottumwa, U.S.A. 80 E8 41 1N 92 25W
Oturkpo, Nigeria 50 G6 7 16N 8 8 E
Otway, B., Chile 96 G2 53 30S 74 0W
Otway, C., Australia 63 F3 38 52S 143 30 E
Otwock, Poland 17 B10 52 5N 21 20 E
Ou →, Laos 38 B4 20 4N 102 13 E
Ou Neua, Laos 38 A3 22 18N 101 48 E
Ou-Sammyaku, Japan 30 E10 39 20N 140 35 E
Ouachita →, U.S.A. 81 K9 31 38N 91 49W
Ouachita, L., U.S.A. 81 H8 34 34N 93 12W
Ouachita Mts., U.S.A. 81 H7 34 40N 94 25W
Ouadâne, Mauritania 50 D2 20 50N 11 40W
Ouadda, C.A.R. 51 G9 8 15N 22 20 E
Ouagadougou, Burkina Faso 50 F4 12 25N 1 30W
Ouahran = Oran, Algeria 50 A4 35 45N 0 39W
Ouallene, Algeria 50 D5 24 41N 1 11 E
Ouanda Djallé, C.A.R. 51 G9 8 55N 22 53 E
Ouango, C.A.R. 52 D4 4 19N 22 30 E
Ouargla, Algeria 50 B6 31 59N 5 16 E
Ouarzazate, Morocco 50 B3 30 55N 6 50W
Oubangi →, Zaïre 52 E3 0 30S 17 50 E
Ouddorp, Neths. 15 C3 51 50N 3 57 E
Oude Rijn →, Neths. 15 B4 52 12N 4 24 E
Oudenaarde, Belgium 15 D3 50 50N 3 37 E
Oudtshoorn, S. Africa 56 E3 33 35S 22 14 E
Ouessant, I. d', France 18 B1 48 28N 5 6W
Ouesso, Congo 52 D3 1 37N 16 5 E
Ouest, Pte., Canada 71 C7 49 52N 64 40W
Ouezzane, Morocco 50 B3 34 51N 5 35W
Ouidah, Benin 50 G5 6 25N 2 0 E
Oujda, Morocco 50 B4 34 41N 1 55W
Oujeft, Mauritania 50 D2 20 2N 13 0W
Ouled Djellal, Algeria 50 B6 34 28N 5 2 E
Oulu, Finland 8 D18 65 1N 25 29 E
Oulu □ = Oulun lääni □, Finland 8 D19 64 36N 27 20 E
Oulujärvi, Finland 8 D19 64 25N 27 15 E
Oulujoki →, Finland 8 D18 65 1N 25 30 E
Oulun lääni □, Finland 8 D19 64 36N 27 20 E
Oum Chalouba, Chad 51 E9 15 48N 20 46 E
Ounguati, Namibia 56 C2 22 0S 15 46 E
Ounianga-Kébir, Chad 51 E9 19 4N 20 29 E
Ounianga Sérir, Chad 51 E9 18 54N 20 51 E
Our →, Lux. 15 E6 49 55N 6 5 E
Ouray, U.S.A. 83 G10 38 1N 107 40W
Ourense = Orense, Spain 19 A2 42 19N 7 55W
Ouricuri, Brazil 93 E10 7 53S 40 5W
Ourinhos, Brazil 95 A6 23 0S 49 54W
Ouro Fino, Brazil 95 A6 22 16S 46 25W
Ouro Prêto, Brazil 95 A7 20 20S 43 30W
Ourthe →, Belgium 15 D5 50 29N 5 35 E
Ouse, Australia 62 G4 42 38S 146 42 E
Ouse →, E. Susx., U.K. 11 G8 50 43N 0 3 E
Ouse →, N. Yorks., U.K. 10 C8 54 3N 0 7 E
Outaouais →, Canada 70 C5 45 27N 74 8W
Outardes →, Canada 71 C6 49 24N 69 30W
Outer Hebrides, U.K. 12 D1 57 30N 7 40W
Outer I., Canada 71 B8 51 10N 58 35W
Outjo, Namibia 56 C2 20 5S 16 7 E
Outlook, Canada 73 C7 51 30N 107 0W
Outlook, U.S.A. 80 A2 48 53N 104 47W
Ouyen, Australia 63 F3 35 1S 142 22 E
Ovalau, Fiji 59 C8 17 40S 178 48 E
Ovalle, Chile 94 C1 30 33S 71 18W
Ovar, Portugal 19 B1 40 51N 8 40W
Overflakkee, Neths. 15 C4 51 44N 4 10 E
Overijssel □, Neths. 15 B6 52 25N 6 35 E
Overpelt, Belgium 15 C5 51 12N 5 20 E
Overton, U.S.A. 85 J12 36 33N 114 27W
Övertorneå, Sweden 8 C17 66 23N 23 38 E
Ovid, U.S.A. 80 E3 40 58N 102 23W
Oviedo, Spain 19 A3 43 25N 5 50W
Övör Hangay □, Mongolia 34 B2 45 0N 102 30 E
Ovruch, Ukraine 25 D3 51 25N 28 45 E
Owaka, N.Z. 59 M2 46 27S 169 40 E
Owase, Japan 31 G8 34 7N 136 12 E
Owatonna, U.S.A. 80 C8 44 5N 93 14W
Owbeh, Afghan. 40 B3 34 28N 63 10 E
Owego, U.S.A. 79 D8 42 6N 76 16W
Owen Falls, Uganda 54 B3 0 30N 33 5 E
Owen Sound, Canada 70 D3 44 35N 80 55W
Owendo, Gabon 52 D1 0 17N 9 30 E
Owens →, U.S.A. 84 J9 36 32N 117 59W
Owens L., U.S.A. 85 J9 36 26N 117 57W
Owensboro, U.S.A. 76 G2 37 46N 87 7W
Owensville, U.S.A. 80 F9 38 21N 91 30W
Owl →, Canada 73 B10 57 51N 92 44W
Owo, Nigeria 50 G6 7 10N 5 39 E
Owosso, U.S.A. 76 D3 43 0N 84 10W
Owyhee, U.S.A. 82 F5 41 57N 116 6W
Owyhee →, U.S.A. 82 E5 43 49N 117 2W
Owyhee, L., U.S.A. 82 E5 43 38N 117 14W
Ox Mts., Ireland 13 B3 54 6N 9 0W
Oxelösund, Sweden 9 G14 58 43N 17 15 E
Oxford, N.Z. 59 K4 43 18S 172 11 E
Oxford, U.K. 11 F6 51 45N 1 15W
Oxford, Miss., U.S.A. 81 H10 34 22N 89 31W
Oxford, N.C., U.S.A. 77 G6 36 19N 78 35W
Oxford, Ohio, U.S.A. 76 F3 39 31N 84 45W
Oxford L., Canada 73 C9 54 51N 95 37W
Oxfordshire □, U.K. 11 F6 51 45N 1 15W
Oxley, Australia 63 E3 34 11S 144 6 E
Oxnard, U.S.A. 85 L7 34 12N 119 11W
Oxus = Amudarya →, Uzbekistan 26 E6 43 40N 59 0 E
Oya, Malaysia 36 D4 2 55N 111 55 E
Oyama, Japan 31 F9 36 18N 139 48 E
Oyem, Gabon 52 D2 1 34N 11 31 E
Oyen, Canada 73 C6 51 22N 110 28W
Oykel →, U.K. 12 D4 57 55N 4 26W
Oymyakon, Russia 27 C15 63 25N 142 44 E
Oyo, Nigeria 50 G5 7 46N 3 56 E
Oyster Bay, U.S.A. 79 F11 40 52N 73 32W
Ōyübari, Japan 30 C11 43 1N 142 5 E
Ozamiz, Phil. 37 C6 8 15N 123 50 E
Ozark, Ala., U.S.A. 77 K3 31 28N 85 39W
Ozark, Ark., U.S.A. 81 H8 35 29N 93 50W
Ozark, Mo., U.S.A. 81 G8 37 1N 93 12W
Ozark Plateau, U.S.A. 81 G9 37 20N 91 40W
Ozarks, L. of the, U.S.A. 80 F8 38 12N 92 38W
Ozette L., U.S.A. 84 B2 48 6N 124 38W
Ozona, U.S.A. 81 K4 30 43N 101 12W
Ozuluama, Mexico 87 C5 21 40N 97 50W

P

P.K. le Roux Dam, S. Africa 56 E3 30 4S 24 40 E
Pa-an, Burma 41 L20 16 51N 97 40 E
Pa Mong Dam, Thailand 38 D4 18 0N 102 22 E
Paamiut = Frederikshåb, Greenland 4 C5 62 0N 49 43W
Paarl, S. Africa 56 E2 33 45S 18 56 E
Paatsi →, Russia 8 B20 68 55N 29 0 E
Paauilo, U.S.A. 74 H17 20 2N 155 22W
Pab Hills, Pakistan 42 F2 26 30N 66 45 E
Pabna, Bangla. 41 G16 24 1N 89 18 E
Pabo, Uganda 54 B3 3 1N 32 10 E
Pacaja →, Brazil 93 D8 1 56S 50 50W
Pacaraima, Sierra, Venezuela 92 C6 4 0N 62 30W
Pacasmayo, Peru 92 E3 7 20S 79 35W
Pachhar, India 42 G7 24 40N 77 42 E
Pachpadra, India 40 G8 25 58N 72 10 E
Pachuca, Mexico 87 C5 20 10N 98 40W
Pacific, Canada 72 C3 54 48N 128 28W
Pacific-Antarctic Ridge, Pac. Oc. 65 M16 43 0S 115 0W
Pacific Grove, U.S.A. 83 H3 36 38N 121 56W
Pacific Ocean, Pac. Oc. 65 G14 10 0N 140 0W
Pacifica, U.S.A. 84 H4 37 36N 122 30W
Pacitan, Indonesia 37 H14 8 12S 111 7 E
Packwood, U.S.A. 84 D5 46 36N 121 40W
Padaido, Kepulauan, Indonesia 37 E9 1 5S 138 0 E
Padang, Indonesia 36 E2 1 0S 100 20 E
Padangpanjang, Indonesia 36 E2 0 40S 100 20 E
Padangsidempuan, Indonesia 36 D1 1 30N 99 15 E
Paddockwood, Canada 73 C7 53 30N 105 30W
Paderborn, Germany 16 C4 51 42N 8 44 E
Padloping Island, Canada 69 B13 67 0N 62 50W
Pádova, Italy 20 B4 45 24N 11 52 E
Padra, India 42 H5 22 15N 73 7 E
Padrauna, India 43 F10 26 54N 83 59 E
Padre I., U.S.A. 81 M6 27 10N 97 25W
Padstow, U.K. 11 G3 50 33N 4 57W
Padua = Pádova, Italy 20 B4 45 24N 11 52 E
Paducah, Ky., U.S.A. 76 G1 37 5N 88 37W
Paducah, Tex., U.S.A. 81 H4 34 1N 100 18W
Paengnyong-do, S. Korea 35 F13 37 57N 124 40 E
Paeroa, N.Z. 59 G5 37 23S 175 41 E
Pafúri, Mozam. 57 C5 22 28S 31 17 E
Pag, Croatia 20 B6 44 30N 14 50 E
Pagadian, Phil. 37 C6 7 55N 123 30 E
Pagai Selatan, P., Indonesia 36 E2 3 0S 100 15 E
Pagai Utara, Indonesia 36 E2 2 35S 100 0 E
Pagalu = Annobón, Atl. Oc. 49 G4 1 25S 5 36 E
Pagastikós Kólpos, Greece 21 E10 39 15N 23 0 E
Pagatan, Indonesia 36 E5 3 33S 115 59 E
Page, Ariz., U.S.A. 83 H8 36 57N 111 27W
Page, N. Dak., U.S.A. 80 B6 47 10N 97 34W
Pago Pago, Amer. Samoa 59 B13 14 16S 170 43W
Pagosa Springs, U.S.A. 83 H10 37 16N 107 1W
Pagwa River, Canada 70 B2 50 2N 85 14W
Pahala, U.S.A. 74 J17 19 12N 155 29W
Pahang →, Malaysia 39 L4 3 30N 103 9 E
Pahiatua, N.Z. 59 J5 40 27S 175 50 E
Pahokee, U.S.A. 77 M5 26 50N 80 40W
Pahrump, U.S.A. 85 J11 36 12N 115 59W
Pahute Mesa, U.S.A. 84 H10 37 20N 116 45W
Pai, Thailand 38 C2 19 19N 98 27 E
Paia, U.S.A. 74 H16 20 54N 156 22W
Paicines, U.S.A. 84 J5 36 44N 121 17W
Paignton, U.K. 11 G4 50 26N 3 33W
Päijänne, Finland 9 F18 61 30N 25 30 E
Painan, Indonesia 36 E2 1 21S 100 34 E
Painesville, U.S.A. 78 E3 41 43N 81 15W
Paint Hills = Nouveau Comptoir, Canada 70 B4 53 0N 78 49W
Paint L., Canada 73 B9 55 28N 97 57W
Paint Rock, U.S.A. 81 K5 31 31N 99 55W
Painted Desert, U.S.A. 83 J8 36 0N 111 0W
Paintsville, U.S.A. 76 G4 37 49N 82 48W
País Vasco □, Spain 19 A4 42 50N 2 45W
Paisley, Canada 78 B3 44 18N 81 16W
Paisley, U.K. 12 F4 55 51N 4 27W
Paisley, U.S.A. 82 E3 42 42N 120 32W
Paita, Peru 92 E2 5 11S 81 9W
Pak Lay, Laos 38 C3 18 15N 101 27 E
Pak Phanang, Thailand 39 H3 8 21N 100 12 E
Pak Sane, Laos 38 C4 18 22N 103 39 E
Pak Song, Laos 38 E6 15 11N 106 14 E
Pak Suong, Laos 38 C4 19 58N 102 15 E
Pakaraima Mts., Guyana 92 B6 6 0N 60 0W
Pákhnes, Greece 23 D6 35 16N 24 4 E
Pakistan ■, Asia 42 E3 30 0N 70 0 E
Pakistan, East = Bangladesh ■, Asia 41 H17 24 0N 90 0 E
Pakkading, Laos 38 C4 18 19N 103 59 E
Pakokku, Burma 41 J19 21 20N 95 0 E
Pakpattan, Pakistan 42 D5 30 25N 73 27 E
Pakse, Laos 38 E5 15 5N 105 52 E
Paktīā □, Afghan. 40 C6 33 0N 69 15 E
Pakwach, Uganda 54 B3 2 28N 31 27 E
Pala, Chad 51 G8 9 25N 15 5 E
Pala, U.S.A. 85 M9 33 22N 117 5W
Pala, Zaïre 54 D2 6 45S 29 30 E
Palabek, Uganda 54 B3 3 22N 32 33 E
Palacios, U.S.A. 81 L6 28 42N 96 13W
Palagruža, Croatia 20 C7 42 24N 16 15 E
Palaiókastron, Greece 23 D8 35 12N 26 15 E
Palaiokhóra, Greece 23 D5 35 16N 23 39 E
Palam, India 40 K10 19 0N 77 0 E
Palamós, Spain 19 B7 41 50N 3 10 E
Palampur, India 42 C7 32 10N 76 30 E
Palana, Australia 62 F4 39 45S 147 55 E
Palana, Russia 27 D16 59 10N 159 59 E
Palanan, Phil. 37 A6 17 8N 122 29 E
Palanan Pt., Phil. 37 A6 17 17N 122 30 E
Palandri, Pakistan 43 C5 33 42N 73 40 E
Palangkaraya, Indonesia 36 E4 2 16S 113 56 E
Palani Hills, India 40 P10 10 14N 77 33 E
Palanpur, India 42 G5 24 10N 72 25 E
Palapye, Botswana 56 C4 22 30S 27 7 E
Palas, Pakistan 43 B5 35 4N 73 14 E
Palatka, Russia 27 C16 60 6N 150 54 E
Palatka, U.S.A. 77 L5 29 39N 81 38W
Palau = Belau ■, Pac. Oc. 64 G5 7 30N 134 30 E
Palawan, Phil. 36 C5 9 30N 118 30 E
Palayankottai, India 40 Q10 8 45N 77 45 E
Paleleh, Indonesia 37 D6 1 10N 121 50 E
Palembang, Indonesia 36 E2 3 0S 104 50 E
Palencia, Spain 19 A3 42 1N 4 34W
Paleokastrítsa, Greece 23 A3 39 40N 19 41 E
Paleometokho, Cyprus 23 D12 35 7N 33 11 E
Palermo, Italy 20 E5 38 8N 13 20 E
Palermo, U.S.A. 82 G3 39 26N 121 33W
Palestine, Asia 47 D4 32 0N 35 0 E
Palestine, U.S.A. 81 K7 31 46N 95 38W
Paletwa, Burma 41 J18 21 10N 92 50 E
Palghat, India 40 P10 10 46N 76 42 E
Palgrave, Mt., Australia 60 D2 23 22S 115 58 E
Pali, India 42 G5 25 50N 73 20 E
Palisade, U.S.A. 80 E4 40 21N 101 7W
Palitana, India 42 J4 21 32N 71 49 E
Palizada, Mexico 87 D6 18 18N 92 8W
Palk Bay, Asia 40 Q11 9 30N 79 15 E
Palk Strait, Asia 40 Q11 10 0N 79 45 E
Palkānah, Iraq 44 C5 35 49N 44 26 E
Palla Road = Dinokwe, Botswana 56 C4 23 29S 26 37 E
Pallisa, Uganda 54 B3 1 12N 33 43 E
Pallu, India 42 E6 28 59N 74 14 E
Palm Beach, U.S.A. 77 M6 26 43N 80 2W
Palm Desert, U.S.A. 85 M10 33 43N 116 22W
Palm Is., Australia 62 B4 18 40S 146 35 E
Palm Springs, U.S.A. 85 M10 33 50N 116 33W
Palma, Mozam. 55 E5 10 46S 40 29 E
Palma →, Brazil 93 F9 12 33S 47 52W
Palma, B. de, Spain 22 B9 39 30N 2 39 E
Palma de Mallorca, Spain 22 B9 39 35N 2 39 E
Palma Soriano, Cuba 88 B4 20 15N 76 0W
Palmares, Brazil 93 E11 8 41S 35 28W
Palmas, Brazil 95 B5 26 29S 52 0W
Palmas, C., Liberia 50 H3 4 27N 7 46W
Pálmas, G. di, Italy 20 E3 39 0N 8 30 E
Palmdale, U.S.A. 85 L8 34 35N 118 7W
Palmeira dos Índios, Brazil 93 E11 9 25S 36 37W
Palmeirinhas, Pta. das, Angola 52 F2 9 2S 12 57 E
Palmer, U.S.A. 68 B5 61 36N 149 7W
Palmer →, Australia 62 B3 16 0S 142 26 E
Palmer Arch., Antarctica 5 C17 64 15S 65 0W
Palmer Lake, U.S.A. 80 F2 39 7N 104 55W
Palmer Land, Antarctica 5 D18 73 0S 63 0W
Palmerston, Canada 78 C4 43 50N 80 51W
Palmerston, N.Z. 59 L3 45 29S 170 43 E
Palmerston North, N.Z. 59 J5 40 21S 175 39 E
Palmerton, U.S.A. 79 F9 40 48N 75 37W
Palmetto, U.S.A. 77 M4 27 31N 82 34W
Palmi, Italy 20 E6 38 21N 15 51 E
Palmira, Argentina 94 C2 32 59S 68 34W
Palmira, Colombia 92 C3 3 32N 76 16W
Palmyra = Tudmur, Syria 44 C3 34 36N 38 15 E
Palmyra, Mo., U.S.A. 80 F9 39 48N 91 32W
Palmyra, N.Y., U.S.A. 78 C7 43 5N 77 18W
Palmyra Is., Pac. Oc. 65 G11 5 52N 162 5W
Palo Alto, U.S.A. 83 H2 37 27N 122 10W
Palo Verde, U.S.A. 85 M12 33 26N 114 44W
Palopo, Indonesia 37 E6 3 0S 120 16 E
Palos, C. de, Spain 19 D5 37 38N 0 40W
Palos Verdes, U.S.A. 85 M8 33 48N 118 23W
Palos Verdes, Pt., U.S.A. 85 M8 33 43N 118 26W
Palouse, U.S.A. 82 C5 46 55N 117 4W
Palparara, Australia 62 C3 24 47S 141 28 E
Palu, Indonesia 37 E5 1 0S 119 52 E
Palu, Turkey 25 G7 38 45N 40 0 E
Paluan, Phil. 37 B6 13 26N 120 29 E
Palwal, India 42 E7 28 8N 77 19 E
Pama, Burkina Faso 50 F5 11 19N 0 44 E
Pamanukan, Indonesia 37 G12 6 16S 107 49 E
Pamekasan, Indonesia 37 G15 7 10S 113 28 E
Pamirs, Tajikistan 26 F8 37 40N 73 0 E
Pamlico →, U.S.A. 77 H7 35 20N 76 28W
Pamlico Sd., U.S.A. 77 H8 35 20N 76 0W
Pampa, U.S.A. 81 H4 35 32N 100 58W
Pampa de las Salinas, Argentina 94 C2 32 1S 66 58W

Pelvoux, Massif de

Pelvoux, Massif de, *France* **18 D7** 44 52N 6 20 E
Pemalang, *Indonesia* **37 G13** 6 53S 109 23 E
Pematangsiantar, *Indonesia* **36 D1** 2 57N 99 5 E
Pemba, *Mozam.* **55 E5** 12 58S 40 30 E
Pemba, *Zambia* **55 F2** 16 30S 27 28 E
Pemba Channel, *Tanzania* **54 D4** 5 0S 39 37 E
Pemba I., *Tanzania* **54 D4** 5 0S 39 45 E
Pemberton, *Australia* **61 F2** 34 30S 116 0 E
Pemberton, *Canada* **72 C4** 50 25N 122 50W
Pembina, *U.S.A.* **73 D9** 48 58N 97 15W
Pembina →, *U.S.A.* **73 D9** 48 58N 97 14W
Pembine, *U.S.A.* **76 C2** 45 38N 87 59W
Pembino, *U.S.A.* **80 A6** 48 58N 97 15W
Pembroke, *Canada* **70 C4** 45 50N 77 7W
Pembroke, *U.K.* **11 F3** 51 41N 4 57W
Pembroke, *U.S.A.* **77 J5** 32 8N 81 37W
Pen-y-Ghent, *U.K.* **10 C5** 54 10N 2 15W
Peña de Francia, Sierra de, *Spain* **19 B2** 40 32N 6 10W
Peñalara, Pico, *Spain* **19 B4** 40 51N 3 57W
Penang = Pinang, *Malaysia* **39 K3** 5 25N 100 15 E
Penápolis, *Brazil* **95 A6** 21 30S 50 0W
Peñarroya-Pueblonuevo, *Spain* **19 C3** 38 19N 5 16W
Peñas, C. de, *Spain* **19 A3** 43 42N 5 52W
Penas, G. de, *Chile* **96 F2** 47 0S 75 0W
Peñas del Chache, *Canary Is.* **22 E6** 29 6N 13 33W
Pench'i = Benxi, *China* **35 D12** 41 20N 123 48 E
Pend Oreille →, *U.S.A.* **82 B5** 49 4N 117 37W
Pend Oreille L., *U.S.A.* **82 C5** 48 10N 116 21W
Pendembu, *S. Leone* **50 G2** 9 7N 12 14W
Pender B., *Australia* **60 C3** 16 45S 122 42 E
Pendleton, *Calif., U.S.A.* **85 M9** 33 16N 117 23W
Pendleton, *Oreg., U.S.A.* **82 D4** 45 40N 118 47W
Penedo, *Brazil* **93 F11** 10 15S 36 36W
Penetanguishene, *Canada* **70 D4** 44 50N 79 55W
Pengalengan, *Indonesia* **37 G12** 7 9S 107 30 E
Penge, *Kasai Or., Zaïre* **54 D1** 5 30S 24 33 E
Penge, *Kivu, Zaïre* **54 C2** 4 27S 28 25 E
Penglai, *China* **35 F11** 37 48N 120 42 E
Penguin, *Australia* **62 G4** 41 8S 146 6 E
Penhalonga, *Zimbabwe* **55 F3** 18 52S 32 40 E
Peniche, *Portugal* **19 C1** 39 19N 9 22W
Penicuik, *U.K.* **12 F5** 55 50N 3 14W
Penida, *Indonesia* **36 F5** 8 45S 115 30 E
Peninsular Malaysia □, *Malaysia* **39 L4** 4 0N 102 0 E
Penmarch, Pte. de, *France* **18 C1** 47 48N 4 22W
Penn Hills, *U.S.A.* **78 F5** 40 28N 79 52W
Penn Yan, *U.S.A.* **78 D7** 42 40N 77 3W
Pennant, *Canada* **73 C7** 50 32N 108 14W
Penner →, *India* **40 M12** 14 35N 80 10 E
Pennines, *U.K.* **10 C5** 54 50N 2 20W
Pennington, *U.S.A.* **84 F5** 39 15N 121 47W
Pennsylvania □, *U.S.A.* **76 E6** 40 45N 77 30W
Penny, *Canada* **72 C4** 53 51N 121 20W
Penola, *Australia* **63 F3** 37 25S 140 48 E
Penong, *Australia* **61 F5** 31 56S 133 1 E
Penonomé, *Panama* **88 E3** 8 31N 80 21W
Penrith, *Australia* **63 E5** 33 43S 150 38 E
Penrith, *U.K.* **10 C5** 54 40N 2 45W
Pensacola, *U.S.A.* **77 K2** 30 25N 87 13W
Pensacola Mts., *Antarctica* **5 E1** 84 0S 40 0W
Pense, *Canada* **73 C8** 50 25N 104 59W
Penshurst, *Australia* **63 F3** 37 49S 142 20 E
Penticton, *Canada* **72 D5** 49 30N 119 38W
Pentland, *Australia* **62 C4** 20 32S 145 25 E
Pentland Firth, *U.K.* **12 C5** 58 43N 3 10W
Pentland Hills, *U.K.* **12 F5** 55 48N 3 25W
Penylan L., *Canada* **73 A7** 61 50N 106 20W
Penza, *Russia* **24 D8** 53 15N 45 5 E
Penzance, *U.K.* **11 G2** 50 7N 5 32W
Penzhino, *Russia* **27 C17** 63 30N 167 55 E
Penzhinskaya Guba, *Russia* **27 C17** 61 30N 163 0 E
Peoria, *Ariz., U.S.A.* **83 K7** 33 35N 112 14W
Peoria, *Ill., U.S.A.* **80 E10** 40 42N 89 36W
Pera Hd., *Australia* **62 A3** 12 55S 141 37 E
Perabumilih, *Indonesia* **36 E2** 3 27S 104 15 E
Pérama, *Kérkira, Greece* **23 A3** 39 34N 19 54 E
Pérama, *Kríti, Greece* **23 D6** 35 20N 24 40 E
Percé, *Canada* **71 C7** 48 31N 64 13W
Perche, *France* **18 B4** 48 31N 1 1 E
Percival Lakes, *Australia* **60 D4** 21 25S 125 0 E
Percy Is., *Australia* **62 C5** 21 39S 150 16 E
Perdido, Mte., *Spain* **19 A6** 42 40N 0 5 E
Perdu, Mt. = Perdido, Mte., *Spain* **19 A6** 42 40N 0 5 E
Pereira, *Colombia* **92 C3** 4 49N 75 43W
Perekerten, *Australia* **63 E3** 34 55S 143 40 E
Perenjori, *Australia* **61 E2** 29 26S 116 16 E
Pereyaslav Khmelnitskiy, *Ukraine* **25 D5** 50 3N 31 28 E
Pérez, I., *Mexico* **87 C7** 22 24N 89 42W
Pergamino, *Argentina* **94 C3** 33 52S 60 30W
Perham, *U.S.A.* **80 B7** 46 36N 95 34W
Perhentian, Kepulauan, *Malaysia* **39 K4** 5 54N 102 42 E
Péribonca →, *Canada* **71 C5** 48 45N 72 5W
Péribonca, L., *Canada* **71 B5** 50 1N 71 10W
Perico, *Argentina* **94 A2** 24 20S 65 5W
Pericos, *Mexico* **86 B3** 25 3N 107 42W
Périgord, *France* **18 D4** 45 0N 0 40 E
Périgueux, *France* **18 D4** 45 10N 0 42 E
Perijá, Sierra de, *Colombia* **92 B4** 9 30N 73 3W
Peristerona →, *Cyprus* **23 D12** 35 8N 33 5 E
Perlas, Arch. de las, *Panama* **88 E4** 8 41N 79 7W
Perlas, Punta de, *Nic.* **88 D3** 12 30N 83 30W
Perm, *Russia* **24 C10** 58 0N 56 10 E
Pernambuco = Recife, *Brazil* **93 E12** 8 0S 35 0W
Pernambuco □, *Brazil* **93 E11** 8 0S 37 0W
Pernatty Lagoon, *Australia* **63 E2** 31 30S 137 12 E
Peron, C., *Australia* **61 E1** 25 30S 113 30 E
Peron Is., *Australia* **60 B5** 13 9S 130 4 E
Peron Pen., *Australia* **61 E1** 26 0S 113 10 E
Perow, *Canada* **72 C3** 54 35N 126 10W
Perpendicular Pt., *Australia* **63 E5** 31 37S 152 52 E
Perpignan, *France* **18 E5** 42 42N 2 53 E
Perris, *U.S.A.* **85 M9** 33 47N 117 14W
Perry, *Fla., U.S.A.* **77 K4** 30 7N 83 35W
Perry, *Ga., U.S.A.* **77 J4** 32 28N 83 44W
Perry, *Iowa, U.S.A.* **80 E7** 41 51N 94 6W
Perry, *Maine, U.S.A.* **77 C12** 44 58N 67 5W
Perry, *Okla., U.S.A.* **81 G6** 36 17N 97 14W
Perryton, *U.S.A.* **81 G4** 36 24N 100 48W
Perryville, *U.S.A.* **81 G10** 37 43N 89 52W
Persia = Iran ■, *Asia* **45 C7** 33 0N 53 0 E
Persian Gulf = Gulf, The, *Asia* **45 E6** 27 0N 50 0 E
Perth, *Australia* **61 F2** 31 57S 115 52 E
Perth, *Canada* **70 D4** 44 55N 76 15W
Perth, *U.K.* **12 E5** 56 24N 3 27W
Perth Amboy, *U.S.A.* **79 F10** 40 31N 74 16W
Peru, *Ill., U.S.A.* **80 E10** 41 20N 89 8W
Peru, *Ind., U.S.A.* **76 E2** 40 45N 86 4W
Peru ■, *S. Amer.* **92 E3** 4 0S 75 0W
Peru-Chile Trench, *Pac. Oc.* **65 K20** 20 0S 72 0W
Perúgia, *Italy* **20 C5** 43 6N 12 24 E
Pervomaysk, *Ukraine* **25 E5** 48 10N 30 46 E
Pervouralsk, *Russia* **24 C10** 56 55N 59 45 E
Pes, Pta. del, *Spain* **22 C7** 38 46N 1 26 E
Pésaro, *Italy* **20 C5** 43 55N 12 53 E
Pescara, *Italy* **20 C6** 42 28N 14 13 E
Peshawar, *Pakistan* **42 B4** 34 2N 71 37 E
Peshtigo, *U.S.A.* **76 C2** 45 4N 87 46W
Pesqueira, *Brazil* **93 E11** 8 20S 36 42W
Petah Tiqwa, *Israel* **47 C3** 32 6N 34 53 E
Petaling Jaya, *Malaysia* **39 L3** 3 4N 101 42 E
Petaloudhes, *Greece* **23 C10** 36 18N 28 5 E
Petaluma, *U.S.A.* **84 G4** 38 14N 122 39W
Petange, *Lux.* **15 E5** 49 33N 5 55 E
Petatlán, *Mexico* **86 D4** 17 31N 101 16W
Petauke, *Zambia* **55 E3** 14 14S 31 20 E
Petawawa, *Canada* **70 C4** 45 54N 77 17W
Petén Itzá, L., *Guatemala* **88 C2** 16 58N 89 50W
Peter I.s Øy, *Antarctica* **5 C16** 69 0S 91 0W
Peter Pond L., *Canada* **73 B7** 55 55N 108 44W
Peterbell, *Canada* **70 C3** 48 36N 83 21W
Peterborough, *Australia* **63 E2** 32 58S 138 51 E
Peterborough, *Canada* **69 D12** 44 20N 78 20W
Peterborough, *U.K.* **11 E7** 52 35N 0 14W
Peterborough, *U.S.A.* **79 D13** 42 53N 71 57W
Peterhead, *U.K.* **12 D7** 57 30N 1 49W
Petermann Bjerg, *Greenland* **66 B17** 73 7N 28 25W
Petersburg, *Alaska, U.S.A.* **72 B2** 56 48N 132 58W
Petersburg, *Ind., U.S.A.* **76 F2** 38 30N 87 17W
Petersburg, *Va., U.S.A.* **76 G7** 37 14N 77 24W
Petersburg, *W. Va., U.S.A.* **76 F6** 39 1N 79 5W
Petford, *Australia* **62 B3** 17 20S 144 58 E
Petit Bois I., *U.S.A.* **77 K1** 30 12N 88 26W
Petit-Cap, *Canada* **71 C7** 49 3N 64 30W
Petit Goâve, *Haiti* **89 C5** 18 27N 72 51W
Petit Lac Manicouagan, *Canada* **71 B6** 51 25N 67 40W
Petitcodiac, *Canada* **71 C6** 45 57N 65 11W
Petite Baleine →, *Canada* **70 A4** 56 0N 76 45W
Petite Saguenay, *Canada* **71 C5** 48 15N 70 4W
Petitsikapau, L., *Canada* **71 B6** 54 37N 66 25W
Petlad, *India* **42 H5** 22 30N 72 45 E
Peto, *Mexico* **87 C7** 20 10N 88 53W
Petone, *N.Z.* **59 J5** 41 13S 174 53 E
Petoskey, *U.S.A.* **76 C3** 45 22N 84 57W
Petra, *Jordan* **47 E4** 30 20N 35 22 E
Petra, *Spain* **22 B10** 39 37N 3 6 E
Petra, Ostrova, *Russia* **4 B13** 76 15N 118 30 E
Petra Velikogo, Zaliv, *Russia* **30 C5** 42 40N 132 0 E
Petrich, *Bulgaria* **21 D10** 41 24N 23 13 E
Petrograd = Sankt-Peterburg, *Russia* **24 C5** 59 55N 30 20 E
Petrolândia, *Brazil* **93 E11** 9 5S 38 20W
Petrolia, *Canada* **70 D3** 42 54N 82 9W
Petrolina, *Brazil* **93 E10** 9 24S 40 30W
Petropavlovsk, *Kazakhstan* **26 D7** 54 53N 69 13 E
Petropavlovsk-Kamchatskiy, *Russia* **27 D16** 53 3N 158 43 E
Petrópolis, *Brazil* **95 A7** 22 33S 43 9W
Petroşeni, *Romania* **17 F11** 45 28N 23 20 E
Petrovaradin, *Serbia, Yug.* **21 B8** 45 16N 19 55 E
Petrovsk, *Russia* **24 D8** 52 22N 45 19 E
Petrovsk-Zabaykalskiy, *Russia* **27 D11** 51 20N 108 55 E
Petrozavodsk, *Russia* **24 B5** 61 41N 34 20 E
Petrus Steyn, *S. Africa* **57 D4** 27 38S 28 8 E
Petrusburg, *S. Africa* **56 D4** 29 4S 25 26 E
Peumo, *Chile* **94 C1** 34 21S 71 12W
Peureulak, *Indonesia* **36 D1** 4 48N 97 45 E
Pevek, *Russia* **27 C18** 69 41N 171 19 E
Pforzheim, *Germany* **16 D4** 48 53N 8 43 E
Phagwara, *India* **40 D9** 31 10N 75 40 E
Phaistós, *Greece* **23 D6** 35 2N 24 50 E
Phala, *Botswana* **56 C4** 23 45S 26 50 E
Phalera = Phulera, *India* **42 F6** 26 52N 75 16 E
Phalodi, *India* **42 F5** 27 12N 72 24 E
Phan, *Thailand* **38 C2** 19 28N 99 43 E
Phan Rang, *Vietnam* **39 G7** 11 34N 109 0 E
Phan Ri = Hoa Da, *Vietnam* **39 G7** 11 16N 108 40 E
Phan Thiet, *Vietnam* **39 G7** 11 1N 108 9 E
Phanat Nikhom, *Thailand* **38 F3** 13 27N 101 11 E
Phangan, Ko, *Thailand* **39 H3** 9 45N 100 0 E
Phangnga, *Thailand* **39 H2** 8 28N 98 30 E
Phanh Bho Ho Chi Minh, *Vietnam* **39 G6** 10 58N 106 40 E
Phanom Sarakham, *Thailand* **38 F3** 13 45N 101 21 E
Pharenda, *India* **43 F10** 27 5N 83 17 E
Phatthalung, *Thailand* **39 J3** 7 39N 100 6 E
Phayao, *Thailand* **38 C2** 19 11N 99 55 E
Phelps, *N.Y., U.S.A.* **78 D7** 42 58N 77 3W
Phelps, *Wis., U.S.A.* **80 B10** 46 4N 89 5W
Phelps L., *Canada* **73 B8** 59 15N 103 15W
Phenix City, *U.S.A.* **77 J3** 32 28N 85 0W
Phet Buri, *Thailand* **38 F2** 13 1N 99 55 E
Phetchabun, *Thailand* **38 D3** 16 25N 101 8 E
Phetchabun, Thiu Khao, *Thailand* **38 E3** 16 0N 101 20 E
Phetchaburi = Phet Buri, *Thailand* **38 F2** 13 1N 99 55 E
Phi Phi, Ko, *Thailand* **39 J2** 7 45N 98 46 E
Phiafay, *Laos* **38 E6** 14 48N 106 0 E
Phibun Mangsahan, *Thailand* **38 E5** 15 14N 105 14 E
Phichai, *Thailand* **38 D3** 17 22N 100 10 E
Phichit, *Thailand* **38 D3** 16 26N 100 22 E
Philadelphia, *Miss., U.S.A.* **81 J10** 32 46N 89 7W
Philadelphia, *N.Y., U.S.A.* **79 B9** 44 9N 75 43W
Philadelphia, *Pa., U.S.A.* **79 F9** 39 57N 75 10W
Philip, *U.S.A.* **80 C4** 44 2N 101 40W
Philippeville, *Belgium* **15 D4** 50 12N 4 33 E
Philippi L., *Australia* **62 C2** 24 20S 138 55 E
Philippines ■, *Asia* **37 B6** 12 0N 123 0 E
Philippolis, *S. Africa* **56 E4** 30 15S 25 16 E
Philippopolis = Plovdiv, *Bulgaria* **21 C11** 42 8N 24 44 E
Philipsburg, *Mont., U.S.A.* **82 C7** 46 20N 113 18W
Philipsburg, *Pa., U.S.A.* **78 F6** 40 54N 78 13W
Philipstown, *S. Africa* **56 E3** 30 28S 24 30 E
Phillip I., *Australia* **63 F4** 38 30S 145 12 E
Phillips, *Tex., U.S.A.* **81 H4** 35 42N 101 22W
Phillips, *Wis., U.S.A.* **80 C9** 45 42N 90 24W
Phillipsburg, *Kans., U.S.A.* **80 F5** 39 45N 99 19W
Phillipsburg, *N.J., U.S.A.* **79 F9** 40 42N 75 12W
Phillott, *Australia* **63 D4** 27 53S 145 50 E
Philmont, *U.S.A.* **79 D11** 42 15N 73 39W
Philomath, *U.S.A.* **82 D2** 44 32N 123 22W
Phimai, *Thailand* **38 E4** 15 13N 102 30 E
Phitsanulok, *Thailand* **38 D3** 16 50N 100 12 E
Phnom Dangrek, *Thailand* **38 E5** 14 20N 104 0 E
Phnom Penh, *Cambodia* **39 G5** 11 33N 104 55 E
Phoenix, *Ariz., U.S.A.* **83 K7** 33 27N 112 4W
Phoenix, *N.Y., U.S.A.* **79 C8** 43 14N 76 18W
Phoenix Is., *Kiribati* **64 H10** 3 30S 172 0W
Phoenixville, *U.S.A.* **79 F9** 40 8N 75 31W
Phon, *Thailand* **38 E4** 15 49N 102 36 E
Phon Tiou, *Laos* **38 D5** 17 53N 104 37 E
Phong →, *Thailand* **38 D4** 16 23N 102 56 E
Phong Saly, *Laos* **38 B4** 21 42N 102 9 E
Phong Tho, *Vietnam* **38 A4** 22 32N 103 21 E
Phonhong, *Laos* **38 C4** 18 30N 102 25 E
Phonum, *Thailand* **39 H2** 8 49N 98 48 E
Phosphate Hill, *Australia* **62 C2** 21 53S 139 58 E
Photharam, *Thailand* **38 F2** 13 41N 99 51 E
Phra Chedi Sam Ong, *Thailand* **38 E2** 15 16N 98 23 E
Phra Nakhon Si Ayutthaya, *Thailand* **38 E3** 14 25N 100 30 E
Phra Thong, Ko, *Thailand* **39 H2** 9 5N 98 17 E
Phrae, *Thailand* **38 C3** 18 7N 100 9 E
Phrom Phiram, *Thailand* **38 D3** 17 2N 100 12 E
Phu Dien, *Vietnam* **38 C5** 18 58N 105 31 E
Phu Loi, *Laos* **38 B4** 20 14N 103 14 E
Phu Ly, *Vietnam* **38 B5** 20 35N 105 50 E
Phu Tho, *Vietnam* **38 B5** 21 24N 105 13 E
Phuc Yen, *Vietnam* **38 B5** 21 16N 105 45 E
Phuket, *Thailand* **39 J2** 7 52N 98 22 E
Phuket, Ko, *Thailand* **39 J2** 8 0N 98 22 E
Phulera, *India* **42 F6** 26 52N 75 16 E
Phumiphon, Khuan, *Thailand* **38 D2** 17 15N 98 58 E
Phun Phin, *Thailand* **39 H2** 9 7N 99 12 E
Piacenza, *Italy* **20 B3** 45 2N 9 42 E
Pialba, *Australia* **63 D5** 25 20S 152 45 E
Pian Cr. →, *Australia* **63 E4** 30 2S 148 12 E
Piapot, *Canada* **73 D7** 49 59N 109 8W
Piatra Neamţ, *Romania* **17 E13** 46 56N 26 21 E
Piauí □, *Brazil* **93 E10** 7 0S 43 0W
Piave →, *Italy* **20 B5** 45 32N 12 44 E
Piazza Ármerina, *Italy* **20 F6** 37 21N 14 20 E
Pibor Post, *Sudan* **51 G11** 6 47N 33 3 E
Pica, *Chile* **92 H5** 20 35S 69 25W
Picardie, *France* **18 B5** 49 50N 3 0 E
Picardy = Picardie, *France* **18 B5** 49 50N 3 0 E
Picayune, *U.S.A.* **81 K10** 30 32N 89 41W
Pichilemu, *Chile* **94 C1** 34 22S 72 0W
Pickerel L., *Canada* **70 C1** 48 40N 91 25W
Pickle Lake, *Canada* **70 B1** 51 30N 90 12W
Pico Truncado, *Argentina* **96 F3** 46 40S 68 0W
Picton, *Australia* **63 E5** 34 12S 150 34 E
Picton, *Canada* **70 D4** 44 1N 77 9W
Picton, *N.Z.* **59 J5** 41 18S 174 3 E
Pictou, *Canada* **71 C7** 45 41N 62 42W
Picture Butte, *Canada* **72 D6** 49 55N 112 45W
Picún Leufú, *Argentina* **96 D3** 39 30S 69 5W
Pidurutalagala, *Sri Lanka* **40 R12** 7 10N 80 50 E
Piedmont = Piemonte □, *Italy* **20 B2** 45 0N 7 30 E
Piedmont, *U.S.A.* **77 J3** 33 55N 85 37W
Piedmont Plateau, *U.S.A.* **77 J5** 34 0N 81 30W
Piedras, R. de las →, *Peru* **92 F5** 12 30S 69 15W
Piedras Negras, *Mexico* **86 B4** 28 35N 100 35W
Piemonte □, *Italy* **20 B2** 45 0N 7 30 E
Pierce, *U.S.A.* **82 C6** 46 30N 115 48W
Piercefield, *U.S.A.* **79 B10** 44 13N 74 35W
Pierre, *U.S.A.* **80 C4** 44 22N 100 21W
Piet Retief, *S. Africa* **57 D5** 27 1S 30 50 E
Pietarsaari = Jakobstad, *Finland* **8 E17** 63 40N 22 43 E
Pietermaritzburg, *S. Africa* **57 D5** 29 35S 30 25 E
Pietersburg, *S. Africa* **57 C4** 23 54S 29 25 E
Pietrosul, *Romania* **17 E12** 47 35N 24 43 E
Pigeon, *U.S.A.* **76 D4** 43 50N 83 16W
Piggott, *U.S.A.* **81 G9** 36 23N 90 11W
Pigüe, *Argentina* **94 D3** 37 36S 62 25W
Pihani, *India* **43 F9** 27 36N 80 15 E
Pikes Peak, *U.S.A.* **80 F2** 38 50N 105 3W
Piketberg, *S. Africa* **56 E2** 32 55S 18 40 E
Pikeville, *U.S.A.* **76 G4** 37 29N 82 31W
Pikou, *China* **35 E12** 39 18N 122 22 E
Pikwitonei, *Canada* **73 B9** 55 35N 97 9W
Pilani, *India* **42 E6** 28 22N 75 33 E
Pilar, *Brazil* **93 E11** 9 36S 35 56W
Pilar, *Paraguay* **94 B4** 26 50S 58 20W
Pilas Group, *Phil.* **37 C6** 6 45N 121 35 E
Pilcomayo →, *Paraguay* **94 B4** 25 21S 57 42W
Pilibhit, *India* **43 E8** 28 40N 79 50 E
Pilica →, *Poland* **17 C10** 51 52N 21 17 E
Pilkhawa, *India* **42 E7** 28 43N 77 42 E
Pílos, *Greece* **21 F9** 36 55N 21 42 E
Pilot Mound, *Canada* **73 D9** 49 15N 98 54W
Pilot Point, *U.S.A.* **81 J6** 33 24N 96 58W
Pilot Rock, *U.S.A.* **82 D4** 45 29N 118 50W
Pilsen = Plzeň, *Czech.* **16 D6** 49 45N 13 22 E
Pima, *U.S.A.* **83 K9** 32 54N 109 50W
Pimba, *Australia* **63 E2** 31 18S 136 46 E
Pimenta Bueno, *Brazil* **92 F6** 11 35S 61 10W
Pimentel, *Peru* **92 E3** 6 45S 79 55W
Pinang, *Malaysia* **39 K3** 5 25N 100 15 E
Pinar, C. del, *Spain* **22 B10** 39 53N 3 12 E
Pinar del Río, *Cuba* **88 B3** 22 26N 83 40W
Pincher Creek, *Canada* **72 D6** 49 30N 113 57W
Pinchi L., *Canada* **72 C4** 54 38N 124 30W
Pinckneyville, *U.S.A.* **80 F10** 38 5N 89 23W
Pińczów, *Poland* **17 C10** 50 32N 20 32 E
Pind Dadan Khan, *Pakistan* **42 C5** 32 36N 73 7 E
Pindar, *Australia* **61 E2** 28 30S 115 47 E
Pindi Gheb, *Pakistan* **42 C5** 33 14N 72 21 E
Pindiga, *Nigeria* **50 G7** 9 58N 10 53 E
Pindos Óros, *Greece* **21 E9** 40 0N 21 0 E
Pindus Mts. = Pindos Óros, *Greece* **21 E9** 40 0N 21 0 E
Pine, *U.S.A.* **83 J8** 34 23N 111 27W
Pine →, *Canada* **73 B7** 58 50N 105 38W
Pine, C., *Canada* **71 C9** 46 37N 53 32W
Pine Bluff, *U.S.A.* **81 H8** 34 13N 92 1W
Pine City, *U.S.A.* **80 C8** 45 50N 92 59W
Pine Falls, *Canada* **73 C9** 50 34N 96 11W
Pine Flat L., *U.S.A.* **84 J7** 36 50N 119 20W
Pine Pass, *Canada* **72 B4** 55 25N 122 42W
Pine Point, *Canada* **72 A6** 60 50N 114 28W
Pine Ridge, *U.S.A.* **80 D3** 43 2N 102 33W
Pine River, *Canada* **73 C8** 51 45N 100 30W
Pine River, *U.S.A.* **80 B7** 46 43N 94 24W
Pine Valley, *U.S.A.* **85 N10** 32 50N 116 32W
Pinecrest, *U.S.A.* **84 G6** 38 12N 120 1W
Pinedale, *U.S.A.* **84 J7** 36 50N 119 48W
Pinega →, *Russia* **24 B8** 64 8N 46 54 E
Pinehill, *Australia* **62 C4** 23 38S 146 57 E
Pinerolo, *Italy* **20 B2** 44 47N 7 21 E
Pinetop, *U.S.A.* **83 J9** 34 8N 109 56W
Pinetown, *S. Africa* **57 D5** 29 48S 30 54 E
Pinetree, *U.S.A.* **82 E11** 43 42N 105 52W
Pineville, *Ky., U.S.A.* **77 G4** 36 46N 83 42W
Pineville, *La., U.S.A.* **81 K8** 31 19N 92 26W
Ping →, *Thailand* **38 E3** 15 42N 100 9 E
Pingaring, *Australia* **61 F2** 32 40S 118 32 E
Pingding, *China* **34 F7** 37 47N 113 38 E
Pingdingshan, *China* **34 H7** 33 43N 113 27 E
Pingdong, *Taiwan* **33 D7** 22 39N 120 30 E
Pingdu, *China* **35 F10** 36 42N 119 59 E
Pingelly, *Australia* **61 F2** 32 32S 117 5 E
Pingliang, *China* **34 G4** 35 35N 106 31 E
Pinglu, *China* **34 E7** 39 31N 112 30 E
Pingluo, *China* **34 E4** 38 52N 106 30 E
Pingquan, *China* **35 D10** 41 1N 118 37 E
Pingrup, *Australia* **61 F2** 33 32S 118 29 E
Pingwu, *China* **34 H3** 32 25N 104 30 E
Pingxiang, *China* **32 D5** 22 6N 106 46 E
Pingyao, *China* **34 F7** 37 12N 112 10 E
Pingyi, *China* **35 G9** 35 30N 117 35 E
Pingyin, *China* **34 F9** 36 20N 116 25 E
Pingyuan, *China* **34 F9** 37 10N 116 22 E
Pinhal, *Brazil* **95 A6** 22 10S 46 46W
Pinhel, *Portugal* **19 B2** 40 50N 7 1W
Pini, *Indonesia* **36 D1** 0 10N 98 40 E
Piniós →, *Greece* **21 E10** 39 55N 22 10 E
Pinjarra, *Australia* **61 F2** 32 37S 115 52 E
Pink →, *Canada* **73 B8** 56 50N 103 50W
Pinnacles, *Australia* **61 E3** 28 12S 120 26 E
Pinnacles, *U.S.A.* **84 J5** 36 33N 121 19W
Pinnaroo, *Australia* **63 F3** 35 17S 140 53 E
Pinon Hills, *U.S.A.* **85 L9** 34 26N 117 39W
Pinos, *Mexico* **86 C4** 22 20N 101 40W
Pinos, Mt., *U.S.A.* **85 L7** 34 49N 119 8W
Pinos Pt., *U.S.A.* **83 H3** 36 38N 121 57W
Pinotepa Nacional, *Mexico* **87 D5** 16 19N 98 3W
Pinrang, *Indonesia* **37 E5** 3 46S 119 41 E
Pinsk, *Belorussia* **24 D4** 52 10N 26 1 E
Pintados, *Chile* **92 H5** 20 35S 69 40W
Pintumba, *Australia* **61 F5** 31 30S 132 12 E
Pinyug, *Russia* **24 B8** 60 5N 48 0 E
Pioche, *U.S.A.* **83 H6** 37 56N 114 27W
Piombino, *Italy* **20 C4** 42 54N 10 30 E
Pioner, Os., *Russia* **27 B10** 79 50N 92 0 E
Piorini, L., *Brazil* **92 D6** 3 15S 62 35W
Piotrków Trybunalski, *Poland* **17 C9** 51 23N 19 43 E
Pip, *Iran* **45 E9** 26 45N 60 10 E
Pipar, *India* **42 F5** 26 25N 73 31 E
Piparia, *India* **42 H8** 22 45N 78 23 E
Pipestone, *U.S.A.* **80 D6** 44 0N 96 19W
Pipestone →, *Canada* **70 B2** 52 53N 89 23W
Pipestone Cr. →, *Canada* **73 D8** 49 38N 100 15W
Pipmuacan, Rés., *Canada* **71 C5** 49 45N 70 30W
Pippingarra, *Australia* **60 D2** 20 27S 118 42 E
Piqua, *U.S.A.* **76 E3** 40 9N 84 15W
Piquiri →, *Brazil* **95 A5** 24 3S 54 14W
Pīr Sohrāb, *Iran* **45 E9** 25 44N 60 54 E
Piracicaba, *Brazil* **95 A6** 22 45S 47 40W
Piracuruca, *Brazil* **93 D10** 3 50S 41 50W

Porto Empédocle

Piræus = Piraiévs, Greece 21 F10 37 57N 23 42 E
Piraiévs, Greece 21 F10 37 57N 23 42 E
Pirajuí, Brazil 95 A6 21 59S 49 29W
Piranè, Argentina 94 B4 25 42S 59 6W
Pirapora, Brazil 93 G10 17 20S 44 56W
Pírgos, Greece 21 F9 37 40N 21 27 E
Piribebuy, Paraguay 94 B4 25 26S 57 2W
Pirin Planina, Bulgaria 21 D10 41 40N 23 30 E
Pirineos, Spain 19 A6 42 40N 1 0 E
Piripiri, Brazil 93 D10 4 15S 41 46W
Pirot, Serbia, Yug. 21 C10 43 9N 22 39 E
Piru, Indonesia 37 E7 3 4S 128 12 E
Piru, U.S.A. 85 L8 34 25N 118 48W
Pisa, Italy 20 C4 43 43N 10 23 E
Pisagua, Chile 92 G4 19 40S 70 15W
Pisco, Peru 92 F3 13 50S 76 12W
Písek, Czech. 16 D7 49 19N 14 10 E
Pishan, China 32 C2 37 30N 78 33 E
Pishin Lora →, Pakistan 42 E1 29 9N 64 5 E
Pising, Indonesia 37 F6 5 8S 121 53 E
Pismo Beach, U.S.A. 85 K6 35 9N 120 38W
Pissouri, Cyprus 23 E11 34 40N 32 42 E
Pistóia, Italy 20 C4 43 57N 10 53 E
Pistol B., Canada 73 A10 62 25N 92 37W
Pisuerga →, Spain 19 B3 41 33N 4 52W
Pitarpunga, L., Australia 63 E3 34 24S 143 30 E
Pitcairn I., Pac. Oc. 65 K14 25 5S 130 5W
Pite älv →, Sweden 8 D16 65 20N 21 25 E
Piteå, Sweden 8 D16 65 20N 21 25 E
Piteşti, Romania 21 B11 44 52N 24 54 E
Pithapuram, India 41 L13 17 10N 82 15 E
Pithara, Australia 61 F2 30 20S 116 35 E
Pitlochry, U.K. 12 E5 56 43N 3 43W
Pitsilia □, Cyprus 23 E12 34 55N 33 0 E
Pitt I., Canada 72 C3 53 30N 129 50W
Pittsburg, Kans., U.S.A. 81 G7 37 25N 94 42W
Pittsburg, Tex., U.S.A. 81 J7 33 0N 94 59W
Pittsburgh, U.S.A. 78 F5 40 26N 80 1W
Pittsfield, Ill., U.S.A. 80 F9 39 36N 90 49W
Pittsfield, Mass., U.S.A. 79 D11 42 27N 73 15W
Pittsfield, N.H., U.S.A. 79 C13 43 18N 71 20W
Pittston, U.S.A. 79 E9 41 19N 75 47W
Pittsworth, Australia 63 D5 27 41S 151 37 E
Pituri →, Australia 62 C2 22 35S 138 30 E
Piura, Peru 92 E2 5 15S 80 38W
Pixley, U.S.A. 84 K7 35 58N 119 18W
Pizzo, Italy 20 E7 38 44N 16 10 E
Placentia, Canada 71 C9 47 20N 54 0W
Placentia B., Canada 71 C9 47 0N 54 40W
Placerville, U.S.A. 84 G6 38 44N 120 48W
Placetas, Cuba 88 B4 22 15N 79 44W
Plain Dealing, U.S.A. 81 J8 32 54N 93 42W
Plainfield, U.S.A. 79 F10 40 37N 74 25W
Plains, Kans., U.S.A. 81 G4 37 16N 100 35W
Plains, Mont., U.S.A. 82 C6 47 28N 114 53W
Plains, Tex., U.S.A. 81 J3 33 11N 102 50W
Plainview, Nebr., U.S.A. 80 D6 42 21N 97 47W
Plainview, Tex., U.S.A. 81 H4 34 11N 101 43W
Plainville, U.S.A. 80 F5 39 14N 99 18W
Plainwell, U.S.A. 76 D3 42 27N 85 38W
Pláka, Ákra, Greece 23 D8 35 11N 26 19 E
Plakhino, Russia 26 C9 67 45N 86 5 E
Plana Cays, Bahamas 89 B5 22 38N 73 30W
Planada, U.S.A. 84 H6 37 16N 120 19W
Plankinton, U.S.A. 80 D5 43 43N 98 29W
Plano, U.S.A. 81 J6 33 1N 96 42W
Plant City, U.S.A. 77 L4 28 1N 82 7W
Plaquemine, U.S.A. 81 K9 30 17N 91 14W
Plasencia, Spain 19 B2 40 3N 6 8W
Plaster City, U.S.A. 85 N11 32 47N 115 51W
Plaster Rock, Canada 71 C6 46 53N 67 22W
Plastun, Russia 30 B8 44 45N 136 19 E
Plata, Río de la, S. Amer. 94 C4 34 45S 57 30W
Platani →, Italy 20 F5 37 23N 13 16 E
Plátanos, Greece 23 D5 35 28N 23 33 E
Plateau du Coteau du Missouri, U.S.A. 80 B4 47 9N 101 5W
Platí, Ákra, Greece 21 D11 40 27N 24 0 E
Plato, Colombia 92 B4 9 47N 74 47W
Platte, U.S.A. 80 D5 43 23N 98 51W
Platte →, U.S.A. 80 F7 39 16N 94 50W
Platteville, U.S.A. 80 E2 40 13N 104 49W
Plattsburgh, U.S.A. 79 B11 44 42N 73 28W
Plattsmouth, U.S.A. 80 E7 41 1N 95 53W
Plauen, Germany 16 C6 50 29N 12 9 E
Playa Blanca, Canary Is. 22 F6 28 55N 13 37W
Playa Blanca Sur, Canary Is. 22 F6 28 51N 13 50W
Playa de las Americas, Canary Is. 22 F3 28 5N 16 43W
Playa de Mogán, Canary Is. 22 G4 27 48N 15 47W
Playa del Inglés, Canary Is. 22 G4 27 45N 15 33W
Playa Esmerelda, Canary Is. 22 F5 28 8N 14 16W
Playgreen L., Canada 73 C9 54 0N 98 15W
Pleasant Bay, Canada 71 C7 46 51N 60 48W
Pleasant Hill, Calif., U.S.A. 84 H4 37 57N 122 4W
Pleasant Hill, Mo., U.S.A. 80 F7 38 47N 94 16W
Pleasanton, U.S.A. 81 L5 28 58N 98 29W
Pleasantville, U.S.A. 76 F8 39 24N 74 32W
Pleiku, Vietnam 38 F7 13 57N 108 0 E
Plenty →, Australia 62 C2 23 25S 136 31 E
Plenty, B. of, N.Z. 59 G6 37 45S 177 0 E
Plentywood, U.S.A. 80 A2 48 47N 104 34W
Plesetsk, Russia 24 B7 62 40N 40 10 E
Plessisville, Canada 71 C5 46 14N 71 47W
Pletipi L., Canada 71 B5 51 44N 70 6W
Pleven, Bulgaria 21 C11 43 26N 24 37 E
Plevlja, Montenegro, Yug. 21 C8 43 21N 19 21 E
Płock, Poland 17 B9 52 32N 19 40 E
Ploieşti, Romania 21 B12 44 57N 26 5 E
Plonge, Lac la, Canada 73 B7 55 8N 107 20W
Plovdiv, Bulgaria 21 C11 42 8N 24 44 E
Plum, U.S.A. 78 F5 40 29N 79 47W
Plum I., U.S.A. 79 E12 41 11N 72 12W
Plumas, U.S.A. 84 F7 39 45N 119 4W
Plummer, U.S.A. 82 C5 47 20N 116 53W
Plumtree, Zimbabwe 55 G2 20 27S 27 55 E
Plymouth, U.K. 11 G3 50 23N 4 9W
Plymouth, Calif., U.S.A. 84 G6 38 29N 120 51W
Plymouth, Ind., U.S.A. 76 E2 41 21N 86 19W
Plymouth, Mass., U.S.A. 79 E14 41 57N 70 40W
Plymouth, N.C., U.S.A. 77 H7 35 52N 76 43W
Plymouth, N.H., U.S.A. 79 C13 43 46N 71 41W
Plymouth, Pa., U.S.A. 79 E9 41 14N 75 57W
Plymouth, Wis., U.S.A. 76 D2 43 45N 87 59W
Plynlimon = Pumlumon Fawr, U.K. 11 E4 52 29N 3 47W
Plzeň, Czech. 16 D6 49 45N 13 22 E
Po →, Italy 20 B5 44 57N 12 4 E
Po Hai = Bo Hai, China 35 E10 39 0N 119 0 E
Pobeda, Russia 27 C15 65 12N 146 12 E
Pobedino, Russia 27 E15 49 51N 142 49 E
Pobedy Pik, Kirghizia 26 E8 40 45N 79 58 E
Pocahontas, Ark., U.S.A. 81 G9 36 16N 90 58W
Pocahontas, Iowa, U.S.A. 80 D7 42 44N 94 40W
Pocatello, U.S.A. 82 E7 42 52N 112 27W
Pochutla, Mexico 87 D5 15 50N 96 31W
Pocito Casas, Mexico 86 B2 28 32N 111 6W
Pocomoke City, U.S.A. 76 F8 38 5N 75 34W
Poços de Caldas, Brazil 95 A6 21 50S 46 33W
Podgorica, Montenegro, Yug. 21 C8 42 30N 19 19 E
Podkamennaya Tunguska →, Russia 27 C10 61 50N 90 13 E
Podolsk, Russia 24 C6 55 25N 37 30 E
Podor, Senegal 50 E1 16 40N 15 2W
Podporozhy, Russia 24 B5 60 55N 34 2 E
Pofadder, S. Africa 56 D2 29 10S 19 22 E
Pogamasing, Canada 70 C3 46 55N 81 50W
Pogranitšnyi, Russia 30 B5 44 25N 131 24 E
Poh, Indonesia 37 E6 0 46S 122 51 E
Pohang, S. Korea 35 F15 36 1N 129 23 E
Pohnpei, Pac. Oc. 64 G7 6 55N 158 10 E
Poinsett, C., Antarctica 5 C8 65 42S 113 18 E
Point Edward, Canada 70 D3 43 0N 82 30W
Point Pedro, Sri Lanka 40 Q12 9 50N 80 15 E
Point Pleasant, N.J., U.S.A. 79 F10 40 5N 74 4W
Point Pleasant, W. Va., U.S.A. 76 F4 38 51N 82 8W
Pointe-à-la Hache, U.S.A. 81 L10 29 35N 89 55W
Pointe-à-Pitre, Guadeloupe 89 C7 16 10N 61 30W
Pointe Noire, Congo 52 E2 4 48S 11 53 E
Poisonbush Ra., Australia 60 D3 22 30S 121 30 E
Poitiers, France 18 C4 46 35N 0 20 E
Pojoaque Valley, U.S.A. 83 J11 35 54N 106 1W
Pokaran, India 40 F7 27 0N 71 50 E
Pokataroo, Australia 63 D4 29 30S 148 36 E
Poko, Zaïre 54 B2 3 7N 26 52 E
Pokrovsk, Russia 24 D8 51 28N 46 6 E
Pokrovsk, Russia 27 C13 61 29N 129 0 E
Polacca, U.S.A. 83 J8 35 50N 110 23W
Polan, Iran 45 E9 25 30N 61 10 E
Poland ■, Europe 17 C10 52 0N 20 0 E
Polatsk = Polotsk, Belorussia 24 C4 55 30N 28 50 E
Polcura, Chile 94 D1 37 17S 71 43W
Polden Hills, U.K. 11 F5 51 7N 2 50W
Polesye, Belorussia 24 D4 52 10N 28 10 E
Polevskoy, Russia 24 C11 56 26N 60 11 E
Polewali, Indonesia 37 E5 3 21S 119 23 E
Pŏlgyo-ri, S. Korea 35 G14 34 51N 127 21 E
Poli, Cameroon 52 C2 8 34N 13 15 E
Polillo Is., Phil. 37 B6 14 56N 122 0 E
Polis, Cyprus 23 D11 35 2N 32 26 E
Políyiros, Greece 21 D10 40 23N 23 25 E
Polk, U.S.A. 78 E5 41 22N 79 56W
Pollachi, India 40 P10 10 35N 77 0 E
Pollensa, Spain 22 B10 39 54N 3 1 E
Pollensa, B. de, Spain 22 B10 39 53N 3 8 E
Pollock, U.S.A. 80 C4 45 55N 100 17W
Polnovat, Russia 26 C7 63 50N 65 54 E
Polo, U.S.A. 80 E10 41 59N 89 35W
Polotsk, Belorussia 24 C4 55 30N 28 50 E
Polson, U.S.A. 82 C6 47 41N 114 9W
Poltava, Ukraine 25 E5 49 35N 34 35 E
Polunochnoye, Russia 24 B11 60 52N 60 25 E
Polyarny, Russia 24 A5 69 8N 33 20 E
Polynesia, Pac. Oc. 65 H11 10 0S 162 0W
Polynésie française ■ = French Polynesia ■, Pac. Oc. 65 J13 20 0S 145 0W
Pomaro, Mexico 86 D4 18 20N 103 18W
Pombal, Brazil 93 E11 6 45S 37 50W
Pombal, Portugal 19 C1 39 55N 8 40W
Pómbia, Greece 23 D6 35 0N 24 51 E
Pomeroy, Ohio, U.S.A. 76 F4 39 2N 82 2W
Pomeroy, Wash., U.S.A. 82 C5 46 28N 117 36W
Pomona, U.S.A. 85 L9 34 4N 117 45W
Pomos, Cyprus 23 D11 35 9N 32 33 E
Pomos, C., Cyprus 23 D11 35 10N 32 33 E
Pompano Beach, U.S.A. 77 M5 26 14N 80 8W
Pompeys Pillar, U.S.A. 82 D10 45 59N 107 57W
Ponape = Pohnpei, Pac. Oc. 64 G7 6 55N 158 10 E
Ponask, L., Canada 70 B1 54 0N 92 41W
Ponass L., Canada 73 C8 52 16N 103 58W
Ponca, U.S.A. 80 D6 42 34N 96 43W
Ponca City, U.S.A. 81 G6 36 42N 97 5W
Ponce, Puerto Rico 89 C6 18 1N 66 37W
Ponchatoula, U.S.A. 81 K9 30 26N 90 26W
Poncheville, L., Canada 70 B4 50 10N 76 55W
Pond, U.S.A. 85 K7 35 43N 119 20W
Pond Inlet, Canada 69 A12 72 40N 77 0W
Pondicherry, India 40 P11 11 59N 79 50 E
Ponds, I. of, Canada 71 B8 53 27N 55 52W
Ponferrada, Spain 19 A2 42 32N 6 35W
Ponnani, India 40 P9 10 45N 75 59 E
Ponnyadaung, Burma 41 J19 22 0N 94 10 E
Ponoi, Russia 24 A7 67 0N 41 0 E
Ponoi →, Russia 24 A7 66 59N 41 17 E
Ponoka, Canada 72 C6 52 42N 113 40W
Ponorogo, Indonesia 37 G14 7 52S 111 27 E
Ponta do Sol, Madeira 22 D2 32 42N 17 7W
Ponta Grossa, Brazil 95 B5 25 7S 50 10W
Ponta Pora, Brazil 95 A4 22 20S 55 35W
Pontarlier, France 18 C7 46 54N 6 20 E
Pontchartrain L., U.S.A. 81 K9 30 5N 90 5W
Ponte do Pungué, Mozam. 55 F3 19 30S 34 33 E
Ponte Nova, Brazil 95 A7 20 25S 42 54W
Pontedera, Italy 20 C4 43 40N 10 37 E
Pontefract, U.K. 10 D6 53 42N 1 19W
Ponteix, Canada 73 D7 49 46N 107 29W
Pontevedra, Spain 19 A1 42 26N 8 40W
Pontiac, Ill., U.S.A. 80 E10 40 53N 88 38W
Pontiac, Mich., U.S.A. 76 D4 42 38N 83 18W
Pontian Kecil, Malaysia 39 M4 1 29N 103 23 E
Pontianak, Indonesia 36 E3 0 3S 109 15 E
Pontic Mts. = Kuzey Anadolu Dağları, Turkey 25 F6 41 30N 35 0 E
Pontine Is. = Ponziane, Isole, Italy 20 D5 40 55N 13 0 E
Pontine Mts. = Kuzey Anadolu Dağları, Turkey 25 F6 41 30N 35 0 E
Ponton →, Canada 72 B5 58 27N 116 11W
Pontypool, Canada 78 B6 44 6N 78 38W
Pontypool, U.K. 11 F4 51 42N 3 1W
Pontypridd, U.K. 11 F4 51 36N 3 21W
Ponziane, Isole, Italy 20 D5 40 55N 13 0 E
Poochera, Australia 63 E1 32 43S 134 51 E
Poole, U.K. 11 G6 50 42N 1 58W
Pooley I., Canada 72 C3 52 45N 128 15W
Poona = Pune, India 40 K8 18 29N 73 57 E
Pooncarie, Australia 63 E3 33 22S 142 31 E
Poopelloe L., Australia 63 E3 31 40S 144 0 E
Poopó, L. de, Bolivia 92 G5 18 30S 67 35W
Popanyinning, Australia 61 F2 32 40S 117 2 E
Popayán, Colombia 92 C3 2 27N 76 36W
Poperinge, Belgium 15 D2 50 51N 2 42 E
Popigay, Russia 27 B12 72 1N 110 39 E
Popilta, L., Australia 63 E3 33 10S 141 42 E
Popio L., Australia 63 E3 33 10S 141 52 E
Poplar, U.S.A. 80 A2 48 7N 105 12W
Poplar →, Man., Canada 73 C9 53 0N 97 19W
Poplar →, N.W.T., Canada 72 A4 61 22N 121 52W
Poplar Bluff, U.S.A. 81 G9 36 46N 90 24W
Poplarville, U.S.A. 81 K10 30 51N 89 32W
Popocatepetl, Mexico 87 D5 19 10N 98 40W
Popokabaka, Zaïre 52 F3 5 41S 16 40 E
Porali →, Pakistan 42 G2 25 35N 66 26 E
Porbandar, India 42 J3 21 44N 69 43 E
Porcher I., Canada 72 C2 53 50N 130 30W
Porcupine →, Canada 73 B8 59 11N 104 46W
Porcupine →, U.S.A. 68 B5 66 34N 145 19W
Pori, Finland 9 F16 61 29N 21 48 E
Porjus, Sweden 8 C15 66 57N 19 50 E
Porkkala, Finland 9 G18 59 59N 24 26 E
Porlamar, Venezuela 92 A6 10 57N 63 51W
Poronaysk, Russia 27 E15 49 13N 143 0 E
Poroshiri-Dake, Japan 30 C11 42 41N 142 52 E
Poroto Mts., Tanzania 55 D3 9 0S 33 30 E
Porpoise B., Antarctica 5 C9 66 0S 127 0 E
Porreras, Spain 22 B10 39 31N 3 2 E
Porretta, Passo di, Italy 20 B4 44 2N 10 56 E
Porsangen, Norway 8 A18 70 40N 25 40 E
Port Adelaide, Australia 63 E2 34 46S 138 30 E
Port Alberni, Canada 72 D4 49 14N 124 50W
Port Alfred, Canada 71 C5 48 18N 70 53W
Port Alfred, S. Africa 56 E4 33 36S 26 55 E
Port Alice, Canada 72 C3 50 20N 127 25W
Port Allegany, U.S.A. 78 E6 41 48N 78 17W
Port Allen, U.S.A. 81 K9 30 27N 91 12W
Port Alma, Australia 62 C5 23 38S 150 53 E
Port Angeles, U.S.A. 84 B3 48 7N 123 27W
Port Antonio, Jamaica 88 C4 18 10N 76 30W
Port Aransas, U.S.A. 81 M6 27 50N 97 4W
Port Arthur = Lüshun, China 35 E11 38 45N 121 15 E
Port Arthur, Australia 62 G4 43 7S 147 50 E
Port Arthur, U.S.A. 81 L8 29 54N 93 56W
Port au Port B., Canada 71 C8 48 40N 58 50W
Port-au-Prince, Haiti 89 C5 18 40N 72 20W
Port Augusta, Australia 63 E2 32 30S 137 50 E
Port Augusta West, Australia 63 E2 32 29S 137 29 E
Port Austin, U.S.A. 78 B2 44 3N 83 1W
Port Bell, Uganda 54 B3 0 18N 32 35 E
Port Bergé Vaovao, Madag. 57 B8 15 33S 47 40 E
Port Blandford, Canada 71 C9 48 20N 54 10W
Port Bradshaw, Australia 62 A2 12 30S 137 20 E
Port Broughton, Australia 63 E2 33 37S 137 56 E
Port Burwell, Canada 70 D3 42 40N 80 48W
Port Canning, India 43 H13 22 23N 88 40 E
Port-Cartier, Canada 71 B6 50 2N 66 50W
Port Chalmers, N.Z. 59 L3 45 49S 170 30 E
Port Chester, U.S.A. 79 F11 41 0N 73 40W
Port Clements, Canada 72 C2 53 40N 132 10W
Port Clinton, U.S.A. 76 E4 41 31N 82 56W
Port Colborne, Canada 70 D4 42 50N 79 10W
Port Coquitlam, Canada 72 D4 49 15N 122 45W
Port Credit, Canada 78 C5 43 33N 79 35W
Port Curtis, Australia 62 C5 23 57S 151 20 E
Port Dalhousie, Canada 78 C5 43 13N 79 16W
Port Darwin, Australia 60 B5 12 24S 130 45 E
Port Darwin, Falk. Is. 96 G5 51 50S 59 0W
Port Davey, Australia 62 G4 43 16S 145 55 E
Port-de-Paix, Haiti 89 C5 19 50N 72 50W
Port Dickson, Malaysia 39 L3 2 30N 101 49 E
Port Douglas, Australia 62 B4 16 30S 145 30 E
Port Dover, Canada 78 D4 42 47N 80 12W
Port Edward, Canada 72 C2 54 12N 130 10W
Port Elgin, Canada 70 D3 44 25N 81 25W
Port Elizabeth, S. Africa 56 E4 33 58S 25 40 E
Port Ellen, U.K. 12 F2 55 38N 6 10W
Port-en-Bessin, France 18 B3 49 21N 0 45W
Port Erin, I. of Man 10 C3 54 5N 4 45W
Port Essington, Australia 60 B5 11 15S 132 10 E
Port Etienne = Nouâdhibou, Mauritania 50 D1 20 54N 17 0W
Port Fairy, Australia 63 F3 38 22S 142 12 E
Port Gamble, U.S.A. 84 C4 47 51N 122 35W
Port-Gentil, Gabon 52 E1 0 40S 8 50 E
Port Gibson, U.S.A. 81 K9 31 58N 90 59W
Port Glasgow, U.K. 12 F4 55 57N 4 40W
Port Harcourt, Nigeria 50 H6 4 40N 7 10 E
Port Hardy, Canada 72 C3 50 41N 127 30W
Port Harrison = Inoucdjouac, Canada 69 C12 58 25N 78 15W
Port Hawkesbury, Canada 71 C7 45 36N 61 22W
Port Hedland, Australia 60 D2 20 25S 118 35 E
Port Henry, U.S.A. 79 B11 44 3N 73 28W
Port Hood, Canada 71 C7 46 0N 61 32W
Port Hope, Canada 70 D4 43 56N 78 20W
Port Hueneme, U.S.A. 85 L7 34 7N 119 12W
Port Huron, U.S.A. 76 D4 42 58N 82 26W
Port Isabel, U.S.A. 81 M6 26 5N 97 12W
Port Jefferson, U.S.A. 79 F11 40 57N 73 3W
Port Jervis, U.S.A. 79 E10 41 22N 74 41W
Port Kelang = Pelabuhan Kelang, Malaysia 39 L3 3 0N 101 23 E
Port Kembla, Australia 63 E5 34 52S 150 49 E
Port Kenny, Australia 63 E1 33 10S 134 41 E
Port-la-Nouvelle, France 18 E5 43 1N 3 3 E
Port Laoise, Ireland 13 C4 53 2N 7 20W
Port Lavaca, U.S.A. 81 L6 28 37N 96 38W
Port Lincoln, Australia 63 E2 34 42S 135 52 E
Port Loko, S. Leone 50 G2 8 48N 12 46W
Port Lyautey = Kenitra, Morocco 50 B3 34 15N 6 40W
Port MacDonnell, Australia 63 F3 38 5S 140 48 E
Port Macquarie, Australia 63 E5 31 25S 152 25 E
Port Maria, Jamaica 88 C4 18 25N 76 55W
Port Mellon, Canada 72 D4 49 32N 123 31W
Port-Menier, Canada 71 C7 49 51N 64 15W
Port Morant, Jamaica 88 C4 17 54N 76 19W
Port Moresby, Papua N. G. 64 H6 9 24S 147 8 E
Port Mouton, Canada 71 D7 43 58N 64 50W
Port Musgrave, Australia 62 A3 11 55S 141 50 E
Port Nelson, Canada 73 B10 57 3N 92 36W
Port Nolloth, S. Africa 56 D2 29 17S 16 52 E
Port Nouveau-Québec, Canada 69 C13 58 30N 65 59W
Port O'Connor, U.S.A. 81 L6 28 26N 96 24W
Port of Spain, Trin. & Tob. 89 D7 10 40N 61 31W
Port Orchard, U.S.A. 84 C4 47 32N 122 38W
Port Orford, U.S.A. 82 E1 42 45N 124 30W
Port Pegasus, N.Z. 59 M1 47 12S 167 41 E
Port Perry, Canada 70 D4 44 6N 78 56W
Port Phillip B., Australia 63 F3 38 10S 144 50 E
Port Pirie, Australia 63 E2 33 10S 138 1 E
Port Radium = Echo Bay, Canada 68 B8 66 5N 117 55W
Port Renfrew, Canada 72 D4 48 30N 124 20W
Port Roper, Australia 62 A2 14 45S 135 25 E
Port Rowan, Canada 78 D4 42 40N 80 30W
Port Safaga = Bûr Safâga, Egypt 51 C11 26 43N 33 57 E
Port Said = Bûr Sa'îd, Egypt 51 B11 31 16N 32 18 E
Port St. Joe, U.S.A. 77 L3 29 49N 85 18W
Port St. Johns, S. Africa 57 E4 31 38S 29 33 E
Port-St.-Louis-du-Rhône, France 18 E6 43 23N 4 49 E
Port Sanilac, U.S.A. 78 C2 43 26N 82 33W
Port Saunders, Canada 71 B8 50 40N 57 18W
Port Severn, Canada 78 B5 44 48N 79 43W
Port Shepstone, S. Africa 57 E5 30 44S 30 28 E
Port Simpson, Canada 72 C2 54 30N 130 20W
Port Stanley = Stanley, Falk. Is. 96 G5 51 40S 59 51W
Port Stanley, Canada 70 D3 42 40N 81 10W
Port Sudan = Bûr Sûdân, Sudan 51 E12 19 32N 37 9 E
Port Talbot, U.K. 11 F4 51 35N 3 48W
Port Townsend, U.S.A. 84 B4 48 7N 122 45W
Port-Vendres, France 18 E5 42 32N 3 8 E
Port Vladimir, Russia 24 A5 69 25N 33 6 E
Port Wakefield, Australia 63 E2 34 12S 138 10 E
Port Washington, U.S.A. 76 D2 43 23N 87 53W
Port Weld, Malaysia 39 K3 4 50N 100 38 E
Portachuelo, Bolivia 92 G6 17 10S 63 20W
Portadown, U.K. 13 B5 54 27N 6 26W
Portage, U.S.A. 80 D10 43 33N 89 28W
Portage La Prairie, Canada 73 D9 49 58N 98 18W
Portageville, U.S.A. 81 G10 36 26N 89 42W
Portalegre, Portugal 19 C2 39 19N 7 25W
Portales, U.S.A. 81 H3 34 11N 103 20W
Portarlington, Ireland 13 C4 53 10N 7 10W
Porter L., N.W.T., Canada 73 A7 61 41N 108 5W
Porter L., Sask., Canada 73 B7 56 20N 107 20W
Porterville, S. Africa 56 E2 33 0S 19 0 E
Porterville, U.S.A. 83 H4 36 4N 119 1W
Porthcawl, U.K. 11 F4 51 28N 3 42W
Porthill, U.S.A. 82 B5 48 59N 116 30W
Portile de Fier, Europe 17 F11 44 42N 22 30 E
Portimão, Portugal 19 D1 37 8N 8 32W
Portland, N.S.W., Australia 63 E4 33 20S 150 0 E
Portland, Vic., Australia 63 F3 38 20S 141 35 E
Portland, Canada 79 B8 44 42N 76 12W
Portland, Conn., U.S.A. 79 E12 41 34N 72 38W
Portland, Maine, U.S.A. 71 D5 43 39N 70 16W
Portland, Mich., U.S.A. 76 D3 42 52N 84 54W
Portland, Oreg., U.S.A. 84 E4 45 32N 122 37W
Portland, I. of, U.K. 11 G5 50 32N 2 25W
Portland B., Australia 63 F3 38 15S 141 45 E
Portland Bill, U.K. 11 G5 50 31N 2 27W
Portland Prom., Canada 69 C12 58 40N 78 33W
Portlands Roads, Australia 62 A3 12 36S 143 25 E
Portneuf, Canada 71 C5 46 43N 71 55W
Porto, Portugal 19 B1 41 8N 8 40W
Pôrto Alegre, Brazil 95 C5 30 5S 51 10W
Porto Amboim = Gunza, Angola 52 G2 10 50S 13 50 E
Porto Cristo, Spain 22 B10 39 33N 3 20 E
Pôrto de Móz, Brazil 93 D8 1 41S 52 13W
Porto Empédocle, Italy 20 F5 37 18N 13 30 E

Ranobe, Helodranon' i

Pweto, *Zaïre* 55 D2 8 25S 28 51 E
Pwllheli, *U.K.* 10 E3 52 54N 4 26W
Pya-ozero, *Russia* 24 A5 66 5N 30 58 E
Pyapon, *Burma* 41 L19 16 20N 95 40 E
Pyasina →, *Russia* 27 B9 73 30N 87 0 E
Pyatigorsk, *Russia* 25 F7 44 2N 43 6 E
Pyè, *Burma* 41 K19 18 49N 95 13 E
Pyinmana, *Burma* 41 K20 19 45N 96 12 E
Pyla, C., *Cyprus* 23 E12 34 56N 33 51 E
Pyŏktong, *N. Korea* 35 D13 40 50N 125 50 E
Pyŏnggang, *N. Korea* 35 E14 38 24N 127 17 E
Pyŏngtaek, *S. Korea* 35 F14 37 1N 127 4 E
P'yŏngyang, *N. Korea* 35 E13 39 0N 125 30 E
Pyote, *U.S.A.* 81 K3 31 32N 103 8W
Pyramid L., *U.S.A.* 82 G4 40 1N 119 35W
Pyramid Pk., *U.S.A.* 85 J10 36 25N 116 37W
Pyrénées, *Europe* 18 E4 42 45N 0 18 E
Pyrénées-Atlantiques □, *France* 18 E3 43 10N 0 50W
Pyrénées-Orientales □, *France* 18 E5 42 35N 2 26 E
Pyu, *Burma* 41 K20 18 30N 96 28 E

Q

Qaanaaq = Thule, *Greenland* 4 B4 77 40N 69 0W
Qachasnek, *S. Africa* 57 E4 30 6S 28 42 E
Qa'el Jafr, *Jordan* 47 E5 30 20N 36 25 E
Qâdib, *Yemen* 46 E5 12 37N 53 57 E
Qa'emābād, *Iran* 45 D9 31 44N 60 2 E
Qā'emshahr, *Iran* 45 B7 36 30N 52 53 E
Qagan Nur, *China* 34 C8 43 30N 114 55 E
Qahar Youyi Zhongqi, *China* 34 D7 41 12N 112 40 E
Qahremānshahr = Bākhtarān, *Iran* 44 C5 34 23N 47 0 E
Qaidam Pendi, *China* 32 C4 37 0N 95 0 E
Qajarīyeh, *Iran* 45 D6 31 1N 48 22 E
Qala, Ras il, *Malta* 23 C1 36 1N 14 20 E
Qala-i-Jadid, *Afghan.* 42 D2 31 1N 66 25 E
Qala Yangi, *Afghan.* 42 B2 34 20N 66 30 E
Qal'at al Akhḑar, *Si. Arabia* 44 E3 28 0N 37 10 E
Qal'at Sukkar, *Iraq* 44 D5 31 51N 46 5 E
Qal'eh Darreh, *Iran* 44 B5 38 47N 47 2 E
Qal'eh Shaharak, *Afghan.* 40 B4 34 10N 64 20 E
Qamar, Ghubbat al, *Yemen* 46 D5 16 20N 52 30 E
Qamdo, *China* 32 C4 31 15N 97 6 E
Qamruddin Karez, *Pakistan* 42 D3 31 45N 68 20 E
Qandahār, *Afghan.* 40 D4 31 32N 65 30 E
Qandahār □, *Afghan.* 40 D4 31 0N 65 0 E
Qapān, *Iran* 45 B7 37 40N 55 47 E
Qaqortoq = Julianehåb, *Greenland* 4 C5 60 43N 46 0W
Qâra, *Egypt* 51 C10 29 38N 26 30 E
Qara Qash →, *India* 43 B8 35 0N 78 30 E
Qaraghandy = Karaganda, *Kazakhstan* 26 E8 49 50N 73 10 E
Qārah, *Si. Arabia* 44 D4 29 55N 40 3 E
Qareh →, *Iran* 44 B5 39 25N 47 22 E
Qareh Tekān, *Iran* 45 B6 36 38N 49 29 E
Qarqan He →, *China* 32 C3 39 30N 88 30 E
Qarshi = Karshi, *Uzbekistan* 26 F7 38 53N 65 48 E
Qartabā, *Lebanon* 47 A4 34 4N 35 50 E
Qaryat al Gharab, *Iraq* 44 D5 31 27N 44 48 E
Qaryat al 'Ulyā, *Si. Arabia* 44 E5 27 33N 47 42 E
Qasr 'Amra, *Jordan* 44 D3 31 48N 36 35 E
Qaşr-e Qand, *Iran* 45 E9 26 15N 60 45 E
Qasr Farâfra, *Egypt* 51 C10 27 0N 28 1 E
Qatanā, *Syria* 47 B5 33 26N 36 4 E
Qatar ■, *Asia* 45 E6 25 30N 51 15 E
Qatlīsh, *Iran* 45 B8 37 50N 57 19 E
Qattâra, Munkhafed el, *Egypt* 51 C10 29 30N 27 30 E
Qattâra Depression = Qattâra, Munkhafed el, *Egypt* 51 C10 29 30N 27 30 E
Qawām al Ḩamzah, *Iraq* 44 D5 31 43N 44 58 E
Qāyen, *Iran* 45 C8 33 40N 59 10 E
Qazaqstan = Kazakhstan ■, *Asia* 26 E7 50 0N 70 0 E
Qazvin, *Iran* 45 B6 36 15N 50 0 E
Qena, *Egypt* 51 C11 26 10N 32 43 E
Qeqertarsuag = Disko, *Greenland* 4 C5 69 45N 53 30W
Qeqertarsuaq = Godhavn, *Greenland* 4 C5 69 15N 53 38W
Qeshlāq, *Iran* 44 C5 34 55N 46 28 E
Qeshm, *Iran* 45 E8 26 55N 56 10 E
Qezi'ot, *Israel* 47 E3 30 52N 34 26 E
Qi Xian, *China* 34 G8 34 40N 114 48 E
Qian Gorlos, *China* 35 B13 45 5N 124 42 E
Qian Xian, *China* 34 G5 34 31N 108 15 E
Qianyang, *China* 34 G4 34 40N 107 8 E
Qibā', *Si. Arabia* 44 E5 27 24N 44 20 E
Qila Safed, *Pakistan* 40 E2 29 0N 61 30 E
Qila Saifullāh, *Pakistan* 42 D3 30 45N 68 17 E
Qilian Shan, *China* 32 C4 38 30N 96 0 E
Qin He →, *China* 34 G7 35 1N 113 22 E
Qin Ling = Qinling Shandi, *China* 34 H5 33 50N 108 10 E
Qin'an, *China* 34 G3 34 48N 105 40 E
Qing Xian, *China* 34 E9 38 35N 116 45 E
Qingcheng, *China* 35 F9 37 15N 117 40 E
Qingdao, *China* 35 F11 36 5N 120 20 E
Qingfeng, *China* 34 G8 35 52N 115 8 E
Qinghai □, *China* 32 C4 36 0N 98 0 E
Qinghai Hu, *China* 32 C5 36 40N 100 10 E
Qinghecheng, *China* 35 D13 41 15N 124 30 E
Qinghemen, *China* 35 D11 41 48N 121 25 E
Qingjian, *China* 34 F6 37 8N 110 8 E
Qingjiang, *China* 35 H10 33 30N 119 2 E
Qingshui, *China* 34 G4 34 48N 106 8 E
Qingshuihe, *China* 34 E6 39 55N 111 35 E
Qingtongxia Shuiku, *China* 34 F3 37 50N 105 58 E
Qingxu, *China* 34 F7 37 34N 112 22 E
Qingyang, *China* 34 F4 36 2N 107 55 E
Qingyuan, *China* 35 C13 42 10N 124 55 E
Qingyun, *China* 35 F9 37 45N 117 20 E
Qinhuangdao, *China* 35 E10 39 56N 119 30 E
Qinling Shandi, *China* 34 H5 33 50N 108 10 E
Qinshui, *China* 34 G7 35 40N 112 8 E
Qinyang, *China* 34 G7 35 7N 112 57 E
Qinyuan, *China* 34 F7 36 29N 112 20 E
Qinzhou, *China* 32 D5 21 58N 108 38 E
Qionghai, *China* 38 C8 19 15N 110 26 E
Qiongshan, *China* 38 C8 19 51N 110 26 E
Qiongzhou Haixia, *China* 38 B8 20 10N 110 15 E
Qiqihar, *China* 27 E13 47 26N 124 0 E
Qirāīya, W. →, *Egypt* 47 E3 30 27N 34 0 E
Qiryat Ata, *Israel* 47 C4 32 47N 35 6 E
Qiryat Gat, *Israel* 47 D3 31 32N 34 46 E
Qiryat Mal'akhi, *Israel* 47 D3 31 44N 34 44 E
Qiryat Shemona, *Israel* 47 B4 33 13N 35 35 E
Qiryat Yam, *Israel* 47 C4 32 51N 35 4 E
Qishan, *China* 34 G4 34 25N 107 38 E
Qitai, *China* 32 B3 44 2N 89 35 E
Qixia, *China* 35 F11 37 17N 120 52 E
Qojūr, *Iran* 44 B5 36 12N 47 55 E
Qom, *Iran* 45 C6 34 40N 51 0 E
Qomsheh, *Iran* 45 D6 32 0N 51 55 E
Qostanay = Kustanay, *Kazakhstan* 26 D7 53 10N 63 35 E
Qu Xian, *China* 33 D6 28 57N 118 54 E
Quairading, *Australia* 61 F2 32 0S 117 21 E
Quakertown, *U.S.A.* 79 F9 40 26N 75 21W
Qualeup, *Australia* 61 F2 33 48S 116 48 E
Quambatook, *Australia* 63 F3 35 49S 143 34 E
Quambone, *Australia* 63 E4 30 57S 147 53 E
Quamby, *Australia* 62 C3 20 22S 140 17 E
Quan Long, *Vietnam* 39 H5 9 7N 105 8 E
Quanah, *U.S.A.* 81 H5 34 18N 99 44W
Quandialla, *Australia* 63 E4 34 1S 147 47 E
Quang Ngai, *Vietnam* 38 E7 15 13N 108 58 E
Quang Yen, *Vietnam* 38 B6 20 56N 106 52 E
Quantock Hills, *U.K.* 11 F4 51 8N 3 10W
Quanzhou, *China* 33 D6 24 55N 118 34 E
Quaraí, *Brazil* 94 C4 30 15S 56 20W
Quartzsite, *U.S.A.* 85 M12 33 40N 114 13W
Quatsino, *Canada* 72 C3 50 30N 127 40W
Quatsino Sd., *Canada* 72 C3 50 25N 127 58W
Qūchān, *Iran* 45 B8 37 10N 58 27 E
Queanbeyan, *Australia* 63 F4 35 17S 149 14 E
Québec, *Canada* 71 C5 46 52N 71 13W
Québec □, *Canada* 71 B6 48 0N 74 0W
Queen Alexandra Ra., *Antarctica* 5 E11 85 0S 170 0 E
Queen Charlotte, *Canada* 72 C2 53 15N 132 2W
Queen Charlotte Is., *Canada* 72 C2 53 20N 132 10W
Queen Charlotte Str., *Canada* 72 C3 51 0N 128 0W
Queen Elizabeth Is., *Canada* 66 B10 76 0N 95 0W
Queen Elizabeth Nat. Park, *Uganda* 54 C3 0 0 30 0 E
Queen Mary Land, *Antarctica* 5 D7 70 0S 95 0 E
Queen Maud G., *Canada* 68 B9 68 15N 102 30W
Queen Maud Land, *Antarctica* 5 D3 72 30S 12 0 E
Queen Maud Mts., *Antarctica* 5 E13 86 0S 160 0W
Queens Chan., *Australia* 60 C4 15 0S 129 30 E
Queenscliff, *Australia* 63 F3 38 16S 144 39 E
Queensland □, *Australia* 62 C3 22 0S 142 0 E
Queenstown, *Australia* 62 G4 42 4S 145 35 E
Queenstown, *N.Z.* 59 L2 45 1S 168 40 E
Queenstown, *S. Africa* 56 E4 31 52S 26 52 E
Queets, *U.S.A.* 84 C2 47 32N 124 20W
Queguay Grande →, *Uruguay* 94 C4 32 9S 58 9W
Queimadas, *Brazil* 93 F11 11 0S 39 38W
Quela, *Angola* 52 F3 9 10S 16 56 E
Quelimane, *Mozam.* 55 F4 17 53S 36 58 E
Quelpart = Cheju Do, *S. Korea* 35 H14 33 29N 126 34 E
Quemado, *N. Mex., U.S.A.* 83 J9 34 20N 108 30W
Quemado, *Tex., U.S.A.* 81 L4 28 58N 100 35W
Quemú-Quemú, *Argentina* 94 D3 36 3S 63 36W
Quequén, *Argentina* 94 D4 38 30S 58 30W
Querétaro, *Mexico* 86 C4 20 40N 100 23W
Querétaro □, *Mexico* 86 C5 20 30N 100 0W
Queshan, *China* 34 H8 32 55N 114 2 E
Quesnel, *Canada* 72 C4 53 0N 122 30W
Quesnel →, *Canada* 72 C4 52 58N 122 29W
Quesnel L., *Canada* 72 C4 52 30N 121 20W
Questa, *U.S.A.* 83 H11 36 42N 105 36W
Quetico Prov. Park, *Canada* 70 C1 48 30N 91 45W
Quetta, *Pakistan* 42 D2 30 15N 66 55 E
Quezaltenango, *Guatemala* 88 D1 14 50N 91 30W
Quezon City, *Phil.* 37 B6 14 38N 121 0 E
Qufār, *Si. Arabia* 44 E4 27 26N 41 37 E
Qui Nhon, *Vietnam* 38 F7 13 40N 109 13 E
Quibaxe, *Angola* 52 F2 8 24S 14 27 E
Quibdo, *Colombia* 92 B3 5 42N 76 40W
Quiberon, *France* 18 C2 47 29N 3 9W
Quick, *Canada* 72 C3 54 36N 126 54W
Quiet L., *Canada* 72 A2 61 5N 133 5W
Quiindy, *Paraguay* 94 B4 25 58S 57 14W
Quila, *Mexico* 86 C3 24 23N 107 13W
Quilán, C., *Chile* 96 E2 43 15S 74 30W
Quilcene, *U.S.A.* 84 C4 47 49N 122 53W
Quilengues, *Angola* 53 G2 14 12S 14 12 E
Quilimarí, *Chile* 94 C1 32 5S 71 30W
Quilino, *Argentina* 94 C3 30 14S 64 29W
Quillabamba, *Peru* 92 F4 12 50S 72 50W
Quillagua, *Chile* 94 A2 21 40S 69 40W
Quillaicillo, *Chile* 94 C1 31 17S 71 40W
Quillota, *Chile* 94 C1 32 54S 71 16W
Quilmes, *Argentina* 94 C4 34 43S 58 15W
Quilon, *India* 40 Q10 8 50N 76 38 E
Quilpie, *Australia* 63 D3 26 35S 144 11 E
Quilpué, *Chile* 94 C1 33 5S 71 33W
Quilua, *Mozam.* 55 F4 16 17S 39 54 E
Quimilí, *Argentina* 94 B3 27 40S 62 30W
Quimper, *France* 18 B1 48 0N 4 9W
Quimperlé, *France* 18 C2 47 53N 3 33W
Quinault →, *U.S.A.* 84 C2 47 21N 124 18W
Quincy, *Calif., U.S.A.* 84 F6 39 56N 120 57W
Quincy, *Fla., U.S.A.* 77 K3 30 35N 84 34W
Quincy, *Ill., U.S.A.* 80 F9 39 56N 91 23W
Quincy, *Mass., U.S.A.* 79 D14 42 15N 71 0W
Quincy, *Wash., U.S.A.* 82 C4 47 22N 119 56W
Quines, *Argentina* 94 C2 32 13S 65 48W
Quinga, *Mozam.* 55 F5 15 49S 40 15 E
Quintana Roo □, *Mexico* 87 D7 19 0N 88 0W
Quintanar de la Orden, *Spain* 19 C4 39 36N 3 5W
Quintanar de la Sierra, *Spain* 19 B4 41 57N 2 55W
Quintero, *Chile* 94 C1 32 45S 71 30W
Quinyambie, *Australia* 63 E3 30 15S 141 0 E
Quipungo, *Angola* 53 G2 14 37S 14 40 E
Quirihue, *Chile* 94 D1 36 15S 72 35W
Quirindi, *Australia* 63 E5 31 28S 150 40 E
Quissanga, *Mozam.* 55 E5 12 24S 40 28 E
Quitilipi, *Argentina* 94 B3 26 50S 60 13W
Quitman, *Ga., U.S.A.* 77 K4 30 47N 83 34W
Quitman, *Miss., U.S.A.* 77 J1 32 2N 88 44W
Quitman, *Tex., U.S.A.* 81 J7 32 48N 95 27W
Quito, *Ecuador* 92 D3 0 15S 78 35W
Quixadá, *Brazil* 93 D11 4 55S 39 0W
Quixaxe, *Mozam.* 55 F5 15 17S 40 4 E
Qumbu, *S. Africa* 57 E4 31 10S 28 48 E
Quneitra, *Syria* 47 B4 33 7N 35 48 E
Quoin I., *Australia* 60 B4 14 54S 129 32 E
Quoin Pt., *S. Africa* 56 E2 34 46S 19 37 E
Quondong, *Australia* 63 E3 33 6S 140 18 E
Quorn, *Australia* 63 E2 32 25S 138 5 E
Quqon = Kokand, *Uzbekistan* 26 E8 40 30N 70 57 E
Qurnat as Sawdā', *Lebanon* 47 A5 34 18N 36 6 E
Qûs, *Egypt* 51 C11 25 55N 32 50 E
Qusaybah, *Iraq* 44 C4 34 24N 40 59 E
Quseir, *Egypt* 51 C11 26 7N 34 16 E
Qūshchī, *Iran* 44 B5 37 59N 45 3 E
Quthing, *Lesotho* 57 E4 30 25S 27 36 E
Qūţīābād, *Iran* 45 C6 35 47N 48 30 E
Quwo, *China* 34 G6 35 38N 111 25 E
Quyang, *China* 34 E8 38 35N 114 40 E
Quynh Nhai, *Vietnam* 38 B4 21 49N 103 33 E
Quzi, *China* 34 F4 36 20N 107 20 E
Qyzylorda = Kzyl-Orda, *Kazakhstan* 26 E7 44 48N 65 28 E

R

Ra, Ko, *Thailand* 39 H2 9 13N 98 16 E
Raahe, *Finland* 8 D18 64 40N 24 28 E
Raasay, *U.K.* 12 D2 57 25N 6 4W
Raasay, Sd. of, *U.K.* 12 D2 57 30N 6 8W
Raba, *Indonesia* 37 F5 8 36S 118 55 E
Rabai, *Kenya* 54 C4 3 50S 39 31 E
Rabat, *Malta* 23 D1 35 53N 14 25 E
Rabat, *Morocco* 50 B3 34 2N 6 48W
Rabaul, *Papua N. G.* 64 H7 4 24S 152 18 E
Rabbit →, *Canada* 72 B3 59 41N 127 12W
Rabbit Lake, *Canada* 73 C7 53 8N 107 46W
Rabbitskin →, *Canada* 72 A4 61 47N 120 42W
Rābor, *Iran* 45 D8 29 17N 56 55 E
Race, C., *Canada* 71 C9 46 40N 53 5W
Rach Gia, *Vietnam* 39 G5 10 5N 105 5 E
Racine, *U.S.A.* 76 D2 42 41N 87 51W
Rackerby, *U.S.A.* 84 F5 39 26N 121 22W
Radama, Nosy, *Madag.* 57 A8 14 0S 47 47 E
Radama, Saikanosy, *Madag.* 57 A8 14 16S 47 53 E
Rădăuţi, *Romania* 17 E12 47 50N 25 59 E
Radford, *U.S.A.* 76 G5 37 8N 80 34W
Radhanpur, *India* 42 H4 23 50N 71 38 E
Radisson, *Canada* 73 C7 52 30N 107 20W
Radium Hot Springs, *Canada* 72 C5 50 35N 116 2W
Radnor Forest, *U.K.* 11 E4 52 17N 3 10W
Radom, *Poland* 17 C10 51 23N 21 12 E
Radomir, *Bulgaria* 21 C10 42 37N 23 4 E
Radomsko, *Poland* 17 C9 51 5N 19 28 E
Radstock, *U.K.* 11 F5 51 17N 2 25W
Radstock, C., *Australia* 63 E1 33 12S 134 20 E
Radville, *Canada* 73 D8 49 30N 104 15W
Rae, *Canada* 72 A5 62 50N 116 3W
Rae Bareli, *India* 43 F9 26 18N 81 20 E
Rae Isthmus, *Canada* 69 B11 66 40N 87 30W
Raeren, *Belgium* 15 D6 50 41N 6 7 E
Raeside, L., *Australia* 61 E3 29 20S 122 0 E
Raetihi, *N.Z.* 59 H5 39 25S 175 17 E
Rafaela, *Argentina* 94 C3 31 10S 61 30W
Rafah, *Egypt* 47 D3 31 18N 34 14 E
Rafai, *C.A.R.* 54 B1 4 59N 23 58 E
Rafḩā, *Si. Arabia* 44 D4 29 35N 43 35 E
Rafsanjān, *Iran* 45 D8 30 30N 56 5 E
Raft Pt., *Australia* 60 C3 16 4S 124 26 E
Ragama, *Sri Lanka* 40 R11 7 0N 79 50 E
Ragged, Mt., *Australia* 61 F3 33 27S 123 25 E
Raglan, *Australia* 62 C5 23 42S 150 49 E
Raglan, *N.Z.* 59 G5 37 55S 174 55 E
Ragusa, *Italy* 20 F6 36 56N 14 42 E
Raha, *Indonesia* 37 E6 4 55S 123 0 E
Rahad al Bardī, *Sudan* 51 F9 11 20N 23 40 E
Rahaeng = Tak, *Thailand* 38 D2 16 52N 99 8 E
Raḩīmah, *Si. Arabia* 45 E6 26 42N 50 4 E
Rahimyar Khan, *Pakistan* 42 E4 28 30N 70 25 E
Rāhjerd, *Iran* 45 C6 34 22N 50 22 E
Raichur, *India* 40 L10 16 10N 77 20 E
Raiganj, *India* 43 G13 25 37N 88 10 E
Raigarh, *India* 41 J13 21 56N 83 25 E
Raijua, *Indonesia* 37 F6 10 37S 121 36 E
Railton, *Australia* 62 G4 41 25S 146 28 E
Rainbow Lake, *Canada* 72 B5 58 30N 119 23W
Rainier, *U.S.A.* 84 D4 46 53N 122 41W
Rainier, Mt., *U.S.A.* 84 D5 46 52N 121 46W
Rainy L., *Canada* 73 D10 48 42N 93 10W
Rainy River, *Canada* 73 D10 48 43N 94 29W
Raipur, *India* 41 J12 21 17N 81 45 E
Raj Nandgaon, *India* 41 J12 21 5N 81 5 E
Raja, Ujung, *Indonesia* 36 D1 3 40N 96 25 E
Raja Ampat, Kepulauan, *Indonesia* 37 E7 0 30S 130 0 E
Rajahmundry, *India* 41 L12 17 1N 81 48 E
Rajajooseppi, *Finland* 8 B20 68 28N 28 29 E
Rajang →, *Malaysia* 36 D4 2 30N 112 0 E
Rajapalaiyam, *India* 40 Q10 9 25N 77 35 E
Rajasthan □, *India* 42 F5 26 45N 73 30 E
Rajasthan Canal, *India* 42 E5 28 0N 72 0 E
Rajauri, *India* 43 C6 33 25N 74 21 E
Rajgarh, *Mad. P., India* 42 G7 24 2N 76 45 E
Rajgarh, *Raj., India* 42 E6 28 40N 75 25 E
Rajkot, *India* 42 H4 22 15N 70 56 E
Rajmahal Hills, *India* 43 G12 24 30N 87 30 E
Rajpipla, *India* 40 J8 21 50N 73 30 E
Rajpura, *India* 42 D7 30 25N 76 32 E
Rajshahi, *Bangla.* 41 G16 24 22N 88 39 E
Rajshahi □, *Bangla.* 43 G13 25 0N 89 0 E
Rakaia, *N.Z.* 59 K4 43 45S 172 1 E
Rakaia →, *N.Z.* 59 K4 43 36S 172 15 E
Rakan, Ra's, *Qatar* 45 E6 26 10N 51 20 E
Rakaposhi, *Pakistan* 43 A6 36 10N 74 25 E
Rakata, Pulau, *Indonesia* 36 F3 6 10S 105 20 E
Rakhni, *Pakistan* 42 D3 30 4N 69 56 E
Rakitnoye, *Russia* 30 B7 45 36N 134 17 E
Rakops, *Botswana* 56 C3 21 1S 24 28 E
Raleigh, *U.S.A.* 77 H6 35 47N 78 39W
Raleigh B., *U.S.A.* 77 H7 34 50N 76 15W
Ralls, *U.S.A.* 81 J4 33 41N 101 24W
Ram →, *Canada* 72 A4 62 1N 123 41W
Rām Allāh, *Jordan* 47 D4 31 55N 35 10 E
Ram Hd., *Australia* 63 F4 37 47S 149 30 E
Rama, *Nic.* 88 D3 12 9N 84 15W
Raman, *Thailand* 39 J3 6 29N 101 18 E
Ramanathapuram, *India* 40 Q11 9 25N 78 55 E
Ramanetaka, B. de, *Madag.* 57 A8 14 13S 47 52 E
Ramat Gan, *Israel* 47 C3 32 4N 34 48 E
Ramatlhabama, *S. Africa* 56 D4 25 37S 25 33 E
Ramban, *India* 43 C6 33 14N 75 12 E
Rambipuji, *Indonesia* 37 H15 8 12S 113 37 E
Ramea, *Canada* 71 C8 47 31N 57 23W
Ramechhap, *Nepal* 43 F12 27 25N 86 10 E
Ramelau, *Indonesia* 37 F7 8 55S 126 22 E
Ramgarh, *Bihar, India* 43 H11 23 40N 85 35 E
Ramgarh, *Raj., India* 42 F6 27 16N 75 14 E
Ramgarh, *Raj., India* 42 F4 27 30N 70 36 E
Rāmhormoz, *Iran* 45 D6 31 15N 49 35 E
Ramīān, *Iran* 45 B7 37 3N 55 16 E
Ramingining, *Australia* 62 A2 12 19S 135 3 E
Ramla, *Israel* 47 D3 31 55N 34 52 E
Ramnad = Ramanathapuram, *India* 40 Q11 9 25N 78 55 E
Ramnagar, *India* 43 C6 32 47N 75 18 E
Ramona, *U.S.A.* 85 M10 33 2N 116 52W
Ramore, *Canada* 70 C3 48 30N 80 25W
Ramotswa, *Botswana* 56 C4 24 50S 25 52 E
Rampur, *H.P., India* 42 D7 31 26N 77 43 E
Rampur, *Mad. P., India* 42 H5 23 25N 73 53 E
Rampur, *Ut. P., India* 43 E8 28 50N 79 5 E
Rampur Hat, *India* 43 G12 24 10N 87 50 E
Rampura, *India* 42 G6 24 30N 75 27 E
Ramree I. = Ramree Kyun, *Burma* 41 K18 19 0N 94 0 E
Ramree Kyun, *Burma* 41 K18 19 0N 94 0 E
Rāmsar, *Iran* 45 B6 36 53N 50 41 E
Ramsey, *Canada* 70 C3 47 25N 82 20W
Ramsey, *U.K.* 10 C3 54 20N 4 21W
Ramsgate, *U.K.* 11 F9 51 20N 1 25 E
Ramtek, *India* 40 J11 21 20N 79 15 E
Ranaghat, *India* 43 H13 23 15N 88 35 E
Ranahu, *Pakistan* 42 G3 25 55N 69 45 E
Ranau, *Malaysia* 36 C5 6 2N 116 40 E
Rancagua, *Chile* 94 C1 34 10S 70 50W
Rancheria →, *Canada* 72 A3 60 13N 129 7W
Ranchester, *U.S.A.* 82 D10 44 54N 107 10W
Ranchi, *India* 43 H11 23 19N 85 27 E
Randers, *Denmark* 9 H11 56 29N 10 1 E
Randfontein, *S. Africa* 57 D4 26 8S 27 45 E
Randle, *U.S.A.* 84 D5 46 32N 121 57W
Randolph, *Mass., U.S.A.* 79 D13 42 10N 71 2W
Randolph, *N.Y., U.S.A.* 78 D6 42 10N 78 59W
Randolph, *Utah, U.S.A.* 82 F8 41 40N 111 11W
Randolph, *Vt., U.S.A.* 79 C12 43 55N 72 40W
Råne älv →, *Sweden* 8 D17 65 50N 22 20 E
Rangae, *Thailand* 39 J3 6 19N 101 44 E
Rangaunu B., *N.Z.* 59 F4 34 51S 173 15 E
Rangeley, *U.S.A.* 79 B14 44 58N 70 39W
Rangely, *U.S.A.* 82 F9 40 5N 108 48W
Ranger, *U.S.A.* 81 J5 32 28N 98 41W
Rangia, *India* 41 F17 26 28N 91 38 E
Rangiora, *N.Z.* 59 K4 43 19S 172 36 E
Rangitaiki →, *N.Z.* 59 G6 37 54S 176 49 E
Rangitata →, *N.Z.* 59 K3 43 45S 171 15 E
Rangkasbitung, *Indonesia* 37 G12 6 21S 106 15 E
Rangon →, *Burma* 41 L20 16 28N 96 40 E
Rangoon, *Burma* 41 L20 16 45N 96 20 E
Rangpur, *Bangla.* 41 G16 25 42N 89 22 E
Rangsit, *Thailand* 38 F3 13 59N 100 37 E
Ranibennur, *India* 40 M9 14 35N 75 30 E
Raniganj, *India* 43 H12 23 40N 87 5 E
Raniwara, *India* 40 G8 24 50N 72 10 E
Rāniyah, *Iraq* 44 B5 36 15N 44 53 E
Ranken →, *Australia* 62 C2 20 31S 137 36 E
Rankin, *U.S.A.* 81 K4 31 13N 101 56W
Rankin Inlet, *Canada* 68 B10 62 30N 93 0W
Rankins Springs, *Australia* 63 E4 33 49S 146 14 E
Rannoch, L., *U.K.* 12 E4 56 41N 4 20W
Rannoch Moor, *U.K.* 12 E4 56 38N 4 48W
Ranobe, Helodranon' i, *Madag.* 57 C7 23 3S 43 33 E

Ranohira

Ranohira, Madag. 57 C8 22 29S 45 24 E
Ranomafana, Toamasina, Madag. 57 B8 18 57S 48 50 E
Ranomafana, Toliara, Madag. 57 C8 24 34S 47 0 E
Ranong, Thailand 39 H2 9 56N 98 40 E
Rānsa, Iran 45 C6 33 39N 48 18 E
Ransiki, Indonesia 37 E8 1 30S 134 10 E
Rantau, Indonesia 36 E5 2 56S 115 9 E
Rantauprapat, Indonesia 36 D1 2 15N 99 50 E
Rantekombola, Indonesia 37 E5 3 15S 119 57 E
Rantoul, U.S.A. 76 E1 40 19N 88 9W
Raoyang, China 34 E8 38 15N 115 45 E
Rapa, Pac. Oc. 65 K13 27 35S 144 20W
Rāpch, Iran 45 E8 25 40N 59 15 E
Rapid →, Canada 72 B3 59 15N 129 5W
Rapid City, U.S.A. 80 D3 44 5N 103 14W
Rapid River, U.S.A. 76 C2 45 55N 86 58W
Rapides des Joachims, Canada 70 C4 46 13N 77 43W
Rarotonga, Cook Is. 65 K12 21 30S 160 0W
Ra's al' Ayn, Syria 44 B4 36 51N 40 4 E
Ra's al Khaymah, U.A.E. 45 E8 25 50N 56 5 E
Ra's al-Unuf, Libya 51 B8 30 25N 18 15 E
Ra's an Naqb, Jordan 47 F4 30 0N 35 29 E
Ras Bânâs, Egypt 51 D12 23 57N 35 59 E
Ras Dashen, Ethiopia 51 F12 13 8N 38 26 E
Râs Timirist, Mauritania 50 E1 19 21N 16 30W
Rasa, Punta, Argentina 96 E4 40 50S 62 15W
Rasca, Pta. de la, Canary Is. 22 G3 27 59N 16 41W
Rashad, Sudan 51 F11 11 55N 31 0 E
Rashîd, Egypt 51 B11 31 21N 30 22 E
Rasht, Iran 45 B6 37 20N 49 40 E
Rasi Salai, Thailand 38 E5 15 20N 104 9 E
Rason L., Australia 61 E3 28 45S 124 25 E
Rasra, India 43 G10 25 50N 83 50 E
Rat Buri, Thailand 38 F2 13 30N 99 54 E
Rat Islands, U.S.A. 68 C1 52 0N 178 0 E
Rat River, Canada 72 A6 61 7N 112 36W
Ratangarh, India 42 E6 28 5N 74 35 E
Raţāwī, Iraq 44 D5 30 38N 47 13 E
Rath, India 43 G8 25 36N 79 37 E
Rath Luirc, Ireland 13 D3 52 21N 8 40W
Rathdrum, Ireland 13 D5 52 57N 6 13W
Rathenow, Germany 16 B6 52 38N 12 23 E
Rathkeale, Ireland 13 D3 52 32N 8 57W
Rathlin, U.K. 13 A5 55 18N 6 14W
Rathlin O'Birne I., Ireland 13 B3 54 40N 8 50W
Ratlam, India 42 H6 23 20N 75 0 E
Ratnagiri, India 40 L8 16 57N 73 18 E
Raton, U.S.A. 81 G2 36 54N 104 24W
Rattaphum, Thailand 39 J3 7 8N 100 16 E
Rattray Hd., U.K. 12 D7 57 38N 1 50W
Ratz, Mt., Canada 72 B2 57 23N 132 12W
Raub, Malaysia 39 L3 3 47N 101 52 E
Rauch, Argentina 94 D4 36 45S 59 5W
Raufarhöfn, Iceland 8 C6 66 27N 15 57W
Raukumara Ra., N.Z. 59 H6 38 5S 177 55 E
Rauma, Finland 9 F16 61 10N 21 30 E
Raurkela, India 43 H11 22 14N 84 50 E
Rausu-Dake, Japan 30 B12 44 4N 145 7 E
Ravānsar, Iran 44 C5 34 43N 46 40 E
Rāvar, Iran 45 D8 31 20N 56 51 E
Ravena, U.S.A. 79 D11 42 28N 73 49W
Ravenna, Italy 20 B5 44 28N 12 15 E
Ravenna, Nebr., U.S.A. 80 E5 41 1N 98 55W
Ravenna, Ohio, U.S.A. 78 E3 41 9N 81 15W
Ravensburg, Germany 16 E4 47 48N 9 38 E
Ravenshoe, Australia 62 B4 17 37S 145 29 E
Ravensthorpe, Australia 61 F3 33 35S 120 2 E
Ravenswood, Australia 62 C4 20 6S 146 54 E
Ravenswood, U.S.A. 76 F5 38 57N 81 46W
Ravi →, Pakistan 42 D4 30 35N 71 49 E
Rawalpindi, Pakistan 42 C5 33 38N 73 8 E
Rawāndūz, Iraq 44 B5 36 40N 44 30 E
Rawang, Malaysia 39 L3 3 20N 101 35 E
Rawdon, Canada 70 C5 46 3N 73 40W
Rawene, N.Z. 59 F4 35 25S 173 32 E
Rawlinna, Australia 61 F4 30 58S 125 28 E
Rawlins, U.S.A. 82 F10 41 47N 107 14W
Rawlinson Ra., Australia 61 D4 24 40S 128 30 E
Rawson, Argentina 96 E3 43 15S 65 5W
Ray, U.S.A. 80 A3 48 21N 103 10W
Ray, C., Canada 71 C8 47 33N 59 15W
Rayadurg, India 40 M10 14 40N 76 50 E
Rayagada, India 41 K13 19 15N 83 20 E
Raychikhinsk, Russia 27 E13 49 46N 129 25 E
Rāyen, Iran 45 D8 29 34N 57 26 E
Raymond, Canada 72 D6 49 30N 112 35W
Raymond, Calif., U.S.A. 84 H7 37 13N 119 54W
Raymond, Wash., U.S.A. 84 D3 46 41N 123 44W
Raymondville, U.S.A. 81 M6 26 29N 97 47W
Raymore, Canada 73 C8 51 25N 104 31W
Rayne, U.S.A. 81 K8 30 14N 92 16W
Rayón, Mexico 86 B2 29 43N 110 35W
Rayong, Thailand 38 F3 12 40N 101 20 E
Rayville, U.S.A. 81 J9 32 29N 91 46W
Raz, Pte. du, France 18 B1 48 2N 4 47W
Razan, Iran 45 C6 35 23N 49 2 E
Razdolnoye, Russia 30 C5 43 30N 131 52 E
Razeh, Iran 45 C6 32 47N 48 9 E
Razgrad, Bulgaria 21 C12 43 33N 26 34 E
Razmak, Pakistan 42 C3 32 45N 69 50 E
Ré, I. de, France 18 C3 46 12N 1 30W
Reading, U.K. 11 F7 51 27N 0 57W
Reading, U.S.A. 79 F9 40 20N 75 56W
Realicó, Argentina 94 D3 35 0S 64 15W
Reata, Mexico 86 B4 26 8N 101 5W
Rebecca, L., Australia 61 F3 30 0S 122 15 E
Rebi, Indonesia 37 F8 6 23S 134 7 E
Rebiana, Libya 51 D9 24 12N 22 10 E
Rebun-Tō, Japan 30 B10 45 23N 141 2 E
Recherche, Arch. of the, Australia 61 F3 34 15S 122 50 E
Recife, Brazil 93 E12 8 0S 35 0W
Recklinghausen, Germany 15 C7 51 36N 7 10 E
Reconquista, Argentina 94 B4 29 10S 59 45W
Recreo, Argentina 94 B2 29 25S 65 10W
Red →, La., U.S.A. 81 K9 31 1N 91 45W
Red →, N. Dak., U.S.A. 80 A6 49 0N 97 15W
Red Bank, U.S.A. 79 F10 40 21N 74 5W
Red Bay, Canada 71 B8 51 44N 56 25W
Red Bluff, U.S.A. 82 F2 40 11N 122 15W
Red Bluff L., U.S.A. 81 K3 31 54N 103 55W
Red Cliffs, Australia 63 E3 34 19S 142 11 E
Red Cloud, U.S.A. 80 E5 40 5N 98 32W
Red Deer, Canada 72 C6 52 20N 113 50W
Red Deer →, Alta., Canada 73 C6 50 58N 110 0W
Red Deer →, Man., Canada 73 C8 52 53N 101 1W
Red Deer L., Canada 73 C8 52 55N 101 20W
Red Indian L., Canada 71 C8 48 35N 57 0W
Red Lake, Canada 73 C10 51 3N 93 49W
Red Lake Falls, U.S.A. 80 B6 47 53N 96 16W
Red Lodge, U.S.A. 82 D9 45 11N 109 15W
Red Mountain, U.S.A. 85 K9 35 37N 117 38W
Red Oak, U.S.A. 80 E7 41 1N 95 14W
Red Rock, Canada 70 C2 48 55N 88 15W
Red Rock, L., U.S.A. 80 E8 41 22N 92 59W
Red Rocks Pt., Australia 61 F4 32 13S 127 32 E
Red Sea, Asia 46 C2 25 0N 36 0 E
Red Slate Mt., U.S.A. 84 H8 37 31N 118 52W
Red Sucker L., Canada 73 C10 54 9N 93 40W
Red Tower Pass = Turnu Rosu Pasul, Romania 17 F12 45 33N 24 17 E
Red Wing, U.S.A. 80 C8 44 34N 92 31W
Redbridge, U.K. 11 F8 51 35N 0 7 E
Redcar, U.K. 10 C6 54 37N 1 4W
Redcliff, Canada 73 C6 50 10N 110 50W
Redcliffe, Australia 63 D5 27 12S 153 0 E
Redcliffe, Mt., Australia 61 E3 28 30S 121 30 E
Reddersburg, S. Africa 56 D4 29 41S 26 10 E
Redding, U.S.A. 82 F2 40 35N 122 24W
Redditch, U.K. 11 E6 52 18N 1 57W
Redfield, U.S.A. 80 C5 44 53N 98 31W
Redknife →, Canada 72 A5 61 14N 119 22W
Redlands, U.S.A. 85 M9 34 4N 117 11W
Redmond, Australia 61 F2 34 55S 117 40 E
Redmond, Oreg., U.S.A. 82 D3 44 17N 121 11W
Redmond, Wash., U.S.A. 84 C4 47 41N 122 7W
Redonda, Antigua 89 C7 16 58N 62 19W
Redondela, Spain 19 A1 42 15N 8 38W
Redondo, Portugal 19 C2 38 39N 7 37W
Redondo Beach, U.S.A. 85 M8 33 50N 118 23W
Redrock Pt., Canada 72 A5 62 11N 115 2W
Redruth, U.K. 11 G2 50 14N 5 14W
Redvers, Canada 73 D8 49 35N 101 40W
Redwater, Canada 72 C6 53 55N 113 6W
Redwood, U.S.A. 79 B9 44 18N 75 48W
Redwood City, U.S.A. 83 H2 37 30N 122 15W
Redwood Falls, U.S.A. 80 C7 44 32N 95 7W
Ree, L., Ireland 13 C4 53 35N 8 0W
Reed, L., Canada 73 C8 54 38N 100 30W
Reed City, U.S.A. 76 D3 43 53N 85 31W
Reeder, U.S.A. 80 B3 46 7N 102 57W
Reedley, U.S.A. 83 H4 36 36N 119 27W
Reedsburg, U.S.A. 80 D9 43 32N 90 0W
Reedsport, U.S.A. 82 E1 43 42N 124 6W
Reefton, N.Z. 59 K3 42 6S 171 51 E
Refugio, U.S.A. 81 L6 28 18N 97 17W
Regensburg, Germany 16 D6 49 1N 12 7 E
Réggio di Calábria, Italy 20 E6 38 7N 15 38 E
Réggio nell' Emilia, Italy 20 B4 44 42N 10 38 E
Regina, Canada 73 C8 50 27N 104 35W
Registro, Brazil 95 A6 24 29S 47 49W
Rehar →, India 43 H10 23 55N 82 40 E
Rehoboth, Namibia 56 C2 23 15S 17 4 E
Rehovot, Israel 47 D3 31 54N 34 48 E
Rei-Bouba, Cameroon 51 G7 8 40N 14 15 E
Reichenbach, Germany 16 C6 50 36N 12 19 E
Reid, Australia 61 F4 30 49S 128 26 E
Reid River, Australia 62 B4 19 40S 146 48 E
Reidsville, U.S.A. 77 G6 36 21N 79 40W
Reigate, U.K. 11 F7 51 14N 0 11W
Reims, France 18 B6 49 15N 4 1 E
Reina Adelaida, Arch., Chile 96 G2 52 20S 74 0W
Reinbeck, U.S.A. 80 D8 42 19N 92 36W
Reindeer →, Canada 73 B8 55 36N 103 11W
Reindeer I., Canada 73 C9 52 30N 98 0W
Reindeer L., Canada 73 B8 57 15N 102 15W
Reinga, C., N.Z. 59 F4 34 25S 172 43 E
Reitz, S. Africa 57 D4 27 48S 28 29 E
Reivilo, S. Africa 56 D3 27 36S 24 8 E
Rekinniki, Russia 27 C17 60 51N 163 40 E
Reliance, Canada 73 A7 63 0N 109 20W
Remarkable, Mt., Australia 63 E2 32 48S 138 10 E
Rembang, Indonesia 37 G14 6 42S 111 21 E
Remedios, Panama 88 E3 8 15N 81 50W
Remeshk, Iran 45 E8 26 55N 58 50 E
Remich, Lux. 15 E6 49 32N 6 22 E
Remscheid, Germany 16 C3 51 11N 7 12 E
Ren Xian, China 34 F8 37 8N 114 40 E
Rendsburg, Germany 16 A4 54 18N 9 41 E
Rene, Russia 27 C19 66 2N 179 25W
Renfrew, Canada 70 C4 45 30N 76 40W
Renfrew, U.K. 12 F4 55 52N 4 24W
Rengat, Indonesia 36 E2 0 30S 102 45 E
Rengo, Chile 94 C1 34 24S 70 50W
Renk, Sudan 51 F11 11 50N 32 50 E
Renkum, Neths. 15 C5 51 58N 5 43 E
Renmark, Australia 63 E3 34 11S 140 43 E
Rennell Sd., Canada 72 C2 53 23N 132 35W
Renner Springs T.O., Australia 62 B1 18 20S 133 47 E
Rennes, France 18 B3 48 7N 1 41W
Reno, U.S.A. 84 F7 39 31N 119 48W
Reno →, Italy 20 B5 44 37N 12 17 E
Renovo, U.S.A. 78 E7 41 20N 77 45W
Renqiu, China 34 E9 38 43N 116 5 E
Rensselaer, Ind., U.S.A. 76 E2 40 57N 87 9W
Rensselaer, N.Y., U.S.A. 79 D11 42 38N 73 45W
Renton, U.S.A. 84 C4 47 29N 122 12W
Reotipur, India 43 G10 25 33N 83 45 E
Republic, Mich., U.S.A. 76 B2 46 25N 87 59W
Republic, Wash., U.S.A. 82 B4 48 39N 118 44W
Republican →, U.S.A. 80 F6 39 4N 96 48W
Republican City, U.S.A. 80 E5 40 6N 99 13W
Repulse Bay, Canada 69 B11 66 30N 86 30W
Requena, Peru 92 E4 5 5S 73 52W
Requena, Spain 19 C5 39 30N 1 4W
Reserve, Canada 73 C8 52 28N 102 39W
Reserve, U.S.A. 83 K9 33 43N 108 45W
Resht = Rasht, Iran 45 B6 37 20N 49 40 E
Resistencia, Argentina 94 B4 27 30S 59 0W
Reşiţa, Romania 17 F10 45 18N 21 53 E
Resolution I., Canada 69 B13 61 30N 65 0W
Resolution I., N.Z. 59 L1 45 40S 166 40 E
Ressano Garcia, Mozam. 57 D5 25 25S 32 0 E
Reston, Canada 73 D8 49 33N 101 6W
Retalhuleu, Guatemala 88 D1 14 33N 91 46W
Retenue, L. de, Zaïre 55 E2 11 0S 27 0 E
Réthímnon, Greece 23 D6 35 18N 24 30 E
Réthímnon □, Greece 23 D6 35 23N 24 28 E
Réunion ■, Ind. Oc. 49 J9 21 0S 56 0 E
Reutlingen, Germany 16 D4 48 28N 9 13 E
Reval = Tallinn, Estonia 24 C3 59 22N 24 48 E
Revda, Russia 24 C10 56 48N 59 57 E
Revelganj, India 43 G11 25 50N 84 40 E
Revelstoke, Canada 72 C5 51 0N 118 10W
Revilla Gigedo, Is., Pac. Oc. 65 F16 18 40N 112 0W
Revillagigedo I., U.S.A. 72 B2 55 50N 131 20W
Revuè →, Mozam. 55 F3 19 50S 34 0 E
Rewa, India 43 G9 24 33N 81 25 E
Rewari, India 42 E7 28 15N 76 40 E
Rexburg, U.S.A. 82 E8 43 49N 111 47W
Rey, Iran 45 C6 35 35N 51 25 E
Rey Malabo, Eq. Guin. 50 H6 3 45N 8 50 E
Reyes, Pt., U.S.A. 84 H3 38 0N 123 0W
Reykjahlíð, Iceland 8 D5 65 40N 16 55W
Reykjanes, Iceland 8 E2 63 48N 22 40W
Reykjavík, Iceland 8 D3 64 10N 21 57W
Reynolds, Canada 73 D9 49 40N 95 55W
Reynolds Ra., Australia 60 D5 22 30S 133 0 E
Reynoldsville, U.S.A. 78 E6 41 5N 78 58W
Reynosa, Mexico 87 B5 26 5N 98 18W
Rezvān, Iran 45 E8 27 34N 56 6 E
Rhayader, U.K. 11 E4 52 19N 3 30W
Rheden, Neths. 15 B6 52 0N 6 3 E
Rhein, Canada 73 C8 51 25N 102 15W
Rhein →, Europe 15 C6 51 52N 6 2 E
Rheine, Germany 16 B3 52 17N 7 25 E
Rheinland-Pfalz □, Germany 16 C3 50 0N 7 0 E
Rhin = Rhein →, Europe 15 C6 51 52N 6 2 E
Rhine = Rhein →, Europe 15 C6 51 52N 6 2 E
Rhineland-Palatinate □ = Rheinland-Pfalz □, Germany 16 C3 50 0N 7 0 E
Rhinelander, U.S.A. 80 C10 45 38N 89 25W
Rhino Camp, Uganda 54 B3 3 0N 31 22 E
Rhode Island □, U.S.A. 79 E13 41 40N 71 30W
Rhodes = Ródhos, Greece 23 C10 36 15N 28 10 E
Rhodesia = Zimbabwe ■, Africa 55 F2 19 0S 30 0 E
Rhodope Mts. = Rhodopi Planina, Bulgaria 21 D11 41 40N 24 20 E
Rhodopi Planina, Bulgaria 21 D11 41 40N 24 20 E
Rhön = Hohe Rhön, Germany 16 C4 50 24N 9 58 E
Rhondda, U.K. 11 F4 51 39N 3 30W
Rhône □, France 18 D6 45 54N 4 35 E
Rhône →, France 18 E6 43 28N 4 42 E
Rhum, U.K. 12 E2 57 0N 6 20W
Rhyl, U.K. 10 D4 53 19N 3 29W
Rhymney, U.K. 11 F4 51 45N 3 17W
Riachão, Brazil 93 E9 7 20S 46 37W
Riasi, India 43 C6 33 10N 74 50 E
Riau □, Indonesia 36 D2 0 0 102 35 E
Riau, Kepulauan, Indonesia 36 D2 0 30N 104 20 E
Riau Arch. = Riau, Kepulauan, Indonesia 36 D2 0 30N 104 20 E
Ribadeo, Spain 19 A2 43 35N 7 5W
Ribatejo, Portugal 19 C1 39 15N 8 30W
Ribble →, U.K. 10 C5 54 13N 2 20W
Ribe, Denmark 9 J10 55 19N 8 44 E
Ribeira Brava, Madeira 22 D2 32 41N 17 4W
Ribeirão Prêto, Brazil 95 A6 21 10S 47 50W
Riberalta, Bolivia 92 F5 11 0S 66 0W
Riccarton, N.Z. 59 K4 43 32S 172 37 E
Rice, U.S.A. 85 L12 34 5N 114 51W
Rice L., Canada 78 B6 44 12N 78 10W
Rice Lake, U.S.A. 80 C9 45 30N 91 44W
Rich Hill, U.S.A. 81 F7 38 6N 94 22W
Richards Bay, S. Africa 57 D5 28 48S 32 6 E
Richards L., Canada 73 B7 59 10N 107 10W
Richardson →, Canada 73 B6 58 25N 111 14W
Richardson Springs, U.S.A. 84 F5 39 51N 121 46W
Richardton, U.S.A. 80 B3 46 53N 102 19W
Riche, C., Australia 61 F2 34 36S 118 47 E
Richey, U.S.A. 80 B2 47 39N 105 4W
Richfield, Idaho, U.S.A. 82 E6 43 3N 114 9W
Richfield, Utah, U.S.A. 83 G8 38 46N 112 5W
Richford, U.S.A. 79 B12 45 0N 72 40W
Richibucto, Canada 71 C7 46 42N 64 54W
Richland, Ga., U.S.A. 77 J3 32 5N 84 40W
Richland, Oreg., U.S.A. 82 D5 44 46N 117 10W
Richland, Wash., U.S.A. 82 C4 46 17N 119 18W
Richland Center, U.S.A. 80 D9 43 21N 90 23W
Richlands, U.S.A. 76 G5 37 6N 81 48W
Richmond, N.S.W., Australia 63 E5 33 35S 150 42 E
Richmond, Queens., Australia 62 C3 20 43S 143 8 E
Richmond, N.Z. 59 J4 41 20S 173 12 E
Richmond, S. Africa 57 D5 29 51S 30 18 E
Richmond, U.K. 10 C6 54 24N 1 43W
Richmond, Calif., U.S.A. 84 H4 37 56N 122 21W
Richmond, Ind., U.S.A. 76 F3 39 50N 84 53W
Richmond, Ky., U.S.A. 76 G3 37 45N 84 18W
Richmond, Mich., U.S.A. 78 D2 42 49N 82 45W
Richmond, Mo., U.S.A. 80 F8 39 17N 93 58W
Richmond, Tex., U.S.A. 81 L7 29 35N 95 46W
Richmond, Utah, U.S.A. 82 F8 41 56N 111 48W
Richmond, Va., U.S.A. 76 G7 37 33N 77 27W
Richmond Ra., Australia 63 D5 29 0S 152 45 E
Richmond-upon-Thames, U.K. 11 F7 51 28N 0 18W
Richton, U.S.A. 77 K1 31 16N 88 56W
Richwood, U.S.A. 76 F5 38 14N 80 32W
Ridder, Kazakhstan 26 D9 50 20N 83 30 E
Ridgecrest, U.S.A. 85 K9 35 38N 117 40W
Ridgedale, Canada 73 C8 53 0N 104 10W
Ridgefield, U.S.A. 84 E4 45 49N 122 45W
Ridgeland, U.S.A. 77 J5 32 29N 80 59W
Ridgelands, Australia 62 C5 23 16S 150 17 E
Ridgetown, Canada 70 D3 42 26N 81 52W
Ridgewood, U.S.A. 79 F10 40 59N 74 7W
Ridgway, U.S.A. 78 E6 41 25N 78 44W
Riding Mountain Nat. Park, Canada 73 C8 50 50N 100 0W
Ridley, Mt., Australia 61 F3 33 12S 122 7 E
Ried, Austria 16 D6 48 14N 13 30 E
Riet →, S. Africa 56 D3 29 0S 23 54 E
Rieti, Italy 20 C5 42 23N 12 50 E
Riffe L., U.S.A. 84 D4 46 32N 122 26W
Rifle, U.S.A. 82 G10 39 32N 107 47W
Rifstangi, Iceland 8 C5 66 32N 16 12W
Rift Valley □, Kenya 54 B4 0 20N 36 0 E
Rig Rig, Chad 51 F7 14 13N 14 25 E
Riga, Latvia 24 C3 56 53N 24 8 E
Riga, G. of = Rīgas Jūras Līcis, Latvia 24 C3 57 40N 23 45 E
Rīgān, Iran 45 D8 28 37N 58 58 E
Rīgas Jūras Līcis, Latvia 24 C3 57 40N 23 45 E
Rigaud, Canada 79 A10 45 29N 74 18W
Rigby, U.S.A. 82 E8 43 40N 111 55W
Rīgestān □, Afghan. 40 D4 30 15N 65 0 E
Riggins, U.S.A. 82 D5 45 25N 116 19W
Rigolet, Canada 71 B8 54 10N 58 23W
Riihimäki, Finland 9 F18 60 45N 24 48 E
Riiser-Larsen-halvøya, Antarctica 5 C4 68 0S 35 0 E
Rijeka, Croatia 20 B6 45 20N 14 21 E
Rijn →, Neths. 15 B4 52 12N 4 21 E
Rijssen, Neths. 15 B6 52 19N 6 31 E
Rijswijk, Neths. 15 B4 52 4N 4 22 E
Rikuzentakada, Japan 30 E10 39 0N 141 40 E
Riley, U.S.A. 82 E4 43 32N 119 28W
Rimah, Wadi ar →, Si. Arabia 44 E4 26 5N 41 30 E
Rimbey, Canada 72 C6 52 35N 114 15W
Rímini, Italy 20 B5 44 3N 12 33 E
Rîmnicu Sărat, Romania 17 F13 45 26N 27 3 E
Rîmnicu Vîlcea, Romania 21 B11 45 9N 24 21 E
Rimouski, Canada 71 C6 48 27N 68 30W
Rimrock, U.S.A. 84 D5 46 38N 121 10W
Rinca, Indonesia 37 F5 8 45S 119 35 E
Rincón de Romos, Mexico 86 C4 22 14N 102 18W
Rinconada, Argentina 94 A2 22 26S 66 10W
Ringkøbing, Denmark 9 H10 56 5N 8 15 E
Ringling, U.S.A. 82 C8 46 16N 110 49W
Ringvassøy, Norway 8 B15 69 56N 19 15 E
Rinía, Greece 21 F11 37 23N 25 13 E
Rinjani, Indonesia 36 F5 8 24S 116 28 E
Rio Branco, Brazil 92 E5 9 58S 67 49W
Río Branco, Uruguay 95 C5 32 40S 53 40W
Rio Brilhante, Brazil 95 A5 21 48S 54 33W
Rio Claro, Brazil 95 A6 22 19S 47 35W
Rio Claro, Trin. & Tob. 89 D7 10 20N 61 25W
Río Colorado, Argentina 96 D4 39 0S 64 0W
Río Cuarto, Argentina 94 C3 33 10S 64 25W
Rio das Pedras, Mozam. 57 C6 23 8S 35 28 E
Rio de Janeiro, Brazil 95 A7 23 0S 43 12W
Rio de Janeiro □, Brazil 95 A7 22 50S 43 0W
Rio do Sul, Brazil 95 B6 27 13S 49 37W
Río Gallegos, Argentina 96 G3 51 35S 69 15W
Río Grande, Argentina 96 G3 53 50S 67 45W
Rio Grande, Brazil 95 C5 32 0S 52 20W
Río Grande, Mexico 86 C4 23 50N 103 2W
Río Grande, Nic. 88 D3 12 54N 83 33W
Rio Grande →, U.S.A. 81 N6 25 57N 97 9W
Rio Grande City, U.S.A. 81 M5 26 23N 98 49W
Río Grande del Norte →, N. Amer. 75 E7 26 0N 97 0W
Rio Grande do Norte □, Brazil 93 E11 5 40S 36 0W
Rio Grande do Sul □, Brazil 95 C5 30 0S 53 0W
Río Hato, Panama 88 E3 8 22N 80 10W
Río Lagartos, Mexico 87 C7 21 36N 88 10W
Rio Largo, Brazil 93 E11 9 28S 35 50W
Río Mulatos, Bolivia 92 G5 19 40S 66 50W
Río Muni = Mbini □, Eq. Guin. 52 D2 1 30N 10 0 E
Rio Negro, Brazil 95 B6 26 0S 49 55W
Rio Pardo, Brazil 95 C5 30 0S 52 30W
Río Segundo, Argentina 94 C3 31 40S 63 59W
Río Tercero, Argentina 94 C3 32 15S 64 8W
Rio Verde, Brazil 93 G8 17 50S 51 0W
Río Verde, Mexico 87 C5 21 56N 99 59W
Rio Vista, U.S.A. 84 G5 38 10N 121 42W
Ríobamba, Ecuador 92 D3 1 50S 78 45W
Ríohacha, Colombia 92 A4 11 33N 72 55W
Riosucio, Caldas, Colombia 92 B3 5 30N 75 40W
Riosucio, Choco, Colombia 92 B3 7 27N 77 7W
Riou L., Canada 73 B7 59 7N 106 25W
Ripley, Canada 78 B3 44 4N 81 35W
Ripley, Calif., U.S.A. 85 M12 33 32N 114 39W
Ripley, N.Y., U.S.A. 78 D5 42 16N 79 43W
Ripley, Tenn., U.S.A. 81 H10 35 45N 89 32W
Ripon, U.K. 10 C6 54 8N 1 31W
Ripon, Calif., U.S.A. 84 H5 37 44N 121 7W
Ripon, Wis., U.S.A. 76 D1 43 51N 88 50W
Risalpur, Pakistan 42 B4 34 3N 71 59 E
Rishā', W. ar →, Si. Arabia 44 E5 25 33N 44 5 E
Rishiri-Tō, Japan 30 B10 45 11N 141 15 E
Rishon le Ziyyon, Israel 47 D3 31 58N 34 48 E
Rison, U.S.A. 81 J8 33 58N 92 11W
Risør, Norway 9 G10 58 43N 9 13 E
Rittman, U.S.A. 78 F3 40 58N 81 47W

Rutledge L.

Ritzville, U.S.A. 82 C4 47 8N 118 23W
Rivadavia, Buenos Aires, Argentina 94 D3 35 29S 62 59W
Rivadavia, Mendoza, Argentina 94 C2 33 13S 68 30W
Rivadavia, Salta, Argentina 94 A3 24 5S 62 54W
Rivadavia, Chile 94 B1 29 57S 70 35W
Rivas, Nic. 88 D2 11 30N 85 50W
Rivera, Uruguay 95 C4 31 0S 55 50W
Riverdale, U.S.A. 84 J7 36 26N 119 52W
Riverhead, U.S.A. 79 F12 40 55N 72 40W
Riverhurst, Canada 73 C7 50 55N 106 50W
Riverina, Australia 61 E3 29 45S 120 40 E
Rivers, Canada 73 C8 50 2N 100 14W
Rivers, L. of the, Canada 73 D7 49 49N 105 44W
Rivers Inlet, Canada 72 C3 51 42N 127 15W
Riversdale, S. Africa 56 E3 34 7S 21 15 E
Riverside, Calif., U.S.A. 85 M9 33 59N 117 22W
Riverside, Wyo., U.S.A. 82 F10 41 13N 106 47W
Riversleigh, Australia 62 B2 19 5S 138 40 E
Riverton, Australia 63 E2 34 10S 138 46 E
Riverton, Canada 73 C9 51 1N 97 0W
Riverton, N.Z. 59 M1 46 21S 168 0 E
Riverton, U.S.A. 82 E9 43 2N 108 23W
Riverton Heights, U.S.A. 84 C4 47 28N 122 17W
Riviera, Europe 16 F4 44 0N 8 30 E
Riviera di Levante, Italy 20 B3 44 23N 9 15 E
Riviera di Ponente, Italy 20 C2 43 50N 7 58 E
Rivière-à-Pierre, Canada 71 C5 46 59N 72 11W
Rivière-au-Renard, Canada 71 C7 48 59N 64 23W
Rivière-du-Loup, Canada 71 C6 47 50N 69 30W
Rivière-Pentecôte, Canada 71 C6 49 57N 67 1W
Rivière-Pilote, Martinique 89 D7 14 26N 60 53W
Rivne = Rovno, Ukraine 25 D4 50 40N 26 10 E
Rivoli B., Australia 63 F3 37 32S 140 3 E
Riyadh = Ar Riyāḑ, Si. Arabia 46 C4 24 41N 46 42 E
Rize, Turkey 25 F7 41 0N 40 30 E
Rizhao, China 35 G10 35 25N 119 30 E
Rizokarpaso, Cyprus 23 D13 35 36N 34 23 E
Rizzuto, C., Italy 20 E7 38 54N 17 5 E
Rjukan, Norway 9 G10 59 54N 8 33 E
Road Town, Virgin Is. 89 C7 18 27N 64 37W
Roag, L., U.K. 12 C2 58 10N 6 55W
Roanne, France 18 C6 46 3N 4 4 E
Roanoke, Ala., U.S.A. 77 J3 33 9N 85 22W
Roanoke, Va., U.S.A. 76 G6 37 16N 79 56W
Roanoke →, U.S.A. 77 H7 35 57N 76 42W
Roanoke I., U.S.A. 77 H8 35 55N 75 40W
Roanoke Rapids, U.S.A. 77 G7 36 28N 77 40W
Roatán, Honduras 88 C2 16 18N 86 35W
Robbins I., Australia 62 G4 40 42S 145 0 E
Robe →, Australia 60 D2 21 42S 116 15 E
Robe →, Ireland 13 C2 53 38N 9 10W
Robert Lee, U.S.A. 81 K4 31 54N 100 29W
Roberts, U.S.A. 82 E7 43 43N 112 8W
Robertsganj, India 43 G10 24 44N 83 4 E
Robertson, S. Africa 56 E2 33 46S 19 50 E
Robertson I., Antarctica 5 C18 65 15S 59 30W
Robertson Ra., Australia 60 D3 23 15S 121 0 E
Robertsport, Liberia 50 G2 6 45N 11 26W
Robertstown, Australia 63 E2 33 58S 139 5 E
Roberval, Canada 71 C5 48 32N 72 15W
Robeson Chan., Greenland 4 A4 82 0N 61 30W
Robinson →, Australia 62 B2 16 3S 137 16 E
Robinson Ra., Australia 61 E2 25 40S 119 0 E
Robinson River, Australia 62 B2 16 45S 136 58 E
Robinvale, Australia 63 E3 34 40S 142 45 E
Roblin, Canada 73 C8 51 14N 101 21W
Roboré, Bolivia 92 G7 18 10S 59 45W
Robson, Mt., Canada 72 C5 53 10N 119 10W
Robstown, U.S.A. 81 M6 27 47N 97 40W
Roca, C. da, Portugal 19 C1 38 40N 9 31W
Roca Partida, I., Mexico 86 D2 19 1N 112 2W
Rocas, I., Brazil 93 D12 4 0S 34 1W
Rocha, Uruguay 95 C5 34 30S 54 25W
Rochdale, U.K. 10 D5 53 36N 2 10W
Rochefort, Belgium 15 D5 50 9N 5 12 E
Rochefort, France 18 D3 45 56N 0 57W
Rochelle, U.S.A. 80 E10 41 56N 89 4W
Rocher River, Canada 72 A6 61 23N 112 44W
Rochester, Canada 72 C6 54 22N 113 27W
Rochester, U.K. 11 F8 51 22N 0 30 E
Rochester, Ind., U.S.A. 76 E2 41 4N 86 13W
Rochester, Minn., U.S.A. 80 C8 44 1N 92 28W
Rochester, N.H., U.S.A. 79 C14 43 18N 70 59W
Rochester, N.Y., U.S.A. 78 C7 43 10N 77 37W
Rock →, Canada 72 A3 60 7N 127 7W
Rock Hill, U.S.A. 77 H5 34 56N 81 1W
Rock Island, U.S.A. 80 E9 41 30N 90 34W
Rock Rapids, U.S.A. 80 D6 43 26N 96 10W
Rock River, U.S.A. 82 F11 41 44N 105 58W
Rock Sound, Bahamas 88 B4 24 54N 76 12W
Rock Springs, Mont., U.S.A. 82 C10 46 49N 106 15W
Rock Springs, Wyo., U.S.A. 82 F9 41 35N 109 14W
Rock Valley, U.S.A. 80 D6 43 12N 96 18W
Rockall, Atl. Oc. 6 D3 57 37N 13 42W
Rockdale, Tex., U.S.A. 81 K6 30 39N 97 0W
Rockdale, Wash., U.S.A. 84 C5 47 22N 121 28W
Rockefeller Plateau, Antarctica 5 E14 80 0S 140 0W
Rockford, U.S.A. 80 D10 42 16N 89 6W
Rockglen, Canada 73 D7 49 11N 105 57W
Rockhampton, Australia 62 C5 23 22S 150 32 E
Rockhampton Downs, Australia 62 B2 18 57S 135 10 E
Rockingham, Australia 61 F2 32 15S 115 38 E
Rockingham B., Australia 62 B4 18 5S 146 10 E
Rockingham Forest, U.K. 11 E7 52 28N 0 42W
Rocklake, U.S.A. 80 A5 48 47N 99 15W
Rockland, Canada 79 A9 45 33N 75 17W
Rockland, Idaho, U.S.A. 82 E7 42 34N 112 53W
Rockland, Maine, U.S.A. 71 D6 44 6N 69 7W
Rockland, Mich., U.S.A. 80 B10 46 44N 89 11W
Rocklin, U.S.A. 84 G5 38 48N 121 14W
Rockmart, U.S.A. 77 H3 34 0N 85 3W
Rockport, Mo., U.S.A. 80 E7 40 25N 95 31W
Rockport, Tex., U.S.A. 81 L6 28 2N 97 3W
Rocksprings, U.S.A. 81 K4 30 1N 100 13W
Rockville, Conn., U.S.A. 79 E12 41 52N 72 28W
Rockville, Md., U.S.A. 76 F7 39 5N 77 9W
Rockwall, U.S.A. 81 J6 32 56N 96 28W
Rockwell City, U.S.A. 80 D7 42 24N 94 38W
Rockwood, U.S.A. 77 H3 35 52N 84 41W
Rocky Ford, U.S.A. 80 F3 38 3N 103 43W
Rocky Gully, Australia 61 F2 34 30S 116 57 E
Rocky Lane, Canada 72 B5 58 31N 116 22W
Rocky Mount, U.S.A. 77 H7 35 57N 77 48W
Rocky Mountain House, Canada 72 C6 52 22N 114 55W
Rocky Mts., N. Amer. 72 C4 55 0N 121 0W
Rockyford, Canada 72 C6 51 14N 113 10W
Rod, Pakistan 40 E3 28 10N 63 5 E
Rødbyhavn, Denmark 9 J11 54 39N 11 22 E
Roddickton, Canada 71 B8 50 51N 56 8W
Roderick I., Canada 72 C3 52 38N 128 22W
Rodez, France 18 D5 44 21N 2 33 E
Rodhopoú, Greece 23 D5 35 34N 23 45 E
Ródhos, Greece 23 C10 36 15N 28 10 E
Rodney, Canada 78 D3 42 34N 81 41W
Rodney, C., N.Z. 59 G5 36 17S 174 50 E
Rodriguez, Ind. Oc. 3 E13 19 45S 63 20 E
Roe →, U.K. 13 A5 55 10N 6 59W
Roebling, U.S.A. 79 F10 40 7N 74 47W
Roebourne, Australia 60 D2 20 44S 117 9 E
Roebuck B., Australia 60 C3 18 5S 122 20 E
Roebuck Plains, Australia 60 C3 17 56S 122 28 E
Roermond, Neths. 15 C5 51 12N 6 0 E
Roes Welcome Sd., Canada 69 B11 65 0N 87 0W
Roeselare, Belgium 15 D3 50 57N 3 7 E
Rogagua, L., Bolivia 92 F5 13 43S 66 50W
Rogaland fylke □, Norway 9 G9 59 12N 6 20 E
Rogdhia, Greece 23 D7 35 22N 25 1 E
Rogers, U.S.A. 81 G7 36 20N 94 7W
Rogers City, U.S.A. 76 C4 45 25N 83 49W
Rogerson, U.S.A. 82 E6 42 13N 114 36W
Rogersville, U.S.A. 77 G4 36 24N 83 1W
Roggan River, Canada 70 B4 54 25N 79 32W
Roggeveldberge, S. Africa 56 E3 32 10S 20 10 E
Rogoaguado, L., Bolivia 92 F5 13 0S 65 30W
Rogue →, U.S.A. 82 E1 42 26N 124 26W
Róhda, Greece 23 A3 39 48N 19 46 E
Rohnert Park, U.S.A. 84 G4 38 16N 122 40W
Rohri, Pakistan 42 F3 27 45N 68 51 E
Rohri Canal, Pakistan 42 F3 26 15N 68 27 E
Rohtak, India 42 E7 28 55N 76 43 E
Roi Et, Thailand 38 D4 16 4N 103 40 E
Rojas, Argentina 94 C3 34 10S 60 45W
Rojo, C., Mexico 87 C5 21 33N 97 20W
Rokan →, Indonesia 36 D2 2 0N 100 50 E
Rokeby, Australia 62 A3 13 39S 142 40 E
Rolândia, Brazil 95 A5 23 18S 51 23W
Rolette, U.S.A. 80 A5 48 40N 99 51W
Rolla, Kans., U.S.A. 81 G4 37 7N 101 38W
Rolla, Mo., U.S.A. 81 G9 37 57N 91 46W
Rolla, N. Dak., U.S.A. 80 A5 48 52N 99 37W
Rolleston, Australia 62 C4 24 28S 148 35 E
Rollingstone, Australia 62 B4 19 2S 146 24 E
Roma, Australia 63 D4 26 32S 148 49 E
Roma, Italy 20 D5 41 54N 12 30 E
Roma, Sweden 9 H15 57 32N 18 26 E
Roman, Romania 17 E13 46 57N 26 55 E
Roman, Russia 27 C12 60 4N 112 14 E
Romang, Indonesia 37 F7 7 30S 127 20 E
Romāni, Egypt 47 E1 30 59N 32 38 E
Romania ■, Europe 21 B11 46 0N 25 0 E
Romano, Cayo, Cuba 88 B4 22 0N 77 30W
Romblon, Phil. 37 B6 12 33N 122 17 E
Rombo □, Tanzania 54 C4 3 10S 37 30 E
Rome = Roma, Italy 20 D5 41 54N 12 30 E
Rome, Ga., U.S.A. 77 H3 34 15N 85 10W
Rome, N.Y., U.S.A. 79 C9 43 13N 75 27W
Romney, U.S.A. 76 F6 39 21N 78 45W
Romney Marsh, U.K. 11 F8 51 4N 0 58 E
Romorantin-Lanthenay, France 18 C4 47 21N 1 45 E
Romsdalen, Norway 8 E10 62 25N 8 0 E
Ron, Vietnam 38 D6 17 53N 106 27 E
Rona, U.K. 12 D3 57 33N 5 57W
Ronan, U.S.A. 82 C6 47 32N 114 6W
Roncador, Cayos, Caribbean 88 D3 13 32N 80 4W
Roncador, Serra do, Brazil 93 F8 12 30S 52 30W
Ronceverte, U.S.A. 76 G5 37 45N 80 28W
Ronda, Spain 19 D3 36 46N 5 12W
Rondane, Norway 9 F10 61 57N 9 50 E
Rondônia □, Brazil 92 F6 11 0S 63 0W
Rondonópolis, Brazil 93 G8 16 28S 54 38W
Ronge, L. la, Canada 73 B7 55 6N 105 17W
Ronne Ice Shelf, Antarctica 5 D18 78 0S 60 0W
Ronsard, C., Australia 61 D1 24 46S 113 10 E
Ronse, Belgium 15 D3 50 45N 3 35 E
Roodepoort, S. Africa 57 D4 26 11S 27 54 E
Roof Butte, U.S.A. 83 H9 36 28N 109 5W
Roorkee, India 42 E7 29 52N 77 59 E
Roosendaal, Neths. 15 C4 51 32N 4 29 E
Roosevelt, Minn., U.S.A. 80 A7 48 48N 95 6W
Roosevelt, Utah, U.S.A. 82 F8 40 18N 109 59W
Roosevelt →, Brazil 92 E6 7 35S 60 20W
Roosevelt, Mt., Canada 72 B3 58 26N 125 20W
Roosevelt I., Antarctica 5 D12 79 30S 162 0W
Roosevelt Res., U.S.A. 83 K8 33 46N 111 0W
Roper →, Australia 62 A2 14 43S 135 27 E
Ropesville, U.S.A. 81 J3 33 26N 102 9W
Roque Pérez, Argentina 94 D4 35 25S 59 24W
Roraima □, Brazil 92 C6 2 0N 61 30W
Roraima, Mt., Venezuela 92 B6 5 10N 60 40W
Rorketon, Canada 73 C9 51 24N 99 35W
Røros, Norway 8 E11 62 35N 11 23 E
Rosa, Zambia 55 D3 9 33S 31 15 E
Rosa, Monte, Europe 16 F3 45 57N 7 53 E
Rosalia, U.S.A. 82 C5 47 14N 117 22W
Rosamond, U.S.A. 85 L8 34 52N 118 10W
Rosario, Argentina 94 C3 33 0S 60 40W
Rosário, Brazil 93 D10 3 0S 44 15W
Rosario, Baja Calif. N., Mexico 86 A1 30 0N 115 50W
Rosario, Sinaloa, Mexico 86 C3 23 0N 105 52W
Rosario, Paraguay 94 A4 24 30S 57 35W
Rosario de la Frontera, Argentina 94 B3 25 50S 65 0W
Rosario de Lerma, Argentina 94 A2 24 59S 65 35W
Rosario del Tala, Argentina 94 C4 32 20S 59 10W
Rosário do Sul, Brazil 95 C5 30 15S 54 55W
Rosarito, U.S.A. 85 N9 32 18N 117 4W
Rosas, Spain 19 A7 42 19N 3 10 E
Rosas, G. de, Spain 19 A7 42 10N 3 15 E
Roscoe, U.S.A. 80 C5 45 27N 99 20W
Roscommon, Ireland 13 C3 53 38N 8 11W
Roscommon, U.S.A. 76 C3 44 30N 84 35W
Roscommon □, Ireland 13 C3 53 40N 8 15W
Roscrea, Ireland 13 D4 52 58N 7 50W
Rose →, Australia 62 A2 14 16S 135 45 E
Rose Blanche, Canada 71 C8 47 38N 58 45W
Rose Harbour, Canada 72 C2 52 15N 131 10W
Rose Pt., Canada 72 C2 54 11N 131 39W
Rose Valley, Canada 73 C8 52 19N 103 49W
Roseau, Domin. 89 C7 15 20N 61 24W
Roseau, U.S.A. 80 A7 48 51N 95 46W
Rosebery, Australia 62 G4 41 46S 145 33 E
Rosebud, U.S.A. 81 K6 31 4N 96 59W
Roseburg, U.S.A. 82 E2 43 13N 123 20W
Rosedale, Australia 62 C5 24 38S 151 53 E
Rosedale, U.S.A. 81 J9 33 51N 91 2W
Roseland, U.S.A. 84 G4 38 25N 122 43W
Rosemary, Canada 72 C6 50 46N 112 5W
Rosenberg, U.S.A. 81 L7 29 34N 95 49W
Rosenheim, Germany 16 E6 47 51N 12 9 E
Rosetown, Canada 73 C7 51 35N 107 59W
Rosetta = Rashîd, Egypt 51 B11 31 21N 30 22 E
Roseville, U.S.A. 84 G5 38 45N 121 17W
Rosewood, N. Terr., Australia 60 C4 16 28S 128 58 E
Rosewood, Queens., Australia 63 D5 27 38S 152 36 E
Roshkhvār, Iran 45 C8 34 58N 59 37 E
Rosignol, Guyana 92 B7 6 15N 57 30W
Roskilde, Denmark 9 J12 55 38N 12 3 E
Roslavl, Russia 24 D5 53 57N 32 55 E
Roslyn, Australia 63 E4 34 29S 149 37 E
Rosmead, S. Africa 56 E4 31 29S 25 8 E
Ross, Australia 62 G4 42 2S 147 30 E
Ross, N.Z. 59 K3 42 53S 170 49 E
Ross I., Antarctica 5 D11 77 30S 168 0 E
Ross Ice Shelf, Antarctica 5 E12 80 0S 180 0 E
Ross L., U.S.A. 82 B3 48 44N 121 4W
Ross-on-Wye, U.K. 11 F5 51 55N 2 34W
Ross Sea, Antarctica 5 D11 74 0S 178 0 E
Rossan Pt., Ireland 13 B3 54 42N 8 47W
Rossburn, Canada 73 C8 50 40N 100 49W
Rosseau, Canada 78 A5 45 16N 79 39W
Rossignol, L., Canada 70 B5 52 43N 73 40W
Rossignol Res., Canada 71 D6 44 12N 65 10W
Rossland, Canada 72 D5 49 6N 117 50W
Rosslare, Ireland 13 D5 52 17N 6 23W
Rosso, Mauritania 50 E1 16 40N 15 45W
Rossosh, Russia 25 D6 50 15N 39 28 E
Rossport, Canada 70 C2 48 50N 87 30W
Røssvatnet, Norway 8 D13 65 45N 14 5 E
Rossville, Australia 62 B4 15 48S 145 15 E
Rosthern, Canada 73 C7 52 40N 106 20W
Rostock, Germany 16 A6 54 4N 12 9 E
Rostov, Russia 24 C6 57 14N 39 25 E
Rostov, Russia 25 E6 47 15N 39 45 E
Roswell, U.S.A. 81 J2 33 24N 104 32W
Rosyth, U.K. 12 E5 56 2N 3 26W
Rotan, U.S.A. 81 J4 32 51N 100 28W
Rothaargebirge, Germany 16 C4 51 0N 8 5 E
Rother →, U.K. 11 G8 50 59N 0 40 E
Rotherham, U.K. 10 D6 53 26N 1 21W
Rothes, U.K. 12 D5 57 31N 3 12W
Rothesay, Canada 71 C6 45 23N 66 0W
Rothesay, U.K. 12 F3 55 50N 5 3W
Roti, Indonesia 37 F6 10 50S 123 0 E
Roto, Australia 63 E4 33 0S 145 30 E
Rotoroa, L., N.Z. 59 J4 41 55S 172 39 E
Rotorua, N.Z. 59 H6 38 9S 176 16 E
Rotorua, L., N.Z. 59 H6 38 5S 176 18 E
Rotterdam, Neths. 15 C4 51 55N 4 30 E
Rottnest I., Australia 61 F2 32 0S 115 27 E
Rottumeroog, Neths. 15 A6 53 33N 6 34 E
Rottweil, Germany 16 D4 48 9N 8 38 E
Rotuma, Fiji 64 J9 12 25S 177 5 E
Roubaix, France 18 A5 50 40N 3 10 E
Rouen, France 18 B4 49 27N 1 4 E
Rouleau, Canada 73 C8 50 10N 104 56W
Round Mt., Australia 63 E5 30 26S 152 16 E
Round Mountain, U.S.A. 82 G5 38 43N 117 4W
Roundup, U.S.A. 82 C9 46 27N 108 33W
Rousay, U.K. 12 B5 59 10N 3 2W
Rouses Point, U.S.A. 79 B11 44 59N 73 22W
Roussillon, France 18 E5 42 30N 2 35 E
Rouxville, S. Africa 56 E4 30 25S 26 50 E
Rouyn, Canada 70 C4 48 20N 79 0W
Rovaniemi, Finland 8 C18 66 29N 25 41 E
Rovereto, Italy 20 B4 45 53N 11 3 E
Rovigo, Italy 20 B4 45 4N 11 48 E
Rovinj, Croatia 20 B5 45 5N 13 40 E
Rovno, Ukraine 25 D4 50 40N 26 10 E
Rovuma →, Tanzania 55 E5 10 29S 40 28 E
Row'ān, Iran 45 C6 35 8N 48 51 E
Rowena, Australia 63 D4 29 48S 148 55 E
Rowley Shoals, Australia 60 C2 17 30S 119 0 E
Roxas, Phil. 37 B6 11 36N 122 49 E
Roxboro, U.S.A. 77 G6 36 24N 78 59W
Roxborough Downs, Australia 62 C2 22 30S 138 45 E
Roxburgh, N.Z. 59 L2 45 33S 169 19 E
Roy, Mont., U.S.A. 82 C9 47 20N 108 58W
Roy, N. Mex., U.S.A. 81 H2 35 57N 104 12W
Roy Hill, Australia 60 D2 22 37S 119 58 E
Royal Leamington Spa, U.K. 11 E6 52 18N 1 32W
Royan, France 18 D3 45 37N 1 2W
Rtishchevo, Russia 24 C7 55 16N 43 50 E
Ruacaná, Angola 56 B1 17 20S 14 12 E
Ruahine Ra., N.Z. 59 H6 39 55S 176 2 E
Ruapehu, N.Z. 59 H5 39 17S 175 35 E
Ruapuke I., N.Z. 59 M2 46 46S 168 31 E
Ruâq, W. →, Egypt 47 F2 30 0N 33 49 E
Rub' al Khali, Si. Arabia 46 D4 18 0N 48 0 E
Rubeho Mts., Tanzania 54 D4 6 50S 36 25 E
Rubh a' Mhail, U.K. 12 F2 55 55N 6 10W
Rubha Hunish, U.K. 12 D2 57 42N 6 20W
Rubicon →, U.S.A. 84 G5 38 53N 121 4W
Rubicone →, Italy 20 B5 44 8N 12 28 E
Rubio, Venezuela 92 B4 7 43N 72 22W
Rubtsovsk, Russia 26 D9 51 30N 81 10 E
Ruby L., U.S.A. 82 F6 40 10N 115 28W
Ruby Mts., U.S.A. 82 F6 40 30N 115 20W
Rūd Sar, Iran 45 B6 37 8N 50 18 E
Rudall, Australia 63 E2 33 43S 136 17 E
Rudall →, Australia 60 D3 22 34S 122 13 E
Rudewa, Tanzania 55 E3 10 7S 34 40 E
Rudnichnyy, Russia 24 C9 59 38N 52 26 E
Rudnogorsk, Russia 27 D11 57 15N 103 42 E
Rudnyy, Kazakhstan 26 D7 52 57N 63 7 E
Rudolf, Ostrov, Russia 26 A6 81 45N 58 30 E
Rudyard, U.S.A. 76 B3 46 14N 84 36W
Rufa'a, Sudan 51 F11 14 44N 33 22 E
Rufiji □, Tanzania 54 D4 8 0S 38 30 E
Rufiji →, Tanzania 54 D4 7 50S 39 15 E
Rufino, Argentina 94 C3 34 20S 62 50W
Rufisque, Senegal 50 F1 14 40N 17 15W
Rufunsa, Zambia 55 F2 15 4S 29 34 E
Rugby, U.K. 11 E6 52 23N 1 16W
Rugby, U.S.A. 80 A5 48 22N 100 0W
Rügen, Germany 16 A6 54 22N 13 25 E
Ruhengeri, Rwanda 54 C2 1 30S 29 36 E
Ruhr →, Germany 16 C3 51 25N 6 44 E
Ruhuhu →, Tanzania 55 E3 10 31S 34 34 E
Ruidosa, U.S.A. 81 L2 29 59N 104 41W
Ruidoso, U.S.A. 83 K11 33 20N 105 41W
Ruivo, Pico, Madeira 22 D3 32 45N 16 56W
Rujm Tal'at al Jamā'ah, Jordan 47 E4 30 24N 35 30 E
Ruk, Pakistan 42 F3 27 50N 68 42 E
Rukwa □, Tanzania 54 D3 7 0S 31 30 E
Rukwa L., Tanzania 54 D3 8 0S 32 20 E
Rulhieres, C., Australia 60 B4 13 56S 127 22 E
Rum Cay, Bahamas 89 B5 23 40N 74 58W
Rum Jungle, Australia 60 B5 13 0S 130 59 E
Rumāḩ, Si. Arabia 44 E5 25 29N 47 10 E
Rumania = Romania ■, Europe 21 B11 46 0N 25 0 E
Rumaylah, Iraq 44 D5 30 47N 47 37 E
Rumbalara, Australia 62 D1 25 20S 134 29 E
Rumbêk, Sudan 51 G10 6 54N 29 37 E
Rumford, U.S.A. 79 B14 44 33N 70 33W
Rumoi, Japan 30 C10 43 56N 141 39 E
Rumonge, Burundi 54 C2 3 59S 29 26 E
Rumsey, Canada 72 C6 51 51N 112 48W
Rumula, Australia 62 B4 16 35S 145 20 E
Rumuruti, Kenya 54 B4 0 17N 36 32 E
Runan, China 34 H8 33 0N 114 30 E
Runanga, N.Z. 59 K3 42 25S 171 15 E
Runaway, C., N.Z. 59 G6 37 32S 177 59 E
Runcorn, U.K. 10 D5 53 20N 2 44W
Rungwa, Tanzania 54 D3 6 55S 33 32 E
Rungwa →, Tanzania 54 D3 7 36S 31 50 E
Rungwe, Tanzania 55 D3 9 11S 33 32 E
Rungwe □, Tanzania 55 D3 9 25S 33 32 E
Runton Ra., Australia 60 D3 23 31S 123 6 E
Ruoqiang, China 32 C3 38 55N 88 10 E
Rupa, India 41 F18 27 15N 92 21 E
Rupar, India 42 D7 31 2N 76 38 E
Rupat, Indonesia 36 D2 1 45N 101 40 E
Rupert →, Canada 70 B4 51 29N 78 45W
Rupert House = Fort Rupert, Canada 70 B4 51 30N 78 40W
Rurrenabaque, Bolivia 92 F5 14 30S 67 32W
Rusambo, Zimbabwe 55 F3 16 30S 32 4 E
Rusape, Zimbabwe 55 F3 18 35S 32 8 E
Ruschuk = Ruse, Bulgaria 21 C11 43 48N 25 59 E
Ruse, Bulgaria 21 C11 43 48N 25 59 E
Rushan, China 35 F11 36 56N 121 30 E
Rushden, U.K. 11 E7 52 17N 0 37W
Rushford, U.S.A. 80 D9 43 49N 91 46W
Rushville, Ill., U.S.A. 80 E9 40 7N 90 34W
Rushville, Ind., U.S.A. 76 F3 39 37N 85 27W
Rushville, Nebr., U.S.A. 80 D3 42 43N 102 28W
Rushworth, Australia 63 F4 36 32S 145 1 E
Russas, Brazil 93 D11 4 55S 37 50W
Russell, Canada 73 C8 50 50N 101 20W
Russell, U.S.A. 80 F5 38 54N 98 52W
Russell L., Man., Canada 73 B8 56 15N 101 30W
Russell L., N.W.T., Canada 72 A5 63 5N 115 44W
Russellkonda, India 41 K14 19 57N 84 42 E
Russellville, Ala., U.S.A. 77 H2 34 30N 87 44W
Russellville, Ark., U.S.A. 81 H8 35 17N 93 8W
Russellville, Ky., U.S.A. 77 G2 36 51N 86 53W
Russia ■, Eurasia 27 C11 62 0N 105 0 E
Russian →, U.S.A. 84 G3 38 27N 123 8W
Russkaya Polyana, Kazakhstan 26 D8 53 47N 73 53 E
Russkoye Ustie, Russia 4 B15 71 0N 149 0 E
Rustam, Pakistan 42 B5 34 25N 72 13 E
Rustam Shahr, Pakistan 42 F2 26 58N 66 6 E
Rustavi, Georgia 25 F8 41 30N 45 0 E
Rustenburg, S. Africa 56 D4 25 41S 27 14 E
Ruston, U.S.A. 81 J8 32 32N 92 38W
Rutana, Burundi 54 C2 3 55S 30 0 E
Ruteng, Indonesia 37 F6 8 35S 120 30 E
Ruth, Mich., U.S.A. 78 C2 43 42N 82 45W
Ruth, Nev., U.S.A. 82 G6 39 17N 114 59W
Rutherford, U.S.A. 84 G4 38 26N 122 24W
Rutherglen, U.K. 12 F4 55 50N 4 11W
Rutland Plains, Australia 62 B3 15 38S 141 43 E
Rutledge →, Canada 73 A6 61 4N 112 0W
Rutledge L., Canada 73 A6 61 33N 110 47W

Rutshuru

Rutshuru, Zaïre **54 C2** 1 13S 29 25 E
Ruurlo, Neths. **15 B6** 52 5N 6 24 E
Ruvu, Tanzania **54 D4** 6 49S 38 43 E
Ruvu →, Tanzania **54 D4** 6 23S 38 52 E
Ruvuma □, Tanzania **55 E4** 10 20S 36 0 E
Ruwais, U.A.E. **45 E7** 24 5N 52 50 E
Ruwenzori, Africa **54 B2** 0 30N 29 55 E
Ruyigi, Burundi **54 C3** 3 29S 30 15 E
Ružomberok, Slovakia **17 D9** 49 3N 19 17 E
Rwanda ■, Africa **54 C3** 2 0S 30 0 E
Ryan, L., U.K. **12 G3** 55 0N 5 2W
Ryazan, Russia **24 D6** 54 40N 39 40 E
Ryazhsk, Russia **24 D7** 53 45N 40 3 E
Rybache, Kazakhstan **26 E9** 46 40N 81 20 E
Rybachiy Poluostrov, Russia **24 A5** 69 43N 32 0 E
Rybinsk, Russia **24 C6** 58 5N 38 50 E
Rybinskoye Vdkhr., Russia **24 C6** 58 30N 38 25 E
Ryde, U.K. **11 G6** 50 44N 1 9W
Ryderwood, U.S.A. **84 D3** 46 23N 123 3W
Rye, U.K. **11 G8** 50 57N 0 46 E
Rye →, U.K. **10 C7** 54 12N 0 53W
Rye Patch Reservoir, U.S.A. **82 F4** 40 28N 118 19W
Ryegate, U.S.A. **82 C9** 46 18N 109 15W
Rylstone, Australia **63 E4** 32 46S 149 58 E
Ryōthu, Japan **30 E9** 38 5N 138 26 E
Rypin, Poland **17 B9** 53 3N 19 25 E
Ryūgasaki, Japan **31 G10** 35 54N 140 11 E
Ryūkyū Is. = Ryūkyū-rettō, Japan **31 M2** 26 0N 126 0 E
Ryūkyū-rettō, Japan **31 M2** 26 0N 126 0 E
Rzeszów, Poland **17 C10** 50 5N 21 58 E
Rzhev, Russia **24 C5** 56 20N 34 20 E

S

Sa, Thailand **38 C3** 18 34N 100 45 E
Sa Dec, Vietnam **39 G5** 10 20N 105 46 E
Sa'ādatābād, Fārs, Iran **45 D7** 30 10N 53 5 E
Sa'ādatābād, Kermān, Iran **45 D7** 28 3N 55 53 E
Saale →, Germany **16 C5** 51 57N 11 56 E
Saar →, Europe **16 D3** 49 41N 6 32 E
Saarbrücken, Germany **16 D3** 49 15N 6 58 E
Saaremaa, Estonia **24 C3** 58 30N 22 30 E
Saariselkä, Finland **8 B20** 68 16N 28 15 E
Saarland □, Germany **15 E7** 49 15N 7 0 E
Sab 'Bi'ār, Syria **44 C3** 33 46N 37 41 E
Saba, W. Indies **89 C7** 17 42N 63 26W
Sabadell, Spain **19 B7** 41 28N 2 7 E
Sabah □, Malaysia **36 C5** 6 0N 117 0 E
Sabak Bernam, Malaysia **39 L3** 3 46N 100 58 E
Sábana de la Mar, Dom. Rep. **89 C6** 19 7N 69 24W
Sábanalarga, Colombia **92 A4** 10 38N 74 55W
Sabang, Indonesia **36 C1** 5 50N 95 15 E
Sabará, Brazil **93 G10** 19 55S 43 46W
Sabarania, Indonesia **37 E9** 2 5S 138 18 E
Sabattis, U.S.A. **79 B10** 44 6N 74 40W
Sabáudia, Italy **20 D5** 41 17N 13 2 E
Sabhah, Libya **51 C7** 27 9N 14 29 E
Sabie, S. Africa **57 D5** 25 10S 30 48 E
Sabinal, Mexico **86 A3** 30 58N 107 25W
Sabinal, U.S.A. **81 L5** 29 19N 99 28W
Sabinas, Mexico **86 B4** 27 50N 101 10W
Sabinas →, Mexico **86 B4** 27 37N 100 42W
Sabinas Hidalgo, Mexico **86 B4** 26 33N 100 10W
Sabine →, U.S.A. **81 L8** 29 59N 93 47W
Sabine L., U.S.A. **81 L8** 29 53N 93 51W
Sabine Pass, U.S.A. **81 L8** 29 44N 93 54W
Sabkhet el Bardawîl, Egypt **47 D2** 31 10N 33 15 E
Sablayan, Phil. **37 B6** 12 50N 120 50 E
Sable, C., U.S.A. **75 E10** 25 9N 81 8W
Sable, C., Canada **71 D6** 43 29N 65 38W
Sable I., Canada **71 D8** 44 0N 60 0W
Sabolev, Russia **27 D16** 54 20N 155 30 E
Sabrina Coast, Antarctica **5 C9** 68 0S 120 0 E
Sabulubek, Indonesia **36 E1** 1 36S 98 40 E
Sabzevār, Iran **45 B8** 36 15N 57 40 E
Sabzvārān, Iran **45 D8** 28 45N 57 50 E
Sac City, U.S.A. **80 D7** 42 25N 95 0W
Sachigo →, Canada **70 A2** 55 6N 88 58W
Sachigo, L., Canada **70 B1** 53 50N 92 12W
Sackets Harbor, U.S.A. **79 C8** 43 57N 76 7W
Saco, Maine, U.S.A. **77 D10** 43 30N 70 27W
Saco, Mont., U.S.A. **82 B10** 48 28N 107 21W
Sacramento, U.S.A. **84 G5** 38 35N 121 29W
Sacramento →, U.S.A. **84 G5** 38 3N 121 56W
Sacramento Mts., U.S.A. **83 K11** 32 30N 105 30W
Sacramento Valley, U.S.A. **84 G5** 39 30N 122 0W
Sádaba, Spain **19 A5** 42 19N 1 12W
Sadani, Tanzania **54 D4** 5 58S 38 35 E
Sadao, Thailand **39 J3** 6 38N 100 26 E
Sadd el Aali, Egypt **51 D11** 23 54N 32 54 E
Saddle Mt., U.S.A. **84 E3** 45 58N 123 41W
Sadimi, Zaïre **55 D1** 9 25S 23 32 E
Sado, Japan **30 E9** 38 0N 138 25 E
Sadon, Burma **41 G20** 25 28N 97 55 E
Saegerstown, U.S.A. **78 E4** 41 43N 80 9W
Şafājah, Si. Arabia **44 E3** 26 25N 39 0 E
Säffle, Sweden **9 G12** 59 8N 12 55 E
Safford, U.S.A. **83 K9** 32 50N 109 43W
Saffron Walden, U.K. **11 E8** 52 2N 0 15 E
Safi, Morocco **50 B3** 32 18N 9 20W
Şafiābād, Iran **45 B8** 36 45N 57 58 E
Safīd Dasht, Iran **45 C6** 33 27N 48 11 E
Safīd Kūh, Afghan. **40 B3** 34 45N 63 0 E
Safwān, Iraq **44 D5** 30 7N 47 43 E
Sag Harbor, U.S.A. **79 F12** 41 0N 72 18W
Saga, Indonesia **37 E8** 2 40S 132 55 E
Saga, Japan **31 H5** 33 15N 130 16 E
Saga □, Japan **31 H5** 33 15N 130 20 E
Sagae, Japan **30 E10** 38 22N 140 17 E
Sagala, Mali **50 F3** 14 9N 6 38W
Sagar, India **40 M9** 14 14N 75 6 E
Sagara, L., Tanzania **54 D3** 5 20S 31 0 E
Saginaw, U.S.A. **76 D4** 43 26N 83 56W
Saginaw B., U.S.A. **76 D4** 43 50N 83 40W
Sagīr, Zab as →, Iraq **44 C4** 35 10N 43 20 E
Saglouc, Canada **69 B12** 62 14N 75 38W
Sagŏ-ri, S. Korea **35 G14** 35 25N 126 49 E
Sagres, Portugal **19 D1** 37 0N 8 58W
Sagua la Grande, Cuba **88 B3** 22 50N 80 10W
Saguache, U.S.A. **83 G10** 38 5N 106 8W
Saguenay →, Canada **71 C5** 48 22N 71 0W
Sagunto, Spain **19 C5** 39 42N 0 18W
Sahagún, Spain **19 A3** 42 18N 5 2W
Saḩam al Jawlān, Syria **47 C4** 32 45N 35 55 E
Sahand, Kūh-e, Iran **44 B5** 37 44N 46 27 E
Sahara, Africa **50 D5** 23 0N 5 0 E
Saharan Atlas = Saharien, Atlas, Algeria **50 B5** 33 30N 1 0 E
Saharanpur, India **42 E7** 29 58N 77 33 E
Saharien, Atlas, Algeria **50 B5** 33 30N 1 0 E
Sahasinaka, Madag. **57 C8** 21 49S 47 49 E
Sahaswan, India **43 E8** 28 5N 78 45 E
Sahibganj, India **43 G12** 25 12N 87 40 E
Sahiwal, Pakistan **42 D5** 30 45N 73 8 E
Şaḩneh, Iran **44 C5** 34 29N 47 41 E
Sahtaneh →, Canada **72 B4** 59 2N 122 28W
Sahuaripa, Mexico **86 B3** 29 0N 109 13W
Sahuarita, U.S.A. **83 L8** 31 57N 110 58W
Sahuayo, Mexico **86 C4** 20 4N 102 43W
Sai Buri, Thailand **39 J3** 6 43N 101 45 E
Sa'id Bundas, Sudan **51 G9** 8 24N 24 48 E
Saïda, Algeria **50 B5** 34 50N 0 11 E
Saīdābād, Kermān, Iran **45 D7** 29 30N 55 45 E
Sa'īdābād, Semnān, Iran **45 B7** 36 8N 54 11 E
Sa'īdīyeh, Iran **45 B6** 36 20N 48 55 E
Saidpur, Bangla. **41 G16** 25 48N 89 0 E
Saidu, Pakistan **43 B5** 34 43N 72 24 E
Saigon = Phanh Bho Ho Chi Minh, Vietnam **39 G6** 10 58N 106 40 E
Saijō, Japan **31 H6** 33 55N 133 11 E
Saikhoa Ghat, India **41 F19** 27 50N 95 40 E
Saiki, Japan **31 H5** 32 58N 131 51 E
Sailolof, Indonesia **37 E8** 1 7S 130 46 E
Şa'īn Dezh, Iran **44 B5** 36 40N 46 25 E
St. Abb's Head, U.K. **12 F6** 55 55N 2 10W
St. Alban's, Canada **71 C8** 47 51N 55 50W
St. Albans, U.K. **11 F7** 51 44N 0 19W
St. Albans, Vt., U.S.A. **79 B11** 44 49N 73 5W
St. Albans, W. Va., U.S.A. **76 F5** 38 23N 81 50W
St. Alban's Head, U.K. **11 G5** 50 34N 2 3W
St. Albert, Canada **72 C6** 53 37N 113 32W
St. Andrew's, Canada **71 C8** 47 45N 59 15W
St. Andrews, U.K. **12 E6** 56 20N 2 48W
St-Anicet, Canada **79 A10** 45 8N 74 22W
St. Ann B., Canada **71 C7** 46 22N 60 25W
St. Ann's Bay, Jamaica **88 C4** 18 26N 77 15W
St. Anthony, Canada **71 B8** 51 22N 55 35W
St. Anthony, U.S.A. **82 E8** 43 58N 111 41W
St. Arnaud, Australia **63 F3** 36 40S 143 16 E
St. Arthur, Canada **71 C6** 47 33N 67 46W
St. Asaph, U.K. **10 D4** 53 15N 3 27W
St-Augustin-Saguenay, Canada **71 B8** 51 13N 58 38W
St. Augustine, U.S.A. **77 L5** 29 54N 81 19W
St. Austell, U.K. **11 G3** 50 20N 4 48W
St.-Barthélemy, I., W. Indies **89 C7** 17 50N 62 50W
St. Bee's Hd., U.K. **10 C4** 54 30N 3 38W
St. Boniface, Canada **73 D9** 49 53N 97 5W
St. Bride's, Canada **71 C9** 46 56N 54 10W
St. Brides B., U.K. **11 F2** 51 48N 5 15W
St.-Brieuc, France **18 B2** 48 30N 2 46W
St. Catharines, Canada **70 D4** 43 10N 79 15W
St. Catherines I., U.S.A. **77 K5** 31 40N 81 10W
St. Catherine's Pt., U.K. **11 G6** 50 34N 1 18W
St. Charles, Ill., U.S.A. **76 E1** 41 54N 88 19W
St. Charles, Mo., U.S.A. **80 F9** 38 47N 90 29W
St. Christopher = St. Kitts, W. Indies **89 C7** 17 20N 62 40W
St. Christopher-Nevis ■, W. Indies **89 C7** 17 20N 62 40W
St. Clair, Mich., U.S.A. **78 D2** 42 50N 82 30W
St. Clair, Pa., U.S.A. **79 F8** 40 43N 76 12W
St. Clair, L., Canada **70 D3** 42 30N 82 45W
St. Clairsville, U.S.A. **78 F4** 40 5N 80 54W
St. Claude, Canada **73 D9** 49 40N 98 20W
St. Cloud, Fla., U.S.A. **77 L5** 28 15N 81 17W
St. Cloud, Minn., U.S.A. **80 C7** 45 34N 94 10W
St-Coeur de Marie, Canada **71 C5** 48 39N 71 43W
St. Cricq, C., Australia **61 E1** 25 17S 113 6 E
St. Croix, Virgin Is. **89 C7** 17 45N 64 45W
St. Croix →, U.S.A. **80 C8** 44 45N 92 48W
St. Croix Falls, U.S.A. **80 C8** 45 24N 92 38W
St. David's, Canada **71 C8** 48 12N 58 52W
St. David's, U.K. **11 F2** 51 54N 5 16W
St. David's Head, U.K. **11 F2** 51 55N 5 16W
St.-Denis, France **18 B5** 48 56N 2 22 E
St. Elias, Mt., U.S.A. **68 B5** 60 18N 140 56W
St. Elias Mts., Canada **72 A1** 60 33N 139 28W
St.-Étienne, France **18 D6** 45 27N 4 22 E
St. Eugène, Canada **79 A10** 45 30N 74 28W
St. Eustatius, W. Indies **89 C7** 17 20N 63 0W
St-Félicien, Canada **70 C5** 48 40N 72 25W
St.-Flour, France **18 D5** 45 2N 3 6 E
St. Francis, U.S.A. **80 F4** 39 47N 101 48W
St. Francis →, U.S.A. **81 H9** 34 38N 90 36W
St. Francis, C., S. Africa **56 E3** 34 14S 24 49 E
St. Francisville, U.S.A. **81 K9** 30 47N 91 23W
St-François, L., Canada **79 A10** 45 10N 74 22W
St-Gabriel-de-Brandon, Canada **70 C5** 46 17N 73 24W
St. Gallen = Sankt Gallen, Switz. **16 E4** 47 26N 9 22 E
St. George, Australia **63 D4** 28 1S 148 30 E
St. George, Canada **71 C6** 45 11N 66 50W
St. George, S.C., U.S.A. **77 J5** 33 11N 80 35W
St. George, Utah, U.S.A. **83 H7** 37 6N 113 35W
St. George, C., Canada **71 C8** 48 30N 59 16W
St. George, C., U.S.A. **77 L3** 29 40N 85 5W
St. George Ra., Australia **60 C4** 18 40S 125 0 E
St-Georges, Belgium **15 D5** 50 37N 5 20 E
St. George's, Canada **71 C8** 48 26N 58 31W
St-Georges, Canada **71 C5** 46 8N 70 40W
St.-Georges, Fr. Guiana **93 C8** 4 0N 52 0W
St. George's, Grenada **89 D7** 12 5N 61 43W
St. George's B., Canada **71 C8** 48 24N 58 53W
St. Georges Basin, Australia **60 C4** 15 23S 125 2 E
St. George's Channel, U.K. **13 E6** 52 0N 6 0W
St. Georges Hd., Australia **63 F5** 35 12S 150 42 E
St. Gotthard P. = San Gottardo, Paso del, Switz. **16 E4** 46 33N 8 33 E
St. Helena, U.S.A. **82 G2** 38 30N 122 28W
St. Helena ■, Atl. Oc. **2 E9** 15 55S 5 44W
St. Helena, Mt., U.S.A. **84 G4** 38 40N 122 36W
St. Helena B., S. Africa **56 E2** 32 40S 18 10 E
St. Helens, Australia **62 G4** 41 20S 148 15 E
St. Helens, U.K. **10 D5** 53 28N 2 44W
St. Helens, U.S.A. **84 E4** 45 52N 122 48W
St. Helens, Mt., U.S.A. **84 D4** 46 12N 122 12W
St. Helier, U.K. **11 H5** 49 11N 2 6W
St-Hubert, Belgium **15 D5** 50 2N 5 23 E
St-Hyacinthe, Canada **70 C5** 45 40N 72 58W
St. Ignace, U.S.A. **76 C3** 45 52N 84 44W
St. Ignace I., Canada **70 C2** 48 45N 88 0W
St. Ignatius, U.S.A. **82 C6** 47 19N 114 6W
St. Ives, Cambs., U.K. **11 E7** 52 20N 0 5W
St. Ives, Corn., U.K. **11 G2** 50 13N 5 29W
St. James, U.S.A. **80 D7** 43 59N 94 38W
St-Jean, Canada **70 C5** 45 20N 73 20W
St-Jean →, Canada **71 B7** 50 17N 64 20W
St-Jean, L., Canada **71 C5** 48 40N 72 0W
St. Jean Baptiste, Canada **73 D9** 49 15N 97 20W
St-Jean-Port-Joli, Canada **71 C5** 47 15N 70 13W
St-Jérôme, Qué., Canada **70 C5** 45 47N 74 0W
St-Jérôme, Qué., Canada **71 C5** 48 26N 71 53W
St. John, Canada **71 C6** 45 20N 66 8W
St. John, Kans., U.S.A. **81 G5** 38 0N 98 46W
St. John, N. Dak., U.S.A. **80 A5** 48 57N 99 43W
St. John →, U.S.A. **71 C6** 45 12N 66 5W
St. John, C., Canada **71 B8** 50 0N 55 32W
St. John's, Antigua **89 C7** 17 6N 61 51W
St. John's, Canada **71 C9** 47 35N 52 40W
St. Johns, Ariz., U.S.A. **83 J9** 34 30N 109 22W
St. Johns, Mich., U.S.A. **76 D3** 43 0N 84 33W
St. Johns →, U.S.A. **77 K5** 30 24N 81 24W
St. Johnsbury, U.S.A. **79 B12** 44 25N 72 1W
St. Johnsville, U.S.A. **79 C10** 43 0N 74 43W
St. Joseph, La., U.S.A. **81 K9** 31 55N 91 14W
St. Joseph, Mich., U.S.A. **76 D2** 42 6N 86 29W
St. Joseph, Mo., U.S.A. **80 F7** 39 46N 94 50W
St. Joseph →, U.S.A. **76 D2** 42 7N 86 29W
St. Joseph, I., Canada **70 C3** 46 12N 83 58W
St. Joseph, L., Canada **70 B1** 51 10N 90 35W
St-Jovite, Canada **70 C5** 46 8N 74 38W
St. Kilda, N.Z. **59 L3** 45 53S 170 31 E
St. Kitts, W. Indies **89 C7** 17 20N 62 40W
St. Kitts-Nevis = St. Christopher-Nevis ■, W. Indies **89 C7** 17 20N 62 40W
St. Laurent, Canada **73 C9** 50 25N 97 58W
St.-Laurent, Fr. Guiana **93 B8** 5 29N 54 3W
St. Lawrence, Australia **62 C4** 22 16S 149 31 E
St. Lawrence, Canada **71 C8** 46 54N 55 23W
St. Lawrence →, Canada **71 C6** 49 30N 66 0W
St. Lawrence, Gulf of, Canada **71 C7** 48 25N 62 0W
St. Lawrence I., U.S.A. **68 B3** 63 30N 170 30W
St. Leonard, Canada **71 C6** 47 12N 67 58W
St. Lewis →, Canada **71 B8** 52 26N 56 11W
St.-Lô, France **18 B3** 49 7N 1 5W
St-Louis, Senegal **50 E1** 16 8N 16 27W
St. Louis, Mich., U.S.A. **76 D3** 43 25N 84 36W
St. Louis, Mo., U.S.A. **80 F9** 38 37N 90 12W
St. Louis →, U.S.A. **80 B8** 47 15N 92 45W
St. Lucia ■, W. Indies **89 D7** 14 0N 60 50W
St. Lucia, L., S. Africa **57 D5** 28 5S 32 30 E
St. Lucia Channel, W. Indies **89 D7** 14 15N 61 0W
St. Lunaire-Griquet, Canada **71 B8** 51 31N 55 28W
St. Maarten, W. Indies **89 C7** 18 0N 63 5W
St.-Malo, France **18 B2** 48 39N 2 1W
St-Marc, Haiti **89 C5** 19 10N 72 41W
St. Maries, U.S.A. **82 C5** 47 19N 116 35W
St-Martin, W. Indies **89 C7** 18 0N 63 0W
St. Martin, L., Canada **73 C9** 51 40N 98 30W
St. Martins, Canada **71 C6** 45 22N 65 34W
St. Martinville, U.S.A. **81 K9** 30 7N 91 50W
St. Mary Pk., Australia **63 E2** 31 32S 138 34 E
St. Marys, Australia **62 G4** 41 35S 148 11 E
St. Marys, Canada **78 C3** 43 20N 81 10W
St. Mary's, U.K. **11 H1** 49 55N 6 18W
St. Marys, U.S.A. **78 E6** 41 26N 78 34W
St. Mary's, C., Canada **71 C9** 46 50N 54 12W
St. Mary's B., Canada **71 C9** 46 50N 53 50W
St. Marys Bay, Canada **71 D6** 44 25N 66 10W
St.-Mathieu, Pte., France **18 B1** 48 20N 4 45W
St. Matthews, I. = Zadetkyi Kyun, Burma **39 H2** 10 0N 98 25 E
St-Maurice →, Canada **70 C5** 46 21N 72 31W
St. Michael's Mount, U.K. **11 G2** 50 7N 5 30W
St.-Nazaire, France **18 C2** 47 17N 2 12W
St. Neots, U.K. **11 E7** 52 14N 0 16W
St. Niklass = Sint Niklaas, Belgium **15 C4** 51 10N 4 9 E
St.-Omer, France **18 A5** 50 45N 2 15 E
St-Pacome, Canada **71 C6** 47 24N 69 58W
St-Pamphile, Canada **71 C6** 46 58N 69 48W
St. Pascal, Canada **71 C6** 47 32N 69 48W
St. Paul, Canada **72 C6** 54 0N 111 17W
St. Paul, Minn., U.S.A. **80 C8** 44 57N 93 6W
St. Paul, Nebr., U.S.A. **80 E5** 41 13N 98 27W
St. Paul, I., Ind. Oc. **3 F13** 38 55S 77 34 E
St. Paul I., Canada **71 C7** 47 12N 60 9W
St. Peter, U.S.A. **80 C8** 44 20N 93 57W
St. Peter Port, Chan. Is. **11 H5** 49 27N 2 31W
St. Peters, N.S., Canada **71 C7** 45 40N 60 53W
St. Peters, P.E.I., Canada **71 C7** 46 25N 62 35W
St. Petersburg = Sankt-Peterburg, Russia **24 C5** 59 55N 30 20 E
St. Petersburg, U.S.A. **77 M4** 27 46N 82 39W
St.-Pierre, St- P. & M. **71 C8** 46 46N 56 12W
St-Pierre, L., Canada **70 C5** 46 12N 72 52W
St.-Pierre et Miquelon □, St- P. & M. **71 C8** 46 55N 56 10W
St.-Quentin, France **18 B5** 49 50N 3 16 E
St. Regis, U.S.A. **82 C6** 47 18N 115 6W
St. Sebastien, Tanjon' i, Madag. **57 A8** 12 26S 48 44 E
St-Siméon, Canada **71 C6** 47 51N 69 54W
St. Stephen, Canada **71 C6** 45 16N 67 17W
St. Thomas, Canada **70 D3** 42 45N 81 10W
St. Thomas I., Virgin Is. **89 C7** 18 20N 64 55W
St-Tite, Canada **70 C5** 46 45N 72 34W
St.-Tropez, France **18 E7** 43 17N 6 38 E
St. Troud = Sint Truiden, Belgium **15 D5** 50 48N 5 10 E
St.-Valéry-sur-Somme, France **18 A4** 50 11N 1 38 E
St. Vincent, W. Indies **89 D7** 13 10N 61 10W
St. Vincent, G., Australia **63 F2** 35 0S 138 0 E
St. Vincent & the Grenadines ■, W. Indies **89 D7** 13 0N 61 10W
St. Vincent Passage, W. Indies **89 D7** 13 30N 61 0W
St-Vith, Belgium **15 D6** 50 17N 6 9 E
Ste-Agathe-des-Monts, Canada **70 C5** 46 3N 74 17W
Ste-Anne de Beaupré, Canada **71 C5** 47 2N 70 58W
Ste-Anne-des-Monts, Canada **71 C6** 49 8N 66 30W
Ste. Genevieve, U.S.A. **80 G9** 37 59N 90 2W
Ste-Marguerite →, Canada **71 B6** 50 9N 66 36W
Ste.-Marie, Martinique **89 D7** 14 48N 61 1W
Ste-Marie de la Madeleine, Canada **71 C5** 46 26N 71 0W
Ste.-Rose, Guadeloupe **89 C7** 16 20N 61 45W
Ste. Rose du Lac, Canada **73 C9** 51 4N 99 30W
Saintes, France **18 D3** 45 45N 0 37W
Saintes, I. des, Guadeloupe **89 C7** 15 50N 61 35W
Saintonge, France **18 D3** 45 40N 0 50W
Saipan, Pac. Oc. **64 F6** 15 12N 145 45 E
Sairang, India **41 H18** 23 50N 92 45 E
Sairecábur, Cerro, Bolivia **94 A2** 22 43S 67 54W
Saitama □, Japan **31 F9** 36 25N 139 30 E
Sajama, Bolivia **92 G5** 18 7S 69 0W
Sajum, India **43 C8** 33 20N 79 0 E
Sak →, S. Africa **56 E3** 30 52S 20 25 E
Sakai, Japan **31 G7** 34 30N 135 30 E
Sakaide, Japan **31 G6** 34 15N 133 50 E
Sakaiminato, Japan **31 G6** 35 38N 133 11 E
Sakākah, Si. Arabia **44 D4** 30 0N 40 8 E
Sakakawea, L., U.S.A. **80 B3** 47 30N 101 25W
Sakami, L., Canada **70 B4** 53 15N 77 0W
Sakania, Zaïre **55 E2** 12 43S 28 30 E
Sakarya = Adapazan, Turkey **25 F5** 40 48N 30 25 E
Sakarya →, Turkey **25 F5** 41 7N 30 39 E
Sakashima-Guntō, Japan **31 M2** 24 46N 124 0 E
Sakata, Japan **30 E9** 38 55N 139 50 E
Sakchu, N. Korea **35 D13** 40 23N 125 2 E
Sakeny →, Madag. **57 C8** 20 0S 45 25 E
Sakha = Yakut Republic □, Russia **27 C13** 62 0N 130 0 E
Sakhalin, Russia **27 D15** 51 0N 143 0 E
Sakhalinskiy Zaliv, Russia **27 D15** 54 0N 141 0 E
Sakon Nakhon, Thailand **38 D5** 17 10N 104 9 E
Sakrand, Pakistan **42 F3** 26 10N 68 15 E
Sakrivier, S. Africa **56 E3** 30 54S 20 28 E
Sakuma, Japan **31 G8** 35 3N 137 49 E
Sakurai, Japan **31 G7** 34 30N 135 51 E
Sala, Sweden **9 G14** 59 58N 16 35 E
Sala-y-Gómez, Pac. Oc. **65 K17** 26 28S 105 28W
Salaberry-de-Valleyfield, Canada **70 C5** 45 15N 74 8W
Saladas, Argentina **94 B4** 28 15S 58 40W
Saladillo, Argentina **94 D4** 35 40S 59 55W
Salado →, Buenos Aires, Argentina **94 D4** 35 44S 57 22W
Salado →, La Pampa, Argentina **96 D3** 37 30S 67 0W
Salado →, Santa Fe, Argentina **94 C3** 31 40S 60 41W
Salado →, Mexico **86 B5** 26 52N 99 19W
Salaga, Ghana **50 G4** 8 31N 0 31W
Sālah, Syria **47 C5** 32 40N 36 45 E
Sálakhos, Greece **23 C9** 36 17N 27 57 E
Salālah, Oman **46 D5** 16 56N 53 59 E
Salamanca, Chile **94 C1** 31 46S 70 59W
Salamanca, Spain **19 B3** 40 58N 5 39W
Salamanca, U.S.A. **78 D6** 42 10N 78 43W
Salāmatābād, Iran **44 C5** 35 39N 47 50 E
Salamis, Cyprus **23 D12** 35 11N 33 54 E
Salamis, Greece **21 F10** 37 56N 23 30 E
Salar de Atacama, Chile **94 A2** 23 30S 68 25W
Salar de Uyuni, Bolivia **92 H5** 20 30S 67 45W
Salatiga, Indonesia **37 G14** 7 19S 110 30 E
Salavat, Russia **24 D10** 53 21N 55 55 E
Salaverry, Peru **92 E3** 8 15S 79 0W
Salawati, Indonesia **37 E8** 1 7S 130 52 E
Salayar, Indonesia **37 F6** 6 7S 120 30 E
Salcombe, U.K. **11 G4** 50 14N 3 47W
Saldaña, Spain **19 A3** 42 32N 4 48W
Saldanha, S. Africa **56 E2** 33 0S 17 58 E
Saldanha B., S. Africa **56 E2** 33 6S 18 0 E
Sale, Australia **63 F4** 38 6S 147 6 E
Salé, Morocco **50 B3** 34 3N 6 48W
Sale, U.K. **10 D5** 53 26N 2 19W
Salekhard, Russia **24 A12** 66 30N 66 35 E
Salem, India **40 P11** 11 40N 78 11 E
Salem, Ind., U.S.A. **76 F2** 38 36N 86 6W
Salem, Mass., U.S.A. **79 D14** 42 31N 70 53W
Salem, Mo., U.S.A. **81 G9** 37 39N 91 32W
Salem, N.J., U.S.A. **76 F8** 39 34N 75 28W

San Ramón de la Nueva Orán

Salem, *Ohio, U.S.A.* **78 F4** 40 54N 80 52W
Salem, *Oreg., U.S.A.* **82 D2** 44 56N 123 2W
Salem, *S. Dak., U.S.A.* **80 D6** 43 44N 97 23W
Salem, *Va., U.S.A.* **76 G5** 37 18N 80 3W
Salerno, *Italy* **20 D6** 40 40N 14 44 E
Salford, *U.K.* **10 D5** 53 30N 2 17W
Salida, *U.S.A.* **74 C5** 38 32N 106 0W
Salima, *Malawi* **53 G6** 13 47S 34 28 E
Salina, *Italy* **20 E6** 38 35N 14 50 E
Salina, *U.S.A.* **80 F6** 38 50N 97 37W
Salina Cruz, *Mexico* **87 D5** 16 10N 95 10W
Salinas, *Brazil* **93 G10** 16 10S 42 10W
Salinas, *Chile* **94 A2** 23 31S 69 29W
Salinas, *Ecuador* **92 D2** 2 10S 80 58W
Salinas, *U.S.A.* **83 H3** 36 40N 121 39W
Salinas →, *Guatemala* **87 D6** 16 28N 90 31W
Salinas →, *U.S.A.* **83 H3** 36 45N 121 48W
Salinas, B. de, *Nic.* **88 D2** 11 4N 85 45W
Salinas, C. de, *Spain* **22 B10** 39 16N 3 4 E
Salinas, Pampa de las, *Argentina* **94 C2** 31 58S 66 42W
Salinas Ambargasta, *Argentina* **94 B3** 29 0S 65 0W
Salinas de Hidalgo, *Mexico* **86 C4** 22 30N 101 40W
Salinas Grandes, *Argentina* **94 B2** 30 0S 65 0W
Saline →, *Ark., U.S.A.* **81 J8** 33 10N 92 8W
Saline →, *Kans., U.S.A.* **80 F6** 38 52N 97 30W
Salines, *Spain* **22 B10** 39 21N 3 3 E
Salinópolis, *Brazil* **93 D9** 0 40S 47 20W
Salisbury = Harare, *Zimbabwe* **55 F3** 17 43S 31 2 E
Salisbury, *Australia* **63 E2** 34 46S 138 40 E
Salisbury, *U.K.* **11 F6** 51 4N 1 48W
Salisbury, *Md., U.S.A.* **76 F8** 38 22N 75 36W
Salisbury, *N.C., U.S.A.* **77 H5** 35 40N 80 29W
Salisbury Plain, *U.K.* **11 F6** 51 13N 1 50W
Salkhad, *Jordan* **47 C5** 32 30N 36 43 E
Sallisaw, *U.S.A.* **81 H7** 35 28N 94 47W
Salmās, *Iran* **44 B5** 38 11N 44 47 E
Salmo, *Canada* **72 D5** 49 10N 117 20W
Salmon, *U.S.A.* **82 D7** 45 11N 113 54W
Salmon →, *Canada* **72 C4** 54 3N 122 40W
Salmon →, *U.S.A.* **82 D5** 45 51N 116 47W
Salmon Arm, *Canada* **72 C5** 50 40N 119 15W
Salmon Falls, *U.S.A.* **82 E6** 42 48N 114 59W
Salmon Gums, *Australia* **61 F3** 32 59S 121 38 E
Salmon Res., *Canada* **71 C8** 48 5N 56 0W
Salmon River Mts., *U.S.A.* **82 D6** 45 0N 114 30W
Salo, *Finland* **9 F17** 60 22N 23 10 E
Salome, *U.S.A.* **85 M13** 33 47N 113 37W
Salonica = Thessaloníki, *Greece* **21 D10** 40 38N 22 58 E
Salonta, *Romania* **17 E10** 46 49N 21 42 E
Salsacate, *Argentina* **94 C2** 31 20S 65 5W
Salsk, *Russia* **25 E7** 46 28N 41 30 E
Salso →, *Italy* **20 F5** 37 6N 13 55 E
Salt →, *Canada* **72 B6** 60 0N 112 25W
Salt →, *U.S.A.* **83 K7** 33 23N 112 19W
Salt Creek, *Australia* **63 F2** 36 8S 139 38 E
Salt Fork Arkansas →, *U.S.A.* **81 G6** 36 36N 97 3W
Salt Lake City, *U.S.A.* **82 F8** 40 45N 111 53W
Salt Range, *Pakistan* **42 C5** 32 30N 72 25 E
Salta, *Argentina* **94 A2** 24 57S 65 25W
Salta □, *Argentina* **94 A2** 24 48S 65 30W
Saltcoats, *U.K.* **12 F4** 55 38N 4 47W
Saltee Is., *Ireland* **13 D5** 52 7N 6 37W
Saltfjorden, *Norway* **8 C13** 67 15N 14 10 E
Salthólmavík, *Iceland* **8 D3** 65 24N 21 57W
Saltillo, *Mexico* **86 B4** 25 30N 100 57W
Salto, *Argentina* **94 C3** 34 20S 60 15W
Salto, *Uruguay* **94 C4** 31 27S 57 50W
Salton City, *U.S.A.* **85 M11** 33 29N 115 51W
Salton Sea, *U.S.A.* **85 M11** 33 15N 115 45W
Saltpond, *Ghana* **50 G4** 5 15N 1 3W
Saltville, *U.S.A.* **76 G5** 36 53N 81 46W
Saluda →, *U.S.A.* **77 H5** 34 1N 81 4W
Salûm, *Egypt* **51 B10** 31 31N 25 7 E
Salûm, Khâlig el, *Egypt* **51 B10** 31 30N 25 9 E
Salur, *India* **41 K13** 18 27N 83 18 E
Saluzzo, *Italy* **20 B2** 44 39N 7 29 E
Salvador, *Brazil* **93 F11** 13 0S 38 30W
Salvador, *Canada* **73 C7** 52 10N 109 32W
Salvador, L., *U.S.A.* **81 L9** 29 43N 90 15W
Salween →, *Burma* **41 L20** 16 31N 97 37 E
Salyany, *Azerbaijan* **25 G8** 39 10N 48 50 E
Salyersville, *U.S.A.* **76 G4** 37 45N 83 4W
Salzburg, *Austria* **16 E6** 47 48N 13 2 E
Salzburg □, *Austria* **16 E6** 47 15N 13 0 E
Salzgitter, *Germany* **16 B5** 52 13N 10 22 E
Sam Neua, *Laos* **38 B5** 20 29N 104 5 E
Sam Ngao, *Thailand* **38 D2** 17 18N 99 0 E
Sam Rayburn Reservoir, *U.S.A.* **81 K7** 31 4N 94 5W
Sam Son, *Vietnam* **38 C5** 19 44N 105 54 E
Sam Teu, *Laos* **38 C5** 19 59N 104 38 E
Sama, *Russia* **26 C7** 60 12N 60 22 E
Sama de Langreo, *Spain* **19 A3** 43 18N 5 40W
Samagaltai, *Russia* **27 D10** 50 36N 95 3 E
Samales Group, *Phil.* **37 C6** 6 0N 122 0 E
Samana, *India* **42 D7** 30 10N 76 13 E
Samana Cay, *Bahamas* **89 B5** 23 3N 73 45W
Samanga, *Tanzania* **55 D4** 8 20S 39 13 E
Samangwa, *Zaïre* **54 C1** 4 23S 24 10 E
Samani, *Japan* **30 C11** 42 7N 142 56 E
Samar, *Phil.* **37 B7** 12 0N 125 0 E
Samara, *Russia* **24 D9** 53 8N 50 6 E
Samaria = Shōmrōn, *Jordan* **47 C4** 32 15N 35 13 E
Samariá, *Greece* **23 D5** 35 17N 23 58 E
Samarinda, *Indonesia* **36 E5** 0 30S 117 9 E
Samarkand, *Uzbekistan* **26 F7** 39 40N 66 55 E
Samarqand = Samarkand, *Uzbekistan* **26 F7** 39 40N 66 55 E
Sāmarrā, *Iraq* **44 C4** 34 12N 43 52 E
Samastipur, *India* **43 G11** 25 50N 85 50 E
Samba, *India* **43 C6** 32 32N 75 10 E
Samba, *Zaïre* **54 C2** 4 38S 26 22 E
Sambalpur, *India* **41 J14** 21 28N 84 4 E
Sambar, Tanjung, *Indonesia* **36 E4** 2 59S 110 19 E
Sambas, *Indonesia* **36 D3** 1 20N 109 20 E
Sambava, *Madag.* **57 A9** 14 16S 50 10 E
Sambawizi, *Zimbabwe* **55 F2** 18 24S 26 13 E
Sambhal, *India* **43 E8** 28 35N 78 37 E
Sambhar, *India* **42 F6** 26 52N 75 6 E
Sambor, *Cambodia* **38 F6** 12 46N 106 0 E
Sambre →, *Europe* **15 D4** 50 27N 4 52 E
Samburu □, *Kenya* **54 B4** 1 10N 37 0 E
Samchŏk, *S. Korea* **35 F15** 37 30N 129 10 E
Samchonpo, *S. Korea* **35 G15** 35 0N 128 6 E
Same, *Tanzania* **54 C4** 4 2S 37 38 E
Samfya, *Zambia* **55 E2** 11 22S 29 31 E
Samnah, *Si. Arabia* **44 E3** 25 10N 37 15 E
Samo Alto, *Chile* **94 C1** 30 22S 71 0W
Samoorombón, B., *Argentina* **94 D4** 36 5S 57 20W
Sámos, *Greece* **21 F12** 37 45N 26 50 E
Samothráki, *Évros, Greece* **21 D11** 40 28N 25 28 E
Samothráki, *Kérkira, Greece* **23 A3** 39 48N 19 31 E
Sampacho, *Argentina* **94 C3** 33 20S 64 50W
Sampang, *Indonesia* **37 G15** 7 11S 113 13 E
Sampit, *Indonesia* **36 E4** 2 34S 113 0 E
Sampit, Teluk, *Indonesia* **36 E4** 3 5S 113 3 E
Samrong, *Cambodia* **38 E4** 14 15N 103 30 E
Samrong, *Thailand* **38 E3** 15 10N 100 40 E
Samsun, *Turkey* **25 F6** 41 15N 36 22 E
Samui, Ko, *Thailand* **39 H3** 9 30N 100 0 E
Samusole, *Zaïre* **55 E1** 10 2S 24 0 E
Samut Prakan, *Thailand* **38 F3** 13 32N 100 40 E
Samut Sakhon, *Thailand* **38 F3** 13 31N 100 13 E
Samut Songkhram →, *Thailand* **38 F3** 13 24N 100 1 E
Samwari, *Pakistan* **42 E2** 28 30N 66 46 E
San, *Mali* **50 F4** 13 15N 4 57W
San →, *Cambodia* **38 F5** 13 32N 105 57 E
San →, *Poland* **17 C10** 50 45N 21 51 E
San Agustin, C., *Phil.* **37 C7** 6 20N 126 13 E
San Agustín de Valle Fértil, *Argentina* **94 C2** 30 35S 67 30W
San Ambrosio, *Pac. Oc.* **65 K20** 26 28S 79 53W
San Andreas, *U.S.A.* **84 G6** 38 12N 120 41W
San Andrés, I. de, *Caribbean* **88 D3** 12 42N 81 46W
San Andres Mts., *U.S.A.* **83 K10** 33 0N 106 30W
San Andrés Tuxtla, *Mexico* **87 D5** 18 30N 95 20W
San Angelo, *U.S.A.* **81 K4** 31 28N 100 26W
San Anselmo, *U.S.A.* **84 H4** 37 59N 122 34W
San Antonio, *Belize* **87 D7** 16 15N 89 2W
San Antonio, *Chile* **94 C1** 33 40S 71 40W
San Antonio, *N. Mex., U.S.A.* **83 K10** 33 55N 106 52W
San Antonio, *Tex., U.S.A.* **81 L5** 29 25N 98 30W
San Antonio →, *U.S.A.* **81 L6** 28 30N 96 54W
San Antonio, C., *Argentina* **94 D4** 36 15S 56 40W
San Antonio, C., *Cuba* **88 B3** 21 50N 84 57W
San Antonio, Mt., *U.S.A.* **85 L9** 34 17N 117 38W
San Antonio Abad, *Spain* **22 C7** 38 59N 1 19 E
San Antonio de los Baños, *Cuba* **88 B3** 22 54N 82 31W
San Antonio de los Cobres, *Argentina* **94 A2** 24 10S 66 17W
San Antonio Oeste, *Argentina* **96 E4** 40 40S 65 0W
San Ardo, *U.S.A.* **84 J6** 36 1N 120 54W
San Augustín, *Canary Is.* **22 G4** 27 47N 15 32W
San Augustine, *U.S.A.* **81 K7** 31 30N 94 7W
San Bartolomé, *Canary Is.* **22 F6** 28 59N 13 37W
San Bartolomé de Tirajana, *Canary Is.* **22 G4** 27 54N 15 34W
San Benedicto, I., *Mexico* **86 D2** 19 18N 110 49W
San Benito, *U.S.A.* **81 M6** 26 8N 97 38W
San Benito →, *U.S.A.* **84 J5** 36 53N 121 34W
San Benito Mt., *U.S.A.* **84 J6** 36 22N 120 37W
San Bernardino, *U.S.A.* **85 L9** 34 7N 117 19W
San Bernardino Mts., *U.S.A.* **85 L10** 34 10N 116 45W
San Bernardino Str., *Phil.* **37 B6** 13 0N 125 0 E
San Bernardo, *Chile* **94 C1** 33 40S 70 50W
San Bernardo, I. de, *Colombia* **92 B3** 9 45N 75 50W
San Blas, *Mexico* **86 B3** 26 4N 108 46W
San Blas, Arch. de, *Panama* **88 E4** 9 50N 78 31W
San Blas, C., *U.S.A.* **77 L3** 29 40N 85 21W
San Borja, *Bolivia* **92 F5** 14 50S 66 52W
San Buenaventura, *Mexico* **86 B4** 27 5N 101 32W
San Carlos, *Argentina* **94 C2** 33 50S 69 0W
San Carlos, *Chile* **94 D1** 36 10S 72 0W
San Carlos, *Mexico* **86 B4** 29 0N 100 54W
San Carlos, *Nic.* **88 D3** 11 12N 84 50W
San Carlos, *Phil.* **37 B6** 10 29N 123 25 E
San Carlos, *Spain* **22 B8** 39 3N 1 34 E
San Carlos, *Uruguay* **95 C5** 34 46S 54 58W
San Carlos, *U.S.A.* **83 K8** 33 21N 110 27W
San Carlos, *Amazonas, Venezuela* **92 C5** 1 55N 67 4W
San Carlos, *Cojedes, Venezuela* **92 B5** 9 40N 68 36W
San Carlos de Bariloche, *Argentina* **96 E2** 41 10S 71 25W
San Carlos del Zulia, *Venezuela* **92 B4** 9 1N 71 55W
San Carlos L., *U.S.A.* **83 K8** 33 11N 110 32W
San Clemente, *Chile* **94 D1** 35 30S 71 29W
San Clemente, *U.S.A.* **85 M9** 33 26N 117 37W
San Clemente I., *U.S.A.* **85 N8** 32 53N 118 29W
San Cristóbal, *Argentina* **94 C3** 30 20S 61 10W
San Cristóbal, *Dom. Rep.* **89 C5** 18 25N 70 6W
San Cristóbal, *Mexico* **87 D6** 16 50N 92 33W
San Cristóbal, *Spain* **22 B11** 39 57N 4 3 E
San Cristóbal, *Venezuela* **92 B4** 7 46N 72 14W
San Diego, *Calif., U.S.A.* **85 N9** 32 43N 117 9W
San Diego, *Tex., U.S.A.* **81 M5** 27 46N 98 14W
San Diego, C., *Argentina* **96 G3** 54 40S 65 10W
San Diego de la Unión, *Mexico* **86 C4** 21 28N 100 52W
San Dimitri, Ras, *Malta* **23 C1** 36 4N 14 11 E
San Estanislao, *Paraguay* **94 A4** 24 39S 56 26W
San Felipe, *Chile* **94 C1** 32 43S 70 42W
San Felipe, *Mexico* **86 A2** 31 0N 114 52W
San Felipe, *Venezuela* **92 A5** 10 20N 68 44W
San Felipe →, *U.S.A.* **85 M11** 33 12N 115 49W
San Feliu de Guíxols, *Spain* **19 B7** 41 45N 3 1 E
San Félix, *Pac. Oc.* **65 K20** 26 23S 80 0W
San Fernando, *Chile* **94 C1** 34 30S 71 0W
San Fernando, *Mexico* **86 B1** 29 55N 115 10W
San Fernando, *La Union, Phil.* **37 A6** 16 40N 120 23 E
San Fernando, *Pampanga, Phil.* **37 A6** 15 5N 120 37 E
San Fernando, *Baleares, Spain* **22 C7** 38 42N 1 28 E
San Fernando, *Cádiz, Spain* **19 D2** 36 28N 6 17W
San Fernando, *Trin. & Tob.* **89 D7** 10 20N 61 30W
San Fernando, *U.S.A.* **85 L8** 34 17N 118 26W
San Fernando →, *Mexico* **86 C5** 24 55N 98 10W
San Fernando de Apure, *Venezuela* **92 B5** 7 54N 67 15W
San Fernando de Atabapo, *Venezuela* **92 C5** 4 3N 67 42W
San Francisco, *Argentina* **94 C3** 31 30S 62 5W
San Francisco, *U.S.A.* **83 H2** 37 47N 122 25W
San Francisco →, *U.S.A.* **83 K9** 32 59N 109 22W
San Francisco, Paso de, *S. Amer.* **94 B2** 27 0S 68 0W
San Francisco de Macorís, *Dom. Rep.* **89 C5** 19 19N 70 15W
San Francisco del Monte de Oro, *Argentina* **94 C2** 32 36S 66 8W
San Francisco del Oro, *Mexico* **86 B3** 26 52N 105 50W
San Francisco Javier, *Spain* **22 C7** 38 42N 1 26 E
San Gil, *Colombia* **92 B4** 6 33N 73 8W
San Gorgonio Mt., *U.S.A.* **85 L10** 34 7N 116 51W
San Gottardo, Paso del, *Switz.* **16 E4** 46 33N 8 33 E
San Gregorio, *Uruguay* **95 C4** 32 37S 55 40W
San Gregorio, *U.S.A.* **84 H4** 37 20N 122 23W
San Ignacio, *Belize* **87 D7** 17 10N 89 0W
San Ignacio, *Bolivia* **92 G6** 16 20S 60 55W
San Ignacio, *Mexico* **86 B2** 27 27N 113 0W
San Ignacio, *Paraguay* **94 B4** 26 52S 57 3W
San Ignacio, L., *Mexico* **86 B2** 26 50N 113 11W
San Ildefonso, C., *Phil.* **37 A6** 16 0N 122 1 E
San Isidro, *Argentina* **94 C4** 34 29S 58 31W
San Jacinto, *U.S.A.* **85 M10** 33 47N 116 57W
San Jaime, *Spain* **22 B11** 39 54N 4 4 E
San Javier, *Misiones, Argentina* **95 B4** 27 55S 55 5W
San Javier, *Santa Fe, Argentina* **94 C4** 30 40S 59 55W
San Javier, *Bolivia* **92 G6** 16 18S 62 30W
San Javier, *Chile* **94 D1** 35 40S 71 45W
San Jeronimo Taviche, *Mexico* **87 D5** 16 38N 96 32W
San Joaquin, *U.S.A.* **84 J6** 36 36N 120 11W
San Joaquin →, *U.S.A.* **83 G3** 38 4N 121 51W
San Joaquin Valley, *U.S.A.* **84 J6** 37 20N 121 0W
San Jordi, *Spain* **22 B9** 39 33N 2 46 E
San Jorge, *Argentina* **94 C3** 31 54S 61 50W
San Jorge, *Spain* **22 C7** 38 54N 1 24 E
San Jorge, B. de, *Mexico* **86 A2** 31 20N 113 20W
San Jorge, G., *Argentina* **96 F3** 46 0S 66 0W
San Jorge, G. de, *Spain* **19 B5** 40 50N 0 55W
San José, *Bolivia* **92 G6** 17 53S 60 50W
San José, *Costa Rica* **88 E3** 9 55N 84 2W
San José, *Guatemala* **88 D1** 14 0N 90 50W
San José, *Mexico* **86 C2** 25 0N 110 50W
San Jose, *Phil.* **37 A6** 15 45N 120 55 E
San José, *Spain* **22 C7** 38 55N 1 18 E
San Jose, *U.S.A.* **83 H3** 37 20N 121 53W
San Jose →, *U.S.A.* **83 J10** 34 25N 106 45W
San Jose de Buenovista, *Phil.* **37 B6** 12 27N 121 4 E
San José de Feliciano, *Argentina* **94 C4** 30 26S 58 46W
San José de Jáchal, *Argentina* **94 C2** 30 15S 68 46W
San José de Mayo, *Uruguay* **94 C4** 34 27S 56 40W
San José de Ocune, *Colombia* **92 C4** 4 15N 70 20W
San José del Cabo, *Mexico* **86 C3** 23 0N 109 40W
San José del Guaviare, *Colombia* **92 C4** 2 35N 72 38W
San Juan, *Argentina* **94 C2** 31 30S 68 30W
San Juan, *Mexico* **86 C4** 21 20N 102 50W
San Juan, *Phil.* **37 C7** 8 25N 126 20 E
San Juan, *Puerto Rico* **89 C6** 18 28N 66 7W
San Juan □, *Argentina* **94 C2** 31 9S 69 0W
San Juan →, *Argentina* **94 C2** 32 20S 67 25W
San Juan →, *Nic.* **88 D3** 10 56N 83 42W
San Juan →, *U.S.A.* **83 H8** 37 16N 110 26W
San Juan, C., *Eq. Guin.* **52 D1** 1 5N 9 20 E
San Juan Bautista, *Paraguay* **94 B4** 26 37S 57 6W
San Juan Bautista, *Spain* **22 B8** 39 5N 1 31 E
San Juan Bautista, *U.S.A.* **83 H3** 36 51N 121 32W
San Juan Bautista Valle Nacional, *Mexico* **87 D5** 17 47N 96 19W
San Juan Capistrano, *U.S.A.* **85 M9** 33 30N 117 40W
San Juan Cr. →, *U.S.A.* **84 J5** 35 40N 120 22W
San Juan de Guadalupe, *Mexico* **86 C4** 24 38N 102 44W
San Juan de los Morros, *Venezuela* **92 B5** 9 55N 67 21W
San Juan del Norte, *Nic.* **88 D3** 10 58N 83 40W
San Juan del Norte, B. de, *Nic.* **88 D3** 11 0N 83 40W
San Juan del Río, *Mexico* **87 C5** 20 25N 100 0W
San Juan del Sur, *Nic.* **88 D2** 11 20N 85 51W
San Juan I., *U.S.A.* **84 B3** 48 32N 123 5W
San Juan Mts., *U.S.A.* **83 H10** 37 30N 107 0W
San Julián, *Argentina* **96 F3** 49 15S 67 45W
San Justo, *Argentina* **94 C3** 30 47S 60 30W
San Kamphaeng, *Thailand* **38 C2** 18 45N 99 8 E
San Lázaro, C., *Mexico* **86 C2** 24 50N 112 18W
San Lázaro, Sa., *Mexico* **86 C3** 23 25N 110 0W
San Leandro, *U.S.A.* **83 H2** 37 44N 122 9W
San Lorenzo, *Argentina* **94 C3** 32 45S 60 45W
San Lorenzo, *Ecuador* **92 C3** 1 15N 78 50W
San Lorenzo, *Paraguay* **94 B4** 25 20S 57 32W
San Lorenzo, *Spain* **22 B10** 39 37N 3 17 E
San Lorenzo →, *Mexico* **86 C3** 24 15N 107 24W
San Lorenzo, I., *Mexico* **86 B2** 28 35N 112 50W
San Lorenzo, I., *Peru* **92 F3** 12 7S 77 15W
San Lorenzo, Mt., *Argentina* **96 F2** 47 40S 72 20W
San Lucas, *Bolivia* **92 H5** 20 5S 65 7W
San Lucas, *Baja Calif. S., Mexico* **86 C3** 22 53N 109 54W
San Lucas, *Baja Calif. S., Mexico* **86 B2** 27 10N 112 14W
San Lucas, *U.S.A.* **84 J5** 36 8N 121 1W
San Lucas, C., *Mexico* **86 C3** 22 50N 110 0W
San Luis, *Argentina* **94 C2** 33 20S 66 20W
San Luis, *Cuba* **88 B3** 22 17N 83 46W
San Luis, *Guatemala* **88 C2** 16 14N 89 27W
San Luis, *U.S.A.* **83 H11** 37 12N 105 25W
San Luis □, *Argentina* **94 C2** 34 0S 66 0W
San Luis, I., *Mexico* **86 B2** 29 58N 114 26W
San Luis, Sierra de, *Argentina* **94 C2** 32 30S 66 10W
San Luis de la Paz, *Mexico* **86 C4** 21 19N 100 32W
San Luis Obispo, *U.S.A.* **85 K6** 35 17N 120 40W
San Luis Potosí, *Mexico* **86 C4** 22 9N 100 59W
San Luis Potosí □, *Mexico* **86 C4** 22 10N 101 0W
San Luis Reservoir, *U.S.A.* **84 H5** 37 4N 121 5W
San Luis Río Colorado, *Mexico* **86 A2** 32 29N 114 58W
San Marcos, *Guatemala* **88 D1** 14 59N 91 52W
San Marcos, *Mexico* **86 B2** 27 13N 112 6W
San Marcos, *U.S.A.* **81 L6** 29 53N 97 56W
San Marino ■, *Europe* **20 C5** 43 56N 12 25 E
San Martín, *Argentina* **94 C2** 33 5S 68 28W
San Martín, L., *Argentina* **96 F2** 48 50S 72 50W
San Mateo, *Spain* **22 B7** 39 3N 1 23 E
San Mateo, *U.S.A.* **83 H2** 37 34N 122 19W
San Matías, *Bolivia* **92 G7** 16 25S 58 20W
San Matías, G., *Argentina* **96 E4** 41 30S 64 0W
San Miguel, *El Salv.* **88 D2** 13 30N 88 12W
San Miguel, *Panama* **88 E4** 8 27N 78 55W
San Miguel, *Spain* **22 B7** 39 3N 1 26 E
San Miguel, *U.S.A.* **83 J3** 35 45N 120 42W
San Miguel →, *Bolivia* **92 F6** 13 52S 63 56W
San Miguel de Tucumán, *Argentina* **94 B2** 26 50S 65 20W
San Miguel del Monte, *Argentina* **94 D4** 35 23S 58 50W
San Miguel I., *U.S.A.* **85 L6** 34 2N 120 23W
San Narciso, *Phil.* **37 A6** 15 2N 120 3 E
San Nicolás, *Canary Is.* **22 G4** 27 58N 15 47W
San Nicolás de los Arroyas, *Argentina* **94 C3** 33 25S 60 10W
San Nicolas I., *U.S.A.* **85 M7** 33 15N 119 30W
San Onofre, *U.S.A.* **85 M9** 33 22N 117 34W
San Pablo, *Bolivia* **94 A2** 21 43S 66 38W
San Pedro, *Buenos Aires, Argentina* **95 B5** 26 30S 54 10W
San Pedro, *Jujuy, Argentina* **94 A3** 24 12S 64 55W
San-Pédro, *Ivory C.* **50 H3** 4 50N 6 33W
San Pedro, *Mexico* **86 C2** 23 55N 110 17W
San Pedro □, *Paraguay* **94 A4** 24 0S 57 0W
San Pedro →, *Chihuahua, Mexico* **86 B3** 28 20N 106 10W
San Pedro →, *Michoacan, Mexico* **86 D4** 19 23N 103 51W
San Pedro →, *Nayarit, Mexico* **86 C3** 21 45N 105 30W
San Pedro →, *U.S.A.* **83 K8** 32 59N 110 47W
San Pedro, Pta., *Chile* **94 B1** 25 30S 70 38W
San Pedro Channel, *U.S.A.* **85 M8** 33 30N 118 25W
San Pedro de Atacama, *Chile* **94 A2** 22 55S 68 15W
San Pedro de Jujuy, *Argentina* **94 A3** 24 12S 64 55W
San Pedro de las Colonias, *Mexico* **86 B4** 25 50N 102 59W
San Pedro de Lloc, *Peru* **92 E3** 7 15S 79 28W
San Pedro de Macorís, *Dom. Rep.* **89 C6** 18 30N 69 18W
San Pedro del Norte, *Nic.* **88 D3** 13 4N 84 33W
San Pedro del Paraná, *Paraguay* **94 B4** 26 43S 56 13W
San Pedro Mártir, Sierra, *Mexico* **86 A1** 31 0N 115 30W
San Pedro Mixtepec, *Mexico* **87 D5** 16 2N 97 7W
San Pedro Ocampo = Melchor Ocampo, *Mexico* **86 C4** 24 52N 101 40W
San Pedro Sula, *Honduras* **88 C2** 15 30N 88 0W
San Quintín, *Mexico* **86 A1** 30 29N 115 57W
San Rafael, *Argentina* **94 C2** 34 40S 68 21W
San Rafael, *Calif., U.S.A.* **84 H4** 37 58N 122 32W
San Rafael, *N. Mex., U.S.A.* **83 J10** 35 7N 107 53W
San Rafael Mt., *U.S.A.* **85 L7** 34 41N 119 52W
San Rafael Mts., *U.S.A.* **85 L7** 34 40N 119 50W
San Ramón de la Nueva Orán, *Argentina* **94 A3** 23 10S 64 20W

San Remo

San Remo, *Italy* **20 C2** 43 48N 7 47 E
San Roque, *Argentina* **94 B4** 28 25S 58 45W
San Rosendo, *Chile* **94 D1** 37 16S 72 43W
San Saba, *U.S.A.* **81 K5** 31 12N 98 43W
San Salvador, *Bahamas* **89 B5** 24 0N 74 40W
San Salvador, *El Salv.* **88 D2** 13 40N 89 10W
San Salvador, *Spain* **22 B10** 39 27N 3 11 E
San Salvador de Jujuy, *Argentina* **94 A3** 24 10S 64 48W
San Salvador I., *Bahamas* **89 B5** 24 0N 74 32W
San Sebastián, *Argentina* **96 G3** 53 10S 68 30W
San Sebastián, *Spain* **19 A5** 43 17N 1 58W
San Sebastian de la Gomera, *Canary Is.* **22 F2** 28 5N 17 7W
San Serra, *Spain* **22 B10** 39 43N 3 13 E
San Simeon, *U.S.A.* **84 K5** 35 39N 121 11W
San Simon, *U.S.A.* **83 K9** 32 16N 109 14W
San Telmo, *Mexico* **86 A1** 30 58N 116 6W
San Telmo, *Spain* **22 B9** 39 35N 2 21 E
San Tiburcio, *Mexico* **86 C4** 24 8N 101 32W
San Valentin, Mte., *Chile* **96 F2** 46 30S 73 30W
San Vicente de la Barquera, *Spain* **19 A3** 43 23N 4 29W
San Ygnacio, *U.S.A.* **81 M5** 27 3N 99 26W
Sana', *Yemen* **46 D3** 15 27N 44 12 E
Sana →, *Bos.-H.* **20 B7** 45 3N 16 23 E
Sanaga →, *Cameroon* **50 H6** 3 35N 9 38 E
Sanaloa, Presa, *Mexico* **86 C3** 24 50N 107 20W
Sanana, *Indonesia* **37 E7** 2 5S 125 59 E
Sanand, *India* **42 H5** 22 59N 72 25 E
Sanandaj, *Iran* **44 C5** 35 18N 47 1 E
Sanandita, *Bolivia* **94 A3** 21 40S 63 45W
Sanawad, *India* **42 H7** 22 11N 76 5 E
Sancellas, *Spain* **22 B9** 39 39N 2 54 E
Sanchahe, *China* **35 B14** 44 50N 126 2 E
Sánchez, *Dom. Rep.* **89 C6** 19 15N 69 36W
Sanchor, *India* **42 G4** 24 45N 71 55 E
Sanco Pt., *Phil.* **37 C7** 8 15N 126 27 E
Sancti-Spíritus, *Cuba* **88 B4** 21 52N 79 33W
Sand →, *S. Africa* **57 C5** 22 25S 30 5 E
Sand Springs, *U.S.A.* **81 G6** 36 9N 96 7W
Sanda, *Japan* **31 G7** 34 53N 135 14 E
Sandakan, *Malaysia* **36 C5** 5 53N 118 4 E
Sandan = Sambor, *Cambodia* **38 F6** 12 46N 106 0 E
Sanday, *U.K.* **12 B6** 59 15N 2 30W
Sanders, *U.S.A.* **83 J9** 35 13N 109 20W
Sanderson, *U.S.A.* **81 K3** 30 9N 102 24W
Sandfly L., *Canada* **73 B7** 55 43N 106 6W
Sandgate, *Australia* **63 D5** 27 18S 153 3 E
Sandía, *Peru* **92 F5** 14 10S 69 30W
Sandnes, *Norway* **9 G8** 58 50N 5 45 E
Sandness, *U.K.* **12 A7** 60 18N 1 38W
Sandoa, *Zaïre* **52 F4** 9 41S 23 0 E
Sandomierz, *Poland* **17 C10** 50 40N 21 43 E
Sandover →, *Australia* **62 C2** 21 43S 136 32 E
Sandoway, *Burma* **41 K19** 18 20N 94 30 E
Sandpoint, *U.S.A.* **82 B5** 48 17N 116 33W
Sandringham, *U.K.* **10 E8** 52 50N 0 30 E
Sandspit, *Canada* **72 C2** 53 14N 131 49W
Sandstone, *Australia* **61 E2** 27 59S 119 16 E
Sandusky, *Mich., U.S.A.* **78 C2** 43 25N 82 50W
Sandusky, *Ohio, U.S.A.* **78 E2** 41 27N 82 42W
Sandviken, *Sweden* **9 F14** 60 38N 16 46 E
Sandwich, C., *Australia* **62 B4** 18 14S 146 18 E
Sandwich B., *Canada* **71 B8** 53 40N 57 15W
Sandwich B., *Namibia* **56 C1** 23 25S 14 20 E
Sandwip Chan., *Bangla.* **41 H17** 22 35N 91 35 E
Sandy, *Nev., U.S.A.* **85 K11** 35 49N 115 36W
Sandy, *Oreg., U.S.A.* **84 E4** 45 24N 122 16W
Sandy Bight, *Australia* **61 F3** 33 50S 123 20 E
Sandy C., *Queens., Australia* **62 C5** 24 42S 153 15 E
Sandy C., *Tas., Australia* **62 G3** 41 25S 144 45 E
Sandy Cay, *Bahamas* **89 B4** 23 13N 75 18W
Sandy Cr. →, *U.S.A.* **82 F9** 41 51N 109 47W
Sandy L., *Canada* **70 B1** 53 2N 93 0W
Sandy Lake, *Canada* **70 B1** 53 0N 93 15W
Sandy Narrows, *Canada* **73 B8** 55 5N 103 4W
Sanford, *Fla., U.S.A.* **77 L5** 28 48N 81 16W
Sanford, *Maine, U.S.A.* **79 C14** 43 27N 70 47W
Sanford, *N.C., U.S.A.* **77 H6** 35 29N 79 10W
Sanford →, *Australia* **61 E2** 27 22S 115 53 E
Sanford, Mt., *U.S.A.* **68 B5** 62 13N 144 8W
Sang-i-Masha, *Afghan.* **42 C2** 33 8N 67 27 E
Sanga, *Mozam.* **55 E4** 12 22S 35 21 E
Sanga →, *Congo* **52 E3** 1 5S 17 0 E
Sanga-Tolon, *Russia* **27 C15** 61 50N 149 40 E
Sangamner, *India* **40 K9** 19 37N 74 15 E
Sangar, *Afghan.* **42 C1** 32 56N 65 30 E
Sangar, *Russia* **27 C13** 64 2N 127 31 E
Sangar Sarai, *Afghan.* **42 B4** 34 27N 70 35 E
Sangasangadalam, *Indonesia* **36 E5** 0 36S 117 13 E
Sange, *Zaïre* **54 D2** 6 58S 28 21 E
Sangeang, *Indonesia* **37 F5** 8 12S 119 6 E
Sanger, *U.S.A.* **83 H4** 36 42N 119 33W
Sanggan He →, *China* **34 E9** 38 12N 117 15 E
Sanggau, *Indonesia* **36 D4** 0 5N 110 30 E
Sangihe, Kepulauan, *Indonesia* **37 D7** 3 0N 126 0 E
Sangihe, P., *Indonesia* **37 D7** 3 45N 125 30 E
Sangju, *S. Korea* **35 F15** 36 25N 128 10 E
Sangkapura, *Indonesia* **36 F4** 5 52S 112 40 E
Sangkhla, *Thailand* **38 E2** 14 57N 98 28 E
Sangli, *India* **40 L9** 16 55N 74 33 E
Sangmélina, *Cameroon* **52 D2** 2 57N 12 1 E
Sangonera →, *Spain* **19 D5** 37 59N 1 4W
Sangre de Cristo Mts., *U.S.A.* **81 G2** 37 0N 105 0W
Sangudo, *Canada* **72 C6** 53 50N 114 54W
Sāniyah, *Iraq* **44 C4** 33 49N 42 43 E
Sanje, *Uganda* **54 C3** 0 49S 31 30 E
Sanjo, *Japan* **30 F9** 37 37N 138 57 E
Sankt Gallen, *Switz.* **16 E4** 47 26N 9 22 E
Sankt Moritz, *Switz.* **16 E4** 46 30N 9 50 E
Sankt-Peterburg, *Russia* **24 C5** 59 55N 30 20 E
Sankuru →, *Zaïre* **52 E4** 4 17S 20 25 E
Sanlúcar de Barrameda, *Spain* **19 D2** 36 46N 6 21W
Sanmenxia, *China* **34 G6** 34 47N 111 12 E
Sanming, *China* **33 D6** 26 13N 117 35 E
Sannaspos, *S. Africa* **56 D4** 29 6S 26 34 E
Sannicandro Gargánico, *Italy* **20 D6** 41 50N 15 34 E
Sannieshof, *S. Africa* **56 D4** 26 30S 25 47 E
Sannīn, J., *Lebanon* **47 B4** 33 57N 35 52 E
Sanok, *Poland* **17 D11** 49 35N 22 10 E
Sanquhar, *U.K.* **12 F5** 55 21N 3 56W
Santa Ana, *Bolivia* **92 F5** 13 50S 65 40W
Santa Ana, *Ecuador* **92 D2** 1 16S 80 20W
Santa Ana, *El Salv.* **88 D2** 14 0N 89 31W
Santa Ana, *Mexico* **86 A2** 30 31N 111 8W
Santa Ana, *U.S.A.* **85 M9** 33 46N 117 52W
Santa Barbara, *Honduras* **88 D2** 14 53N 88 14W
Santa Bárbara, *Mexico* **86 B3** 26 48N 105 50W
Santa Barbara, *U.S.A.* **85 L7** 34 25N 119 42W
Santa Barbara Channel, *U.S.A.* **85 L7** 34 15N 120 0W
Santa Barbara I., *U.S.A.* **85 M7** 33 29N 119 2W
Santa Catalina, *Mexico* **86 B2** 25 40N 110 50W
Santa Catalina, Gulf of, *U.S.A.* **85 N9** 33 10N 117 50W
Santa Catalina I., *U.S.A.* **85 M8** 33 23N 118 25W
Santa Catarina □, *Brazil* **95 B6** 27 25S 48 30W
Santa Catarina, I. de, *Brazil* **95 B6** 27 30S 48 40W
Santa Cecília, *Brazil* **95 B5** 26 56S 50 18W
Santa Clara, *Cuba* **88 B4** 22 20N 80 0W
Santa Clara, *Calif., U.S.A.* **83 H3** 37 21N 121 57W
Santa Clara, *Utah, U.S.A.* **83 H7** 37 8N 113 39W
Santa Clara de Olimar, *Uruguay* **95 C5** 32 50S 54 54W
Santa Clotilde, *Peru* **92 D4** 2 33S 73 45W
Santa Coloma de Gramanet, *Spain* **19 B7** 41 27N 2 13 E
Santa Cruz, *Argentina* **96 G3** 50 0S 68 32W
Santa Cruz, *Bolivia* **92 G6** 17 43S 63 10W
Santa Cruz, *Chile* **94 C1** 34 38S 71 27W
Santa Cruz, *Costa Rica* **88 D2** 10 15N 85 35W
Santa Cruz, *Madeira* **22 D3** 32 42N 16 46W
Santa Cruz, *Phil.* **37 B6** 14 20N 121 24 E
Santa Cruz, *U.S.A.* **83 H2** 36 58N 122 1W
Santa Cruz →, *Argentina* **96 G3** 50 10S 68 20W
Santa Cruz de la Palma, *Canary Is.* **22 F2** 28 41N 17 46W
Santa Cruz de Tenerife, *Canary Is.* **22 F3** 28 28N 16 15W
Santa Cruz del Norte, *Cuba* **88 B3** 23 9N 81 55W
Santa Cruz del Sur, *Cuba* **88 B4** 20 44N 78 0W
Santa Cruz do Rio Pardo, *Brazil* **95 A6** 22 54S 49 37W
Santa Cruz do Sul, *Brazil* **95 B5** 29 42S 52 25W
Santa Cruz I., *Solomon Is.* **64 J8** 10 30S 166 0 E
Santa Cruz I., *U.S.A.* **85 M7** 34 1N 119 43W
Santa Domingo, Cay, *Bahamas* **88 B4** 21 25N 75 15W
Santa Elena, *Argentina* **94 C4** 30 58S 59 47W
Santa Elena, *Ecuador* **92 D1** 2 16S 80 52W
Santa Elena, C., *Costa Rica* **88 D2** 10 54N 85 56W
Santa Eugenia, Pta., *Mexico* **86 B1** 27 50N 115 5W
Santa Eulalia, *Spain* **22 C8** 38 59N 1 32 E
Santa Fe, *Argentina* **94 C3** 31 35S 60 41W
Santa Fe, *U.S.A.* **83 J11** 35 41N 105 57W
Santa Fé □, *Argentina* **94 C3** 31 50S 60 55W
Santa Filomena, *Brazil* **93 E9** 9 6S 45 50W
Santa Galdana, *Spain* **22 B10** 39 56N 3 58 E
Santa Gertrudis, *Spain* **22 B7** 39 0N 1 26 E
Santa Inés, *Spain* **22 B7** 39 3N 1 21 E
Santa Inés, I., *Chile* **96 G2** 54 0S 73 0W
Santa Isabel = Rey Malabo, *Eq. Guin.* **50 H6** 3 45N 8 50 E
Santa Isabel, *Argentina* **94 D2** 36 10S 66 54W
Santa Isabel, *Brazil* **93 F8** 11 45S 51 30W
Santa Lucía, *Corrientes, Argentina* **94 B4** 28 58S 59 5W
Santa Lucía, *San Juan, Argentina* **94 C2** 31 30S 68 30W
Santa Lucia, *Uruguay* **94 C4** 34 27S 56 24W
Santa Lucia Range, *U.S.A.* **83 J3** 36 0N 121 20W
Santa Magdalena, I., *Mexico* **86 C2** 24 40N 112 15W
Santa Margarita, *Argentina* **94 D3** 38 28S 61 35W
Santa Margarita, *Mexico* **86 C2** 24 30N 111 50W
Santa Margarita, *Spain* **22 B10** 39 42N 3 6 E
Santa Margarita, *U.S.A.* **84 K6** 35 23N 120 37W
Santa Margarita →, *U.S.A.* **85 M9** 33 13N 117 23W
Santa María, *Argentina* **94 B2** 26 40S 66 0W
Santa Maria, *Brazil* **95 B5** 29 40S 53 48W
Santa María, *Spain* **22 B9** 39 38N 2 47 E
Santa Maria, *U.S.A.* **85 L6** 34 57N 120 26W
Santa María →, *Mexico* **86 A3** 31 0N 107 14W
Santa María, B. de, *Mexico* **86 B3** 25 10N 108 40W
Santa Maria da Vitória, *Brazil* **93 F10** 13 24S 44 12W
Santa Maria di Leuca, C., *Italy* **21 E8** 39 48N 18 20 E
Santa Marta, *Colombia* **92 A4** 11 15N 74 13W
Santa Marta, Sierra Nevada de, *Colombia* **92 A4** 10 55N 73 50W
Santa Marta Grande, C., *Brazil* **95 B6** 28 43S 48 50W
Santa Maura = Levkás, *Greece* **21 E9** 38 40N 20 43 E
Santa Monica, *U.S.A.* **85 M8** 34 1N 118 29W
Santa Ponsa, *Spain* **22 B9** 39 30N 2 28 E
Santa Rita, *U.S.A.* **83 K10** 32 48N 108 4W
Santa Rosa, *La Pampa, Argentina* **94 D3** 36 40S 64 17W
Santa Rosa, *San Luis, Argentina* **94 C2** 32 21S 65 10W
Santa Rosa, *Bolivia* **92 F5** 10 36S 67 20W
Santa Rosa, *Brazil* **95 B5** 27 52S 54 29W
Santa Rosa, *Calif., U.S.A.* **84 G4** 38 26N 122 43W
Santa Rosa, *N. Mex., U.S.A.* **81 H2** 34 57N 104 41W
Santa Rosa de Copán, *Honduras* **88 D2** 14 47N 88 46W
Santa Rosa de Río Primero, *Argentina* **94 C3** 31 8S 63 20W
Santa Rosa I., *Calif., U.S.A.* **85 M6** 33 58N 120 6W
Santa Rosa I., *Fla., U.S.A.* **77 K2** 30 20N 86 50W
Santa Rosa Range, *U.S.A.* **82 F5** 41 45N 117 40W
Santa Rosalía, *Mexico* **86 B2** 27 20N 112 20W
Santa Sylvina, *Argentina* **94 B3** 27 50S 61 10W
Santa Tecla = Nueva San Salvador, *El Salv.* **88 D2** 13 40N 89 18W
Santa Teresa, *Argentina* **94 C3** 33 25S 60 47W
Santa Teresa, *Mexico* **87 B5** 25 17N 97 51W
Santa Vitória do Palmar, *Brazil* **95 C5** 33 32S 53 25W
Santa Ynez →, *U.S.A.* **85 L6** 35 41N 120 36W
Santa Ynez Mts., *U.S.A.* **85 L6** 34 30N 120 0W
Santa Ysabel, *U.S.A.* **85 M10** 33 7N 116 40W
Santai, *China* **32 C5** 31 5N 104 58 E
Santana, *Madeira* **22 D3** 32 48N 16 52W
Santana, Coxilha de, *Brazil* **95 C4** 30 50S 55 35W
Santana do Livramento, *Brazil* **95 C4** 30 55S 55 30W
Santanayi, *Spain* **22 B10** 39 20N 3 5 E
Santander, *Spain* **19 A4** 43 27N 3 51W
Santander Jiménez, *Mexico* **87 C5** 24 11N 98 29W
Santaquin, *U.S.A.* **82 G8** 39 59N 111 47W
Santarém, *Brazil* **93 D8** 2 25S 54 42W
Santarém, *Portugal* **19 C1** 39 12N 8 42W
Santaren Channel, *W. Indies* **88 B4** 24 0N 79 30W
Santee, *U.S.A.* **85 N10** 32 50N 116 58W
Santiago, *Brazil* **95 B5** 29 11S 54 52W
Santiago, *Chile* **94 C1** 33 24S 70 40W
Santiago, *Panama* **88 E3** 8 0N 81 0W
Santiago □, *Chile* **94 C1** 33 30S 70 50W
Santiago →, *Peru* **92 D3** 4 27S 77 38W
Santiago de Compostela, *Spain* **19 A1** 42 52N 8 37W
Santiago de Cuba, *Cuba* **88 C4** 20 0N 75 49W
Santiago de los Cabelleros, *Dom. Rep.* **89 C5** 19 30N 70 40W
Santiago del Estero, *Argentina* **94 B3** 27 50S 64 15W
Santiago del Estero □, *Argentina* **94 B3** 27 40S 63 15W
Santiago del Teide, *Canary Is.* **22 F3** 28 17N 16 48W
Santiago Ixcuintla, *Mexico* **86 C3** 21 50N 105 11W
Santiago Papasquiaro, *Mexico* **86 B3** 25 0N 105 20W
Santiaguillo, L. de, *Mexico* **86 C4** 24 50N 104 50W
Santo Amaro, *Brazil* **93 F11** 12 30S 38 43W
Santo Anastácio, *Brazil* **95 A5** 21 58S 51 39W
Santo André, *Brazil* **95 A6** 23 39S 46 29W
Santo Ângelo, *Brazil* **95 B5** 28 15S 54 15W
Santo Antonio, *Brazil* **93 G7** 15 50S 56 0W
Santo Corazón, *Bolivia* **92 G7** 18 0S 58 45W
Santo Domingo, *Dom. Rep.* **89 C6** 18 30N 69 59W
Santo Domingo, *Baja Calif. N., Mexico* **86 A1** 30 43N 116 2W
Santo Domingo, *Baja Calif. S., Mexico* **86 B2** 25 32N 112 2W
Santo Domingo, *Nic.* **88 D3** 12 14N 84 59W
Santo Tomás, *Mexico* **86 A1** 31 33N 116 24W
Santo Tomás, *Peru* **92 F4** 14 26S 72 8W
Santo Tomé, *Argentina* **95 B4** 28 40S 56 5W
Santo Tomé de Guayana = Ciudad Guayana, *Venezuela* **92 B6** 8 0N 62 30W
Santoña, *Spain* **19 A4** 43 29N 3 27W
Santos, *Brazil* **95 A6** 24 0S 46 20W
Santos Dumont, *Brazil* **95 A7** 22 55S 43 10W
Sanyuan, *China* **34 G5** 34 35N 108 58 E
Sanza Pombo, *Angola* **52 F3** 7 18S 15 56 E
São Anastácio, *Brazil* **95 A5** 22 0S 51 40W
São Bernado de Campo, *Brazil* **95 A6** 23 45S 46 34W
São Borja, *Brazil* **95 B4** 28 39S 56 0W
São Carlos, *Brazil* **95 A6** 22 0S 47 50W
São Cristóvão, *Brazil* **93 F11** 11 1S 37 15W
São Domingos, *Brazil* **93 F9** 13 25S 46 19W
São Francisco, *Brazil* **93 G10** 16 0S 44 50W
São Francisco →, *Brazil* **93 F11** 10 30S 36 24W
São Francisco do Sul, *Brazil* **95 B6** 26 15S 48 36W
São Gabriel, *Brazil* **95 C5** 30 20S 54 20W
São Gonçalo, *Brazil* **95 A7** 22 48S 43 5W
Sao Hill, *Tanzania* **55 D4** 8 20S 35 12 E
São João da Boa Vista, *Brazil* **95 A6** 22 0S 46 52W
São João del Rei, *Brazil* **95 A7** 21 8S 44 15W
São João do Araguaia, *Brazil* **93 E9** 5 23S 48 46W
São João do Piauí, *Brazil* **93 E10** 8 21S 42 15W
São Jorge, Pta. de, *Madeira* **22 D3** 32 50N 16 53W
São José do Rio Prêto, *Brazil* **95 A6** 20 50S 49 20W
São José dos Campos, *Brazil* **95 A6** 23 7S 45 52W
São Leopoldo, *Brazil* **95 B5** 29 50S 51 10W
São Lourenço, *Brazil* **95 A6** 22 7S 45 3W
São Lourenço →, *Brazil* **93 G7** 17 53S 57 27W
São Lourenço, Pta. de, *Madeira* **22 D3** 32 44N 16 39W
São Luís, *Brazil* **93 D10** 2 39S 44 15W
São Luís Gonzaga, *Brazil* **95 B5** 28 25S 55 0W
São Marcos →, *Brazil* **93 G9** 18 15S 47 37W
São Marcos, B. de, *Brazil* **93 D10** 2 0S 44 0W
São Mateus, *Brazil* **93 G11** 18 44S 39 50W
São Paulo, *Brazil* **95 A6** 23 32S 46 37W
São Paulo □, *Brazil* **95 A6** 22 0S 49 0W
São Paulo, I., *Atl. Oc.* **2 D8** 0 50N 31 40W
São Roque, *Madeira* **22 D3** 32 46N 16 48W
São Roque, C. de, *Brazil* **93 E11** 5 30S 35 16W
São Sebastião, I. de, *Brazil* **95 A6** 23 50S 45 18W
São Sebastião do Paraíso, *Brazil* **95 A6** 20 54S 46 59W
São Tomé, *Atl. Oc.* **48 F4** 0 10N 6 39 E
São Tomé, C. de, *Brazil* **95 A7** 22 0S 40 59W
São Tomé & Principe ■, *Africa* **49 F4** 0 12N 6 39 E
São Vicente, *Brazil* **95 A6** 23 57S 46 23W
São Vicente, *Madeira* **22 D2** 32 48N 17 3W
São Vicente, C. de, *Portugal* **19 D1** 37 0N 9 0W
Saona, I., *Dom. Rep.* **89 C6** 18 10N 68 40W
Saône →, *France* **18 D6** 45 44N 4 50 E
Saône-et-Loire □, *France* **18 C6** 46 30N 4 50 E
Saonek, *Indonesia* **37 E8** 0 22S 130 55 E
Saparua, *Indonesia* **37 E7** 3 33S 128 40 E
Sapele, *Nigeria* **50 G6** 5 50N 5 40 E
Sapelo I., *U.S.A.* **77 K5** 31 25N 81 12W
Saposoa, *Peru* **92 E3** 6 55S 76 45W
Sappho, *U.S.A.* **84 B2** 48 4N 124 16W
Sapporo, *Japan* **30 C10** 43 0N 141 21 E
Sapudi, *Indonesia* **37 G16** 7 6S 114 20 E
Sapulpa, *U.S.A.* **81 G7** 35 59N 96 5W
Saqqez, *Iran* **44 B5** 36 15N 46 20 E
Sar Dasht, *Iran* **45 C6** 32 32N 48 52 E
Sar Gachīneh, *Iran* **45 D6** 30 31N 51 31 E
Sar Planina, *Macedonia* **21 C9** 42 10N 21 0 E
Sara Buri, *Thailand* **38 E3** 14 30N 100 55 E
Sarāb, *Iran* **44 B5** 38 0N 47 30 E
Sarabadi, *Iraq* **44 C5** 33 1N 44 48 E
Sarada →, *India* **41 F12** 27 21N 81 23 E
Saragossa = Zaragoza, *Spain* **19 B5** 41 39N 0 53W
Saraguro, *Ecuador* **92 D3** 3 35S 79 16W
Sarajevo, *Bos.-H.* **21 C8** 43 52N 18 26 E
Saran, G., *Indonesia* **36 E4** 0 30S 111 25 E
Saranac Lake, *U.S.A.* **79 B10** 44 20N 74 8W
Saranda, *Tanzania* **54 D3** 5 45S 34 59 E
Sarandí del Yi, *Uruguay* **95 C4** 33 18S 55 38W
Sarandí Grande, *Uruguay* **94 C4** 33 44S 56 20W
Sarangani B., *Phil.* **37 C7** 6 0N 125 13 E
Sarangani Is., *Phil.* **37 C7** 5 25N 125 25 E
Sarangarh, *India* **41 J13** 21 30N 83 5 E
Saransk, *Russia* **24 D8** 54 10N 45 10 E
Sarapul, *Russia* **24 C9** 56 28N 53 48 E
Sarasota, *U.S.A.* **77 M4** 27 20N 82 32W
Saratoga, *Calif., U.S.A.* **84 H4** 37 16N 122 2W
Saratoga, *Wyo., U.S.A.* **82 F10** 41 27N 106 49W
Saratoga Springs, *U.S.A.* **79 C11** 43 5N 73 47W
Saratov, *Russia* **24 D8** 51 30N 46 2 E
Saravane, *Laos* **38 E6** 15 43N 106 25 E
Sarawak □, *Malaysia* **36 D4** 2 0N 113 0 E
Sarbāz, *Iran* **45 E9** 26 38N 61 19 E
Sarbīsheh, *Iran* **45 C8** 32 30N 59 40 E
Sarda → = Sarada →, *India* **41 F12** 27 21N 81 23 E
Sardalas, *Libya* **50 C7** 25 50N 10 34 E
Sardarshahr, *India* **42 E6** 28 30N 74 29 E
Sardegna, *Italy* **20 E3** 39 57N 9 0 E
Sardhana, *India* **42 E7** 29 9N 77 39 E
Sardina, Pta., *Canary Is.* **22 F4** 28 9N 15 44W
Sardinia = Sardegna, *Italy* **20 E3** 39 57N 9 0 E
Sārdūīyeh = Dar Mazār, *Iran* **45 D8** 29 14N 57 20 E
Sargent, *U.S.A.* **80 E5** 41 39N 99 22W
Sargodha, *Pakistan* **42 C5** 32 10N 72 40 E
Sarh, *Chad* **51 G8** 9 5N 18 23 E
Sārī, *Iran* **45 B7** 36 30N 53 4 E
Sarikei, *Malaysia* **36 D4** 2 8N 111 30 E
Sarina, *Australia* **62 C4** 21 22S 149 13 E
Sarita, *U.S.A.* **81 M6** 27 13N 97 47W
Sariwŏn, *N. Korea* **35 E13** 38 31N 125 46 E
Sark, *Chan. Is.* **11 H5** 49 25N 2 20W
Sarlat-la-Canéda, *France* **18 D4** 44 54N 1 13 E
Sarles, *U.S.A.* **80 A5** 48 58N 99 0W
Sarmi, *Indonesia* **37 E9** 1 49S 138 44 E
Sarmiento, *Argentina* **96 F3** 45 35S 69 5W
Sarnia, *Canada* **70 D3** 42 58N 82 23W
Sarny, *Ukraine* **24 D4** 51 17N 26 40 E
Sarolangun, *Indonesia* **36 E2** 2 19S 102 42 E
Saronikós Kólpos, *Greece* **21 F10** 37 45N 23 45 E
Saros Körfezi, *Turkey* **21 D12** 40 30N 26 15 E
Sarpsborg, *Norway* **9 G11** 59 16N 11 12 E
Sarre = Saar →, *Europe* **16 D3** 49 41N 6 32 E
Sarro, *Mali* **50 F3** 13 40N 5 15W
Sartène, *France* **20 D3** 41 38N 8 58 E
Sarthe □, *France* **18 C4** 47 58N 0 10 E
Sarthe →, *France* **18 C3** 47 33N 0 31W
Sartynya, *Russia* **26 C7** 63 22N 63 11 E
Sarvestān, *Iran* **45 D7** 29 20N 53 10 E
Sary-Tash, *Kirghizia* **26 F8** 39 44N 73 15 E
Saryshagan, *Kazakhstan* **26 E8** 46 12N 73 38 E
Sasabeneh, *Ethiopia* **46 F3** 7 59N 44 43 E
Sasaram, *India* **43 G11** 24 57N 84 5 E
Sasebo, *Japan* **31 H4** 33 10N 129 43 E
Saser, *India* **43 B7** 34 50N 77 50 E
Saskatchewan □, *Canada* **73 C7** 54 40N 106 0W
Saskatchewan →, *Canada* **73 C8** 53 37N 100 40W
Saskatoon, *Canada* **73 C7** 52 10N 106 38W
Saskylakh, *Russia* **27 B12** 71 55N 114 1 E
Sasolburg, *S. Africa* **57 D4** 26 46S 27 49 E
Sasovo, *Russia* **24 D7** 54 25N 41 55 E
Sassandra, *Ivory C.* **50 H3** 4 55N 6 8W
Sassandra →, *Ivory C.* **50 H3** 4 58N 6 5W
Sássari, *Italy* **20 D3** 40 44N 8 33 E
Sassnitz, *Germany* **16 A6** 54 29N 13 39 E
Sasumua Dam, *Kenya* **54 C4** 0 45S 36 40 E
Sata-Misaki, *Japan* **31 J5** 31 0N 130 40 E
Satadougou, *Mali* **50 F2** 12 25N 11 25W
Satanta, *U.S.A.* **81 G4** 37 26N 100 59W
Satara, *India* **40 L8** 17 44N 73 58 E
Satilla →, *U.S.A.* **77 K5** 30 59N 81 29W
Satka, *Russia* **24 C10** 55 3N 59 1 E
Satmala Hills, *India* **40 J9** 20 15N 74 40 E

Shackleton Inlet

Satna, India 43 G9 24 35N 80 50 E
Sátoraljaújhely, Hungary 17 D10 48 25N 21 41 E
Satpura Ra., India 42 J7 21 25N 76 10 E
Satsuna-Shotō, Japan 31 K5 30 0N 130 0 E
Sattahip, Thailand 38 F3 12 41N 100 54 E
Satu Mare, Romania 17 E11 47 46N 22 55 E
Satui, Indonesia 36 E5 3 50S 115 27 E
Satun, Thailand 39 J3 6 43N 100 2 E
Saturnina →, Brazil 92 F7 12 15S 58 10W
Sauce, Argentina 94 C4 30 5S 58 46W
Sauceda, Mexico 86 B4 25 55N 101 18W
Saucillo, Mexico 86 B3 28 1N 105 17W
Sauda, Norway 9 G9 59 40N 6 20 E
Sauðarkrókur, Iceland 8 D4 65 45N 19 40W
Saudi Arabia ■, Asia 46 B3 26 0N 44 0 E
Sauer →, Germany 15 E6 49 44N 6 31 E
Saugeen →, Canada 78 B3 44 30N 81 22W
Saugerties, U.S.A. 79 D11 42 5N 73 57W
Sauk Centre, U.S.A. 80 C7 45 44N 94 57W
Sauk Rapids, U.S.A. 80 C7 45 35N 94 10W
Sault Ste. Marie, Canada 70 C3 46 30N 84 20W
Sault Ste. Marie, U.S.A. 76 B3 46 30N 84 21W
Saumlaki, Indonesia 37 F8 7 55S 131 20 E
Saumur, France 18 C3 47 15N 0 5W
Saunders C., N.Z. 59 L3 45 53S 170 45 E
Saunders I., Antarctica 5 B1 57 48S 26 28W
Saunders Point, Australia 61 E4 27 52S 125 38 E
Saurbær, Borgarfjarðarsýsla, Iceland 8 D3 64 24N 21 35W
Saurbær, Eyjafjarðarsýsla, Iceland 8 D4 65 27N 18 13W
Sauri, Nigeria 50 F6 11 42N 6 44 E
Saurimo, Angola 52 F4 9 40S 20 12 E
Sausalito, U.S.A. 84 H4 37 51N 122 29W
Savá, Honduras 88 C2 15 32N 86 15W
Sava →, Serbia, Yug. 21 B9 44 50N 20 26 E
Savage, U.S.A. 80 B2 47 27N 104 21W
Savage I. = Niue, Cook Is. 65 J11 19 2S 169 54W
Savai'i, W. Samoa 59 A12 13 28S 172 24W
Savalou, Benin 50 G5 7 57N 1 58 E
Savane, Mozam. 55 F4 19 37S 35 8 E
Savanna, U.S.A. 80 D9 42 5N 90 8W
Savanna la Mar, Jamaica 88 C4 18 10N 78 10W
Savannah, Ga., U.S.A. 77 J5 32 5N 81 6W
Savannah, Mo., U.S.A. 80 F7 39 56N 94 50W
Savannah, Tenn., U.S.A. 77 H1 35 14N 88 15W
Savannah →, U.S.A. 77 J5 32 2N 80 53W
Savannakhet, Laos 38 D5 16 30N 104 49 E
Savant L., Canada 70 B1 50 16N 90 44W
Savant Lake, Canada 70 B1 50 14N 90 40W
Savanur, India 40 M9 14 59N 75 21 E
Savé, Benin 50 G5 8 2N 2 29 E
Save →, Mozam. 57 C5 21 16S 34 0 E
Sāveh, Iran 45 C6 35 2N 50 20 E
Savelugu, Ghana 50 G4 9 38N 0 54W
Savoie □, France 18 D7 45 26N 6 25 E
Savona, Italy 20 B3 44 19N 8 29 E
Savonlinna, Finland 24 B4 61 52N 28 53 E
Sawahlunto, Indonesia 36 E2 0 40S 100 52 E
Sawai, Indonesia 37 E7 3 0S 129 5 E
Sawai Madhopur, India 42 F7 26 0N 76 25 E
Sawang Daen Din, Thailand 38 D4 17 28N 103 28 E
Sawankhalok, Thailand 38 D2 17 19N 99 50 E
Sawara, Japan 31 G10 35 55N 140 30 E
Sawatch Mts., U.S.A. 83 G10 38 30N 106 30W
Sawel, U.K. 13 B4 54 48N 7 5W
Sawi, Thailand 39 G2 10 14N 99 5 E
Sawmills, Zimbabwe 55 F2 19 30S 28 2 E
Sawu, Indonesia 37 F6 9 35S 121 50 E
Sawu Sea, Indonesia 37 F6 9 30S 121 50 E
Saxby →, Australia 62 B3 18 25S 140 53 E
Saxton, U.S.A. 78 F6 40 13N 78 15W
Say, Niger 50 F5 13 8N 2 22 E
Sayabec, Canada 71 C6 48 35N 67 41W
Sayaboury, Laos 38 C3 19 15N 101 45 E
Sayán, Peru 92 F3 11 8S 77 12W
Sayan, Vostochnyy, Russia 27 D10 54 0N 96 0 E
Sayan, Zapadnyy, Russia 27 D10 52 30N 94 0 E
Saydā, Lebanon 47 B4 33 35N 35 25 E
Sayhan-Ovoo, Mongolia 34 B2 45 27N 103 54 E
Sayhandulaan, Mongolia 34 B5 44 40N 109 1 E
Sayḩut, Yemen 46 D5 15 12N 51 10 E
Saynshand, Mongolia 34 B6 44 55N 110 11 E
Sayre, Okla., U.S.A. 81 H5 35 18N 99 38W
Sayre, Pa., U.S.A. 79 E8 41 59N 76 32W
Sayula, Mexico 86 D4 19 50N 103 40W
Sazan, Albania 21 D8 40 30N 19 20 E
Sázava →, Czech. 16 D7 49 53N 14 24 E
Sazin, Pakistan 43 B5 35 35N 73 30 E
Scafell Pikes, U.K. 10 C4 54 26N 3 14W
Scalpay, U.K. 12 D2 57 51N 6 40W
Scandia, Canada 72 C6 50 20N 112 0W
Scandinavia, Europe 8 E12 64 0N 12 0 E
Scapa Flow, U.K. 12 C5 58 52N 3 6W
Scappoose, U.S.A. 84 E4 45 45N 122 53W
Scarborough, Trin. & Tob. 89 D7 11 11N 60 42W
Scarborough, U.K. 10 C7 54 17N 0 24W
Scebeli, Wabi →, Somali Rep. 46 G3 2 0N 44 0 E
Scenic, U.S.A. 80 D3 43 47N 102 33W
Schaffhausen, Switz. 16 E4 47 42N 8 39 E
Schagen, Neths. 15 B4 52 49N 4 48 E
Schefferville, Canada 71 B6 54 48N 66 50W
Schelde →, Belgium 15 C4 51 15N 4 16 E
Schell Creek Ra., U.S.A. 82 G6 39 15N 114 30W
Schenectady, U.S.A. 79 D11 42 49N 73 57W
Scheveningen, Neths. 15 B4 52 6N 4 16 E
Schiedam, Neths. 15 C4 51 55N 4 25 E
Schiermonnikoog, Neths. 15 A6 53 30N 6 15 E
Schio, Italy 20 B4 45 42N 11 21 E
Schleswig, Germany 16 A4 54 32N 9 34 E
Schleswig-Holstein □, Germany 16 A4 54 10N 9 40 E
Schofield, U.S.A. 80 C10 44 54N 89 36W
Scholls, U.S.A. 84 E4 45 24N 122 56W
Schouten I., Australia 62 G4 42 20S 148 20 E
Schouten Is. = Supriori, Kepulauan, Indonesia 37 E9 1 0S 136 0 E
Schouwen, Neths. 15 C3 51 43N 3 45 E
Schreiber, Canada 70 C2 48 45N 87 20W
Schuler, Canada 73 C6 50 20N 110 6W
Schumacher, Canada 70 C3 48 30N 81 16W
Schurz, U.S.A. 82 G4 38 57N 118 49W
Schuyler, U.S.A. 80 E6 41 27N 97 4W
Schuylkill Haven, U.S.A. 79 F8 40 37N 76 11W
Schwäbische Alb, Germany 16 D4 48 30N 9 30 E
Schwaner, Pegunungan, Indonesia 36 E4 1 0S 112 30 E
Schwarzwald, Germany 16 E4 48 0N 8 0 E
Schweinfurt, Germany 16 C5 50 3N 10 12 E
Schweizer-Reneke, S. Africa 56 D4 27 11S 25 18 E
Schwerin, Germany 16 B5 53 37N 11 22 E
Schwyz, Switz. 16 E4 47 2N 8 39 E
Sciacca, Italy 20 F5 37 30N 13 3 E
Scilla, Italy 20 E6 38 18N 15 44 E
Scilly, Isles of, U.K. 11 H1 49 55N 6 15W
Scioto →, U.S.A. 76 F4 38 44N 83 1W
Scobey, U.S.A. 80 A2 48 47N 105 25W
Scone, Australia 63 E5 32 5S 150 52 E
Scoresbysund, Greenland 4 B6 70 20N 23 0W
Scotia, Calif., U.S.A. 82 F1 40 29N 124 6W
Scotia, N.Y., U.S.A. 79 D11 42 50N 73 58W
Scotia Sea, Antarctica 5 B18 56 5S 56 0W
Scotland, U.S.A. 80 D6 43 9N 97 43W
Scotland □, U.K. 12 E5 57 0N 4 0W
Scotland Neck, U.S.A. 77 G7 36 8N 77 25W
Scott, C., Australia 60 B4 13 30S 129 49 E
Scott City, U.S.A. 80 F4 38 29N 100 54W
Scott Glacier, Antarctica 5 C8 66 15S 100 5 E
Scott I., Antarctica 5 C11 67 0S 179 0 E
Scott Inlet, Canada 69 A12 71 0N 71 0W
Scott Is., Canada 72 C3 50 48N 128 40W
Scott L., Canada 73 B7 59 55N 106 18W
Scott Reef, Australia 60 B3 14 0S 121 50 E
Scottburgh, S. Africa 57 E5 30 15S 30 47 E
Scottdale, U.S.A. 78 F5 40 6N 79 35W
Scottsbluff, U.S.A. 80 E3 41 52N 103 40W
Scottsboro, U.S.A. 77 H2 34 40N 86 2W
Scottsburg, U.S.A. 76 F3 38 41N 85 47W
Scottsdale, Australia 62 G4 41 9S 147 31 E
Scottsville, Ky., U.S.A. 77 G2 36 45N 86 11W
Scottsville, N.Y., U.S.A. 78 C7 43 2N 77 47W
Scottville, U.S.A. 76 D2 43 58N 86 17W
Scranton, U.S.A. 79 E9 41 25N 75 40W
Scugog, L., Canada 78 B6 44 10N 78 55W
Scunthorpe, U.K. 10 D7 53 35N 0 38W
Scusciuban, Somali Rep. 46 E5 10 18N 50 12 E
Scutari = Üsküdar, Turkey 21 D13 41 0N 29 5 E
Seabrook, L., Australia 61 F2 30 55S 119 40 E
Seaford, U.S.A. 76 F8 38 39N 75 37W
Seaforth, Canada 70 D3 43 35N 81 25W
Seagraves, U.S.A. 81 J3 32 57N 102 34W
Seal →, Canada 73 B10 59 4N 94 48W
Seal Cove, Canada 71 C8 49 57N 56 22W
Seal L., Canada 71 B7 54 20N 61 30W
Sealy, U.S.A. 81 L6 29 47N 96 9W
Searchlight, U.S.A. 85 K12 35 28N 114 55W
Searcy, U.S.A. 81 H9 35 15N 91 44W
Searles L., U.S.A. 85 K9 35 44N 117 21W
Seaside, Calif., U.S.A. 84 J5 36 37N 121 50W
Seaside, Oreg., U.S.A. 84 E3 46 0N 123 56W
Seaspray, Australia 63 F4 38 25S 147 15 E
Seattle, U.S.A. 84 C4 47 36N 122 20W
Seaview Ra., Australia 62 B4 18 40S 145 45 E
Sebastián Vizcaíno, B., Mexico 86 B2 28 0N 114 30W
Sebastopol = Sevastopol, Ukraine 25 F5 44 35N 33 30 E
Sebastopol, U.S.A. 84 G4 38 24N 122 49W
Sebewaing, U.S.A. 76 D4 43 44N 83 27W
Sebha = Sabhah, Libya 51 C7 27 9N 14 29 E
Sebring, Fla., U.S.A. 77 M5 27 30N 81 27W
Sebring, Ohio, U.S.A. 78 F3 40 55N 81 2W
Sebringville, Canada 78 C3 43 24N 81 4W
Sebta = Ceuta, Morocco 50 A3 35 52N 5 18W
Sebuku, Indonesia 36 E5 3 30S 116 25 E
Sebuku, Teluk, Malaysia 36 D5 4 0N 118 10 E
Sechelt, Canada 72 D4 49 25N 123 42W
Sechura, Desierto de, Peru 92 E2 6 0S 80 30W
Secretary I., N.Z. 59 L1 45 15S 166 56 E
Secunderabad, India 40 L11 17 28N 78 30 E
Sedalia, U.S.A. 80 F8 38 42N 93 14W
Sedan, Australia 63 E2 34 34S 139 19 E
Sedan, France 18 B6 49 43N 4 57 E
Sedan, U.S.A. 81 G6 37 8N 96 11W
Seddon, N.Z. 59 J5 41 40S 174 7 E
Seddonville, N.Z. 59 J4 41 33S 172 1 E
Sedeh, Fārs, Iran 45 D7 30 45N 52 11 E
Sedeh, Khorāsān, Iran 45 C8 33 20N 59 14 E
Sederot, Israel 47 D3 31 32N 34 37 E
Sedgewick, Canada 72 C6 52 48N 111 41W
Sedhiou, Senegal 50 F1 12 44N 15 30W
Sedley, Canada 73 C8 50 10N 104 0W
Sedova, Pik, Russia 26 B6 73 29N 54 58 E
Sedro-Woolley, U.S.A. 84 B4 48 30N 122 14W
Seeheim, Namibia 56 D2 26 50S 17 45 E
Seekoei →, S. Africa 56 E4 30 18S 25 1 E
Seg-ozero, Russia 24 B5 63 0N 33 10 E
Segamat, Malaysia 39 L4 2 30N 102 50 E
Seget, Indonesia 37 E8 1 24S 130 58 E
Segezha, Russia 24 B5 63 44N 34 19 E
Ségou, Mali 50 F3 13 30N 6 16W
Segovia = Coco →, Cent. Amer. 88 D3 15 0N 83 8W
Segovia, Spain 19 B3 40 57N 4 10W
Segre →, Spain 19 B6 41 40N 0 43 E
Séguéla, Ivory C. 50 G3 7 55N 6 40W
Seguin, U.S.A. 81 L6 29 34N 97 58W
Segundo →, Argentina 94 C3 30 53S 62 44W
Segura →, Spain 19 C5 38 6N 0 54W
Seh Qal'eh, Iran 45 C8 33 40N 58 24 E
Sehitwa, Botswana 56 C3 20 30S 22 30 E
Sehore, India 42 H7 23 10N 77 5 E
Sehwan, Pakistan 42 F2 26 28N 67 53 E
Seiland, Norway 8 A17 70 25N 23 15 E
Seiling, U.S.A. 81 G5 36 9N 98 56W
Seinäjoki →, Finland 8 E17 62 40N 22 45 E
Seine →, France 18 B4 49 26N 0 26 E
Seine-et-Marne □, France 18 B5 48 45N 3 0 E
Seine-Maritime □, France 18 B4 49 40N 1 0 E
Seine-St.-Denis □, France 18 B5 48 58N 2 24 E
Seistan, Iran 45 D9 30 50N 61 0 E
Sekayu, Indonesia 36 E2 2 51S 103 51 E
Seke, Tanzania 54 C3 3 20S 33 31 E
Sekenke, Tanzania 54 C3 4 18S 34 11 E
Sekondi-Takoradi, Ghana 50 H4 4 58N 1 45W
Sekuma, Botswana 56 C3 24 36S 23 50 E
Selah, U.S.A. 82 C3 46 39N 120 32W
Selama, Malaysia 39 K3 5 12N 100 42 E
Selaru, Indonesia 37 F8 8 9S 131 0 E
Selby, U.K. 10 D6 53 47N 1 5W
Selby, U.S.A. 80 C4 45 31N 100 2W
Selden, U.S.A. 80 F4 39 33N 100 34W
Sele →, Italy 20 D6 40 27N 14 58 E
Selemdzha →, Russia 27 D13 51 42N 128 53 E
Selenga = Selenge Mörön →, Asia 32 A5 52 16N 106 16 E
Selenge Mörön →, Asia 32 A5 52 16N 106 16 E
Seletan, Tg., Indonesia 36 E4 4 10S 114 40 E
Selfridge, U.S.A. 80 B4 46 2N 100 56W
Sélibabi, Mauritania 50 E2 15 10N 12 15W
Seligman, U.S.A. 83 J7 35 20N 112 53W
Selîma, El Wâhât el, Sudan 51 D10 21 22N 29 19 E
Selinda Spillway, Botswana 56 B3 18 35S 23 10 E
Selkirk, Canada 73 C9 50 10N 96 55W
Selkirk, U.K. 12 F6 55 33N 2 50W
Selkirk I., Canada 73 C9 53 20N 99 6W
Selkirk Mts., Canada 72 C5 51 15N 117 40W
Selliá, Greece 23 D6 35 12N 24 23 E
Sells, U.S.A. 83 L8 31 55N 111 53W
Selma, Ala., U.S.A. 77 J2 32 25N 87 1W
Selma, Calif., U.S.A. 83 H4 36 34N 119 37W
Selma, N.C., U.S.A. 77 H6 35 32N 78 17W
Selmer, U.S.A. 77 H1 35 10N 88 36W
Selowandoma Falls, Zimbabwe 55 G3 21 15S 31 50 E
Selpele, Indonesia 37 E8 0 1S 130 5 E
Selsey Bill, U.K. 11 G7 50 44N 0 47W
Selu, Indonesia 37 F8 7 32S 130 55 E
Selva, Argentina 94 B3 29 50S 62 0W
Selvas, Brazil 92 E5 6 30S 67 0W
Selwyn, Australia 62 C3 21 32S 140 30 E
Selwyn L., Canada 73 A8 60 0N 104 30W
Selwyn Ra., Australia 62 C3 21 10S 140 0 E
Seman →, Albania 21 D8 40 45N 19 50 E
Semarang, Indonesia 37 G14 7 0S 110 26 E
Semau, Indonesia 37 F6 10 13S 123 22 E
Semabule, Uganda 54 C3 0 4S 31 25 E
Semeru, Indonesia 37 H15 8 4S 112 55 E
Semey = Semipalatinsk, Kazakhstan 26 D9 50 30N 80 10 E
Seminoe Reservoir, U.S.A. 82 E10 42 9N 106 55W
Seminole, Okla., U.S.A. 81 H6 35 14N 96 41W
Seminole, Tex., U.S.A. 81 J3 32 43N 102 39W
Semiozernoye, Kazakhstan 26 D7 52 22N 64 8 E
Semipalatinsk, Kazakhstan 26 D9 50 30N 80 10 E
Semirara Is., Phil. 37 B6 12 0N 121 20 E
Semisopochnoi I., U.S.A. 68 C2 51 55N 179 36 E
Semitau, Indonesia 36 D4 0 29N 111 57 E
Semiyarskoye, Kazakhstan 26 D8 50 55N 78 23 E
Semmering Pass, Austria 16 E7 47 41N 15 45 E
Semnān, Iran 45 C7 35 55N 53 25 E
Semnān □, Iran 45 C7 36 0N 54 0 E
Semois →, Europe 15 E4 49 53N 4 44 E
Semporna, Malaysia 37 D5 4 30N 118 33 E
Semuda, Indonesia 36 E4 2 51S 112 58 E
Senā, Iran 45 D6 28 27N 51 36 E
Sena, Mozam. 55 F3 17 25S 35 0 E
Sena Madureira, Brazil 92 E5 9 5S 68 45W
Senador Pompeu, Brazil 93 E11 5 40S 39 20W
Senaja, Malaysia 36 C5 6 45N 117 3 E
Senanga, Zambia 56 B3 16 2S 23 14 E
Senatobia, U.S.A. 81 H10 34 37N 89 58W
Sendai, Kagoshima, Japan 31 J5 31 50N 130 20 E
Sendai, Miyagi, Japan 30 E10 38 15N 140 53 E
Sendai-Wan, Japan 30 E10 38 15N 141 0 E
Seneca, Oreg., U.S.A. 82 D4 44 8N 118 58W
Seneca, S.C., U.S.A. 77 H4 34 41N 82 57W
Seneca Falls, U.S.A. 79 D8 42 55N 76 48W
Seneca L., U.S.A. 78 D8 42 40N 76 54W
Senegal ■, W. Afr. 50 F2 14 30N 14 30W
Senegal →, W. Afr. 50 E1 15 48N 16 32W
Senegambia, Africa 48 E2 12 45N 12 0W
Senekal, S. Africa 57 D4 28 20S 27 36 E
Senga Hill, Zambia 55 D3 9 19S 31 11 E
Senge Khambab = Indus →, Pakistan 42 G2 24 20N 67 47 E
Sengerema □, Tanzania 54 C3 2 10S 32 20 E
Sengkang, Indonesia 37 E6 4 8S 120 1 E
Sengua →, Zimbabwe 55 F2 17 7S 28 5 E
Senhor-do-Bonfim, Brazil 93 F10 10 30S 40 10W
Senigállia, Italy 20 C5 43 42N 13 12 E
Senj, Croatia 20 B6 45 0N 14 58 E
Senja, Norway 8 B14 69 25N 17 30 E
Senlis, France 18 B5 49 13N 2 35 E
Senmonorom, Cambodia 38 F6 12 27N 107 12 E
Sennâr, Sudan 51 F11 13 30N 33 35 E
Senneterre, Canada 70 C4 48 25N 77 15W
Seno, Laos 38 D5 16 35N 104 50 E
Sens, France 18 B5 48 11N 3 15 E
Senta, Serbia, Yug. 21 B9 45 55N 20 3 E
Sentery, Zaïre 54 D2 5 17S 25 42 E
Sentinel, U.S.A. 83 K7 32 52N 113 13W
Sentolo, Indonesia 37 G14 7 55S 110 13 E
Seo de Urgel, Spain 19 A6 42 22N 1 23 E
Seohara, India 43 E8 29 15N 78 33 E
Seoni, India 43 H8 22 5N 79 30 E
Seoul = Sŏul, S. Korea 35 F14 37 31N 126 58 E
Separation Point, Canada 71 B8 53 37N 57 25W
Sepīdān, Iran 45 D7 30 20N 52 5 E
Sepo-ri, N. Korea 35 E14 38 57N 127 25 E
Sepone, Laos 38 D6 16 45N 106 13 E
Sept-Îles, Canada 71 B6 50 13N 66 22W
Sequim, U.S.A. 84 B3 48 5N 123 6W
Sequoia National Park, U.S.A. 83 H4 36 30N 118 30W
Seraing, Belgium 15 D5 50 35N 5 32 E
Seraja, Indonesia 39 L7 2 41N 108 35 E
Serakhis →, Cyprus 23 D11 35 13N 32 55 E
Seram, Indonesia 37 E7 3 10S 129 0 E
Seram Laut, Kepulauan, Indonesia 37 E8 4 5S 131 25 E
Seram Sea, Indonesia 37 E7 2 30S 128 30 E
Serang, Indonesia 37 G12 6 8S 106 10 E
Serasan, Indonesia 39 L7 2 29N 109 4 E
Serbia □, Yugoslavia 21 C9 43 30N 21 0 E
Serdobsk, Russia 24 D7 52 28N 44 10 E
Seremban, Malaysia 39 L3 2 43N 101 53 E
Serengeti □, Tanzania 54 C3 2 0S 34 30 E
Serengeti Plain, Tanzania 54 C3 2 40S 35 0 E
Serenje, Zambia 55 E3 13 14S 30 15 E
Sereth = Siret →, Romania 17 F14 45 24N 28 1 E
Sergino, Russia 26 C7 62 30N 65 38 E
Sergipe □, Brazil 93 F11 10 30S 37 30W
Sergiyev Posad, Russia 24 C6 56 20N 38 10 E
Seria, Brunei 36 D4 4 37N 114 23 E
Serian, Malaysia 36 D4 1 10N 110 31 E
Seribu, Kepulauan, Indonesia 36 F3 5 36S 106 33 E
Seringapatam Reef, Australia 60 B3 13 38S 122 5 E
Sermata, Indonesia 37 F7 8 15S 128 50 E
Serny Zavod, Turkmenistan 26 F6 39 59N 58 50 E
Serov, Russia 24 C11 59 29N 60 35 E
Serowe, Botswana 56 C4 22 25S 26 43 E
Serpentine, Australia 61 F2 32 23S 115 58 E
Serpentine Lakes, Australia 61 E4 28 30S 129 10 E
Serpukhov, Russia 24 D6 54 55N 37 28 E
Sérrai, Greece 21 D10 41 5N 23 31 E
Serrezuela, Argentina 94 C2 30 40S 65 20W
Serrinha, Brazil 93 F11 11 39S 39 0W
Sertânia, Brazil 93 E11 8 5S 37 20W
Sertanópolis, Brazil 95 A5 23 4S 51 2W
Serua, Indonesia 37 F8 6 18S 130 1 E
Serui, Indonesia 37 E9 1 53S 136 10 E
Serule, Botswana 56 C4 21 57S 27 20 E
Sese Is., Uganda 54 C3 0 20S 32 20 E
Sesepe, Indonesia 37 E7 1 30S 127 59 E
Sesfontein, Namibia 56 B1 19 7S 13 39 E
Sesheke, Zambia 56 B3 17 29S 24 13 E
S'estañol, Spain 22 B9 39 22N 2 54 E
Setana, Japan 30 C9 42 26N 139 51 E
Sète, France 18 E5 43 25N 3 42 E
Sete Lagôas, Brazil 93 G10 19 27S 44 16W
Sétif, Algeria 50 A6 36 9N 5 26 E
Seto, Japan 31 G8 35 14N 137 6 E
Setonaikai, Japan 31 G6 34 20N 133 30 E
Settat, Morocco 50 B3 33 0N 7 40W
Setté-Cama, Gabon 52 E1 2 32S 9 45 E
Setting L., Canada 73 B9 55 0N 98 38W
Settle, U.K. 10 C5 54 5N 2 18W
Settlement Pt., Bahamas 77 M6 26 40N 79 0W
Setúbal, Portugal 19 C1 38 30N 8 58W
Setúbal, B. de, Portugal 19 C1 38 40N 8 56W
Seulimeum, Indonesia 36 C1 5 27N 95 15 E
Sevan, Ozero, Armenia 25 F8 40 30N 45 20 E
Sevastopol, Ukraine 25 F5 44 35N 33 30 E
Seven Emu, Australia 62 B2 16 20S 137 8 E
Seven Sisters, Canada 72 C3 54 56N 128 10W
Severn →, Canada 70 A2 56 2N 87 36W
Severn →, U.K. 11 F5 51 35N 2 38W
Severn L., Canada 70 B1 53 54N 90 48W
Severnaya Zemlya, Russia 27 B10 79 0N 100 0 E
Severnyye Uvaly, Russia 24 C8 60 0N 50 0 E
Severo-Kurilsk, Russia 27 D16 50 40N 156 8 E
Severo-Yeniseyskiy, Russia 27 C10 60 22N 93 1 E
Severodvinsk, Russia 24 B6 64 27N 39 58 E
Severomorsk, Russia 24 A5 69 5N 33 27 E
Severouralsk, Russia 24 B10 60 9N 59 57 E
Sevier, U.S.A. 83 G7 38 39N 112 11W
Sevier →, U.S.A. 83 G7 39 4N 113 6W
Sevier L., U.S.A. 82 G7 38 54N 113 9W
Sevilla, Spain 19 D3 37 23N 6 0W
Seville = Sevilla, Spain 19 D3 37 23N 6 0W
Seward, Alaska, U.S.A. 68 B5 60 7N 149 27W
Seward, Nebr., U.S.A. 80 E6 40 55N 97 6W
Seward Pen., U.S.A. 68 B3 65 0N 164 0W
Sewell, Chile 94 C1 34 10S 70 23W
Sewer, Indonesia 37 F8 5 53S 134 40 E
Sewickley, U.S.A. 78 F4 40 32N 80 12W
Sexsmith, Canada 72 B5 55 21N 118 47W
Seychelles ■, Ind. Oc. 29 K9 5 0S 56 0 E
Seyðisfjörður, Iceland 8 D7 65 16N 13 57W
Seydvān, Iran 44 B5 38 34N 45 2 E
Seymchan, Russia 27 C16 62 54N 152 30 E
Seymour, Australia 63 F4 37 0S 145 10 E
Seymour, S. Africa 57 E4 32 33S 26 46 E
Seymour, Conn., U.S.A. 79 E11 41 24N 73 4W
Seymour, Ind., U.S.A. 76 F3 38 58N 85 53W
Seymour, Tex., U.S.A. 81 J5 33 35N 99 16W
Seymour, Wis., U.S.A. 76 C1 44 31N 88 20W
Sfax, Tunisia 51 B7 34 49N 10 48 E
Sfîntu Gheorghe, Romania 21 B11 45 52N 25 48 E
Shaanxi □, China 34 G5 35 0N 109 0 E
Shaba □, Zaïre 54 D2 8 0S 25 0 E
Shabunda, Zaïre 54 C2 2 40S 27 16 E
Shache, China 32 C2 38 20N 77 10 E
Shackleton Ice Shelf, Antarctica 5 C8 66 0S 100 0 E
Shackleton Inlet, Antarctica 5 E11 83 0S 160 0 E

Shādegān

Shādegān, *Iran* **45 D6** 30 40N 48 38 E
Shadi, *India* **43 C7** 33 24N 77 14 E
Shadrinsk, *Russia* **26 D7** 56 5N 63 32 E
Shafter, *Calif., U.S.A.* **85 K7** 35 30N 119 16W
Shafter, *Tex., U.S.A.* **81 L2** 29 49N 104 18W
Shaftesbury, *U.K.* **11 F5** 51 0N 2 12W
Shagram, *Pakistan* **43 A5** 36 24N 72 20 E
Shah Bunder, *Pakistan* **42 G2** 24 13N 67 56 E
Shahabad, *Punjab, India* **42 D7** 30 10N 76 55 E
Shahabad, *Raj., India* **42 G7** 25 15N 77 11 E
Shahabad, *Ut. P., India* **43 F8** 27 36N 79 56 E
Shahadpur, *Pakistan* **42 G3** 25 55N 68 35 E
Shahba, *Syria* **47 C5** 32 52N 36 38 E
Shahdād, *Iran* **45 D8** 30 30N 57 40 E
Shahdadkot, *Pakistan* **42 F2** 27 50N 67 55 E
Shahe, *China* **34 F8** 37 0N 114 32 E
Shahganj, *India* **43 F10** 26 3N 82 44 E
Shahgarh, *India* **40 F6** 27 15N 69 50 E
Shaḩḩāt, *Libya* **51 B9** 32 48N 21 54 E
Shahjahanpur, *India* **43 F8** 27 54N 79 57 E
Shahpur, *India* **42 H7** 22 12N 77 58 E
Shahpur, *Pakistan* **42 E3** 28 46N 68 27 E
Shahpura, *India* **43 H9** 23 10N 80 45 E
Shahr Kord, *Iran* **45 C6** 32 15N 50 55 E
Shāhrakht, *Iran* **45 C9** 33 38N 60 16 E
Shahrig, *Pakistan* **42 D2** 30 15N 67 40 E
Shahukou, *China* **34 D7** 40 20N 112 18 E
Shaikhabad, *Afghan.* **42 B3** 34 2N 68 45 E
Shajapur, *India* **42 H7** 23 27N 76 21 E
Shakargarh, *Pakistan* **42 C6** 32 17N 75 10 E
Shakawe, *Botswana* **56 B3** 18 28S 21 49 E
Shaker Heights, *U.S.A.* **78 E3** 41 29N 81 32W
Shakhty, *Russia* **25 E7** 47 40N 40 16 E
Shakhunya, *Russia* **24 C8** 57 40N 46 46 E
Shaki, *Nigeria* **50 G5** 8 41N 3 21 E
Shakopee, *U.S.A.* **80 C8** 44 48N 93 32W
Shala, L., *Ethiopia* **51 G12** 7 30N 38 30 E
Shallow Lake, *Canada* **78 B3** 44 36N 81 5W
Shaluli Shan, *China* **32 C4** 30 40N 99 55 E
Shām, *Iran* **45 E8** 26 39N 57 21 E
Shamâl Dârfûr □, *Sudan* **51 E10** 15 0N 25 0 E
Shamâl Kordofân □, *Sudan* **51 F10** 15 0N 30 0 E
Shamattawa, *Canada* **73 B10** 55 51N 92 5W
Shamattawa →, *Canada* **70 A2** 55 1N 85 23W
Shamīl, *Iran* **45 E8** 27 30N 56 55 E
Shāmkūh, *Iran* **45 C8** 35 47N 57 50 E
Shamli, *India* **42 E7** 29 32N 77 18 E
Shamo = Gobi, *Asia* **34 C5** 44 0N 111 0 E
Shamo, L., *Ethiopia* **51 G12** 5 45N 37 30 E
Shamokin, *U.S.A.* **79 F8** 40 47N 76 34W
Shamrock, *U.S.A.* **81 H4** 35 13N 100 15W
Shamva, *Zimbabwe* **55 F3** 17 20S 31 32 E
Shan □, *Burma* **41 J21** 21 30N 98 30 E
Shan Xian, *China* **34 G9** 34 50N 116 5 E
Shanchengzhen, *China* **35 C13** 42 20N 125 20 E
Shāndak, *Iran* **45 D9** 28 28N 60 27 E
Shandon, *U.S.A.* **84 K6** 35 39N 120 23W
Shandong □, *China* **35 F10** 36 0N 118 0 E
Shandong Bandao, *China* **35 F11** 37 0N 121 0 E
Shang Xian, *China* **34 H5** 33 50N 109 58 E
Shangalowe, *Zaïre* **55 E2** 10 50S 26 30 E
Shangani →, *Zimbabwe* **55 F2** 18 41S 27 10 E
Shangbancheng, *China* **35 D10** 40 50N 118 1 E
Shangdu, *China* **34 D7** 41 30N 113 30 E
Shanghai, *China* **33 C7** 31 15N 121 26 E
Shanghe, *China* **35 F9** 37 20N 117 10 E
Shangnan, *China* **34 H6** 33 32N 110 50 E
Shangqiu, *China* **34 G8** 34 26N 115 36 E
Shangrao, *China* **33 D6** 28 25N 117 59 E
Shangshui, *China* **34 H8** 33 42N 114 35 E
Shangzhi, *China* **35 B14** 45 22N 127 56 E
Shanhetun, *China* **35 B14** 44 33N 127 15 E
Shaniko, *U.S.A.* **82 D3** 45 0N 120 45W
Shannon, *Greenland* **4 B7** 75 10N 18 30W
Shannon, *N.Z.* **59 J5** 40 33S 175 25 E
Shannon →, *Ireland* **13 D2** 52 35N 9 30W
Shansi = Shanxi □, *China* **34 F7** 37 0N 112 0 E
Shantar, Ostrov Bolshoy, *Russia* **27 D14** 55 9N 137 40 E
Shantipur, *India* **43 H13** 23 17N 88 25 E
Shantou, *China* **33 D6** 23 18N 116 40 E
Shantung = Shandong □, *China* **35 F10** 36 0N 118 0 E
Shanxi □, *China* **34 F7** 37 0N 112 0 E
Shanyang, *China* **34 H5** 33 31N 109 55 E
Shanyin, *China* **34 E7** 39 25N 112 56 E
Shaoguan, *China* **33 D6** 24 48N 113 35 E
Shaoxing, *China* **33 C7** 30 0N 120 35 E
Shaoyang, *China* **33 D6** 27 14N 111 25 E
Shapinsay, *U.K.* **12 B6** 59 2N 2 50W
Shaqra', *Si. Arabia* **44 E5** 25 15N 45 16 E
Shaqrā', *Yemen* **46 E4** 13 22N 45 44 E
Sharbot Lake, *Canada* **79 B8** 44 46N 76 41W
Shari, *Japan* **30 C12** 43 55N 144 40 E
Sharjah = Ash Shāriqah, *U.A.E.* **45 E7** 25 23N 55 26 E
Shark B., *Australia* **61 E1** 25 30S 113 32 E
Sharon, *Mass., U.S.A.* **79 D13** 42 7N 71 11W
Sharon, *Pa., U.S.A.* **78 E4** 41 14N 80 31W
Sharon Springs, *U.S.A.* **80 F4** 38 54N 101 45W
Sharp Pt., *Australia* **62 A3** 10 58S 142 43 E
Sharpe L., *Canada* **73 C10** 54 24N 93 40W
Sharpsville, *U.S.A.* **78 E4** 41 15N 80 29W
Sharq el Istiwa'iya □, *Sudan* **51 G11** 5 0N 33 0 E
Sharya, *Russia* **24 C8** 58 22N 45 20 E
Shashi, *Botswana* **57 C4** 21 15S 27 27 E
Shashi, *China* **33 C6** 30 25N 112 14 E
Shashi →, *Africa* **55 G2** 21 14S 29 20 E
Shasta, Mt., *U.S.A.* **82 F2** 41 25N 122 12W
Shasta L., *U.S.A.* **82 F2** 40 43N 122 25W
Shatt al'Arab →, *Iraq* **45 D6** 29 57N 48 34 E
Shattuck, *U.S.A.* **81 G5** 36 16N 99 53W
Shaunavon, *Canada* **73 D7** 49 35N 108 25W
Shaver L., *U.S.A.* **84 H7** 37 9N 119 18W
Shaw →, *Australia* **60 D2** 20 21S 119 17 E
Shaw I., *Australia* **62 C4** 20 30S 149 2 E
Shawanaga, *Canada* **78 A4** 45 31N 80 17W
Shawano, *U.S.A.* **76 C1** 44 47N 88 36W
Shawinigan, *Canada* **70 C5** 46 35N 72 50W
Shawnee, *U.S.A.* **81 H6** 35 20N 96 55W
Shaybārā, *Si. Arabia* **44 E3** 25 26N 36 47 E
Shaykh Sa'īd, *Iraq* **44 C5** 32 34N 46 17 E
Shcherbakov = Rybinsk, *Russia* **24 C6** 58 5N 38 50 E
Shchuchiosk, *Kazakhstan* **26 D8** 52 56N 70 12 E
She Xian, *China* **34 F7** 36 30N 113 40 E
Shebele = Scebeli, Wabi →, *Somali Rep.* **46 G3** 2 0N 44 0 E
Sheboygan, *U.S.A.* **76 D2** 43 46N 87 45W
Shediac, *Canada* **71 C7** 46 14N 64 32W
Sheelin, L., *Ireland* **13 C4** 53 48N 7 20W
Sheep Haven, *Ireland* **13 A4** 55 12N 7 55W
Sheerness, *U.K.* **11 F8** 51 26N 0 47 E
Sheet Harbour, *Canada* **71 D7** 44 56N 62 31W
Sheffield, *U.K.* **10 D6** 53 23N 1 28W
Sheffield, *Ala., U.S.A.* **77 H2** 34 46N 87 41W
Sheffield, *Mass., U.S.A.* **79 D11** 42 5N 73 21W
Sheffield, *Pa., U.S.A.* **78 E5** 41 42N 79 3W
Sheffield, *Tex., U.S.A.* **81 K4** 30 41N 101 49W
Sheho, *Canada* **73 C8** 51 35N 103 13W
Sheikhpura, *India* **43 G11** 25 9N 85 53 E
Shekhupura, *Pakistan* **42 D5** 31 42N 73 58 E
Shelburne, *N.S., Canada* **71 D6** 43 47N 65 20W
Shelburne, *Ont., Canada* **70 D3** 44 4N 80 15W
Shelburne, *U.S.A.* **79 B11** 44 23N 73 14W
Shelburne B., *Australia* **62 A3** 11 50S 142 50 E
Shelburne Falls, *U.S.A.* **79 D12** 42 36N 72 45W
Shelby, *Mich., U.S.A.* **76 D2** 43 37N 86 22W
Shelby, *Mont., U.S.A.* **82 B8** 48 30N 111 51W
Shelby, *N.C., U.S.A.* **77 H5** 35 17N 81 32W
Shelby, *Ohio, U.S.A.* **78 F2** 40 53N 82 40W
Shelbyville, *Ill., U.S.A.* **80 F10** 39 24N 88 48W
Shelbyville, *Ind., U.S.A.* **76 F3** 39 31N 85 47W
Shelbyville, *Tenn., U.S.A.* **77 H2** 35 29N 86 28W
Sheldon, *U.S.A.* **80 D7** 43 11N 95 51W
Sheldrake, *Canada* **71 B7** 50 20N 64 51W
Shelikhova, Zaliv, *Russia* **27 D16** 59 30N 157 0 E
Shell Lake, *Canada* **73 C7** 53 19N 107 2W
Shell Lakes, *Australia* **61 E4** 29 20S 127 30 E
Shellbrook, *Canada* **73 C7** 53 13N 106 24W
Shellharbour, *Australia* **63 E5** 34 31S 150 51 E
Shelling Rocks, *Ireland* **13 E1** 51 45N 10 35W
Shelton, *Conn., U.S.A.* **79 E11** 41 19N 73 5W
Shelton, *Wash., U.S.A.* **84 C3** 47 13N 123 6W
Shen Xian, *China* **34 F8** 36 15N 115 40 E
Shenandoah, *Iowa, U.S.A.* **80 E7** 40 46N 95 22W
Shenandoah, *Pa., U.S.A.* **79 F8** 40 49N 76 12W
Shenandoah, *Va., U.S.A.* **76 F6** 38 29N 78 37W
Shenandoah →, *U.S.A.* **76 F7** 39 19N 77 44W
Shenchi, *China* **34 E7** 39 8N 112 10 E
Shendam, *Nigeria* **50 G6** 8 49N 9 30 E
Shendî, *Sudan* **51 E11** 16 46N 33 22 E
Shengfang, *China* **34 E9** 39 3N 116 42 E
Shenjingzi, *China* **35 B13** 44 40N 124 30 E
Shenmu, *China* **34 E6** 38 50N 110 29 E
Shenqiu, *China* **34 H8** 33 25N 115 5 E
Shenqiucheng, *China* **34 H8** 33 24N 115 2 E
Shensi = Shaanxi □, *China* **34 G5** 35 0N 109 0 E
Shenyang, *China* **35 D12** 41 48N 123 27 E
Sheopur Kalan, *India* **40 G10** 25 40N 76 40 E
Shepparton, *Australia* **63 F4** 36 23S 145 26 E
Sheqi, *China* **34 H7** 33 12N 112 57 E
Sher Qila, *Pakistan* **43 A6** 36 7N 74 2 E
Sherborne, *U.K.* **11 G5** 50 56N 2 31W
Sherbro I., *S. Leone* **50 G2** 7 30N 12 40W
Sherbrooke, *Canada* **71 C5** 45 28N 71 57W
Sheridan, *Ark., U.S.A.* **81 H8** 34 19N 92 24W
Sheridan, *Wyo., U.S.A.* **82 D10** 44 48N 106 58W
Sherkot, *India* **43 E8** 29 22N 78 35 E
Sherman, *U.S.A.* **81 J6** 33 40N 96 35W
Sherridon, *Canada* **73 B8** 55 8N 101 5W
Sherwood, *N. Dak., U.S.A.* **80 A4** 48 57N 101 38W
Sherwood, *Tex., U.S.A.* **81 K4** 31 18N 100 45W
Sherwood Forest, *U.K.* **10 D6** 53 5N 1 5W
Sheslay, *Canada* **72 B2** 58 17N 131 52W
Sheslay →, *Canada* **72 B2** 58 48N 132 5W
Shethanei L., *Canada* **73 B9** 58 48N 97 50W
Shetland □, *U.K.* **12 A7** 60 30N 1 30W
Shetland Is., *U.K.* **12 A7** 60 30N 1 30W
Sheyenne, *U.S.A.* **80 B5** 47 50N 99 7W
Sheyenne →, *U.S.A.* **80 B6** 47 2N 96 50W
Shibām, *Yemen* **46 D4** 16 0N 48 36 E
Shibata, *Japan* **30 F9** 37 57N 139 20 E
Shibecha, *Japan* **30 C12** 43 17N 144 36 E
Shibetsu, *Japan* **30 B11** 44 10N 142 23 E
Shibogama L., *Canada* **70 B2** 53 35N 88 15W
Shibushi, *Japan* **31 J5** 31 25N 131 8 E
Shickshock Mts. = Chic-Chocs, Mts., *Canada* **71 C6** 48 55N 66 0W
Shidao, *China* **35 F12** 36 50N 122 25 E
Shido, *Japan* **31 G7** 34 19N 134 10 E
Shiel, L., *U.K.* **12 E3** 56 48N 5 32W
Shield, C., *Australia* **62 A2** 13 20S 136 20 E
Shiga □, *Japan* **31 G8** 35 20N 136 0 E
Shigaib, *Sudan* **51 E9** 15 5N 23 35 E
Shiguaigou, *China* **34 D6** 40 52N 110 15 E
Shihchiachuangi = Shijiazhuang, *China* **34 E8** 38 2N 114 28 E
Shijiazhuang, *China* **34 E8** 38 2N 114 28 E
Shikarpur, *India* **42 E8** 28 17N 78 7 E
Shikarpur, *Pakistan* **42 F3** 27 57N 68 39 E
Shikoku □, *Japan* **31 H6** 33 30N 133 30 E
Shikoku-Sanchi, *Japan* **31 H6** 33 30N 133 30 E
Shilabo, *Ethiopia* **46 F3** 6 22N 44 32 E
Shiliguri, *India* **41 F16** 26 45N 88 25 E
Shilka, *Russia* **27 D12** 52 0N 115 55 E
Shilka →, *Russia* **27 D13** 53 20N 121 26 E
Shillelagh, *Ireland* **13 D5** 52 46N 6 32W
Shillong, *India* **41 G17** 25 35N 91 53 E
Shilo, *Jordan* **47 C4** 32 4N 35 18 E
Shilou, *China* **34 F6** 37 0N 110 48 E
Shimabara, *Japan* **31 H5** 32 48N 130 20 E
Shimada, *Japan* **31 G9** 34 49N 138 10 E
Shimane □, *Japan* **31 G6** 35 0N 132 30 E
Shimanovsk, *Russia* **27 D13** 52 15N 127 30 E
Shimizu, *Japan* **31 G9** 35 0N 138 30 E
Shimodate, *Japan* **31 F9** 36 20N 139 55 E
Shimoga, *India* **40 N9** 13 57N 75 32 E
Shimoni, *Kenya* **54 C4** 4 38S 39 20 E
Shimonoseki, *Japan* **31 H5** 33 58N 130 55 E
Shimpuru Rapids, *Angola* **56 B2** 17 45S 19 55 E
Shin, L., *U.K.* **12 C4** 58 7N 4 30W
Shin-Tone →, *Japan* **31 G10** 35 44N 140 51 E
Shinano →, *Japan* **31 F9** 36 50N 138 30 E
Shīndand, *Afghan.* **40 C3** 33 12N 62 8 E
Shingleton, *U.S.A.* **70 C2** 46 21N 86 28W
Shingū, *Japan* **31 H7** 33 40N 135 55 E
Shinjō, *Japan* **30 E10** 38 46N 140 18 E
Shinshār, *Syria* **47 A5** 34 36N 36 43 E
Shinyanga, *Tanzania* **54 C3** 3 45S 33 27 E
Shinyanga □, *Tanzania* **54 C3** 3 50S 34 0 E
Shiogama, *Japan* **30 E10** 38 19N 141 1 E
Shiojiri, *Japan* **31 F8** 36 6N 137 58 E
Ship I., *U.S.A.* **81 K10** 30 13N 88 55W
Shiping, *China* **32 D5** 23 45N 102 23 E
Shipki La, *India* **40 D11** 31 45N 78 40 E
Shippegan, *Canada* **71 C7** 47 45N 64 45W
Shippensburg, *U.S.A.* **78 F7** 40 3N 77 31W
Shiprock, *U.S.A.* **83 H9** 36 47N 108 41W
Shiqma, N. →, *Israel* **47 D3** 31 37N 34 30 E
Shiquan, *China* **34 H5** 33 5N 108 15 E
Shīr Kūh, *Iran* **45 D7** 31 39N 54 3 E
Shiragami-Misaki, *Japan* **30 D10** 41 24N 140 12 E
Shirakawa, *Fukushima, Japan* **31 F10** 37 7N 140 13 E
Shirakawa, *Gifu, Japan* **31 F8** 36 17N 136 56 E
Shirane-San, *Gumma, Japan* **31 F9** 36 48N 139 22 E
Shirane-San, *Yamanashi, Japan* **31 G9** 35 42N 138 9 E
Shiraoi, *Japan* **30 C10** 42 33N 141 21 E
Shīrāz, *Iran* **45 D7** 29 42N 52 30 E
Shire →, *Africa* **55 F4** 17 42S 35 19 E
Shiretoko-Misaki, *Japan* **30 B12** 44 21N 145 20 E
Shirinab →, *Pakistan* **42 D2** 30 15N 66 28 E
Shiriya-Zaki, *Japan* **30 D10** 41 25N 141 30 E
Shiroishi, *Japan* **30 E10** 38 0N 140 37 E
Shīrvān, *Iran* **45 B8** 37 30N 57 50 E
Shirwa, L. = Chilwa, L., *Malawi* **55 F4** 15 15S 35 40 E
Shivpuri, *India* **42 G7** 25 26N 77 42 E
Shixian, *China* **35 C15** 43 5N 129 50 E
Shizuishan, *China* **34 E4** 39 15N 106 50 E
Shizuoka, *Japan* **31 G9** 34 57N 138 24 E
Shizuoka □, *Japan* **31 G9** 35 15N 138 40 E
Shkoder = Shkodra, *Albania* **21 C8** 42 6N 19 20 E
Shkodra, *Albania* **21 C8** 42 6N 19 20 E
Shkumbini →, *Albania* **21 D8** 41 5N 19 50 E
Shmidta, O., *Russia* **27 A10** 81 0N 91 0 E
Shō-Gawa →, *Japan* **31 F8** 36 47N 137 4 E
Shoal Lake, *Canada* **73 C8** 50 30N 100 35W
Shōdo-Shima, *Japan* **31 G7** 34 30N 134 15 E
Shoeburyness, *U.K.* **11 F8** 51 31N 0 49 E
Sholapur = Solapur, *India* **40 L9** 17 43N 75 56 E
Shologontsy, *Russia* **27 C12** 66 13N 114 0 E
Shōmrōn, *Jordan* **47 C4** 32 15N 35 13 E
Shoshone, *Calif., U.S.A.* **85 K10** 35 58N 116 16W
Shoshone, *Idaho, U.S.A.* **82 E6** 42 56N 114 25W
Shoshone L., *U.S.A.* **82 D8** 44 22N 110 43W
Shoshone Mts., *U.S.A.* **82 G5** 39 20N 117 25W
Shoshong, *Botswana* **56 C4** 22 56S 26 31 E
Shoshoni, *U.S.A.* **82 E9** 43 14N 108 7W
Shouguang, *China* **35 F10** 37 52N 118 45 E
Shouyang, *China* **34 F7** 37 54N 113 8 E
Show Low, *U.S.A.* **83 J9** 34 15N 110 2W
Shreveport, *U.S.A.* **81 J8** 32 31N 93 45W
Shrewsbury, *U.K.* **10 E5** 52 42N 2 45W
Shrirampur, *India* **43 H13** 22 44N 88 21 E
Shropshire □, *U.K.* **11 E5** 52 36N 2 45W
Shuangcheng, *China* **35 B14** 45 20N 126 15 E
Shuanggou, *China* **35 G9** 34 2N 117 30 E
Shuangliao, *China* **35 C12** 43 29N 123 30 E
Shuangshanzi, *China* **35 D10** 40 20N 119 8 E
Shuangyang, *China* **35 C13** 43 28N 125 40 E
Shuangyashan, *China* **33 B8** 46 28N 131 5 E
Shuguri Falls, *Tanzania* **55 D4** 8 33S 37 22 E
Shuicheng, *China* **32 D5** 26 38N 104 48 E
Shuiye, *China* **34 F8** 36 7N 114 8 E
Shujalpur, *India* **42 H7** 23 18N 76 46 E
Shukpa Kunzang, *India* **43 B8** 34 22N 78 22 E
Shulan, *China* **35 B14** 44 28N 127 0 E
Shule, *China* **32 C2** 39 25N 76 3 E
Shumagin Is., *U.S.A.* **68 C4** 55 7N 159 45W
Shumikha, *Russia* **26 D7** 55 10N 63 15 E
Shungnak, *U.S.A.* **68 B4** 66 52N 157 9W
Shuo Xian, *China* **34 E7** 39 20N 112 33 E
Shūr →, *Iran* **45 D7** 28 30N 55 0 E
Shūr Āb, *Iran* **45 C6** 34 23N 51 11 E
Shūr Gaz, *Iran* **45 D8** 29 10N 59 20 E
Shūrāb, *Iran* **45 C8** 33 43N 56 29 E
Shūrjestān, *Iran* **45 D7** 31 24N 52 25 E
Shurugwi, *Zimbabwe* **55 F3** 19 40S 30 0 E
Shūsf, *Iran* **45 D9** 31 50N 60 5 E
Shūshtar, *Iran* **45 D6** 32 0N 48 50 E
Shuswap L., *Canada* **72 C5** 50 55N 119 3W
Shuyang, *China* **35 G10** 34 10N 118 42 E
Shūzū, *Iran* **45 D7** 29 52N 54 30 E
Shwebo, *Burma* **41 H19** 22 30N 95 45 E
Shwegu, *Burma* **41 G20** 24 15N 96 26 E
Shweli →, *Burma* **41 H20** 23 45N 96 45 E
Shymkent = Chimkent, *Kazakhstan* **26 E7** 42 18N 69 36 E
Shyok, *India* **43 B8** 34 15N 78 12 E
Shyok →, *Pakistan* **43 B6** 35 13N 75 53 E
Si Chon, *Thailand* **39 H2** 9 0N 99 54 E
Si Kiang = Xi Jiang →, *China* **33 D6** 22 5N 113 20 E
Si-ngan = Xi'an, *China* **34 G5** 34 15N 109 0 E
Si Prachan, *Thailand* **38 E3** 14 37N 100 9 E
Si Racha, *Thailand* **38 F3** 13 10N 100 48 E
Si Xian, *China* **35 H9** 33 30N 117 50 E
Siahan Range, *Pakistan* **40 F4** 27 30N 64 40 E
Siaksrindrapura, *Indonesia* **36 D2** 0 51N 102 0 E
Sialkot, *Pakistan* **42 C6** 32 32N 74 30 E
Siam = Thailand ■, *Asia* **38 E4** 16 0N 102 0 E
Siantan, P., *Indonesia* **39 L6** 3 10N 106 15 E
Sīāreh, *Iran* **45 D9** 28 5N 60 14 E
Siargao, *Phil.* **37 C7** 9 52N 126 3 E
Siari, *Pakistan* **43 B7** 34 55N 76 40 E
Siasi, *Phil.* **37 C6** 5 34N 120 50 E
Siau, *Indonesia* **37 D7** 2 50N 125 25 E
Šiauliai, *Lithuania* **24 C3** 55 56N 23 15 E
Siaya □, *Kenya* **54 B3** 0 0 34 20 E
Sibasa, *S. Africa* **57 C5** 22 53S 30 33 E
Sibay, *Russia* **24 D10** 52 42N 58 39 E
Sibayi, L., *S. Africa* **57 D5** 27 20S 32 45 E
Šibenik, *Croatia* **20 C6** 43 48N 15 54 E
Siberia, *Russia* **4 D13** 60 0N 100 0 E
Siberut, *Indonesia* **36 E1** 1 30S 99 0 E
Sibi, *Pakistan* **42 E2** 29 30N 67 54 E
Sibil, *Indonesia* **37 E10** 4 59S 140 35 E
Sibiti, *Congo* **52 E2** 3 38S 13 19 E
Sibiu, *Romania* **21 B11** 45 45N 24 9 E
Sibley, *Iowa, U.S.A.* **80 D7** 43 24N 95 45W
Sibley, *La., U.S.A.* **81 J8** 32 33N 93 18W
Sibolga, *Indonesia* **36 D1** 1 42N 98 45 E
Sibsagar, *India* **41 F19** 27 0N 94 36 E
Sibu, *Malaysia* **36 D4** 2 18N 111 49 E
Sibuco, *Phil.* **37 C6** 7 20N 122 10 E
Sibuguey B., *Phil.* **37 C6** 7 50N 122 45 E
Sibutu, *Phil.* **37 D5** 4 45N 119 30 E
Sibutu Passage, *E. Indies* **37 D5** 4 50N 120 0 E
Sibuyan, *Phil.* **37 B6** 12 25N 122 40 E
Sibuyan Sea, *Phil.* **37 B6** 12 30N 122 20 E
Sicamous, *Canada* **72 C5** 50 49N 119 0W
Sichuan □, *China* **32 C5** 31 0N 104 0 E
Sicilia, *Italy* **20 F6** 37 30N 14 30 E
Sicilia □, *Italy* **20 F6** 37 30N 14 30 E
Sicily = Sicilia, *Italy* **20 F6** 37 30N 14 30 E
Sicuani, *Peru* **92 F4** 14 21S 71 10W
Sidári, *Greece* **23 A3** 39 47N 19 41 E
Siddhapur, *India* **42 H5** 23 56N 72 25 E
Siddipet, *India* **40 K11** 18 5N 78 51 E
Sidéradougou, *Burkina Faso* **50 F4** 10 42N 4 12W
Sídheros, Ákra, *Greece* **23 D8** 35 19N 26 19 E
Sîdi Barrâni, *Egypt* **51 B10** 31 38N 25 58 E
Sidi-bel-Abbès, *Algeria* **50 A4** 35 13N 0 39W
Sidlaw Hills, *U.K.* **12 E5** 56 32N 3 10W
Sidley, Mt., *Antarctica* **5 D14** 77 2S 126 2W
Sidmouth, *U.K.* **11 G4** 50 40N 3 13W
Sidmouth, C., *Australia* **62 A3** 13 25S 143 36 E
Sidney, *Canada* **72 D4** 48 39N 123 24W
Sidney, *Mont., U.S.A.* **80 B2** 47 43N 104 9W
Sidney, *N.Y., U.S.A.* **79 D9** 42 19N 75 24W
Sidney, *Nebr., U.S.A.* **80 E3** 41 8N 102 59W
Sidney, *Ohio, U.S.A.* **76 E3** 40 17N 84 9W
Sidoarjo, *Indonesia* **37 G15** 7 27S 112 43 E
Sidon = Saydā, *Lebanon* **47 B4** 33 35N 35 25 E
Sidra, G. of = Surt, Khalīj, *Libya* **51 B8** 31 40N 18 30 E
Siedlce, *Poland* **17 B11** 52 10N 22 20 E
Siegen, *Germany* **16 C4** 50 52N 8 2 E
Siem Pang, *Cambodia* **38 E6** 14 7N 106 23 E
Siem Reap, *Cambodia* **38 F4** 13 20N 103 52 E
Siena, *Italy* **20 C4** 43 20N 11 20 E
Sierra Blanca, *U.S.A.* **83 L11** 31 11N 105 22W
Sierra Blanca Peak, *U.S.A.* **83 K11** 33 23N 105 49W
Sierra City, *U.S.A.* **84 F6** 39 34N 120 38W
Sierra Colorada, *Argentina* **96 E3** 40 35S 67 50W
Sierra Gorda, *Chile* **94 A2** 22 50S 69 15W
Sierra Leone ■, *W. Afr.* **50 G2** 9 0N 12 0W
Sierra Madre, *Mexico* **87 D6** 16 0N 93 0W
Sierra Mojada, *Mexico* **86 B4** 27 19N 103 42W
Sierraville, *U.S.A.* **84 F6** 39 36N 120 22W
Sífnos, *Greece* **21 F11** 37 0N 24 45 E
Sifton, *Canada* **73 C8** 51 21N 100 8W
Sifton Pass, *Canada* **72 B3** 57 52N 126 15W
Sighetu-Marmatiei, *Romania* **17 E11** 47 57N 23 52 E
Sighişoara, *Romania* **17 E12** 46 12N 24 50 E
Sigli, *Indonesia* **36 C1** 5 25N 96 0 E
Siglufjörður, *Iceland* **8 C4** 66 12N 18 55W
Signal, *U.S.A.* **85 L13** 34 30N 113 38W
Signal Pk., *U.S.A.* **85 M12** 33 20N 114 2W
Sigsig, *Ecuador* **92 D3** 3 0S 78 50W
Sigtuna, *Sweden* **9 G14** 59 36N 17 44 E
Sigüenza, *Spain* **19 B4** 41 3N 2 40W
Siguiri, *Guinea* **50 F3** 11 31N 9 10W
Sigurd, *U.S.A.* **83 G8** 38 50N 111 58W
Sihanoukville = Kompong Som, *Cambodia* **39 G4** 10 38N 103 30 E
Sijarira Ra., *Zimbabwe* **55 F2** 17 36S 27 45 E
Sikao, *Thailand* **39 J2** 7 34N 99 21 E
Sikar, *India* **42 F6** 27 33N 75 10 E
Sikasso, *Mali* **50 F3** 11 18N 5 35W
Sikeston, *U.S.A.* **81 G10** 36 53N 89 35W
Sikhote Alin, Khrebet, *Russia* **27 E14** 45 0N 136 0 E
Sikhote Alin Ra. = Sikhote Alin, Khrebet, *Russia* **27 E14** 45 0N 136 0 E
Síkinos, *Greece* **21 F11** 36 40N 25 8 E
Sikkani Chief →, *Canada* **72 B4** 57 47N 122 15W
Sikkim □, *India* **41 F16** 27 50N 88 30 E
Sikotu-Ko, *Japan* **30 C10** 42 45N 141 25 E
Sil →, *Spain* **19 A2** 42 27N 7 43W
Silacayoapan, *Mexico* **87 D5** 17 30N 98 9W
Silchar, *India* **41 G18** 24 49N 92 48 E
Silcox, *Canada* **73 B10** 57 12N 94 10W
Siler City, *U.S.A.* **77 H6** 35 44N 79 28W
Silesia = Śląsk, *Poland* **16 C8** 51 0N 16 30 E
Silgarhi Doti, *Nepal* **43 E9** 29 15N 81 0 E
Silghat, *India* **41 F18** 26 35N 93 0 E
Silifke, *Turkey* **25 G5** 36 22N 33 58 E
Siliguri = Shiliguri, *India* **41 F16** 26 45N 88 25 E
Siling Co, *China* **32 C3** 31 50N 89 20 E
Silistra, *Bulgaria* **21 B12** 44 6N 27 19 E
Siljan, *Sweden* **9 F13** 60 55N 14 45 E
Silkeborg, *Denmark* **9 H10** 56 10N 9 32 E
Sillajhuay, Cordillera, *Chile* **92 G5** 19 46S 68 40W

Son Ha

Son Ha, *Vietnam* 38 E7 15 3N 108 34 E
Son Hoa, *Vietnam* 38 F7 13 2N 108 58 E
Son La, *Vietnam* 38 B4 21 20N 103 50 E
Son Tay, *Vietnam* 38 B5 21 8N 105 30 E
Soná, *Panama* 88 E3 8 0N 81 20W
Sonamarg, *India* 43 B6 34 18N 75 21 E
Sonamukhi, *India* 43 H12 23 18N 87 27 E
Sŏnchŏn, *N. Korea* 35 E13 39 48N 124 55 E
Sondags →, *S. Africa* 56 E4 33 44S 25 51 E
Sondar, *India* 43 C6 33 28N 75 56 E
Sønderborg, *Denmark* 9 J10 54 55N 9 49 E
Søndre Strømfjord, *Greenland* 69 B14 66 59N 50 40W
Sone, *Mozam.* 55 F3 17 23S 34 55 E
Sonepur, *India* 41 J13 20 55N 83 50 E
Song, *Thailand* 38 C3 18 28N 100 11 E
Song Cau, *Vietnam* 38 F7 13 27N 109 18 E
Song Xian, *China* 34 G7 34 12N 112 8 E
Songchŏn, *N. Korea* 35 E14 39 12N 126 15 E
Songea, *Tanzania* 55 E4 10 40S 35 40 E
Songea □, *Tanzania* 55 E4 10 30S 36 0 E
Songhua Hu, *China* 35 C14 43 35N 126 50 E
Songhua Jiang →, *China* 33 B8 47 45N 132 30 E
Songjin, *N. Korea* 35 D15 40 40N 129 10 E
Songjŏng-ni, *S. Korea* 35 G14 35 8N 126 47 E
Songkhla, *Thailand* 39 J3 7 13N 100 37 E
Songnim, *N. Korea* 35 E13 38 45N 125 39 E
Songpan, *China* 32 C5 32 40N 103 30 E
Songwe, *Zaïre* 54 C2 3 20S 26 16 E
Songwe →, *Africa* 55 D3 9 44S 33 58 E
Sonid Youqi, *China* 34 C7 42 45N 112 48 E
Sonipat, *India* 42 E7 29 0N 77 5 E
Sonmiani, *Pakistan* 42 G2 25 25N 66 40 E
Sono →, *Brazil* 93 E9 9 58S 48 11W
Sonora, *Calif., U.S.A.* 83 H3 37 59N 120 23W
Sonora, *Tex., U.S.A.* 81 K4 30 34N 100 39W
Sonora □, *Mexico* 86 B2 29 0N 111 0W
Sonora →, *Mexico* 86 B2 28 50N 111 33W
Sonora Desert, *U.S.A.* 85 M12 33 40N 114 15W
Sonoyta, *Mexico* 86 A2 31 51N 112 50W
Sŏnsan, *S. Korea* 35 F15 36 14N 128 17 E
Sonsonate, *El Salv.* 88 D2 13 43N 89 44W
Soochow = Suzhou, *China* 33 C7 31 19N 120 38 E
Sop Hao, *Laos* 38 B5 20 33N 104 27 E
Sop Prap, *Thailand* 38 D2 17 53N 99 20 E
Sopi, *Indonesia* 37 D7 2 34N 128 28 E
Sopot, *Poland* 17 A9 54 27N 18 31 E
Sop's Arm, *Canada* 71 C8 49 46N 56 56W
Sopur, *India* 43 B6 34 18N 74 27 E
Sør-Rondane, *Antarctica* 5 D4 72 0S 25 0 E
Sør-Trøndelag fylke □, *Norway* 8 E10 63 0N 9 30 E
Sorah, *Pakistan* 42 F3 27 13N 68 56 E
Sorata, *Bolivia* 92 G5 15 50S 68 40W
Sorel, *Canada* 70 C5 46 0N 73 10W
Soreq, N. →, *Israel* 47 D3 31 57N 34 43 E
Soria, *Spain* 19 B4 41 43N 2 32W
Soriano, *Uruguay* 94 C4 33 24S 58 19W
Sorkh, Kuh-e, *Iran* 45 C8 35 40N 58 30 E
Sorocaba, *Brazil* 95 A6 23 31S 47 27W
Sorochinsk, *Russia* 24 D9 52 26N 53 10 E
Soron, *India* 43 F8 27 55N 78 45 E
Sorong, *Indonesia* 37 E8 0 55S 131 15 E
Soroni, *Greece* 23 C10 36 21N 28 1 E
Soroti, *Uganda* 54 B3 1 43N 33 35 E
Sørøya, *Norway* 8 A17 70 40N 22 30 E
Sørøysundet, *Norway* 8 A17 70 25N 23 0 E
Sorrento, *Australia* 63 F3 38 22S 144 47 E
Sorrento, *Italy* 20 D6 40 38N 14 23 E
Sorsele, *Sweden* 8 D14 65 31N 17 30 E
Sorsogon, *Phil.* 37 B6 13 0N 124 0 E
Sortavala, *Russia* 24 B5 61 42N 30 41 E
Sŏsan, *S. Korea* 35 F14 36 47N 126 27 E
Soscumica, L., *Canada* 70 B4 50 15N 77 27W
Sosnogorsk, *Russia* 24 B9 63 37N 53 51 E
Sosnovka, *Russia* 27 D11 54 9N 109 35 E
Sosnowiec, *Poland* 17 C9 50 20N 19 10 E
Sŏsura, *N. Korea* 35 C16 42 16N 130 36 E
Sosva, *Russia* 24 C11 59 10N 61 50 E
Soto la Marina →, *Mexico* 87 C5 23 40N 97 40W
Sotuta, *Mexico* 87 C7 20 29N 89 43W
Souanké, *Congo* 52 D2 2 10N 14 3 E
Soúdha, *Greece* 23 D6 35 29N 24 4 E
Soúdhas, Kólpos, *Greece* 23 D6 35 25N 24 10 E
Soukhouma, *Laos* 38 E5 14 38N 105 48 E
Sŏul, *S. Korea* 35 F14 37 31N 126 58 E
Sound, The = Øresund, *Europe* 9 H12 55 45N 12 45 E
Sound, The, *U.K.* 11 G3 50 20N 4 10W
Sources, Mt. aux, *Lesotho* 57 D4 28 45S 28 50 E
Soure, *Brazil* 93 D9 0 35S 48 30W
Souris, *Man., Canada* 73 D8 49 40N 100 20W
Souris, *P.E.I., Canada* 71 C7 46 21N 62 15W
Souris →, *Canada* 80 A5 49 40N 99 34W
Sousa, *Brazil* 93 E11 6 45S 38 10W
Sousel, *Brazil* 93 D8 2 38S 52 29W
Sousse, *Tunisia* 51 A7 35 50N 10 38 E
South Africa ■, *Africa* 56 E3 32 0S 23 0 E
South Aulatsivik I., *Canada* 71 A7 56 45N 61 30W
South Australia □, *Australia* 63 E2 32 0S 139 0 E
South Baldy, *U.S.A.* 83 J10 33 59N 107 11W
South Bend, *Ind., U.S.A.* 76 E2 41 41N 86 15W
South Bend, *Wash., U.S.A.* 84 D3 46 40N 123 48W
South Boston, *U.S.A.* 77 G6 36 42N 78 54W
South Branch, *Canada* 71 C8 47 55N 59 2W
South Brook, *Canada* 71 C8 49 26N 56 5W
South Buganda □, *Uganda* 54 C3 0 15S 31 30 E
South Carolina □, *U.S.A.* 77 J5 34 0N 81 0W
South Charleston, *U.S.A.* 76 F5 38 22N 81 44W
South China Sea, *Asia* 36 C4 10 0N 113 0 E
South Dakota □, *U.S.A.* 80 C5 44 15N 100 0W
South Downs, *U.K.* 11 G7 50 53N 0 10W
South East C., *Australia* 62 G4 43 40S 146 50 E
South East Is., *Australia* 61 F3 34 17S 123 30 E
South Esk →, *U.K.* 12 E5 56 44N 3 3W
South Foreland, *U.K.* 11 F9 51 7N 1 23 E
South Fork →, *U.S.A.* 82 C7 47 54N 113 15W
South Fork, American →, *U.S.A.* 84 G5 38 45N 121 5W
South Fork, Feather →, *U.S.A.* 84 F5 39 17N 121 36W
South Georgia, *Antarctica* 5 B1 54 30S 37 0W
South Glamorgan □, *U.K.* 11 F4 51 30N 3 20W
South Haven, *U.S.A.* 76 D2 42 24N 86 16W
South Henik, L., *Canada* 73 A9 61 30N 97 30W
South Honshu Ridge, *Pac. Oc.* 64 E6 23 0N 143 0 E
South Horr, *Kenya* 54 B4 2 12N 36 56 E
South I., *Kenya* 54 B4 2 35N 36 35 E
South I., *N.Z.* 59 L3 44 0S 170 0 E
South Invercargill, *N.Z.* 59 M2 46 26S 168 23 E
South Knife →, *Canada* 73 B10 58 55N 94 37W
South Korea ■, *Asia* 35 F15 36 0N 128 0 E
South Lake Tahoe, *U.S.A.* 84 G6 38 57N 119 59W
South Loup →, *U.S.A.* 80 E5 41 4N 98 39W
South Magnetic Pole, *Antarctica* 5 C9 64 8S 138 8 E
South Milwaukee, *U.S.A.* 76 D2 42 55N 87 52W
South Molton, *U.K.* 11 F4 51 1N 3 50W
South Nahanni →, *Canada* 72 A4 61 3N 123 21W
South Natuna Is. = Natuna Selatan, Kepulauan, *Indonesia* 39 L7 2 45N 109 0 E
South Negril Pt., *Jamaica* 88 C4 18 14N 78 30W
South Orkney Is., *Antarctica* 5 C18 63 0S 45 0W
South Pagai, I. = Pagai Selatan, P., *Indonesia* 36 E2 3 0S 100 15 E
South Pass, *U.S.A.* 82 E9 42 20N 108 58W
South Pittsburg, *U.S.A.* 77 H3 35 1N 85 42W
South Platte →, *U.S.A.* 80 E4 41 7N 100 42W
South Pole, *Antarctica* 5 E 90 0S 0 0 E
South Porcupine, *Canada* 70 C3 48 30N 81 12W
South River, *Canada* 70 C4 45 52N 79 23W
South River, *U.S.A.* 79 F10 40 27N 74 23W
South Ronaldsay, *U.K.* 12 C6 58 46N 2 58W
South Sandwich Is., *Antarctica* 5 B1 57 0S 27 0W
South Saskatchewan →, *Canada* 73 C7 53 15N 105 5W
South Seal →, *Canada* 73 B9 58 48N 98 8W
South Shetland Is., *Antarctica* 5 C18 62 0S 59 0W
South Shields, *U.K.* 10 C6 54 59N 1 26W
South Sioux City, *U.S.A.* 80 D6 42 28N 96 24W
South Taranaki Bight, *N.Z.* 59 H5 39 40S 174 5 E
South Thompson →, *Canada* 72 C4 50 40N 120 20W
South Twin I., *Canada* 70 B4 53 7N 79 52W
South Tyne →, *U.K.* 10 C5 54 46N 2 25W
South Uist, *U.K.* 12 D1 57 20N 7 15W
South West Africa = Namibia ■, *Africa* 56 C2 22 0S 18 9 E
South West C., *Australia* 62 G4 43 34S 146 3 E
South Yorkshire □, *U.K.* 10 D6 53 30N 1 20W
Southampton, *Canada* 70 D3 44 30N 81 25W
Southampton, *U.K.* 11 G6 50 54N 1 23W
Southampton, *U.S.A.* 79 F12 40 53N 72 23W
Southampton I., *Canada* 69 B11 64 30N 84 0W
Southbridge, *N.Z.* 59 K4 43 48S 172 16 E
Southbridge, *U.S.A.* 79 D12 42 5N 72 2W
Southend, *Canada* 73 B8 56 19N 103 22W
Southend-on-Sea, *U.K.* 11 F8 51 32N 0 42 E
Southern □, *Malawi* 55 F4 15 0S 35 0 E
Southern □, *Zambia* 55 F2 16 20S 26 20 E
Southern Alps, *N.Z.* 59 K3 43 41S 170 11 E
Southern Cross, *Australia* 61 F2 31 12S 119 15 E
Southern Hills, *Australia* 61 F3 32 15S 122 40 E
Southern Indian L., *Canada* 73 B9 57 10N 98 30W
Southern Ocean, *Antarctica* 5 C6 62 0S 60 0 E
Southern Pines, *U.S.A.* 77 H6 35 11N 79 24W
Southern Uplands, *U.K.* 12 F5 55 30N 3 3W
Southington, *U.S.A.* 79 E12 41 36N 72 53W
Southold, *U.S.A.* 79 E12 41 4N 72 26W
Southport, *Australia* 63 D5 27 58S 153 25 E
Southport, *U.K.* 10 D4 53 38N 3 1W
Southport, *U.S.A.* 77 J6 33 55N 78 1W
Southwest C., *N.Z.* 59 M1 47 17S 167 28 E
Southwold, *U.K.* 11 E9 52 19N 1 41 E
Soutpansberg, *S. Africa* 57 C4 23 0S 29 30 E
Sovetsk, *Russia* 24 C3 55 6N 21 50 E
Sovetsk, *Russia* 24 C8 57 38N 48 53 E
Sovetskaya Gavan, *Russia* 27 E15 48 50N 140 5 E
Soviet Union = Commonwealth of Independent States ■, *Eurasia* 27 D11 60 0N 100 0 E
Soweto, *S. Africa* 57 D4 26 14S 27 54 E
Sôya-Kaikyô = La Perouse Str., *Asia* 30 B11 45 40N 142 0 E
Sôya-Misaki, *Japan* 30 B10 45 30N 141 55 E
Soyo, *Angola* 52 F2 6 13S 12 20 E
Sozh →, *Belorussia* 24 D5 51 57N 30 48 E
Spa, *Belgium* 15 D5 50 29N 5 53 E
Spain ■, *Europe* 19 C4 39 0N 4 0W
Spalding, *Australia* 63 E2 33 30S 138 37 E
Spalding, *U.K.* 10 E7 52 47N 0 9W
Spalding, *U.S.A.* 80 E5 41 42N 98 22W
Spangler, *U.S.A.* 78 F6 40 39N 78 48W
Spaniard's Bay, *Canada* 71 C9 47 38N 53 20W
Spanish, *Canada* 70 C3 46 12N 82 20W
Spanish Fork, *U.S.A.* 82 F8 40 7N 111 39W
Spanish Town, *Jamaica* 88 C4 18 0N 76 57W
Sparks, *U.S.A.* 84 F7 39 32N 119 45W
Sparta = Spárti, *Greece* 21 F10 37 5N 22 25 E
Sparta, *Ga., U.S.A.* 77 J4 33 17N 82 58W
Sparta, *Wis., U.S.A.* 80 D9 43 56N 90 49W
Spartanburg, *U.S.A.* 77 H4 34 56N 81 57W
Spartansburg, *U.S.A.* 78 E5 41 49N 79 41W
Spárti, *Greece* 21 F10 37 5N 22 25 E
Spartivento, C., *Calabria, Italy* 20 F7 37 56N 16 4 E
Spartivento, C., *Sard., Italy* 20 E3 38 52N 8 50 E
Spassk-Dalniy, *Russia* 27 E14 44 40N 132 48 E
Spátha, Ákra, *Greece* 23 D5 35 42N 23 43 E
Spatsizi →, *Canada* 72 B3 57 42N 128 7W
Spearfish, *U.S.A.* 80 C3 44 30N 103 52W
Spearman, *U.S.A.* 81 G4 36 12N 101 12W
Speers, *Canada* 73 C7 52 43N 107 34W
Speightstown, *Barbados* 89 D8 13 15N 59 39W
Speke Gulf, *Tanzania* 54 C3 2 20S 32 50 E
Spence Bay, *Canada* 68 B10 69 32N 93 32W
Spencer, *Idaho, U.S.A.* 82 D7 44 22N 112 11W
Spencer, *Iowa, U.S.A.* 80 D7 43 9N 95 9W
Spencer, *N.Y., U.S.A.* 79 D8 42 13N 76 30W
Spencer, *Nebr., U.S.A.* 80 D5 42 53N 98 42W
Spencer, *W. Va., U.S.A.* 76 F5 38 48N 81 21W
Spencer, C., *Australia* 63 F2 35 20S 136 53 E
Spencer B., *Namibia* 56 D1 25 30S 14 47 E
Spencer G., *Australia* 63 E2 34 0S 137 20 E
Spencerville, *Canada* 79 B9 44 51N 75 33W
Spences Bridge, *Canada* 72 C4 50 25N 121 20W
Spenser Mts., *N.Z.* 59 K4 42 15S 172 45 E
Sperrin Mts., *U.K.* 13 B5 54 50N 7 0W
Spessart, *Germany* 16 C4 50 10N 9 20 E
Spey →, *U.K.* 12 D5 57 26N 3 25W
Speyer, *Germany* 16 D4 49 19N 8 26 E
Spili, *Greece* 23 D6 35 13N 24 31 E
Spin Baldak = Qala-i-Jadid, *Afghan.* 42 D2 31 1N 66 25 E
Spinalónga, *Greece* 23 D7 35 18N 25 44 E
Spinazzola, *Italy* 20 D7 40 58N 16 5 E
Spirit Lake, *Idaho, U.S.A.* 82 C5 47 58N 116 52W
Spirit Lake, *Wash., U.S.A.* 84 D4 46 15N 122 9W
Spirit River, *Canada* 72 B5 55 45N 118 50W
Spiritwood, *Canada* 73 C7 53 24N 107 33W
Spithead, *U.K.* 11 G6 50 43N 1 5W
Spitzbergen = Svalbard, *Arctic* 4 B8 78 0N 17 0 E
Split, *Croatia* 20 C7 43 31N 16 26 E
Split L., *Canada* 73 B9 56 8N 96 15W
Splügenpass, *Switz.* 16 E4 46 30N 9 20 E
Spofford, *U.S.A.* 81 L4 29 10N 100 25W
Spokane, *U.S.A.* 82 C5 47 40N 117 24W
Spoleto, *Italy* 20 C5 42 46N 12 47 E
Spooner, *U.S.A.* 80 C9 45 50N 91 53W
Sporyy Navolok, Mys, *Russia* 26 B7 75 50N 68 40 E
Spragge, *Canada* 70 C3 46 15N 82 40W
Sprague, *U.S.A.* 82 C5 47 18N 117 59W
Sprague River, *U.S.A.* 82 E3 42 27N 121 30W
Spratly I., *S. China Sea* 36 C4 8 20N 112 0 E
Spray, *U.S.A.* 82 D4 44 50N 119 48W
Spree →, *Germany* 16 B6 52 32N 13 13 E
Spremberg, *Germany* 16 C7 51 33N 14 21 E
Spring City, *U.S.A.* 82 G8 39 29N 111 30W
Spring Garden, *U.S.A.* 84 F6 39 52N 120 47W
Spring Mts., *U.S.A.* 83 H6 36 0N 115 45W
Spring Valley, *Calif., U.S.A.* 85 N10 32 45N 117 5W
Spring Valley, *Minn., U.S.A.* 80 D8 43 41N 92 23W
Springbok, *S. Africa* 56 D2 29 42S 17 54 E
Springdale, *Canada* 71 C8 49 30N 56 6W
Springdale, *Ark., U.S.A.* 81 G7 36 11N 94 8W
Springdale, *Wash., U.S.A.* 82 B5 48 4N 117 45W
Springer, *U.S.A.* 81 G2 36 22N 104 36W
Springerville, *U.S.A.* 83 J9 34 8N 109 17W
Springfield, *Canada* 78 D4 42 50N 80 56W
Springfield, *N.Z.* 59 K3 43 19S 171 56 E
Springfield, *Colo., U.S.A.* 81 G3 37 24N 102 37W
Springfield, *Ill., U.S.A.* 80 F10 39 48N 89 39W
Springfield, *Mass., U.S.A.* 79 D12 42 6N 72 35W
Springfield, *Mo., U.S.A.* 81 G8 37 13N 93 17W
Springfield, *Ohio, U.S.A.* 76 F4 39 55N 83 49W
Springfield, *Oreg., U.S.A.* 82 D2 44 3N 123 1W
Springfield, *Tenn., U.S.A.* 77 G2 36 31N 86 53W
Springfield, *Vt., U.S.A.* 79 C12 43 18N 72 29W
Springfontein, *S. Africa* 56 E4 30 15S 25 40 E
Springhill, *Canada* 71 C7 45 40N 64 4W
Springhouse, *Canada* 72 C4 51 56N 122 7W
Springhurst, *Australia* 63 F4 36 10S 146 31 E
Springs, *S. Africa* 57 D4 26 13S 28 25 E
Springsure, *Australia* 62 C4 24 8S 148 6 E
Springvale, *Queens., Australia* 62 C3 23 33S 140 42 E
Springvale, *W. Austral., Australia* 60 C4 17 48S 127 41 E
Springvale, *U.S.A.* 79 C14 43 28N 70 48W
Springville, *Calif., U.S.A.* 84 J8 36 8N 118 49W
Springville, *N.Y., U.S.A.* 78 D6 42 31N 78 40W
Springville, *Utah, U.S.A.* 82 F8 40 10N 111 37W
Springwater, *Canada* 73 C7 51 58N 108 23W
Spruce-Creek, *U.S.A.* 78 F6 40 36N 78 9W
Spur, *U.S.A.* 81 J4 33 28N 100 52W
Spurn Hd., *U.K.* 10 D8 53 34N 0 8 E
Spuzzum, *Canada* 72 D4 49 37N 121 23W
Squam L., *U.S.A.* 79 C13 43 45N 71 32W
Squamish, *Canada* 72 D4 49 45N 123 10W
Square Islands, *Canada* 71 B8 52 47N 55 47W
Squires, Mt., *Australia* 61 E4 26 14S 127 28 E
Sragen, *Indonesia* 37 G14 7 26S 111 2 E
Srbija = Serbia □, *Yugoslavia* 21 C9 43 30N 21 0 E
Sre Khtum, *Cambodia* 39 F6 12 10N 106 52 E
Sre Umbell, *Cambodia* 39 G4 11 8N 103 46 E
Sredinny Ra. = Sredinnyy Khrebet, *Russia* 27 D16 57 0N 160 0 E
Sredinnyy Khrebet, *Russia* 27 D16 57 0N 160 0 E
Sredne Tambovskoye, *Russia* 27 D14 50 55N 137 45 E
Srednekolymsk, *Russia* 27 C16 67 27N 153 40 E
Srednevilyuysk, *Russia* 27 C13 63 50N 123 5 E
Srepok →, *Cambodia* 38 F6 13 33N 106 16 E
Sretensk, *Russia* 27 D12 52 10N 117 40 E
Sri Lanka ■, *Asia* 40 R12 7 30N 80 50 E
Srikakulam, *India* 41 K13 18 14N 83 58 E
Srinagar, *India* 43 B6 34 5N 74 50 E
Staaten →, *Australia* 62 B3 16 24S 141 17 E
Staðarhólskirkja, *Iceland* 8 D3 65 23N 21 58W
Stadlandet, *Norway* 8 E8 62 10N 5 10 E
Stadskanaal, *Neths.* 15 A6 53 4N 6 55 E
Stafafell, *Iceland* 8 D6 64 25N 14 52W
Staffa, *U.K.* 12 E2 56 26N 6 21W
Stafford, *U.K.* 10 E5 52 49N 2 9W
Stafford, *U.S.A.* 81 G5 37 58N 98 36W
Stafford Springs, *U.S.A.* 79 E12 41 57N 72 18W
Staffordshire □, *U.K.* 10 E5 52 53N 2 10W
Staines, *U.K.* 11 F7 51 26N 0 30W
Stakhanov = Kadiyevka, *Ukraine* 25 E6 48 35N 38 40 E
Stalingrad = Volgograd, *Russia* 25 E7 48 40N 44 25 E
Staliniri = Tskhinvali, *Georgia* 25 F7 42 14N 44 1 E
Stalino = Donetsk, *Ukraine* 25 E6 48 0N 37 45 E
Stalinogorsk = Novomoskovsk, *Russia* 24 D6 54 5N 38 15 E
Stalis, *Greece* 23 D7 35 17N 25 25 E
Stalybridge, *U.K.* 10 D5 53 29N 2 4W
Stamford, *Australia* 62 C3 21 15S 143 46 E
Stamford, *U.K.* 11 E7 52 39N 0 29W
Stamford, *Conn., U.S.A.* 79 E11 41 3N 73 32W
Stamford, *Tex., U.S.A.* 81 J5 32 57N 99 48W
Stamps, *U.S.A.* 81 J8 33 22N 93 30W
Stanberry, *U.S.A.* 80 E7 40 13N 94 35W
Standerton, *S. Africa* 57 D4 26 55S 29 7 E
Standish, *U.S.A.* 76 D4 43 59N 83 57W
Stanford, *U.S.A.* 82 C8 47 9N 110 13W
Stanger, *S. Africa* 57 D5 29 27S 31 14 E
Stanislaus →, *U.S.A.* 84 H5 37 40N 121 14W
Stanislav = Ivano-Frankovsk, *Ukraine* 25 E3 48 40N 24 40 E
Stanke Dimitrov, *Bulgaria* 21 C10 42 17N 23 9 E
Stanley, *Australia* 62 G4 40 46S 145 19 E
Stanley, *N.B., Canada* 71 C6 46 20N 66 44W
Stanley, *Sask., Canada* 73 B8 55 24N 104 22W
Stanley, *Falk. Is.* 96 G5 51 40S 59 51W
Stanley, *Idaho, U.S.A.* 82 D6 44 13N 114 56W
Stanley, *N. Dak., U.S.A.* 80 A3 48 19N 102 23W
Stanley, *N.Y., U.S.A.* 78 D7 42 48N 77 6W
Stanley, *Wis., U.S.A.* 80 C9 44 58N 90 56W
Stanovoy Khrebet, *Russia* 27 D13 55 0N 130 0 E
Stanovoy Ra. = Stanovoy Khrebet, *Russia* 27 D13 55 0N 130 0 E
Stansmore Ra., *Australia* 60 D4 21 23S 128 33 E
Stanthorpe, *Australia* 63 D5 28 36S 151 59 E
Stanton, *U.S.A.* 81 J4 32 8N 101 48W
Stanwood, *U.S.A.* 84 B4 48 15N 122 23W
Staples, *U.S.A.* 80 B7 46 21N 94 48W
Stapleton, *U.S.A.* 80 E4 41 29N 100 31W
Star City, *Canada* 73 C8 52 50N 104 20W
Stara Planina, *Bulgaria* 21 C10 43 15N 23 0 E
Stara Zagora, *Bulgaria* 21 C11 42 26N 25 39 E
Staraya Russa, *Russia* 24 C5 57 58N 31 23 E
Starbuck I., *Kiribati* 65 H12 5 37S 155 55W
Staritsa, *Russia* 24 C5 56 33N 34 55 E
Starke, *U.S.A.* 77 K4 29 57N 82 7W
Starkville, *Colo., U.S.A.* 81 G2 37 8N 104 30W
Starkville, *Miss., U.S.A.* 77 J1 33 28N 88 49W
Starogard, *Poland* 17 B9 53 59N 18 30 E
Start Pt., *U.K.* 11 G4 50 13N 3 38W
Staryy Kheydzhan, *Russia* 27 C15 60 0N 144 50 E
Staryy Oskol, *Russia* 24 D6 51 19N 37 55 E
State College, *U.S.A.* 78 F7 40 48N 77 52W
Stateline, *U.S.A.* 84 G7 38 57N 119 56W
Staten, I. = Estados, I. de Los, *Argentina* 96 G4 54 40S 64 30W
Staten I., *U.S.A.* 79 F10 40 35N 74 9W
Statesboro, *U.S.A.* 77 J5 32 27N 81 47W
Statesville, *U.S.A.* 77 H5 35 47N 80 53W
Stauffer, *U.S.A.* 85 L7 34 45N 119 3W
Staunton, *Ill., U.S.A.* 80 F10 39 1N 89 47W
Staunton, *Va., U.S.A.* 76 F6 38 9N 79 4W
Stavanger, *Norway* 9 G8 58 57N 5 40 E
Staveley, *N.Z.* 59 K3 43 40S 171 32 E
Stavelot, *Belgium* 15 D5 50 23N 5 55 E
Staveren, *Neths.* 15 B5 52 53N 5 22 E
Stavern, *Norway* 9 G11 59 0N 10 1 E
Stavropol, *Russia* 25 E7 45 5N 42 0 E
Stavros, *Cyprus* 23 D11 35 1N 32 38 E
Stavrós, *Greece* 23 D6 35 12N 24 45 E
Stavros, Ákra, *Greece* 23 D6 35 26N 24 58 E
Stawell, *Australia* 63 F3 37 5S 142 47 E
Stawell →, *Australia* 62 C3 20 20S 142 55 E
Stayner, *Canada* 78 B4 44 25N 80 5W
Steamboat Springs, *U.S.A.* 82 F10 40 29N 106 50W
Steele, *U.S.A.* 80 B5 46 51N 99 55W
Steelton, *U.S.A.* 78 F8 40 14N 76 50W
Steelville, *U.S.A.* 81 G9 37 58N 91 22W
Steen River, *Canada* 72 B5 59 40N 117 12W
Steenkool = Bintuni, *Indonesia* 37 E8 2 7S 133 32 E
Steenwijk, *Neths.* 15 B6 52 47N 6 7 E
Steep Pt., *Australia* 61 E1 26 8S 113 8 E
Steep Rock, *Canada* 73 C9 51 30N 98 48W
Stefanie L. = Chew Bahir, *Ethiopia* 51 H12 4 40N 36 50 E
Stefansson Bay, *Antarctica* 5 C5 67 20S 59 8 E
Steiermark □, *Austria* 16 E7 47 26N 15 0 E
Steilacoom, *U.S.A.* 84 C4 47 10N 122 36W
Steinbach, *Canada* 73 D9 49 32N 96 40W
Steinfort, *Lux.* 15 E5 49 39N 5 55 E
Steinkjer, *Norway* 8 E11 63 59N 11 31 E
Steinkopf, *S. Africa* 56 D2 29 18S 17 43 E
Stellarton, *Canada* 71 C7 45 32N 62 30W
Stellenbosch, *S. Africa* 56 E2 33 58S 18 50 E
Stendal, *Germany* 16 B5 52 36N 11 50 E
Stensele, *Sweden* 8 D14 65 3N 17 8 E
Stepanakert = Khankendy, *Azerbaijan* 25 G8 39 40N 46 25 E
Stephen, *U.S.A.* 80 A6 48 27N 96 53W
Stephens Creek, *Australia* 63 E3 31 50S 141 30 E
Stephens I., *Canada* 72 C2 54 10N 130 45W
Stephenville, *Canada* 71 C8 48 31N 58 35W

Swainsboro

Swainsboro, *U.S.A.* **77 J4** 32 36N 82 20W
Swakopmund, *Namibia* **56 C1** 22 37S 14 30 E
Swale →, *U.K.* **10 C6** 54 5N 1 20W
Swan Hill, *Australia* **63 F3** 35 20S 143 33 E
Swan Hills, *Canada* **72 C5** 54 42N 115 24W
Swan Is., *W. Indies* **88 C3** 17 22N 83 57W
Swan L., *Canada* **73 C8** 52 30N 100 40W
Swan River, *Canada* **73 C8** 52 10N 101 16W
Swanage, *U.K.* **11 G6** 50 36N 1 59W
Swansea, *Australia* **63 E5** 33 3S 151 35 E
Swansea, *U.K.* **11 F4** 51 37N 3 57W
Swar →, *Pakistan* **43 B5** 34 40N 72 5 E
Swartberge, *S. Africa* **56 E3** 33 20S 22 0 E
Swartmodder, *S. Africa* **56 D3** 28 1S 20 32 E
Swartruggens, *S. Africa* **56 D4** 25 39S 26 42 E
Swastika, *Canada* **70 C3** 48 7N 80 6W
Swatow = Shantou, *China* **33 D6** 23 18N 116 40 E
Swaziland ■, *Africa* **57 D5** 26 30S 31 30 E
Sweden ■, *Europe* **9 H13** 57 0N 15 0 E
Sweet Home, *U.S.A.* **82 D2** 44 24N 122 44W
Sweetwater, *Nev., U.S.A.* **84 G7** 38 27N 119 9W
Sweetwater, *Tex., U.S.A.* **81 J4** 32 28N 100 25W
Sweetwater →, *U.S.A.* **82 E10** 42 31N 107 2W
Swellendam, *S. Africa* **56 E3** 34 1S 20 26 E
Świdnica, *Poland* **16 C8** 50 50N 16 30 E
Świebodzin, *Poland* **16 B7** 52 15N 15 31 E
Swift Current, *Canada* **73 C7** 50 20N 107 45W
Swiftcurrent →, *Canada* **73 C7** 50 38N 107 44W
Swilly, L., *Ireland* **13 A4** 55 12N 7 35W
Swindle, I., *Canada* **72 C3** 52 30N 128 35W
Swindon, *U.K.* **11 F6** 51 33N 1 47W
Swinemünde = Świnoujście, *Poland* **16 B7** 53 54N 14 16 E
Świnoujście, *Poland* **16 B7** 53 54N 14 16 E
Switzerland ■, *Europe* **16 E4** 46 30N 8 0 E
Swords, *Ireland* **13 C5** 53 27N 6 15W
Sydney, *Australia* **63 E5** 33 53S 151 10 E
Sydney, *Canada* **71 C7** 46 7N 60 7W
Sydney Mines, *Canada* **71 C7** 46 18N 60 15W
Sydprøven, *Greenland* **4 C5** 60 30N 45 35W
Sydra G. of = Surt, Khalīj, *Libya* **51 B8** 31 40N 18 30 E
Syktyvkar, *Russia* **24 B9** 61 45N 50 40 E
Sylacauga, *U.S.A.* **77 J2** 33 10N 86 15W
Sylarna, *Sweden* **8 E12** 63 2N 12 13 E
Sylhet, *Bangla.* **41 G17** 24 54N 91 52 E
Sylvan Lake, *Canada* **72 C6** 52 20N 114 3W
Sylvania, *U.S.A.* **77 J5** 32 45N 81 38W
Sylvester, *U.S.A.* **77 K4** 31 32N 83 50W
Sym, *Russia* **26 C9** 60 20N 88 18 E
Symón, *Mexico* **86 C4** 24 42N 102 35W
Synnott Ra., *Australia* **60 C4** 16 30S 125 20 E
Syracuse, *Kans., U.S.A.* **81 F4** 37 59N 101 45W
Syracuse, *N.Y., U.S.A.* **79 C8** 43 3N 76 9W
Syrdarya →, *Kazakhstan* **26 E7** 46 3N 61 0 E
Syria ■, *Asia* **44 C3** 35 0N 38 0 E
Syul'dzhyukyor, *Russia* **27 C12** 63 14N 113 32 E
Syzran, *Russia* **24 D8** 53 12N 48 30 E
Szczecin, *Poland* **16 B7** 53 27N 14 27 E
Szczecinek, *Poland* **16 B8** 53 43N 16 41 E
Szechwan = Sichuan □, *China* **32 C5** 31 0N 104 0 E
Szeged, *Hungary* **17 E10** 46 16N 20 10 E
Székesfehérvár, *Hungary* **17 E9** 47 15N 18 25 E
Szekszárd, *Hungary* **17 E9** 46 22N 18 42 E
Szentes, *Hungary* **17 E10** 46 39N 20 21 E
Szolnok, *Hungary* **17 E10** 47 10N 20 15 E
Szombathely, *Hungary* **16 E8** 47 14N 16 38 E

T

Ta Khli Khok, *Thailand* **38 E3** 15 18N 100 20 E
Ta Lai, *Vietnam* **39 G6** 11 24N 107 23 E
Tabacal, *Argentina* **94 A3** 23 15S 64 15W
Tabaco, *Phil.* **37 B6** 13 22N 123 44 E
Ţābah, *Si. Arabia* **44 E4** 26 55N 42 38 E
Tabarka, *Tunisia* **50 A6** 36 56N 8 46 E
Ţabas, *Khorāsān, Iran* **45 C9** 32 48N 60 12 E
Ţabas, *Khorāsān, Iran* **45 C8** 33 35N 56 55 E
Tabasará, Serranía de, *Panama* **88 E3** 8 35N 81 40W
Tabasco □, *Mexico* **87 D6** 17 45N 93 30W
Tabatinga, Serra da, *Brazil* **93 F10** 10 30S 44 0W
Tabāzīn, *Iran* **45 D8** 31 12N 57 54 E
Taber, *Canada* **72 D6** 49 47N 112 8W
Tablas, *Phil.* **37 B6** 12 25N 122 2 E
Table B. = Tafelbaai, *S. Africa* **56 E2** 33 35S 18 25 E
Table B., *Canada* **71 B8** 53 40N 56 25W
Table Mt., *S. Africa* **56 E2** 34 0S 18 22 E
Tableland, *Australia* **60 C4** 17 16S 126 51 E
Tabletop, Mt., *Australia* **62 C4** 23 24S 147 11 E
Tábor, *Czech.* **16 D7** 49 25N 14 39 E
Tabora, *Tanzania* **54 D3** 5 2S 32 50 E
Tabora □, *Tanzania* **54 D3** 5 0S 33 0 E
Tabou, *Ivory C.* **50 H3** 4 30N 7 20W
Tabrīz, *Iran* **44 B5** 38 7N 46 20 E
Tabuaeran, *Pac. Oc.* **65 G12** 3 51N 159 22W
Tabūk, *Si. Arabia* **44 D3** 28 23N 36 36 E
Tacámbaro de Codallos, *Mexico* **86 D4** 19 14N 101 28W
Tacheng, *China* **32 B3** 46 40N 82 58 E
Tach'ing Shan = Daqing Shan, *China* **34 D6** 40 40N 111 0 E
Tacloban, *Phil.* **37 B6** 11 15N 124 58 E
Tacna, *Peru* **92 G4** 18 0S 70 20W
Tacoma, *U.S.A.* **84 C4** 47 14N 122 26W
Tacuarembó, *Uruguay* **95 C4** 31 45S 56 0W
Tademaït, Plateau du, *Algeria* **50 C5** 28 30N 2 30 E
Tadjoura, *Djibouti* **46 E3** 11 50N 42 55 E
Tadmor, *N.Z.* **59 J4** 41 27S 172 45 E
Tadoule, L., *Canada* **73 B9** 58 36N 98 20W
Tadoussac, *Canada* **71 C6** 48 11N 69 42W
Tadzhikistan = Tajikistan ■, *Asia* **26 F8** 38 30N 70 0 E
Taechŏn-ni, *S. Korea* **35 F14** 36 21N 126 36 E
Taegu, *S. Korea* **35 G15** 35 50N 128 37 E
Taegwan, *N. Korea* **35 D13** 40 13N 125 12 E
Taejŏn, *S. Korea* **35 F14** 36 20N 127 28 E
Tafalla, *Spain* **19 A5** 42 30N 1 41W
Tafelbaai, *S. Africa* **56 E2** 33 35S 18 25 E
Tafermaar, *Indonesia* **37 F8** 6 47S 134 10 E
Tafí Viejo, *Argentina* **94 B2** 26 43S 65 17W
Tafīhān, *Iran* **45 D7** 29 25N 52 39 E
Taft, *Iran* **45 D7** 31 45N 54 14 E
Taft, *Phil.* **37 B7** 11 57N 125 30 E
Taft, *Calif., U.S.A.* **85 K7** 35 8N 119 28W
Taft, *Tex., U.S.A.* **81 M6** 27 59N 97 24W
Taga Dzong, *Bhutan* **41 F16** 27 5N 89 55 E
Taganrog, *Russia* **25 E6** 47 12N 38 50 E
Tagbilaran, *Phil.* **37 C6** 9 39N 123 51 E
Tagish, *Canada* **72 A2** 60 19N 134 16W
Tagish L., *Canada* **72 A2** 60 10N 134 20W
Tagliamento →, *Italy* **20 B5** 45 38N 13 5 E
Tagomago, I. de, *Spain* **22 B8** 39 2N 1 39 E
Taguatinga, *Brazil* **93 F10** 12 16S 42 26W
Tagum, *Phil.* **37 C7** 7 33N 125 53 E
Tagus = Tejo →, *Europe* **19 C1** 38 40N 9 24W
Tahakopa, *N.Z.* **59 M2** 46 30S 169 23 E
Tahan, Gunong, *Malaysia* **39 K4** 4 34N 102 17 E
Tahat, *Algeria* **50 D6** 23 18N 5 33 E
Tāherī, *Iran* **45 E7** 27 43N 52 20 E
Tahiti, *Pac. Oc.* **65 J13** 17 37S 149 27W
Tahoe, L., *U.S.A.* **84 G6** 39 6N 120 2W
Tahoe City, *U.S.A.* **84 F6** 39 10N 120 9W
Taholah, *U.S.A.* **84 C2** 47 21N 124 17W
Tahoua, *Niger* **50 F6** 14 57N 5 16 E
Tahta, *Egypt* **51 C11** 26 44N 31 32 E
Tahulandang, *Indonesia* **37 D7** 2 27N 125 23 E
Tahuna, *Indonesia* **37 D7** 3 38N 125 30 E
Taï, *Ivory C.* **50 G3** 5 55N 7 30W
Tai Shan, *China* **35 F9** 36 25N 117 20 E
Tai'an, *China* **35 F9** 36 12N 117 8 E
Taibei, *Taiwan* **33 D7** 25 4N 121 29 E
Taibique, *Canary Is.* **22 G2** 27 42N 17 58W
Taibus Qi, *China* **34 D8** 41 54N 115 22 E
T'aichung = Taizhong, *Taiwan* **33 D7** 24 12N 120 35 E
Taieri →, *N.Z.* **59 M3** 46 3S 170 12 E
Taigu, *China* **34 F7** 37 28N 112 30 E
Taihang Shan, *China* **34 G7** 36 0N 113 30 E
Taihape, *N.Z.* **59 H5** 39 41S 175 48 E
Taihe, *China* **34 H8** 33 20N 115 42 E
Taikang, *China* **34 G8** 34 5N 114 50 E
Tailem Bend, *Australia* **63 F2** 35 12S 139 29 E
Taimyr Peninsula = Taymyr, Poluostrov, *Russia* **27 B11** 75 0N 100 0 E
Tain, *U.K.* **12 D4** 57 49N 4 4W
Tainan, *Taiwan* **33 D7** 23 17N 120 18 E
Taínaron, Ákra, *Greece* **21 F10** 36 22N 22 27 E
T'aipei = Taibei, *Taiwan* **33 D7** 25 4N 121 29 E
T'aipei, *Taiwan* **33 D7** 25 2N 121 30 E
Taiping, *Malaysia* **39 K3** 4 51N 100 44 E
Taipingzhen, *China* **34 H6** 33 35N 111 42 E
Taita □, *Kenya* **54 C4** 4 0S 38 30 E
Taita Hills, *Kenya* **54 C4** 3 25S 38 15 E
Taitao, Pen. de, *Chile* **96 F2** 46 30S 75 0W
Taivalkoski, *Finland* **8 D20** 65 33N 28 12 E
Taiwan ■, *Asia* **33 D7** 23 30N 121 0 E
Taïyetos Óros, *Greece* **21 F10** 37 0N 22 23 E
Taiyiba, *Israel* **47 C4** 32 36N 35 27 E
Taiyuan, *China* **34 F7** 37 52N 112 33 E
Taizhong, *Taiwan* **33 D7** 24 12N 120 35 E
Ta'izz, *Yemen* **46 E3** 13 35N 44 2 E
Tājābād, *Iran* **45 D7** 30 2N 54 24 E
Tajikistan ■, *Asia* **26 F8** 38 30N 70 0 E
Tajima, *Japan* **31 F9** 37 12N 139 46 E
Tajo = Tejo →, *Europe* **19 C1** 38 40N 9 24W
Tajrīsh, *Iran* **45 C6** 35 48N 51 25 E
Tājūrā, *Libya* **51 B7** 32 51N 13 21 E
Tak, *Thailand* **38 D2** 16 52N 99 8 E
Takāb, *Iran* **44 B5** 36 24N 47 7 E
Takachiho, *Japan* **31 H5** 32 42N 131 18 E
Takada, *Japan* **31 F9** 37 7N 138 15 E
Takahagi, *Japan* **31 F10** 36 43N 140 45 E
Takaka, *N.Z.* **59 J4** 40 51S 172 50 E
Takamatsu, *Japan* **31 G7** 34 20N 134 5 E
Takaoka, *Japan* **31 F8** 36 47N 137 0 E
Takapuna, *N.Z.* **59 G5** 36 47S 174 47 E
Takasaki, *Japan* **31 F9** 36 20N 139 0 E
Takatsuki, *Japan* **31 G7** 34 51N 135 37 E
Takaungu, *Kenya* **54 C4** 3 38S 39 52 E
Takayama, *Japan* **31 F8** 36 18N 137 11 E
Take-Shima, *Japan* **31 J5** 30 49N 130 26 E
Takefu, *Japan* **31 G8** 35 50N 136 10 E
Takengon, *Indonesia* **36 D1** 4 45N 96 50 E
Takeo, *Cambodia* **39 G5** 10 59N 104 47 E
Takeo, *Japan* **31 H5** 33 12N 130 1 E
Tākestān, *Iran* **45 C6** 36 0N 49 40 E
Taketa, *Japan* **31 H5** 32 58N 131 24 E
Takh, *India* **43 C7** 33 6N 77 32 E
Takhman, *Cambodia* **39 G5** 11 29N 104 57 E
Takikawa, *Japan* **30 C10** 43 33N 141 54 E
Takla L., *Canada* **72 B3** 55 15N 125 45W
Takla Landing, *Canada* **72 B3** 55 30N 125 50W
Takla Makan = Taklamakan Shamo, *China* **28 F12** 38 0N 83 0 E
Taklamakan Shamo, *China* **28 F12** 38 0N 83 0 E
Taku →, *Canada* **72 B2** 58 30N 133 50W
Takum, *Nigeria* **50 G6** 7 18N 9 36 E
Tal Halāl, *Iran* **45 D7** 28 54N 55 1 E
Tala, *Uruguay* **95 C4** 34 21S 55 46W
Talagante, *Chile* **94 C1** 33 40S 70 50W
Talamanca, Cordillera de, *Cent. Amer.* **88 E3** 9 20N 83 20W
Talara, *Peru* **92 D2** 4 38S 81 18W
Talas, *Kirghizia* **26 E8** 42 30N 72 13 E
Talâta, *Egypt* **47 E1** 30 36N 32 20 E
Talaud, Kepulauan, *Indonesia* **37 D7** 4 30N 127 10 E
Talaud Is. = Talaud, Kepulauan, *Indonesia* **37 D7** 4 30N 127 10 E
Talavera de la Reina, *Spain* **19 C3** 39 55N 4 46W
Talawana, *Australia* **60 D3** 22 51S 121 9 E
Talayan, *Phil.* **37 C6** 6 52N 124 24 E
Talbot, C., *Australia* **60 B4** 13 48S 126 43 E
Talbragar →, *Australia* **63 E4** 32 12S 148 37 E
Talca, *Chile* **94 D1** 35 28S 71 40W
Talca □, *Chile* **94 D1** 35 20S 71 46W
Talcahuano, *Chile* **94 D1** 36 40S 73 10W
Talcher, *India* **41 J14** 21 0N 85 18 E
Taldy Kurgan, *Kazakhstan* **26 E8** 45 10N 78 45 E
Taldyqorghan = Taldy Kurgan, *Kazakhstan* **26 E8** 45 10N 78 45 E
Talesh, *Iran* **45 B6** 37 58N 48 58 E
Ţalesh, Kūhhā-ye, *Iran* **45 B6** 39 0N 48 30 E
Tali Post, *Sudan* **51 G11** 5 55N 30 44 E
Taliabu, *Indonesia* **37 E6** 1 45S 124 55 E
Talibon, *Phil.* **37 B6** 10 9N 124 20 E
Talibong, Ko, *Thailand* **39 J2** 7 15N 99 23 E
Talihina, *U.S.A.* **81 H7** 34 45N 95 3W
Taliwang, *Indonesia* **36 F5** 8 50S 116 55 E
Tall 'Asūr, *Jordan* **47 D4** 31 59N 35 17 E
Tall Kalakh, *Syria* **47 A5** 34 41N 36 15 E
Talladega, *U.S.A.* **77 J2** 33 26N 86 6W
Tallahassee, *U.S.A.* **77 K3** 30 27N 84 17W
Tallangatta, *Australia* **63 F4** 36 15S 147 19 E
Tallarook, *Australia* **63 F4** 37 5S 145 6 E
Tallering Pk., *Australia* **61 E2** 28 6S 115 37 E
Tallinn, *Estonia* **24 C3** 59 22N 24 48 E
Tallulah, *U.S.A.* **81 J9** 32 25N 91 11W
Talodi, *Sudan* **51 F11** 10 35N 30 22 E
Talpa de Allende, *Mexico* **86 C4** 20 23N 104 51W
Taltal, *Chile* **94 B1** 25 23S 70 33W
Taltson →, *Canada* **72 A6** 61 24N 112 46W
Talwood, *Australia* **63 D4** 28 29S 149 29 E
Talyawalka Cr. →, *Australia* **63 E3** 32 28S 142 22 E
Tam Chau, *Vietnam* **39 G5** 10 48N 105 12 E
Tam Ky, *Vietnam* **38 E7** 15 34N 108 29 E
Tam Quan, *Vietnam* **38 E7** 14 35N 109 3 E
Tama, *U.S.A.* **80 E8** 41 58N 92 35W
Tamala, *Australia* **61 E1** 26 42S 113 47 E
Tamale, *Ghana* **50 G4** 9 22N 0 50W
Tamano, *Japan* **31 G6** 34 29N 133 59 E
Tamanrasset, *Algeria* **50 D6** 22 50N 5 30 E
Tamaqua, *U.S.A.* **79 F9** 40 48N 75 58W
Tamar →, *U.K.* **11 G3** 50 33N 4 15W
Tamarang, *Australia* **63 E5** 31 27S 150 5 E
Tamarinda, *Spain* **22 B10** 39 55N 3 49 E
Tamashima, *Japan* **31 G6** 34 32N 133 40 E
Tamaské, *Niger* **50 F6** 14 49N 5 43 E
Tamaulipas □, *Mexico* **87 C5** 24 0N 99 0W
Tamaulipas, Sierra de, *Mexico* **87 C5** 23 30N 98 20W
Tamazula, *Mexico* **86 C3** 24 55N 106 58W
Tamazunchale, *Mexico* **87 C5** 21 16N 98 47W
Tambacounda, *Senegal* **50 F2** 13 45N 13 40W
Tambelan, Kepulauan, *Indonesia* **36 D3** 1 0N 107 30 E
Tambellup, *Australia* **61 F2** 34 4S 117 37 E
Tambo, *Australia* **62 C4** 24 54S 146 14 E
Tambo de Mora, *Peru* **92 F3** 13 30S 76 8W
Tambohorano, *Madag.* **57 B7** 17 30S 43 58 E
Tambora, *Indonesia* **36 F5** 8 12S 118 5 E
Tambov, *Russia* **24 D7** 52 45N 41 28 E
Tambuku, *Indonesia* **37 G15** 7 8S 113 40 E
Tamburâ, *Sudan* **51 G10** 5 40N 27 25 E
Tâmchekket, *Mauritania* **50 E2** 17 25N 10 40W
Tamega →, *Portugal* **19 B1** 41 5N 8 21W
Tamenglong, *India* **41 G18** 25 0N 93 35 E
Tamgak, Mts., *Niger* **50 E6** 19 12N 8 35 E
Tamiahua, L. de, *Mexico* **87 C5** 21 30N 97 30W
Tamil Nadu □, *India* **40 P10** 11 0N 77 0 E
Tamluk, *India* **43 H12** 22 18N 87 58 E
Tammerfors = Tampere, *Finland* **9 F17** 61 30N 23 50 E
Tammisaari, *Finland* **9 F17** 60 0N 23 26 E
Tamo Abu, Pegunungan, *Malaysia* **36 D5** 3 10N 115 5 E
Tampa, *U.S.A.* **77 M4** 27 57N 82 27W
Tampa B., *U.S.A.* **77 M4** 27 50N 82 30W
Tampere, *Finland* **9 F17** 61 30N 23 50 E
Tampico, *Mexico* **87 C5** 22 20N 97 50W
Tampin, *Malaysia* **39 L4** 2 28N 102 13 E
Tamrida = Qādib, *Yemen* **46 E5** 12 37N 53 57 E
Tamu, *Burma* **41 G19** 24 13N 94 12 E
Tamworth, *Australia* **63 E5** 31 7S 150 58 E
Tamworth, *U.K.* **11 E6** 52 38N 1 41W
Tamyang, *S. Korea* **35 G14** 35 19N 126 59 E
Tan An, *Vietnam* **39 G6** 10 32N 106 25 E
Tana, *Norway* **8 A20** 70 26N 28 14 E
Tana →, *Kenya* **54 C5** 2 32S 40 31 E
Tana →, *Norway* **8 A20** 70 30N 28 23 E
Tana, L., *Ethiopia* **51 F12** 13 5N 37 30 E
Tana River, *Kenya* **54 C4** 2 0S 39 30 E
Tanabe, *Japan* **31 H7** 33 44N 135 22 E
Tanafjorden, *Norway* **8 A20** 70 45N 28 25 E
Tanaga, Pta., *Canary Is.* **22 G1** 27 42N 18 10W
Tanahbala, *Indonesia* **36 E1** 0 30S 98 30 E
Tanahgrogot, *Indonesia* **36 E5** 1 55S 116 15 E
Tanahjampea, *Indonesia* **37 F6** 7 10S 120 35 E
Tanahmasa, *Indonesia* **36 E1** 0 12S 98 39 E
Tanahmerah, *Indonesia* **37 F10** 6 5S 140 16 E
Tanakura, *Japan* **31 F10** 37 10N 140 20 E
Tanami, *Australia* **60 C4** 19 59S 129 43 E
Tanami Desert, *Australia* **60 C5** 18 50S 132 0 E
Tanana, *U.S.A.* **68 B4** 65 10N 152 4W
Tanana →, *U.S.A.* **68 B4** 65 10N 151 58W
Tananarive = Antananarivo, *Madag.* **57 B8** 18 55S 47 31 E
Tánaro →, *Italy* **20 B3** 45 1N 8 47 E
Tanbar, *Australia* **62 D3** 25 51S 141 55 E
Tancheng, *China* **35 G10** 34 25N 118 20 E
Tanchŏn, *N. Korea* **35 D15** 40 27N 128 54 E
Tanda, *Ut. P., India* **43 F10** 26 33N 82 35 E
Tanda, *Ut. P., India* **43 E8** 28 57N 78 56 E
Tandag, *Phil.* **37 C7** 9 4N 126 9 E
Tandaia, *Tanzania* **55 D3** 9 25S 34 15 E
Tandaué, *Angola* **56 B2** 16 58S 18 5 E
Tandil, *Argentina* **94 D4** 37 15S 59 6W
Tandil, Sa. del, *Argentina* **94 D4** 37 30S 59 0W
Tandlianwala, *Pakistan* **42 D5** 31 3N 73 9 E
Tando Adam, *Pakistan* **42 G3** 25 45N 68 40 E
Tandou L., *Australia* **63 E3** 32 40S 142 5 E
Tane-ga-Shima, *Japan* **31 J5** 30 30N 131 0 E
Taneatua, *N.Z.* **59 H6** 38 4S 177 1 E
Tanen Tong Dan, *Burma* **38 D2** 16 30N 98 30 E
Tanezrouft, *Algeria* **50 D5** 23 9N 0 11 E
Tang, Koh, *Cambodia* **39 G4** 10 16N 103 7 E
Tang Krasang, *Cambodia* **38 F5** 12 34N 105 3 E
Tanga, *Tanzania* **54 D4** 5 5S 39 2 E
Tanga □, *Tanzania* **54 D4** 5 20S 38 0 E
Tanganyika, L., *Africa* **54 D2** 6 40S 30 0 E
Tanger, *Morocco* **50 A3** 35 50N 5 49W
Tangerang, *Indonesia* **37 G12** 6 11S 106 37 E
Tanggu, *China* **35 E9** 39 2N 117 40 E
Tanggula Shan, *China* **32 C4** 32 40N 92 10 E
Tanghe, *China* **34 H7** 32 47N 112 50 E
Tangier = Tanger, *Morocco* **50 A3** 35 50N 5 49W
Tangorin P.O., *Australia* **62 C3** 21 47S 144 12 E
Tangshan, *China* **35 E10** 39 38N 118 10 E
Tangtou, *China* **35 G10** 35 28N 118 30 E
Tanimbar, Kepulauan, *Indonesia* **37 F8** 7 30S 131 30 E
Tanimbar Is. = Tanimbar, Kepulauan, *Indonesia* **37 F8** 7 30S 131 30 E
Tanjay, *Phil.* **37 C6** 9 30N 123 5 E
Tanjong Malim, *Malaysia* **39 L3** 3 42N 101 31 E
Tanjore = Thanjavur, *India* **40 P11** 10 48N 79 12 E
Tanjung, *Indonesia* **36 E5** 2 10S 115 25 E
Tanjungbalai, *Indonesia* **36 D1** 2 55N 99 44 E
Tanjungbatu, *Indonesia* **36 D5** 2 23N 118 3 E
Tanjungkarang Telukbetung, *Indonesia* **36 F3** 5 20S 105 10 E
Tanjungpandan, *Indonesia* **36 E3** 2 43S 107 38 E
Tanjungpinang, *Indonesia* **36 D2** 1 5N 104 30 E
Tanjungpriok, *Indonesia* **37 G12** 6 8S 106 55 E
Tanjungredeb, *Indonesia* **36 D5** 2 9N 117 29 E
Tanjungselor, *Indonesia* **36 D5** 2 55N 117 25 E
Tank, *Pakistan* **42 C4** 32 14N 70 25 E
Tannu-Ola, *Russia* **27 D10** 51 0N 94 0 E
Tanout, *Niger* **50 F6** 14 50N 8 55 E
Tanta, *Egypt* **51 B11** 30 45N 30 57 E
Tantoyuca, *Mexico* **87 C5** 21 21N 98 10W
Tantung = Dandong, *China* **35 D13** 40 10N 124 20 E
Tanunda, *Australia* **63 E2** 34 30S 139 0 E
Tanzania ■, *Africa* **54 D3** 6 0S 34 0 E
Tanzilla →, *Canada* **72 B2** 58 8N 130 43W
Tao Ko, *Thailand* **39 G2** 10 5N 99 52 E
Tao'an, *China* **35 B12** 45 22N 122 40 E
Tao'er He →, *China* **35 B13** 45 45N 124 5 E
Taolanaro, *Madag.* **57 D8** 25 2S 47 0 E
Taole, *China* **34 E4** 38 48N 106 40 E
Taos, *U.S.A.* **83 H11** 36 24N 105 35W
Taoudenni, *Mali* **50 D4** 22 40N 3 55W
Taourirt, *Morocco* **50 B4** 34 25N 2 53W
Tapa Shan = Daba Shan, *China* **33 C5** 32 0N 109 0 E
Tapachula, *Mexico* **87 E6** 14 54N 92 17W
Tapah, *Malaysia* **39 K3** 4 12N 101 15 E
Tapajós →, *Brazil* **93 D8** 2 24S 54 41W
Tapaktuan, *Indonesia* **36 D1** 3 15N 97 10 E
Tapanui, *N.Z.* **59 L2** 45 56S 169 18 E
Tapauá →, *Brazil* **92 E6** 5 40S 64 21W
Tapeta, *Liberia* **50 G3** 6 29N 8 52W
Taphan Hin, *Thailand* **38 D3** 16 13N 100 26 E
Tapi →, *India* **40 J8** 21 8N 72 41 E
Tapirapecó, Serra, *Venezuela* **92 C6** 1 10N 65 0W
Tappahannock, *U.S.A.* **76 G7** 37 56N 76 52W
Tapuaenuku, Mt., *N.Z.* **59 J4** 42 0S 173 39 E
Tapul Group, *Phil.* **37 C6** 5 35N 120 50 E
Taqīābād, *Iran* **45 C8** 35 33N 59 11 E
Ţaqţaq, *Iraq* **44 C5** 35 53N 44 35 E
Taquara, *Brazil* **95 B5** 29 36S 50 46W
Taquari →, *Brazil* **92 G7** 19 15S 57 17W
Tara, *Australia* **63 D5** 27 17S 150 31 E
Tara, *Canada* **78 B3** 44 28N 81 9W
Tara, *Russia* **26 D8** 56 55N 74 24 E
Tara, *Zambia* **55 F2** 16 58S 26 45 E
Tara →, *Montenegro, Yug.* **21 C8** 43 21N 18 51 E
Tara →, *Russia* **26 D8** 56 42N 74 36 E
Tarabagatay, Khrebet, *Kazakhstan* **26 E9** 48 0N 83 0 E
Tarābulus, *Lebanon* **47 A4** 34 31N 35 50 E
Tarābulus, *Libya* **51 B7** 32 49N 13 7 E
Tarajalejo, *Canary Is.* **22 F5** 28 12N 14 7W
Tarakan, *Indonesia* **36 D5** 3 20N 117 35 E
Tarakit, Mt., *Kenya* **54 B4** 2 2N 35 10 E
Taralga, *Australia* **63 E4** 34 26S 149 52 E
Tarama-Jima, *Japan* **31 M2** 24 39N 124 42 E
Taranagar, *India* **42 E6** 28 43N 74 50 E
Taranaki □, *N.Z.* **59 H5** 39 25S 174 30 E
Taranga, *India* **42 H5** 23 56N 72 43 E
Taranga Hill, *India* **42 H5** 24 0N 72 40 E
Táranto, *Italy* **20 D7** 40 30N 17 11 E
Táranto, G. di, *Italy* **20 D7** 40 0N 17 15 E
Tarapacá, *Colombia* **92 D5** 2 56S 69 46W
Tarapacá □, *Chile* **94 A2** 20 45S 69 30W
Tararua Ra., *N.Z.* **59 J5** 40 45S 175 25 E
Tarauacá, *Brazil* **92 E4** 8 6S 70 48W
Tarauacá →, *Brazil* **92 E5** 6 42S 69 48W
Tarawera, *N.Z.* **59 H6** 39 2S 176 36 E
Tarawera L., *N.Z.* **59 H6** 38 13S 176 27 E
Tarbat Ness, *U.K.* **12 D5** 57 52N 3 48W
Tarbela Dam, *Pakistan* **42 B5** 34 8N 72 52 E
Tarbert, *Strath., U.K.* **12 F3** 55 55N 5 25W
Tarbert, *W. Isles, U.K.* **12 D2** 57 54N 6 49W
Tarbes, *France* **18 E4** 43 15N 0 3 E
Tarboro, *U.S.A.* **77 H7** 35 54N 77 32W
Tarbrax, *Australia* **62 C3** 21 7S 142 26 E
Tarcoola, *Australia* **63 E1** 30 44S 134 36 E
Tarcoon, *Australia* **63 E4** 30 15S 146 43 E
Taree, *Australia* **63 E5** 31 50S 152 30 E
Tarentaise, *France* **18 D7** 45 30N 6 35 E
Tarfaya, *Morocco* **50 C2** 27 55N 12 55W
Tarifa, *Spain* **19 D3** 36 1N 5 36W

Thekulthili L.

Tarija, Bolivia 94 A3 21 30S 64 40W
Tarija □, Bolivia 94 A3 21 30S 63 30W
Tariku →, Indonesia 37 E9 2 55S 138 26 E
Tarim Basin = Tarim Pendi, China 32 B3 40 0N 84 0 E
Tarim He →, China 32 C3 39 30N 88 30 E
Tarim Pendi, China 32 B3 40 0N 84 0 E
Tarime □, Tanzania 54 C3 1 15S 34 0 E
Taritatu →, Indonesia 37 E9 2 54S 138 27 E
Tarka →, S. Africa 56 E4 32 10S 26 0 E
Tarkastad, S. Africa 56 E4 32 0S 26 16 E
Tarkhankut, Mys, Ukraine 25 E5 45 25N 32 30 E
Tarko Sale, Russia 26 C8 64 55N 77 50 E
Tarkwa, Ghana 50 G4 5 20N 2 0W
Tarlac, Phil. 37 A6 15 29N 120 35 E
Tarlton Downs, Australia 62 C2 22 40S 136 45 E
Tarma, Peru 92 F3 11 25S 75 45W
Tarn □, France 18 E5 43 49N 2 8 E
Tarn →, France 18 D4 44 5N 1 6 E
Tarn-et-Garonne □, France 18 D4 44 8N 1 20 E
Tarnów, Poland 17 C10 50 3N 21 0 E
Tarnowskie Góry, Poland 17 C9 50 27N 18 54 E
Taroom, Australia 63 D4 25 36S 149 48 E
Taroudannt, Morocco 50 B3 30 30N 8 52W
Tarpon Springs, U.S.A. 77 L4 28 9N 82 45W
Tarragona, Spain 19 B6 41 5N 1 17 E
Tarrasa, Spain 19 B7 41 34N 2 1 E
Tarrytown, U.S.A. 79 E11 41 4N 73 52W
Tarshiha = Me'ona, Israel 47 B4 33 1N 35 15 E
Tarso Emissi, Chad 51 D8 21 27N 18 36 E
Tarsus, Turkey 25 G5 36 58N 34 55 E
Tartagal, Argentina 94 A3 22 30S 63 50W
Tartu, Estonia 24 C4 58 20N 26 44 E
Tarţūs, Syria 44 C2 34 55N 35 55 E
Tarumizu, Japan 31 J5 31 29N 130 42 E
Tarutao, Ko, Thailand 39 J2 6 33N 99 40 E
Tarutung, Indonesia 36 D1 2 0N 98 54 E
Tasāwah, Libya 51 C7 26 0N 13 30 E
Taschereau, Canada 70 C4 48 40N 78 40W
Taseko →, Canada 72 C4 52 8N 123 45W
Tash-Kumyr, Kirghizia 26 E8 41 40N 72 10 E
Tashauz, Turkmenistan 26 E6 41 49N 59 58 E
Tashi Chho Dzong = Thimphu, Bhutan 41 F16 27 31N 89 45 E
Tashkent, Uzbekistan 26 E7 41 20N 69 10 E
Tashtagol, Russia 26 D9 52 47N 87 53 E
Tasikmalaya, Indonesia 37 G13 7 18S 108 12 E
Tåsjön, Sweden 8 D13 64 15N 15 40 E
Taskan, Russia 27 C16 62 59N 150 20 E
Tasman B., N.Z. 59 J4 40 59S 173 25 E
Tasman Mts., N.Z. 59 J4 41 3S 172 25 E
Tasman Pen., Australia 62 G4 43 10S 148 0 E
Tasman Sea, Pac. Oc. 64 L8 36 0S 160 0 E
Tasmania □, Australia 62 G4 42 0S 146 30 E
Tasu Sd., Canada 72 C2 52 47N 132 2W
Tatar Republic □, Russia 24 C9 55 30N 51 30 E
Tatarsk, Russia 26 D8 55 14N 76 0 E
Tateyama, Japan 31 G9 35 0N 139 50 E
Tathlina L., Canada 72 A5 60 33N 117 39W
Tathra, Australia 63 F4 36 44S 149 59 E
Tatinnai L., Canada 73 A9 60 55N 97 40W
Tatnam, C., Canada 73 B10 57 16N 91 0W
Tatra = Tatry, Slovakia 17 D10 49 20N 20 0 E
Tatry, Slovakia 17 D10 49 20N 20 0 E
Tatsuno, Japan 31 G7 34 52N 134 33 E
Tatta, Pakistan 42 G2 24 42N 67 55 E
Tatuí, Brazil 95 A6 23 25S 47 53W
Tatum, U.S.A. 81 J3 33 16N 103 19W
Tat'ung = Datong, China 34 D7 40 6N 113 18 E
Tatvan, Turkey 25 G7 38 31N 42 15 E
Taubaté, Brazil 95 A6 23 0S 45 36W
Tauern, Austria 16 E6 47 15N 12 40 E
Taumarunui, N.Z. 59 H5 38 53S 175 15 E
Taumaturgo, Brazil 92 E4 8 54S 72 51W
Taung, S. Africa 56 D3 27 33S 24 47 E
Taungdwingyi, Burma 41 J19 20 1N 95 40 E
Taunggyi, Burma 41 J20 20 50N 97 0 E
Taungup, Burma 41 K19 18 51N 94 14 E
Taungup Pass, Burma 41 K19 18 40N 94 45 E
Taungup Taunggya, Burma 41 K18 18 20N 93 40 E
Taunsa Barrage, Pakistan 42 D4 30 42N 70 50 E
Taunton, U.K. 11 F4 51 1N 3 7W
Taunton, U.S.A. 79 E13 41 54N 71 6W
Taunus, Germany 16 C4 50 15N 8 20 E
Taupo, N.Z. 59 H6 38 41S 176 7 E
Taupo, L., N.Z. 59 H5 38 46S 175 55 E
Tauranga, N.Z. 59 G6 37 42S 176 11 E
Tauranga Harb., N.Z. 59 G6 37 30S 176 5 E
Taurianova, Italy 20 E7 38 22N 16 1 E
Taurus Mts. = Toros Dağlan, Turkey 25 G5 37 0N 32 30 E
Tavda, Russia 26 D7 58 7N 65 8 E
Tavda →, Russia 26 D7 59 20N 63 28 E
Taveta, Tanzania 54 C4 3 23S 37 37 E
Taveuni, Fiji 59 C9 16 51S 179 58W
Tavira, Portugal 19 D2 37 8N 7 40W
Tavistock, Canada 78 C4 43 19N 80 50W
Tavistock, U.K. 11 G3 50 33N 4 9W
Tavoy, Burma 38 E2 14 2N 98 12 E
Taw →, U.K. 11 F3 51 4N 4 11W
Tawas City, U.S.A. 76 C4 44 16N 83 31W
Tawau, Malaysia 36 D5 4 20N 117 55 E
Tawitawi, Phil. 37 B6 5 10N 120 0 E
Taxila, Pakistan 42 C5 33 42N 72 52 E
Tay →, U.K. 12 E5 56 37N 3 38W
Tay, Firth of, U.K. 12 E5 56 25N 3 8W
Tay, L., Australia 61 F3 32 55S 120 48 E
Tay, L., U.K. 12 E4 56 30N 4 10W
Tay Ninh, Vietnam 39 G6 11 20N 106 5 E
Tayabamba, Peru 92 E3 8 15S 77 16W
Taylakovy, Russia 26 D8 59 13N 74 0 E
Taylor, Canada 72 B4 56 13N 120 40W
Taylor, Nebr., U.S.A. 80 E5 41 46N 99 23W
Taylor, Pa., U.S.A. 79 E9 41 23N 75 43W
Taylor, Tex., U.S.A. 81 K6 30 34N 97 25W
Taylor, Mt., U.S.A. 83 J10 35 14N 107 37W
Taylorville, U.S.A. 80 F10 39 33N 89 18W
Taymā, Si. Arabia 44 E3 27 35N 38 45 E
Taymyr, Oz., Russia 27 B11 74 20N 102 0 E
Taymyr, Poluostrov, Russia 27 B11 75 0N 100 0 E
Tayport, U.K. 12 E6 56 27N 2 52W
Tayshet, Russia 27 D10 55 58N 98 1 E
Tayside □, U.K. 12 E5 56 25N 3 30W
Taytay, Phil. 37 B5 10 45N 119 30 E
Taz →, Russia 26 C8 67 32N 78 40 E
Taza, Morocco 50 B4 34 16N 4 6W
Tāzah Khurmātū, Iraq 44 C5 35 18N 44 20 E
Tazawa-Ko, Japan 30 E10 39 43N 140 40 E
Tazin L., Canada 73 B7 59 44N 108 42W
Tazovskiy, Russia 26 C8 67 30N 78 44 E
Tbilisi, Georgia 25 F7 41 43N 44 50 E
Tchad = Chad ■, Africa 51 E8 15 0N 17 15 E
Tchad, L., Chad 51 F7 13 30N 14 30 E
Tch'eng-tou = Chengdu, China 32 C5 30 38N 104 2 E
Tchentlo L., Canada 72 B4 55 15N 125 0W
Tchibanga, Gabon 52 E2 2 45S 11 0 E
Tch'ong-k'ing = Chongqing, China 32 D5 29 35N 106 25 E
Te Anau, L., N.Z. 59 L1 45 15S 167 45 E
Te Aroha, N.Z. 59 G5 37 32S 175 44 E
Te Awamutu, N.Z. 59 H5 38 1S 175 20 E
Te Kuiti, N.Z. 59 H5 38 20S 175 11 E
Te Puke, N.Z. 59 G6 37 46S 176 22 E
Te Waewae B., N.Z. 59 M1 46 13S 167 33 E
Tea Tree, Australia 62 C1 22 5S 133 22 E
Teague, U.S.A. 81 K6 31 38N 96 17W
Teapa, Mexico 87 D6 18 35N 92 56W
Tebakang, Malaysia 36 D4 1 6N 110 30 E
Tébessa, Algeria 50 A6 35 22N 8 8 E
Tebicuary →, Paraguay 94 B4 26 36S 58 16W
Tebingtinggi, Indonesia 36 D1 3 20N 99 9 E
Tecate, Mexico 85 N10 32 34N 116 38W
Tecomán, Mexico 86 D4 18 55N 103 53W
Tecopa, U.S.A. 85 K10 35 51N 116 13W
Tecoripa, Mexico 86 B3 28 37N 109 57W
Tecuala, Mexico 86 C3 22 23N 105 27W
Tecuci, Romania 17 F13 45 51N 27 27 E
Tecumseh, U.S.A. 76 D4 42 0N 83 57W
Tedzhen, Turkmenistan 26 F7 37 23N 60 31 E
Tees →, U.K. 10 C6 54 36N 1 25W
Teesside, U.K. 10 C6 54 37N 1 13W
Teeswater, Canada 78 C3 43 59N 81 17W
Tefé, Brazil 92 D6 3 25S 64 50W
Tegal, Indonesia 37 G13 6 52S 109 8 E
Tegelen, Neths. 15 C6 51 20N 6 9 E
Teghra, India 43 G11 25 30N 85 34 E
Tegid, L. = Bala, L., U.K. 10 E4 52 53N 3 38W
Tegina, Nigeria 50 F6 10 5N 6 11 E
Tegucigalpa, Honduras 88 D2 14 5N 87 14W
Tehachapi, U.S.A. 85 K8 35 8N 118 27W
Tehachapi Mts., U.S.A. 85 L8 35 0N 118 30W
Tehrān, Iran 45 C6 35 44N 51 30 E
Tehuacán, Mexico 87 D5 18 30N 97 30W
Tehuantepec, Mexico 87 D5 16 21N 95 13W
Tehuantepec, G. de, Mexico 87 D5 15 50N 95 12W
Tehuantepec, Istmo de, Mexico 87 D6 17 0N 94 30W
Teide, Canary Is. 22 F3 28 15N 16 38W
Teifi →, U.K. 11 E3 52 4N 4 14W
Teign →, U.K. 11 G4 50 41N 3 42W
Teignmouth, U.K. 11 G4 50 33N 3 30W
Tejo →, Europe 19 C1 38 40N 9 24W
Tejon Pass, U.S.A. 85 L8 34 49N 118 53W
Tekamah, U.S.A. 80 E6 41 47N 96 13W
Tekapo, L., N.Z. 59 K3 43 53S 170 33 E
Tekax, Mexico 87 C7 20 11N 89 18W
Tekeli, Kazakhstan 26 E8 44 50N 79 0 E
Tekirdağ, Turkey 21 D12 40 58N 27 30 E
Tekkali, India 41 K14 18 37N 84 15 E
Tekoa, U.S.A. 82 C5 47 14N 117 4W
Tel Aviv-Yafo, Israel 47 C3 32 4N 34 48 E
Tel Lakhish, Israel 47 D3 31 34N 34 51 E
Tel Megiddo, Israel 47 C4 32 35N 35 11 E
Tela, Honduras 88 C2 15 40N 87 28W
Telanaipura = Jambi, Indonesia 36 E2 1 38S 103 30 E
Telavi, Georgia 25 F8 42 0N 45 30 E
Telde, Canary Is. 22 G4 27 59N 15 25W
Telegraph Creek, Canada 72 B2 58 0N 131 10W
Telemark fylke □, Norway 9 G10 59 25N 8 30 E
Telén, Argentina 94 D2 36 15S 65 31W
Teleng, Iran 45 E9 25 47N 61 3 E
Teles Pires →, Brazil 92 E7 7 21S 58 3W
Telescope Pk., U.S.A. 85 J9 36 10N 117 5W
Telford, U.K. 10 E5 52 42N 2 31W
Télimélé, Guinea 50 F2 10 54N 13 2W
Telkwa, Canada 72 C3 54 41N 127 5W
Tell City, U.S.A. 76 G2 37 57N 86 46W
Tellicherry, India 40 P9 11 45N 75 30 E
Telluride, U.S.A. 83 H10 37 56N 107 49W
Teloloapán, Mexico 87 D5 18 21N 99 51W
Telpos Iz, Russia 24 B10 63 35N 57 30 E
Telsen, Argentina 96 E3 42 30S 66 50W
Teluk Anson, Malaysia 39 K3 4 3N 101 0 E
Teluk Betung = Tanjungkarang Telukbetung, Indonesia 36 F3 5 20S 105 10 E
Teluk Intan = Teluk Anson, Malaysia 39 K3 4 3N 101 0 E
Telukbutun, Indonesia 39 K7 4 13N 108 12 E
Telukdalem, Indonesia 36 D1 0 33N 97 50 E
Tema, Ghana 50 G5 5 41N 0 0 E
Temanggung, Indonesia 37 G14 7 18S 110 10 E
Temapache, Mexico 87 C5 21 4N 97 38W
Temax, Mexico 87 C7 21 10N 88 50W
Temba, S. Africa 57 D4 25 20S 28 17 E
Tembe, Zaïre 54 C2 0 16S 28 14 E
Temblor Range, U.S.A. 85 K7 35 20N 119 50W
Teme →, U.K. 11 E5 52 23N 2 15W
Temecula, U.S.A. 85 M9 33 30N 117 9W
Temerloh, Malaysia 39 L4 3 27N 102 25 E
Temir, Kazakhstan 26 E6 49 21N 57 3 E
Temirtau, Kazakhstan 26 D8 50 5N 72 56 E
Temirtau, Russia 26 D9 53 10N 87 30 E
Témiscaming, Canada 70 C4 46 44N 79 5W
Temma, Australia 62 G3 41 12S 144 48 E
Temora, Australia 63 E4 34 30S 147 30 E
Temosachic, Mexico 86 B3 28 58N 107 50W
Tempe, U.S.A. 83 K8 33 25N 111 56W
Tempe Downs, Australia 60 D5 24 22S 132 24 E
Tempiute, U.S.A. 84 H11 37 39N 115 38W
Temple, U.S.A. 81 K6 31 6N 97 21W
Temple B., Australia 62 A3 12 15S 143 3 E
Templemore, Ireland 13 D4 52 48N 7 50W
Templeton, U.S.A. 84 K6 35 33N 120 42W
Templeton →, Australia 62 C2 21 0S 138 40 E
Tempoal, Mexico 87 C5 21 31N 98 23W
Temuco, Chile 96 D2 38 45S 72 40W
Temuka, N.Z. 59 L3 44 14S 171 17 E
Tenabo, Mexico 87 C6 20 2N 90 12W
Tenaha, U.S.A. 81 K7 31 57N 94 15W
Tenali, India 40 L12 16 15N 80 35 E
Tenancingo, Mexico 87 D5 19 0N 99 33W
Tenango, Mexico 87 D5 19 7N 99 33W
Tenasserim, Burma 39 F2 12 6N 99 3 E
Tenasserim □, Burma 38 F2 14 0N 98 30 E
Tenby, U.K. 11 F3 51 40N 4 42W
Tendaho, Ethiopia 46 E3 11 48N 40 54 E
Tenerife, Canary Is. 22 F3 28 15N 16 35W
Tenerife, Pico, Canary Is. 22 G1 27 43N 18 1W
Teng Xian, China 35 G9 35 5N 117 10 E
Tengah □, Indonesia 37 E6 2 0S 122 0 E
Tengah Kepulauan, Indonesia 36 F5 7 5S 118 15 E
Tengchong, China 32 D4 25 0N 98 28 E
Tengchowfu = Penglai, China 35 F11 37 48N 120 42 E
Tenggara □, Indonesia 37 E6 3 0S 122 0 E
Tenggarong, Indonesia 36 E5 0 24S 116 58 E
Tenggol, P., Malaysia 39 K4 4 48N 103 41 E
Tengiz, Ozero, Kazakhstan 26 D7 50 30N 69 0 E
Tenino, U.S.A. 84 D4 46 51N 122 51W
Tenkasi, India 40 Q10 8 55N 77 20 E
Tenke, Shaba, Zaïre 55 E2 11 22S 26 40 E
Tenke, Shaba, Zaïre 55 E2 10 32S 26 7 E
Tenkodogo, Burkina Faso 50 F4 11 54N 0 19W
Tennant Creek, Australia 62 B1 19 30S 134 15 E
Tennessee □, U.S.A. 77 H2 36 0N 86 30W
Tennessee →, U.S.A. 76 G1 37 4N 88 34W
Tennille, U.S.A. 77 J4 32 56N 82 48W
Teno, Pta. de, Canary Is. 22 F3 28 21N 16 55W
Tenom, Malaysia 36 C5 5 4N 115 57 E
Tenosique, Mexico 87 D6 17 30N 91 24W
Tenryū-Gawa →, Japan 31 G8 35 39N 137 48 E
Tent L., Canada 73 A7 62 25N 107 54W
Tenterfield, Australia 63 D5 29 0S 152 0 E
Teófilo Otoni, Brazil 93 G10 17 50S 41 30W
Teotihuacán, Mexico 87 D5 19 44N 98 50W
Tepa, Indonesia 37 F7 7 52S 129 31 E
Tepalcatepec →, Mexico 86 D4 18 35N 101 59W
Tepehuanes, Mexico 86 B3 25 21N 105 44W
Tepetongo, Mexico 86 C4 22 28N 103 9W
Tepic, Mexico 86 C4 21 30N 104 54W
Teplice, Czech. 16 C6 50 40N 13 48 E
Tepoca, C., Mexico 86 A2 30 20N 112 25W
Tequila, Mexico 86 C4 20 54N 103 47W
Ter →, Spain 19 A7 42 2N 3 12 E
Ter Apel, Neths. 15 B7 52 53N 7 5 E
Téra, Niger 50 F5 14 0N 0 45 E
Teraina, Kiribati 65 G11 4 43N 160 25W
Téramo, Italy 20 C5 42 40N 13 40 E
Terang, Australia 63 F3 38 15S 142 55 E
Tercero →, Argentina 94 C3 32 58S 61 47W
Terek →, Russia 25 F8 44 0N 47 30 E
Teresina, Brazil 93 E10 5 9S 42 45W
Terewah, L., Australia 63 D4 29 52S 147 35 E
Terhazza, Mali 50 D3 23 38N 5 22W
Teridgerie Cr. →, Australia 63 E4 30 25S 148 50 E
Termez, Uzbekistan 26 F7 37 15N 67 15 E
Términos, L. de, Mexico 87 D6 18 35N 91 30W
Térmoli, Italy 20 D6 42 0N 15 0 E
Ternate, Indonesia 37 D7 0 45N 127 25 E
Terneuzen, Neths. 15 C3 51 20N 3 50 E
Terney, Russia 27 E14 45 3N 136 37 E
Terni, Italy 20 C5 42 34N 12 38 E
Ternopol, Ukraine 25 E4 49 30N 25 40 E
Terowie, N.S.W., Australia 63 E4 32 27S 147 52 E
Terowie, S. Austral., Australia 63 E2 33 8S 138 55 E
Terra Bella, U.S.A. 85 K7 35 58N 119 3W
Terrace, Canada 72 C3 54 30N 128 35W
Terrace Bay, Canada 70 C2 48 47N 87 5W
Terracina, Italy 20 D5 41 17N 13 12 E
Terralba, Italy 20 E3 39 42N 8 38 E
Terranova = Ólbia, Italy 20 D3 40 55N 9 30 E
Terrassa = Tarrasa, Spain 19 B7 41 34N 2 1 E
Terre Haute, U.S.A. 76 F2 39 28N 87 25W
Terrebonne B., U.S.A. 81 L9 29 5N 90 35W
Terrell, U.S.A. 81 J6 32 44N 96 17W
Terrenceville, Canada 71 C9 47 40N 54 44W
Terrick Terrick, Australia 62 C4 24 44S 145 5 E
Terry, U.S.A. 80 B2 46 47N 105 19W
Terschelling, Neths. 15 A5 53 25N 5 20 E
Teruel, Spain 19 B5 40 22N 1 8W
Tervola, Finland 8 C18 66 6N 24 49 E
Teryaweyna L., Australia 63 E3 32 18S 143 22 E
Tešanj, Bos.-H. 21 B7 44 38N 18 1 E
Teshio, Japan 30 B10 44 53N 141 44 E
Teshio-Gawa →, Japan 30 B10 44 53N 141 45 E
Tesiyn Gol →, Mongolia 32 A4 50 40N 93 20 E
Teslin, Canada 72 A2 60 10N 132 43W
Teslin →, Canada 72 A2 61 34N 134 35W
Teslin L., Canada 72 A2 60 15N 132 57W
Tessalit, Mali 50 D5 20 12N 1 0 E
Tessaoua, Niger 50 F6 13 47N 7 56 E
Test →, U.K. 11 F6 51 7N 1 30W
Tetachuck L., Canada 72 C3 53 18N 125 55W
Tetas, Pta., Chile 94 A1 23 31S 70 38W
Tete, Mozam. 55 F3 16 13S 33 33 E
Tete □, Mozam. 55 F3 15 15S 32 40 E
Teteven, Bulgaria 21 C11 42 58N 24 17 E
Tethul →, Canada 72 A6 60 35N 112 12W
Teton →, U.S.A. 82 C8 47 56N 110 31W
Tétouan, Morocco 50 A3 35 35N 5 21W
Tetovo, Macedonia 21 C9 42 1N 21 2 E
Tetuán = Tétouan, Morocco 50 A3 35 35N 5 21W
Tetyukhe Pristan, Russia 30 B7 44 22N 135 48 E
Teuco →, Argentina 94 B3 25 35S 60 11W
Teulon, Canada 73 C9 50 23N 97 16W
Teun, Indonesia 37 F7 6 59S 129 8 E
Teutoburger Wald, Germany 16 B4 52 5N 8 20 E
Tevere →, Italy 20 D5 41 44N 12 14 E
Teverya, Israel 47 C4 32 47N 35 32 E
Teviot →, U.K. 12 F6 55 21N 2 51W
Tewantin, Australia 63 D5 26 27S 153 3 E
Tewkesbury, U.K. 11 F5 51 59N 2 8W
Texada I., Canada 72 D4 49 40N 124 25W
Texarkana, Ark., U.S.A. 81 J8 33 26N 94 2W
Texarkana, Tex., U.S.A. 81 J7 33 26N 94 3W
Texas, Australia 63 D5 28 49S 151 9 E
Texas □, U.S.A. 81 K5 31 40N 98 30W
Texas City, U.S.A. 81 L7 29 24N 94 54W
Texel, Neths. 15 A4 53 5N 4 50 E
Texhoma, U.S.A. 81 G4 36 30N 101 47W
Texline, U.S.A. 81 G3 36 23N 103 2W
Texoma, L., U.S.A. 81 J6 33 50N 96 34W
Tezin, Afghan. 42 B3 34 24N 69 30 E
Teziutlán, Mexico 87 D5 19 50N 97 22W
Tezpur, India 41 F18 26 40N 92 45 E
Tezzeron L., Canada 72 C4 54 43N 124 30W
Tha-anne →, Canada 73 A10 60 31N 94 37W
Tha Deua, Laos 38 D4 17 57N 102 53 E
Tha Deua, Laos 38 C3 19 26N 101 50 E
Tha Pla, Thailand 38 D3 17 48N 100 32 E
Tha Rua, Thailand 38 E3 14 34N 100 44 E
Tha Sala, Thailand 39 H2 8 40N 99 56 E
Tha Song Yang, Thailand 38 D1 17 34N 97 55 E
Thaba Nchu, S. Africa 56 D4 29 17S 26 52 E
Thaba Putsoa, Lesotho 57 D4 29 45S 28 0 E
Thabana Ntlenyana, Lesotho 57 D4 29 30S 29 16 E
Thabazimbi, S. Africa 57 C4 24 40S 27 21 E
Thai Binh, Vietnam 38 B6 20 35N 106 1 E
Thai Hoa, Vietnam 38 C5 19 20N 105 20 E
Thai Muang, Thailand 39 H2 8 24N 98 16 E
Thai Nguyen, Vietnam 38 B5 21 35N 105 55 E
Thailand ■, Asia 38 E4 16 0N 102 0 E
Thailand, G. of, Asia 39 G3 11 30N 101 0 E
Thakhek, Laos 38 D5 17 25N 104 45 E
Thal, Pakistan 42 C4 33 28N 70 33 E
Thal Desert, Pakistan 42 D4 31 10N 71 30 E
Thala La, Burma 41 E20 28 25N 97 23 E
Thalabarivat, Cambodia 38 F5 13 33N 105 57 E
Thallon, Australia 63 D4 28 39S 148 49 E
Thame →, U.K. 11 F6 51 35N 1 8W
Thames, N.Z. 59 G5 37 7S 175 34 E
Thames →, Canada 70 D3 42 20N 82 25W
Thames →, U.K. 11 F8 51 30N 0 35 E
Thames →, U.S.A. 79 E12 41 18N 72 5W
Thamesford, Canada 78 C3 43 4N 81 0W
Thamesville, Canada 78 D3 42 33N 81 59W
Than Uyen, Vietnam 38 B4 22 0N 103 54 E
Thane, India 40 K8 19 12N 72 59 E
Thanesar, India 42 D7 30 1N 76 52 E
Thanet, I. of, U.K. 11 F9 51 21N 1 20 E
Thangoo, Australia 60 C3 18 10S 122 22 E
Thangool, Australia 62 C5 24 38S 150 42 E
Thanh Hoa, Vietnam 38 C5 19 48N 105 46 E
Thanh Hung, Vietnam 39 H5 9 55N 105 43 E
Thanh Pho Ho Chi Minh = Phanh Bho Ho Chi Minh, Vietnam 39 G6 10 58N 106 40 E
Thanh Thuy, Vietnam 38 A5 22 55N 104 51 E
Thanjavur, India 40 P11 10 48N 79 12 E
Thap Sakae, Thailand 39 G2 11 30N 99 37 E
Thap Than, Thailand 38 E2 15 27N 99 54 E
Thar Desert, India 42 F4 28 0N 72 0 E
Tharad, India 42 G4 24 30N 71 44 E
Thargomindah, Australia 63 D3 27 58S 143 46 E
Tharrawaddy, Burma 41 L19 17 38N 95 48 E
Tharthar, W. →, Iraq 44 C4 33 59N 43 12 E
Thásos, Greece 21 D11 40 40N 24 40 E
That Khe, Vietnam 38 A6 22 16N 106 28 E
Thatcher, Ariz., U.S.A. 83 K9 32 51N 109 46W
Thatcher, Colo., U.S.A. 81 G2 37 33N 104 7W
Thaton, Burma 41 L20 16 55N 97 22 E
Thaungdut, Burma 41 G19 24 30N 94 40 E
Thayer, U.S.A. 81 G9 36 31N 91 33W
Thayetmyo, Burma 41 K19 19 20N 95 10 E
Thazi, Burma 41 J20 21 0N 96 5 E
The Alberga →, Australia 63 D2 27 6S 135 33 E
The Bight, Bahamas 89 B4 24 19N 75 24W
The Coorong, Australia 63 F2 35 50S 139 20 E
The Dalles, U.S.A. 82 D3 45 36N 121 10W
The English Company's Is., Australia 62 A2 11 50S 136 32 E
The Frome →, Australia 63 D2 29 8S 137 54 E
The Grampians, Australia 63 F3 37 0S 142 20 E
The Great Divide = Great Dividing Ra., Australia 62 C4 23 0S 146 0 E
The Hague = 's-Gravenhage, Neths. 15 B4 52 7N 4 17 E
The Hamilton →, Australia 63 D2 26 40S 135 19 E
The Macumba →, Australia 63 D2 27 52S 137 12 E
The Neales →, Australia 63 D2 28 8S 136 47 E
The Officer →, Australia 61 E5 27 46S 132 30 E
The Pas, Canada 73 C8 53 45N 101 15W
The Range, Zimbabwe 55 F3 19 2S 31 2 E
The Rock, Australia 63 F4 35 15S 147 2 E
The Salt L., Australia 63 E3 30 6S 142 8 E
The Stevenson →, Australia 63 D2 27 6S 135 33 E
The Warburton →, Australia 63 D2 28 4S 137 28 E
Thebes = Thívai, Greece 21 E10 38 19N 23 19 E
Thedford, Canada 78 C3 43 9N 81 51W
Thedford, U.S.A. 80 E4 41 59N 100 35W
Theebine, Australia 63 D5 25 57S 152 34 E
Thekulthili L., Canada 73 A7 61 3N 110 0W

Tungnafellsjökull

Toppenish, *U.S.A.* **82 C3** 46 23N 120 19W
Toraka Vestale, *Madag.* . **57 B7** 16 20S 43 58 E
Torata, *Peru* **92 G4** 17 23S 70 1W
Torbay, *Canada* **71 C9** 47 40N 52 42W
Torbay, *U.K.* **11 G4** 50 26N 3 31W
Tordesillas, *Spain* **19 B3** 41 30N 5 0W
Torfajökull, *Iceland* **8 E4** 63 54N 19 0W
Torgau, *Germany* ...... **16 C6** 51 32N 13 0 E
Torhout, *Belgium* ...... **15 C3** 51 5N 3 7 E
Tori-Shima, *Japan* ..... **31 J10** 30 29N 140 19 E
Torin, *Mexico* **86 B2** 27 33N 110 15W
Torino, *Italy* .... **20 B2** 45 4N 7 40 E
Torit, *Sudan* .... **51 H11** 4 27N 32 31 E
Tormes →, *Spain* **19 B2** 41 18N 6 29W
Tornado Mt., *Canada* ... **72 D6** 49 55N 114 40W
Torne älv →, *Sweden* . **8 D18** 65 50N 24 12 E
Torneå = Tornio, *Finland* **8 D18** 65 50N 24 12 E
Torneträsk, *Sweden* .... **8 B15** 68 24N 19 15 E
Tornio, *Finland* .... **8 D18** 65 50N 24 12 E
Tornionjoki →, *Finland* **8 D18** 65 50N 24 12 E
Tornquist, *Argentina* ... **94 D3** 38 8S 62 15W
Toro, *Spain* .... **22 B11** 39 59N 4 8 E
Toro, Cerro del, *Chile* .. **94 B2** 29 10S 69 50W
Toro Pk., *U.S.A.* **85 M10** 33 34N 116 24W
Toronìos Kólpos, *Greece* **21 D10** 40 5N 23 30 E
Toronto, *Australia* ..... **63 E5** 33 0S 151 30 E
Toronto, *Canada* **70 D4** 43 39N 79 20W
Toronto, *U.S.A.* **78 F4** 40 28N 80 36W
Toropets, *Russia* .... **24 C5** 56 30N 31 40 E
Tororo, *Uganda* .... **54 B3** 0 45N 34 12 E
Toros Dağlan, *Turkey* .. **25 G5** 37 0N 32 30 E
Torquay, *Canada* **73 D8** 49 9N 103 30W
Torquay, *U.K.* **11 G4** 50 27N 3 31W
Torrance, *U.S.A.* **85 M8** 33 50N 118 19W
Tôrre de Moncorvo,
Portugal **19 B2** 41 12N 7 8W
Torre del Greco, *Italy* ... **20 D6** 40 47N 14 22 E
Torrelavega, *Spain* **19 A3** 43 20N 4 5W
Torremolinos, *Spain* ... **19 D3** 36 38N 4 30W
Torrens, L., *Australia* .. **63 E2** 31 0S 137 50 E
Torrens Cr. →, *Australia* **62 C4** 22 23S 145 9 E
Torrens Creek, *Australia* **62 C4** 20 48S 145 3 E
Torréon, *Mexico* **86 B4** 25 33N 103 25W
Torres, *Mexico* **86 B2** 28 46N 110 47W
Torres Strait, *Australia* . **64 H6** 9 50S 142 20 E
Torres Vedras, *Portugal* . **19 C1** 39 5N 9 15W
Torrevieja, *Spain* **19 D5** 37 59N 0 42W
Torrey, *U.S.A.* **83 G8** 38 18N 111 25W
Torridge →, *U.K.* **11 G3** 50 51N 4 10W
Torridon, L., *U.K.* **12 D3** 57 35N 5 50W
Torrington, *Conn., U.S.A.* **79 E11** 41 48N 73 7W
Torrington, *Wyo., U.S.A.* **80 D2** 42 4N 104 11W
Tortola, *Virgin Is.* **89 C7** 18 19N 64 45W
Tortosa, *Spain* .... **19 B6** 40 49N 0 31 E
Tortosa, C., *Spain* ...... **19 B6** 40 41N 0 52 E
Tortue, I. de la, *Haiti* ... **89 B5** 20 5N 72 57W
Torūd, *Iran* .... **45 C7** 35 25N 55 5 E
Toruń, *Poland* .... **17 B9** 53 2N 18 39 E
Tory I., *Ireland* **13 A3** 55 17N 8 12W
Tosa, *Japan* .... **31 H6** 33 24N 133 23 E
Tosa-Shimizu, *Japan* ... **31 H6** 32 52N 132 58 E
Tosa-Wan, *Japan* ...... **31 H6** 33 15N 133 30 E
Toscana, *Italy* .... **20 C4** 43 30N 11 5 E
Toshkent = Tashkent,
Uzbekistan .... **26 E7** 41 20N 69 10 E
Tostado, *Argentina* **94 B3** 29 15S 61 50W
Tostón, Pta. de,
Canary Is. **22 F5** 28 42N 14 2W
Tosu, *Japan* .... **31 H5** 33 22N 130 31 E
Toteng, *Botswana* ..... **56 C3** 20 22S 22 58 E
Totma, *Russia* .... **24 C7** 60 0N 42 40 E
Totnes, *U.K.* **11 G4** 50 26N 3 41W
Totonicapán, *Guatemala* **88 D1** 14 58N 91 12W
Totten Glacier, *Antarctica* **5 C8** 66 45S 116 10 E
Tottenham, *Australia* ... **63 E4** 32 14S 147 21 E
Tottenham, *Canada* **78 B5** 44 1N 79 49W
Tottori, *Japan* .... **31 G7** 35 30N 134 15 E
Tottori □, *Japan* .... **31 G7** 35 30N 134 12 E
Touba, *Ivory C.* **50 G3** 8 22N 7 40W
Toubkal, Djebel, *Morocco* **50 B3** 31 0N 8 0W
Tougan, *Burkina Faso* .. **50 F4** 13 11N 2 58W
Touggourt, *Algeria* ..... **50 B6** 33 6N 6 4 E
Tougué, *Guinea* **50 F2** 11 25N 11 50W
Toul, *France* .... **18 B6** 48 40N 5 53 E
Toulepleu, *Ivory C.* **50 G3** 6 32N 8 24W
Toulon, *France* .... **18 E6** 43 10N 5 55 E
Toulouse, *France* ...... **18 E4** 43 37N 1 27 E
Toummo, *Niger* .... **51 D7** 22 45N 14 8 E
Toungoo, *Burma* .... **41 K20** 19 0N 96 30 E
Touraine, *France* ...... **18 C4** 47 20N 0 30 E
Tourane = Da Nang,
Vietnam .... **38 D7** 16 4N 108 13 E
Tourcoing, *France* ..... **18 A5** 50 42N 3 10 E
Tournai, *Belgium* ...... **15 D3** 50 35N 3 25 E
Tournon, *France* ...... **18 D6** 45 4N 4 50 E
Tours, *France* .... **18 C4** 47 22N 0 40 E
Touwsrivier, *S. Africa* .. **56 E3** 33 20S 20 2 E
Towada, *Japan* .... **30 D10** 40 37N 141 13 E
Towada-Ko, *Japan* ..... **30 D10** 40 28N 140 55 E
Towamba, *Australia* .... **63 F4** 37 6S 149 43 E
Towanda, *U.S.A.* **79 E8** 41 46N 76 27W
Towang, *India* .... **41 F17** 27 37N 91 50 E
Tower, *U.S.A.* **80 B8** 47 48N 92 17W
Towerhill Cr. →,
Australia .... **62 C3** 22 28S 144 35 E
Towner, *U.S.A.* **80 A4** 48 21N 100 25W
Townsend, *U.S.A.* **82 C8** 46 19N 111 31W
Townshend I., *Australia* . **62 C5** 22 10S 150 31 E
Townsville, *Australia* ... **62 B4** 19 15S 146 45 E
Towson, *U.S.A.* **76 F7** 39 24N 76 36W
Toya-Ko, *Japan* .... **30 C10** 42 35N 140 51 E
Toyah, *U.S.A.* **81 K3** 31 19N 103 48W
Toyahvale, *U.S.A.* **81 K3** 30 57N 103 47W
Toyama, *Japan* .... **31 F8** 36 40N 137 15 E
Toyama □, *Japan* ...... **31 F8** 36 45N 137 30 E
Toyama-Wan, *Japan* .. **31 F8** 37 0N 137 30 E
Toyohashi, *Japan* ...... **31 G8** 34 45N 137 25 E
Toyokawa, *Japan* ...... **31 G8** 34 48N 137 27 E
Toyonaka, *Japan* ...... **31 G7** 34 50N 135 28 E
Toyooka, *Japan* .... **31 G7** 35 35N 134 48 E
Toyota, *Japan* .... **31 G8** 35 3N 137 7 E
Tozeur, *Tunisia* .... **50 B6** 33 56N 8 8 E
Tra On, *Vietnam* .... **39 H5** 9 58N 105 55 E
Trabzon, *Turkey* .... **25 F6** 41 0N 39 45 E
Tracadie, *Canada* **71 C7** 47 30N 64 55W
Tracy, *Calif., U.S.A.* **83 H3** 37 44N 121 26W
Tracy, *Minn., U.S.A.* **80 C7** 44 14N 95 37W
Tradovoye, *Russia* ..... **30 C6** 43 17N 132 5 E
Trafalgar, C., *Spain* **19 D2** 36 10N 6 2W
Trail, *Canada* **72 D5** 49 5N 117 40W
Trainor L., *Canada* **72 A4** 60 24N 120 17W
Trákhonas, *Cyprus* ..... **23 D12** 35 12N 33 21 E
Tralee, *Ireland* **13 D2** 52 16N 9 42W
Tralee B., *Ireland* **13 D2** 52 17N 9 55W
Tramore, *Ireland* **13 D4** 52 10N 7 10W
Tran Ninh, Cao Nguyen,
Laos .... **38 C4** 19 30N 103 10 E
Tranås, *Sweden* .... **9 G13** 58 3N 14 59 E
Trancas, *Argentina* **94 B2** 26 11S 65 20W
Trang, *Thailand* .... **39 J2** 7 33N 99 38 E
Trangahy, *Madag.* ..... **57 B7** 19 7S 44 31 E
Trangan, *Indonesia* .... **37 F8** 6 40S 134 20 E
Trangie, *Australia* ...... **63 E4** 32 4S 148 0 E
Trani, *Italy* .... **20 D7** 41 17N 16 24 E
Tranoroa, *Madag.* ..... **57 C8** 24 42S 45 4 E
Tranqueras, *Uruguay* ... **95 C4** 31 13S 55 45W
Trans Nzoia □, *Kenya* .. **54 B3** 1 0N 35 0 E
Transantarctic Mts.,
Antarctica **5 E12** 85 0S 170 0W
Transcaucasia =
Zakavkazye, *Asia* .... **25 F7** 42 0N 44 0 E
Transcona, *Canada* **73 D9** 49 55N 97 0W
Transilvania, *Romania* .. **21 B11** 45 19N 25 0 E
Transkei □, *S. Africa* .. **57 E4** 32 15S 28 15 E
Transvaal □, *S. Africa* .. **56 D4** 25 0S 29 0 E
Transylvania =
Transilvania, *Romania* **21 B11** 45 19N 25 0 E
Transylvanian Alps,
Romania .... **6 F10** 45 30N 25 0 E
Trápani, *Italy* .... **20 E5** 38 1N 12 30 E
Trapper Pk., *U.S.A.* **82 D6** 45 54N 114 18W
Traralgon, *Australia* .... **63 F4** 38 12S 146 34 E
Tras os Montes e Alto
Douro, *Portugal* **19 B2** 41 25N 7 20W
Trasimeno, L., *Italy* .... **20 C5** 43 10N 12 5 E
Trat, *Thailand* .... **39 F4** 12 14N 102 33 E
Traveller's L., *Australia* . **63 E3** 33 20S 142 0 E
Travers, Mt., *N.Z.* ..... **59 K4** 42 1S 172 45 E
Traverse City, *U.S.A.* ... **76 C3** 44 46N 85 38W
Travnik, *Bos.-H.* .... **21 B7** 44 17N 17 39 E
Trayning, *Australia* ..... **61 F2** 31 7S 117 40 E
Trébbia →, *Italy* ...... **20 B3** 45 4N 9 41 E
Trebinje, *Bos.-H.* ...... **21 C8** 42 44N 18 22 E
Třeboň, *Czech.* .... **16 D7** 48 59N 14 48 E
Tredegar, *U.K.* **11 F4** 51 47N 3 16W
Tregaron, *U.K.* **11 E4** 52 14N 3 56W
Tregrosse Is., *Australia* . **62 B5** 17 41S 150 43 E
Tréguier, *France* **18 B2** 48 47N 3 16W
Treherne, *Canada* **73 D9** 49 38N 98 42W
Treinta y Tres, *Uruguay* . **95 C5** 33 16S 54 17W
Trekveld, *S. Africa* ..... **56 E2** 30 35S 19 45 E
Trelew, *Argentina* **96 E3** 43 10S 65 20W
Trelleborg, *Sweden* ..... **9 J12** 55 20N 13 10 E
Tremonton, *U.S.A.* **82 F7** 41 43N 112 10W
Tremp, *Spain* .... **19 A6** 42 10N 0 52 E
Trenche →, *Canada* ... **70 C5** 47 46N 72 53W
Trenggalek, *Indonesia* .. **37 H14** 8 3S 111 43 E
Trenque Lauquen,
Argentina **94 D3** 36 5S 62 45W
Trent →, *U.K.* **10 D7** 53 33N 0 44W
Trentino-Alto Adige □,
Italy .... **20 A4** 46 30N 11 0 E
Trento, *Italy* .... **20 A4** 46 5N 11 8 E
Trenton, *Canada* **70 D4** 44 10N 77 34W
Trenton, *Mo., U.S.A.* ... **80 E8** 40 5N 93 37W
Trenton, *N.J., U.S.A.* ... **79 F10** 40 14N 74 46W
Trenton, *Nebr., U.S.A.* .. **80 E4** 40 11N 101 1W
Trenton, *Tenn., U.S.A.* .. **81 H10** 35 59N 88 56W
Trepassey, *Canada* **71 C9** 46 43N 53 25W
Tres Arroyos, *Argentina* **94 D3** 38 26S 60 20W
Três Corações, *Brazil* ... **95 A6** 21 44S 45 15W
Três Lagoas, *Brazil* **93 H8** 20 50S 51 43W
Tres Marías, *Mexico* **86 C3** 21 25N 106 28W
Tres Montes, C., *Chile* .. **96 F1** 46 50S 75 30W
Tres Pinos, *U.S.A.* **84 J5** 36 48N 121 19W
Três Pontas, *Brazil* **95 A6** 21 23S 45 29W
Tres Puentes, *Chile* **94 B1** 27 50S 70 15W
Tres Puntas, C.,
Argentina **96 F3** 47 0S 66 0W
Três Rios, *Brazil* **95 A7** 22 6S 43 15W
Tres Valles, *Mexico* **87 D5** 18 15N 96 8W
Treungen, *Norway* ..... **9 G10** 59 1N 8 31 E
Treviso, *Italy* .... **20 B5** 45 40N 12 15 E
Triabunna, *Australia* ... **62 G4** 42 30S 147 55 E
Triánda, *Greece* .... **23 C10** 36 25N 28 10 E
Triang, *Malaysia* .... **39 L4** 3 15N 102 26 E
Tribulation, C., *Australia* **62 B4** 16 5S 145 29 E
Tribune, *U.S.A.* **80 F4** 38 28N 101 45W
Trichinopoly =
Tiruchchirappalli, *India* **40 P11** 10 45N 78 45 E
Trichur, *India* .... **40 P10** 10 30N 76 18 E
Trida, *Australia* .... **63 E4** 33 1S 145 1 E
Trier, *Germany* .... **16 D3** 49 45N 6 37 E
Trieste, *Italy* .... **20 B5** 45 39N 13 45 E
Triglav, *Slovenia* ...... **20 A5** 46 21N 13 50 E
Tríkkala, *Greece* .... **21 E9** 39 34N 21 47 E
Trikomo, *Cyprus* .... **23 D12** 35 17N 33 52 E
Trikora, Puncak,
Indonesia .... **37 E9** 4 15S 138 45 E
Trim, *Ireland* **13 C5** 53 34N 6 48W
Trincomalee, *Sri Lanka* . **40 Q12** 8 38N 81 15 E
Trindade, I., *Atl. Oc.* **2 F8** 20 20S 29 50W
Trinidad, *Bolivia* **92 F6** 14 46S 64 50W
Trinidad, *Colombia* **92 B4** 5 25N 71 40W
Trinidad, *Cuba* **88 B3** 21 48N 80 0W
Trinidad, *Uruguay* **94 C4** 33 30S 56 50W
Trinidad, *U.S.A.* **81 G2** 37 10N 104 31W
Trinidad, *W. Indies* **89 D7** 10 30N 61 15W
Trinidad →, *Mexico* ... **87 D5** 17 49N 95 9W
Trinidad, I., *Argentina* .. **96 D4** 39 10S 62 0W
Trinidad & Tobago ■,
W. Indies **89 D7** 10 30N 61 20W
Trinity, *Canada* **71 C9** 48 59N 53 55W
Trinity, *U.S.A.* **81 K7** 30 57N 95 22W
Trinity →, *Calif., U.S.A.* **82 F2** 41 11N 123 42W
Trinity →, *Tex., U.S.A.* . **81 L7** 29 45N 94 43W
Trinity B., *Canada* **71 C9** 48 20N 53 10W
Trinity Range, *U.S.A.* ... **82 F4** 40 15N 118 45W
Trinkitat, *Sudan* .... **51 E12** 18 45N 37 51 E
Trion, *U.S.A.* **77 H3** 34 33N 85 19W
Tripoli = Tarābulus,
Lebanon .... **47 A4** 34 31N 35 50 E
Tripoli = Tarābulus,
Libya .... **51 B7** 32 49N 13 7 E
Trípolis, *Greece* .... **21 F10** 37 31N 22 25 E
Tripp, *U.S.A.* **80 D6** 43 13N 97 58W
Tripura □, *India* .... **41 H17** 24 0N 92 0 E
Tripylos, *Cyprus* .... **23 E11** 34 59N 32 41 E
Tristan da Cunha, *Atl. Oc.* **2 F9** 37 6S 12 20W
Trivandrum, *India* ...... **40 Q10** 8 41N 77 0 E
Trnava, *Slovakia* .... **17 D8** 48 23N 17 35 E
Trochu, *Canada* **72 C6** 51 50N 113 13W
Trodely I., *Canada* **70 B4** 52 15N 79 26W
Troglav, *Croatia* .... **20 C7** 43 56N 16 36 E
Troilus, L., *Canada* **70 B5** 50 50N 74 35W
Trois-Pistoles, *Canada* .. **71 C6** 48 5N 69 10W
Trois-Rivières, *Canada* .. **70 C5** 46 25N 72 34W
Troitsk, *Russia* .... **26 D7** 54 10N 61 35 E
Troitsko Pechorsk, *Russia* **24 B10** 62 40N 56 10 E
Trölladyngja, *Iceland* ... **8 D5** 64 54N 17 16W
Trollhättan, *Sweden* .... **9 G12** 58 17N 12 20 E
Troms fylke □, *Norway* . **8 B15** 68 56N 19 0 E
Tromsø, *Norway* .... **8 B15** 69 40N 18 56 E
Trona, *U.S.A.* **85 K9** 35 46N 117 23W
Tronador, *Argentina* **96 E2** 41 10S 71 50W
Trondheim, *Norway* .... **8 E11** 63 36N 10 25 E
Trondheimsfjorden,
Norway .... **8 E11** 63 35N 10 30 E
Troodos, *Cyprus* .... **23 E11** 34 55N 32 52 E
Troon, *U.K.* **12 F4** 55 33N 4 40W
Tropic, *U.S.A.* **83 H7** 37 37N 112 5W
Trossachs, The, *U.K.* **12 E4** 56 14N 4 24W
Trostan, *U.K.* **13 A5** 55 4N 6 10W
Trotternish, *U.K.* **12 D2** 57 32N 6 15W
Troup, *U.S.A.* **81 J7** 32 9N 95 7W
Trout →, *Canada* **72 A5** 61 19N 119 51W
Trout L., *N.W.T., Canada* **72 A4** 60 40N 121 14W
Trout L., *Ont., Canada* .. **73 C10** 51 20N 93 15W
Trout Lake, *Mich., U.S.A.* **70 C2** 46 12N 85 1W
Trout Lake, *Wash., U.S.A.* **84 E5** 46 0N 121 32W
Trout River, *Canada* **71 C8** 49 29N 58 8W
Trouville-sur-Mer, *France* **18 B4** 49 21N 0 5 E
Trowbridge, *U.K.* **11 F5** 51 18N 2 12W
Troy, *Ala., U.S.A.* **77 K3** 31 48N 85 58W
Troy, *Idaho, U.S.A.* **82 C5** 46 44N 116 46W
Troy, *Kans., U.S.A.* **80 F7** 39 47N 95 5W
Troy, *Mo., U.S.A.* **80 F9** 38 59N 90 59W
Troy, *Mont., U.S.A.* **82 B6** 48 28N 115 53W
Troy, *N.Y., U.S.A.* **79 D11** 42 44N 73 41W
Troy, *Ohio, U.S.A.* **76 E3** 40 2N 84 12W
Troyes, *France* .... **18 B6** 48 19N 4 3 E
Trucial States = United
Arab Emirates ■, *Asia* **45 F7** 23 50N 54 0 E
Truckee, *U.S.A.* **84 F6** 39 20N 120 11W
Trujillo, *Honduras* **88 C2** 16 0N 86 0W
Trujillo, *Peru* **92 E3** 8 6S 79 0W
Trujillo, *Spain* **19 C3** 39 28N 5 55W
Trujillo, *U.S.A.* **81 H2** 35 32N 104 42W
Trujillo, *Venezuela* **92 B4** 9 22N 70 38W
Truk, *Pac. Oc.* .... **64 G7** 7 25N 151 46 E
Trumann, *U.S.A.* **81 H9** 35 41N 90 31W
Trumbull, Mt., *U.S.A.* ... **83 H7** 36 25N 113 8W
Trundle, *Australia* ...... **63 E4** 32 53S 147 35 E
Trung-Phan, *Vietnam* ... **38 E7** 16 0N 108 0 E
Truro, *Canada* **71 C7** 45 21N 63 14W
Truro, *U.K.* **11 G2** 50 17N 5 2W
Truslove, *Australia* ..... **61 F3** 33 20S 121 45 E
Truth or Consequences,
U.S.A. **83 K10** 33 8N 107 15W
Trutnov, *Czech.* .... **16 C7** 50 37N 15 54 E
Tryon, *U.S.A.* **77 H4** 35 13N 82 14W
Tryonville, *U.S.A.* **78 E5** 41 42N 79 48W
Tsaratanana, *Madag.* ... **57 B8** 16 47S 47 39 E
Tsaratanana, Mt. de,
Madag. .... **57 A8** 14 0S 49 0 E
Tsau, *Botswana* .... **56 C3** 20 8S 22 22 E
Tselinograd, *Kazakhstan* **26 D8** 51 10N 71 30 E
Tsetserleg, *Mongolia* ... **32 B5** 47 36N 101 32 E
Tshabong, *Botswana* .... **56 D3** 26 2S 22 29 E
Tshane, *Botswana* ..... **56 C3** 24 5S 21 54 E
Tshela, *Zaïre* .... **52 E2** 4 57S 13 4 E
Tshesebe, *Botswana* .... **57 C4** 21 51S 27 32 E
Tshibeke, *Zaïre* .... **54 C2** 2 40S 28 35 E
Tshibinda, *Zaïre* .... **54 C2** 2 23S 28 43 E
Tshikapa, *Zaïre* .... **52 F4** 6 28S 20 48 E
Tshilenge, *Zaïre* .... **54 D1** 6 17S 23 48 E
Tshinsenda, *Zaïre* ...... **55 E2** 12 20S 28 0 E
Tshofa, *Zaïre* .... **54 D2** 5 13S 25 16 E
Tshwane, *Botswana* .... **56 C3** 22 24S 22 1 E
Tsigara, *Botswana* ...... **56 C4** 20 22S 25 54 E
Tsihombe, *Madag.* ..... **57 D8** 25 10S 45 41 E
Tsimlyansk Res. =
Tsimlyanskoye Vdkhr.,
Russia .... **25 E7** 48 0N 43 0 E
Tsimlyanskoye Vdkhr.,
Russia .... **25 E7** 48 0N 43 0 E
Tsinan = Jinan, *China* .. **34 F9** 36 38N 117 1 E
Tsineng, *S. Africa* ...... **56 D3** 27 5S 23 5 E
Tsinghai = Qinghai □,
China .... **32 C4** 36 0N 98 0 E
Tsingtao = Qingdao,
China .... **35 F11** 36 5N 120 20 E
Tsinjomitondraka,
Madag. .... **57 B8** 15 40S 47 8 E
Tsiroanomandidy,
Madag. .... **57 B8** 18 46S 46 2 E
Tsivory, *Madag.* .... **57 C8** 24 4S 46 5 E
Tskhinvali, *Georgia* ..... **25 F7** 42 14N 44 1 E
Tsna →, *Russia* .... **24 D7** 54 55N 41 58 E
Tso Moriri, L., *India* .... **43 C8** 32 50N 78 20 E
Tsodilo Hill, *Botswana* .. **56 B3** 18 49S 21 43 E
Tsogttsetsiy, *Mongolia* .. **34 C3** 43 43N 105 35 E
Tsolo, *S. Africa* .... **57 E4** 31 18S 28 37 E
Tsomo, *S. Africa* .... **57 E4** 32 0S 27 42 E
Tsu, *Japan* .... **31 G8** 34 45N 136 25 E
Tsu L., *Canada* **72 A6** 60 40N 111 52W
Tsuchiura, *Japan* ...... **31 F10** 36 5N 140 15 E
Tsugaru-Kaikyō, *Japan* . **30 D10** 41 35N 141 0 E
Tsumeb, *Namibia* ...... **56 B2** 19 9S 17 44 E
Tsumis, *Namibia* ...... **56 C2** 23 39S 17 29 E
Tsuruga, *Japan* .... **31 G8** 35 45N 136 2 E
Tsurugi-San, *Japan* .... **31 H7** 33 51N 134 6 E
Tsuruoka, *Japan* .... **30 E9** 38 44N 139 50 E
Tsushima, *Gifu, Japan* .. **31 G8** 35 10N 136 43 E
Tsushima, *Nagasaki,*
Japan .... **31 G4** 34 20N 129 20 E
Tual, *Indonesia* .... **37 F8** 5 38S 132 44 E
Tuam, *Ireland* **13 C3** 53 30N 8 50W
Tuamotu Arch. =
Tuamotu Is., *Pac. Oc.* . **65 J13** 17 0S 144 0W
Tuamotu Is., *Pac. Oc.* .. **65 J13** 17 0S 144 0W
Tuamotu Ridge, *Pac. Oc.* **65 K14** 20 0S 138 0W
Tuao, *Phil.* .... **37 A6** 17 55N 121 22 E
Tuapse, *Russia* .... **25 F6** 44 5N 39 10 E
Tuatapere, *N.Z.* .... **59 M1** 46 8S 167 41 E
Tuba City, *U.S.A.* **83 H8** 36 8N 111 14W
Tuban, *Indonesia* ...... **37 G15** 6 54S 112 3 E
Tubarão, *Brazil* **95 B6** 28 30S 49 0W
Tūbās, *Jordan* .... **47 C4** 32 20N 35 22 E
Tubau, *Malaysia* .... **36 D4** 3 10N 113 40 E
Tübingen, *Germany* .... **16 D4** 48 31N 9 4 E
Tubruq, *Libya* .... **51 B9** 32 7N 23 55 E
Tubuai Is., *Pac. Oc.* **65 K12** 25 0S 150 0W
Tuc Trung, *Vietnam* .... **39 G6** 11 1N 107 12 E
Tucacas, *Venezuela* **92 A5** 10 48N 68 19W
Tuchodi →, *Canada* ... **72 B4** 58 17N 123 42W
Tucson, *U.S.A.* **83 K8** 32 13N 110 58W
Tucumán □, *Argentina* . **94 B2** 26 48S 66 2W
Tucumcari, *U.S.A.* **81 H3** 35 10N 103 44W
Tucupita, *Venezuela* **92 B6** 9 2N 62 3W
Tucuruí, *Brazil* **93 D9** 3 42S 49 44W
Tudela, *Spain* **19 A5** 42 4N 1 39W
Tudmur, *Syria* .... **44 C3** 34 36N 38 15 E
Tudor, L., *Canada* **71 A6** 55 50N 65 25W
Tuen, *Australia* .... **63 D4** 28 33S 145 37 E
Tugela →, *S. Africa* ... **57 D5** 29 14S 31 30 E
Tuguegarao, *Phil.* ...... **37 A6** 17 35N 121 42 E
Tugur, *Russia* .... **27 D14** 53 44N 136 45 E
Tuineje, *Canary Is.* **22 F5** 28 19N 14 3W
Tukangbesi, Kepulauan,
Indonesia .... **37 F6** 6 0S 124 0 E
Tukarak I., *Canada* **70 A4** 56 15N 78 45W
Tukayyid, *Iraq* .... **44 D5** 29 47N 45 36 E
Tūkrah, *Libya* .... **51 B9** 32 30N 20 37 E
Tuktoyaktuk, *Canada* ... **68 B6** 69 27N 133 2W
Tukuyu, *Tanzania* ...... **55 D3** 9 17S 33 35 E
Tula, *Hidalgo, Mexico* ... **87 C5** 20 5N 99 20W
Tula, *Tamaulipas, Mexico* **87 C5** 23 0N 99 40W
Tula, *Russia* .... **24 D6** 54 13N 37 38 E
Tulancingo, *Mexico* **87 C5** 20 5N 99 22W
Tulare, *U.S.A.* **83 H4** 36 13N 119 21W
Tulare Lake Bed, *U.S.A.* . **83 J4** 36 0N 119 48W
Tularosa, *U.S.A.* **83 K10** 33 5N 106 1W
Tulbagh, *S. Africa* ...... **56 E2** 33 16S 19 6 E
Tulcán, *Ecuador* **92 C3** 0 48N 77 43W
Tulcea, *Romania* .... **21 B13** 45 13N 28 46 E
Tūleh, *Iran* .... **45 C7** 34 35N 52 33 E
Tulemalu L., *Canada* **73 A9** 62 58N 99 25W
Tuli, *Indonesia* .... **37 E6** 1 24S 122 26 E
Tuli, *Zimbabwe* .... **55 G2** 21 58S 29 13 E
Tulia, *U.S.A.* **81 H4** 34 32N 101 46W
Ţulkarm, *Jordan* .... **47 C4** 32 19N 35 2 E
Tullahoma, *U.S.A.* **77 H2** 35 22N 86 13W
Tullamore, *Australia* .... **63 E4** 32 39S 147 36 E
Tullamore, *Ireland* **13 C4** 53 17N 7 30W
Tulle, *France* .... **18 D4** 45 16N 1 46 E
Tullibigeal, *Australia* ... **63 E4** 33 25S 146 44 E
Tullow, *Ireland* **13 D4** 52 48N 6 45W
Tully, *Australia* .... **62 B4** 17 56S 145 55 E
Ţulmaythah, *Libya* ..... **51 B9** 32 40N 20 55 E
Tulmur, *Australia* ...... **62 C3** 22 40S 142 20 E
Tulsa, *U.S.A.* **81 G7** 36 10N 95 55W
Tulsequah, *Canada* **72 B2** 58 39N 133 35W
Tulua, *Colombia* **92 C3** 4 6N 76 11W
Tulun, *Russia* .... **27 D11** 54 32N 100 35 E
Tulungagung, *Indonesia* **36 F4** 8 5S 111 54 E
Tum, *Indonesia* .... **37 E8** 3 36S 130 21 E
Tuma →, *Nic.* **88 D3** 13 6N 84 35W
Tumaco, *Colombia* **92 C3** 1 50N 78 45W
Tumatumari, *Guyana* ... **92 B7** 5 20N 58 55W
Tumba, L., *Zaïre* ...... **52 E3** 0 50S 18 0 E
Tumbarumba, *Australia* . **63 F4** 35 44S 148 0 E
Tumbaya, *Argentina* **94 A2** 23 50S 65 26W
Túmbes, *Peru* **92 D2** 3 37S 80 27W
Tumbwe, *Zaïre* .... **55 E2** 11 25S 27 15 E
Tumby Bay, *Australia* ... **63 E2** 34 21S 136 8 E
Tumd Youqi, *China* ..... **34 D6** 40 30N 110 30 E
Tumen, *China* .... **35 C15** 43 0N 129 50 E
Tumen Jiang →, *China* **35 C16** 42 20N 130 35 E
Tumeremo, *Venezuela* .. **92 B6** 7 18N 61 30W
Tumkur, *India* .... **40 N10** 13 18N 77 6 E
Tummel, L., *U.K.* **12 E5** 56 43N 3 55W
Tump, *Pakistan* .... **40 F3** 26 7N 62 16 E
Tumpat, *Malaysia* ...... **39 J4** 6 11N 102 10 E
Tumu, *Ghana* **50 F4** 10 56N 1 56W
Tumucumaque, Serra,
Brazil **93 C8** 2 0N 55 0W
Tumut, *Australia* .... **63 F4** 35 16S 148 13 E
Tumwater, *U.S.A.* **82 C2** 47 1N 122 54W
Tunas de Zaza, *Cuba* ... **88 B4** 21 39N 79 34W
Tunbridge Wells, *U.K.* .. **11 F8** 51 7N 0 16 E
Tuncurry, *Australia* .... **63 E5** 32 17S 152 29 E
Tunduru, *Tanzania* ..... **55 E4** 11 8S 37 25 E
Tunduru □, *Tanzania* ... **55 E4** 11 5S 37 22 E
Tundzha →, *Bulgaria* .. **21 D12** 41 40N 26 35 E
Tunga Pass, *India* ..... **41 E19** 29 0N 94 14 E
Tungabhadra →, *India* . **40 M11** 15 57N 78 15 E
Tungaru, *Sudan* .... **51 F11** 10 9N 30 52 E
Tungla, *Nic.* **88 D3** 13 24N 84 21W
Tungnafellsjökull, *Iceland* **8 D5** 64 45N 17 55W

Tungsten

Tungsten, Canada 72 A3 61 57N 128 16W
Tunguska, Nizhnyaya →, Russia 27 C9 65 48N 88 4 E
Tunica, U.S.A. 81 H9 34 41N 90 23W
Tunis, Tunisia 50 A7 36 50N 10 11 E
Tunisia ■, Africa 50 B6 33 30N 9 10 E
Tunja, Colombia 92 B4 5 33N 73 25W
Tunkhannock, U.S.A. 79 E9 41 32N 75 57W
Tunliu, China 34 F7 36 13N 112 52 E
Tunnsjøen, Norway 8 D12 64 45N 13 25 E
Tunungayualok I., Canada 71 A7 56 0N 61 0W
Tunuyán, Argentina 94 C2 33 35S 69 0W
Tunuyán →, Argentina 94 C2 33 33S 67 30W
Tunxi, China 33 D6 29 42N 118 25 E
Tuolumne, U.S.A. 83 H3 37 58N 120 15W
Tuolumne →, U.S.A. 84 H5 37 36N 121 13W
Tuoy-Khaya, Russia 27 C12 62 32N 111 25 E
Tūp Āghāj, Iran 44 B5 36 3N 47 50 E
Tupã, Brazil 95 A5 21 57S 50 28W
Tupelo, U.S.A. 77 H1 34 16N 88 43W
Tupik, Russia 27 D12 54 26N 119 57 E
Tupinambaranas, Brazil 92 D7 3 0S 58 0W
Tupiza, Bolivia 94 A2 21 30S 65 40W
Tupman, U.S.A. 85 K7 35 18N 119 21W
Tupper, Canada 72 B4 55 32N 120 1W
Tupper Lake, U.S.A. 79 B10 44 14N 74 28W
Tupungato, Cerro, S. Amer. 94 C2 33 15S 69 50W
Tuquan, China 35 B11 45 18N 121 38 E
Túquerres, Colombia 92 C3 1 5N 77 37W
Tura, Russia 27 C11 64 20N 100 17 E
Turabah, Si. Arabia 44 D4 28 20N 43 15 E
Tūrān, Iran 45 C8 35 39N 56 42 E
Turan, Russia 27 D10 51 55N 95 0 E
Turayf, Si. Arabia 44 D3 31 41N 38 39 E
Turda, Romania 17 E11 46 34N 23 47 E
Turek, Poland 17 B9 52 3N 18 30 E
Turfan = Turpan, China 32 B3 43 58N 89 10 E
Turgutlu, Turkey 25 G4 38 30N 27 48 E
Turia →, Spain 19 C5 39 27N 0 19W
Turiaçu, Brazil 93 D9 1 40S 45 19W
Turiaçu →, Brazil 93 D9 1 36S 45 19W
Turin = Torino, Italy 20 B2 45 4N 7 40 E
Turin, Canada 72 D6 49 58N 112 31W
Turkana □, Kenya 54 B4 3 0N 35 30 E
Turkana, L., Africa 54 B4 3 30N 36 5 E
Turkestan, Kazakhstan 26 E7 43 17N 68 16 E
Turkey ■, Eurasia 25 G6 39 0N 36 0 E
Turkey Creek, Australia 60 C4 17 2S 128 12 E
Turkmenistan ■, Asia 26 F6 39 0N 59 0 E
Turks & Caicos Is. ■, W. Indies 89 B5 21 20N 71 20W
Turks Island Passage, W. Indies 89 B5 21 30N 71 30W
Turku, Finland 9 F17 60 30N 22 19 E
Turkwe →, Kenya 54 B4 3 6N 36 6 E
Turlock, U.S.A. 83 H3 37 30N 120 51W
Turnagain →, Canada 72 B3 59 12N 127 35W
Turnagain, C., N.Z. 59 J6 40 28S 176 38 E
Turneffe Is., Belize 87 D7 17 20N 87 50W
Turner, Australia 60 C4 17 52S 128 16 E
Turner, U.S.A. 82 B9 48 51N 108 24W
Turner Pt., Australia 62 A1 11 47S 133 32 E
Turner Valley, Canada 72 C6 50 40N 114 17W
Turners Falls, U.S.A. 79 D12 42 36N 72 33W
Turnhout, Belgium 15 C4 51 19N 4 57 E
Turnor L., Canada 73 B7 56 35N 108 35W
Tŭrnovo, Bulgaria 21 C11 43 5N 25 41 E
Turnu Măgurele, Romania 17 G12 43 46N 24 56 E
Turnu Rosu Pasul, Romania 17 F12 45 33N 24 17 E
Turon, U.S.A. 81 G5 37 48N 98 26W
Turpan, China 32 B3 43 58N 89 10 E
Turriff, U.K. 12 D6 57 32N 2 28W
Tursāq, Iraq 44 C5 33 27N 45 47 E
Turtle Head I., Australia 62 A3 10 56S 142 37 E
Turtle L., Canada 73 C7 53 36N 108 38W
Turtle Lake, N. Dak., U.S.A. 80 B4 47 31N 100 53W
Turtle Lake, Wis., U.S.A. 80 C8 45 24N 92 8W
Turtleford, Canada 73 C7 53 23N 108 57W
Turukhansk, Russia 27 C9 65 21N 88 5 E
Turun ja Porin lääni □, Finland 9 F17 60 27N 22 15 E
Tuscaloosa, U.S.A. 77 J2 33 12N 87 34W
Tuscany = Toscana, Italy 20 C4 43 30N 11 5 E
Tuscola, Ill., U.S.A. 76 F1 39 48N 88 17W
Tuscola, Tex., U.S.A. 81 J5 32 12N 99 48W
Tuscumbia, U.S.A. 77 H2 34 44N 87 42W
Tuskar Rock, Ireland 13 D5 52 12N 6 10W
Tuskegee, U.S.A. 77 J3 32 25N 85 42W
Tuticorin, India 40 Q11 8 50N 78 12 E
Tutóia, Brazil 93 D10 2 45S 42 20W
Tutong, Brunei 36 D4 4 47N 114 40 E
Tutrakan, Bulgaria 21 B12 44 2N 26 40 E
Tutshi L., Canada 72 B2 59 56N 134 30W
Tuttle, U.S.A. 80 B5 47 9N 100 0W
Tuttlingen, Germany 16 E4 47 59N 8 50 E
Tutuala, Indonesia 37 F7 8 25S 127 15 E
Tutuila, Amer. Samoa 59 B13 14 19S 170 50W
Tututepec, Mexico 87 D5 16 9N 97 38W
Tuva Republic □, Russia 27 D10 51 30N 95 0 E
Tuvalu ■, Pac. Oc. 64 H9 8 0S 178 0 E
Tuxpan, Mexico 87 C5 20 58N 97 23W
Tuxtla Gutiérrez, Mexico 87 D6 16 50N 93 10W
Tuy, Spain 19 A1 42 3N 8 39W
Tuy An, Vietnam 38 F7 13 17N 109 16 E
Tuy Duc, Vietnam 39 F6 12 15N 107 27 E
Tuy Hoa, Vietnam 38 F7 13 5N 109 10 E
Tuy Phong, Vietnam 39 G7 11 14N 108 43 E
Tuya L., Canada 72 B2 59 7N 130 35W
Tuyen Hoa, Vietnam 38 D6 17 50N 106 10 E
Tuyen Quang, Vietnam 38 B5 21 50N 105 10 E
Tūysarkān, Iran 45 C6 34 33N 48 27 E
Tuz Gölü, Turkey 25 G5 38 45N 33 30 E
Ṭūz Khurmātū, Iraq 44 C5 34 56N 44 38 E
Tuzla, Bos.-H. 21 B8 44 34N 18 41 E
Tver, Russia 24 C6 56 55N 35 55 E
Twain, U.S.A. 84 E5 40 1N 121 3W
Twain Harte, U.S.A. 84 G6 38 2N 120 14W
Tweed, Canada 78 B7 44 29N 77 19W
Tweed →, U.K. 12 F7 55 42N 1 59W
Tweed Heads, Australia 63 D5 28 10S 153 31 E
Tweedsmuir Prov. Park, Canada 72 C3 53 0N 126 20W
Twentynine Palms, U.S.A. 85 L10 34 8N 116 3W
Twillingate, Canada 71 C9 49 42N 54 45W
Twin Bridges, U.S.A. 82 D7 45 33N 112 20W
Twin Falls, U.S.A. 82 E6 42 34N 114 28W
Twin Valley, U.S.A. 80 B6 47 16N 96 16W
Twisp, U.S.A. 82 B3 48 22N 120 7W
Two Harbors, U.S.A. 80 B9 47 2N 91 40W
Two Hills, Canada 72 C6 53 43N 111 52W
Two Rivers, U.S.A. 76 C2 44 9N 87 34W
Twofold B., Australia 63 F4 37 8S 149 59 E
Tychy, Poland 17 C9 50 9N 18 59 E
Tyler, U.S.A. 75 D7 32 18N 95 17W
Tyler, Minn., U.S.A. 80 C6 44 18N 96 8W
Tyler, Tex., U.S.A. 81 J7 32 21N 95 18W
Tynda, Russia 27 D13 55 10N 124 43 E
Tyne →, U.K. 10 C6 54 58N 1 28W
Tyne & Wear □, U.K. 10 C6 54 55N 1 35W
Tynemouth, U.K. 10 B6 55 1N 1 27W
Tyre = Sūr, Lebanon 47 B4 33 19N 35 16 E
Tyrifjorden, Norway 9 F11 60 2N 10 8 E
Tyrol = Tirol □, Austria 16 E5 47 3N 10 43 E
Tyrone, U.S.A. 78 F6 40 40N 78 14W
Tyrrell →, Australia 63 F3 35 26S 142 51 E
Tyrrell, L., Australia 63 F3 35 20S 142 50 E
Tyrrell Arm, Canada 73 A9 62 27N 97 30W
Tyrrell L., Canada 73 A7 63 7N 105 27W
Tyrrhenian Sea, Europe 20 E5 40 0N 12 30 E
Tysfjorden, Norway 8 B14 68 7N 16 25 E
Tyulgan, Russia 24 D10 52 22N 56 12 E
Tyumen, Russia 26 D7 57 11N 65 29 E
Tywi →, U.K. 11 F3 51 48N 4 20W
Tywyn, U.K. 11 E3 52 36N 4 5W
Tzaneen, S. Africa 57 C5 23 47S 30 9 E
Tzermiádhes, Greece 23 D7 35 12N 25 29 E
Tzukong = Zigong, China 32 D5 29 15N 104 48 E

U

U Taphao, Thailand 38 F3 12 35N 101 0 E
U.S.A. = United States of America ■, N. Amer. 74 C7 37 0N 96 0W
Uanda, Australia 62 C3 21 37S 144 55 E
Uarsciek, Somali Rep. 46 G4 2 28N 45 55 E
Uasin □, Kenya 54 B4 0 30N 35 20 E
Uato-Udo, Indonesia 37 F7 9 7S 125 36 E
Uatumã →, Brazil 92 D7 2 26S 57 37W
Uaupés, Brazil 92 D5 0 8S 67 5W
Uaupés →, Brazil 92 C5 0 2N 67 16W
Uaxactún, Guatemala 88 C2 17 25N 89 29W
Ubá, Brazil 95 A7 21 8S 43 0W
Ubaitaba, Brazil 93 F11 14 18S 39 20W
Ubangi = Oubangi →, Zaïre 52 E3 0 30S 17 50 E
Ubauro, Pakistan 42 E3 28 15N 69 45 E
Ube, Japan 31 H5 33 56N 131 15 E
Ubeda, Spain 19 C4 38 3N 3 23W
Uberaba, Brazil 93 G9 19 50S 47 55W
Uberlândia, Brazil 93 G9 19 0S 48 20W
Ubolratna Res., Thailand 38 D4 16 45N 102 30 E
Ubombo, S. Africa 57 D5 27 31S 32 4 E
Ubon Ratchathani, Thailand 38 E5 15 15N 104 50 E
Ubondo, Zaïre 54 C2 0 55S 25 42 E
Ubundu, Zaïre 54 C2 0 22S 25 30 E
Ucayali →, Peru 92 D4 4 30S 73 30W
Uchi Lake, Canada 73 C10 51 5N 92 35W
Uchiura-Wan, Japan 30 C10 42 25N 140 40 E
Uchur →, Russia 27 D14 58 48N 130 35 E
Ucluelet, Canada 72 D3 48 57N 125 32W
Uda →, Russia 27 D14 54 42N 135 14 E
Udaipur, India 42 G5 24 36N 73 44 E
Udaipur Garhi, Nepal 43 F12 27 0N 86 35 E
Uddevalla, Sweden 9 G11 58 21N 11 55 E
Uddjaur, Sweden 8 D16 65 25N 21 15 E
Udgir, India 40 K10 18 25N 77 5 E
Udhampur, India 43 C6 33 0N 75 5 E
Udi, Nigeria 50 G6 6 17N 7 21 E
Údine, Italy 20 A5 46 5N 13 10 E
Udmurt Republic □, Russia 24 C9 57 30N 52 30 E
Udon Thani, Thailand 38 D4 17 29N 102 46 E
Udupi, India 40 N9 13 25N 74 42 E
Udzungwa Range, Tanzania 55 D4 9 30S 35 10 E
Ueda, Japan 31 F9 36 24N 138 16 E
Uedineniya, Os., Russia 4 B12 78 0N 85 0 E
Uele →, Zaïre 52 D4 3 45N 24 45 E
Uelen, Russia 27 C19 66 10N 170 0W
Uelzen, Germany 16 B5 53 0N 10 33 E
Ufa, Russia 24 D10 54 45N 55 55 E
Ufa →, Russia 24 D10 54 40N 56 0 E
Ugab →, Namibia 56 C1 20 55S 13 30 E
Ugalla →, Tanzania 54 D3 5 8S 30 42 E
Uganda ■, Africa 54 B3 2 0N 32 0 E
Ugie, S. Africa 57 E4 31 10S 28 13 E
Uglegorsk, Russia 27 E15 49 5N 142 2 E
Ugolyak, Russia 27 C13 64 33N 120 30 E
Ugûn Mûsa, Egypt 47 F1 29 53N 32 40 E
Uhrichsville, U.S.A. 78 F3 40 24N 81 21W
Uíge, Angola 52 F2 7 30S 14 40 E
Uijŏngbu, S. Korea 35 F14 37 48N 127 0 E
Ŭiju, N. Korea 35 D13 40 15N 124 35 E
Uinta Mts., U.S.A. 82 F8 40 45N 110 30W
Uitenhage, S. Africa 56 E4 33 40S 25 28 E
Uithuizen, Neths. 15 A6 53 24N 6 41 E
Ujhani, India 43 F8 28 0N 79 6 E
Uji-guntō, Japan 31 J4 31 15N 129 25 E
Ujjain, India 42 H6 23 9N 75 43 E
Újpest, Hungary 17 E9 47 32N 19 6 E
Ujung Pandang, Indonesia 37 F5 5 10S 119 20 E
Uka, Russia 27 D17 57 50N 162 0 E
Ukara I., Tanzania 54 C3 1 50S 33 0 E
Uke-Shima, Japan 31 K4 28 2N 129 14 E
Ukerewe □, Tanzania 54 C3 2 0S 32 30 E
Ukerewe I., Tanzania 54 C3 2 0S 33 0 E
Ukhrul, India 41 G19 25 10N 94 25 E
Ukhta, Russia 24 B9 63 55N 54 0 E
Ukiah, U.S.A. 84 F3 39 9N 123 13W
Ukki Fort, India 43 C7 33 28N 76 54 E
Ukraine ■, Europe 25 E5 49 0N 32 0 E
Ukwi, Botswana 56 C3 23 29S 20 30 E
Ulaanbaatar, Mongolia 27 E11 47 55N 106 53 E
Ulaangom, Mongolia 32 A4 50 5N 92 10 E
Ulamba, Zaïre 55 D1 9 3S 23 38 E
Ulan Bator = Ulaanbaatar, Mongolia 27 E11 47 55N 106 53 E
Ulan Ude, Russia 27 D11 51 45N 107 40 E
Ulanga □, Tanzania 55 D4 8 40S 36 50 E
Ulaya, Morogoro, Tanzania 54 D4 7 3S 36 55 E
Ulaya, Tabora, Tanzania 54 C3 4 25S 33 30 E
Ulcinj, Montenegro, Yug. 21 D8 41 58N 19 10 E
Ulco, S. Africa 56 D3 28 21S 24 15 E
Ulhasnagar, India 40 K8 19 15N 73 10 E
Ulladulla, Australia 63 F5 35 21S 150 29 E
Ullapool, U.K. 12 D3 57 54N 5 10W
Ullswater, U.K. 10 C5 54 35N 2 52W
Ullung-do, S. Korea 35 F16 37 30N 130 30 E
Ulm, Germany 16 D4 48 23N 10 0 E
Ulmarra, Australia 63 D5 29 37S 153 4 E
Ulonguè, Mozam. 55 E3 14 37S 34 19 E
Ulricehamn, Sweden 9 H12 57 46N 13 26 E
Ulsan, S. Korea 35 G15 35 20N 129 15 E
Ulster □, U.K. 13 B5 54 35N 6 30W
Ulubaria, India 43 H13 22 31N 88 4 E
Uluguru Mts., Tanzania 54 D4 7 15S 37 40 E
Ulungur He →, China 32 B3 47 1N 87 24 E
Ulutau, Kazakhstan 26 E7 48 39N 67 1 E
Ulverston, U.K. 10 C4 54 13N 3 7W
Ulverstone, Australia 62 G4 41 11S 146 11 E
Ulya, Russia 27 D15 59 10N 142 0 E
Ulyanovsk = Simbirsk, Russia 24 D8 54 20N 48 25 E
Ulyasutay, Mongolia 32 B4 47 56N 97 28 E
Ulysses, U.S.A. 81 G4 37 35N 101 22W
Umala, Bolivia 92 G5 17 25S 68 5W
Uman, Ukraine 25 E5 48 40N 30 12 E
Umaria, India 41 H12 23 35N 80 50 E
Umarkot, Pakistan 40 G6 25 15N 69 40 E
Umatilla, U.S.A. 82 D4 45 55N 119 21W
Umba, Russia 24 A5 66 50N 34 20 E
Umbrella Mts., N.Z. 59 L2 45 35S 169 5 E
Umbria □, Italy 20 C5 42 53N 12 30 E
Ume älv →, Sweden 8 E16 63 45N 20 20 E
Umeå, Sweden 8 E16 63 45N 20 20 E
Umera, Indonesia 37 E7 0 12S 129 37 E
Umfuli →, Zimbabwe 55 F2 17 30S 29 23 E
Umgusa, Zimbabwe 55 F2 19 29S 27 52 E
Umkomaas, S. Africa 57 E5 30 13S 30 48 E
Umm ad Daraj, J., Jordan 47 C4 32 18N 35 48 E
Umm al Qaywayn, U.A.E. 45 E7 25 30N 55 35 E
Umm al Qittayn, Jordan 47 C5 32 18N 36 40 E
Umm Bāb, Qatar 45 E6 25 12N 50 48 E
Umm Bel, Sudan 51 F10 13 35N 28 0 E
Umm el Fahm, Israel 47 C4 32 31N 35 9 E
Umm Lajj, Si. Arabia 44 E3 25 0N 37 23 E
Umm Ruwaba, Sudan 51 F11 12 50N 31 20 E
Umnak I., U.S.A. 68 C3 53 15N 168 20W
Umniati →, Zimbabwe 55 F2 16 49S 28 45 E
Umpqua →, U.S.A. 82 E1 43 40N 124 12W
Umreth, India 42 H5 22 41N 73 4 E
Umtata, S. Africa 57 E4 31 36S 28 49 E
Umuarama, Brazil 95 A5 23 45S 53 20W
Umvukwe Ra., Zimbabwe 55 F3 16 45S 30 45 E
Umzimvubu = Port St. Johns, S. Africa 57 E4 31 38S 29 33 E
Umzingwane →, Zimbabwe 55 G2 22 12S 29 56 E
Umzinto, S. Africa 57 E5 30 15S 30 45 E
Una, India 42 J4 20 46N 71 8 E
Unac →, Bos.-H. 20 B7 44 30N 16 9 E
Unadilla, U.S.A. 79 D9 42 20N 75 19W
Unalaska, U.S.A. 68 C3 53 53N 166 32W
Uncía, Bolivia 92 G5 18 25S 66 40W
Uncompahgre Peak, U.S.A. 83 G10 38 4N 107 28W
Underberg, S. Africa 57 D4 29 50S 29 22 E
Underbool, Australia 63 F3 35 10S 141 51 E
Ungarie, Australia 63 E4 33 38S 146 56 E
Ungarra, Australia 63 E2 34 12S 136 2 E
Ungava B., Canada 69 C13 59 30N 67 30W
Ungava Pen., Canada 69 C12 60 0N 74 0W
Unggi, N. Korea 35 C16 42 16N 130 28 E
União da Vitória, Brazil 95 B5 26 13S 51 5W
Unimak I., U.S.A. 68 C3 54 45N 164 0W
Union, Miss., U.S.A. 81 J10 32 34N 89 7W
Union, Mo., U.S.A. 80 F9 38 27N 91 0W
Union, S.C., U.S.A. 77 H5 34 43N 81 37W
Union, Mt., U.S.A. 83 J7 34 34N 112 21W
Union City, Calif., U.S.A. 84 H4 37 36N 122 1W
Union City, N.J., U.S.A. 79 F10 40 45N 74 2W
Union City, Pa., U.S.A. 78 E5 41 54N 79 51W
Union City, Tenn., U.S.A. 81 G10 36 26N 89 3W
Union Gap, U.S.A. 82 C3 46 33N 120 28W
Union of Soviet Socialist Republics = Commonwealth of Independent States ■, Eurasia 27 D11 60 0N 100 0 E
Union Springs, U.S.A. 77 J3 32 9N 85 43W
Uniondale, S. Africa 56 E3 33 39S 23 7 E
Uniontown, U.S.A. 76 F6 39 54N 79 44W
Unionville, U.S.A. 80 E8 40 29N 93 1W
United Arab Emirates ■, Asia 45 F7 23 50N 54 0 E
United Kingdom ■, Europe 7 E5 53 0N 2 0W
United States of America ■, N. Amer. 74 C7 37 0N 96 0W
Unity, Canada 73 C7 52 30N 109 5W
Unjha, India 42 H5 23 46N 72 24 E
Unnao, India 43 F9 26 35N 80 30 E
Unst, U.K. 12 A8 60 50N 0 55W
Unuk →, Canada 72 B2 56 5N 131 3W
Uozu, Japan 31 F8 36 48N 137 24 E
Upata, Venezuela 92 B6 8 1N 62 24W
Upemba, L., Zaïre 55 D2 8 30S 26 20 E
Upernavik, Greenland 4 B5 72 49N 56 20W
Upington, S. Africa 56 D3 28 25S 21 15 E
Upleta, India 42 J4 21 46N 70 16 E
Upolu, W. Samoa 59 A13 13 58S 172 0W
Upper Alkali Lake, U.S.A. 82 F3 41 47N 120 8W
Upper Arrow L., Canada 72 C5 50 30N 117 50W
Upper Foster L., Canada 73 B7 56 47N 105 20W
Upper Hutt, N.Z. 59 J5 41 8S 175 5 E
Upper Klamath L., U.S.A. 82 E3 42 25N 121 55W
Upper Lake, U.S.A. 84 F4 39 10N 122 54W
Upper Musquodoboit, Canada 71 C7 45 10N 62 58W
Upper Red L., U.S.A. 80 A7 48 8N 94 45W
Upper Sandusky, U.S.A. 76 E4 40 50N 83 17W
Upper Volta = Burkina Faso ■, Africa 50 F4 12 0N 1 0W
Uppsala, Sweden 9 G14 59 53N 17 38 E
Uppsala län □, Sweden 9 G14 60 0N 17 30 E
Upshi, India 43 C7 33 48N 77 52 E
Upstart, C., Australia 62 B4 19 41S 147 45 E
Upton, U.S.A. 80 C2 44 6N 104 38W
Ur, Iraq 44 D5 30 55N 46 25 E
Uracara, Brazil 92 D7 2 20S 57 50W
Urad Qianqi, China 34 D5 40 40N 108 30 E
Urakawa, Japan 30 C11 42 9N 142 47 E
Ural, Australia 63 E4 33 21S 146 12 E
Ural →, Kazakhstan 25 E9 47 0N 51 48 E
Ural Mts. = Uralskie Gory, Russia 24 C10 60 0N 59 0 E
Uralla, Australia 63 E5 30 37S 151 29 E
Uralsk, Kazakhstan 24 D9 51 20N 51 20 E
Uralskie Gory, Russia 24 C10 60 0N 59 0 E
Urambo, Tanzania 54 D3 5 4S 32 0 E
Urambo □, Tanzania 54 D3 5 0S 32 0 E
Urandangi, Australia 62 C2 21 32S 138 14 E
Uranium City, Canada 73 B7 59 34N 108 37W
Uranquinty, Australia 63 F4 35 10S 147 12 E
Urawa, Japan 31 G9 35 50N 139 40 E
Uray, Russia 26 C7 60 5N 65 15 E
'Uray'irah, Si. Arabia 45 E6 25 57N 48 53 E
Urbana, Ill., U.S.A. 76 E1 40 7N 88 12W
Urbana, Ohio, U.S.A. 76 E4 40 7N 83 45W
Urbino, Italy 20 C5 43 43N 12 38 E
Urbión, Picos de, Spain 19 A4 42 1N 2 52W
Urcos, Peru 92 F4 13 40S 71 38W
Urda, Kazakhstan 25 E8 48 52N 47 23 E
Urdinarrain, Argentina 94 C4 32 37S 58 52W
Urdzhar, Kazakhstan 26 E9 47 5N 81 38 E
Ure →, U.K. 10 C6 54 20N 1 25W
Ures, Mexico 86 B2 29 30N 110 30W
Urfa, Turkey 25 G6 37 12N 38 50 E
Urfahr, Austria 16 D7 48 19N 14 17 E
Urganch = Urgench, Uzbekistan 26 E7 41 40N 60 41 E
Urgench, Uzbekistan 26 E7 41 40N 60 41 E
Uri, India 43 B6 34 8N 74 2 E
Uribia, Colombia 92 A4 11 43N 72 16W
Uriondo, Bolivia 94 A3 21 41S 64 41W
Urique, Mexico 86 B3 27 13N 107 55W
Urique →, Mexico 86 B3 26 29N 107 58W
Urk, Neths. 15 B5 52 39N 5 36 E
Urmia = Orūmīyeh, Iran 44 B5 37 40N 45 0 E
Urmia, L. = Orūmīyeh, Daryācheh-ye, Iran 44 B5 37 50N 45 30 E
Uruana, Brazil 93 G9 15 30S 49 41W
Uruapan, Mexico 86 D4 19 30N 102 0W
Urubamba, Peru 92 F4 13 20S 72 10W
Urubamba →, Peru 92 F4 10 43S 73 48W
Uruçuí, Brazil 93 E10 7 20S 44 28W
Uruguai →, Brazil 95 B5 26 0S 53 30W
Uruguaiana, Brazil 94 B4 29 50S 57 0W
Uruguay ■, S. Amer. 94 C4 32 30S 56 30W
Uruguay →, S. Amer. 94 C4 34 12S 58 18W
Urumchi = Ürümqi, China 26 E9 43 45N 87 45 E
Ürümqi, China 26 E9 43 45N 87 45 E
Urup, Os., Russia 27 E16 46 0N 151 0 E
Uryung-Khaya, Russia 27 B12 72 48N 113 23 E
Usa →, Russia 24 A10 65 57N 56 55 E
Uşak, Turkey 25 G4 38 43N 29 28 E
Usakos, Namibia 56 C2 21 54S 15 31 E
Usedom, Germany 16 B6 53 50N 13 55 E
Ush-Tobe, Kazakhstan 26 E8 45 16N 78 0 E
Ushakova, Os., Russia 4 A12 82 0N 80 0 E
Ushant = Ouessant, I. d', France 18 B1 48 28N 5 6W
Ushashi, Tanzania 54 C3 1 59S 33 57 E
Ushibuka, Japan 31 H5 32 11N 130 1 E
Ushuaia, Argentina 96 G3 54 50S 68 23W
Ushumun, Russia 27 D13 52 47N 126 32 E
Usk →, U.K. 11 F5 51 37N 2 56W
Üsküdar, Turkey 21 D13 41 0N 29 5 E
Usman, Russia 24 D6 52 5N 39 48 E
Usoke, Tanzania 54 D3 5 8S 32 24 E
Usolye Sibirskoye, Russia 27 D11 52 48N 103 40 E
Uspallata, P. de, Argentina 94 C2 32 37S 69 22W
Uspenskiy, Kazakhstan 26 E8 48 41N 72 43 E
Ussuri →, Asia 30 A7 48 27N 135 0 E
Ussuriysk, Russia 27 E14 43 48N 131 59 E
Ussurka, Russia 30 B6 45 12N 133 31 E
Ust-Aldan = Batamay, Russia 27 C13 63 30N 129 15 E
Ust Amginskoye = Khandyga, Russia 27 C14 62 42N 135 35 E
Ust-Bolsheretsk, Russia 27 D16 52 50N 156 15 E
Ust Chaun, Russia 27 C18 68 47N 170 30 E
Ust'-Ilga, Russia 27 D11 55 5N 104 55 E
Ust Ilimpeya = Yukti, Russia 27 C11 63 26N 105 42 E
Ust-Ilimsk, Russia 27 D11 58 3N 102 39 E
Ust Ishim, Russia 26 D8 57 45N 71 10 E
Ust-Kamchatsk, Russia 27 D17 56 10N 162 28 E

Victoria I.

Ust-Kamenogorsk, Kazakhstan **26 E9** 50 0N 82 36 E
Ust-Karenga, Russia **27 D12** 54 25N 116 30 E
Ust Khayryuzova, Russia **27 D16** 57 15N 156 45 E
Ust-Kut, Russia **27 D11** 56 50N 105 42 E
Ust Kuyga, Russia **27 B14** 70 1N 135 43 E
Ust Maya, Russia **27 C14** 60 30N 134 28 E
Ust-Mil, Russia **27 D14** 59 40N 133 11 E
Ust-Nera, Russia **27 C15** 64 35N 143 15 E
Ust-Nyukzha, Russia **27 D13** 56 34N 121 37 E
Ust Olenek, Russia **27 B12** 73 0N 120 5 E
Ust-Omchug, Russia **27 C15** 61 9N 149 38 E
Ust Port, Russia **26 C9** 69 40N 84 26 E
Ust Tsilma, Russia **24 A9** 65 25N 52 0 E
Ust-Tungir, Russia **27 D13** 55 25N 120 36 E
Ust Urt = Ustyurt, Plato, Kazakhstan **26 E6** 44 0N 55 0 E
Ust Usa, Russia **24 A10** 66 0N 56 30 E
Ust Vorkuta, Russia **26 C7** 67 24N 64 0 E
Ústí nad Labem, Czech. **16 C7** 50 41N 14 3 E
Ustica, Italy **20 E5** 38 42N 13 10 E
Ustinov = Izhevsk, Russia **24 C9** 56 51N 53 14 E
Ustye, Russia **27 D10** 57 46N 94 37 E
Ustyurt, Plato, Kazakhstan **26 E6** 44 0N 55 0 E
Usu, China **32 B3** 44 27N 84 40 E
Usuki, Japan **31 H5** 33 8N 131 49 E
Usulután, El Salv. **88 D2** 13 25N 88 28W
Usumacinta →, Mexico **87 D6** 17 0N 91 0W
Usumbura = Bujumbura, Burundi **54 C2** 3 16S 29 18 E
Usure, Tanzania **54 C3** 4 40S 34 22 E
Uta, Indonesia **37 E9** 4 33S 136 0 E
Utah □, U.S.A. **82 G8** 39 20N 111 30W
Utah, L., U.S.A. **82 F8** 40 10N 111 58W
Ute Creek →, U.S.A. **81 H3** 35 21N 103 50W
Utete, Tanzania **54 D4** 8 0S 38 45 E
Uthai Thani, Thailand **38 E3** 15 22N 100 3 E
Uthal, Pakistan **42 G2** 25 44N 66 40 E
Utiariti, Brazil **92 F7** 13 0S 58 10W
Utica, N.Y., U.S.A. **79 C9** 43 6N 75 14W
Utica, Ohio, U.S.A. **78 F2** 40 14N 82 27W
Utik L., Canada **73 B9** 55 15N 96 0W
Utikuma L., Canada **72 B5** 55 50N 115 30W
Utrecht, Neths. **15 B5** 52 5N 5 8 E
Utrecht, S. Africa **57 D5** 27 38S 30 20 E
Utrecht □, Neths. **15 B5** 52 6N 5 7 E
Utrera, Spain **19 D3** 37 12N 5 48W
Utsjoki, Finland **8 B19** 69 51N 26 59 E
Utsunomiya, Japan **31 F9** 36 30N 139 50 E
Uttar Pradesh □, India **43 F9** 27 0N 80 0 E
Uttaradit, Thailand **38 D3** 17 36N 100 5 E
Uttoxeter, U.K. **10 E6** 52 53N 1 50W
Uudenmaan lääni □, Finland **9 F18** 60 25N 25 0 E
Uusikaarlepyy, Finland **8 E17** 63 32N 22 31 E
Uusikaupunki, Finland **9 F16** 60 47N 21 25 E
Uva, Russia **24 C9** 56 59N 52 13 E
Uvalde, U.S.A. **81 L5** 29 13N 99 47W
Uvat, Russia **26 D7** 59 5N 68 50 E
Uvinza, Tanzania **54 D3** 5 5S 30 24 E
Uvira, Zaïre **54 C2** 3 22S 29 3 E
Uvs Nuur, Mongolia **32 A4** 50 20N 92 30 E
Uwajima, Japan **31 H6** 33 10N 132 35 E
Uxbridge, Canada **78 B5** 44 6N 79 7W
Uxin Qi, China **34 E5** 38 50N 109 5 E
Uxmal, Mexico **87 C7** 20 22N 89 46W
Uyandi, Russia **27 C15** 69 19N 141 0 E
Uyuni, Bolivia **92 H5** 20 28S 66 47W
Uzbekistan ■, Asia **26 E7** 41 30N 65 0 E
Uzen, Kazakhstan **25 F9** 43 27N 53 10 E
Uzerche, France **18 D4** 45 25N 1 34 E

V

Vaal →, S. Africa **56 D3** 29 4S 23 38 E
Vaal Dam, S. Africa **57 D4** 27 0S 28 14 E
Vaalwater, S. Africa **57 C4** 24 15S 28 8 E
Vaasa, Finland **8 E16** 63 6N 21 38 E
Vaasan lääni □, Finland **8 E17** 63 2N 22 50 E
Vác, Hungary **17 E9** 47 49N 19 10 E
Vacaria, Brazil **95 B5** 28 31S 50 52W
Vacaville, U.S.A. **84 G5** 38 21N 121 59W
Vach →, Russia **26 C8** 60 45N 76 45 E
Vache, I.-à-, Haiti **89 C5** 18 2N 73 35W
Vadnagar, India **42 H5** 23 47N 72 40 E
Vadodara, India **42 H5** 22 20N 73 10 E
Vadsø, Norway **8 A20** 70 3N 29 50 E
Værøy, Norway **8 C12** 67 40N 12 40 E
Váh →, Slovakia **17 E9** 47 43N 18 7 E
Vahsel B., Antarctica **5 D1** 75 0S 35 0W
Vái, Greece **23 D8** 35 15N 26 18 E
Vaigach, Russia **26 B6** 70 10N 59 0 E
Val-de-Marne □, France **18 B5** 48 45N 2 28 E
Val-d'Oise □, France **18 B5** 49 5N 2 10 E
Val d'Or, Canada **70 C4** 48 7N 77 47W
Val Marie, Canada **73 D7** 49 15N 107 45W
Valahia, Romania **21 B11** 44 35N 25 0 E
Valcheta, Argentina **96 E3** 40 40S 66 8W
Valdayskaya Vozvyshennost, Russia **24 C5** 57 0N 33 30 E
Valdepeñas, Spain **19 C4** 38 43N 3 25W
Valdés, Pen., Argentina **96 E4** 42 30S 63 45W
Valdez, U.S.A. **68 B5** 61 7N 146 16W
Valdivia, Chile **96 D2** 39 50S 73 14W
Valdosta, U.S.A. **77 K4** 30 50N 83 17W
Vale, U.S.A. **82 E5** 43 59N 117 15W
Valença, Brazil **93 F11** 13 20S 39 5W
Valença do Piauí, Brazil **93 E10** 6 20S 41 45W
Valence, France **18 D6** 44 57N 4 54 E
Valencia, Spain **19 C5** 39 27N 0 23W
Valencia, Venezuela **92 A5** 10 11N 68 0W
Valencia □, Spain **19 C5** 39 20N 0 40W
Valencia, Albufera de, Spain **19 C5** 39 20N 0 27W
Valencia, G. de, Spain **19 C6** 39 30N 0 20 E
Valencia de Alcántara, Spain **19 C2** 39 25N 7 14W
Valenciennes, France **18 A5** 50 20N 3 34 E
Valentia Harbour, Ireland **13 E1** 51 56N 10 17W
Valentia I., Ireland **13 E1** 51 54N 10 22W
Valentim, Sa. do, Brazil **93 E10** 6 0S 43 30W
Valentin, Russia **30 C7** 43 8N 134 17 E
Valentine, Nebr., U.S.A. **80 D4** 42 52N 100 33W
Valentine, Tex., U.S.A. **81 K2** 30 35N 104 30W
Valera, Venezuela **92 B4** 9 19N 70 37W
Valier, U.S.A. **82 B7** 48 18N 112 16W
Valjevo, Serbia, Yug. **21 B8** 44 18N 19 53 E
Valkeakoski, Finland **9 F18** 61 16N 24 2 E
Valkenswaard, Neths. **15 C5** 51 21N 5 29 E
Valladolid, Mexico **87 C7** 20 40N 88 11W
Valladolid, Spain **19 B3** 41 38N 4 43W
Valldemosa, Spain **22 B9** 39 43N 2 37 E
Valle d'Aosta □, Italy **20 B2** 45 45N 7 22 E
Valle de la Pascua, Venezuela **92 B5** 9 13N 66 0W
Valle de las Palmas, Mexico **85 N10** 32 20N 116 43W
Valle de Santiago, Mexico **86 C4** 20 25N 101 15W
Valle de Suchil, Mexico **86 C4** 23 38N 103 55W
Valle de Zaragoza, Mexico **86 B3** 27 28N 105 49W
Valle Fértil, Sierra del, Argentina **94 C2** 30 20S 68 0W
Valle Hermoso, Mexico **87 B5** 25 35N 97 40W
Vallecas, Spain **19 B4** 40 23N 3 41W
Valledupar, Colombia **92 A4** 10 29N 73 15W
Vallehermoso, Canary Is. **22 F2** 28 10N 17 15W
Vallejo, U.S.A. **84 G4** 38 7N 122 14W
Vallenar, Chile **94 B1** 28 30S 70 50W
Valletta, Malta **23 D2** 35 54N 14 31 E
Valley Center, U.S.A. **85 M9** 33 13N 117 2W
Valley City, U.S.A. **80 B6** 46 55N 98 0W
Valley Falls, U.S.A. **82 E3** 42 29N 120 17W
Valley Springs, U.S.A. **84 G6** 38 12N 120 50W
Valley Wells, U.S.A. **85 K11** 35 27N 115 46W
Valleyview, Canada **72 B5** 55 5N 117 17W
Vallimanca, Arroyo, Argentina **94 D4** 35 40S 59 10W
Valls, Spain **19 B6** 41 18N 1 15 E
Valognes, France **18 B3** 49 30N 1 28W
Valona = Vlóra, Albania **21 D8** 40 32N 19 28 E
Valparaíso, Chile **94 C1** 33 2S 71 40W
Valparaíso, Mexico **86 C4** 22 50N 103 32W
Valparaiso, U.S.A. **76 E2** 41 28N 87 4W
Valparaíso □, Chile **94 C1** 33 2S 71 40W
Vals →, S. Africa **56 D4** 27 23S 26 30 E
Vals, Tanjung, Indonesia **37 F9** 8 26S 137 25 E
Valsad, India **40 J8** 20 40N 72 58 E
Valverde, Canary Is. **22 G2** 27 48N 17 55W
Valverde del Camino, Spain **19 D2** 37 35N 6 47W
Vámos, Greece **23 D6** 35 24N 24 13 E
Van, Turkey **25 G7** 38 30N 43 20 E
Van, L. = Van Gölü, Turkey **25 G7** 38 30N 43 0 E
Van Alstyne, U.S.A. **81 J6** 33 25N 96 35W
Van Bruyssel, Canada **71 C5** 47 56N 72 9W
Van Buren, Canada **71 C6** 47 10N 67 55W
Van Buren, Ark., U.S.A. **81 H7** 35 26N 94 21W
Van Buren, Maine, U.S.A. **77 B11** 47 10N 67 58W
Van Buren, Mo., U.S.A. **81 G9** 37 0N 91 1W
Van Canh, Vietnam **38 F7** 13 37N 109 0 E
Van Diemen, C., N. Terr., Australia **60 B5** 11 9S 130 24 E
Van Diemen, C., Queens., Australia **62 B2** 16 30S 139 46 E
Van Diemen G., Australia **60 B5** 11 45S 132 0 E
Van Gölü, Turkey **25 G7** 38 30N 43 0 E
Van Horn, U.S.A. **81 K2** 31 3N 104 50W
Van Ninh, Vietnam **38 F7** 12 42N 109 14 E
Van Reenen P., S. Africa **57 D4** 28 22S 29 27 E
Van Rees, Pegunungan, Indonesia **37 E9** 2 35S 138 15 E
Van Tassell, U.S.A. **80 D2** 42 40N 104 5W
Van Wert, U.S.A. **76 E3** 40 52N 84 35W
Van Yen, Vietnam **38 B5** 21 4N 104 42 E
Vanavara, Russia **27 C11** 60 22N 102 16 E
Vancouver, Canada **72 D4** 49 15N 123 10W
Vancouver, U.S.A. **84 E4** 45 38N 122 40W
Vancouver, C., Australia **61 G2** 35 2S 118 11 E
Vancouver I., Canada **72 D3** 49 50N 126 0W
Vandalia, Ill., U.S.A. **80 F10** 38 58N 89 6W
Vandalia, Mo., U.S.A. **80 F9** 39 19N 91 29W
Vandenburg, U.S.A. **85 L6** 34 35N 120 33W
Vanderbijlpark, S. Africa **57 D4** 26 42S 27 54 E
Vandergrift, U.S.A. **78 F5** 40 36N 79 34W
Vanderhoof, Canada **72 C4** 54 0N 124 0W
Vanderlin I., Australia **62 B2** 15 44S 137 2 E
Vandyke, Australia **62 C4** 24 10S 147 51 E
Vänern, Sweden **9 G12** 58 47N 13 30 E
Vänersborg, Sweden **9 G12** 58 26N 12 19 E
Vang Vieng, Laos **38 C4** 18 58N 102 32 E
Vanga, Kenya **54 C4** 4 35S 39 12 E
Vangaindrano, Madag. **57 C8** 23 21S 47 36 E
Vanguard, Canada **73 D7** 49 55N 107 20W
Vanier, Canada **70 C4** 45 27N 75 40W
Vankarem, Russia **27 C17** 67 51N 175 50W
Vankleek Hill, Canada **70 C5** 45 32N 74 40W
Vanna, Norway **8 A15** 70 6N 19 50 E
Vännäs, Sweden **8 E15** 63 58N 19 48 E
Vannes, France **18 C2** 47 40N 2 47W
Vanrhynsdorp, S. Africa **56 E2** 31 36S 18 44 E
Vanrook, Australia **62 B3** 16 57S 141 57 E
Vansbro, Sweden **9 F13** 60 32N 14 15 E
Vansittart B., Australia **60 B4** 14 3S 126 17 E
Vanthli, India **42 J4** 21 28N 70 25 E
Vanua Levu, Fiji **59 C8** 16 33S 179 15 E
Vanua Mbalavu, Fiji **59 C9** 17 40S 178 57W
Vanuatu ■, Pac. Oc. **64 J8** 15 0S 168 0 E
Vanwyksvlei, S. Africa **56 E3** 30 18S 21 49 E
Vanzylsrus, S. Africa **56 D3** 26 52S 22 4 E
Var □, France **18 E7** 43 27N 6 18 E
Varanasi, India **43 G10** 25 22N 83 0 E
Varangerfjorden, Norway **8 A20** 70 3N 29 25 E
Varaždin, Croatia **20 A7** 46 20N 16 20 E
Varberg, Sweden **9 H12** 57 6N 12 20 E
Vardar →, Macedonia **21 D10** 41 15N 22 33 E
Varella, Mui, Vietnam **38 F7** 12 54N 109 26 E
Varese, Italy **20 B3** 45 49N 8 50 E
Varginha, Brazil **95 A6** 21 33S 45 25W
Variadero, U.S.A. **81 H2** 35 43N 104 17W
Varillas, Chile **94 A1** 24 0S 70 10W
Värmlands län □, Sweden **9 G12** 60 0N 13 20 E
Varna, Bulgaria **21 C12** 43 13N 27 56 E
Värnamo, Sweden **9 H13** 57 10N 14 3 E
Vars, Canada **79 A9** 45 21N 75 21W
Varzaneh, Iran **45 C7** 32 25N 52 40 E
Vasa, Finland **8 E16** 63 6N 21 38 E
Vasa Barris →, Brazil **93 F11** 11 10S 37 10W
Vascongadas = País Vasco □, Spain **19 A4** 42 50N 2 45W
Vasht = Khāsh, Iran **40 E2** 28 15N 61 15 E
Vaslui, Romania **17 E13** 46 38N 27 42 E
Vassar, Canada **73 D9** 49 10N 95 55W
Vassar, U.S.A. **76 D4** 43 22N 83 35W
Västerås, Sweden **9 G14** 59 37N 16 38 E
Västerbottens län □, Sweden **8 D14** 64 58N 18 0 E
Västernorrlands län □, Sweden **8 E14** 63 30N 17 30 E
Västervik, Sweden **9 H14** 57 43N 16 43 E
Västmanlands län □, Sweden **9 G14** 59 45N 16 20 E
Vasto, Italy **20 C6** 42 8N 14 40 E
Vatili, Cyprus **23 D12** 35 6N 33 40 E
Vatnajökull, Iceland **8 D5** 64 30N 16 48W
Vatneyri, Iceland **8 D2** 65 35N 24 0W
Vatoa, Fiji **59 D9** 19 50S 178 13W
Vatólakkos, Greece **23 D5** 35 27N 23 53 E
Vatoloha, Madag. **57 B8** 17 52S 47 48 E
Vatomandry, Madag. **57 B8** 19 20S 48 59 E
Vatra-Dornei, Romania **17 E12** 47 22N 25 22 E
Vättern, Sweden **9 G13** 58 25N 14 30 E
Vaucluse □, France **18 E6** 43 50N 5 20 E
Vaughn, Mont., U.S.A. **82 C8** 47 33N 111 33W
Vaughn, N. Mex., U.S.A. **83 J11** 34 36N 105 13W
Vaupés = Uaupés →, Brazil **92 C5** 0 2N 67 16W
Vauxhall, Canada **72 C6** 50 5N 112 9W
Vava'u, Tonga **59 D11** 18 36S 174 0W
Växjö, Sweden **9 H13** 56 52N 14 50 E
Vaygach, Ostrov, Russia **26 C6** 70 0N 60 0 E
Váyia, Ákra, Greece **23 C10** 36 15N 28 11 E
Vechte →, Neths. **15 B6** 52 34N 6 6 E
Vedea →, Romania **17 G12** 43 53N 25 59 E
Vedia, Argentina **94 C3** 34 30S 61 31W
Vedra, I. del, Spain **22 C7** 38 52N 1 12 E
Veendam, Neths. **15 A6** 53 5N 6 52 E
Veenendaal, Neths. **15 B5** 52 2N 5 34 E
Vefsna →, Norway **8 D12** 65 48N 13 10 E
Vega, Norway **8 D11** 65 40N 11 55 E
Vega, U.S.A. **81 H3** 35 15N 102 26W
Vegafjorden, Norway **8 D12** 65 37N 12 0 E
Veghel, Neths. **15 C5** 51 37N 5 32 E
Vegreville, Canada **72 C6** 53 30N 112 5W
Vejer de la Frontera, Spain **19 D3** 36 15N 5 59W
Vejle, Denmark **9 J10** 55 43N 9 30 E
Velas, C., Costa Rica **88 D2** 10 21N 85 52W
Velasco, Sierra de, Argentina **94 B2** 29 20S 67 10W
Velay, Mts. du, France **18 D5** 45 0N 3 40 E
Velddrif, S. Africa **56 E2** 32 42S 18 11 E
Velebit Planina, Croatia **20 B6** 44 50N 15 20 E
Vélez, Colombia **92 B4** 6 1N 73 41W
Vélez Málaga, Spain **19 D3** 36 48N 4 5W
Vélez Rubio, Spain **19 D4** 37 41N 2 5W
Velhas →, Brazil **93 G10** 17 13S 44 49W
Velikaya →, Russia **24 C4** 57 48N 28 10 E
Velikaya Kema, Russia **30 B8** 45 30N 137 12 E
Veliki Ustyug, Russia **24 B8** 60 47N 46 20 E
Velikiye Luki, Russia **24 C5** 56 25N 30 32 E
Velikonda Range, India **40 M11** 14 45N 79 10 E
Velletri, Italy **20 D5** 41 43N 12 43 E
Vellore, India **40 N11** 12 57N 79 10 E
Velsen-Noord, Neths. **15 B4** 52 27N 4 40 E
Velsk, Russia **24 B7** 61 10N 42 5 E
Velva, U.S.A. **80 A4** 48 4N 100 56W
Venado Tuerto, Argentina **94 C3** 33 50S 62 0W
Venda □, S. Africa **57 C5** 22 40S 30 35 E
Vendée □, France **18 C3** 46 50N 1 35W
Véneto □, Italy **20 B5** 45 40N 12 0 E
Venézia, Italy **20 B5** 45 27N 12 20 E
Venézia, G. di, Italy **20 B5** 45 20N 13 0 E
Venezuela ■, S. Amer. **92 B5** 8 0N 66 0W
Venezuela, G. de, Venezuela **92 A4** 11 30N 71 0W
Vengurla, India **40 M8** 15 53N 73 45 E
Venice = Venézia, Italy **20 B5** 45 27N 12 20 E
Venkatapuram, India **41 K12** 18 20N 80 30 E
Venlo, Neths. **15 C6** 51 22N 6 11 E
Venraij, Neths. **15 C5** 51 31N 6 0 E
Ventana, Punta de la, Mexico **86 C3** 24 4N 109 48W
Ventana, Sa. de la, Argentina **94 D3** 38 0S 62 30W
Ventersburg, S. Africa **56 D4** 28 7S 27 9 E
Venterstad, S. Africa **56 E4** 30 47S 25 48 E
Ventnor, U.K. **11 G6** 50 35N 1 12W
Ventspils, Latvia **9 H16** 57 25N 21 32 E
Ventuarí →, Venezuela **92 C5** 3 58N 67 2W
Ventucopa, U.S.A. **85 L7** 34 50N 119 29W
Ventura, U.S.A. **85 L7** 34 17N 119 18W
Venus B., Australia **63 F4** 38 40S 145 42 E
Vera, Argentina **94 B3** 29 30S 60 20W
Vera, Spain **19 D5** 37 15N 1 51W
Veracruz, Mexico **87 D5** 19 10N 96 10W
Veracruz □, Mexico **87 D5** 19 0N 96 15W
Veraval, India **42 J4** 20 53N 70 27 E
Vercelli, Italy **20 B3** 45 19N 8 25 E
Verdalsøra, Norway **8 E11** 63 48N 11 30 E
Verde →, Argentina **96 E3** 41 56S 65 5W
Verde →, Chihuahua, Mexico **86 B3** 26 29N 107 58W
Verde →, Oaxaca, Mexico **87 D5** 15 59N 97 50W
Verde →, Veracruz, Mexico **86 C4** 21 10N 102 50W
Verde →, Paraguay **94 A4** 23 9S 57 37W
Verde, Cay, Bahamas **88 B4** 23 0N 75 5W
Verden, Germany **16 B4** 52 58N 9 18 E
Verdi, U.S.A. **84 F7** 39 31N 119 59W
Verdigre, U.S.A. **80 D5** 42 36N 98 2W
Verdun, France **18 B6** 49 9N 5 24 E
Vereeniging, S. Africa **57 D4** 26 38S 27 57 E
Vérendrye, Parc Prov. de la, Canada **70 C4** 47 20N 76 40W
Verga, C., Guinea **50 F2** 10 30N 14 10W
Vergemont, Australia **62 C3** 23 33S 143 1 E
Vergemont Cr. →, Australia **62 C3** 24 16S 143 16 E
Vergennes, U.S.A. **79 B11** 44 10N 73 15W
Verkhnevilyuysk, Russia **27 C13** 63 27N 120 18 E
Verkhneye Kalinino, Russia **27 D11** 59 54N 108 8 E
Verkhniy Baskunchak, Russia **25 E8** 48 14N 46 44 E
Verkhoyansk, Russia **27 C14** 67 35N 133 25 E
Verkhoyansk Ra. = Verkhoyanskiy Khrebet, Russia **27 C13** 66 0N 129 0 E
Verkhoyanskiy Khrebet, Russia **27 C13** 66 0N 129 0 E
Verlo, Canada **73 C7** 50 19N 108 35W
Vermilion, Canada **73 C6** 53 20N 110 50W
Vermilion →, Alta., Canada **73 C6** 53 22N 110 51W
Vermilion →, Qué., Canada **70 C5** 47 38N 72 56W
Vermilion, B., U.S.A. **81 L9** 29 45N 91 55W
Vermilion Bay, Canada **73 D10** 49 51N 93 34W
Vermilion Chutes, Canada **72 B6** 58 22N 114 51W
Vermilion L., U.S.A. **80 B8** 47 53N 92 26W
Vermillion, U.S.A. **80 D6** 42 47N 96 56W
Vermont □, U.S.A. **79 C12** 44 0N 73 0W
Vernal, U.S.A. **82 F9** 40 27N 109 32W
Vernalis, U.S.A. **84 H5** 37 36N 121 17W
Verner, Canada **70 C3** 46 25N 80 8W
Verneukpan, S. Africa **56 D3** 30 0S 21 0 E
Vernon, Canada **72 C5** 50 20N 119 15W
Vernon, U.S.A. **81 H5** 34 9N 99 17W
Vernonia, U.S.A. **84 E3** 45 52N 123 11W
Vero Beach, U.S.A. **77 M5** 27 38N 80 24W
Véroia, Greece **21 D10** 40 34N 22 12 E
Verona, Italy **20 B4** 45 27N 11 0 E
Veropol, Russia **27 C17** 65 15N 168 40 E
Versailles, France **18 B5** 48 48N 2 8 E
Vert, C., Senegal **50 F1** 14 45N 17 30W
Verulam, S. Africa **57 D5** 29 38S 31 2 E
Verviers, Belgium **15 D5** 50 37N 5 52 E
Veselovskoye Vdkhr., Russia **25 E7** 47 0N 41 0 E
Vesoul, France **18 C7** 47 40N 6 11 E
Vest-Agder fylke □, Norway **9 G9** 58 30N 7 15 E
Vesterålen, Norway **8 B13** 68 45N 15 0 E
Vestfjorden, Norway **8 C13** 67 55N 14 0 E
Vestfold fylke □, Norway **9 G11** 59 15N 10 0 E
Vestmannaeyjar, Iceland **8 E3** 63 27N 20 15W
Vestspitsbergen, Svalbard **4 B8** 78 40N 17 0 E
Vestvågøy, Norway **8 B12** 68 18N 13 50 E
Vesuvio, Italy **20 D6** 40 50N 14 22 E
Vesuvius, Mt. = Vesuvio, Italy **20 D6** 40 50N 14 22 E
Veszprém, Hungary **17 E8** 47 8N 17 57 E
Vetlanda, Sweden **9 H13** 57 24N 15 3 E
Vetlugu →, Russia **26 D5** 56 18N 46 24 E
Veurne, Belgium **15 C2** 51 5N 2 40 E
Veys, Iran **45 D6** 31 30N 49 0 E
Vezhen, Bulgaria **21 C11** 42 50N 24 20 E
Vi Thanh, Vietnam **39 H5** 9 42N 105 26 E
Viacha, Bolivia **92 G5** 16 39S 68 18W
Viamão, Brazil **95 C5** 30 5S 51 0W
Viana, Brazil **93 D10** 3 13S 44 55W
Viana, Portugal **19 C2** 38 20N 8 0W
Viana do Castelo, Portugal **19 B1** 41 42N 8 50W
Vianópolis, Brazil **93 G9** 16 40S 48 35W
Vibank, Canada **73 C8** 50 20N 103 56W
Viborg, Denmark **9 H10** 56 27N 9 23 E
Vicenza, Italy **20 B4** 45 32N 11 31 E
Vich, Spain **19 B7** 41 58N 2 19 E
Vichy, France **18 C5** 46 9N 3 26 E
Vicksburg, Ariz., U.S.A. **85 M13** 33 45N 113 45W
Vicksburg, Mich., U.S.A. **76 D3** 42 7N 85 32W
Vicksburg, Miss., U.S.A. **81 J9** 32 21N 90 53W
Viçosa, Brazil **93 E11** 9 28S 36 14W
Victor, India **42 J4** 21 0N 71 30 E
Victor, Colo., U.S.A. **80 F2** 38 43N 105 9W
Victor, N.Y., U.S.A. **78 D7** 42 58N 77 24W
Victor Harbor, Australia **63 F2** 35 30S 138 37 E
Victoria, Argentina **94 C3** 32 40S 60 10W
Victoria, Canada **72 D4** 48 30N 123 25W
Victoria, Chile **96 D2** 38 13S 72 20W
Victoria, Guinea **50 F2** 10 50N 14 32W
Victoria, Malaysia **36 C5** 5 20N 115 14 E
Victoria, Malta **23 C1** 36 2N 14 14 E
Victoria, Kans., U.S.A. **80 F5** 38 52N 99 9W
Victoria, Tex., U.S.A. **81 L6** 28 48N 97 0W
Victoria □, Australia **63 F3** 37 0S 144 0 E
Victoria →, Australia **60 C4** 15 10S 129 40 E
Victoria, Grand L., Canada **70 C4** 47 31N 77 30W
Victoria, L., Africa **54 C3** 1 0S 33 0 E
Victoria, L., Australia **63 E3** 33 57S 141 15 E
Victoria Beach, Canada **73 C9** 50 40N 96 35W
Victoria de Durango, Mexico **86 C4** 24 3N 104 39W
Victoria de las Tunas, Cuba **88 B4** 20 58N 76 59W
Victoria Falls, Zimbabwe **55 F2** 17 58S 25 52 E
Victoria Harbour, Canada **78 B5** 44 45N 79 45W
Victoria I., Canada **68 A8** 71 0N 111 0W

Westall Pt.

Wallowa, *U.S.A.* **82 D5** 45 34N 117 32W
Wallowa Mts., *U.S.A.* **82 D5** 45 20N 117 30W
Wallsend, *Australia* **63 E5** 32 55S 151 40 E
Wallsend, *U.K.* **10 C6** 54 59N 1 30W
Wallula, *U.S.A.* **82 C4** 46 5N 118 54W
Wallumbilla, *Australia* **63 D4** 26 33S 149 9 E
Walmsley, L., *Canada* **73 A7** 63 25N 108 36W
Walney, I. of, *U.K.* **10 C4** 54 5N 3 15W
Walnut Creek, *U.S.A.* **84 H4** 37 54N 122 4W
Walnut Ridge, *U.S.A.* **81 G9** 36 4N 90 57W
Walsall, *U.K.* **11 E6** 52 36N 1 59W
Walsenburg, *U.S.A.* **81 G2** 37 38N 104 47W
Walsh, *U.S.A.* **81 G3** 37 23N 102 17W
Walsh →, *Australia* **62 B3** 16 31S 143 42 E
Walsh P.O., *Australia* **62 B3** 16 40S 144 0 E
Walterboro, *U.S.A.* **77 J5** 32 55N 80 40W
Walters, *U.S.A.* **81 H5** 34 22N 98 19W
Waltham, *U.S.A.* **79 D13** 42 23N 71 14W
Waltham Station, *Canada* **70 C4** 45 57N 76 57W
Waltman, *U.S.A.* **82 E10** 43 4N 107 12W
Walton, *U.S.A.* **79 D9** 42 10N 75 8W
Walvisbaai, *S. Africa* **56 C1** 23 0S 14 28 E
Wamba, *Kenya* **54 B4** 0 58N 37 19 E
Wamba, *Zaïre* **54 B2** 2 10N 27 57 E
Wamego, *U.S.A.* **80 F6** 39 12N 96 18W
Wamena, *Indonesia* **37 E9** 4 4S 138 57 E
Wamsasi, *Indonesia* **37 E7** 3 27S 126 7 E
Wan Xian, *China* **34 E8** 38 47N 115 7 E
Wana, *Pakistan* **42 C3** 32 20N 69 32 E
Wanaaring, *Australia* **63 D3** 29 38S 144 9 E
Wanaka, *N.Z.* **59 L2** 44 42S 169 9 E
Wanaka L., *N.Z.* **59 L2** 44 33S 169 7 E
Wanapiri, *Indonesia* **37 E9** 4 30S 135 59 E
Wanapitei L., *Canada* **70 C3** 46 45N 80 40W
Wanbi, *Australia* **63 E3** 34 46S 140 17 E
Wandarrie, *Australia* **61 E2** 27 50S 117 52 E
Wanderer, *Zimbabwe* **55 F3** 19 36S 30 1 E
Wandoan, *Australia* **63 D4** 26 5S 149 55 E
Wanfu, *China* **35 D12** 40 8N 122 38 E
Wang →, *Thailand* **38 D2** 17 8N 99 2 E
Wang Noi, *Thailand* **38 E3** 14 13N 100 44 E
Wang Saphung, *Thailand* **38 D3** 17 18N 101 46 E
Wang Thong, *Thailand* **38 D3** 16 50N 100 26 E
Wanga, *Zaïre* **54 B2** 2 58N 29 12 E
Wangal, *Indonesia* **37 F8** 6 8S 134 9 E
Wanganella, *Australia* **63 F3** 35 6S 144 49 E
Wanganui, *N.Z.* **59 H5** 39 56S 175 3 E
Wangaratta, *Australia* **63 F4** 36 21S 146 19 E
Wangary, *Australia* **63 E2** 34 35S 135 29 E
Wangdu, *China* **34 E8** 38 40N 115 7 E
Wangerooge, *Germany* **16 B3** 53 47N 7 52 E
Wangi, *Kenya* **54 C5** 1 58S 40 58 E
Wangiwangi, *Indonesia* **37 F6** 5 22S 123 37 E
Wangqing, *China* **35 C15** 43 12N 129 42 E
Wankaner, *India* **42 H4** 22 35N 71 0 E
Wanless, *Canada* **73 C8** 54 11N 101 21W
Wanon Niwat, *Thailand* **38 D4** 17 38N 103 46 E
Wanquan, *China* **34 D8** 40 50N 114 40 E
Wanrong, *China* **34 G6** 35 25N 110 50 E
Wanxian, *China* **33 C5** 30 42N 108 20 E
Wapakoneta, *U.S.A.* **76 E3** 40 34N 84 12W
Wapato, *U.S.A.* **82 C3** 46 27N 120 25W
Wapawekka L., *Canada* **73 C8** 54 55N 104 40W
Wapikopa L., *Canada* **70 B2** 52 56N 87 53W
Wappingers Falls, *U.S.A.* **79 E11** 41 36N 73 55W
Wapsipinicon →, *U.S.A.* **80 E9** 41 44N 90 19W
Warangal, *India* **40 L11** 17 58N 79 35 E
Waratah, *Australia* **62 G4** 41 30S 145 30 E
Waratah B., *Australia* **63 F4** 38 54S 146 5 E
Warburton, *Vic., Australia* **63 F4** 37 47S 145 42 E
Warburton, *W. Austral., Australia* **61 E4** 26 8S 126 35 E
Warburton Ra., *Australia* **61 E4** 25 55S 126 28 E
Ward, *N.Z.* **59 J5** 41 49S 174 11 E
Ward →, *Australia* **63 D4** 26 28S 146 6 E
Ward Cove, *U.S.A.* **72 B2** 55 25N 132 43W
Ward Mt., *U.S.A.* **84 H8** 37 12N 118 54W
Warden, *S. Africa* **57 D4** 27 50S 29 0 E
Wardha, *India* **40 J11** 20 45N 78 39 E
Wardha →, *India* **40 K11** 19 57N 79 11 E
Wardlow, *Canada* **72 C6** 50 56N 111 31W
Ware, *Canada* **72 B3** 57 26N 125 41W
Ware, *U.S.A.* **79 D12** 42 16N 72 14W
Wareham, *U.S.A.* **79 E14** 41 46N 70 43W
Warialda, *Australia* **63 D5** 29 29S 150 33 E
Wariap, *Indonesia* **37 E8** 1 30S 134 5 E
Warin Chamrap, *Thailand* **38 E5** 15 12N 104 53 E
Warkopi, *Indonesia* **37 E8** 1 12S 134 9 E
Warley, *U.K.* **11 E6** 52 30N 1 58W
Warm Springs, *U.S.A.* **83 G5** 38 10N 116 20W
Warman, *Canada* **73 C7** 52 19N 106 30W
Warmbad, *Namibia* **56 D2** 28 25S 18 42 E
Warmbad, *S. Africa* **57 C4** 24 51S 28 19 E
Warnambool Downs, *Australia* **62 C3** 22 48S 142 52 E
Warnemünde, *Germany* **16 A6** 54 9N 12 5 E
Warner, *Canada* **72 D6** 49 17N 112 12W
Warner Mts., *U.S.A.* **82 F3** 41 40N 120 15W
Warner Robins, *U.S.A.* **77 J4** 32 37N 83 36W
Waroona, *Australia* **61 F2** 32 50S 115 58 E
Warracknabeal, *Australia* **63 F3** 36 9S 142 26 E
Warragul, *Australia* **63 F4** 38 10S 145 58 E
Warrawagine, *Australia* **60 D3** 20 51S 120 42 E
Warrego →, *Australia* **63 E4** 30 24S 145 21 E
Warrego Ra., *Australia* **62 C4** 24 58S 146 0 E
Warren, *Australia* **63 E4** 31 42S 147 51 E
Warren, *Ark., U.S.A.* **81 J8** 33 37N 92 4W
Warren, *Mich., U.S.A.* **76 D4** 42 30N 83 0W
Warren, *Minn., U.S.A.* **80 A6** 48 12N 96 46W
Warren, *Ohio, U.S.A.* **78 E4** 41 14N 80 49W
Warren, *Pa., U.S.A.* **78 E5** 41 51N 79 9W
Warrenpoint, *U.K.* **13 B5** 54 7N 6 15W
Warrensburg, *U.S.A.* **80 F8** 38 46N 93 44W
Warrenton, *S. Africa* **56 D3** 28 9S 24 47 E
Warrenton, *U.S.A.* **84 D3** 46 10N 123 56W
Warrenville, *Australia* **63 D4** 25 48S 147 22 E
Warri, *Nigeria* **50 G6** 5 30N 5 41 E
Warrina, *Australia* **63 D2** 28 12S 135 50 E
Warrington, *U.K.* **10 D5** 53 25N 2 38W
Warrington, *U.S.A.* **77 K2** 30 23N 87 17W
Warrnambool, *Australia* **63 F3** 38 25S 142 30 E
Warroad, *U.S.A.* **80 A7** 48 54N 95 19W
Warsa, *Indonesia* **37 E9** 0 47S 135 55 E
Warsaw = Warszawa, *Poland* **17 B10** 52 13N 21 0 E
Warsaw, *Ind., U.S.A.* **76 E3** 41 14N 85 51W
Warsaw, *N.Y., U.S.A.* **78 D6** 42 45N 78 8W
Warsaw, *Ohio, U.S.A.* **78 F2** 40 20N 82 0W
Warszawa, *Poland* **17 B10** 52 13N 21 0 E
Warta →, *Poland* **16 B7** 52 35N 14 39 E
Warthe = Warta →, *Poland* **16 B7** 52 35N 14 39 E
Waru, *Indonesia* **37 E8** 3 30S 130 36 E
Warwick, *Australia* **63 D5** 28 10S 152 1 E
Warwick, *U.K.* **11 E6** 52 17N 1 36W
Warwick, *U.S.A.* **79 E13** 41 42N 71 28W
Warwickshire □, *U.K.* **11 E6** 52 20N 1 30W
Wasaga Beach, *Canada* **78 B4** 44 31N 80 1W
Wasatch Ra., *U.S.A.* **82 F8** 40 30N 111 15W
Wasbank, *S. Africa* **57 D5** 28 15S 30 9 E
Wasco, *Calif., U.S.A.* **85 K7** 35 36N 119 20W
Wasco, *Oreg., U.S.A.* **82 D3** 45 36N 120 42W
Waseca, *U.S.A.* **80 C8** 44 5N 93 30W
Wasekamio L., *Canada* **73 B7** 56 45N 108 45W
Wash, The, *U.K.* **10 E8** 52 58N 0 20 E
Washago, *Canada* **78 B5** 44 45N 79 20W
Washburn, *N. Dak., U.S.A.* **80 B4** 47 17N 101 2W
Washburn, *Wis., U.S.A.* **80 B9** 46 40N 90 54W
Washim, *India* **40 J10** 20 3N 77 0 E
Washington, *D.C., U.S.A.* **76 F7** 38 54N 77 2W
Washington, *Ga., U.S.A.* **77 J4** 33 44N 82 44W
Washington, *Ind., U.S.A.* **76 F2** 38 40N 87 10W
Washington, *Iowa, U.S.A.* **80 E9** 41 18N 91 42W
Washington, *Mo., U.S.A.* **80 F9** 38 33N 91 1W
Washington, *N.C., U.S.A.* **77 H7** 35 33N 77 3W
Washington, *N.J., U.S.A.* **79 F10** 40 46N 74 59W
Washington, *Pa., U.S.A.* **78 F4** 40 10N 80 15W
Washington, *Utah, U.S.A.* **83 H7** 37 8N 113 31W
Washington □, *U.S.A.* **82 C3** 47 30N 120 30W
Washington, Mt., *U.S.A.* **79 B13** 44 16N 71 18W
Washington I., *U.S.A.* **76 C2** 45 23N 86 54W
Washougal, *U.S.A.* **84 E4** 45 35N 122 21W
Wasian, *Indonesia* **37 E8** 1 47S 133 19 E
Wasior, *Indonesia* **37 E8** 2 43S 134 30 E
Waskaiowaka, L., *Canada* **73 B9** 56 33N 96 23W
Waskesiu Lake, *Canada* **73 C7** 53 55N 106 5W
Wassenaar, *Neths.* **15 B4** 52 8N 4 24 E
Waswanipi, *Canada* **70 C4** 49 40N 76 29W
Waswanipi, L., *Canada* **70 C4** 49 35N 76 40W
Watangpone, *Indonesia* **37 E6** 4 29S 120 25 E
Water Park Pt., *Australia* **62 C5** 22 56S 150 47 E
Water Valley, *U.S.A.* **81 H10** 34 10N 89 38W
Waterberge, *S. Africa* **57 C4** 24 10S 28 0 E
Waterbury, *Conn., U.S.A.* **79 E11** 41 33N 73 3W
Waterbury, *Vt., U.S.A.* **79 B12** 44 20N 72 46W
Waterbury L., *Canada* **73 B8** 58 10N 104 22W
Waterdown, *Canada* **78 C5** 43 20N 79 53W
Waterford, *Canada* **78 D4** 42 56N 80 17W
Waterford, *Ireland* **13 D4** 52 16N 7 8W
Waterford, *U.S.A.* **84 H6** 37 38N 120 46W
Waterford □, *Ireland* **13 D4** 52 10N 7 40W
Waterford Harbour, *Ireland* **13 D5** 52 10N 6 58W
Waterhen L., *Man., Canada* **73 C9** 52 10N 99 40W
Waterhen L., *Sask., Canada* **73 C7** 54 28N 108 25W
Waterloo, *Belgium* **15 D4** 50 43N 4 25 E
Waterloo, *Ont., Canada* **70 D3** 43 30N 80 32W
Waterloo, *Qué., Canada* **79 A12** 45 22N 72 32W
Waterloo, *S. Leone* **50 G2** 8 26N 13 8W
Waterloo, *Ill., U.S.A.* **80 F9** 38 20N 90 9W
Waterloo, *Iowa, U.S.A.* **80 D8** 42 30N 92 21W
Waterloo, *N.Y., U.S.A.* **78 D8** 42 54N 76 52W
Watersmeet, *U.S.A.* **80 B10** 46 16N 89 11W
Waterton-Glacier International Peace Park, *U.S.A.* **82 B7** 48 45N 115 0W
Watertown, *Conn., U.S.A.* **79 E11** 41 36N 73 7W
Watertown, *N.Y., U.S.A.* **79 C9** 43 59N 75 55W
Watertown, *S. Dak., U.S.A.* **80 C6** 44 54N 97 7W
Watertown, *Wis., U.S.A.* **80 D10** 43 12N 88 43W
Waterval-Boven, *S. Africa* **57 D5** 25 40S 30 18 E
Waterville, *Canada* **79 A13** 45 16N 71 54W
Waterville, *Maine, U.S.A.* **71 D6** 44 33N 69 38W
Waterville, *N.Y., U.S.A.* **79 D9** 42 56N 75 23W
Waterville, *Pa., U.S.A.* **78 E7** 41 19N 77 21W
Waterville, *Wash., U.S.A.* **82 C3** 47 39N 120 4W
Watervliet, *U.S.A.* **79 D11** 42 44N 73 42W
Wates, *Indonesia* **37 G14** 7 51S 110 10 E
Watford, *Canada* **78 D3** 42 57N 81 53W
Watford, *U.K.* **11 F7** 51 38N 0 23W
Watford City, *U.S.A.* **80 B3** 47 48N 103 17W
Wathaman →, *Canada* **73 B8** 57 16N 102 59W
Watheroo, *Australia* **61 F2** 30 15S 116 0 E
Wating, *China* **34 G4** 35 40N 106 38 E
Watkins Glen, *U.S.A.* **78 D8** 42 23N 76 52W
Watling I. = San Salvador, *Bahamas* **89 B5** 24 0N 74 40W
Watonga, *U.S.A.* **81 H5** 35 51N 98 25W
Watrous, *Canada* **73 C7** 51 40N 105 25W
Watrous, *U.S.A.* **81 H2** 35 48N 104 59W
Watsa, *Zaïre* **54 B2** 3 4N 29 30 E
Watseka, *U.S.A.* **76 E2** 40 47N 87 44W
Watson, *Australia* **61 F5** 30 29S 131 31 E
Watson, *Canada* **73 C8** 52 10N 104 30W
Watson Lake, *Canada* **72 A3** 60 6N 128 49W
Watsonville, *U.S.A.* **83 H3** 36 55N 121 45W
Wattiwarriganna Cr. →, *Australia* **63 D2** 28 57S 136 10 E
Watuata = Batuata, *Indonesia* **37 F6** 6 12S 122 42 E
Watubela, Kepulauan, *Indonesia* **37 E8** 4 28S 131 35 E
Watubela Is. = Watubela, Kepulauan, *Indonesia* **37 E8** 4 28S 131 35 E
Waubamik, *Canada* **78 A4** 45 27N 80 1W
Waubay, *U.S.A.* **80 C6** 45 20N 97 18W
Waubra, *Australia* **63 F3** 37 21S 143 39 E
Wauchope, *Australia* **63 E5** 31 28S 152 45 E
Wauchula, *U.S.A.* **77 M5** 27 33N 81 49W
Waugh, *Canada* **73 D9** 49 40N 95 11W
Waukarlycarly, L., *Australia* **60 D3** 21 18S 121 56 E
Waukegan, *U.S.A.* **76 D2** 42 22N 87 50W
Waukesha, *U.S.A.* **76 D1** 43 1N 88 14W
Waukon, *U.S.A.* **80 D9** 43 16N 91 29W
Wauneta, *U.S.A.* **80 E4** 40 25N 101 23W
Waupaca, *U.S.A.* **80 C10** 44 21N 89 5W
Waupun, *U.S.A.* **80 D10** 43 38N 88 44W
Waurika, *U.S.A.* **81 H6** 34 10N 98 0W
Wausau, *U.S.A.* **80 C10** 44 58N 89 38W
Wautoma, *U.S.A.* **80 C10** 44 4N 89 18W
Wauwatosa, *U.S.A.* **76 D2** 43 3N 88 0W
Wave Hill, *Australia* **60 C5** 17 32S 131 0 E
Waveney →, *U.K.* **11 E9** 52 24N 1 20 E
Waverley, *N.Z.* **59 H5** 39 46S 174 37 E
Waverly, *Iowa, U.S.A.* **80 D8** 42 44N 92 29W
Waverly, *N.Y., U.S.A.* **79 D8** 42 1N 76 32W
Wavre, *Belgium* **15 D4** 50 43N 4 38 E
Wâw, *Sudan* **51 G10** 7 45N 28 1 E
Wāw al Kabīr, *Libya* **51 C8** 25 20N 16 43 E
Wawa, *Canada* **70 C3** 47 59N 84 47W
Wawanesa, *Canada* **73 D9** 49 36N 99 40W
Wawona, *U.S.A.* **84 H7** 37 32N 119 39W
Waxahachie, *U.S.A.* **81 J6** 32 24N 96 51W
Way, L., *Australia* **61 E3** 26 45S 120 16 E
Wayabula Rau, *Indonesia* **37 D7** 2 29N 128 17 E
Wayatinah, *Australia* **62 G4** 42 19S 146 27 E
Waycross, *U.S.A.* **77 K4** 31 13N 82 21W
Wayne, *Nebr., U.S.A.* **80 D6** 42 14N 97 1W
Wayne, *W. Va., U.S.A.* **76 F4** 38 13N 82 27W
Waynesboro, *Ga., U.S.A.* **77 J4** 33 6N 82 1W
Waynesboro, *Miss., U.S.A.* **77 K1** 31 40N 88 39W
Waynesboro, *Pa., U.S.A.* **76 F7** 39 45N 77 35W
Waynesboro, *Va., U.S.A.* **76 F6** 38 4N 78 53W
Waynesburg, *U.S.A.* **76 F5** 39 54N 80 11W
Waynesville, *U.S.A.* **77 H4** 35 28N 82 58W
Waynoka, *U.S.A.* **81 G5** 36 35N 98 53W
Wazirabad, *Pakistan* **42 C6** 32 30N 74 8 E
We, *Indonesia* **36 C1** 5 51N 95 18 E
Weald, The, *U.K.* **11 F8** 51 7N 0 29 E
Wear →, *U.K.* **10 C6** 54 55N 1 22W
Weatherford, *Okla., U.S.A.* **81 H5** 35 32N 98 43W
Weatherford, *Tex., U.S.A.* **81 J6** 32 46N 97 48W
Weaverville, *U.S.A.* **82 F2** 40 44N 122 56W
Webb City, *U.S.A.* **81 G7** 37 9N 94 28W
Webster, *Mass., U.S.A.* **79 D13** 42 3N 71 53W
Webster, *N.Y., U.S.A.* **78 C7** 43 13N 77 26W
Webster, *S. Dak., U.S.A.* **80 C6** 45 20N 97 31W
Webster, *Wis., U.S.A.* **80 C8** 45 53N 92 22W
Webster City, *U.S.A.* **80 D8** 42 28N 93 49W
Webster Green, *U.S.A.* **80 F9** 38 38N 90 20W
Webster Springs, *U.S.A.* **76 F5** 38 29N 80 25W
Weda, *Indonesia* **37 D7** 0 21N 127 50 E
Weda, Teluk, *Indonesia* **37 D7** 0 30N 127 50 E
Weddell I., *Falk. Is.* **96 G4** 51 50S 61 0W
Weddell Sea, *Antarctica* **5 D1** 72 30S 40 0W
Wedderburn, *Australia* **63 F3** 36 26S 143 33 E
Wedgeport, *Canada* **71 D6** 43 44N 65 59W
Wedza, *Zimbabwe* **55 F3** 18 40S 31 33 E
Wee Waa, *Australia* **63 E4** 30 11S 149 26 E
Weed, *U.S.A.* **82 F2** 41 25N 122 23W
Weed Heights, *U.S.A.* **84 G7** 38 59N 119 13W
Weedsport, *U.S.A.* **79 C8** 43 3N 76 35W
Weedville, *U.S.A.* **78 E6** 41 17N 78 30W
Weemelah, *Australia* **63 D4** 29 2S 149 15 E
Weenen, *S. Africa* **57 D5** 28 48S 30 7 E
Weert, *Neths.* **15 C5** 51 15N 5 43 E
Wei He →, *Hebei, China* **34 F8** 36 10N 115 45 E
Wei He →, *Shaanxi, China* **34 G6** 34 38N 110 15 E
Weichang, *China* **35 D9** 41 58N 117 49 E
Weichuan, *China* **34 G7** 34 20N 113 59 E
Weifang, *China* **33 C6** 36 47N 119 10 E
Weifang, *Shandong, China* **35 F10** 36 44N 119 7 E
Weihai, *China* **35 F12** 37 30N 122 6 E
Weimar, *Germany* **16 C5** 51 0N 11 20 E
Weinan, *China* **34 G5** 34 31N 109 29 E
Weipa, *Australia* **62 A3** 12 40S 141 50 E
Weir →, *Australia* **63 D4** 28 20S 149 50 E
Weir →, *Canada* **73 B10** 56 54N 93 21W
Weir River, *Canada* **73 B10** 56 49N 94 6W
Weirton, *U.S.A.* **78 F4** 40 24N 80 35W
Weiser, *U.S.A.* **82 D5** 44 10N 117 0W
Weishan, *China* **35 G9** 34 47N 117 5 E
Weiyuan, *China* **34 G3** 35 7N 104 10 E
Wejherowo, *Poland* **17 A9** 54 35N 18 12 E
Wekusko L., *Canada* **73 C9** 54 40N 99 50W
Welbourn Hill, *Australia* **63 D1** 27 21S 134 6 E
Welch, *U.S.A.* **76 G5** 37 26N 81 35W
Welkom, *S. Africa* **56 D4** 28 0S 26 46 E
Welland, *Canada* **70 D4** 43 0N 79 15W
Welland →, *U.K.* **10 E7** 52 43N 0 10W
Wellesley Is., *Australia* **62 B2** 16 42S 139 30 E
Wellin, *Belgium* **15 D5** 50 5N 5 6 E
Wellingborough, *U.K.* **11 E7** 52 18N 0 41W
Wellington, *Australia* **63 E4** 32 35S 148 59 E
Wellington, *Canada* **70 D4** 43 57N 77 20W
Wellington, *N.Z.* **59 J5** 41 19S 174 46 E
Wellington, *S. Africa* **56 E2** 33 38S 19 1 E
Wellington, *Shrops., U.K.* **10 E5** 52 42N 2 31W
Wellington, *Somst., U.K.* **11 G4** 50 58N 3 13W
Wellington, *Colo., U.S.A.* **80 E2** 40 42N 105 0W
Wellington, *Kans., U.S.A.* **81 G6** 37 16N 97 24W
Wellington, *Nev., U.S.A.* **84 G7** 38 45N 119 23W
Wellington, *Ohio, U.S.A.* **78 E2** 41 10N 82 13W
Wellington, *Tex., U.S.A.* **81 H4** 34 51N 100 13W
Wellington, I., *Chile* **96 F1** 49 30S 75 0W
Wellington, L., *Australia* **63 F4** 38 6S 147 20 E
Wells, *Norfolk, U.K.* **10 E8** 52 57N 0 51 E
Wells, *Somst., U.K.* **11 F5** 51 12N 2 39W
Wells, *Maine, U.S.A.* **79 C14** 43 20N 70 35W
Wells, *Minn., U.S.A.* **80 D8** 43 45N 93 44W
Wells, *Nev., U.S.A.* **82 F6** 41 7N 114 58W
Wells, L., *Australia* **61 E3** 26 44S 123 15 E
Wells Gray Prov. Park, *Canada* **72 C4** 52 30N 120 15W
Wells River, *U.S.A.* **79 B12** 44 9N 72 4W
Wellsboro, *U.S.A.* **78 E7** 41 45N 77 18W
Wellsburg, *U.S.A.* **78 F4** 40 16N 80 37W
Wellsville, *Mo., U.S.A.* **80 F9** 39 4N 91 34W
Wellsville, *N.Y., U.S.A.* **78 D7** 42 7N 77 57W
Wellsville, *Ohio, U.S.A.* **78 F4** 40 36N 80 39W
Wellsville, *Utah, U.S.A.* **82 F8** 41 38N 111 56W
Wellton, *U.S.A.* **83 K6** 32 40N 114 8W
Wels, *Austria* **16 D7** 48 9N 14 1 E
Welshpool, *U.K.* **11 E4** 52 40N 3 9W
Wem, *U.K.* **10 E5** 52 52N 2 45W
Wembere →, *Tanzania* **54 C3** 4 10S 34 15 E
Wen Xian, *Gansu, China* **34 H3** 32 43N 104 36 E
Wen Xian, *Henan, China* **34 G7** 34 55N 113 5 E
Wenatchee, *U.S.A.* **82 C3** 47 25N 120 19W
Wenchang, *China* **38 C8** 19 38N 110 42 E
Wenchi, *Ghana* **50 G4** 7 46N 2 8W
Wenchow = Wenzhou, *China* **33 D7** 28 0N 120 38 E
Wendell, *U.S.A.* **82 E6** 42 47N 114 42W
Wenden, *U.S.A.* **85 M13** 33 49N 113 33W
Wendeng, *China* **35 F12** 37 15N 122 5 E
Wendesi, *Indonesia* **37 E8** 2 30S 134 17 E
Wendover, *U.S.A.* **82 F6** 40 44N 114 2W
Wenlock →, *Australia* **62 A3** 12 2S 141 55 E
Wenshan, *China* **32 D5** 23 20N 104 18 E
Wenshang, *China* **34 G9** 35 45N 116 30 E
Wenshui, *China* **34 F7** 37 26N 112 1 E
Wensu, *China* **32 B3** 41 15N 80 10 E
Wentworth, *Australia* **63 E3** 34 2S 141 54 E
Wenut, *Indonesia* **37 E8** 3 11S 133 19 E
Wenxi, *China* **34 G6** 35 20N 111 10 E
Wenzhou, *China* **33 D7** 28 0N 120 38 E
Weott, *U.S.A.* **82 F2** 40 20N 123 55W
Wepener, *S. Africa* **56 D4** 29 42S 27 3 E
Werda, *Botswana* **56 D3** 25 24S 23 15 E
Werder, *Ethiopia* **46 F4** 6 58N 45 1 E
Weri, *Indonesia* **37 E8** 3 10S 132 38 E
Werribee, *Australia* **63 F3** 37 54S 144 40 E
Werrimull, *Australia* **63 E3** 34 25S 141 38 E
Werris Creek, *Australia* **63 E5** 31 18S 150 38 E
Wersar, *Indonesia* **37 E8** 1 30S 131 55 E
Weser →, *Germany* **16 B4** 53 33N 8 30 E
Wesiri, *Indonesia* **37 F7** 7 30S 126 30 E
Wesley Vale, *U.S.A.* **83 J10** 35 3N 106 2W
Wesleyville, *Canada* **71 C9** 49 8N 53 36W
Wesleyville, *U.S.A.* **78 D4** 42 9N 80 0W
Wessel, C., *Australia* **62 A2** 10 59S 136 46 E
Wessel Is., *Australia* **62 A2** 11 10S 136 45 E
Wessington, *U.S.A.* **80 C5** 44 27N 98 42W
Wessington Springs, *U.S.A.* **80 C5** 44 5N 98 34W
West, *U.S.A.* **81 K6** 31 48N 97 6W
West B., *U.S.A.* **81 L10** 29 3N 89 22W
West Baines →, *Australia* **60 C4** 15 38S 129 59 E
West Bend, *U.S.A.* **76 D1** 43 25N 88 11W
West Bengal □, *India* **43 H12** 23 0N 88 0 E
West Beskids = Západné Beskydy, *Europe* **17 D9** 49 30N 19 0 E
West Branch, *U.S.A.* **76 C3** 44 17N 84 14W
West Bromwich, *U.K.* **11 E5** 52 32N 2 1W
West Cape Howe, *Australia* **61 G2** 35 8S 117 36 E
West Chazy, *U.S.A.* **79 B11** 44 49N 73 28W
West Chester, *U.S.A.* **76 F8** 39 58N 75 36W
West Columbia, *U.S.A.* **81 L7** 29 9N 95 39W
West Covina, *U.S.A.* **85 L9** 34 4N 117 54W
West Des Moines, *U.S.A.* **80 E8** 41 35N 93 43W
West End, *Bahamas* **88 A4** 26 41N 78 58W
West Falkland, *Falk. Is.* **96 G4** 51 40S 60 0W
West Fjord = Vestfjorden, *Norway* **8 C13** 67 55N 14 0 E
West Frankfort, *U.S.A.* **80 G10** 37 54N 88 55W
West Glamorgan □, *U.K.* **11 F4** 51 40N 3 55W
West Hartford, *U.S.A.* **79 E12** 41 45N 72 44W
West Haven, *U.S.A.* **79 E12** 41 17N 72 57W
West Helena, *U.S.A.* **81 H9** 34 33N 90 38W
West Ice Shelf, *Antarctica* **5 C7** 67 0S 85 0 E
West Indies, *Cent. Amer.* **89 C7** 15 0N 65 0W
West Lorne, *Canada* **78 D3** 42 36N 81 36W
West Lunga →, *Zambia* **55 E1** 13 6S 24 39 E
West Memphis, *U.S.A.* **81 H9** 35 9N 90 11W
West Midlands □, *U.K.* **11 E6** 52 30N 1 55W
West Mifflin, *U.S.A.* **78 F5** 40 22N 79 52W
West Monroe, *U.S.A.* **81 J8** 32 31N 92 9W
West Newton, *U.S.A.* **78 F5** 40 14N 79 46W
West Nicholson, *Zimbabwe* **55 G2** 21 2S 29 20 E
West Palm Beach, *U.S.A.* **77 M5** 26 43N 80 3W
West Plains, *U.S.A.* **81 G9** 36 44N 91 51W
West Pt. = Ouest, Pte., *Canada* **71 C7** 49 52N 64 40W
West Pt., *Australia* **63 F2** 35 1S 135 56 E
West Point, *Ga., U.S.A.* **77 J3** 32 53N 85 11W
West Point, *Miss., U.S.A.* **77 J1** 33 36N 88 39W
West Point, *Nebr., U.S.A.* **80 E6** 41 51N 96 43W
West Point, *Va., U.S.A.* **76 G7** 37 32N 76 48W
West Pokot □, *Kenya* **54 B4** 1 30N 35 15 E
West Road →, *Canada* **72 C4** 53 18N 122 53W
West Rutland, *U.S.A.* **79 C11** 43 38N 73 5W
West Schelde = Westerschelde →, *Neths.* **15 C3** 51 25N 3 25 E
West Seneca, *U.S.A.* **78 D6** 42 51N 78 48W
West Siberian Plain, *Russia* **28 C11** 62 0N 75 0 E
West Sussex □, *U.K.* **11 G7** 50 55N 0 30W
West-Terschelling, *Neths.* **15 A5** 53 22N 5 13 E
West Virginia □, *U.S.A.* **76 F5** 38 45N 80 30W
West-Vlaanderen □, *Belgium* **15 D3** 51 0N 3 0 E
West Walker →, *U.S.A.* **84 G7** 38 54N 119 9W
West Wyalong, *Australia* **63 E4** 33 56S 147 10 E
West Yellowstone, *U.S.A.* **82 D8** 44 40N 111 6W
West Yorkshire □, *U.K.* **10 D6** 53 45N 1 40W
Westall Pt., *Australia* **63 E1** 32 55S 134 4 E

Westbrook

Westbrook, *Maine, U.S.A.* **77 D10** 43 41N 70 22W
Westbrook, *Tex., U.S.A.* **81 J4** 32 21N 101 1W
Westbury, *Australia* **62 G4** 41 30S 146 51 E
Westby, *U.S.A.* **80 A2** 48 52N 104 3W
Westend, *U.S.A.* **85 K9** 35 42N 117 24W
Western □, *Kenya* **54 B3** 0 30N 34 30 E
Western □, *Uganda* **54 B3** 1 45N 31 30 E
Western □, *Zambia* **55 F1** 15 15S 24 30 E
Western Australia □, *Australia* **61 E2** 25 0S 118 0 E
Western Ghats, *India* **40 N9** 14 0N 75 0 E
Western Isles □, *U.K.* **12 D1** 57 30N 7 10W
Western Sahara ■, *Africa* **50 D2** 25 0N 13 0W
Western Samoa ■, *Pac. Oc.* **59 A13** 14 0S 172 0W
Westernport, *U.S.A.* **76 F6** 39 29N 79 3W
Westerschelde →, *Neths.* **15 C3** 51 25N 3 25 E
Westerwald, *Germany* **16 C3** 50 39N 8 0 E
Westfield, *Mass., U.S.A.* **79 D12** 42 7N 72 45W
Westfield, *N.Y., U.S.A.* **78 D5** 42 20N 79 35W
Westfield, *Pa., U.S.A.* **78 E7** 41 55N 77 32W
Westhope, *U.S.A.* **80 A4** 48 55N 101 1W
Westland Bight, *N.Z.* **59 K3** 42 55S 170 5 E
Westlock, *Canada* **72 C6** 54 9N 113 55W
Westmeath □, *Ireland* **13 C4** 53 30N 7 30W
Westminster, *U.S.A.* **76 F7** 39 34N 76 59W
Westmorland, *U.S.A.* **83 K6** 33 2N 115 37W
Weston, *Malaysia* **36 C5** 5 10N 115 35 E
Weston, *Oreg., U.S.A.* **82 D4** 45 49N 118 26W
Weston, *W. Va., U.S.A.* **76 F5** 39 2N 80 28W
Weston I., *Canada* **70 B4** 52 33N 79 36W
Weston-super-Mare, *U.K.* **11 F5** 51 20N 2 59W
Westport, *Canada* **79 B8** 44 40N 76 25W
Westport, *Ireland* **13 C2** 53 44N 9 31W
Westport, *N.Z.* **59 J3** 41 46S 171 37 E
Westport, *Oreg., U.S.A.* **84 D3** 46 8N 123 23W
Westport, *Wash., U.S.A.* **82 C1** 46 53N 124 6W
Westray, *Canada* **73 C8** 53 36N 101 24W
Westray, *U.K.* **12 B6** 59 18N 3 0W
Westree, *Canada* **70 C3** 47 26N 81 34W
Westville, *Calif., U.S.A.* **84 F6** 39 8N 120 42W
Westville, *Ill., U.S.A.* **76 E2** 40 2N 87 38W
Westville, *Okla., U.S.A.* **81 G7** 35 58N 94 40W
Westwood, *U.S.A.* **82 F3** 40 18N 121 0W
Wetar, *Indonesia* **37 F7** 7 30S 126 30 E
Wetaskiwin, *Canada* **72 C6** 52 55N 113 24W
Wethersfield, *U.S.A.* **79 E12** 41 42N 72 40W
Wetteren, *Belgium* **15 D3** 51 0N 3 52 E
Wetzlar, *Germany* **16 C4** 50 33N 8 30 E
Wewoka, *U.S.A.* **81 H6** 35 9N 96 30W
Wexford, *Ireland* **13 D5** 52 20N 6 28W
Wexford □, *Ireland* **13 D5** 52 20N 6 25W
Wexford Harbour, *Ireland* **13 D5** 52 20N 6 25W
Weyburn, *Canada* **73 D8** 49 40N 103 50W
Weyburn L., *Canada* **72 A5** 63 0N 117 59W
Weymouth, *Canada* **71 D6** 44 30N 66 1W
Weymouth, *U.K.* **11 G5** 50 36N 2 28W
Weymouth, *U.S.A.* **79 D14** 42 13N 70 58W
Weymouth, C., *Australia* **62 A3** 12 37S 143 27 E
Whakatane, *N.Z.* **59 G6** 37 57S 177 1 E
Whale →, *Canada* **71 A6** 58 15N 67 40W
Whale Cove, *Canada* **73 A10** 62 11N 92 36W
Whales, B. of, *Antarctica* **5 D12** 78 0S 165 0W
Whalsay, *U.K.* **12 A7** 60 22N 1 0W
Whangamomona, *N.Z.* **59 H5** 39 8S 174 44 E
Whangarei, *N.Z.* **59 F5** 35 43S 174 21 E
Whangarei Harb., *N.Z.* **59 F5** 35 45S 174 28 E
Wharfe →, *U.K.* **10 D6** 53 55N 1 30W
Wharfedale, *U.K.* **10 C5** 54 7N 2 4W
Wharton, *N.J., U.S.A.* **79 F10** 40 54N 74 35W
Wharton, *Pa., U.S.A.* **78 E6** 41 31N 78 1W
Wharton, *Tex., U.S.A.* **81 L6** 29 19N 96 6W
Wheatland, *Calif., U.S.A.* **84 F5** 39 1N 121 25W
Wheatland, *Wyo., U.S.A.* **80 D2** 42 3N 104 58W
Wheatley, *Canada* **78 D2** 42 6N 82 27W
Wheaton, *U.S.A.* **80 C6** 45 48N 96 30W
Wheelbarrow Pk., *U.S.A.* **84 H10** 37 26N 116 5W
Wheeler, *Oreg., U.S.A.* **82 D2** 45 41N 123 53W
Wheeler, *Tex., U.S.A.* **81 H4** 35 27N 100 16W
Wheeler →, *Canada* **73 B7** 57 25N 105 30W
Wheeler Pk., *N. Mex., U.S.A.* **83 H11** 36 34N 105 25W
Wheeler Pk., *Nev., U.S.A.* **83 G6** 38 57N 114 15W
Wheeler Ridge, *U.S.A.* **85 L8** 35 0N 118 57W
Wheeling, *U.S.A.* **78 F4** 40 4N 80 43W
Whernside, *U.K.* **10 C5** 54 14N 2 24W
Whidbey I., *U.S.A.* **72 D4** 48 12N 122 17W
Whiskey Gap, *Canada* **72 D6** 49 0N 113 3W
Whiskey Jack L., *Canada* **73 B8** 58 23N 101 55W
Whistleduck Cr. →, *Australia* **62 C2** 20 15S 135 18 E
Whitby, *Canada* **78 C6** 43 52N 78 56W
Whitby, *U.K.* **10 C7** 54 29N 0 37W
White →, *Ark., U.S.A.* **81 J9** 33 57N 91 5W
White →, *Ind., U.S.A.* **76 F2** 38 25N 87 45W
White →, *S. Dak., U.S.A.* **80 D5** 43 42N 99 27W
White →, *Utah, U.S.A.* **82 F9** 40 4N 109 41W
White →, *Wash., U.S.A.* **84 C4** 47 12N 122 15W
White, L., *Australia* **60 D4** 21 9S 128 56 E
White B., *Canada* **71 B8** 50 0N 56 35W
White Bear Res., *Canada* **71 C8** 48 10N 57 5W
White Bird, *U.S.A.* **82 D5** 45 46N 116 18W
White Butte, *U.S.A.* **80 B3** 46 23N 103 18W
White City, *U.S.A.* **80 F6** 38 48N 96 44W
White Cliffs, *Australia* **63 E3** 30 50S 143 10 E
White Deer, *U.S.A.* **81 H4** 35 26N 101 10W
White Hall, *U.S.A.* **80 F9** 39 26N 90 24W
White Haven, *U.S.A.* **79 E9** 41 4N 75 47W
White I., *N.Z.* **59 G6** 37 30S 177 13 E
White L., *Canada* **79 A8** 45 18N 76 31W
White L., *U.S.A.* **81 L8** 29 44N 92 30W
White Mts., *Calif., U.S.A.* **83 H4** 37 30N 118 15W
White Mts., *N.H., U.S.A.* **75 B12** 44 15N 71 15W
White Nile = Nîl el Abyad →, *Sudan* **51 E11** 15 38N 32 31 E
White Otter L., *Canada* **70 C1** 49 5N 91 55W
White Pass, *Canada* **72 B1** 59 40N 135 3W
White Pass, *U.S.A.* **84 D5** 46 38N 121 24W
White Plains, *U.S.A.* **79 E11** 41 2N 73 46W
White River, *Canada* **70 C2** 48 35N 85 20W
White River, *S. Africa* **57 D5** 25 20S 31 0 E
White River, *U.S.A.* **80 D4** 43 34N 100 45W
White Russia = Belorussia ■, *Europe* **24 D4** 53 30N 27 0 E
White Sea = Beloye More, *Russia* **24 A6** 66 30N 38 0 E
White Sulphur Springs, *Mont., U.S.A.* **82 C8** 46 33N 110 54W
White Sulphur Springs, *W. Va., U.S.A.* **76 G5** 37 48N 80 18W
White Swan, *U.S.A.* **84 D6** 46 23N 120 44W
Whitecliffs, *N.Z.* **59 K3** 43 26S 171 55 E
Whitecourt, *Canada* **72 C5** 54 10N 115 45W
Whiteface, *U.S.A.* **81 J3** 33 36N 102 37W
Whitefield, *U.S.A.* **79 B13** 44 23N 71 37W
Whitefish, *U.S.A.* **82 B6** 48 25N 114 20W
Whitefish L., *Canada* **73 A7** 62 41N 106 48W
Whitefish Point, *U.S.A.* **76 B3** 46 45N 84 59W
Whitegull, L., *Canada* **71 A7** 55 27N 64 17W
Whitehall, *Mich., U.S.A.* **76 D2** 43 24N 86 21W
Whitehall, *Mont., U.S.A.* **82 D7** 45 52N 112 6W
Whitehall, *N.Y., U.S.A.* **79 C11** 43 33N 73 24W
Whitehall, *Wis., U.S.A.* **80 C9** 44 22N 91 19W
Whitehaven, *U.K.* **10 C4** 54 33N 3 35W
Whitehorse, *Canada* **72 A1** 60 43N 135 3W
Whitehorse, Vale of, *U.K.* **11 F6** 51 37N 1 30W
Whitemark, *Australia* **62 G4** 40 7S 148 3 E
Whitemouth, *Canada* **73 D9** 49 57N 95 58W
Whitesboro, *N.Y., U.S.A.* **79 C9** 43 7N 75 18W
Whitesboro, *Tex., U.S.A.* **81 J6** 33 39N 96 54W
Whiteshell Prov. Park, *Canada* **73 C9** 50 0N 95 40W
Whitetail, *U.S.A.* **80 A2** 48 54N 105 10W
Whiteville, *U.S.A.* **77 H6** 34 20N 78 42W
Whitewater, *U.S.A.* **76 D1** 42 50N 88 44W
Whitewater Baldy, *U.S.A.* **83 K9** 33 20N 108 39W
Whitewater L., *Canada* **70 B2** 50 50N 89 10W
Whitewood, *Australia* **62 C3** 21 28S 143 30 E
Whitewood, *Canada* **73 C8** 50 20N 102 20W
Whitfield, *Australia* **63 F4** 36 42S 146 24 E
Whithorn, *U.K.* **12 G4** 54 44N 4 25W
Whitianga, *N.Z.* **59 G5** 36 47S 175 41 E
Whitman, *U.S.A.* **79 D14** 42 5N 70 56W
Whitmire, *U.S.A.* **77 H5** 34 30N 81 37W
Whitney, *Canada* **78 A6** 45 31N 78 14W
Whitney, Mt., *U.S.A.* **83 H4** 36 35N 118 18W
Whitney Point, *U.S.A.* **79 D9** 42 20N 75 58W
Whitstable, *U.K.* **11 F9** 51 21N 1 2 E
Whitsunday I., *Australia* **62 C4** 20 15S 149 4 E
Whittier, *U.S.A.* **85 M8** 33 58N 118 3W
Whittlesea, *Australia* **63 F4** 37 27S 145 9 E
Whitwell, *U.S.A.* **77 H3** 35 12N 85 31W
Wholdaia L., *Canada* **73 A8** 60 43N 104 20W
Whyalla, *Australia* **63 E2** 33 2S 137 30 E
Whyjonta, *Australia* **63 D3** 29 41S 142 28 E
Wiarton, *Canada* **78 B3** 44 40N 81 10W
Wibaux, *U.S.A.* **80 B2** 46 59N 104 11W
Wichian Buri, *Thailand* **38 E3** 15 39N 101 7 E
Wichita, *U.S.A.* **81 G6** 37 42N 97 20W
Wichita Falls, *U.S.A.* **81 J5** 33 54N 98 30W
Wick, *U.K.* **12 C5** 58 26N 3 5W
Wickenburg, *U.S.A.* **83 K7** 33 58N 112 44W
Wickepin, *Australia* **61 F2** 32 50S 117 30 E
Wickham, C., *Australia* **62 F3** 39 35S 143 57 E
Wickliffe, *U.S.A.* **78 E3** 41 36N 81 28W
Wicklow, *Ireland* **13 D5** 53 0N 6 2W
Wicklow □, *Ireland* **13 D5** 52 59N 6 25W
Wicklow Hd., *Ireland* **13 D5** 52 59N 6 3W
Widgiemooltha, *Australia* **61 F3** 31 30S 121 34 E
Widnes, *U.K.* **10 D5** 53 22N 2 44W
Wieliczka, *Poland* **17 D10** 50 0N 20 5 E
Wieluń, *Poland* **17 C9** 51 15N 18 34 E
Wien, *Austria* **16 D8** 48 12N 16 22 E
Wiener Neustadt, *Austria* **16 E8** 47 49N 16 16 E
Wierden, *Neths.* **15 B6** 52 22N 6 35 E
Wiesbaden, *Germany* **16 C4** 50 7N 8 17 E
Wigan, *U.K.* **10 D5** 53 33N 2 38W
Wiggins, *Colo., U.S.A.* **80 E2** 40 14N 104 4W
Wiggins, *Miss., U.S.A.* **81 K10** 30 51N 89 8W
Wight, I. of □, *U.K.* **11 G6** 50 40N 1 20W
Wigtown, *U.K.* **12 G4** 54 52N 4 27W
Wigtown B., *U.K.* **12 G4** 54 46N 4 15W
Wilber, *U.S.A.* **80 E6** 40 29N 96 58W
Wilberforce, *Canada* **78 A6** 45 2N 78 13W
Wilberforce, C., *Australia* **62 A2** 11 54S 136 35 E
Wilburton, *U.S.A.* **81 H7** 34 55N 95 19W
Wilcannia, *Australia* **63 E3** 31 30S 143 26 E
Wilcox, *U.S.A.* **78 E6** 41 35N 78 41W
Wildrose, *Calif., U.S.A.* **85 J9** 36 14N 117 11W
Wildrose, *N. Dak., U.S.A.* **80 A3** 48 38N 103 11W
Wildwood, *U.S.A.* **76 F8** 38 59N 74 50W
Wilge →, *S. Africa* **57 D4** 27 3S 28 20 E
Wilhelm II Coast, *Antarctica* **5 C7** 68 0S 90 0 E
Wilhelmshaven, *Germany* **16 B4** 53 30N 8 9 E
Wilhelmstal, *Namibia* **56 C2** 21 58S 16 21 E
Wilkes-Barre, *U.S.A.* **79 E9** 41 15N 75 53W
Wilkesboro, *U.S.A.* **77 G5** 36 9N 81 10W
Wilkie, *Canada* **73 C7** 52 27N 108 42W
Wilkinsburg, *U.S.A.* **78 F5** 40 26N 79 53W
Wilkinson Lakes, *Australia* **61 E5** 29 40S 132 39 E
Willamina, *U.S.A.* **82 D2** 45 5N 123 29W
Willandra Billabong Creek →, *Australia* **63 E4** 33 22S 145 52 E
Willapa B., *U.S.A.* **82 C2** 46 40N 124 0W
Willapa Hills, *U.S.A.* **84 D3** 46 35N 123 25W
Willard, *N. Mex., U.S.A.* **83 J10** 34 36N 106 2W
Willard, *Utah, U.S.A.* **82 F7** 41 25N 112 2W
Willcox, *U.S.A.* **83 K9** 32 15N 109 50W
Willemstad, *Neth. Ant.* **89 D6** 12 5N 69 0W
Willeroo, *Australia* **60 C5** 15 14S 131 37 E
William →, *Canada* **73 B7** 59 8N 109 19W
William Creek, *Australia* **63 D2** 28 58S 136 22 E
Williambury, *Australia* **61 D2** 23 45S 115 12 E
Williams, *Australia* **61 F2** 33 2S 116 52 E
Williams, *Ariz., U.S.A.* **83 J7** 35 15N 112 11W
Williams, *Calif., U.S.A.* **84 F4** 39 9N 122 9W
Williams Lake, *Canada* **72 C4** 52 10N 122 10W
Williamsburg, *Ky., U.S.A.* **77 G3** 36 44N 84 10W
Williamsburg, *Pa., U.S.A.* **78 F6** 40 28N 78 12W
Williamsburg, *Va., U.S.A.* **76 G7** 37 17N 76 44W
Williamson, *N.Y., U.S.A.* **78 C7** 43 14N 77 11W
Williamson, *W. Va., U.S.A.* **76 G4** 37 41N 82 17W
Williamsport, *U.S.A.* **78 E7** 41 15N 77 0W
Williamston, *U.S.A.* **77 H7** 35 51N 77 4W
Williamstown, *Australia* **63 F3** 37 51S 144 52 E
Williamstown, *Mass., U.S.A.* **79 D11** 42 41N 73 12W
Williamstown, *N.Y., U.S.A.* **79 C9** 43 26N 75 53W
Williamsville, *U.S.A.* **81 G9** 36 58N 90 33W
Willimantic, *U.S.A.* **79 E12** 41 43N 72 13W
Willis Group, *Australia* **62 B5** 16 18S 150 0 E
Williston, *S. Africa* **56 E3** 31 20S 20 53 E
Williston, *Fla., U.S.A.* **77 L4** 29 23N 82 27W
Williston, *N. Dak., U.S.A.* **80 A3** 48 9N 103 37W
Williston L., *Canada* **72 B4** 56 0N 124 0W
Willits, *U.S.A.* **82 G2** 39 25N 123 21W
Willmar, *U.S.A.* **80 C7** 45 7N 95 3W
Willoughby, *U.S.A.* **78 E3** 41 39N 81 24W
Willow Bunch, *Canada* **73 D7** 49 20N 105 35W
Willow L., *Canada* **72 A5** 62 10N 119 8W
Willow Lake, *U.S.A.* **80 C6** 44 38N 97 38W
Willow Springs, *U.S.A.* **81 G8** 37 0N 91 58W
Willow Wall, The, *China* **35 C12** 42 10N 122 0 E
Willowlake →, *Canada* **72 A4** 62 42N 123 8W
Willowmore, *S. Africa* **56 E3** 33 15S 23 30 E
Willows, *Australia* **62 C4** 23 39S 147 25 E
Willows, *U.S.A.* **84 F4** 39 31N 122 12W
Willowvale = Gatyana, *S. Africa* **57 E4** 32 16S 28 31 E
Wills, L., *Australia* **60 D4** 21 25S 128 51 E
Wills Cr. →, *Australia* **62 C3** 22 43S 140 2 E
Wills Point, *U.S.A.* **81 J7** 32 43N 96 1W
Willunga, *Australia* **63 F2** 35 15S 138 30 E
Wilmette, *U.S.A.* **76 D2** 42 5N 87 42W
Wilmington, *Australia* **63 E2** 32 39S 138 7 E
Wilmington, *Del., U.S.A.* **76 F8** 39 45N 75 33W
Wilmington, *Ill., U.S.A.* **76 E1** 41 18N 88 9W
Wilmington, *N.C., U.S.A.* **77 H7** 34 14N 77 55W
Wilmington, *Ohio, U.S.A.* **76 F4** 39 27N 83 50W
Wilpena Cr. →, *Australia* **63 E2** 31 25S 139 29 E
Wilsall, *U.S.A.* **82 D8** 45 59N 110 38W
Wilson, *U.S.A.* **77 H7** 35 44N 77 55W
Wilson →, *Queens., Australia* **63 D3** 27 38S 141 24 E
Wilson →, *W. Austral., Australia* **60 C4** 16 48S 128 16 E
Wilson Bluff, *Australia* **61 F4** 31 41S 129 0 E
Wilsons Promontory, *Australia* **63 F4** 38 55S 146 25 E
Wilton, *U.K.* **11 F6** 51 5N 1 52W
Wilton, *U.S.A.* **80 B4** 47 10N 100 47W
Wilton →, *Australia* **62 A1** 14 45S 134 33 E
Wiltshire □, *U.K.* **11 F6** 51 20N 2 0W
Wiltz, *Lux.* **15 E5** 49 57N 5 55 E
Wiluna, *Australia* **61 E3** 26 36S 120 14 E
Wimmera →, *Australia* **63 F3** 36 8S 141 56 E
Winam G., *Kenya* **54 C3** 0 20S 34 15 E
Winburg, *S. Africa* **56 D4** 28 30S 27 2 E
Winchendon, *U.S.A.* **79 D12** 42 41N 72 3W
Winchester, *U.K.* **11 F6** 51 4N 1 19W
Winchester, *Conn., U.S.A.* **79 E11** 41 53N 73 9W
Winchester, *Idaho, U.S.A.* **82 C5** 46 14N 116 38W
Winchester, *Ind., U.S.A.* **76 E3** 40 10N 84 59W
Winchester, *Ky., U.S.A.* **76 G3** 38 0N 84 11W
Winchester, *N.H., U.S.A.* **79 D12** 42 46N 72 23W
Winchester, *Nev., U.S.A.* **85 J11** 36 6N 115 10W
Winchester, *Tenn., U.S.A.* **77 H2** 35 11N 86 7W
Winchester, *Va., U.S.A.* **76 F6** 39 11N 78 10W
Wind →, *U.S.A.* **82 E9** 43 12N 108 12W
Wind River Range, *U.S.A.* **82 E9** 43 0N 109 30W
Windau = Ventspils, *Latvia* **9 H16** 57 25N 21 32 E
Windber, *U.S.A.* **78 F6** 40 14N 78 50W
Windermere, L., *U.K.* **10 C5** 54 20N 2 57W
Windfall, *Canada* **72 C5** 54 12N 116 13W
Windflower L., *Canada* **72 A5** 62 52N 118 30W
Windhoek, *Namibia* **56 C2** 22 35S 17 4 E
Windom, *U.S.A.* **80 D7** 43 52N 95 7W
Windorah, *Australia* **62 D3** 25 24S 142 36 E
Window Rock, *U.S.A.* **83 J9** 35 41N 109 3W
Windrush →, *U.K.* **11 F6** 51 48N 1 35W
Windsor, *Australia* **63 E5** 33 37S 150 50 E
Windsor, *N.S., Canada* **71 D7** 44 59N 64 5W
Windsor, *Nfld., Canada* **71 C8** 48 57N 55 40W
Windsor, *Ont., Canada* **70 D3** 42 18N 83 0W
Windsor, *U.K.* **11 F7** 51 28N 0 36W
Windsor, *Colo., U.S.A.* **80 E2** 40 29N 104 54W
Windsor, *Conn., U.S.A.* **79 E12** 41 50N 72 39W
Windsor, *Mo., U.S.A.* **80 F8** 38 32N 93 31W
Windsor, *N.Y., U.S.A.* **79 D9** 42 5N 75 37W
Windsor, *Vt., U.S.A.* **79 C12** 43 29N 72 24W
Windsorton, *S. Africa* **56 D3** 28 16S 24 44 E
Windward Is., *W. Indies* **89 D7** 13 0N 61 0W
Windward Passage = Vientos, Paso de los, *Caribbean* **89 C5** 20 0N 74 0W
Windy L., *Canada* **73 A8** 60 20N 100 2W
Winefred L., *Canada* **73 B6** 55 30N 110 30W
Winfield, *U.S.A.* **81 G6** 37 15N 96 59W
Wingate Mts., *Australia* **60 B5** 14 25S 130 40 E
Wingen, *Australia* **63 E5** 31 54S 150 54 E
Wingham, *Australia* **63 E5** 31 48S 152 22 E
Wingham, *Canada* **70 D3** 43 55N 81 20W
Winifred, *U.S.A.* **82 C9** 47 34N 109 23W
Winisk, *Canada* **70 A2** 55 20N 85 15W
Winisk →, *Canada* **70 A2** 55 17N 85 5W
Winisk L., *Canada* **70 B2** 52 55N 87 22W
Wink, *U.S.A.* **81 K3** 31 45N 103 9W
Winkler, *Canada* **73 D9** 49 10N 97 56W
Winlock, *U.S.A.* **84 D4** 46 30N 122 56W
Winneba, *Ghana* **50 G4** 5 25N 0 36W
Winnebago, *U.S.A.* **80 D7** 43 46N 94 10W
Winnebago, L., *U.S.A.* **76 D1** 44 0N 88 26W
Winnecke Cr. →, *Australia* **60 C5** 18 35S 131 34 E
Winnemucca, *U.S.A.* **82 F5** 40 58N 117 44W
Winnemucca L., *U.S.A.* **82 F4** 40 7N 119 21W
Winner, *U.S.A.* **80 D5** 43 22N 99 52W
Winnett, *U.S.A.* **82 C9** 47 0N 108 21W
Winnfield, *U.S.A.* **81 K8** 31 56N 92 38W
Winnibigoshish, L., *U.S.A.* **80 B7** 47 27N 94 13W
Winning, *Australia* **60 D1** 23 9S 114 30 E
Winnipeg, *Canada* **73 D9** 49 54N 97 9W
Winnipeg →, *Canada* **73 C9** 50 38N 96 19W
Winnipeg, L., *Canada* **73 C9** 52 0N 97 0W
Winnipeg Beach, *Canada* **73 C9** 50 30N 96 58W
Winnipegosis, *Canada* **73 C9** 51 39N 99 55W
Winnipegosis L., *Canada* **73 C9** 52 30N 100 0W
Winnipesaukee, L., *U.S.A.* **79 C13** 43 38N 71 21W
Winnsboro, *La., U.S.A.* **81 J9** 32 10N 91 43W
Winnsboro, *S.C., U.S.A.* **77 H5** 34 23N 81 5W
Winnsboro, *Tex., U.S.A.* **81 J7** 32 58N 95 17W
Winokapau, L., *Canada* **71 B7** 53 15N 62 50W
Winona, *Minn., U.S.A.* **80 C9** 44 3N 91 39W
Winona, *Miss., U.S.A.* **81 J10** 33 29N 89 44W
Winooski, *U.S.A.* **79 B11** 44 29N 73 11W
Winschoten, *Neths.* **15 A7** 53 9N 7 3 E
Winslow, *Ariz., U.S.A.* **83 J8** 35 2N 110 42W
Winslow, *Wash., U.S.A.* **84 C4** 47 38N 122 31W
Winsted, *U.S.A.* **79 E11** 41 55N 73 4W
Winston-Salem, *U.S.A.* **77 G5** 36 6N 80 15W
Winter Garden, *U.S.A.* **77 L5** 28 34N 81 35W
Winter Haven, *U.S.A.* **77 M5** 28 1N 81 44W
Winter Park, *U.S.A.* **77 L5** 28 36N 81 20W
Winterhaven, *U.S.A.* **85 N12** 32 47N 114 39W
Winters, *Calif., U.S.A.* **84 G5** 38 32N 121 58W
Winters, *Tex., U.S.A.* **81 K5** 31 58N 99 58W
Winterset, *U.S.A.* **80 E7** 41 20N 94 1W
Wintersville, *U.S.A.* **78 F4** 40 23N 80 42W
Winterswijk, *Neths.* **15 C6** 51 58N 6 43 E
Winterthur, *Switz.* **16 E4** 47 30N 8 44 E
Winthrop, *Minn., U.S.A.* **80 C7** 44 32N 94 22W
Winthrop, *Wash., U.S.A.* **82 B3** 48 28N 120 10W
Winton, *Australia* **62 C3** 22 24S 143 3 E
Winton, *N.Z.* **59 M2** 46 8S 168 20 E
Winton, *U.S.A.* **77 G7** 36 24N 76 56W
Wirral, *U.K.* **10 D4** 53 25N 3 0W
Wirrulla, *Australia* **63 E1** 32 24S 134 31 E
Wisbech, *U.K.* **10 E8** 52 39N 0 10 E
Wisconsin □, *U.S.A.* **80 C10** 44 45N 89 30W
Wisconsin →, *U.S.A.* **80 D9** 43 0N 91 15W
Wisconsin Dells, *U.S.A.* **80 D10** 43 38N 89 46W
Wisconsin Rapids, *U.S.A.* **80 C10** 44 23N 89 49W
Wisdom, *U.S.A.* **82 D7** 45 37N 113 27W
Wishaw, *U.K.* **12 F5** 55 46N 3 55W
Wishek, *U.S.A.* **80 B5** 46 16N 99 33W
Wisła →, *Poland* **17 A9** 54 22N 18 55 E
Wismar, *Germany* **16 B5** 53 53N 11 23 E
Wisner, *U.S.A.* **80 E6** 41 59N 96 55W
Witbank, *S. Africa* **57 D4** 25 51S 29 14 E
Witdraai, *S. Africa* **56 D3** 26 58S 20 48 E
Witham →, *U.K.* **10 D7** 53 3N 0 8W
Withernsea, *U.K.* **10 D8** 53 43N 0 2 E
Witney, *U.K.* **11 F6** 51 47N 1 29W
Witnossob →, *Namibia* **56 D3** 26 55S 20 37 E
Witten, *Germany* **15 C7** 51 26N 7 19 E
Wittenberg, *Germany* **16 C6** 51 51N 12 39 E
Wittenberge, *Germany* **16 B5** 53 0N 11 44 E
Wittenoom, *Australia* **60 D2** 22 15S 118 20 E
Wkra →, *Poland* **17 B10** 52 27N 20 44 E
Wlingi, *Indonesia* **37 H15** 8 5S 112 25 E
Włocławek, *Poland* **17 B9** 52 40N 19 3 E
Woburn, *U.S.A.* **79 D13** 42 29N 71 9W
Wodian, *China* **34 H7** 32 50N 112 35 E
Wodonga, *Australia* **63 F4** 36 5S 146 50 E
Woëvre, *France* **18 B6** 49 15N 5 45 E
Wokam, *Indonesia* **37 F8** 5 45S 134 28 E
Wolf →, *Canada* **72 A2** 60 17N 132 33W
Wolf Creek, *U.S.A.* **82 C7** 47 0N 112 4W
Wolf L., *Canada* **72 A2** 60 24N 131 40W
Wolf Point, *U.S.A.* **80 A2** 48 5N 105 39W
Wolfe I., *Canada* **70 D4** 44 7N 76 20W
Wolfsburg, *Germany* **16 B5** 52 27N 10 49 E
Wolin, *Poland* **16 B7** 53 50N 14 37 E
Wollaston, Is., *Chile* **96 H3** 55 40S 67 30W
Wollaston L., *Canada* **73 B8** 58 7N 103 10W
Wollaston Pen., *Canada* **68 B8** 69 30N 115 0W
Wollogorang, *Australia* **62 B2** 17 13S 137 57 E
Wollongong, *Australia* **63 E5** 34 25S 150 54 E
Wolmaransstad, *S. Africa* **56 D4** 27 12S 25 59 E
Wolseley, *Australia* **63 F3** 36 23S 140 54 E
Wolseley, *Canada* **73 C8** 50 25N 103 15W
Wolseley, *S. Africa* **56 E2** 33 26S 19 7 E
Wolstenholme, C., *Canada* **66 C12** 62 35N 77 30W
Wolvega, *Neths.* **15 B6** 52 52N 6 0 E
Wolverhampton, *U.K.* **11 E5** 52 35N 2 6W
Wonarah, *Australia* **62 B2** 19 55S 136 20 E
Wondai, *Australia* **63 D5** 26 20S 151 49 E
Wongalarroo L., *Australia* **63 E3** 31 32S 144 0 E
Wongan Hills, *Australia* **61 F2** 30 51S 116 37 E
Wongawol, *Australia* **61 E3** 26 5S 121 55 E
Wŏnju, *S. Korea* **35 F14** 37 22N 127 58 E
Wonosari, *Indonesia* **37 G14** 7 58S 110 36 E
Wŏnsan, *N. Korea* **35 E14** 39 11N 127 27 E
Wonthaggi, *Australia* **63 F4** 38 37S 145 37 E
Woocalla, *Australia* **63 E2** 31 42S 137 12 E
Wood Buffalo Nat. Park, *Canada* **72 B6** 59 0N 113 41W
Wood Is., *Australia* **60 C3** 16 24S 123 19 E
Wood L., *Canada* **73 B8** 55 17N 103 17W
Wood Lake, *U.S.A.* **80 D4** 42 38N 100 14W
Woodah I., *Australia* **62 A2** 13 27S 136 10 E
Woodanilling, *Australia* **61 F2** 33 31S 117 24 E
Woodbridge, *Canada* **78 C5** 43 47N 79 36W
Woodburn, *Australia* **63 D5** 29 6S 153 23 E
Woodenbong, *Australia* **63 D5** 28 24S 152 39 E
Woodend, *Australia* **63 F3** 37 20S 144 33 E
Woodfords, *U.S.A.* **84 G7** 38 47N 119 50W
Woodgreen, *Australia* **62 C1** 22 26S 134 12 E
Woodlake, *U.S.A.* **84 J7** 36 25N 119 6W
Woodland, *U.S.A.* **84 G5** 38 41N 121 46W
Woodlands, *Australia* **60 D2** 24 46S 118 8 E
Woodpecker, *Canada* **72 C4** 53 30N 122 40W

Yermakovo

Woodridge, *Canada* **73 D9** 49 20N 96 9W
Woodroffe, Mt., *Australia* **61 E5** 26 20S 131 45 E
Woodruff, *Ariz., U.S.A.* . **83 J8** 34 51N 110 1W
Woodruff, *Utah, U.S.A.* . **82 F8** 41 31N 111 10W
Woods, L., *Australia* .... **62 B1** 17 50S 133 30 E
Woods, L., *Canada* **71 B6** 54 30N 65 13W
Woods, L. of the, *Canada* **73 D10** 49 15N 94 45W
Woodstock, *Queens., Australia* .... **62 B4** 19 35S 146 50 E
Woodstock, *W. Austral., Australia* .... **60 D2** 21 41S 118 57 E
Woodstock, *N.B., Canada* **71 C6** 46 11N 67 37W
Woodstock, *Ont., Canada* **70 D3** 43 10N 80 45W
Woodstock, *U.K.* **11 F6** 51 51N 1 20W
Woodstock, *Ill., U.S.A.* .. **80 D10** 42 19N 88 27W
Woodstock, *Vt., U.S.A.* . **79 C12** 43 37N 72 31W
Woodsville, *U.S.A.* **79 B13** 44 9N 72 2W
Woodville, *N.Z.* .... **59 J5** 40 20S 175 53 E
Woodville, *U.S.A.* **81 K7** 30 47N 94 25W
Woodward, *U.S.A.* **81 G5** 36 26N 99 24W
Woody, *U.S.A.* **85 K8** 35 42N 118 50W
Woolamai, C., *Australia* . **63 F4** 38 30S 145 23 E
Woolgoolga, *Australia* .. **63 E5** 30 6S 153 11 E
Woombye, *Australia* ... **63 D5** 26 40S 152 55 E
Woomera, *Australia* .... **63 E2** 31 5S 136 50 E
Woonsocket, *R.I., U.S.A.* **79 D13** 42 0N 71 31W
Woonsocket, *S. Dak., U.S.A.* **80 C5** 44 3N 98 17W
Wooramel, *Australia* ... **61 E1** 25 45S 114 17 E
Wooramel →, *Australia* **61 E1** 25 47S 114 10 E
Wooroloo, *Australia* .... **61 F2** 31 48S 116 18 E
Wooster, *U.S.A.* **78 F3** 40 48N 81 56W
Worcester, *S. Africa* .... **56 E2** 33 39S 19 27 E
Worcester, *U.K.* **11 E5** 52 12N 2 12W
Worcester, *Mass., U.S.A.* **79 D13** 42 16N 71 48W
Worcester, *N.Y., U.S.A.* . **79 D10** 42 36N 74 45W
Workington, *U.K.* **10 C4** 54 39N 3 34W
Worksop, *U.K.* **10 D6** 53 19N 1 9W
Workum, *Neths.* .... **15 B5** 52 59N 5 26 E
Worland, *U.S.A.* **82 D10** 44 1N 107 57W
Worms, *Germany* ...... **16 D4** 49 37N 8 21 E
Wortham, *U.S.A.* **81 K6** 31 47N 96 28W
Worthing, *U.K.* **11 G7** 50 49N 0 21W
Worthington, *U.S.A.* **80 D7** 43 37N 95 36W
Wosi, *Indonesia* .... **37 E7** 0 15S 128 0 E
Wou-han = Wuhan, *China* .... **33 C6** 30 31N 114 18 E
Wour, *Chad* .... **51 D8** 21 14N 16 0 E
Wousi = Wuxi, *China* .. **33 C7** 31 33N 120 18 E
Wowoni, *Indonesia* .... **37 E6** 4 5S 123 5 E
Woy Woy, *Australia* .... **63 E5** 33 30S 151 19 E
Wrangel I. = Vrangelya, Ostrov, *Russia* ...... **27 B19** 71 0N 180 0 E
Wrangell, *U.S.A.* **68 C6** 56 28N 132 23W
Wrangell I., *U.S.A.* **72 B2** 56 16N 132 12W
Wrangell Mts., *U.S.A.* .. **68 B5** 61 30N 142 0W
Wrath, C., *U.K.* **12 C3** 58 38N 5 0W
Wray, *U.S.A.* **80 E3** 40 5N 102 13W
Wrekin, The, *U.K.* **10 E5** 52 41N 2 35W
Wrens, *U.S.A.* **77 J4** 33 12N 82 23W
Wrexham, *U.K.* **10 D4** 53 5N 3 0W
Wright, *Canada* **72 C4** 51 52N 121 40W
Wright, *Phil.* .... **37 B7** 11 42N 125 2 E
Wrightson Mt., *U.S.A.* .. **83 L8** 31 42N 110 51W
Wrightwood, *U.S.A.* **85 L9** 34 21N 117 38W
Wrigley, *Canada* **68 B7** 63 16N 123 37W
Wrocław, *Poland* ...... **16 C8** 51 5N 17 5 E
Września, *Poland* ...... **17 B8** 52 21N 17 36 E
Wu Jiang →, *China* ... **32 D5** 29 40N 107 20 E
Wu'an, *China* .... **34 F8** 36 40N 114 15 E
Wubin, *Australia* ...... **61 F2** 30 6S 116 37 E
Wubu, *China* .... **34 F6** 37 28N 110 42 E
Wuchang, *China* ...... **35 B14** 44 55N 127 5 E
Wucheng, *China* ...... **34 F9** 37 12N 116 20 E
Wuchuan, *China* ...... **34 D6** 41 5N 111 28 E
Wudi, *China* .... **35 F9** 37 40N 117 35 E
Wuding He →, *China* .. **34 F6** 37 2N 110 23 E
Wudu, *China* .... **34 H3** 33 22N 104 54 E
Wuhan, *China* .... **33 C6** 30 31N 114 18 E
Wuhe, *China* .... **35 H9** 33 10N 117 50 E
Wuhsi = Wuxi, *China* .. **33 C7** 31 33N 120 18 E
Wuhu, *China* .... **33 C6** 31 22N 118 21 E
Wukari, *Nigeria* .... **50 G6** 7 51N 9 42 E
Wulajie, *China* .... **35 B14** 44 6N 126 33 E
Wulanbulang, *China* ... **34 D6** 41 5N 110 55 E
Wulian, *China* .... **35 G10** 35 40N 119 12 E
Wuliaru, *Indonesia* .... **37 F8** 7 27S 131 0 E
Wuluk'umushih Ling, *China* .... **32 C3** 36 25N 87 25 E
Wulumuchi = Ürümqi, *China* .... **26 E9** 43 45N 87 45 E
Wum, *Cameroon* ...... **50 G7** 6 24N 10 2 E
Wunnummin L., *Canada* **70 B2** 52 55N 89 10W
Wuntho, *Burma* .... **41 H19** 23 55N 95 45 E
Wuppertal, *Germany* ... **16 C3** 51 15N 7 8 E
Wuppertal, *S. Africa* .... **56 E2** 32 13S 19 12 E
Wuqing, *China* .... **35 E9** 39 23N 117 4 E
Wurung, *Australia* ..... **62 B3** 19 13S 140 38 E
Würzburg, *Germany* ... **16 D4** 49 46N 9 55 E
Wushan, *China* .... **34 G3** 34 43N 104 53 E
Wusuli Jiang = Ussuri →, *Asia* ..... **30 A7** 48 27N 135 0 E
Wutai, *China* .... **34 E7** 38 40N 113 12 E
Wuting = Huimin, *China* **35 F9** 37 27N 117 28 E
Wutonghaolai, *China* ... **35 C11** 42 50N 120 5 E
Wutongqiao, *China* .... **32 D5** 29 22N 103 50 E
Wuwei, *China* .... **32 C5** 37 57N 102 34 E
Wuxi, *China* .... **33 C7** 31 33N 120 18 E
Wuxiang, *China* ...... **34 F7** 36 49N 112 50 E
Wuxing, *China* .... **33 C7** 30 51N 120 8 E
Wuyang, *China* .... **34 H7** 33 25N 113 35 E
Wuyi, *China* .... **34 F8** 37 46N 115 56 E
Wuyi Shan, *China* ..... **33 D6** 27 0N 117 0 E
Wuyuan, *China* .... **34 D5** 41 2N 108 20 E
Wuzhai, *China* .... **34 E6** 38 54N 111 48 E
Wuzhi Shan, *China* .... **38 C7** 18 45N 109 45 E
Wuzhong, *China* ...... **34 E4** 38 2N 106 12 E
Wuzhou, *China* .... **33 D6** 23 30N 111 18 E
Wyaaba Cr. →, *Australia* **62 B3** 16 27S 141 35 E
Wyalkatchem, *Australia* . **61 F2** 31 8S 117 22 E
Wyalusing, *U.S.A.* **79 E8** 41 40N 76 16W
Wyandotte, *U.S.A.* **76 D4** 42 12N 83 9W
Wyandra, *Australia* .... **63 D4** 27 12S 145 56 E
Wyangala Res., *Australia* **63 E4** 33 54S 149 0 E
Wyara, L., *Australia* .... **63 D3** 28 42S 144 14 E
Wycheproof, *Australia* .. **63 F3** 36 5S 143 17 E
Wye →, *U.K.* **11 F5** 51 36N 2 40W
Wyemandoo, *Australia* . **61 E2** 28 28S 118 29 E
Wymondham, *U.K.* **11 E7** 52 45N 0 42W
Wymore, *U.S.A.* **80 E6** 40 7N 96 40W
Wynbring, *Australia* ... **63 E1** 30 33S 133 32 E
Wyndham, *Australia* ... **60 C4** 15 33S 128 3 E
Wyndham, *N.Z.* .... **59 M2** 46 20S 168 51 E
Wyndmere, *U.S.A.* **80 B6** 46 16N 97 8W
Wynne, *U.S.A.* **81 H9** 35 14N 90 47W
Wynnum, *Australia* .... **63 D5** 27 27S 153 9 E
Wynyard, *Australia* .... **62 G4** 41 5S 145 44 E
Wynyard, *Canada* **73 C8** 51 45N 104 10W
Wyola, L., *Australia* .... **61 E5** 29 8S 130 17 E
Wyoming □, *U.S.A.* **82 E10** 43 0N 107 30W
Wyong, *Australia* ..... **63 E5** 33 14S 151 24 E
Wytheville, *U.S.A.* **76 G5** 36 57N 81 5W

X

Xai-Xai, *Mozam.* .... **57 D5** 25 6S 33 31 E
Xainza, *China* .... **32 C3** 30 58N 88 35 E
Xangongo, *Angola* ..... **56 B2** 16 45S 15 5 E
Xánthi, *Greece* .... **21 D11** 41 10N 24 58 E
Xapuri, *Brazil* **92 F5** 10 35S 68 35W
Xar Moron He →, *China* **35 C11** 43 25N 120 35 E
Xau, L., *Botswana* ..... **56 C3** 21 15S 24 44 E
Xavantina, *Brazil* **95 A5** 21 15S 52 48W
Xenia, *U.S.A.* **76 F4** 39 41N 83 56W
Xeropotamos →, *Cyprus* **23 E11** 34 42N 32 33 E
Xhora, *S. Africa* ...... **57 E4** 31 55S 28 38 E
Xhumo, *Botswana* ..... **56 C3** 21 7S 24 35 E
Xi Jiang →, *China* .... **33 D6** 22 5N 113 20 E
Xi Xian, *China* .... **34 F6** 36 41N 110 58 E
Xia Xian, *China* .... **34 G6** 35 8N 111 12 E
Xiachengzi, *China* ..... **35 B16** 44 40N 130 18 E
Xiaguan, *China* .... **32 D5** 25 32N 100 16 E
Xiajin, *China* .... **34 F8** 36 56N 116 0 E
Xiamen, *China* .... **33 D6** 24 25N 118 4 E
Xi'an, *China* .... **34 G5** 34 15N 109 0 E
Xian Xian, *China* ...... **34 E9** 38 12N 116 6 E
Xiang Jiang →, *China* . **33 D6** 28 55N 112 50 E
Xiangcheng, *Henan, China* .... **34 H8** 33 29N 114 52 E
Xiangcheng, *Henan, China* .... **34 H7** 33 50N 113 27 E
Xiangfan, *China* ...... **33 C6** 32 2N 112 8 E
Xianghuang Qi, *China* .. **34 C7** 42 2N 113 50 E
Xiangning, *China* ..... **34 G6** 35 58N 110 50 E
Xiangquan, *China* ..... **34 F7** 36 30N 113 1 E
Xiangshui, *China* ..... **35 G10** 34 12N 119 33 E
Xiangtan, *China* ...... **33 D6** 27 51N 112 54 E
Xianyang, *China* ...... **34 G5** 34 20N 108 40 E
Xiao Hinggan Ling, *China* **33 B7** 49 0N 127 0 E
Xiao Xian, *China* ..... **34 G9** 34 15N 116 55 E
Xiaoyi, *China* .... **34 F6** 37 8N 111 48 E
Xiawa, *China* .... **35 C11** 42 35N 120 38 E
Xiayi, *China* .... **34 G9** 34 15N 116 10 E
Xichang, *China* ...... **32 D5** 27 51N 102 19 E
Xichuan, *China* ...... **34 H6** 33 0N 111 30 E
Xieng Khouang, *Laos* .. **38 C4** 19 17N 103 25 E
Xifei He →, *China* .... **34 H9** 32 45N 116 40 E
Xifeng, *China* .... **35 C13** 42 42N 124 45 E
Xifengzhen, *China* .... **34 G4** 35 40N 107 40 E
Xigazê, *China* .... **32 D3** 29 5N 88 45 E
Xihe, *China* .... **34 G3** 34 2N 105 20 E
Xihua, *China* .... **34 H8** 33 45N 114 30 E
Xiliao He →, *China* ... **35 C12** 43 32N 123 35 E
Xin Xian, *China* ...... **34 E7** 38 22N 112 46 E
Xinavane, *Mozam.* .... **57 D5** 25 2S 32 47 E
Xinbin, *China* .... **35 D13** 41 40N 125 2 E
Xing Xian, *China* ..... **34 E6** 38 27N 111 7 E
Xing'an, *China* .... **33 D6** 25 38N 110 40 E
Xingcheng, *China* .... **35 D11** 40 40N 120 45 E
Xinghe, *China* .... **34 D7** 40 55N 113 55 E
Xinghua, *China* ...... **35 H10** 32 58N 119 48 E
Xinglong, *China* ...... **35 D9** 40 25N 117 30 E
Xingping, *China* ...... **34 G5** 34 20N 108 28 E
Xingtai, *China* .... **34 F8** 37 3N 114 32 E
Xingu →, *Brazil* **93 D8** 1 30S 51 53W
Xingyang, *China* ..... **34 G7** 34 45N 112 52 E
Xinhe, *China* .... **34 F8** 37 30N 115 15 E
Xining, *China* .... **32 C5** 36 34N 101 40 E
Xinjiang, *China* ...... **34 G6** 35 34N 111 11 E
Xinjiang Uygur Zizhiqu □, *China* ..... **32 B3** 42 0N 86 0 E
Xinjin, *China* .... **35 E11** 39 25N 121 58 E
Xinkai He →, *China* ... **35 C12** 43 32N 123 35 E
Xinle, *China* .... **34 E8** 38 25N 114 40 E
Xinlitun, *China* ...... **35 D12** 42 0N 122 8 E
Xinmin, *China* .... **35 D12** 41 59N 122 50 E
Xintai, *China* .... **35 G9** 35 55N 117 45 E
Xinxiang, *China* ...... **34 G7** 35 18N 113 50 E
Xinzhan, *China* ...... **35 C14** 43 50N 127 18 E
Xinzheng, *China* ..... **34 G7** 34 20N 113 45 E
Xiong Xian, *China* .... **34 E9** 38 59N 116 8 E
Xiongyuecheng, *China* . **35 D12** 40 12N 122 5 E
Xiping, *Henan, China* .. **34 H8** 33 22N 114 5 E
Xiping, *Henan, China* .. **34 H6** 33 25N 111 8 E
Xique-Xique, *Brazil* **93 F10** 10 50S 42 40W
Xisha Qundao = Hsisha Chuntao, *Pac. Oc.* ..... **36 A4** 15 50N 112 0 E
Xiuyan, *China* .... **35 D12** 40 18N 123 11 E
Xixabangma Feng, *China* **41 E14** 28 20N 85 40 E
Xixia, *China* .... **34 H6** 33 25N 111 29 E
Xixiang, *China* .... **34 H4** 33 0N 107 44 E
Xiyang, *China* .... **34 F7** 37 38N 113 38 E
Xizang □, *China* ...... **32 C3** 32 0N 88 0 E
Xlendi, *Malta* .... **23 C1** 36 1N 14 12 E
Xuan Loc, *Vietnam* .... **39 G6** 10 56N 107 14 E
Xuanhua, *China* ...... **34 D8** 40 40N 115 2 E
Xuchang, *China* ...... **34 G7** 34 2N 113 48 E
Xun Xian, *China* ...... **34 G8** 35 42N 114 33 E
Xunyang, *China* ...... **34 H5** 32 48N 109 22 E
Xunyi, *China* .... **34 G5** 35 8N 108 20 E
Xushui, *China* .... **34 E8** 39 2N 115 40 E
Xuyen Moc, *Vietnam* ... **39 G6** 10 34N 107 25 E
Xuzhou, *China* .... **35 G9** 34 18N 117 10 E
Xylophagou, *Cyprus* ... **23 E12** 34 54N 33 51 E

Y

Ya Xian, *China* .... **38 C7** 18 14N 109 29 E
Yaamba, *Australia* .... **62 C5** 23 8S 150 22 E
Yaapeet, *Australia* .... **63 F3** 35 45S 142 3 E
Yabelo, *Ethiopia* ...... **51 H12** 4 50N 38 8 E
Yablonovy Khrebet, *Russia* .... **27 D12** 53 0N 114 0 E
Yablonovy Ra. = Yablonovy Khrebet, *Russia* .... **27 D12** 53 0N 114 0 E
Yabrai Shan, *China* .... **34 E2** 39 40N 103 0 E
Yabrūd, *Syria* .... **47 B5** 33 58N 36 39 E
Yacheng, *China* ...... **33 E5** 18 22N 109 6 E
Yacuiba, *Bolivia* **94 A3** 22 0S 63 43W
Yadgir, *India* .... **40 L10** 16 45N 77 5 E
Yadkin →, *U.S.A.* **77 H5** 35 29N 80 9W
Yagodnoye, *Russia* .... **27 C15** 62 33N 149 40 E
Yagoua, *Cameroon* .... **52 B3** 10 20N 15 13 E
Yaha, *Thailand* .... **39 J3** 6 29N 101 8 E
Yahila, *Zaïre* .... **54 B1** 0 13N 24 28 E
Yahk, *Canada* **72 D5** 49 6N 116 10W
Yahuma, *Zaïre* .... **52 D4** 1 0N 23 10 E
Yaita, *Japan* .... **31 F9** 36 48N 139 56 E
Yaiza, *Canary Is.* **22 F6** 28 57N 13 46W
Yakima, *U.S.A.* **82 C3** 46 36N 120 31W
Yakima →, *U.S.A.* **82 C3** 47 0N 120 30W
Yakovlevka, *Russia* .... **30 B6** 44 26N 133 28 E
Yaku-Shima, *Japan* .... **31 J5** 30 20N 130 30 E
Yakut Republic □, *Russia* **27 C13** 62 0N 130 0 E
Yakutat, *U.S.A.* **68 C6** 59 33N 139 44W
Yakutsk, *Russia* ...... **27 C13** 62 5N 129 50 E
Yala, *Thailand* .... **39 J3** 6 33N 101 18 E
Yalbalgo, *Australia* .... **61 E1** 25 10S 114 45 E
Yalboroo, *Australia* .... **62 C4** 20 50S 148 40 E
Yale, *U.S.A.* **78 C2** 43 8N 82 48W
Yalgoo, *Australia* ..... **61 E2** 28 16S 116 39 E
Yalinga, *C.A.R.* .... **51 G9** 6 33N 23 10 E
Yalkubul, Punta, *Mexico* **87 C7** 21 32N 88 37W
Yalleroi, *Australia* .... **62 C4** 24 3S 145 42 E
Yalobusha →, *U.S.A.* .. **81 J9** 33 33N 90 10W
Yalong Jiang →, *China* **32 D5** 26 40N 101 55 E
Yalta, *Ukraine* .... **25 F5** 44 30N 34 10 E
Yalu Jiang →, *China* .. **35 E13** 40 0N 124 22 E
Yalutorovsk, *Russia* ... **26 D7** 56 41N 66 12 E
Yam Ha Melaḥ = Dead Sea, *Asia* .... **47 D4** 31 30N 35 30 E
Yam Kinneret, *Israel* ... **47 C4** 32 45N 35 35 E
Yamada, *Japan* .... **31 H5** 33 33N 130 49 E
Yamagata, *Japan* ..... **30 E10** 38 15N 140 15 E
Yamagata □, *Japan* ... **30 E10** 38 30N 140 0 E
Yamaguchi, *Japan* .... **31 G5** 34 10N 131 32 E
Yamaguchi □, *Japan* .. **31 G5** 34 20N 131 40 E
Yamal, Poluostrov, *Russia* .... **26 B8** 71 0N 70 0 E
Yamal Pen. = Yamal, Poluostrov, *Russia* ... **26 B8** 71 0N 70 0 E
Yamanashi □, *Japan* .. **31 G9** 35 40N 138 40 E
Yamantau, Gora, *Russia* **24 D10** 54 15N 58 6 E
Yamba, *N.S.W., Australia* **63 D5** 29 26S 153 23 E
Yamba, *S. Austral., Australia* .... **63 E3** 34 10S 140 52 E
Yambah, *Australia* .... **62 C1** 23 10S 133 50 E
Yambarran Ra., *Australia* **60 C5** 15 10S 130 25 E
Yâmbiô, *Sudan* .... **51 H10** 4 35N 28 16 E
Yambol, *Bulgaria* ..... **21 C12** 42 30N 26 36 E
Yamdena, *Indonesia* ... **37 F8** 7 45S 131 20 E
Yame, *Japan* .... **31 H5** 33 13N 130 35 E
Yamethin, *Burma* ..... **41 J20** 20 29N 96 18 E
Yamma-Yamma, L., *Australia* .... **63 D3** 26 16S 141 20 E
Yamoussoukro, *Ivory C.* **50 G3** 6 49N 5 17W
Yampa →, *U.S.A.* **82 F9** 40 32N 108 59W
Yampi Sd., *Australia* ... **60 C3** 16 8S 123 38 E
Yamuna →, *India* .... **43 G9** 25 30N 81 53 E
Yamzho Yumco, *China* . **32 D4** 28 48N 90 35 E
Yana →, *Russia* ..... **27 B14** 71 30N 136 0 E
Yanac, *Australia* ..... **63 F3** 36 8S 141 25 E
Yanagawa, *Japan* .... **31 H5** 33 10N 130 24 E
Yanai, *Japan* .... **31 H6** 33 58N 132 7 E
Yan'an, *China* .... **34 F5** 36 35N 109 26 E
Yanaul, *Russia* ...... **24 C10** 56 25N 55 0 E
Yanbu 'al Baḥr, *Si. Arabia* **44 F3** 24 0N 38 5 E
Yancannia, *Australia* ... **63 E3** 30 12S 142 35 E
Yanchang, *China* ..... **34 F6** 36 43N 110 1 E
Yancheng, *Henan, China* **34 H7** 33 35N 114 0 E
Yancheng, *Jiangsu, China* **35 H11** 33 23N 120 8 E
Yanchi, *China* .... **34 F4** 37 48N 107 20 E
Yanchuan, *China* ..... **34 F6** 36 51N 110 10 E
Yanco Cr. →, *Australia* **63 F4** 35 14S 145 35 E
Yandal, *Australia* ..... **61 E3** 27 35S 121 10 E
Yandanooka, *Australia* . **61 E2** 29 18S 115 29 E
Yandaran, *Australia* ... **62 C5** 24 43S 152 6 E
Yandoon, *Burma* ..... **41 L19** 17 0N 95 40 E
Yang Xian, *China* ..... **34 H4** 33 15N 107 30 E
Yangambi, *Zaïre* ..... **54 B1** 0 47N 24 20 E
Yangcheng, *China* .... **34 G7** 35 28N 112 22 E
Yangch'ü = Taiyuan, *China* .... **34 F7** 37 52N 112 33 E
Yanggao, *China* ...... **34 D7** 40 21N 113 55 E
Yanggu, *China* .... **34 F8** 36 8N 115 43 E
Yangi-Yer, *Kazakhstan* . **26 E7** 43 50N 68 48 E
Yangliuqing, *China* .... **35 E9** 39 2N 117 5 E
Yangon = Rangoon, *Burma* .... **41 L20** 16 45N 96 20 E
Yangpingguan, *China* .. **34 H4** 32 58N 106 5 E
Yangquan, *China* ..... **34 F7** 37 58N 113 31 E
Yangtze Kiang = Chang Jiang →, *China* ..... **33 C7** 31 48N 121 10 E
Yangyang, *S. Korea* .... **35 E15** 38 4N 128 38 E
Yangyuan, *China* ..... **34 D8** 40 1N 114 10 E
Yangzhou, *China* ..... **33 C6** 32 21N 119 26 E
Yanji, *China* .... **35 C15** 42 59N 129 30 E
Yankton, *U.S.A.* **80 D6** 42 53N 97 23W
Yanna, *Australia* ..... **63 D4** 26 58S 146 0 E
Yanonge, *Zaïre* .... **54 B1** 0 35N 24 38 E
Yanqi, *China* .... **32 B3** 42 5N 86 35 E
Yanqing, *China* ...... **34 D8** 40 30N 115 58 E
Yanshan, *China* ...... **35 E9** 38 4N 117 22 E
Yanshou, *China* ...... **35 B15** 45 28N 128 22 E
Yantabulla, *Australia* ... **63 D4** 29 21S 145 0 E
Yantai, *China* .... **35 F11** 37 34N 121 22 E
Yanzhou, *China* ...... **34 G9** 35 35N 116 49 E
Yao, *Chad* .... **51 F8** 12 56N 17 33 E
Yao Xian, *China* ..... **34 G5** 34 55N 108 59 E
Yao Yai, Ko, *Thailand* .. **39 J2** 8 0N 98 35 E
Yaoundé, *Cameroon* ... **50 H7** 3 50N 11 35 E
Yaowan, *China* ...... **35 G10** 34 15N 118 3 E
Yap I., *Pac. Oc.* ...... **64 G5** 9 30N 138 10 E
Yapen, *Indonesia* ..... **37 E9** 1 50S 136 0 E
Yapen, Selat, *Indonesia* . **37 E9** 1 20S 136 10 E
Yappar →, *Australia* ... **62 B3** 18 22S 141 16 E
Yaqui →, *Mexico* **86 B2** 27 37N 110 39W
Yar-Sale, *Russia* ..... **26 C8** 66 50N 70 50 E
Yaraka, *Australia* ..... **62 C3** 24 53S 144 3 E
Yaransk, *Russia* ...... **24 C8** 57 22N 47 49 E
Yardea P.O., *Australia* .. **63 E2** 32 23S 135 32 E
Yare →, *U.K.* .... **11 E9** 52 36N 1 28 E
Yarensk, *Russia* ...... **24 B8** 61 10N 49 8 E
Yarí →, *Colombia* **92 D4** 0 20S 72 20W
Yarkand = Shache, *China* **32 C2** 38 20N 77 10 E
Yarker, *Canada* **79 B8** 44 23N 76 46W
Yarkhun →, *Pakistan* .. **43 A5** 36 17N 72 30 E
Yarmouth, *Canada* **71 D6** 43 50N 66 7W
Yarmūk →, *Syria* ..... **47 C4** 32 42N 35 40 E
Yaroslavl, *Russia* ..... **24 C6** 57 35N 39 55 E
Yarqa, W. →, *Egypt* ... **47 F2** 30 0N 33 49 E
Yarra Yarra Lakes, *Australia* .... **61 E2** 29 40S 115 45 E
Yarraden, *Australia* ... **62 A3** 14 17S 143 15 E
Yarraloola, *Australia* ... **60 D2** 21 33S 115 52 E
Yarram, *Australia* .... **63 F4** 38 29S 146 39 E
Yarraman, *Australia* ... **63 D5** 26 50S 152 0 E
Yarranvale, *Australia* .. **63 D4** 26 50S 145 20 E
Yarras, *Australia* ..... **63 E5** 31 25S 152 20 E
Yarrowmere, *Australia* . **62 C4** 21 27S 145 53 E
Yartsevo, *Russia* ..... **27 C10** 60 20N 90 0 E
Yasawa Group, *Fiji* .... **59 C7** 17 0S 177 23 E
Yasin, *Pakistan* ...... **43 A5** 36 24N 73 23 E
Yasinski, L., *Canada* ... **70 B4** 53 16N 77 35W
Yasothon, *Thailand* ... **38 E5** 15 50N 104 10 E
Yass, *Australia* ...... **63 E4** 34 49S 148 54 E
Yates Center, *U.S.A.* .. **81 G7** 37 53N 95 44W
Yathkyed L., *Canada* ... **73 A9** 62 40N 98 0W
Yatsushiro, *Japan* .... **31 H5** 32 30N 130 40 E
Yatta Plateau, *Kenya* .. **54 C4** 2 0S 38 0 E
Yauyos, *Peru* **92 F3** 12 19S 75 50W
Yavari →, *Peru* **92 D4** 4 21S 70 2W
Yavatmal, *India* ...... **40 J11** 20 20N 78 15 E
Yavne, *Israel* .... **47 D3** 31 52N 34 45 E
Yawatahama, *Japan* ... **31 H6** 33 27N 132 24 E
Yayama-Rettō, *Japan* .. **31 M1** 24 30N 123 40 E
Yazd, *Iran* .... **45 D7** 31 55N 54 27 E
Yazd □, *Iran* .... **45 D7** 32 0N 55 0 E
Yazoo →, *U.S.A.* **81 J9** 32 22N 90 54W
Yazoo City, *U.S.A.* **81 J9** 32 51N 90 25W
Yding Skovhøj, *Denmark* **9 J10** 55 59N 9 46 E
Ye Xian, *Henan, China* .. **34 H7** 33 35N 113 25 E
Ye Xian, *Shandong, China* .... **35 F10** 37 8N 119 57 E
Yealering, *Australia* ... **61 F2** 32 36S 117 36 E
Yebyu, *Burma* .... **41 M21** 14 15N 98 13 E
Yechŏn, *S. Korea* ..... **35 F15** 36 39N 128 27 E
Yecla, *Spain* **19 C5** 38 35N 1 5W
Yécora, *Mexico* **86 B3** 28 20N 108 58W
Yeeda, *Australia* ..... **60 C3** 17 31S 123 38 E
Yeelanna, *Australia* ... **63 E2** 34 9S 135 45 E
Yegros, *Paraguay* **94 B4** 26 20S 56 25W
Yehuda, Midbar, *Israel* . **47 D4** 31 35N 35 15 E
Yei, *Sudan* .... **51 H11** 4 9N 30 40 E
Yekaterinburg, *Russia* .. **24 C11** 56 50N 60 30 E
Yekaterinodar = Krasnodar, *Russia* .... **25 E6** 45 5N 39 0 E
Yelanskoye, *Russia* ... **27 C13** 61 25N 128 0 E
Yelarbon, *Australia* ... **63 D5** 28 33S 150 38 E
Yelets, *Russia* .... **24 D6** 52 40N 38 30 E
Yelizavetgrad = Kirovograd, *Ukraine* ... **25 E5** 48 35N 32 20 E
Yell, *U.K.* **12 A7** 60 35N 1 5W
Yell Sd., *U.K.* **12 A7** 60 33N 1 15W
Yellow Sea, *China* .... **35 G12** 35 0N 123 0 E
Yellowhead Pass, *Canada* **72 C5** 52 53N 118 25W
Yellowknife, *Canada* ... **72 A6** 62 27N 114 29W
Yellowknife →, *Canada* **72 A6** 62 31N 114 19W
Yellowstone →, *U.S.A.* . **80 B3** 47 59N 103 59W
Yellowstone L., *U.S.A.* . **82 D8** 44 27N 110 22W
Yellowstone National Park, *U.S.A.* **82 D8** 44 40N 110 30W
Yellowtail Res., *U.S.A.* .. **82 D9** 45 6N 108 8W
Yelvertoft, *Australia* ... **62 C2** 20 13S 138 45 E
Yemen ■, *Asia* ...... **46 E3** 15 0N 44 0 E
Yen Bai, *Vietnam* ..... **38 B5** 21 42N 104 52 E
Yenangyaung, *Burma* .. **41 J19** 20 30N 95 0 E
Yenbo = Yanbu 'al Baḥr, *Si. Arabia* .... **44 F3** 24 0N 38 5 E
Yenda, *Australia* ..... **63 E4** 34 13S 146 14 E
Yenisey →, *Russia* ... **26 B9** 71 50N 82 40 E
Yeniseysk, *Russia* .... **27 D10** 58 27N 92 13 E
Yeniseyskiy Zaliv, *Russia* **26 B9** 72 20N 81 0 E
Yennádhi, *Greece* ..... **23 C9** 36 2N 27 56 E
Yenyuka, *Russia* ..... **27 D13** 57 57N 121 15 E
Yeo, L., *Australia* ..... **61 E3** 28 0S 124 30 E
Yeola, *India* .... **40 J9** 20 2N 74 30 E
Yeoryioúpolis, *Greece* .. **23 D6** 35 20N 24 15 E
Yeovil, *U.K.* **11 G5** 50 57N 2 38W
Yeppoon, *Australia* ... **62 C5** 23 5S 150 47 E
Yerbent, *Turkmenistan* . **26 F6** 39 30N 58 50 E
Yerbogachen, *Russia* .. **27 C11** 61 16N 108 0 E
Yerevan, *Armenia* .... **25 F7** 40 10N 44 31 E
Yerilla, *Australia* ..... **61 E3** 29 24S 121 47 E
Yermak, *Kazakhstan* ... **26 D8** 52 2N 76 55 E
Yermakovo, *Russia* .... **27 D13** 52 25N 126 20 E

Z